Adult Neurogenesis 2

Adult Neurogenesis 2

Stem Cells and Neuronal Development in the Adult Brain

Gerd Kempermann, PhD
CRTD – Center for Regenerative Therapies Dresden
Dresden, Germany

UNIVERSITY PRESS
2011

OXFORD
UNIVERSITY PRESS

Oxford University Press, Inc., publishes works that further
Oxford University's objective of excellence
in research, scholarship, and education.

Oxford New York
Auckland   Cape Town   Dar es Salaam   Hong Kong   Karachi
Kuala Lumpur   Madrid   Melbourne   Mexico City   Nairobi
New Delhi   Shanghai   Taipei   Toronto

With offices in
Argentina   Austria   Brazil   Chile   Czech Republic   France   Greece
Guatemala   Hungary   Italy   Japan   Poland   Portugal   Singapore
South Korea   Switzerland   Thailand   Turkey   Ukraine   Vietnam

Published by Oxford University Press, Inc.
198 Madison Avenue, New York, New York 10016
www.oup.com

Oxford is a registered trademark of Oxford University Press

Library of Congress Cataloging-in-Publication Data

Kempermann, Gerd.
    Adult neurogenesis 2/Gerd Kempermann.—2nd ed.
    p.; cm.
    Includes bibliographical references and index.
    ISBN 978-0-19-972969-2
1. Developmental neurobiology. 2. Neural stem cells. I. Title.
[DNLM: 1. Neurons—physiology. 2. Adult. 3. Brain—physiology.
4. Nerve Regeneration—physiology. 5. Stem Cells—physiology. WL 102.5 K32a 2011]
    QP363.5.K466 2011
    612.6'4018—dc22

                                                    2010012908

The science of medicine is a rapidly changing field. As new research and clinical
experience broaden our knowledge, changes in treatment and drug therapy occur.
The author and publisher of this work have checked with sources believed to be
reliable in their efforts to provide information that is accurate and complete, and
in accordance with the standards accepted at the time of publication. However,
in light of the possibility of human error or changes in the practice of medicine,
neither the author, nor the publisher, nor any other party who has been involved
in the preparation or publication of this work warrants that the information
contained herein is in every respect accurate or complete. Readers are encouraged
to confirm the information contained herein with other reliable sources, and are
strongly advised to check the product information sheet provided by the
pharmaceutical company for each drug they plan to administer.

9 8 7 6 5 4 3 2 1
Printed in China
on acid-free paper

As before: for Uta, Georg, and Rusty

Cartoon drawn by Wayne Stayskal for the Tampa Tribune on occasion of the publication of the report that physical exercise increases adult hippocampal neurogenesis (see Chapter 9, p. 343). The figure was kindly provided by Wayne Stayskal. Reprinted with permission from the Tampa Tribune.

Preface

ADULT NEUROGENESIS—that is, the generation of new neurons in the adult brain—has progressed through various stages of study and appreciation in the rather short period of 45 years between 1965 and today. First it was regarded as an impossibility, then as a curiosity, later as an exception to otherwise steadfast rules, and finally as an accepted phenomenon of mammalian neurobiology. This book is meant to provide an introduction to this field, which still appears to be in the log phase of expansion (see also Fig. 1–1 in the introductory Chapter 1). Consequently, a great deal has happened in the mere five years since the first edition of this book. Approximately 1,000 new publications had to be covered. It is safe to say that the field has now left its infancy, has passed the period of being just the "flavor of the month", and has matured to an established area of neurobiology with ever-increasing influence on general concepts. Can we understand the cellular bases of "learning" without including adult neurogenesis? I am convinced: whoever wants to understand the brain and how it works can now longer ignore that there are new neurons even in adulthood.

Nevertheless, at a time when even scientists use obscure and nonscientific bibliometric measures to decide what is important in science and what is not, so that where a new result is published trumps the actual content, a book might seem to be placed into a void and thus wasted time on the part of the author. Books are hardly ever cited, and their "impact" remains unestimated. At the same time, the (laudable) move towards open access publishing and wiki-like knowledge bases and structures on the Internet might increasingly challenge the value of any book. I am convinced, however, that not despite but because of all these cultural changes there will be an important, albeit very specific, role for books, not least because of their haptic qualities. The brain is rather slow compared to "the cloud". It has evolved not in response to virtual worlds but challenged by the demands of a physical environment, and for proper functioning still appreciates tangible anchor points. A book might be (or—what psychologically would amount to the same thing—at least seem to be) the best assurance against the experience of helplessness when facing the vast depth of the knowledge space behind the computer screen.

The web and its contents are constantly in flux, which promises to the user that developments might be captured in "real time." But in the end that is an illusion, and science remains an art of abstraction. To go in-depth we have to start from "freezes" of information and gain a solid place to stand before we move on. This book is meant to provide such a foundation, from which the reader might embark on the discovery of the newer exciting developments not yet covered.

I am happy and proud to see that "the white Adult Neurogenesis book" has become what it was intended to be: an accepted entrance gate to one of the most exciting areas of neuroscience. Students and more advanced scholars alike have been using the first edition to support their navigation through the ever-increasing number of publications on new neurons in the adult brain and to dig down to the underlying concepts. I hope that they will do the same with the second edition.

Just adding new literature was not sufficient to bring the book up to date. The entire work was essentially rewritten, and many aspects not discussed in the first edition are now covered. Errors have been corrected, and many new figures have been added. But, most important, much conceptual progress has been made, in particular with respect to the molecular basis and the possible function of adult neurogenesis. With the exception of an additional handful of publications from early 2010, every attempt has been made to adequately cover the literature as of the end of 2009.

As before, everything presented here is meant to be an invitation for discussion. Please let me know your thoughts: gerd.kempermann@crt-dresden.de. Future editions will incorporate the results of any ensuing interaction, so that the community's views, rather than a personal opinion, will take shape and be taught to the next generation of neuroscientists.

Gerd Kempermann
Dresden, January 1, 2010

From the Preface to the First Edition

WHAT YOU WILL FIND on these pages is certainly not all that could or perhaps should be said about adult neurogenesis. The ideal was to draw a coherent picture and identify the common threads and evolving ideas, but as anybody knows who has embarked on such an endeavor, this effort must occasionally lead to a biased view. But bias was not intended, and in fact, every attempt was made to avoid the unfair treatment of other people's results. On the other hand, honest subjectivity is better than pseudo-objectivity. Science is not as objective as we would like it to be: even today it is made by people, a fact that the use of computers and the Internet tend to obscure. A book has one identified author whom you can hold responsible. The arguments laid down in this book have been tested in discussions with many colleagues, most of whom are, hopefully, mentioned in the Acknowledgements. If I have stepped on anyone's toes, please accept my apologies. The goal was certainly not to hurt toes, but to provide a critical and scientifically sound argument. It is, of course, the proverbial shoulders of giants that I have stood on. By now, the giants should, with all due respect, be rather used to this treatment. But I am well aware that I have also used the less trained shoulders of many students and postdoctoral researchers, whose work has contributed to what we now know about adult neurogenesis. They might rightfully feel more injustice if I have misrepresented their work. By no means do I intend to suggest that I have delivered the definitive interpretation of the field. But what I do hope to convey is my enthusiasm for this field of research. So what the reader will find here is one scientist's take on the world of adult neurogenesis. I hope this view comes equipped with good-enough arguments to withstand the critique.

Acknowledgments

MY FIRST THANK-YOU goes to the two people who brought me into the field of adult neurogenesis. I would like to thank Fred H. Gage for his continued support and the many challenging discussions over the past 15 years, and H. Georg Kuhn, with whom I did my first experiments on adult neurogenesis. H. Georg Kuhn also commented on the manuscript of the first edition and helped shape its perspective. So did Theo D. Palmer, my desk neighbor at the Salk Institute, whose contributions during many early discussions have been extremely valuable in developing the ideas on neural stem cells presented here and with whom, along with Georg Kuhn and Philipp Horner, I founded "Route 28—Summits in Neurobiology," a series of workshops that between 1999 and 2006 invited graduate and postdoctoral students to discover adult neurogenesis and neural stem cell biology. I would also like to thank former Gage lab colleague Daniel Peterson, whose advice on many methodological issues has been unsurpassed. I had an inspiring exchange of emails with Joseph Altman, who also commented on parts of the first edition. I also benefited greatly from several challenging discussions with Pasko Rakic. Particular thanks also goes to Robert W. Williams, who shaped my thinking about genetics, genomics, and the "systems" versions of those two. From him one can learn not to be afraid of complexity. And adult neurogenesis is clearly becoming an increasingly complex topic.

I thank the following colleagues who contributed images to this book and allowed their reproduction: Joseph Altman, Harish Babu, Monika Brilt, Michael Brand, Heather Cameron, Christiana Cooper-Kuhn, Peter Eriksson (*deceased*), Klaus Fabel, Fred H. Gage, Alexander Garthe, Rainer Glass, Magdalena Götz, Sebastian Jessberger, Helmut Kettenmann, Zaal Kokaia, David Kornack, Monika Kott, Golo Kronenberg, H. Georg Kuhn, Olle Lindvall, Gudrun Lutsch, Jeffrey D. Macklis, Pasko Rakic, Constanze Scharff, Sophie Scotto-Lomassese, Barbara Steiner, Li Ping Wang, and Andreas Wodarz. In addition, I received very useful comments on particular questions from Harish Babu, Alexander Garthe, Sophie Scotto-Lomassese, Sam Pleasure, and Tracey Shors.

I would also like to thank the current and former colleagues in my laboratory in Dresden (and previously Berlin). Needless to say, without such a wonderful team the endeavor would not have been possible.

I am grateful to Craig Panner at Oxford University Press, who took over the project *Adult Neurogenesis* and greatly supported the second edition, also by suggesting a few important changes in format and layout and by introducing a more extensive color usage. I would also like to thank Astrid Poppenhusen for her good advice as literary agent.

Finally, but actually foremost, I thank my wife, Uta, for her love, patience, and support.

Contents

Adult Neurogenesis 2

1

Introduction

NEW NEURONS FOR OLD BRAINS! A thought of great suggestive power—like the fountain of youth.

Most of our body is constantly renewed. On average, the age of our cells might be as low as 7 to 10 years (Spalding et al., 2005; Spalding et al., 2008). It should pose a considerable challenge to philosophy that we do not seem to be the same physical person throughout our life.

Many cell types in our body show lifelong turnover

The basis of our constant rejuvenation is that bone marrow, skin, and intestines turn over all the time and at highest rate. Our skin constantly wears off its outermost layer, which is replaced from the innermost zone. Pasko Rakic from Yale University phrased it: "With respect to my skin I am a new man every year." With respect to the most superficial layer, he is even a new man every day. Hairdressers make a living from the activity of the stem cells in our hair follicles. Intestinal epithelia do not live to experience very many warm meals before they are replaced. Blood donation is a safe experience because we can rely on our bone marrow to replace our donation within a short period of time. Blood, skin, and intestines are the classic regenerative tissues.

(Re-)discovery of adult neurogenesis was one of the scientific breakthroughs of the 1990s

Intuitively we are not always particularly appreciative about the often-lifesaving cellular turnover. But when it became clear in the 1990s that something similar is possible in the brain, even if it occurs on a minute scale, the finding was hailed as a fundamental breakthrough. Peter Eriksson's 1998 study (Eriksson et al., 1998), in which he reported that new neurons are generated in the hippocampus of adult humans, moved research on adult neurogenesis from heresy to where it is now: not quite in orthodoxy but at least firmly established. The *New York Times* hailed the discovery of ongoing neurogenesis in the adult brain as the most important research result from the "Decade of the Brain," which spanned the years 1990 to 1999 (Blakeslee, 2000). That decade saw an impressive surge in scientific and public awareness of the phenomenon of neurogenesis in the adult mammalian brain, and the wealth of studies published certainly constituted a major scientific progress (Fig. 1–1). Now at the end of the following decade, publications on adult neurogenesis soar, and it seems that we are still in the logarithmic phase of growth.

Despite adult neurogenesis, the brain regenerates poorly

The journalistic blessing by the *New York Times*, a long feature article in the *New Yorker* (Specter, 2001), and a growing number of articles in newspapers and magazines during the last decade also contributed, however, to some widely held misconceptions about this exciting field of science.

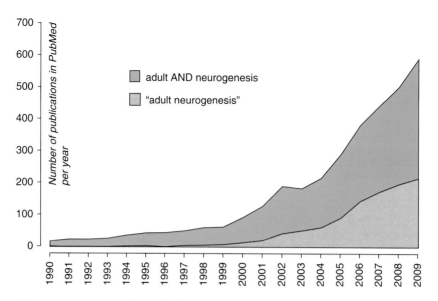

FIGURE 1–1 The ever-increasing numbers of publications on adult neurogenesis. The numbers are derived from simple online PubMed searches (performed on February 4, 2010). The upper green curve gives numbers for the less stringent search "adult AND neurogenesis," the lower gray curve for the very stringent search "adult neurogenesis." While the first search generates many false positive results, and the latter many false negatives (and would, for example already miss "adult hippocampal neurogenesis"), so "the truth" will lie somewhere in between. But in any case, the growth is impressive and reveals "adult neurogenesis" as an extremely lively research area attracting more and more interest.

Some of these misconceptions appear to have—because scientists read newspapers as well—found their way back into the scientific literature.

Most important, and contrary to ideas prevailing in the public as well as among some scientists, the existence of adult neurogenesis unfortunately does not suddenly turn the brain into a regenerative organ. The brain still regenerates poorly, and the call "Brain, heal thyself" that was frequently heard in the early years of the field (Lowenstein and Parent, 1999) is not so straightforward to follow as the presence of new neurons in the adult brain might suggest.

Also, adult neurogenesis was discovered much earlier than the 1990s and came out of the blue much less than one might be tempted to think. Research on adult neurogenesis was firmly rooted in general progress throughout the twentieth century in developmental neurobiology, stem cell biology, and research on brain plasticity. Nonetheless, few scientific developments have caught the imagination of the scientific community and the public as much as the so-called stem cell revolution. Adult neurogenesis is a highly visible and intriguing part of this revolution. It is the stem cells of the adult brain that drive adult neurogenesis. Publications about neural stem cells and adult neurogenesis make up roughly 5% to 10% of all publications on stem cells. This is a remarkable number, given that hematology, where stem cell therapy is clinical routine, comes to about 20% to 25%. These are only very rough estimates, and a more sophisticated bibliometrical analysis might change actual counts but will probably not affect their relative preponderance.

The fundamental change that modern stem cell biology has introduced to science is similar to the paradigm shift caused by the sequencing of the genome: our view changed from a deductive perspective, which we have become accustomed to in science, to an essentially open situation. Genes and stem cells are to a large degree characterized by their potential.

||||||||||||||||||||||||||||||||||||||||||||||||||||||||||||||||||

Rediscovery of adult neurogenesis was part of the stem cell revolution

||||||||||||||||||||||||||||||||||||||||||||||||||||||||||||||||||

||||||||||||||||||||||||||||||||||||||||||||||||||||||||||||||||||

Stem cell research leads to a change in perspective

||||||||||||||||||||||||||||||||||||||||||||||||||||||||||||||||||

This causes a reversed perspective that is much more difficult to deal with than the classical method, working backwards along a chain of causes. Within the realm of stem cell research, adult neurogenesis appears to be something like an island of classical perspective in a sea of contexts that fall apart. Understanding of this circumscribed process will allow insights into a strange new world. But we should not deceive ourselves—even the old view was not straightforward. High-dimensional regulatory networks with seemingly fuzzy or chaotic properties within a cell and on a systems level have always made it impossible to speak in a strict and simple sense of "causes." Therefore, what on the surface we perceive as a paradigm shift is actually a shift in perception itself and in awareness. And while it is true that adult neurogenesis is a scientific question very much at the heart of this type of new scientific problem, it is also obvious that what seemed to be a footnote to neurobiology has rightfully become one of the most intriguing questions and hottest topics in science today.

What Is Adult Neurogenesis?

- Adult neurogenesis is the production of new, functional neurons in the adult brain.
- Adult neurogenesis is neuronal development under the otherwise non-neurogenic conditions of the adult brain.
- Adult neurogenesis the exception, not the rule.
- Adult neurogenesis is a process, not an event.

In the older literature, "neurogenesis" was often equated with precursor cell division leading to neuronal development only

The most general definition of adult neurogenesis is: Adult neurogenesis is the production of new functional neurons in the adult brain (Fig. 1–2). The term comprises a complex process, beginning with the division of a precursor cell (or potentially even before) and ending with the existence of a new, functioning neuron (Fig. 1–3). Neurogenesis is thus much more than just the division of a precursor cell. In the literature on embryonic and fetal neurogenesis, one still finds the term *neurogenesis* used for an event on the level of the precursor cell, most notably its division. Because under the conditions of the adult brain the division of the precursor cell alone is not at all indicative of whether in the end a new neuron will exist, and because regulation of the result largely occurs well after the division of the precursor cell, this use of the term has been deemed too narrow (Kempermann et al., 1997).

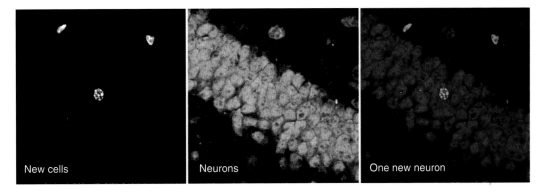

New cells

Neurons

One new neuron

FIGURE 1–2 New neurons in the adult brain. Both newly generated cells (first image) and neurons (second image) can be visualized with immunohistochemistry. The third image shows the co-localization of both markers in one granule cell of the adult dentate gyrus. The method of "birth-dating" cells and thereby marking them as new is based on the incorporation of a tagged false base (bromodeoxyuridine, BrdU) into the DNA of dividing stem or progenitor cells (see Chap. 7 for further explanation). Image kindly provided by Klaus Fabel, Dresden.

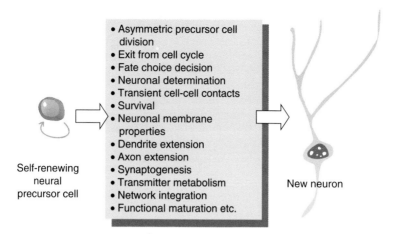

- Asymmetric precursor cell division
- Exit from cell cycle
- Fate choice decision
- Neuronal determination
- Transient cell-cell contacts
- Survival
- Neuronal membrane properties
- Dendrite extension
- Axon extension
- Synaptogenesis
- Transmitter metabolism
- Network integration
- Functional maturation etc.

Self-renewing neural precursor cell

New neuron

FIGURE 1–3 Neurogenesis is a process, not an event. In the context of neuronal development in the adult brain, the term *neurogenesis* has undergone an important shift in meaning. Whereas the old usage of the word, influenced by the conditions in the developing nervous system, tended to equate proliferation of neural precursor cells with neurogenesis, this limited use of the word is not helpful in the context of adult neurogenesis. Adult neurogenesis is regulated at many different stages of cell development, most of the newly generated cells die, and gliogenesis appears to originate from the same proliferating precursor cells. Here the term *neurogenesis* thus comprises all necessary steps, starting with division of a precursor cell and resulting in the existence of a functionally fully integrated new neuron.

Adult neurogenesis needs to be seen as a complex process

Adult neurogenesis as we understand it here is thus a process, not an event. It involves decisions at the precursor cell level, such as whether symmetrical divisions (with identical daughter cells) or asymmetrical divisions (with two different daughter cells) occur. After asymmetrical division, various factors determine when the progeny is to become a neuron. Neurogenesis includes securing the survival of the new cell because only part of the daughter cells survives. Neurogenesis involves migration of the differentiating progeny and the differentiation process itself—that is, the turning on and off of genes according to a neuronal profile. It entails sending out the cellular processes: dendrites and axons that make connections and form synapses. It involves the fulfillment of electrophysiological criteria of neuronal function: the presence of sodium currents and the ability to generate action potentials. And there are even more steps and criteria we could rightfully subsume to "neurogenesis." In everyday research and in most contexts, all of these criteria cannot and need not be demonstrated. It is nonetheless important to realize that in our discussion of neurogenesis, all of these steps and criteria are meant.

Neurogenesis in vitro is an abstraction from the in vivo perspective

For the purpose of this book, adult neurogenesis is essentially considered a process *in vivo*. This is in accordance with how the term is widely used, but of course the development of new neurons can be studied in cell culture, and this has been done very successfully. Hynek Wichterle, Tom Jessell, and coworkers (Wichterle et al., 2002; Wichterle and Peljto, 2008), for example, have been able to recapitulate all of the many regulatory steps that guide embryonic stem cells toward becoming motor neurons. Similar strategies have been used to generate dopamine-producing neurons from embryonic stem cells, and considerable hope is placed on the idea that successfully differentiated cells might be used for cell replacement therapies (Lee et al., 2000; Kim et al., 2002; Martinat et al., 2006).

In some sense, such experiments are perhaps the most straightforward way to demonstrate "neurogenesis." The approach taken by these groups is a prime example of what is possible in controllable and reductionistic experimental systems. But of thousands of neuronal cell types, at best a handful has yet been generated in vitro. And the question always remains: does the cell "created" in vitro truly correspond to the desired cell type in vivo? It has been extremely hard, for example, to define what constitutes a dopaminergic neuron, the desired target cell for cell therapy of Parkinson disease.

Adult brain regions capable of neurogenesis are called "neurogenic" and provide stem cells plus a permissive microenvironment

A brain area that can generate neurons is called a "neurogenic" region. *Neurogenic* implies two things: first, the presence of immature precursor cells from which new neurons can develop, and second, a microenvironment that permits neurogenesis to occur. The latter is important, and we will later see the strengths and limitations of this definition.

In the adult brain of rodents and primates, there are two known neurogenic regions, the hippocampus and the olfactory system. In these two regions a sizable amount of adult neurogenesis can be found (Fig. 1–4). In the hippocampal dentate gyrus, new granule cells are generated, and in the olfactory bulb, new interneurons are produced in the granule cell layer and the periglomerular region (Fig. 1–5). We refer to the rest of the brain as "non-neurogenic," although this categorization might ultimately turn out to be premature (Fig. 8–1). As we will discuss, there is an increasing number of publications that have quite successfully challenged the view that most of the brain is non-neurogenic even in rodents (Gould et al., 1999; Arvidsson et al., 2002; Parent et al., 2002; Dayer et al., 2005; Ohira et al., 2010). Although *grosso modo* the distinction still holds, several issues remain to be resolved. And the quite categorical concept deserves to be constantly challenged. There are considerable species differences, and even among mammals large differences exist (see Chap. 11, p. 496). The distinction of *neurogenic regions*, in which adult neurogenesis occurs physiologically and at a measurable level, from *regions*, in which neurogenesis might not be outright impossible (dependent on species, situation, age, and other factors) is still useful, but is a construct only. We thus refer to the two neurogenic regions described for rodents and primates as "canonical" or "conventional" to elucidate this fact.

The hippocampus and olfactory bulb are the neurogenic regions in rodents and primates

Neurogenic permissiveness is the umbrella term for all cell-extrinsic factors necessary for adult neurogenesis to occur

The key idea of the current concept of a neurogenic versus non-neurogeneic region is that there is presence or absence of "neurogenic permissiveness"; that is, whether or not the local microenvironment physiologically allows an endogenous or implanted neural precursor cell to develop into a neuron. A neurogenic region provides a neurogenic environment. The presence or absence of precursor cells alone is not sufficient, in this view. So far, transplantation into the adult brain has yielded neurons only in the olfactory system and the hippocampus, although the same precursor cells could give rise to neurons under appropriate in vitro conditions (Gage et al., 1995; Suhonen et al., 1996; Herrera et al., 1999; Shihabuddin et al., 2000; Lie et al., 2002). But overall, only a few transplantation experiments have been done.

Many claims of adult neurogenesis in non-neurogenic regions raise methodological concerns

There are numerous single reports of adult neurogenesis occurring outside the hippocampus and the olfactory system, but none of these reports has passed the stage of being anecdotal. One study that stirred excitement during the decade of the brain, reporting neurogenesis in the neocortex of adult primates (Gould et al., 1999; Gould et al., 2001), was not confirmed by other laboratories (Kornack and Rakic, 2001a; Rakic, 2002; Koketsu et al., 2003). Similarly, no evidence for (neo-)cortical neurogenesis has been found in adult mice (Magavi et al., 2000; Ehninger and Kempermann, 2003). Nevertheless, there is convincing evidence that neurogenesis in Layer VI of the cortex persists for relatively long periods postnatally (Inta et al., 2008) and might be found on an extremely low level in adult and aged animals (Dayer et al., 2005). These cells supposedly originate from the SVZ. There are also suggestive findings of ischemia-induced neurogenesis of neocortical interneurons originating from precursor cells in Layer I (Ohira et al., 2010), which still need to be confirmed but might fundamentally challenge the current concepts.

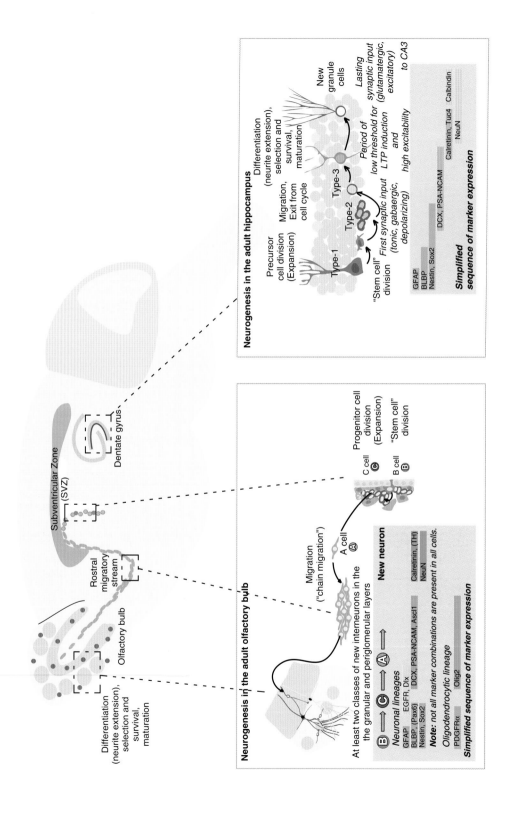

Neurogenesis in the adult hippocampus

Precursor cell division (Expansion)

Differentiation (neurite extension), selection and survival, maturation

"Stem cell" division

Migration, Exit from cell cycle

New granule cells

Type-1

Type-2

Type-3

First synaptic input (tonic, gabaergic, depolarizing)

Period of low threshold for LTP induction and high excitability

Lasting synaptic input (glutamatergic, excitatory) to CA3

GFAP
BLBP
Nestin, Sox2

DCX, PSA-NCAM

Calretinin, Tuc4   Calbindin
NeuN

Simplified sequence of marker expression

Subventricular Zone (SVZ)

Dentate gyrus

Rostral migratory stream

Olfactory bulb

Differentiation (neurite extension), selection and survival, maturation

Progenitor cell division (Expansion)

"Stem cell" division

C cell (C)

B cell (B)

Migration ("chain migration")

A cell (A)

Neurogenesis in the adult olfactory bulb

At least two classes of new interneurons in the the granular and periglomerular layers

(B) ⟷ (C) ⟷ (A)   **New neuron**

Neuronal lineages
GFAP   EGFR, Dlx   Calretinin, (TH)
BLBP, (Pax6)   DCX, PSA-NCAM, Ascl1   NeuN
Nestin, Sox2

Note: *not all marker combinations are present in all cells.*

Oligodendrocytic lineage
PDGFRα   Olig2

Simplified sequence of marker expression

8

FIGURE 1–4 The two neurogenic regions in the adult brain. Only two regions of the adult mammalian brain appear to be neurogenic under physiological conditions. Precursor cells residing in the lateral walls of the lateral ventricle give rise to new interneurons in the olfactory bulb. Neurogenesis in the adult hippocampus generates new excitatory granule cells throughout life. Both forms of adult neurogenesis originate from different precursor cell populations, are independently regulated, and serve entirely different functions. See Chapter 8 for details on the distinction between neurogenic and non-neurogenic regions, the key criteria for this distinction, and also their limitations.

Right side: Adult hippocampal neurogenesis. Neural precursor cells reside in the border zone between the granule cell layer and the hilus, the subgranular zone (SGZ). The SGZ is a neurogenic niche that, in addition to containing neural precursor cells, also provides the microenvironment in which neuronal development can occur. Neurogenesis progresses from division of a radial glia–like stem cell over rapid divisions of intermediate progenitor cells to production of post-mitotic immature and mature granule cells. All cues for all stages of development must be present and accessible for the developing neurons. The mature new granule cells extend their dendrites into the molecular layer, where they, like all other granule cells, receive input from the entorhinal cortex and send their axon along the mossy fiber pathway to hippocampal region CA3.

Left side: Adult neurogenesis in the olfactory bulb. Neural precursor cells reside in the subventricular zone, the lateral wall of the lateral ventricles. Stem cells generate transiently amplifying progenitor cells, which generate migratory neural progenitor cells, which travel along the rostral migratory stream to the olfactory bulb, where they terminally differentiate and integrate functionally into the local network. Neurogenesis in the adult olfactory bulb and that in the hippocampus are completely independent processes.

9

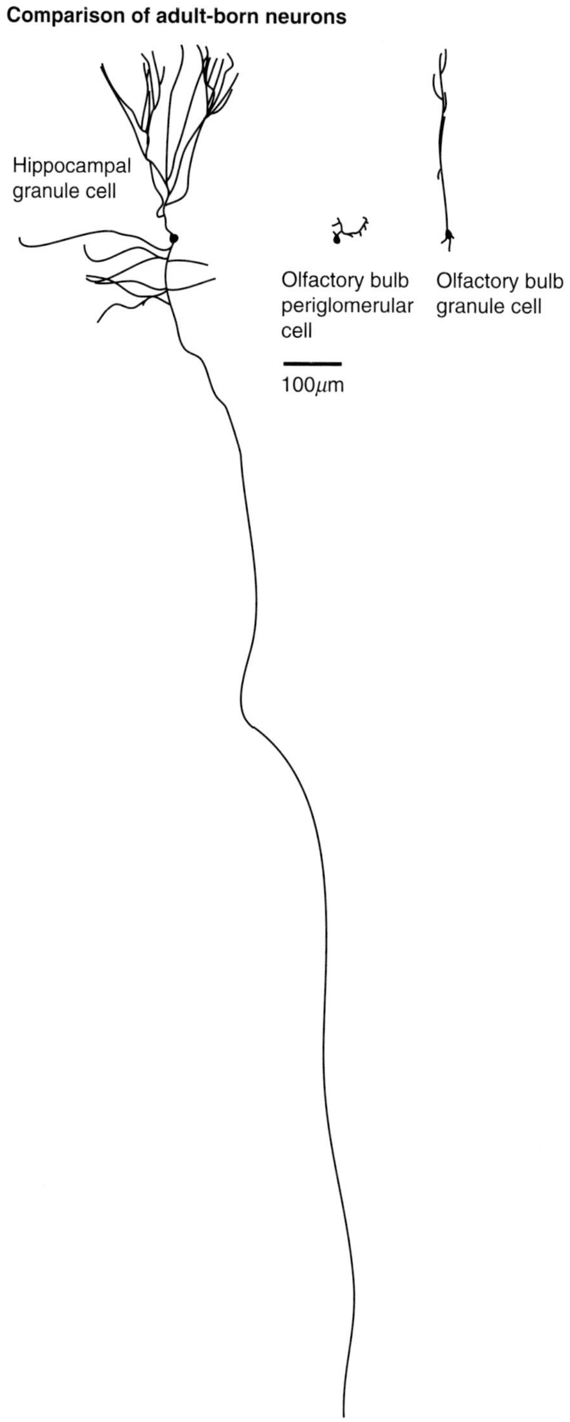

Comparison of adult-born neurons

Hippocampal
granule cell

Olfactory bulb
periglomerular
cell

Olfactory bulb
granule cell

100μm

FIGURE 1–5 The three major types of neurons generated in adult neurogenesis. In adult neurogenesis, excitatory hippocampal granule cells and inhibitory interneurons are produced in the olfactory bulb, here drawn to scale. New interneurons in the olfactory bulb come in a total of seven known types and are found in the granule cell layer of the bulb (95%) and in the periglomerular regions (5%). The periglomerular interneurons of the olfactory bulb show three neurotransmitter phenotypes (GABA, GABA and dopamine, and glutamate).

To fundamentally prove the
complete absence of adult
neurogenesis in a region is
impossible

The best advice might be that because the adult brain has been good for so many surprises with respect to adult neurogenesis, one should not dismiss such observations prematurely. Technical concerns or extremely low numbers of observations are nonetheless very serious issues. One important thing to consider, however: it seems that under pathological conditions, more is possible than normally. There is reasonable evidence of induced adult neurogenesis in neurogenic and non-neurogenic regions as a response to brain damage, most notably ischemia (Arvidsson et al., 2002; Parent et al., 2002; Thored et al., 2006; Thored et al., 2007; Ohira et al., 2010). But this reactive neurogenesis, which will be our topic in Chapters 8 and 12, seems to occur on a minute scale. The term non-neurogenic should be taken with a grain of salt: non-neurogenic regions are brain areas that under normal conditions do not show overt signs of neurogenesis, as in the hippocampus and olfactory system, but might do so under special conditions. The litmus test remains the implantation experiment.

This book is mostly on adult
neurogenesis in mammals

Thus, to refine the definition for this book, adult neurogenesis is the generation of new neurons in the central nervous system of adult rodents and primates. We will touch only briefly on neurogenesis in invertebrates and lower vertebrates, including birds, although adult neurogenesis in songbirds has played a particular role in getting the field started (see Chap. 2, p. 39). These topics will be dealt with according to their relevance to the mammalian situation, not with the intent to provide extensive coverage. For many decades, preceding the first reports of adult neurogenesis in mammals, adult neurogenesis in non-mammalian species, including birds, has hardly been controversial. It is not the phenomenon of new neurons in the adult brain that has stirred controversy, but the question of to which extent the same could occur in mammals, and especially, humans.

In the experimental context,
"adult" often means "adolescent"

Adult here means at the earliest stage that mammals can reproduce. In female mice, this is around day 28 to 30 postnatally (Safranski et al., 1993). Consequently, adulthood includes puberty! This is also the most wide-spread understanding of the word across cultures, but nevertheless is not identical to how many (including most parents when dealing with their children) would like the word to be used. For them, adulthood is defined as the phase in life after growth has ended and the organism is "mature." By its very nature, however, adult neurogenesis defies such definition of adulthood because adult neurogenesis proves that in some sense development never ends.

Postnatal neurogenesis might be
distinct from both embryonic
and adult neurogenesis

The definition of adulthood as beginning with puberty is important because a large part of brain development occurs postnatally but before puberty. Although *adult* is of course after birth, the term *postnatal* in this context implies "after birth but before the ability to sexually reproduce." The neurons of the cerebellum, for example, are generated almost entirely after birth—in humans, as late as 12 years of age. The hippocampus, too, is to a large degree formed postnatally. But this postnatal neurogenesis can be clearly set apart from adult neurogenesis. Postnatal neurogenesis is a direct continuation of embryonic and fetal neurogenesis. Adult neurogenesis is a very restricted process that occurs in a cellular environment that otherwise has terminated its development. Postnatal neurogenesis is the rule for many brain parts; adult neurogenesis is the absolute exception.

Adult neurogenesis in the
peripheral nervous system
comprises the olfactory
epithelium, which is tightly
linked to neurogenesis in the
olfactory bulb

Irrespective of these definitions, it seems that in terms of the structural requirements in the neurogenic niches (see Chap. 8) "adult" neurogenesis starts well before "adulthood"; namely, when postnatal neurogenesis with its specific conditions has given way to the new structures allowing adult neurogenesis. It seems that in mice this change occurs between two and three weeks after birth.

Adult neurogenesis also occurs in the peripheral nervous system, most notably in the olfactory epithelium, which will be covered in Chapter 8, p. 304. The olfactory epithelium lies outside the CNS, but the axons of the receptor neurons, which routinely are generated anew throughout life, project to the olfactory bulb inside the CNS. Adult neurogenesis of the olfactory epithelium might thus constitute a functional unit with adult neurogenesis in the olfactory bulb. Also, adult neurogenesis of the olfactory epithelium shares many similarities with cerebral adult neurogenesis and it generates massive numbers of new neurons, many more than in any other neurogenic region. This feature alone makes it interesting as a model system for adult neurogenesis in general.

Adult Neurogenesis in the Hippocampus and Olfactory System

Adult hippocampal neurogenesis originates from precursor cells in the subgranular zone and generates excitatory granule cells

In the adult hippocampus, a population of precursor cells resides in the subgranular zone (SGZ) (Fig. 1–4). For earlier anatomists, the area that today is referred to as SGZ was simply the border between the granule cell layer of the dentate gyrus and the hilus (or plexiforme layer, or CA4). The term *subgranular zone* was coined by Joseph Altman in a study on neurogenesis in the cat in 1975 (Altman, 1975). The progeny of these continuously dividing cells migrate varying distances into the granule cell layer. They then extend their dendrites, as do all other granule cells, into the molecular layer and send an axon along the mossy fiber tract to area CA3, the projection area of the granule cells. Within a couple of weeks and by all known standards, the new granule cells become virtually indistinguishable from their older siblings. The entire process of development is regulated in an activity-dependent fashion, and numerous individual factors have been identified that can influence adult neurogenesis. Adult hippocampal neurogenesis has been plausibly linked to the plasticity of hippocampal learning and memory processes across the lifespan.

In adult olfactory bulb neurogenesis, the precursor cells reside in the lateral wall of the lateral ventricles and produce various types of interneurons in the bulb

In the adult olfactory system, the population of precursor cells is found in the subventricular zone (SVZ) in the temporal walls of the lateral ventricles (Fig. 1–4). Some authors refer to the SVZ, arguably more correctly, as the "subependymal zone" (SEZ). Note that in adult neurogenesis the term *SVZ*, which is very precisely defined as the zone below the ependymal layer in the context of fetal corticogenesis, is now largely used as name for the entire neurogenic region of the lateral walls of the lateral ventricles, including the ependymal layer. In neurogenesis originating from the SVZ, progeny migrate over a long distance along the *rostral migratory stream* to the olfactory bulb, where they differentiate into interneurons—one population in the granule cell layer of the bulb and two populations with different neurotransmitter phenotypes in the periglomerular layer. Manipulation of the olfactory system influences neurogenesis in the olfactory bulb, and first theoretical considerations have tried to place neurogenesis in this system into a functional context.

The two regions with adult neurogenesis show many differences

Even at this superficial level of description, numerous differences between neurogenesis in the hippocampus and that in the SVZ become obvious (see Tables 8–2 and 10–1 for more details). Most dramatic is the quantitative difference. In rodents, adult neurogenesis is many-fold higher in the olfactory bulb than in the dentate gyrus. In humans it might be the other way around (Sanai et al., 2004), possibly due to the lesser functional relevance of olfaction in humans. But another study suggests that adult human olfactory neurogenesis might not be as rare as assumed (Curtis et al., 2007). In any case, qualitative differences might be more interesting; functionally, very distinct types of neurons are produced.

Although many questions on adult neurogenesis have been answered, many remain open

Open Questions in Adult Neurogenesis

Despite thousands of papers and more than 15 years of increasingly intense research, a great number of questions on adult neurogenesis remain open.

It is no wonder that so many people have become interested in adult neurogenesis. That the mammalian brain can make new neurons, but under normal conditions does so only in a few privileged regions and otherwise apparently neglects this potential for regeneration, is a stunning phenomenon. This is clearly still a blank spot on the map of neuroscience— *"Hic sunt leones"* ("Here are the lions"), as the medieval cartographers wrote when they did not know how to fill in the map. Today it is often "Here are the stem cells" when we do not know either. But at least we have become much better at asking the questions. The fact that we still do not know that much does not imply that we will not know anything. The field is literally exploding.

Adult neurogenesis is also a model system for cell and tissue engineering

Adult neurogenesis is scientifically interesting for two reasons that to some degree are independent of each other: as phenomenon itself and as model for something else. Increasingly, further progress in research on adult neurogenesis will require interdisciplinary approaches. It was only a decade ago that everybody stood in awe, realizing that there are indeed new neurons in the adult brain. But the days of sheer phenomenology are clearly over. To address the full range of questions will require contributions from numerous disciplines. In the context of brain function in health and disease, the ability of the brain to make new neurons is important. At the same time, adult neurogenesis can be used as a model system for investigating central questions in (applied) stem cell biology.

The focus, then, is to study factors controlling the development of new neurons under conditions prevailing in the adult brain and discern what has to be achieved if stem cells or immature neurons are implanted in the brain. How much of adult neurogenesis lies in the precursor cells, and how much is dictated by the cellular microenvironment? What is the nature of the stem and progenitor cells underlying adult neurogenesis? And how large is the contribution of the local milieu, the neurogenic "niche"? What defines a neurogenic region as neurogenic? For most of the young history of neurobiology, the adult brain was essentially considered to be non-neurogenic, if not anti-neurogenic. But the adult brain routinely does exactly what tissue engineers intend to do in (stem) cell-based therapy for the brain: make new neurons.

Adult neurogenesis is a late-discovered neurobiological phenomenon of surprising importance

But how does the brain do it? And why does it go to all the trouble of maintaining this complex machinery? What is adult neurogenesis good for? What is special about the new neurons besides their later birth? How are they integrated into the existing networks? How could do they contribute to brain function? Why is adult neurogenesis restricted to only few brain regions? Why would the neurogenic regions of the brain rely on a mechanism to produce new neurons, whereas the rest of the brain seems to do perfectly fine without it? How does the brain "know" that more neurons are needed, and how is this demand translated into signals a precursor cell can "understand"? How is adult neurogenesis regulated on the various levels from behavioral down to molecular? Which genes govern adult neurogenesis, and are any of them specific to *adult* neurogenesis? What is the relationship between adult neurogenesis and embryonic neurogenesis anyway? How do the many messenger systems and signaling cascades interact to control neurogenesis?

Has adult neurogenesis anything to do with regeneration? Or is it at least a futile attempt at this? And vice versa: what happens if adult neurogenesis fails? Could the failure of adult neurogenesis have anything to do with degeneration? Are there neuroplastic and other stem cell disorders of the brain? Can adult neurogenesis be stimulated and employed to improve regeneration?

Adult neurogenesis calls for inter- and transdisciplinary approaches

Adult neurogenesis is difficult to grasp, because it draws from many areas of neuroscience. This interdisciplinary nature of the field makes research on adult neurogenesis fascinating because it enables researchers to take a fresh look at many long-known phenomena of neurobiology. How, for example, does the existence of adult hippocampal neurogenesis influence concepts of how the hippocampus functions? None of the older theories of learning and memory (and other hippocampal functions) had to consider the fact that the underlying network could be altered by adult neurogenesis. So how did pioneer of research on learning and memory and Nobel laureate Eric Kandel perceive the appearance of "adult neurogenesis" on the stage of his field? In an interview he gave at the 2008 Neuroforum organized by the Hertie Foundation in Frankfurt, Germany, he answered that question:

> So I think the role of it [adult neurogenesis] in cognitive function is unclear but . . . it would be interesting to incorporate [adult neurogenesis]. . . . I did not see this as challenging what I thought. I thought of it as an important new dimension but even if it did challenge what I did, I would rather like to know than to close my eyes and not know. I think that knowledge is the best we can do at any given time and it will be superseded without question. (Kandel, 2008)

We will see (mostly in Chap. 9) whether it is really true that adult neurogenesis does not challenge some assumptions in the learning and memory field. But it is also evident that, although it might not appear so first, adult neurogenesis is very well compatible with what we know about how the brain works. And it is better to know than not to know.

A Revolution in our Heads?

Adult neurogenesis does not stand outside other fields of neurobiology

The impending integration of new neurons into many older and well-established concepts does not mean that everything in the field will fundamentally change. After all, the adult brain does regenerate poorly and does not seem to make use of the regenerative potential resting in its more or less abundant stem cells. Does it have a good reason for this poor performance, or is this just the path evolution coincidentally took? There is a risk in any new and burgeoning field that it will overestimate its relevance. We should aim for a careful and realistic consideration. The overview given in this book should help with this consideration, to learn what adult neurogenesis is about and, perhaps even more important, what it is not about. I believe, though, that research on adult neurogenesis could make some critical and relevant contributions to a number of fields of neurobiology because it touches on a line of thought that previously was absent: the possibility of plasticity on the level of neuronal numbers. This type of phenomenon simply was not supposed to occur in a good mammalian brain (Rakic, 1985). But in a way, this argument is similar to a famous line from a poem by Christian Morgenstern: "*daß nicht sein kann, was nicht sein darf*" ("that which must not be, cannot be").

In 1965, Joseph Altman was the first to describe adult neurogenesis

With today's knowledge it is easy to wonder about some previous views prevalent in neurobiology that disregarded the mere possibility of plasticity on the level of neurons. Joseph Altman first described adult neurogenesis in the rodent hippocampus in 1965 (Altman and Das, 1965). His landmark study was preceded by two other articles, two and three years earlier, in which he suggested the possibility of adult neurogenesis with somewhat weaker evidence (Altman, 1962; Altman, 1963). He also first described adult neurogenesis in the olfactory bulb (Altman, 1969), preceded by a study by James Hinds, who had presented postnatal neurogenesis in the olfactory bulb until the age of postnatal day 20 (Hinds, 1968). Michael Kaplan picked up on neurogenesis in the adult olfactory bulb in 1977, adding electron-microscopic evidence (Kaplan and Hinds, 1977). Even Fernando Nottebohm's exciting work in the 1980s on adult neurogenesis in songbirds (Goldman and Nottebohm, 1983) did not fundamentally change the general perception of adult neurogenesis as some sort of oddity or perhaps evolutionary atavism. Nottebohm stated in a 2001 interview in *The New Yorker*: "The view that neurons in the adult brain come and go was considered the view of a lunatic" (Specter, 2001). This seems to be only partially true, however, because Nottebohm's work on adult neurogenesis in songbirds received widespread attention and appraisal. Still, with respect to the situation in mammals, his remark might capture something of the spirit of that time. For adult mammalian neurogenesis it took more than 30 years after the initial description until, along with the discovery of neural stem cells in the adult SVZ by Brent Reynolds and Sam Weiss (Reynolds and Weiss, 1992) and by Perry Bartlett and colleagues (Richards et al., 1992), as well as their discovery in the hippocampus by Fred H. Gage and his colleagues Jasodarah Ray and Theo D. Palmer (Palmer et al., 1995; Ray et al., 1995), research on adult neurogenesis in mammals found its plausible basis, became widely accepted, and was steered into the mainstream.

Despite its importance and the growing wealth of information, adult neurogenesis remains an exception to the rule

Today, however, the principal and somewhat stubborn skepticism of the earlier days has often given way to the unspoken belief that essentially everything is possible. Out of this enthusiasm many observations are dangerously taken at face value. Not all supposedly new neurons, detected immunohistochemically after ischemia or some other type of lesion, make this a finding of reactive neurogenesis. Such images alone are also not the proof of an exploitable potential for self-repair and regeneration only waiting to be stimulated by some clever drug, leading to restoration of function. Before we can really make such a claim, a number of important questions will have to be answered. And these questions are the same ones that come to mind with respect to physiological adult neurogenesis.

The Myth of the "No New Neurons" Dogma

The reluctance of the scientific community to accept adult neurogenesis is often linked to a "no-new-neurons" dogma, which is, however, of questionable existence

The reluctance of the scientific world to embrace the idea of new neurons in the adult brain is often referred to as obedience to the so-called no-new-neurons dogma. Literally, a dogma is teaching *ex cathedra*; the final word on a fundamental issue about which, when all arguments have been heard, no logical decision can be made. The body of dogmata is the catalog of fundamental sentences in theology, upon which everything else is built, but which by definition can neither be proven nor falsified.

Whereas dogma is from the domain of theology, the natural sciences know axioms. Unlike dogma, axioms do not involve a decision and a binding statement made by an authority. Since the Renaissance, natural science has been built on empiricism, not on a body of thought from the authorities.

In the history of science, many discoveries have taken a long time to become accepted. In hindsight it is tempting to mock the ignorant contemporaries who did not recognize the obvious truth. But this is a dangerous attitude that blocks a clear view of how science works, and it is unfair to criticize those who quarreled and remained skeptical; science is based on skepticism. This does not exclude, unfortunately, the possibility that generally justified skepticism could be misused to suppress the inconvenient truth.

Joseph Altman described adult neurogenesis decades before the scientific world was ripe for this discovery

In some sense, Altman's discovery of adult neurogenesis was a discovery made before its time. Some of the reasons for its lukewarm welcome in the world of science are obvious. The evidence then for new neurons was based on autoradiographically detected grains from tritiated thymidine over cells that had to be morphologically identified as neurons. These were not pyramidal neurons in the cortex or Purkinje cells in the cerebellum, which are clearly identifiable neurons even by plain morphological standards, but "microneurons" and "granule cells." Nevertheless, in the dentate gyrus, which consists essentially only of granule cells, it is hard to argue against the morphology of a granule cell neuron, even though their polarity and other neuronal features are less prominent than in other neuronal cell types. Generally, however, modern glial biology has shown how complex and neuron-like astrocytes might appear. In maintaining the widely accepted standard that extraordinary claims require extraordinary evidence, it was quite in order that even Altman's meticulous studies were at first greeted with healthy skepticism. Nevertheless, the editors of *Science, Nature,* and *The Journal of Comparative Neurology* did publish his work. The influence of his studies remained limited, partially because the finding could not be reconciled with the otherwise available concepts of neurobiology of the time, not because they were suppressed even before they were published. The great anatomists, from His, to Cajal, and Spielmeyer, along with many others, had asked whether neurons could divide, but they had never found evidence of it. And they were right—even today we remain convinced that neurons do not divide. And at the time of Altman's pioneering work, the concept of stem cells in the adult brain did not yet exist. Without stem cells, the idea of adult neurogenesis conflicted with the central principles of neuroscience known in the 1960s as much as they do with those known today. Altman was well aware of this discrepancy and reasoned that "precursor cells" might exist in the brain. There was no authority that proclaimed the dogma of "no new neurons." Rather, the finding was at odds with the knowledge of the time, and even those who found it interesting lacked the technical means to take the issue further.

The pioneers of adult neurogenesis research felt marginalized, and factual critique seems to have been used against them personally

There is, however, a very disturbing aspect to this story. Even though nobody was in the scientific and formal position to issue a "No new neurons" dogma, some people took a dogmatic position and made political use of the fact that neurogenesis could not be proven. In 2001, Michael Kaplan, who in the late 1970s had published a series of studies on adult neurogenesis in which he used electron microscopy to prove the neuronal nature of the newly generated cells, wrote a commentary on his personal history in the field of adult neurogenesis: "But in any revolution, whether political or scientific, there are crusades and battles: not all are winners. In the midst of a revolution one must choose allegiance, and during the 1960s and 1970s, those who chose to support the notion of neurogenesis in the adult brain were ignored or silenced" (Kaplan, 2001). These are strong accusations but there seems to be truth to them. However, a scholarly historical investigation of this difficult phase of neurogenesis research has yet to be done. The historical account that follows in Chapter 2 of this book is meant to track the development of evidence for adult neurogenesis and cannot do fully justice to this aspect of its history. In any case, the impression remains that the two leading figures of the field at the time, Altman and Kaplan, received too many discouraging signals from the scientific community to continue their exciting work. Something obviously went terribly wrong here.

Pasko Rakic's influential 1985 article, "Limits of Neurogenesis in Adult Primates," also published by *Science*, is usually considered the key ex cathedra writing that, as one commenter phrased it, "single-handedly held the field of neurogenesis back by at least a decade" (Specter, 2001). But although Rakic has been and still is the most outspoken and most influential skeptic of adult neurogenesis, the issue cannot and should not be tied just to his person and this one paper. The arguments discussed in Rakic's paper had to be raised against far-reaching conclusions from Altman's, Kaplan's, and even Nottebohm's data. If such generally valid skepticism, which is no proof of the opposite anyway, was used against the people, not their data, (and this is what seems to have happened) then this is certainly not in order. The stem cell field of today, however, with its disturbingly quick pushes into clinical applications, clearly shows how easy it is to get carried away by suggestive ideas derived from premature extrapolation of single scientific observations. Tellingly, when new tools became available, even Rakic was ready to admit to the idea that adult neurogenesis occurs in the neurogenic zones of the adult brain, despite remaining very skeptical about its functional relevance (Kornack and Rakic, 1999; Kornack and Rakic, 2001b). Many arguments raised in 1985 remain valid and important. In 2005 there is still no convincing evidence of physiological neurogenesis in most parts of the brain, and the dilemma between stability and plasticity applied by Rakic in his comment on neuronal development and plasticity remains one of the fundamental issues in neuronal network theory.

High standards of critical thinking need to be maintained

Finally, and thereby returning to the quote from Eric Kandel, science is not about being 100% right every time. To the contrary, it is a responsibility of the community to raise arguments against the impressions elicited by a new observation, even if one might turn out to be wrong in this skepticism. Opportunism in science is as bad as it is everywhere else. Enthusiasm about a scientific topic should never lead to a watering down of the standards. It is no coincidence that research on adult neurogenesis took off only when the link to neural stem cells could be made and when immunohistochemistry, confocal microscopy, and stereological quantification tools enabled a clearer picture of what was going on in the neurogenic zones.

Around the year 2000, we saw a trend in neural stem cell biology that illustrates the issue. When it was reported that adult stem cells might be able to cross organ and germ layer borders so that "blood" could make "brain" and "brain" could make "blood," these findings were received with skepticism by some (Anderson et al., 2001) but not most investigators. Quite mysteriously, these findings were hardly questioned, by and large. They immediately found their way into scientific and folkloristic visions of the medicine of the future: with stem cells, everything would be possible. However, later data strongly supported the critics, and although it cannot be categorically ruled out that transdifferentiation is principally possible (just as it is impossible to categorically rule out some neurogenesis in non-neurogenic regions), it is not what it first seemed to be. The rather blind surrender to a suggestive idea on the basis of limited evidence turned out to be the wrong choice.

The scientific community's opinion on adult neurogenesis took the slow path that many new ideas in science must take. There is nothing inherently wrong with this. One must argue about style here and there, and there appears to have been misuse of scientific arguments to influence research politics, but the general course along which science progressed in this case is not refutable. Some ideas, such as the double helix structure of the DNA, find their terrain well prepared. When they are first brought up, their explanatory power is so extensive that many open questions are immediately resolved; other important discoveries, however, are made well ahead of the questions to which they provide the answers.

The myth of the no-new-neurons dogma should be abandoned

We should thus bury the no-new-neurons dogma because it is misleading: nobody spoke *ex cathedra* that there are no new neurons, and the apparent lack of neurogenesis in the adult brain has never been an axiom. It simply reflected the state of knowledge at the time. The rest of the story was human nature—moving too slowly in some places, too quickly in others.

What to Expect in This Book

Chapter 2: History

Chapter 2 gives a historic account and summarizes the development of the concept of *plasticity*, the fundamental capacity of the brain to alter its structure in response to function. This discussion might at first glance seem to be a typical historical digression, but this is hardly the case. Adult neurogenesis is a plastic event that takes place in the context of other signs of plasticity. Like other instances of plasticity, neurogenesis is intricately linked to function; thus, it does not make sense to view adult neurogenesis in isolation. The concept of plasticity sets the stage for the description of how adult neurogenesis was discovered and how it was discussed.

Chapter 3: Stem cells

In Chapter 3 we describe the *conditio sine qua non* of adult neurogenesis: stem or progenitor cells with the potential to generate new neurons. The term *stem cell* has suffered inflationary use in recent years. A clear picture of what is actually meant by "stemness" is important to the understanding of some fundamental principles underlying adult neurogenesis. This endeavor is not at all trivial, as the discussion of numerous conflicting concepts of stem cells in the adult brain will show. Although many important things can be said about them, stem cells of the brain still remain rather elusive creatures.

Chapter 4: Neuronal development

Chapter 4 reviews the more general aspects of brain development. Neural stem cell biology is essentially developmental biology, but developmental neurobiology is also to a large degree stem cell biology. Adult neurogenesis is neuronal development under the conditions of the adult brain. To understand to what degree adult neurogenesis is a recapitulation of embryonic and fetal neurogenesis and is an independent process with its own rules and mechanisms, a review of some principles of neural stem cell biology in brain development is necessary. We will place adult neurogenesis in the context of neuronal development in the embryonic and early postnatal brain. What is similar and what is different between these two forms of neuronal development? Whereas embryonic development occurs in a microenvironment that is itself developing, adult neurogenesis has to proceed in a cellular milieu that is in general hostile to neuronal development. Adult neurogenesis has to implement and maintain its own permissive microenvironment. Also, embryonic neuronal development is massively parallel, allowing us to determine with relative precision which brain parts are generated at which gestational time point. In the adult, however, all stages of neuronal development can be found next to each other. How can regulatory cues address the correct target cells only?

Chapters 5 and 6: Neurogenesis in the adult hippocampus and olfactory bulb

Chapter 7: Methods

Chapters 5 and 6 will give a detailed description of what is known about neurogenesis in the adult mammalian hippocampus and olfactory system. The emphasis here is on analysis of the "naked" process itself, not yet its regulation or function, which will be covered extensively in Chapters 9 and 10. Still, thoughts about plasticity (from Chap. 2) and stem cells (from Chap. 3) allow determination of where regulation sets in.

Chapter 7 deals with the experimental techniques used to investigate adult neurogenesis. There are several principal issues and technical pitfalls that have to be considered when designing experiments on adult neurogenesis and analyzing the data. The selection of marker proteins to identify different populations of cells is crucial, as is the proper use of confocal microscopy to examine the expression of these marker antigens in vivo. Other important questions are how adult neurogenesis can be quantified and how it can be compared between different experimental conditions. Many of the problems in interpreting seemingly contradictory findings on adult neurogenesis arise at least in part from methodological problems.

Chapter 8: Neurogenic and non-neurogenic regions

With that in mind, Chapter 8 will turn to a discussion of adult neurogenesis outside the "canonical" neurogenic regions. For this we will revisit the concept of the *stem cell niche* as the microenvironment that enables precursor cell activity and neurogenesis. How do neurogenic regions differ from non-neurogenic ones? Does adult neurogenesis physiologically occur in the cortex and elsewhere outside the hippocampus and the olfactory system? This issue has been discussed controversially and is to some degree a debate on methods. But there is more to the discussion than methodological issues. The previous chapters feed into hypotheses on what might be happening in the adult cortex.

Chapter 9: Regulation

Chapter 9 discusses the regulation of adult neurogenesis. Factors that have been shown to affect adult neurogenesis are discussed here. The years since the first edition of this book have yielded a large body of literature on the topic, and we can dare now to attempt some first syntheses. The chapter will discuss the deduction of some very fundamental principles and the development of a number of testable hypotheses.

Chapter 10: Function

Chapter 10 finally addresses the function of adult neurogenesis. Function can be considered on different conceptual levels. *Function* here can mean the function of the individual new neuron or of that same new neuron but within the context of its neuronal network, as well as the relevance of the entire process of neurogenesis for brain function, cognition, and behavior.

Chapter 11: Evolution

Chapter 11 is on the comparative biology of adult neurogenesis. How much can we learn about adult neurogenesis as it is relevant to humans by studying the fruit fly and zebra fish? Adult neurogenesis is well preserved throughout evolution, but there are some fundamental changes between adult neurogenesis in rodents and that in other vertebrates. Why can lizards regenerate entire brain parts, but "higher animals" cannot? Is adult neurogenesis advantageous or disadvantageous? Is it just an alternative mechanism, the use of some principle originally developed for something else, or something that has been shaped under direct evolutionary pressure? We will discuss adult neurogenesis in songbirds in some detail and consider how much we know about adult neurogenesis in humans.

Chapter 12: Medicine

The final chapter (12) deals with medical implications of adult neurogenesis. What happens if adult neurogenesis fails? What role does adult neurogenesis play in the pathogenesis of neurological disorders? We will also discuss the therapeutic consequences that might arise from such a role and the lessons that can be learned from adult neurogenesis for neurotransplantation.

A Hypothesis

||||||||||||||||||||||||||||||||||||||||||||||||||||||||||

Adult neurogenesis proves that brain development never ends

||||||||||||||||||||||||||||||||||||||||||||||||||||||||||

The general hypothesis underlying this book is that adult neurogenesis is a particularly prominent manifestation of a far more general principle of neurobiology: the idea that brain development never ends and that plasticity can be taken as continued development. As we will see, quantitatively, adult neurogenesis is minute over the largest periods of adulthood. Most of what we call "adult neurogenesis" actually occurs fairly early in life. But the point is that it never seems to end and that it remains regulated by activity and many other factors. Because neurons are involved, adult neurogenesis receives and deserves particular interest. But adult neurogenesis also needs to be seen in the context of other aspects of cell genesis throughout life and in connection with the many other aspects of plasticity—for example, on the level of neurites and synapses or in glial cells. The new neurons themselves serve a specific purpose and fulfill a particular functional need that could not be met by means of other types of plasticity. New neurons might be an exception, but they still contribute to higher brain function in a very significant way. Adult neurogenesis is not so much about regeneration as about looking at brain function across the lifespan with a new perspective.

References

Altman J (1962). Are new neurons formed in the brains of adult mammals? *Science* 135:1128–1129.

Altman J (1963). Autoradiographic investigation of cell proliferation in the brains of rats and cats. *Anat Rec* 145:573–591.

Altman J (1969). Autoradiographic and histological studies of postnatal neurogenesis. IV. Cell proliferation and migration in the anterior forebrain, with special reference to persisting neurogenesis in the olfactory bulb. *J Comp Neurol* 137:433–457.

Altman J (1975). Postnatal development of the hippocampal dentate gyrus under normal and experimental conditions. In: *The Hippocampus* (Isaacson RL, Pribram KH, eds.), pp. 95–122. New York: Plenum Press.

Altman J, Das GD (1965). Autoradiographic and histological evidence of postnatal hippocampal neurogenesis in rats. *J Comp Neurol* 124:319–335.

Anderson DJ, Gage FH, Weissman IL (2001). Can stem cells cross lineage boundaries? *Nat Med* 7:393–395.

Arvidsson A, Collin T, Kirik D, Kokaia Z, Lindvall O (2002). Neuronal replacement from endogenous precursors in the adult brain after stroke. *Nat Med* 8:963-970.

Blakeslee S (2000). A decade of discovery yields a shock about the brain. In: *New York Times*, p 1.

Curtis MA, Kam M, Nannmark U, Anderson MF, Axell MZ, Wikkelso C, Holtas S, van Roon-Mom WM, Bjork-Eriksson T, Nordborg C, Frisen J, Dragunow M, Faull RL, Eriksson PS (2007). Human neuroblasts migrate to the olfactory bulb via a lateral ventricular extension. *Science* 315:1243–1249.

Dayer AG, Cleaver KM, Abouantoun T, Cameron HA (2005). New GABAergic interneurons in the adult neocortex and striatum are generated from different precursors. *J Cell Biol* 168:415–427.

Ehninger D, Kempermann G (2003). Regional effects of wheel running and environmental enrichment on cell genesis and microglia proliferation in the adult murine neocortex. *Cereb Cortex* 13:845–851.

Eriksson PS, Perfilieva E, Bjork-Eriksson T, Alborn AM, Nordborg C, Peterson DA, Gage FH (1998). Neurogenesis in the adult human hippocampus. *Nat Med* 4:1313–1317.

Gage FH, Coates PW, Palmer TD, Kuhn HG, Fisher LJ, Suhonen JO, Peterson DA, Suhr ST, Ray J (1995). Survival and differentiation of adult neuronal progenitor cells transplanted to the adult brain. *Proc Natl Acad Sci USA* 92:11879–11883.

Goldman SA, Nottebohm F (1983). Neuronal production, migration and differentiation in a vocal control nucleus of the adult female canary brain. *Proc Natl Acad Sci USA* 80:2390–2394.

Gould E, Reeves AJ, Graziano MS, Gross CG (1999). Neurogenesis in the neocortex of adult primates. *Science* 286:548–552.

Gould E, Vail N, Wagers M, Gross CG (2001). Adult-generated hippocampal and neocortical neurons in macaques have a transient existence. *Proc Natl Acad Sci* USA 98:10910–10917.

Herrera DG, Garcia-Verdugo JM, Alvarez-Buylla A (1999). Adult-derived neural precursors transplanted into multiple regions in the adult brain. *Ann Neurol* 46:867–877.

Hinds JW (1968). Autoradiographic study of histogenesis in the mouse olfactory bulb. I. Time of origin of neurons and neuroglia. *J Comp Neurol* 134:287–304.

Inta D, Alfonso J, von Engelhardt J, Kreuzberg MM, Meyer AH, van Hooft JA, Monyer H (2008). Neurogenesis and widespread forebrain migration of distinct GABAergic neurons from the postnatal subventricular zone. *Proc Natl Acad Sci* USA 105:20994–20999.

Kandel ER (2008). Interview with Eric R. Kandel: From memory, free will, and the problem with Freud to fortunate decisions. *J Vis Exp.* 24:pii 762

Kaplan MS (2001). Environment complexity stimulates visual cortex neurogenesis: Death of a dogma and a research career. *Trends Neurosci* 24:617–620.

Kaplan MS, Hinds JW (1977). Neurogenesis in the adult rat: Electron microscopic analysis of light radioautographs. *Science* 197:1092–1094.

Kempermann G, Kuhn HG, Gage FH (1997). Genetic influence on neurogenesis in the dentate gyrus of adult mice. *Proc Natl Acad Sci USA* 94:10409–10414.

Kim JH, Auerbach JM, Rodriguez-Gomez JA, Velasco I, Gavin D, Lumelsky N, Lee SH, Nguyen J, Sanchez-Pernaute R, Bankiewicz K, McKay R (2002). Dopamine neurons derived from embryonic stem cells function in an animal model of Parkinson's disease. *Nature* 418:50–56.

Koketsu D, Mikami A, Miyamoto Y, Hisatsune T (2003). Nonrenewal of neurons in the cerebral neocortex of adult macaque monkeys. *J Neurosci* 23:937–942.

Kornack DR, Rakic P (1999). Continuation of neurogenesis in the hippocampus of the macaque monkey. *Proc Natl Acad Sci USA* 96:5768–5773.

Kornack DR, Rakic P (2001a). Cell proliferation without neurogenesis in adult primate neocortex. *Science* 294:2127–2130.

Kornack DR, Rakic P (2001b). The generation, migration, and differentiation of olfactory neurons in the adult primate brain. *Proc Natl Acad Sci USA* 98:4752–4757.

Lee SH, Lumelsky N, Studer L, Auerbach JM, McKay RD (2000). Efficient generation of midbrain and hindbrain neurons from mouse embryonic stem cells. *Nat Biotechnol* 18:675–679.

Lie DC, Dziewczapolski G, Willhoite AR, Kaspar BK, Shults CW, Gage FH (2002). The adult substantia nigra contains progenitor cells with neurogenic potential. *J Neurosci* 22:6639–6649.

Lowenstein DH, Parent JM (1999). Brain, heal thyself. *Science* 283:1126–1127.

Magavi S, Leavitt B, Macklis J (2000). Induction of neurogenesis in the neocortex of adult mice. *Nature* 405: 951–955.

Martinat C, Bacci JJ, Leete T, Kim J, Vanti WB, Newman AH, Cha JH, Gether U, Wang H, Abeliovich A (2006). Cooperative transcription activation by Nurr1 and Pitx3 induces embryonic stem cell maturation to the midbrain dopamine neuron phenotype. *Proc Natl Acad Sci USA* 103:2874–2879.

Ohira K, Furuta T, Hioki H, Nakamura KC, Kuramoto E, Tanaka Y, Funatsu N, Shimizu K, Oishi T, Hayashi M, Miyakawa T, Kaneko T, Nakamura S (2010). Ischemia-induced neurogenesis of neocortical layer 1 progenitor cells. *Nat Neurosci* 13:173-179.

Palmer TD, Ray J, Gage FH (1995). FGF-2-responsive neuronal progenitors reside in proliferative and quiescent regions of the adult rodent brain. *Mol Cell Neurosci* 6:474–486.

Parent JM, Vexler ZS, Gong C, Derugin N, Ferriero DM (2002). Rat forebrain neurogenesis and striatal neuron replacement after focal stroke. *Ann Neurol* 52:802-813.

Rakic P (1985). Limits of neurogenesis in primates. *Science* 227:1054–1056.

Rakic P (2002). Neurogenesis in adult primate neocortex: An evaluation of the evidence. *Nat Rev Neurosci* 3:65–71.

Ray J, Raymon HK, Gage FH (1995). Generation and culturing of precursor cells and neuroblasts from embryonic and adult central nervous system. *Methods in enzymology* 254:20–37.

Reynolds BA, Weiss S (1992). Generation of neurons and astrocytes from isolated cells of the adult mammalian central nervous system. *Science* 255:1707–1710.

Richards LJ, Kilpatrick TJ, Bartlett PF (1992). De novo generation of neuronal cells from the adult mouse brain. *Proc Natl Acad Sci USA* 89:8591–8595.

Safranski TJ, Lamberson WR, Keisler DH (1993). Correlations among three measures of puberty in mice and relationships with estradiol concentration and ovulation. *Biol Reprod* 48:669–673.

Sanai N, Tramontin AD, Quinones-Hinojosa A, Barbaro NM, Gupta N, Kunwar S, Lawton MT, McDermott MW, Parsa AT, Manuel-Garcia Verdugo J, Berger MS, Alvarez-Buylla A (2004). Unique astrocyte ribbon in adult human brain contains neural stem cells but lacks chain migration. *Nature* 427:740–744.

Shihabuddin LS, Horner PJ, Ray J, Gage FH (2000). Adult spinal cord stem cells generate neurons after transplantation in the adult dentate gyrus. *J Neurosci* 20:8727–8735.

Spalding KL, Bhardwaj RD, Buchholz BA, Druid H, Frisen J (2005). Retrospective birth dating of cells in humans. *Cell* 122:133–143.

Spalding KL, Arner E, Westermark PO, Bernard S, Buchholz BA, Bergmann O, Blomqvist L, Hoffstedt J, Naslund E, Britton T, Concha H, Hassan M, Ryden M, Frisen J, Arner P (2008). Dynamics of fat cell turnover in humans. *Nature* 453:783–787.

Specter M (2001). Rethinking the brain. In: *The New Yorker*, pp. 42–53.

Thored P, Wood J, Arvidsson A, Cammenga J, Kokaia Z, Lindvall O (2007). Long-term neuroblast migration along blood vessels in an area with transient angiogenesis and increased vascularization after stroke. *Stroke* 38:3032–3039.

Thored P, Arvidsson A, Cacci E, Ahlenius H, Kallur T, Darsalia V, Ekdahl CT, Kokaia Z, Lindvall O (2006). Persistent production of neurons from adult brain stem cells during recovery after stroke. *Stem Cells* 24: 739–747.

Suhonen JO, Peterson DA, Ray J, Gage FH (1996). Differentiation of adult hippocampus-derived progenitors into olfactory neurons in vivo. *Nature* 383:624–627.

Wichterle H, Peljto M (2008). Differentiation of mouse embryonic stem cells to spinal motor neurons. In *Current Protocols in Stem Cell Biology*, Chap. 1: Unit 1H 1 1–1H 1 9.

Wichterle H, Lieberam I, Porter JA, Jessell TM (2002). Directed differentiation of embryonic stem cells into motor neurons. *Cell* 110:385–397.

2

History

Joseph Altman discovered adult neurogenesis in the early 1960s; the groundbreaking paper was published in 1965

ADULT NEUROGENESIS was discovered relatively late in the twentieth century, and research on new neurons in the adult brain took off only amidst a surge in public awareness of brain research in the mid- to late 1990s. The first reports on neurogenesis in the adult hippocampus, however, had come from Joseph Altman in 1962, 1963, and 1965 (Altman, 1962b; Altman, 1963; Altman and Das, 1965a; Altman and Das, 1965b). The first report on neurogenesis in the "truly" adult olfactory bulb (at the age of three months in the rat) was from Michael Kaplan in 1977 (Kaplan and Hinds, 1977) and this is the paper most often cited. But it was again Joseph Altman who had published two studies on postnatal and young-adult neurogenesis in the olfactory bulb in 1965 and 1969 (Altman and Das, 1965a; Altman, 1969). Altman had labeled cells at postnatal day 30 and examined the brains 120 days later.

The discovery of neural stem cells in the early 1990s promoted renewed interest in adult neurogenesis

Despite the relatively large number of publications from Altman and Kaplan as well as a few later, and not as widely known, studies from Maxwell Cowan and his colleagues (Boss et al., 1985; Crespo et al., 1986), and a rather amazing solitary report by Stanfield and Trice on the projection of adult-born neurons in the hippocampus (Stanfield and Trice, 1988), it was only in the early 1990s, beginning with the first studies identifying stem cells in the adult brain (Reynolds and Weiss, 1992; Kilpatrick and Bartlett, 1993; Palmer et al., 1995), that research on adult neurogenesis did gain momentum.

As neurons do not divide, adult neurogenesis seemed implausible without neural precursor cells in the adult brain

To appreciate the phenomenon of adult mammalian neurogenesis, one must know the essentials of neurobiology. Without a working knowledge of cells and their ability to divide or nerve cells and their inability to divide, one can not grasp the excitement that arose about adult neurogenesis. When cell divisions were discovered in the nineteenth century, naturally related questions arose. Can all cells divide? Is division a characteristic feature of cells? Consequently, can neurons divide as well? Because it was soon found out that they cannot, adult neurogenesis was simply declared out of the question. When adult neurogenesis was first reported, it just appeared to be mysterious, to put it mildly. Given the general state of biological knowledge at the time, adult neurogenesis, if the reports on it were to be taken seriously at all, must have seemed an oddity, a strange exception to a plausible rule.

Today, when stem cells are the talk of the town and adult neurogenesis has gone mainstream, it is important not to forget that these phenomena have their roots in something that was long considered an oddity. Carl Sagan has aphoristically defined science as "the marriage of skepticism and wonder" (Sagan, 1995). In the case of any scientific topic and adult neurogenesis in particular, it is important not to lose the ability to marvel at what we see and the detachment to question what we believe we have seen. A historical perspective makes this dual responsibility easier.

Cells as the Structuring Principle of All Living Things

Biology has not always been a science of cells

Today it is difficult to think of biology other than as a science of cells. In some sense, cells are the basic unit of biology. But cells can only be seen under the microscope and thus were completely unknown until optical lenses and microscopes were invented. Ancient biology was a biology of material, forces, and spirits. For Aristotle, the most important authority on ancient biology, there was uninterrupted continuity between living and nonliving matter. Earth and sea could "spontaneously" generate organisms (for a comprehensive survey of the history of cell biology, see Mazzarello, 1999). This idea was not refuted until the eighteenth century, when Lazzaro Spallanzani and others showed that only organisms could produce other organisms. Where the antique philosopher-biologists had been convinced that "nature does not jump" (*natura non saltat*), these empiricists claimed that there is in fact a gap between living and nonliving things. With the invention of the microscope, a whole new universe was made accessible to human curiosity. Researchers began to look for the smallest units of life. In 1676, Antoni van Leeuwenhoek, the inventor of the microscope, reported to the Royal Society that moving, and thus presumably living, particles ("animalcules") could be seen in pond water (for an insightful account, see Dobell, 1960).

Robert Hooke was the first to describe cells in 1665

Robert Hooke was the English Leonardo da Vinci. A true Renaissance man, born in 1635, he made contributions to almost all fields of science, from physics (Hooke's law, the equation describing the elasticity of a spring) and chemistry, to astronomy, architecture, and engineering. His greatest biological achievement was the first description of the "cell" as the constituent of plants. He used his own improvement of Antony van Leeuwenhoek's microscope. Studying a piece of cork, he could "*exceedingly plainly perceive it to be all perforated and porous, much like a honey-comb . . . these pores or cells . . . were indeed the first microscopical pores I ever saw, and perhaps, that were ever seen, for I had not met with any writer or person that had made any mention of them before this*" (*Micrographia*, Observation XVIII, 1665). The "pores" reminded him of the cells of monks in their monastery, and the analogy stuck. Preceding this discovery, however, had been the first description of red blood "corpuscles" by Jan Swammerdam (Winsor, 1980).

That all living things consist of cells was realized only in the nineteenth century, and first stated by Theodor Schwann

After Hooke it took almost 200 years until scientists recognized that all living things are entirely composed of cells and, even more important, that development depends on cells, because cells give rise to cells. It is amazing that none of the first describers of cells developed a theory about this discovery. The significance of the observation was not appreciated until much later (see Wolpert, 1995). The first speculative steps toward a cell theory were made in 1805 by Lorenz Oken, who proposed that plants and animals are conglomerates of tiny living "infusoria," which, however, he did not equate with cells but rather with the animalcules of Leeuwenhoek. In 1832, Christian Gottfried Ehrenberg recognized that fungi and "infusoria" propagated by division (Ehrenberg, 1832). Again, the link to multicellular organisms was not made. An important next milestone was the discovery of the cell nucleus by Robert Brown in 1833 (see references in Watermann, 1982).

Around 1838–1839, the botanist Matthias Jacob and the physician Theodor Schwann (who today is mostly known for the "Schwann cells" of the peripheral nervous system) met in the laboratory of famous physiologist Johannes Müller in Berlin, where they discussed their view that organisms are made up of cells. They did not publish their work together, but are jointly recognized today as the fathers of cell biology. Schwann stated that "the elementary parts of all tissues are formed of cells." Even more important, he suggested that the "universal principle of development . . . is the formation of cells." Neither Schleiden nor Schwann, however, came to terms with the origin of cells. They adhered to the view, influentially propagated in Schwann's major written works, that cells could generate *de novo* and arise within or even outside other cells (Watermann, 1982).

Rudolf Virchow claimed that pathology originates from events on the level of cells and that cells originate from cells

In the late nineteenth century, the idea gained acceptance that cell divisions are at the heart of development

In 1858, the German pathologist Rudolf Virchow coined the term *cellular pathology*. Key to his work was the statement, "Every animal appears as a sum of vital units, each of which bears in itself the complete characteristics of life." His famous dictum *omnis cellula e cellula* ("every cell is from a cell"), and its implications for medicine made his theory one of the most influential ideas in biology. Virchow implied that "cells make cells," in health and disease, and that therefore the brain (or any other organ) could not be taken as static, but as a product of development (Virchow, 1859).

In 1766, Abraham Trembley was the first to report the direct observation that "animalcules" could reproduce, but there is not one single name to attach to the discovery that cell divisions underlie growth and development. This concept became increasingly accepted towards the end of the nineteenth century. Another scientist from Berlin, Robert Remak, is credited with the first explicit proposal of a theory of cell propagation by cell divisions: "The cells . . . multiply by continuous division, which starts at the nucleus" (Remak, 1852, 1855). Rudolf Albrecht von Kölliker reported cell divisions in the early vertebrate embryo that yielded cells that went on to differentiate into the many different types of cells found in the mature organism. In some sense, this is the birth of the idea of stem cells. In the 1860s, von Kölliker and others began to interpret embryology in light of the new cell theory.

The discovery of mitosis as the mechanism of cell division was extremely protracted (for more complete review of the topic, see Gourret, 1995). It is said to have begun with the work of Hermann Fol, Otto Bütschli, and Eduard Strasburger, who around 1873 first described intranuclear figures that correspond to the *mitotic apparatus*, as it is called today.

Cells of the Nervous System

Neurons and Glia

It turned out to be tremendously difficult to study cells in the brain. The major breakthrough came in 1873, when Camillo Golgi announced his now-famous staining method, a silver impregnation that miraculously (the reason is still not completely known) stained only selected neurons, thus distinguishing them from the overwhelmingly dense background. The structure of the nervous tissue became accessible.

Rudolf Virchow and Ramon y Cajal were the first to distinguish neurons from glia

As far as we know, Rudolf Virchow was the first to distinguish between neurons and glia. He coined the term *glia*, a Greek term for *Nervenkitt*, or "nerve glue" (Virchow, 1846). Camillo Golgi's silver impregnation methods allowed the visualization of individual glial cells and the description of astrocytes. Golgi was also the first to describe radial fibers that extended between the ventricular and pial surface of the neural tube (Golgi, 1886). However, another new staining method, introduced by Weigert in 1895, led to the dismissal of the idea that glial cells were individual entities. Rather, this became the period of the "glial syncytium," some sort of continuous cytoplasm with interspersed nuclei, engulfing all nerve cells (Hardesty, 1904). The idea gained much influence but was never fully accepted. In 1913, Ramon y Cajal used his gold-chloride-sublimate staining to prove that glial cells were indeed individual cells. He distinguished neurons (the "first element") from astrocytes (the "second element").

A group with small, rounded nuclei did not fall into either category and became the so-called third element of Cajal. Pío Del Rio-Hortega discovered that this third group of cells consisted of oligoden-drocytes and another class of cells, for which he coined the term *microglia* (Rio-Hortega, 1919). He made the case that microglia alone should be considered the third element and oligodendrocytes grouped with astrocytes. Thus, by 1921, the major cell classes of the central nervous system had been described, and the general pattern of lineage relationships as we know it today was in place.

Dividing Cells in the Central Nervous System

Mitoses in the course of brain development were first mapped by Ludwig Merk in 1896

To understand growth of the embryo and the development of organs, mapping of mitotic figures became an important tool. In 1881, Altmann described mitotic figures in the walls of the neural tubes (Altmann, 1881), and in 1896, Ludwig Merk from Graz wrote *Mitosen im Centralnervensysteme—ein Beitrag zur Lehre vom Wachstum desselben*. This large, comparative study of many species contained the observation that the numbers of cell divisions change over time, and secondary centers of cell division might take the place of the proliferative plane near the ventricle. He also described the characteristic pattern of cerebellar development: after an initial phase of cell division near the ventricle, the granule cell layer is formed in an outside-inside fashion (Merk, 1896).

Wilhelm His discovered what today would be considered the neuroepithelial stem cells

The first description of germinal cells (*Keimzellen*) in the developing nervous system goes back to Wilhelm His (His, 1889), who, along with the Spaniard Santiago Ramon y Cajal and the Italian Camillo Golgi, was the greatest neuroanatomist of the time. These three laid the foundation of modern neurobiology. Wilhelm His particularly emphasized developmental aspects (many of them studied in salmon). These early investigations were based on the light-microscopic identification of mitoses and meticulous anatomical analyses.

His described how neuroepithelial (NE) cells of the neural tube (the "medullary canal" in his nomenclature) produced "germinal cells" and differentiated cells, including the "spongioblasts." The idea was that germinal cells produced neuroblasts, which generated nerve cells. The "spongioblasts," in contrast, were equivalent to what today are called *radial glia* and were thought to produce glia only. In 1897, Alfred Schaper reported that, not only the epithelial cells of the tube, but also the germinal cells could give rise to spongioblasts. Schaper also introduced the "indifferent cell," or "medulloblast," a sort of parenchymal germinal cell, separated from the ventricular epithelium. He proposed that these cells retained the capabilities of the spongioblasts after migration into the tissue (Schaper, 1897).

With Wilder Penfield's work in the 1930s, the neural lineage tree reached a first completion

Wilder Penfield's seminal work of 1932, *Neuroglia*, contains a schematic drawing that summarizes the knowledge about lineage relationships at that time (Penfield, 1932; Fig. 2–1). Today it seems that, *cum grano salis*, both spongioblastic and medulloblastic lineages, which were thought to be mutually exclusive in Penfield's time, can be found in the adult brain. A plausible hypothesis is that spongioblasts would relate to the precursor cells of the dorsal brain, generating principal neurons and astrocytes; the "medulloblasts," to the precursor cells of the ventral brain, giving rise to interneurons and oligodendrocytes. This analogy is somewhat deceptive and remains problematic (not the least because the distinction between ventral and dorsal precursor cells is not as firm any more as it appeared to be only a few years ago) and the term *medulloblast* has additional controversial connotations. In 1938, John Kershman argued that medulloblasts would only exist in the cerebellum, where they could give rise to medulloblastomas.

Today we know the tumor entities of medulloblastomas as primitive neuroectodermal tumors in the roof of the fourth ventricle, but the medulloblast as a cell entity has disappeared altogether. However, the proposal of an "indifferent" apolar parenchymal cell as a "bipotential undifferentiated element" (Kershman, 1938) as suggested by Schaper has regained appeal with the discovery of

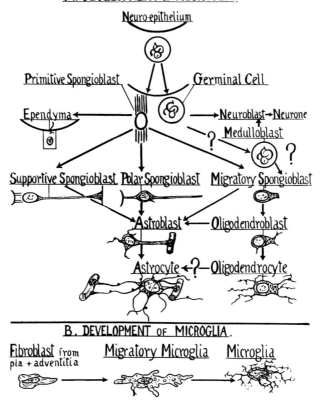

FIGURE 2–1 Penfield's scheme of cellular origin in the central nervous system. In 1932 Wilder Penfield published this drawing in his work on neuroglia (Penfield, 1932). Several types of cells in this scheme, for example, the "medulloblast" and the "spongioblast," did not appear in most later systematics. Modern precursor cell biology, however, leads to a new appreciation of the lineage relationships proposed during Penfield's time. The "spongioblast," for example, first described by Schaper in 1897, shows intriguing similarities to the proposed precursor cells of the brain tissue outside the neurogenic regions ("NG2 cells"). Such historical links remain speculation at present, but they indicate that adult precursor cell biology has its roots in the work of classical early twentieth-century neuroanatomy.

NG2-expressing parenchymal precursor cells. NG2 cells linger between the lineages, and it is not fully clear whether they truly compose one category of cell. But some NG2 cells have even been found to be multipotent ex vivo, and in vivo can express markers of multiple lineages (Belachew et al., 2003; Tamura et al., 2007) although the matter remains far from clear (Buffo et al., 2008). We will come back to this issue in greater detail in Chapters 3 (p. 71) and 7 (p. 230). Remarkably, however, already Schaper had suggested that some of the indifferent cells remained in a bipotent state until late in life and might perhaps provide the material for "regeneration processes" in the brain (Kershman, 1938).

Much later than in the context of the rare medulloblastoma and despite an early theory about the origin of "subependymomas" (Globus and Kuhlenbeck, 1944), it was recognized that proliferative cells in the ventricular walls of adult primates might be the origin of gliomas (Lewis, 1968). Today, the idea is gaining ground that quite generally preserved precursor cells in the adult brain can become the origin of tumors (Canoll and Goldman, 2008). We will come back to "cancer stem cells" of the brain in Chapters 3 (p. 83) and 12 (p. 524).

Can Neurons Divide?

Walther Spielmeyer established the postmitotic nature of neurons

Independently of the evolving understanding of the course of brain development and the lineage relationship between the major classes of cells in the brain, it was generally accepted that cell genesis almost completely ceases after embryogenesis and is absent in the adult. Walter Spielmeyer, in his *Histopathologie des Nervensystems* (Spielmeyer, 1922), wrote, "in general, the ganglion cells of the postembryonic period do not have a capability for cell division." But he did mention the possible exception of retinal ganglion cells, in which Schreiber and Wengler had induced mitoses by injections of *Scharlachöl* ("scarlet oil"; i.e., an antiseptic drug in veterinary medicine containing mineral oil, etheric oils, alcohols, methyl salicylate, and others, together with Scarlet Red dye). Nevertheless, Schreiber and Wengler themselves had remained skeptical of whether they had observed complete mitoses, and they carefully weighed the evidence (Schreiber and Wengler, 1908, 1910).

The brain's fundamental lack of regenerative capability was known by the beginning of the twentieth century

Regeneration in the central nervous system of amphibians had long been known and described by Müller, Masius, van Lair, and others. Reports about similar observations in mammals by Brown-Séguard and Dentans were already disputed at the time. Some authors claimed that among the proliferative responses found after needle-stick injuries to the adult mammalian cortex, dividing ganglion cells could be detected. Others argued against these conclusions. Vitzous's bold claims of complete regeneration of the monkey visual system after ablation of large areas of the visual cortex were greeted with great skepticism (see references in Schreiber and Wengler, 1910). Thus, at the end of the nineteenth century, the topic of neuronal regeneration had already been extensively discussed. Intriguingly, many of the arguments have since remained the same: How can we prove the neuronal nature of a cell that has undergone division? How can we prove lineage relationships?

Ramón y Cajal's pessimistic statement "everything may die, nothing regenerate" summarizes the prevailing opinion of his time

An alternate theory, the idea of amitotic nuclear divisions in nerve cells, was discussed with fervor but refuted early on. Cell division became generally accepted as the only path of growth in animals. It was also recognized that, with demonstration of a true mitosis of ganglion cells, proof of possible neuronal regeneration would be made, the implications of which did not go unnoticed (Schreiber and Wengler, 1910). Regarding this possibility, however, Ramón y Cajal made the following rather pessimistic statement:

> Once development was ended, the fonts of growth and regeneration of the axons and dendrites dried up irrevocably. In the adult centers, the nerve paths are something fixed, and immutable: everything may die, nothing may be regenerated. *(Una vez que el desarrollo ha concluido, las fuentes de regeneración de los axones y dendritas se agotan irrevocablemente. Preciso es reconocer que, en los centros adultos, las vías nerviosas son algo fijo, acabado, inmutable. Todo puede morir, nada renacer.)* (Ramon y Cajal, 1928).

This statement, found in the introduction to many reviews on adult neurogenesis, is considered the origin of the "no-new-neurons dogma." The problematic implications of the term *dogma* in this context have been discussed in Chapter 1. In addition, even this short historical outline underscores that Cajal's statement was not a lonely decision *ex cathedra* but one that reflected his conclusions from ongoing scientific discussions of that time. He found that there was not enough evidence to support the idea that neurons could divide.

Induction of cell division in neurons causes cell death

In the context of "adult neurogenesis," two side notes on the topic "Can neurons divide?" are interesting. First, as discussed in more detail in Chapter 12, in Alzheimer disease, neurons erroneously try to enter the cell cycle, which precedes their death (Herrup and Yang, 2001; Yang et al., 2001; Yang et al., 2003; Herrup et al., 2004; see also Chap. 12, p. 534).

Herrup and Yang have argued that this stress effect that leads to the induction of cell cycle–related factors can be separated from the actual initiation of cell death and have described the resulting state as "undead." Such undead cells might be part of the disease problem rather than "innocent victims" (Herrup and Yang, 2007). Herrup and Yang made the point that one should not be led into false security about the fact that "cell cycle regulation in neurons is an oxymoron." Neurons do not divide, but if they try, this nevertheless causes problems.

||||||||||||||||||||||||||||||||||||||||||||||||||||||||||

Dividing neurons exist only as immortalized cell lines

||||||||||||||||||||||||||||||||||||||||||||||||||||||||||

Second, in order to study neuronal cells, researchers have developed "immortalized" neuronal cell lines. The best known example is the PC12 line; another the "mouse motor neuron like hybrid cell line" NSC-34 by the company Tebu-Bio; or the hNT line from Stratagene. Invariably, such lines are based on tumor cells—pheochromocytoma in the case of PC12, neuroblastoma in the case of NSC-34, and teratocarcinoma in the case of hNT. Nobody has yet succeeded in immortalizing a true neuron: at best, a fusion between a more mature phenotype and the tumor cell was achieved. Neurons just do not divide and are extremely resistant to learning it. Enter neural stem cells.

The Origins of Stem Cell Biology

||||||||||||||||||||||||||||||||||||||||||||||||||||||||||||||||||

Modern stem cell biology has is origins in the first attempts at in vitro fertilization in the 1960s

||||||||||||||||||||||||||||||||||||||||||||||||||||||||||||||||||

The term *stem cell* first appeared in the scientific literature of the nineteenth century, but its use was not defined, and conflicting terms for what we would consider a stem cell today were common. Although a history of stem cell biology remains to be written, one aspect of it is important for our context.

The modern, multifaceted idea of stem cells arose largely in the context of the first work on human in vitro fertilization in the early 1960s (Edwards, 2001). Robert G. Edwards, who became the father of in vitro fertilization, began to study cells isolated from mammalian embryos and blastocysts and their development in vitro. Explants of inner cell masses from blastocysts derived from mice showed outgrowth into many cellular lineages in culture (Cole et al., 1966). Implantation of such "embryonic stem cells" (which did not bear that name at the time) into blastocysts originating from mice of a different strain produced the first mouse chimeras, which were obvious because of their mixes of coat-color patterns (Gardner, 1968). Despite its relevance for the field, this research was not stem cell research in any stricter sense of the word, because the cells themselves could not be handled, expanded, and studied. In the 1970s, experimental "stem cell" research largely consisted of work on teratoma and teratocarcinoma cells, which could be readily expanded and showed many but not all characteristics of embryonic stem cells. Most important, it was not possible to generate chimeras.

||||||||||||||||||||||||||||||||||||||||||||||||||||||||||||||||||

Embryonic stem cells of mice (1981) and humans (1998) became the quintessential stem cells triggering a change in scientific perspective

||||||||||||||||||||||||||||||||||||||||||||||||||||||||||||||||||

Only in 1981 did Matthew Kaufman and Martin Evans (Evans and Kaufman, 1981) and Gail R. Martin (Martin, 1981) succeed in culturing embryonic stem cells from mice, thereby allowing us to tap the theoretically unlimited expansion of the stem cells. Twenty-seven more years passed before James ("Jamie") Thomson reported the same for human embryonic stem cells (Thomson et al., 1998), thereby initiating the heated "stem cell debate" about the use of human embryos for research that has raged for several years in many countries.

Quite amazingly, the report on murine embryonic stem cells did not trigger the visions for therapeutic applications that would later follow Thomson's report on the same cells for humans. Six years after Evans and Kaufman, Peter Hollands used embryonic stem cells to repopulate the bone marrow of irradiated mice (Hollands, 1987). This report bridged the work on embryonic stem cells with E.D. Thomas's pioneering work on cell transplantation in hematology. Today, for certain types of leukemia and other disorders, bone marrow transplantation or the transplantation of bone marrow stem cells has become clinical routine (Thomas, 1999). The rise of embryonic stem cells, with their unlimited capacity for self-renewal and differentiation into various cell types, has sparked the hope that embryonic stem cells might be used for cell replacement beyond the hematopoetic system.

The Need for New Cells: Cell Therapy in the Brain

That cell therapy in the brain is possible has been shown for Parkinson disease

In the nervous system, cell transplantation had been tried with varying but mostly limited success. A severe setback to such treatment strategy was the widespread but pointless use of adrenal transplants to treat Parkinson disease (Madrazo et al., 1987; Madrazo et al., 1988). Around the same time, however, Olle Lindvall, Anders Björklund, and colleagues first succeeded in treating Parkinson's patients with striatal implants of fetal mesencephalic tissue (Lindvall et al., 1987). The ensuing series of open-label trials suggested that cell therapy in Parkinson's is possible and promising (Wenning et al., 1997; Piccini et al., 1999), but these success stories were not matched by the results from two placebo-controlled double-blind trials in the United States (Freed et al., 2001; Olanow et al., 2003), where dyskinesias as potential side effects of the treatment became apparent (Hagell et al., 2002; Freed et al., 2003). In some sense, even the reports on side effects can be taken as proof that neuronal replacement therapy is possible in humans, albeit not yet in the best interests of patients. Because one of the major obstacles in realizing neural cell replacement therapy is the availability of graftable cells, stem cells, with their theoretically unlimited expandability, would provide an ideal source for cells used in neural transplantation.

Recruitment of endogenous neurogenesis for regeneration might be an attractive alternative to extrinsic cell replacement strategies

Historically, the greatest hopes in neural transplantation coincided with the reemerging interest in adult neurogenesis and the discovery of neural precursor cells in the adult brain. Stem cells in the adult brain have never been considered solely an interesting biological fact, and adult neurogenesis was readily seen as nature's example of what cell engineers were trying to achieve: new neurons for the adult brain.

As predicted by Schreiber and Wengler in 1910, the evidence of successful mitoses in the brain had been immediately regarded as evidence of regeneration and became linked to important medical implications. The discovery of actively proliferative neural stem cells in the early 1990s and of adult neurogenesis is thus inseparable from the general rise in stem cell–based medicine. In many reports on adult neurogenesis, the two views on stem cells intermingle—the perspective of developmental biology on one hand, and the perspective of regenerative medicine on the other. There are very many reports on adult neurogenesis in vivo that start with a reference to the presumed fact that studies on adult neurogenesis are important because of their relevance for regeneration and the potential treatment of neurodegenerative disease.

There is nothing wrong with either of these perspectives, but the two different conceptual frameworks in stem cell biology can cause profound misunderstandings (see also Anderson, 2001). Research on adult neurogenesis might benefit from the booming interest in stem cells, but despite the many important and interesting medical implications of adult neurogenesis, it remains, foremost, an area of developmental neurobiology and physiological brain function.

The Idea of Plasticity

Plasticity is the reciprocal relationship between structure and function

The impact of adult neurogenesis on the neurosciences can be only appreciated within the conceptual framework of "plasticity." The idea of plasticity as we know it today in neuroscience goes back to pioneer psychologist William James, who stated, "Plasticity, then, in the wide sense of the word, means the possession of a structure weak enough to yield to an influence, but strong enough not to yield all at once" (James, 1890). Broadly speaking, *plasticity* means something like "adaptability and malleability." But there is no single true definition of plasticity; many disciplines claim the term as theirs and use it differently, and even more often the term is used only vaguely. In the sense of "neuroplasticity," the term represents the observation that the brain alters its structure, depending on its activity. Plasticity is how form follows function in the brain. As we will see, adult neurogenesis is a particular case of brain plasticity, because it affects brain structure on the level of entire neuronal cells, not just synapses and neurites. Sometimes structural plasticity is distinguished from functional plasticity, but the very concept of neuroplasticity implies that these two are so closely related that they are actually one phenomenon.

Despite Cajal's pessimistic statement routinely found in review articles on adult neurogenesis ("Everything may die, nothing may be regenerated. It is for the science of the future to change, if possible, this harsh decree"—Ramon y Cajal, 1928), the early neuroanatomists, and Ramon y Cajal in particular, did not consider the brain to be static, even if they saw how poorly it regenerated. But they did not have the tools to see the reciprocal link between function and structure that is characteristic for the brain. Today we know that a static brain would be dead. Brain function depends on plasticity. Ironically, then, it was not the anatomists but the psychologists who (despite James' introduction of the concept of plasticity) first favored a rigid, hardwired brain structure. The behaviorists who dominated psychology in the first decades of the twentieth century exemplify how ideology allows one to ignore obvious facts, silence opposing arguments through sheer influence and power, and thereby prevent progress. The behaviorists declared that all learning was based on a simple relationship between input and output. A reflex, such as the patellar tendon reflex, is the prototype of this relationship. All behavior could be reduced to units of such reflex-like responses. Learning was essentially Pavlovian and based only on conditioning; that is, stimuli of different kinds were linked to motor outputs. This reduction of learning to motor behavior should have made investigators skeptical, even during the heyday of the theory, and it is only one of the many bizarre features of behaviorism. Behaviorism began to be questioned when psychiatrist Hans Berger from Jena demonstrated the electroencephalogram (EEG) in 1924. The complexity and restlessness of continuous and widespread brain activity were impossible to explain in terms of behaviorism. In the behavioristic mindset, brain activity between input and output did not exist, nor did activity independent of input or motoric output. The behavioristic brain could be thought of as a hardwired black box, and of course the concept failed to explain signs of recovery after damage and functional plasticity.

Donald Hebb made plasticity a key aspect of neurobiology by postulating that synaptic plasticity would underlie learning processes

The great thinker in the neurosciences whose work marks the transition from this narrow perspective to modern cognitive neurobiology is Donald Hebb, known mostly for his famous "neurophysiological postulate." The key sentence in his book *The organization of behavior: A Neurophysiological theory* (Hebb, 1949), is: "When an axon of cell A is near enough to excite cell B and repeatedly or persistently takes part in firing it, some growth process or metabolic change takes place in one or both cells such that A's efficiency, as one of the cells firing B, is increased." In essence, what Hebb implied was a basis for learning on the level of synapses. If activity in two cells was closely enough related in time, this activity would strengthen their connection. Contiguity between pre- and postsynaptic activity leads to permanent structural changes, which is plasticity. Hebb postulated that plasticity was fundamental to learning and long-term memory and thereby essentially the basis for cognition and brain function in general.

Long-term potentiation (LTP), the presumed electrophysiological correlate of learning, is the realization of Hebb's postulate

It was later found that, for this principle to work, it had to account for both strengthening and weakening of connections with equal specificity. In 1973, Tim Bliss and Terje Lomo discovered long-term potentiation (LTP), which became generally accepted as the electrophysiological correlate to learning (Bliss and Lomo, 1973). The induction of LTP is indeed Hebbian (Kelso et al., 1986). But so is its counterpart, long-term depression (LTD), the long-lasting weakening of a neuronal connection. Whether LTP or LTD occurred could depend solely on the temporal spacing of the stimuli. If cell A fired shortly before B, LTP was found; if cell A fired shortly after

B, LTD resulted (Markram et al., 1997). This is called the "temporally asymmetric" form of Hebb's rule and helps explain the constant formation and dismantling of connections in the brain that underlie cognitive processes. Hebb postulated plasticity as the fundamental principle of how the brain works.

While "Hebb's synapse" became the most influential part of his work, Hebb did not stop at the level of single cell–cell interactions. He postulated further that information was not stored in single synapses but that the synapses formed the basis of autonomously "reverberating activity" in what Hebb called "cell assemblies." This not only introduced the idea of neuronal networks but also proposed that self-sustained activity in such networks, independent of concurrent input, underlay cognition. Intrinsic activity as much as extrinsic input could feed into the network activity. An implicit consequence of this idea is that continuous brain activity, as seen on the EEG, is paralleled by an equally continuous structural plasticity.

Hebb already reasoned that cellular changes might underlie plasticity

The oft-quoted thesis of Hebb's postulate is immediately preceded by an equally interesting sentence: "Let us assume that the persistence or repetition of reverberatory activity (or trace) tends to induce lasting cellular changes that add to its stability." Literally, there is no reason to believe that these changes should be restricted to synaptic changes. Consequently, adult neurogenesis relates to this aspect of Hebb's idea on how the brain and mind work. New cells obviously add new synapses, but they also provide new knots in the network, and they might change, not only the size (quantity), but also the quality of neuronal cell assemblies. Adult neurogenesis might thereby be a matter more of qualitative change than of just adding greater quantities of cells, analogous to increasing the processing power in our computer by putting in extra working memory.

Hebb also performed the first enriched environment experiment

Hebb was interested in more than the basic rules of learning and their underlying neurophysiological principles. He also stimulated the field of developmental psychology by studying experimentally how early experience influenced cognitive abilities later in life. In his most influential investigation in this field, he compared rats kept in laboratory cages with rats he had reared at home (supposedly as pets for his children) and found long-lasting positive effects on learning and memory (Hebb, 1947). He concluded that experience shapes neuronal development and, consequently, cognition. The finding was reported in a 43-line abstract in *American Psychologist* and is indeed not much more than an anecdote: group sizes were small and no proper controls had been included; no exact data can be found in the text. However, in 1952, Bernard Hymovitch, a doctoral student working with Hebb, published a replication and confirmation of Hebb's observation with larger groups and better controls and thus proved the initial claim to be correct (Hymovitch, 1952).

With this work, Hebb became the father of research on the effects of "enriched environments," an experimental setup that in the late 1950s to the mid-1970s became a central paradigm in developmental psychology. It allowed investigators to address experimentally the fundamental question of how we become what we are. How much of us is inherited and how much is acquired by education? Or in terms of biology, how much of an individual is determined by his or her genes and how much by interaction with the environment? Especially in pedagogical debates during this period, exact percentages were confidently declared concerning how much of one's personal character is inherited and how much comes from education. Depending on the individual position in this debate, the percentage could range between 0% and 100%. Today we know that development is a continued interaction between genes and environment, and that at every single moment, any living organism is 100% "genes" and 100% "environment," with "environment" beginning essentially on the subcellular level and constituting everything outside the genome itself. This insight originated partly from the studies of enriched environments, because in the laboratory situation one could keep the genetic influence constant by using inbred strains of rodents. Twin studies in humans complemented this line of research and allowed investigators to address issues that were more specifically human. With the rise of molecular biology, the molecular basis of the constant interaction between genes and environment was discovered, providing another crucial insight into what constitutes plasticity.

Rosenzweig and Bennett showed that living in an enriched environment alters many aspects of brain anatomy and physiology

Mark Rosenzweig, a psychologist at the University of California, Berkeley, built an amazing and influential scientific empire on the initial observation of effects of environmental enrichment on behavior as Hebb had reported it. Together with E.L. Bennett and colleagues, Rosenzweig published a large series of studies in the 1960s in which he showed manifold and dynamic changes, after environmental enrichment, in numerous measures describing the brain (Rosenzweig and Bennett, 1996; Rosenzweig, 2003; Bennett, 1976). Most notably, they made the conceptual link between "activity" or experience, and gross alterations in brain structure. They showed that environmental enrichment profoundly influences the anatomy and physiology of the brain.

Methodology was not yet advanced enough to answer the question of whether environmental stimuli could change neuron numbers

The question was even asked whether environmental enrichment could influence the amount of brain cells, neurons, and glia. One of the early studies by Joseph Altman, who later was the first to describe adult neurogenesis, addressed the question of whether environmental enrichment could stimulate the production of new neurons in the adult brain (Altman and Das, 1964). He, and later others (Diamond et al., 1966), reported more glia in some brain regions, but no increased neuronal numbers. The question was timely and important, but the methods of that time did not provide reliable quantification, the basis of discovering a significant increase in cell numbers.

An enriched environment experiment is more than the reversion of impoverished laboratory conditions

An enriched environment is "the complex combination of social and inanimate stimulation" (Rosenzweig and Bennett, 1996), and the experimental model usually consists of a large group of animals living in a large cage with toys and exchangeable tunnels, bridges, and other equipment. These mice are compared with animals living under the usual, rather Spartan conditions of laboratory housing. This simple manipulation has far-reaching effects on the brain and its function. Because of the special laboratory situation, however, it has often been claimed that enriched environments do not really reflect "enrichment" but actually bring "impoverished" animals back to normal (Cummins et al., 1977). Although there is some truth to this critique, for research rodents, laboratory caging has become the "natural" habitat for hundreds of generations. They do not survive when left out in the wild, and there is good evidence that they experience enrichment as rather stressful and the regulatory cages as normal. Besides this qualification, in addressing the many questions about learning, the differences that can be seen between enriched and control animals are much more relevant than determining how the finding relates to feral conditions.

The most important general conclusion to be drawn from this wealth of studies is that activity can regulate many aspects of neuronal development and function in the adult brain. The magnitude of the effect and its detailed relationship to specific stimuli or stressors are important but secondary to the result that such regulation is occurring at all, and which mechanisms underlie the observed change. Related to the experimental paradigm of environmental enrichment are studies on voluntary physical activity, another paradigm with numerous and impressive effects on the brain. Here the findings from active rodents compared to those from animals in standard cages might intuitively seem more directly applicable to the situation of modern humans with their sedentary lifestyle. Both environmental enrichment and physical activity up-regulate the production of new neurons (Kempermann et al., 1997; Van Praag et al., 1999; Kempermann, 2002; Kronenberg et al., 2003), demonstrating that the regulation of adult neurogenesis as a prime example of plasticity is in fact tied to behavior.

Not only does brain development not end, it is also profoundly shaped by activity and experience

The effects of environmental enrichment and physical activity on brain morphology and function have shown that the brain is not static and its development is never finished. There is a large body of literature (not reviewed here) on its structural reorganization after brain lesions. Many clinical observations confirm the idea of activity-induced plasticity (Wilson et al., 2002; Colcombe and Kramer, 2003; Abbott et al., 2004; Colcombe et al., 2004; Andel et al., 2008). The studies of animals in enriched environments delivered the systems and behavioral framework for research on synaptic and cellular plasticity. Adult neurogenesis and its activity- and experience-dependent regulation is therefore not an island of plasticity in a sea of static connections. Rather, adult neurogenesis fits seamlessly into a larger vision of the brain as an ever-changing structure that is constantly developing and refining its structure in response to functional demands.

The History of Research on Adult Neurogenesis

Prehistory

The first report of cell divisions in the adult ventricular wall is from 1912 (Ezra Allen)

A study reported in 1912 by Ezra Allen from McGill University is considered to be the first account of cell divisions in the adult rodent brain. Allen showed mitotic figures in the wall of the lateral ventricles of albino rats up to 120 days of age and depicted the germinal and mantle layers (roughly corresponding to ventricular and subventricular zones) of the ventricular wall at up to two years of age (Fig. 2–2) (Allen, 1912).

An extensive characterization of the ventricular wall during late human embryonic development came from Swedish pathologist Erik Rydberg in 1932 (Rydberg, 1932). He adhered to the prevailing concept of the time and described spongioblasts and germinal cells as well as "migratory spongioblasts," which were thought to produce oligodendrocytes or even to be bipotent and generate oligodendrocytes and astrocytes. His book is a neuropathological account of birth defects. From today's perspective it is amazing to see how elaborate the descriptive approach to brain anatomy and pathology and thus to neurobiology of the time had become. The 1930s were a turning point in that the classical neuroanatomical studies of, for example, Korbinian Brodman, Oskar Vogt, and their contemporaries had reached their limits. On the level of what was accessible by the mid-1930s, the brain had essentially been charted. There were nevertheless still many gaps in descriptions of the germinative zones of the adult brain.

Neuro-oncological questions stimulated the first studies on germinative matrices in the adult brain

In John Kershman's (1938) work on the "medulloblast and the medulloblastoma," referring to Schaper's "medulloblast," Kershman introduced the term *subependymal layer* for the cell layers immediately below the *ependymal zone* (Kershman, 1938). Work by Globus and Kuhlenbeck (1944) indicated that the human subependyma persisted into adult life and might be the occasional origin of tumors (the entity of subependymomas is

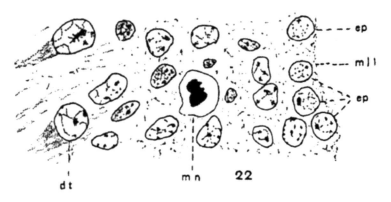

FIGURE 2–2 Dividing cell in the adult rodent subependymal layer. The first known depiction of cell divisions in the wall of the lateral ventricles is found in the work of Ezra Allen (1912) and shows mitotic figures in the subependymal layer of a 120-day-old rat.

indeed recognized by the modern World Health Organization classification of brain tumors: Globus and Kuhlenbeck, 1944). The hippocampus as a germinative zone of the adult brain, however, had gone completely unnoticed and did not enter the picture until work by Joseph Altman in 1963.

Revision of Nomenclature by the Boulder Committee

In 1970, a committee of 11 distinguished neuroscientists, called the "Boulder Committee" after their meeting place in Boulder, Colorado, published a consensus declaration for a revised nomenclature of the ventricular wall in the embryonic central nervous system (Committee, 1970). This nomenclature replaced the original nomenclature of His (His, 1889, 1904) and its many modifications over the following 80 years. The description by the Boulder Committee contains an interesting paragraph summarizing the state of knowledge at the beginning of the 1970s:

> No basis exists today for assigning more specific names to cells of the innermost two zones than is indicated by the terms *ventricular cell* and *subventricular cell*. All ventricular cells are identical in structure and behavior by every available criterion, though it is quite possible that different clones will be shown to exist among them. Subventricular cells fall into two classes based on cell size and nuclear morphology; cells of both classes proliferate and the possible relationships remain to be worked out.

The Boulder Committee thus canonized the term subventricular zone (SVZ) as we know it today and recognized different populations of dividing (precursor) cells in the SVZ. They did not explicitly apply this knowledge to the situation in the adult (but rather attempted to devise a new nomenclature that would avoid confusion with structures existing in the adult), although they referred to the older studies mentioned above that had suggested the persistence of a germinative matrix into adulthood. Ironically, thus, the text stated that "The layers and cells of the early developing central nervous system lack direct counterparts in the adult and must be designated by a special terminology," whereas today we know that the developing structures do indeed have (partial) counterparts in the adult to which we consequently now apply the terms coined for the developing nervous system. An update based on the progress since 1970 has been proposed (Bystron et al., 2008), but still the situation in the adult has not been explicitly incorporated.

From Descriptive Anatomy to Biological Experiments

The first study in which dividing cells were not only identified by the presence of mitotic figures but actively labeled came from Charles-Philippe Leblond and colleagues, also at McGill University. Leblond pioneered the use of tritiated thymidine that was incorporated into the DNA of dividing cells and detected by autoradiography (see the interesting biographical account in Bennet, 2008). They reported that "in the cerebrum, reactions were most frequent in clusters of cells underlying the ependyma of the lateral ventricle." Labeled cells were also found in other brain regions—for example, as satellite cells of neurons—and the question was raised of whether occasional neurons that showed the label were in fact dividing (Messier et al., 1958).

In 1959, William Bryans used colchicine injections to arrest mitoses and thereby increase the likelihood of finding them in the adult brain. He was able to confirm earlier reports and concluded that "it is conceivable that division of cells within the sub-ependymal region may represent a continuous source of glia in the adult rat" (Bryans, 1959).

In a number of studies, I. Smart continued the work with tritiated thymidine, and in 1961 he published the first extensive survey on DNA synthesis in the adult mouse brain, concluding that glial cells might divide throughout the parenchyma (Smart and Leblond, 1961). Smart also studied the subependymal layer of 35-day-old mice in greater detail and saw DNA synthesis and mitotic figures. He estimated the number of newly generated cells and wondered whether cell death might balance this production of cells. He reported that "the number of pycnotic nuclei correspond with the number of metaphases" and concluded that "in any case, the evidence was that subependymal cells do not add a significant number of new cells to the cell population of the adult brain" (Smart, 1961).

The first study on the adult mouse brain, including the SVZ, came from Smart in 1961

Leblond and Smart clearly had the methods to find adult neurogenesis. But no new neurons arise in the subependymal layer of the ventricular wall, and consequently they could not find them there. They underestimated cell migration away from the site of cell division and did not study the olfactory bulb. And they apparently never studied the hippocampus.

Joseph Altman

Despite the fact that essentially all previous evidence hinting at the possibility of adult neurogenesis was obtained in the SVZ, adult neurogenesis was first discovered in the hippocampus. Joseph Altman deserves credit for a remarkably complete and insightful initial description.

The father of adult neurogenesis research is Joseph Altman

In 1962, Joseph Altman was a scientist in the Psychophysiological Laboratory at Massachusetts Institute of Technology (Fig. 2–3). He applied the tritiated thymidine method to study cell divisions (DNA synthesis) in the adult brain. The initial

FIGURE 2–3 Joseph Altman in 1965, around the time of his seminal work on adult neurogenesis in Boston. The photograph was kindly provided by Joseph Altman.

results were interesting but far from clear. In 1962, Altman injected tritiated thymidine directly into the brain and found [3H]-thymidine incorporation into glial cells primarily around the lesion site. These cells survived up to two months after the injury (Altman, 1962a).

Altman had no bias for the SVZ. Like Smart and Leblond, he found occasional labeling in parenchymal cells along with the morphology of neurons. Could these indeed be neurons? In 1962 he published an article in *Science*, asking, "Are new neurons formed in the brains of adult mammals?" Remarkably, for an article published in *Science*, his answer was only "perhaps." Altman wrote that the absence of mitoses in the adult neurons "does not definitely rule out neogenesis of neurons in the adult, for new neurons might arise from non-differentiated precursors such as ependymal cells" (Altman, 1962b). He went on to describe the following: "In addition to the numerous labeled glia cells, which presumably underwent proliferation in response to the lesions, a few labeled glia cells, some labeled neuroblasts, and also labeled nuclei of some neurons were observed in brain regions not necessarily associated with the lesion area." An example of a labeled neuron shown in the article might look suspiciously like a satellite cell (see Chap. 7), and the term *neuroblast* in this context is rather elusive. But the study introduced the appropriate methods, asked the right questions, and proposed a clear idea that precursor cells might generate new neurons. Despite its provocative title, the 1962 *Science* article did not contain proof of adult neurogenesis and today appears rather as a remarkable indication of the things to come.

> Altman took a careful, unbiased approach and asked the central question of from which cells new neurons could originate

The next big step was a study published in 1963 in *Anatomical Record* (Altman, 1963). Here, tritiated thymidine was given systemically to rats and cats. Again, labeled glia were found and also some possible neurons in the cortex. Most strikingly, however, "a proliferative region of granule cells was identified in the dentate gyrus of the hippocampus." Figure 2–4 is taken from this article and shows the first depiction of adult neurogenesis in history. The study also contains proliferating cells in the SVZ and their migration below the corpus callosum.

> The first "new neuron" in the hippocampus was shown in 1963

As discussed in later chapters, adult cortical neurogenesis under physiological conditions has not yet been convincingly replicated. Satellite glia and methodological factors might account for these false-positive results. However, Altman recognized the problem of satellite cells and reported labeled cortical cells that he thought were unambiguously neurons with no satellite cells identifiable. It is probably understandable that this claim was not unanimously accepted. The convincing power of a few autoradiographic grains was not great enough. That neurogenesis appeared to be a rather widespread phenomenon in the adult brain directly contradicted the preceding work by Smart, Leblond, and Messier, and this discrepancy certainly did not help acceptance of the work.

Although the 1963 study contained the first images of newly generated granule cells, Altman's 1965 study published in the *Journal of Comparative Neurology*, "Autoradiographic and Histological Evidence of Postnatal Neurogenesis in Rats," is widely considered to be the inaugural article of the field of adult neurogenesis (Altman and Das, 1965a). The coauthor of this fundamental paper was Gopal D. Das. It was a remarkably complete study that did not stop with the mere demonstration of tritiated thymidine in granule cells. The pictorial evidence of adult hippocampal neurogenesis is similar to that in the 1963 study. Altman and Das showed that neurogenesis was restricted to the granule cell layer and that no new neurons were found in the hilus and the CA fields. They also reported that new cells survived at least two months after labeling, that neurogenesis showed a steep age-dependent decline with very high levels soon after birth, and that adult hippocampal neurogenesis could be detected at least up to the age of eight months in the rat. In addition to the labeled granule cells, they also found smaller labeled cells with dark nuclei and hypothesized that these cells might be the local precursor cell population. One strength of this study was that it sought evidence of neuronal development and did not deliver a mere snapshot in time. Another strength was that adult neurogenesis was demonstrated with a second method that was independent of the use of tritiated thymidine.

> The inaugural paper of the field is Altman's and Das's 1965 report of adult hippocampal neurogenesis in rats

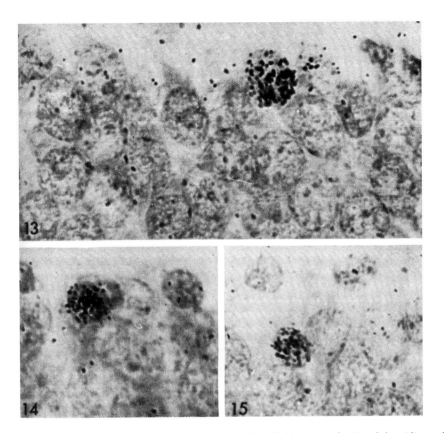

FIGURE 2–4 Altman's first image of an adult-generated neuron. Joseph Altman used tritiated thymidine to label proliferating cells and visualized the incorporation by autoradiography. In 1963, he published a study (from which this image is taken) providing the first known depiction of adult neurogenesis (Altman, 1963). The picture shows grains from the blackened photoemulsion over hippocampal granule cells in an adult rat. Similar images and a wealth of related data are found in second study done two years later, which is generally taken as the first report on adult neurogenesis (Altman and Das, 1965). Reprinted with kind permission of Joseph Altman and John Wiley & Sons Inc.

The second half of the article gave a careful analysis of changes in the anatomy of the dentate gyrus over time and the changes in its cellular composition. Here, Altman and Das showed that postnatal and adult hippocampal neurogenesis caused a sixfold increase in the number of granule cells between six days after birth and the age of three months. They found that the decrease in the number of undifferentiated cells in the granule cell layer gradually gave way to increasing numbers of neurons. This independent confirmation based on absolute cell counts set the study apart from the earlier and most of the later studies. It also made the evidence for neurogenesis in the adult hippocampus considerably stronger than that in the case of supposed signs of adult cortical neurogenesis.

Altman also first described the rostral migratory stream to the olfactory bulb

In addition to the data on hippocampal neurogenesis, the 1965 article also contained more information on cell proliferation in the subependymal zone. Here a strong decrease in labeled cells was observed over time after the injection of tritiated thymidine. The authors concluded "these facts would suggest a high rate of cell proliferation in adult rats in this region and the migration of labeled cells to as yet undetermined in regions." In a second 1965 study, the authors published first evidence that cells that were postnatally generated in the subependymal zone would migrate to the olfactory bulb and mature into neurons there (Altman and Das, 1965b). In 1969, Altman was the first to describe in detail the "rostral migratory stream," the route of migration between the SVZ and olfactory bulb (Altman, 1969).

In summary, the first report on adult hippocampal neurogenesis was an unusual piece of scientific work. Its true value could only be recognized much later, but even for its time, the evidence for adult neurogenesis seemed to be strong. With today's knowledge it is amazing to see how many important questions were asked and answered in this first study. One would think that these arguments in favor of adult neurogenesis would be hard to argue against, even at the time they were published. But despite all this evidence, adult neurogenesis was not accepted for almost another 30 years. This lack of acceptance turned out to be tragic for Altman, who saw his career stifled, possibly not despite, but rather because of, making a "discovery of a century."

Although not all of Altman's initial observations on adult neurogenesis have been replicated by others, much more than just the core set of data has been confirmed. Some possible reasons for the striking lack of acceptance of Altman's discovery were mentioned in Chapter 1, but given the quality of this piece of work, the reluctance to acknowledge his achievement is still not defensible. For Altman, the response from the scientific community, characterized by Fernando Nottebohm as "stiff resistance" (Nottebohm, 2002), turned into an unfortunate mix of justifiable skepticism and valid critique on one hand and ignorance and sometimes even outright hostility on the other. This part of the history of research on adult neurogenesis remains to be investigated. Although Altman made many more important contributions to research on brain development, he stopped publishing on adult neurogenesis in the 1970s. His wife, Shirley Bayer, however, published a number of reports that built on the initial observations and, for example, described the net growth of the rodent dentate gyrus (Bayer et al., 1982; Bayer, 1985).

Michael Kaplan

Fully 15 years passed before another researcher took up studies on adult neurogenesis with the same dedication as Altman's. Michael Kaplan published an extensive series of studies on adult neurogenesis in the hippocampus, the olfactory bulb, and the cortex—most notably the visual cortex. Kaplan's great contribution was the use of electron microscopy to prove the neuronal nature of cells labeled with tritiated thymidine. His most important study, published with James Hinds in 1977, showed adult neurogenesis in the hippocampus and the olfactory bulb of three-month-old rats (Kaplan and Hinds, 1977). In his electron-microscopic investigations, Kaplan described microtubules in long processes, synapses, and the characteristic karyoplasm of the labeled granule cells. In 1984, Kaplan identified the proliferative "neuroblast" in the subgranular zone and found that, a few hours after labeling with radioactive thymidine, these cells had somatic synapses and small neurites (Kaplan and Bell, 1984). Despite calling them *neuroblasts*, however, what Kaplan seemed to suggest was that they were dividing neurons. In the absence of stem cells in the adult brain, and consequently a clear concept of neuroblasts in adulthood, his assumption was not greeted with much enthusiasm and support. Kaplan also published a careful and extensive characterization of neurogenesis in the olfactory bulb at different ages, demonstrating the long-term survival of newly generated neurons (Kaplan, 1985).

In 1983 Kaplan used tritiated thymidine to label dividing cells in the subependymal layer of an almost five-year-old rhesus monkey, but had to conclude that the cells incorporated only minute amounts of radioactivity, preventing further conclusions (Kaplan, 1983).

Kaplan's greatest interest, however, was in neurogenesis in the visual cortex. As an undergraduate he investigated the effects of environmental enrichment on cell generation in the adult rat and found Layer IV neurons labeled. Although the study was never published, Kaplan later gave an account of it in a 2001 article on the current paradigm shift in neurogenesis research (Kaplan, 2001). Kaplan also did a study giving electron-microscopic evidence of neurogenesis in the visual cortex of adult rats (Kaplan, 1981). The evidence was perhaps not as strong as in the case of the hippocampus and olfactory bulb, because the cells have a less distinctive morphology, but this alone could not explain the discrepancy with the extensive studies published by Pasko Rakic on cell genesis in the adult monkey cortex. Rakic found no evidence of neurogenesis in the adult monkey brain, as reflected in the title of his article "Limits of Neurogenesis in the Adult Primate" (Rakic, 1985). Many later studies could not confirm the existence of adult cortical neurogenesis (Magavi et al., 2000; Kornack and Rakic, 2001b; Ehninger and Kempermann, 2003; Koketsu et al., 2003), although the claim has been made by others as well (Gould et al., 1999a). Today we know that neurogenesis persists for a long time postnatally in Layer VI of the cortex (Inta et al., 2008) and might even be present at minute levels in later adulthood (Dayer et al., 2005).

Despite this rare exception (which will be discussed in greater detail in Chap. 8, p. 293), what Kaplan might have detected were, in hindsight, dividing parenchymal precursor cells in the non-neurogenic regions of the adult brain, many of which express the proteoglycan NG2 and show both glial and some immature neuronal features (see also Chap. 8, p. 292). They often have a stellate or bipolar appearance. Synapses have not yet been reported on them, but the characterization of this peculiar cell population is still in flux. In the electron-microscopic image, Kaplan might thus have detected neuronal features on a cell that light-microscopically does not appear truly neuronal and according to current knowledge does not qualify as a functioning neuron. He would therefore have been both right and wrong at the same time. Determining the existence of a tissue precursor cell showing both glial and neuronal characteristics was clearly beyond the reach of the techniques of that time.

Like Altman, Kaplan was disappointed by the lack of appreciation of his discoveries by the neuroscientific mainstream

Kaplan left the field of neurogenesis disappointed. Justified skepticism toward his data from cortex studies seems to have spoiled the recognition of his achievements in research on adult neurogenesis in the hippocampus and olfactory bulb. Kaplan felt marginalized and excluded. Thus he became "both excited and jealous" (Kaplan, 2001) when, in 1998, Peter Eriksson published his report on neurogenesis in the adult human brain (Eriksson et al., 1998). Kaplan claims to have suggested a similar project to the ethics community of his university in 1982, but said he could not carry it out (Kaplan, 2001). Because he had intended to use radioactively labeled thymidine in brain cancer patients and did not have the immunohistochemical and microscopic tools of the Eriksson study, success would have been questionable and the data were likely to have remained controversial. His idea counts nonetheless, and reflects the many innovative lines along which Kaplan thought to increase knowledge about adult neurogenesis.

Fernando Nottebohm

Goldman and Nottebohm discovered adult neurogenesis in the canary brain related to song-learning

The next major step in research on adult neurogenesis was taken in 1983 with Fernando Nottebohm's work on neuronal replacement in the brains of songbirds. In canaries the males have elaborate and complex song repertoires, whereas the females sing much less. Treatment with testosterone, however, increases the females' singing (Nottebohm, 1980). In the 1970s it was found that gender difference in song behavior paralleled sexually dimorphic anatomy. Some of the nuclei of the canary song system are much larger in males than in females. When females receive testosterone, their altered singing behavior is associated with a growth of certain nuclei in the song system, including the archistriatum and the high vocal center (HVC) (Nottebohm, 1980). In males, these same nuclei grow and shrink during the course of the year, with their greatest volume correlating with the seasons in which the birds depend on the results of their singing (Nottebohm, 1981). In the spring, with the beginning of the breeding season and when the birds learn their songs, the volume of the HVC is largest. Nottebohm first assumed that these changes were due to fluctuations in the neuropil and primarily reflected synaptic growth and pruning. But was this sufficient to explain the large volume changes? Nottebohm applied the method of birthmarking dividing cells with tritiated thymidine and discovered a large number of labeled cells of neuronal appearance in the HVC (Goldman and Nottebohm, 1983). Somewhat against the initial hypothesis, however, was the finding that the ratio of labeled cells among the neurons of the HVC was the same for males and females. Nottebohm and his colleague Steven Goldman reasoned that these marked neurons might reflect a neuronal turnover.

Nottebohm introduced electrophysiology and targeted manipulations to do research on adult neurogenesis

Like Altman and Kaplan, Nottebohm first used microscopic techniques to identify labeled neurons. With electron microscopy, synapses were detected on labeled neurons (Burd and Nottebohm, 1985). The breakthrough experiment, however, was a truly revolutionary functional study and probably marked the turning point in the acceptance of adult neurogenesis by a wider audience. In canaries treated with tritiated thymidine four weeks earlier, random electrophysiological recordings were made in the HVC. The chance of hitting a labeled cell was about one in ten. After recording, the cells were filled with horseradish peroxidase, which could be detected histochemically after autoradiography for the radioactive thymidine. Indeed, about 10% of the filled cells were marked with silver grains and thus identified as new (Paton and Nottebohm, 1984). Nottebohm later wrote, "I believe this experiment, combining autoradiography, fine anatomical description, and neurophysiological recordings established the credibility of past and future claims that new neurons continued to be added to the adult vertebrate brain" (Nottebohm, 2002).

In adult birds, neurogenesis occurs throughout the telencephalon and is not restricted to a few neurogenic zones as in mammals. Like in mammals, however, cell proliferation in the adult avian brain is largely concentrated on the walls of the ventricles. Here, a great number of radial glia-like cells were found whose long processes projected into the gray matter and provided a guidance structure along which the new neurons could migrate (Alvarez-Buylla et al., 1987; Alvarez-Buylla and Nottebohm, 1988). Arturo Alvarez-Buylla, who later applied these insights from the avian brain to mammalian neurogenic regions, found that the radial glial cells even divided asymmetrically, with their long process remaining in place (Alvarez-Buylla et al., 1990). Strikingly, the HVC is not penetrated by radial fibers, leaving open the possibility that another population of precursor cells causes the production of the projection neurons in this region.

Thus, by 1990, the work on songbirds had not only shown activity-dependent regulation of adult neurogenesis and functional integration of new neurons but had also prepared the groundwork for studies on the identity (and possible heterogeneity) of precursor cells in the adult brain.

Elizabeth Gould, Heather Cameron, and Bruce McEwen

Elizabeth Gould and colleagues rediscovered adult neurogenesis within the context of stress-related changes in the hippocampus

The third rediscovery of adult hippocampal neurogenesis (after Altman's and Kaplan's) was made by Heather Cameron and Elizabeth Gould, working with Bruce McEwen at Rockefeller University in New York (Gould et al., 1992; Cameron et al., 1993). These researchers were interested in the effects of stress and the accompanying hormonal changes on the brain(Gould et al., 1991a, b). They wondered why the dentate gyrus did not seem to be sensitive to stress-induced cell death as found in other hippocampal regions. In line with Altman's observations and the findings of Nottebohm and coworkers, they hypothesized that adult hippocampal neurogenesis might result in regeneration of the dentate gyrus, balancing the amount of cell death. They combined the classical thymidine method with immunohistochemistry for neuronal marker neuron-specific enolase (NSE), thereby applying a straightforward double-labeling method (Cameron et al., 1993). Adult hippocampal neurogenesis was confirmed by the new method.

In the following years, the group published a series of studies on the negative regulation of adult hippocampal neurogenesis by stress and increased corticosterone levels. In contrast to the initial idea, stress, with its detrimental consequences for some hippocampal neuronal populations, also decreased adult neurogenesis and did not induce regenerative neurogenesis. Adrenalectomy and thus the removal of all stress hormones caused adult neurogenesis to increase (Cameron and Gould, 1994). The advance was thus not only methodical, it also included the demonstration that adult hippocampal neurogenesis in mammals could be regulated, albeit in these cases mainly down-regulated. These studies coincided with the discovery of neural precursor cells in adult mammals (Reynolds and Weiss, 1992), which provided a broader conceptual basis for the acceptance of adult neurogenesis in the research community.

Quite remarkably, Bruce McEwen, the principal investigator, in whose laboratory Elizabeth Gould and Heather Cameron were working at the time, and senior author on all of their initial studies, has never publicly claimed adult neurogenesis as his topic.

Tatsunori Seki

In 1991, Japanese neuroscientist Tatsunori Seki reported that newborn granule cells in the dentate gyrus would express PSA-NCAM (Seki and Arai, 1991). He did not actually do the experiment to confirm this in this first paper (but did so two years later, roughly at the same time of the much wider known description by Cameron and Gould [Cameron et al., 1993]). Over the years he made several important contributions; for example, by describing the structural relationship between radial glia-like cells and newborn, PSA-NCAM-positive cells in the SGZ at a time, when the concept of radial glia in adult neurogenesis had hardly been touched (Seki and Arai, 1999).

Rediscovery of Neurogenesis in the Adult Olfactory Bulb

Around the same time, several groups, most notably Marla Luskin's from Emory University, rediscovered new neurons in the adult olfactory bulb

Parallel to Cameron and Gould's work on adult hippocampal neurogenesis were two independent rediscoveries of neurogenesis in the adult olfactory bulb (Corotto et al., 1993; Luskin, 1993). At the University of Missouri at Columbia, Frank Corotto and Joel Maruniak were the first to use the bromodeoxyuridine (BrdU) method, albeit without confocal microscopy, to show that proliferating cells from the SVZ migrated into the olfactory bulb, where they expressed NSE and Calretinin (Corotto et al., 1993). The group also reported that odor deprivation reduced neurogenesis in the olfactory bulb (Corotto et al., 1994), thereby providing the first example of regulation for the SVZ.

Marla Luskin, from Emory University, used a retrovirus carrying the β-galactosidase gene to mark the progeny of proliferative precursor cells in the SVZ and reported how the cells use a narrow migratory pathway to reach the olfactory bulb, where they differentiate into two types of interneurons (Luskin, 1993). Using a similar method, Steven Levison and James Goldman at Columbia University showed that SVZ precursor cells postnatally also give rise to astrocytes and oligodendrocytes in striatum and cortex (Levison and Goldman, 1993).

Arturo Alvarez-Buylla

Arturo Alvarez-Buylla came from studying adult neurogenesis in songbirds and worked out the detailed anatomy of the SVZ and the rostral migratory stream

By that same time, Arturo Alvarez-Buylla applied his extensive experience from his work with Fernando Nottebohm on adult neurogenesis in birds to the adult mouse brain. In 1993, Carlos Lois and Alvarez-Buylla reported the neurogenic potential of murine SVZ precursor cells in explant cultures, thereby directly linking adult neurogenesis with stem cell biology (Lois and Alvarez-Buylla, 1993). The same authors published their account of olfactory bulb neurogenesis in vivo in 1994 (Lois and Alvarez-Buylla, 1994) and in the following years went on to characterize the migration of the precursor cells. They identified the particular mechanism of *chain migration,* by which the newborn cells move along each other in the rostral migratory stream (RMS) of rodents to reach the olfactory bulb (Lois et al., 1996).

Alvarez-Buylla and colleague Fiona Doetsch later characterized the microanatomical structure of the SVZ and identified the different precursor cell types in this region (Doetsch et al., 1997; Doetsch et al., 1999; Doetsch et al., 2002a; Doetsch et al., 2002b). Those years saw a strangely intense debate over whether only astrocytes of the SVZ or the ependymal lining of the ventricles might represent the "true" stem cells of this area (we will come back to the issue in Chaps. 3 and 6) and Alvarez-Buylla and colleagues made a strong case for the astrocytes, named "B cells" in the nomenclature they introduced. More recently Alvarez-Buylla further pushed the limits of anatomical analysis by demonstrating the rosette-like arrangement of cells in the ventricular wall (Mirzadeh et al., 2008). Overall he has made numerous contributions to the field and set the standards for many investigations.

Fred H. Gage

Fred H. Gage integrated neural stem cell biology and adult neurogenesis research

The scientist in whose laboratory a first synthesis of the field was attempted and who from the beginning saw adult neurogenesis from a perspective of cell and molecular biology was Fred H. Gage, at that time at the University of California–San Diego, since 1995 at the Salk Institute for Biological Studies in La Jolla, California. Gage came from neurotransplantation research and had worked with Anders Björklund in Lund, Sweden (Gage et al., 1988). Lund was a hot spot in the neurotransplantation field, and there also the first fetal cell transplants for Parkinson disease had taken place. At that time the idea of promoting survival and integration of implanted cells with growth or neurotrophic factors was much en vogue (Fischer et al., 1991). When experimenting with basic fibroblast growth factor (bFGF or FGF2), Gage and coworker Jasodarah Ray discovered cells in their neuronal cell cultures from the hippocampus that were massively responsive to the growth factor. It turned out that these cells were progenitor cells (Ray et al., 1993; Ray et al., 1995). Ray's colleague Theo D. Palmer took up the challenge to search for the same cells in adult tissue. The first report on neural stem cells from the adult hippocampus was published in 1995 in *Molecular and Cellular Neurosciences* (Palmer et al., 1995). That the "big" journals let this opportunity pass might indicate that, even three years after Reynolds and Weiss, the scientific world was still not ready to fully embrace the idea of stem cells in the brain. But in that same year, Gage published a review in the *Annual Review of Neurosciences*, which essentially set the stage for what was going to come and is the first synthetic overview of the nascent field (Gage et al., 1995).

The BrdU method combined with confocal microscopy paved the path for Peter Eriksson's 1998 report on adult human neurogenesis

Two more methodological milestones from Gage's laboratory brought the field to where it is today. In 1982, Gratzner and colleagues had been the first to replace tritiated thymidine with the immunohistochemically detectable thymidine analog bromodeoxyuridine to label proliferating cells (Gratzner, 1982). This method had been applied to brain tissue by Nowakowski and colleagues in 1989 and had been first used to investigate adult olfactory neurogenesis by Frank Corotto and colleagues in 1993 (see above; Corotto et al., 1993). But Georg Kuhn now combined the BrdU-method with confocal microscopy and double- and triple-labeling paradigms to investigate the newly born cells, thereby greatly expanding the range of information to be gained. The facilitated identification of new neurons also made quantification more reliable and feasible and prepared the ground for many later studies on the regulation of adult neurogenesis (Kuhn et al., 1996). In 1998, Peter Eriksson, together with Gage, applied that same method to report neurogenesis in the adult human hippocampus (Eriksson et al., 1998; Fig. 2–5).

In 2001, Henriette van Praag and coworkers, also at Fred H. Gage's laboratory, used a retrovirus labeled with the green fluorescent protein (GFP) to visualize newly generated cells in living tissue slices and were thus able to confirm that newborn neurons in the adult rodent hippocampus become functional granule cells (van Praag et al., 2002).

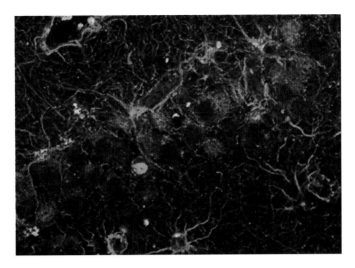

FIGURE 2–5 Neurogenesis in the adult human hippocampus. Peter Eriksson, neurologist in Gothenburg, Sweden, identified terminally ill patients who had received injections of bromodeoxyuridine (BrdU) for tumor staging purposes. He received their informed consent and after their death was able to examine their brains. He applied the same histological methods that are routinely used to visualize adult neurogenesis in rodents. The confocal microscopic image shows the human dentate gyrus; BrdU is *green*, NeuN is *red*, astrocytic marker GFAP is *blue*. One granule cell is double-labeled for BrdU and NeuN and thereby identified as a new neuron (compare with Fig. 1–1). Reprinted from Eriksson et al. (1998), with permission of the authors and Macmillan Publishers Ltd (*Nature Medicine*), copyright 1998.

> The discovery that adult neurogenesis is regulated by physical and cognitive activity marks a turning point in the appreciation of adult neurogenesis as a physiologically relevant phenomenon

Finally, and not without effect on the public perception of adult neurogenesis, from Gage's lab came the discovery that exposure to an enriched environment would stimulate the production of neurogenesis in the adult and aging hippocampus. These reports gave the first examples that adult neurogenesis in rodents is positively regulated in a context of plausible behavioral and functional relevance (Kempermann et al., 1997, 1998).

Today Gage is the most productive and best-cited researcher in the field of adult neurogenesis. His spectrum spans from meticulous anatomical analyses based on retroviral labeling techniques (Zhao et al., 2006; Toni et al., 2008) over classical molecular studies in vivo and in vitro to a sophisticated computational model of the function of newborn neurons in the adult hippocampus (Aimone et al., 2009) and a broad range of other aspects, including novel ideas such as the molecular control of neurogenesis by retrotransposons (Muotri et al., 2005). Gage did not discover adult neurogenesis as some people believe, but he did discover the width of the topic. His impact on the field has thus been extraordinarily comprehensive.

Adult Neurogenesis in Nonhuman Primates and Humans

> By the end of the 1990s, adult neurogenesis had also been demonstrated for non-human primates and humans

General acceptance of adult neurogenesis was aided by the fact that the late 1990s saw a series of publications in which adult hippocampal neurogenesis climbed up the evolutionary ladder and was first described in primates (Gould et al., 1997; Gould et al., 1999b; Kornack and Rakic, 1999). The work in nonhuman primates was followed in 1998 by Peter Eriksson's study (Eriksson et al., 1998) showing evidence of adult hippocampal neurogenesis in humans (Fig. 2–5; see also Chap. 11, p. 499).

Neurogenesis in the adult primate olfactory bulb was first demonstrated by David Kornack and Pasko Rakic of Yale University (Kornack and Rakic, 2001a). Some indirect evidence for neurogenesis in the adult human olfactory bulb was reported in 2004 (Bedard and Parent, 2004), but one other study seemed to suggest that the structure of the SVZ in humans differed considerably from that of rodents and no rostral migratory stream to the olfactory bulb existed (Sanai et al., 2004). But an extensive study, again from Peter Eriksson, soon revealed the structure of the human rostral migratory stream and could confirm adult neurogenesis in the human olfactory bulb (Curtis et al., 2007). Although some discrepancies between the two studies remain unresolved (because *both* showed an extraordinarily high level of technical quality) and only more studies with different methodology will give ultimate clarity, the question has lost much of its nature as a white spot on the map.

Adult Neurogenesis in the Twenty-First Century

In the first decade of the twenty-first century, adult neurogenesis was described in great anatomical detail, many molecular mechanisms were unraveled, and its function was successfully studied

By the year 2000, adult neurogenesis had become generally accepted as a phenomenon, but with respect to some key issues the field still had more questions than answers. Most important, the question of what the new neurons are good for had barely been touched. This changed in the course of the following decade. As discussed in greater detail in Chapter 10, we today have a fairly good idea of how and why new neurons might be beneficial for the brain. Between the first ablation experiment by Tracey Shors and Elizabeth Gould (Shors et al., 2001) and the first truly comprehensive theoretical construct by Brad Aimone and Fred H. Gage (Aimone et al., 2006), there was a wide range of publications from several groups, tackling the problem from different angles. More and mechanistically different strategies to selectively ablate the newborn neurons and better and more sensitive behavioral tests were developed (see overview and references in Chap. 10).

As of the time of this writing, there might still not be a grand unifying theory of the function of new neurons in the adult brain (although a quite far-reaching synthesis is attempted in Chap. 10), but many key questions have been thoroughly investigated, and a core idea has emerged. There are obviously some specific functional and computational needs in the adult brain that are best addressed by adding new neurons rather than just improving synaptic connectivity. These functions have to do with the particular network properties of the dentate gyrus and the olfactory bulb, and they seem to provide an effective solution to the so-called stability-plasticity dilemma that both the hippocampus and the olfactory bulb face. In both cases adult neurogenesis occurs in a highly convergent network structure—a true bottleneck through which much information has to pass. New neurons are long-term investments, although they also respond to short-term computational needs. Ultimately, they adapt the network to the level of challenges experienced by the individual and allow an optimization presumably not achievable just by the means of synaptic plasticity.

The increased specificity of our ideas of what new neurons do in the adult brain has also led to the reconsideration of the medical relevance of adult neurogenesis. One of the most stimulating ideas (which also greatly enhanced public awareness of adult neurogenesis) has been the hypothesis that failing adult hippocampal neurogenesis might be linked to the pathogenesis of depression (Jacobs et al., 2000; Kempermann and Kronenberg, 2003; Sahay and Hen, 2007). In the strongest variant, this hypothesis is almost certainly wrong, but testing its less absolute forms continues to allow important insights into the connection between affective behavior and learning.

References

Abbott RD, White LR, Ross GW, Masaki KH, Curb JD, Petrovitch H (2004). Walking and dementia in physically capable elderly men. *JAMA* 292:1447–1453.

Aimone JB, Wiles J, Gage FH (2006). Potential role for adult neurogenesis in the encoding of time in new memories. *Nat Neurosci* 9:723–727.

Aimone JB, Wiles J, Gage FH (2009). Computational influence of adult neurogenesis on memory encoding. *Neuron* 61:187–202.

Allen E (1912). The cessation of mitosis in the central nervous system of the albino rat. *J Comp Neurol* 22: 547–568.

Altman J (1962a). Autoradiographic study of degenerative and regenerative proliferation of neuroglia cells with tritiated thymidine. *Exp Neurol* 5:302–318.

Altman J (1962b). Are new neurons formed in the brains of adult mammals? *Science* 135:1128–1129.

Altman J (1963). Autoradiographic investigation of cell proliferation in the brains of rats and cats. *Anat Rec* 145:573–591.

Altman J (1969). Autoradiographic and histological studies of postnatal neurogenesis. IV: Cell proliferation and migration in the anterior forebrain, with special reference to persisting neurogenesis in the olfactory bulb. *J Comp Neurol* 137:433–457.

Altman J, Das GD (1964). Autoradiographic examination of the effects of enriched environment on the rate of glial muliplication in the adult rat brain. *Nature* 204:1161–1163.

Altman J, Das GD (1965a). Autoradiographic and histological evidence of postnatal hippocampal neurogenesis in rats. *J Comp Neurol* 124:319–335.

Altman J, Das GD (1965b). Post-natal origin of microneurons in the rat brain. *Nature* 207:953–956.

Altmann R (1881). *Über Embryonales Wachstum.* Leipzig.

Alvarez-Buylla A, Buskirk DR, Nottebohm F (1987). Monoclonal antibody reveals radial glia in adult avian brain. *J Comp Neurol* 264:159–170.

Alvarez-Buylla A, Nottebohm F (1988). Migration of young neurons in adult avian brain. *Nature* 335:353–354.

Alvarez-Buylla A, Theelen M, Nottebohm F (1990). Proliferation "hot spots" in adult avian ventricular zone reveal radial cell division. *Neuron* 5:101–109.

Andel R, Crowe M, Pedersen NL, Fratiglioni L, Johansson B, Gatz M (2008). Physical exercise at midlife and risk of dementia three decades later: A population-based study of Swedish twins. *J Gerontol* 63:62–66.

Anderson DJ (2001). Stem cells and pattern formation in the nervous system: The possible versus the actual. *Neuron* 30:19–35.

Bayer SA (1985). Neuron production in the hippocampus and olfactory bulb of the adult rat brain: Addition or replacement? *Ann NY Acad Sci* 457:163–172.

Bayer SA, Yackel JW, Puri PS (1982). Neurons in the rat dentate gyrus granular layer substantially increase during juvenile and adult life. *Science* 216:890–892.

Bedard A, Parent A (2004). Evidence of newly generated neurons in the human olfactory bulb. *Brain Res Dev Brain Res* 151:159–168.

Belachew S, Chittajallu R, Aguirre AA, Yuan X, Kirby M, Anderson S, Gallo V (2003). Postnatal NG2 proteoglycan-expressing progenitor cells are intrinsically multipotent and generate functional neurons. *J Cell Biol* 161:169–186.

Bennett EL (1976). Cerebral effects of differential experience and training. In: Neural mechanisms of learning and memory (Rosenzweig MR, Bennett EL, eds), pp 279-287. Cambridge: MIT Press.

Bennett G (2008). Charles Philippe Leblond. 5 February 1910 - 10 April 2007. *Biogr Mems Fell R Soc* 54: 175–191.

Bliss TV, Lomo T (1973). Long-lasting potentiation of synaptic transmission in the dentate area of the anaesthetized rabbit following stimulation of the perforant path. *J Physiol* 232:331–356.

Boss BD, Peterson GM, Cowan WM (1985). On the number of neurons in the dentate gyrus of the rat. *Brain Res* 338:144–150.

Bryans WA (1959). Mitotic activity in the brain of the adult rat. *Anat Rec* 133:65–71.

Buffo A, Rite I, Tripathi P, Lepier A, Colak D, Horn AP, Mori T, Gotz M (2008). Origin and progeny of reactive gliosis: A source of multipotent cells in the injured brain. *Proc Natl Acad Sci USA* 105:3581–3586.

Burd GD, Nottebohm F (1985). Ultrastructural characterization of synaptic terminals formed on newly generated neurons in a song control nucleus of the adult canary forebrain. *J Comp Neurol* 240:143–152.

Bystron I, Blakemore C, Rakic P (2008). Development of the human cerebral cortex: Boulder Committee revisited. *Nat Rev Neurosci* 9:110–122.

Cameron HA, Gould E (1994). Adult neurogenesis is regulated by adrenal steroids in the dentate gyrus. *Neuroscience* 61:203–209.

Cameron HA, Woolley CS, McEwen BS, Gould E (1993). Differentiation of newly born neurons and glia in the dentate gyrus of the adult rat. *Neuroscience* 56:337–344.

Canoll P, Goldman JE (2008). The interface between glial progenitors and gliomas. *Acta Neuropathol* 116: 465–477.

Colcombe S, Kramer AF (2003). Fitness effects on the cognitive function of older adults: A meta-analytic study. *Psychol Sci* 14:125–130.

Colcombe SJ, Kramer AF, Erickson KI, Scalf P, McAuley E, Cohen NJ, Webb A, Jerome GJ, Marquez DX, Elavsky S (2004). Cardiovascular fitness, cortical plasticity, and aging. *Proc Natl Acad Sci USA* 101:3316–3321.

Cole RJ, Edwards RG, Paul J (1966). Cytodifferentiation and embryogenesis in cell colonies and tissue cultures derived from ova and blastocysts of the rabbit. *Dev Biol* 13:385–407.

Committee B (1970). Embryonic vertebrate central nervous system: Revised terminology. The Boulder Committee. *Anat Rec* 166:257–261.

Corotto FS, Henegar JA, Maruniak JA (1993). Neurogenesis persists in the subependymal layer of the adult mouse brain. *Neurosci Lett* 149:111–114.

Corotto FS, Henegar JR, Maruniak JA (1994). Odor deprivation leads to reduced neurogenesis and reduced neuronal survival in the olfactory bulb of the adult mouse. *Neuroscience* 61:739–744.

Crespo D, Stanfield BB, Cowan WM (1986). Evidence that late-generated granule cells do not simply replace earlier formed neurons in the rat dentate gyrus. *Exp Brain Res* 62:541–548.

Cummins RA, Livesey PJ, Evans JG (1977). A developmental theory of environmental enrichment. *Science* 197:692–694.

Curtis MA, Kam M, Nannmark U, Anderson MF, Axell MZ, Wikkelso C, Holtas S, van Roon-Mom WM, Bjork-Eriksson T, Nordborg C, Frisen J, Dragunow M, Faull RL, Eriksson PS (2007). Human neuroblasts migrate to the olfactory bulb via a lateral ventricular extension. *Science* 315:1243–1249.

Dayer AG, Cleaver KM, Abouantoun T, Cameron HA (2005). New GABAergic interneurons in the adult neocortex and striatum are generated from different precursors. *J Cell Biol* 168:415–427.

Diamond MC, Law F, Rhodes H, Lindner B, Rosenzweig MR, Krech D, Bennett EL (1966). Increases in cortical depth and glia numbers in rats subjected to enriched environment. *J Comp Neurol* 128:117–126.

Dobell C (1960). *Antony van Leeuwenhoek and His "Little Animals."* New York: Dover.

Doetsch F, Garcia-Verdugo JM, Alvarez-Buylla A (1997). Cellular composition and three-dimensional organization of the subventricular germinal zone in the adult mammalian brain. *J Neurosci* 17:5046–5061.

Doetsch F, Caille I, Lim DA, Garcia-Verdugo JM, Alvarez-Buylla A (1999). Subventricular zone astrocytes are neural stem cells in the adult mammalian brain. *Cell* 97:703–716.

Doetsch F, Petreanu L, Caille I, Garcia-Verdugo JM, Alvarez-Buylla A (2002a). EGF converts transit-amplifying neurogenic precursors in the adult brain into multipotent stem cells. *Neuron* 36:1021–1034.

Doetsch F, Verdugo JM, Caille I, Alvarez-Buylla A, Chao MV, Casaccia-Bonnefil P (2002b). Lack of the cell-cycle inhibitor p27Kip1 results in selective increase of transit-amplifying cells for adult neurogenesis. *J Neurosci* 22:2255–2264.

Edwards RG (2001). IVF and the history of stem cells. *Nature* 413:349–351.

Ehninger D, Kempermann G (2003). Regional effects of wheel running and environmental enrichment on cell genesis and microglia proliferation in the adult murine neocortex. *Cereb Cortex* 13:845–851.

Ehrenberg CG (1832). Über das Entstehen des Organischen aus einfacher Materie, und über die organischen Molecüle und Atome, insbesondere als Erfahrungsgegenstände. *Poggendorff's Annalen der Physik und Chemie* 24:1–48.

Eriksson PS, Perfilieva E, Bjork-Eriksson T, Alborn AM, Nordborg C, Peterson DA, Gage FH (1998). Neurogenesis in the adult human hippocampus. *Nat Med* 4:1313–1317.

Evans MJ, Kaufman MH (1981). Establishment in culture of pluripotential cells from mouse embryos. *Nature* 292:154–156.

Fischer W, Bjorklund A, Chen K, Gage FH (1991). NGF improves spatial memory in aged rodents as a function of age. *J Neurosci* 11:1889–1906.

Freed CR, Leehey MA, Zawada M, Bjugstad K, Thompson L, Breeze RE (2003). Do patients with Parkinson's disease benefit from embryonic dopamine cell transplantation? *J Neurol* 250 Suppl 3:44–46.

Freed CR, Greene PE, Breeze RE, Tsai WY, DuMouchel W, Kao R, Dillon S, Winfield H, Culver S, Trojanowski JQ, Eidelberg D, Fahn S (2001). Transplantation of embryonic dopamine neurons for severe Parkinson's disease. *N Engl J Med* 344:710–719.

Gage FH, Ray J, Fisher LJ (1995). Isolation, characterization, and use of stem cells from the CNS. *Annu Rev Neurosci* 18:159–192.

Gage FH, Brundin P, Strecker R, Dunnett SB, Isacson O, Bjorklund A (1988). Intracerebral neuronal grafting in experimental animal models of age-related motor dysfunction. *Ann NY Acad Sci* 515:383–394.

Gardner RL (1968). Mouse chimeras obtained by the injection of cells into the blastocyst. *Nature* 220:596–597.

Globus JH, Kuhlenbeck H (1944). The subependymal cell plate (matrix) and its relationship to brain tumors of the ependymal type. *J Neuropathol Exp Neurol* 3:1–35.

Goldman SA, Nottebohm F (1983). Neuronal production, migration and differentiation in a vocal control nucleus of the adult female canary brain. *Proc Natl Acad Sci USA* 80:2390–2394.

Golgi C (1886). *Sulla fina anatomia degli organi centrali des sisterna nervoso*. Milan: Hoepli.

Gould E, Cameron HA, Daniels DC, Woolley CS, McEwen BS (1992). Adrenal hormones suppress cell division in the adult rat dentate gyrus. *J Neurosci* 12:3642–3650.

Gould E, McEwen BS, Tanapat P, Galea LA, Fuchs E (1997). Neurogenesis in the dentate gyrus of the adult tree shrew is regulated by psychosocial stress and NMDA receptor activation. *J Neurosci* 17:2492–2498.

Gould E, Reeves AJ, Graziano MS, Gross CG (1999a). Neurogenesis in the neocortex of adult primates. *Science* 286:548–552.

Gould E, Reeves AJ, Fallah M, Tanapat P, Gross CG, Fuchs E (1999b). Hippocampal neurogenesis in adult Old World primates. *Proc Natl Acad Sci USA* 96:5263–5267.

Gould E, Woolley CS, McEwen BS (1991a). Naturally occurring cell death in the developing dentate gyrus of the rat. *J Comp Neurol* 304:408–418.

Gould E, Woolley CS, McEwen BS (1991b). Adrenal steroids regulate postnatal development of the rat dentate gyrus: I. Effects of glucocorticoids on cell death. *J Comp Neurol* 313:479–485.

Gourret JP (1995). Modelling the mitotic apparatus. From the discovery of the bipolar spindle to modern concepts. *Acta Biotheor* 43:127–142.

Gratzner HG (1982). Monoclonal antibody to 5-bromo- and 5-iododeoxyuridine: A new reagent for detection of DNA replication. *Science* 218:474–475.

Hagell P, Piccini P, Bjorklund A, Brundin P, Rehncrona S, Widner H, Crabb L, Pavese N, Oertel WH, Quinn N, Brooks DJ, Lindvall O (2002). Dyskinesias following neural transplantation in Parkinson's disease. *Nat Neurosci* 5:627–628.

Hardesty I (1904). On the development and nature of the neuroglia. *Am J Anat* 3:229–268.

Hebb DO (1947). The effects of early experience on problem-solving at maturity. *Am Psychol* 2:306–307.

Hebb DO (1949). *The Organization of Behavior*. New York: John Wiley & Sons.

Herrup K, Neve R, Ackerman SL, Copani A (2004). Divide and die: Cell cycle events as triggers of nerve cell death. *J Neurosci* 24:9232–9239.

Herrup K, Yang Y (2001). Pictures in molecular medicine: Contemplating Alzheimer's disease as cancer: a loss of cell-cycle control. *Trends Mol Med* 7:527.

Herrup K, Yang Y (2007). Cell cycle regulation in the postmitotic neuron: Oxymoron or new biology? *Nat Rev Neurosci* 8:368–378.

His W (1889). Die Neuroblasten und deren Entstehung im embryonalen Mark. *Arch Anat Physiol Anat*. Abt.: 249–300.

His W (1904). Die Entwicklung des menschlichen Gehirns während der ersten Monate. Leipzig: S. Hirzel.

Hollands P (1987). Differentiation and grafting of haemopoietic stem cells from early postimplantation mouse embryos. *Development* 99:69–76.

Hymovitch B (1952). The effects of experimental variation on problem solving in the rat. *J Comp Physiol Psychol* 45:313–321.

Inta D, Alfonso J, von Engelhardt J, Kreuzberg MM, Meyer AH, van Hooft JA, Monyer H (2008). Neurogenesis and widespread forebrain migration of distinct GABAergic neurons from the postnatal subventricular zone. *Proc Natl Acad Sci USA* 105:20994–20999.

Jacobs BL, Praag H, Gage FH (2000). Adult brain neurogenesis and psychiatry: A novel theory of depression. *Mol Psychiatry* 5:262–269.

James W (1890). *The Principles of Psychology*. New York: Holt and Macmillan.

Kaplan MS (1981). Neurogenesis in the 3-month-old rat visual cortex. *J Comp Neurol* 195:323–338.

Kaplan MS (1983). Proliferation of subependymal cells in the adult primate CNS: Differential uptake of DNA labelled precursors. *J Hirnforsch* 24:23–33.

Kaplan MS (1985). Formation and turnover of neurons in young and senescent animals: An electron-microscopic and morphometric analysis. *Ann NY Acad Sci* 457:173–192.

Kaplan MS (2001). Environment complexity stimulates visual cortex neurogenesis: Death of a dogma and a research career. *Trends Neurosci* 24:617–620.

Kaplan MS, Hinds JW (1977). Neurogenesis in the adult rat: Electron microscopic analysis of light radioautographs. *Science* 197:1092–1094.

Kaplan MS, Bell DH (1984). Mitotic neuroblasts in the 9-day-old and 11-month-old rodent hippocampus. *J Neurosci* 4:1429–1441.

Kelso SR, Ganong AH, Brown TH (1986). Hebbian synapses in hippocampus. *Proc Natl Acad Sci USA* 83:5326–5330.

Kempermann G (2002). Regulation of adult hippocampal neurogenesis—implications for novel theories of major depression. *Bipolar Disord* 4:17–33.

Kempermann G, Kronenberg G (2003). Depressed new neurons—adult hippocampal neurogenesis and a cellular plasticity hypothesis of major depression. *Biol Psychiatry* 54:499–503.

Kempermann G, Kuhn HG, Gage FH (1997). More hippocampal neurons in adult mice living in an enriched environment. *Nature* 386:493–495.

Kempermann G, Kuhn HG, Gage FH (1998). Experience-induced neurogenesis in the senescent dentate gyrus. *J Neurosci* 18:3206–3212.

Kershman J (1938). The medulloblast and the medulloblastoma. *Arch Neurol Psychiat* 40:937–967.

Kilpatrick TJ, Bartlett PF (1993). Cloning and growth of multipotential neural precursors: Requirements for proliferation and differentiation. *Neuron* 10:255–265.

Koketsu D, Mikami A, Miyamoto Y, Hisatsune T (2003). Nonrenewal of neurons in the cerebral neocortex of adult macaque monkeys. *J Neurosci* 23:937–942.

Kornack DR, Rakic P (1999). Continuation of neurogenesis in the hippocampus of the macaque monkey. *Proc Natl Acad Sci USA* 96:5768–5773.

Kornack DR, Rakic P (2001a). The generation, migration, and differentiation of olfactory neurons in the adult primate brain. *Proc Natl Acad Sci USA* 98:4752–4757.

Kornack DR, Rakic P (2001b). Cell proliferation without neurogenesis in adult primate neocortex. *Science* 294:2127–2130.

Kronenberg G, Reuter K, Steiner B, Brandt MD, Jessberger S, Yamaguchi M, Kempermann G (2003). Subpopulations of proliferating cells of the adult hippocampus respond differently to physiologic neurogenic stimuli. *J Comp Neurol* 467:455–463.

Kuhn HG, Dickinson-Anson H, Gage FH (1996). Neurogenesis in the dentate gyrus of the adult rat: Age-related decrease of neuronal progenitor proliferation. *J Neurosci* 16:2027–2033.

Levison SW, Goldman JE (1993). Both oligodendrocytes and astrocytes develop from progenitors in the subventricular zone of postnatal rat forebrain. *Neuron* 10:201–212.

Lewis PD (1968). Mitotic activity in the primate subependymal layer and the genesis of gliomas. *Nature* 217:974–975.

Lindvall O, Backlund EO, Farde L, Sedvall G, Freedman R, Hoffer B, Nobin A, Seiger A, Olson L (1987). Transplantation in Parkinson's disease: Two cases of adrenal medullary grafts to the putamen. *Ann Neurol* 22:457–468.

Lois C, Alvarez-Buylla A (1993). Proliferating subventricular zone cells in the adult mammalian forebrain can differentiate into neurons and glia. *Proc Natl Acad Sci USA* 90:2074–2077.

Lois C, Alvarez-Buylla A (1994). Long-distance neuronal migration in the adult mammalian brain. *Science* 264:1145–1148.

Lois C, Garcia-Verdugo J-M, Alvarez-Buylla A (1996). Chain migration of neuronal precursors. *Science* 271:978–981.

Luskin MB (1993). Restricted proliferation and migration of postnatally generated neurons derived from the forebrain subventricular zone. *Neuron* 11:173–189.

Madrazo I, Drucker-Colin R, Diaz V, Martinez-Mata J, Torres C, Becerril JJ (1987). Open microsurgical autograft of adrenal medulla to the right caudate nucleus in two patients with intractable Parkinson's disease. *N Engl J Med* 316:831–834.

Madrazo I, Leon V, Torres C, Aguilera MC, Varela G, Alvarez F, Fraga A, Drucker-Colin R, Ostrosky F, Skurovich M, et al. (1988). Transplantation of fetal substantia nigra and adrenal medulla to the caudate nucleus in two patients with Parkinson's disease. *N Engl J Med* 318:51.

Magavi S, Leavitt B, Macklis J (2000). Induction of neurogenesis in the neocortex of adult mice. *Nature* 405:951–955.

Markram H, Lubke J, Frotscher M, Sakmann B (1997). Regulation of synaptic efficacy by coincidence of postsynaptic APs and EPSPs. *Science* 275:213–215.

Martin GR (1981). Isolation of a pluripotent cell line from early mouse embryos cultured in medium conditioned by teratocarcinoma stem cells. *Proc Natl Acad Sci USA* 78:7634–7638.

Mazzarello P (1999). A unifying concept: The history of cell theory. *Nat Cell Biol* 1:E13–15.

Merk L (1896). Die Mitosen im Centralnervensysteme. *Denkschriften der königlichen Akademie der Wissenschaften, math-naturwissenschaftliche Classe* 53:79–118.

Messier B, Leblond CP, Smart I (1958). Presence of DNA synthesis and mitosis in the brain of young adult mice. *Exp Cell Res* 14:224–226.

Mirzadeh Z, Merkle FT, Soriano-Navarro M, Garcia-Verdugo JM, Alvarez-Buylla A (2008). Neural stem cells confer unique pinwheel architecture to the ventricular surface in neurogenic regions of the adult brain. *Cell Stem Cell* 3:265–278.

Muotri AR, Chu VT, Marchetto MC, Deng W, Moran JV, Gage FH (2005). Somatic mosaicism in neuronal precursor cells mediated by L1 retrotransposition. *Nature* 435:903–910.

Nottebohm F (1980). Testosterone triggers growth of brain vocal control nuclei in adult female canaries. *Brain Res* 189:429–436.

Nottebohm F (1981). A brain for all seasons: Cyclical anatomical changes in song control nuclei of the canary brain. *Science* 214:1368–1370.

Nottebohm F (2002). Neuronal replacement in the adult brain. *Brain Res Bull* 57:737–749.

Olanow CW, Goetz CG, Kordower JH, Stoessl AJ, Sossi V, Brin MF, Shannon KM, Nauert GM, Perl DP, Godbold J, Freeman TB (2003). A double-blind controlled trial of bilateral fetal nigral transplantation in Parkinson's disease. *Ann Neurol* 54:403–414.

Palmer TD, Ray J, Gage FH (1995). FGF-2–responsive neuronal progenitors reside in proliferative and quiescent regions of the adult rodent brain. *Mol Cell Neurosci* 6:474–486.

Paton JA, Nottebohm FN (1984). Neurons generated in the adult brain are recruited into functional circuits. *Science* 225:1046–1048.

Penfield W (1932). Neuroglia: normal and pathological. In: *Cytology and Cellular Pathology of the Nervous System* (Penfield W, ed.), pp. 423–479. New York: Hoeber.

Piccini P, Brooks DJ, Bjorklund A, Gunn RN, Grasby PM, Rimoldi O, Brundin P, Hagell P, Rehncrona S, Widner H, Lindvall O (1999). Dopamine release from nigral transplants visualized in vivo in a Parkinson's patient. *Nat Neurosci* 2:1137–1140.

Rakic P (1985). Limits of Neurogenesis in Primates. *Science* 227:1054–1056.

Ramony Cajal S (1928). *Degeneration and Regeneration of the Nervous System*. New York: Hafner.

Ray J, Peterson DA, Schinstine M, Gage FH (1993). Proliferation, differentiation, and long-term culture of primary hippocampal neurons. *Proc Natl Acad Sci USA* 90:3602–3606.

Ray J, Raymon HK, Gage FH (1995). Generation and culturing of precursor cells and neuroblasts from embryonic and adult central nervous system. *Methods Enzymol* 254:20–37.

Remak R (1852). Über extracelluläre Entehung thierischer Zellen und über die Vermehrung derselben durch Theilung. *Archiv für Anatomie, Physiologie und wissenschaftliche Medicin*:137–176.

Remak R (1855). *Untersuchungen über die Entwicklung der Wirbelthiere*. Berlin: G. Reimer.

Reynolds BA, Weiss S (1992). Generation of neurons and astrocytes from isolated cells of the adult mammalian central nervous system. *Science* 255:1707–1710.

Rio-Hortega P (1919). El tercer elemento de los centros nerviosos. *Bol Soc Esp de Biol* 9:69.

Rosenzweig MR (2003). Effects of differential experience on the brain and behavior. *Dev Neuropsychol* 24: 523–540.

Rosenzweig MR, Bennett EL (1996). Psychobiology of plasticity: effects of training and experience on brain and behavior. *Behav Brain Res* 78:57–65.

Rydberg E (1932). Cerebral injury in newborn children consequent to birth trauma. *Acta Pathol Microbiol Scand* Suppl. 10:1–247.

Sagan C (1995). *The Demon-Haunted World*. New York: Random House.

Sahay A, Hen R (2007). Adult hippocampal neurogenesis in depression. *Nat Neurosci* 10:1110–1115.

Sanai N, Tramontin AD, Quinones-Hinojosa A, Barbaro NM, Gupta N, Kunwar S, Lawton MT, McDermott MW, Parsa AT, Manuel-Garcia Verdugo J, Berger MS, Alvarez-Buylla A (2004). Unique astrocyte ribbon in adult human brain contains neural stem cells but lacks chain migration. *Nature* 427:740–744.

Schaper A (1897). Die frühesten Differenzierungsvorgänge im Zentralnervensystem. *Arch f Entwicklungsmech* 5:81.

Schreiber L, Wengler F (1908). Über die Wirkungen des Scharlachöls auf die Netzhaut. Mitosenbildung in Ganglienzellen. *Centralblatt für allgemeine Pathologie* 19:529.

Schreiber L, Wengler F (1910). Über die Wirkungen des Scharlachöls auf die Netzhaut. Mitosenbildung in Ganglienzellen. *Graefes Archiv für Ophthalmologie* 74.

Seki T, Arai Y (1991). The persistent expression of a highly polysialylated NCAM in the dentate gyrus of the adult rat. *Neurosci Res* 12:503–513.

Seki T, Arai Y (1993). Highly polysialylated neural cell adhesion molecule (NCAM-H) is expressed by newly generated granule cells in the dentate gyrus of the adult rat. *J Neurosci* 13:2351–2358.

Seki T, Arai Y (1999). Temporal and spatial relationships between PSA-NCAM-expressing, newly generated granule cells, and radial glia–like cells in the adult dentate gyrus. *J Comp Neurol* 410:503–513.

Shors TJ, Miesegaes G, Beylin A, Zhao M, Rydel T, Gould E (2001). Neurogenesis in the adult is involved in the formation of trace memories. *Nature* 410:372–376.

Smart I (1961). The subependymal layer of the mouse brain and its cell production as shown by radiography after thymidine-H3 injection. *J Comp Neurol* 116:325–347.

Smart I, Leblond CP (1961). Evidence for division and transformation of neuroglia cells in the mouse brain as derived from radioautography after injection of thymidine-H3. *J Comp Neurol* 116:349–367.

Spielmeyer W (1922). *Histopathologie des Nervensystems*. Berlin: Julius Springer.

Stanfield BB, Trice JE (1988). Evidence that granule cells generated in the dentate gyrus of adult rats extend axonal projections. *Exp Brain Res* 72:399–406.

Tamura Y, Kataoka Y, Cui Y, Takamori Y, Watanabe Y, Yamada H (2007). Multi-directional differentiation of doublecortin- and NG2-immunopositive progenitor cells in the adult rat neocortex in vivo. *Eur J Neurosci* 25:3489–3498.

Thomas ED (1999). Bone marrow transplantation: A review. *Semin Hematol* 36:95–103.

Thomson JA, Itskovitz-Eldor J, Shapiro SS, Waknitz MA, Swiergiel JJ, Marshall VS, Jones JM (1998). Embryonic stem cell lines derived from human blastocysts. *Science* 282:1145–1147.

Toni N, Laplagne DA, Zhao C, Lombardi G, Ribak CE, Gage FH, Schinder AF (2008). Neurons born in the adult dentate gyrus form functional synapses with target cells. *Nat Neurosci* 11:901–907.

van Praag H, Kempermann G, Gage FH (1999). Running increases cell proliferation and neurogenesis in the adult mouse dentate gyrus. *Nat Neurosci* 2:266–270.

van Praag H, Schinder AF, Christie BR, Toni N, Palmer TD, Gage FH (2002). Functional neurogenesis in the adult hippocampus. *Nature* 415:1030–1034.

Virchow R (1846). Über das granulierte Aussehen der Wandungen der Gehirnventrikel. *Allgemeine Zeitschrift für Psychiatrie* 3:242–250.

Virchow R (1859). *Die Cellularpathologie in ihrer Begründung auf physiologische und pathologische Gewebelehre.* Berlin: Verlag von August Hirschfeld.

Watermann R (1982). [Schwann's microscopic research and his cell theory. Contribution on his 100th anniversary.] *Z Mikrosk Anat Forsch* 96:1032–1043.

Wenning GK, Odin P, Morrish P, Rehncrona S, Widner H, Brundin P, Rothwell JC, Brown R, Gustavii B, Hagell P, Jahanshahi M, Sawle G, Bjorklund A, Brooks DJ, Marsden CD, Quinn NP, Lindvall O (1997). Short- and long-term survival and function of unilateral intrastriatal dopaminergic grafts in Parkinson's disease. *Ann Neurol* 42:95–107.

Wilson RS, Mendes De Leon CF, Barnes LL, Schneider JA, Bienias JL, Evans DA, Bennett DA (2002). Participation in cognitively stimulating activities and risk of incident Alzheimer disease. *JAMA* 287:742–748.

Winsor MP (1980). "Swammerdam, Jan." In: *Dictionary of Scientific Biography*, vol. 13 (Gillespie C, ed.), pp. 168–175. New York: Scribner.

Wolpert L (1995). Evolution of cell theory. *Phil Trans R Soc Lond B* 349:227–233.

Yang Y, Geldmacher DS, Herrup K (2001). DNA replication precedes neuronal cell death in Alzheimer's disease. *J Neurosci* 21:2661–2668.

Yang Y, Mufson EJ, Herrup K (2003). Neuronal cell death is preceded by cell cycle events at all stages of Alzheimer's disease. *J Neurosci* 23:2557–2563.

Zhao C, Teng EM, Summers RG, Jr., Ming GL, Gage FH (2006). Distinct morphological stages of dentate granule neuron maturation in the adult mouse hippocampus. *J Neurosci* 26:3–11.

3

Neural Stem Cells

The insight that stem cells might be central to brain plasticity, including but not limited to adult neurogenesis, is a recent idea

The discovery of adult neural stem cells solved the question of the origin of adult neurogenesis

Somatic stem cells are presumably found in all organs

Adult neural stem cells are heterogeneous

ADULT NEUROGENESIS originates from precursor cells in the adult brain. Therefore neural stem cell biology is the basis of the development of new neurons in the adult brain just like it is during embryonic and fetal brain development. The idea that stem cells are an indispensable prerequisite of adult neurogenesis might today seem self-evident. But for decades, concepts of brain plasticity fared well without accommodating stem cells, and in fact, even today, stem cells, cell-based plasticity, and adult neurogenesis have hardly found entry into most "mainstream" concepts of how the brain works.

Although even the earliest reports on adult neurogenesis pondered the idea of precursor cells' giving rise to the neurons, at the time there was no direct experimental evidence in support of this assumption (Altman and Das, 1965; Kaplan, 1983). It is thus no coincidence that research on adult neurogenesis and its acceptance in the scientific community gained ground only when neural stem cells were discovered in the early 1990s (Reynolds and Weiss, 1992; Richards et al., 1992; Kilpatrick and Bartlett, 1993, 1995; Palmer et al., 1995; Ray et al., 1995; Palmer et al., 1997). The existence of stem cells solved the problem of the origin of the new neurons and allowed adult neurogenesis to be seen within generally accepted frameworks of developmental neurobiology.

Historically, it was thought that stem cells would be active only during embryogenesis and would, once development had ceased, persist only in organs with inherent regenerative ability: bone marrow, skin, and the intestines. Only with the demonstration of precursor cells in organs considered to be non-regenerating, such as the lung, the liver, and finally the heart and the brain, were stem cells viewed increasingly as a ubiquitous phenomenon in the body. Visions of exploiting the regenerative potential of these cells for new therapeutic approaches for incurable disorders stimulated the exploration of the identity and nature of stem cells in the adult brain.

Whereas in the adult hematopoietic system all different mature cell types originate from one single stem cell (Fig. 3–1), this is apparently not the case in the adult brain, where precursor cells in the subventricular zone differ from those in the hippocampus and from the (glial) precursor cells in the white matter (Fig. 3–2). In the adult brain we do not seem to find an

51

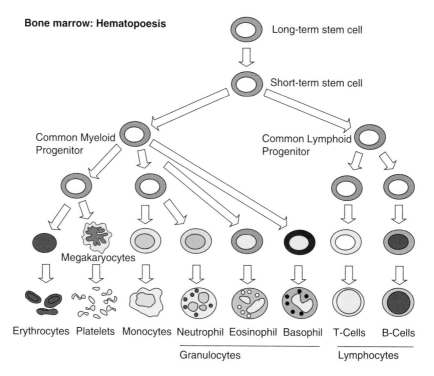

Bone marrow: Hematopoesis

Long-term stem cell

Short-term stem cell

Common Myeloid Progenitor

Common Lymphoid Progenitor

Megakaryocytes

Erythrocytes Platelets Monocytes Neutrophil Eosinophil Basophil T-Cells B-Cells

Granulocytes Lymphocytes

FIGURE 3–1 Stem cell hierarchies. In hematology, there is a detailed hierarchy of the various mature cell types and the progenitor cells from which the former are derived. Although many questions remain open, the individual cell types can be comparatively well defined by sets of key surface markers.

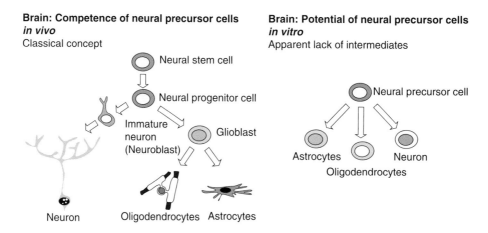

Brain: Competence of neural precursor cells
in vivo
Classical concept

Neural stem cell

Neural progenitor cell

Immature neuron (Neuroblast)

Glioblast

Neuron Oligodendrocytes Astrocytes

Brain: Potential of neural precursor cells
in vitro
Apparent lack of intermediates

Neural precursor cell

Astrocytes Neuron

Oligodendrocytes

FIGURE 3–2 Compared to the situation in hematology (Fig. 3–1), a widely used schema of precursor cell hierarchy in the adult brain (*left*) appears to be comparatively crude. In contrast to the situation in hematology, where all adult-generated lineages trace back to one type of bone marrow precursor cell, the highest ranking stem cells in the hierarchy, comparable to the neuroepithelial cell from brain development does not persist into adulthood. Even more problematic, most in vitro systems (*right*) do not reflect the proposed situation in vivo. Lineages in vitro are often assessed with one marker only.

equivalent to the neuroepithelial cell, from which essentially the entire nervous system is derived and which exists in the walls of the neural tube, a few days to weeks after conception, depending on the species.

If one looks at putative "stem cell" activity solely from the perspective of cell proliferation in the adult brain, neurogenesis contributes only a minor fraction to it. Both astrocytes and oligodendrocytes are produced in considerable numbers in the adult brain. Microglia is also made in the adult brain but supposedly originates from blood monocytes that enter the brain. Their origins notwithstanding, microglia can proliferate in situ, under both normal and pathological conditions. The minority opinion that microglia and macroglia (astrocytes and oligodendrocytes) might share a common precursor has lost ground because microglia has never been found in neural stem cell cultures.

Most precursor cells in the adult brain do not seem to be neurogenic

The rise of neural stem cell biology has made investigators wonder whether mature macroglia—astrocytes and oligodendrocytes—can divide at all or whether all their apparent proliferation in fact reflects precursor cell activity. Along these lines, new concepts in neuro-oncology hypothesize that gliomas—that is, brain tumors with characteristics of glia—in fact originate from precursor cells. Similarly, in the reactive gliosis that occurs after brain injury, a phenomenon long considered to originate from mature astrocytes, local precursor cells might generate the activated astrocytes.

The ultimate specific potential of adult neural precursor cells is unknown

There are hundreds of known types of neurons in the adult brain, and it is actually difficult to establish what would constitute the identity any two neurons. To date it is not known whether all neural and glial cell types might at least in theory be generated from precursor cells that are still present in the adult brain. Nor do we know whether the cellular environment in the adult brain could support new neurons or glial cells of any desirable phenotype. This seems unlikely, given the high specialization of adult neurogenesis. However, it remains to be shown to what degree this specificity is a function of precursor cell properties or of the permissive microenvironment (Chap. 8).

The Two Cardinal Features of Stem Cells: Self-Renewal and Multipotency

Neural stem cells are undifferentiated cells that divide and from which development of the nervous system (or parts of it) originates. Adult neurogenesis is thus a concrete manifestation of stem cell biology (Fig. 3–3).

Stem cells are characterized by two cardinal properties: their ability to undergo unlimited self-renewal by division and the potency to generate at least two different cell types (McKay, 1997; van der Kooy and Weiss, 2000; Weissman et al., 2001; Parker et al., 2005). The ability to generate a variety of differentiated cells is called *multipotency*. The term has a second, more restricted meaning in stem cell biology, referring to the concrete range of developmental potential of somatic, or "adult," stem cells (see below). Many researchers in the field consider the multipotency criterion unnecessary and base their definition and assessment of "stemness" solely on self-renewal. In that concept, *unipotent* stem cells can exist, giving rise to only one type of differentiated cell.

In practice, "stem cell" is a rather generic label, and the theoretically defining criteria have usually not been tested

"Self-renewal plus multipotency" is a minimal definition, but in most concrete instances it remains unclear whether even these most basic criteria are met. Most of the time, *stem cell* is a bona fide label whose concrete meaning and implications depend on the context. Particularly in the popular media, the term *stem cell* often represents some medical silver bullet and has thus accrued so many everyday connotations that talking about stem cells scientifically has sometimes become difficult. Fortunately, in most contexts the damage due to the superficial usage of *stem cell* is limited, as long as it is understood that no specific statements about these cells are made. *Stem cell* in the popular

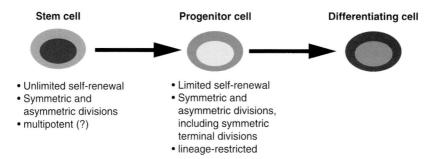

Stem cell   **Progenitor cell**   **Differentiating cell**

- Unlimited self-renewal
- Symmetric and asymmetric divisions
- multipotent (?)

- Limited self-renewal
- Symmetric and asymmetric divisions, including symmetric terminal divisions
- lineage-restricted

FIGURE 3–3 Definitions of stem and progenitor cells. Stem cells produce progenitor cells, not differentiating cells. Progenitor cells generate differentiating cells. Stem and progenitor cells are summarized under the umbrella term *precursor cells.*

sense is a generic term for one cell that can "make" a complex piece of bodily tissue. This meaning is not wrong, but in the end it is insufficient.

Self-Renewal

A self-renewing stem cell generates at least one daughter cell that is an identical copy of the mother

Self-renewal is the most important criterion of stemness and means that cell division generates at least one identical copy of the mother cell. If the division yields two identical copies, the division is called *symmetric* (Fig. 3–4). *Asymmetric* divisions produce one new stem cell that is identical to the mother cell and one cell that is more determined for a certain lineage of cellular differentiation than the stem cell. These daughter cells have reduced stem cell properties and are referred to as *progenitor cells.*

Symmetric division does not have to generate two cells identical to the mother cell. A division would still be symmetric if the two daughter cells were identical to each other but different from the mother cell. This is, for example, the case with the "neurogenic division" at later stages of neuronal development, where a progenitor cell divides terminally to give rise to two neurons. However, when "stem cells" are defined, it is implied that daughter and mother cells are identical. If a stem cell were to divide in such a way that none of the daughter cells was an identical copy of the mother cell, another mandatory criterion of stem cells would be violated: the unlimited nature of its self-renewal. In theory, unlimited self-renewal makes stem cells eternal; in reality, however, it is difficult to prove unlimited self-renewal even for the life span of the organism.

The adult brain contains precursor cells fulfilling the self-renewal criterion

Self-renewal within the limits of the applicable test has been shown for precursor cells from both the adult dentate gyrus and the subventricular zone (Palmer et al., 1997; Seaberg and van der Kooy, 2002; Babu et al., 2007). Details on the methodology how this is achieved is found in Chapter 7, page 221. The concept of symmetric vs. asymmetrical divisions is revisited in the next Chapter, page 115.

Progenitor cells are the progeny of stem cells and are incapable of unlimited self-renewal but can proliferate and expand their population

Progenitor Cells and Precursor Cells

The unlimited self-renewal of stem cells is conceptually set against a limited self-renewal of the progenitor cells, the immediate progeny of true stem cells (Weissman et al., 2001; Seaberg and van der Kooy, 2003). Irrespective of their limited self-renewal, progenitor cells are usually more proliferative than stem cells and can vastly expand the number of new cells.

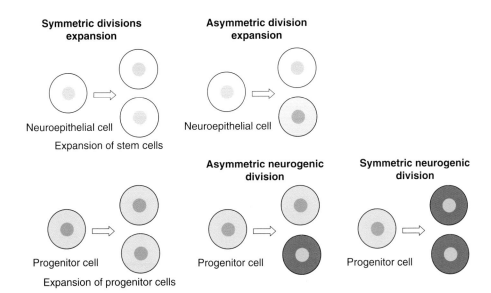

Symmetric divisions expansion

Neuroepithelial cell

Expansion of stem cells

Asymmetric division expansion

Neuroepithelial cell

Progenitor cell

Expansion of progenitor cells

Asymmetric neurogenic division

Progenitor cell

Symmetric neurogenic division

Progenitor cell

FIGURE 3–4 Asymmetrical divisions. Asymmetrical division is one hallmark of stem cells. The key criterion for stem cells is self-renewal. Every stem cell thus has to generate one new stem cell during its division. To produce differentiating progeny at the same time, asymmetrical divisions have to occur. Progenitor cells are the progeny of stem cells and have limited stem cell properties, most notably limited self-renewal. Symmetrical division of progenitor cells can produce two daughter cells that are different from the mother cell. The schema presented here is based primarily on the work of Wieland Huttner and colleagues (Wodarz and Huttner, 2003; Kosodo et al., 2004).

They are thus referred to as *transit-amplifying* or *transiently amplifying progenitor cells*. The fact that the number of divisions is limited in such highly proliferative cells might act as a means of protection against uncontrolled growth (Fig. 3–3).

Because self-renewal of progenitor cells is by definition not unlimited, a terminal symmetric division of progenitor cells can generate two daughter cells that are identical to each other but different from the mother cell (Takahashi et al., 1996; Cai et al., 2002; Huttner and Kosodo, 2005; Zhong and Chia, 2008). The neuroepithelial cells, the stem cells in the wall of the neural tube in embryonic development, divide symmetrically during the initial expansion phase (Fig. 3–4). They divide asymmetrically to produce neuronally determined progenitor cells in asymmetrical neurogenic divisions. Such progenitor cells, detached from the surface of the ventricle, can divide symmetrically into two cells that are both differentiating into neurons. This constitutes a symmetric neurogenic division (Kosodo et al., 2004; Huttner and Kosodo, 2005). Consequently, these progenitor cells are not self-renewing and thus are not considered stem cells in the strict sense of the definition. Precursor cells that are not stem cells but have stemness properties are also referred to as "intermediate progenitor cells" (see also below, p. 60) as they stand between stem cells and mature cells.

The conceptual distinction between stem and progenitor cells occasionally has concrete relevance, because many conclusions in stem cell biology and research on adult neurogenesis are based on the inferred properties and the assumed potential of the precursor cells (Seaberg and van der Kooy, 2003). By definition, the immediate progeny of stem cells are progenitor cells, not differentiated cells. Progenitor cells in turn generate differentiated cells and have limited stem cell properties. Consequently, the regulatory principles controlling development at the two different stages of stem cells vs. progenitor cells will differ and so will the consequences of the regulatory event. Regulation of stem cells might be slow but persistent, whereas regulation of precursor cells allows a swift but transient response.

Despite the usefulness of the concept of distinguishing between stem and progenitor cells, it is extremely difficult if not impossible to practically distinguish the two classes, particularly in vivo. Experiments can be designed that allow the classification in vitro (Seaberg and van der Kooy, 2002; Kim and Morshead, 2003; Seaberg and van der Kooy, 2003), but culture conditions influence the outcome. The sphere-forming assay alone, if not combined with other analyses, has its limitations, which we will discuss in Chapter 8.

Precursor cells is the preferred umbrella term for cells whose stemness properties were not or could not be practically determined

Cells with different degrees of stemness (self-renewal and multipotency) can be identified in the brain (and elsewhere), but in a concrete situation it is usually not possible to categorize a cell as either a stem cell or a progenitor cell. Sometimes the term *progenitor cell* is used as the more parsimonious label, implying that it remains undetermined whether the given cell is in fact capable of unlimited self-renewal and has the potential to generate multiple cell lineages. However, to avoid confusion, the term *precursor cell* should be favored as an umbrella term encompassing both stem and progenitor cells.

Levels of Potency

Stem cells differ by their potential to generate different cell lineages

The second defining criterion of stem cells besides self-renewal is multipotency. Based on the expression of different marker sets and the lack of an overlap between other markers, cell lineages can be identified in vivo. We will look at the two canonical neurogenic regions in more detail in Chapters 5 and 6. That the proliferatively active cells are precursor cells, however, is usually a conclusion by analogy. When isolated from the brain and propagated in cell culture, many cells with antigen profiles equivalent to those of the putative precursor cells in vivo are self-renewing and multipotent. It has not yet been possible, however, to recapitulate the hypothesized in vivo precursor cell hierarchy in vitro, or vice versa. Some authors thus a make distinction between the "potential" of precursor cells in vitro and their realized "competence" in vivo (Stiles and Rowitch, 2008). Multipotency can be studied in vivo by clonal retroviral labeling (Walsh and Cepko, 1992), but no successful attempt in the adult brain has been reported so far. Non-clonal analyses based on retroviruses or conditional transgenesis, however, have been performed in the two neurogenic zones and support multipotency (Suh et al., 2007).

The zygote has the highest degree of developmental potential, called totipotency

The potential for differentiation into different cell types can be subdivided into a number of categories of potency, or potential (Fig. 3–5). We call the differentiation potential of the fertilized egg *totipotent*, the one of embryonic stem cells *pluripotent*, and the one of all somatic stem cells *multipotent*.

The fertilized egg, the zygote, is considered totipotent because it can produce an entire organism, including the trophoblast, the placenta. The zygote is thus often considered the ultimate stem cell. However, the zygote violates the stem cell definition of unlimited self-renewal because it loses its totipotency after two or three divisions. There are no totipotent cells beyond these earliest stages of development.

Further divisions lead to the morula and the blastocyst stage, which begins at approximately the 64–cell stage. A cavity forms and the wall of the little cyst thickens at one side. This local accumulation of cells is called the *inner cell mass*.

The potency level of ES cells excludes only the trophoblast and is called pluripotency

Embryonic Stem Cells

Embryonic stem (ES) cells are pluripotent and are found in the inner cell mass of the blastocyst. Murine ES cell cultures were first described independently by Evans and Kaufman and by Martin in 1981 (Evans and Kaufman, 1981;

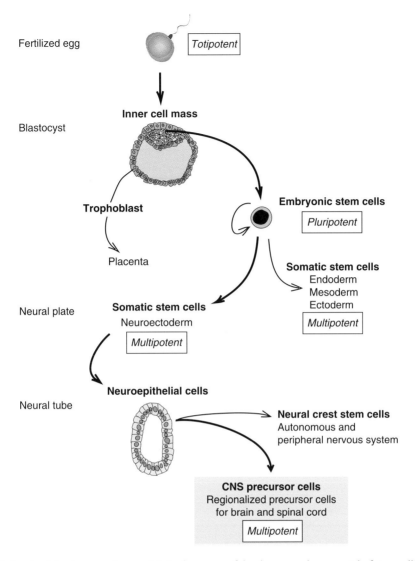

Fertilized egg — *Totipotent*

Blastocyst

Inner cell mass

Trophoblast

Placenta

Embryonic stem cells
Pluripotent

Somatic stem cells
Endoderm
Mesoderm
Ectoderm
Multipotent

Neural plate

Somatic stem cells
Neuroectoderm
Multipotent

Neural tube

Neuroepithelial cells

Neural crest stem cells
Autonomous and
peripheral nervous system

CNS precursor cells
Regionalized precursor cells
for brain and spinal cord
Multipotent

FIGURE 3–5 Levels of developmental potential. In the course of development, the potential of stem cells to generate different types of cells becomes increasingly restricted. Adult neurogenesis originates from multipotent somatic precursor cells.

Martin, 1981). The first ES cell cultures from humans were presented by Thomson in 1998 (Thomson et al., 1998), thereby initiating a gigantic boom in stem cell research as well as the worldwide debates about the ethical aspects of research on human embryos.

The trophoblast, from which the placenta originates, is not part of the inner cell mass but surrounds it. *Pluripotency* is consequently defined as the ability to generate cells of all body tissues except the trophoblast. Or more simply: Pluripotency is totipotency minus placenta. If an ES cell is implanted into the uterus, it cannot give rise to a complete new organism because it cannot generate the placenta. Because ES cells are pluripotent and can be propagated in vitro with relative ease, they hold great promise for regenerative medicine. In theory they might provide a source for implantable cells in cell replacement therapies, possibly a strategy to treat some incurable diseases.

The potential therapeutic use of human ES cells has stirred stem cell debates in many countries. Embryonic stem cells are not controversial because of their pluripotency, and the use of ES cells is not controversial per se; what is controversial is the use of *human* ES cells or of pluripotent cells, as long as these have to be derived from human embryos. The stem cell debate has largely been a debate not over stem cells but over the ethical status of the human embryo.

‖‖

Pluripotency can be induced artificially in somatic cells

‖‖

Since 2008, with the report of "induced pluripotency" (see below), this emphasis on therapy has somewhat waned because induced pluripotent cells have none of the controversial ethical implications the use of human ES cells has. If "induced pluripotency" holds its promises, ES cell research will largely withdraw to basic research and further lose its therapeutic appeal. In theory, induced pluripotent cells can be made from cells with any genotype, opening new avenues for "personalized medicine," something that in the case of regular ES cells would be only possible by means of somatic cell nuclear transfer (SCNT, the so-called therapeutic cloning), a technique raising additional and different ethical and technical concerns.

Another important goal is to generate cell lines from pluripotent cells that carry specific mutations to develop better cell models of human disease. Such cell lines would be ideally suited to test new drugs and learn about disease-specific signaling mechanisms (see Chap. 12).

Neural Differentiation from Pluripotent Cells In Vitro

ES cells can be differentiated into neurons in vitro, and this has been shown for various species, including humans. Mainly three different core strategies have been used:

- To culture ES cells as monolayers and use sequentially applied growth factors or other signaling molecules such as noggin to obtain neurons of different specification (Reubinoff et al., 2001; Perrier et al., 2004; Gerrard et al., 2005; Itsykson et al., 2005; Keirstead et al., 2005).
- To co-culture ES cells with stromal cells or astrocytes to induce the differentiation into neurons (Kawasaki et al., 2000; Morizane et al., 2006; Yue et al., 2006).
- To let the ES cells form embryoid bodies in the presence of serum (Carpenter et al., 2001; Tropepe et al., 2001; Zhang et al., 2001) or defined media (Zhou et al., 2008) and subsequent selection of neural precursor cells.

‖‖

ES cells in vitro can be guided to develop into neurons (and other neural cells)

‖‖
‖‖

The generation of defined neural precursor cells rather than generic types of neurons posed a major challenge

‖‖

With such protocols, neurons of different subspecification have been obtained, most notably dopaminergic neurons (Kawasaki et al., 2000; Zeng et al., 2004; Yue et al., 2006) and motoneurons (Li et al., 2005; Wichterle and Peljto, 2008). In some animal models of human disease, ES cell–based cell replacement has been promising (for review, see, for example, Glaser et al., 2008; Li et al., 2008; Srivastava et al., 2008).

The level of confidence with which one can state that the cells that were obtained by the differentiation protocols were truly neurons or neural precursor cells varies between studies, so not all success stories should be taken at face value. What turned out to be particularly challenging—more challenging than the generation of more differentiated neurons, anyway—was the generation of specified neural precursor cell populations. This is of foremost importance to the obtaining of defined cell sources that could be safely expanded to generate desired neuronal phenotypes without the necessity of handling pluripotent cells. The question is also very interesting from a basic-research point of view. Ontogenetically, ES cells precede the more specified neural precursor cells. For the initial transition from pluripotency to the more restricted neural stem cell state, a suppression of Oct4 by germ cell nuclear factor (GCNF) is required (Akamatsu et al., 2009).

Much effort has thus focused on the development of neuroepithelial cells from ES cells in vivo. ES cells could be developed into an NE-like cell that forms rosette-like structures in vitro, intriguingly suggestive of their attempts to build a neural tube (Karki et al., 2006; Elkabetz et al., 2008; Koch et al., 2009). These cells could be prospectively identified on the basis of their reaction with Forse1 antibody (Elkabetz et al., 2008), which recognizes an epitope that is found on early neuronal precursor cells (Tole et al., 1995) and is possibly related to LeX/SSEA1 (Allendoerfer et al., 1995).

ES cells can be differentiated into neuroepithelial-like cells

Their pluripotency makes ES cells challenging to use in vivo, as ES cells have a tumorigenic potential and can generate dysontogenetic tumors, teratomas, or teratocarcinomas (Langa et al., 2000; Reubinoff et al., 2001; Asano et al., 2003; Blum and Benvenisty, 2008). These tumors are characterized by the presence of more or less differentiated cell types of all three germ layers. In that sense, the ability to form teratomas is characteristic of pluripotent cells. But ES cells have further tumorigenic potential in that they might adapt to the culture conditions, which would promote karyotypic instabilities and cancerous growth (Baker et al., 2007).

ES cells can by definition generate teratomas; that is, tumors containing all three germ layers

Neural Precursor Cells Are Multipotent

Stem cells within each germ layer are called *multipotent*. Because neural stem cells are part of the neuroectoderm, they are by definition multipotent within the boundaries of neuroectoderm. The use of the term in this context is more concrete and restricted than when the same term is used to describe the principle that stem cells by definition can produce mature cells of two or more lineages (multipotency).

The term *somatic stem cells* should be preferred to *adult stem cells*

In many, especially popular, contexts, the terms *adult stem cells* and *somatic stem cells* are used synonymously for "multipotent stem cells," which are in contrast to the pluripotent ES cells. In this sense, *adult* stem cells can be derived from an embryo, a fetus, or early postnatally. In contrast, in the scientific literature, *adult stem cells* normally implies that the cells in question are found in or isolated from the organism after general development has ceased.

Multipotent adult precursor cells (MAPC) are rare cells in vitro that can develop beyond the boundary of one germ layer

Matters became further complicated around the year 2000 by the identification of adult precursor cells that seem to take a position between multipotency and pluripotency. In bone marrow, Catherine Verfaillie and her coworkers identified a rare precursor cell that can differentiate into cells across germ layer boundaries (Jiang et al., 2002b). Confusingly for many, although for these cells pluripotency was suggested, they were still (and in the end presumably more appropriately) named "multipotent adult precursor cells" (MAPC). They have been related to endothelial progenitor cells, but their exact nature remains to be clarified. Besides similar findings from bone marrow by others (Sauerzweig et al., 2009), related observations have been made for cells from skin (Toma et al., 2001; Toma et al., 2005) and postnatal and adult brains (Rietze et al., 2001; Jiang et al., 2002a). Ratajczak and colleagues have called such cells "very small embryonic-like stem cell[s]" and described various conditions under which such cells might be found (Ratajczak et al., 2008); they even reported that these cells were released into the blood stream in greater numbers after stroke (Kucia et al., 2008; Paczkowska et al., 2009).

In any case, these cells are exquisitely rare and thus far have only been identified in cell cultures and after a great number of passages. Although they might thus be a product of culture conditions, their rareness might indicate the differential susceptibility of various cells in the cultures to such inducing conditions and thus still support the idea of a pre-existing precursor-cell heterogeneity. Apparently, MAPC can generate differentiating cells across the limits of one germ layer, but they are not fully pluripotent in the sense in which embryonic stem cells are. Whereas MAPC have been studied for years now, and a successful clinical application has even been claimed (Kovacsovics-Bankowski et al., 2009), there have not been many follow-ups on the analogous cell types from other tissues, and, most important for our context, there has been no additional evidence for more-than-multipotent stem cells in the brain. MAPC are today considered to be a particular type of mesenchymal stem cells that will be discussed in more detail below (p. 87).

Precursor cells in the adult neurogenic zones are multipotent

Tripotency for neurons, astrocytes and oligodendrocytes has been established in vitro but not in vivo

For most purposes, it is sufficient to distinguish between the three levels of totipotency, pluripotency, and multipotency. In adult neurogenesis we only encounter multipotent precursor cells. These cells could be stem or progenitor cells with respect to their degree of self-renewal and multipotency, but their exact identity in a specific situation in vivo is often unknown. In theory, these cells are actually "tri-potent" in that they might generate neurons, astrocytes, and oligodendrocytes.

Indeed, ex vivo, precursor cells from the neurogenic regions of the adult brain exhibit stem cell properties (Reynolds and Weiss, 1992, 1996; Palmer et al., 1997; Seaberg and van der Kooy, 2002; Reynolds and Rietze, 2005; Babu et al., 2007) and can generate cells from the three neural lineages. But neither for the adult nor for the embryonic brain has the existence of a tripotent precursor cell been confirmed in vivo. To conduct such proof, one must rely on ultimately clonal retroviral lineage tracing, which is challenging to perform (compare Suh et al., 2007). Support of the idea that no tripotent progenitor cells exist in vivo has come from another side by the demonstration that fibroblast growth factor (FGF) might non-physiologically induce the tripotent nature of the cultured cells (Gabay et al., 2003). In one of the key publications of this particular research area, Connie Cepko concluded from such diverging results that the "potential" of a cell in vitro might not be identical to its "competence" in vivo (Cepko et al., 1996).

Even within one neurogenic region there are often intriguing differences in precursor cell properties and potential, depending on their origin (Golmohammadi et al., 2008). Clones of different composition are found ex vivo, and observation of cell-cultures with video-microscopy has revealed that lineage determination in vitro appears to manifest itself sometimes days before any fate markers are expressed (Ravin et al., 2008). Again, this might tell us more about the culture conditions than about the "true" potential of neural precursor cells and the homogeneity of their response.

Neuroblasts are migratory progenitor cells with limited proliferative activity

Hierarchically below the multipotent cells we find more-limited precursor cells that can generate only two cell lineages or even only one. These are referred to as *bipotent* and *unipotent*. Unipotent progenitor cells are cells that can expand the population of immediate precursors of differentiated cells, the last proliferative stage before differentiation and a post-mitotic stage is reached. Unipotent progenitor cells are sometimes called "blasts," but this term, widely used in the hematopoietic system, remains ambiguous in the context of neural stem cells. The type-3 cells of the dentate gyrus (p. 199) and the A cells of the SVZ (p. 167) might be referred to as *neuroblasts*.

The Basic Model

A simple model describes development as the sequence stem cell, progenitor cell, neuroblast, neuron

Most work on stem cell–based development is founded on a general basic model (Fig. 3–6). In this (sometimes implicit) model, development originates from a self-renewing multipotent stem cell that gives rise to transiently amplifying progenitor cells. Progenitor cells generate lineage-determined blasts, which via a facultative terminal division produce post-mitotic progeny. This basic model has been extended, taking into account insights from hematopoesis, by introducing a long-term stem cell at the beginning of this process. This long-term stem cell serves as a more or less

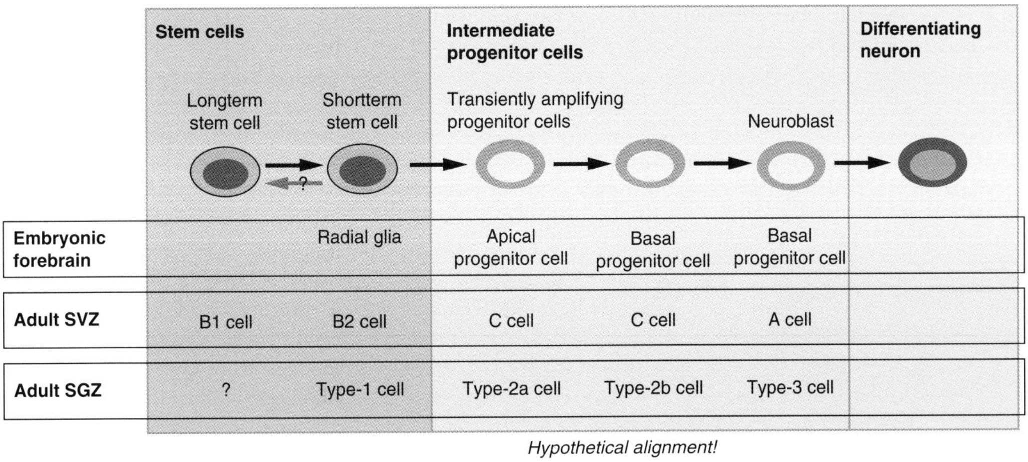

	Stem cells		Intermediate progenitor cells			Differentiating neuron
	Longterm stem cell	Shortterm stem cell	Transiently amplifying progenitor cells		Neuroblast	
Embryonic forebrain		Radial glia	Apical progenitor cell	Basal progenitor cell	Basal progenitor cell	
Adult SVZ	B1 cell	B2 cell	C cell	C cell	A cell	
Adult SGZ	?	Type-1 cell	Type-2a cell	Type-2b cell	Type-3 cell	

Hypothetical alignment!

FIGURE 3–6 The basic model. The basic model of stem cell biology consists of the core sequence stem cell – progenitor cell – blast – differentiating cell and is found in most systems. There are, however, certain variations, and often the exact alignment of model and reality is still speculative. Here, the situation during brain development (cortex) and in the adult neurogenic zones are shown.

quiescent reservoir that can replenish the population of stem cells that is now referred to as *short-term stem cells.*

According to most evidence this model is unidirectional

This linear model is extremely useful, and we will adhere to it throughout this book, but it is relatively simplistic and does not do justice to all observations that have been made. The question of directionality is an especially important issue. Can cells return to a level that their predecessors have previously passed through? Not only has the work on induced pluripotency added a new quality to the question of whether cells can "de-differentiate": we may also wonder, "Is the entire process strictly unidirectional?" Another issue is whether all steps are mandatory. This applies to the "blast" stage of lineage-determined progenitor cells. Are shortcuts to cell-cycle-exit possible and entry into maturation possible?

Figure 3–6 aligns the proposed sequence of precursor cells in adult neurogenesis. Details of these sequences are discussed in Chapters 5 and 6. The point here is the fact that, despite the general usefulness of the basic model, it is not a perfect fit, and it remains rather hypothetical in nature. The second point is that, even at this relatively low level of descriptive sophistication, differences between the two neurogenic regions of the adult brain become apparent.

Neural Stem Cells in the Developing Brain

Neuroepithelial cells in the wall of the neural tube are the primary stem cells of the developing nervous system

Development of the nervous system begins with the induction of neuroepithelial (NE) cells from ES cells in the blastocyst (Fig. 3–5). NE cells are the primary neural stem cells. The first stage of differentiation is induction of the primitive ectoderm from which all ectodermal cells arise. The primitive ectoderm gives way to the two ectodermal lineages: surface ectoderm and neuroectoderm.

The earliest neuroepithelial cells are found along the forming neural axis. Later, the primary population of stem cells resides in the wall of the neural tube. These cells are multipotent and characterized by expression of markers such as Sox1 and Sox2, nestin, and neural cell adhesion molecule (NCAM). From these earliest neural stem cells the central nervous system is generated. A subpopulation of stem cells migrates laterally and forms the neural crest, which becomes the origin of the peripheral and autonomous nervous systems.

The ventricular zone is the primary germinative center of the forebrain; the subventricular zone is the site of massive expansion

The primary germinative zone for forebrain development is the ventricular plate, the wall of this primordial ventricle. Here stem cells divide in symmetric and asymmetrical divisions and give rise to progenitor cell populations that reside in the subventricular zone, just below the ventricular zone. The subventricular progenitor cells proliferate massively and produce, first, neurons and later, glia.

Precursor cells from the ventricular plate in the dorsal brain produce the principal neurons of the cortex and astrocytes (Fig. 3–7). Principal neurons are excitatory neurons: for example, the pyramidal cells. Precursor cells from defined regions in the ventricular zone of the ventral brain (the "ganglionic eminences") generate neurons in the ventral brain (e.g., the basal ganglia), oligodendrocytes, and the complex diversity of inhibitory interneurons of both the ventral and dorsal brain. Mid- and hindbrain structures, as well as specialized structures such as the hypothalamus, neurophyphophysis, and others, originate from yet other stem cell populations whose relationship to those of the ventricular plate and the ganglionic eminences is not yet exactly known. Primary sensory epithelia such as retina, hair cells of the inner ear, and olfactory epithelium are derived from the optic, auditory, or olfactory placodes. Initiation of their development precedes cortical development.

Radial Glia as Stem Cells

After the neuroepithelial cells, radial glia act as key stem cell populations and double in function as guidance structure

Radial glia is the conventional name for a characteristic type of cell that plays several important roles during brain development. Radial glia's function is dual: it acts as a precursor cell in the developing brain and serves as a guidance structure that leads new neurons from the site of division near the ventricular surface to their final position in the cortex (Fig. 3–8). Cortical neurons originate from radial glia (Cameron and Rakic, 1991; Gaiano et al., 2000; Malatesta et al., 2000; Noctor et al., 2001; Noctor et al., 2002; Hartfuss et al., 2003; Malatesta et al., 2003; Tramontin et al., 2003), and the glial mother cells guide their daughter cells along their processes into their appropriate target region in the developing cortex (Noctor et al., 2001). Accordingly, radial glia is most abundant during the peak time of neurogenesis in the developing brain and afterwards transforms into astrocytes (Rakic, 1971; Eckenhoff and Rakic, 1984; Voigt, 1989; Misson et al., 1991). It was thought that radial glia would thereby disappear entirely with the cessation of brain development; it is now obvious, however, that adult neurogenesis is tied to an at least highly similar cell type. In the neurogenic regions, morphologically characteristic radial glia persists into adulthood (Alvarez-Buylla et al., 2001; Garcia-Verdugo et al., 2002; Filippov et al., 2003; Steiner et al., 2006). The transformation of a subset of radial glia into radial glia–like precursor cells of the adult neurogenic regions is in line with other types of specialized development, such as into Bergmann glia of the cerebellum or Müller cells of the retina.

Besides their distinct morphology, radial glia have a particular distinct marker profile including BLBP, Nestin, GLAST and Sox2

In the walls of the neural tube, the predecessors of radial glia originate from the neuroepithelial cells. Cell bodies of radial glia are found in the ventricular zone (VZ) of the developing cortex, with a long process reaching the pial surface. The main distinction between the neuroepithelial cells and radial glia is the development of glial features in the latter, such as the expression of glial filaments; marker proteins such as tenascin C, GLAST, RC2, and brain-lipid-binding protein (BLBP); as well as the occurrence of glycogen inclusions (Hartfuss et al., 2001, Kriegstein and Gotz, 2003). Radial glia express the bona-fide precursor cell gene nestin (Hartfuss et al., 2001) and Sox2 throughout development (Ellis et al., 2004; Ferri et al., 2004; Bani-Yaghoub et al., 2006). Overexpression of Sox2 resulted in increased gliogenesis and inhibited neurogenesis in a Notch-dependent manner (Bani-Yaghoub et al., 2006).

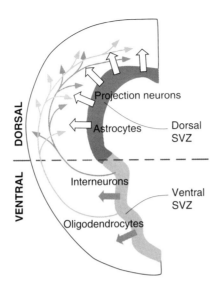

FIGURE 3–7 Regionalization of precursor cells. Brain development originates from precursor cells in the walls of the neural tube, which later become the walls of the ventricular system, the ventricular zone (VZ), and the cell layer below, the subventricular zone (SVZ). Different parts of the VZ and SVZ contribute differently to brain development. The precursor cells are regionalized. Here, specification in ventral and dorsal precursor cells is depicted. In addition to ventral neurons, ventral precursor cells in the ganglionic eminences generate interneurons in the dorsal brain. The precursor cells in the adult brain appear to maintain this developmental regionalization (see also Fig. 4–5).

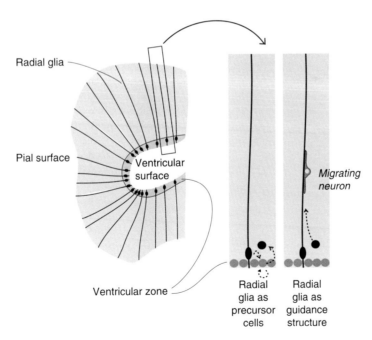

FIGURE 3–8 Dual functions of radial glia during development. Radial glia serves as precursor cells and as a guidance structure for the migration of newly generated neurons to their appropriate position in the cortex. Expansion of the precursor pool occurs mainly on the level of intermediate progenitor cells that lack the characteristic morphology. In both neurogenic regions of the adult brain, cells with radial glia–like properties persist and seem to function as the highest-ranking precursor cells of these regions.

Radial glia show regional differences

Most brain regions contain radial glia, but despite obvious similarities, radial glia are a considerably heterogeneous class of cells (Kriegstein and Gotz, 2003; Malatesta et al., 2003; Hack et al., 2004; Pinto and Gotz, 2007; Alvarez-Buylla et al., 2008). Radial glia from different regions of the developing brain differ in the expression of distinct sets of neurogenic transcription factors known to control patterning of the developing brain (Stoykova et al., 1996; Corbin et al., 2000; Yun et al., 2001; Pinto et al., 2008). In fact, several lineages of glial and neuronal precursor cells exist (Malatesta et al., 2000; McCarthy et al., 2001; Pinto and Gotz, 2007; Alvarez-Buylla et al., 2008). Conclusions drawn from one brain region in one species and one point of development thus cannot be generalized. In the adult SVZ, radial glia–like precursor cells from different parts of the ventricular wall give rise to different types of interneurons in the olfactory bulb (De Marchis et al., 2007; Lledo et al., 2008), further supporting the idea of precursor-cell heterogeneity at this level.

Neural stem cells in the adult brain have some but not all properties of radial glia

In both the adult SGZ and SVZ, astrocyte-like, radial glia–like cells retain stem cell properties and are thought to link adult neurogenesis with embryonic brain development (Alvarez-Buylla et al., 2001; Gotz and Huttner, 2005; Ihrie and Alvarez-Buylla, 2008; Malatesta et al., 2008; Gubert et al., 2009). In the hippocampus, radial glia cells of the adult dentate gyrus are a secondary population that are displaced from the original ventricular location and no longer extend between the ventricular wall and the pial surface of the brain (Rickmann et al., 1987). But these cells show many characteristics of astrocytes (Filippov et al., 2003). In the SVZ, the original location near the ventricular surface is maintained.

In adult neurogenesis of canary birds, BLBP expression has been noted (Rousselot et al., 1997), but in mammalian neurogenic zones not all radial cells showed BLPB expression (Sundholm-Peters et al., 2004; Steiner et al., 2006). GLAST and RC2 protein does not seem to be noticeably expressed in adult neurogenic zones, but the GLAST promotor in conjunction with a regulatable construct has been successfully used to mark all progeny of radial glia so that the gene must be active (Ninkovic et al., 2007).

A key transcription factor of radial glia in the dorsal forebrain is Pax6, which is also found in some adult precursor cells

Radial glia of the dorsal forebrain express Pax6. Accordingly, hippocampal radial glia–like cells in the adult hippocampus might express Pax6 as well, and Pax6 is down-regulated in the progenitor cells. In the adult SVZ, however, which is of dorsal origin, the radial glia–derived B cells are Pax6–negative, whereas migratory progeny of these cells, producing periglomerular interneurons in the olfactory bulb, do express Pax6. Consequently, the pattern of Pax6 and thus its presumed functional role seem to differ between the fetal and adult period and also between the two neurogenic regions. In Pax6–deficient mice, however, radial glia from the dorsal forebrain act like radial glia of the ventral forebrain and instead of cortical-projection neurons generates mainly glia and few neurons (Gotz et al., 1998). Intriguingly, these neurons were interneurons of the olfactory bulb. Radial glia in the lateral ganglionic eminence express Gsh2 and normally give rise to interneurons in the olfactory bulb (Stenman et al., 2003); radial glia in the medial ganglionic eminence express *Olig2*, generating cortical interneurons and oligodendrocytes (Malatesta et al., 2003). Ex vivo, forced expression of Pax6 in astrocytes was sufficient to impose radial glia–like properties in them and effectively reprogram them to a precursor-cell state (Moon et al., 2008).

Intermediate Progenitor Cells

Radial glia–like precursor cells generate intermediate progenitor cells

Radial glia is not the only cell type with stem cell properties in neurogenesis, and the picture is thus somewhat blurred (Sawamoto et al., 2001; Gal et al., 2006; Mizutani et al., 2007). According to the standard model of stem cell biology, outlined above, stem cells by definition give rise, not to differentiated progeny, but to intermediate progenitor cells. Such intermediate progenitor cells do not extend processes to the pial or ventricular surface and appear to be "between the lineages" in that they show neither (radial) glial features nor (yet) neuronal determination (Doetsch, 2003; Steiner et al., 2006; Spella et al., 2007; Bello et al., 2008). In most cases they will be the progeny of radial glial cells, so that essentially all neurons would indirectly derive from radial glia (Anthony et al., 2004). It cannot yet be excluded, though, that truly independent neural-precursor cell populations exist. The best candidates for such independent populations are a subset of the NG2 cells that are discussed below. They are not involved in adult neurogenesis in the two canonical neurogenic regions. The intermediate progenitor cells are also referred to as "transient amplifying progenitor cells" as they tend to have much greater proliferative activity than the "true" stem cells and carry the load of expanding the progenitor cell pool during development.

Lineage progression from radial glia (Pax6–positive) to intermediate progenitor cells (Tbr2–positive) and neurons (Tbr1–positive) could be demonstrated for cortical development in the embryo (Englund et al., 2005) and is related to similar patterns in the adult neurogenic zones (Hodge et al., 2008).

Neural Precursor Cells in the Adult Brain

Precursor cells in the adult brain comprise many distinct populations

The adult mammalian brain contains a variety of precursor cells, including but not limited to the neurogenic precursor cells in the SVZ and the SGZ as well as their progeny (Table 3–1). These cell types are not all distinct entities but rather reflect the current state of phenomenology. For example, identity between cells with stemness properties ex vivo does not necessarily correspond strictly to precursor cells found in vivo.

Some of these precursor cells are multipotent ex vivo, but in vivo their potential appears to be more limited to bi- or unipotency.

The term *neural* does not necessarily imply that neurons are generated

The initial reports on precursor cells isolated from the adult brain focused on the two neurogenic regions, the hippocampus and the SVZ/olfactory bulb. The in vivo and ex vivo lines of research on "adult neurogenesis" became mutually reinforcing, once this conceptual link was made (see Chap. 2, page 42). The existence of stem cells in the regions of supposed neurogenesis made this proposition all the more realistic, while adult neurogenesis gave meaning to the presence of stem cells in the adult brain. In somewhat neuronocentric terminology, the precursor cells of the neurogenic regions in the adult brain are sometimes referred to as "neuronal precursor cells." For these regions, the term is equivalent, however, to *neural precursor cells*. As Table 3–1 shows, there are other, possibly non-neurogenic precursor cells in the adult brain as well. Based on the present stage of knowledge, to name, for example, the NG2 cells of the brain parenchyma "neuronal" would be inappropriate, although they might be neural precursor cells in that they produce glia.

Most experimental data on *adult* precursor cells reflect a situation in rather early adulthood. In rodents, which are used in most studies, sexual maturity and thus adulthood begin at approximately four weeks of age, the time (or just a few weeks thereafter) when cells are isolated in most studies (see also Chap. 1 on a practical definition of adulthood, and Chap. 10). In truly adult animals, the yield is very low; adult neural stem cells are rare.

Precursor cell counts in the brain decrease with increasing age

Prevalence of precursor cells in the adult brain appears to decline with increasing age, although so far robust data are available only for the neurogenic zones. SVZ precursor cells have been isolated even from the oldest brain tissue investigated, up to about two years in mice (Bailey et al., 2004; Maslov et al., 2004), consistent with the observed age-related changes in vivo (Luo et al., 2006; Quinones-Hinojosa et al., 2006). To date, the oldest age at which adult SVZ precursor cells were obtained from the human brain is around 40 years (Sanai et al., 2004; and personal communication by Arturo Alvarez-Buylla).

Table 3–1 Precursor cell populations in and from the adult mammalian brain

Brain region	Cell		Progeny	Reference
Ventricle walls				
SVZ (Forebrain)	B	In vivo	C, astrocytes	(Doetsch et al., 1999b; Ciccolini et al., 2005)
	C	In vivo	A	
	A	In vivo	Interneurons, oligodendrocytes	
	NG2-cell	Ex vivo	?; neurogenic potential only found in one model and not confirmed by other studies	(Belachew et al., 2003; Aguirre and Gallo, 2004; Aguirre et al., 2004)
	"Neural stem cells" (NSC)	Ex vivo	Neurons, astrocytes, oligodendrocytes	(Reynolds and Weiss, 1992)
Ependyma	Ependymal cell (E cell)	In vivo	Neurogenic potential only after pathology	(Carlen et al., 2009)
Periventricular organs	?	Ex vivo	Neurons, astrocytes, oligodendrocytes	(Bennett et al., 2009)
		In vivo	Astrocytes. Very few neurons only in Area postrema, not fully characterized	(Bennett et al., 2009)
SVZ (3rd ventricle)	"Ependymal cell", tanycytes	In vivo	Orexin-positive neurons in the hypothalamus (see Chap. 8, p. 296)	(Xu et al., 2005)
SVZ (4th ventricle)	?	In vivo	Astrocytes, oligodendrocytes	(Martens et al., 2002)
Central canal (spinal cord)	Ependymal cell	In vivo	Astrocytes, oligodendrocytes	(Meletis et al., 2008)
	NG2 cells	In vivo	NG2 cells	(Horner et al., 2000)
	Neural precursor cells	Ex vivo	Neurons, astrocytes, oligodendrocytes	(Shihabuddin et al., 2000)
Parenchyma				
Rostral migratory stream	A	Ex vivo	Neurons, astrocytes, oligodendrocytes	(Liu and Martin, 2003)

Region	Cell type		Progeny	References
Olfactory bulb	A	Ex vivo	Neurons, astrocytes, oligodendrocytes	(Liu and Martin, 2003; Vicario-Abejon et al., 2003)
Hippocampus	Type-1	In vivo	Type-2, astrocytes	(Seri et al., 2001; Filippov et al., 2003; Steiner et al., 2006; Suh et al., 2007)
	Type-2	In vivo	Type-3, oligodendrocytes (?), granule cell neurons	(Seri et al., 2001; van Praag et al., 2002; Filippov et al., 2003; Steiner et al., 2006; Suh et al., 2007; Jessberger et al., 2008)
	Type-3	In vivo	Granule cells	(Filippov et al., 2003; Jessberger et al., 2005; Steiner et al., 2006)
	Neural precursor cells	In vitro	Granule-cell like cells, astrocytes, oligodendrocytes	(Palmer et al., 1997; Babu et al., 2007; Suh et al., 2007)
Fiber tracts	WMPC, Neuronal precursor cell	Ex vivo		(Palmer et al., 1999; Kondo and Raff, 2000)
	NG2 cells	In vivo	Layer VI interneurons (up to adolescence);	(Dayer et al., 2005; Inta et al., 2008)
Gray matter	A-like cells (DCX-positive)	In vivo	Neurons (after ischemia); unknown	(Arvidsson et al., 2002; Parent et al., 2002; Glass et al., 2005)
	NG2 cells	In vivo	NG2 cells,	(Kronenberg et al., 2007)

Note that the different categories have been derived from many different reports in the literature and might not necessarily in all cases truly represent separate entities. This is particularly true for the NG2 cells. The identity of precursor cells in vivo and ex vivo has not been demonstrated in most cases.

The heterogeneity among precursor cells that is summarized in Table 3–1 is further enhanced by the finding that, even within the same area, here most notably the SVZ of the forebrain, precursor cells show considerable diversity (De Marchis et al., 2007; Merkle et al., 2007; Young et al., 2007). We will come back to this issue in Chapter 5, page 173. But it needs to be noted that there is no such thing as "the neural stem cell." In many publications, an ominous, apparently generic "Neural Stem Cells" term appears and is introduced by its acronym "NSC." Although such abbreviation might occasionally be useful, the term requires qualification and specification under nearly all circumstances. "NSC" suggests an entity where there is actually none, and true stemness of these cells has hardly ever been proven. The probably best generic term is simply "neural precursor cells," often abbreviated to "NPC."

Additional precursor cell heterogeneity arises from sub-regionalization

Glial Progenitor Cells

The literature on glial progenitor cells uses a distinct nomenclature that is not always intuitively compatible with neural stem cell biology

Glial progenitor cells are neural precursor cells that produce only glia. Historically, research on glial precursor cell biology has preceded the rise of neural stem cell biology by more than a decade. The O2A progenitor cell, first described by Martin Raff and Marc Noble in 1983 as a precursor giving rise to oligodendrocytes and one type of astrocyte (type 2 astrocyte) in vitro (Raff et al., 1983), has over the years been characterized and (in vitro) detailed lineage relationships have been revealed. However, the O2A precursor cell has never been unambiguously identified in vivo. O2A-like progenitor cells might be found among the so-called NG2 cells of the brain (Nishiyama et al., 1996).

Despite this head-start of the gliologists, neural stem cell biology evolved to a large degree independently of glial precursor biology (Fig. 3–9): thus the terminology, concepts, and conclusions can differ considerably between the two fields (see, for example, Liu and Rao, 2004, Rao, 2004). Many studies on glial precursor cell biology have been done on the spinal cord, whereas most studies in neural stem cell biology have favored the SVZ of the forebrain. Consequently, it is not clear, for example, whether the glial-restricted precursor cells (GRPs), which have been studied extensively (Noble et al., 2004), are identical to the glial progenitor cells of neural stem cell biology. In some sense the migratory doublecortin-expressing progenitor cells of the adult hippocampus and SVZ might be considered neural-restricted precursor cells (NRP), described as the neuronal counterpart to the GRPs (Herrera et al., 2001; Gregori et al., 2002; Liu et al., 2002). Such equalizations make intuitive sense, but no experimental data exist to really bridge the two worlds. Although the wording seems similar, the evidence to support the similarity of the differently evolved concepts is still lacking. This is not a trivial matter, because stem cells of the adult brain actually represent a type of astrocyte, blurring the boundaries of seemingly well-defined cell populations on several conceptual levels.

Astrocytes as Stem Cells in the Adult Brain and Stem Cells with Astrocytic Features

Precursor cells in the adult brain have astrocytic properties and originate from radial glia after cortical development

Related to the insights on radial glia as precursor cells, Arturo Alvarez-Buylla and coworkers found evidence that glial fibrillary acidic protein (GFAP)–positive cells of the SVZ (for the olfactory system) and the SGZ (for the hippocampus) could be the stem cells of these neurogenic regions in the adult. After temporary ablation of all proliferating cells with the cytostatic agent cytosine-arabinoside (Ara-C), GFAP-expressing cells were the first to reappear (Doetsch et al., 1999b; Seri et al., 2001). Also, a

Neural precursor cells *in vivo*
Glial precursor cell perspective

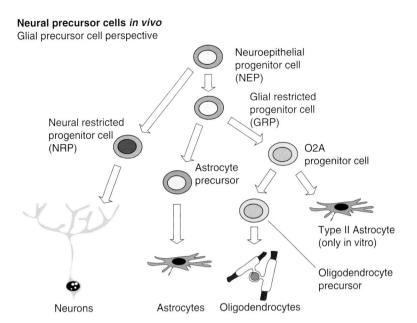

FIGURE 3–9 Hierarchy of glial precursor cells. Historically, the rise of glial precursor cell biology preceded neural precursor cell biology by almost a decade. Most data in this field have been obtained in fetal spinal cord. The resulting schema is similar to the pedigree depicted in Figure 3–2 (*left*) but the nomenclature is different. The precursor cell function of radial glia and astrocyte-like cells is not reflected in this scheme.

reporter gene transmitted by a virus that could only infect GFAP-expressing cells that had been genetically engineered to be susceptible for the virus was later found in neurons and gave evidence that neurons could originate from GFAP-positive cells in this region (Doetsch et al., 1999a). These astrocytes in turn were the progeny of radial glia (Merkle et al., 2004). From these data, Alvarez-Buylla derived a theory on the lineage relationship between glia and neurons during embryonic and adult neural development (Alvarez-Buylla et al., 2001; Kriegstein and Alvarez-Buylla, 2009). According to this model, astrocytes with stem cell qualities are derived from radial glia, which had ultimately originated from the early ventricular zone (Kriegstein and Gotz, 2003). Alvarez-Buylla proposed that "neural stem cells are contained within the developmental lineage that proceeds from neuroepithelial cells, through radial glia, to astrocytes" (Alvarez-Buylla et al., 2001). In this "grand unifying theory," it is essentially a question of time during development whether the stem cells will display neuroepithelial, radial glial, or astrocytic characteristics. Consequently, the nature of stem cells during neural development shifts. Given the heterogeneity of precursor cell types in the adult SVZ, reality is more complex than this model, and Alvarez-Buylla elaborated on this issue in later publications, refining his original "unifying" concept (Lledo et al., 2006; Alvarez-Buylla et al., 2008; Kriegstein and Alvarez-Buylla, 2009).

Even back in 2001, experimental findings by Jonas Frisén's group (Johansson et al., 1999) and the prospective isolation of SVZ precursor cells by Parry Bartlett and coworkers (Rietze et al., 2001) seemed to argue against the idea of a single stem cell population. Frisén had, for example, shown that sorting based on both DiI incorporation or Notch expression allowed the isolation of multipotent cells from the adult ventricular wall, suggesting that ependymal cells could act as stem cells in the adult brain (Johansson et al., 1999). Under normal conditions, ependymal cells, however, are post-mitotic and do not seem to be contributing to adult neurogenesis in vivo (Spassky et al., 2005). During embryonic development, ependymal cells are derived from radial glia (Spassky et al., 2005).

Ependymal cells physiologically do not exert stem cell functions

The idea that GFAP-positive cells in the SVZ rather than the GFAP-negative ependymal cells act as precursor cells was further supported by the finding that killing GFAP-expressing cells with a genetic construct that made them sensitive to ganciclovir resulted in a twenty-fold reduction in the number of self-renewing and multipotent precursor cells that could be isolated from the brain (Morshead et al., 2003). In a later study using a transgenic approach, a similar result was obtained. That study even suggested that morphologically the neurogenic astrocytes might differ from non-neurogenic astrocytes by having fewer processes (Garcia et al., 2004). A direct comparison of the neurogenic potential ex vivo had also favored the astrocytes-only hypothesis, also supporting the idea that astrocytes in neurogenic regions have particular properties (Laywell et al., 2000).

All attempts to close the case, however, turned out to be premature. Prominin 1 (CD133) is present on ependymal cells but is not expressed by the SVZ astrocytes (Pfenninger et al., 2007). Nevertheless, sorting for Prominin1 expression allowed the prospective isolation of precursor cells from the SVZ (Coskun et al., 2008). The particular fine structure of the ventricular wall, however, makes such distinctions very difficult, and the Prominin1 and cilia-bearing process might still belong to the underlying B cell (Fig. 5–4). But other lines of evidence suggest that ependymal cells might act as precursor cells after brain injury (Colak et al., 2008; Carlen et al., 2009). Together, these latter observations reopened the old debate about whether a subset of ependymal cells might act as precursor cells for the SVZ (Nakafuku et al., 2008). In the end, only very detailed lineage-tracing experiments will allow us to settle the issue. While there is little doubt that astrocyte-like cells do carry the main load, the question is now whether ependymal cells act as backup stem cell reservoir for the SVZ (Fig. 3–10 and Fig. 5–4). This idea is interesting because it might suggest a bi-directional exchange between the ependyma and the B cells in the SVZ that would act as long- versus short-term stem cells of the SVZ.

In any case, despite the detection of astroctyic marker GFAP in precursor cells and the possible presence of other astrocytic features, such as vascular endfeet and distinct electrophysiological characteristics (Filippov et al., 2003), it will not be wise to bluntly apply the term *astrocytes*, which is associated with a differentiated, albeit heterogeneous, type of glial cell, to

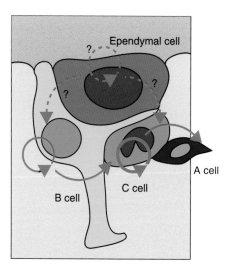

FIGURE 3–10 The proposed relationship between ependymal cells and astrocyte-like precursor cells in the SVZ (B cells), in particular after injury. The schematic is based on (Nakafuku, 2008) and those cited therein. The idea is that, in cases of pathology, ependymal cells might (re-)gain stem cell properties and contribute to cell genesis in the SVZ by producing B or C cells.

precursor cells. Within the hierarchy of stem cell biology, this kind of "astrocyte" would be found on a level that is conceptually super-ordinate to astrocytes in the classical sense. Also, not all astrocytes are stem cells, even if (for example, by means of Pax6 over-expression) it might be possible to effectively reprogram them to multipotency (Moon et al., 2008).

What constitutes an astrocyte versus a precursor cell with astrocytic properties needs to be established

On the other hand, GFAP immunoreactivity alone is certainly not a sufficient criterion to identify either differentiated "true" astrocytes or stem cells. Hence the description of precursor cells as "astrocyte-like" or "radial glia–like" is useful but incomplete. For some glial biologists, this issue serves as one further argument to abandon the idea of astrocytes as being one single cell type; rather, *astrocytes* might be an umbrella term for all neuroepithelial brain cells that are neither neurons nor oligodendrocytes (and NG2 cells). In this sense, stem cells might be astrocytes, but the reverse will still not always be true (Morshead and van der Kooy, 2004).

White Matter Precursor Cells and NG2 Cells

Precursor cells expressing proteoglycan NG2 and lacking astrocytic properties are found predominantly in white but also in gray matter

The non-neurogenic regions of the adult brain harbor precursor cells, but at least under physiological conditions these cells are unrelated to neurogenesis. Because of their relative abundance in the white matter tracts, they have been loosely referred to as "white matter precursor cells" (WMPC). From a theoretical perspective they might relate to the "spongioblasts" of the neuroanatomists in the late nineteenth century and Wilder Penfield's schema as depicted in Figure 2–1. They were largely considered to be progenitors for oligodendrocytes (hence their abundance in the fiber tract) and many express the proteoglycan NG2 and many also Nestin. Because most WMPC are NG2–positive and NG2 cells also exist in gray matter, the term "NG2 cells" for the entire class of cells has become commonplace, although there is considerable heterogeneity among NG2 cells. Most notably, most proliferating NG2 cells generate only NG2 cells. NG2 cells comprise by far the most abundant population of proliferative cells in the adult brain (Dawson et al., 2003). It has been estimated that they comprise up to 4% of all cells in the white matter (Canoll and Goldman, 2008). In gray matter they are almost evenly spaced out and show an intriguing non-overlapping territorial pattern (Fig. 3–11).

NG2 cells somehow stand between neuronal and glial lineages

NG2 itself is a chondroitin-sulfate proteoglycan transmembrane protein containing extracellular laminin-like domains, which is suggestive of a function in cell adhesion, recognition, and migration. In vitro, a blocking antibody against NG2 inhibits migration of NG2–expressing cells (Niehaus et al., 1999). The mouse homologue was originally called "AN2" (Schneider et al., 2001) but the name NG2 is now generally used for all species.

NG2 cells are regarded as somewhat mysterious cell type between the domains of glia and neurons, with a generally ill-defined precursor cell potential (Berry et al., 2002; Horner et al., 2002; Stallcup, 2002). "NG" actually stands for "neuron glia" to underscore that fact. The NG2/Nestin-positive cells of the cortex, for example, receive excitatory GABAergic input and are depolarized by GABA (Tanaka et al., 2009). Their depolarization in turn resulted in the release of brain-derived neurotrophic factor (BDNF)—a mechanism known from neurons (Obrietan et al., 2002)—suggesting a role of these cells in cellular plasticity that goes beyond or is independent of their presumed nature as precursor cells. NG2 cells react to injury and ischemia with proliferation, and a related population in the CA1 region of the adult hippocampus has also responded to voluntary physical activity (Kronenberg et al., 2007).

Morphologically, there are also different types of NG2 cells. One type is bipolar and has a small nucleus, whereas the other has a ramified appearance. It is not clear how these anatomical phenotypes relate to the functional distinction.

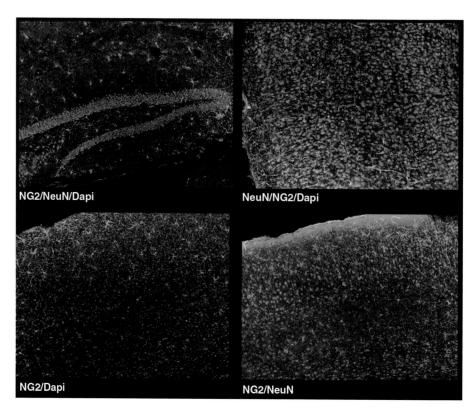

FIGURE 3–11 NG2 cells. *Upper left panel*: NG2 cells (*green*), here in the molecular layer of CA1, have an extensive tree of processes, with which they cover hardly overlapping territories. *Other panels*: This same territorial pattern is also found in the neocortex (NG2 *green*, except for upper right panel, where NG2 is *red*). Figure by Klaus Fabel, Dresden.

NG2 cells comprise at least two major functional sub-types but are likely to be even more heterogeneous

A current working hypothesis holds that there are two types of NG2–positive cells in the developing and adult CNS (Mallon et al., 2002; Nishiyama et al., 2002). First, NG2 is indeed thought to identify precursor cells in the oligodendrocytic lineage (Keirstead et al., 1998; Chari and Blakemore, 2002; Watanabe et al., 2002). Most, but not all, NG2 cells express Olig2, which is necessary to initiating the development of oligodendrocytes (Ligon et al., 2006). Ex vivo, NG2–expressing cells were initially described as giving rise to oligodendrocytes and one type of astrocyte (Stallcup and Beasley, 1987). Appropriately, NG2 cells can generate O2A progenitor cells, suggesting that they might in fact be at least bipotent (Baracskay et al., 2007). During development, NG2 cells characteristically express platelet-derived growth factor (PDGF) receptor alpha (Nishiyama et al., 1999). PDGF is secreted by neurons and astrocytes and maintains oligodendrocyte progenitor cells in a proliferative stage (Hart et al., 1989). In vivo lineage-tracing by means of a double transgenic mouse expressing Cre recombinase under the NG2 promoter crossed into a Cre reporter mouse line expressing EGFP (see Chap. 7, p. 257, for a detailed description of this methodology) has confirmed that at least a subset of NG2 cells can develop into both oligodendrocytes and protoplasmic astrocytes within the gray matter (Zhu et al., 2008).

Second, NG2–positive glia might be a highly specific type of astrocyte, closely related to synapses and nodes of Ranvier (Butt et al., 1999; Ong and Levine, 1999; Butt et al., 2002). In astrocytic differentiation, NG2 cells are consistently negative for GFAP, but they might express low levels of S100β. Reporter gene mice with the S100β promoter showed enhanced GFP expression in both astrocytes and NG2–positive cells of the hippocampus (Lin and Bergles, 2002). On the basis of electrophysiological examinations, two types of astrocytes can be distinguished: one with classical passive properties, the other called *complex*, with a very high input resistance, lower resting conductance to potassium, and lower expression of glutamate transporters (Steinhauser et al., 1992; Steinhauser et al., 1994). This pattern was originally described for O2A precursor cells in vitro (Steinhauser et al., 1992), later for one type of astrocyte (Akopian et al., 1997), NG2–expressing hippocampal cells (Bergles et al., 2000), and finally for type-2 precursor cells in the adult hippocampus (Filippov et al., 2003). This characteristic functional measure further supports the idea of some identity between these cell types that would all share some involvement in cellular or synaptic plasticity.

There might be more common identity beyond the shared expression of NG2

The existence of a link between these two types of NG2 cells is supported by a study that looked at the co-expression of NG2 with DCX in cortical perineuronal satellite cells; i.e., glial cells closely attached to neurons (Fig. 7–7). Cells with this marker combination can be found throughout the brain. Perineuronal NG2 cells were never DCX-negative. Based on a BrdU-labeling and the analysis of immediately adjacent pairs of NG2/ DCX double-positive cells, it was concluded that NG2/DCX-double-positive cells could either become DCX- or NG2–single-positive cells or replicate to again double-positive progeny. The single-positive cells showed further signs of pre-neuronal (Tuc4) or oligodendrocytic (glutathione-S-transferase-pi, or GST-pi) lineage (Tamura et al., 2007). Suggestive as this might be, there is clearly a need for thorough lineage studies in vivo to resolve this issue.

NG2 is expressed by some pericytes, which also might have certain precursor cell properties

In addition to these cell types, pericytes (which by definition are not part of the brain parenchyma but are separated from it by a basal membrane) might also express NG2 and can show some precursor cell properties ex vivo (Dore-Duffy et al., 2006; Dore-Duffy, 2008). There is also a possible relationship to a monocytic/microglial cell type (Fiedorowicz et al., 2008).

There is only limited evidence for exceptional neuronal differentiation from NG2 cells

Despite the fact that NG2–positive cells are negative for oligodendrocyte marker CNP, the CNP promoter could be used to isolate NG2–expressing multipotent cells (Belachew et al., 2003; Aguirre et al., 2004), which were also able to generate interneurons. NG2 cells have also been postulated as a source for the minute population of adult-generated Layer VI interneurons in the cortex (Dayer et al., 2005), and these NG2 cells appear to originate from the SVZ, presumably the C progenitor cells (Aguirre et al., 2004; Inta et al., 2008). Whereas thus NG2–positive precursor cells that are bipotent for oligodendrocytes and interneurons might exist in the adult SVZ, such cells are conspicuously absent from the SGZ of the dentate gyrus (Steiner et al., 2004).

The fate choice of NG2 cells to the different lineages, including the neuronal lineage, is dependent on epigenetic control (Liu et al., 2007). In that view, NG2 cells have an epigenetic memory for the oligodendrocyte lineage that allows them to remain quiescent during phases of neurogenesis and the generation of astrocytes. If this memory was experimentally erased by treating the cells with histone deacetylase (HDAC) inhibitor valproic acid, the NG2 cells became responsive to cues that steered their differentiation to the astroglial and neuronal phenotype (Liu et al., 2007).

Dividing NG2 cells mostly produces only new NG2 cells

Interestingly, however, and in some contrast to these results, even over longer time periods, most dividing NG2 cells in vivo seem to generate primarily other NG2 cells, not differentiating oligodendroglia or other cell types (Horner et al., 2002). If it exists at all in vivo, the realized neurogenic potential of NG2 cells is extremely low.

NG2 does not unanimously identify precursor cells

Consequently, neither the sensitivity nor the specificity of NG2 as a precursor cell marker are known. When precursor cells were isolated on the basis of Sox2 expression, in the resulting neurospheres the presence of GFAP and NG2 turned out to be mutually exclusive, further suggesting separate entities but without clarifying the exact lineage relationships (Brazel et al., 2005). It is not clear whether all NG2 cells can have precursor-cell properties, and it is even less clear whether this potential includes neuronal differentiation in all cases. In one study, NG2 cells have been reported as multipotent ex vivo (Belachew et al., 2003), but this has not been confirmed by others (Buffo et al., 2008). NG2 cells from the SVZ that had been isolated on the basis of CNPase-promoter activity and were transplanted into the dentate gyrus exhibited a neurogenic potential there as well (Aguirre et al., 2004). These findings do not prove that all NG2 cells are multipotent but further indicate heterogeneity and lineage plasticity. It is also not known whether the NG2 protein itself plays a role in inducing or maintaining any presumed precursor cell identity. Consequently, NG2 remains an interesting but still problematic marker for identifying precursor cells in the brain.

NG2 cells might rightfully become considered a major brain cell class of their own

Because of their extraordinary nature partaking of all categories, it has been suggested that NG2 cells should represent an entity of their own, a fifth neural lineage (Greenwood and Butt, 2003; Butt et al., 2005). Various new designations have been given to these cells: for example "synantocytes" (primarily for the astrocyte-like type [Butt et al., 2002] but later generalized to all NG2 cells [Butt et al., 2005]) and "polydendrocytes" (Nishiyama, 2007; Nishiyama et al., 2009). This might be helpful or not, and neither of the names has stuck with the community that calls the cells "NG2 cells." Given the heterogeneity of the NG2–positive population, this might be a wise choice.

NG2 cells respond to injury

NG2 expression and the proliferation of NG2–expressing cells are up-regulated after many different types of injury (Levine et al., 2001), and NG2 cells are part of the glial scar. This response is not limited to mechanical damage. Demyelination induces a very focal response of NG2–positive cells, which initiate differentiation into new oligodendrocytes (Keirstead et al., 1998; Redwine and Armstrong, 1998). However, remyelination from these cells is not always successful, especially not in locations of chronic damage (Reynolds et al., 2002). We will return to NG2 cells in detail in Chapter 8, p. 230.

Neural Precursor Cells in Non-Neurogenic Regions

Precursor cells with neurogenic potential ex vivo could be isolated from several brain regions

Despite the limited realized potential of NG2 cells in vivo, precursor cells with a neurogenic potential can be isolated from many brain regions other than the neurogenic regions. Neural precursor cells may be found in very many brain areas, and lineage-determined progenitor cells for the glial lineage possibly in all. The exact relationship to NG2 cells remains to be established. Overall, the situation is far from clear. The question of how cells with precursor cell properties and a neurogenic potential are distributed in the brain is extremely difficult to address. The only feasible available method to determine stemness is to isolate the cells and demonstrate self-renewal and multipotency in vitro. This approach, however, imposes culture conditions on the cells, to which they might be differentially sensitive.

FGF2 might be critical for eliciting the neurogenic potential

The first systematic study on neurogenic precursor cells outside the SGZ and SVZ came from Theo D. Palmer, then with Fred H. Gage at the Salk Institute in La Jolla. Palmer found that fibroblast growth factor-2 (FGF-2) can elicit a neurogenic potential in cells isolated from the neocortex, septum, striatum, corpus callosum, and optic nerve (Palmer et al., 1999). He applied a culture method that had proven effective on hippocampal precursor cells (Palmer et al., 1995; Palmer et al., 1997). The yield from the non-neurogenic regions was generally much lower (about three orders of magnitude) than that from neurogenic regions. The nature of these cells in vivo is not clear. At the time, NG2 had not been studied.

Most brain precursor cells are found in the walls of the ventricular system

The most plausible location for persistent precursor cells in adulthood is along the ventricular system, and indeed it seems that cells with precursor cell properties can be derived from entire ventricular walls, not only from the lateral ventricles but also from the fourth ventricle (Martens et al., 2002) and the central canal of the spinal cord (Shihabuddin et al., 1997). The periventricular organs, areas of an open blood-brain barrier, including the pineal gland, area postrema, subcommissural organ, and choroids plexus, express many markers also found in neurogenic zones. In fact, tri-potent neurospheres (see below) could be isolated from these regions (Bennett et al., 2009). In vivo, mainly astrocytes were generated. In the Area postrema a few BrdU/NeuN-positive cells were seen but not yet further characterized. See Chapter 7 for the applicable caveats.

Neurogenic precursor cells alone do not make a brain region neurogenic

Despite the existence of potentially neurogenic cells, under normal conditions, the adult brain outside the SGZ and the SVZ appears to be essentially non-neurogenic, although the picture is somewhat blurred. Reports about physiological neurogenesis in other brain regions of mammals have remained unconfirmed or controversial at best, mostly because of technical issues (Chaps. 7 and 8). The case is less problematic, however, for genesis of a specific population of interneurons in Layer VI of the neocortex that occurs postnatally in rodents and might be found in extremely small numbers well into adulthood, but these cells are ultimately of SVZ origin (Dayer et al., 2005; Inta et al., 2008).

Similarly, in the cases of regenerative or reactive neurogenesis that will be described in Chapter 8 in detail, the response did not seem to originate from local precursor cells but rather to be based on precursor cells recruited from the SVZ. Jeffrey Macklis and coworkers from Harvard University have shown that under the condition of a highly targeted, induced cell death in the neocortex of mice, new neurons can be generated in these regions (Magavi et al., 2000; Chen et al., 2004). In response to ischemia, reactive neurogenesis seems to be possible in at least some non-neurogenic regions (Arvidsson et al., 2002; Parent et al., 2002).

Isolating Neural Precursor Cells

Precursor cell cultures are a valuable tool, but they are a model system

The developmental potential and many other aspects of neural stem cell biology usually cannot be studied in vivo; they have to be isolated and investigated in cell cultures. In fact, most conclusions about "stemness," self-renewal, and multipotency are inferred from ex vivo results. Cell culture work aims at studying individual precursor cells and their properties as representatives of a class of cell. In the past, biotechnological stem cell research has often made only limited reference to in vivo situations. The isolated cells were the object of the studies, and it was only at the endpoint of the studies—for example, of the generation of a transplantable cell—where the compatibility with the in vivo situation came into play. This has partly changed and is going to change even further. For basic research and thereby also in the context of adult neurogenesis, this has always been considerably different. Neural precursor cells are studied in vitro to gain insight about functions and properties of relevance for the in vivo situation. In the organism, however, stem cells never occur in isolation. Cultured neural precursor cells, like all cultured stem cells, are an abstraction. In some sense, isolated stem cells are a useful but quite artificial construct, and their relationship to the in vivo situation and their presumed in vivo counterpart still has to be constantly questioned.

Prospective isolation means that you know beforehand the selectable criteria for stem cells

Like for other cell cultures, neural precursor cells are isolated by first generating a single cell suspension from brain tissue. Ideally, from this suspension, the precursor cells can be isolated prospectively. Prospective isolation means that a set of markers is known and allows the researcher to identify cells with defined properties. For prospective isolation methods, stem cells are thus operationally defined as cells with a particular antigen profile. This prospective definition per se does not reveal anything about the two cardinal stemness criteria of self-renewal and multipotency.

Surface markers allow the
sorting of precursor cell types
from cell suspensions

In the hematopoietic system, different precursor cells can be unambiguously identified based on the binding of specific antibodies to surface antigens. Practically, this can be achieved, for example, through fluorescence-activated cell sorting (FACS): in a fine stream of single cells, the fluorescent antibody that has bound to the surface antigen of the precursor cells (or the intrinsic expression of, for example, green fluorescent protein, GFP) is recognized, and an electric field is applied to direct each individual identified cell into a receptacle that takes up the cells of the desired antigen pattern. For smaller volumes, one can alternatively (and more cheaply) couple iron beads to antibodies against precursor cell antigens and pull the binding cells from the solution by means of magnetic force.

Several strategies for prospective
isolation of neural precursor cells
have been developed

For prospective isolation of precursor cells from adult brain Sox2, high expression of EGF receptor, nestin, as well as GFAP in conjunction with LeX or SSEA-1 (CD15) have been used to isolated neural precursor cells from the adult forebrain. Table 3–2 summarizes these results. What is almost entirely lacking are comparisons between these different techniques and other standard protocols in order to find out whether they are actually the same cells.

Besides the markers exploited for the isolation of neural precursor cells from the adult, prospective isolation of embryonic and fetal precursor cells has also been achieved based on the expression of syndecan-1, Notch-1, and integrin-beta1 (Nagato et al., 2005). Integrin-beta1 and integrin-alpha1 were also successfully used to isolate precursor cells from fetal human tissue (Hall et al., 2006). The expression of Sox2 in both embryonic neural precursor cells with ES cells has been exploited to compare these two cell types (D'Amour and Gage, 2003). Isolation based on surface antigens requires

Table 3–2 Prospective isolation of neurogenic precursor cells from adult brain				
Marker	**Reference**	**Region**	**Type of culture**	**Remark**
Sox2-GFP	(Brazel et al., 2005)	Forebrain	Neurospheres	Not all cells neurogenic; distinction of NG2-pos./GFAP-neg. and NG2-neg./GFAP-pos. cells
	(Barraud et al., 2005)	Forebrain	Neurospheres	Not all cells neurogenic in the adult
Nestin-GFP	(Mignone et al., 2004)	Forebrain and hippocampus	Neurospheres	Distinction of GFP-bright GFAP-pos./blll-tubulin-neg. cells and GFP-dim GFAP-neg./blll-tubulin-pos. cells
Doublecortin-GFP	(Walker et al., 2007)	Forebrain and hippocampus	Neurospheres	DCX-low cells show precursor cell properties but High-high do not
Prominin	(Corti et al., 2007)	Forebrain	Neurospheres	Co-expression with Sox1/2, nestin, and Musashi. Differentiation into neurons and glia after transplantation.
EGFR	(Ciccolini et al., 2005)	Forebrain		Co-expression with GFAP and LeX/SSEA-1 in adult tissue
GFAP, LeX/SSEA-1	(Imura et al., 2006)	Forebrain	Neurospheres, Monolayers	Neurogenic cells only in the LeX/SSEA-1-positive fraction

optimal preservation of the antigens during the preparation of the single cell suspension (Panchision et al., 2007). In a study on fetal murine precursor cells, an optimized protocol revealed that neural precursor cells expressed higher levels of Prominin1 (CD133) and LeX/SSEA-1 (CD15) than earlier multipotent precursor cells. In addition, CD24 turned out to be a useful additional marker to further select within these populations (Panchision et al., 2007).

Stem cells, including neural stem cells, enrich in "side populations," which separate in the FACS due to fast dye efflux

In 1996, Margaret A. Goodell and colleagues reported that hematopoietic stem cells could also be isolated by FACS on the basis of the efflux of fluorescent dyes from these cells (Goodell et al., 1996). The dye Hoechst 33342 binds to histone proteins of all cells, but different classes of cells have different abilities to actively pump out the dye because they differ in their expression of a membrane-bound ABC transporter (Abcg2) (Challen and Little, 2006). The physiological function seems to be linked to protection from xenobiotics (Robey et al., 2009). A related protein, Abcb1, is also present in adult brain cells (Mouthon et al., 2006).

Side populations containing neural precursor cells could be isolated from the adult brain (Murayama et al., 2002; Kim and Morshead, 2003). But side populations in the mouse brain greatly varied depending on the developmental stage (Murayama et al., 2002). At the same time, whereas one report did not find differences in precursor cell activity between side and main populations from E14.5 mouse embryos in the first place (Nagato et al., 2005), one report questioned whether neural stem cells are contained in the side population at all (Mouthon et al., 2006). As a consequence, using dye efflux to isolate neural stem cells from the adult brain will still require additional validation work. It seems that side populations from the adult brain contain precursor cells, but it is not clear how sensitive or specific this approach is.

Using side populations has also been widely used to identify neural precursor cells from brain tumors. We will discuss tumor stem cells below.

Culturing Neural Precursor Cells

Neural precursor cells are cultured as so-called neurospheres or monolayers

Neural stem cells can be cultured in two different ways: as floating aggregates, called *neurospheres*, or as adherent monolayer cultures (Fig. 3–12, Table 3–3). See Chapter 7 for more details.

Neurospheres

The neurosphere technique dominates the literature but has many shortcomings

Overall, the neurosphere method is by far the most widely used technique. It was originally described by Reynolds and Weiss in 1992 and further characterized by Angelo Viscovi and Derek van der Kooy and colleagues (see Table 3–3 for references). In cell suspension with no adherent substrates but containing growth factors, the putative stem cells form floating clusters of cells, the *neurospheres*. These neurospheres might contain a few percent stem cells and are thus not per se identical to stem cells. In addition, the detection of sphere-forming cells is not necessarily equivalent to the detection of stem cells. Because of some exaggerated conclusions from neurosphere experiments, the assay was rightfully criticized (Jensen and Parmar, 2006; Singec et al., 2006; Jessberger et al., 2007; Marshall et al., 2007). Several attempts have been made to respond to this critique and improve the neurosphere assay (Reynolds and Rietze, 2005; Rietze and Reynolds, 2006). In any case, the neurosphere technique has also clear advantages: the cells retain cell–cell contacts and show more realistic patterns of differentiation. The neighborhood relationships in the sphere replace the niche in which the cells would be found in vivo. In many experiments, though, this advantage is not exploited, and the sphere assay is used for what it might not be optimal: quantifying "stem cell activity" and studying the behavior of individual precursor cells.

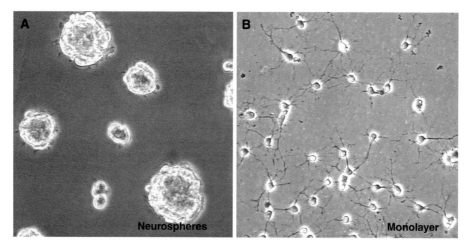

FIGURE 3–12 The two forms of precursor cell cultures from the adult brain. Neural precursor cells can be cultured as suspension or aggregate cultures, called "neurospheres," or as adherent monolayers. The two strategies do not yield exactly identical results. Note that neurospheres might contain stem cells but are not identical to stem cells. Only clonal analysis (Fig. 7–1) can ultimately determine whether a cell culture truly contains stem cells. See Table 3–3 for a comparison of the two cell culture strategies, and Figure 7–2. Images by Harish Babu.

Table 3–3 Comparison of neurosphere and monolayer cultures of adult neural precursor cells		
	Neurospheres	**Monolayers**
Synonym	Aggregate cultures	Adherent cultures
Description	Cells grown as agglomerates floating in the culture medium	Cells grown individualized on surfaces coated with adhesives such as laminin, poly-D-lysin, or fibronectin
Growth factors		
Mouse	EGF and / or FGF2	EGF and FGF2
Rat	EGF or FGF2	FGF2
Human	EGF and / or FGF2 and / or LIF	EGF, FGF2, LIF
Pros and cons:	• Technically simple, including tests of clonality • High yield • Heterogeneous • Individual cells normally not accessible • More realistic with respect to cell-cell interactions	• Technically demanding, especially clonal analyses • Lower yield • Relatively homogeneous • Individual cells accessible (e.g. for electrophysiology) • More stringent approach with respect to intrinsic precursor cell properties
References	(Reynolds et al., 1992; Reynolds and Weiss, 1992; Morshead et al., 1994; Gritti et al., 1995; Craig et al., 1996; Gritti et al., 1996; Morshead et al., 1998; Pincus et al., 1998; Gritti et al., 1999; Reynolds and Rietze, 2005)	(Palmer et al., 1995; Palmer et al., 1997; Palmer et al., 1999; Palmer et al., 2001; Schwartz et al., 2003; Babu et al., 2007)

EGF—epidermal growth factor; FGF2—fibroblast growth factor 2; LIF—leucocyte inhibitory factor.

The formation of a neurosphere is not sufficient to prove stemness

Many researchers prefer use of neurosphere cultures to the adherent cultures described below because neurospheres allow production of a greater number of cells and easier preparation of single-cell suspensions for clonal analysis. However, this apparently greater ease of use might be deceptive. The ability to form neurospheres alone does not prove the stemness of the cells. One aspect is that usually self-renewal *and* multipotency must be demonstrated (Fig. 3–13). Clonal analysis is a means of establishing the self-renewal of putative stem cells; multipotency of the clones remains to be established as well. Another aspect, however, is that clonal assays are only meaningful if the "clones" truly originate from single cells. This is not a given in most circumstances, because spheres tend to fuse in culture (Singec et al., 2006; Jessberger et al., 2007).

Neurospheres are cellularly heterogenous

The aggregates of neurospheres resemble the embryoid bodies found when ES cells are propagated in vitro. Within the spheres, the cells form three-dimensional structures; there are often inside-out gradients for certain characteristics. For example, most proliferating cells might be found on the outside, whereas the differentiating cells are found in the core of the sphere. If spheres grow too large, cells in the core might die.

Monolayer Cultures

Monolayer cultures are more radical in separating precursor cells and thus allow greater control of cell-extrinsic influences

The monolayer method, in which the cells grow adherent on coated surfaces was introduced to the field by Theo D. Palmer, Fred H. Gage, and colleagues (Palmer et al., 1995; Palmer et al., 1997; Palmer et al., 1999). The advantage is that the cells remain more isolated and their development can be studied individually. The reduced number of cell–cell contacts and the more even exposure to the culture medium might be an advantage or a disadvantage, depending on the research question.

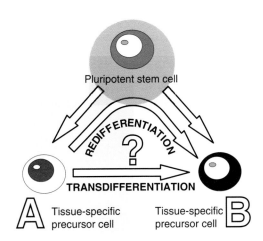

FIGURE 3–13 Can stem cells cross tissue boundaries? To explain the phenomenon that in some experiments tissue-specific precursor cells apparently generate progeny belonging to a different germ layer, the two concepts of transdifferentiation and redifferentiation were developed, the latter involving a more-than-multipotent intermediate stage. Most supposed cases of transdifferentiation or redifferentiation have turned out to be caused by cell fusion, but this does not rule out the possibility that transdifferentiation or redifferentiation might actually exist.

Comparison

Both approaches share the use of growth factors and strictly serum-free conditions and yield similar, though not always identical, results. Clive Svendsen, from the University of Madison, Wisconsin, has compared the two ways of culturing neural stem cells with the two philosophies in the world of computing, PC and Macintosh. This image nicely illustrates the combination of generally identical possibilities with at times still limited compatibility. The hope is perhaps to develop a kind of Linux of stem cell research, an open and stable code to produce neurons from defined stem cells. Table 3–3 outlines the comparison. In Chapter 7, p. 217, we will return to the technical aspects and the critique in greater detail.

Independent of the method used to culture the cells: In the absence of sensitive and specific precursor cell markers, the question remains as to how precursor cells can be identified in the cultures and, consequently, also how candidate stemness markers could be validated. In many contexts of brain research, the deliberate use of the term *stem cell* is somewhat misleading, because an analysis of stemness has actually not been performed.

The Question of Transdifferentiation

Around the year 2000, "transdifferentiation" seemed to question the limits in the potential for differentiation of somatic precursor cells

The traditional concept of stem cell biology is that stem cells of the adult organism are tissue-specific—that is, they are only able to generate cell types of the tissue they are found in. From an evolutionary perspective, this is beneficial, because for regenerating the intestinal epithelium, for example, oligodendrocytes or pancreatic islet cells are not only not needed but also not wanted. The lineage restriction reduces the risk of malformation tumors, such as teratomas or teratocarcinomas. Around the year 2000, however, a number of studies challenged this view by suggesting that precursor cells from one organ could cross limits of tissue specificity and generate differentiated cells in other organs (Fig. 3–14).

For example, studies suggested there was generation of blood from neural stem or progenitor cells and of neurons and glia from hematopoietic stem cells (Brazelton et al., 2000). Most surprisingly, one year after bone marrow transplantation, Purkinje neurons carrying the markers of the implanted bone marrow were found in the cerebellum (Priller et al., 2001; Weimann et al., 2003). Similarly, in the brains of female patients who had received a bone marrow graft from male donors, brain autopsy results revealed neurons carrying the Y chromosome (Mezey and Chandross, 2000; Cogle et al., 2004). Because females do not have Y chromosomes, these neurons had to originate from the donor bone marrow cells.

Other reports apparently demonstrated the generation of hepatocytes (Lagasse et al., 2000) and skeletal muscle from bone marrow cells (Ferrari et al., 1998; Gussoni et al., 1999), bone marrow repopulation (Bjornson et al., 1999), and muscle cell development (Galli et al., 2000) from neural precursor cells. These data were regarded as evidence of transdifferentiation, in that under certain conditions cells of one germ layer might develop into cells from another. As yet unidentified cues *in vitro* or from the local microenvironment after implantation appeared to be responsible for this process, and plausible candidate mechanisms included epigenetic changes at the chromatin level and in DNA methylation and acetylation—altering the accessibility of genes for regulatory cues. The findings suggestive of transdifferentiation put into question the validity of many ideas on stem cells and stirred tremendous attention. As of 2008, four reports from this era were among the 10 most frequently cited papers in the entire stem cell field, making it by far the most-debated issue in stem cell research.

Most observations of "transdifferentiation" could be explained by cell fusion

As part of this response, a number of serious concerns had early been raised against the plausibility of transdifferentiation (Anderson et al., 2001; Wagers et al., 2002; Wagers and Weissman, 2004; Joseph and Morrison, 2005; Jaenisch and Young, 2008). Most notably, cell fusion could underlie these findings (Terada et al., 2002; Ying et al., 2002). "Cell fusion" means

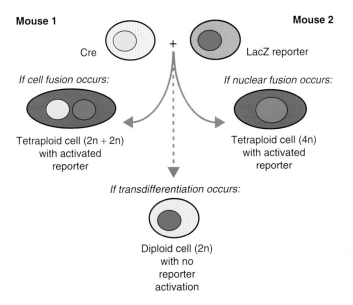

FIGURE 3–14 Fusion scheme. Most examples of "transdifferentiation" appear to be readily explainable by cell fusion. That fusion occurs was shown by the use of two transgenic lines. In the case of transdifferentiation, only the reporter gene from the grafted donor would be present in the supposedly transdifferentiated cells. In the case of fusion, both reporters appear in the same cell, which then shows various degrees of aneuploidy. The schematic reflects the work in Alvarez-Dolado, 2003.

that the grafted cell would not reprogram its developmental program but would fuse with a host neuron, thus generating a cell with both donor and neuron markers. One way to detect cell fusion is very straightforward: fused cells should have more than the two chromosome sets of normal cells—they should be tetraploid, not diploid. Indeed, tetraploid cells can be found in some but not all circumstances of supposed transdifferentiation; both hepatocytes and Purkinje cells, two of the supposedly transdifferentiated cell types, are known to be aneuploid. Nevertheless, in vitro, fusion has been found to be a very rare event (Terada et al., 2002; Ying et al., 2002).

Arguably, detection of aneuploidy in vivo is technically challenging. But the Purkinje neurons and hepatocytes supposedly derived from bone marrow could indeed be explained by cell fusion (Alvarez-Dolado et al., 2003; Wang et al., 2003), as suggested by studies in transgenic reporter gene mice (Fig. 3–14). Purkinje cells and hepatocytes might have a particular propensity to fuse. If cell fusion occurs, the fused cells would be tetraploid, unless there was an unprecedented merging and consecutive reduction of the genome (for which there is no evidence) or a fusion after which the invading nucleus threw out the host nucleus. Thus tetraploidy can be visualized with relative ease when the two fusing cells each contain a reporter gene that allows their unambiguous identification (Alvarez-Dolado et al., 2003). In the case of the "bone marrow–derived" Purkinje cell neurons and hepatocytes, exactly this was found. The reporter gene of the transplanted bone marrow was expressed in the same cell as the host reporter gene. If real transdifferentiation had occurred, only the graft reporter gene would have been detectable.

Whether "transdifferentiation" is categorically impossible cannot be stated with confidence but it is not a common event

If fusion has been established as a serious confounding factor in the theory of transdifferentiation, does this rule out the possibility of transdifferentiation? Proof of definite absence is generally difficult to establish. In fact, transdifferentiation of neural precursor cells into endothelial cells induced by cell–cell contact with other endothelial cells was shown not to depend on cell fusion (Wurmser et al., 2004). It is not clear to what degree this finding can be applied in principle to other examples of transdifferentiation, and this particular example has not been studied further. But the hope that the phenomenon is so common and robust that it can be used as basis for novel stem cell–based therapeutic approaches is not justified, given the available evidence.

Intriguingly, in the beginning of 2010 it was reported that a successful reprogramming of fibroblasts into neurons was possible in vitro. The transition was based on the overexpression of only three genes, Ascl1/Mash1, Brn2/Pou3f2 and Myt1l (Vierbuchen et al., 2010). The conversion went much further than in other previous attempts before but it remains to be seen how far.

Redifferentiation stands for the idea that differentiated cells might revert to a stem cell state

A concept related to transdifferentiation is *redifferentiation*. Whereas in transdifferentiation the differentiating cells develop directly into a cell type from another germ layer, it is thought that in redifferentiation an intermediate step exists at which the cells revert to a more immature cell type common to the two germ layers. At present it is not known whether redifferentiation can occur. Most of the evidence put forth against transdifferentiation applies to redifferentiation as well. The perception of "transdifferentiation" or "redifferentiation" changed fundamentally when it was reported that somatic cells could be effectively reprogrammed into a pluripotent state.

Induced Pluripotent Stem (IPS) Cells

Pluripotency can be induced in somatic cells with only four or fewer factors

The most important event in stem cell biology in the past few years has been the report by Takahashi and Yamanaka as well as Wernig und Jaenisch and their colleagues that with a set of only four genes to be manipulated, somatic cells of mice (Takahashi and Yamanaka, 2006) and humans could be induced to revert to a pluripotent state (Takahashi et al., 2007; Wernig et al., 2007).

Basis for the search for factors that might reprogram a somatic cell into a pluripotent state was the observation that if the nucleus of a somatic cell was exposed to the cytoplasm of a zygote ("somatic cell nuclear transfer", SCNT; or in popular parlance, "therapeutic cloning"), in some cases the resulting chimeric cell was pluripotent. Dolly, the first cloned sheep, was conceived by a similar process, then called "reproductive cloning" (Wilmut et al., 1997). This process is terribly inefficient, but in principle, functions. The question thus arose as to which constituents of the cytoplasm of the zygote had elicited this effect and which changes in the genome had to be induced to achieve the reversion to pluripotency. Surprisingly, overexpression of four genes turned out to be sufficient to induce pluripotency in skin fibroblasts. These genes were Oct-3/4, Sox2, c-Myc, and Klf4. Oct4 together with Nanog is the key gene to maintain ES cells in a pluripotent state. Sox2 is similarly essential for pluripotency. The role of oncogenes c-myc and Klf4, however, seemed less clear. In fact, it was first shown that c-myc is dispensable, although the already low efficiency of the reprogramming processes is thereby further reduced (Yu et al., 2007; Nakagawa et al., 2008; Wernig et al., 2008a). Efficiency of the induction of pluripotency by these means is very low, too, but the entire protocol is much more straightforward than in SCNT, so that this low success rate is less of a problem. In several studies published so far, the efficiency was at approximately 0.015% of the transfected cells (Okita et al., 2007; Kim et al., 2009).

Because neural precursor cells already express Sox2, using them as a source of cells to start the reprogramming process further reduced the number of factors that had to be manipulated (Duinsbergen et al., 2008; Kim et al., 2008). Besides, Sox2–negative neural progenitor cells could also be reprogrammed to pluripotency (Eminli et al., 2008).

Of the four factors, only Oct4 is necessary to induce pluripotency, but the efficiency is extremely low

Finally, because of the preexisting stem cell properties of the neural precursor cells, the oncogene, too, turned out to be superfluous: Oct4 alone elicited the reprogramming (Kim et al., 2009). When implanted to a blastocyst, these neural stem cell–derived pluripotent cells contributed to the germline, one of the most rigorous tests for pluripotency, although it cannot rule out epigenetic alterations that might prevent proper development. Oct4 is usually restricted to ES cells (and germline cells), and its overexpression in somatic cells results in severe dysplasias (Hochedlinger et al., 2005; Lengner et al., 2008). In the neural stem cells this was not the case.

To prove pluripotency requires stringent tests, the strongest of which is "tetraploid complementation"

The most rigorous test for pluripotency would be by so-called tetraploid complementation (Jaenisch and Young, 2008). By applying an electric current to two-cell-stage embryos, a fusion into one single diploid cell is achieved, which after one round of division generates tetraploid embryo, which can be taken to the blastocyst stage (Kubiak and Tarkowski, 1985). The tetraploid cells cannot contribute to further development of the embryo but can generate the trophoblast (placenta). If pluripotent cells, including iPS cells, are injected into the 4n blastocysts, only the injected cells can thus generate the embryo. If this occurs, the proof of full pluripotency has been achieved. In some of the reports of iPS cells this has most stringent test has been passed (Wernig et al., 2007), germ-line contribution in the others. This rigor sets induced pluripotency apart from the earlier claims of pluripotency during the "transdifferentiation" wave around the year 2000, where these criteria had generally not been tested for.

The best of the induced cells, selected on the basis of their expression of endogenous Oct4 and Nanog, the key regulators of pluripotency, thus behaved like ES cells. On the level of gene expression and epigenetic states, they were essentially indistinguishable (Maherali et al., 2007). Nevertheless, all manipulations and assays involve extensive in vitro work, which will alienate the cells from the in vivo situations. Initially described types of iPS cells were generated by means of retroviral vectors, making the cells inapplicable for any therapeutic use. Later protocols used small molecules to achieve reprogramming (Soldner et al., 2009).

iPS cells might offer new avenues for cell replacement strategies but above all allow the development of patient- and disease-specific cell models

iPS cells have been applied in a couple of proof-of-principle animal studies for cell replacement, among them prominently the derivation of neuronal cells that alleviated symptoms in a rat model of Parkinson disease (Wernig et al., 2008b). IPS cells from patients with amyotrophic lateral sclerosis (ALS) were successfully differentiated into motoneurons in vitro, following similar principles previously established for ES cells (Dimos et al., 2008).

The immediately even more relevant application of iPS cells, however, lies in the development of cellular models of complex disease (see also Fig. 12–1). As first proof of principle, besides the mentioned example of iPS cells from ALS patients, patient-specific pluripotent cell lines were isolated from patients with Parkinson disease (Soldner et al., 2009).

Neural Stem Cells as the Origin of Brain Tumors

Brain tumors might arise from neural stem cells

Where there is light, there is shadow. An unexpected dark side of neural stem cell biology became apparent with the hypothesis that brain tumors might originate from neural precursor cells (Noble and Dietrich, 2004). Quite general, the "stem cell hypothesis" of oncology has profoundly transformed concepts of how cancer develops. Not unexpectedly, the hypothesis has its roots in hematology with its relative clear lineage relationships and the existence of distinct types of cancerous transformations that fit these lineages (Reya et al., 2001). For example, the expression of the fusion gene characteristic for chronic myelogenous leucemias associated with a t(9;22) chromosomal translocation was sufficient to induce myeloid leukemia in mice, especially if a second hit impaired the control of apoptosis (Jaiswal et al., 2003). The stem cell hypothesis might also help to explain why many oncological therapies ultimately fail: they might be simply directed against the wrong cell type.

Historically, the stem cell hypothesis of brain tumors has first been proposed for subependymomas

The first brain tumors for which a connection between precursor cell activity and tumorigenesis was proposed were subependymomas, relatively rare tumors of the ventricular wall. At the time when this idea was first phrased (Globus and Kuhlenbeck, 1944), precursor cells in the adult brain had not yet been discovered, but a hypothetical link to persistently dividing cells in the walls of the ventricles could be made. Thus, a neurooncological report became one of the first comments on continued cell division in the adult brain. Classical stem cell tumors, in particular teratomas and teratocarcinomas, can affect the brain as well, but were always regarded as a special case. Nevertheless, given the fact that these tumors are otherwise predominantly found in the gonads, their existence in brain (and other tissues) at least raised some questions (Sano, 1995).

In some, admittedly very rare, tumors, the link between tumor cells and some sort of precursor cell seems obvious. Olfactory neuroblastoma (esthesioneuroblastoma), originating from the highly proliferative yet neurogenic olfactory epithelium is such an example (although a tumor of the peripheral nervous system).

For primitive, dysontogenetic tumors of the brain, a link to precursor cells has long been suspected

This kind of tumors often have a characteristically uncharacteristic morphology, known in neuropathology as "small round cell tumors." A whole group of so-called dysontogenetic tumors shares such appearance, including the primitive neuroectodermal tumors (PNET), an entity for which an origin from precursor cells (or "blasts") had been suggested for a long time. The best-known example of these tumors is the medullablastoma, a childhood tumor of the cerebellum and brain stem. The tumor cells respond to Notch, sonic hedgehog (Shh), or Wnt, as do neural precursor cells, underscoring a potential identity (for review, see Ross and Spengler, 2007; Fan and Eberhart, 2008). In about 25% of cases, medulloblastomas arise due to a mutation in Shh signaling that normally governs the generation of cerebellar granule cells. Medulloblastomas are extremely rare in adulthood, consistent with the idea that they arise from a precursor cell, and if they occur late, their origin from a precancerous alteration in childhood cannot be excluded. Despite the names "neuroblastoma" or "medulloblastoma" and the speculation that such tumors are of precursor cell origin, no actual "neuroblast" or "medulloblast" has been identified for decades. There might be no such cell as a medulloblast, but the transformed cell might still be a neuroepithelial precursor cell.

Isolated tumor stem cells can propagate tumors after transplantation

The perception changed with the rise of neural stem cell biology. Today the existence of so-called tumor-propagating cells (TPC) or tumor-initiating cells (TIC) that share many features with stem cells of the tissue in question can be isolated from many tumors. If transplanted into a host tissue, tumor stem cells can give rise to new tumors, hence the designation as "tumor propagating cells." These terms, TPC or TIC, are used to avoid premature judgments about stemness in these cells, but the term *tumor stem cells* stuck and is more generally used despite these problems. As in most other contexts, the term *stem cell* has to be taken with a grain of salt here as well.

Tumor stem cells express stem cell markers

Surface marker Prominin-1/CD133 has been particularly useful for isolating TPC in many tumors, but the relationship between TPC and intrinsic stem cells is not quite clear (see also below; and Bidlingmaier et al., 2008; Mizrak et al., 2008; and Zhu et al., 2009). In intestine, for example (see below), Prominin 1/CD133 expression identified stem cells that were susceptible to neoplastic transformation (here through activation of aberrant Wnt signaling: Zhu et al., 2009). The key question is whether TPCs originate from normal-tissue stem cells or from later-stage progenitor or even mature cells that are reverted to a more stem-like phenotype (Lowry and Temple, 2009). In the case of medulloblastoma, it turned out that the TPC in a mouse model of the disease (based on the mutated Shh-receptor Patched) were distinct from the CD133–positive population but expressing other progenitor cell markers, CD15/SSEA and Math1, thereby simultaneously challenging and supporting the stem cell hypothesis of brain tumors (Read et al., 2009). The bottom line of such studies is that the hypothesis of stem cells' being the origin of tumors has great appeal, but the actual data that are available defy an easy identification of the culprit cells.

Gliomas might arise from the precursor cell populations of the adult brain

If somatic precursor cells are the origin of brain tumors, the question is whether tumors might arise from multipotent neural precursor cells as they continue to exist in the neurogenic zones (and only possibly, and in very low numbers, beyond), or from restricted glial progenitor cells, such as NG2 cells, residing throughout the brain parenchyma but favoring the white matter, or from both.

Gliomas, tumors of apparently glial origin and with certain glial differentiation, can occur throughout life but show a greater prevalence in older age. Gliomas show a large variability in neuropathological appearance, and their most malignant variant, glioblastoma multiforme (GBM), is with 15% the most frequent brain tumor, accounting for roughly half of all gliomas, and has an incidence of about three to five cases per 100,000 person-years. As the name multiforme indicates, GBM are characterized by a mix of differently differentiated cell types, vascularization, necrosis, and lymphocytic reaction. Most GBM are dominated by astrocyte-like differentiation, but forms with intermingled oligodendrocytic and mesodermal differentiation exist. In addition, some signs of neuronal differentiation have been found in tumor cells (Labrakakis et al., 1998). The notion that these neuronal features originate from resident neurons that become enclosed by growing tumor has been largely abandoned. This range of differentiation patterns within one tumor is indeed best explained if one assumes that the origin is from a cell with a wide potential for differentiation. Gene defects known to result in tumorigenic transformation concern genes that also often play prominent roles in stem cell biology and can be used to transform neural precursor cells experimentally (Holland et al., 2000; Uhrbom et al., 2002; Seoane et al., 2004). From human glioma samples, cells could be isolated that behaved as stem cells in vitro and caused tumor formation in a xenograft model (Singh et al., 2004). Currently this evidence from the transplantation of stemlike cells is the strongest argument in favor of the stem cell hypothesis of tumorigenesis.

The location of glioblastomas might also be suggestive of a link to regions rich in precursor cells. Most glioblastomas seem to arise from anterior subcortical structures and the temporal lobes (Larjavaara et al., 2007).

Stiles and Rowitch have reviewed the evidence in favor of the existence of glioma stem cells and have come to the conclusion that "it is only Ockham's razor, applied to the data in toto, that currently favors a multipotent stem cell origin for malignant glioma," but they also raise the questions that remain to be addressed (Stiles and Rowitch, 2008).

White matter precursor cells have attracted particular attention as potential tumor stem cells

Many presently available data concern the possible role of oligodendrocyte precursor cells in the genesis of gliomas (Noble, 1997). GBM show varying degrees of oligodendrocytic differentiation, but oligodendroglioma as well as mixed gliomas with lower degrees of malignancy than GBM exist. Consequently, human gliomas can express many genes that are active in the oligodendrocyte lineage (Kashima et al., 1993), and from human gliomas a CD133–negative population of tumor stem cells can be identified that reacts with the A2B5 antibody, which identifies O2A progenitor cells (Ogden et al., 2008). Transforming putative O2A progenitor cells with oncogenes *c-myc* and *Ras* resulted in gliomas after implantation in the mouse brain (Barnett et al., 1998). Similarly, activation of Akt and Ras in neural precursor cells led to the development of glioblastoma-like tumors in mice (Holland et al., 2000). White-matter precursor cells in the adult rat brain could be driven to form glioblastoma-like tumors by the retrovirally mediated overexpression of platelet-derived growth factor (Assanah et al., 2006). PDGF has a been suspected as key tumorigenic factor since it was discovered that the cancerogenic gene in the simian sarcoma virus is PDGF β (Wong-Staal et al., 1981). In fact, PDGF turned out to be sufficient to induce glioma growth (Uhrbom et al., 1998). PDGF is secreted from tumor cells and acts in a paracrine and autocrine fashion on cell proliferation, migration, and differentiation as well as on angiogenesis (Shih and Holland, 2006). As mentioned before, PDGF receptor α is also expressed on precursor cells in the SVZ.

The occurrence of tumors with neuronal differentiation can best be explained with a stem cell hypothesis rather than a de-differentiation process

Tumors in which neuronal differentiation goes beyond the expression of some marker proteins, such as in gangliocytomas or gangliogliomas, are particularly interesting in this context. Both are very rare entities but are found relatively more often in the temporal lobe. Gangliogliomas show a mix of neoplastic neuronal and glial cells; in gangliocytomas, only neoplastic neurons are present. Even rarer are central neurocytomas, an intraventricular, rather benign tumor, in which the cells appear to be arrested at the stage of neuronal immaturity. Because differentiated neurons cannot divide, the idea of a neuronal origin of neurocytomas and gangliocytomas seemed improbable. The as-yet-unproven hypothesis that they develop from a type of precursor cell is more plausible but requires further molecular analysis.

Markers allow the prospective isolation of tumor stem cells

Putative tumor stem cells have been isolated from tumor samples on the basis of marker expression. Nestin, for example, is expressed in neuroectodermal tumors and provided an early hint at potential precursor cells in brain tumors (Tohyama et al., 1992; Kashima et al., 1993). However, the specificity of this finding remained low, since germ cell tumors of the CNS, an exceedingly rare group of tumors, also express nestin (Sakurada et al., 2008), and after injury, nestin expression in the adult brain is widespread in normal astrocytes (Krum and Rosenstein, 1999; Sahin Kaya et al., 1999). Thus the observation of canonical stem cell gene Sox2 in gliomas (Phi et al., 2008) as well as in teratomas in the CNS (Phi et al., 2007) might be somewhat more relevant, but as discussed above, Sox2 expression alone is also insufficient to define stemness, and it is also not specific enough.

The most interesting "stem cell marker" in tumors is Prominin1, or CD133. On the basis of CD133–expression, cells with stem cell properties could be isolated from human tumor samples and formed tumors after implantation in nude mice (Galli et al., 2004; Singh et al., 2004; Yuan et al., 2004; Ogden et al., 2008). The biology of Prominin1 and its putative role in neural stem cells will be discussed in Chapter 4. As mentioned before, on the basis of CD133 expression, precursor cells can be isolated from the adult mammalian brain (as well as from other organs), but their relationship to precursor cells isolated by other means has not yet been fully established (Bidlingmaier et al., 2008; Dell'Albani, 2008). Tumor propagation can also be achieved with the CD133–negative fraction, suggesting that CD133–positive cells are only a subset of potential tumor stem cells (Beier et al., 2007).

CD133 shows an overlap with white-matter precursor cell markers such as A2B5 and PDGF receptor alpha in gliomas, and the expression appears to increase with increasing malignancy (Rebetz et al., 2008). This suggests that the lines of evidence are beginning to converge. On the other hand, there are considerable differences between the putative tumor stem cells and precursor cells isolated from non-cancerous adult human brain; e.g., with respect to cell proliferation and differentiation and, not surprisingly, the ability to form tumors (Varghese et al., 2008).

Tumor stem cells can also be found in side populations

In addition, side populations (see above) also turned out to be rich in tumor stem cells. Initially from neuroblastoma samples and glioblastoma cell lines side populations have been isolated by fluorescence-activated cell sorting, and these cells showed stem cell properties (Hirschmann-Jax et al., 2004). When side population cells were extracted from solid tumors (e.g., from lung), these stemlike cells were more malignant after implantation into nude mice than the cells from the corresponding main population (Ho et al., 2007). This observation was not consistent between different studies, however (Patrawala et al., 2005). Side population cells were also found in murine glioma models (Harris et al., 2008), opening up the opportunity for more experimental research. On the other hand, tumor cell lines might be particular in that they contain unrealistically high numbers of tumor stem cells (Zheng et al., 2007). Bleau and colleagues derived side population cells from murine gliomas, induced by PDGF, as well as from human glioma samples (Bleau et al., 2009). When neurospheres from these cells were subcloned, the side populations from secondary spheres exhibited even greater sphere-forming capabilities, suggesting that the ABCG2 transporter activity enriched for stem-like cells. When the gene PTEN was deleted, the side population nearly doubled, and the side population also increased after treatment with the cytostatic agent of choice for gliomas, temozolomide (Bleau et al., 2009). The latter finding points to the counterintuitive and possibly paradoxical implications that the stem cell hypothesis of brain tumors might have for therapy (Persson and Weiss, 2009). The efflux of incorporated substances through the membrane transporter ABCG2 not only allows us to identify tumor precursor cells but also helps explain how these tumorigenic cells escape from drug treatment. Chemoresistance of stem cells is also mediated via ABCG2 in that the drug, just like the dye, is effectively pumped out of the cell.

Mesenchymal Stem Cells

Mesenchymal stem cells first seemed particularly suitable cells for "transdifferentiation"

Mesenchymal stem cells are found in the bone marrow (and other sites) and generate connective tissues

Mesenchymal stroma cells can show limited signs of neuronal differentiation

The potential for partial neuronal differentiation might be explainable by the origin of mesenchymal stem cells from the neural crest

Mesenchymal stem cells (MSC) became interesting in neurobiological contexts when experiments suggested that they might be a promising cellular source for neuronal replacement therapies (Liu et al., 2009). This euphoria peaked during the high days of "transdifferentiation" (Woodbury et al., 2000) and has waned since, but mesenchymal stem cells remain a particularly interesting cell type in neural precursor cell biology. The MAPCs, discussed above, presumably belong to the category MSCs.

Mesenchymal stem cells are found in the bone marrow, where they are part of the stroma. They are thus also referred to as "mesenchymal stroma cells." Their lineage is independent of the hematopoietic lineage. However, the stromal stem cells are part of the stem cell niche of the bone marrow, contributing to the microenvironment sustaining precursor cell activity in the hematopoietic system. In addition, mesenchymal stem cells are found in the adipose tissue and cartilage as well as in cord blood. Like hematopoietic stem cells, mesenchymal stem cells can be identified prospectively by marker combinations (Gronthos et al., 1994), and under appropriate conditions, the cells differentiate into the osteoblast, adipocyte, and chondrocyte lineages (Pittenger et al., 1999). Little is known about heterogeneity among mesenchymal stem cells. They have largely been characterized ex vivo, and hardly any knowledge exists about their nature and possible positional identity in vivo (da Silva Meirelles et al., 2008).

Under appropriate culture conditions, mesenchymal stem cells can show some signs of neuronal differentiation and express markers associated with neuronal lineage (Levy et al., 2003). These conditions include, for example, exposure to DMSO (Woodbury et al., 2002) or conditioned media (Fu et al., 2004). The bone marrow–derived cells lose their mesenchymal markers, grow in agglomerates resembling neurospheres, and can be induced to differentiate into cells expressing neuronal, astrocytic, and oligodendrocytic markers in vitro (Hermann et al., 2004). But neuronal differentiation has turned out to be been rather limited. Under DMSO an incomplete morphological change toward a neuronal shape was associated with the expression of some but not other neuronal markers (Neuhuber et al., 2004). Similar morphological alterations could also be induced in fibroblasts. The expression of neuronal features has also been interpreted as a stress response of the cells, and as such, more or less as an artifact (Croft and Przyborski, 2006). In vitro, only three genes, Ascl1/Mash1, Brn2/Pou3f2, and Myt1l, were required to converse fibroblasts into neurons (Vierbuchen et al., 2010). Here more advanced neuronal phenotypes were achieved, but it remains to be seen which in vivo standards the neurons will ultimately meet. The Vierbuchen report might lead to a reconsideration of the entire question of "neurons" from mesodermal cell types, but this is too early to decide.

The most parsimonious interpretation of the available data from the majority of studies is that, under specific culture conditions, mesenchymal precursor cells can show signs of incomplete neuronal development. This at-first-counterintuitive observation becomes somewhat less surprising if one considers the hypothesis that mesenchymal stem cells are actually related to the neural crest (Morikawa et al., 2009). In an elegant series of experiments, Takashima and colleagues have demonstrated that a population of Sox1–positive cells in the early embryo gives rise to MSCs, which in part involves an intermediate step in the neural crest (Takashima et al., 2007). As expression of Wnt1 is specific to the neural crest, Nagoshi and colleagues could use a Wnt1–Cre/Floxed-EGFP construct to follow the progeny of neural crest precursor cells. They detected "neural crest-derived stem cells" in various tissues, including the bone marrow (Nagoshi et al., 2008). These cells also showed some potential for mesodermal differentiation. Interestingly, at a young age it seemed that such cells were wandering with the bloodstream.

So far no signs of full neuronal
maturation have been found in
differentiated mesenchymal
stem cells

At present it is not clear whether mesenchymal precursor cells can truly produce fully functional neurons under any circumstances. Most studies on the neurogenic potential of mesenchymal stem cells have been in vitro studies, so the end point of neuronal maturation and functionality is very difficult to determine. As discussed in Chapter 10, function on a single-cell level is different from that on a system level in vivo. Studies showing signs of neuronal differentiation in mesenchymal stem cell cultures were not able to show appropriate electrophysiological responses (Padovan et al., 2003), but the cells did respond to NMDA and showed voltage-gated channels. In other studies, the differentiated cells revealed basic physiological properties of neurons, such as, for example, delayed rectifier potassium currents (Mareschi et al., 2006) and even single-action potentials have been observed (Wislet-Gendebien et al., 2005).

Because they appear to be "somewhat more multipotent" than other precursor cell types in the adult, MSCs are sometimes incorrectly referred to as "pluripotent" (Porada et al., 2006; Dore-Duffy, 2008; Koyama et al., 2009), which should be avoided and the term be reserved to the potency level of embryonic stem cells.

Mesenschymal stem cells might
secrete trophic factors, which
could explain the therapeutic
effects of implanted cells in the
absence of "transdifferentiation"

MSCs can secrete many factors that affect other cells in a paracrine fashion (Sze et al., 2007). Most of the beneficial observations in clinical studies that used MSCs, for example after stroke, have been linked to such effects (Caplan and Dennis, 2006). Among this plethora of factors, MSCs can also release factors that promote neuronal differentiation from neural precursor cells in vitro (Hermann et al., 2006). Effects on bona fide neural precursor cells in vivo have been reported in a stroke model (Yoo et al., 2008).

Another interesting link exists between MSCs and NG2–positive pericytes of the adult brain. Pericytes might be multipotent ex vivo (Dore-Duffy, 2008) and have been brought into connection with MSCs (Crisan et al., 2008) but the potential origin from the neural crest suggests that pericytes and MSCs might share an ancestor rather than being identical.

Germinal Niches in the Adult Brain

At the end of the stem cell chapter, we need to turn briefly to the fact that stem cells are not islands and do not exist by themselves. Bone marrow stem cells might be found in the peripheral bloodstream, but they, too, originate from a tissue in which they form close neighborhood relationships with other cell types. In fact, one might consider the entire idea of stem cells as an intrinsic property as an artifice, given the fact that stem cells depend so much on environmental cues to determine their state. We will consider this issue further in the following paragraphs. While it does not make too much sense to abandon the concept of intrinsic stemness altogether, it seems wiser to include the relationship to the environment as part of the concept. This is done in the so-called stem cell niche.

At the core of this concept is the idea that precursor cells in the neurogenic zones (as well as in other organs) are embedded in a microenvironment with which they form a functional unit, the stem cell or germinative (or germinal) niche (Fig. 8–1). In the brain, these niches consist of the precursor cell proper, astrocytes, endothelial cells, microglia or macrophages, extracellular matrix, and close contact with the basal membrane. This particular structure does not seem to be found outside the classical neurogenic zones in the hippocampus and olfactory bulb. Therefore, the germinative niche, rather than the presence of precursor cells alone, might provide the basis for adult neurogenesis. Precursor cells with neurogenic potential ex vivo have been isolated from outside the neurogenic zones. Germinative niches serve two functions: to maintain the stem and progenitor cell activity, and to permit the initial stages of differentiation to occur. This is a somewhat paradoxical situation, because maintaining the stem cell state requires an active suppression of the tendency to differentiate.

In non-neurogenic regions, precursor cells appear to be maintained in a different, niche-independent manner: it might not be a coincidence that no neuronal development occurs from niche-free precursor cells. On the other hand, the idea that tumor angiogenesis might provide tumor stem cells with an abnormal niche might call for a broader take on the niche concept. In Chapter 8 we will return to the question of neurogenic permissiveness and the niche.

Still a Mystery: What Is a Stem Cell?

Stemness is dependent on cell-extrinsic cues

The example of the MSCs aptly demonstrates how culture conditions can determine the properties of precursor cells and even induce stemness, with different sets of culture conditions yielding incommensurable results. This is also true for neural precursor cells. The key role of FGF2 in releasing the stemness properties of cells in the adult brain is one key example (Palmer et al., 1999); the potential of Pax6 to convert bona fide astrocytes into neuronal progenitor cells, another (Heins et al., 2002). This does not mean that precursor cell identities are culture artifacts but that properties of neural precursor cells can be changed by external cues. This is in some contrast to the view that stemness is an intrinsic property that is only dependent on preceding developmental steps. In any case, there seem to be differences in sensitivity of various cell types to particular sets of environmental cues or genetic manipulations. The determinants of such differences in susceptibility might be interpreted as preconditions for stemness as well, but this rather stretches the concept.

Temporospatial contexts in development determine identity and potential of precursor cells

On the other hand, all cells of an individual share the same genetic information. Genetically, any cell is a potential stem cell. Whether this potential is realized in vivo depends on temporospatial contexts. If environment is everything except the gene itself, epigenetic and environmental conditions must exist that allow stemness. The question is no longer whether these conditions exist at all (it has been proven by induced pluripotency that they can be identified and successfully manipulated), but how exactly to change conditions adequately in time and space to obtain the desired types of cells. This has far-reaching implications for our ideas of what stem cells are and how they contribute to development, plasticity, and regeneration.

The radical view is that stem cells are entirely a product of the environment

At the crest of the findings suggestive of transdifferentiation, Helen Blau from Stanford University proposed in an influential and important comment that stem cells might not be a distinct type of cell at all, but rather reflect a functional state that can be acquired by different types of cells under different circumstances (Blau et al., 2001). In the extreme variant of that idea, the "highway model of stem cell biology" (Fig. 3–15), cells susceptible to local stemness-inducing cues would circulate throughout the body and context-dependently perform different stem cell functions. This radical interpretation has been "great food for thought," because it turned the previous concept on its head. As a provable scientific hypothesis it faced considerable difficulties, because in this view stemness is inherently transient—to prove the cellular identity over time and through different states of stemness and differentiation would be challenging. One main argument against this concept is that stem cells can in fact be prospectively isolated according to fixed sets of criteria. In turn, one could then still argue that these criteria define a transient state that can at times be fulfilled by different cells. But the question is whether such a condition could still be aptly named a functional state only. While the radical version of the model was therefore rightly abandoned with the collapse of the transdifferentiation hype, the idea touched on a fundamental and rather new aspect of stem cell biology: In principle any cell can be made into stem cells. This might not happen under physiological conditions in vivo, as the "highway model" suggested, but the necessary manipulations in vitro to induce even pluripotency are actually quite modest.

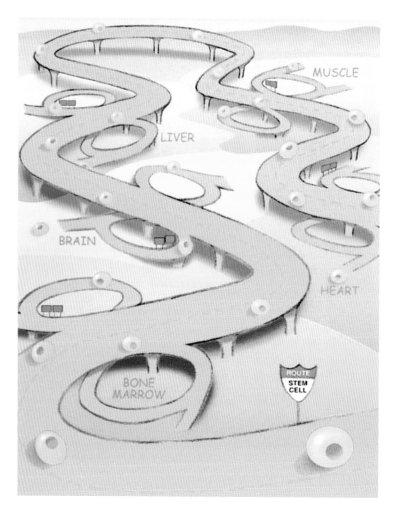

FIGURE 3–15 The Highway Model of stem cells. A famous radical take on the issue of transdifferentiation was taken by Helen Blau of Stanford University in an influential article (Blau. 2001) that proposed that stemness is a functional state of mobile cells, so that depending on the tissue, different stem cell properties emerge. Although in its literal meaning this hypothesis turned out to be incorrect, the concept stimulated discussion because it turned the old idea of stemness solely as a cell-intrinsic property literally on its head. To some degree, the same hypothesis can still be maintained for the fusion of circulating cells with selected tissue cells. Artwork by N. Gewertz and B. Colyear, kindly provided by Helen Blau. Reprinted from the cited publication with permission from Elsevier, copyright 2001.

Single genetic switches can exert great influence on stem cell properties

There are other examples of single molecular switches that exert a large effect on precursor cells' activity. The transcriptional repressor element 1–silencing transcription factor (REST) is the prime example (Kagalwala et al., 2009). Besides containing a DNA-binding domain, which qualifies REST as a transcription factor, REST also has two binding sites for co-repressors. REST is highly expressed in ES cells, moderately in neural stem cells, and is absent from differentiated neural cells (Fig. 3–16). REST appears to maintain "stemness" via micro-RNA-mediated mechanisms, and disinhibition of REST activates neuronal target genes (Immaneni et al., 2000), including, for example, BDNF (Zuccato et al., 2003), and induces neuronal differentiation from adult hippocampal precursor cells (Kuwabara et al., 2004). It has even been reported that a neuronal differentiation of myoblasts could be induced

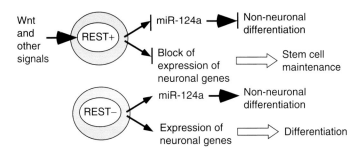

FIGURE 3–16 REST as a master regulator of adult neurogenesis. REST expression, which is up-regulated by many signaling molecules, including the Wnt pathway, directly blocks the expression of neuronal genes and indirectly (through miRNA-124a) non-neuronal differentiation, effectively promoting precursor cell maintenance.

via REST (Watanabe et al., 2004). In the presence of particular binding partners, the repressor turns into an activator of neurogenic processes (Kuwabara et al., 2004). We will come back to REST in Chapter 9, p. 405 and 408. The relevant conclusion at this point is that the precursor cell state is more flexible than previously appreciated, in that the presence or absence of one factor, its subcellular distribution, and its variable interaction with numerous binding partners is sufficient to determine the actual "stemness" of the cell in question. Stem cells do not just generate plasticity—they are, in some sense, plastic themselves.

The best-known image of this context-dependency is the "epigenetic landscape"

A classical visualization of the now-historic view on "development" is Conrad Hal Waddington's "epigenetic landscape" model (Fig. 3–17, inset), in which a ball rolls down a slope and can go down any of into several valleys, but every "decision" at every bifurcation restricts future decisions (Wadington, 1957). This description is equivalent to the concept that development means increasing limitation of potential in a progressively less permissive environment. The ball cannot roll uphill and go down into a new valley. What is less known is Waddington's glimpse of the underside of the epigenetic landscape (Fig. 3–17), in which he shows how guy lines leading to complexes of "genes" pull on the surface and thereby form the landscape (Slack, 2002). In that view the landscape appears, not "rock-solid" as it might appear at first, but malleable by the forces of the genome (and the environment). By that it turns out that the "epigenetic landscape" model is much more modern than often perceived and might even be reconciled with induced pluripotency and similar observations that contradict strict directionality in stem cell biology. It also shows why "stem cells" escape a final definition. Stem cells, too, are context-dependent.

Stem cell biology introduces openness and potential into biology

Stem cells are at the center of a scientific revolution that profoundly changes the principles on which medicine is based. Not unlike the parallel genomic revolution, stem cell biology forces humanity to start looking at biology from within rather than deductively from afar. Stem cells are characterized by a potential, and this potential is almost by definition inherently open. The world seen from the perspective of the stem cell (or the genome) is less deterministic than most deductive concepts of classical biology. This openness stimulates fantasy, and far-reaching consequences of stem cell biology for the treatment of incurable disease have been prophesied. At the same time, seemingly boundless potential raises fears because of the impending dangers to what people consider to be essentially human. This brings forth the old nature-versus-nurture debate in new clothing. How much of us is in our stem cells?

The potential of stem cells has stimulated fantasies for biomedical application that are detached from the realities of developmental biology

There is an obvious discrepancy between the public perception of stem cell biology and the science itself. This gap is the extension of what David Anderson has aptly named the "possible versus the actual" in stem cell research. In an influential review, Anderson placed the two fundamentally different types of stem cell research in opposition to each other, depending on whether the underlying biology or the potential application was the focus of interest (Anderson, 2001). He found only limited transfer of knowledge between the two groups and suspected a further divergence of

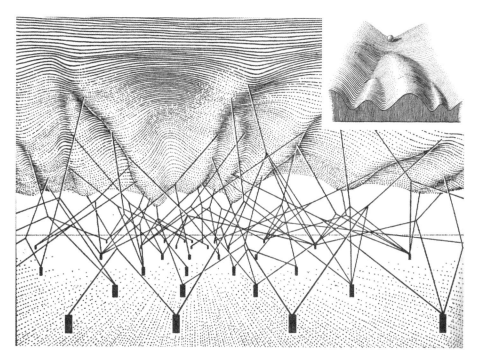

FIGURE 3–17 Waddington's famous "epigenetic landscape" (*inset*) has an underside that is far less well known. It shows the guy lines and strings that pull the landscape into place and suggest its plasticity. Both figures are taken from E. H. Waddington, *The Strategy of the Genes*, George Allen & Unwin, London, 1957. (The current rights-holder of the figure could not be identified.)

evolving concepts. Applied science sees stem cell biology essentially as a toolbox for developing new cell-based therapies. It often uses nomenclature taken from the different context of basic research, but with different meanings. Basic research tries to understand the fundamentals of cellular development and the parameters defining stemness. The term *lineage restriction*, for example, has solely descriptive and neutral meaning in basic stem cell biology, whereas in the field of applied stem cell research, it is viewed as either an imposed limitation or a desired goal.

But both fields could and should benefit from each other. And while they are occasionally in the hand of one and the same researcher, both the public and the scientific debates have suffered from a lack of clearly communicated concepts. The excitement over the great therapeutic potential of stem cells has often occluded the simple fact that it is not at all clear what is meant by *stem cell*.

Even today stem cells escape clear definitions

There is currently no single sufficiently precise definition of "stem cells." In a 1997 review on neural stem cells, Morrison, Shah, and Anderson coined an often quoted aphorism, comparing the difficulties in defining stem cells to the remark made by U.S. Supreme Court Justice Potter Stewart on pornography (in the paper erroneously ascribed to his colleague Byron White): "It is hard to define, but I know it when I see it" (Morrison et al., 1997). The analogy stuck, because it reveals the important fact that, despite the lack of an articulated definition, subconscious, operational definitions of stem cells are very much in place. The consequence is twofold. On the positive side, the tacit understanding that stem cells are somehow sufficiently definable enables everyday research. On the negative side, the lack of a matured concept obviously breeds errors and missed opportunities. This is particularly important when it comes to political and legal decisions affecting stem cell biology and its application in medicine.

References

Aguirre A, Gallo V (2004). Postnatal neurogenesis and gliogenesis in the olfactory bulb from NG2-expressing progenitors of the subventricular zone. *J Neurosci* 24:10530–10541.

Aguirre AA, Chittajallu R, Belachew S, Gallo V (2004). NG2-expressing cells in the subventricular zone are type C–like cells and contribute to interneuron generation in the postnatal hippocampus. *J Cell Biol* 165: 575–589.

Akamatsu W, DeVeale B, Okano H, Cooney AJ, van der Kooy D (2009). Suppression of Oct4 by germ cell nuclear factor restricts pluripotency and promotes neural stem cell development in the early neural lineage. *J Neurosci* 29:2113–2124.

Akopian G, Kuprijanova E, Kressin K, Steinhuser C (1997). Analysis of ion channel expression by astrocytes in red nucleus brain stem slices of the rat. *Glia* 19:234–246.

Allendoerfer KL, Magnani JL, Patterson PH (1995). FORSE-1, an antibody that labels regionally restricted sub-populations of progenitor cells in the embryonic central nervous system, recognizes the Le(x) carbohydrate on a proteoglycan and two glycolipid antigens. *Mol Cell Neurosci* 6:381–395.

Altman J, Das GD (1965). Autoradiographic and histological evidence of postnatal hippocampal neurogenesis in rats. *J Comp Neurol* 124:319–335.

Alvarez-Buylla A, Garcia-Verdugo JM, Tramontin AD (2001). A unified hypothesis on the lineage of neural stem cells. *Nat Rev Neurosci* 2:287–293.

Alvarez-Buylla A, Kohwi M, Nguyen TM, Merkle FT (2008). The heterogeneity of adult neural stem cells and the emerging complexity of their niche. *Cold Spring Harbor Symposia on Quantitative Biology* 73:357–365.

Alvarez-Dolado M, Pardal R, Garcia-Verdugo JM, Fike JR, Lee HO, Pfeffer K, Lois C, Morrison SJ, Alvarez-Buylla A (2003). Fusion of bone-marrow–derived cells with Purkinje neurons, cardiomyocytes and hepatocytes. *Nature* 425:968–973.

Anderson DJ (2001). Stem cells and pattern formation in the nervous system: The possible versus the actual. *Neuron* 30:19–35.

Anderson DJ, Gage FH, Weissman IL (2001). Can stem cells cross lineage boundaries? *Nat Med* 7:393–395.

Anthony TE, Klein C, Fishell G, Heintz N (2004). Radial glia serve as neuronal progenitors in all regions of the central nervous system. *Neuron* 41:881–890.

Arvidsson A, Collin T, Kirik D, Kokaia Z, Lindvall O (2002). Neuronal replacement from endogenous precursors in the adult brain after stroke. *Nat Med* 8:963–970.

Asano T, Ageyama N, Takeuchi K, Momoeda M, Kitano Y, Sasaki K, Ueda Y, Suzuki Y, Kondo Y, Torii R, Hasegawa M, Ookawara S, Harii K, Terao K, Ozawa K, Hanazono Y (2003). Engraftment and tumor formation after allogeneic in utero transplantation of primate embryonic stem cells. *Transplantation* 76: 1061–1067.

Assanah M, Lochhead R, Ogden A, Bruce J, Goldman J, Canoll P (2006). Glial progenitors in adult white matter are driven to form malignant gliomas by platelet-derived growth factor–expressing retroviruses. *J Neurosci* 26:6781–6790.

Babu H, Cheung G, Kettenmann H, Palmer TD, Kempermann G (2007). Enriched monolayer precursor cell cultures from micro-dissected adult mouse dentate gyrus yield functional granule cell–like neurons. *PLoS ONE* 2:e388.

Bailey KJ, Maslov AY, Pruitt SC (2004). Accumulation of mutations and somatic selection in aging neural stem/ progenitor cells. *Aging Cell* 3:391–397.

Baker DE, Harrison NJ, Maltby E, Smith K, Moore HD, Shaw PJ, Heath PR, Holden H, Andrews PW (2007). Adaptation to culture of human embryonic stem cells and oncogenesis in vivo. *Nat Biotechnol* 25:207–215.

Bani-Yaghoub M, Tremblay RG, Lei JX, Zhang D, Zurakowski B, Sandhu JK, Smith B, Ribecco-Lutkiewicz M, Kennedy J, Walker PR, Sikorska M (2006). Role of Sox2 in the development of the mouse neocortex. *Dev Biol* 295:52–66.

Baracskay KL, Kidd GJ, Miller RH, Trapp BD (2007). NG2-positive cells generate A2B5-positive oligodendro-cyte precursor cells. *Glia* 55:1001–1010.

Barnett SC, Robertson L, Graham D, Allan D, Rampling R (1998). Oligodendrocyte-type-2 astrocyte (O-2A) progenitor cells transformed with c-myc and H-ras form high-grade glioma after stereotactic injection into the rat brain. *Carcinogenesis* 19:1529–1537.

Barraud P, Thompson L, Kirik D, Bjorklund A, Parmar M (2005). Isolation and characterization of neural precursor cells from the Sox1–GFP reporter mouse. *Eur J Neurosci* 22:1555–1569.

Beier D, Hau P, Proescholdt M, Lohmeier A, Wischhusen J, Oefner PJ, Aigner L, Brawanski A, Bogdahn U, Beier CP (2007). CD133(+) and CD133(–) glioblastoma-derived cancer stem cells show differential growth characteristics and molecular profiles. *Cancer Res* 67:4010–4015.

Belachew S, Chittajallu R, Aguirre AA, Yuan X, Kirby M, Anderson S, Gallo V (2003). Postnatal NG2 proteoglycan-expressing progenitor cells are intrinsically multipotent and generate functional neurons. *J Cell Biol* 161:169–186.

Bello BC, Izergina N, Caussinus E, Reichert H (2008). Amplification of neural stem cell proliferation by intermediate progenitor cells in *Drosophila* brain development. *Neural Dev* 3:5.

Bennett L, Yang M, Enikolopov G, Iacovitti L (2009). Circumventricular organs: A novel site of neural stem cells in the adult brain. *Mol Cell Neurosci* 41:337–347.

Bergles DE, Roberts JD, Somogyi P, Jahr CE (2000). Glutamatergic synapses on oligodendrocyte precursor cells in the hippocampus. *Nature* 405:187–191.

Berry M, Hubbard P, Butt AM (2002). Cytology and lineage of NG2–positive glia. *J Neurocytol* 31:457–467.

Bidlingmaier S, Zhu X, Liu B (2008). The utility and limitations of glycosylated human CD133 epitopes in defining cancer stem cells. *J Mol Med* 86:1025–1032.

Bjornson CR, Rietze RL, Reynolds BA, Magli MC, Vescovi AL (1999). Turning brain into blood: A hematopoetic fate adopted by adult neural stem cells in vivo. *Science* 283:534–537.

Blau HM, Brazelton TR, Weimann JM (2001). The evolving concept of a stem cell: Entity or function? *Cell* 105:829–841.

Bleau AM, Hambardzumyan D, Ozawa T, Fomchenko EI, Huse JT, Brennan CW, Holland EC (2009). PTEN/PI3K/Akt pathway regulates the side population phenotype and ABCG2 activity in glioma tumor stem-like cells. *Cell Stem Cell* 4:226–235.

Blum B, Benvenisty N (2008). The tumorigenicity of human embryonic stem cells. *Adv Cancer Res* 100: 133–158.

Brazel CY, Limke TL, Osborne JK, Miura T, Cai J, Pevny L, Rao MS (2005). Sox2 expression defines a heterogeneous population of neurosphere-forming cells in the adult murine brain. *Aging Cell* 4:197–207.

Brazelton TR, Rossi FMV, Keshet GI, Blau HM (2000). From marrow to brain: Expression of neuronal phenotypes in adult mice. *Science* 290:1775–1779.

Buffo A, Rite I, Tripathi P, Lepier A, Colak D, Horn AP, Mori T, Gotz M (2008). Origin and progeny of reactive gliosis: A source of multipotent cells in the injured brain. *Proc Natl Acad Sci USA* 105:3581–3586.

Butt AM, Duncan A, Hornby MF, Kirvell SL, Hunter A, Levine JM, Berry M (1999). Cells expressing the NG2 antigen contact nodes of Ranvier in adult CNS white matter. *Glia* 26:84–91.

Butt AM, Kiff J, Hubbard P, Berry M (2002). Synantocytes: New functions for novel NG2-expressing glia. *J Neurocytol* 31:551–565.

Butt AM, Hamilton N, Hubbard P, Pugh M, Ibrahim M (2005). Synantocytes: The fifth element. *J Anat* 207: 695–706.

Cai L, Hayes NL, Takahashi T, Caviness VS, Jr., Nowakowski RS (2002). Size distribution of retrovirally marked lineages matches prediction from population measurements of cell cycle behavior. *J Neurosci Res* 69:731–744.

Cameron RS, Rakic P (1991). Glial cell lineage in the cerebral cortex: A review and synthesis. *Glia* 4:124–137.

Canoll P, Goldman JE (2008). The interface between glial progenitors and gliomas. *Acta Neuropathol* 116:465–477.

Caplan AI, Dennis JE (2006). Mesenchymal stem cells as trophic mediators. *J Cell Biochem* 98:1076–1084.

Carlen M, Meletis K, Goritz C, Darsalia V, Evergren E, Tanigaki K, Amendola M, Barnabe-Heider F, Yeung MS, Naldini L, Honjo T, Kokaia Z, Shupliakov O, Cassidy RM, Lindvall O, Frisen J (2009). Forebrain ependymal cells are Notch-dependent and generate neuroblasts and astrocytes after stroke. *Nat Neurosci* 12:259–267.

Carpenter MK, Inokuma MS, Denham J, Mujtaba T, Chiu CP, Rao MS (2001). Enrichment of neurons and neural precursors from human embryonic stem cells. *Exp Neurol* 172:383–397.

Cepko CL, Austin CP, Yang X, Alexiades M, Ezzeddine D (1996). Cell fate determination in the vertebrate retina. *Proc Natl Acad Sci USA* 93:589–595.

Challen GA, Little MH (2006). A side order of stem cells: The SP phenotype. *Stem Cells* 24:3–12.

Chari DM, Blakemore WF (2002). Efficient recolonisation of progenitor-depleted areas of the CNS by adult oligodendrocyte progenitor cells. *Glia* 37:307–313.

Chen J, Magavi SS, Macklis JD (2004). Neurogenesis of corticospinal motor neurons extending spinal projections in adult mice. *Proc Natl Acad Sci USA* 101:16357–16362.

Ciccolini F, Mandl C, Holzl-Wenig G, Kehlenbach A, Hellwig A (2005). Prospective isolation of late development multipotent precursors whose migration is promoted by EGFR. *Dev Biol* 284:112–125.

Cogle CR, Yachnis AT, Laywell ED, Zander DS, Wingard JR, Steindler DA, Scott EW (2004). Bone marrow transdifferentiation in brain after transplantation: A retrospective study. *Lancet* 363:1432–1437.

Colak D, Mori T, Brill MS, Pfeifer A, Falk S, Deng C, Monteiro R, Mummery C, Sommer L, Gotz M (2008). Adult neurogenesis requires Smad4–mediated bone morphogenic protein signaling in stem cells. *J Neurosci* 28:434–446.

Corbin JG, Gaiano N, Machold RP, Langston A, Fishell G (2000). The Gsh2 homeodomain gene controls multiple aspects of telencephalic development. *Development* 127:5007–5020.

Corti S, Nizzardo M, Nardini M, Donadoni C, Locatelli F, Papadimitriou D, Salani S, Del Bo R, Ghezzi S, Strazzer S, Bresolin N, Comi GP (2007). Isolation and characterization of murine neural stem/progenitor cells based on Prominin-1 expression. *Exp Neurol* 205:547–562.

Coskun V, Wu H, Blanchi B, Tsao S, Kim K, Zhao J, Biancotti JC, Hutnick L, Krueger RC, Jr., Fan G, de Vellis J, Sun YE (2008). CD133+ neural stem cells in the ependyma of mammalian postnatal forebrain. *Proc Natl Acad Sci USA* 105:1026–1031.

Craig CG, Tropepe V, Morshead CM, Reynolds BA, Weiss S, van der Kooy D (1996). In vivo growth factor expansion of endogenous subependymal neural precursor cell populations in the adult mouse brain. *J Neurosci* 16:2649–2658.

Crisan M et al. (2008). A perivascular origin for mesenchymal stem cells in multiple human organs. *Cell Stem Cell* 3:301–313.

Croft AP, Przyborski SA (2006). Formation of neurons by non-neural adult stem cells: Potential mechanism implicates an artifact of growth in culture. *Stem Cells* 24:1841–1851.

D'Amour KA, Gage FH (2003). Genetic and functional differences between multipotent neural and pluripotent embryonic stem cells. *Proc Natl Acad Sci USA* 100 Suppl 1:11866–11872.

da Silva Meirelles L, Caplan AI, Nardi NB (2008). In search of the in vivo identity of mesenchymal stem cells. *Stem Cells* 26:2287–2299.

Dawson MR, Polito A, Levine JM, Reynolds R (2003). NG2-expressing glial progenitor cells: An abundant and widespread population of cycling cells in the adult rat CNS. *Mol Cell Neurosci* 24:476–488.

Dayer AG, Cleaver KM, Abouantoun T, Cameron HA (2005). New GABAergic interneurons in the adult neocortex and striatum are generated from different precursors. *J Cell Biol* 168:415–427.

De Marchis S, Bovetti S, Carletti B, Hsieh YC, Garzotto D, Peretto P, Fasolo A, Puche AC, Rossi F (2007). Generation of distinct types of periglomerular olfactory bulb interneurons during development and in adult mice: Implication for intrinsic properties of the subventricular zone progenitor population. *J Neurosci* 27: 657–664.

Dell'Albani P (2008). Stem cell markers in gliomas. *Neurochem Res* 33:2407–2415.

Dimos JT, Rodolfa KT, Niakan KK, Weisenthal LM, Mitsumoto H, Chung W, Croft GF, Saphier G, Leibel R, Goland R, Wichterle H, Henderson CE, Eggan K (2008). Induced pluripotent stem cells generated from patients with ALS can be differentiated into motor neurons. *Science* 321:1218–1221.

Doetsch F (2003). The glial identity of neural stem cells. *Nat Neurosci* 6:1127–1134.

Doetsch F, Garcia-Verdugo JM, Alvarez-Buylla A (1999a) Regeneration of a germinal layer in the adult mammalian brain. *Proc Natl Acad Sci USA* 96:11619–11624.

Doetsch F, Caille I, Lim DA, Garcia-Verdugo JM, Alvarez-Buylla A (1999b) Subventricular zone astrocytes are neural stem cells in the adult mammalian brain. *Cell* 97:703–716.

Dore-Duffy P (2008). Pericytes: Pluripotent cells of the blood brain barrier. *Curr Pharm Des* 14:1581–1593.

Dore-Duffy P, Katychev A, Wang X, Van Buren E (2006). CNS microvascular pericytes exhibit multipotential stem cell activity. *J Cereb Blood Flow Metab* 26:613–624.

Duinsbergen D, Eriksson M, t Hoen PA, Frisen J, Mikkers H (2008). Induced pluripotency with endogenous and inducible genes. *Exp Cell Res* 314:3255–3263.

Eckenhoff MF, Rakic P (1984). Radial organization of the hippocampal dentate gyrus: A Golgi, ultrastructural, and immunocytochemical analysis in the developing rhesus monkey. *J Comp Neurol* 223:1–21.

Elkabetz Y, Panagiotakos G, Al Shamy G, Socci ND, Tabar V, Studer L (2008). Human ES cell–derived neural rosettes reveal a functionally distinct early neural stem cell stage. *Genes Dev* 22:152–165.

Ellis P, Fagan BM, Magness ST, Hutton S, Taranova O, Hayashi S, McMahon A, Rao M, Pevny L (2004). Sox2, a persistent marker for multipotential neural stem cells derived from embryonic stem cells, the embryo or the adult. *Dev Neurosci* 26:148–165.

Eminli S, Utikal J, Arnold K, Jaenisch R, Hochedlinger K (2008). Reprogramming of neural progenitor cells into induced pluripotent stem cells in the absence of exogenous Sox2 expression. *Stem Cells* 26:2467–2474.

Englund C, Fink A, Lau C, Pham D, Daza RA, Bulfone A, Kowalczyk T, Hevner RF (2005). Pax6, Tbr2, and Tbr1 are expressed sequentially by radial glia, intermediate progenitor cells, and postmitotic neurons in developing neocortex. *J Neurosci* 25:247–251.

Evans MJ, Kaufman MH (1981). Establishment in culture of pluripotential cells from mouse embryos. *Nature* 292:154–156.

Fan X, Eberhart CG (2008). Medulloblastoma stem cells. *J Clin Oncol* 26:2821–2827.

Ferrari G, Cusella-De Angelis G, Coletta M, Paolucci E, Stornaiuolo A, Cossu G, Mavilio F (1998). Muscle regeneration by bone marrow–derived myogenic progenitors. *Science* 279:1528–1530.

Ferri AL, Cavallaro M, Braida D, Di Cristofano A, Canta A, Vezzani A, Ottolenghi S, Pandolfi PP, Sala M, DeBiasi S, Nicolis SK (2004). Sox2 deficiency causes neurodegeneration and impaired neurogenesis in the adult mouse brain. *Development* 131:3805–3819.

Fiedorowicz A, Figiel I, Zaremba M, Dzwonek K, Oderfeld-Nowak B (2008). The ameboid phenotype of NG2 (+) cells in the region of apoptotic dentate granule neurons in trimethyltin intoxicated mice shares antigen properties with microglia/macrophages. *Glia* 56:209–222.

Filippov V, Kronenberg G, Pivneva T, Reuter K, Steiner B, Wang LP, Yamaguchi M, Kettenmann H, Kempermann G (2003). Subpopulation of nestin-expressing progenitor cells in the adult murine hippocampus shows electro-physiological and morphological characteristics of astrocytes. *Mol Cell Neurosci* 23:373–382.

Fu YS, Shih YT, Cheng YC, Min MY (2004). Transformation of human umbilical mesenchymal cells into neurons in vitro. *J Biomed Sci* 11:652–660.

Gabay L, Lowell S, Rubin LL, Anderson DJ (2003). Deregulation of dorsoventral patterning by FGF confers trilineage differentiation capacity on CNS stem cells in vitro. *Neuron* 40:485–499.

Gaiano N, Nye JS, Fishell G (2000). Radial glial identity is promoted by Notch1 signaling in the murine forebrain. *Neuron* 26:395–404.

Gal JS, Morozov YM, Ayoub AE, Chatterjee M, Rakic P, Haydar TF (2006). Molecular and morphological heterogeneity of neural precursors in the mouse neocortical proliferative zones. *J Neurosci* 26:1045–1056.

Galli R, Borello U, Gritti A, Minasi MG, Bjornson C, Coletta M, Mora M, De Angelis MG, Fiocco R, Cossu G, Vescovi AL (2000). Skeletal myogenic potential of human and mouse neural stem cells. *Nat Neurosci* 3:986–991.

Galli R, Binda E, Orfanelli U, Cipelletti B, Gritti A, De Vitis S, Fiocco R, Foroni C, Dimeco F, Vescovi A (2004). Isolation and characterization of tumorigenic, stem-like neural precursors from human glioblastoma. *Cancer Res* 64:7011–7021.

Garcia A, Steiner B, Kronenberg G, Bick-Sander A, Kempermann G (2004). Age-dependent expression of gluco-corticoid and mineralocorticoid receptors on neural precursor cell populations in the adult murine hippocampus. *Aging Cell* 3:363–371.

Garcia-Verdugo JM, Ferron S, Flames N, Collado L, Desfilis E, Font E (2002). The proliferative ventricular zone in adult vertebrates: A comparative study using reptiles, birds, and mammals. *Brain Res Bull* 57:765–775.

Gerrard L, Rodgers L, Cui W (2005). Differentiation of human embryonic stem cells to neural lineages in adher-ent culture by blocking bone morphogenetic protein signaling. *Stem Cells* 23:1234–1241.

Glaser T, Schmandt T, Brustle O (2008). Generation and potential biomedical applications of embryonic stem cell–derived glial precursors. *J Neurol Sci* 265:47–58.

Glass R, Synowitz M, Kronenberg G, Walzlein JH, Markovic DS, Wang LP, Gast D, Kiwit J, Kempermann G, Kettenmann H (2005). Glioblastoma-induced attraction of endogenous neural precursor cells is associated with improved survival. *J Neurosci* 25:2637–2646.

Globus JH, Kuhlenbeck H (1944). The subependymal cell plate (matrix) and its relationship to brain tumors of the ependymal type. *J Neuropathol Exp Neurol* 3:1–35.

Golmohammadi MG, Blackmore DG, Large B, Azari H, Esfandiary E, Paxinos G, Franklin KB, Reynolds BA, Rietze RL (2008). Comparative analysis of the frequency and distribution of stem and progenitor cells in the adult mouse brain. *Stem Cells* 26:979–987.

Goodell MA, Brose K, Paradis G, Conner AS, Mulligan RC (1996). Isolation and functional properties of murine hematopoietic stem cells that are replicating in vivo. *J Exp Med* 183:1797–1806.

Gotz M, Stoykova A, Gruss P (1998). Pax6 controls radial glia differentiation in the cerebral cortex. *Neuron* 21:1031–1044.

Gotz M, Huttner WB (2005). The cell biology of neurogenesis. *Nat Rev Mol Cell Biol* 6:777–788.

Greenwood K, Butt AM (2003). Evidence that perinatal and adult NG2-glia are not conventional oligodendro-cyte progenitors and do not depend on axons for their survival. *Mol Cell Neurosci* 23:544–558.

Gregori N, Proschel C, Noble M, Mayer-Proschel M (2002). The tripotential glial-restricted precursor (GRP) cell and glial development in the spinal cord: Generation of bipotential oligodendrocyte-type-2 astrocyte progeni-tor cells and dorsal-ventral differences in GRP cell function. *J Neurosci* 22:248–256.

Gritti A, Cova L, Parati EA, Galli R, Vescovi AL (1995). Basic fibroblast growth factor supports the proliferation of epidermal growth factor–generated neuronal precursor cells of the adult mouse CNS. *Neurosci Lett* 185:151–154.

Gritti A, Parati EA, Cova L, Frolichsthal P, Galli R, Wanke E, Faravelli L, Morassutti DJ, Roisen F, Nickel DD, Vescovi AL (1996). Multipotential stem cells from the adult mouse brain proliferate and self-renew in response to basic fibroblast growth factor. *J Neurosci* 16:1091–1100.

Gritti A, Frolichsthal-Schoeller P, Galli R, Parati EA, Cova L, Pagano SF, Bjornson CR, Vescovi AL (1999). Epidermal and fibroblast growth factors behave as mitogenic regulators for a single multipotent stem cell–like population from the subventricular region of the adult mouse forebrain. *J Neurosci* 19:3287–3297.

Gronthos S, Graves SE, Ohta S, Simmons PJ (1994). The STRO-1+ fraction of adult human bone marrow contains the osteogenic precursors. *Blood* 84:4164–4173.

Gubert F, Zaverucha-do-Valle C, Pimentel-Coelho PM, Mendez-Otero R, Santiago MF (2009). Radial glia-like cells persist in the adult rat brain. *Brain Res* 1258:43–52.

Gussoni E, Soneoka Y, Strickland CD, Buzney EA, Khan MK, Flint AF, Kunkel LM, Mulligan RC (1999). Dystrophin expression in the mdx mouse restored by stem cell transplantation. *Nature* 401:390–394.

Hack MA, Sugimori M, Lundberg C, Nakafuku M, Gotz M (2004). Regionalization and fate specification in neurospheres: The role of Olig2 and Pax6. *Mol Cell Neurosci* 25:664–678.

Hall PE, Lathia JD, Miller NG, Caldwell MA, ffrench-Constant C (2006). Integrins are markers of human neural stem cells. *Stem Cells* 24:2078–2084.

Harris MA, Yang H, Low BE, Mukherje J, Guha A, Bronson RT, Shultz LD, Israel MA, Yun K (2008). Cancer stem cells are enriched in the side population cells in a mouse model of glioma. *Cancer Res* 68:10051–10059.

Hart IK, Richardson WD, Heldin CH, Westermark B, Raff MC (1989). PDGF receptors on cells of the oligoden-drocyte-type-2 astrocyte (O-2A) cell lineage. *Development* 105:595–603.

Hartfuss E, Galli R, Heins N, Gotz M (2001). Characterization of CNS precursor subtypes and radial glia. *Dev Biol* 229:15–30.

Hartfuss E, Forster E, Bock HH, Hack MA, Leprince P, Luque JM, Herz J, Frotscher M, Gotz M (2003). Reelin signaling directly affects radial glia morphology and biochemical maturation. *Development* 130:4597–4609.

Heins N, Malatesta P, Cecconi F, Nakafuku M, Tucker KL, Hack MA, Chapouton P, Barde YA, Gotz M (2002). Glial cells generate neurons: The role of the transcription factor Pax6. *Nat Neurosci* 5:308–315.

Hermann A, Gastl R, Liebau S, Popa MO, Fiedler J, Boehm BO, Maisel M, Lerche H, Schwarz J, Brenner R, Storch A (2004). Efficient generation of neural stem cell–like cells from adult human bone marrow stromal cells. *J Cell Sci* 117:4411–4422.

Hermann A, Maisel M, Liebau S, Gerlach M, Kleger A, Schwarz J, Kim KS, Antoniadis G, Lerche H, Storch A (2006). Mesodermal cell types induce neurogenesis from adult human hippocampal progenitor cells. *J Neurochem* 98:629–640.

Herrera J, Yang H, Zhang SC, Proschel C, Tresco P, Duncan ID, Luskin M, Mayer-Proschel M (2001). Embryonic-derived glial-restricted precursor cells (GRP cells) can differentiate into astrocytes and oligodendrocytes in vivo. *Exp Neurol* 171:11–21.

Hirschmann-Jax C, Foster AE, Wulf GG, Nuchtern JG, Jax TW, Gobel U, Goodell MA, Brenner MK (2004). A distinct "side population" of cells with high drug efflux capacity in human tumor cells. *Proc Natl Acad Sci USA* 101:14228–14233.

Ho MM, Ng AV, Lam S, Hung JY (2007). Side population in human lung cancer cell lines and tumors is enriched with stem-like cancer cells. *Cancer Res* 67:4827–4833.

Hochedlinger K, Yamada Y, Beard C, Jaenisch R (2005). Ectopic expression of Oct-4 blocks progenitor-cell differentiation and causes dysplasia in epithelial tissues. *Cell* 121:465–477.

Hodge RD, Kowalczyk TD, Wolf SA, Encinas JM, Rippey C, Enikolopov G, Kempermann G, Hevner RF (2008). Intermediate progenitors in adult hippocampal neurogenesis: Tbr2 expression and coordinate regulation of neuronal output. *J Neurosci* 28:3707–3717.

Holland EC, Celestino J, Dai C, Schaefer L, Sawaya RE, Fuller GN (2000). Combined activation of Ras and Akt in neural progenitors induces glioblastoma formation in mice. *Nat Genet* 25:55–57.

Horner PJ, Power AE, Kempermann G, Kuhn HG, Palmer TD, Winkler J, Thal LJ, Gage FH (2000). Proliferation and differentiation of progenitor cells throughout the intact adult rat spinal cord. *J Neurosci* 20:2218–2228.

Horner PJ, Thallmair M, Gage FH (2002). Defining the NG2-expressing cell of the adult CNS. *J Neurocytol* 31:469–480.

Huttner WB, Kosodo Y (2005). Symmetric versus asymmetric cell division during neurogenesis in the developing vertebrate central nervous system. *Curr Opin Cell Biol* 17:648–657.

Ihrie RA, Alvarez-Buylla A (2008). Cells in the astroglial lineage are neural stem cells. *Cell Tissue Res* 331:179–191.

Immaneni A, Lawinger P, Zhao Z, Lu W, Rastelli L, Morris JH, Majumder S (2000). REST-VP16 activates multiple neuronal differentiation genes in human NT2 cells. *Nucleic Acids Res* 28:3403–3410.

Imura T, Nakano I, Kornblum HI, Sofroniew MV (2006). Phenotypic and functional heterogeneity of GFAP-expressing cells in vitro: Differential expression of LeX/CD15 by GFAP-expressing multipotent neural stem cells and non-neurogenic astrocytes. *Glia* 53:277–293.

Inta D, Alfonso J, von Engelhardt J, Kreuzberg MM, Meyer AH, van Hooft JA, Monyer H (2008). Neurogenesis and widespread forebrain migration of distinct GABAergic neurons from the postnatal subventricular zone. *Proc Natl Acad Sci USA* 105:20994–20999.

Itsykson P, Ilouz N, Turetsky T, Goldstein RS, Pera MF, Fishbein I, Segal M, Reubinoff BE (2005). Derivation of neural precursors from human embryonic stem cells in the presence of noggin. *Mol Cell Neurosci* 30:24–36.

Jaenisch R, Young R (2008). Stem cells, the molecular circuitry of pluripotency and nuclear reprogramming. *Cell* 132:567–582.

Jaiswal S, Traver D, Miyamoto T, Akashi K, Lagasse E, Weissman IL (2003). Expression of BCR/ABL and BCL-2 in myeloid progenitors leads to myeloid leukemias. *Proc Natl Acad Sci USA* 100:10002–10007.

Jensen JB, Parmar M (2006). Strengths and limitations of the neurosphere culture system. *Mol Neurobiol* 34: 153–161.

Jessberger S, Romer B, Babu H, Kempermann G (2005). Seizures induce proliferation and dispersion of doublecortin-positive hippocampal progenitor cells. *Exp Neurol* 196:342–351.

Jessberger S, Clemenson GD, Jr., Gage FH (2007). Spontaneous fusion and nonclonal growth of adult neural stem cells. *Stem Cells* 25:871–874.

Jessberger S, Toni N, Clemenson Jr GD, Ray J, Gage FH (2008). Directed differentiation of hippocampal stem/progenitor cells in the adult brain. *Nat Neurosci* 11:888–893.

Jiang Y, Vaessen B, Lenvik T, Blackstad M, Reyes M, Verfaillie CM (2002a) Multipotent progenitor cells can be isolated from postnatal murine bone marrow, muscle, and brain. *Exp Hematol* 30:896–904.

Jiang Y, Jahagirdar BN, Reinhardt RL, Schwartz RE, Keene CD, Ortiz-Gonzalez XR, Reyes M, Lenvik T, Lund T, Blackstad M, Du J, Aldrich S, Lisberg A, Low WC, Largaespada DA, Verfaillie CM (2002b) Pluripotency of mesenchymal stem cells derived from adult marrow. *Nature* 418:41–49.

Johansson CB, Momma S, Clarke DL, Risling M, Lendahl U, Frisen J (1999). Identification of a neural stem cell in the adult mammalian central nervous system. *Cell* 96:25–34.

Joseph NM, Morrison SJ (2005). Toward an understanding of the physiological function of mammalian stem cells. *Developmental Cell* 9:173–183.

Kagalwala MN, Singh SK, Majumder S (2009). Stemness is only a state of the cell. Cold Spring Harb Symp Quant Biol 73:227–34.

Kaplan MS (1983). Proliferation of subependymal cells in the adult primate CNS: Differential uptake of DNA-labelled precursors. *J Hirnforsch* 24:23–33.

Karki S, Pruszak J, Isacson O, Sonntag KC (2006). ES cell–derived neuroepithelial cell cultures. *J Vis Exp* (1) :118.

Kashima T, Tiu SN, Merrill JE, Vinters HV, Dawson G, Campagnoni AT (1993). Expression of oligodendrocyte-associated genes in cell lines derived from human gliomas and neuroblastomas. *Cancer Res* 53:170–175.

Kawasaki H, Mizuseki K, Nishikawa S, Kaneko S, Kuwana Y, Nakanishi S, Nishikawa SI, Sasai Y (2000). Induction of midbrain dopaminergic neurons from ES cells by stromal cell–derived inducing activity. *Neuron* 28:31–40.

Keirstead HS, Levine JM, Blakemore WF (1998). Response of the oligodendrocyte progenitor cell population (defined by NG2 labelling) to demyelination of the adult spinal cord. *Glia* 22:161–170.

Keirstead HS, Nistor G, Bernal G, Totoiu M, Cloutier F, Sharp K, Steward O (2005). Human embryonic stem cell-derived oligodendrocyte progenitor cell transplants remyelinate and restore locomotion after spinal cord injury. *J Neurosci* 25:4694–4705.

Kilpatrick TJ, Bartlett PF (1993). Cloning and growth of multipotential neural precursors: Requirements for proliferation and differentiation. *Neuron* 10:255–265.

Kilpatrick TJ, Bartlett PF (1995). Cloned multipotential precursors from the mouse cerebrum require FGF-2, whereas glial restricted precursors are stimulated with either FGF-2 or EGF. *J Neurosci* 15:3653–3661.

Kim JB, Zaehres H, Wu G, Gentile L, Ko K, Sebastiano V, Arauzo-Bravo MJ, Ruau D, Han DW, Zenke M, Scholer HR (2008). Pluripotent stem cells induced from adult neural stem cells by reprogramming with two factors. *Nature* 454:646–650.

Kim JB, Sebastiano V, Wu G, Arauzo-Bravo MJ, Sasse P, Gentile L, Ko K, Ruau D, Ehrich M, van den Boom D, Meyer J, Hubner K, Bernemann C, Ortmeier C, Zenke M, Fleischmann BK, Zaehres H, Scholer HR (2009). Oct4-induced pluripotency in adult neural stem cells. *Cell* 136:411–419.

Kim M, Morshead CM (2003). Distinct populations of forebrain neural stem and progenitor cells can be isolated using side-population analysis. *J Neurosci* 23:10703–10709.

Koch P, Opitz T, Steinbeck JA, Ladewig J, Brustle O (2009). A rosette-type, self-renewing human ES cell-derived neural stem cell with potential for in vitro instruction and synaptic integration. *Proc Natl Acad Sci USA* 106:3225–30.

Kondo T, Raff M (2000). Oligodendrocyte precursor cells reprogrammed to become multipotential CNS stem cells. *Science* 289:1754–1757.

Kosodo Y, Roper K, Haubensak W, Marzesco AM, Corbeil D, Huttner WB (2004). Asymmetric distribution of the apical plasma membrane during neurogenic divisions of mammalian neuroepithelial cells. *Embo J* 23: 2314–2324.

Kovacsovics-Bankowski M, Streeter PR, Mauch KA, Frey MR, Raber A, van't Hof W, Deans R, Maziarz RT (2009). Clinical scale expanded adult pluripotent stem cells prevent graft-versus-host disease. *Cell Immunol* 255:55–60.

Koyama N, Okubo Y, Nakao K, Bessho K (2009). Evaluation of pluripotency in human dental pulp cells. *J Oral Maxillofac Surg* 67:501–506.

Kriegstein A, Alvarez-Buylla A (2009). The glial nature of embryonic and adult neural stem cells. *Annu Rev Neurosci* 32:149–184.

Kriegstein AR, Gotz M (2003). Radial glia diversity: A matter of cell fate. *Glia* 43:37–43.

Kronenberg G, Wang LP, Geraerts M, Babu H, Synowitz M, Vicens P, Lutsch G, Glass R, Yamaguchi M, Baekelandt V, Debyser Z, Kettenmann H, Kempermann G (2007). Local origin and activity-dependent generation of nestin-expressing protoplasmic astrocytes in CA1. *Brain Structure & Function* 212:19–35.

Krum JM, Rosenstein JM (1999). Transient coexpression of nestin, GFAP, and vascular endothelial growth factor in mature reactive astroglia following neural grafting or brain wounds. *Exp Neurol* 160:348–360.

Kubiak JZ, Tarkowski AK (1985). Electrofusion of mouse blastomeres. *Exp Cell Res* 157:561–566.

Kucia MJ, Wysoczynski M, Wu W, Zuba-Surma EK, Ratajczak J, Ratajczak MZ (2008). Evidence that very small embryonic-like stem cells are mobilized into peripheral blood. *Stem Cells* 26:2083–2092.

Kuwabara T, Hsieh J, Nakashima K, Taira K, Gage FH (2004). A small modulatory dsRNA specifies the fate of adult neural stem cells. *Cell* 116:779–793.

Labrakakis C, Patt S, Hartmann J, Kettenmann H (1998). Functional GABA(A) receptors on human glioma cells. *Eur J Neurosci* 10:231–238.

Lagasse E, Connors H, Al-Dhalimy M, Reitsma M, Dohse M, Osborne L, Wang X, Finegold M, Weissman IL, Grompe M (2000). Purified hematopoietic stem cells can differentiate into hepatocytes in vivo. *Nat Med* 6:1229–1234.

Langa F, Kress C, Colucci-Guyon E, Khun H, Vandormael-Pournin S, Huerre M, Babinet C (2000). Teratocarcinomas induced by embryonic stem (ES) cells lacking vimentin: An approach to study the role of vimentin in tumorigenesis. *J Cell Sci* 113 Pt 19:3463–3472.

Larjavaara S, Mantyla R, Salminen T, Haapasalo H, Raitanen J, Jaaskelainen J, Auvinen A (2007). Incidence of gliomas by anatomic location. *Neuro-oncology* 9:319–325.

Laywell ED, Rakic P, Kukekov VG, Holland EC, Steindler DA (2000). Identification of a multipotent astrocytic stem cell in the immature and adult mouse brain. *Proc Natl Acad Sci USA* 97:13883–13888.

Lengner CJ, Welstead GG, Jaenisch R (2008). The pluripotency regulator Oct4: A role in somatic stem cells? *Cell Cycle* 7:725–728.

Levine JM, Reynolds R, Fawcett JW (2001). The oligodendrocyte precursor cell in health and disease. *Trends Neurosci* 24:39–47.

Levy YS, Merims D, Panet H, Barhum Y, Melamed E, Offen D (2003). Induction of neuron-specific enolase promoter and neuronal markers in differentiated mouse bone marrow stromal cells. *J Mol Neurosci* 21:121–132.

Li JY, Christophersen NS, Hall V, Soulet D, Brundin P (2008). Critical issues of clinical human embryonic stem cell therapy for brain repair. *Trends Neurosci* 31:146–153.

Li XJ, Du ZW, Zarnowska ED, Pankratz M, Hansen LO, Pearce RA, Zhang SC (2005). Specification of motoneurons from human embryonic stem cells. *Nat Biotechnol* 23:215–221.

Ligon KL, Kesari S, Kitada M, Sun T, Arnett HA, Alberta JA, Anderson DJ, Stiles CD, Rowitch DH (2006). Development of NG2 neural progenitor cells requires Olig gene function. *Proc Natl Acad Sci USA* 103:7853–7858.

Lin SC, Bergles DE (2002). Physiological characteristics of NG2-expressing glial cells. *J Neurocytol* 31:537–549.

Liu A, Han YR, Li J, Sun D, Ouyang M, Plummer MR, Casaccia-Bonnefil P (2007). The glial or neuronal fate choice of oligodendrocyte progenitors is modulated by their ability to acquire an epigenetic memory. *J Neurosci* 27:7339–7343.

Liu Y, Wu Y, Lee JC, Xue H, Pevny LH, Kaprielian Z, Rao MS (2002). Oligodendrocyte and astrocyte development in rodents: An in situ and immunohistological analysis during embryonic development. *Glia* 40:25–43.

Liu Y, Rao MS (2004). Glial progenitors in the CNS and possible lineage relationships among them. *Biol Cell* 96:279–290.

Liu YP, Lang BT, Baskaya MK, Dempsey RJ, Vemuganti R (2009). The potential of neural stem cells to repair stroke-induced brain damage. *Acta Neuropathol* 117:469–80.

Liu Z, Martin LJ (2003). Olfactory bulb core is a rich source of neural progenitor and stem cells in adult rodent and human. *J Comp Neurol* 459:368–391.

Lledo PM, Alonso M, Grubb MS (2006). Adult neurogenesis and functional plasticity in neuronal circuits. *Nat Rev Neurosci* 7:179–193.

Lledo PM, Merkle FT, Alvarez-Buylla A (2008). Origin and function of olfactory bulb interneuron diversity. *Trends Neurosci* 31:392–400.

Lowry NA, Temple S (2009). Identifying the perpetrator in medulloblastoma: Dorian Gray versus Benjamin Button. *Cancer Cell* 15:83–85.

Luo J, Daniels SB, Lennington JB, Notti RQ, Conover JC (2006). The aging neurogenic subventricular zone. *Aging Cell* 5:139–152.

Magavi S, Leavitt B, Macklis J (2000). Induction of neurogenesis in the neocortex of adult mice. *Nature* 405: 951–955.

Maherali N, Sridharan R, Xie W, Utikal J, Eminli S, Arnold K, Stadtfeld M, Yachechko R, Tchieu J, Jaenisch R, Plath K, Hochedlinger K (2007). Directly reprogrammed fibroblasts show global epigenetic remodeling and widespread tissue contribution. *Cell Stem Cell* 1:55–70.

Malatesta P, Hartfuss E, Gotz M (2000). Isolation of radial glial cells by fluorescent-activated cell sorting reveals a neuronal lineage. *Development* 127:5253–5263.

Malatesta P, Hack MA, Hartfuss E, Kettenmann H, Klinkert W, Kirchhoff F, Gotz M (2003). Neuronal or glial progeny: Regional differences in radial glia fate. *Neuron* 37:751–764.

Malatesta P, Appolloni I, Calzolari F (2008). Radial glia and neural stem cells. *Cell Tissue Res* 331:165–178.

Mallon BS, Shick HE, Kidd GJ, Macklin WB (2002). Proteolipid promoter activity distinguishes two populations of NG2–positive cells throughout neonatal cortical development. *J Neurosci* 22:876–885.

Mareschi K, Novara M, Rustichelli D, Ferrero I, Guido D, Carbone E, Medico E, Madon E, Vercelli A, Fagioli F (2006). Neural differentiation of human mesenchymal stem cells: Evidence for expression of neural markers and eag K+ channel types. *Exp Hematol* 34:1563–1572.

Marshall GP, 2nd, Reynolds BA, Laywell ED (2007). Using the neurosphere assay to quantify neural stem cells in vivo. *Current Pharmaceutical Biotechnology* 8:141–145.

Martens DJ, Seaberg RM, van der Kooy D (2002). In vivo infusions of exogenous growth factors into the fourth ventricle of the adult mouse brain increase the proliferation of neural progenitors around the fourth ventricle and the central canal of the spinal cord. *Eur J Neurosci* 16:1045–1057.

Martin GR (1981). Isolation of a pluripotent cell line from early mouse embryos cultured in medium conditioned by teratocarcinoma stem cells. *Proc Natl Acad Sci USA* 78:7634–7638.

Maslov AY, Barone TA, Plunkett RJ, Pruitt SC (2004). Neural stem cell detection, characterization, and age-related changes in the subventricular zone of mice. *J Neurosci* 24:1726–1733.

McCarthy M, Turnbull DH, Walsh CA, Fishell G (2001). Telencephalic neural progenitors appear to be restricted to regional and glial fates before the onset of neurogenesis. *J Neurosci* 21:6772–6781.

McKay R (1997). Stem cells in the central nervous system. *Science* 276:66–71.

Meletis K, Barnabe-Heider F, Carlen M, Evergren E, Tomilin N, Shupliakov O, Frisen J (2008). Spinal cord injury reveals multilineage differentiation of ependymal cells. *Public Library of Science Biology* 6:e182.

Merkle FT, Tramontin AD, Garcia-Verdugo JM, Alvarez-Buylla A (2004). Radial glia give rise to adult neural stem cells in the subventricular zone. *Proc Natl Acad Sci USA* 101:17528–17532.

Merkle FT, Mirzadeh Z, Alvarez-Buylla A (2007). Mosaic organization of neural stem cells in the adult brain. *Science* 317:381–384.

Mezey E, Chandross KJ (2000). Bone marrow: A possible alternative source of cells in the adult nervous system [In Process Citation]. *Eur J Pharmacol* 405:297–302.

Mignone JL, Kukekov V, Chiang AS, Steindler D, Enikolopov G (2004). Neural stem and progenitor cells in nestin-GFP transgenic mice. *J Comp Neurol* 469:311–324.

Misson JP, Austin CP, Takahashi T, Cepko CL, Caviness VS, Jr. (1991). The alignment of migrating neural cells in relation to the murine neopallial radial glial fiber system. *Cereb Cortex* 1:221–229.

Mizrak D, Brittan M, Alison MR (2008). CD133: Molecule of the moment. *J Pathol* 214:3–9.

Mizutani K, Yoon K, Dang L, Tokunaga A, Gaiano N (2007). Differential Notch signalling distinguishes neural stem cells from intermediate progenitors. *Nature* 449:351–355.

Moon JH, Yoon BS, Kim B, Park G, Jung HY, Maeng I, Jun EK, Yoo SJ, Kim A, Oh S, Whang KY, Kim H, Kim DW, Kim KD, You S (2008). Induction of neural stem cell–like cells (NSCLCs) from mouse astrocytes by Bmi1. *Biochem Biophys Res Commun* 371:267–272.

Morikawa S, Mabuchi Y, Niibe K, Suzuki S, Nagoshi N, Sunabori T, Shimmura S, Nagai Y, Nakagawa T, Okano H, Matsuzaki Y (2009). Development of mesenchymal stem cells partially originates from the neural crest. *Biochem Biophys Res Commun* 379:1114–1119.

Morizane A, Takahashi J, Shinoyama M, Ideguchi M, Takagi Y, Fukuda H, Koyanagi M, Sasai Y, Hashimoto N (2006). Generation of graftable dopaminergic neuron progenitors from mouse ES cells by a combination of coculture and neurosphere methods. *J Neurosci Res* 83:1015–1027.

Morrison SJ, Shah NM, Anderson DJ (1997). Regulatory mechanisms in stem cell biology. *Cell* 88:287–298.

Morshead CM, Reynolds BA, Craig CG, McBurney MW, Staines WA, Morassutti D, Weiss S, van der Kooy D (1994). Neural stem cells in the adult mammalian forebrain: A relatively quiescent subpopulation of subependymal cells. *Neuron* 13:1071–1082.

Morshead CM, Craig CG, van der Kooy D (1998). In vivo clonal analyses reveal the properties of endogenous neural stem cell proliferation in the adult mammalian forebrain. *Development* 125:2251–2261.

Morshead CM, Garcia AD, Sofroniew MV, van Der Kooy D (2003). The ablation of glial fibrillary acidic protein-positive cells from the adult central nervous system results in the loss of forebrain neural stem cells but not retinal stem cells. *Eur J Neurosci* 18:76–84.

Morshead CM, van der Kooy D (2004). Disguising adult neural stem cells. *Curr Opin Neurobiol* 14:125–131.

Mouthon MA, Fouchet P, Mathieu C, Sii-Felice K, Etienne O, Lages CS, Boussin FD (2006). Neural stem cells from mouse forebrain are contained in a population distinct from the "side population." *J Neurochem* 99:807–817.

Murayama A, Matsuzaki Y, Kawaguchi A, Shimazaki T, Okano H (2002). Flow cytometric analysis of neural stem cells in the developing and adult mouse brain. *J Neurosci Res* 69:837–847.

Nagato M, Heike T, Kato T, Yamanaka Y, Yoshimoto M, Shimazaki T, Okano H, Nakahata T (2005). Prospective characterization of neural stem cells by flow cytometry analysis using a combination of surface markers. *J Neurosci Res* 80:456–466.

Nagoshi N, Shibata S, Kubota Y, Nakamura M, Nagai Y, Satoh E, Morikawa S, Okada Y, Mabuchi Y, Katoh H, Okada S, Fukuda K, Suda T, Matsuzaki Y, Toyama Y, Okano H (2008). Ontogeny and multipotency of neural crest-derived stem cells in mouse bone marrow, dorsal root ganglia, and whisker pad. *Cell Stem Cell* 2:392–403.

Nakafuku M, Nagao M, Grande A, Cancelliere A (2008). Revisiting neural stem cell identity. *Proc Natl Acad Sci USA* 105:829–830.

Nakagawa M, Koyanagi M, Tanabe K, Takahashi K, Ichisaka T, Aoi T, Okita K, Mochiduki Y, Takizawa N, Yamanaka S (2008). Generation of induced pluripotent stem cells without Myc from mouse and human fibroblasts. *Nat Biotechnol* 26:101–106.

Neuhuber B, Gallo G, Howard L, Kostura L, Mackay A, Fischer I (2004). Reevaluation of in vitro differentiation protocols for bone marrow stromal cells: Disruption of actin cytoskeleton induces rapid morphological changes and mimics neuronal phenotype. *J Neurosci Res* 77:192–204.

Niehaus A, Stegmuller J, Diers-Fenger M, Trotter J (1999). Cell-surface glycoprotein of oligodendrocyte progenitors involved in migration. *J Neurosci* 19:4948–4961.

Ninkovic J, Mori T, Gotz M (2007). Distinct modes of neuron addition in adult mouse neurogenesis. *J Neurosci* 27:10906–10911.

Nishiyama A (2007). Polydendrocytes: NG2 cells with many roles in development and repair of the CNS. *Neuroscientist* 13:62–76.

Nishiyama A, Lin XH, Giese N, Heldin CH, Stallcup WB (1996). Co-localization of NG2 proteoglycan and PDGF alpha-receptor on O2A progenitor cells in the developing rat brain. *J Neurosci Res* 43:299–314.

Nishiyama A, Chang A, Trapp BD (1999). NG2+ glial cells: A novel glial cell population in the adult brain. *J Neuropathol Exp Neurol* 58:1113–1124.

Nishiyama A, Watanabe M, Yang Z, Bu J (2002). Identity, distribution, and development of polydendrocytes: NG2-expressing glial cells. *J Neurocytol* 31:437–455.

Nishiyama A, Komitova M, Suzuki R, Zhu X (2009). Polydendrocytes (NG2 cells): multifunctional cells with lineage plasticity. *Nat Rev Neurosci* 10:9–22.

Noble M (1997). The oligodendrocyte-type-2 astrocyte lineage: In vitro and in vivo studies on development, tissue repair and neoplasia. In: *Isolation, Characterization and Utilization of CNS Stem Cells* (Gage FH, Christen Y, eds.), pp. 101–128. Berlin, Heidelberg: Springer.

Noble M, Dietrich J (2004). The complex identity of brain tumors: Emerging concerns regarding origin, diversity and plasticity. *Trends Neurosci* 27:148–154.

Noble M, Proschel C, Mayer-Proschel M (2004). Getting a GR(i)P on oligodendrocyte development. *Dev Biol* 265:33–52.

Noctor SC, Flint AC, Weissman TA, Dammerman RS, Kriegstein AR (2001). Neurons derived from radial glial cells establish radial units in neocortex. *Nature* 409:714–720.

Noctor SC, Flint AC, Weissman TA, Wong WS, Clinton BK, Kriegstein AR (2002). Dividing precursor cells of the embryonic cortical ventricular zone have morphological and molecular characteristics of radial glia. *J Neurosci* 22:3161–3173.

Obrietan K, Gao XB, Van Den Pol AN (2002). Excitatory actions of GABA increase BDNF expression via a MAPK-CREB-dependent mechanism—a positive feedback circuit in developing neurons. *J Neurophysiol* 88:1005–1015.

Ogden AT, Waziri AE, Lochhead RA, Fusco D, Lopez K, Ellis JA, Kang J, Assanah M, McKhann GM, Sisti MB, McCormick PC, Canoll P, Bruce JN (2008). Identification of A2B5+CD133– tumor-initiating cells in adult human gliomas. *Neurosurgery* 62:505–514.

Okita K, Ichisaka T, Yamanaka S (2007). Generation of germline-competent induced pluripotent stem cells. *Nature* 448:313–317.

Ong WY, Levine JM (1999). A light and electron microscopic study of NG2 chondroitin sulfate proteoglycan-positive oligodendrocyte precursor cells in the normal and kainate-lesioned rat hippocampus. *Neuroscience* 92:83–95.

Paczkowska E, Kucia M, Koziarska D, Halasa M, Safranow K, Masiuk M, Karbicka A, Nowik M, Nowacki P, Ratajczak MZ, Machalinski B (2009). Clinical evidence that very small embryonic-like stem cells are mobilized into peripheral blood in patients after stroke. *Stroke* 40:1237–44.

Padovan CS, Jahn K, Birnbaum T, Reich P, Sostak P, Strupp M, Straube A (2003). Expression of neuronal markers in differentiated marrow stromal cells and CD133+ stem-like cells. *Cell Transplant* 12:839–848.

Palmer TD, Ray J, Gage FH (1995). FGF-2-responsive neuronal progenitors reside in proliferative and quiescent regions of the adult rodent brain. *Mol Cell Neurosci* 6:474–486.

Palmer TD, Takahashi J, Gage FH (1997). The adult rat hippocampus contains premordial neural stem cells. *Mol Cell Neurosci* 8:389–404.

Palmer TD, Markakis EA, Willhoite AR, Safar F, Gage FH (1999). Fibroblast growth factor-2 activates a latent neurogenic program in neural stem cells from diverse regions of the adult CNS. *J Neurosci* 19:8487–8497.

Palmer TD, Schwartz PH, Taupin P, Kaspar B, Stein SA, Gage FH (2001). Cell culture. Progenitor cells from human brain after death. *Nature* 411:42–43.

Panchision DM, Chen HL, Pistollato F, Papini D, Ni HT, Hawley TS (2007). Optimized flow cytometric analysis of central nervous system tissue reveals novel functional relationships among cells expressing CD133, CD15, and CD24. *Stem Cells* 25:1560–1570.

Parent JM, Vexler ZS, Gong C, Derugin N, Ferriero DM (2002). Rat forebrain neurogenesis and striatal neuron replacement after focal stroke. *Ann Neurol* 52:802–813.

Parker MA, Anderson JK, Corliss DA, Abraria VE, Sidman RL, Park KI, Teng YD, Cotanche DA, Snyder EY (2005). Expression profile of an operationally defined neural stem cell clone. *Exp Neurol* 194:320–332.

Patrawala L, Calhoun T, Schneider-Broussard R, Zhou J, Claypool K, Tang DG (2005). Side population is enriched in tumorigenic, stem-like cancer cells, whereas ABCG2+ and ABCG2– cancer cells are similarly tumorigenic. *Cancer Res* 65:6207–6219.

Perrier AL, Tabar V, Barberi T, Rubio ME, Bruses J, Topf N, Harrison NL, Studer L (2004). Derivation of midbrain dopamine neurons from human embryonic stem cells. *Proc Natl Acad Sci USA* 101:12543–12548.

Persson AI, Weiss WA (2009). The side story of stem-like glioma cells. *Cell Stem Cell* 4:191–192.

Pfenninger CV, Roschupkina T, Hertwig F, Kottwitz D, Englund E, Bengzon J, Jacobsen SE, Nuber UA (2007). CD133 is not present on neurogenic astrocytes in the adult subventricular zone, but on embryonic neural stem cells, ependymal cells, and glioblastoma cells. *Cancer Res* 67:5727–5736.

Phi JH, Park SH, Paek SH, Kim SK, Lee YJ, Park CK, Cho BK, Lee DH, Wang KC (2007). Expression of Sox2 in mature and immature teratomas of central nervous system. *Mod Pathol* 20:742–748.

Phi JH, Park SH, Kim SK, Paek SH, Kim JH, Lee YJ, Cho BK, Park CK, Lee DH, Wang KC (2008). Sox2 expression in brain tumors: A reflection of the neuroglial differentiation pathway. *Am J Surg Pathol* 32:103–112.

Pincus DW, Keyoung HM, Harrison-Restelli C, Goodman RR, Fraser RA, Edgar M, Sakakibara S, Okano H, Nedergaard M, Goldman SA (1998). Fibroblast growth factor-2/brain-derived neurotrophic factor-associated maturation of new neurons generated from adult human subependymal cells. *Ann Neurol* 43:576–585.

Pinto L, Gotz M (2007). Radial glial cell heterogeneity—the source of diverse progeny in the CNS. *Prog Neurobiol* 83:2–23.

Pinto L, Mader MT, Irmler M, Gentilini M, Santoni F, Drechsel D, Blum R, Stahl R, Bulfone A, Malatesta P, Beckers J, Gotz M (2008). Prospective isolation of functionally distinct radial glial subtypes—lineage and transcriptome analysis. *Mol Cell Neurosci* 38:15–42.

Pittenger MF, Mackay AM, Beck SC, Jaiswal RK, Douglas R, Mosca JD, Moorman MA, Simonetti DW, Craig S, Marshak DR (1999). Multilineage potential of adult human mesenchymal stem cells. *Science* 284:143–147.

Porada CD, Zanjani ED, Almeida-Porad G (2006). Adult mesenchymal stem cells: A pluripotent population with multiple applications. *Current Stem Cell Research & Therapy* 1:365–369.

Priller J, Persons DA, Klett FF, Kempermann G, Kreutzberg GW, Dirnagl U (2001). Neogenesis of cerebellar Purkinje neurons from gene-marked bone marrow cells in vivo. *J Cell Biol* 155:733–738.

Quinones-Hinojosa A, Sanai N, Soriano-Navarro M, Gonzalez-Perez O, Mirzadeh Z, Gil-Perotin S, Romero-Rodriguez R, Berger MS, Garcia-Verdugo JM, Alvarez-Buylla A (2006). Cellular composition and cytoarchitecture of the adult human subventricular zone: A niche of neural stem cells. *J Comp Neurol* 494: 415–434.

Raff MC, Miller RH, Noble M (1983). A glial progenitor cell that develops in vitro into an astrocyte or an oligodendrocyte depending on culture medium. *Nature* 303:390–396.

Rakic P (1971). Neuron–glia relationship during granule cell migration in developing cerebellar cortex. A Golgi and electronmicroscopic study in *Macacus rhesus*. *J Comp Neurol* 141:283–312.

Rao M (2004). Stem and precursor cells in the nervous system. *J Neurotrauma* 21:415–427.

Ratajczak MZ, Zuba-Surma EK, Machalinski B, Ratajczak J, Kucia M (2008). Very small embryonic-like (VSEL) stem cells: Purification from adult organs, characterization, and biological significance. *Stem Cell Reviews* 4:89–99.

Ravin R, Hoeppner DJ, Munno DM, Carmel L, Sullivan J, Levitt DL, Miller JL, Athaide C, Panchision DM, McKay RD (2008). Potency and fate specification in CNS stem cell populations in vitro. *Cell Stem Cell* 3: 670–680.

Ray J, Raymon HK, Gage FH (1995). Generation and culturing of precursor cells and neuroblasts from embryonic and adult central nervous system. *Methods Enzymol* 254:20–37.

Read TA, Fogarty MP, Markant SL, McLendon RE, Wei Z, Ellison DW, Febbo PG, Wechsler-Reya RJ (2009). Identification of CD15 as a marker for tumor-propagating cells in a mouse model of medulloblastoma. *Cancer Cell* 15:135–147.

Rebetz J, Tian D, Persson A, Widegren B, Salford LG, Englund E, Gisselsson D, Fan X (2008). Glial progenitor-like phenotype in low-grade glioma and enhanced CD133-expression and neuronal lineage differentiation potential in high-grade glioma. *Public Library of Science ONE* 3:e1936.

Redwine JM, Armstrong RC (1998). In vivo proliferation of oligodendrocyte progenitors expressing PDGFalphaR during early remyelination. *J Neurobiol* 37:413–428.

Reubinoff BE, Itsykson P, Turetsky T, Pera MF, Reinhartz E, Itzik A, Ben-Hur T (2001). Neural progenitors from human embryonic stem cells. *Nat Biotechnol* 19:1134–1140.

Reya T, Morrison SJ, Clarke MF, Weissman IL (2001). Stem cells, cancer, and cancer stem cells. *Nature* 414: 105–111.

Reynolds BA, Weiss S (1992). Generation of neurons and astrocytes from isolated cells of the adult mammalian central nervous system. *Science* 255:1707–1710.

Reynolds BA, Tetzlaff W, Weiss S (1992). A multipotent EGF-responsive striatal embryonic progenitor cell produces neurons and astrocytes. *J Neurosci* 12:4565–4574.

Reynolds BA, Weiss S (1996). Clonal and population analyses demonstrate that an EGF-responsive mammalian embryonic CNS precursor is a stem cell. *Dev Biol* 175:1–13.

Reynolds BA, Rietze RL (2005). Neural stem cells and neurospheres—re-evaluating the relationship. *Nat Methods* 2:333–336.

Reynolds R, Dawson M, Papadopoulos D, Polito A, Di Bello IC, Pham-Dinh D, Levine J (2002). The response of NG2-expressing oligodendrocyte progenitors to demyelination in MOG-EAE and MS. *J Neurocytol* 31:523–536.

Richards LJ, Kilpatrick TJ, Bartlett PF (1992). De novo generation of neuronal cells from the adult mouse brain. *Proc Natl Acad Sci USA* 89:8591–8595.

Rickmann M, Amaral DG, Cowan WM (1987). Organization of radial glial cells during the development of the rat dentate gyrus. *J Comp Neurol* 264:449–479.

Rietze RL, Valcanis H, Brooker GF, Thomas T, Voss AK, Bartlett PF (2001). Purification of a pluripotent neural stem cell from the adult mouse brain. *Nature* 412:736–739.

Rietze RL, Reynolds BA (2006). Neural stem cell isolation and characterization. *Methods Enzymol* 419:3–23.

Robey RW, To KK, Polgar O, Dohse M, Fetsch P, Dean M, Bates SE (2009). ABCG2: A perspective. *Adv Drug Deliv Rev* 61:3–13.

Ross RA, Spengler BA (2007). Human neuroblastoma stem cells. *Semin Cancer Biol* 17:241–247.

Rousselot P, Heintz N, Nottebohm F (1997). Expression of brain lipid binding protein in the brain of the adult canary and its implications for adult neurogenesis. *J Comp Neurol* 385:415–426.

Sahin Kaya S, Mahmood A, Li Y, Yavuz E, Chopp M (1999). Expression of nestin after traumatic brain injury in rat brain. *Brain Res* 840:153–157.

Sakurada K, Saino M, Mouri W, Sato A, Kitanaka C, Kayama T (2008). Nestin expression in central nervous system germ cell tumors. *Neurosurg Rev* 31:173–176; discussion 176–177.

Sanai N, Tramontin AD, Quinones-Hinojosa A, Barbaro NM, Gupta N, Kunwar S, Lawton MT, McDermott MW, Parsa AT, Manuel-Garcia Verdugo J, Berger MS, Alvarez-Buylla A (2004). Unique astrocyte ribbon in adult human brain contains neural stem cells but lacks chain migration. *Nature* 427:740–744.

Sano K (1995). So-called intracranial germ cell tumours: Are they really of germ cell origin? *Br J Neurosurg* 9:391–401.

Sauerzweig S, Munsch T, Lessmann V, Reymann KG, Braun H (2009). A population of serum deprivation–induced bone marrow stem cells (SD-BMSC) expresses marker typical for embryonic and neural stem cells. *Exp Cell Res* 315:50–66.

Sawamoto K, Yamamoto A, Kawaguchi A, Yamaguchi M, Mori K, Goldman SA, Okano H (2001). Direct isolation of committed neuronal progenitor cells from transgenic mice coexpressing spectrally distinct fluorescent proteins regulated by stage-specific neural promoters. *J Neurosci Res* 65:220–227.

Schneider S, Bosse F, D'Urso D, Muller H, Sereda MW, Nave K, Niehaus A, Kempf T, Schnolzer M, Trotter J (2001). The AN2 protein is a novel marker for the Schwann cell lineage expressed by immature and nonmyelinating Schwann cells. *J Neurosci* 21:920–933.

Schwartz PH, Bryant PJ, Fuja TJ, Su H, O'Dowd DK, Klassen H (2003). Isolation and characterization of neural progenitor cells from post-mortem human cortex. *J Neurosci Res* 74:838–851.

Seaberg RM, van der Kooy D (2002). Adult rodent neurogenic regions: The ventricular subependyma contains neural stem cells, but the dentate gyrus contains restricted progenitors. *J Neurosci* 22:1784–1793.

Seaberg RM, van der Kooy D (2003). Stem and progenitor cells: The premature desertion of rigorous definitions. *Trends Neurosci* 26:125–131.

Seoane J, Le HV, Shen L, Anderson SA, Massague J (2004). Integration of Smad and forkhead pathways in the control of neuroepithelial and glioblastoma cell proliferation. *Cell* 117:211–223.

Seri B, Garcia-Verdugo JM, McEwen BS, Alvarez-Buylla A (2001). Astrocytes give rise to new neurons in the adult mammalian hippocampus. *J Neurosci* 21:7153–7160.

Shih AH, Holland EC (2006). Platelet-derived growth factor (PDGF) and glial tumorigenesis. *Cancer Lett* 232:139–147.

Shihabuddin LS, Ray J, Gage FH (1997). FGF-2 is sufficient to isolate progenitors found in the adult mammalian spinal cord. *Exp Neurol* 148:577–586.

Shihabuddin LS, Horner PJ, Ray J, Gage FH (2000). Adult spinal cord stem cells generate neurons after transplantation in the adult dentate gyrus. *J Neurosci* 20:8727–8735.

Singec I, Knoth R, Meyer RP, Maciaczyk J, Volk B, Nikkhah G, Frotscher M, Snyder EY (2006). Defining the actual sensitivity and specificity of the neurosphere assay in stem cell biology. *Nat Methods* 3:801–806.

Singh SK, Hawkins C, Clarke ID, Squire JA, Bayani J, Hide T, Henkelman RM, Cusimano MD, Dirks PB (2004). Identification of human brain tumour initiating cells. *Nature* 432:396–401.

Slack JM (2002). Conrad Hal Waddington: The last Renaissance biologist? *Nat Rev Genet* 3:889–895.

Soldner F, Hockemeyer D, Beard C, Gao Q, Bell GW, Cook EG, Hargus G, Blak A, Cooper O, Mitalipova M, Isacson O, Jaenisch R (2009). Parkinson's disease patient-derived induced pluripotent stem cells free of viral reprogramming factors. *Cell* 136:964–977.

Spassky N, Merkle FT, Flames N, Tramontin AD, Garcia-Verdugo JM, Alvarez-Buylla A (2005). Adult ependymal cells are postmitotic and are derived from radial glial cells during embryogenesis. *J Neurosci* 25:10–18.

Spella M, Britz O, Kotantaki P, Lygerou Z, Nishitani H, Ramsay RG, Flordellis C, Guillemot F, Mantamadiotis T, Taraviras S (2007). Licensing regulators Geminin and Cdt1 identify progenitor cells of the mouse CNS in a specific phase of the cell cycle. *Neuroscience* 147:373–387.

Srivastava AS, Malhotra R, Sharp J, Berggren T (2008). Potentials of ES cell therapy in neurodegenerative diseases. *Curr Pharm Des* 14:3873–3879.

Stallcup WB (2002). The NG2 proteoglycan: Past insights and future prospects. *J Neurocytol* 31:423–435.

Stallcup WB, Beasley L (1987). Bipotential glial precursor cells of the optic nerve express the NG2 proteoglycan. *J Neurosci* 7:2737–2744.

Steiner B, Kronenberg G, Jessberger S, Brandt MD, Reuter K, Kempermann G (2004). Differential regulation of gliogenesis in the context of adult hippocampal neurogenesis in mice. *Glia* 46:41–52.

Steiner B, Klempin F, Wang L, Kott M, Kettenmann H, Kempermann G (2006). Type-2 cells as link between glial and neuronal lineage in adult hippocampal neurogenesis. *Glia* 54:805–814.

Steinhauser C, Berger T, Frotscher M, Kettenmann H (1992). Heterogeneity in the membrane current pattern of identified glial cells in the hippocampal slice. *Eur J Neurosci* 4:472–484.

Steinhauser C, Kressin K, Kuprijanova E, Weber M, Seifert G (1994). Properties of voltage-activated Na$^+$ and K$^+$ currents in mouse hippocampal glial cells in situ and after acute isolation from tissue slices. *Pflugers Arch* 428:610–620.

Stenman J, Toresson H, Campbell K (2003). Identification of two distinct progenitor populations in the lateral ganglionic eminence: Implications for striatal and olfactory bulb neurogenesis. *J Neurosci* 23:167–174.

Stiles CD, Rowitch DH (2008). Glioma stem cells: A midterm exam. *Neuron* 58:832–846.

Stoykova A, Fritsch R, Walther C, Gruss P (1996). Forebrain patterning defects in Small eye mutant mice. *Development* 122:3453–3465.

Suh H, Consiglio A, Ray J, Sawai T, D'Amour KA, Gage FH (2007). In vivo fate analysis reveals the multipotent and self-renewal capacities of Sox2+ neural stem cells in the adult hippocampus. *Cell Stem Cell* 1:515–528.

Sundholm-Peters NL, Yang HK, Goings GE, Walker AS, Szele FG (2004). Radial glia-like cells at the base of the lateral ventricles in adult mice. *J Neurocytol* 33:153–164.

Sze SK, de Kleijn DP, Lai RC, Khia Way Tan E, Zhao H, Yeo KS, Low TY, Lian Q, Lee CN, Mitchell W, El Oakley RM, Lim SK (2007). Elucidating the secretion proteome of human embryonic stem cell–derived mesenchymal stem cells. *Mol Cell Proteomics* 6:1680–1689.

Takahashi T, Nowakowski RS, Caviness VS, Jr. (1996). The leaving or Q fraction of the murine cerebral proliferative epithelium: A general model of neocortical neuronogenesis. *J Neurosci* 16:6183–6196.

Takahashi K, Yamanaka S (2006). Induction of pluripotent stem cells from mouse embryonic and adult fibroblast cultures by defined factors. *Cell* 126:663–676.

Takahashi K, Tanabe K, Ohnuki M, Narita M, Ichisaka T, Tomoda K, Yamanaka S (2007). Induction of pluripotent stem cells from adult human fibroblasts by defined factors. *Cell* 131:861–872.

Takashima Y, Era T, Nakao K, Kondo S, Kasuga M, Smith AG, Nishikawa S (2007). Neuroepithelial cells supply an initial transient wave of MSC differentiation. *Cell* 129:1377–1388.

Tamura Y, Kataoka Y, Cui Y, Takamori Y, Watanabe Y, Yamada H (2007). Multi-directional differentiation of doublecortin- and NG2-immunopositive progenitor cells in the adult rat neocortex in vivo. *Eur J Neurosci* 25:3489–3498.

Tanaka Y, Tozuka Y, Takata T, Shimazu N, Matsumura N, Ohta A, Hisatsune T (2009). Excitatory GABAergic activation of cortical dividing glial cells. *Cereb Cortex* 19:2181–95.

Terada N, Hamazaki T, Oka M, Hoki M, Mastalerz DM, Nakano Y, Meyer EM, Morel L, Petersen BE, Scott EW (2002). Bone marrow cells adopt the phenotype of other cells by spontaneous cell fusion. *Nature* 416: 542–545.

Thomson JA, Itskovitz-Eldor J, Shapiro SS, Waknitz MA, Swiergiel JJ, Marshall VS, Jones JM (1998). Embryonic stem cell lines derived from human blastocysts. *Science* 282:1145–1147.

Tohyama T, Lee VM, Rorke LB, Marvin M, McKay RD, Trojanowski JQ (1992). Nestin expression in embryonic human neuroepithelium and in human neuroepithelial tumor cells. *Lab Invest* 66:303–313.

Tole S, Kaprielian Z, Ou SK, Patterson PH (1995). FORSE-1: A positionally regulated epitope in the developing rat central nervous system. *J Neurosci* 15:957–969.

Toma JG, Akhavan M, Fernandes KJ, Barnabe-Heider F, Sadikot A, Kaplan DR, Miller FD (2001). Isolation of multipotent adult stem cells from the dermis of mammalian skin. *Nat Cell Biol* 3:778–784.

Toma JG, McKenzie IA, Bagli D, Miller FD (2005). Isolation and characterization of multipotent skin-derived precursors from human skin. *Stem Cells* 23:727–737.

Tramontin AD, Garcia-Verdugo JM, Lim DA, Alvarez-Buylla A (2003). Postnatal development of radial glia and the ventricular zone (VZ): A continuum of the neural stem cell compartment. *Cereb Cortex* 13:580–587.

Tropepe V, Hitoshi S, Sirard C, Mak TW, Rossant J, van der Kooy D (2001). Direct neural fate specification from embryonic stem cells: A primitive mammalian neural stem cell stage acquired through a default mechanism. *Neuron* 30:65–78.

Uhrbom L, Hesselager G, Nister M, Westermark B (1998). Induction of brain tumors in mice using a recombinant platelet-derived growth factor B-chain retrovirus. *Cancer Res* 58:5275–5279.

Uhrbom L, Dai C, Celestino JC, Rosenblum MK, Fuller GN, Holland EC (2002). Ink4a-Arf loss cooperates with KRas activation in astrocytes and neural progenitors to generate glioblastomas of various morphologies depending on activated Akt. *Cancer Res* 62:5551–5558.

van der Kooy D, Weiss S (2000). Why stem cells? *Science* 287:1439–1441.

van Praag H, Schinder AF, Christie BR, Toni N, Palmer TD, Gage FH (2002). Functional neurogenesis in the adult hippocampus. *Nature* 415:1030–1034.

Varghese M, Olstorn H, Sandberg C, Vik-Mo EO, Noordhuis P, Nister M, Berg-Johnsen J, Moe MC, Langmoen IA (2008). A comparison between stem cells from the adult human brain and from brain tumors. *Neurosurgery* 63:1022–1033; discussion 1033–1024.

Vicario-Abejon C, Yusta-Boyo MJ, Fernandez-Moreno C, de Pablo F (2003). Locally born olfactory bulb stem cells proliferate in response to insulin-related factors and require endogenous insulin-like growth factor-I for differentiation into neurons and glia. *J Neurosci* 23:895–906.

Vierbuchen T, Ostermeier A, Pang ZP, Kokubu Y, Sudhof TC, Wernig M (2010). Direct conversion of fibroblasts to functional neurons by defined factors. *Nature* 463:1035–41.

Voigt T (1989). *Development* of glial cells in the cerebral wall of ferrets: direct tracing of their transformation from radial glia into astrocytes. *J Comp Neurol* 289:74–88.

Wadington CH (1957). *The Strategy of the Genes.* London: Geo. Allen & Unwin.

Wagers AJ, Sherwood RI, Christensen JL, Weissman IL (2002). Little evidence for developmental plasticity of adult hematopoietic stem cells. *Science* 297:2256–2259.

Wagers AJ, Weissman IL (2004). Plasticity of adult stem cells. *Cell* 116:639–648.

Walker TL, Yasuda T, Adams DJ, Bartlett PF (2007). The doublecortin-expressing population in the developing and adult brain contains multipotential precursors in addition to neuronal-lineage cells. *J Neurosci* 27: 3734–3742.

Walsh C, Cepko CL (1992). Widespread dispersion of neuronal clones across functional regions of the cerebral cortex. *Science* 255:434–440.

Wang X, Willenbring H, Akkari Y, Torimaru Y, Foster M, Al-Dhalimy M, Lagasse E, Finegold M, Olson S, Grompe M (2003). Cell fusion is the principal source of bone-marrow–derived hepatocytes. *Nature* 422: 897–901.

Watanabe M, Toyama Y, Nishiyama A (2002). Differentiation of proliferated NG2-positive glial progenitor cells in a remyelinating lesion. *J Neurosci Res* 69:826–836.

Watanabe Y, Kameoka S, Gopalakrishnan V, Aldape KD, Pan ZZ, Lang FF, Majumder S (2004). Conversion of myoblasts to physiologically active neuronal phenotype. *Genes Dev* 18:889–900.

Weimann JM, Charlton CA, Brazelton TR, Hackman RC, Blau HM (2003). Contribution of transplanted bone marrow cells to Purkinje neurons in human adult brains. *Proc Natl Acad Sci USA* 100:2088–2093.

Weissman IL, Anderson DJ, Gage F (2001). Stem and progenitor cells: Origins, phenotypes, lineage commitments, and transdifferentiations. *Annu Rev Cell Dev Biol* 17:387–403.

Wernig M, Meissner A, Foreman R, Brambrink T, Ku M, Hochedlinger K, Bernstein BE, Jaenisch R (2007). In vitro reprogramming of fibroblasts into a pluripotent ES-cell-like state. *Nature* 448:318–324.

Wernig M, Meissner A, Cassady JP, Jaenisch R (2008a) c-Myc is dispensable for direct reprogramming of mouse fibroblasts. *Cell Stem Cell* 2:10–12.

Wernig M, Zhao JP, Pruszak J, Hedlund E, Fu D, Soldner F, Broccoli V, Constantine-Paton M, Isacson O, Jaenisch R (2008b) Neurons derived from reprogrammed fibroblasts functionally integrate into the fetal brain and improve symptoms of rats with Parkinson's disease. *Proc Natl Acad Sci USA* 105:5856–5861.

Wichterle H, Peljto M (2008). Differentiation of mouse embryonic stem cells to spinal motor neurons. *Current Protocols in Stem Cell Biology*, Chapter 1: Unit 1H.1.1–1H 1.9.

Wilmut I, Schnieke AE, McWhir J, Kind AJ, Campbell KH (1997). Viable offspring derived from fetal and adult mammalian cells. *Nature* 385:810–813.

Wislet-Gendebien S, Hans G, Leprince P, Rigo JM, Moonen G, Rogister B (2005). Plasticity of cultured mesenchymal stem cells: Switch from nestin-positive to excitable neuron-like phenotype. *Stem Cells* 23:392–402.

Wong-Staal F, Dalla-Favera R, Gelmann EP, Manzari V, Szala S, Josephs SF, Gallo RC (1981). The v-sis transforming gene of simian sarcoma virus is a new onc. gene of primate origin. *Nature* 294:273–275.

Woodbury D, Schwarz EJ, Prockop DJ, Black IB (2000). Adult rat and human bone marrow stromal cells differentiate into neurons. *J Neurosci Res* 61:364–370.

Woodbury D, Reynolds K, Black IB (2002). Adult bone marrow stromal stem cells express germline, ectodermal, endodermal, and mesodermal genes prior to neurogenesis. *J Neurosci Res* 69:908–917.

Wurmser AE, Nakashima K, Summers RG, Toni N, D'Amour KA, Lie DC, Gage FH (2004). Cell fusion–independent differentiation of neural stem cells to the endothelial lineage. *Nature* 430:350–356.

Xu Y, Tamamaki N, Noda T, Kimura K, Itokazu Y, Matsumoto N, Dezawa M, Ide C (2005). Neurogenesis in the ependymal layer of the adult rat 3rd ventricle. *Exp Neurol* 192:251–264.

Ying QL, Nichols J, Evans EP, Smith AG (2002). Changing potency by spontaneous fusion. *Nature* 416:545–548.

Yoo SW, Kim SS, Lee SY, Lee HS, Kim HS, Lee YD, Suh-Kim H (2008). Mesenchymal stem cells promote proliferation of endogenous neural stem cells and survival of newborn cells in a rat stroke model. *Exp Mol Med* 40:387–397.

Young KM, Fogarty M, Kessaris N, Richardson WD (2007). Subventricular zone stem cells are heterogeneous with respect to their embryonic origins and neurogenic fates in the adult olfactory bulb. *J Neurosci* 27:8286–8296.

Yu J, Vodyanik MA, Smuga-Otto K, Antosiewicz-Bourget J, Frane JL, Tian S, Nie J, Jonsdottir GA, Ruotti V, Stewart R, Slukvin, II, Thomson JA (2007). Induced pluripotent stem cell lines derived from human somatic cells. *Science* 318:1917–1920.

Yuan X, Curtin J, Xiong Y, Liu G, Waschsmann-Hogiu S, Farkas DL, Black KL, Yu JS (2004). Isolation of cancer stem cells from adult glioblastoma multiforme. *Oncogene* 23:9392–9400.

Yue F, Cui L, Johkura K, Ogiwara N, Sasaki K (2006). Induction of midbrain dopaminergic neurons from primate embryonic stem cells by coculture with sertoli cells. *Stem Cells* 24:1695–1706.

Yun K, Potter S, Rubenstein JL (2001). Gsh2 and Pax6 play complementary roles in dorsoventral patterning of the mammalian telencephalon. *Development* 128:193–205.

Zeng X, Cai J, Chen J, Luo Y, You ZB, Fotter E, Wang Y, Harvey B, Miura T, Backman C, Chen GJ, Rao MS, Freed WJ (2004). Dopaminergic differentiation of human embryonic stem cells. *Stem Cells* 22:925–940.

Zhang SC, Wernig M, Duncan ID, Brustle O, Thomson JA (2001). In vitro differentiation of transplantable neural precursors from human embryonic stem cells. *Nat Biotechnol* 19:1129–1133.

Zheng X, Shen G, Yang X, Liu W (2007). Most C6 cells are cancer stem cells: Evidence from clonal and population analyses. *Cancer Res* 67:3691–3697.

Zhong W, Chia W (2008). Neurogenesis and asymmetric cell division. *Curr Opin Neurobiol* 18:4–11.

Zhou JM, Chu JX, Chen XJ (2008). An improved protocol that induces human embryonic stem cells to differentiate into neural cells in vitro. *Cell Biol Int* 32:80–85.

Zhu L, Gibson P, Currle DS, Tong Y, Richardson RJ, Bayazitov IT, Poppleton H, Zakharenko S, Ellison DW, Gilbertson RJ (2009). Prominin 1 marks intestinal stem cells that are susceptible to neoplastic transformation. *Nature* 457:603–607.

Zhu X, Bergles DE, Nishiyama A (2008). NG2 cells generate both oligodendrocytes and gray matter astrocytes. *Development* 135:145–157.

Zuccato C, Tartari M, Crotti A, Goffredo D, Valenza M, Conti L, Cataudella T, Leavitt BR, Hayden MR, Timmusk T, Rigamonti D, Cattaneo E (2003). Huntingtin interacts with REST/NRSF to modulate the transcription of NRSE-controlled neuronal genes. *Nat Genet* 35:76–83.

4

Neuronal Development

Adult neurogenesis needs to be
seen in the context of
neurogenesis during brain
development

NEURAL STEM CELL BIOLOGY of the adult organism and the study of adult neurogenesis are essentially neural developmental biology under the conditions of the adult brain. To understand the particularities of adult neurogenesis, it is necessary to see adult neurogenesis in the context of neural development in general. In this chapter we will take up some of the thoughts about precursor cells in the developing brain that were discussed in the previous chapter. We will take a closer look at the development of the brain and the different roles that precursor cells play within it. Although the overview provided here must be limited and selective to those key concepts most relevant to adult neurogenesis, it is important that adult neurogenesis be seen against the backdrop of neurogenesis in a more general sense. The developmental perspective is particularly to be kept mind in for all attempts to manipulate stem cells in order to manipulate adult neurogenesis.

If neural stem cell biology is developmental biology, the reverse is also true: brain development is to a large degree a matter of stem cells. All development originates from stem cells, which are the cells in which the genome is in its most accessible form. *Development* is the realization of the potentials the genome harbors. By tampering with the origins of development, we might do more than we intend. The inherent openness of stem cell function, always facing the future rather than present and past, and the impact of activity and environment on development reduce predictability of the results.

Stem Cells Are the Origin of Development

Stem cells are not only the
source but also the consequence
of development

The fact that stem cells are the origin of development is based on their particular genetic program, which determines their self-renewal behavior and their developmental potential. There will, however, be as many stemness programs as there are different types of stem cells, and the shared core genes that define all stem cells have not yet been identified. Except for the zygote, which stands at the beginning of the development of the organism, all stem cells and hence their genetic program result from preceding development. Even in the zygote, two lines of development from the parental germ cells come together and form something new. In the end, one cannot talk about stem cells without development, and vice versa.

From the very first moment the egg becomes fertilized, its further development is governed by an interaction between the DNA and the environment; that is, everything that is not the DNA. One might even argue that for any given gene, *environment* includes all other genes as well. The very first steps of embryonic development are still governed by maternal genes whose transcripts have been stored in the cytoplasm of the oocyte. Control of their biological activity takes place on the translational level. Only with the first divisions of the fertilized eggs gene transcription sets in the embryo and developmental control shifts from maternal factors to the embryo.

Independent development starts with the first divisions of the zygote

To split the consideration into only the genes on one side and everything else as the rest on the other is not without problems. The DNA can to some degree be interpreted as a text, and many metaphors speak to the fact that this is how most people see it: "deciphering the book of life," "open reading frame," etc. If the genome and its function seem complicated, the "rest" is by orders of magnitude more complex. The text metaphor of the DNA actually does not carry us very far. It suggests a linearity that is hardly there, and it misleads us to apply all the techniques we are used to in dealing with "texts" to the genome. Epigenetics—that is, all heritable factors except for the DNA sequence that influence development (for example, DNA methylation and chromatin remodeling)—begins to shake the grounds of these implicit assumptions, but still more or less remains in the same frame of mind. But the fields of systems biology and systems genetics presently give us the first glimpse at what lies behind the "text" and how its relationship to the "rest" can possibly be grasped and understood. Such understanding will often require new kinds of mathematics, much like in the transition from Newton's mechanics to the physics of Einstein and Planck. Present-day biology has a molecular bias, and the journals are the graveyards for innumerable reports on the supposed "functions" of individual genes:

The key to development lies in the complex interaction between genome and shifting environmental contexts

> We speak piously of taking measurements and making small studies that will add another brick
> to the temple of science. Most such bricks just lie around in the brickyard.
>
> —J. R. Platt (1964)

There is an obvious need for and relevance in studying individual genes, so conventional molecular biology cannot be questioned in general; but the severe limitations of such studies in illuminating the greater picture need to be taken into account. In the following paragraphs (and especially in Chap. 9), many individual genes will be mentioned, but the goal is usually only to highlight the contribution of a few singers stepping out from the chorus line for their solo. In the end we are interested in the entire piece, and this piece cannot be understood from the perspectives of the soloists alone—even if one day we know them all.

The most direct molecular interactors with DNA are transcription factors, which are regulatory molecules that have a DNA binding motif. Alone but often together with cofactors a transcription factor can bind to regulatory regions of a specific gene and induce or inhibit its transcription. Because transcription factors interact directly with the genome and its potentials, they are at the center of studies on mechanisms of development and stem cell biology. Networks of intracellular signaling cascades converge on transcription factors and lead to their activation or inhibition. Transcription factors are thus like the home stretch in a complex cascade of regulation.

Regulatory intracellular signaling cascades converge on transcription factors

The human organism is built from a mere 23,000 genes. This efficiency is only possible because genes have different meanings in different contexts. Many structural proteins and even RNA can double as enzymes,

The effect of regulatory factors is combinatorial and context-dependent

for example. Different organs contain the same basic structural elements but use them toward different ends. The same principle applies to transcription factors and other intracellular signaling molecules. During brain development the same repertoire of regulatory molecules and transcription factors is active at different stages of development. The exact function, however, differs with location and time and is thus context-sensitive. *Development* is a temporally unidirectional shift of such contexts. Stem cells are the site where this shift takes place and where it unfolds its effect on development. As a consequence, transcription factors ultimately induce the shifts in their own meaning. Very few factors are specific for only one developmental situation. An understanding of the molecular bases of adult neurogenesis requires, not only identification of the factors involved, but also their position in a regulatory network. Although there is a transition phase shortly after birth it might be months and years of development that separate adult neurogenesis from embryonic brain development.

How to Make a Neuron

Development proceeds from stem cell to mature cells over mitotic and post-mitotic intermediate steps, all of which might be regulated

Neurogenesis follows a rough general pattern. A multipotent stem cell gives rise to intermediate progeny. At the level of these progenitors, expansion occurs by massive proliferation, and the first signs of neuronal determination are found. Lineage-determined progenitor cells might continue to divide and give rise to migratory cells, sometimes referred to as "neuroblasts," which reach the target localization of the new neuron. Neuroblasts might still divide but at a much lower level than the preceding progenitors. At the latest when cells have reached that stage exit from the cell cycle occurs. The cells go through a post-mitotic maturation stage durging which connectivity is established. Most new neurons do not survive this stage. Only those that become integrated into functional circuits are recruited for long-term survival. Preceding all this might lie a quiescent long-term stem cell that provides reconstitution of the stem cell pool. This sequence— long-term stem cell, stem cell, intermediate progenitor cell, lineage-determined progenitor cells, "blast," post-mitotic immature cell, mature new cell—is best exemplified in hematopoietic development. It is a rather coarse but useful scheme that, however, is no end in itself (Fig. 3–6).

The abstract basic model of development can also be applied to adult neurogenesis

The best-studied cases of neurogenesis are probably the development of the cortex and the spinal cord. We also have considerable insight into the development of the cerebellum and the events at the midbrain/hindbrain boundary. Even a rough comparison between these examples of neurogenesis reveals how diverse neurogenesis can be and how much lies in the details. The mentioned model is a an abstraction that cannot deliver the necessary level of sophistication appropriate to each situation. As both neurogenic regions of the adult brain relate to cortical structures, we will largely refer to cortical development in the following paragraphs. Table 4–1 aligns the theoretical concept with what is known from cortical development with the lineage schemes as they have been developed for the adult neurogenic regions (which will be covered in greater detail in the next two chapters). This alignment is to some degree hypothetical but is implicit in many current statements about adult neurogenesis. We use it as a framework, knowing that in reality there will be many relevant deviations.

From the Blastocyst to the Neural Tube

Brain development originates from the embryonic stem cells in the blastocyst

Brain development begins at the blastocyst stage (Fig. 3–5). The blastocyst contains three populations of stem cells: embryonic stem (ES) cells, trophoblast stem (TS) cells, and extraembryonic endoderm (XEN) cells. We begin to understand the molecular events governing this lineage distinction. Single transcription factors exert massive effects at this early stage of development.

Table 4–1 Embryonic versus adult neurogenesis

	Embryo / fetus	Adult
Abundance	Rule	Exception
Spatial distribution	Generalized	Spatially highly restricted to essentially two neurogenic zones
Differentiation spectrum	All types of neurons	2 types of interneurons in the olfactory bulb, 1 type of excitatory neuron in the dentate gyrus
Temporal pattern	Often in waves and synchronous	Asynchronous
Yield	Very large numbers	Very low numbers
Relation to behavior	Dependent on intrauterine sensory input but essentially independent of behavior of the individual; partly dependent on maternal behavior	Dependent on activity and behavior of the individual
Key role in	Development	Plasticity

Transcription factor Sall4, for example, is critically involved in the specification of these three lineages (Lim et al., 2008). All tissues, including the nervous system, derive from the ES cells.

Diversification already sets in at the level of ES cells

The pluripotent ES cells are, among other markers, characterized by their expression of Oct4, which is down-regulated when the cells lose their pluripotency (Scholer et al., 1990; Rathjen et al., 2002). Oct4 does not work alone: a complex regulatory network around Oct4 and Nanog controls pluripotency (Loh et al., 2006). In such a network, Oct4 and Sox2, for example, stand in a reciprocally regulated relationship (Chew et al., 2005). Neural stem cells do not express Oct4 but maintain expression of Sox2 even into adulthood. Given the nature of the transcription networks that define ES cells it seems increasingly unlikely that even for one brief period all embryonic stem cells are equal in their potential. Two models compete: differences from the zygote onward and among the ES cells might either be introduced by stochastic processes or be the product of a sequence of symmetrical versus asymmetrical divisions (see below). Probably both mechanisms contribute. Early development goes through a series of symmetrical divisions followed by an immediate step of differentiation that makes the two daughter cells different from the mother cell. Interaction among these rapidly expanding cells would at once introduce further differences, through the simple fact that some cells find themselves in the center of the cell cluster and are thus surrounded by many others of their kind, whereas other cells are exposed to the outer environment.

At the ES cell level neural development does not need to be actively induced

For embryonic stem cells, development into neuroepithelial precursor cells appears to be something like a default state. Neurons also readily develop from embryonic stem cell cultures, although their neuronal nature is quite generic. The non-neuronal differentiation into epidermis, in contrast, requires active signaling through the bone morphogenic proteins (BMPs). BMP 4, for example, antagonizes neural induction. A key transcription factor that comes on at the level of ES cells and marks the initiation of the induction of the neuroectoderm is Sox1, which, however, maintains the cells in an undifferentiated state. Sox1 is down-regulated when Pax6 is switched on. Pax6 in turn appears to be the key transcription factor that determines the development of radial glia (Suter et al., 2008). Experimental overexpression of Pax6 in ES cells can induce the transition to the radial glial fate. Without Pax6 no normal radial glia forms, and cortical neurogenesis is impaired (Gotz et al., 1998). Besides Sox1 and Pax6, early neural precursor cells are characterized by their expression of Musashi1, another factor that is also found in neural precursor cells at much later stage of development.

Neural development is a specialization within the neuroectodermal germ layer

Within the ectoderm, neural induction leads to specification of the neuroectoderm. The early neuroectodermal tissue forms a layer along the anterior midline of the egg cylinder, the *neural plate*. The neural plate initially contains a single layer of neural precursor cells, called *neuroepithelial cells*. The entire structure is called the *neuroepithelium*. During formation of the neural plate, Sox1 and genes such as Gbx2, which initiate early differentiation steps, are up-regulated (Wassarman et al., 1997; Pevny et al., 1998; Li et al., 2005). Induced by so-called organizer regions, the neural plate next invaginates along the midline. Known factors involved in this initial neural induction are noggin, chordin, and follistatin. The longitudinal groove closes on the top and thereby forms the neural tube, whose walls contain the neuroepithelial cells. The anterior part of the neural tube will develop into the brain; the more caudal parts into the spinal cord. Above (i.e., dorsal of) the neural tube, once the tissue has thickened, the neural crest appears, from which the peripheral and autonomous nervous system are derived.

Neuroepithelial precursor cells reside in the walls of the primordial ventricular system

The neural tube encloses a cavity that becomes the primordial ventricular system of the brain. Throughout life, the walls of the ventricles remain a location of cell division and precursor cell action. This also holds true for parts of the ventricular system that become obliterated in the adult. The rostral migratory stream, for example, which provides the route of migration during neurogenesis in the adult olfactory system, surrounds the remnants of the olfactory ventricle (see Chap. 11, p. 504, for a more detailed discussion).

Neural Precursor Cells Are Regionalized

Precursor cells have a temporospatial identity, defined by key morphogens

While the neural tube develops, the precursor cells become *regionalized*; that is, they acquire a potential that is based on their spatial position. Given that development progresses with time in an essentially unidirectional manner, this positional identity is also a spatio-temporal identity. The concept of "positional identity" is of foremost importance for understanding how, in the adult brain, neural precursor cells might continue to generate new neurons. Positional identity means that for a stem cell, it matters where in the brain and at what time of development it is found (Fig. 3–7). *Development* implies increasingly restricted potential. Precursor cells at different locations in the adult brain differ in their positional identity. They originate from a differential development that begins at the level of the neuroepithelial cells in the wall of the neural tube.

First patterning occurs along the antero-posterior axis

It is believed that the originally very similar precursor cells begin to distinguish themselves from their neighbors by divisions, analogously to the model situation that has been found in the neuroblasts of *Drosophila*, one of the best-studied model systems. The first patterning of this kind occurs along the anterior–posterior axis of the embryo. From the posterior regions, the spinal cord develops; the anterior part generates the brain. One key molecule relevant for the transition to neural patterning is retinoic acid. In the developing nervous system, a posterior–anterior gradient of retinoic acid expression forms, which in turn leads to a graded induction of a class of genes called Hox genes in the posterior part of the neural tube that will later become spinal cord and hindbrain. Making use of this role, retinoic acid is added in vitro when growth factors are withdrawn to induce differentiation in neural precursor cell cultures.

Hox genes are involved in the definition of segments along the anterior-posterior axis

The different Hox genes regulate further subdivisions of the future spinal cord. In the more anterior regions, multiple other so-called homeodomain genes are involved in specification of the precursor cells from which the telencephalic structures will develop. A particularly clear boundary is the midbrain–hindbrain border, which is subject to a comparatively well-understood molecular control (Brand et al., 1996; Raible and Brand, 2004). For example, the midbrain/hindbrain boundary is characterized by a sharply defined expression of two transcription factors, Wnt1 and engrailed-1 (En1), and the growth factor FGF-8. Overexpression of FGF-8 moves the midbrain–hindbrain border posteriorly (Brodski et al., 2003). In the anterior parts of the brain, the anterior–posterior axis is defined by growth factors FGF-2, -4, and -8, Wnt 3A and 8, ligands of nuclear RA receptors, and other so-called morphogens (Altmann and Brivanlou, 2001). Transcription factor Emx1 defines the anterior half of the hemisphere in the telencephalon, and Emx2 defines the posterior half. Emx2 is required for the development of a normally proportioned neocortex (Bishop et al., 2000; Mallamaci et al., 2000).

Pax genes are found at many decision points of development

The Pax gene family consists of nine members with numerous effects on brain development and is still present in the adult brain, both in radial glia–like precursor cells as well as in some mature cells (Blake et al., 2008). The current idea is that Pax genes are instrumental in orchestrating the interaction between intrinsic mechanisms of development and cues from the cellular environment. In the adult brain, such interaction would underlie the many forms of "plasticity," including adult neurogenesis. *Pax6* in particular plays key roles at several restriction points of development. One of these is that Pax6 is expressed in a gradient that is the reverse of Emx2 expression (Fig. 4–1). Such opposing gradients are a frequent mechanism of determining positional identity. Between the two sources of the regulatory molecules, individual positions are characterized by two coordinates defined by graded expression of each of the two factors. Moving from one source toward the other is reflected in decreasing expression of the first factor while the other increases. Higher dimensional systems with more than two gradients probably exist. Although it is not clear how a cell can sense subtle gradients and respond specifically, the spatial resolution of such a system is extremely high. The only theoretical alternative would be a system in which every single position is defined by individual molecules.

Pax6 and Emx2 define the anterior–posterior patterning of the hemispheres

Emx2 is most strongly expressed in the posterior part of the hemispheres; Pax6 in the anterior part. In the mature cortex, various specialized areas are found along the anterior–posterior axis—the frontal, cingular, somatosensory, motor, insular, and visual cortices, to name just the largest. In a Pax6 mutant, the posterior areas, the visual cortex, become enlarged because Emx2 activity dominates. In an Emx2 mutant, by contrast, the more anterior areas, such as the motor cortex, are bigger because the anterior Pax6 effects dominate (Bishop et al., 2000; Mallamaci et al., 2000). FGF-8 seems to up-regulate Emx2 expression, and thus FGF-8 overexpression has a similar effect to a Pax6 knockout. Pax6 is a

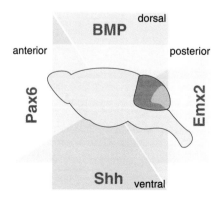

FIGURE 4–1 Oppositional gradients as key principle for regionalization. The schematic shows (in a simplified way) the anterior-posterior gradients build by Pax6 and Emx2 and the dorsal-ventral gradient build by Shh and BMP. In theory, in two dimensions. each point of the brain is thereby defined by a unique combination of factor concentrations.

regulator of cortical cell numbers in an at least partly cell-autonomous way (Estivill-Torrus et al., 2002; Quinn et al., 2007).

Parallel to anterior–posterior regionalization, dorsal–ventral specification occurs (Fig. 4–1). In the neural tube, the most dorsal cells (the roof plate) specialize in secreting bone morphogenic proteins (BMPs), whereas the most ventral cells (the floor plate) secrete sonic hedgehog (Shh), thereby forming another reciprocal gradient. BMP signaling is also involved in the most dorsal part of the neural tube when a population of precursor cells sequester and migrate laterally to form the neural crest, the origin of the autonomic nervous system. Different Pax genes, for example, respond to either BMP or Shh, thereby linking the patterning along the different axes (Jostes et al., 1990; Stoykova et al., 1996).

The fact that transcriptional regulation is context-sensitive gives precursor cells their positional identity: their spatially and temporally determined nature depending on transcription factor patterns. This determination is more flexible in terms of space than in terms of time. Precursor cells from the spinal cord can be transplanted into the hippocampus and behave region-specifically. Precursor cells from the hippocampus can participate in the production of retinal cells. Embryonic precursor cells also readily integrate after implantation into later developmental stages. But the reverse is not necessarily true: adult precursor cells cannot generally regain functions of earlier developmental stages because development is essentially unidirectional and characterized by an increasing restriction of developmental potentials. This does not mean that under no circumstances could this pattern be overcome. As discussed in Chapter 3, mature cells have effectively been reprogrammed to become stem and progenitor cells. The question, however, is if this process occurs naturally at all.

Neurogenesis in the adult brain originates from two distinct pools of precursor cells with different positional identities. The hippocampus is a cortical structure arising from the dorsal telencephalon, together with the olfactory cortex and the neocortex. Neural precursor cells here generate glutamatergic principal neurons. Neurogenesis in the adult SVZ and olfactory bulb mostly originates from the ventral telencephalon, where precursor cells for interneurons and oligodendrocytes reside in the so-called ganglionic eminences. The two populations of precursor cells are differentially regionalized. Adult neurogenesis is in continuity with embryonic brain development. There is no evidence that completely different principles govern embryonic and adult neurogenesis. However, precursor cells in adult neurogenesis have a strong temporal and positional identity with a limited developmental potential in vivo. The roots for this specification lie in embryonic development. Precursor cells in the neurogenic regions of the adult brain have intrinsic properties that are region-dependent. On the other hand, as demonstrated by ex vivo experiments and manipulation of the in vivo conditions, the realization of their potential is highly dependent on the cellular microenvironment, the stem cell niche that provides neurogenic permissiveness.

General Patterns of Mammalian Brain Development

Brain development in mammals is characterized by three major stages. The first phase is an expansion phase, during which stem cells produce growth by massive cell divisions. During this expansion phase, different types of divisions occur in different types of precursor cells, leading to a highly orchestrated expansion of regionalized precursor cells. In species with a gyrified cortex, the basis for the later gyri and sulci is laid at this level by a differential expansion (Kriegstein et al., 2006). There is less expansion in sectors underlying later sulci than those underlying gyri. Differential expansion is achieved by symmetrical and asymmetrical division of so-called founder stem cells (Kriegstein et al., 2006) and their intermediate progeny.

The expansion phase is followed by the phases of first, neurogenesis, and then gliogenesis, during which the expanded precursor cells give rise to differentiating cells. At birth, neurogenesis has ceased in most parts of the brain. Radial glia, from which neurogenesis originates, is largely converted into astrocytes but remains present in the neurogenic zones of the adult brain (Fig. 4–2; see also Chapter 3, p. 62). Neurogenesis and gliogenesis are characterized by the expansion of the precursor cells and the co-occurrence of cell death, by which a surplus of newly generated cells that presumably do not make appropriate connections is eliminated. It has been estimated that more than ten times more neurons are produced during development than are present in the mature brain. This pattern is also found in adult neurogenesis, where many more immature neurons are produced than reach maturity and functional integration. Within neurogenesis, excitatory principal or projecting neurons and inhibitory interneurons are derived from different sources and at different times.

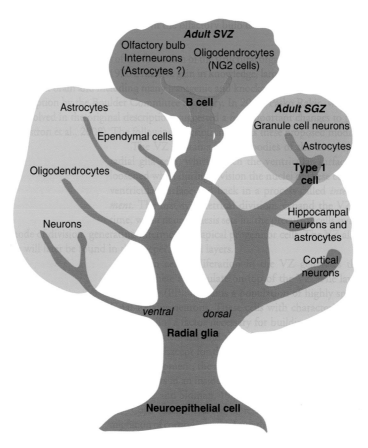

FIGURE 4–2 The two neurogenic regions of the adult brain are both ultimately derived from a common precursor cell, the neuroepithelial cell. However, they are the product of two differently regionalized lineages in the dorsal (hippocampus) and ventral brain (SVZ, olfactory bulb). Radial glia and radial glia–like precursor cells embody these lineages. The separation is not strict, however. The dorsal branch returning to the olfactory bulb represents the glutamatergic neurons in the adult olfactory bulb (Brill et al. 2009).

Gliogenesis again occurs in two waves. First, astrocytes are produced after neurogenesis has begun to cease. Then oligodendrocytes are generated, most of them postnatally. The links between neurogenesis and the generation of astrocytes and of oligodendrocytes are not equivalent. In the adult hippocampus, for example, normally almost no oligodendrocytes are produced (Steiner et al., 2004) but they can be induced by overexpression of Ascl1 in intermediate progenitor cells (Jessberger et al., 2008a). In the SVZ, in contrast, many Olig2-expressing progenitor cells are found, and oligodendrocytes are generated alongside the interneurons (Menn et al., 2006).

In gliogenesis, oligodendrogenesis follows astrogenesis

The shifting potential of precursor cells over time can be recapitulated in vitro when precursor cells are isolated at different time points of development (Qian et al., 2000). Acutely isolated cells will tend to differentiate into neurons during the peak of neurogenesis in vivo: into astrocytes when cells are taken, while astrogenesis is high; and into oligodendrocytes, if they are isolated, when the generation of oligodendrocytes dominates in vivo. The acutely isolated cells reflect the stage of brain development in vivo.

Acutely isolated precursor cells retain some of their positional identity in vivo

More surprisingly, the temporal specification also becomes apparent when precursors isolated at early embryonic stages (embryonic day 12 [E12] of a mouse) are followed longitudinally in culture (Qian et al., 2000). These cells sequentially go through phases of predominant neurogenesis, astrogenesis, and oligodendrogenesis. The earlier cells are still multipotent; the later cells are destined to a glial fate. This implies that there must be a precursor cell–immanent program that determines the changes in developmental potential. Positional identity is thus not entirely separable from temporal identity. To some degree, a certain time is equivalent to a specific place, and vice versa.

Asymmetrical Division as a Fundamental Mechanism of Stem Cell Biology

The switch between symmetrical and asymmetrical divisions is key to the early steps of development

Besides determining positional identity, two other key regulations have to occur. First, within a precursor cell's positional identity, the cell must be instructed as to whether it should expand by symmetrical or asymmetrical divisions, be quiescent, or die. Second, and finally, cellular differentiation must be initiated. Brain size is to a large degree determined by the balance between expansion and differentiation. The switch between symmetrical and asymmetrical divisions of neuroepithelial and radial precursor cells and of intermediate progenitor cells is a key mechanism controlling this balance.

By convention, the apical surface of the neuroepithelium is depicted at the bottom

Neuroepithelial cells and radial glia touch the ventricular surface. Their cell bodies lie in the ventricular zone, immediately adjacent to the ventricle. By convention this orientation is called *apical*, although it is usually depicted at the bottom. Progenitor cells that delaminate from the ventricular surface and move towards the pial surface are called *basal*, although they are usually shown above the apical cells. While we today tend to depict the developing brain with the ventricular surface at the bottom (in the depth of the tissue) and the pial surface on top, the classical embryologists, whose nomenclature persists here, looked at the expansion of the brain as occurring from the ventricular wall downwards. The modern perspective, however, is found in the numbering of the cortical layers, where Layer I is up and Layer VI is down. So both perspectives more or less peacefully coexist.

In principle, the pattern of cortical development also applies to the hippocampus

In the hippocampus, the situation is similar but less obvious, because with reference to the ventricular wall, adult neurogenesis takes place in an ectopic location. The precursor cells that drive adult neurogenesis in rodents still face the ventricular surface (albeit at a distance) and migrate towards the pial surface (which is partly hidden in the hippocampal fissure). The terminology of *apical* (i.e. in the SGZ) vs. *basal* progenitor cells (more into the granule cell layer) is not in use for adult neurogenesis, but the logic of the nomenclature would be applicable. What is confusing is that, with respect to the neurons, the basal dendrites are those that lie in the SGZ and thus where the apical precursor cells would be located, and vice versa.

In cortical development, the stem cells lie at the ventricular surface and generate apical intermediate progenitor cells

The current model of cortical development is as follows: the neuroepithelial cell or radial glial cell is the highest-ranking stem cell in the system. The hallmark of stem cells is their ability to divide asymmetrically. Thereby the cells can self-renew and give rise to an intermediate progeny. These intermediates are referred to as *apical progenitor cells*, remain close to the ventricular surface, and can divide either symmetrically or asymmetrically. Their asymmetrical division yields a progenitor cell that migrates away from the ventricular wall and becomes a basal progenitor cell. Basal progenitor cells can divide symmetrically into two daughter cells, either as self-renewing division or generating two daughter cells that turn into neurons (Haubensak et al., 2004). This latter type of division represents a terminal "neurogenic division," a special case of symmetrical division, in which both daughter cells become different from the mother cell (Kriegstein et al., 2006; Farkas and Huttner, 2008). The sequence of symmetrical vs. asymmetrical divisions is orchestrated in a very precise way, and a few key molecules have been identified that govern the switch between the two types of division or between division and differentiation. Prominin1, for example, is involved in this decision in neuroepithelial cells and apical progenitor cells (Corbeil et al., 1999). Tis21 comes on in basal progenitor cells and is associated with the terminal neurogenic divisions (Iacopetti et al., 1999).

The sequence of apical and basal divisions has been observed with live microscopy

The described sequence of events has been largely observed by live imaging. Symmetric and asymmetrical divisions in the single layer of proliferating neuroepithelial cells in the developing mouse brain were visualized by time-lapse microscopy in slice preparations, and their progeny was followed (Qian et al., 1998; Qian et al., 2000; Haubensak et al., 2004). Similarly, radial glia was observed to divide asymmetrically, giving rise to another radial glia cell and one cell entering neuronal differentiation (Miyata et al., 2001; Noctor et al., 2001; Fishell and Kriegstein, 2003). Tracking basal and apical precursor cells it was also established that in basal progenitor cells symmetrical and rapid up-regulation of neuronal reporter genes occurs, visualizing neurogenic divisions in these cells (Attardo et al., 2008).

Drosophila has been the primary model organism to study asymmetrical and symmetrical divisions

Insight into the molecular bases of neuronal development and in particular the role of symmetrical vs. asymmetrical divisions came initially from work on the dew fly *Drosophila melanogaster*. From *Drosophila* (as well as in nematodes and zebra fishes) many genes and mechanisms regulating neurogenesis are highly conserved. The homologues in the different animals can play different roles, however, and mammals often have more and other isoforms of a given factor than nematodes, insects, or lower vertebrates. The contributions of individual molecules to identifiable steps of neuronal development can nonetheless be successfully studied in these less complex organisms. The divisions of neuroblasts in *Drosophila* contain some essential concepts of neural stem cell biology in nucleo. In theory, only through asymmetrical divisions can self-renewal be combined with the ability to differentiate. Asymmetrical divisions generate cells with different fates. It is still possible that in some contexts, asymmetrical division might not be a characteristic of individual precursor cells, but its consequence (daughter cells with different fates) could be achieved in a stochastic manner within a population of differentiating precursor cells.

Symmetric versus asymmetrical divisions are linked to the polarity of the epithelial cells

In *Drosophila* larvae, the neuroblasts individually move from their initial position in the band of neuroepithelial cells into the embryo. The neuroblasts divide asymmetrically, producing one neuroblast and one so-called ganglion mother cell. This ganglion mother cell divides exactly one more time, resulting in the presence of two glial or neuronal cells (plus the neuroblast from the first, asymmetrical division). As shown in Figure 4–3, this asymmetrical division occurs in a highly organized spatial pattern. The neuroblasts are polarized cells, just like most other epithelial (surface) cells. *Polarity* means, for

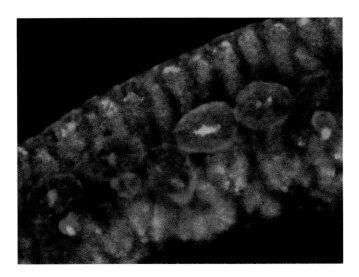

FIGURE 4–3 Asymmetrical cell division in *Drosophila* larvae. Embryonic *Drosophila* neuroblasts delaminate from the single-layered neuroectodermal epithelium (*top*) and subsequently divide repeatedly in an asymmetrical fashion. Both the epithelium and the neuroblasts show a pronounced apical–basal polarity. In both cell types, the Bazooka/PAR-3 protein (*red*) is localized in the apical cytocortex. In contrast, the Miranda protein (*blue*), which functions as an adaptor protein for the cell fate determinant Prospero, is restricted to the basal cortex of dividing neuroblasts and segregates exclusively into the budding ganglion mother cell. DNA is stained in green. Apical is the top of the figure. Image by Andreas Wodarz, Göttingen.

example, that with reference to the apical surface, cellular contents are not randomly distributed within the cell but show a spatial preference. In *Drosophila* neuroblasts, a number of transcription factors, such as Numb, Prospero, Miranda, Staufen, Inscutable, and Bazooka, show a polar distribution. Inscutable and Bazooka are located apically; Numb, Prospero, and Miranda are located basally. This distribution is an active, cytoskeleton-dependent process.

Segregation of proteins before division results in asymmetrically equipped daughter cells

The neuroblasts divide along a horizontal plane parallel to the apical surface (Fig. 4–4). The metaphase plane of the mitotic spindle separates the cell into two compartments, one apical, one basal, that differ in their contents of transcription factors. Consequently, when the division is completed, the two daughter cells carry different transcriptional signals within them. For example, Prospero, which is located in the basal compartment of the neuroblast and is thus later found in the ganglion mother cell, ultimately induces the exit from the cell cycle and is involved in determining that a ganglion mother cell divides only one more time. The mammalian homologue to Prospero is Prox1, which seems to play an important and potentially similar role in adult hippocampal neurogenesis (Pleasure et al., 2000a; Steiner et al., 2008). Numb, on the other hand, which co-segregates with Prospero to the basal compartment of the neuroblast, can again be distributed asymmetrically before the second division, allowing the development of cells with two different fates (two neuronal phenotypes) from one ganglion mother cell.

The mammalian Numb homologue is linked to asymmetrical division in mouse cortical development (Shen et al., 2002). Notch1 is another example of a key regulatory transcription factor that is distributed asymmetrically during early mammalian neurogenesis (Chenn and McConnell, 1995). In rodents, it turned out that Numb, which also functions as an inhibitor of Notch, is necessary for the polarity of the precursor cells, in that Numb is critical for maintaining adherens junctions between radial glial precursor cells (Rasin et al., 2007). Rodent Numb thus appears to be involved in the control of asymmetrical divisions in a manner somewhat different from fly Numb (Kim and Walsh, 2007).

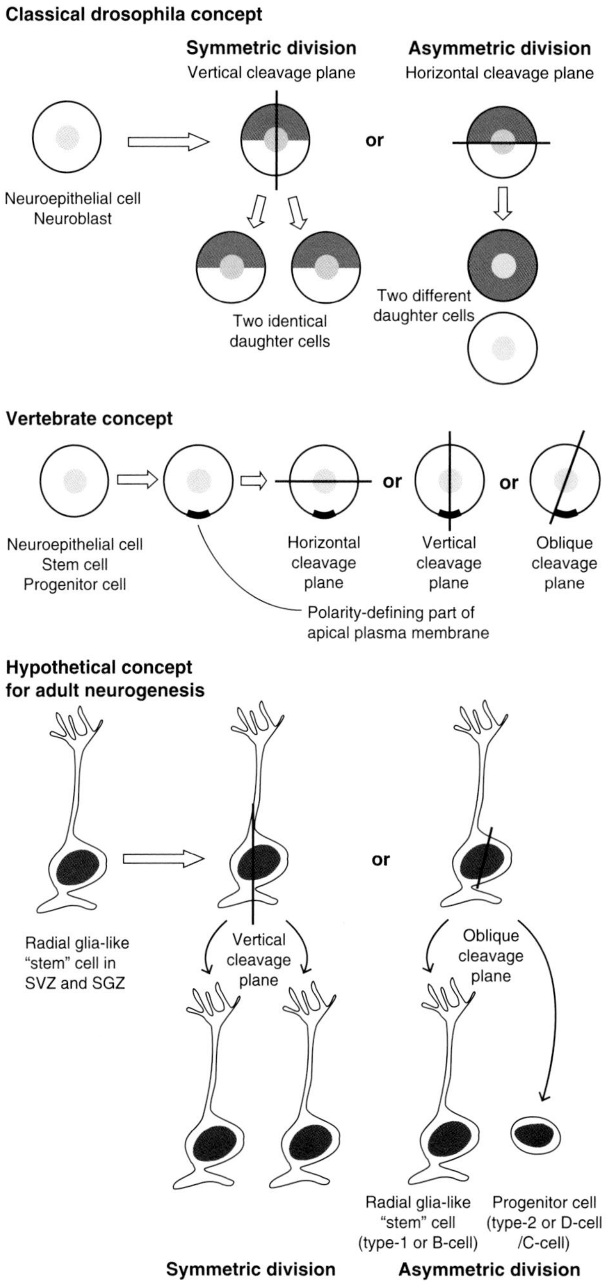

Classical drosophila concept

Symmetric division
Vertical cleavage plane

Asymmetric division
Horizontal cleavage plane

Neuroepithelial cell
Neuroblast

Two identical
daughter cells

Two different
daughter cells

Vertebrate concept

Neuroepithelial cell
Stem cell
Progenitor cell

Horizontal
cleavage
plane

Vertical
cleavage
plane

Oblique
cleavage
plane

Polarity-defining part of
apical plasma membrane

**Hypothetical concept
for adult neurogenesis**

Radial glia-like
"stem" cell in
SVZ and SGZ

Vertical
cleavage
plane

Oblique
cleavage
plane

Radial glia-like
"stem" cell
(type-1 or B-cell)

Progenitor cell
(type-2 or D-cell
/C-cell)

Symmetric division

Asymmetric division

FIGURE 4–4 Asymmetrical cell division in the mammalian brain. Mammalian stem cells show asymmetrical cell divisions but depart from the pattern seen in *Drosophila* (see Fig. 4–3). In neuroepithelial cells, orientation of a small, highly specialized membrane region, characterized by expression of Prominin/ CD133, in relation to the cleavage plane determines whether the division is symmetrical or asymmetrical. The application of this concept to adult neuro-genesis, as shown in the lower half of the figure remains to be fully confirmed. Prominin 1 is expressed by B1 cells of the SVZ (and ependymal cells) but expression in the hippocampus has not yet been unambiguously detected. The concept is based on the work of Wieland Huttner and colleagues (Wodarz and Huttner, 2003; Kosodo et al., 2004).

Mammalian asymmetrical versus
symmetrical divisions are less
dependent on cleavage plane
orientation than on cilia-bearing
membrane parts

In *Drosophila*, symmetrical divisions are associated with a vertical cleavage plane; asymmetrical divisions with a horizontal cleavage plane. In mammals, no such strict spatial orientation of the mitotic plane with a mandatory 90° turn of the cleavage plane between symmetrical and asymmetrical division is found. In some cases, most notably the initiation of cortical development and the pattern was initially thought to be suggestively similar to the situation in *Drosophila* (Chenn and McConnell, 1995; Noctor et al., 2008), but this impression appears not to be fully generalizable (Farkas and Huttner, 2008). Rather, in neuroepithelial cells and their progeny, both symmetrical and asymmetrical divisions can involve a vertical cleavage plane (Kosodo et al., 2004). Only inclusion or exclusion of a small but highly specialized portion of the apical cell membrane, characterized by the expression of Prominin1/CD133 and the presence of a cilium, is sufficient to distinguish symmetrical from asymmetrical divisions. The "horizontal" division thus differs only by a few degrees from a true vertical division, which would cut exactly through this small apical spot (Kosodo et al., 2004). This apical part of the membrane contains a primary cilium that protrudes into the ventricular lumen. If both daughter cells inherit the cilium, they will stay at the luminal surface. A daughter cell without cilium will delaminate and move towards the basal (i.e., pial surface). For the astrocyte-like B cells of the adult SVZ, the presence of a cilium has been convincingly demonstrated (Mirzadeh et al., 2008), suggesting that a similar mechanism might be in place there. For the postnatal and adult hippocampus, the presence of a cilium on the radial glia–like type-1 precursor cells has been demonstrated, and a link between the presence of this cilium and Shh-dependent signaling has been found (Breunig et al., 2008). In both cases the role in asymmetrical vs. symmetrical divisions remains to be shown.

The Immortal Strand Hypothesis

Are DNA strands differentially
segregated during asymmetrical
divisions?

Due to their "indefinite" self-renewal, stem cells could be particularly prone to accumulating mutations and thereby cause harm to the organism. In 1975, John Cairns proposed the "immortal strand hypothesis" as a mechanism by which stem cells would avoid accumulating mutations acquired during mitosis (Cairns, 1975). His idea was that during asymmetrical divisions, the daughter cell that remained a stem cell would always get the original DNA strand, whereas the differentiating daughter cell would receive the copy. This elegant idea is actually backed by relatively little evidence but has stimulated debate ever since (Lansdorp, 2007; Rando, 2007). Arguments against the hypothesis include that replication is far from the only, and certainly not the most relevant, source of mutations, and that ongoing strand repair causes exchanges between sister chromatids, which would counteract the immortal strand mechanism (Lansdorp, 2007). Due to technical limitations, no experiment presented in support of the immortal strand hypothesis has convinced onlookers. Intriguingly, none of the existing studies has dealt with germline cells or ES cells, and there is no evidence of non-random sister chromatid segregation in these cells. This raises the question of why there should be more protection needed in somatic stem cells than in stem cells of early development. The asymmetrical segregation could also not be demonstrated for hematopoeitic stem cells (Kiel et al., 2007). But for other stem cells, including adult neural stem cells, an asymmetry has been described (Karpowicz et al., 2005; Conboy et al., 2007), although again technical caveats need to be taken into account.

If the immortal strand hypothesis were true, an as-yet-unknown mechanism would have to be in place that would allow the directed, non-random segregation of the chromatids. One idea is that two distinct spindle poles (as they have already been described for several species) would exist and somehow connect to two matching and again different centromeres (Lew et al., 2008). In fact, the two strands of DNA are replicated by different mechanisms: the leading strand apparatus and the lagging strand apparatus, in which the lagging-strand mechanism is way more complicated and hence might attract more and different epigenetic marks than the simpler sister strand. But even without assuming that this mechanism would allow the cells to avoid collecting mutations, the epigenetic differences between sister chromatids might play an important role in the asymmetrical divisions in the first place (Lansdorp, 2007). The two sister chromatids would differ in their epigenetic status, thereby determining which genes would be expressed in the two different daughter cells after completion of the asymmetrical division.

Expansion Through Precursor Cell Proliferation

Many morphogens also double as mitogens that regulate the expansion phase

In the course of brain development, the morphogens stage, responsible for positional identity, is followed by the mitogen phase (Fig. 4–5). Positioned precursor cell populations are adequately expanded by means of the described coordination of symmetrical and asymmetrical divisions. Cell-intrinsic as well as cell-extrinsic mechanisms have to interact to bring about this pattern. Interestingly, many of the morphogens also serve as inducers of proliferation. This additional function might be a consequence of positional identity with its specific patterns of gene expression. Retinoic acid, for example, is required to maintain sufficient levels of FGF-8 and Shh expression. FGF-8 in turn appears to maintain proliferative activity. In the dorsal–ventral gradient of Wnt signaling, Wnt causes precursor cell self-renewal near the roof plate (the source of Wnt) but induces exit from the cell cycle at greater distances (Megason and McMahon, 2002). Consequently, the Wnt gradient is inversely related to neuronal

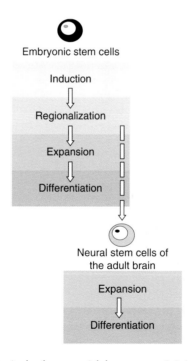

FIGURE 4–5 Sequence of events in brain development. Adult neurogenesis is in continuity with embryonic brain development. The neurogenic zones of the adult brain are formed during brain development (see also Fig. 3–7). Adult neurogenesis recapitulates only parts of embryonic and fetal neurogenesis. The precursor cells for the two neurogenic regions are differently regionalized. This explains the many differences that exist between neurogenesis in the adult hippocampus and olfactory bulb.

differentiation, which begins ventrally. In adult neurogenesis, Wnt is involved in maintaining precursor cell activity (Lie et al., 2005). As a final example, Emx2 and Pax6, both involved in patterning of the cortex, are also necessary to maintain a sufficient level of cell proliferation in the cortical germinative zones.

Control of cell cycle length and exit are key switches for neurogenesis

In the presence of the appropriate growth factors, the precursor cells follow a cell autonomous program that is strictly timed. Intrinsic control of proliferation relies on the regulation of cell cycle in a two-fold way: first with respect to entry into the cell cycle or progression through the cell cycle, second with respect to exit from cell cycle. Neither function is independent (Nguyen et al., 2006; Farkas and Huttner, 2008). P27(kip1) is one example of a molecule that might exert influence on both proliferation itself and differentiation (Nguyen et al., 2006). Consistent data on a role of p27(kip1) have also been obtained for neurogenesis in the adult SVZ (Li et al., 2009b). A second example is BM88/Cend1, which links cell cycle exit with neuronal differentiation (Politis et al., 2007). In the adult SVZ, too, BM88/Cend1 is expressed in progenitor cells undergoing their terminal neuronal division (Koutmani et al., 2004).

How precise control of cell cycle-related events is also becomes obvious with respect to the influence of cell cycle length on neurogenesis. With progression from apical to basal and from symmetrical to asymmetrical to symmetrical divisions, the cell cycle lengthens. Intriguingly, it was found that slowing down the cell cycle extrinsically stimulated neurogenesis by increasing the number of neurogenic divisions (Calegari et al., 2005). Conversely, speeding up the cell cycle in the precursor cells reduced the proliferation of basal progenitor cells and the generation of neurons (Lange et al., 2009).

The expansion phase is tightly controlled

Proliferative activity is an intrinsic property of the precursor cells but is not independent of cell-extrinsic factors. The niche concept (see Chap. 3, p. 88, and Chap. 8, p. 190) aims to explain the microenvironmental conditions governing precursor cell activity in a tissue. Although in embryonic brain development neurogenesis is massively parallel (whereas it is spread out over much larger time spans and is locally much more limited in the adult), expansion is far from being disorganized and irregular. Even though a large surplus of neurons is generated, the entire process is tightly controlled. In the end, brain size is a trait that is highly dependent on inheritance. The gigantic differences in brain size between vertebrate species, all essentially dependent on very similar mechanisms of brain development, speak to the fact. How exactly this determination is achieved is still largely unknown. In mice, neuron number maps to a locus on Chromosome 11 containing the genes for the receptors for retinoic acid, thyroxine, and neuregulin (Williams et al., 1998). Far from being conclusive at present, these data still suggest the existence of master regulators of brain development that could underlie the striking natural variation.

Growth factors are cell-extrinsic pro-proliferative signals

Expansion of precursor cells is dependent on growth factors. EGF (epidermal growth factor) is the strongest of the mitogens involved. Mice lacking the EGF receptor B show dysgenesis of the cortex. This malformation might not be due only to the mitogenic effects of EGF, though. EGF also has effects on many other aspects of development as well as on the maturation of radial glia (Patten et al., 2003; Schmid et al., 2003), so its influence is complex. EGF acts through a family of four receptors, which act in various combinations (Fox and Kornblum, 2005). EGF is ubiquitously present in the brain and produced in many cell types, including neurons. EGF receptors are expressed on precursor cells during brain development (Kornblum et al., 2000). In the adult SVZ, the intermediate progenitor cells (C cells) can be identified by their expression of the EGF receptor (Ciccolini et al., 2005). The EGF receptor belongs to the proteins that are differentially distributed during asymmetrical division (Sun et al., 2005b). Ex vivo, the sensitivity of neural precursor cells to EGF is used to stimulate proliferation in culture.

The second growth factor of particular relevance is FGF2 (fibroblast growth factor 2). Without FGF2, precursor-cell proliferation decreases (Raballo et al., 2000), and the mutant mice have a much reduced number of cortical neurons and astrocytes (Vaccarino et al., 1999; Chen et al., 2008). FGF2 is also used to grow neural precursor cells ex vivo.

Basic Helix-Loop-Helix Factors and the Induction of Neuronal Differentiation

bHLH transcription factors control neuronal differentiation

Whereas differentiation of embryonic stem cells into neural precursor cells might be something like a default state, differentiation into neurons from later neural precursor cells requires active induction. Possibly all genes directly involved in this induction, called *proneural genes*, belong to the family of basic helix-loop-helix (bHLH) genes. There seems to be a core program of bHLH activity that is involved in neurogenesis, no matter when and where the given neurons develop (Kintner, 2002; Ross et al., 2003). This principle is highly relevant for adult neurogenesis, although our present knowledge of bHLH genes in adult neurogenesis is still relatively limited (see also Chap. 9). Unless otherwise noted, the following information relates to cortical development in rodents.

bHLH transcription factor effects are balanced to achieve tight control of development

The bHLH genes Id and Hes precede the activity of proneural genes and maintain precursor cells, whereas the proneural genes themselves, Ascl1/Mash1, Neurogenin1, and Neurogenin2, are primarily involved in the step from proliferation to neural determination, which is a reversible stage and different from terminal differentiation (Fig. 4–6). Hes1 and Hes5 actively inhibit neuronal differentiation and keep precursor cells in cell cycle (Ishibashi et al., 1995). So do the Id genes, albeit by different mechanisms, introducing redundancy to the system. Id factors lack the DNA binding motif. They rather sequester the E protein, a co-factor required for other bHLH factors to be active, thereby controlling their activity (Norton, 2000). Action of bHLH factors is such that they might physically interact at the binding site of the promoters, thereby creating a very direct but interdependent control.

Hes1 expression oscillates

Hes1 is an unusual gene in that it suppresses its own expression by binding to its own promoter (Takebayashi et al., 1994). Because Hes1 mRNA and protein are very unstable, this suppression leads to the absence of Hes1 activity in the cell. Consequently, after approximately two hours, the Hes1 promoter becomes de-repressed and Hes1 is expressed again. By

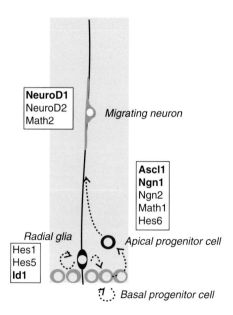

FIGURE 4–6 The family of Hes (and a few other) basic helix loop helix transcription factors play a prominent role at the beginning of neurogenesis in that they control precursor cell maintenance versus differentiation.

this mechanism Hes1 shows an oscillating expression in many cell types, including neural stem cells (Shimojo et al., 2008). It is assumed that these oscillations underlie the cell-autonomous timing of developmental processes. Dependent on Hes1 oscillations, other bHLH factors down the developmental line show oscillations as well (Shimojo et al., 2008).

bHLH transcription factors control both neuronal determination and terminal differentiation

Upon initiation of neuronal differentiation, Id and Hes genes are down-regulated and the proneural genes Ascl1/Mash1 and Neurogenin1 and Neurogenin2 come on. Activity of the proneural genes leads to a state of a predetermined neuroblast. The bHLH genes NeuroD1, NeuroD2, and Math2, in turn, are involved in inducing terminal neuronal differentiation (Farah et al., 2000). The activity of the latter is a consequence of the activity of the first, so that neurogenesis appears as tightly bound to the sequence of patterned bHLH activity. A cascade of different bHLH genes consecutively regulates neuronal determination and differentiation.

bHLH patterns define precursor cell identities

Ascl1/Mash1, NeuroD, Neurogenin 1 and 2, and Math 1 and 2 are all involved in determining positional identity and something like a generic neuronal identity. The expression of the various pro-neural bHLH factors is largely not overlapping, suggesting that they have further functions in initiating the specification of neuronal subtypes (Farah et al., 2000; Tomita et al., 2000; Mizuguchi et al., 2001; Nieto et al., 2001).

Proneuronal factors are anti-gliogenic

All of these proneural factors also prevent gliogenesis at this stage. Hes genes, in contrast, promote the production of astrocytes—at least in spinal cord and retina (Hojo et al., 2000; Wu et al., 2003). To make matters more complicated, however, one related molecule, Hes6, does the opposite and promotes neurogenesis while preventing gliogenesis (Jhas et al., 2006).

bHLH transcription factors also control adult neurogenesis

An important question is whether such a cascade of bHLH activity is relevant to adult neurogenesis and serves the same functions there. Id1, for example, identifies stem cells in both SVZ and SGZ (Nam and Benezra, 2009). Id2 is involved in determining the dopaminergic lineage among the periglomerular interneurons in the olfactory bulb (Havrda et al., 2008). Nevertheless, in the two neurogenic systems of the adult brain, bHLH genes might have somewhat different functions than during development. A good example is Ascl1/Mash1, which primarily promotes the generation of interneurons in the SVZ but of oligodendrocytes in the SGZ (Jessberger et al., 2008a). Nevertheless, there is a Ascl1/Mash1-dependent line of oligodendrocyte differentiation in the SVZ as well. bHLH transcription factor Olig2, in contrast, is not present in the hippocampus but specifies the development of oligodendrocytes and astrocytes at the expense of neuronal differentiation in the SVZ (Marshall et al., 2005; Ono et al., 2008). Transcription factor Dlx suppresses Olig2, and is necessary for the interneuron lineage (Petryniak et al., 2007). In the adult SVZ, the Olig2-positive lineage contributes to myelinating and non-myelinating NG2 cells and oligodendrocytes in the striatum, the corpus callosum, and the cortex (Menn et al., 2006). Both neuronal and glial lineage thus originate from the radial glial cells of the adult SVZ, and bHLH factors control the fate choice.

Tlx is not a transcription factor but a cell-autonomous determinant of precursor-cell maintenance versus differentiation

Other genes play a role in determining the choice between proliferation and differentiation as well. Tailless (Tlx) is an orphan receptor (i.e., a receptor for which no ligand is known) that is exclusively expressed in the forebrain. Tlx expression prevents premature maturation in the precursor cells, especially in the subventricular zone (Roy et al., 2004). Tlx keeps precursor cells cycling; down-regulation leads to cell-cycle exit and precocious differentiation (Li et al., 2008). Null-mutants for *Tlx*, however, have no severe developmental phenotype, suggesting redundant mechanisms (Shi et al., 2004). In adult precursor cells, however, this redundancy appears to be lost and the null-mutation of the gene results in a severe impairment of neurogenesis in both neurogenic regions of the adult brain (Shi et al., 2004; Liu et al., 2008). Tlx is expressed in the radial glia–like stem cells (type-1 and B cells, respectively).

Paracrine Signaling: Notch, Shh, BMP, and Wnt Pathways

Short-range signaling induces
cell diversity

One key property of development is that cells must become different from each other. Asymmetrical divisions are the quintessential example of a cell-autonomous mechanism to achieve diversity. The second key mechanism involves the action of short-range secreted molecules, so-called paracrine factors, which signal between neighboring cells. Key examples of factors in this category are Notch, Shh, BMP, and Wnt (Fig. 4–7). If the factors act back on the cell they were released from, the mechanism is called *autocrine*. Paracrine signaling that leads to cellular diversity is based on a principle named *lateral inhibition* (Beatus and Lendahl, 1998). Lateral inhibition literally means that a cell inhibits other cells at its sides.

The model mechanism for
inducing differences is "lateral
inhibition" through Notch
signaling

The prime example of such molecule is Notch, a receptor molecule that makes cells maintain an epidermal fate. Among the main target genes of Notch signaling are the Hes genes, and the classic example for "lateral inhibition" is a function of Notch. When Notch is knocked out in *Drosophila*, essentially all ectodermal precursor cells become neuroblasts. *Achaete-scute* induces the expression of Delta, one of the ligands to Notch. Even within a group of neighboring identical ectodermal cells, the expression levels of individual proteins, including Delta, varies slightly in a stochastic manner. An increased level of Delta in one cell, call it *A*, will lead to increased binding of Delta to Notch on the neighboring cell, call it *B*, and the Notch activation will down-regulate *achaete-scute* and consequently Delta in *B*. Consequently, the neighboring cells *A* and *B* are developmentally driven apart. Lowered Delta expression in *B* will mean less Notch activation on *A*, further supporting neuroblast differentiation of *A*. Vertebrates have *achaete-scute* homologues such as Mash1 and have at least four Notch genes plus four Notch ligands (two Delta genes and two *jagged* genes, which are homologues to the second *Drosophila* Notch-ligand besides Delta, serrate). Unfortunately, in vertebrates the intuitive system of lateral inhibition is not found in the same form and clarity, but the principle seems to hold. Newborn neurons that migrate away from the ventricular and subventricular zone of the developing neocortex up-regulate Delta, and Delta activates Hes5 activity via the Notch receptor on neighboring progenitor cells. Notch-mediated Hes5 activity is a property of defined subpopulation of neural precursor cells that are self-renewing and multipotent ex vivo but do not co-express the marker of basal progenitor cells, Tbr2, suggesting that the mechanism primarily applies to apical progenitor cells (Basak and Taylor, 2007). Appropriately, Notch and the EGF receptor collaborate in neural precursor cells (Campos et al., 2006). Hes activity directly suppresses proneural bHLH genes in the progenitor cells (Giagtzoglou et al., 2003) and maintains them in their precursor cell until the new neuron is far enough away to make place for new differentiating progeny. Notch also favors the development of astrocytes from neural precursor cells (Tanigaki et al., 2001).

Expression patterns of BMP
receptors determine BMP effects
on proliferation versus
differentiation

Other locally secreted factors are also involved in inducing the shift from proliferation to differentiation. Again, some of the usual suspects reappear with different functions. BMP activity, which played an important role during regionalization of the developing nervous system, for example, induces, not only proliferation via activation of the BMP-1A receptor, but also expression of another receptor, the BMP-1B receptor, which has an entirely different function. Activation of BMP-1B up-regulates p21^{kip1}, an inhibitor of the cell cycle (Panchision et al., 2001). After dorsalized cells have been primed by BMP-1A activation, BMP-1B signaling leads to neuronal differentiation. Consequently, there will be a point in development at which there is competition between the activity of the two receptors, and the balance can shift either way. There are other examples of such competitive actions, and thus more or less stochastic mechanisms might be involved in managing transitions between two developmental stages.

BMPs maintain adult SVZ
precursor cells

BMP also counteracts neurogenesis in the adult SVZ and is antagonized by locally secreted noggin to achieve neurogenic permissiveness (Lim et al., 2000). Downstream of BMP/Noggin is a factor named Smad4. Deletion of Smad4 in adult SVZ stem (but not progenitor cells) resulted in more

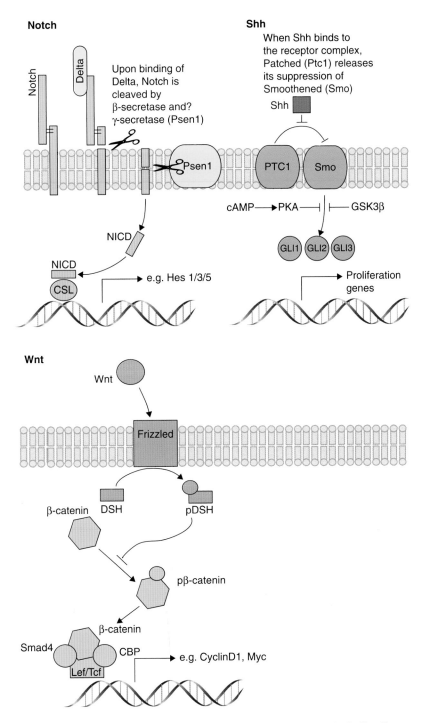

FIGURE 4–7 Paracrine signaling molecules control neuronal development at the level of cell–cell communication. Here, the intracellular cascades for Notch, Shh and Wnt signaling are depicted (in a simplified way). Notch is particular in that part of the receptor is cleaved under ligand binding and translocates to the nucleus. Wnt signaling is particularly pleiotropic due to the fact that its main transcription factor β-catenin accepts numerous co-factors, which allows integration of numerous signals. Compare also with Figure 9–7.

Olig2-positive progeny to be generated, so that a greater number of cells migrated into the corpus callosum and into the cortex to become oligodendrocytes (Colak et al., 2008). In the dentate gyrus, BMP function might be different. Here, BMP is expressed by precursor cells and noggin maintains and expands the pool of radial glia–like type-1 cells (Bonaguidi et al., 2008).

Combinatorial use of a limited number of factors allows a wide range of effects

In the tissues of the dorsal brain that expand late, such as the cerebellum and the cortex, Shh acts as a mitogen counteracting BMP signals that induce differentiation (Rios et al., 2004). In this case, the mediator is Smad5. Again, a balance of two antagonistic factors, here two ligands at different receptors, in contrast to the situation with one ligand at two receptors discussed above, controls expansion of cell numbers and exit from the cell cycle. The bottom line of this spectrum of observations is that brain development in all its complexity makes combinatorial use of a rather limited set of tools (Aboitiz and Montiel, 2007). The same signaling pathways are reused in different contexts, and by shifts in their interaction, novel determinants are created. Adult neurogenesis is again a new context. It does not come as a surprise to find the same players again. Depending on the particular context, however, the specific functions of individual factors might differ. Observations from embryonic brain development cannot necessarily be extrapolated to adult neurogenesis in a straightforward way.

Wnt is the quintessential pleiotropic regulator of development

Wnt signaling might be the best example of a pleiotropic regulatory system, with manifold functions across brain development and with great importance for adult neurogenesis as well. Wnt signaling is involved in processes as diverse as the self-renewal and maintenance of neural precursor cells (Kalani et al., 2008) (as well as in many other systems), proliferation (mostly up-regulatory), regionalization (dorsalization), fate choice (generally pro-neuronal), determining neuronal polarity (Nishimura et al., 2004), dendrite formation, axon guidance and remodeling, and synapse formation. Wnt also plays a major role in the fate decision of neural crest stem cells (Barembaum and Bronner-Fraser, 2005).

Wnt proteins, of which about 19 are known in vertebrates, are highly conserved. They are secreted molecules binding to the Frizzled receptor and acting through at least three different signaling pathways: the canonical pathway, which involves β-catenin, the Frizzled/planar cell polarity pathway (Strutt, 2003), and the Ca^{2+} pathway (Kuhl et al., 2000). Wnt signaling is modulated by numerous other factors both at the receptor level and beyond.

Factors like Wnt can link between many different developmental mechanisms

This plethora of functions, the three different signaling pathways, and the myriad interaction partners indicate that Wnt is not so much involved in controlling single cellular events as in orchestrating development. This is not to say that the loss of particular Wnt functions might not result in specific deficits. Mutation of Wnt3a, for example, leads to loss of the dentate gyrus with other aspects of brain development left undisturbed (Lee et al., 2000; Roelink, 2000). Similarly, mutating the downstream targets of Wnt signaling, transcription factors LEF1/TCF, again prevents the formation of the dentate gyrus (Galceran et al., 2000). Wnt family members are also expressed in the SVZ (Morris et al., 2007), and Wnt3a is found in the neurogenic niche of the adult hippocampus, where it positively regulates adult neurogenesis on the precursor cell level (Lie et al., 2005). But this apparent specificity still proves the point that the Wnt system as a whole links developmental events rather than being responsible for individual ones (see also Michaelidis and Lie, 2008, for an overview of the possible regulators, and Chap. 9, p. 373).

Principles of Cortical Layer Formation

Developing neurons grow into a regionalized identity

The neocortex is characterized by a highly ordered, six-layered structure parallel to the ventricular and outer surface, and a subdivision in cortical areas in the tangential dimension. The hippocampus is also a cortical region but belongs to the evolutionary older archicortex and has a simplified layered structure with only three strata. Neuronal function is specified within these spatial coordinates. Areas differ in their input and output pathways and in details of their network structure. Layers contain different neuronal populations. The two major classes are principal or projection neurons, the large pyramidal neurons of Layer II being the most obvious examples, and numerous types of interneurons. The cortical projection neurons are derived from the (dorsal) ventricular zone. Most interneurons, however, originate from the ganglionic eminences in the ventral ventricular zone and thus have to migrate in from a greater distance and cross the paths of radially migrating neurons from the VZ. All neurons have to find their place within the enfolding structure of the developing cortex (Figs. 4–8 and 4–9).

Positional identity is determined by the combination of gradients of secreted signaling factors

Molecularly, areas and layers are not defined by the expression of unique genes. Rather, positional information is derived from distinguishing expression patterns of many different genes, most notably transcription factors, which are not unique to any layer or region. The expression of the same gene can differ in different layers and different areas. Therefore, in the early VZ, where no layers yet exist, no strict molecular spatial pre-specification that would represent the later areas *and* layers is possible. Arealization and layer formation are intricately linked and interdependent over development. The details of how this ordered diversity is achieved on a molecular level are still largely unknown. The degree to which cortical specification of areas and layers occurs is dependent on intrinsic versus extrinsic influences, and is the subject of the debate between the "protomap" theory (which favors intrinsic regulation) and the "protocortex" theory (which favors dependence on extrinsic cues). As often in science, no extreme position can explain all aspects of a phenomenon, but how exactly cell-intrinsic and cell-extrinsic mechanisms interact remains to be unveiled.

The Boulder committee set the framework for the description of cortical development

In 1970, a meeting of eminent neuroembryologists in Boulder, Colorado, ended with a consensus description of the basic principles that govern the development of the cortex (see also). Although the data used to arrive at this consensus were largely obtained in studies of the human and monkey brain, it turned out that the proposed model was applicable to other vertebrate species as well. Over the years, it became the standard concept of

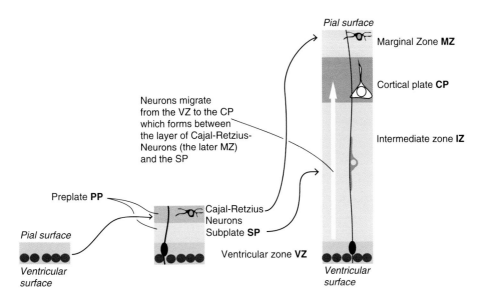

FIGURE 4–8 Layer formation during cortex development. The six-layered neocortex of the mammalian brain is built up from expanded progenitor cell pools in the ventricular and subventricular zones. (See text for details.) Cajal Retzius neurons are pioneer neurons in the outermost layer. Radial glia serves as a guidance structure for the positioning of new neurons. Adult hippocampal neurogenesis takes place in an archicortical region and reflects some of the characteristics of this pattern. These are absent in adult olfactory bulb neurogenesis.

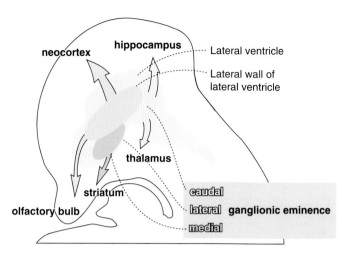

FIGURE 4–9 The ganglionic eminences. Hippocampal projection neurons, including granule cells, are derived from the dorsal subventricular zone, but hippocampal interneurons originate from the medial ganglionic eminence. Olfactory bulb interneurons are derived from the lateral ganglionic eminence. *Grosso modo* adult hippocampal neurogenesis is dorsal, whereas adult neurogenesis in the olfactory bulb is ventral.

brain development in terms of the formation of the structural organization, affecting ideas far beyond the cortex. Nevertheless, since 1970 a massive gain in knowledge, largely coming from research on the developing rodent brain and including many transgenic and knockout models, has made some adaptations of the description by the Boulder Committee necessary. In 2008, Pasko Rakic, who had already been centrally involved in the original description, suggested a few important changes to the original nomenclature (Bystron et al., 2008). The following description reflects these proposed modifications.

The ventricular zone (VZ) is the layer of the stem cells

The VZ contains the cell bodies of the neuroepithelial cells and the radial glia cells, which touch the ventricular surface. This contact is not loosened when during division the nuclei of these cells move away from the ventricular surface and back in a process called *interkinetic nuclear movement*. The first symmetrical divisions expand the VZ horizontally. At the time, when neurogenesis sets in, the precursor cells switch to an asymmetrical mode of division, generating intermediate, apical progenitor cells. The VZ also produces neurons, which will later be found in the deeper cortical layers.

Cajal-Retzius cells are pioneer neurons originating from the VZ

When cell proliferation in the VZ generates the first post-mitotic neurons, these accumulate on top of the VZ. The layer that forms, called the *preplate* (PP), contains a population of highly specialized neurons, the Cajal-Retzius neurons. These cells with characteristic horizontal processes secrete reelin, a factor necessary for building the glial scaffold along which the cortical neurons find their final positions, and for regulating neuronal migration itself (Forster et al., 2002). Except for the Cajal-Retzius neurons that remain near the pial surface of the brain throughout development, the neuronal populations that form first are those of the deep layers. Cortical maturation occurs in an inside-out fashion. The later cells have to pass those that were generated previously (Angevine and Sidman, 1961; Rakic, 1974).

Reelin signaling from Cajal-Retzius cells regulates the development of radial guidance structures for the following migration of neurons

Cajal-Retzius neurons express reelin, which is involved in regulating the migration of cortical neurons from the ventricular zone outward along the radial glial fibers (D'Arcangelo et al., 1995; Rice and Curran, 2001). In the absence of reelin, no radial glia scaffolding forms in the hippocampus (Forster et al., 2002) and cortex (Hartfuss et al., 2003), suggesting a close interdependency between Cajal-Retzius neurons and precursor cells. On the other hand, selective ablation of Cajal-Retzius neurons during neocortical development did not severely impair layering, suggesting the existence of a redundant mechanism.

Between VZ and PP a thickening layer of intermediate progenitor cells forms, the subventricular zone. These SVZ progenitor cells are basal progenitors that do not maintain contact to the ventricular surface any more, lack radial glia features, have down-regulated Pax6, and have up-regulated Tbr2 and Ngn2. They can show terminal neurogenic divisions, producing two neurons in asymmetrical divisions, which is preceded by Tis21 expression.

Neurons are generated from
both basal and apical progenitor
cells

At this stage of cortical development, neurons are thus generated from two cellular sources: from radial glia and apical intermediate progenitor cells in the VZ, and from the basal progenitor cells in the SVZ (Noctor et al., 2004). New neurons use the radial processes to migrate toward the pial surface and thereby pass those that were produced earlier (Rakic, 1974). During this process they show a complex pattern of movements, including an initial retrograde migration toward the ventricle, followed by a reversal of their bipolar shape and migration toward the pial surface (Noctor et al., 2004). The migrating cells accumulate within the PP and split it into an outer marginal zone (MZ), which contains the Cajal-Retzius neurons, and an inner deep subplate (SP). The new zone between MZ and SP is called the *cortical plate* (CP). The CP widens while more cells migrate in and over time form the cortical layers in an inside-out fashion. The MZ later transforms into Layer I; the VZ resolves in the ependymal layer and the underlying SVZ. The SVZ is called a *secondary germinative matrix*, because it is derived from the primary germinative zone of the VZ.

The adult SVZ is only partially a
remnant of the SVZ underlying
cortical development

What is called SVZ in the adult brain and in the context of adult neurogenesis is not a direct remnant of this SVZ of the developing cortex. Rather, neurogenesis in the adult SVZ largely corresponds to the much-less-well-studied neurogenic processes in the ventral telencephalon. Interneurons and oligodendrocytes have their origin in the ganglionic eminences, which are ventral structures (Fig. 4–9). The described pattern primarily applies to dorsal structures such as neocortex and hippocampus. Independent of this regionalization, the adult SVZ does not lie below a VZ anymore. Consequently, the older literature more appropriately speaks of the "subependymal" layer. Nevertheless, the term "SVZ" for this germinative zone in the adult brain stuck and is unlikely to be changed again. But adult SVZ and the SVZ of the Boulder Committee are not the same.

The subplate neurons form the
first long-distance connections,
which are activity-dependently
pruned

The SP below the CP is very homogenous across the entire cortex. The earliest-developing projection neurons within the SP form very extensive and excessive immature connections. Parallel to layer formation, this far-reaching connectivity is later trimmed back to the appropriate distribution patterns. Just as neurons are generated in excess and then selected for function, the same happens with neurites of the new neurons. These selection and trimming processes are activity-dependent. The major input to the neocortex arises from the dorsal thalamus and mediates the somatic, auditory, and visual sensory input. The sensory input provides the strongest structure-forming signals at a time when the sensory experience of the outer world is limited to the amniotic sac. Ex vivo data suggest that this input might already affect very early, proliferative stages of neurogenesis (Dehay et al., 2001).

Input from the thalamus
becomes critical for driving
further specification in the CP

However, thalamocortical input of the various modalities becomes instrumental in defining the formation of areas within the CP. Increasing specification of neurons within the CP is thereby tightly linked to the growing complexity of thalamocortical projections during development. The CP becomes regionalized in an input-dependent manner.

Layering factors like COUP-
TF1 are dependent on
thalamocortical input

Transcription factors have been identified that are linked to layer formation. Similar to the anterior–posterior gradient build by Emx2 and Pax6 expression, the orphan nuclear receptor COUP-TF1 (official gene symbol: Nr2f1) is expressed low anteriorly and high posteriorly. This gradient is probably dependent on the Emx2/Pax6 gradient, because in contrast to other factors, the distribution of Emx2 and Pax6 expression is normal in COUP-TF1 knockouts (Zhou et al., 2001). COUP-TF1 mutant mice lack

SP and Layer IV neurons (Zhou et al., 1999). Because normally COUP-TF1 is highly expressed in the dorsal thalamus, it might well be that these losses are due to a disturbance in thalamocortical input (Liu et al., 2000).

There are molecular mechanisms of arealization that are to some degree independent of thalamocortical input. Mice that lack the homeodomain gene Gbx2 or bHLH gene Ascl1/Mash1 fail to develop a normal prenatal thalamocortical input but have normal expression patterns for many genes in the hemispheres, a finding suggesting that these are independent of thalamocortical projections (Miyashita-Lin et al., 1999; Tuttle et al., 1999). In general, however, thalamocortical afferents are a key regulator of cortical development, as much as the development of proper thalamocortical input is dependent on preceding intrinsic molecular patterns.

Migration

Projection neurons migrate radially, interneurons migrate largely tangentially

Two distinct types of cell migration are involved in cortical development. Newborn neurons from the VZ and SVZ use the radial glial processes to migrate towards the pial surface until they have reached their target position. Cortical interneurons, in contrast, migrate in from the ventral brain and move tangentially to the radial fibers. This textbook dichotomy does not fully hold up against reality: cortical interneurons, too, move radially once they have reached their appropriate columnar location (Nadarajah and Parnavelas, 2002; Poluch and Juliano, 2007). The layered structure is primarily dependent on the coordinated migration of the cortical progenitor cells, and a number of distinct developmental disorders exist in which mutation of a particular gene leads to a characteristic defect in layer formation. Such defects, in turn, have given us evidence about the potential role of the underlying genes. Overall, however, the exact control of neuronal migration during development remains far from understood.

Both forms of migration are also found in adult neurogenesis, plus the characteristic chain migration between SVZ and olfactory bulb

In adult neurogenesis, both types of migration are found again. In addition, at least in rodents and monkeys, a characteristic third way of migration can be observed: progenitor cells from the adult SVZ destined to become interneurons in the olfactory bulb use a mode of locomotion called homotypic migration or "chain migration" to wander along the rostral migratory stream to the bulb, where they switch to a conventional radial migration (Fig. 5–5). Chain migration is covered in greater detail in Chapter 5, p. 168.

Most migration is dependent on (radial) guidance structures

In radial migration during cortical development, two sub-forms can be distinguished. The first precursor cells do not freely migrate through the tissue but find their final position by translocation of their soma. Later-stage progenitor cells, in contrast, move along the radial processes (Nadarajah and Parnavelas, 2002; Nadarajah et al., 2003). The interneurons migrating in tangentially from the ganglionic eminences require the radial scaffolds to reach their target as well (Poluch and Juliano, 2007).

Reelin is a key secreted factor for cortical layer formation and hippocampal neurogenesis

A number of genetic factors involved in controlling the migration that leads to cortical layer formation have been identified on the basis of mutations that lead to cortical malformations (Guerrini and Marini, 2006; Vaillend et al., 2008). Mutations of Filamin A or ADP-ribosylation factor guanine exchange factor 2 (ARFGEF2), for example, prevent the onset of migration, resulting in what is called *periventricular heterotopia*, in which neurons are displaced alongside the ventricular wall (Fox et al., 1998; Ferland et al., 2009). The Reeler phenotype, elicited by the lack or the mutation of Reelin, for example, is characterized by an inability of the migrating neurons to enter the PP. Reelin is often referred to as a stop signal to migrating neurons, but this is only partially true. The Reeler mutants show a disorganized cortex (D'Arcangelo et al., 1995), in that the cells accumulate beneath the PP, resulting in an inverted cortex. Mutation of the Reelin receptors, VLDLR (low-density lipoprotein receptor) and

ApoE receptor 2, as well as of Disabled 1, which is the downstream effector molecule of Reelin signaling, result in a similar phenotype (Hack et al., 2007). Such a phenotype is also found in the dentate gyrus (Drakew et al., 2002). In all of these, the formation of the radial glial scaffold, not only in the cortex but also the dentate gyrus, is impaired, suggesting that Reelin-related regulation is not only relevant at the level of the migrating cells (Weiss et al., 2003; Zhao et al., 2007). Adult neurogenesis is disturbed in mice with Reelin deficiency as well (Won et al., 2006). Again, the phenotype was related to the migration of precursor cell (Gong et al., 2007) as well as the link between gliogenesis and neurogenesis and the existence of the appropriate radial scaffold (Zhao et al., 2007). The role of Reelin in the hippocampus and its relevance for adult neurogenesis has received particular attention because of the fact that Reelin mutations are a candidate for the pathogenesis of schizophrenia (Reif et al., 2007; Toro and Deakin, 2007; Gregorio et al., 2009).

Migratory neurons characteristically express doublecortin

The mutation of Lis1 and of doublecortin (Dcx) result in "lissencephaly" or "smooth brain." In the case of Lis1, a four-layered cortex is found. Adult hippocampal neurogenesis was also disturbed in mice with Lis1 haploinsufficiency and the germinative matrix appeared abnormal (Wang and Baraban, 2008). Proliferation, migration, and differentiation were impaired. Granule cells were dispersed, presumably as a result of ectopic neurogenesis (Wang and Baraban, 2007). DCX is expressed by precursor cells and early post-mitotic neurons in the course of adult neurogenesis in both neurogenic regions. Given its wide use as a marker for intermediate stages of adult neurogenesis, comparatively little is known about the biology of DCX itself. But DCX is necessary for the translocation of the nucleus in the migrating neurons and for maintaining their bipolar morphology during migration (Koizumi et al., 2006). Knockdown of Dcx in the developing rat neocortex effectively blocked the radial migration of newborn cells (Bai et al., 2003). Dcx is discussed in more detail in Chapter 7, p. 229.

Mutations in six different glycosyltransferases (POMT1, POMT2, POMGnT1, Fukutin, FKRP and LARGE) all result in overshooting migration of newborn neurons, even as far as into the meningeal spaces. The resulting syndromes are known as *dystroglycanopathies* and all involve malfunction of a glycoprotein complex, which is also relevant for Duchenne muscular atrophy (Muntoni et al., 2008). As yet, no particular links to adult neurogenesis have been established for these genes.

Neurite Extension

Neuronal Polarity

Cell polarity is a hallmark of neurons

Neurons are polar structures with two types of neurites: the (usually) one axon and multiple dendrites differ structurally and functionally. Dendrites receive signals from other (presynaptic) neurons, and axons transmit signals to other (postsynaptic) neurons. Whereas the axon does not allow that much room for individuality, the dendritic tree of a given neuronal population is something like its morphological fingerprint. Primarily by virtue of its dendritic tree, every neuron has this distinct structural identity. To achieve neurite development, neurons must first establish polarity and then translate this into determining the origins of the axon and dendrites on the soma. They must physically extend the neurites in an organized way and in tune with neighboring cells.

First signs of polarity might appear with asymmetrical precursor cell divisions

The complex of signaling molecules Par3, Par6, and atypical protein kinase (aPKC), which is highly conserved throughout evolution, is involved in determining polarity in epithelial cells. To some degree, neurons inherit their polarity from precursor cells in asymmetrical divisions, and a signal transduction cascade involving the Par3-Par6-aPKC complex takes part in establishing early neuronal polarity. If the complex is blocked in neuronal cell cultures, no axon forms among the earliest undetermined neurites (Shi et al., 2003). Beyond this internal switch, changes in cytoskeleton assembly (Inagaki et al., 2001) as well as extracellular cues determine axon selection (Esch et al., 1999). However, in vivo, axonal growth might precede dendritic growth, so that a sequential instruction rather than a selection among undetermined neurites leads to axonal growth.

Despite their fundamental differences, axons and dendrites share many similarities, and many factors affect their development (Goldberg, 2003). In particular, some extracellular factors such as semaphorin 3A and slit affect both axon and dendrites. Slit and its receptors robo, for example, promote both dendrite and axon growth (Wang et al., 1999; Whitford et al., 2002). Semaphorin 3A, in contrast, attracts dendrites and repulses axons and thus has a differential effect requiring a different action on the receptor side (Song et al., 1998; Polleux et al., 2000). One hypothesis that attempts to make sense of the seemingly limited specificity of factors regulating axon and dendrite development is that neurons switch between a dendritic and axonal growth mode (Goldberg et al., 2002a). In addition, for example, the neurotrophins nerve growth factor (NGF), brain-derived neurotrophic factor, and NT3 influence dendritic growth, but different types of neurons respond with a different growth and branching pattern (McAllister et al., 1997; Goldberg et al., 2002b). Taken together, this implies that there are intrinsic differences both between different types of neurons and within a given neuron over time that allow specification of dendritic and axonal morphology.

Axonal Pathfinding

Neurites grow by sending out a highly plastic and sensitive structure called a "growth cone" that travels toward the target and trails behind the elongating neurite. Most neurons have only one axon, which can be extremely long; motor neurons in the lumbar spinal cord of humans are more than one meter in length. The growth cone does not have to cover this entire distance because the already correctly positioned axon extends with the growing organism.

Axonal growth cones have to choose a path to follow, and they have to decide the direction to go on this path. The pathways are defined by cell–cell interactions and extracellular matrix molecules (Hynes and Lander, 1992; Tessier-Lavigne and Goodman, 1996) and diffusible repulsive cues, to which the growth cones react. These can act as chemoattractants or repellants. Slit, for example, first discovered in *Drosophila*, is expressed in the midline and acts as a repulsant (Simpson et al., 2000a). The different receptors for slit—robo, robo2, and robo3—mediate this effect. Axons from different cells express different combinations of robo receptors. Ectopic expression of robo2 and robo3 causes medial axons to go laterally; deletion of robo and robo2 makes lateral axons divert medially (Rajagopalan et al., 2000; Simpson et al., 2000b). Slit chemorepulsion is also active in the adult SVZ (Koizumi et al., 2006). Slit continues to be expressed in the septum, and the robos are expressed on precursor cells. In mice lacking Slit, groups of newborn cells migrate caudally rather than into the rostral migratory stream and towards the olfactory bulb (Nguyen-Ba-Charvet et al., 2004). Intriguingly, Slit was also expressed by the precursor cells (A and C cells) themselves, and a knockdown in vitro resulted in impaired migration (Nguyen-Ba-Charvet et al., 2004). Semaphorin 3 also acts as a repellent of axonal growth and leads to a concentration of axonal projections. When Sema3a or its receptors are knocked out, axons tend to have wider, more diffuse target fields (Kitsukawa et al., 1997; Taniguchi et al., 1997). In the adult brain, Sema3a also modulates synaptic plasticity (Bouzioukh et al., 2006).

Directional information and polarity of axonal growth are mediated by long-range–acting diffusible cues, which can act as chemoattractants or repellents. Some axons (the commissural fibers) have to cross the midline to reach their target structure, which poses a particular problem because axons somehow have to "know" which side of the midline they are on in order to draw opposite conclusions from the cues (Tessier-Lavigne et al., 1988).

In spinal cord development, chemoattractants secreted by the floor plate direct commissural growth (Placzek et al., 1990). Extracellular matrix molecules called *Netrins* are candidate mediators of this effect (Kennedy et al., 1994; Shirasaki et al., 1995; Serafini et al., 1996; Shirasaki et al., 1996).

The floor plate (and presumably its equivalent in rostral brain parts) is also thought to secrete a repellent factor that prevents those axons that should not cross the midline from doing so (Colamarino and Tessier-Lavigne, 1995; Tamada et al., 1995). Interestingly, one of these cues might be identical to one of the chemoattractants. Netrin 1 seems to act as a chemorepellent on some axons (Varela-Echavarria et al., 1997). It thus seems that, depending on the type of neurons, growth cones can draw different, even opposite conclusions from the same available cues. Seizures, which strongly up-regulate adult hippocampal neurogenesis, are also associated with an up-regulation of Netrin 1 (Yang et al., 2008) and a displacement of migratory progenitor cells (type-3 cells) (Jessberger et al., 2005).

Dendrite Development

Dendrites acquire a complex morphology covering large territories

Because dendrites are so numerous and cover such large territories, dendritic growth is even more complex than axon development and much less understood. Different classes of neurons have the signature architectures of the dendritic trees, but within this framework dendritic development is highly individual. How is this balance between shared pattern and individual form achieved? How do the dendrites know where to branch? Some factors with other functions during neuronal development play a role in the shaping of dendrites—BMPs, for example (Lein et al., 1995), and the neurotrophins (McAllister et al., 1997) are involved in this step. The slit–robo system also takes part in shaping dendritic morphology (Wang et al., 1999; Whitford et al., 2002).

A number of interesting regulators of dendrite morphology have as yet only been described in *Drosophila,* although the respective genes are involved in mammalian neurogenesis as well. Mutation or overexpression of homeodomain transcription factor *cut,* for example, which also plays an earlier role in neuronal development, causes cell class–specific changes in dendritic morphology (Grueber et al., 2003). The mammalian homologue Cux1 is involved in producing Cajal-Retzius neurons during development (Nieto et al., 2004), whereas Cux2 is associated with the expansion of apical cortical progenitor cells (Cubelos et al., 2008) and interneuron specification (Cobos et al., 2006). Loss of both Cux1 and Cux2 leads to the absence of Reelin (Nieto et al., 2004).

Neurite Development in Adult Neurogenesis

Control of neurite growth is linked to other neurodevelopmental events like migration and synaptogenesis

Neuritogenesis cannot be neatly separated from other developmental processes, most notably migration and synaptic formation. The diverse functions of some classes of transcription factors reach well into this stage, and some new genes participate in this immensely complex part of development. Like essentially all other parts of neuronal development, neurite extension is based on the interaction of innate genetic programs and environmental cues. Neurite formation is activity-dependent, and some factors such as CPG15, whose expression responds to activity, take part in mediating this aspect of development (Nedivi et al., 1998).

Information about the regulation of neuritogenesis in adult neurogenesis is limited but is likely to involve factors already relevant during cortical development

At present, very limited specific information is available about the mechanisms underlying neurite extension during adult neurogenesis. Cell-specific overexpression of Cyclin-dependent kinase 5, Cdk5, in adult hippocampal precursor cells in vivo leads to an aberrant growth of basal dendrites, ectopic migration towards the hilus, and the establishment of incorrect connections onto hilar neurons (Jessberger et al., 2008b). Adult hippocampal neurogenesis produces excitatory granule cells with a large dendritic field in the molecular layer of the dentate gyrus and a relatively long axon along the mossy fiber tract to hippocampal region CA3.

Neurogenesis in the adult olfactory bulb, in contrast, produces two classes of interneurons with only local connections. The profoundly different patterns of connectivity of new neurons in the two neurogenic regions make it likely that various developmental mechanisms will make different contributions to the process.

Activity-Dependent Development of Neuronal Circuits

In utero, the brain generates activity that shapes connections

In the developing brain, two forms of activity shape the evolving structure. First, spontaneous activity is generated by the tissue itself in the absence of external stimuli (Katz and Shatz, 1996). This activity spreads through excitatory synapses but also through gap junctions. Second, postnatal and adult refinement of networks is heavily dependent on sensory stimuli and behavior and is mediated by synaptic activity as well as extrasynaptic mediators.

Postnatally, activity gains increasing influence on neuronal development

Developing neurons respond to electrical activity and fine-tune their development accordingly. Embryonic development proper—that is, development up to the existence of the brain anlage—appears to be relatively independent of electrical activity. Consequently, the overall idea is that embryonic and fetal brain development are largely controlled by genetic determinants, whereas postnatally, environmental influences and activity play an ever greater role. In the adult, precursor cells can directly respond to electrical activity, and such activity may be one of the factors determining the course of precursor cell proliferation and neuronal differentiation (Deisseroth et al., 2004, Babu, 2009 #5447). Adult neurogenesis is obviously still dependent on genetic conditions but at the same time is extremely sensitive to all kinds of environmental events and the behavior of the individual. In the adult, activity-dependent processes regulate neurogenesis on several stages of neuronal development.

Embryonic brain development is relatively more controlled by genetic programs and chemical gradients but not independent of activity

For the developing brain, in contrast, most data on activity-dependent regulation relate to post-mitotic stages and the formation of synapses and circuits. It was long thought that axon and dendrite elongation were independent of electrical activity and only guided by chemical cues. However, electrical activity is, for example, necessary for the establishment of thalamocortical projections (Catalano and Shatz, 1998). In vitro, electrical stimulation of *Xenopus* neurons leads to an increase in cAMP, which modulates the response of growth cones to guidance cues, from being repulsed to being attracted (Ming et al., 2001). In general, the response of axonal growth cones to electrical activity is cell type–dependent, making it difficult to generalize how activity influences neurite extension. Synapse formation and the establishment of neuronal networks, however, are clearly activity-dependent. Activity has influences even before synapses are present, and activity determines where synapses form. Thus the presence of spots of membrane specialization that allow the exchange of information between neurons precedes the existence of morphologically definable synapses. After the axon has reached the target zone, the filopodia of the arriving growth cone seek contact. Immediately before synapses are established, filopodia on both the future pre- and postsynaptic side become highly mobile. Their contact marks the initiation of synapse formation (Alsina et al., 2001). The presynaptic side goes well prepared into this encounter and contains preassembled packages of structural elements such as receptors that will be used in the building of the synapse (Washbourne et al., 2002). At least in hippocampal neurons, presynaptic development occurs before postsynaptic differentiation (Okabe et al., 2001). The filopodia on the dendritic side are soon replaced by spines, or thorn-like protrusions on the dendrite, on which the synapses sit. The spines first appear as immature protospines and over time develop into mature spines. Even in established networks, however, spines and synapses remain in some flux; in the brain, connections are not welded together. Maintenance is activity-dependent as well.

Network formation can be
thought of as being dependent
on learning

Intracellularly, calcium ions mediate activity-dependent effects by acting on various intracellular messenger systems. Calcium oscillations can be found in many developing neuronal circuits and affect fiber outgrowth and differentiation (Spitzer et al., 2000). *N*-methyl-d-aspartate (NMDA) receptors are involved in this response. When NMDA receptors were blocked in *Xenopus*, dendritic branching was inhibited (Rajan et al., 1999). There is thus a link between axon and dendrite formation and synapse development. Young synapses might be silent until they are depolarized for the first time; their maturation depends on NMDA receptor activation together with strong and concurrent afferent inputs. This situation is similar if not identical to that of long-term potentiation (LTP), the electrophysiological and structural substrate of learning. In fact, one can imagine the entire process of network formation as "learning." Activity within the network determines which connections are maintained and which are eliminated. Many more synapses are initiated than will be maintained. Neurons that fail to make useful connections in this process are removed.

During network formation, the building and dismantling of synapses is a very fast process that takes only a few hours for a single synapse (Alsina et al., 2001). After a synapse is removed, the branch of the axon will retract. In this sense, the activity-dependent shaping of neuronal networks is a highly dynamic process that occurs on the level of single axons and dendrites by means of a rapid fine-tuning of the synaptic connections (Antonini and Stryker, 1993; Cline, 2001; Trachtenberg and Stryker, 2001).

However, activity-dependent effects go beyond the individual synapse. Long-term potentiation is quite specific and might only spread to adjacent synapses. However, LTP might also have effects on local filopodia motility and dendritic sprouting, resulting in new spines and possibly new synapses (Engert and Bonhoeffer, 1999). In acute hippocampal slice models, repetitive prolonged electrical stimulation not only induced short-lasting early-phase LTP but also long-lasting late LTP and synaptogenesis (Tominaga-Yoshino et al., 2008).

Synaptic activity is associated
with the expression of many
plasticity genes

Electrical activity triggers the expression of many plasticity-related genes (Sun et al., 2005a; Ploski et al., 2006), including many factors involved in neurotrophic signaling (Altar et al., 2004). Calcium influx increases cAMP levels, which induce the phosphorylation of cAMP response element binding protein (Creb). Creb is a transcription factor that is involved not only in mediating the structural changes underlying synapse maturation but also in many other intracellular correlates of plasticity. In adult neurogenesis, for example, Creb phosphorylation is associated with neuronal differentiation (Giachino et al., 2005). Preventing Creb activation effectively blocked reactive adult hippocampal neurogenesis in a stroke model, which up-regulates Creb phosphorylation (Zhu et al., 2004).

The activity-dependent up-regulation of CPG15 discussed above underscores the relationship between neurite and synapse formation. Prominent among the genes induced by electrical activity are the neurotrophins, which are produced and secreted in response to electrical activity (Schinder and Poo, 2000). Neurotrophins have many effects on synapse formation, maturation, and maintenance (Poo, 2001) as well as on neurite and spine stability (McAllister et al., 1997). For example, BDNF can be released from axons (Hartmann et al., 2001) and affects dendrites (Aakalu et al., 2001). At least in the retina, BDNF is also a potent modulator of axonal development, indicative of pleiotropic and complex functions of the different signaling mechanisms (Lom et al., 2002; Hu et al., 2005).

In co-cultures of adult hippocampal precursor cells with primary hippocampal neurons, excitation of the neurons in an LTP-like fashion resulted in calcium oscillations, which in an NMDA-dependent fashion evoked the activity-dependent release of BDNF from the neurons that in turn triggered neuronal development from the precursor cells (Babu et al., 2009).

Neurotrophic Factors Promote Cell Survival or Death

The brain is built by selecting needed cells from a surplus

In embryonic and adult neurogenesis, a wave of cell death occurs shortly after the new neurons have begun to make functional connections. The usefulness of the connection they make appears to be the selection principle. The neuronal network is not built up neuron by neuron; instead, a vast number of neurons are generated, of which only a small functional subset survives. It has been estimated that five times as many projection neurons are generated than will survive (Oppenheim, 1991). This subtractive method is somewhat similar to that of a sculptor who chisels Venus of Milo out of a block of marble. The central difference is that in the brain the sculptor is the sculpture itself; the network liberates itself from excess parts in an activity-dependent way. The first victims are synapses and branches of neurites, but if activity in a given cell falls below a particular threshold, the cell is eliminated. This occurs via programmed cell death, or apoptosis. *Development* is thus inseparably interwoven with cell death. Paradoxically, apoptosis is a principle of maturation. Activity-dependent fate is determined by secreted factors that promote either survival or cell death.

Neuronal survival is dependent on trophic support

The neurotrophic hypothesis developed by Rita Levi-Montalcini states that neuronal survival is dependent on trophic support from the target zones. In a classic experiment, she showed that administration of antibodies against NGF when the developing neurons innervated their targets would lead to the death of the neurons, whereas exogenous NGF would rescue neurons that otherwise would have died (Levi-Montalcini, 1987). When NGF or its receptor TrkA is lacking, cell survival similarly decreases (Lewin and Barde, 1996). These initial findings have been extended to the large family of neurotrophic factors, including the neurotrophins proper (NGF, BDNF, NT3, NT4), neurotrophic cytokines (CNTF, LIF-1, interleukin-6, and others), and the families of glia-derived neurotrophic factor (GDNF)- and hepatocyte growth factor (HGF)-related proteins (Davis and Murphey, 1994; Lewin and Barde, 1996; Airaksinen and Saarma, 2002). Different factors act differently on different populations of neurons, and it must be noted that the majority of publications on neurotrophins deals with sensory neurons. How well these principles transfer to other systems is not always known. Quite generally, though, neurons read and integrate over a spectrum of neurotrophic signals. Among other parameters, this pattern depends on the distribution of the different receptors that mediate neurotrophic factor action. Receptor expression not only differs between different cells but also over time in the same cell (Buchman and Davies, 1993; Ninkina et al., 1996).

A key neuronal survival factor is BDNF

Neurotrophic factors have many functions during neuronal development, but only relatively late neurotrophic signals become a matter of life and death. The acquired responsiveness to BDNF, for example, is mirrored in the delayed expression of the BDNF receptor Trk-B (Vogel and Davies, 1991; Robinson et al., 1996). BDNF is also transported in anterograde form. It is not only involved in stimulating dendritic growth but also affects the survival of the post-mitotic neuron (Altar et al., 1997; Caleo et al., 2000). An activity-dependent autocrine signaling by neurotrophins has been shown for hippocampal pyramidal cells in vitro (Boukhaddaoui et al., 2001). In some populations of sensory neurons, the preference of different growth factors can change over time (Buchman and Davies, 1993; Enokido et al., 1999), and sometimes extracellular signals are required to mediate this switch (Paul and Davies, 1995). It is likely that similar activity-dependent switches (albeit perhaps very subtle) are more the rule than the exception and are found in other neuronal cell types as well. The means by which sequential expression of receptors is regulated is not yet known. Some brain regions are patterned independently of complete input signals, which implies that there must be an intrinsic program that makes them independent of external neurotrophic cues (Lopez-Mascaraque and de Castro, 2002).

Neurotrophins bind, not only to the Trk receptors, but also to the p75 receptor, which, among many other functions (which include both pro-proliferative and survival-promoting effects), can also induce cell death via an intracellular cell death domain (Coulson et al., 2004). The co-activation of the antagonistically acting receptors is modulated by context and activity. When Trk receptors are absent in sensory neurons, the pro-apoptotic action of p75 activation tips the balance toward cell death (Barrett and Bartlett, 1994). Active Trk-A, in contrast, stimulates signaling along the PI3K/Akt pathway, which suppresses apoptotic signals mediated by p75. P53, a central molecule in the induction of cell death, is inhibited and the expression of trk-A itself is increased (Kaplan and Miller, 2000).

Numerous neurotrophins have been investigated in the context of adult neurogenesis (Hagg, 2009). They are discussed in greater detail in Chapter 9, p. 386. The best-studied neurotrophin in adult neurogenesis is again BDNF, which was found to promote neurogenesis in the SVZ and olfactory bulb in vivo (Pencea et al., 2001). Also, BDNF heterozygous null mutants had reduced adult hippocampal neurogenesis (Lee et al., 2002). Because BDNF expression and secretion from neurons are regulated in an activity-dependent way (Kuczewski et al., 2009), BDNF is a strong candidate for mediating pro-neurogenic stimuli from the surrounding network to the developing new neurons (Babu et al., 2009). On the other hand, in the adult SVZ both TrkB and BDNF (Galvao et al., 2008) as well as p75 (Bath et al., 2008) seemed unnecessary to maintaining neurogenesis. Nevertheless, a valine to methionine substitution at position 66 (Val66Met) impaired adult neurogenesis in the olfactory bulb (Bath et al., 2008), and ex vivo experiments revealed a deficit of neuronal migration (Chiaramello et al., 2007). In summary, the role of neurotrophins and BDNF in particular in adult neurogenesis require further investigation, but their general involvement is rather undisputed.

Development of the Neurogenic Zones of the Adult Brain

The Subventricular Zone

In mice, the SVZ first appears in the ventral brain around E11 (Sturrock and Smart, 1980). During embryonic development, an SVZ can be found in all ventricles, but not below the fourth ventricle. The dividing SVZ cells, the apical progenitor cells, do not show the characteristic interkinetic nuclear movements during the cell cycle described for the VZ (Takahashi et al., 1995). In mice, at E15 cell divisions in the SVZ have surpassed proliferation in the VZ and lead to a massive thickening of the SVZ. Around birth the VZ is no longer recognizable as a germinative layer (Bayer and Altman, 1991).

The embryonic SVZ can be divided into two distinct regions, each of which gives rise to specific neuronal populations (Figs. 3–7 and 4–9). As part of the dorsal forebrain, as described above, the neocortical SVZ gives rise to the layered cortex, including the principal neurons of the hippocampus; as part of the ventral forebrain, the ganglionic eminences produce interneurons, astrocytes, and oligodendrocytes in many brain regions in the ventral and dorsal brain (Pleasure et al., 2000b; He et al., 2001; Marshall and Goldman, 2002). Whereas in the dorsal brain, radial glial processes span from the ventricular to the pial surface, in the ganglionic eminences they only reach into the *anlage* of the basal ganglia (Edwards et al., 1990).

The term SVZ is here not used in the same strict sense as in the dorsal telencephalon. No Boulder Committee has yet proposed a consensus model for the ganglionic eminences.

The ganglionic eminences
correspond to the ventral SVZ
and produce interneurons and
oligodendrocytes

The "ganglionic eminences" got their name from the protrusions they form in the ventricular wall and can be divided into the medial ganglionic eminence (MGE), the lateral ganglionic eminence (LGE), and the caudal ganglionic eminence (CGE) (Nery et al., 2002). The three ganglionic eminences are distinguished by the expression of marker transcription factors (Fig. 4–9). The expression of homeobox genes Nkx2.1, Dlx1, and Dlx2 distinguishes the ganglionic eminences from the dorsal SVZ (Parnavelas, 2000; Wonders and Anderson, 2006). Most notably, Dlx identifies progeny from the ganglionic eminences but is not expressed in progeny of dorsal precursor cells. However, Dlx is not expressed in all ventral precursor cells, which suggests that Dlx-negative cells might give rise to Dlx-positive cells (He et al., 2001).

In some sense the adult SVZ is a
remnant of the ganglionic
eminences, interspersed,
however, with dorsal cells

Migration from the MGE precedes migration from the LGE, and at different times interneurons with different properties are generated in the ganglionic eminences (Butt et al., 2005). The MGE rather diffusely produces interneurons in the neocortex and hippocampus. It is not clear whether the pioneer interneurons, the Cajal-Retzius cells, which are found in the marginal zone of the cortex and hippocampus, are derived from the MGE (Anderson et al., 1999; Wichterle et al., 2001). The cells from the ganglionic eminences migrate tangential to the radially migrating cells in the neocortical SVZ and thus have to cross their path. The LGE might contribute to cortical interneurogenesis as well, but this scenario is not yet fully resolved (Tamamaki et al., 1997; Anderson et al., 1999). However, with some contributions from the MGE, the LGE produces many neurons of the striatum, including the DARPP32-positive spiny interneurons (Deacon et al., 1994), as well as neurons in many other regions, such as the septum and thalamus (Wichterle et al., 1999; Wichterle et al., 2001). The anterior part of the LGE produces interneurons in the olfactory bulb, the posterior part in the striatum (Stenman et al., 2003). Throughout adulthood, astrocyte-like stem cells (B cells) of the adult SVZ produce new interneurons for the olfactory bulb. Additionally, after ischemia, new striatal interneurons, including DARPP32-positive cells (which are of ventral origin as well), can be generated from SVZ precursor cells (Arvidsson et al., 2002). B cells of the adult SVZ that are Dlx negative are derived from Gsh2-expressing cells of the LGE but do not express transcription factor Er81, which is found in the anterior LGE, in migrating cells of the adult SVZ and in olfactory bulb interneurons (see also Fig. 4–9). Like the situation during embryonic development, it thus seems that Dlx-negative precursor cells generate Dlx-positive intermediates, which then give rise to adult-born interneurons. Consequently, all currently available data suggest that the system of adult neurogenesis in the adult olfactory bulb is of predominantly ventral origin. In contrast, embryonic dorsal SVZ precursor cells give rise to the later precursor cells of the adult SGZ and glutamatergic projection neurons including the dentate gyrus granule cells. There is at least one lineage in adult olfactory bulb neurogenesis that is of dorsal origin (Brill et al., 2009).

The caudal ganglionic eminence is characterized by the expression of COUP-TFII (Nr2f2) (Kanatani et al., 2008). As determined around E13.5 (in mice), interneurons migrate from the CGE in the caudal migratory stream primarily to the hippocampus and the most caudal telencephalon (Yozu et al., 2005; Miyoshi et al., 2010).

The lateral wall of the adult SVZ can be divided into two parts: a cell-rich spandril below the corpus callosum, and the ventricular wall adjacent to the striatum. It has been proposed that the latter are the remnants of the lateral ganglionic eminence (Brazel et al., 2003; Kohwi et al., 2007). Transplanting LGE precursor cells heterochronically into the adult SVZ produced the appropriate cell types (Kohwi et al., 2007). In this theory, the roof of the lateral ventricles would then represent the remnants of the dorsal SVZ. The presence of "dorsal" factors such as Pax6 in apparently ventral parts further blurs the definitions. One unconfirmed study has claimed that after ischemia and induction by growth factors EGF and FGF, new CA1 projection neurons arise in the hippocampus, presumably originating from a dorsal SVZ population (Nakatomi et al., 2002). The finding has not yet been replicated by others, however, and remains controversial. Under normal conditions, no migration from the adult dorsal SVZ into the hippocampus occurs (Kronenberg et al., 2007).

Migratory chains of precursor cells crisscross the SVZ towards the entry to the RMS

In some early publications, one finds a division of the SVZ of the postnatal brain into an anterior part (designated "aSVZ"; Luskin et al., 1997) as the source of the cells migrating into the rostral migratory stream to the olfactory bulb, and the dorsolateral part as a main source of glia. However, migrating SVZ cells forming a network of crisscrossing chains can be found throughout the lateral ventricular walls of the lateral ventricles in rodents (Doetsch et al., 1997), and thus the distinction of an "aSVZ" has not been adopted. Migrating precursor cells from many parts of the SVZ converging toward the entrance to the rostral migratory stream might give the impression of an increased neurogenic potential by resident precursor cells of this region. In the postnatal SVZ, Dlx identifies the transiently amplifying progenitor cells (C cells) and the migratory A cells (Doetsch et al., 2002). The expression pattern shows a large overlap with polysialated neural cell adhesion molecule (PSA-NCAM) and doublecortin. The surrounding cells are astrocyte-like cells, including the local radial glia–like precursor cells (B cells), classical astrocytes without precursor cell function, and ependyma. Consequently, in the adult SVZ, the Dlx-positive cells are interspersed with other cell types; anterior–ventrally they concentrate and form a core in the spandril leading into the rostral migratory stream. This pattern gives the impression of a different cellular composition at this location than that in the SVZ proper (Rothstein and Levison, 2002; Brazel et al., 2003).

The specific function of the medial SVZ is not yet known, but it contributes to neurogenesis in the olfactory bulb

The medial walls of the lateral ventricles show an SVZ as well, but their function in the developing and adult brain is not known. It might well be that they form a continuity with the SVZ in the lateral wall. Heterochronic transplantation of embryonic precursor cells from outside the lateral wall into the adult lateral wall still resulted in the production of new interneurons in the olfactory bulb (Kohwi et al., 2007).

Nevertheless, the adult SVZ is not homogenous, and cells from different locations within the SVZ produce different types of cells (Merkle et al., 2007; Lledo et al., 2008; and see Chap. 5, p. 175).

Development of the Olfactory Bulb

The olfactory bulb is the first relay station in the olfactory pathway

The olfactory bulb contains the first central neuron of the olfactory system, to which the olfactory receptor neurons of the olfactory epithelium in the nasal cavity project. The olfactory bulb shows a rather simple layered structure and is otherwise organized in glomeruli, consisting of the principal projecting neurons, mitral cells, tufted cells, and several types of interneurons within the glomeruli and in the periglomerular layer (Fig. 4–10).

The olfactory epithelium develops first

Development of the olfactory system begins with formation of the olfactory epithelium from the olfactory placode, and the olfactory bulb from the olfactory primordium in the rostral telencephalon. This specific part of the ventricular wall later becomes the (obliterated) olfactory ventricle. Even in the adult, the olfactory bulb contains neural precursor cells. It is not clear whether these cells are direct descendants of this primordium or represent the precursor cells migrating in from the adult SVZ. From the ventricular primordium, the projecting neurons, the mitral cells, and the tufted cells are generated. The interneurons are derived from the LGE (Stenman et al., 2003; Kohwi et al., 2007; Lledo et al., 2008).

Axons from the olfactory epithelium induce outgrowth from the ventricular wall

The first pioneer axons arriving from the olfactory epithelium (the peripheral receptor field in the olfactory system) reach the ventricular wall and induce an increased number of cells to exit from the cell cycle (Gong and Shipley, 1995). This causes a bulging of the olfactory bulb (De Carlos et al., 1995, 1996). Although induction by the pioneer axons seems

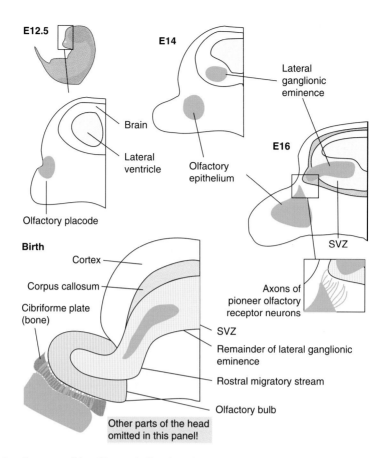

E12.5

E14

Brain

Lateral ventricle

Olfactory placode

Olfactory epithelium

Lateral ganglionic eminence

E16

SVZ

Birth

Cortex

Corpus callosum

Cibriforme plate (bone)

Axons of pioneer olfactory receptor neurons

SVZ

Remainder of lateral ganglionic eminence

Rostral migratory stream

Olfactory bulb

Other parts of the head omitted in this panel!

FIGURE 4–10 Development of the olfactory bulb. The olfactory epithelium is derived from the olfactory placode, whereas the olfactory bulb is of ventral subventricular zone origin. Early input from axons of pioneer olfactory receptor neurons influences development of the olfactory bulb.

important, even the Pax6 null-mutant, which lacks an olfactory epithelium and consequently olfactory axons, develops a rudimentary olfactory bulb (Jimenez et al., 2000).

Olfactory bulb development is only partly dependent on the olfactory epithelium

Like the hypotheses for cortical development, a *protomap* hypothesis (the patterning occurs before migration, and the migrating cells are heterogeneous and follow intrinsic control) and a *protocortex* hypothesis (the cells are homogeneous and are patterned according to specific local stimuli at later time points) have been formulated. As in the cortex, it seems that both hypotheses are true, yet neither is sufficient for a full explanation of olfactory bulb development. *Development* of the olfactory bulb is influenced by the olfactory epithelium but not entirely dependent on it. Particularly the late stages of neuronal maturation in the olfactory bulb might be influenced by input from the olfactory epithelium (Stout and Graziadei, 1980; Matsutani and Yamamoto, 2000). But olfactory bulb neurons can survive without input from the olfactory epithelium (see Chap. 5, p. 171).

The projection neurons in the olfactory bulb form before the interneurons migrate in

The earliest projection neurons form before the bulb itself becomes recognizable. *Development* of the olfactory bulb follows a rather rigid sequence (Hinds, 1968a, 1968b). Neurogenesis in the accessory olfactory bulb precedes that in the main bulb. The projection neurons arise first; the various types of interneurons are generated last. At the time the interneurons are migrating in, a morphologically recognizable olfactory bulb has formed. The interneurons have to migrate over increasing distances from the lateral ganglionic eminence (and later the adult SVZ) into the bulb. Their migratory route becomes ensheathed by glial cells and turns into the rostral migratory stream that persists into adulthood. Only neurogenesis of interneurons originates from precursor cells in the adult SVZ and can be found throughout life.

The olfactory bulb principal neurons project to the olfactory cortex, olfactory tubercle, anterior olfactory nucleus, piriform cortex, entorhinal cortex, and amygdala (reviewed in Lopez-Mascaraque and de Castro, 2002). The earliest cells project to the most posterior target structures, the later-born neurons to the more anterior nuclei (Derer et al., 1977; Bayer, 1983). The projections of the olfactory bulb are not topographic, because projection neurons from all over the bulb can project to the same neuronal populations in the olfactory cortex and amygdala (Scott et al., 1980).

Afferents from the olfactory epithelium originating from the same type of odor receptor converge on very few olfactory glomeruli (Treloar et al., 2002), which in turn project to clusters of target neurons in the olfactory cortex (Fig. 4–11). The pattern of glomeruli representing specific receptors is highly characteristic (Ressler et al., 1994; Vassar et al., 1994; Mombaerts et al., 1996). Thus, although the olfactory bulb is not a topographical map of the olfactory epithelium, it is, much more astonishingly, an anatomical map of the approximately 1,500 receptor types. This complex and specific patterning seems to be dependent on the olfactory receptor proteins themselves.

Efferents from the olfactory bulb target the olfactory cortex, the piriform cortex, the amygdala, and other regions

The olfactory bulb is organized into separate domains, the glomeruli, for the different receptor types

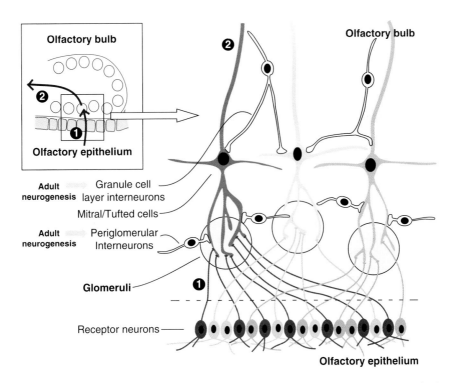

FIGURE 4–11 Principal network structure of the olfactory bulb. The olfactory receptor neurons are the first neurons of the olfactory tract. The mitral/tufted cells, located in the glomeruli of the olfactory bulb are the second neuron, which project to the olfactory tubercle and other structures of the rhinencephalon. Adult neurogenesis generates interneurons in the granule cell layer and in the periglomerular region. No mitral or tufted cells are produced in adulthood. The olfactory bulb is not topologically structured like the visual system. Rather, olfactory neurons of the same type project to the same glomeruli in the olfactory bulb.

The individual olfactory proteins are characteristic of the different receptor types. If one region of the receptor protein is genetically replaced by another, this causes a targeting of the projection to other glomeruli (Bozza et al., 2002).

Development of the olfactory bulb projections involves several candidate guidance mechanisms. PSA-NCAM as well as other NCAM forms are associated with development of the lateral olfactory tract, the main output structure of the olfactory bulb (Seki and Arai, 1991; Stoeckli and Landmesser, 1995). The Slit–Robo system is also prominently involved in the formation of the projections (Li et al., 1999). Interestingly, both systems also seem to play a role in guiding adult-generated neurons into the olfactory bulb—in the opposite direction from that of the axons during development.

Development of Hippocampal Dentate Gyrus and Subgranular Zone

The hippocampus is a special case of cortical development characterized by a displaced precursor cell population

The dentate gyrus is unusual in that a population of primary precursor cells becomes displaced from the ventricular wall and drives development throughout prenatal, postnatal, and adult neurogenesis.

The hippocampus is part of the archicortex. Despite the simple network structure of archicortex, with three layers instead of six, the sequence of gross development in the dentate gyrus is more complicated than in the rest of the hippocampus and in the neocortex. Whereas development of the CA fields follows the general pattern of other cortical regions, in the dentate gyrus, three germinative matrices follow each other. During embryonic development, the hippocampus becomes displaced by the much stronger growing cortical regions and is rolled into the characteristic shape that provoked its name, *hippocampus*, meaning "sea horse." The sequence of germinative centers distinguishes neurogenesis in the dentate gyrus from the olfactory system, where the germinative zone of the adult SVZ is at least locally the more or less direct successor of the matrix active during embryonic brain development, functionally representing a remnant of the ganglionic eminences. With respect to its ultimately ventricular origin, development of the dentate gyrus thus originates from an ectopic precursor cell pool, and so does adult neurogenesis.

Hippocampal development originates from a specialized region in the ventricular wall, the cortical hem

Like the other cortical regions, development of the hippocampus originates from the VZ in the dorsal forebrain. The hippocampus is part of the medial cortex. Patterning signals for this regions are provided by a small region called the "cortical hem," at the most dorsal tip of the telencephalon. This area is rich in the expression of BMPs (Furuta et al., 1997) and Wnt factors (Grove et al., 1998). The hippocampus originates from the region dorsal to the cortical hem. This region is characterized by the expression of Wnt receptor Frizzled (Kim et al., 2001). From E10 on, Wnt3a is the only Wnt whose expression is exclusive to the cortical hem. Wnt3a null mutants lack the hippocampus (Lee et al., 2000). In contrast, mutants for Lef1, which is a transcription factor in the Wnt signaling pathway, lack the dentate gyrus only (Galceran et al., 2000). With respect to their origin from the ventricular wall, neocortical and hippocampal development are not contiguous. Rather, a section of transitional cortex, which will later form the subiculum, is interspersed between neocortex and hippocampus (Fig. 4–12). In humans the massive growth of the neocortex displaces the hippocampus from its dorsal to a ventral position. In rodents the hippocampus remains dorsal.

Like in cortical development, first expansion occurs in a subventricular zone

Initially, the region from which the dentate gyrus will develop is not clearly demarcated, but by E14.5 in the mouse, cell proliferation in this zone decreases, while the neighboring regions that give rise to ammon's horn and the fimbrial structures maintain their proliferation. A subventricular proliferative zone, the secondary germinative matrix, builds up during this time. From this matrix, cells migrate to the future site of the dentate gyrus (Altman and Bayer, 1990a, b). Cells from the secondary matrix later form the outer shell of the dentate gyrus (Fig. 4–13).

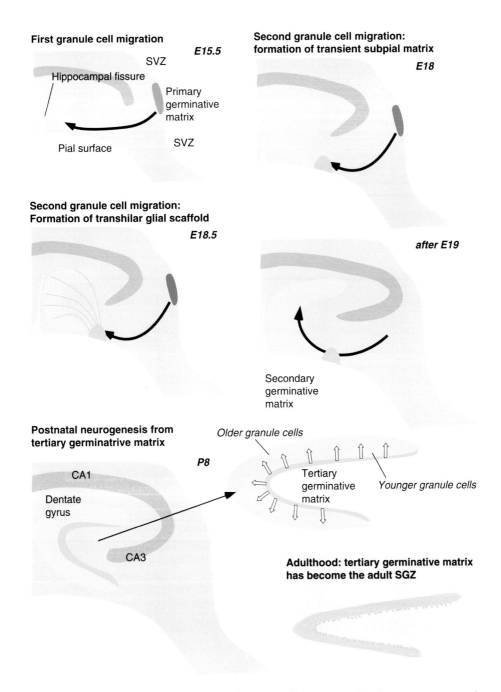

First granule cell migration

SVZ *E15.5*

Hippocampal fissure

Primary germinative matrix

Pial surface SVZ

Second granule cell migration: formation of transient subpial matrix

E18

Second granule cell migration: Formation of transhilar glial scaffold

E18.5

after E19

Secondary germinative matrix

Postnatal neurogenesis from tertiary germinatrive matrix

Older granule cells

P8

CA1

Tertiary germinative matrix

Younger granule cells

Dentate gyrus

CA3

Adulthood: tertiary germinative matrix has become the adult SGZ

FIGURE 4–12 Development of the dentate gyrus. In the course of dentate gyrus development, two waves of migration of precursor cells follow each other. From a primary germinative region in the wall of the lateral ventricle, these cells find a secondary matrix that gives way to a third matrix, which persists into adulthood as the precursor cell population in the subgranular zone. The granule cell layer is built in an outside–inside fashion. The first granule cell migration forms the outer shell; the second migration and the secondary and tertiary germinative matrices produce the inner layers. The second wave of migration establishes a "base camp" at the pial surface, from which trans-hilar radial fibers emerge, and the precursor cells translocate to the border between hilus and granule cell layer, the later "subgranular zone." This simplified compilation is based on works by Rickmann et al. (1987), Altman and Bayer (1990a, 1990b) and Li and Pleasure (2009).

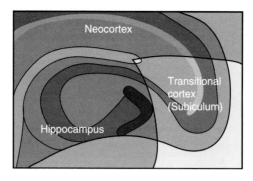

FIGURE 4–13 Neocortical and hippocampal development are not in continuity. There is a piece of "transitional cortex," the later subiculum, interspersed, which shows yet a distinct (but less well described) development.

Whereas the secondary matrix retained contact with the ventricular wall, the tertiary matrix forms at what will later become the hilus

Early postnatally, the secondary matrix is almost resolved, and a tertiary matrix that is most active during early postnatal development, from postnatal day 3 (P3) to P10, takes over. The tertiary matrix is derived from the secondary matrix, which splits around E18. One part of the secondary matrix in a first migration gives rise to the outer granule cells, the other to the tertiary matrix in what later will be referred to as the hilus, or plexiform layer, of the dentate gyrus. The tertiary matrix produces the inner shell of the granule cell layer and thereby the greater part of the total granule cell number. In the granule cell layer, the oldest granule cells are on the outside; the youngest are on the inside. The tertiary germinative matrix becomes increasingly confined to the SGZ, from which the new neurons of adult hippocampal neurogenesis are generated. At present it is not clear whether this process is only an increasing spatial confinement or represents a transition to a qualitatively distinct fourth germinative matrix. Hippocampal interneurons, like other interneurons in the cortex, originate from the ganglionic eminences (Pleasure et al., 2000b; Yozu et al., 2005).

Between secondary and tertiary matrix, a transient neurogenic matrix at the subpial surface forms

The transition from secondary to tertiary matrix is not yet fully understood, but much progress has been made. Before E18 an intermediate structure, something like a base camp, forms at the pial surface. From this transient matrix, a transhilar glial scaffold forms, and the precursor cells reach their position in the hilus and the later subgranular zone (Li et al., 2009a).

Radial glia guides development of the dentate gyrus

Dentate gyrus development is dependent on Cajal-Retzius neurons, reelin signaling, and radial glia, as in the neocortical regions. At least on the level of the tertiary matrix, but presumably also earlier, radial glia has a dual function in this process. Function of radial glia–like astrocytes as precursor cells has only been shown for adult neurogenesis (Seri et al., 2001; Seri et al., 2004), but the data from neocortical development suggest a similar function at earlier stages as well. In any case, the link between the two functions has not yet been fully elucidated, and thus it is not clear whether all radial glial cells do in fact serve both functions and whether all radial glial cells have the potential to do so. In the dentate anlage radial glia formation takes place around E13 and thus precedes granule cell development, and in all of the subsequent germinative matrices, radial glia is discernible. In mice with a defect in Wnt signaling, radial glia scaffolding was disorganized, and the precursor pool did not develop properly (Zhou et al., 2004).

Radial glia forms two bundles,
one of which turns
perpendicular to the future
granule cell layer

The original radial glia, which spans between the ventricular and the pial surfaces as in other cortical regions, becomes separated into two bundles. One is the supragranular bundle, which runs parallel to the hippocampal fissure; the other is the fimbrial bundle, which co-localizes with the secondary germinative matrix. Radial glia in the secondary matrix first runs tangential to what later becomes the granule cell layer, but a subset leads to the future hilus (Rickmann et al., 1987; Sievers et al., 1992; Li et al., 2009a). These latter fibers originate from the transient subpial cluster of precursor cells (Li et al., 2009a). Consequently, radial glia of the tertiary matrix (in the hilus) is then found to be almost perpendicular to the supragranular glia and crosses the granule cell layer radially. Development of the dentate gyrus is particularly dependent on a subpopulation of radial glia, characterized by the expression of Nuclear factor 1B (Nfib), which is necessary for the supragranular bundle to form but is not required for the fimbrial bundle (Barry et al., 2008). Consequently, an aberrant migration of the dentate gyrus precursor cells was found. This suggests that different types of radial glia contribute to the development of the dentate gyrus.

The tertiary matrix transforms
into a persisting adult matrix
with an associated radial scaffold

Late postnatally, the cell bodies of the radial glia become restricted to the SGZ; their processes terminate in the molecular layer. This condition persists into adulthood. A subset of these radial glial cells functions as precursor cells in the adult dentate gyrus. It is therefore possible that each germinative matrix in the developing dentate gyrus has its own distinctive set of radial glia. The first set would have the typical orientation of a base at the ventricular surface and apical contact at the pia. The secondary, tertiary, and quaternary radial glia would lack these contacts and only maintain an end foot on the basal membrane of a blood vessel (just as in the SVZ the B cell touches the ventricular surface). The radial process would be independent of surface contact. Alternatively, other blood vessels could provide the second contact as they seem to do in the SVZ. This hypothetical structure has to be confirmed in detailed anatomical studies. It seems, however, that radial glia–like cells of the adult SGZ terminate in the molecular layer without mandatory contact with blood vessels.

Comparison of Neuronal Development in the Adult and in the Fetus

Despite many shared features,
adult neurogenesis differs from
embryonic and fetal brain
development

Neuronal development in the adult is no exact replication or continuation of embryonic and fetal brain development. Conditions, starting points, time course, the cells involved, activity-dependency, and the purpose the entire process serves in the larger picture are too different (Table 4–1). On the other hand, evolution tends to reuse established systems, which are then modified according their new function. Apparent similarities might thus be misleading; apparent differences less relevant than they might appear at first. The further away from the core principles of transcriptional control we move, the larger the differences between the two types of neurogenesis become. Conversely, this implies that, in order to understand adult neurogenesis, one needs to relate and compare all findings to events during embryonic and fetal brain development.

Despite many differences, adult neurogenesis has all its roots in embryonic brain development and shares many principles with it. Adult and embryonic/fetal neurogenesis both originate from radial glia–like precursor cells and intermediate progenitor cells. They follow the same gross scheme of events. In both cases a great surplus of neurons is produced, from which only a subset is recruited for terminal integration and function. In both cases this recruitment is activity-dependent, but only in the adult brain it is due to actual behavior of the individual, which is rather limited in utero. In the end, the type and extent of activity-dependency might be the most important difference between the two types of neurogenesis. Consequently, the role of adult neurogenesis lies in plastic remodeling rather than "development" in the sense of building a nervous structure in the first place. One might, however, rightfully consider plasticity as a form of continued development under the conditions of the adult brain. The key issue is the selectivity. If development means the reduction of potential, adult neurogenesis reflects the maintenance of an extremely reduced potential compared to the opportunities at the beginning of brain development, where the generation of hundreds of neuron types lies ahead.

Adult neurogenesis also does not have to recapitulate all aspects of brain development: it is highly selective and starts from a vantage point. Positional information and patterning are advanced and remain fixed; the developmental potential in vivo is strictly limited to only few specific neuronal populations. Ironically, thus, a large part of regulation in the neurogenic zones of the adult brain is actually antineurogenic, in the sense that it limits neurogenesis in order to achieve it within a physiologically desirable range.

References

Aakalu G, Smith WB, Nguyen N, Jiang C, Schuman EM (2001). Dynamic visualization of local protein synthesis in hippocampal neurons. *Neuron* 30:489–502.

Aboitiz F, Montiel J (2007). Co-option of signaling mechanisms from neural induction to telencephalic patterning. *Rev Neurosci* 18:311–342.

Airaksinen MS, Saarma M (2002). The GDNF family: Signalling, biological functions and therapeutic value. *Nat Rev Neurosci* 3:383–394.

Alsina B, Vu T, Cohen-Cory S (2001). Visualizing synapse formation in arborizing optic axons in vivo: Dynamics and modulation by BDNF. *Nat Neurosci* 4:1093–1101.

Altar CA, Cai N, Bliven T, Juhasz M, Conner JM, Acheson AL, Lindsay RM, Wiegand SJ (1997). Anterograde transport of brain-derived neurotrophic factor and its role in the brain. *Nature* 389:856–860.

Altar CA, Laeng P, Jurata LW, Brockman JA, Lemire A, Bullard J, Bukhman YV, Young TA, Charles V, Palfreyman MG (2004). Electroconvulsive seizures regulate gene expression of distinct neurotrophic signaling pathways. *J Neurosci* 24:2667–2677.

Altman J, Bayer SA (1990a) Migration and distribution of two populations of hippocampal progenitors during the perinatal and postnatal periods. *J Comp Neurol* 301:365–381.

Altman J, Bayer SA (1990b) Mosaic organization of the hippocampal neuroepithelium and the multiple germinal sources of dentate granule cells. *J Comp Neurol* 301:325–342.

Altmann CR, Brivanlou AH (2001). Neural patterning in the vertebrate embryo. *Int Rev Cytol* 203:447–482.

Anderson S, Mione M, Yun K, Rubenstein JL (1999). Differential origins of neocortical projection and local circuit neurons: Role of Dlx genes in neocortical interneuronogenesis. *Cereb Cortex* 9:646–654.

Angevine J, Sidman R (1961). Autoradiographic study of cell migration during histogenesis of the cerebral cortex of the mouse. *Nature* 192:266–268.

Antonini A, Stryker MP (1993). Rapid remodeling of axonal arbors in the visual cortex. *Science* 260: 1819–1821.

Arvidsson A, Collin T, Kirik D, Kokaia Z, Lindvall O (2002). Neuronal replacement from endogenous precursors in the adult brain after stroke. *Nat Med* 8:963–970.

Attardo A, Calegari F, Haubensak W, Wilsch-Brauninger M, Huttner WB (2008). Live imaging at the onset of cortical neurogenesis reveals differential appearance of the neuronal phenotype in apical versus basal progenitor progeny. *PLoS ONE* 3:e2388.

Babu H, Ramirez-Rodriguez G, Fabel K, Bischofberger J, Kempermann G (2009). Synaptic network activity induces neuronal differentiation of adult hippocampal precursor cells through BDNF signaling. Front. *Neurosci.* 3:49.

Bai J, Ramos RL, Ackman JB, Thomas AM, Lee RV, LoTurco JJ (2003). RNAi reveals doublecortin is required for radial migration in rat neocortex. *Nat Neurosci* 6:1277–1283.

Barembaum M, Bronner-Fraser M (2005). Early steps in neural crest specification. *Semin Cell Dev Biol* 16:642–646.

Barrett GL, Bartlett PF (1994). The p75 nerve growth factor receptor mediates survival or death depending on the stage of sensory neuron development. *Proc Natl Acad Sci USA* 91:6501–6505.

Barry G, Piper M, Lindwall C, Moldrich R, Mason S, Little E, Sarkar A, Tole S, Gronostajski RM, Richards LJ (2008). Specific glial populations regulate hippocampal morphogenesis. *J Neurosci* 28:12328–12340.

Basak O, Taylor V (2007). Identification of self-replicating multipotent progenitors in the embryonic nervous system by high Notch activity and Hes5 expression. *Eur J Neurosci* 25:1006–1022.

Bath KG, Mandairon N, Jing D, Rajagopal R, Kapoor R, Chen ZY, Khan T, Proenca CC, Kraemer R, Cleland TA, Hempstead BL, Chao MV, Lee FS (2008). Variant brain-derived neurotrophic factor (Val66Met) alters adult olfactory bulb neurogenesis and spontaneous olfactory discrimination. *J Neurosci* 28:2383–2393.

Bayer SA (1983). 3H-thymidine-radiographic studies of neurogenesis in the rat olfactory bulb. *Exp Brain Res* 50:329–340.

Bayer SA, Altman J (1991). *Neocortical Development*. New York: Raven Press.

Beatus P, Lendahl U (1998). Notch and neurogenesis. *J Neurosci Res* 54:125–136.

Bishop KM, Goudreau G, O'Leary DD (2000). Regulation of area identity in the mammalian neocortex by Emx2 and Pax6. *Science* 288:344–349.

Blake JA, Thomas M, Thompson JA, White R, Ziman M (2008). Perplexing Pax: From puzzle to paradigm. *Dev Dyn* 237:2791–2803.

Bonaguidi MA, Peng CY, McGuire T, Falciglia G, Gobeske KT, Czeisler C, Kessler JA (2008). Noggin expands neural stem cells in the adult hippocampus. *J Neurosci* 28:9194–9204.

Boukhaddaoui H, Sieso V, Scamps F, Valmier J (2001). An activity-dependent neurotrophin-3 autocrine loop regulates the phenotype of developing hippocampal pyramidal neurons before target contact. *J Neurosci* 21:8789–8797.

Bouzioukh F, Daoudal G, Falk J, Debanne D, Rougon G, Castellani V (2006). Semaphorin3A regulates synaptic function of differentiated hippocampal neurons. *Eur J Neurosci* 23:2247–2254.

Bozza T, Feinstein P, Zheng C, Mombaerts P (2002). Odorant receptor expression defines functional units in the mouse olfactory system. *J Neurosci* 22:3033–3043.

Brand M, Heisenberg CP, Jiang YJ, Beuchle D, Lun K, Furutani-Seiki M, Granato M, Haffter P, Hammerschmidt M, Kane DA, Kelsh RN, Mullins MC, Odenthal J, van Eeden FJ, Nusslein-Volhard C (1996). Mutations in zebrafish genes affecting the formation of the boundary between midbrain and hindbrain. *Development* 123:179–190.

Brazel CY, Romanko MJ, Rothstein RP, Levison SW (2003). Roles of the mammalian subventricular zone in brain development. *Prog Neurobiol* 69:49–69.

Breunig JJ, Sarkisian MR, Arellano JI, Morozov YM, Ayoub AE, Sojitra S, Wang B, Flavell RA, Rakic P, Town T (2008). Primary cilia regulate hippocampal neurogenesis by mediating sonic hedgehog signaling. *Proc Natl Acad Sci USA* 105:13127–13132.

Brill MS, Ninkovic J, Winpenny E, Hodge RD, Ozen I, Yang R, Lepier A, Gascon S, Erdelyi F, Szabo G, Parras C, Guillemot F, Frotscher M, Berninger B, Hevner RF, Raineteau O, Gotz M (2009). Adult generation of glutamatergic olfactory bulb interneurons. *Nat Neurosci* 12:1524–33.

Brodski C, Weisenhorn DM, Signore M, Sillaber I, Oesterheld M, Broccoli V, Acampora D, Simeone A, Wurst W (2003). Location and size of dopaminergic and serotonergic cell populations are controlled by the position of the midbrain-hindbrain organizer. *J Neurosci* 23:4199–4207.

Buchman VL, Davies AM (1993). Different neurotrophins are expressed and act in a developmental sequence to promote the survival of embryonic sensory neurons. *Development* 118:989–1001.

Butt SJ, Fuccillo M, Nery S, Noctor S, Kriegstein A, Corbin JG, Fishell G (2005). The temporal and spatial origins of cortical interneurons predict their physiological subtype. *Neuron* 48:591–604.

Bystron I, Blakemore C, Rakic P (2008). Development of the human cerebral cortex: Boulder Committee revisited. *Nat Rev Neurosci* 9:110–122.

Cairns J (1975). Mutation selection and the natural history of cancer. *Nature* 255:197–200.

Calegari F, Haubensak W, Haffner C, Huttner WB (2005). Selective lengthening of the cell cycle in the neurogenic subpopulation of neural progenitor cells during mouse brain development. *J Neurosci* 25:6533–6538.

Caleo M, Menna E, Chierzi S, Cenni MC, Maffei L (2000). Brain-derived neurotrophic factor is an anterograde survival factor in the rat visual system. *Curr Biol* 10:1155–1161.

Campos LS, Decker L, Taylor V, Skarnes W (2006). Notch, epidermal growth factor receptor, and beta1-integrin pathways are coordinated in neural stem cells. *J Biol Chem* 281:5300–5309.

Catalano SM, Shatz CJ (1998). Activity-dependent cortical target selection by thalamic axons. *Science* 281:559–562.

Chen K, Ohkubo Y, Shin D, Doetschman T, Sanford LP, Li H, Vaccarino FM (2008). Decrease in excitatory neurons, astrocytes and proliferating progenitors in the cerebral cortex of mice lacking exon 3 from the Fgf2 gene. *BMC Neurosci* 9:94.

Chenn A, McConnell S (1995). Cleavage orientation and the asymmetric inheritance of Notch1 immunoreactivity in mammalian neurogenesis. *Cell* 82:631–641.

Chew JL, Loh YH, Zhang W, Chen X, Tam WL, Yeap LS, Li P, Ang YS, Lim B, Robson P, Ng HH (2005). Reciprocal transcriptional regulation of Pou5f1 and Sox2 via the Oct4/Sox2 complex in embryonic stem cells. *Mol Cell Biol* 25:6031–6046.

Chiaramello S, Dalmasso G, Bezin L, Marcel D, Jourdan F, Peretto P, Fasolo A, De Marchis S (2007). BDNF/TrkB interaction regulates migration of SVZ precursor cells via PI3-K and MAP-K signalling pathways. *Eur J Neurosci* 26:1780–1790.

Ciccolini F, Mandl C, Holzl-Wenig G, Kehlenbach A, Hellwig A (2005). Prospective isolation of late development multipotent precursors whose migration is promoted by EGFR. *Dev Biol* 284:112–125.

Cline HT (2001). Dendritic arbor development and synaptogenesis. *Curr Opin Neurobiol* 11:118–126.

Cobos I, Long JE, Thwin MT, Rubenstein JL (2006). Cellular patterns of transcription factor expression in developing cortical interneurons. *Cereb Cortex* 16 Suppl 1:i82–88.

Colak D, Mori T, Brill MS, Pfeifer A, Falk S, Deng C, Monteiro R, Mummery C, Sommer L, Gotz M (2008). Adult neurogenesis requires Smad4–mediated bone morphogenic protein signaling in stem cells. *J Neurosci* 28:434–446.

Colamarino SA, Tessier-Lavigne M (1995). The axonal chemoattractant netrin-1 is also a chemorepellent for trochlear motor axons. *Cell* 81:621–629.

Committee B (1970). Embryonic vertebrate central nervous system: Revised terminology. The Boulder Committee. *Anat Rec* 166:257–261.

Conboy MJ, Karasov AO, Rando TA (2007). High incidence of non-random template strand segregation and asymmetric fate determination in dividing stem cells and their progeny. *PLoS Biol* 5:e102.

Corbeil D, Roper K, Hannah MJ, Hellwig A, Huttner WB (1999). Selective localization of the polytopic membrane protein prominin in microvilli of epithelial cells - a combination of apical sorting and retention in plasma membrane protrusions. *J Cell Sci* 112 (Pt 7):1023–1033.

Coulson EJ, Reid K, Shipham KM, Morley S, Kilpatrick TJ, Bartlett PF (2004). The role of neurotransmission and the Chopper domain in p75 neurotrophin receptor death signaling. *Prog Brain Res* 146:41–62.

Cubelos B, Sebastian-Serrano A, Kim S, Moreno-Ortiz C, Redondo JM, Walsh CA, Nieto M (2008). Cux-2 controls the proliferation of neuronal intermediate precursors of the cortical subventricular zone. *Cereb Cortex* 18:1758–1770.

D'Arcangelo G, Miao GG, Chen SC, Soares HD, Morgan JI, Curran T (1995). A protein related to extracellular matrix proteins deleted in the mouse mutant reeler. *Nature* 374:719–723.

Davis GW, Murphey RK (1994). Long-term regulation of short-term transmitter release properties: Retrograde signaling and synaptic development. *Trends Neurosci* 17:9–13.

De Carlos JA, Lopez-Mascaraque L, Valverde F (1995). The telencephalic vesicles are innervated by olfactory placode-derived cells: A possible mechanism to induce neocortical development. *Neuroscience* 68:1167–1178.

De Carlos JA, Lopez-Mascaraque L, Valverde F (1996). Early olfactory fiber projections and cell migration into the rat telencephalon. *Int J Dev Neurosci* 14:853–866.

Deacon TW, Pakzaban P, Isacson O (1994). The lateral ganglionic eminence is the origin of cells committed to striatal phenotypes: Neural transplantation and developmental evidence. *Brain Res* 668:211–219.

Dehay C, Savatier P, Cortay V, Kennedy H (2001). Cell-cycle kinetics of neocortical precursors are influenced by embryonic thalamic axons. *J Neurosci* 21:201–214.

Deisseroth K, Singla S, Toda H, Monje M, Palmer TD, Malenka RC (2004). Excitation-neurogenesis coupling in adult neural stem/progenitor cells. *Neuron* 42:535–552.

Derer P, Caviness VS, Jr., Sidman RL (1977). Early cortical histogenesis in the primary olfactory cortex of the mouse. *Brain Res* 123:27–40.

Doetsch F, Garcia-Verdugo JM, Alvarez-Buylla A (1997). Cellular composition and three-dimensional organization of the subventricular germinal zone in the adult mammalian brain. *J Neurosci* 17:5046–5061.

Doetsch F, Petreanu L, Caille I, Garcia-Verdugo JM, Alvarez-Buylla A (2002). EGF converts transit-amplifying neurogenic precursors in the adult brain into multipotent stem cells. *Neuron* 36:1021–1034.

Drakew A, Deller T, Heimrich B, Gebhardt C, Del Turco D, Tielsch A, Forster E, Herz J, Frotscher M (2002). Dentate granule cells in reeler mutants and VLDLR and ApoER2 knockout mice. *Exp Neurol* 176:12–24.

Edwards MA, Yamamoto M, Caviness VS, Jr. (1990). Organization of radial glia and related cells in the developing murine CNS. An analysis based upon a new monoclonal antibody marker. *Neuroscience* 36:121–144.

Engert F, Bonhoeffer T (1999). Dendritic spine changes associated with hippocampal long-term synaptic plasticity. *Nature* 399:66–70.

Enokido Y, Wyatt S, Davies AM (1999). Developmental changes in the response of trigeminal neurons to neurotrophins: Influence of birthdate and the ganglion environment. *Development* 126:4365–4373.

Esch T, Lemmon V, Banker G (1999). Local presentation of substrate molecules directs axon specification by cultured hippocampal neurons. *J Neurosci* 19:6417–6426.

Estivill-Torrus G, Pearson H, van Heyningen V, Price DJ, Rashbass P (2002). Pax6 is required to regulate the cell cycle and the rate of progression from symmetrical to asymmetrical division in mammalian cortical progenitors. *Development* 129:455–466.

Farah MH, Olson JM, Sucic HB, Hume RI, Tapscott SJ, Turner DL (2000). Generation of neurons by transient expression of neural bHLH proteins in mammalian cells. *Development* 127:693–702.

Farkas LM, Huttner WB (2008). The cell biology of neural stem and progenitor cells and its significance for their proliferation versus differentiation during mammalian brain development. *Curr Opin Cell Biol* 20:707–715.

Ferland RJ, Batiz LF, Neal J, Lian G, Bundock E, Lu J, Hsiao YC, Diamond R, Mei D, Banham AH, Brown PJ, Vanderburg CR, Joseph J, Hecht JL, Folkerth R, Guerrini R, Walsh CA, Rodriguez EM, Sheen VL (2009). Disruption of neural progenitors along the ventricular and subventricular zones in periventricular heterotopia. *Hum Mol Genet* 18:497–516.

Fishell G, Kriegstein AR (2003). Neurons from radial glia: The consequences of asymmetric inheritance. *Curr Opin Neurobiol* 13:34–41.

Forster E, Tielsch A, Saum B, Weiss KH, Johanssen C, Graus-Porta D, Muller U, Frotscher M (2002). Reelin, Disabled 1, and beta 1 integrins are required for the formation of the radial glial scaffold in the hippocampus. *Proc Natl Acad Sci USA* 99:13178–13183.

Fox IJ, Kornblum HI (2005). Developmental profile of ErbB receptors in murine central nervous system: Implications for functional interactions. *J Neurosci Res* 79:584–597.

Fox JW, Lamperti ED, Eksioglu YZ, Hong SE, Feng Y, Graham DA, Scheffer IE, Dobyns WB, Hirsch BA, Radtke RA, Berkovic SF, Huttenlocher PR, Walsh CA (1998). Mutations in filamin 1 prevent migration of cerebral cortical neurons in human periventricular heterotopia. *Neuron* 21:1315–1325.

Furuta Y, Piston DW, Hogan BL (1997). Bone morphogenetic proteins (BMPs) as regulators of dorsal forebrain development. *Development* 124:2203–2212.

Galceran J, Miyashita-Lin EM, Devaney E, Rubenstein JL, Grosschedl R (2000). Hippocampus development and generation of dentate gyrus granule cells is regulated by LEF1. *Development* 127:469–482.

Galvao RP, Garcia-Verdugo JM, Alvarez-Buylla A (2008). Brain-derived neurotrophic factor signaling does not stimulate subventricular zone neurogenesis in adult mice and rats. *J Neurosci* 28:13368–13383.

Giachino C, De Marchis S, Giampietro C, Parlato R, Perroteau I, Schutz G, Fasolo A, Peretto P (2005). cAMP response element-binding protein regulates differentiation and survival of newborn neurons in the olfactory bulb. *J Neurosci* 25:10105–10118.

Giagtzoglou N, Alifragis P, Koumbanakis KA, Delidakis C (2003). Two modes of recruitment of E(spl) repressors onto target genes. *Development* 130:259–270.

Goldberg JL (2003). How does an axon grow? *Genes Dev* 17:941–958.

Goldberg JL, Klassen MP, Hua Y, Barres BA (2002a) Amacrine-signaled loss of intrinsic axon growth ability by retinal ganglion cells. *Science* 296:1860–1864.

Goldberg JL, Espinosa JS, Xu Y, Davidson N, Kovacs GT, Barres BA (2002b) Retinal ganglion cells do not extend axons by default: Promotion by neurotrophic signaling and electrical activity. *Neuron* 33:689–702.

Gong C, Wang TW, Huang HS, Parent JM (2007). Reelin regulates neuronal progenitor migration in intact and epileptic hippocampus. *J Neurosci* 27:1803–1811.

Gong Q, Shipley MT (1995). Evidence that pioneer olfactory axons regulate telencephalon cell cycle kinetics to induce the formation of the olfactory bulb. *Neuron* 14:91–101.

Gotz M, Stoykova A, Gruss P (1998). Pax6 controls radial glia differentiation in the cerebral cortex. *Neuron* 21:1031–1044.

Gregorio SP, Sallet PC, Do KA, Lin E, Gattaz WF, Dias-Neto E (2009). Polymorphisms in genes involved in neurodevelopment may be associated with altered brain morphology in schizophrenia: Preliminary evidence. *Psychiatr Res* 165:1–9.

Grove EA, Tole S, Limon J, Yip L, Ragsdale CW (1998). The hem of the embryonic cerebral cortex is defined by the expression of multiple Wnt genes and is compromised in Gli3–deficient mice. *Development* 125:2315–2325.

Grueber WB, Jan LY, Jan YN (2003). Different levels of the homeodomain protein cut regulate distinct dendrite branching patterns of *Drosophila* multidendritic neurons. *Cell* 112:805–818.

Guerrini R, Marini C (2006). Genetic malformations of cortical development. *Exp Brain Res* 173:322–333.

Hack I, Hellwig S, Junghans D, Brunne B, Bock HH, Zhao S, Frotscher M (2007). Divergent roles of ApoER2 and Vldlr in the migration of cortical neurons. *Development* 134:3883–3891.

Hagg T (2009). From neurotransmitters to neurotrophic factors to neurogenesis. *Neuroscientist* 15:20–27.

Hartfuss E, Forster E, Bock HH, Hack MA, Leprince P, Luque JM, Herz J, Frotscher M, Gotz M (2003). Reelin signaling directly affects radial glia morphology and biochemical maturation. *Development* 130:4597–4609.

Hartmann M, Heumann R, Lessmann V (2001). Synaptic secretion of BDNF after high-frequency stimulation of glutamatergic synapses. *Embo J* 20:5887–5897.

Haubensak W, Attardo A, Denk W, Huttner WB (2004). Neurons arise in the basal neuroepithelium of the early mammalian telencephalon: A major site of neurogenesis. *Proc Natl Acad Sci USA* 101:3196–3201.

Havrda MC, Harris BT, Mantani A, Ward NM, Paolella BR, Cuzon VC, Yeh HH, Israel MA (2008). Id2 is required for specification of dopaminergic neurons during adult olfactory neurogenesis. *J Neurosci* 28: 14074–14086.

He W, Ingraham C, Rising L, Goderie S, Temple S (2001). Multipotent stem cells from the mouse basal forebrain contribute GABAergic neurons and oligodendrocytes to the cerebral cortex during embryogenesis. *J Neurosci* 21:8854–8862.

Hinds JW (1968a) Autoradiographic study of histogenesis in the mouse olfactory bulb. II. *Cell* proliferation and migration. *J Comp Neurol* 134:305–322.

Hinds JW (1968b) Autoradiographic study of histogenesis in the mouse olfactory bulb. I. Time of origin of neurons and neuroglia. *J Comp Neurol* 134:287–304.

Hojo M, Ohtsuka T, Hashimoto N, Gradwohl G, Guillemot F, Kageyama R (2000). Glial cell fate specification modulated by the bHLH gene Hes5 in mouse retina. *Development* 127:2515–2522.

Hu B, Nikolakopoulou AM, Cohen-Cory S (2005). BDNF stabilizes synapses and maintains the structural complexity of optic axons in vivo. *Development* 132:4285–4298.

Hynes RO, Lander AD (1992). Contact and adhesive specificities in the associations, migrations, and targeting of cells and axons. *Cell* 68:303–322.

Iacopetti P, Michelini M, Stuckmann I, Oback B, Aaku-Saraste E, Huttner WB (1999). Expression of the antiproliferative gene TIS21 at the onset of neurogenesis identifies single neuroepithelial cells that switch from proliferative to neuron-generating division. *Proc Natl Acad Sci USA* 96:4639–4644.

Inagaki N, Chihara K, Arimura N, Menager C, Kawano Y, Matsuo N, Nishimura T, Amano M, Kaibuchi K (2001). CRMP-2 induces axons in cultured hippocampal neurons. *Nat Neurosci* 4:781–782.

Ishibashi M, Ang SL, Shiota K, Nakanishi S, Kageyama R, Guillemot F (1995). Targeted disruption of mammalian hairy and Enhancer of split homolog-1 (HES-1). leads to up-regulation of neural helix-loop-helix factors, premature neurogenesis, and severe neural tube defects. *Genes Dev* 9:3136–3148.

Jessberger S, Romer B, Babu H, Kempermann G (2005). Seizures induce proliferation and dispersion of doublecortin-positive hippocampal progenitor cells. *Exp Neurol* 196:342–351.

Jessberger S, Toni N, Clemenson Jr GD, Ray J, Gage FH (2008a) Directed differentiation of hippocampal stem/progenitor cells in the adult brain. *Nat Neurosci* 11:888–893.

Jessberger S, Aigner S, Clemenson GD, Toni N, Lie DC, Karalay O, Overall R, Kempermann G, Gage FH (2008b) Cdk5 Regulates Accurate Maturation of Newborn Granule Cells in the Adult Hippocampus. *PLoS Biol* 6:e272.

Jhas S, Ciura S, Belanger-Jasmin S, Dong Z, Llamosas E, Theriault FM, Joachim K, Tang Y, Liu L, Liu J, Stifani S (2006). Hes6 inhibits astrocyte differentiation and promotes neurogenesis through different mechanisms. *J Neurosci* 26:11061–11071.

Jimenez D, Garcia C, de Castro F, Chedotal A, Sotelo C, de Carlos JA, Valverde F, Lopez-Mascaraque L (2000). Evidence for intrinsic development of olfactory structures in Pax-6 mutant mice. *J Comp Neurol* 428:511–526.

Jostes B, Walther C, Gruss P (1990). The murine paired box gene, Pax7, is expressed specifically during the development of the nervous and muscular system. *Mech Dev* 33:27–37.

Kalani MY, Cheshier SH, Cord BJ, Bababeygy SR, Vogel H, Weissman IL, Palmer TD, Nusse R (2008). Wnt-mediated self-renewal of neural stem/progenitor cells. *Proc Natl Acad Sci USA* 105:16970–16975.

Kanatani S, Yozu M, Tabata H, Nakajima K (2008). COUP-TFII is preferentially expressed in the caudal ganglionic eminence and is involved in the caudal migratory stream. *J Neurosci* 28:13582–13591.

Kaplan DR, Miller FD (2000). Neurotrophin signal transduction in the nervous system. *Curr Opin Neurobiol* 10:381–391.

Karpowicz P, Morshead C, Kam A, Jervis E, Ramunas J, Cheng V, van der Kooy D (2005). Support for the immortal strand hypothesis: Neural stem cells partition DNA asymmetrically in vitro. *J Cell Biol* 170:721–732.

Katz LC, Shatz CJ (1996). Synaptic activity and the construction of cortical circuits. *Science* 274:1133–1138.

Kennedy TE, Serafini T, de la Torre JR, Tessier-Lavigne M (1994). Netrins are diffusible chemotropic factors for commissural axons in the embryonic spinal cord. *Cell* 78:425–435.

Kiel MJ, He S, Ashkenazi R, Gentry SN, Teta M, Kushner JA, Jackson TL, Morrison SJ (2007). Haematopoietic stem cells do not asymmetrically segregate chromosomes or retain BrdU. *Nature* 449:238–242.

Kim AS, Lowenstein DH, Pleasure SJ (2001). Wnt receptors and Wnt inhibitors are expressed in gradients in the developing telencephalon. *Mech Dev* 103:167–172.

Kim S, Walsh CA (2007). Numb, neurogenesis and epithelial polarity. *Nat Neurosci* 10:812–813.

Kintner C (2002). Neurogenesis in embryos and in adult neural stem cells. *J Neurosci* 22:639–643.

Kitsukawa T, Shimizu M, Sanbo M, Hirata T, Taniguchi M, Bekku Y, Yagi T, Fujisawa H (1997). Neuropilin-semaphorin III/D-mediated chemorepulsive signals play a crucial role in peripheral nerve projection in mice. *Neuron* 19:995–1005.

Kohwi M, Petryniak MA, Long JE, Ekker M, Obata K, Yanagawa Y, Rubenstein JL, Alvarez-Buylla A (2007). A subpopulation of olfactory bulb GABAergic interneurons is derived from Emx1– and Dlx5/6–expressing progenitors. *J Neurosci* 27:6878–6891.

Koizumi H, Higginbotham H, Poon T, Tanaka T, Brinkman BC, Gleeson JG (2006). Doublecortin maintains bipolar shape and nuclear translocation during migration in the adult forebrain. *Nat Neurosci* 9:779–786.

Kornblum HI, Yanni DS, Easterday MC, Seroogy KB (2000). Expression of the EGF receptor family members ErbB2, ErbB3, and ErbB4 in germinal zones of the developing brain and in neurosphere cultures containing CNS stem cells. *Dev Neurosci* 22:16–24.

Kosodo Y, Roper K, Haubensak W, Marzesco AM, Corbeil D, Huttner WB (2004). Asymmetric distribution of the apical plasma membrane during neurogenic divisions of mammalian neuroepithelial cells. *Embo J* 23:2314–2324.

Koutmani Y, Hurel C, Patsavoudi E, Hack M, Gotz M, Thomaidou D, Matsas R (2004). BM88 is an early marker of proliferating precursor cells that will differentiate into the neuronal lineage. *Eur J Neurosci* 20:2509–2523.

Kriegstein A, Noctor S, Martinez-Cerdeno V (2006). Patterns of neural stem and progenitor cell division may underlie evolutionary cortical expansion. *Nat Rev Neurosci* 7:883–890.

Kronenberg G, Wang LP, Geraerts M, Babu H, Synowitz M, Vicens P, Lutsch G, Glass R, Yamaguchi M, Baekelandt V, Debyser Z, Kettenmann H, Kempermann G (2007). Local origin and activity-dependent generation of nestin-expressing protoplasmic astrocytes in CA1. *Brain Structure & Function* 212:19–35.

Kuczewski N, Porcher C, Lessmann V, Medina I, Gaiarsa JL (2009). Activity-dependent dendritic release of BDNF and biological consequences. *Mol Neurobiol* 39:37–49.

Kuhl M, Sheldahl LC, Park M, Miller JR, Moon RT (2000). The Wnt/Ca2+ pathway: A new vertebrate Wnt signaling pathway takes shape. *Trends Genet* 16:279–283.

Lange C, Huttner WB, Calegari F (2009). Cdk4/cyclinD1 overexpression in neural stem cells shortens G1, delays neurogenesis, and promotes the generation and expansion of basal progenitors. *Cell Stem Cell* 5:320–331.

Lansdorp PM (2007). Immortal strands? Give me a break. *Cell* 129:1244–1247.

Lee J, Duan W, Mattson MP (2002). Evidence that brain-derived neurotrophic factor is required for basal neurogenesis and mediates, in part, the enhancement of neurogenesis by dietary restriction in the hippocampus of adult mice. *J Neurochem* 82:1367–1375.

Lee SM, Tole S, Grove E, McMahon AP (2000). A local Wnt-3a signal is required for development of the mammalian hippocampus. *Development* 127:457–467.

Lein P, Johnson M, Guo X, Rueger D, Higgins D (1995). Osteogenic protein-1 induces dendritic growth in rat sympathetic neurons. *Neuron* 15:597–605.

Levi-Montalcini R (1987). The nerve growth factor 35 years later. *Science* 237:1154–1162.

Lew DJ, Burke DJ, Dutta A (2008). The immortal strand hypothesis: How could it work? *Cell* 133:21–23.

Lewin GR, Barde YA (1996). Physiology of the neurotrophins. *Annu Rev Neurosci* 19:289–317.

Li G, Kataoka H, Coughlin SR, Pleasure SJ (2009a) Identification of a transient subpial neurogenic zone in the developing dentate gyrus and its regulation by Cxcl12 and reelin signaling. *Development* 136:327–335.

Li HS, Chen JH, Wu W, Fagaly T, Zhou L, Yuan W, Dupuis S, Jiang ZH, Nash W, Gick C, Ornitz DM, Wu JY, Rao Y (1999). Vertebrate slit, a secreted ligand for the transmembrane protein roundabout, is a repellent for olfactory bulb axons. *Cell* 96:807–818.

Li JY, Lao Z, Joyner AL (2005). New regulatory interactions and cellular responses in the isthmic organizer region revealed by altering Gbx2 expression. *Development* 132:1971–1981.

Li W, Sun G, Yang S, Qu Q, Nakashima K, Shi Y (2008). Nuclear receptor TLX regulates cell cycle progression in neural stem cells of the developing brain. *Mol Endocrinol* 22:56–64.

Li X, Tang X, Jablonska B, Aguirre A, Gallo V, Luskin MB (2009b) p27(KIP1). regulates neurogenesis in the rostral migratory stream and olfactory bulb of the postnatal mouse. *J Neurosci* 29:2902–2914.

Lie DC, Colamarino SA, Song HJ, Desire L, Mira H, Consiglio A, Lein ES, Jessberger S, Lansford H, Dearie AR, Gage FH (2005). Wnt signalling regulates adult hippocampal neurogenesis. *Nature* 437:1370–1375.

Lim CY, Tam WL, Zhang J, Ang HS, Jia H, Lipovich L, Ng HH, Wei CL, Sung WK, Robson P, Yang H, Lim B (2008). Sall4 regulates distinct transcription circuitries in different blastocyst-derived stem cell lineages. *Cell Stem Cell* 3:543–554.

Lim DA, Tramontin AD, Trevejo JM, Herrera DG, Garcia-Verdugo JM, Alvarez-Buylla A (2000). Noggin antagonizes BMP signaling to create a niche for adult neurogenesis. *Neuron* 28:713–726.

Liu HK, Belz T, Bock D, Takacs A, Wu H, Lichter P, Chai M, Schutz G (2008). The nuclear receptor tailless is required for neurogenesis in the adult subventricular zone. *Genes Dev* 22:2473–2478.

Liu Q, Dwyer ND, O'Leary DD (2000). Differential expression of COUP-TFI, CHL1, and two novel genes in developing neocortex identified by differential display PCR. *J Neurosci* 20:7682–7690.

Lledo PM, Merkle FT, Alvarez-Buylla A (2008). Origin and function of olfactory bulb interneuron diversity. *Trends Neurosci* 31:392–400.

Loh YH et al. (2006). The Oct4 and Nanog transcription network regulates pluripotency in mouse embryonic stem cells. *Nat Genet* 38:431–440.

Lom B, Cogen J, Sanchez AL, Vu T, Cohen-Cory S (2002). Local and target-derived brain-derived neurotrophic factor exert opposing effects on the dendritic arborization of retinal ganglion cells in vivo. *J Neurosci* 22: 7639–7649.

Lopez-Mascaraque L, de Castro F (2002). The olfactory bulb as an independent developmental domain. *Cell Death Differ* 9:1279–1286.

Luskin MB, Zigova T, Soteres BJ, Stewart RR (1997). Neuronal progenitor cells derived from the anterior subventricular zone of the neonatal rat forebrain continue to proliferate in vitro and express a neuronal phenotype. *Mol Cell Neurosci* 8:351–366.

Mallamaci A, Muzio L, Chan CH, Parnavelas J, Boncinelli E (2000). Area identity shifts in the early cerebral cortex of Emx2–/- mutant mice. *Nat Neurosci* 3:679–686.

Marshall CA, Goldman JE (2002). Subpallial dlx2–expressing cells give rise to astrocytes and oligodendrocytes in the cerebral cortex and white matter. *J Neurosci* 22:9821–9830.

Marshall CA, Novitch BG, Goldman JE (2005). Olig2 directs astrocyte and oligodendrocyte formation in postnatal subventricular zone cells. *J Neurosci* 25:7289–7298.

Matsutani S, Yamamoto N (2000). Differentiation of mitral cell dendrites in the developing main olfactory bulbs of normal and naris-occluded rats. *J Comp Neurol* 418:402–410.

McAllister AK, Katz LC, Lo DC (1997). Opposing roles for endogenous BDNF and NT-3 in regulating cortical dendritic growth. *Neuron* 18:767–778.

Megason SG, McMahon AP (2002). A mitogen gradient of dorsal midline Wnts organizes growth in the CNS. *Development* 129:2087–2098.

Menn B, Garcia-Verdugo JM, Yaschine C, Gonzalez-Perez O, Rowitch D, Alvarez-Buylla A (2006). Origin of oligodendrocytes in the subventricular zone of the adult brain. *J Neurosci* 26:7907–7918.

Merkle FT, Mirzadeh Z, Alvarez-Buylla A (2007). Mosaic organization of neural stem cells in the adult brain. *Science* 317:381–384.

Michaelidis TM, Lie DC (2008). Wnt signaling and neural stem cells: Caught in the Wnt web. *Cell Tissue Res* 331:193–210.

Ming G, Henley J, Tessier-Lavigne M, Song H, Poo M (2001). Electrical activity modulates growth cone guidance by diffusible factors. *Neuron* 29:441–452.

Mirzadeh Z, Merkle FT, Soriano-Navarro M, Garcia-Verdugo JM, Alvarez-Buylla A (2008). Neural stem cells confer unique pinwheel architecture to the ventricular surface in neurogenic regions of the adult brain. *Cell Stem Cell* 3:265–278.

Miyashita-Lin EM, Hevner R, Wassarman KM, Martinez S, Rubenstein JL (1999). Early neocortical regionalization in the absence of thalamic innervation. *Science* 285:906–909.

Miyata T, Kawaguchi A, Okano H, Ogawa M (2001). Asymmetric inheritance of radial glial fibers by cortical neurons. *Neuron* 31:727–741.

Miyoshi G, Hjerling-Leffler J, Karayannis T, Sousa VH, Butt SJ, Battiste J, Johnson JE, Machold RP, Fishell G (2010). Genetic fate mapping reveals that the caudal ganglionic eminence produces a large and diverse population of superficial cortical interneurons. *J Neurosci* 30:1582–1594.

Mizuguchi R, Sugimori M, Takebayashi H, Kosako H, Nagao M, Yoshida S, Nabeshima Y, Shimamura K, Nakafuku M (2001). Combinatorial roles of olig2 and neurogenin2 in the coordinated induction of pan-neuronal and subtype-specific properties of motoneurons. *Neuron* 31:757–771.

Mombaerts P, Wang F, Dulac C, Chao SK, Nemes A, Mendelsohn M, Edmondson J, Axel R (1996). Visualizing an olfactory sensory map. *Cell* 87:675–686.

Morris DC, Zhang ZG, Wang Y, Zhang RL, Gregg S, Liu XS, Chopp M (2007). Wnt expression in the adult rat subventricular zone after stroke. *Neurosci Lett* 418:170–174.

Muntoni F, Torelli S, Brockington M (2008). Muscular dystrophies due to glycosylation defects. *Neurotherapeutics* 5:627–632.

Nadarajah B, Parnavelas JG (2002). Modes of neuronal migration in the developing cerebral cortex. *Nat Rev Neurosci* 3:423–432.

Nadarajah B, Alifragis P, Wong RO, Parnavelas JG (2003). Neuronal migration in the developing cerebral cortex: Observations based on real-time imaging. *Cereb Cortex* 13:607–611.

Nakatomi H, Kuriu T, Okabe S, Yamamoto S, Hatano O, Kawahara N, Tamura A, Kirino T, Nakafuku M (2002). Regeneration of hippocampal pyramidal neurons after ischemic brain injury by recruitment of endogenous neural progenitors. *Cell* 110:429–441.

Nam HS, Benezra R (2009). High levels of Id1 expression define B1 type adult neural stem cells. *Cell Stem Cell* 5:515–526.

Nedivi E, Wu GY, Cline HT (1998). Promotion of dendritic growth by CPG15, an activity-induced signaling molecule. *Science* 281:1863–1866.

Nery S, Fishell G, Corbin JG (2002). The caudal ganglionic eminence is a source of distinct cortical and subcortical cell populations. *Nat Neurosci* 5:1279–1287.

Nguyen L, Besson A, Roberts JM, Guillemot F (2006). Coupling cell cycle exit, neuronal differentiation and migration in cortical neurogenesis. *Cell Cycle* 5:2314–2318.

Nguyen-Ba-Charvet KT, Picard-Riera N, Tessier-Lavigne M, Baron-Van Evercooren A, Sotelo C, Chedotal A (2004). Multiple roles for slits in the control of cell migration in the rostral migratory stream. *J Neurosci* 24:1497–1506.

Nieto M, Schuurmans C, Britz O, Guillemot F (2001). Neural bHLH genes control the neuronal versus glial fate decision in cortical progenitors. *Neuron* 29:401–413.

Nieto M, Monuki ES, Tang H, Imitola J, Haubst N, Khoury SJ, Cunningham J, Gotz M, Walsh CA (2004). Expression of Cux-1 and Cux-2 in the subventricular zone and upper layers II-IV of the cerebral cortex. *J Comp Neurol* 479:168–180.

Ninkina N, Adu J, Fischer A, Pinon LG, Buchman VL, Davies AM (1996). Expression and function of TrkB variants in developing sensory neurons. *Embo J* 15:6385–6393.

Nishimura T, Kato K, Yamaguchi T, Fukata Y, Ohno S, Kaibuchi K (2004). Role of the PAR-3–KIF3 complex in the establishment of neuronal polarity. *Nat Cell Biol* 6:328–334.

Noctor SC, Flint AC, Weissman TA, Dammerman RS, Kriegstein AR (2001). Neurons derived from radial glial cells establish radial units in neocortex. *Nature* 409:714–720.

Noctor SC, Martinez-Cerdeno V, Ivic L, Kriegstein AR (2004). Cortical neurons arise in symmetric and asymmetric division zones and migrate through specific phases. *Nat Neurosci* 7:136–144.

Noctor SC, Martinez-Cerdeno V, Kriegstein AR (2008). Distinct behaviors of neural stem and progenitor cells underlie cortical neurogenesis. *J Comp Neurol* 508:28–44.

Norton JD (2000). ID helix-loop-helix proteins in cell growth, differentiation and tumorigenesis. *J Cell Sci* 113 (Pt 22):3897–3905.

Okabe S, Miwa A, Okado H (2001). Spine formation and correlated assembly of presynaptic and postsynaptic molecules. *J Neurosci* 21:6105–6114.

Ono K, Takebayashi H, Ikeda K, Furusho M, Nishizawa T, Watanabe K, Ikenaka K (2008). Regional- and temporal-dependent changes in the differentiation of Olig2 progenitors in the forebrain, and the impact on astrocyte development in the dorsal pallium. *Dev Biol* 320:456–468.

Oppenheim RW (1991). Cell death during development of the nervous system. *Annu Rev Neurosci* 14:453–501.

Panchision DM, Pickel JM, Studer L, Lee SH, Turner PA, Hazel TG, McKay RD (2001). Sequential actions of BMP receptors control neural precursor cell production and fate. *Genes Dev* 15:2094–2110.

Parnavelas JG (2000). The origin and migration of cortical neurones: New vistas. *Trends Neurosci* 23:126–131.

Patten BA, Peyrin JM, Weinmaster G, Corfas G (2003). Sequential signaling through Notch1 and erbB receptors mediates radial glia differentiation. *J Neurosci* 23:6132–6140.

Paul G, Davies AM (1995). Trigeminal sensory neurons require extrinsic signals to switch neurotrophin dependence during the early stages of target field innervation. *Dev Biol* 171:590–605.

Pencea V, Bingaman KD, Freedman LJ, Luskin MB (2001). Neurogenesis in the subventricular zone and rostral migratory stream of the neonatal and adult primate forebrain. *Exp Neurol* 172:1–16.

Petryniak MA, Potter GB, Rowitch DH, Rubenstein JL (2007). Dlx1 and Dlx2 control neuronal versus oligodendroglial cell fate acquisition in the developing forebrain. *Neuron* 55:417–433.

Pevny LH, Sockanathan S, Placzek M, Lovell-Badge R (1998). A role for SOX1 in neural determination. *Development* 125:1967–1978.

Placzek M, Tessier-Lavigne M, Jessell T, Dodd J (1990). Orientation of commissural axons in vitro in response to a floor plate-derived chemoattractant. *Development* 110:19–30.

Platt JR (1964). Strong Inference: Certain systematic methods of scientific thinking may produce much more rapid progress than others. *Science* 146:347–353.

Pleasure SJ, Collins AE, Lowenstein DH (2000a) Unique expression patterns of cell fate molecules delineate sequential stages of dentate gyrus development. *J Neurosci* 20:6095–6105.

Pleasure SJ, Anderson S, Hevner R, Bagri A, Marin O, Lowenstein DH, Rubenstein JL (2000b) Cell migration from the ganglionic eminences is required for the development of hippocampal GABAergic interneurons. *Neuron* 28:727–740.

Ploski JE, Newton SS, Duman RS (2006). Electroconvulsive seizure-induced gene expression profile of the hippocampus dentate gyrus granule cell layer. *J Neurochem* 99:1122–1132.

Politis PK, Makri G, Thomaidou D, Geissen M, Rohrer H, Matsas R (2007). BM88/CEND1 coordinates cell cycle exit and differentiation of neuronal precursors. *Proc Natl Acad Sci USA* 104:17861–17866.

Polleux F, Morrow T, Ghosh A (2000). Semaphorin 3A is a chemoattractant for cortical apical dendrites. *Nature* 404:567–573.

Poluch S, Juliano SL (2007). A normal radial glial scaffold is necessary for migration of interneurons during neocortical development. *Glia* 55:822–830.

Poo MM (2001). Neurotrophins as synaptic modulators. *Nat Rev Neurosci* 2:24–32.

Qian X, Goderie SK, Shen Q, Stern JH, Temple S (1998). Intrinsic programs of patterned cell lineages in isolated vertebrate CNS ventricular zone cells. *Development* 125:3143–3152.

Qian X, Shen Q, Goderie SK, He W, Capela A, Davis AA, Temple S (2000). Timing of CNS cell generation: A programmed sequence of neuron and glial cell production from isolated murine cortical stem cells. *Neuron* 28:69–80.

Quinn JC, Molinek M, Martynoga BS, Zaki PA, Faedo A, Bulfone A, Hevner RF, West JD, Price DJ (2007). Pax6 controls cerebral cortical cell number by regulating exit from the cell cycle and specifies cortical cell identity by a cell autonomous mechanism. *Dev Biol* 302:50–65.

Raballo R, Rhee J, Lyn-Cook R, Leckman JF, Schwartz ML, Vaccarino FM (2000). Basic fibroblast growth factor (Fgf2). is necessary for cell proliferation and neurogenesis in the developing cerebral cortex. *J Neurosci* 20:5012–5023.

Raible F, Brand M (2004). Divide et Impera–the midbrain-hindbrain boundary and its organizer. *Trends Neurosci* 27:727–734.

Rajagopalan S, Vivancos V, Nicolas E, Dickson BJ (2000). Selecting a longitudinal pathway: Robo receptors specify the lateral position of axons in the *Drosophila* CNS. *Cell* 103:1033–1045.

Rajan I, Witte S, Cline HT (1999). NMDA receptor activity stabilizes presynaptic retinotectal axons and postsynaptic optic tectal cell dendrites in vivo. *J Neurobiol* 38:357–368.

Rakic P (1974). Neurons in rhesus monkey visual cortex: Systematic relation between time of origin and eventual disposition. *Science* 183:425–427.

Rando TA (2007). The immortal strand hypothesis: Segregation and reconstruction. *Cell* 129:1239–1243.

Rasin MR, Gazula VR, Breunig JJ, Kwan KY, Johnson MB, Liu-Chen S, Li HS, Jan LY, Jan YN, Rakic P, Sestan N (2007). Numb and Numbl are required for maintenance of cadherin-based adhesion and polarity of neural progenitors. *Nat Neurosci* 10:819–827.

Rathjen J, Haines BP, Hudson KM, Nesci A, Dunn S, Rathjen PD (2002). Directed differentiation of pluripotent cells to neural lineages: Homogeneous formation and differentiation of a neurectoderm population. *Development* 129:2649–2661.

Reif A, Schmitt A, Fritzen S, Lesch KP (2007). Neurogenesis and schizophrenia: Dividing neurons in a divided mind? *Eur Arch Psychiatry Clin Neurosci* 257:290–299.

Ressler KJ, Sullivan SL, Buck LB (1994). Information coding in the olfactory system: Evidence for a stereotyped and highly organized epitope map in the olfactory bulb. *Cell* 79:1245–1255.

Rice DS, Curran T (2001). Role of the reelin signaling pathway in central nervous system development. *Annu Rev Neurosci* 24:1005–1039.

Rickmann M, Amaral DG, Cowan WM (1987). Organization of radial glial cells during the development of the rat dentate gyrus. *J Comp Neurol* 264:449–479.

Rios I, Alvarez-Rodriguez R, Marti E, Pons S (2004). Bmp2 antagonizes sonic hedgehog-mediated proliferation of cerebellar granule neurones through Smad5 signalling. *Development* 131:3159–3168.

Robinson M, Adu J, Davies AM (1996). Timing and regulation of trkB and BDNF mRNA expression in placode-derived sensory neurons and their targets. *Eur J Neurosci* 8:2399–2406.

Roelink H (2000). Hippocampus formation: An intriguing collaboration. *Curr Biol* 10:R279–281.

Ross SE, Greenberg ME, Stiles CD (2003). Basic helix-loop-helix factors in cortical development. *Neuron* 39:13–25.

Rothstein RP, Levison SW (2002). Damage to the choroid plexus, ependyma and subependyma as a consequence of perinatal hypoxia/ischemia. *Dev Neurosci* 24:426–436.

Roy K, Kuznicki K, Wu Q, Sun Z, Bock D, Schutz G, Vranich N, Monaghan AP (2004). The Tlx gene regulates the timing of neurogenesis in the cortex. *J Neurosci* 24:8333–8345.

Schinder AF, Poo M (2000). The neurotrophin hypothesis for synaptic plasticity. *Trends Neurosci* 23:639–645.

Schmid RS, McGrath B, Berechid BE, Boyles B, Marchionni M, Sestan N, Anton ES (2003). Neuregulin 1–erbB2 signaling is required for the establishment of radial glia and their transformation into astrocytes in cerebral cortex. *Proc Natl Acad Sci USA* 100:4251–4256.

Scholer HR, Ruppert S, Suzuki N, Chowdhury K, Gruss P (1990). New type of POU domain in germ line-specific protein Oct-4. *Nature* 344:435–439.

Scott JW, McBride RL, Schneider SP (1980). The organization of projections from the olfactory bulb to the piriform cortex and olfactory tubercle in the rat. *J Comp Neurol* 194:519–534.

Seki T, Arai Y (1991). Expression of highly polysialylated NCAM in the neocortex and piriform cortex of the developing and the adult rat. *Anat Embryol* 184:395–401.

Serafini T, Colamarino SA, Leonardo ED, Wang H, Beddington R, Skarnes WC, Tessier-Lavigne M (1996). Netrin-1 is required for commissural axon guidance in the developing vertebrate nervous system. *Cell* 87:1001–1014.

Seri B, Garcia-Verdugo JM, McEwen BS, Alvarez-Buylla A (2001). Astrocytes give rise to new neurons in the adult mammalian hippocampus. *J Neurosci* 21:7153–7160.

Seri B, Garcia-Verdugo JM, Collado-Morente L, McEwen BS, Alvarez-Buylla A (2004). *Cell* types, lineage, and architecture of the germinal zone in the adult dentate gyrus. *J Comp Neurol* 478:359.

Shen Q, Zhong W, Jan YN, Temple S (2002). Asymmetric Numb distribution is critical for asymmetric cell division of mouse cerebral cortical stem cells and neuroblasts. *Development* 129:4843–4853.

Shi SH, Jan LY, Jan YN (2003). Hippocampal neuronal polarity specified by spatially localized mPar3/mPar6 and PI 3–kinase activity. *Cell* 112:63–75.

Shi Y, Chichung Lie D, Taupin P, Nakashima K, Ray J, Yu RT, Gage FH, Evans RM (2004). Expression and function of orphan nuclear receptor TLX in adult neural stem cells. *Nature* 427:78–83.

Shimojo H, Ohtsuka T, Kageyama R (2008). Oscillations in Notch signaling regulate maintenance of neural progenitors. *Neuron* 58:52–64.

Shirasaki R, Tamada A, Katsumata R, Murakami F (1995). Guidance of cerebellofugal axons in the rat embryo: Directed growth toward the floor plate and subsequent elongation along the longitudinal axis. *Neuron* 14: 961–972.

Shirasaki R, Mirzayan C, Tessier-Lavigne M, Murakami F (1996). Guidance of circumferentially growing axons by netrin-dependent and -independent floor plate chemotropism in the vertebrate brain. *Neuron* 17:1079–1088.

Sievers J, Hartmann D, Pehlemann FW, Berry M (1992). Development of astroglial cells in the proliferative matrices, the granule cell layer, and the hippocampal fissure of the hamster dentate gyrus. *J Comp Neurol* 320:1–32.

Simpson JH, Bland KS, Fetter RD, Goodman CS (2000a) Short-range and long-range guidance by Slit and its Robo receptors: A combinatorial code of Robo receptors controls lateral position. *Cell* 103:1019–1032.

Simpson JH, Kidd T, Bland KS, Goodman CS (2000b) Short-range and long-range guidance by slit and its Robo receptors. Robo and Robo2 play distinct roles in midline guidance. *Neuron* 28:753–766.

Song H, Ming G, He Z, Lehmann M, McKerracher L, Tessier-Lavigne M, Poo M (1998). Conversion of neuronal growth cone responses from repulsion to attraction by cyclic nucleotides. *Science* 281:1515–1518.

Spitzer NC, Lautermilch NJ, Smith RD, Gomez TM (2000). Coding of neuronal differentiation by calcium transients. *Bioessays* 22:811–817.

Steiner B, Kronenberg G, Jessberger S, Brandt MD, Reuter K, Kempermann G (2004). Differential regulation of gliogenesis in the context of adult hippocampal neurogenesis in mice. *Glia* 46:41–52.

Steiner B, Zurborg S, Horster H, Fabel K, Kempermann G (2008). Differential 24 h responsiveness of Prox1–expressing precursor cells in adult hippocampal neurogenesis to physical activity, environmental enrichment, and kainic acid-induced seizures. *Neuroscience* 154:521–529.

Stenman J, Toresson H, Campbell K (2003). Identification of two distinct progenitor populations in the lateral ganglionic eminence: Implications for striatal and olfactory bulb neurogenesis. *J Neurosci* 23:167–174.

Stoeckli ET, Landmesser LT (1995). Axonin-1, Nr-CAM, and Ng-CAM play different roles in the in vivo guidance of chick commissural neurons. *Neuron* 14:1165–1179.

Stout RP, Graziadei PP (1980). Influence of the olfactory placode on the development of the brain in Xenopus laevis (Daudin). I. Axonal growth and connections of the transplanted olfactory placode. *Neuroscience* 5: 2175–2186.

Stoykova A, Fritsch R, Walther C, Gruss P (1996). Forebrain patterning defects in Small eye mutant mice. *Development* 122:3453–3465.

Strutt D (2003). Frizzled signalling and cell polarisation in *Drosophila* and vertebrates. *Development* 130: 4501–4513.

Sturrock RR, Smart IH (1980). A morphological study of the mouse subependymal layer from embryonic life to old age. *J Anat* 130:391–415.

Sun W, Park KW, Choe J, Rhyu IJ, Kim IH, Park SK, Choi B, Choi SH, Park SH, Kim H (2005a) Identification of novel electroconvulsive shock-induced and activity-dependent genes in the rat brain. *Biochem Biophys Res Commun* 327:848–856.

Sun Y, Goderie SK, Temple S (2005b) Asymmetric distribution of EGFR receptor during mitosis generates diverse CNS progenitor cells. *Neuron* 45:873–886.

Suter DM, Tirefort D, Julien S, Krause KH (2008). A Sox1 to Pax6 switch drives neuroectoderm to radial glia progression during differentiation of mouse embryonic stem cells. *Stem Cells* 27(1):49–58.

Takahashi T, Nowakowski RS, Caviness VS, Jr. (1995). The cell cycle of the pseudostratified ventricular epithelium of the embryonic murine cerebral wall. *J Neurosci* 15:6046–6057.

Takebayashi K, Sasai Y, Sakai Y, Watanabe T, Nakanishi S, Kageyama R (1994). Structure, chromosomal locus, and promoter analysis of the gene encoding the mouse helix-loop-helix factor HES-1. Negative autoregulation through the multiple N box elements. *J Biol Chem* 269:5150–5156.

Tamada A, Shirasaki R, Murakami F (1995). Floor plate chemoattracts crossed axons and chemorepels uncrossed axons in the vertebrate brain. *Neuron* 14:1083–1093.

Tamamaki N, Fujimori KE, Takauji R (1997). Origin and route of tangentially migrating neurons in the developing neocortical intermediate zone. *J Neurosci* 17:8313–8323.

Tanigaki K, Nogaki F, Takahashi J, Tashiro K, Kurooka H, Honjo T (2001). Notch1 and Notch3 instructively restrict bFGF-responsive multipotent neural progenitor cells to an astroglial fate. *Neuron* 29:45–55.

Taniguchi M, Yuasa S, Fujisawa H, Naruse I, Saga S, Mishina M, Yagi T (1997). Disruption of semaphorin III/D gene causes severe abnormality in peripheral nerve projection. *Neuron* 19:519–530.

Tessier-Lavigne M, Placzek M, Lumsden AG, Dodd J, Jessell TM (1988). Chemotropic guidance of developing axons in the mammalian central nervous system. *Nature* 336:775–778.

Tessier-Lavigne M, Goodman CS (1996). The molecular biology of axon guidance. *Science* 274:1123–1133.

Tominaga-Yoshino K, Urakubo T, Okada M, Matsuda H, Ogura A (2008). Repetitive induction of late-phase LTP produces long-lasting synaptic enhancement accompanied by synaptogenesis in cultured hippocampal slices. *Hippocampus* 18:281–293.

Tomita K, Moriyoshi K, Nakanishi S, Guillemot F, Kageyama R (2000). Mammalian achaete-scute and atonal homologs regulate neuronal versus glial fate determination in the central nervous system. *Embo J* 19: 5460–5472.

Toro CT, Deakin JF (2007). Adult neurogenesis and schizophrenia: A window on abnormal early brain development? *Schizophr Res* 90:1–14.

Trachtenberg JT, Stryker MP (2001). Rapid anatomical plasticity of horizontal connections in the developing visual cortex. *J Neurosci* 21:3476–3482.

Treloar HB, Feinstein P, Mombaerts P, Greer CA (2002). Specificity of glomerular targeting by olfactory sensory axons. *J Neurosci* 22:2469–2477.

Tuttle R, Nakagawa Y, Johnson JE, O'Leary DD (1999). Defects in thalamocortical axon pathfinding correlate with altered cell domains in Mash-1–deficient mice. *Development* 126:1903–1916.

Vaccarino FM, Schwartz ML, Raballo R, Nilsen J, Rhee J, Zhou M, Doetschman T, Coffin JD, Wyland JJ, Hung YT (1999). Changes in cerebral cortex size are governed by fibroblast growth factor during embryogenesis. *Nat Neurosci* 2:246–253.

Vaillend C, Poirier R, Laroche S (2008). Genes, plasticity and mental retardation. *Behav Brain Res* 192:88–105.

Varela-Echavarria A, Tucker A, Puschel AW, Guthrie S (1997). Motor axon subpopulations respond differentially to the chemorepellents netrin-1 and semaphorin D. *Neuron* 18:193–207.

Vassar R, Chao SK, Sitcheran R, Nunez JM, Vosshall LB, Axel R (1994). Topographic organization of sensory projections to the olfactory bulb. *Cell* 79:981–991.

Vogel KS, Davies AM (1991). The duration of neurotrophic factor independence in early sensory neurons is matched to the time course of target field innervation. *Neuron* 7:819–830.

Wang KH, Brose K, Arnott D, Kidd T, Goodman CS, Henzel W, Tessier-Lavigne M (1999). Biochemical purification of a mammalian slit protein as a positive regulator of sensory axon elongation and branching. *Cell* 96:771–784.

Wang Y, Baraban SC (2007). Granule cell dispersion and aberrant neurogenesis in the adult hippocampus of an LIS1 mutant mouse. *Dev Neurosci* 29:91–98.

Wang Y, Baraban SC (2008). Aberrant dentate gyrus cytoarchitecture and fiber lamination in Lis1 mutant mice. *Hippocampus* 18:758–765.

Washbourne P, Bennett JE, McAllister AK (2002). Rapid recruitment of NMDA receptor transport packets to nascent synapses. *Nat Neurosci* 5:751–759.

Wassarman KM, Lewandoski M, Campbell K, Joyner AL, Rubenstein JL, Martinez S, Martin GR (1997). Specification of the anterior hindbrain and establishment of a normal mid/hindbrain organizer is dependent on Gbx2 gene function. *Development* 124:2923–2934.

Weiss KH, Johanssen C, Tielsch A, Herz J, Deller T, Frotscher M, Forster E (2003). Malformation of the radial glial scaffold in the dentate gyrus of reeler mice, scrambler mice, and ApoER2/VLDLR-deficient mice. *J Comp Neurol* 460:56–65.

Whitford KL, Marillat V, Stein E, Goodman CS, Tessier-Lavigne M, Chedotal A, Ghosh A (2002). Regulation of cortical dendrite development by Slit-Robo interactions. *Neuron* 33:47–61.

Wichterle H, Garcia-Verdugo JM, Herrera DG, Alvarez-Buylla A (1999). Young neurons from medial ganglionic eminence disperse in adult and embryonic brain. *Nat Neurosci* 2:461–466.

Wichterle H, Turnbull DH, Nery S, Fishell G, Alvarez-Buylla A (2001). In utero fate mapping reveals distinct migratory pathways and fates of neurons born in the mammalian basal forebrain. *Development* 128: 3759–3771.

Williams RW, Strom RC, Goldowitz D (1998). Natural variation in neuron number in mice is linked to a major quantitative trait locus on Chr 11. *J Neurosci* 18:138–146.

Won SJ, Kim SH, Xie L, Wang Y, Mao XO, Jin K, Greenberg DA (2006). Reelin-deficient mice show impaired neurogenesis and increased stroke size. *Exp Neurol* 198:250–259.

Wonders CP, Anderson SA (2006). The origin and specification of cortical interneurons. *Nat Rev Neurosci* 7: 687–696.

Wu Y, Liu Y, Levine EM, Rao MS (2003). Hes1 but not Hes5 regulates an astrocyte versus oligodendrocyte fate choice in glial restricted precursors. *Dev Dyn* 226:675–689.

Yang F, Wang JC, Han JL, Zhao G, Jiang W (2008). Different effects of mild and severe seizures on hippocampal neurogenesis in adult rats. *Hippocampus* 18:460–468.

Yozu M, Tabata H, Nakajima K (2005). The caudal migratory stream: A novel migratory stream of interneurons derived from the caudal ganglionic eminence in the developing mouse forebrain. *J Neurosci* 25:7268–7277.

Zhao S, Chai X, Frotscher M (2007). Balance between neurogenesis and gliogenesis in the adult hippocampus: Role for reelin. *Dev Neurosci* 29:84–90.

Zhou C, Tsai SY, Tsai MJ (2001). COUP-TFI: An intrinsic factor for early regionalization of the neocortex. *Genes Dev* 15:2054–2059.

Zhou C, Qiu Y, Pereira FA, Crair MC, Tsai SY, Tsai MJ (1999). The nuclear orphan receptor COUP-TFI is required for differentiation of subplate neurons and guidance of thalamocortical axons. *Neuron* 24:847–859.

Zhou CJ, Zhao C, Pleasure SJ (2004). Wnt signaling mutants have decreased dentate granule cell production and radial glial scaffolding abnormalities. *J Neurosci* 24:121–126.

Zhu DY, Lau L, Liu SH, Wei JS, Lu YM (2004). Activation of cAMP-response-element-binding protein (CREB) after focal cerebral ischemia stimulates neurogenesis in the adult dentate gyrus. *Proc Natl Acad Sci USA* 101:9453–9457.

5

Neurogenesis in the Adult Subventricular Zone and Olfactory Bulb

Neurogenesis in the adult olfactory bulb generates several types of interneurons; neurogenesis in the adult olfactory epithelium replaces receptor neurons

THE ADULT OLFACTORY SYSTEM CONTAINS, not one, but two neurogenic regions: a turnover of olfactory receptor neurons in the olfactory epithelium and adult neurogenesis in the olfactory bulb, in which several types of interneurons at two major locations are generated. Historically, the exploration of both processes was done parallel to and to a large degree independently of each other. Functionally, however, both forms of adult neurogenesis might be intricately linked. And despite the different precursor cell populations, many features are shared between neurogenesis in the olfactory epithelium and within the CNS. Neurogenesis in adult olfactory epithelium is described and discussed in Chapter 8, p. 304.

Smelling is the evolutionary conserved detection of chemicals in the air and has strong links to emotional behavior

Smelling is our sense that detects chemicals in the air. Our olfactory system can distinguish an estimated 10,000 smells and thus thousands of chemical compounds. This task is accomplished by the integration of signals from about ten million olfactory receptor neurons specialized in the detection of individual chemical compounds. Roughly 2% of the human genome is needed to encode for the variety of hundreds of odor receptors.

It is not necessary to consciously identify and name the chemical compounds underlying a particular smell to react appropriately. Smelling is a powerful alarm system that not only warns of impending dangers but also attracts animals to food and mates. Rodents and canines rely on the olfactory system as much as humans rely on the visual system to form an inner representation of the world. The olfactory memory is one of the strongest forms of memory we have. We cannot actively evoke olfactory memory, but a tiny olfactory cue is sufficient to kick off a wealth of associations and profound memories, often closely associated with emotions. Marcel Proust's monumental novel *Remembrance of Things Past* unfolds from such minimal cues and buttresses the point. The tight link between smell and emotions allows intuitive responses to potentially dangerous situations, so that smell, much more than the other senses, relates to an intrinsic alarm system.

Neurogenesis in the adult olfactory epithelium takes place in the peripheral nervous system from local precursor cells

Olfactory receptor neurons are not nearly so long-lasting as the memories they help generate. Throughout one's life these neurons are continuously exchanged through neurogenesis in the adult olfactory epithelium. We will discuss adult neurogenesis in the olfactory epithelium, together with other examples of neurogenesis outside the CNS, in Chapter 8. The olfactory receptor neurons project to the olfactory bulb where they form connections to the mitral and tufted cells within distinct clusters called *glomeruli*. The mitral and tufted cells project out of the olfactory bulb to the olfactory tubercle.

The second and more widely appreciated form of adult neurogenesis in the olfactory system occurs within the boundaries of the CNS and generates interneurons in the granule cell layer and the periglomerular region of the olfactory bulb. Based on the combined expression of different markers (e.g., calretinin, calbindin, tyrosine hydroxylase, glutamate, etc.; as well as numerous transcription factors) various "types" of adult-born interneurons are identifiable, but it is not yet clear how many of these really constitute distinct functional classes (Lledo et al., 2008; Brill et al., 2009). To date, seven types of new interneurons have been identified (see Table 5–1).

The precursor cells for adult neurogenesis in the adult olfactory bulb reside in the walls of the lateral ventricle. Most proliferation occurs in the lateral walls, but precursor cell activity is not limited to this location. The germinative matrix is called the *subventricular zone*, usually abbreviated to SVZ. From the adult SVZ the developing new neurons migrate over a relatively long distance into the olfactory bulb. They do so in a particular form of migration called "chain migration." The anatomical structure along which this migration occurs is called the "rostral migratory stream" (RMS) or, less popular, "rostral migratory path" (RMP).

The receptor neurons in the olfactory epithelium project to the olfactory bulb, where they form synapses with the second neuron of the olfactory pathway, the mitral and tufted cells. The synapses cluster in distinctive structures, the glomeruli of the olfactory bulb. The newborn interneurons interact with mitral and tufted cells. Thus, whereas the receptor neurons in the periphery are constantly replaced, the mitral and tufted cells, the second neuron in the olfactory system, are stable: only interneurons relating between them are replaced or added. The projection neurons themselves are generated prenatally. It is important to note, however, that not all olfactory bulb interneurons are

Table 5–1 Adult-generated interneurons in the olfactory bulb					
Localization	**Marker**	**Connect to …**	**%**	**Origin**	**Transcription factors**
Granule cell layer					
• superficial	GABA	Tufted cells	95%	Dorsal SVZ	Emx1
• deep	GABA, CamK4	Mitral cells		Ventral SVZ	Nkx2.1
Periglomerular layer	Calretinin, GABA	?	4.5%	Septal SVZ	Dlx5/6, Gsh2
	Calbindin, GABA	?		Ventral SVZ	Dlx5/6, Gsh2, Nkx2.1
	Glutamate	?	?	Dorsal SVZ	Neurog2, Tbr2
	Tyrosine hydroxylase (TH), GABA, Er81		0.5%	Dorsal SVZ	Dlx5/6, Gsh2, Emx1, Pax6, Er81
Plexiform layer	Parvalbumin (?)	?	< 0.1%	?	?

born postnatally or in the adult; many develop before birth (Batista-Brito et al., 2008). But calretinin-expressing periglomerular interneurons, for example, form almost exclusively postnatally (Batista-Brito et al., 2008).

||||||||||||||||||||||||||||||||||||||||||||||||||||||||||

Neuronal development is spread out over a long distance between the SVZ and olfactory bulb

||||||||||||||||||||||||||||||||||||||||||||||||||||||||||

Adult neurogenesis of interneurons in the olfactory bulb (OB) is often simply referred to as "adult olfactory bulb neurogenesis" because in the olfactory bulb the terminally differentiated new neurons are found. Alternatively, and actually by now more prevalently, one finds the term *adult SVZ neurogenesis* or, combining the two: *adult SVZ/OB neurogenesis*. The latter might be most appropriate as the entire process of neuronal development spreads out over a large brain area between the ventricle walls and the olfactory bulb (Fig. 1–4 and 5–1). Neurogenesis in the adult SVZ/OB persists lifelong, at least in rodents (Kuhn et al., 1996; Luo et al., 2006; Molofsky et al., 2006) and is intricately linked with the genesis of oligodendrocytes arising from the same precursor cells as the neurons (Nait-Oumesmar et al., 1999; Hack et al., 2004; Menn et al., 2006). Strictly speaking, the bipotentiality of the precursor cells for neurons and oligodendrocytes has not yet been proven in vivo, but the ex vivo evidence is there (Menn et al., 2006).

||||||||||||||||||||||||||||||||||||||||||||||||||||||||||

Adult neurogenesis results in a turnover of interneurons in the olfactory bulb

||||||||||||||||||||||||||||||||||||||||||||||||||||||||||

In the adult olfactory bulb of young adult rodents, tens of thousands of new neurons are generated each day (Lois and Alvarez-Buylla, 1994). As many as half of these cells die within days and weeks (Winner et al., 2002), but over the lifespan of a mouse or rat, there might be a modest increase in the size of the olfactory bulb. This net growth, however, has not yet been firmly established by stereological means. In 1977, Hinds and McNelly

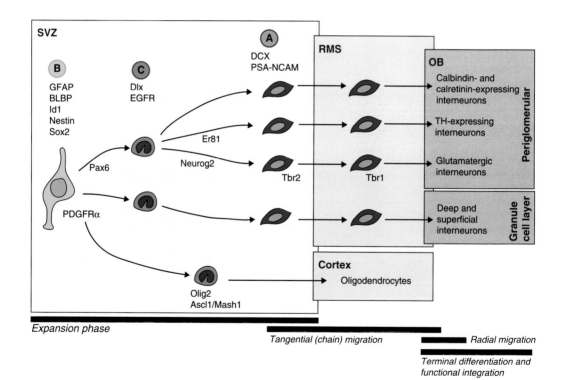

FIGURE 5–1 Developmental stages in adult olfactory bulb neurogenesis. Neuronal development in the subventricular zone (SVZ), rostral migratory stream (RMS), and olfactory bulb proceeds through a series of stages that can be identified immunohistochemically. This pattern shows similarities to, but also important differences from, the situation in the hippocampus (Fig. 6–6).

wrote that the rodent olfactory bulb does not grow over the rodent life span and in rats shows a stable size until two years of age, after which there is a decrease in size (Hinds and McNelly, 1977). The study focused, however, on the mitral and tufted cells. Kaplan and Hinds later reported that the number of granule cells in the rat olfactory bulb almost doubled between four and 31 months of age (Kaplan, 1985). New neurons in the olfactory bulb have long survival periods after an initial peak in the number of cells at about two weeks after cell division (Kato et al., 2000). After labeling dividing cells with BrdU, the number of new neurons in the olfactory bulb did not change between four weeks and two years after labeling (Winner et al., 2002). A cumulative genetic ablation study eliminating the adult-generated progeny of nestin-expressing cells, however, indicated that in contrast to the dentate gyrus, the lack of new neurons over time resulted in a decrease in granule cells in the olfactory bulb (Imayoshi et al., 2008). This finding argues in favor of the idea that adult neurogenesis in the olfactory bulb is, especially in younger age, cumulative and later leads to a substantial neuronal turnover. This is in contrast to the cellular turnover in the olfactory epithelium, where adult neurogenesis maintains a constant receptor cell number under the pressure of constant damage to the olfactory neurons.

The Cellular Basis of the Adult SVZ/Olfactory Bulb Neurogenesis

The Adult Subventricular Zone (SVZ)

The term SVZ for the adult germinative niche grew historically and is problematic

It was long thought that the neurogenic SVZ would be limited to the lateral walls of the lateral ventricles; and in fact this is the region where most of the precursor cells reside. In addition, however, precursor cells that contribute to neurogenesis in the olfactory bulb are also found in the dorsal SVZ (i.e., the roof of the lateral ventricle) and the medial (i.e., septal) wall (De Marchis et al., 2007; Kohwi et al., 2007; Merkle et al., 2007; Young et al., 2007). In rodents the SVZ is defined as a one- or two-cell body–wide zone below the ependyma, whereas in humans a conspicuous cell-free gap exists between the ependyma and the band of the putative precursor cells (Sanai et al., 2004; Curtis et al., 2007). The in some respects more appropriate term *subependymal layer* is sometimes used synonymously, but the literature overwhelmingly prefers the term *subventricular zone*. In fact, and as discussed further below, neither term seems to be fully precise in that the true highest-ranking precursor cells with their cell bodies in the SVZ are in physical contact with the ventricular surface next to the ependymal cells. This and a number of other inconsistencies have led Andrew K. Chojnacki, Gloria M. Mak, and Samuel Weiss to question whether the ventricular zone has indeed disappeared postnatally and if the "SVZ" should not be better referred to as the "periventricular" region (Chojnacki et al., 2009). We will here continue to use the widely accepted and established term "SVZ," but it needs to be acknowledged that it is a historical and operational term, not identical to the SVZ during embryonic cortical development. Neurogenic precursor cells do not only reside periventricularly but also within the RMS and OB.

Most but not all of the adult SVZ relates to the ganglionic eminences in the embryonic brain

The adult SVZ as we encounter it in reports on adult neurogenesis in the olfactory bulb is a remnant of the SVZ in the ganglionic eminences of the embryonic forebrain, presumably primarily the lateral ganglionic eminence. When LGE tissue was transplanted into the postnatal SVZ, many types of periglomerular interneurons formed; but not the calretinin-expressing interneurons of the granule cell layer (Kohwi et al., 2007). A population of Emx1-expressing cells from outside the LGE produced the Calretinin-positive as well as other OB interneurons (Kohwi et al., 2007). According to one study by Monika Brill, Magdalena Götz, and colleagues, precursor cells in the dorsal part of the adult SVZ gave rise to a population of glutamatergic juxtaglomerular interneurons in the olfactory bulb, which based on the transcription factors involved was a "dorsal" process (Brill et al., 2009).

The adult SVZ is thus presumably largely not entirely of ventral origin, whereas the "embryonic SVZ," as it was defined by the Boulder Committee (see p. 127), referred explicitly only to the development of the cortex in the dorsal forebrain (Committee, 1970). Nevertheless, in the ganglionic eminences, too, precursor cell activity in a ventricular zone precedes proliferation in a subventricular layer, and many other features that characterize neurogenesis in the (dorsal) embryonic SVZ are also applicable to the (originally ventral) adult SVZ. The adult SVZ as it appears in the publications about adult neurogenesis is nevertheless not identical to "the" SVZ of the embryonic brain and even less so to the SVZ of the Boulder Committee. Normally, the imprecise use of the term does not cause much harm, but occasionally unjustified extrapolations are made if something is described as "for the SVZ", although a specific developmental time or location was meant.

During ventral forebrain development, neuroepithelial cells directly lining the ventricle expand, first thickening the ventricular layer, then building up a distinct SVZ. In mice, the SVZ of the ganglionic eminences first becomes visible around embryonic day 11 (E11) (Smart, 1976). For more details see also Chapter 4, p. 126. By E15, proliferation in the SVZ surpasses cell divisions in the VZ. Radial glia not only provides guidance structure to newly generated cells but also acts as precursor cells. Postnatally, radial glia transforms into a type of astrocyte located in the SVZ while still retaining some characteristics of radial glia (Alvarez-Buylla et al., 2001; Chojnacki et al., 2009). These radial glia–like astrocytes no longer span the entire cortical thickness but reach from the ventricular surface to Virchow-Robin spaces around the deep blood vessels of the cortex, thereby still maintaining contact with the pial surface (Fig. 5–2). In the classical view, in the postnatal and adult brain, the ventricular zone has transformed into the ependyma, the epithelial lining of the ventricles. In this process, the pseudostratified epithelium of the ventricular zone turns into a simple, columnar epithelium (Sturrock and Smart, 1980). Through gaps between the ependymal cells, the precursor cells of the adult SVZ (B1 cells) reach the ventricular surface.

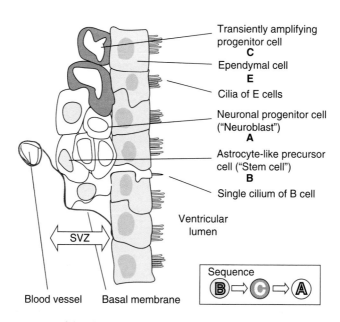

FIGURE 5–2 Cellular composition of the adult subventricular zone. The original concept and nomenclature were based on the work of Fiona Doetsch, Arturo Alvarez-Buylla, and colleagues (Doetsch et al., 1999a, 1999b). See also Figure 5–3.

B Cells Are the Astrocyte-Like Stem Cells of the Subventricular Zone and Contact the Ventricle

||||||||||||||||||||||||||||||||||||||||||||||||||||||||||||||

Stem cells in the adult SVZ are often referred to as B cells

||||||||||||||||||||||||||||||||||||||||||||||||||||||||||||||

The question of the identity of the precursor cells in the adult SVZ that is the ultimate origin of adult neurogenesis has remained a matter of debate since the initial report by Reynolds and Weiss that "stem cells" could be isolated from this brain region (Reynolds and Weiss, 1992). The key issue is the relationship between these precursor cells and the ependyma. To some degree the question is semantic and taxonomic. But the difficulties in defining the identity of particular cell types lead to uncertainties about the precise lineage relationships.

||||||||||||||||||||||||||||||||||||||||||||||||||||||||||||||

B cells express GFAP and generate intermediate progenitor cells called C cells

||||||||||||||||||||||||||||||||||||||||||||||||||||||||||||||

In 1997, Fiona Doetsch, Arturo Alvarez-Buylla, and colleagues first showed that glial fibrillary acidic protein–expressing astrocytic cells act as the primary stem cells in the adult SVZ (Doetsch et al., 1997). They named these astrocytes "B cells" and identified in vivo a transiently amplifying progenitor cell (C cell) originating from the B cell and in turn giving rise to immature neurons or neuroblasts (A cell; Figs. 5–1, 5–2, and 5–3). Ultrastructurally, two types of SVZ astrocytes could be distinguished; B1 cells close to the ependyma, and B2 cells toward the striatum. B2 cells ensheathe migratory A cells (Doetsch et al., 1997). B1 cells express transcription factor Id1, which is responsible for their maintenance (Nam and Benezra, 2009). B cells are considered to represent direct successors of radial glial cells

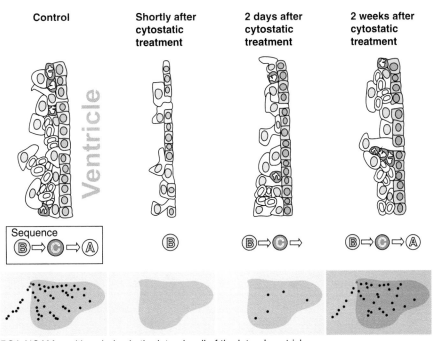

PSA-NCAM-positive chains in the lateral wall of the lateral ventricle

FIGURE 5–3 Cytostatic drugs disrupt integrity of the subventricular zone. Cytostatic drugs wipe out the highly proliferative progenitor cell populations (C and A) but spare some B cells. These repopulate the germinative matrix over time after the chemical assault. These findings have provided evidence of the hypothesis that astrocyte-like B cells are the stem cells of this region. Note, however, the B cells within the band of ependymal cells early after drug administration. The question of whether ependymal cells might transform into B cells remains open. Schematic drawing is based on a figure in Doetsch et al. (1999a).

(Merkle et al., 2004) and in fact show a rudimentary radial morphology. They have a long basal process that runs parallel to the ventricular surface (Mirzadeh et al., 2008). Whether one adheres to the designation of B cells as "astrocyte-like" or "radial glia–like" will ultimately not matter. In any case the suffix "-like" deserves attention (see also Chap. 3, p. 68). Electrophysiologically, postnatal B cells stand somewhere between what has been described for astrocytes and for radial glia (Liu et al., 2006).

There is little doubt today that the stem cell of the adult SVZ is a GFAP-expressing cell and by that standard can be referred to as an astrocyte. The selective elimination of dividing cells in the SVZ with cytostatic agent cytosine arabinoside (Ara-C) resulted in a complete loss of all signs of adult neurogenesis (Fig. 5–3). In particular, the rapidly dividing progenitor cells (C cells) disappeared. Over time, after the treatment, neurogenesis reappeared (Doetsch et al., 1999a). Cells that expressed glial fibrillary acidic protein reappeared first. Other proliferative cells followed, and this sequence of events was interpreted as a genealogy: later cells originated from the early GFAP-positive cells.

GFAP-expressing B cells show stem cell properties ex vivo

An isolation protocol based on the expression of GFAP allowed the prospective culture of cells with stem cell properties from the adult SVZ (Laywell et al., 2000). Laywell and colleagues examined the precursor cell properties of astrocytes from various brain regions and found that astrocyte monolayers, including those derived from the SVZ, showed self-renewal and multipotency, whereas ependymal cells showed growth in agglomerates but were unipotent only (Laywell et al., 2000). Derek van der Kooy's group at the University of Toronto addressed the same issue by microdissecting the SVZ and the ependymal layer. They found that ependymal cells were unipotent and produced only glial cells in their neurosphere assay. Cells in the SVZ, by contrast, were multipotent ex vivo (Chiasson et al., 1999).

To address at least the question of aberrant GFAP expression in vivo, Doetsch and colleagues made use of a transgenic mouse (GFAP-TVA, which had been developed by Eric Holland and Harold Varmus) in which the receptor for an avian virus not infectious for mammalian cells is expressed under the GFAP promoter. The avian virus was coupled to β-galactosidase as a reporter gene and injected into the ventricle. Because the GFAP promoter drives the receptor expression, only GFAP-positive cells could be infected by the virus and thus acquire the reporter gene. Because the reporter gene itself was expressed under a ubiquitous promoter, the expression of the reporter gene product was maintained after infection and after the GFAP promoter had become inactive. Thus, cells that at later time points expressed β-galactosidase could be identified as originating from a cell that must have expressed GFAP (and therefore the avian virus receptor) when the virus was injected. No isolation step that might induce other properties in the cells was involved. This method confirmed that lineage-restricted precursor cells and new neurons were derived from an originally GFAP-positive cell in vivo (Doetsch et al., 1999b). Similarly, elimination of all proliferating GFAP-positive cells by expressing tymidine kinase under the GFAP promoter, which made them sensitive to the deadly action of the drug ganclicovir, resulted in a loss of stem cells that could be isolated from the adult SVZ (Imura et al., 2003). This stem cell depletion was specific to the SVZ (and probably the hippocampus) but did not affect the retina, a region that does contain neural precursor cells but undergoes no adult neurogenesis (Morshead et al., 2003).

Whether ependymal cells are neural stem cells, too, has been hotly debated

Because ependymal cells can in principle express GFAP, although they do not do so under normal conditions, the possibility that ependymal cells might act as stem cells could not be excluded from these experiments. Dye-labeling experiments indeed suggested that newborn neurons in the OB were of ependymal origin (Johansson et al., 1999). In addition, the process of isolating cells from the tissue might also have induced GFAP expression in ependymal cells. In many non-mammalian species, both ependymal and non-ependymal cells act as precursor cells, so the exact identity of neural stem cells as primarily ependymal or sub-ependymal and astrocytic remained debatable (Morshead and van der Kooy, 2004; Chojnacki et al., 2009). The situation might be different in cases of pathology. After ischemia, ependymal cells in fact could produce new neuroblasts and neurons, but this potential is normally suppressed by Notch signaling. Inhibiting Notch allowed neurogenesis from ependymal cells (Fig. 3–10) (Carlen et al., 2009). "Normal" ependymal cells do not proliferate in vivo (or hardly do so: Spassky et al., 2005; Coskun et al., 2008). They also do not behave as multipotent precursor cells ex vivo (Chiasson et al., 1999; Laywell et al., 2000), but for the studies to prove this point, the absence of GFAP had been used as a criterion, which just turned the argument on its head. The question could not be terminally decided so long as no independent additional defining characteristic of ependymal vs. subventricular astrocytes could be used.

||||||||||||||||||||||||||||||||||||||||||||||||||||||||||||||||

B1 cells contact the ventricular surface with a cilia-bearing process

||||||||||||||||||||||||||||||||||||||||||||||||||||||||||||||||

In summary, the stem cell (that is, here, the precursor cell highest in the local precursor cell hierarchy) of the adult ventricular wall appears to be a cell with astrocytic features whose cell body resides in the SVZ but extends a cellular process between the ependymal cells and thus maintains direct contact with the ventricular surface (Fig. 5–2). Unlike common ependymal cells, which carry a number of long cilia, B1 cells present one single cilium on their ventricular (i.e., apical) surface, similar to the one cilium of neural precursor cells during embryonic development. That cilium of embryonic precursor cells from the VZ originates from a specialized patch of cell membrane that is also characterized by the expression of prominin1/CD133 and presumably plays a critical role in the process of asymmetrical divisions (see p. 115, and Weigmann et al., 1997). In at least two studies, however, Prominin1 was detectable not on the subventricular astrocytes but on ependymal cells (Pfenninger et al., 2007; Coskun et al., 2008). But were these "real" ependymal cells or the apical parts of otherwise subventricular astrocytes? Mirzadeh et al. described Prominin1 on the apical process touching the ventricular surface of a little fewer than one third of the B1 cells (Mirzadeh et al., 2008).

||||||||||||||||||||||||||||||||||||||||||||||||||||||||||||||||

Ependymal cells surround B1 cells in a rosette-like pattern

||||||||||||||||||||||||||||||||||||||||||||||||||||||||||||||||

The ependymal cells are arranged in rosette- or pinwheel-like arrangements around the protruding B1 cells, creating a delicate anatomical fine structure (Fig. 5–4). The ependymal cells immediately around the opening carry cilia in large numbers (now referred to as *E1 cells*), whereas the other ependymal cells carry only two cilia (now E2) (Mirzadeh et al., 2008). This organization presumably represents an important aspect of the stem cell niche in the adult SVZ (Currle and Gilbertson, 2008), and the close contact between ependyma and astrocyte-like precursor cells is likely to have important regulatory functions. It does not solve all relationship problems between the two cell types. A similar but not identical model of the fine structure of the SVZ has been proposed by Shen et al. and emphasized the contact of SVZ astrocytes with

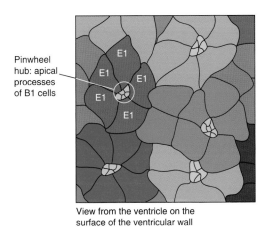
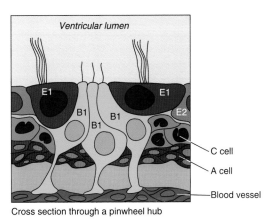

View from the ventricle on the surface of the ventricular wall

Cross section through a pinwheel hub

FIGURE 5–4 Fine structure of the ventricular wall. The processes of the B1 cells, the stem cells of the SVZ, whose cell soma lies in the subependymal layer, reach the ventricular surface surrounded by a rosette of ependymal cells (E1). E2 cells are other ependymal cells outside these pinwheels. B1 cells also have basal processes reaching the wall of blood vessels. B1 cells give rise to C cells, which in turn generate migratory A cells (compare Figs. 3–10, 5–1, 5–2 and 5–3). The figure has been freely drawn after the results presented in Mirzadeh et al. (2008).

the local vasculature. Through the laminin receptor α6β1-integrin, the astrocytes linked to the laminin-rich extracellular matrix around the vessels, and blockade of α6β1-integrin led to their detachment from the vasculature and reduced proliferation (Shen et al., 2008). Despite their similarities, the two models are not identical and cannot simply be merged.

||||||||||||||||||||||||||||||||||||||||||||||||||||||||||

Precursor cell identities in the adult SVZ have not yet been fully resolved

||||||||||||||||||||||||||||||||||||||||||||||||||||||||||

As a summary, despite much progress in the identification of the stem cell of the adult ventricular wall, the issue is far from resolved (Chojnacki et al., 2009). There is no question that GFAP-expressing cells drive adult neurogenesis, but GFAP is not specific to astrocytes. The complex, delicate spatial relationship between the processes of periventricular astrocytes in their B1 and B2 varieties (or in their two forms touching the ventricle or expanding parallel to its surface, according to Shen et al., 2008), common ependymal cells and the more specialized E2 form, and tanycytes (see below, p. 168) further complicates the exact determination of which cells contribute, and how, to the process of adult neurogenesis. In the following paragraphs, we thus operationally continue to use the term *B cell* for the highest-order precursor cell of the adult SVZ (and ventricular region) as introduced by Fiona Doetsch, Arturo Alvarez-Buylla and colleagues, independent of the obvious fact that its exact nature still remains a matter of further exploration.

For more details on the niche concept please refer to Chapter 3, p. 88, and for the question of factors determining "neurogenic permissiveness" within the niche to Chapter 8, p. 277.

C Cells Are the Transiently Amplifying Progenitor Cells of the Adult SVZ

||||||||||||||||||||||||||||||||||||||||||||||||||||||||||

C cells represent the largest population of proliferating SVZ progenitor cells

||||||||||||||||||||||||||||||||||||||||||||||||||||||||||

In addition to the GFAP-positive B cells and the ependymal cells, Doetsch and Alvarez-Buylla identified by light and electron microscopy the transiently amplifying progenitor cell they called the "C cell" (Doetsch, 2003). After injection of proliferation markers such as bromodeoxyuridine (BrdU) or tritiated thymidine, C cells are most frequently labeled; they represent the largest pool of dividing cells in the adult SVZ. C cells maintain close proximity to the radial elements provided by B cells, leading to highly proliferative clusters of C cells.

||||||||||||||||||||||||||||||||||||||||||||||||||||||||||

C cells are EGF-responsive

||||||||||||||||||||||||||||||||||||||||||||||||||||||||||

Ex vivo, C cells are multipotent (Doetsch et al., 2002a), but whether this is a constituent intrinsic property or a case of reprogramming by the culture conditions is not yet clear. C cells express the EGF receptor (Ciccolini et al., 2005), and extrinsic application of EGF to the ventricle induces massive cell proliferation (Kuhn et al., 1997). Intriguingly, the EGF infusions did not increase neurogenesis in the OB but rather reduced it, whereas the production of new astrocytes in the bulb was increased (Kuhn et al., 1997). Intracerebroventricular infusion of FGF2 in contrast stimulated neurogenesis. This implies that the EGF-responsive cells are presumably bipotent for the neuronal and astrocytic lineage.

||||||||||||||||||||||||||||||||||||||||||||||||||||||||||

Precursor cells ex vivo presumably largely correspond to C cells in vivo

||||||||||||||||||||||||||||||||||||||||||||||||||||||||||

Most precursor cells isolated from the SVZ by means of any protocol seem to correspond to C cells in vivo, indicating that upon exposure to EGF, C cells reveal a latent potential for true stem cell behavior, self-renewal, and within-tissue multipotency (Doetsch et al., 2002a). Prospective isolation of an SVZ precursor cell with strong sphere-forming capacity has been based on the expression of nestin and the absence of GFAP, heat-stable antigen, and peanut agglutinin (two surface molecules; Rietze et al., 2001), presumably also corresponding to C cells.

||||||||||||||||||||||||||||||||||||||||||||||||||||||||||

C cells express transcription factor Dlx2

||||||||||||||||||||||||||||||||||||||||||||||||||||||||||

C cells express homeobox transcription factor Dlx2 and can be isolated on the basis of this expression (Doetsch et al., 2002a). In the developing brain, progenitor cells in the medial and lateral ganglionic eminences that express Dlx2 generate interneurons and oligodendrocytes throughout the brain (Pleasure et al., 2000; He et al., 2001), including the hippocampus and the olfactory bulb. In null-mutant mice for Dlx1/2, interneuron development in the olfactory bulb was massively disturbed (Bulfone et al., 1998). In addition, Dlx transcription factors play a specific role in dopaminergic development (Saino-Saito et al., 2003), which is presumably of relevance for the development of TH-positive periglomerular OB interneurons (see below). Dlx2 plays a dual role in that it is generally required for neurogenesis in the adult olfactory bulb, but together with Pax6 also directs, specifically, differentiation into dopaminergic periglomerular interneurons (Brill et al., 2008). It seems that Dlx2 both facilitates the transition from the stem cell to the amplifying progenitor cell stage and the responsiveness of the C cells to mitogenic stimulation by EGF (Suh et al., 2009).

Some C cells express "dorsal" transcription factor Pax6

Many C cells also express Pax6 (Hack et al., 2004). During embryonic development, Pax6 is expressed in radial glia of the dorsal brain, and null mutants of Pax6 lack normal radial glia in the developing cortex and consequently show abortive cortex development (Gotz et al., 1998). Pax6, however, is expressed neither in radial glia of the lateral ganglionic eminence that gives rise to olfactory bulb interneurons, nor in the adult B cells.

Proliferation of C cells is controlled through a p27kip1-dependent mechanism

Cell cycle length is an important parameter in controlling the rate of cell division. Regulation of cell cycle length occurs mainly by varying the duration of the G1 phase, bringing dividing cells in a state of transient cell cycle arrest. This state is also called *G1 transition* and is associated with the expression of p27kip1. P27kip1 binds to cell cycle–dependent kinases (CDK) whose enzymatic activity controls the length of the cell cycle and specifically causes a lengthening of the G1 phase. Accordingly, both dividing and terminally post-mitotic cells are p27kip1 negative. Increased p27kip1 activity provides a mechanism for arresting proliferative cells in the cell cycle and thereby effectively controlling the expansion of the population. Loss of p27kip1 function in null-mutant mice resulted in an increase in BrdU incorporation in the SVZ by increasing the population of C cells and, consequently, A cells. The number of B cells, in contrast, was unaffected (Doetsch et al., 2002b). This did not lead to increased neurogenesis in the bulb, however. Rather, the RMS was increased in diameter, and a higher rate of apoptosis was detected. Growth of the olfactory bulb was delayed (Li et al., 2009). This suggests a function of p27kip1 that goes beyond the expansion of the progenitor cells.

A Cells Are the Migrating Neuroblasts

From the transiently amplifying progenitor cells the young migratory "neuroblasts," or A cells, are derived. Most of the A cells express the polysialylated form of the neural cell adhesion molecule (PSA-NCAM) and doublecortin (DCX), two molecules associated with neuronal migration. All three cell types are considered to be self-renewing, although to a different degree (Doetsch et al., 2002a).

Heterogeneity of A cells reflects diversity of new interneurons

A cells are not homogenous and express a variety of different markers. These patterns presumably indicate the types of interneuron in the bulb they are destined to become. As discussed below, the site of origin within the SVZ largely determines the fate of the newborn cells. A cells from different domains in the SVZ feed the production of different populations of interneurons (Table 5–1).

During migration, A cells continue to divide and initiate neuronal maturation (Menezes et al., 1995). Proliferation, however, is much slower than in the SVZ, and the cell cycle time is lengthened (Smith and Luskin, 1998). Migrating A cells express immature neuronal markers such as DCX, PSA-NCAM, and β-III-tubulin. Like in the adult hippocampus, DCX expression in the SVZ and RMS is associated with the most critical period of neuronal development, ranging from a progenitor cell stage to the existence of a differentiated neuron that is integrated into the circuitry (Brown et al., 2003; Ambrogini et al., 2004; Rao and Shetty, 2004).

The sequence of developmental stages in the adult SVZ, RMS, and olfactory bulb is B–C–A, with the E cells (for "ependyma") standing outside this order (Fig. 5–1). Note that in this scheme there are currently no D cells.

Tanycytes: Where Are the D cells?

The role of D cells (tanycytes) is not fully clear

The initial model also contained a D cell, the tanycyte (Doetsch et al., 1997). Tanycytes are unusual cells that maintain contact with both the ventricular and the pial surface of the brain. They are the most numerous macroglial cell type in lower vertebrates but in adult mammals are limited to the periventricular organs and a few other places. Their functional role as a mediators between vasculature, ventricle, and neural tissue, especially neuro-endocrine tissue, is well described for the hypothalamus (Rodriguez et al., 2005). Their morphology shows some resemblance to that of radial glia, and in fact they are thought to be derivatives of radial glia. They are also related to astrocytes by being able to express GFAP under certain conditions (Roessmann et al., 1980) and at least in certain locations (Redecker, 1989). The more recent models of the anatomical fine-structure of the adult ventricular wall proposed by the group of Arturo Alvarez-Buylla (Mirzadeh et al., 2008) did not contain D cells and tanycytes any more, and they were also not mentioned in the model from Sally Temple's group (Shen et al., 2008). B cells and tanycytes share many properties, however, so they cannot be ignored in this context (Chojnacki et al., 2009). Being secretory cells, tanycytes might exert signaling functions or even represent precursor cells themselves, as suggested for other brain regions and species (Chetverukhin and Polenov, 1993). In the end the question boils down to the discrimination power between the different cell types and the validity of our knowledge about the properties of these different cells.

Chain Migration of Neuroblasts to the Olfactory Bulb

Migration of A cells is called "homotypic" and occurs in characteristic chains of cells

Precursor cells can be found in the entire lateral wall of the lateral ventricle. Neural precursor cells and A cells in particular can be isolated from distant caudal parts of the SVZ—in rats, millimeters away from the beginning of the RMS. The RMS itself is several millimeters long in rats, but many centimeters long in primates. Migration over these distances occurs in a unique pattern, or chain migration (Lois and Alvarez-Buylla, 1994; Lois et al., 1996). The migrating cells use each other as guides for the next step. Because they slide along other cells of their own kind, this type of migration is called *homotypic*. Small groups (chains) of cells thereby lead each other on the way. One cell will extend a prominent growth cone that serves as a guidance structure for the migration of a few following cells. These in turn leave behind a tail process that the first cell of the next chain can use to direct its migration.

As described in the previous chapter (p. 126), during development, immature neurons migrate primarily in two distinct forms (Fig. 5–5). Movement of newly generated neurons from the ventricular plate out to their terminal position within the layers of the cortex is radial migration and depends on radial glia as a guiding structure. Tangential migration, in contrast, is typical for the integration of interneurons from the ganglionic eminences. It occurs parallel to the germinative matrix and does not rely on radial glia as a guiding structure. Positioning of interneurons is still impaired, if the appropriate radial guidance structure is not present, because the cells that move in tangentially switch to a radial path once they are below their target location. This is also the case in the bulb, where the A cells switch from chain migration to radial migration that brings them to their terminal location in the granule cell layer or the periglomerular areas. The RMS terminates in the core of the olfactory bulb. The incoming immature neurons switch the direction of their migration and start to move radially into the granule cell layer and toward the periglomerular layer. It seems that this type of radial migration is dependent on a scaffold provided by blood vessels (Bovetti et al., 2007c).

The rodent RMS is ensheathed by a glial tube of unknown function

Chain migration itself has features of both radial and tangential migration. In rodents, as the route along which the developing neurons migrate, the RMS is shielded against the rest of the brain by glial cells, which form a tube-like structure. These astrocytes are slowly dividing and have been equated to the B cells (B2) of the SVZ. They may perform precursor cell

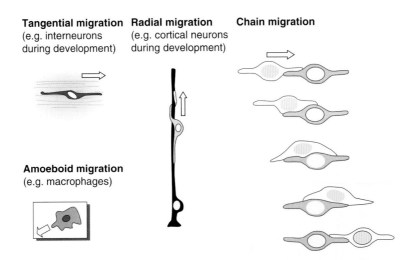

Tangential migration
(e.g. interneurons
during development)

Radial migration
(e.g. cortical neurons
during development)

Chain migration

Amoeboid migration
(e.g. macrophages)

FIGURE 5–5 Chain migration. Migrating neuronally determined progenitor cells (A cells) travel between the subventricular zone and olfactory bulb in a particular form of migration that is independent of a glial guidance structure. In chain migration, the migrating cells guide each other in what is called "homotypic" migration. This mechanism is distinct from the radial and tangential migration found during brain development. In the brain, amoeboid (free) migration is found in microglia and macrophages.

functions in the RMS and olfactory bulb as well (Gritti et al., 2002). Otherwise the exact function of this glial tube is unknown, because chain migration can occur in the absence of astrocytes. Disrupting the integrity of the glial sheath by manipulating the Beta1 integrin (which is the receptor for laminin) resulted in cells moving away from the stream (Belvindrah et al., 2007). One other idea is that the surrounding glia might regulate the speed of migration by tritrating ambient levels of GABA (Bolteus and Bordey, 2004). In rabbits, migration toward the olfactory bulb occurs in chains but independently of a glial ensheathing (Luzzati et al., 2003). In humans, it has been difficult to identify the anatomical structure of an RMS for similar reasons and detect migratory chains (Sanai et al., 2004). Extensive microscopic analyses revealed the migration of human neuroblasts along the rudiments of the obliterated olfactory ventricle in a fashion quite dissimilar to that of rodents in that neither the glial tube nor the characteristic clusters of migrating cells were found (Curtis et al., 2007). Chain migration has been described for monkeys, however (Kornack and Rakic, 2001; see also Chap. 11, p. 500, and Fig. 11–4). Whether or not the cells migrate in chains or more individually might be a matter of cell numbers in relation to the distance to be covered.

The glial tube does not seem to be necessary for migration to the bulb, and many species, including humans, do not have one. Removal of the rodent olfactory bulb causes an almost complete disappearance of the glia ensheathing the RMS, but only a reduction in the migration toward the site of the olfactory bulb, which also does not occur in a disorderly fashion (Jankovski et al., 1998). The underlying mechanism might be found in particular interactions with the extracellular matrix in that individual migrating cells secrete metalloproteinases (Bovetti et al., 2007a). Migration within the glial tube was independent of metalloproteinase activity; e.g., ADAM2 (Bovetti et al., 2007a; Murase et al., 2008). This mechanism would not explain, however, how the cells find their way in the absence of a glial guidance structure.

The structure of the RMS relates to the olfactory ventricle, which is usually obliterated in adulthood

Pathways of this kind can also be found as a network crisscrossing the lateral walls of the ventricles, becoming denser rostrally and merging toward the characteristic dorsoanterior spandril under the corpus callosum. All of these paths feed into the RMS, which essentially represents a tissue volume immediately around the obliterated olfactory ventricle connecting the lateral ventricle with the olfactory bulb. In this way, there is continuity between the SVZ, the RMS, and the core tissue of the olfactory bulb, where the arriving new cells disperse and migrate toward the granule cell layer and the periglomerular layers.

A cells express migratory molecules PSA-NCAM and DCX

The network of migratory chains crisscrossing the SVZ have a predominant orientation toward the RMS (Sawamoto et al., 2006). Migration is generally unidirectional toward the olfactory bulb, although many individual pieces of the paths can have other directions. Because no clear anatomical guidance structure exists, differentially expressed guidance molecules might be responsible for steering migration. PSA-NCAM, which is expressed on the migrating cells, enables migration but it is neither specific nor sufficient (Hu et al., 1996; Chazal et al., 2000). Nevertheless, interfering with PSA-NCAM, for example by enzymatically removing the polysialization, reduced migration and resulted in smaller olfactory bulbs (Tomasiewicz et al., 1993; Cremer et al., 1994; Ono et al., 1994). Concomitantly, the migrating cells express DCX, which is thought to facilitate cytoskeletal flexibility. In vitro, DCX was necessary to allow migration of cells away from neurospheres, and overexpression after knockdown could rescue the effect (Ocbina et al., 2006).

Several other factors have been identified that control normal migration in the SVZ and the RMS but no complete picture has emerged yet. Migrating cells, for example, show calcium transients depending on L-type calcium channels, suggesting conceptual links with other types of plasticity, but the exact function of these transients is not yet known. Their manipulation did not alter migration (Darcy and Isaacson, 2009).

Ephrin signaling plays diverse roles during migration and maturation

The different SVZ cells express both Ephrin receptor tyrosine kinases and their ligands in varying patterns. Besides controlling precursor cell proliferation in a negative feedback mechanism (Holmberg et al., 2005; Jiao et al., 2008), Ephrin activity is involved in additional aspects of plasticity such as in axon guidance (Drescher, 1997; Orioli and Klein, 1997) and pruning (Xu and Henkemeyer, 2009), synaptic maturation (Lim et al., 2008), and especially neuronal migration (Santiago and Erickson, 2002; Takeuchi et al., 2009). Interference with Ephrin signaling in vivo inhibited the migration of A cells toward the olfactory bulb and increased cell proliferation in the SVZ (Conover et al., 2000). Ephrin-dependent migration is achieved through matrix metalloproteinases, which alter the cell's interaction with the extracellular matrix (Lin et al., 2008). By that both EphrinA5 and B2 act repulsive (Lin et al., 2008; Zimmer et al., 2008), but neither of these details has yet been shown for adult neurogenesis in the SVZ/OB. Ephrin signaling is also relevant for adult hippocampal neurogenesis (Ricard et al., 2006; Chumley et al., 2007).

Neuregulins might be necessary for migration

The receptor tyrosine kinases ErbB and their ligands, the neuregulins, are expressed in the SVZ and RMS (Perroteau et al., 1999). The neuregulins have functional properties similar to those of growth factor EGF and the ErbB/neuregulin systems play a major role during brain development (Britsch, 2007). Conditional ErbB4 mutants targeted to nestin- or GFAP-expressing cells showed impaired chain migration of A cells and disturbed neuronal differentiation (Anton et al., 2004). Expression of ErbB4 in periglomerular cells and mitral cells, as well as of Neuregulin1 is dependent on synaptic input from the olfactory epithelium (Perroteau et al., 1999; Oberto et al., 2001). Here, the link to adult neurogenesis and the effects on the incoming A cells has not yet been established.

The roles of chemoattractants and -repellants is not yet clear

There is as yet no strong evidence that an attractant factor in the olfactory bulb would direct migrating A cells toward the bulb (but see below). The cells migrated in the right direction even when the olfactory bulb was ablated (Jankovski et al., 1998; Kirschenbaum et al., 1999). The pcd/pcd mutant mouse is characterized by a postnatal loss of mitral cells in the olfactory bulb. This quite-severe alteration of the network integrity in the bulb also did not exert a strong effect on adult neurogenesis, and in particular, migration was normal. Due to the lack of mitral cells, only the terminal integration was altered (Valero et al., 2007).

Others have proposed that it is less an attractant in the olfactory bulb and more a repellent from the more caudal parts of the SVZ that directs cell migration. Part of this driving force is Slit, a repulsant factor that is present in the adult striatum and choroid plexus and repels neural precursor cells both in vivo and in vitro (Wu et al., 1999). In Slit mutants, migration in the SVZ and RMS is reduced and occurs laterally and backwards, so Slit might normally prevent migrating SVZ cells from wandering into other brain regions, somewhat like putting a fence around the SVZ. Migrating A cells express the receptor for Slit: Robo (Marillat et al., 2002). Migrating chains were present in adult Slit1 null-mutants, but the cells were oriented towards the corpus callosum and not the olfactory bulb (Nguyen-Ba-Charvet et al., 2004). The repulsant theory of migration in the RMS had already been dismissed because Slit is effective over a distance of a millimeter at most. If Slit were secreted in an autocrine or paracrine fashion, however, the migrating neuroblasts themselves would help distribute the factor necessary for migration. If Slit acted solely as a repulsant, this would not make sense; the cells would prevent their own migration and that of the following cells. One still-debated idea about how this discrepancy might be resolved is that the migrating cells would follow the rostro-caudal flow of the cerebrospinal fluid, which carries Slit protein from the choroid plexus with it (Sawamoto et al., 2006).

Guofa Liu and Yi Rao from Washington University in St. Louis revisited the chemoattractant hypothesis and found that when the olfactory bulb was removed, migration toward the olfactory bulb strongly decreased but was not completely stopped (Liu and Rao, 2003). This decrease could be averted by transplanting a piece of olfactory bulb. The same result did not occur when a piece of neocortex was implanted or when some known chemoattractants were applied. These results argue in favor of some chemoattractant in the olfactory bulb. One candidate is Sonic hedgehog (Shh), which has numerous effects on precursor cell activity in the SVZ (Ahn and Joyner, 2005; Palma et al., 2005). Migrating A cells express Shh receptor Patched and were attracted by experimentally manipulated sources of Shh (Angot et al., 2008). There is also a feedback mechanism increasing Patched expression in response to elevated Shh levels (Charytoniuk et al., 2002). According to the *Allen Brain Atlas*, Shh mRNA is found in the periglomerular layer of the olfactory bulb (http:// mouse.brain-map.org) (Lein et al., 2007), but its role there has not yet been explored further.

After removal of the olfactory bulb, new post-mitotic neurons accumulate at the lesion where they might leave remnants of the RMS and direct their further migration specifically toward the frontal cortex and the anterior olfactory nucleus. The promotion of cell migration was also maintained for implanted SVZ precursor cells (Jankovski et al., 1998), which suggests that the olfactory bulb is not required for many aspects of neuronal migration and differentiation.

Not all progeny of cell proliferation in the adult SVZ migrate to the bulb anyway. Rather, a subset of cells diverges towards the cortex. Here, two large streams can be identified: cells from the SVZ are found both directly below the corpus callosum and more diffusely in the cortex above the corpus callosum. The latter migration shows resemblance to migration from the ganglionic eminences during embryonic brain development. Because many of the migrating cells express DCX (and other supposedly neuronal markers), this process has sometimes been misidentified as "cortical neurogenesis." Under physiological conditions, however, neither of these populations generates neurons in the adult brain. This is still different early postnatally, when substantial numbers of interneurons for cortical Layer VI are derived from the SVZ (Inta et al., 2008). Highly circumscript photothrombotic lesions, however, not only caused attraction of precursor cells from the SVZ to the cortex, but also resulted in regenerative neurogenesis (Magavi et al., 2000).

An experimental tumor in the cortex converted Nestin-positive proliferative cells destined to the OB towards the tumor site and thereby effectively reduced neurogenesis in the OB. Nevertheless, the cells did not contribute to neurogenesis near the tumor, but most cells retained an undifferentiated yet DCX- and NG2-positive phenotype (Glass et al.,

2005; Walzlein et al., 2008). Ischemia in the cortex elicited migration from the SVZ directly into the striatum, where under these conditions even a low amount of neurogenesis could be found (Arvidsson et al., 2002).

Molecular Control of Neurogenesis in the Adult SVZ/OB

The expression patterns of key transcription factors have shed some light on the potential heterogeneity and lineage relationship of precursor cells in the adult SVZ. These patterns relate to molecular events that are found during embryonic brain development, but important differences might occur. Also, the mechanisms are not necessarily shared between the two neurogenic regions of the adult brain. (See also Chapter 8, p. 287.)

Stem cell transcription factor Sox2 is expressed in B cells and suppresses neurogenesis

Transcription factor Sox2 identifies multipotent precursor cells throughout brain development and also in the adult (Ellis et al., 2004), but Sox2-positive precursor cells are heterogeneous (Brazel et al., 2005). In rodents as well as in humans, Sox2 is expressed by but not limited to B cells (Baer et al., 2007). Not surprisingly, knockout mice for Sox2, which is strongly expressed in the adult SVZ, also show severely impaired adult neurogenesis, among very many other problems (Ferri et al., 2004). More detailed analyses for the consequences of Sox2 knockdown are available for neurogenesis in the adult hippocampus (p. 407). Besides Sox2, Sox3 is expressed in the adult neurogenic zones. All Sox genes 1–3 suppress neurogenesis and thus allow the maintenance of stem cells (Bylund et al., 2003).

Bmi1 allows self-renewal by suppressing Cdkn2a

Relatively little is known about the transcriptional control of self-renewal, but some interesting molecules have been identified. Cyclin-dependent kinase inhibitor 2A (Cdkn2a), also known as p16INK4a/Arf, needs to be repressed in order to allow self-renewal of SVZ stem cells (Molofsky et al., 2003; Bruggeman et al., 2005). This repression is achieved by transcription factor Bmi1 (Molofsky et al., 2005) and is lost with age, presumably contributing to the senescence of SVZ precursor cells in old age (Molofsky et al., 2006). Interestingly, p19ink4d/Arf shows lower expression in the SVZ than in the RMS (Coskun and Luskin, 2001). The Bmi1-dependent effect was also stronger ex vivo than in vivo (He et al., 2009), so that redundant mechanisms that also differ between different progenitor cells might be in use.

Tlx maintains precursor cells and prevents astrocytic differentiation

Tlx is in the SVZ expressed solely by all B cells, and is necessary both for the transition from the radial glia of the embryonic brain to the adult B cells as well as for self-renewal of the B cells (Liu et al., 2008). Tlx also prevents the (further) astrocytic differentiation of the astrocyte-like precursor cells, thereby maintaining the precursor cell pool (Shi et al., 2004). Tlx is an orphan receptor, not a transcription factor, and its mechanism of action is not yet known.

Ascl1/Mash1 induces neuronal specification in C cells but is also involved in oligodendrogenesis

All types of neural precursor cells in the SVZ can express Ascl1/Mash1, especially transiently amplifying progenitor cells (Parras et al., 2007). Ascl1/Mash1 is required for lineage specification of GABAergic interneurons and induces Dlx genes (Fode et al., 2000), which in turn are a hallmark of C cells. But Ascl1/Mash1 is also involved in the generation of oligodendrocytes, presumably in cells that have already taken this fate choice. For more details on oligodendrogenesis in the SVZ, see below, p. 175.

Emx2, Vax1, and Gli1 are also maintenance factors in the SVZ

Transcription factor Gli1 is the intracellular target of Sonic hedgehog (Shh) signaling and is expressed by SVZ precursor cells in vivo (Ahn and Joyner, 2005; Palma et al., 2005), presumably an important niche factor in the control of precursor cell maintenance. Most notably, Gli1 was found also in quiescent precursor cells, which allowed the reconstitution of the SVZ after an ablation of the intermediate progenitor cells (C and A cells) and contributed to this reconstitution for over a year after the damage (Ahn and Joyner, 2005).

In addition, proliferation of SVZ precursor cells is also controlled by Emx2 (Gangemi et al., 2001; Galli et al., 2002) and Vax1(Soria et al., 2004; Taglialatela et al., 2004), the ablation of both of which impairs precursor cell activity.

Diversity of New Interneurons in the Adult Olfactory Bulb

Different adult-generated interneurons affect all aspects of olfactory bulb circuitry

In migrating A cells, expression of signs of a more mature neuronal phenotype is delayed until the cells have reached the olfactory bulb. Only here do they become positive for their neurotransmitter and show electrophysiological signs of neuronal maturity (Carleton et al., 2003). By the standard of localization and key marker expression, seven different types of adult-generated neurons can be found in the olfactory bulb (Table 5–1) (Bovetti et al., 2007b; Lledo et al., 2008). As it is not clear how these categories relate to functional properties, we do not yet know how many truly different types of neurons are made during adulthood. It has been noted, though, that all known neuronal microcircuits in the olfactory bulb are affected by the addition of new neurons (Lledo et al., 2008). Both temporal and spatial determinants are involved in specifying the neuronal subtype differentiation. Transplantation studies revealed that both the type of interneuron as well as its dendritic morphology are largely dependent on cell-autonomous mechanisms (De Marchis et al., 2007; Kelsch et al., 2007; Kohwi et al., 2007; Merkle et al., 2007).

Most new interneurons are found in the granule cell layer

The new granule cells account for about 95% of the new neurons and are GABAergic (Fig. 5–6). The second population of newly generated interneurons is found in the periglomerular layer and accounts for 5% of the total number of new neurons. These neurons are GABAergic as well, but 10% of them (0.5% of the total population of new neurons) also have dopamine as neurotransmitter (Gall et al., 1987; Kosaka et al., 1987; Winner et al., 2002). In addition, there is one type of glutamatergic interneuron (Brill et al., 2009). Very rarely, new neurons are seen in the external plexiform layer of the olfactory bulb (Winner et al., 2002; Yang, 2008).

Dopaminergic subtypes are specified early

The dopaminergic neurons can be recognized by their expression of tyrosine hydroxylase (TH). A few migrating cells start expressing calretinin during their migration (Jankovski and Sotelo, 1996). Calretinin identifies interneurons in the glomeruli and the granule cell layer of the olfactory bulb but is not found in the tyrosine hydroxylase–positive periglomerular interneurons (Jacobowitz and Winsky, 1991; Rogers, 1992; Li et al., 2002). The first signs of mRNA for TH are found in migrating cells in the RMS as well; TH protein, however, is only in the mature new cells in the olfactory bulb (Baker et al., 2001). A similar result has been obtained based on TH-EGFP reporter gene mice (Saino-Saito et al., 2004).

Many, if not most, of the new olfactory bulb neurons are eliminated by apoptosis

A large proportion of new cells arriving in the olfactory bulb die (Biebl et al., 2000; Winner et al., 2002). Consequently, a surplus of neurons reaches the olfactory bulb and is eliminated thereafter (Fig. 5–6). Among the new cells in the granule cell layer of the rat, about 40% survive for long periods (at least up to 19 months, the latest time-point investigated: Winner et al., 2002). There is a high density of cell death in the SVZ and RMS as well, but in absolute terms, most of the elimination occurs in the olfactory bulb. The fact that cell death is largely delayed to this late point of development suggests that functional connections are required to decide whether a new cell will survive or not. Survival would depend on a Hebbian mechanism of recruiting new cells into function.

Local distribution of neurogenesis in the olfactory bulb changes with time

Addition of new neurons to the different regions of the bulb does not occur uniformly over time. With respect to the granule cell layer, for example, there appears to be an outside-in gradient with more superficial cells tending to become integrated first (Lemasson et al., 2005). It has also been proposed that the progenitor cells that divide latest during migration in the

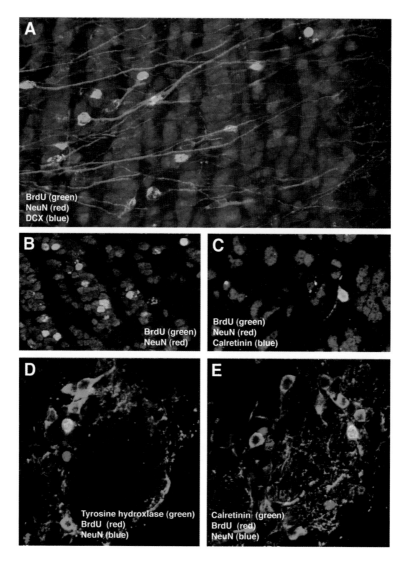

FIGURE 5–6 Adult-generated neurons in the olfactory bulb. Confocal-microscopic depiction of key antigens is used to identify newly generated cells in the course of adult neurogenesis in the olfactory bulb. (A and B): Demonstration of neurogenesis by BrdU/NeuN colocalization. (C and E): New neurons in the granule cell layer are calretinin positive, whereas periglomerular interneurons are not. (D): A small percentage of the new periglomerular interneurons expresses tyrosine hydroxylase, the key enzyme of the dopamine synthesis pathway. Images by H. Georg Kuhn and Christiana Cooper-Kuhn, Gothenburg.

SVZ produce the periglomerular neurons (Betarbet et al., 1996). Transplanting precursor cells from the LGE and adult SVZ heterochronically into the neonatal or adult SVZ revealed that integration into the OB relied more on the age of the donor than of the recipient, and that interneurons for the granule cell layer tend to be produced before the periglomerular cells (De Marchis et al., 2007). Combining a genetic approach based on the Dlx-enhancer to identify the Dlx-derived lineage with a temporal analysis, Batista-Brito and colleagues found that CR-positive periglomerular cells preceded TH-positive cells with the calbindin-positive population peaking in between (Batista-Brito et al., 2008). This result was in some contrast to the findings from the transplantation study by De Marchis and colleagues (De Marchis et al., 2007), but in any case, the time of birth influences the specification of the new interneurons.

Precursor cells in the RMS might still have a multi-lineage potential

When the RMS was transected near its "knee," Sox2-positive proliferating cells that also express DCX were detected in the remainder of the RMS, although no new cells could migrate in from the SVZ. These cells primarily contributed to the periglomerular layer (Mendoza-Torreblanca et al., 2008), largely confirming an earlier, less-detailed study (Fukushima et al., 2002). Nevertheless, the limited neurogenesis that persisted in the bulb affected both the granule cell layer and the periglomerular zone, suggesting that proliferative cells within the RMS and the bulb itself can generate interneurons for both regions. No further sub-specification (e.g., on the basis of transcription factor profiles) has yet been obtained, so it is not clear if in fact the full range of adult neurogenesis can originate from precursor cells that reside in the RMS and the bulb.

Neural precursor cells can also be isolated from the olfactory bulb. After killing dividing cells with a cytostatic agent, large numbers of new C cells appeared in the olfactory bulb as early as two days after the treatment—too soon for migration from the SVZ (Gritti et al., 2002). C cells are normally rare in the olfactory bulb. In this experiment, however, no new TH-expressing cells were found during the recovery of adult neurogenesis in the olfactory bulb.

Precursor cells in the SVZ are regionalized with respect to the interneurons they can generate

Several studies have confirmed that the generation of the different types of interneurons depends on the localization of the primary precursor cell in the SVZ (Merkle et al., 2007). In a study that used an inducible transgenic model that allowed the specific targeting of nestin-expressing cells, only the newly generated periglomerular neurons were labeled but not new interneurons in the granule cell layer. This result suggests a specified developmental path, possibly originating from a separate population of precursor cells (Beech et al., 2004). Interestingly, the different types of new interneurons in the periglomerular zone, TH-positive or calretinin-positive, appeared to originate from the precursor cell pools in the dorsal SVZ, which is separate from the cells that produce new interneurons in the granule cell layer. A Tbr2- and Neurogenin1-expressing population of precursor cells in the dorsal SVZ gives rise to a small number of glutamatergic interneurons in the periglomerular zone (Fig. 5–7) (Brill et al., 2009).

Gsh2 expression allows the distinction of LGE precursor cells that give rise to olfactory bulb interneurons vs. striatal projection neurons (Parmar et al., 2003; Stenman et al., 2003). In Gsh2 mutants, the number of Er81-positive cells was drastically reduced (Stenman et al., 2003).

Pax6 expression in C cells particularly promotes the generation of new dopaminergic interneurons in the adult OB. Blockade of Pax6 expression reduced predominantly this population, not the non-dopaminergic cells, which are dependent on transcription factors Sp8 (Waclaw et al., 2006) and Er81 (Stenman et al., 2003). Pax6 interacts with Dlx2 to specify the dopaminergic periglomerular cell type (Brill et al., 2008). For Sp8, a concrete function in the specification of OB interneurons has been described, and Sp8 is inversely balanced with Pax6 expression (Kohwi et al., 2005). Er81 is present in the SVZ and persists in the OB in subsets of cells, largely limited to the periglomerular layers and weak in granule cells (Kohwi et al., 2005). Er81 identifies the dopaminergic interneurons of the periglomerular layer. In the OB, Er81 shows an inverse distribution with CamK4, which is found in the deep granule cell layer (Saino-Saito et al., 2007).

Oligodendrocytes from the Adult SVZ

The adult SVZ also produces oligodendrocytes, mostly in the cortex

To integrate the oligodendrocytic lineage into the described concept has posed considerable problems. Oligodendrocyte precursor cells from the adult SVZ allow some remyelination after demyelinating lesions and also contribute to baseline production of oligodendrocytes (Nait-Oumesmar et al., 1999; Menn et al., 2006).

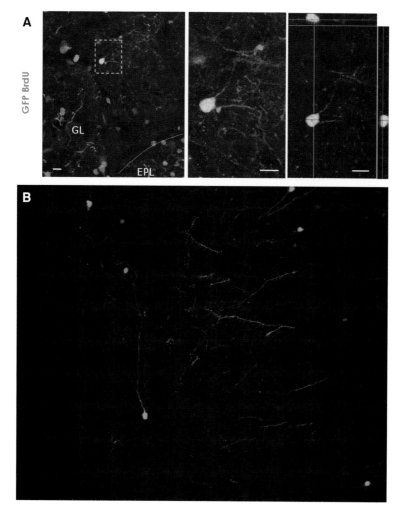

FIGURE 5–7 New glutamatergic interneurons in the adult olfactory bulb. A small fraction of adult-generated neurons in the olfactory bulb is in the glutamatergic lineage and is found in the periglomerular (juxtaglomerular) region. (A): BrdU label in juxtaglomerular neurons expressing GFP derived from Neurog2 expression (E1–Neurog2–cre). BrdU was applied through the drinking water for three weeks, followed by a three-week BrdU-free period. The boxed area is shown at higher magnification as a Z projection (*middle*) and as a single optical section (*right*). Scale bars, 20μm. (B): Lentiviral labeling of periglomerular cells, six weeks after virus injection (many of these cells express vGluT2, not shown here). Images kindly provided by Monika Brill and Magdalena Götz, Munich, and reprinted from Brill et al. (2009) with permission from Macmillan Publishers (*Nature Medicine*), copyright 2009.

SVZ precursor cells ex vivo can generate oligodendrocytes, but whether the precursor cells are tri-potent in vivo was not clear. The embryonic/fetal stem cell antigen LeX/Ssea1/CD15 is expressed by both astrocytic and non-astrocytic cells in the SVZ (Capela and Temple, 2002: Imura, 2006 #4052), which has been taken as indicative of their multipotency or tripotency.

Oligodendrocyte precursor cells are found within the adult SVZ in vivo (Nait-Oumesmar et al., 1999), but the these cells normally do not seem to migrate into the olfactory bulb, but rather contribute to cell genesis in cortical areas. The key marker for the lineage is Olig2 (Hack et al., 2004), but putative oligodendrocyte precursor cells are often identified by their expression of NG2 (see also Chap. 3, p. 71). Their density in the SVZ and RMS is low compared to the brain parenchyma. NG2 cells are also found in the OB, where they can produce oligodendrocytes and astrocytes but no neurons (Komitova et al., 2009). Nevertheless, NG2 cells and oligodendrocytes of SVZ origin can be found in other brain regions (Levison and Goldman, 1993; Marshall and Goldman, 2002; Menn et al., 2006). There, often two types of oligodendrocyte precursor cells are distinguished: myelinating (which express the full range of oligodendrocyte markers and show a mature morphology) and non-myelinating (which essentially retain characteristics of the precursor cells). About the latter, little is known besides the fact that they share markers such as Olig2 and NG2 with the former. Within the adult SVZ, the parenchymal oligodendrocyte precursor cells originate from a separate lineage of development than the neurons in the OB (Fig. 5–1). The NG2-positive oligodendrocyte precursor cells within the SVZ have essentially stopped dividing at this stage. Only one in 200 proliferating cells of the adult murine SVZ is NG2 positive (Komitova et al., 2009). This suggests that the proliferative precursor cell must be found before.

PDGF receptor α-positive B cells generate Olig2-expressing progenitor cells

A subset of B cells is positive for the PDGF receptor alpha and produces progeny that express Olig2 (Menn et al., 2006). Olig2 represses neuronal differentiation and directs SVZ progenitors toward astrocytic and oligodendrocytic fates (Marshall et al., 2005). Olig2 expression overlaps with Nkx2.2, another transcription factor (Kuhlmann et al., 2008). In vitro, overexpression of Olig2 resulted in oligodendrocytic differentiation of embryonic neural precursor cells from mice (Copray et al., 2006). In the developing forebrain, Dlx1/2 acts upstream of Olig2. Dlx1/2 promotes the generation of interneurons, whereas Mash1/Ascl1 promotes the production of oligodendrocytes (Petryniak et al., 2007). Reportedly, PDGF receptor alpha expression is maintained in a few C cells (Menn et al., 2006) but if one uses Dlx1/2 as defining criterion for C cells, the lineage split would have to occur at the level of B cells. Given that PDGF receptor alpha only identifies a subset of B cells, it seems also plausible that the oligodendrocytic lineage might be entirely separate from the neuronal lineage. At a transcriptional level, Olig2 and Mash1/Ascl1 interact to define the oligodendrocytic lineage (Parras et al., 2007). Transplanted cells expressing NG2, Olig2, and Mash1/Ascl1 also contributed to remyelination (Aguirre et al., 2007).

The Olig2-positive cells also express PSA-NCAM (and presumably Doublecortin) but are negative for βIII-tubulin (Menn et al., 2006) and interneuron marker Er81 (Aguirre et al., 2004). On a side note, this observation again underscores that neither Doublecortin nor PSA-NCAM suffice as markers of "neurogenesis." The relationship to NG2 cells in the adult SVZ is not yet entirely clear.

Might NG2 cells in the SVZ retain multilineage potential?

Early postnatally, the picture seems somewhat different. Some EGF-responsive cells also express NG2 (although the majority do not) (Cesetti et al., 2009). In studies using a reporter gene (EGFP) under the CNPase-promoter to identify cells within the oligodendrocytic lineage, Adan Aguirre, Vittorio Gallo, and coworkers at the Children's National Medical Center in Washington, D.C., found that EGFP-positive cells were also NG2-positive and expressed markers of both an interneuronal (including Er81) and oligodendrocytic differentiation (Aguirre and Gallo, 2004). They behaved like C cells ex vivo (but note that "C cell" is a term related to adult neurogenesis in vivo) and were multipotent (Aguirre et al., 2004). What is more, upon implantation in the developing hippocampus, the NG2-positive cells also developed into interneurons (Aguirre et al., 2004). This might imply that only postnatally and with the transition to adult neurogenesis do the two lineages separate.

References

Aguirre A, Gallo V (2004). Postnatal neurogenesis and gliogenesis in the olfactory bulb from NG2-expressing progenitors of the subventricular zone. *J Neurosci* 24:10530–10541.

Aguirre A, Dupree JL, Mangin JM, Gallo V (2007). A functional role for EGFR signaling in myelination and remyelination. *Nat Neurosci* 10:990–1002.

Aguirre AA, Chittajallu R, Belachew S, Gallo V (2004). NG2-expressing cells in the subventricular zone are type C-like cells and contribute to interneuron generation in the postnatal hippocampus. *J Cell Biol* 165:575–589.

Ahn S, Joyner AL (2005). In vivo analysis of quiescent adult neural stem cells responding to Sonic hedgehog. *Nature* 437:894–897.

Alvarez-Buylla A, Garcia-Verdugo JM, Tramontin AD (2001). A unified hypothesis on the lineage of neural stem cells. *Nat Rev Neurosci* 2:287–293.

Ambrogini P, Lattanzi D, Ciuffoli S, Agostini D, Bertini L, Stocchi V, Santi S, Cuppini R (2004). Morpho-functional characterization of neuronal cells at different stages of maturation in granule cell layer of adult rat dentate gyrus. *Brain Res* 1017:21–31.

Angot E, Loulier K, Nguyen-Ba-Charvet KT, Gadeau AP, Ruat M, Traiffort E (2008). Chemoattractive activity of sonic hedgehog in the adult subventricular zone modulates the number of neural precursors reaching the olfactory bulb. *Stem Cells* 26:2311–2320.

Anton ES, Ghashghaei HT, Weber JL, McCann C, Fischer TM, Cheung ID, Gassmann M, Messing A, Klein R, Schwab MH, Lloyd KC, Lai C (2004). Receptor tyrosine kinase ErbB4 modulates neuroblast migration and placement in the adult forebrain. *Nat Neurosci* 7:1319–1328.

Arvidsson A, Collin T, Kirik D, Kokaia Z, Lindvall O (2002). Neuronal replacement from endogenous precursors in the adult brain after stroke. *Nat Med* 8:963–970.

Baer K, Eriksson PS, Faull RL, Rees MI, Curtis MA (2007). Sox-2 is expressed by glial and progenitor cells and Pax-6 is expressed by neuroblasts in the human subventricular zone. *Exp Neurol* 204:828–831.

Baker H, Liu N, Chun HS, Saino S, Berlin R, Volpe B, Son JH (2001). Phenotypic differentiation during migration of dopaminergic progenitor cells to the olfactory bulb. *J Neurosci* 21:8505–8513.

Batista-Brito R, Close J, Machold R, Fishell G (2008). The distinct temporal origins of olfactory bulb interneuron subtypes. *J Neurosci* 28:3966–3975.

Beech RD, Cleary MA, Treloar HB, Eisch AJ, Harrist AV, Zhong W, Greer CA, Duman RS, Picciotto MR (2004). Nestin promoter/enhancer directs transgene expression to precursors of adult generated periglomerular neurons. *J Comp Neurol* 475:128–141.

Belvindrah R, Hankel S, Walker J, Patton BL, Muller U (2007). Beta1 integrins control the formation of cell chains in the adult rostral migratory stream. *J Neurosci* 27:2704–2717.

Betarbet R, Zigova T, Bakay RA, Luskin MB (1996). Dopaminergic and GABAergic interneurons of the olfactory bulb are derived from the neonatal subventricular zone. *Int J Dev Neurosci* 14:921–930.

Biebl M, Cooper CM, Winkler J, Kuhn HG (2000). Analysis of neurogenesis and programmed cell death reveals a self- renewing capacity in the adult rat brain. *Neurosci Lett* 291:17–20.

Bolteus AJ, Bordey A (2004). GABA release and uptake regulate neuronal precursor migration in the postnatal subventricular zone. *J Neurosci* 24:7623–7631.

Bovetti S, Bovolin P, Perroteau I, Puche AC (2007a). Subventricular zone-derived neuroblast migration to the olfactory bulb is modulated by matrix remodelling. *Eur J Neurosci* 25:2021–2033.

Bovetti S, Peretto P, Fasolo A, De Marchis S (2007b). Spatio-temporal specification of olfactory bulb interneurons. *J Mol Histol* 38:563–569.

Bovetti S, Hsieh YC, Bovolin P, Perroteau I, Kazunori T, Puche AC (2007c). Blood vessels form a scaffold for neuroblast migration in the adult olfactory bulb. *J Neurosci* 27:5976–5980.

Brazel CY, Limke TL, Osborne JK, Miura T, Cai J, Pevny L, Rao MS (2005). Sox2 expression defines a heterogeneous population of neurosphere-forming cells in the adult murine brain. *Aging Cell* 4:197–207.

Brill MS, Snapyan M, Wohlfrom H, Ninkovic J, Jawerka M, Mastick GS, Ashery-Padan R, Saghatelyan A, Berninger B, Gotz M (2008). A dlx2- and pax6-dependent transcriptional code for periglomerular neuron specification in the adult olfactory bulb. *J Neurosci* 28:6439–6452.

Brill MS, Ninkovic J, Winpenny E, Hodge RD, Ozen I, Yang R, Lepier A, Gascon S, Erdelyi F, Szabo G, Parras C, Guillemot F, Frotscher M, Berninger B, Hevner RF, Raineteau O, Gotz M (2009). Adult generation of glutamatergic olfactory bulb interneurons. *Nat Neurosci* 12:1524–33.

Britsch S (2007). The neuregulin-I/ErbB signaling system in development and disease. *Adv Anat Embryol Cell Biol* 190:1–65.

Brown JP, Couillard-Despres S, Cooper-Kuhn CM, Winkler J, Aigner L, Kuhn HG (2003). Transient expression of doublecortin during adult neurogenesis. *J Comp Neurol* 467:1–10.

Bruggeman SW, Valk-Lingbeek ME, van der Stoop PP, Jacobs JJ, Kieboom K, Tanger E, Hulsman D, Leung C, Arsenijevic Y, Marino S, van Lohuizen M (2005). Ink4a and Arf differentially affect cell proliferation and neural stem cell self-renewal in Bmi1–deficient mice. *Genes Dev* 19:1438–1443.

Bulfone A, Wang F, Hevner R, Anderson S, Cutforth T, Chen S, Meneses J, Pedersen R, Axel R, Rubenstein JL (1998). An olfactory sensory map develops in the absence of normal projection neurons or GABAergic interneurons. *Neuron* 21:1273–1282.

Bylund M, Andersson E, Novitch BG, Muhr J (2003). Vertebrate neurogenesis is counteracted by Sox1-3 activity. *Nat Neurosci* 6:1162–1168.

Capela A, Temple S (2002). LeX/ssea-1 is expressed by adult mouse CNS stem cells, identifying them as non-ependymal. *Neuron* 35:865–875.

Carlen M, Meletis K, Goritz C, Darsalia V, Evergren E, Tanigaki K, Amendola M, Barnabe-Heider F, Yeung MS, Naldini L, Honjo T, Kokaia Z, Shupliakov O, Cassidy RM, Lindvall O, Frisen J (2009). Forebrain ependymal cells are Notch-dependent and generate neuroblasts and astrocytes after stroke. *Nat Neurosci* 12:259–267.

Carleton A, Petreanu LT, Lansford R, Alvarez-Buylla A, Lledo PM (2003). Becoming a new neuron in the adult olfactory bulb. *Nat Neurosci* 6:507–518.

Cesetti T, Obernier K, Bengtson CP, Fila T, Mandl C, Holzl-Wenig G, Worner K, Eckstein V, Ciccolini F (2009). Analysis of stem cell lineage progression in the neonatal subventricular zone identifies EGFR+/NG2– cells as transit-amplifying precursors. *Stem Cells* 27:1443–1454.

Charytoniuk D, Traiffort E, Hantraye P, Hermel JM, Galdes A, Ruat M (2002). Intrastriatal sonic hedgehog injection increases Patched transcript levels in the adult rat subventricular zone. *Eur J Neurosci* 16:2351–2357.

Chazal G, Durbec P, Jankovski A, Rougon G, Cremer H (2000). Consequences of neural cell adhesion molecule deficiency on cell migration in the rostral migratory stream of the mouse. *J Neurosci* 20:1446–1457.

Chetverukhin VK, Polenov AL (1993). Ultrastructural radioautographic analysis of neurogenesis in the hypothalamus of the adult frog, *Rana temporaria*, with special reference to physiological regeneration of the preoptic nucleus. I. Ventricular zone cell proliferation. *Cell Tissue Res* 271:341–350.

Chiasson BJ, Tropepe V, Morshead CM, van der Kooy D (1999). Adult mammalian forebrain ependymal and subependymal cells demonstrate proliferative potential, but only subependymal cells have neural stem cell characteristics. *J Neurosci* 19:4462–4471.

Chojnacki AK, Mak GK, Weiss S (2009). Identity crisis for adult periventricular neural stem cells: Subventricular zone astrocytes, ependymal cells or both? *Nat Rev Neurosci* 10:153–163.

Chumley MJ, Catchpole T, Silvany RE, Kernie SG, Henkemeyer M (2007). EphB receptors regulate stem/progenitor cell proliferation, migration, and polarity during hippocampal neurogenesis. *J Neurosci* 27: 13481–13490.

Ciccolini F, Mandl C, Holzl-Wenig G, Kehlenbach A, Hellwig A (2005). Prospective isolation of late development multipotent precursors whose migration is promoted by EGFR. *Dev Biol* 284:112–125.

Committee B (1970). Embryonic vertebrate central nervous system: Revised terminology. The Boulder Committee. *Anat Rec* 166:257–261.

Conover JC, Doetsch F, Garcia-Verdugo JM, Gale NW, Yancopoulos GD, Alvarez-Buylla A (2000). Disruption of Eph/ephrin signaling affects migration and proliferation in the adult subventricular zone. *Nat Neurosci* 3:1091–1097.

Copray S, Balasubramaniyan V, Levenga J, de Bruijn J, Liem R, Boddeke E (2006). Olig2 overexpression induces the in vitro differentiation of neural stem cells into mature oligodendrocytes. *Stem Cells* 24: 1001–1010.

Coskun V, Luskin MB (2001). The expression pattern of the cell cycle inhibitor p19(INK4d). by progenitor cells of the rat embryonic telencephalon and neonatal anterior subventricular zone. *J Neurosci* 21:3092–3103.

Coskun V, Wu H, Blanchi B, Tsao S, Kim K, Zhao J, Biancotti JC, Hutnick L, Krueger RC, Jr., Fan G, de Vellis J, Sun YE (2008). CD133+ neural stem cells in the ependyma of mammalian postnatal forebrain. *Proc Natl Acad Sci USA* 105:1026–1031.

Cremer H, Lange R, Christoph A, Plomann M, Vopper G, Roes J, Brown R, Baldwin S, Kraemer P, Scheff S, et al. (1994). Inactivation of the N-CAM gene in mice results in size reduction of the olfactory bulb and deficits in spatial learning. *Nature* 367:455–459.

Currle DS, Gilbertson RJ (2008). The niche revealed. *Cell Stem Cell* 3:234–236.

Curtis MA, Kam M, Nannmark U, Anderson MF, Axell MZ, Wikkelso C, Holtas S, van Roon-Mom WM, Bjork-Eriksson T, Nordborg C, Frisen J, Dragunow M, Faull RL, Eriksson PS (2007). Human neuroblasts migrate to the olfactory bulb via a lateral ventricular extension. *Science* 315:1243–1249.

Darcy DP, Isaacson JS (2009). L-type calcium channels govern calcium signaling in migrating newborn neurons in the postnatal olfactory bulb. *J Neurosci* 29:2510–2518.

De Marchis S, Bovetti S, Carletti B, Hsieh YC, Garzotto D, Peretto P, Fasolo A, Puche AC, Rossi F (2007). Generation of distinct types of periglomerular olfactory bulb interneurons during development and in adult mice: Implication for intrinsic properties of the subventricular zone progenitor population. *J Neurosci* 27:657–664.

Doetsch F (2003). The glial identity of neural stem cells. *Nat Neurosci* 6:1127–1134.

Doetsch F, Garcia-Verdugo JM, Alvarez-Buylla A (1997). Cellular composition and three-dimensional organization of the subventricular germinal zone in the adult mammalian brain. *J Neurosci* 17:5046–5061.

Doetsch F, Garcia-Verdugo JM, Alvarez-Buylla A (1999a). Regeneration of a germinal layer in the adult mammalian brain. *Proc Natl Acad Sci USA* 96:11619–11624.

Doetsch F, Caille I, Lim DA, Garcia-Verdugo JM, Alvarez-Buylla A (1999b). Subventricular zone astrocytes are neural stem cells in the adult mammalian brain. *Cell* 97:703–716.

Doetsch F, Petreanu L, Caille I, Garcia-Verdugo JM, Alvarez-Buylla A (2002a). EGF converts transit-amplifying neurogenic precursors in the adult brain into multipotent stem cells. *Neuron* 36:1021–1034.

Doetsch F, Verdugo JM, Caille I, Alvarez-Buylla A, Chao MV, Casaccia-Bonnefil P (2002b). Lack of the cell-cycle inhibitor p27Kip1 results in selective increase of transit-amplifying cells for adult neurogenesis. *J Neurosci* 22:2255–2264.

Drescher U (1997). The Eph family in the patterning of neural development. *Curr Biol* 7:R799–807.

Ellis P, Fagan BM, Magness ST, Hutton S, Taranova O, Hayashi S, McMahon A, Rao M, Pevny L (2004). Sox2, a persistent marker for multipotential neural stem cells derived from embryonic stem cells, the embryo or the adult. *Dev Neurosci* 26:148–165.

Ferri AL, Cavallaro M, Braida D, Di Cristofano A, Canta A, Vezzani A, Ottolenghi S, Pandolfi PP, Sala M, DeBiasi S, Nicolis SK (2004). Sox2 deficiency causes neurodegeneration and impaired neurogenesis in the adult mouse brain. *Development* 131:3805–3819.

Fode C, Ma Q, Casarosa S, Ang SL, Anderson DJ, Guillemot F (2000). A role for neural determination genes in specifying the dorsoventral identity of telencephalic neurons. *Genes Dev* 14:67–80.

Fukushima N, Yokouchi K, Kawagishi K, Moriizumi T (2002). Differential neurogenesis and gliogenesis by local and migrating neural stem cells in the olfactory bulb. *Neurosci Res* 44:467–473.

Gall CM, Hendry SH, Seroogy KB, Jones EG, Haycock JW (1987). Evidence for coexistence of GABA and dopamine in neurons of the rat olfactory bulb. *J Comp Neurol* 266:307–318.

Galli R, Fiocco R, De Filippis L, Muzio L, Gritti A, Mercurio S, Broccoli V, Pellegrini M, Mallamaci A, Vescovi AL (2002). Emx2 regulates the proliferation of stem cells of the adult mammalian central nervous system. *Development* 129:1633–1644.

Gangemi RM, Daga A, Marubbi D, Rosatto N, Capra MC, Corte G (2001). Emx2 in adult neural precursor cells. *Mech Dev* 109:323–329.

Glass R, Synowitz M, Kronenberg G, Walzlein JH, Markovic DS, Wang LP, Gast D, Kiwit J, Kempermann G, Kettenmann H (2005). Glioblastoma-induced attraction of endogenous neural precursor cells is associated with improved survival. *J Neurosci* 25:2637–2646.

Gotz M, Stoykova A, Gruss P (1998). Pax6 controls radial glia differentiation in the cerebral cortex. *Neuron* 21:1031–1044.

Gritti A, Bonfanti L, Doetsch F, Caille I, Alvarez-Buylla A, Lim DA, Galli R, Verdugo JM, Herrera DG, Vescovi AL (2002). Multipotent neural stem cells reside into the rostral extension and olfactory bulb of adult rodents. *J Neurosci* 22:437–445.

Hack MA, Sugimori M, Lundberg C, Nakafuku M, Gotz M (2004). Regionalization and fate specification in neurospheres: The role of Olig2 and Pax6. *Mol Cell Neurosci* 25:664–678.

He S, Iwashita T, Buchstaller J, Molofsky AV, Thomas D, Morrison SJ (2009). Bmi-1 over-expression in neural stem/progenitor cells increases proliferation and neurogenesis in culture but has little effect on these functions in vivo. *Dev Biol* 328:257–272.

He W, Ingraham C, Rising L, Goderie S, Temple S (2001). Multipotent stem cells from the mouse basal forebrain contribute GABAergic neurons and oligodendrocytes to the cerebral cortex during embryogenesis. *J Neurosci* 21:8854–8862.

Hinds JW, McNelly NA (1977). Aging of the rat olfactory bulb: Growth and atrophy of constituent layers and changes in size and number of mitral cells. *J Comp Neurol* 72:345–367.

Holmberg J, Armulik A, Senti KA, Edoff K, Spalding K, Momma S, Cassidy R, Flanagan JG, Frisen J (2005). Ephrin-A2 reverse signaling negatively regulates neural progenitor proliferation and neurogenesis. *Genes Dev* 19:462–471.

Hu H, Tomasiewicz H, Magnuson T, Rutishauser U (1996). The role of polysialic acid in migration of olfactory bulb interneuron precursors in the subventricular zone. *Neuron* 16:735–743.

Imayoshi I, Sakamoto M, Ohtsuka T, Takao K, Miyakawa T, Yamaguchi M, Mori K, Ikeda T, Itohara S, Kageyama R (2008). Roles of continuous neurogenesis in the structural and functional integrity of the adult forebrain. *Nat Neurosci* 11:1153–1161.

Imura T, Kornblum HI, Sofroniew MV (2003). The predominant neural stem cell isolated from postnatal and adult forebrain but not early embryonic forebrain expresses GFAP. *J Neurosci* 23:2824–2832.

Inta D, Alfonso J, von Engelhardt J, Kreuzberg MM, Meyer AH, van Hooft JA, Monyer H (2008). Neurogenesis and widespread forebrain migration of distinct GABAergic neurons from the postnatal subventricular zone. *Proc Natl Acad Sci USA* 105:20994–20999.

Jacobowitz DM, Winsky L (1991). Immunocytochemical localization of calretinin in the forebrain of the rat. *J Comp Neurol* 304:198–218.

Jankovski A, Sotelo C (1996). Subventricular zone-olfactory bulb migratory pathway in the adult mouse: Cellular composition and specificity as determined by heterochronic and heterotopic transplantation. *J Comp Neurol* 371:376–396.

Jankovski A, Garcia C, Soriano E, Sotelo C (1998). Proliferation, migration and differentiation of neuronal progenitor cells in the adult mouse subventricular zone surgically separated from its olfactory bulb. *Eur J Neurosci* 10:3853–3868.

Jiao JW, Feldheim DA, Chen DF (2008). Ephrins as negative regulators of adult neurogenesis in diverse regions of the central nervous system. *Proc Natl Acad Sci USA* 105:8778–8783.

Johansson CB, Momma S, Clarke DL, Risling M, Lendahl U, Frisen J (1999). Identification of a neural stem cell in the adult mammalian central nervous system. *Cell* 96:25–34.

Kaplan MS (1985). Formation and turnover of neurons in young and senescent animals: An electronmicroscopic and morphometric analysis. *Ann NY Acad Sci* 457:173–192.

Kato T, Yokouchi K, Kawagishi K, Fukushima N, Miwa T, Moriizumi T (2000). Fate of newly formed periglomerular cells in the olfactory bulb. *Acta Otolaryngol* 120:876–879.

Kelsch W, Mosley CP, Lin CW, Lois C (2007). Distinct mammalian precursors are committed to generate neurons with defined dendritic projection patterns. *PLoS Biol* 5:e300.

Kirschenbaum B, Doetsch F, Lois C, Alvarez-Buylla A (1999). Adult subventricular zone neuronal precursors continue to proliferate and migrate in the absence of the olfactory bulb. *J Neurosci* 19:2171–2180.

Kohwi M, Osumi N, Rubenstein JL, Alvarez-Buylla A (2005). Pax6 is required for making specific subpopulations of granule and periglomerular neurons in the olfactory bulb. *J Neurosci* 25:6997–7003.

Kohwi M, Petryniak MA, Long JE, Ekker M, Obata K, Yanagawa Y, Rubenstein JL, Alvarez-Buylla A (2007). A subpopulation of olfactory bulb GABAergic interneurons is derived from Emx1– and Dlx5/6-expressing progenitors. *J Neurosci* 27:6878–6891.

Komitova M, Zhu X, Serwanski DR, Nishiyama A (2009). NG2 cells are distinct from neurogenic cells in the postnatal mouse subventricular zone. *J Comp Neurol* 512:702–716.

Kornack DR, Rakic P (2001). The generation, migration, and differentiation of olfactory neurons in the adult primate brain. *Proc Natl Acad Sci USA* 98:4752–4757.

Kosaka T, Kosaka K, Heizmann CW, Nagatsu I, Wu JY, Yanaihara N, Hama K (1987). An aspect of the organization of the GABAergic system in the rat main olfactory bulb: Laminar distribution of immunohistochemically defined subpopulations of GABAergic neurons. *Brain Res* 411:373–378.

Kuhlmann T, Miron V, Cuo Q, Wegner C, Antel J, Bruck W (2008). Differentiation block of oligodendroglial progenitor cells as a cause for remyelination failure in chronic multiple sclerosis. *Brain* 131:1749–1758.

Kuhn HG, Dickinson-Anson H, Gage FH (1996). Neurogenesis in the dentate gyrus of the adult rat: Age-related decrease of neuronal progenitor proliferation. *J Neurosci* 16:2027–2033.

Kuhn HG, Winkler J, Kempermann G, Thal LJ, Gage FH (1997). Epidermal growth factor and fibroblast growth factor-2 have different effects on neural progenitors in the adult rat brain. *J Neurosci* 17:5820–5829.

Laywell ED, Rakic P, Kukekov VG, Holland EC, Steindler DA (2000). Identification of a multipotent astrocytic stem cell in the immature and adult mouse brain. *Proc Natl Acad Sci USA* 97:13883–13888.

Lein ES et al. (2007). Genome-wide atlas of gene expression in the adult mouse brain. *Nature* 445:168–176.

Lemasson M, Saghatelyan A, Olivo-Marin JC, Lledo PM (2005). Neonatal and adult neurogenesis provide two distinct populations of newborn neurons to the mouse olfactory bulb. *J Neurosci* 25:6816–6825.

Levison SW, Goldman JE (1993). Both oligodendrocytes and astrocytes develop from progenitors in the subventricular zone of postnatal rat forebrain. *Neuron* 10:201–212.

Li X, Tang X, Jablonska B, Aguirre A, Gallo V, Luskin MB (2009). p27(KIP1). regulates neurogenesis in the rostral migratory stream and olfactory bulb of the postnatal mouse. *J Neurosci* 29:2902–2914.

Li Z, Kato T, Kawagishi K, Fukushima N, Yokouchi K, Moriizumi T (2002). Cell dynamics of calretinin-immunoreactive neurons in the rostral migratory stream after ibotenate-induced lesions in the forebrain. *Neurosci Res* 42:123–132.

Lim BK, Matsuda N, Poo MM (2008). Ephrin-B reverse signaling promotes structural and functional synaptic maturation in vivo. *Nat Neurosci* 11:160–169.

Lin KT, Sloniowski S, Ethell DW, Ethell IM (2008). Ephrin-B2–induced cleavage of EphB2 receptor is mediated by matrix metalloproteinases to trigger cell repulsion. *J Biol Chem* 283:28969–28979.

Liu G, Rao Y (2003). Neuronal migration from the forebrain to the olfactory bulb requires a new attractant persistent in the olfactory bulb. *J Neurosci* 23:6651–6659.

Liu HK, Belz T, Bock D, Takacs A, Wu H, Lichter P, Chai M, Schutz G (2008). The nuclear receptor tailless is required for neurogenesis in the adult subventricular zone. *Genes Dev* 22:2473–2478.

Liu X, Bolteus AJ, Balkin DM, Henschel O, Bordey A (2006). GFAP-expressing cells in the postnatal subventricular zone display a unique glial phenotype intermediate between radial glia and astrocytes. *Glia* 54:394–410.

Lledo PM, Merkle FT, Alvarez-Buylla A (2008). Origin and function of olfactory bulb interneuron diversity. *Trends Neurosci* 31:392–400.

Lois C, Alvarez-Buylla A (1994). Long-distance neuronal migration in the adult mammalian brain. *Science* 264:1145–1148.

Lois C, Garcia-Verdugo J-M, Alvarez-Buylla A (1996). Chain migration of neuronal precursors. *Science* 271: 978–981.

Luo J, Daniels SB, Lennington JB, Notti RQ, Conover JC (2006). The aging neurogenic subventricular zone. *Aging Cell* 5:139–152.

Luzzati F, Peretto P, Aimar P, Ponti G, Fasolo A, Bonfanti L (2003). Glia-independent chains of neuroblasts through the subcortical parenchyma of the adult rabbit brain. *Proc Natl Acad Sci USA* 100:13036–13041.

Magavi S, Leavitt B, Macklis J (2000). Induction of neurogenesis in the neocortex of adult mice. *Nature* 405: 951–955.

Marillat V, Cases O, Nguyen-Ba-Charvet KT, Tessier-Lavigne M, Sotelo C, Chedotal A (2002). Spatiotemporal expression patterns of slit and robo genes in the rat brain. *J Comp Neurol* 442:130–155.

Marshall CA, Goldman JE (2002). Subpallial dlx2-expressing cells give rise to astrocytes and oligodendrocytes in the cerebral cortex and white matter. *J Neurosci* 22:9821–9830.

Marshall CA, Novitch BG, Goldman JE (2005). Olig2 directs astrocyte and oligodendrocyte formation in post-natal subventricular zone cells. *J Neurosci* 25:7289–7298.

Mendoza-Torreblanca JG, Martinez-Martinez E, Tapia-Rodriguez M, Ramirez-Hernandez R, Gutierrez-Ospina G (2008). The rostral migratory stream is a neurogenic niche that predominantly engenders periglomerular cells: In vivo evidence in the adult rat brain. *Neurosci Res* 60:289–299.

Menezes JRL, Smith CM, Nelson KC, Luskin MB (1995). The division of neuronal progenitor cells during migration in the neonatal mammalian forebrain. *Mol Cell Neurosci* 6:496–508.

Menn B, Garcia-Verdugo JM, Yaschine C, Gonzalez-Perez O, Rowitch D, Alvarez-Buylla A (2006). Origin of oligodendrocytes in the subventricular zone of the adult brain. *J Neurosci* 26:7907–7918.

Merkle FT, Tramontin AD, Garcia-Verdugo JM, Alvarez-Buylla A (2004). Radial glia give rise to adult neural stem cells in the subventricular zone. *Proc Natl Acad Sci USA* 101:17528–17532.

Merkle FT, Mirzadeh Z, Alvarez-Buylla A (2007). Mosaic organization of neural stem cells in the adult brain. *Science* 317:381–384.

Mirzadeh Z, Merkle FT, Soriano-Navarro M, Garcia-Verdugo JM, Alvarez-Buylla A (2008). Neural stem cells confer unique pinwheel architecture to the ventricular surface in neurogenic regions of the adult brain. *Cell Stem Cell* 3:265–278.

Molofsky AV, Pardal R, Iwashita T, Park IK, Clarke MF, Morrison SJ (2003). Bmi-1 dependence distinguishes neural stem cell self-renewal from progenitor proliferation. *Nature* 425:962–967.

Molofsky AV, He S, Bydon M, Morrison SJ, Pardal R (2005). Bmi-1 promotes neural stem cell self-renewal and neural development but not mouse growth and survival by repressing the p16Ink4a and p19Arf senescence pathways. *Genes Dev* 19:1432–1437.

Molofsky AV, Slutsky SG, Joseph NM, He S, Pardal R, Krishnamurthy J, Sharpless NE, Morrison SJ (2006). Increasing p16INK4a expression decreases forebrain progenitors and neurogenesis during ageing. *Nature* 443:448–452.

Morshead CM, van der Kooy D (2004). Disguising adult neural stem cells. *Curr Opin Neurobiol* 14:125–131.

Morshead CM, Garcia AD, Sofroniew MV, van Der Kooy D (2003). The ablation of glial fibrillary acidic protein-positive cells from the adult central nervous system results in the loss of forebrain neural stem cells but not retinal stem cells. *Eur J Neurosci* 18:76–84.

Murase S, Cho C, White JM, Horwitz AF (2008). ADAM2 promotes migration of neuroblasts in the rostral migratory stream to the olfactory bulb. *Eur J Neurosci* 27:1585–1595.

Nait-Oumesmar B, Decker L, Lachapelle F, Avellana-Adalid V, Bachelin C, Van Evercooren AB (1999). Progenitor cells of the adult mouse subventricular zone proliferate, migrate and differentiate into oligodendrocytes after demyelination. *Eur J Neurosci* 11:4357–4366.

Nam HS, Benezra R (2009). High levels of Id1 expression define B1 type adult neural stem cells. *Cell Stem Cell* 5:515–526.

Nguyen-Ba-Charvet KT, Picard-Riera N, Tessier-Lavigne M, Baron-Van Evercooren A, Sotelo C, Chedotal A (2004). Multiple roles for slits in the control of cell migration in the rostral migratory stream. *J Neurosci* 24:1497–1506.

Oberto M, Soncin I, Bovolin P, Voyron S, De Bortoli M, Dati C, Fasolo A, Perroteau I (2001). ErbB-4 and neuregulin expression in the adult mouse olfactory bulb after peripheral denervation. *Eur J Neurosci* 14: 513–521.

Ocbina PJ, Dizon ML, Shin L, Szele FG (2006). Doublecortin is necessary for the migration of adult subventricular zone cells from neurospheres. *Mol Cell Neurosci* 33:126–135.

Ono K, Tomasiewicz H, Magnuson T, Rutishauser U (1994). N-CAM mutation inhibits tangential neuronal migration and is phenocopied by enzymatic removal of polysialic acid. *Neuron* 13:595–609.

Orioli D, Klein R (1997). The Eph receptor family: Axonal guidance by contact repulsion. *Trends Genet* 13: 354–359.

Palma V, Lim DA, Dahmane N, Sanchez P, Brionne TC, Herzberg CD, Gitton Y, Carleton A, Alvarez-Buylla A, Ruiz i Altaba A (2005). Sonic hedgehog controls stem cell behavior in the postnatal and adult brain. *Development* 132:335–344.

Parmar M, Sjoberg A, Bjorklund A, Kokaia Z (2003). Phenotypic and molecular identity of cells in the adult subventricular zone. In vivo and after expansion in vitro. *Mol Cell Neurosci* 24:741–752.

Parras CM, Hunt C, Sugimori M, Nakafuku M, Rowitch D, Guillemot F (2007). The proneural gene Mash1 specifies an early population of telencephalic oligodendrocytes. *J Neurosci* 27:4233–4242.

Perroteau I, Oberto M, Soncin I, Voyron S, De Bortoli M, Bovolin P, Fasolo A (1999). Transregulation of erbB expression in the mouse olfactory bulb. *Cell Mol Biol* (Noisy-le-Grand, France). 45:293–301.

Petryniak MA, Potter GB, Rowitch DH, Rubenstein JL (2007). Dlx1 and Dlx2 control neuronal versus oligoden-droglial cell fate acquisition in the developing forebrain. *Neuron* 55:417–433.

Pfenninger CV, Roschupkina T, Hertwig F, Kottwitz D, Englund E, Bengzon J, Jacobsen SE, Nuber UA (2007). CD133 is not present on neurogenic astrocytes in the adult subventricular zone, but on embryonic neural stem cells, ependymal cells, and glioblastoma cells. *Cancer Res* 67:5727–5736.

Pleasure SJ, Anderson S, Hevner R, Bagri A, Marin O, Lowenstein DH, Rubenstein JL (2000). Cell migration from the ganglionic eminences is required for the development of hippocampal GABAergic interneurons. *Neuron* 28:727–740.

Rao MS, Shetty AK (2004). Efficacy of doublecortin as a marker to analyse the absolute number and dendritic growth of newly generated neurons in the adult dentate gyrus. *Eur J Neurosci* 19:234–246.

Redecker P (1989). Immunogold electron microscopic localization of glial fibrillary acidic protein (GFAP). in neurohypophyseal pituicytes and tanycytes of the Mongolian gerbil (Meriones unguiculatus). *Histochemistry* 91:333–337.

Reynolds BA, Weiss S (1992). Generation of neurons and astrocytes from isolated cells of the adult mammalian central nervous system. *Science* 255:1707–1710.

Ricard J, Salinas J, Garcia L, Liebl DJ (2006). EphrinB3 regulates cell proliferation and survival in adult neuro-genesis. *Mol Cell Neurosci* 31:713–722.

Rietze RL, Valcanis H, Brooker GF, Thomas T, Voss AK, Bartlett PF (2001). Purification of a pluripotent neural stem cell from the adult mouse brain. *Nature* 412:736–739.

Rodriguez EM, Blazquez JL, Pastor FE, Pelaez B, Pena P, Peruzzo B, Amat P (2005). Hypothalamic tanycytes: A key component of brain-endocrine interaction. *Int Rev Cytol* 247:89–164.

Roessmann U, Velasco ME, Sindely SD, Gambetti P (1980). Glial fibrillary acidic protein (GFAP). in ependymal cells during development. An immunocytochemical study. *Brain Res* 200:13–21.

Rogers JH (1992). Immunohistochemical markers in rat brain: Colocalization of calretinin and calbindin-D28k with tyrosine hydroxylase. *Brain Res* 587:203–210.

Saino-Saito S, Berlin R, Baker H (2003). Dlx-1 and Dlx-2 expression in the adult mouse brain: Relationship to dopaminergic phenotypic regulation. *J Comp Neurol* 461:18–30.

Saino-Saito S, Sasaki H, Volpe BT, Kobayashi K, Berlin R, Baker H (2004). Differentiation of the dopaminergic phenotype in the olfactory system of neonatal and adult mice. *J Comp Neurol* 479:389–398.

Saino-Saito S, Cave JW, Akiba Y, Sasaki H, Goto K, Kobayashi K, Berlin R, Baker H (2007). ER81 and CaMKIV identify anatomically and phenotypically defined subsets of mouse olfactory bulb interneurons. *J Comp Neurol* 502:485–496.

Sanai N, Tramontin AD, Quinones-Hinojosa A, Barbaro NM, Gupta N, Kunwar S, Lawton MT, McDermott MW, Parsa AT, Manuel-Garcia Verdugo J, Berger MS, Alvarez-Buylla A (2004). Unique astrocyte ribbon in adult human brain contains neural stem cells but lacks chain migration. *Nature* 427:740–744.

Santiago A, Erickson CA (2002). Ephrin-B ligands play a dual role in the control of neural crest cell migration. *Development* 129:3621–3632.

Sawamoto K, Wichterle H, Gonzalez-Perez O, Cholfin JA, Yamada M, Spassky N, Murcia NS, Garcia-Verdugo JM, Marin O, Rubenstein JL, Tessier-Lavigne M, Okano H, Alvarez-Buylla A (2006). New neurons follow the flow of cerebrospinal fluid in the adult brain. *Science* 311:629–632.

Shen Q, Wang Y, Kokovay E, Lin G, Chuang SM, Goderie SK, Roysam B, Temple S (2008). Adult SVZ stem cells lie in a vascular niche: A quantitative analysis of niche cell–cell interactions. *Cell Stem Cell* 3:289–300.

Shi Y, Chichung Lie D, Taupin P, Nakashima K, Ray J, Yu RT, Gage FH, Evans RM (2004). Expression and func-tion of orphan nuclear receptor TLX in adult neural stem cells. *Nature* 427:78–83.

Smart IH (1976). A pilot study of cell production by the ganglionic eminences of the developing mouse brain. *J Anat* 121:71–84.

Smith CM, Luskin MB (1998). Cell cycle length of olfactory bulb neuronal progenitors in the rostral migratory stream. *Dev Dyn* 213:220–227.

Soria JM, Taglialatela P, Gil-Perotin S, Galli R, Gritti A, Verdugo JM, Bertuzzi S (2004). Defective postnatal neurogenesis and disorganization of the rostral migratory stream in absence of the Vax1 homeobox gene. *J Neurosci* 24:11171–11181.

Spassky N, Merkle FT, Flames N, Tramontin AD, Garcia-Verdugo JM, Alvarez-Buylla A (2005). Adult ependymal cells are postmitotic and are derived from radial glial cells during embryogenesis. *J Neurosci* 25:10–18.

Stenman J, Toresson H, Campbell K (2003). Identification of two distinct progenitor populations in the lateral ganglionic eminence: Implications for striatal and olfactory bulb neurogenesis. *J Neurosci* 23:167–174.

Sturrock RR, Smart IH (1980). A morphological study of the mouse subependymal layer from embryonic life to old age. *J Anat* 130:391–415.

Suh Y, Obernier K, Holzl-Wenig G, Mandl C, Herrmann A, Worner K, Eckstein V, Ciccolini F (2009). Interaction between DLX2 and EGFR regulates proliferation and neurogenesis of SVZ precursors. *Mol Cell Neurosci* 42:308–314.

Taglialatela P, Soria JM, Caironi V, Moiana A, Bertuzzi S (2004). Compromised generation of GABAergic interneurons in the brains of Vax1–/– mice. *Development* 131:4239–4249.

Takeuchi S, Yamaki N, Iwasato T, Negishi M, Katoh H (2009). Beta2–chimaerin binds to EphA receptors and regulates cell migration. *FEBS Lett* 583:1237–1242.

Tomasiewicz H, Ono K, Yee D, Thompson C, Goridis C, Rutishauser U, Magnuson T (1993). Genetic deletion of a neural cell adhesion molecule variant (N-CAM-180). produces distinct defects in the central nervous system. *Neuron* 11:1163–1174.

Valero J, Weruaga E, Murias AR, Recio JS, Curto GG, Gomez C, Alonso JR (2007). Changes in cell migration and survival in the olfactory bulb of the pcd/pcd mouse. *Dev Neurobiol* 67:839–859.

Waclaw RR, Allen ZJ, 2nd, Bell SM, Erdelyi F, Szabo G, Potter SS, Campbell K (2006). The zinc finger transcription factor Sp8 regulates the generation and diversity of olfactory bulb interneurons. *Neuron* 49:503–516.

Walzlein JH, Synowitz M, Engels B, Markovic DS, Gabrusiewicz K, Nikolaev E, Yoshikawa K, Kaminska B, Kempermann G, Uckert W, Kaczmarek L, Kettenmann H, Glass R (2008). The antitumorigenic response of neural precursors depends on subventricular proliferation and age. *Stem Cells* 26:2945–2954.

Weigmann A, Corbeil D, Hellwig A, Huttner WB (1997). Prominin, a novel microvilli-specific polytopic membrane protein of the apical surface of epithelial cells, is targeted to plasmalemmal protrusions of non-epithelial cells. *Proc Natl Acad Sci USA* 94:12425–12430.

Winner B, Cooper-Kuhn CM, Aigner R, Winkler J, Kuhn HG (2002). Long-term survival and cell death of newly generated neurons in the adult rat olfactory bulb. *Eur J Neurosci* 16:1681–1689.

Wu W, Wong K, Chen J, Jiang Z, Dupuis S, Wu JY, Rao Y (1999). Directional guidance of neuronal migration in the olfactory system by the protein Slit. *Nature* 400:331–336.

Xu NJ, Henkemeyer M (2009). Ephrin-B3 reverse signaling through Grb4 and cytoskeletal regulators mediates axon pruning. *Nat Neurosci* 12:268–276.

Yang Z (2008). Postnatal subventricular zone progenitors give rise not only to granular and periglomerular interneurons but also to interneurons in the external plexiform layer of the rat olfactory bulb. *J Comp Neurol* 506:347–358.

Young KM, Fogarty M, Kessaris N, Richardson WD (2007). Subventricular zone stem cells are heterogeneous with respect to their embryonic origins and neurogenic fates in the adult olfactory bulb. *J Neurosci* 27: 8286–8296.

Zimmer G, Garcez P, Rudolph J, Niehage R, Weth F, Lent R, Bolz J (2008). Ephrin-A5 acts as a repulsive cue for migrating cortical interneurons. *Eur J Neurosci* 28:62–73.

6

Adult Hippocampal Neurogenesis

Adult hippocampal neurogenesis is found in the dentate gyrus only and generates excitatory granule cells

ADULT HIPPOCAMPAL neurogenesis produces new excitatory granule cells in the dentate gyrus, the first relay station of the trisynaptic network backbone of the hippocampus (Fig. 6–1). Although we speak of neurogenesis *in the hippocampus*, neurogenesis actually takes place only in one subregion, the dentate gyrus, not other sub-areas. Confusingly, in historic nomenclatures, the "dentate gyrus" is not even considered part of the hippocampus proper. In fact, there are developmental and structural reasons to subsume the dentate gyrus only to the hippocampal formation and not the hippocampus proper. From some functional perspectives this distinction might seem somewhat outdated and artificial (but see Chap. 10, p. 449). When we talk about adult hippocampal neurogenesis, we mean neurogenesis in the adult dentate gyrus.

The axons of the new granule cells add to the mossy fiber connection between the dentate gyrus and CA3. Neurogenesis in the adult hippocampus is locally much more confined than in the subventricular zone and olfactory bulb. In mice, proliferation of the precursor cells, migration, and differentiation into mature neurons occur within a radius of about 100 μm; only a few cells migrate close to the molecular layer (Kempermann et al., 2003) and only under pathological conditions such as epileptic seizures is a greater areal coverage found (Jessberger et al., 2005). Quantitatively the two neurogenic regions differ dramatically: against the thousands of neurons generated for the rodent olfactory bulb, hippocampal neurogenesis produces only few new cells. In a six-week-old C57BL/6 mouse about 200–300 new neurons that have a lasting existence are born per day, and this number further declines with age. Whereas neurogenesis in the olfactory bulb contributes primarily to a turnover, adult neurogenesis in the hippocampus is cumulative (Ninkovic et al., 2007; Imayoshi et al., 2008).

Adult hippocampal neurogenesis is linked to learning

Extreme local restriction and scarcity of the new cells does not prevent adult hippocampal neurogenesis from exerting a particular fascination. The scientific studies on adult hippocampal neurogenesis by far outnumber those on adult olfactory bulb neurogenesis; in 2008 they were roughly three times as numerous. One reason for this preference is that, from an anthropocentric point of view, the potential functional relevance of new hippocampal neurons appears greater than that of those in the olfactory system. Dogs and rodents, no doubt, would have an opposite opinion and differently structure their research programs. But the hippocampus plays a central role in the concepts of how the human brain and mind work. The hippocampus is called the "gateway to memory," and although this does apply to rodents as well, adult

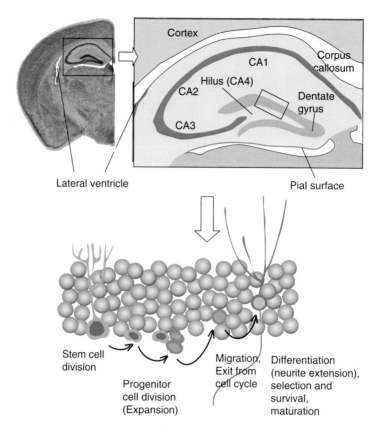

FIGURE 6–1 Summary scheme of adult hippocampal neurogenesis in mice. See text for details.

neurogenesis here might be linked to many processes that we consider essentially human. All animals learn and all mammals require their hippocampus for certain types of learning. The hippocampus prepares and evaluates information before long-term storage takes place in cortical regions. The hippocampus is thus not the primary storage area itself but processes and consolidates so-called declarative information, the knowledge of facts and events. One aspect of how the hippocampus processes declarative contents is by relating items in space and time. The hippocampus enables the positioning of information in coordinate systems, thus making memory "episodic." Autobiographical memory is a special type of episodic memory and is of particular importance to our human self-awareness. Human long-term memory of declarative and episodic contents is thus considered an essential prerequisite for consciousness and our sense of a personal history that can be experienced and communicated. From this insight it is only a small step to the consideration of adult hippocampal neurogenesis in the context of hippocampal malfunctioning as we encounter it in two of the most pressing fields of modern medicine: dementia and depression.

Adult hippocampal neurogenesis is very likely important for human cognition

Thus it is not the hippocampus itself that makes us human—rather, its function contributes to preparing the foundation for the cognitive abilities that set us apart from other animals. Adult hippocampal neurogenesis cannot be the factor that explains these important differences because we share it with all mammals. Adult neurogenesis certainly is in no sense uniquely human; quite the contrary, it seems that we have less of it than other animals. On the other hand, only mammals developed a dentate gyrus at all (see Chap. 11). The new neurons in the hippocampus will not directly solve the puzzle of human cognition and consciousness, but if we attempt to really understand the hippocampus, it will not be possible to leave out adult hippocampal neurogenesis.

Adult hippocampal neurogenesis can be conceptually divided into four phases: (1) the precursor cell phase, during which an expansion of the precursor cell pool occurs; (2) the early survival phase, in which the majority of the newborn cells are eliminated even before they make synaptic contacts with their target regions; (3) the post-mitotic maturation phase, during which functional integration into the network is established; and (4) the late survival phase, in which a final selection presumably based on the new neuron's functionality is achieved.

Anatomy

The rodent hippocampus does not look like a seahorse

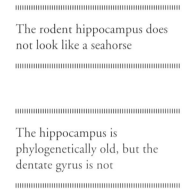

The hippocampus is phylogenetically old, but the dentate gyrus is not

The hippocampus is a bilateral structure found within the temporal lobes of the hemispheres. *Hippocampus* stands for the Latin word for "seahorse," and a prepared human hippocampus indeed resembles the elegantly curved appearance of a seahorse. The area dentata of the hippocampus even mirrors the characteristic serrated back of the seahorse. The rodent hippocampus, on which most of our observations are based, lacks this imaginative resemblance.

The hippocampus is part of the evolutionarily "old" part of the cortex: the *allocortex* with its subdivision called *archicortex*. This region of the cortex is also called the *limbic cortex*; the hippocampus is part of the limbic system and is thereby involved in the processing of emotions. In contrast to the six-layered neocortex, the allocortex has a three-layered structure with a deep plexiforme layer, a band of principal neurons, and a superficial fiber layer. Phylogenetically, the dentate gyrus appears to be younger (see Chap. 10, p. 450).

The hippocampus (or strictly speaking the "hippocampal formation") consists of four parts: the dentate gyrus (also area dentata, or fascia dentata); the cornu ammonis (Ammon's horn); the presubiculum; and the subiculum (Amaral and Witter, 1989). In all modern descriptions of the hippocampus, we encounter the term *CA*, which originally stood for *cornu ammonis* but has taken on a life of its own. In 1933, the anatomist Lorente de Nò, disciple of Ramón y Cajal in Madrid, subdivided the cornu ammonis into four parts, which have been numbered CA1, CA2, CA3, and CA4. Both the patterns of fiber connections and the expression patterns of marker genes have confirmed this anatomical description (Lein et al., 2004). Intriguingly, this even applied to area CA2, which is skipped in the classical trisynaptic scheme of the flow of information.

The mammalian hippocampus has a trisynaptic backbone structure

Axons from the granule cells constitute the mossy fiber tract between dentate gyrus and CA3

The essentially trisynaptic core circuit of the hippocampus (Figs. 6–2 and 10–5) is constituted by the dentate gyrus, CA3 and CA1. Afferents from the entorhinal cortex (EC) via the perforant path reach the outer segments of the dendrites of the granule cells in the dentate gyrus (in the outer molecular layer), where they form the first synapse. The perforant path crosses the hippocampal fissure.

The axons of the granule cells build the mossy fiber tract that reaches the pyramidal neurons in CA3, where they terminate in unusually large and complex structures, the mossy fiber boutons. These are interspersed with interneurons and contain the second synapse with the neurons of CA3. The pyramidal neurons of CA3 project via the Shaffer collateral pathway to CA1, where the third synapse is located. CA1 pyramidal neurons have an axon to the subiculum (from which a projection returns to the entorhinal cortex). This "trisynaptic circuit" is an important and useful concept, but it is a simplification. For example, the EC also directly projects to CA3 and CA2, bypassing the dentate gyrus (Fig. 10–5), and there are fibers from CA3 that project back to the dentate gyrus (Scharfman, 2007).

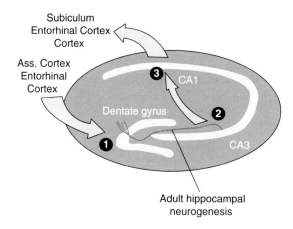

FIGURE 6–2 The trisynaptic backbone of hippocampal circuitry. Adult hippocampal neurogenesis generates only new granule cells in the dentate gyrus and thus only modifies the mossy fiber connection between the dentate gyrus and CA3. No other subsystems incorporate new neurons.

Main input is from the entorhinal cortex

The perforant path can be subdivided into medial and lateral sections, which corresponds to their origin from the medial and lateral entorhinal cortex, and despite their superficially similar appearance show a remarkable number of differences at the biochemical level (Hjorth-Simonsen, 1972; Hjorth-Simonsen and Jeune, 1972; Fredens et al., 1984). The fibers of the lateral EC project to the outermost third of the granule cell dendrites, whereas the fibers from the medial EC terminate at the middle third of the dendrites. Commissural association fibers from the contralateral hippocampus make synapses with the inner segment of the granule cell dendrites in the inner molecular layer (Forster et al., 2006).

Neuronal activity in the dentate gyrus is strongly inhibited by local interneurons

Numerous types of interneurons are the primary modulators of neuronal activity in the dentate gyrus and help in processing the information that is coming from the cortical regions; e.g., the sensory fields, via the EC (Freund and Buzsaki, 1996; Houser, 2007). These are broadly classified into those that form synapses on or near the soma (such as, for example, the basket cells) and those that make connections to the dendrites of the granule cells. Interneurons are inhibitory and express GABA and either or both of the GABA-synthesizing enzymes, GABA decarboxylase, GAD65 or GAD67. They show a large variety of morphological and neurochemical properties that allows their subclassification. Nevertheless, no complete picture has emerged yet, and details about their full connectivity are scarce. By their sheer size, the most prominent interneurons of the dentate gyrus are the basket cells, of which alone at least five subtypes exist (Ribak and Seress, 1983). Whereas most classical interneurons are inhibitory, the hilus also contains mossy cells, a class of excitatory neurons that receives input from numerous sources including collaterals of the mossy fibers and projects back to the inner molecular layer of the dentate gyrus (Soltesz et al., 1993; Soriano and Frotscher, 1994).

Many brain regions project to the dentate gyrus using a wide range of neurotransmitters

In addition, a number of other neurotransmitter systems ("extrinsic afferent systems") project to the hippocampus and the dentate gyrus (Leranth and Hajszan, 2007). They are primarily of subcortical and commissural origin. Serotonergic input from the raphe nuclei, acetylcholinergic and GABAergic projections from the medial septum and diagonal band of Broca, noradrenergic fibers from pontine nuclei and the locus coeruleus, and finally rather weak dopaminergic fibers from the ventral tegmental area terminate in the dentate gyrus. The commissural fibers largely originate from hilar neurons on the contralateral side and mostly end to the inner third of the dendrites of the granule cells. Together all input systems link the hippocampus with essentially the entire brain, and the extrinsic afferent systems exert a very powerful control over the network activity in the dentate gyrus (Leranth and Hajszan, 2007). Their similarly impressive regulatory effects on adult hippocampal neurogenesis are described in greater detail in Chapter 9, p. 354 (Fig. 6–3).

The different cell types that are found in the SGZ have not yet been fully characterized in terms of the types of neuronal input they might receive and how they interact to integrate this spectrum of signals that reaches the neurogenic zone. Although the exact position of the SGZ within the network of the hippocampus is not yet clear, the SGZ is a brain region with unusually complex and diverse innervation. This situation provides a particular backdrop for adult neurogenesis and certainly does not seem to be without relevance for both the function and the regulation of adult neurogenesis.

The dentate gyrus has an upper and a lower blade

The dentate gyrus itself is composed of three laminae: the hilus or polymorphic layer, the granule cell layer with the densely packed band of granule cells, and the molecular layer with its inner and outer sub-lamination. In rostral coronal sections the dentate gyrus appears as a lying "V." The two blades of the V are often referred to as dorsal (or "upper"; i.e., closer to the overarching CA1 but above the pyramidal cells of CA3, and hence also called "suprapyramidal") and ventral (or lower; i.e., opposite to the CA fields, or "infrapyramidal"). This informal nomenclature tends to cause confusion because, as we move more caudally, the dentate gyrus curves around, becomes more and more C-shaped, and what was dorsal now becomes ventral. To overcome this problem, Helen Scharfman and Menno Witter have proposed naming the blade that, independently of being rostral or caudal, remains covered by the CA fields as "enclosed" blade, versus the other blade as "exposed" (Scharfman and Witter, 2007).

The dentate gyrus has a dorsal (or anterior or medial) and a ventral (or posterior or temporal) part

The confusion about the terminology for the blades also arises because of the fact that, functionally, an important distinction can be made between the dorsal and ventral hippocampus as a whole (Fig. 6–4) (Bannerman et al., 2004). The "dorsal hippocampus" plays a more prominent role in learning and memory (Moser et al., 1993), whereas the ventral hippocampus is more involved in affective behavior (Trivedi and Coover, 2004; Czerniawski et al., 2009). In rodents, "dorsal" largely corresponds to the more rostral parts, which because of the curved nature of the dentate gyrus are also closer to the midline and hence the septum. The "ventral" part, in contrast, corresponds to the more caudal section, which is also referred to as "temporal," because it is found at a greater

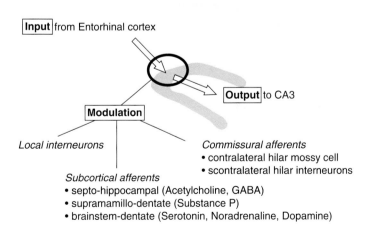

FIGURE 6–3 Input to the dentate gyrus. Several fiber systems, characterized by their neurotransmitter, reach the dentate gyrus and influence the regulation of adult neurogenesis.

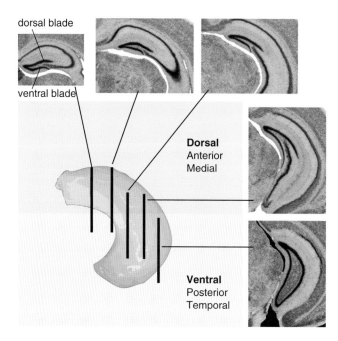

dorsal blade

ventral blade

Dorsal
Anterior
Medial

Ventral
Posterior
Temporal

FIGURE 6–4 Distinction of the dorsal and ventral hippocampus. The dentate gyrus can be divided into two regions, a dorsal (or anterior or medial) part, presumably more involved in cognition, and a ventral (or posterior or temporal) part with greater putative relevance in affective behavior. This distinction is independent from the usage of "dorsal/ventral" to describe the two blades of the granule cell layer.

distance from the midline. The main rostro-caudal axis in rodents is thus also referred to as "septo-temporal." This distinction is independent of the two blades.

Under physiological conditions, adult hippocampal neurogenesis is found only in the dentate gyrus, where it generates new granule cells. There is no conclusive evidence that new neurons are formed in other hippocampal regions, although single reports have proposed otherwise (Rietze et al., 2000). The question of whether inducible restorative neurogenesis is possible in CA1 will be discussed in Chapter 12, p. 553. In an ex vivo study on neurogenic precursor cells from CA1, the separation from the neighboring periventricular area was not warranted so that the interpretation is confounded (Becq et al., 2005), and only gliogenesis was found in CA1 in vivo (Kronenberg et al., 2007).

The Subgranular Zone (SGZ)

The SGZ is the germinative matrix of the dentate gyrus

In vivo, the precursor cells of the dentate gyrus reside in a narrow band of tissue, the subgranular zone (SGZ). The term was coined by Joseph Altman in 1975 (Altman, 1975). For earlier anatomists, this zone was not recognized as a structure that required a name and was nothing more than the border between the granule cell layer and the hilus (or plexiforme layer, or CA4). Other than the SVZ, before 1965 the SGZ had not been identified as a germinative matrix. Descriptively, the SGZ is usually defined as a layer about three cell nuclei wide (20 to 25 μm), including the most basal cell band of the granule cell layer and a two nucleus–wide zone into the hilus (Fig. 6–5). In rodents, the transition between the granule cell layer and the hilus is abrupt, whereas in the human dentate gyrus, no such sharp delineation is found, and the dentate gyrus shows a rather serrated border with the hilus.

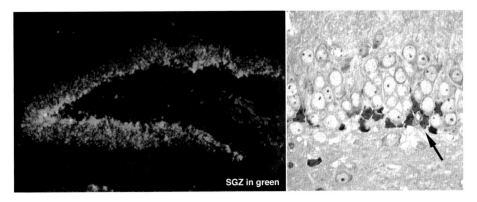

FIGURE 6–5 The subgranular zone. The SGZ contains the neurogenic niche in the adult dentate gyrus. The SGZ can be visualized with nuclear staining that contrasts it against the neurons of the granule cell layer (*left*). In the semi-thin-section, SGZ cells, including the radial glia–like structures, can be distinguished, even in the absence of immunohistochemical labeling (*right*). Note how the dark cells, the putative precursor cells, tightly surround a blood vessel (*arrow*). Figure kindly provided by Gudrun Lutsch. Compare also with Figure 6–7.

The neurogenic niche is composed of precursor cells, immature neurons, and other niche cells

The SGZ provides a neurogenic microenvironment that is permissive for neuronal development throughout life. This microenvironment is often referred to as a "neurogenic niche." The SGZ contains astrocytes resembling radial glial elements (Seri et al., 2001; Filippov et al., 2003; Fukuda et al., 2003; Breunig et al., 2008) and represent the putative stem cells of this region. In addition, a number of different types of intermediate neuronal and glial progenitor cells can be identified (Filippov et al., 2003; Fukuda et al., 2003; Kronenberg et al., 2003; Seri et al., 2004; Encinas et al., 2006), as well as neurons in all stages of differentiation and maturation (Brandt et al., 2003; Ambrogini et al., 2004; Plumpe et al., 2006).

Astrocytes are important niche cells, and precursor cells share astrocytic properties

Astrocytes play a particular role in defining the stem cell niche in the dentate gyrus. Not only have the radial glia–like cells many astrocytic properties (Filippov et al., 2003), astrocytes also "cradle" the developing new neurons (Seri et al., 2001; Shapiro et al., 2005b; Plumpe et al., 2006) and secrete supportive factors that promote precursor-cell maintenance and neuronal development (Song et al., 2002). Some of the astrocyte-derived factors that might mediate these effects have been identified and include two interleukins, IL-1beta and IL-6 (Barkho et al., 2006).

The neurogenic niche is also a vascular niche

The SGZ is also a highly vascularized region, and proliferative hotspots are found close to the blood vessels. Theo Palmer of Stanford University has established the concept of the "vascular niche" as the enabling microenvironment for the precursor cells of the SGZ (Palmer et al., 2000; Palmer, 2002; Fabel et al., 2003a). His theory assumes a close interaction between vascular structures and precursor cell activity and makes reference to similar concepts for bone marrow (Kiel and Morrison, 2006; Sugiyama et al., 2006). For adult neurogenesis in songbirds, a close co-regulation between neurogenesis and angiogenesis was described first (Louissaint et al., 2002). The finding that vascular endothelial growth factor (VEGF) is a potent regulator of angiogenesis on one hand and precursor cell activity and adult neurogenesis on the other supports this idea (Jin et al., 2002; Cao et al., 2004; Jin et al., 2004; Schanzer et al., 2004). Blockade of VEGF signaling even blocked the up-regulating effect of exercise on adult neurogenesis (Fabel et al., 2003b).

Despite the contact between precursor cells and blood vessels in the SGZ, there is no support for the speculation that the hippocampal precursor cells themselves might be actually blood-borne.

In experiments with genetically marked bone marrow, no evidence of integration of bone marrow–derived precursor cells into the SGZ has been found (Priller et al., 2001). However, one study suggested that after ischemia, adventitia cells of local blood vessels might function as precursor cells (Yamashima et al., 2004). The issue of the possible precursor cell potential of pericytes from the microvasculature in the brain has been mentioned in the context of mesenchymal stem cells and NG2 cells (Chap. 3, p. 71). As with other reports on the neurogenic potential of NG2 cells (or MSCs), there is no evidence yet that these cells can mature into functional neurons (Dore-Duffy et al., 2006).

In careful microanatomical analyses, Mercier, Kitasako, and Hatton (2002) described how the entire structural unit, including precursor cells, blood vessels, and microglia, is enclosed by a shared basal membrane (Mercier et al., 2002). Again, this is in line with reports from other stem cell systems, suggesting a prominent role for the basal membrane in maintaining tissue-specific stem cell niches (reviewed in Nikolova et al., 2007). The concept of the stem cell niche and of neurogenic permissiveness will be discussed in greater detail in Chapter 8.

The Precursor Cell Phase of Adult Hippocampal Neurogenesis

The sequence of stem cell, progenitor cell, and "neuroblast" is also found in the adult dentate gyrus

A central concept of stem cell biology is that different degrees of stemness can be distinguished with respect to both the extent of self-renewal and the degree of multipotency. A simplified version of this idea is the pattern of stem cell–progenitor cell–blast that can be found repeatedly in the context of many stem cell systems and that has been discussed in Chapter 3, p. 87 (Fig. 3–6). We have seen that this pattern is also applicable to the SVZ, where astrocyte-like B cells as the highest-ranking precursor cells (stem cells) give rise to transiently amplifying progenitor cells, the C cells, which in turn produce migratory neuroblasts, the A cells. We find this scheme is also realized in the dentate gyrus.

The stem cell of the SGZ expresses GFAP

Bettina Seri and Arturo Alvarez-Buylla identified the astrocyte-like nature of neural precursor cells in the adult hippocampus. In analogy to the previous work by Fiona Doetsch on the SVZ (Doetsch et al., 1999), the proliferative hippocampal cells were ablated with an anti-mitotic agent, and the first cell cohort to reappear expressed astrocytic marker GFAP (Seri et al., 2001). In another set of experiments, avian virus receptor was targeted to glial cells (receptor expression was controlled by the GFAP or nestin promoter). This approach allowed the selective introduction of a reporter gene into avian virus receptor–expressing cells, in this case only the GFAP- or nestin-expressing cells. The fact that the reporter gene first appeared in glial cells and only later was found in neurons further demonstrated the developmental potential of some hippocampal astrocytes (Seri et al., 2001; Seri et al., 2004).

The SGZ contains cells that fulfill the stemness criteria self-renewal and multipotency

In 1995, Jasodarah Ray, Theo D. Palmer, and Fred H. Gage from San Diego reported that fibroblast growth factor-2 (FGF-2) –responsive precursor cells could be isolated from the rodent hippocampus (Ray et al., 1993; Palmer et al., 1995). Palmer and coworkers reported in 1997 that "primordial stem cells" existed in the adult rodent hippocampus (Palmer et al., 1997). The use of the term *primordial* underscored the finding that these cells fulfilled the criteria of true stem cells with unlimited self-renewal and multipotency: they produced neurons, astrocytes and oligodendrocytes in vitro. The study introduced a protocol for adherent hippocampal precursor cell cultures, as described in Chapter 3 (Fig. 3–12 and 7–1).

The report of "stem cells" in the adult hippocampus was later disputed (Seaberg and van der Kooy, 2002; Bull and Bartlett, 2005), and it was argued that hippocampal precursor cells, in contrast to cells from the SVZ, were lineage-restricted progenitors with only limited self-renewal. Under the conditions of the neurosphere assay, no secondary spheres were found when single cells from the original spheres were plated out, and the cells did not show signs of multipotency in that no spheres displayed both glial and neuronal differentiation (Seaberg and van der Kooy, 2002, 2003). It was

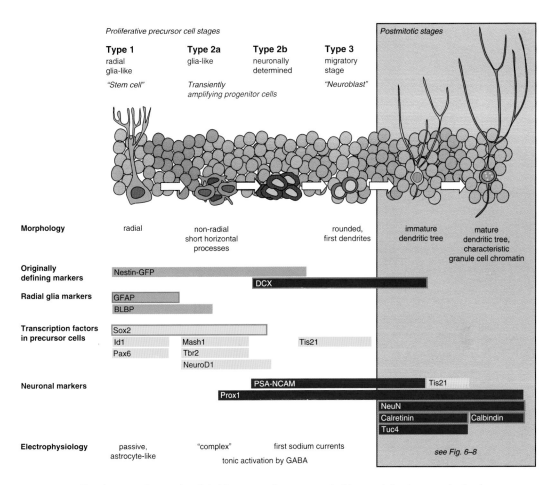

Proliferative precursor cell stages / Postmitotic stages

	Type 1 radial glia-like	**Type 2a** glia-like	**Type 2b** neuronally determined	**Type 3** migratory stage		
	"Stem cell"	Transiently amplifying progenitor cells		"Neuroblast"		

Morphology	radial	non-radial short horizontal processes		rounded, first dendrites	immature dendritic tree	mature dendritic tree, characteristic granule cell chromatin
Originally defining markers	Nestin-GFP		DCX			
Radial glia markers	GFAP					
	BLBP					
Transcription factors in precursor cells	Sox2					
	Id1	Mash1		Tis21		
	Pax6	Tbr2				
		NeuroD1				
Neuronal markers			PSA-NCAM		Tis21	
		Prox1				
				NeuN		
				Calretinin		Calbindin
				Tuc4		
Electrophysiology	passive, astrocyte-like	"complex"	first sodium currents		see Fig. 6–8	
		tonic activation by GABA				

FIGURE 6–6 Developmental stages in adult hippocampal neurogenesis. Neuronal development in the dentate gyrus proceeds through a series of stages that can be identified immunohistochemically. This pattern shows similarities to, but also relevant differences from, the situation in the olfactory bulb (Fig. 5–1).

hypothesized that the detection of stem cells in hippocampal precursor cell cultures could in fact be due to a contamination with stem cells from the ventricular wall. But by using a microdissection protocol in combination with an enrichment procedure that overcame the issue of sub-critical culture densities and an improved culture protocol, the existence of self-renewing stem cells with multi-lineage potential was confirmed for the adult murine dentate gyrus (Babu et al., 2007). In the end the discrepancy thus presumably resulted from differences in cell culture and isolation protocols, leading to the enrichment of different cell populations. This explanation would further strengthen the theory that stem cell properties are not entirely intrinsic and can be modulated by environmental cues. Independently of these issues, however, the problem of precursor cell heterogeneity between SVZ and SGZ persists.

The basic cellular model of cortical development is in principle also applicable to adult neurogenesis

All currently available data suggest that neuronal development in the adult dentate gyrus obeys the canonical scheme of "stem cell," "apical intermediate progenitor cell," and "basal progenitor cell" that we have encountered as a key concept describing the role of stem cells during brain development (Fig. 3–6). More specifically, and like in the adult SVZ, we find the sequence "radial glia–like precursor cell," "glia-like transient amplifying progenitor cell," "neuronally determined transient amplifying progenitor cell," and "migratory neuroblast," which is similar but not identical to the

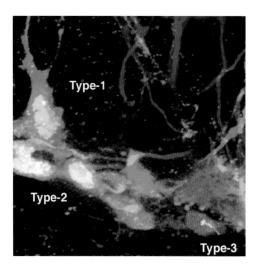

FIGURE 6–7 Precursor cell types in the adult subgranular zone. Radial glia–like type-1 cells give rise to transiently amplifying type-2 cells that in turn generate type-3 cells (*left*). See Figure 6–4 for details (nestin-GFP, *green*; DCX, *blue*; BrdU, *red*). Reprinted from Kempermann et al. (2004) with kind permission from Elsevier.

first description. All of the terms are constructs based on a priori assumptions, and markers have to be used to identify the different cells in most experimental settings. To facilitate discussion and the comparison of data between different research groups, operational nomenclatures based on markers are in use. For the adult SVZ, this is the B-C-A scheme described on p. 162. For the adult dentate gyrus, a provisional nomenclature that does not forestalls specific cellular identities is based on numbering the precursor cell stages (type-1, type-2a, type-2b, and type-3), which are identified by marker proteins and morphology (see below), and has found some acceptance (Fig. 6–6) (Kempermann et al., 2004). In a cell-cultural *tour de force,* this approach has been recapitulated in vitro with a modified neurosphere assay, for once taking advantage of the heterogeneity of neurospheres (Imbeault et al., 2009).

Applying the classical construct of stem cell biology to these cells, type-1 cells are the stem cells, type-2 cells are the transiently amplifying progenitor cells, and type-3 cells are the (migrating) neuroblasts that mark the transition to post-mitotic immature neurons. These cells differ in their morphology, proliferative activity, migratory behavior, and expression of key marker antigens. All of these cells can divide, and after a single injection of BrdU, over time the distribution of BrdU-labeled cells progresses through these four types of cells. This unidirectional redistribution among the BrdU-labeled cells of the SGZ suggests sequential marker progression and therefore development.

The equivalence between this terminology and the supposedly underlying concept is largely inferred, however, and necessarily remains subject to discussion. The model was initially based on morphological criteria, the expression of nestin-GFP and DCX, and on proliferative activity (Kempermann et al., 2004). Meanwhile the expression of additional markers Sox2, Blbp, Prox1, Neurod1, and Tis21 has been used to describe the early developmental stages in the course of adult neurogenesis (Steiner et al., 2006; Steiner et al., 2008; Attardo et al., 2009). Others have introduced markers such as HOP (De Toni et al., 2008), Id1 (Nam and Benezra, 2009), Tbr2 (Hodge et al., 2008), and Ascl1 (Kim et al., 2008). Although in general the new markers have confirmed the initial description, the proposed operational nomenclature will remain subject to future refinement. This scheme is no end in itself but provides some common ground. The "types" of cells do not actually represent truly distinct cellular entities but rather milestones of neurogenesis.

Type 1 Cells Are the Radial Glia–like Precursor Cells of the Subgranular Zone

The putative stem cells of the adult SGZ show a characteristic morphology resembling radial glia and have astrocytic properties (Seri et al., 2001; Filippov et al., 2003). The cells have a triangular-shaped soma that is somewhat larger than the surrounding granule cells (Fig. 6–7). Their resemblance to radial glia is due to a long and strong apical process that reaches into the granule cell layer, where it branches sparsely into the outer third of the granule cell band (Filippov et al., 2003; Fukuda et al., 2003; Mignone et al., 2004). When the processes reach the inner molecular layer, they spread out in numerous small branches, which gives the cells a treelike appearance. The existence of radial glia–like cells in the developing and adult dentate gyrus has been known for a long time (Eckenhoff and Rakic, 1984; Kosaka and Hama, 1986; Rickmann et al., 1987). In the developing hippocampus, radial glia–like cells provide a scaffold is necessary for the normal formation of the dentate gyrus (Caviness, 1973; Stanfield and Cowan, 1979; Forster et al., 2002; Li and Pleasure, 2005; Barry et al., 2008). Radial glia cells also function as precursor cells during brain and especially cortical development (Chaps. 3 and 4). Hippocampus and dentate gyrus are of dorsal origin, but interneurons are derived from ventral sources. Presumably, thus, radial glia in the adult dentate gyrus should be a remnant of dorsal radial glia from development. This is supported by the expression of "dorsal" transcription factor Pax6 in these cells and a consecutive progression through marker stages that are characteristic of embryonic neocortical development (Nacher et al., 2005; Hevner et al., 2006; Hodge et al., 2008). Nevertheless, in the context of the adult hippocampus, the term radial glia–like is often mainly descriptive, because in actual experiments the markers do not show complete overlap, leaving open the possibility that the population of radial elements is not homogenous (Steiner et al., 2006).

Initially, the radial glia–like cells that express precursor cell–marker nestin were called "type-1 cells" (Fig. 6–7), but the definition has since been extended to include additional markers of either radial glia or neural precursor cells, such as Sox2, BLBP, Id1, and Hopx (Filippov et al., 2003; Steiner et al., 2006; De Toni et al., 2008; Nam and Benezra, 2009). In contrast to the corresponding astrocytic cells of the SVZ (B cells), type-1 cells not only have astrocytic properties but also a suggestive radial glia–like morphology (Filippov et al., 2003; Fukuda et al., 2003; Mignone et al., 2004). They consistently express astrocytic marker protein GFAP and are negative for S100β, another astrocytic protein (Seri et al., 2004; Steiner et al., 2004). Many of them express radial glia marker BLBP (Steiner et al., 2006). Type-1 cells have a cilium, and this cilium is necessary for the maintenance of the precursor cell pool in response to Sonic hedgehog signaling, a primary niche factor of neurogenic permissiveness (Breunig et al., 2008; Han et al., 2008). A similar dependence is also known for other areas of neurogenesis; for example, the cerebellum (Spassky et al., 2008).

Type-1 cells have vascular endfeet in the SGZ with a process attached to the basal membrane of blood vessels, another hallmark of astrocytes (Filippov et al., 2003). Electrophysiologically, type-1 cells have classical astrocytic properties in that they have passive membrane properties and potassium currents (Filippov et al., 2003; Fukuda et al., 2003). Their astrocytic nature is thus firmly established. Analogous to cells in the SVZ, the GFAP-positive cells with radial glia–like morphology in the dentate gyrus were first designated B cells by Bettina Seri and Arturo Alvarez-Buylla (Seri et al., 2001), but the nomenclature was abandoned in a later study, and the cells would only be designated "vertical astrocytes" (Seri et al., 2004).

Radial glia–like precursor cells in the SGZ are coupled by gap junctions, and this coupling is important for their function as precursor cells (Kunze et al., 2009).

At least a subset of the GFAP-positive radial glia–like processes is also nestin-positive but the degree of this overlap varies between studies (Filippov et al., 2003; Steiner et al., 2006; Lagace et al., 2007). It is not clear whether Nestin expression marks an activational state of these cells or whether it identifies a constitutive subpopulation. Depending on the species, the model, and the methodology, the overlap seems to vary. Nestin expression is associated with precursor cell activity in vitro but its sensitivity and specificity in vivo are largely undetermined (Lendahl et al., 1990). On the basis of nestin-GFP expression, nestin-positive cells were extracted from the brain with fluorescence-activated cell sorting. These cells formed primary neurospheres in vitro, but unfortunately, no clonal analysis or generation of secondary spheres was performed in that experiment (Mignone et al., 2004). Cumulative lineage tracing based on nestin-CreER(T2)/R26R-YFP, which is explained in detail in Chapter 7, p. 257 (Fig. 7–10), revealed that nestin-expressing cells are the source of adult-born neurons (Lagace et al., 2007; Ninkovic et al., 2007). Nestin, which is an intermediate filament, has largely been abandoned as reference marker for the "stem cells" of the dentate gyrus, and this role has been taken over by Sox2, although for many practical purposes, Nestin remains in use. Using Sox2-GFP reporter gene mice together with lineage-tracing based on a retrovirus-mediated construct, Suh et al. demonstrated the self-renewal of Sox2-positive cells of the adult dentate gyrus and the generation of new neurons and glia from these cells (Suh et al., 2007). Converging evidence thus indicates that GFAP-, Nestin, and Sox2-positive cells are the origin of adult hippocampal neurogenesis. Not all cells with this marker combination have an obvious radial morphology, though, so an intermediate cell, here named *type-2a* (see below), which shows radial glia–like properties but lacks the radial morphology, is proposed (Steiner et al., 2006). At present it is not clear how far radial morphology is a prerequisite of "stemness" in the adult dentate gyrus. The studies on the role of primary cilia in hippocampal precursor cells do not yet allow a conclusion concerning whether cells with non-radial morphology might bear cilia (Han et al., 2008).

Morphologically and with regard to spatial orientation in the tissue, the radial glia–like precursor cells of the SGZ differ from subventricular B cells, although they presumably serve an analogous function. Precursor cells in the SGZ represent an ectopic pool that has been displaced from the ventricular wall during development (see Chap. 4, p. 142, and Fig. 4–13). Whereas SVZ precursor cells generate inhibitory interneurons in the olfactory bulb and oligodendrocytes, SGZ precursor cells produce one type of excitatory neuron. Ex vivo, precursor cells from the SVZ and the SGZ behave differently.

Type-1 cells account for approximately two-thirds of the nestin-expressing cells in the SGZ of adult mice. In contrast, only 5% of the cell divisions among the nestin-expressing cells of the SGZ are in type-1 cells with radial glia morphology (Kronenberg et al., 2003). The number of type-1 cells identified as labeled after a single injection of proliferation marker bromodeoxyuridine remains very constant over long periods of time (Kronenberg et al., 2003). Under pathological conditions, however, such as the induction of experimental seizures (Huttmann et al., 2003; Steiner et al., 2008) or an ischemic lesion (Kunze et al., 2005), proliferation of radial glia–like cells was increased. Also, memantine, an NMDA-receptor antagonist, increased proliferation of radial glia–like cells in the SGZ (Namba et al., 2008). It might seem improbable that type-1 cells, with their treelike morphology and extension between the blood vessels of the SGZ and fibers of the inner molecular layer, could undergo division at all. But type-1 cells would divide asymmetrically. As asymmetrical division is a hallmark of stem cells, this feature would further support the idea that type-1 cells are the highest-ranking precursor cells in the SGZ. The orientation in the granule cell layer and the gross morphology would be maintained during this division, and a daughter cell would bud off at the base of the type-1 cell (see also Fig. 4–4). A few type-1 cells have been caught during division, so that a few pictures of the presumably asymmetrically dividing type-1 cells exist (Namba et al., 2008).

> **Type-1 cells are rarely dividing and can divide asymmetrically**

Type 2 Are Transiently Amplifying Progenitor Cells

> **Type-2 cells are highly proliferative and make the transition from a glial to a neuronal phenotype**

The daughter cells of type-1 cells lack a radial morphology and have been named *type-2 cells* (Filippov et al., 2003; Fukuda et al., 2003). They are highly proliferative and constitute the largest fraction of proliferating cells in the adult SGZ. In nestin-GFP transgenic reporter mice, type-2 cells appear as nestin-expressing cells with plump, short processes oriented more or less parallel to the SGZ (Figs. 6–6 and 6–7). Immunohistochemistry and light microscopy suggest that they do not express GFAP, but in a GFAP-GFP reporter mouse, GFP-labeled type-2 cells were seen (Steiner et al., 2006; Liu et al., 2010). This might be indicative either of GFAP-promoter activity in some type-2 cells or an inheritance of GFP from the GFAP-expressing mother cell ("poor man's lineage tracing"). Type-2 cells have an irregularly shaped nucleus with dense chromatin.

Many proliferating cells cluster near blood vessels and the radial scaffold

Cell division in the SGZ occurs in clusters, which is somewhat more obvious in rats than in mice. These clusters form around central blood vessels and contain the radial glia–like cells. Within days after the initial division, the newly generated cells might spread out along the SGZ (Kuhn et al., 1996). The morphology of type-2 cells supports the possibility of tangential migration at this stage but has not yet been specifically investigated. In any case, this tangential migration is spatially very limited in comparison to migration in the SVZ and rostral migratory stream (RMS). At least a subset of type-2 cells remains in close proximity to the radial scaffold. Type-1 cells often "cradle" type-2 cells in that they might show a basket-like structure shielding the type-2 cells towards the SGZ (Fig. 8–3) (Shapiro et al., 2005b; Plumpe et al., 2006). At later stages, mostly when first signs of neuronal differentiation have appeared, the progenitor cells can be found at a greater distance into the granule cell layer, while still retaining this contact. As astrocytes have been identified as important niche factors (Song et al., 2002), it seems reasonable to assume that the radial glia–like cells serve both as precursor cells and as neurogenesis-promoting guidance structure. Nevertheless, most precursor cell proliferation occurs in the SGZ itself, and migration at this early stage is very limited. More than 75% of dividing cells are found in the SGZ, and even among the long-term progeny, fewer than 40% migrate into the upper two-thirds of the granule cell layer (Kempermann et al., 2003). There is no indication of cell cycle–dependent nuclear movements ("interkinetic nuclear migration") or the existence of layered sites of proliferation for different precursor cell types such as the VZ and SVZ during embryonic cortex development.

Early type-2 cells (type-2a) express stem cell and radial glial markers

Type-2 cells can express BLBP and Sox2, which is suggestive of their close relationship to the radial glia–like type-1 cells and supports the idea that they represent a type of radial glia–like cell without radial morphology (Steiner et al., 2006). This hypothesis was confirmed by a study based on retrovirus-mediated lineage tracing (Suh et al., 2007) and by an experiment in a reporter gene mouse with the murine GFAP-promoter (Liu et al., 2010). The retrovirus experiment also revealed an overlapping expression with Musashi1, which is associated with fetal neural stem cells (Sakakibara et al., 2002) and whose expression in brain co-varies with adult hippocampal neurogenesis (Kempermann et al., 2006). The GFAP reporter study confirmed the previously proposed expression of Ascl1/Mash1 in type-2a cells (Kim et al., 2007; Liu et al., 2010). Type-2 cells might also express Tbr2, which according to the criteria from embryonic cortical neurogenesis would identify them as "apical progenitor cells" (Hodge et al., 2008). So far this term is not in use in the context of adult neurogenesis, however.

Many type-2a express Ascl1/Mash1

Some type-2a cells (as well as some type-1 cells) also express Mash1/Ascl1 (Kim et al., 2007; Uda et al., 2007). In the developing forebrain, Mash1/Ascl1 cooperates with Olig2 in suppressing the astrocytic lineage and favors the generation of oligodendrocytes and interneurons. Mash1/Ascl1 there promotes the generation of interneurons (Yung et al., 2002), and this also applies to the developing (Long et al., 2007) and adult olfactory bulb (Kim et al., 2008). As a general rule, for all brain regions investigated at early developmental stages, precursor cells expressing Mash1/Ascl1 preferentially become neurons, whereas at later stages they become oligodendrocytes (Battiste et al., 2007; Kim et al., 2008). Appropriately, in the adult SGZ, retrovirally mediated overexpression of Mash1/Ascl1 resulted in an increased production of new oligodendrocytes (Jessberger et al., 2008a), also consistent with the claim that SGZ precursor cell are tripotent for all three major neural lineages (Suh et al., 2007). Physiologically, however, only extremely few newborn oligodendrocytes can be found in the adult dentate gyrus (Steiner et al., 2004) and no interneurons are produced in the adult SGZ (Brandt et al., 2003). Lineage tracing based on a Ascl1-CreErT2 construct (see Chap. 7, p. 257) confirmed that in the adult SGZ the Ascl1-positive lineage does not produce oligodendrocytes but, to some degree unexpectedly, generates the newborn granule cells (Kim et al., 2008). This result remains to be explained, but in spinal cord, for example, Mash1/Ascl1 is also involved in fate choice decisions between inhibitory and excitatory lineages, so that an additional function is conceivable (Mizuguchi et al., 2006).

Late type-2 cells (type-2b) are neuronally determined but still highly proliferative

At the stage of type-2 cells, the first signs of neuronal determination or differentiation are found. Originally, this distinction was made on the basis of the expression of doublecortin in nestin-GFP expressing cells, so that type-2 cells come in two subtypes, one negative (type-2a) and one positive (type-2b) for DCX (Kronenberg et al., 2003). Type-2b cells thus combine signs of precursor cell identity (e.g., the expression of BLBP and Sox2, albeit to a lesser extent than type-2a) and neuronal lineage determination (such as DCX) (Steiner et al., 2006). Type-2a cells express Neurogenin 2, whereas NeuroD comes on only at the type-2b stage (Roybon et al., 2009). At the level of type-2 cells, thus, a transition for a glial to a neuronal phenotype takes place. This could happen by asymmetrical or symmetrical neurogenic division or could simply represent further development without a new round of division. Consequently, it is not clear how far the glial vs. the neuronal stage truly corresponds to two different cell types (hence the weaker distinction by the letters *a* and *b* among the type-2 cells has been chosen). Type-2b cells are only slightly less proliferative than type-2a cells. Together they essentially constitute the reservoir of cycling precursor cells. But roughly half of the dividing cells in the SGZ show clear signs of neuronal lineage determination.

Migration and neuritogenesis markers DCX and PSA-NCAM are expressed from the type-2b stage onwards

DCX shows an almost complete overlap with the polysialylated form of the neural cell adhesion molecule, whereas type-1 cells are always negative for DCX and PSA-NCAM. DCX expression is associated with both the initiation of neuronal differentiation and migration (Francis et al., 1999). Developing granule cells go through a transient stage of DCX and PSA-NCAM expression (Seki and Arai, 1993; Brandt et al., 2003; Rao and Shetty, 2004). Although the signaling upstream of DCX expression is not yet known in detail, the DCX promoter contains a binding motif for NeuroD that becomes expressed in type-2 cells. NeuroD expression precedes PSA-NCAM expression during adult hippocampal neurogenesis (Seki, 2002).

Early neuronal markers on the type-2b stage also comprise NeuroD, Prox1, Tbr2, and Tis21

The type-2 stage thus comprises the presumably unidirectional transition from a glia-like precursor cell to neuronal determination. At no later stage has an overlap between glial and neuronal markers or properties been found, which suggests that if the precursor cells of the SGZ can give rise to both neurons and glia, the fate choice should occur on the level of type-1 or type-2a cells. At least in the SGZ, DCX is strictly associated with neuronal lineage, and DCX shows a large overlap with other neuronal markers such as NeuroD1 and β-III-tubulin as well as Prox1 (Steiner et al., 2008), which in the brain is specific for hippocampal granule cells (Pleasure et al., 2000). In the course of development, the new cells sequentially express Pax6, Ngn2, Tbr2 and Tbr1 (Hevner et al., 2006; Raineteau et al., 2006; Ozen et al., 2007; Hodge et al., 2008), a pattern also found in the development of cortical glutamatergic neurons. A subset of type-2b cells (and type-3 cells that are discussed in the next paragraph) expresses Tis21, which in the developing cortex marks basal progenitor cells, in which Tis21 is up-regulated before the terminal neurogenic division (Attardo et al., 2009). This function has not yet been confirmed for adult neurogenesis, but the pattern of marker expression thus strongly resembles the sequence found during cortical development.

Type-2 cells receive GABAergic synaptic input

At the level of type-2 cells, the progenitor cells receive first synaptic input, which is GABAergic (Wang et al., 2005). In addition, ambient GABA affects the cells (Ge et al., 2006). At this stage GABA leads to an excitatory depolarization rather than the inhibitory hyperpolarization that is characteristic of GABA action on mature neurons. This polarity of GABA effects largely depends on intracellular chloride concentrations (Ben-Ari, 2002). In the course of neuronal development, chloride importer NKCCl is expressed first and only replaced by chloride exporter KCC2 when the neurons have received glutamatergic input. This pattern has also been described for adult-born granule cells (Ge et al., 2006). GABA promotes neuronal differentiation in the adult dentate gyrus and does so between the precursor cell stage and the synaptic integration of the newborn neurons (Tozuka et al., 2005; Ge et al., 2006).

Type 3 Are the Migratory Neuroblast of the Adult SGZ

Type-3 cells show characteristics of migratory cells but do not migrate far

Type-3 cells are positive for DCX, PSA-NCAM and other neuronal markers, but are negative not only for all glial markers but also for precursor cell markers such as Sox2. Depending on the markers used, the distinction from type-2b cells can be difficult; the original definition had been based on the absence of nestin-GFP expression in DCX-positive cells. To some degree the transition between the two states is fluent. Although there are thus some good reasons to collapse type-2b and type-3 cells into one category (Seri et al., 2004), there are also arguments in favor of distinguishing the neuroblast-like stage. One argument is that this phase is associated with large morphological changes and with radial migration into the granule cell layer. The distance of radial migration into the granule cell layer, however, is limited. Most new cells remain in the SGZ and the inner third of the granule cell layer; few reach the outer third (Kempermann et al., 2003). The cells arrive at their final position early, presumably still during a progenitor cell stage. Afterwards the distribution of new cells does not substantially change. Most of the cells remain in the lower third of the granule cell layer (Kempermann et al., 2003).

Type-3 cells can still divide and respond to seizures

The nucleus of type-3 cells is rounded, and many cells show a first apical dendrite (Plumpe et al., 2006). Besides horizontally oriented type-3 cells that are similar to the orientation of type-2 cells, but often have longer horizontal processes, we find many type-3 cells with a more vertical orientation. Even cells with a dendrite that reaches a length of approximately half the width of the granule cell layer can still divide (Plumpe et al., 2006).

Few dividing cells can be found within the granule cell layer above the SGZ at any given time, and these cells generally seem to be DCX positive. Occasionally, however, these cells can be nestin-positive as well. Therefore, radial migration, at least in some cases, might be initiated on the level of type-2b cells, but the migrating cells readily lose their nestin expression when they reach the granule cell layer. It is not clear whether radial migration needs to have ended before the cells terminally exit the cell cycle. It is also not clear whether the final exit from the cell cycle can only occur on the level of type-3 cells. The electrophysiological data and the expression of neuronal transcription factors such as NeuroD in type-2 cells suggest that signals for terminal neuronal differentiation can successfully reach the cells earlier. Type-2b and type-3 cells also express transcription factor Prox1, which remains expressed throughout granule cell development and is found in all adult granule cells (Kronenberg et al., 2003; Steiner et al., 2008). The type-3 stage thus comprises the transition from a potentially proliferative state to the post-mitotic immature neuron.

Experimental seizures lead to a disproportional increase in proliferation of type-3 progenitor cells and increase their migration (Jessberger et al., 2005). This suggests that both proliferation and migration are sensitive to changes in neuronal excitation.

In DCX-expressing cells the electrophysiological transition from immature to mature neuronal properties occurs (Ambrogini et al., 2004). DCX- or PSA-NCAM-positive cells with a more mature dendritic morphology suggestive of receiving physiological input (and thus representing a stage beyond the type-3 level) showed increased synaptic plasticity compared to that of older granule cells (see below; also Wang et al., 2000; Snyder et al., 2001; and Schmidt-Hieber et al., 2004).

The Early Phase of Survival or Cell Death

Most newborn cells are again eliminated by apoptotic cell death

Like during embryonic, fetal, and postnatal neurogenesis and as in neurogenesis of the SVZ/olfactory bulb system (Winner et al., 2002), the precursor cells of the adult SGZ produce a vast surplus of progenitor cells of neuronal lineage and a surplus of immature neurons (Kempermann et al., 2003).

The majority of dividing cells in the SGZ are type-2 cells. By one day later, the progeny of these divisions has doubled in the number of cells labeled with proliferation marker BrdU (see p. 232 for details). This result is expected because every cell division generates two labeled daughter cells, and one cell cycle apparently lasts about one day. The maximum number of labeled cells is reached before three days after a single injection of BrdU (Kronenberg et al., 2003). It is estimated that many of the dividing cells go through two to three cell cycles. By this time, most of the BrdU-labeled cells have already turned into type-3 cells. The number of labeled cells is higher than at day after BrdU injection, but not eight times as high (2^3), the hypothetical number of cells that could be reached if all new cells survived (compare also Dayer et al., 2003). The dilution of BrdU below the level of detection leads to some false-negative cells (Hayes and Nowakowski, 2002).

Variation in survival is strain-dependent

The cells not recruited into function are eliminated by apoptotic cell death (Biebl et al., 2000; Kuhn et al., 2005; Plumpe et al., 2006; Vandenbosch et al., 2007), although the exact mechanism by which apoptosis is induced remains unknown. Inhibiting caspase activity in the adult dentate gyrus accordingly resulted in greater numbers of newborn neurons (Ekdahl et al., 2001; Gemma et al., 2007). Similarly, in mice overexpressing anti-apoptosis gene Bcl2, adult neurogenesis was dramatically increased (Kuhn et al., 2005; Sasaki et al., 2006). The same was found in null-mutant mice for pro-apoptotic gene BAX (Sun et al., 2004). This elimination is highly dependent on the genetic background: whereas in CD1 mice almost 75% of the newborn cells survive, in 129/SvJ mice almost 75% die (Kempermann et al., 1997a). Absolute numbers of newborn neurons are also greatly affected by age: there is a strong decline in adult neurogenesis with increasing age (Kuhn et al., 1996; Kempermann et al., 1998). The survival rate—i.e., the number of surviving cells in relation to the number of proliferating cells—is less affected. For the young-adult C57BL/6 mice used in most studies, the survival rate is about one third (Kempermann et al., 1997a); for Fisher344 or Sprague Dawley rats, approximately 50% (Dayer et al., 2003). Because the denominator in this ratio, the number of proliferating cells, is dependent on the methodology used to identify these cells, these values have to be taken with a grain of salt, though. In most cases the expansion of intermediate progenitor cells that were generated between the two time-points of investigation is not assessed and thus not taken into account. Both net production of cells and cell death might therefore be substantially higher than in these estimates.

Selective activity-dependent promotion of survival largely takes place at an early post-mitotic stage

Most cell death seems to occur once the cells have left the cell cycle, have become post-mitotic, and express both DCX and Calretinin. The elimination of cells is rapid, so that in less than one to two weeks after cell cycle exit, the survival curve flattens out. The number of fragmented nuclei, a morphological indication of cell death, is highest at three days after an injection with BrdU (Seki, 2002). From this dynamic one is tempted to hypothesize that elimination might be the default pathway, which is counteracted by activity-dependent, survival-promoting rescue effects specifically for the cells that are useful. Such rescue effects are indeed elicited by "cognitive" stimuli such as exposure to an enriched environment or more specific learning stimuli and including the putative electrophysiological mechanism of learning, induced long-term potentiation (Kempermann et al., 1997b, 1998; Gould et al., 1999; Bruel-Jungerman et al., 2006; Trouche et al., 2009). Learning stimuli induce the expression of immediate early genes in granule cells and also in newborn neurons, once they have reached a certain level of maturity (Jessberger and Kempermann, 2003). This underscores that the newborn cells can indeed respond to an external stimulus. When mice were trained in the Morris water maze, a classical test of hippocampal function, maturing newborn cells were activated by the learning stimulus with greater likelihood (Kee et al., 2007; Trouche et al., 2009). It seems that the survival-promoting effect of experience and learning is greatest at about one to three weeks after the cells were born (Gould et al., 1999; Greenough et al., 1999; Tashiro et al., 2007; Trouche et al., 2009). In vivo, a critical window between two and three weeks was identified (Fig. 6–8), during which newborn neurons are engaged in competitive survival that involves activation of NMDA receptor subtype NR1 (Tashiro et al., 2006).

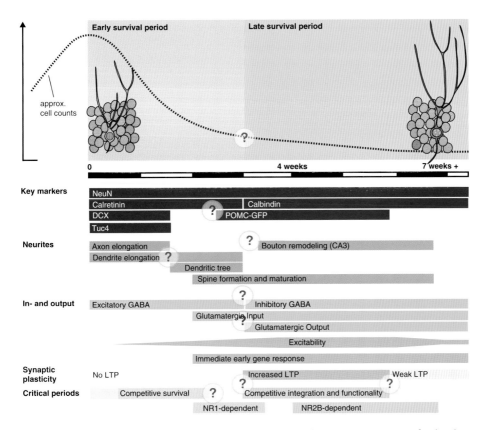

FIGURE 6–8 The phase of post-mitotic maturation in adult hippocampal neurogenesis. See text for details.

Most cells are eliminated before they could have possibly been fully integrated into the network

Consequently, these time windows do not match well: cell death is greatest before the best responsiveness of the cells is reached. Most newborn cells are eliminated within a few days after they are born; they are long gone before they could have been truly integrated into the neuronal network of the dentate gyrus. Different mechanisms controlling the elimination of new cells must be considered, but as yet it is not clear how the earliest phase of elimination is regulated. The lack of complete functional integration and maturation does not mean that the cells could not respond to excitatory stimuli. This has even been shown for hippocampal precursor cells in vitro. Precursor cells can sense neuronal activity and translate it into a program of neuronal differentiation (Deisseroth et al., 2004; Babu et al., 2009). A single bout of experimental seizures one day after the precursor cells were labeled with BrdU increased the number of the labeled neurons four weeks later (Steiner et al., 2008), further supporting the existence of very early pro-survival mechanisms.

The not yet fully integrated cell might serve a specific function during the time-window critical for their survival

In slice preparations with stimulation of the entorhinal cortex, roughly three-week-old immature neurons could already be efficiently activated (Mongiat et al., 2009), but even younger cells have not yet been tested. Coinciding with that result, synaptic plasticity of newborn cells is greatest at about four weeks after they are born (Schmidt-Hieber et al., 2004), and at about that time they also are preferentially recruited (Kee et al., 2007) and seem to make critical contributions to hippocampus-dependent learning (Farioli-Vecchioli et al., 2008). The current hypothesis about the functional relevance of newborn neurons in the hippocampus ascribes a benefit to the partial turnover of immature newborn neurons, which might or might not be independent of the function after long-term integration (see Chap. 10, p. 470).

The early phase of activity-dependent survival thus appears to be relatively non-specific, even if it is elicited in the context of a specific task. There are thus two more or less independent survival effects: one early, one late during the development of the cell. In a given situation the two effects might be exerted by the same learning stimulus that affects different cells at different stages of development. The idea that one stimulus might at the same time have different results in the course of neurogenesis is supported by the finding that the recruitment of new neurons also has a secondary effect on progenitor cell proliferation, as if the recruitment would signal to the precursor cells that there is increased demand for potentially recruitable cells (Dupret et al., 2007).

If cells survive the critical time-window, they are likely to survive for long times

The survival-promoting effects are long-lasting (Kempermann and Gage, 1999). Two to three weeks after exit from the cell cycle, the number of new cells remains very stable (Dayer et al., 2003; Kempermann et al., 2003). Even at one year after the BrdU injection, the latest time-point investigated in the mouse hippocampus, the number is essentially unchanged (Kempermann et al., 2003). Nevertheless, even in the oldest animal investigated and in the absence of any change in living conditions or obvious learning stimuli, a very low number of new neurons is found (Kempermann et al., 1998; Kempermann et al., 2002), which might indicate a low constitutive baseline rate of adult neurogenesis.

Proliferation of precursor cells is a poor predictor of net neurogenesis

One consequence of this developmental pattern is that proliferation is a poor predictor of net neurogenesis, explaining only approximately 20% of the variance in adult hippocampal neurogenesis in an inbred panel of 26 mouse strains, whereas variability in the number of surviving cells explained already 85% (Kempermann et al., 2006). Adult hippocampal neurogenesis is largely regulated on the level of cell survival and neuronal differentiation.

In some studies, especially older ones, proliferation was measured as the only parameter describing neurogenesis. Cell division in the SGZ is correlated with net neurogenesis but not identical to it. In addition, assessment of proliferation is also influenced by the labeling paradigm (Hayes and Nowakowski, 2002). The same applies to the number of DCX-expressing cells as sole indicator of "neurogenesis," because the DCX-population comprises a large number of proliferating cells. Nevertheless, survival requires preceding expansion, and the size of the available progenitor cell population to some degree determines that appropriate numbers of cells are selectable (Fabel et al., 2009).

The Phase of Post-Mitotic Maturation

After cell cycle exit, new cells might rapidly express NeuN

Type-3 cells (and possibly some type-2b cells) exit from the cell cycle and begin the terminal post-mitotic differentiation of granule cells (Fig. 6–8). Very early after cell cycle exit, they might express neuronal marker NeuN. After Prox1, NeuN is the second known mature marker that persists in the new neurons. Accordingly, one day after a single injection of BrdU, a small number of BrdU/NeuN-positive cells can be found; after three days their number is already quite high. The early presence of NeuN in BrdU-labeled cells cannot be taken as evidence of dividing neurons. To the contrary, NeuN-positive cells have never been found to express cell cycle markers (Ki67, PCNA, pH3, etc.) or to contain BrdU within minutes after the injection, confirming that NeuN-positive cells are post-mitotic. However, because a NeuN-positive young neuron is derived from a predifferentiated type-2b/3 cell, it might be detected by BrdU/NeuN-immunoreaction as early as one or two days after the last division. The "mature" marker NeuN consequently also detects also very immature neurons. In nestin-GFP reporter gene mice, a low but consistent number of GFP/NeuN-doublepositive cells is found at one day after labeling with BrdU, suggesting that a proliferating cell has generated a NeuN-positive neuron that inherited GFP from the nestin-expressing progenitor cell (Petrus et al., 2008).

The post-mitotic phase is
characterized by the transient
expression of calretinin

In mice, the early post-mitotic stage of neuronal development is also characterized by the transient expression of calretinin, whereas mature granule cells express calbindin (Brandt et al., 2003). This switch of calcium-binding proteins must occur very rapidly and in mice takes place approximately three to four weeks after the cells have become post-mitotic. One as-yet-unproven idea is that the change in calcium-binding protein is a response to the onset of glutamatergic innervation that sets in approximately at that time.

Neuritogenesis and
morphological and functional
maturation take place at this
time

The phase of DCX and calretinin expression is the period of dendrite and axon formation and functional maturation (Figs. 6–9 and 6–10). The largest part of the dendritic tree is built during this period (Brandt et al., 2003; Ambrogini et al., 2004; Rao and Shetty, 2004; Plumpe et al., 2006; Zhao et al., 2006), including the transient appearance of a basal dendrite (Ribak et al., 2004). Overexpression of Cdk5, a cyclin-dependent kinase with functions largely outside cell cycle control, in newborn neurons resulted in ectopic migration of the cells and the extension of abnormal basal dendrites (Jessberger et al., 2008b). Persistent basal dendrites are a neuropathological hallmark in models of temporal lobe epilepsy (Ribak et al., 2000) and have been found on the adult-generated granule cells (Shapiro et al., 2005a).

During the post-mitotic phase of neurite extension, the cells keep the vertical morphology of the late type-3 cells. Their nucleus is rounded or slightly triangular, and an apical dendrite becomes clearly visible on DCX or PSA-NCAM staining. At the stage when Calretinin becomes strongly expressed, the dendrite has already acquired relatively mature morphology and can reach far up into the inner molecular layer (Fig. 6–9). Within the granule cell layer proper, it often has a close spatial relationship to the apical process of vertical astrocytes and type-1 cells (Seki and Arai, 1999).

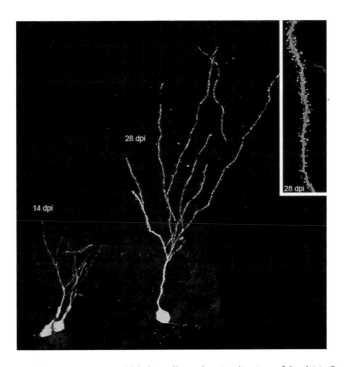

FIGURE 6–9 Dendritic development. Retroviral labeling allows the visualization of dendritic fine morphology, including spines, of the newborn neurons. The time-course of dendritic maturation can be followed. Figure kindly provided by Sebastian Jessberger, Zürich.

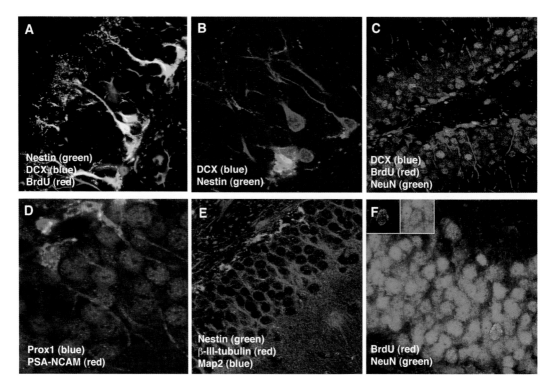

FIGURE 6–10 Phenotypes of adult-generated cells in the dentate gyrus. Confocal-microscopic depiction of key antigens is used to identify newly generated cells in the course of adult neurogenesis in the hippocampus. (A): Nestin-GFP-expressing radial type-1 cells and type-2 progenitor cells. (B and C): DCX expression characterizes a range of cells from rounded proliferative cells to immature neurons with dendritic trees. (D): Transcription factor Prox1 is expressed early in the course of development in the adult hippocampus and here coincides with the expression of PSA-NCAM. (E): β-III tubulin and Map2 are neuronal markers expressed in the adult dentate gyrus, but because of their cytoplasmic expression they are more difficult to use than nuclear antigen to demonstrate co-localization. β-III-tubulin is expressed transiently in the SGZ; Map2 labels all neuronal processes. (F): NeuN is the standard marker used to identify new neurons in the adult brain. Panel A from Filippov et al. (2003) with kind permission of Elsevier. Images by Barbara Steiner (A–E) and Golo Kronenberg (F), Berlin.

This might indicate that the radial glia–like process serves as a guidance structure for the forming dendrites. The time-course of dendrite extension has been found to be remarkably stable across different situations that regulate adult neurogenesis, including experimental seizures (Plumpe et al., 2006), but qualitative differences might be found in cases of pathology (Jessberger et al., 2007).

Dendritic growth initially often occurs in two or more branches, whereas in mature granule cells, only one dendritic stem reaches into the outer granule cell layer and inner molecular layer, where branching occurs. Roughly four weeks after cell cycle exit, the cells have matured to show dendritic spines (Fig. 6–9, inset).

||||||||||||||||||||||||||||||||||||||||||||||||||||||||||

Axonal connectivity to CA3 and hilar interneurons is fully established within about two to 10 weeks

||||||||||||||||||||||||||||||||||||||||||||||||||||||||||

Axon elongation occurs rapidly after the cells have become post-mitotic and reach their target in CA3 within approximately 10 days (Stanfield and Trice, 1988; Hastings and Gould, 1999; Markakis and Gage, 1999; Zhao et al., 2006). Note that the first description of this newly established connectivity in the adult brain was given by Stanfield and Trice in 1988, well before the rediscovery of adult neurogenesis in the mid-1990s and any thorough characterization of neuronal development in the adult dentate gyrus (Stanfield and Trice, 1988). TUC4, associated with growth cone activity, is expressed during this phase (see Chap. 7). Combining retrovirus-mediated expression of GFP in newborn cells with confocal microscopy and immuno-electron-microscopy, Nicolas Toni and colleagues from Fred H. Gage's and Alejandro Schinder's groups determined that the new cells make in fact appropriate synaptic contacts both in the hilus (where axon collaterals target interneurons) and on pyramidal cells of CA3 (Toni et al., 2008). In addition contacts were made to interneurons in CA3, which is consistent with the physiological pattern (Acsady et al., 1998). They found that initially, i.e., at two to three weeks after labeling, the contacts were only adjacent to thorny excrescences of the pyramidal cells; at one month they might share a thorny excrescence with an unlabeled fiber; and at 75 days, the latest time-point investigated, most labeled boutons possessed their own individual thorny excrescence (Toni et al., 2008). This pattern, which was confirmed in a competitor's study showing that knockdown of schizophrenia-related gene Disc1 impaired this maturation process (Faulkner et al., 2008), suggests that functional connectivity is established early but needs to mature over a prolonged period of time. To demonstrate the actual functionality of the connectivity in the target area, Toni and colleagues made use of a technology allowing the activation of a neuron by light. To achieve this, newborn cells were transduced with a vector expressing channelrhodopsin 2. Light shone onto the slice preparation elicited action potentials only in the newborn neurons. A response could then be measured in postsynaptic target neurons, confirming that the synaptic integration was effective (Toni et al., 2008). Previously, Marie Carlén, Jonas Frisén, and colleagues had demonstrated that an infection of CA1 with a GFP-vector based on pseudorabies virus, which is transmitted only trans-synaptically, resulted in GFP-labeled cells in CA3 and in newborn neurons in the dentate gyrus, which had been pre-marked with BrdU (Carlen et al., 2002).

Dendritic spines begin to appear after the axon has reached CA3

At approximately 10 to 14 days, that is when the axon has reached CA3, the cells might begin to have the appearance of mature granule cells, although functionally they are still immature (Zhao et al., 2006). Only at this time, the first spines form on the dendrites, and the dendrites reach their targets in the molecular layer. During the following weeks, dendrites and spines mature, and the gross morphology has been established at approximately four weeks. Neither aging nor exercise, for example, which strongly influence adult neurogenesis, affected spine density and maturation on adult-born cells (van Praag et al., 2005), whereas seizures profoundly disturbed dendrite morphology (Jessberger et al., 2007). At the interesting time-point of about two weeks after cell cycle exit, the newborn cells can also be visualized with EGFP as reporter gene under a truncated version of the promoter of proopiomelanocortin (POMC) (Overstreet et al., 2004). POMC is a precursor protein for many bioactive factors, including β-endorphin, but is not endogenously expressed in granule cells, so the biological meaning of this transient promoter activity is not clear.

The first GABAergic input is excitatory and promotes further development

Functional maturation of the newborn granule cells relatively closely recapitulates neuronal development in the embryo (Esposito et al., 2005). During the first week (after labeling proliferating cells with a GFP-expressing retrovirus to identify the newborn cells under the microscope of the electrophysiology setup), the cells were synaptically silent and showed only immature spikes (Esposito et al., 2005). In experiments based on nestin-GFP reporter gene mice, synaptic responses to GABA were recorded even from the precursor cells (presumably type-2) (Ambrogini et al., 2004; Wang et al., 2005). In any case, the first GABA response is slow and occurs well before spines have formed on the dendrites and connectivity to CA3 has been established (Esposito et al., 2005). GABA drives the functional maturation and synaptic integration (Ge et al., 2006; Ge et al., 2007a). Maturation was aborted when the effects of chloride importer NKCCl (resulting in depolarization; that is, excitation of the newborn cells) were overridden by gene transfer of the hyperpolarizing chloride exporter KCC2, resulting in a premature conversion of GABA-dependent excitation to inhibition (Ge et al., 2006).

Once glutamatergic input is established, a phase of increased synaptic plasticity begins

The first glutamatergic input becomes detectable at approximately four weeks (Esposito et al., 2005), coinciding with the period of increased synaptic plasticity (Schmidt-Hieber et al., 2004) and the existence of mature spines on the dendrites (Zhao et al., 2006). At around that same time, there are synapses on the dendrites, the dendritic spines, and the soma of the new neurons (Toni et al., 2007). New synapses primarily form in the vicinity of already existing synapses, and dendritic filopodia precede synapse formation, consistent with their role in synaptogenesis (Toni et al., 2007). It thus appears that a functional connectivity as it is characteristic for older granule cells has been established within approximately one month after the cells were born.

Dentate gyrus LTP is largely contributed by the synaptic plasticity of newborn neurons

The latest functional feature that forms during the following three to four weeks is the inhibitory response to GABA (Esposito et al., 2005). Mature granule cells are massively inhibited by local interneurons, which is the reason why, in electrophysiological recordings of long-term potentiation in the adult dentate gyrus (and in the absence of a pharmacological blockade of this inhibition), only the response of the as-yet-uninhibited newborn neurons becomes visible (Fig. 10–4) (Wang et al., 2000; Snyder et al., 2001; Schmidt-Hieber et al., 2004; Saxe et al., 2006; Garthe et al., 2009).

Full maturation takes about two months

Although fine-tuning of dendritic morphology and of the connectivity continues at least for two months after the appearance of the first protrusions on the dendrites (Toni et al., 2007), with respect to their overall functionality the new cells can be considered "mature" at approximately seven to eight weeks after they are born (van Praag et al., 2002; Jessberger and Kempermann, 2003). The newborn granule cells become functionally indistinguishable from older granule cells, so that the entire granule cell layer represents a functionally homogenous population with the exception of the interspersed immature cells (Laplagne et al., 2006; Laplagne et al., 2007). Compared to neurogenesis at neonatal age, the entire process of functional maturation is somewhat slower in the adult (Overstreet-Wadiche et al., 2006). The duration of depolarizing GABA is longer in the adult, which might indicate that the period during which the cells are open for activity-dependent recruitment is adjusted to the situation of a freely moving and constantly learning and experiencing animal. The critical time window of increased synaptic plasticity is also dependent on the expression of the NR2B subtype of the NMDA receptor on the newborn cells (Ge et al., 2007b).

The Late Survival Phase

A second wave of activity-dependent survival takes place after functional integration and the phase of increased synaptic plasticity

Based on the total number of newborn cells, most changes occur before the two identified critical time windows (Fig. 6–8). But during both periods, before and during the phase of increased synaptic plasticity, promotion of survival of newborn cells has been demonstrated (Bruel-Jungerman et al., 2006; Tashiro et al., 2006; Ge et al., 2007b; Kee et al., 2007; Tashiro et al., 2007). It appears that the two periods differ in the contribution two NMDA receptor subtypes make to the plastic effect: NR1 to the early period, NR2B to the late period (Tashiro et al., 2006; Ge et al., 2007b). The first period would also be primarily dependent on GABA signaling, the second on glutamate. As discussed above, most cells are even eliminated before they could possibly be rescued by these mechanisms. Nothing is known about the mechanisms underlying the selective elimination of newborn cells during the phase of increased synaptic plasticity. Effects on total cell numbers are low. The likelihood for a newborn cell that has survived the first selection phase to also survive the second phase is very high.

Long-term integrated cells function like older granule cells and lead to persistent network changes

The late survival phase is thus one of stability. The new cells are fully integrated into the network and do not show any more properties that distinguish them from older granule cells. They blend into the crowd and therefore presumably function with the other granule cells in a homogenous way (Laplagne et al., 2006). But they have changed the network. Because there is no evidence of any meaningful turnover of neurons in the adult dentate gyrus, adult neurogenesis is cumulative. Presumably, the network becomes increasingly complex over time.

Consequently, when only young, immature cells are dying, adult neurogenesis must lead to a growing dentate gyrus. Rats have approximately 1.2 million granule cells per dentate gyrus, mice have 400,000, and humans have 15 million (West and Gundersen, 1990). Even the earliest reports on adult hippocampal neurogenesis concluded that new neurons are added to the granule cell layer and do not replace dying older cells (Altman and Das, 1965; Bayer et al., 1982; Bayer, 1985; Crespo et al., 1986). In 1982, Shirley Bayer estimated a 35% to 43% net growth of the dentate gyrus in rats (Bayer, 1982; Bayer et al., 1982; Bayer, 1985). According to stereological analysis, however, one later study found no significant change in the total granule cell number of rats between five months and 24 months of age (Merrill et al., 2003), possibly because there was considerable inter-individual and inter-strain variation. In mice there was a 40% increase in the volume of the dentate gyrus in roughly the first year of life (Peirce et al., 2003). Also in mice, and based on a genetically mediated permanent labeling of adult-born neurons, after the age of four months a relatively constant ratio of newborn to total neurons of approximately 7% was estimated (Ninkovic et al., 2007). That would correspond to a total number of about 30,000 adult-born granule cells in C57BL/6 mice. From this it follows that adult hippocampal neurogenesis is essentially a qualitative phenomenon. There does not seem to be a need for the integration of many new cells across the life span, but there is a clear need for adult neurogenesis to happen at all.

Genetic ablation of new cells over longer periods causes a decrease in granule cells but not below a certain baseline, supporting the idea that in the end, adult neurogenesis is cumulative (Imayoshi et al., 2008). The turnover of immature neurons plus the cumulative network integration of mature cells has many implications for functional theories on adult hippocampal neurogenesis (see Chap. 10).

Gliogenesis in the Adult Dentate Gyrus

Precursor cells isolated from the adult hippocampus are multipotent in vitro in that they can produce all three neural lineages—neurons, astrocytes, and oligodendrocytes (Fig. 6–10) (Palmer et al., 1997; Babu et al., 2007). Whether the precursor cells in vivo are tri-potent as well has not been fully resolved. Some suggestive evidence in favor of this idea comes from an in vivo lineage study (Suh et al., 2007). In the dentate gyrus, gliogenesis is largely equivalent to the generation of new astrocytes (Cameron et al., 1993; Kempermann et al., 2003).

Astrocytes

Astrocyte-like precursor cells produce both neurons and new astrocytes (Seri et al., 2001). Morphologically, newly generated astrocytes in the adult dentate gyrus differ from astrocyte-like precursor cells. The radial glia–like type-1 cells have also been called "vertical astrocytes," whereas the newly generated astrocytes of the adult SGZ are "horizontal astrocytes" (Seri et al., 2001; Steiner et al., 2004). Both types of astrocytes express GFAP, but in contrast to type-1 cells, horizontal astrocytes also express S100β. In the murine SGZ, S100β-positive cells have never been found in cell cycle (Steiner et al., 2004). The immediate precursor of the S100β-positive astrocytes is a progenitor cell within the glial lineage. This glial-determined progenitor is characterized by GFAP expression and the absence of S100β, nestin, and radial glia morphology (Steiner et al., 2004). Like neurogenesis, astrogenesis is regulated by numerous stimuli, with many stimuli affecting both lineages similarly. Nevertheless, regulation in both lineages appears to be independent (Steiner et al., 2004). The ratio between neurogenesis and gliogenesis differs greatly between strains (Kempermann and Gage, 2002). In terms of absolute cell numbers generated, there is no correlation between neurogenesis and astrogenesis, and different sets of regulatory candidate genes are associated with either of them (Kempermann et al., 2006). One of the few genes explicitly studied in this context is reelin. Mutated reelin caused neurogenesis to decrease and gliogenesis to increase (Zhao et al., 2007).

It is not clear whether intermediate progenitor cells can (re-)generate type-1 cells

Whether non-radial astrocytes or type-2a cells can generate radial glia–like type-1 cells has not yet been fully clarified. If so, this might imply that the radial status of the cells may serve as a relatively quiescent reservoir that is both generating intermediate progenitor cells and being replenished by them. Although this idea has some charm, it has not been experimentally demonstrated. At face value, retroviral labeling supported the idea (Suh et al., 2007) but in that study it was not fully established whether the new "radial" cells had acquired full radial morphology or might still have belonged to the non-radial type-2a cells.

Populations of astrocytes throughout the adult hippocampus can be produced throughout adulthood. Protoplasmic astrocytes in CA1, for example, were even produced in response to environmental enrichment and running, as do astrocytes within the dentate gyrus (Kronenberg et al., 2007). This process, however, is independent of the astrogenesis in the SGZ that is directly related to adult neurogenesis.

Oligodendrocytes

Only a very few oligodendrocytes are generated in the adult dentate gyrus

With respect to the generation of new oligodendrocytes, the results are conflicting, possibly reflecting strain and species differences. Very low numbers of new oligodendrocytes can be detected (Eckenhoff and Rakic, 1988; Kempermann et al., 2003; Steiner et al., 2004). NG2-expressing cells as the putative precursor cells in the oligodendrocytic lineage can be found in the SGZ, but their rate of division is extremely low (Steiner et al., 2004). Given that the mossy fibers are a non-myelinated tract, this local lack of oligodendrogenesis might not come as a complete surprise. Generally, the origin of oligodendrocytes is from ganglionic eminences in the ventral SVZ, whereas the neural precursor cells of the adult SGZ derive from the dorsal ventricular wall. Nevertheless, overexpression of Ascl1/Mash1 in the SGZ was sufficient to induce oligodendrogenesis in the dentate gyrus (Jessberger et al., 2008a) and Mash1/Ascl1 is normally expressed by some type-2a cells as well as a few type-1 cells (Kim et al., 2008).

References

Acsady L, Kamondi A, Sik A, Freund T, Buzsaki G (1998). GABAergic cells are the major postsynaptic targets of mossy fibers in the rat hippocampus. *J Neurosci* 18:3386–3403.

Altman J (1975). Postnatal development of the hippocampal dentate gyrus under normal and experimental conditions. In: *The Hippocampus* (Isaacson RL, Pribram KH, eds.), pp. 95–122. New York: Plenum Press.

Altman J, Das GD (1965). Autoradiographic and histological evidence of postnatal hippocampal neurogenesis in rats. *J Comp Neurol* 124:319–335.

Amaral DG, Witter MP (1989). The three-dimensional organization of the hippocampal formation: A review of anatomical data. *Neuroscience* 31:571–591.

Ambrogini P, Lattanzi D, Ciuffoli S, Agostini D, Bertini L, Stocchi V, Santi S, Cuppini R (2004). Morpho-functional characterization of neuronal cells at different stages of maturation in granule cell layer of adult rat dentate gyrus. *Brain Res* 1017:21–31.

Attardo A, Fabel K, Krebs J, Haubensak W, Huttner WB, Kempermann G (2009). Tis21 expression marks not only populations of neurogenic precursor cells but also new postmitotic neurons in adult hippocampal neurogenesis. *Cereb Cortex* 20:304–14.

Babu H, Cheung G, Kettenmann H, Palmer TD, Kempermann G (2007). Enriched monolayer precursor cell cultures from micro-dissected adult mouse dentate gyrus yield functional granule cell-like neurons. *PLoS ONE* 2:e388.

Babu H, Ramirez-Rodriguez G, Fabel K, Bischofberger J, Kempermann G (2009). Synaptic network activity induces neuronal differentiation of adult hippocampal precursor cells through BDNF signaling. *Front. Neurosci* 3:49.

Bannerman DM, Rawlins JN, McHugh SB, Deacon RM, Yee BK, Bast T, Zhang WN, Pothuizen HH, Feldon J (2004). Regional dissociations within the hippocampus—memory and anxiety. *Neurosci Biobehav Rev* 28:273–283.

Barkho BZ, Song H, Aimone JB, Smrt RD, Kuwabara T, Nakashima K, Gage FH, Zhao X (2006). Identification of astrocyte-expressed factors that modulate neural stem/progenitor cell differentiation. *Stem Cells Dev* 15: 407–421.

Barry G, Piper M, Lindwall C, Moldrich R, Mason S, Little E, Sarkar A, Tole S, Gronostajski RM, Richards LJ (2008). Specific glial populations regulate hippocampal morphogenesis. *J Neurosci* 28:12328–12340.

Battiste J, Helms AW, Kim EJ, Savage TK, Lagace DC, Mandyam CD, Eisch AJ, Miyoshi G, Johnson JE (2007). Ascl1 defines sequentially generated lineage-restricted neuronal and oligodendrocyte precursor cells in the spinal cord. *Development* 134:285–293.

Bayer SA (1982). Changes in the total number of dentate granule cells in juvenile and adult rats: A correlated volumetric and 3H-thymidine autoradiographic study. *Exp Brain Res* 46:315–323.

Bayer SA (1985). Neuron production in the hippocampus and olfactory bulb of the adult rat brain: Addition or replacement? *Ann NY Acad Sci* 457:163–172.

Bayer SA, Yackel JW, Puri PS (1982). Neurons in the rat dentate gyrus granular layer substantially increase during juvenile and adult life. *Science* 216:890–892.

Becq H, Jorquera I, Ben-Ari Y, Weiss S, Represa A (2005). Differential properties of dentate gyrus and CA1 neural precursors. *J Neurobiol* 62:243–261.

Ben-Ari Y (2002). Excitatory actions of gaba during development: The nature of the nurture. *Nat Rev Neurosci* 3:728–739.

Biebl M, Cooper CM, Winkler J, Kuhn HG (2000). Analysis of neurogenesis and programmed cell death reveals a self-renewing capacity in the adult rat brain. *Neurosci Lett* 291:17–20.

Brandt MD, Jessberger S, Steiner B, Kronenberg G, Reuter K, Bick-Sander A, von der Behrens W, Kempermann G (2003). Transient calretinin expression defines early postmitotic step of neuronal differentiation in adult hippocampal neurogenesis of mice. *Mol Cell Neurosci* 24:603–613.

Breunig JJ, Sarkisian MR, Arellano JI, Morozov YM, Ayoub AE, Sojitra S, Wang B, Flavell RA, Rakic P, Town T (2008). Primary cilia regulate hippocampal neurogenesis by mediating Sonic hedgehog signaling. *Proc Natl Acad Sci USA* 105:13127–13132.

Bruel-Jungerman E, Davis S, Rampon C, Laroche S (2006). Long-term potentiation enhances neurogenesis in the adult dentate gyrus. *J Neurosci* 26:5888–5893.

Bull ND, Bartlett PF (2005). The adult mouse hippocampal progenitor is neurogenic but not a stem cell. *J Neurosci* 25:10815–10821.

Cameron HA, Woolley CS, McEwen BS, Gould E (1993). Differentiation of newly born neurons and glia in the dentate gyrus of the adult rat. *Neuroscience* 56:337–344.

Cao L, Jiao X, Zuzga DS, Liu Y, Fong DM, Young D, During MJ (2004). VEGF links hippocampal activity with neurogenesis, learning and memory. *Nat Genet* 36:827–835.

Carlen M, Cassidy RM, Brismar H, Smith GA, Enquist LW, Frisen J (2002). Functional integration of adult-born neurons. *Curr Biol* 12:606–608.

Caviness VS, Jr. (1973). Time of neuron origin in the hippocampus and dentate gyrus of normal and reeler mutant mice: An autoradiographic analysis. *J Comp Neurol* 151:113–120.

Crespo D, Stanfield BB, Cowan WM (1986). Evidence that late-generated granule cells do not simply replace earlier formed neurons in the rat dentate gyrus. *Exp Brain Res* 62:541–548.

Czerniawski J, Yoon T, Otto T (2009). Dissociating space and trace in dorsal and ventral hippocampus. *Hippocampus* 19:20–32.

Dayer AG, Ford AA, Cleaver KM, Yassaee M, Cameron HA (2003). Short-term and long-term survival of new neurons in the rat dentate gyrus. *J Comp Neurol* 460:563–572.

De Toni A, Zbinden M, Epstein JA, Ruiz i Altaba A, Prochiantz A, Caille I (2008). Regulation of survival in adult hippocampal and glioblastoma stem cell lineages by the homeodomain-only protein HOP. *Neural Dev* 3:13.

Deisseroth K, Singla S, Toda H, Monje M, Palmer TD, Malenka RC (2004). Excitation-neurogenesis coupling in adult neural stem/progenitor cells. *Neuron* 42:535–552.

Doetsch F, Garcia-Verdugo JM, Alvarez-Buylla A (1999). Regeneration of a germinal layer in the adult mammalian brain. *Proc Natl Acad Sci USA* 96:11619–11624.

Dore-Duffy P, Katychev A, Wang X, Van Buren E (2006). CNS microvascular pericytes exhibit multipotential stem cell activity. *J Cereb Blood Flow Metab* 26:613–624.

Dupret D, Fabre A, Dobrossy MD, Panatier A, Rodriguez JJ, Lamarque S, Lemaire V, Oliet SH, Piazza PV, Abrous DN (2007). Spatial learning depends on both the addition and removal of new hippocampal neurons. *PLoS Biol* 5:e214.

Eckenhoff MF, Rakic P (1984). Radial organization of the hippocampal dentate gyrus: A Golgi, ultrastructural, and immunocytochemical analysis in the developing rhesus monkey. *J Comp Neurol* 223:1–21.

Eckenhoff MF, Rakic P (1988). Nature and fate of proliferative cells in the hippocampal dentate gyrus during the life span of the rhesus monkey. *J Neurosci* 8:2729–2747.

Ekdahl CT, Mohapel P, Elmer E, Lindvall O (2001). Caspase inhibitors increase short-term survival of progenitor-cell progeny in the adult rat dentate gyrus following status epilepticus. *Eur J Neurosci* 14:937–945.

Encinas JM, Vaahtokari A, Enikolopov G (2006). Fluoxetine targets early progenitor cells in the adult brain. *Proc Natl Acad Sci USA* 103:8233–8238.

Esposito MS, Piatti VC, Laplagne DA, Morgenstern NA, Ferrari CC, Pitossi FJ, Schinder AF (2005). Neuronal differentiation in the adult hippocampus recapitulates embryonic development. *J Neurosci* 25:10074–10086.

Fabel K, Toda H, Palmer T (2003a). Copernican stem cells: Regulatory constellations in adult hippocampal neurogenesis. *J Cell Biochem* 88:41–50.

Fabel K, Fabel K, Tam B, Kaufer D, Baiker A, Simmons N, Kuo CJ, Palmer TD (2003b). VEGF is necessary for exercise-induced adult hippocampal neurogenesis. *Eur J Neurosci* 18:2803–2812.

Fabel K, Wolf SA, Ehninger D, Babu H, Galicia PL, Kempermann G (2009). Additive effects of physical exercise and environmental enrichment on adult hippocampal neurogenesis in mice. *Front. Neurosci* 3:50.

Farioli-Vecchioli S, Saraulli D, Costanzi M, Pacioni S, Cina I, Aceti M, Micheli L, Bacci A, Cestari V, Tirone F (2008). The timing of differentiation of adult hippocampal neurons is crucial for spatial memory. *PLoS Biol* 6:e246.

Faulkner RL, Jang MH, Liu XB, Duan X, Sailor KA, Kim JY, Ge S, Jones EG, Ming GL, Song H, Cheng HJ (2008). Development of hippocampal mossy fiber synaptic outputs by new neurons in the adult brain. *Proc Natl Acad Sci USA* 105:14157–14162.

Filippov V, Kronenberg G, Pivneva T, Reuter K, Steiner B, Wang LP, Yamaguchi M, Kettenmann H, Kempermann G (2003). Subpopulation of nestin-expressing progenitor cells in the adult murine hippocampus shows electrophysiological and morphological characteristics of astrocytes. *Mol Cell Neurosci* 23:373–382.

Forster E, Tielsch A, Saum B, Weiss KH, Johanssen C, Graus-Porta D, Muller U, Frotscher M (2002). Reelin, Disabled 1, and beta 1 integrins are required for the formation of the radial glial scaffold in the hippocampus. *Proc Natl Acad Sci USA* 99:13178–13183.

Forster E, Zhao S, Frotscher M (2006). Laminating the hippocampus. *Nat Rev Neurosci* 7:259–267.

Francis F, Koulakoff A, Boucher D, Chafey P, Schaar B, Vinet MC, Friocourt G, McDonnell N, Reiner O, Kahn A, McConnell SK, Berwald-Netter Y, Denoulet P, Chelly J (1999). Doublecortin is a developmentally regulated, microtubule-associated protein expressed in migrating and differentiating neurons. *Neuron* 23: 247–256.

Fredens K, Stengaard-Pedersen K, Larsson LI (1984). Localization of enkephalin and cholecystokinin immunoreactivities in the perforant path terminal fields of the rat hippocampal formation. *Brain Res* 304:255–263.

Freund TF, Buzsaki G (1996). Interneurons of the hippocampus. *Hippocampus* 6:347–470.

Fukuda S, Kato F, Tozuka Y, Yamaguchi M, Miyamoto Y, Hisatsune T (2003). Two distinct subpopulations of nestin-positive cells in adult mouse dentate gyrus. *J Neurosci* 23:9357–9366.

Garthe A, Behr J, Kempermann G (2009). Adult-generated hippocampal neurons allow the flexible use of spatially precise learning strategies. *PLoS ONE* 4:e5464.

Ge S, Goh EL, Sailor KA, Kitabatake Y, Ming GL, Song H (2006). GABA regulates synaptic integration of newly generated neurons in the adult brain. *Nature* 439:589–593.

Ge S, Pradhan DA, Ming GL, Song H (2007a). GABA sets the tempo for activity-dependent adult neurogenesis. *Trends Neurosci* 30:1–8.

Ge S, Yang CH, Hsu KS, Ming GL, Song H (2007b). A critical period for enhanced synaptic plasticity in newly generated neurons of the adult brain. *Neuron* 54:559–566.

Gemma C, Bachstetter AD, Cole MJ, Fister M, Hudson C, Bickford PC (2007). Blockade of caspase-1 increases neurogenesis in the aged hippocampus. *Eur J Neurosci* 26:2795–2803.

Gould E, Beylin A, Tanapat P, Reeves A, Shors TJ (1999). Learning enhances adult neurogenesis in the hippocampal formation. *Nat Neurosci* 2:260–265.

Greenough WT, Cohen NJ, Juraska JM (1999). New neurons in old brains: Learning to survive? *Nat Neurosci* 2:203–205.

Han YG, Spassky N, Romaguera-Ros M, Garcia-Verdugo JM, Aguilar A, Schneider-Maunoury S, Alvarez-Buylla A (2008). Hedgehog signaling and primary cilia are required for the formation of adult neural stem cells. *Nat Neurosci* 11:277–284.

Hastings NB, Gould E (1999). Rapid extension of axons into the CA3 region by adult-generated granule cells. *J Comp Neurol* 413:146–154.

Hayes NL, Nowakowski RS (2002). Dynamics of cell proliferation in the adult dentate gyrus of two inbred strains of mice. *Brain Res Dev Brain Res* 134:77–85.

Hevner RF, Hodge RD, Daza RA, Englund C (2006). Transcription factors in glutamatergic neurogenesis: Conserved programs in neocortex, cerebellum, and adult hippocampus. *Neurosci Res* 55:223–233.

Hjorth-Simonsen A (1972). Projection of the lateral part of the entorhinal area to the hippocampus and fascia dentata. *J Comp Neurol* 146:219–232.

Hjorth-Simonsen A, Jeune B (1972). Origin and termination of the hippocampal perforant path in the rat studied by silver impregnation. *J Comp Neurol* 144:215–232.

Hodge RD, Kowalczyk TD, Wolf SA, Encinas JM, Rippey C, Enikolopov G, Kempermann G, Hevner RF (2008). Intermediate progenitors in adult hippocampal neurogenesis: Tbr2 expression and coordinate regulation of neuronal output. *J Neurosci* 28:3707–3717.

Houser CR (2007). Interneurons of the dentate gyrus: An overview of cell types, terminal fields and neurochemical identity. *Prog Brain Res* 163:217–232.

Huttmann K, Sadgrove M, Wallraff A, Hinterkeuser S, Kirchhoff F, Steinhauser C, Gray WP (2003). Seizures preferentially stimulate proliferation of radial glia–like astrocytes in the adult dentate gyrus: Functional and immunocytochemical analysis. *Eur J Neurosci* 18:2769–2778.

Imayoshi I, Sakamoto M, Ohtsuka T, Takao K, Miyakawa T, Yamaguchi M, Mori K, Ikeda T, Itohara S, Kageyama R (2008). Roles of continuous neurogenesis in the structural and functional integrity of the adult forebrain. *Nat Neurosci* 11:1153–1161.

Imbeault S, Gauvin LG, Toeg HD, Pettit A, Sorbara CD, Migahed L, DesRoches R, Menzies AS, Nishii K, Paul DL, Simon AM, Bennett SA (2009). The extracellular matrix controls gap junction protein expression and function in postnatal hippocampal neural progenitor cells. *BMC Neurosci* 10:13.

Jessberger S, Kempermann G (2003). Adult-born hippocampal neurons mature into activity-dependent responsiveness. *Eur J Neurosci* 18:2707–2712.

Jessberger S, Romer B, Babu H, Kempermann G (2005). Seizures induce proliferation and dispersion of doublecortin-positive hippocampal progenitor cells. *Exp Neurol* 196:342–351.

Jessberger S, Zhao C, Toni N, Clemenson GD, Jr., Li Y, Gage FH (2007). Seizure-associated, aberrant neurogenesis in adult rats characterized with retrovirus-mediated cell labeling. *J Neurosci* 27:9400–9407.

Jessberger S, Toni N, Clemenson Jr GD, Ray J, Gage FH (2008a). Directed differentiation of hippocampal stem/progenitor cells in the adult brain. *Nat Neurosci* 11:888–893.

Jessberger S, Aigner S, Clemenson GD, Toni N, Lie DC, Karalay O, Overall R, Kempermann G, Gage FH (2008b). Cdk5 regulates accurate maturation of newborn granule cells in the adult hippocampus. *PLoS Biol* 6:e272.

Jin K, Zhu Y, Sun Y, Mao XO, Xie L, Greenberg DA (2002). Vascular endothelial growth factor (VEGF) stimulates neurogenesis in vitro and in vivo. *Proc Natl Acad Sci USA* 99:11946–11950.

Jin K, Galvan V, Xie L, Mao XO, Gorostiza OF, Bredesen DE, Greenberg DA (2004). Enhanced neurogenesis in Alzheimer's disease transgenic (PDGF-APPSw,Ind) mice. *Proc Natl Acad Sci USA* 101:13363–13367.

Kee N, Teixeira CM, Wang AH, Frankland PW (2007). Preferential incorporation of adult-generated granule cells into spatial memory networks in the dentate gyrus. *Nat Neurosci* 10:355–362.

Kempermann G, Gage FH (1999). Experience-dependent regulation of adult hippocampal neurogenesis: Effects of long-term stimulation and stimulus withdrawal. *Hippocampus* 9:321–332.

Kempermann G, Gage FH (2002). Genetic influence on phenotypic differentiation in adult hippocampal neurogenesis. *Brain Res Dev Brain Res* 134:1–12.

Kempermann G, Kuhn HG, Gage FH (1997a). Genetic influence on neurogenesis in the dentate gyrus of adult mice. *Proc Natl Acad Sci USA* 94:10409–10414.

Kempermann G, Kuhn HG, Gage FH (1997b). More hippocampal neurons in adult mice living in an enriched environment. *Nature* 386:493–495.

Kempermann G, Kuhn HG, Gage FH (1998). Experience-induced neurogenesis in the senescent dentate gyrus. *J Neurosci* 18:3206–3212.

Kempermann G, Gast D, Gage FH (2002). Neuroplasticity in old age: Sustained fivefold induction of hippocampal neurogenesis by long-term environmental enrichment. *Ann Neurol* 52:135–143.

Kempermann G, Gast D, Kronenberg G, Yamaguchi M, Gage FH (2003). Early determination and long-term persistence of adult-generated new neurons in the hippocampus of mice. *Development* 130:391–399.

Kempermann G, Jessberger S, Steiner B, Kronenberg G (2004). Milestones of neuronal development in the adult hippocampus. *Trends Neurosci* 27:447–452.

Kempermann G, Chesler EJ, Lu L, Williams RW, Gage FH (2006). Natural variation and genetic covariance in adult hippocampal neurogenesis. *Proc Natl Acad Sci USA* 103:780–785.

Kiel MJ, Morrison SJ (2006). Maintaining hematopoietic stem cells in the vascular niche. *Immunity* 25: 862–864.

Kim EJ, Leung CT, Reed RR, Johnson JE (2007). In vivo analysis of Ascl1 defined progenitors reveals distinct developmental dynamics during adult neurogenesis and gliogenesis. *J Neurosci* 27:12764–12774.

Kim EJ, Battiste J, Nakagawa Y, Johnson JE (2008). Ascl1 (Mash1) lineage cells contribute to discrete cell populations in CNS architecture. *Mol Cell Neurosci* 38:595–606.

Kosaka T, Hama K (1986). Three-dimensional structure of astrocytes in the rat dentate gyrus. *J Comp Neurol* 249:242–260.

Kronenberg G, Reuter K, Steiner B, Brandt MD, Jessberger S, Yamaguchi M, Kempermann G (2003). Subpopulations of proliferating cells of the adult hippocampus respond differently to physiologic neurogenic stimuli. *J Comp Neurol* 467:455–463.

Kronenberg G, Wang LP, Geraerts M, Babu H, Synowitz M, Vicens P, Lutsch G, Glass R, Yamaguchi M, Baekelandt V, Debyser Z, Kettenmann H, Kempermann G (2007). Local origin and activity-dependent generation of nestin-expressing protoplasmic astrocytes in CA1. *Brain Structure & Function* 212:19–35.

Kuhn HG, Dickinson-Anson H, Gage FH (1996). Neurogenesis in the dentate gyrus of the adult rat: Age-related decrease of neuronal progenitor proliferation. *J Neurosci* 16:2027–2033.

Kuhn HG, Biebl M, Wilhelm D, Li M, Friedlander RM, Winkler J (2005). Increased generation of granule cells in adult Bcl-2o-verexpressing mice: A role for cell death during continued hippocampal neurogenesis. *Eur J Neurosci* 22:1907–1915.

Kunze A, Grass S, Witte OW, Yamaguchi M, Kempermann G, Redecker C (2005). Proliferative response of distinct hippocampal progenitor cell populations after cortical infarcts in the adult brain. *Neurobiol Dis* 21:324–32.

Kunze A, Congreso MR, Hartmann C, Wallraff-Beck A, Huttmann K, Bedner P, Requardt R, Seifert G, Redecker C, Willecke K, Hofmann A, Pfeifer A, Theis M, Steinhauser C (2009). Connexin expression by radial glia–like cells is required for neurogenesis in the adult dentate gyrus. *Proc Natl Acad Sci USA.*

Lagace DC, Whitman MC, Noonan MA, Ables JL, DeCarolis NA, Arguello AA, Donovan MH, Fischer SJ, Farnbauch LA, Beech RD, DiLeone RJ, Greer CA, Mandyam CD, Eisch AJ (2007). Dynamic contribution of nestin-expressing stem cells to adult neurogenesis. *J Neurosci* 27:12623–12629.

Laplagne DA, Esposito MS, Piatti VC, Morgenstern NA, Zhao C, van Praag H, Gage FH, Schinder AF (2006). Functional convergence of neurons generated in the developing and adult hippocampus. *PLoS Biol* 4:e409.

Laplagne DA, Kamienkowski JE, Esposito MS, Piatti VC, Zhao C, Gage FH, Schinder AF (2007). Similar GABAergic inputs in dentate granule cells born during embryonic and adult neurogenesis. *Eur J Neurosci* 25:2973–2981.

Lein ES, Zhao X, Gage FH (2004). Defining a molecular atlas of the hippocampus using DNA microarrays and high-throughput in situ hybridization. *J Neurosci* 24:3879–3889.

Lendahl U, Zimmerman LB, McKay RDG (1990). CNS Stem cells express a new class of intermediate filament protein. *Cell* 60:585–595.

Leranth C, Hajszan T (2007). Extrinsic afferent systems to the dentate gyrus. *Prog Brain Res* 163:63–84.

Li G, Pleasure SJ (2005). Morphogenesis of the dentate gyrus: What we are learning from mouse mutants. *Dev Neurosci* 27:93–99.

Liu Y, Namba T, Liu J, Suzuki R, Shioda S, Seki T (2010). Glial fibrillary acidic protein-expressing neural progenitors give rise to immature neurons via early intermediate progenitors expressing both glial fibrillary acidic protein and neuronal markers in the adult hippocampus. *Neuroscience* 166:241–251.

Long JE, Garel S, Alvarez-Dolado M, Yoshikawa K, Osumi N, Alvarez-Buylla A, Rubenstein JL (2007). Dlx-dependent and -independent regulation of olfactory bulb interneuron differentiation. *J Neurosci* 27:3230–3243.

Louissaint A, Jr., Rao S, Leventhal C, Goldman SA (2002). Coordinated interaction of neurogenesis and angiogenesis in the adult songbird brain. *Neuron* 34:945–960.

Markakis E, Gage FH (1999). Adult-generated neurons in the dentate gyrus send axonal projections to the field CA3 and are surrounded by synaptic vesicles. *J Comp Neurol* 406:449–460.

Mercier F, Kitasako JT, Hatton GI (2002). Anatomy of the brain neurogenic zones revisited: Fractones and the fibroblast/macrophage network. *J Comp Neurol* 451:170–188.

Merrill DA, Karim R, Darraq M, Chiba AA, Tuszynski MH (2003). Hippocampal cell genesis does not correlate with spatial learning ability in aged rats. *J Comp Neurol* 459:201–207.

Mignone JL, Kukekov V, Chiang AS, Steindler D, Enikolopov G (2004). Neural stem and progenitor cells in nestin-GFP transgenic mice. *J Comp Neurol* 469:311–324.

Mizuguchi R, Kriks S, Cordes R, Gossler A, Ma Q, Goulding M (2006). Ascl1 and Gsh1/2 control inhibitory and excitatory cell fate in spinal sensory interneurons. *Nat Neurosci* 9:770–778.

Mongiat LA, Esposito MS, Lombardi G, Schinder AF (2009). Reliable activation of immature neurons in the adult hippocampus. *PLoS ONE* 4:e5320.

Moser E, Moser MB, Andersen P (1993). Spatial learning impairment parallels the magnitude of dorsal hippocampal lesions, but is hardly present following ventral lesions. *J Neurosci* 13:3916–3925.

Nacher J, Varea E, Blasco-Ibanez JM, Castillo-Gomez E, Crespo C, Martinez-Guijarro FJ, McEwen BS (2005). Expression of the transcription factor Pax6 in the adult rat dentate gyrus. *J Neurosci Res* 81:753–761.

Nam HS, Benezra R (2009). High levels of Id1 expression define B1 type adult neural stem cells. *Cell Stem Cell* 5:515–526.

Namba T, Maekawa M, Yuasa S, Kohsaka S, Uchino S (2008). The Alzheimer's disease drug memantine increases the number of radial glia–like progenitor cells in adult hippocampus. *Glia* 57:1082–90.

Nikolova G, Strilic B, Lammert E (2007). The vascular niche and its basement membrane. *Trends Cell Biol* 17:19–25.

Ninkovic J, Mori T, Gotz M (2007). Distinct modes of neuron addition in adult mouse neurogenesis. *J Neurosci* 27:10906–10911.

Overstreet LS, Hentges ST, Bumaschny VF, de Souza FS, Smart JL, Santangelo AM, Low MJ, Westbrook GL, Rubinstein M (2004). A transgenic marker for newly born granule cells in dentate gyrus. *J Neurosci* 24:3251–3259.

Overstreet-Wadiche LS, Bensen AL, Westbrook GL (2006). Delayed development of adult-generated granule cells in dentate gyrus. *J Neurosci* 26:2326–2334.

Ozen I, Galichet C, Watts C, Parras C, Guillemot F, Raineteau O (2007). Proliferating neuronal progenitors in the postnatal hippocampus transiently express the proneural gene Ngn2. *Eur J Neurosci* 25:2591–2603.

Palmer TD (2002). Adult neurogenesis and the vascular Nietzsche. *Neuron* 34:856–858.

Palmer TD, Ray J, Gage FH (1995). FGF-2r-esponsive neuronal progenitors reside in proliferative and quiescent regions of the adult rodent brain. *Mol Cell Neurosci* 6:474–486.

Palmer TD, Takahashi J, Gage FH (1997). The adult rat hippocampus contains premordial neural stem cells. *Mol Cell Neurosci* 8:389–404.

Palmer TD, Willhoite AR, Gage FH (2000). Vascular niche for adult hippocampal neurogenesis. *J Comp Neurol* 425:479–494.

Peirce JL, Chesler EJ, Williams RW, Lu L (2003). Genetic architecture of the mouse hippocampus: Identification of gene loci with selective regional effects. *Genes Brain Behav* 2:238–252.

Petrus DS, Fabel K, Kronenberg G, Winter C, Steiner B, Kempermann G (2008). NMDA and benzodiazepine receptors have synergistic and antagonistic effects on precursor cells in adult hippocampal neurogenesis. *Eur J Neurosci* 29:244–52.

Pleasure SJ, Collins AE, Lowenstein DH (2000). Unique expression patterns of cell fate molecules delineate sequential stages of dentate gyrus development. *J Neurosci* 20:6095–6105.

Plumpe T, Ehninger D, Steiner B, Klempin F, Jessberger S, Brandt M, Romer B, Rodriguez GR, Kronenberg G, Kempermann G (2006). Variability of doublecortin-associated dendrite maturation in adult hippocampal neurogenesis is independent of the regulation of precursor cell proliferation. *BMC Neurosci* 7:77.

Priller J, Persons DA, Klett FF, Kempermann G, Kreutzberg GW, Dirnagl U (2001). Neogenesis of cerebellar Purkinje neurons from gene-marked bone marrow cells in vivo. *J Cell Biol* 155:733–738.

Raineteau O, Hugel S, Ozen I, Rietschin L, Sigrist M, Arber S, Gahwiler BH (2006). Conditional labeling of newborn granule cells to visualize their integration into established circuits in hippocampal slice cultures. *Mol Cell Neurosci* 32:344–355.

Rao MS, Shetty AK (2004). Efficacy of doublecortin as a marker to analyse the absolute number and dendritic growth of newly generated neurons in the adult dentate gyrus. *Eur J Neurosci* 19:234–246.

Ray J, Peterson DA, Schinstine M, Gage FH (1993). Proliferation, differentiation, and long-term culture of primary hippocampal neurons. *Proc Natl Acad Sci USA* 90:3602–3606.

Ribak CE, Seress L (1983). Five types of basket cell in the hippocampal dentate gyrus: A combined Golgi and electron microscopic study. *J Neurocytol* 12:577–597.

Ribak CE, Tran PH, Spigelman I, Okazaki MM, Nadler JV (2000). Status epilepticus-induced hilar basal dendrites on rodent granule cells contribute to recurrent excitatory circuitry. *J Comp Neurol* 428:240–253.

Ribak CE, Korn MJ, Shan Z, Obenaus A (2004). Dendritic growth cones and recurrent basal dendrites are typical features of newly generated dentate granule cells in the adult hippocampus. *Brain Res* 1000:195–199.

Rickmann M, Amaral DG, Cowan WM (1987). Organization of radial glial cells during the development of the rat dentate gyrus. *J Comp Neurol* 264:449–479.

Rietze R, Poulin P, Weiss S (2000). Mitotically active cells that generate neurons and astrocytes are present in multiple regions of the adult mouse hippocampus. *J Comp Neurol* 424:397–408.

Roybon L, Deierborg T, Brundin P, Li JY (2009). Involvement of Ngn2, Tbr and NeuroD proteins during postnatal olfactory bulb neurogenesis. *Eur J Neurosci* 29:232–243.

Sakakibara S, Nakamura Y, Yoshida T, Shibata S, Koike M, Takano H, Ueda S, Uchiyama Y, Noda T, Okano H (2002). RNA-binding protein Musashi family: Roles for CNS stem cells and a subpopulation of ependymal cells revealed by targeted disruption and antisense ablation. *Proc Natl Acad Sci USA* 99:15194–15199.

Sasaki T, Kitagawa K, Yagita Y, Sugiura S, Omura-Matsuoka E, Tanaka S, Matsushita K, Okano H, Tsujimoto Y, Hori M (2006). Bcl2 enhances survival of newborn neurons in the normal and ischemic hippocampus. *J Neurosci Res* 84:1187–1196.

Saxe MD, Battaglia F, Wang JW, Malleret G, David DJ, Monckton JE, Garcia AD, Sofroniew MV, Kandel ER, Santarelli L, Hen R, Drew MR (2006). Ablation of hippocampal neurogenesis impairs contextual fear conditioning and synaptic plasticity in the dentate gyrus. *Proc Natl Acad Sci USA* 103:17501–17506.

Schanzer A, Wachs FP, Wilhelm D, Acker T, Cooper-Kuhn C, Beck H, Winkler J, Aigner L, Plate KH, Kuhn HG (2004). Direct stimulation of adult neural stem cells in vitro and neurogenesis in vivo by vascular endothelial growth factor. *Brain Pathol* 14:237–248.

Scharfman H, Witter MP (2007). Preface. *Prog Brain Res* 163:IX-XIII.

Scharfman HE (2007). The CA3 "backprojection" to the dentate gyrus. *Prog Brain Res* 163:627–637.

Schmidt-Hieber C, Jonas P, Bischofberger J (2004). Enhanced synaptic plasticity in newly generated granule cells of the adult hippocampus. *Nature* 429:184–187.

Seaberg RM, van der Kooy D (2002). Adult rodent neurogenic regions: The ventricular subependyma contains neural stem cells, but the dentate gyrus contains restricted progenitors. *J Neurosci* 22:1784–1793.

Seaberg RM, van der Kooy D (2003). Stem and progenitor cells: The premature desertion of rigorous definitions. *Trends Neurosci* 26:125–131.

Seki T (2002). Expression patterns of immature neuronal markers PSA-NCAM, CRMP-4 and NeuroD in the hippocampus of young adult and aged rodents. *J Neurosci Res* 70:327–334.

Seki T, Arai Y (1993). Highly polysialylated neural cell adhesion molecule (NCAM-H) is expressed by newly generated granule cells in the dentate gyrus of the adult rat. *J Neurosci* 13:2351–2358.

Seki T, Arai Y (1999). Temporal and spacial relationships between PSA-NCAM-expressing, newly generated granule cells, and radial glia–like cells in the adult dentate gyrus. *J Comp Neurol* 410:503–513.

Seri B, Garcia-Verdugo JM, McEwen BS, Alvarez-Buylla A (2001). Astrocytes give rise to new neurons in the adult mammalian hippocampus. *J Neurosci* 21:7153–7160.

Seri B, Garcia-Verdugo JM, Collado-Morente L, McEwen BS, Alvarez-Buylla A (2004). Cell types, lineage, and architecture of the germinal zone in the adult dentate gyrus. *J Comp Neurol* 478:359.

Shapiro LA, Korn MJ, Ribak CE (2005a). Newly generated dentate granule cells from epileptic rats exhibit elongated hilar basal dendrites that align along GFAP-immunolabeled processes. *Neuroscience* 136:823–831.

Shapiro LA, Korn MJ, Shan Z, Ribak CE (2005b). GFAP-expressing radial glia–like cell bodies are involved in a one-to-one relationship with doublecortin-immunolabeled newborn neurons in the adult dentate gyrus. *Brain Res* 1040:81–91.

Snyder JS, Kee N, Wojtowicz JM (2001). Effects of adult neurogenesis on synaptic plasticity in the rat dentate gyrus. *J Neurophysiol* 85:2423–2431.

Soltesz I, Bourassa J, Deschenes M (1993). The behavior of mossy cells of the rat dentate gyrus during theta oscillations in vivo. *Neuroscience* 57:555–564.

Song H, Stevens CF, Gage FH (2002). Astroglia induce neurogenesis from adult neural stem cells. *Nature* 417:39–44.

Soriano E, Frotscher M (1994). Mossy cells of the rat fascia dentata are glutamate-immunoreactive. *Hippocampus* 4:65–69.

Spassky N, Han YG, Aguilar A, Strehl L, Besse L, Laclef C, Ros MR, Garcia-Verdugo JM, Alvarez-Buylla A (2008). Primary cilia are required for cerebellar development and Shh-dependent expansion of progenitor pool. *Dev Biol* 317:246–259.

Stanfield BB, Cowan WM (1979). The development of the hippocampus and dentate gyrus in normal and reeler mice. *J Comp Neurol* 185:423–459.

Stanfield BB, Trice JE (1988). Evidence that granule cells generated in the dentate gyrus of adult rats extend axonal projections. *Exp Brain Res* 72:399–406.

Steiner B, Kronenberg G, Jessberger S, Brandt MD, Reuter K, Kempermann G (2004). Differential regulation of gliogenesis in the context of adult hippocampal neurogenesis in mice. *Glia* 46:41–52.

Steiner B, Klempin F, Wang L, Kott M, Kettenmann H, Kempermann G (2006). Type-2 cells as link between glial and neuronal lineage in adult hippocampal neurogenesis. *Glia* 54:805–814.

Steiner B, Zurborg S, Horster H, Fabel K, Kempermann G (2008). Differential 24 h responsiveness of Prox1-expressing precursor cells in adult hippocampal neurogenesis to physical activity, environmental enrichment, and kainic acid-induced seizures. *Neuroscience* 154:521–529.

Sugiyama T, Kohara H, Noda M, Nagasawa T (2006). Maintenance of the hematopoietic stem cell pool by CXCL12–CXCR4 chemokine signaling in bone marrow stromal cell niches. *Immunity* 25:977–988.

Suh H, Consiglio A, Ray J, Sawai T, D'Amour KA, Gage FH (2007). In vivo fate analysis reveals the multipotent and self-renewal capacities of Sox2+ neural stem cells in the adult hippocampus. *Cell Stem Cell* 1:515–528.

Sun W, Winseck A, Vinsant S, Park OH, Kim H, Oppenheim RW (2004). Programmed cell death of adult-generated hippocampal neurons is mediated by the proapoptotic gene Bax. *J Neurosci* 24:11205–11213.

Tashiro A, Sandler VM, Toni N, Zhao C, Gage FH (2006). NMDA-receptor-mediated, cell-specific integration of new neurons in adult dentate gyrus. *Nature* 442:929–933.

Tashiro A, Makino H, Gage FH (2007). Experience-specific functional modification of the dentate gyrus through adult neurogenesis: A critical period during an immature stage. *J Neurosci* 27:3252–3259.

Toni N, Teng EM, Bushong EA, Aimone JB, Zhao C, Consiglio A, van Praag H, Martone ME, Ellisman MH, Gage FH (2007). Synapse formation on neurons born in the adult hippocampus. *Nat Neurosci* 10:727–734.

Toni N, Laplagne DA, Zhao C, Lombardi G, Ribak CE, Gage FH, Schinder AF (2008). Neurons born in the adult dentate gyrus form functional synapses with target cells. *Nat Neurosci* 11:901–907.

Tozuka Y, Fukuda S, Namba T, Seki T, Hisatsune T (2005). GABAergic excitation promotes neuronal differentiation in adult hippocampal progenitor cells. *Neuron* 47:803–815.

Trivedi MA, Coover GD (2004). Lesions of the ventral hippocampus, but not the dorsal hippocampus, impair conditioned fear expression and inhibitory avoidance on the elevated T-maze. *Neurobiol Learn Mem* 81: 172–184.

Trouche S, Bontempi B, Roullet P, Rampon C (2009). Recruitment of adult-generated neurons into functional hippocampal networks contributes to updating and strengthening of spatial memory. *Proc Natl Acad Sci USA* 106:5919–5924.

Uda M, Ishido M, Kami K (2007). Features and a possible role of Mash1-immunoreactive cells in the dentate gyrus of the hippocampus in the adult rat. *Brain Res* 1171:9–17.

van Praag H, Shubert T, Zhao C, Gage FH (2005). Exercise enhances learning and hippocampal neurogenesis in aged mice. *J Neurosci* 25:8680–8685.

van Praag H, Schinder AF, Christie BR, Toni N, Palmer TD, Gage FH (2002). Functional neurogenesis in the adult hippocampus. *Nature* 415:1030–1034.

Vandenbosch R, Borgs L, Beukelaers P, Foidart A, Nguyen L, Moonen G, Berthet C, Kaldis P, Gallo V, Belachew S, Malgrange B (2007). CDK2 is dispensable for adult hippocampal neurogenesis. *Cell Cycle* 6:3065–3069.

Wang LP, Kempermann G, Kettenmann H (2005). A subpopulation of precursor cells in the mouse dentate gyrus receives synaptic GABAergic input. *Mol Cell Neurosci* 29:181–189.

Wang S, Scott BW, Wojtowicz JM (2000). Heterogenous properties of dentate granule neurons in the adult rat. *J Neurobiol* 42:248–257.

West MJ, Gundersen HJ (1990). Unbiased stereological estimation of the number of neurons in the human hippocampus. *J Comp Neurol* 296:1–22.

Winner B, Cooper-Kuhn CM, Aigner R, Winkler J, Kuhn HG (2002). Long-term survival and cell death of newly generated neurons in the adult rat olfactory bulb. *Eur J Neurosci* 16:1681–1689.

Yamashima T, Tonchev AB, Vachkov IH, Popivanova BK, Seki T, Sawamoto K, Okano H (2004). Vascular adventitia generates neuronal progenitors in the monkey hippocampus after ischemia. *Hippocampus* 14:861–875.

Yung SY, Gokhan S, Jurcsak J, Molero AE, Abrajano JJ, Mehler MF (2002). Differential modulation of BMP signaling promotes the elaboration of cerebral cortical GABAergic neurons or oligodendrocytes from a common Sonic hedgehog-responsive ventral forebrain progenitor species. *Proc Natl Acad Sci USA* 99:16273–16278.

Zhao C, Teng EM, Summers RG, Jr., Ming GL, Gage FH (2006). Distinct morphological stages of dentate granule neuron maturation in the adult mouse hippocampus. *J Neurosci* 26:3–11.

Zhao S, Chai X, Frotscher M (2007). Balance between neurogenesis and gliogenesis in the adult hippocampus: Role for reelin. *Dev Neurosci* 29:84–90.

7

Technical Notes

Research on adult neurogenesis
still largely relies on a relatively
narrow spectrum of core methods

NOVEL GENETIC TOOLS have greatly expanded the spectrum of methods used to study adult neurogenesis in the past few years, but a large and quite central part of the present knowledge about adult neurogenesis still relies on a rather limited repertoire of techniques, in particular on the birthdating of newborn neurons with bromodeoxyuridine. This makes the body of information particularly sensitive to confounding technical problems. Consequently, critical evaluation of the methods has been an important part of interpreting the data on adult neurogenesis. In the peer review process, the sweat of both authors and referees is spent on disagreement over methodology, and although there can and should never be ironclad standards (which would prevent any progress), the field has matured enough to avoid obvious and rather common pitfalls and confounds.

This chapter cannot cover all methods relevant to research on adult neurogenesis. But for both in vitro and in vivo techniques, a few fundamentals should be discussed. The goal is less to introduce the reader to how to actually perform the methods than to provide information that facilitates interpretation of published results.

Methodological issues that were
solved long ago might again
confound present-day study

The first three decades after Altman's initial description of adult neurogenesis in the 1960s were dominated by fundamental skepticism about this phenomenon. Evidence of adult neurogenesis was not yet good enough. Since the mid-1990s, the mere existence of neurogenesis in the adult mammalian brain is no longer questioned. Consequently, today many of the methodological discussions are not as fervent as they were in the early years, and not every author is required to go the extra mile to prove again that adult neurogenesis occurs. Nevertheless, one should know about the relevant methodological issues, and not just for historical reasons. In fact, science sometimes goes full circle, and old problems and their solutions slip from awareness because we have started to take things for granted, which is almost never justifiable in science. Clearly, one reason for the increasing acceptance of adult neurogenesis, besides the rise of neural stem cell biology, has been technical progress. But now-widely-used techniques such as confocal microscopy, stereology, conditional mutagenesis, and cell-specific genetic tools are not without problems. The methodological issues around adult neurogenesis essentially fall into four categories.

- *Problems related to the identification of a cell as neural precursor cell.* These problems are related to the in vitro experiments, in which the proof of "stemness" is performed, and the question is how well they represent the in vivo conditions.

216

- *Problems related to the identification of newly generated cells.* A key question in adult neurogenesis research in vivo is: Is this cell, is this neuron new? Practically, these issues circle around the use of markers such as BrdU as persistent markers of proliferating cells. Do these methods reliably mark only dividing cells, or is a contamination by DNA repair or cell death possible? And also: How can these markers be used to identify precursor cells in vivo? Similar issues affect the use of retroviruses to permanently label dividing cells.
- *Problems related to identifying the phenotype of a cell.* Here difficulties arise primarily from the challenges of multi-channel immunofluorescence and the potential ambiguities of the markers used in these experiments. The core problem is how to demonstrate the neuronal or precursor identity of a cell that has been generated in the adult brain. What are the minimal criteria for a neuron? Can they be addressed in vivo? What is a reasonable compromise between feasibility and validity? Related are issues with the new genetic tools for lineage-tracing in vivo, for cell-specific gene targeting and conditional mutagenesis, or cell-specific ablation.
- *Problems related to quantification.* This issue has to do with the not-so-trivial task of obtaining valid estimates of total cell numbers from counts of marked cells in the available sample. The solution to this problem is stereology, a solution that some researchers fear will pose greater problems than the pitfall it supposedly prevents. What are the consequences of sloppy quantification?

Some additional difficulties in interpreting data about adult neurogenesis, not covered here in detail, have to do with conclusions about a dynamic process being derived from essentially static data. How are our data influenced by the dynamics of the system?

In Vitro Methods

Practically, "stemness" is largely a property studied and defined in vitro

Although the majority of research reports on adult neurogenesis deal with the situation in vivo, an increasing number supports in vivo results by in vitro evidence, if they are not in vitro studies from the beginning. As it will have become clear in Chapters 3 through 6, many research questions in the context of adult neurogenesis involve issues concerning the properties and the identity of the precursor cells from which the new neurons originate. In vivo demonstration of "stemness" is possible, but it is technically challenging (see, for example, Suh et al., 2007), and the interpretation is of course not without problems of its own. Studies exploring detailed molecular mechanisms of neurogenesis and research on the cellular biology of neural stem cells heavily rely on cell culture experiments anyway.

Isolating Precursor Cells

Only cultures from microdissected brain regions can reflect the regionalization of precursor cells

Neural precursor cell cultures can be obtained from whole brain homogenate with relative ease using any of the available protocols for neurospheres or adherent cultures. Using large parts of the brain as source material, however, gives away spatial information and potential insight into regional differences. As outlined in Chapter 3, there is no such thing as a generic neural stem cell in the adult brain. From the neuroepithelial cells onwards, precursor cells are regionalized (Fig. 4–5). Thus in most cases the tissue has to be dissected to a varying degree to reduce the number of potential precursor cell populations that could show up in the cultures. Microdissection of specific regions, such as parts of the SVZ (Merkle et al., 2007) or only the dentate gyrus (Babu et al., 2007), provide additional challenges but improve regional specificity.

Culture protocols start from single-cell suspensions

The tissue samples have to be homogenized in media to obtain a single-cell suspension. Besides mechanical trituration, the partial digestion with enzymes such as trypsin or papain is necessary. The digestion step is crucial for the success of the establishment of the culture and might determine the exact culture composition. Digestion removes many of the surface epitopes, especially the carbohydrates that play an important role in cell-to-cell signaling. Cell processes, including the radial processes of the primary stem cells of SVZ and SGZ, are presumably torn during this treatment, reducing the cell's chance for survival in culture.

In most protocols the precursor cells are selected for by conditions not suitable for other cell types

Most neural precursor-cell culture protocols are "survival of the fittest" protocols. The culture conditions are such that differentiated cells do not survive, and only the precursor cells are enriched over the course of a few days. Contrary to common perception, stem cells are quite robust. This approach is not particularly effective, as in the beginning, the cultures contain much irrelevant cellular material, which might even counteract the successful establishment of the cultures. Myelin, for example, has to be removed because it inhibits the precursor cells. This can be achieved by centrifugation and aided by the addition of beads that have a high affinity for myelin.

Enrichment of the precursor cells can be enhanced by gradient centrifugation (Palmer et al., 1995; Palmer et al., 1999; Babu et al., 2007). The specific buoyancy of stem and progenitor cells in a Percoll gradient is used to enrich putative stem cells by separating them from blood and other neural cells and getting rid of myelin and cell debris. The method is effective but rather cumbersome.

Prospective isolation with surface antigens or transgene-driven reporter genes is possible but has not been extensively validated

The first prospective isolation of neural precursor cells, albeit from the fetal brain, was achieved in 2000 by Uchida, Gage, Weissmann and colleagues, who used fluorescence-activated cell sorting and surface antigens but did not have any follow-ups. Phenotypically, the stem cells had the signature CD133-positive, CD34-negative, CD45-negative, and CD24-negative or -low (Uchida et al., 2000). The isolation of "neural stem cells" based on the magnetic separation of Prominin1/CD133 is marketed by companies and works well for the embryonic brain but has not been fully validated for the adult brain, and especially not for the hippocampus (see below for Prominin1/CD133 as stem cell marker). Table 3–2 (p. 76) lists examples of markers that have been successfully used for prospective isolation of (adult) neural precursor cells. The list includes many examples, which are based on transgenic reporter gene constructs, in which the fluorescent marker used for selection is expressed under the promoter of a candidate precursor cell gene such as Sox2. These approaches are not applicable for naïve wildtype animals, unless a transient infection with a viral vector is chosen to express the reporter gene (Roy et al., 1999; Roy et al., 2000a; Roy et al., 2000b). What is still lacking is a side-by-side comparison of the precursor cell population obtained with the various methods. Presumably, the populations are not identical.

Side populations have not yet been widely used to isolate neural precursor cells

In hematology, the analysis of so-called side populations has become an important tool. Cells differ in their ability to bind or actively exclude dyes with affinity to their DNA (see also Chap. 3, p. 77). During FACS one can identify a side population characterized by low binding to the fluorescent dye and high outflux of the dye from the cells. Kim and Morshead have applied this method to neural stem cell biology and found that the main and side populations differed in their level of stemness; nearly all cells with a sphere-forming capacity in their type of the assay were found in the side population (Kim and Morshead, 2003). GFAP, however, was found in both populations. Combining the side population criteria with positivity for the marker LeX/SSEA1 (Lewis X) further enriched for sphere-forming cells, although LeX/SSEA1 is expressed in both GFAP-expressing and non-expressing putative precursor cells (Capela and Temple, 2002).

Neurospheres

"Neurospheres" have largely become a standard tool in neural stem cell biology (Figs. 3–12 and 7–1). As neurospheres are free-floating aggregate cultures, they have many advantages; most important, they are easy to use. After obtaining the cell suspension, the procedure requires little more than using the appropriate medium and regular passaging (see Chap. 3, p. 77, for more details on the cell biology of neurospheres). Neurosphere cultures represent the simplest and most straightforward way to bring precursor cells from the adult brain in vitro. But quite often the term *neurospheres* is incorrectly used synonymously for *neural stem cells*, which is not justified. The presence of aggregates in the cell culture does not constitute a proof of "stemness." Despite its widespread use, the classical neurosphere assay has a few disadvantages and is unfortunately confounded with respect to some key aspects (Reynolds and Weiss, 1992; Reynolds and Rietze, 2005; Jensen and Parmar, 2006; Singec et al., 2006; Jessberger et al., 2007; Marshall et al., 2007). Measures can be taken to avoid these problems, but neurospheres do not meet all needs, and the interpretation of published data requires us to look into the details of both the protocol and the results.

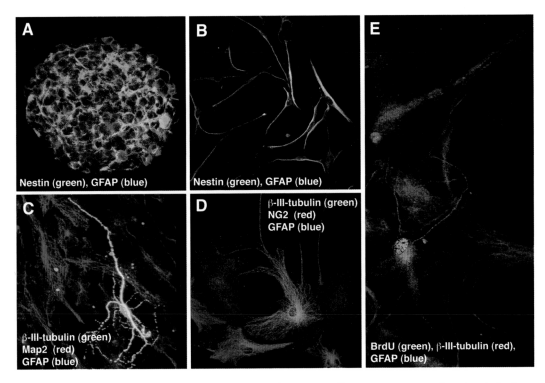

FIGURE 7–1 Neural precursor cell markers in vitro. Neural precursor cells express precursor cell markers in both types of precursor cell cultures–neurospheres (A) and monolayers (B). Neurospheres from the adult hippocampus (A) express both nestin and GFAP. Even under proliferation conditions, precursor cell cultures show heterogeneity. This is particularly obvious in neurospheres. Upon transition to differentiation conditions (C–E), precursor cells differentiate into cells with characteristics of the three neural lineages–neurons (here shown by immunoreactivity against β-III-tubulin), astrocytes (GFAP), and oligodendrocytes (NG2). Note that the specificity of these markers is not undisputed (see Chap. 8). In (E), a BrdU-labeled cell has acquired a neuronal phenotype, evidence of neurogenesis in vitro. Images by Harish Babu.

The main arguments against the unconditioned and sometimes unreflected use of neurospheres have been: (1) the heterogeneity of the cellular aggregates, in which precursor cells amounting to an unknown quantity of only a few percent; (2) the problem that individual cells are differently exposed to the culture medium and its factors, depending on where in the cluster they are located; (3) the variable extent of cell–cell contact exposes the precursor cells and their progeny to different levels of cell–cell signaling and tissue integration; (4) the problems with quantifying neurospheres, because spheres tend to fuse; and (5) that the growth of the cells in a compact cluster prevents a straightforward investigation of individual cells.

In addition, a number of other issues have been raised with respect to the attempts to use the formation of neurospheres as a quantitative measure of neurogenic potential or stem cell counts. These have been criticized because individual spheres show a variable propensity to fuse, so that neither number of spheres nor sphere composition is representative of the putative cell of origin (Singec et al., 2006; Jessberger et al., 2007). Several solutions have been proposed to cope with these issues, and they will be discussed below (p. 225).

Adherent Monolayer Cultures

Stem cells are not static elements but defined by transitions or the controlled absence thereof. They are in a fluent state, called *potency* or *potential*, that requires delicate molecular control. This control is an active process. Cell–cell interactions within the spheres counteract the maintenance of the precursor cells. This is the reason why neurospheres contain so few true precursor cells and are thus not ideally suited to study stemness (Reynolds and Rietze, 2005; Babu et al., 2007; Jessberger et al., 2007). While it is true that even from the first divisions of an isolated stem cells onward, small stochastic differences between cells exist that drive the cells apart; there appears to be a restriction point, after which true differentiation sets in. Usually this restriction point is associated with cell cycle exit. This process can (at least in theory) be better controlled in monolayer cultures (Figs. 3–12 and 7–1).

In this method, the isolated cell suspension is seeded on coated surfaces, and the cells are then grown in the presence of mitogens as adherent monolayer cultures. Historically, enrichment by centrifugation gradients has been part of these protocols (Palmer et al., 1995; Palmer et al., 1999) but is not mandatory. The enrichment step increases the yield, which is particularly relevant with the monolayer protocol, which tends to give cells less support at early growth stages. The cells can be grown on substrates such as laminin, fibronectin, or poly-D-lysin, depending on the exact protocol. Success of the culture is greatly dependent on the chosen substrates. The initial steps of establishing monolayer cultures are much more cumbersome and time-consuming than with neurospheres.

Cell-to-cell contacts are less prominent in adherent cultures compared to neurospheres, which probably influences survival and differentiation because the cells receive fewer niche-like stimuli that could be provided by other cells. While on one side this is a disadvantage, it leads to more homogenous cultures in the longer run. Adherent cultures allow better visualization of single-cell morphology, cell migration, differentiation, and maturation than neurospheres, even though the initial yield is lower. Given the actual low abundance of stem cells in the cell-rich neurospheres, that impression of low efficiency might be misleading. A direct quantitative comparison of both methods is still lacking, however. Importantly, the monolayers can be easily superseded onto layers of other cell types, allowing, for example, the relatively homogenous co-culture between neurons and precursor cells (Babu et al., 2009). Also, the transduction of the precursor cells grown as monolayers is more efficient and easier to monitor than in neurospheres. Finally, electrophysiology is much easier to investigate in the monolayer cultures with their direct access to individual cells.

Assessing the Properties of Neural Precursor Cells In Vitro

Precursor cells ex vivo do not yet fully reflect their in vivo counterparts

Today's neural stem cell cultures do not yet adequately reflect the situation that seems to prevail in vivo. Many regional identities are lost when the cells are brought into culture (Santa-Olalla et al., 2003; Hack et al., 2004). Neither self-renewal nor multipotency, however, can be feasibly and routinely addressed in vivo, so an ex vivo approach to assess these aspects of stem cell identity and nature appears to be inevitable. But to put cells in vitro means that they will undergo a radical and complete change in their environment. The spatiotemporal signaling cues these cells are exposed to in vivo change into a highly controlled environment created by the researcher. Many factors that are rather scarce in vivo are now delivered at high concentrations. The precursor cells only show a certain and relatively specific, intrinsically determined behavior in vitro that allows one to distinguish them from other cell populations brought into the same culture conditions and that would not show self-renewal and multipotency. The degree to which the culture conditions induce or disinhibit this behavior in the isolated cells is not known. Consequently, the potential of precursor cells ex vivo is not necessarily identical to their realized potential in vivo. In vitro conditions do not exactly replicate in vivo conditions, although progress has been made toward a better understanding of this issue. More selective isolation protocols will help to make cultures truer to their in vivo correspondence, but the main issue is the identification of the best media compositions that only bring out the "natural" competence while preventing the emergence of artificial properties. Overall, the results have been both encouraging and disturbing, because ex vivo, the cells often tend to show a greater potential than that assumed from the in vivo data. Essentially, precursor cells ex vivo have greater capacity for self-renewal and multipotency than that expected from their putative in vivo counterparts.

Assessing Self-Renewal

Self-renewal is demonstrated by showing that single cells can repeatedly reconstitute the full lineage potential

Self-renewal is tested by a clonal analysis, in which single cells isolated from spheres or adherent cultures can clonally generate new secondary spheres or colonies with multi-lineage differentiation (Fig. 7–2). This step needs to be repeated to serial propagation in order to fully establish self-renewal. In hematology the ultimate proof of stemness is based on serial transplantation: the putative stem cells are implanted into the host and re-isolated after integration to establish a new generation of precursor cell cultures. This has so far been rarely attempted in neural stem cell biology; in the context of one study without success (Marshall et al., 2006).

Practically, experiments need to go down to the single cell level, which creates a highly reductionistic situation

A suspension of putative precursor cells is diluted to such a degree that when pipetted onto a microtiter plate, only one single cell is found in each individual well. Preferably, the cell suspensions are simply given onto a microwell plate with densely packed cavities, in which the cavity size is optimized for the cell size and usually only one cell ends up per well (Cordey et al., 2008). This can be confirmed under the microscope. Most standard protocols only rely on the a high dilution of the cell suspension ("clonal density"), however. It is then examined whether the single cells can again form neurospheres or clusters of adherent cells, that under differentiation conditions show signs of multipotency. By this standard, the yield of stem cells is low in both types of precursor cell cultures but higher in monolayers than in neurospheres where it usually does not surpass a few percent. To test for so-called "sphere-forming units" as a measure of stemness, rather than equaling spheres and stem cells 1:1 marked a clear progress (Reynolds and Rietze, 2005) but neither spheres nor sphere-forming units have an unambiguous in vivo counterpart. In any case the assay exploits a highly artificial situation. We do not have any deep insight into how much of a stem cell's ability to self-renew is dependent on the anatomy of the niche. We might be able to mimic the composition of the humeral environment in the culture medium to a reasonable degree, but in vivo stem cells are never alone. The niche is a three-dimensional anatomical structure that presumably forms a functional unit with the

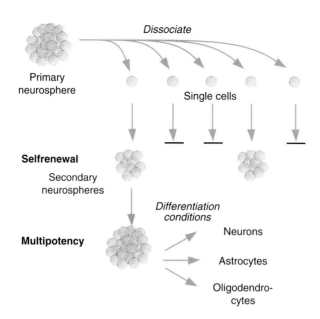

FIGURE 7–2 Stemness assay. To determine whether neurospheres in fact contain stem cells, they are dissociated and plated as single cells (clonal density). A varying percentage of the individualized cells form secondary neurospheres. This step can be repeated to support the conclusion that cells from the primary neurosphere are indeed self-renewing and thus fulfill the cardinal criterion of stem cells. The secondary sphere can also be transferred into differentiation conditions to assess its ability to form different cell types. Multipotency is the second stem cell criterion. In the eyes of some experts, this criterion is not necessary if true self-renewal has been demonstrated. Assessment of multipotency on the level of primary neurospheres is almost meaningless because it cannot be determined whether the primary sphere indeed originated from one single stem cell.

stem cells themselves (see also Chap. 3, p. 88 and Chap. 8, p. 277). Reducing the niche to the humeral factors that can be provided by the culture medium assesses self-renewal under conditions, under which no stem cell in vivo ever has to perform. From similar studies in hematology we might learn how to deal with this problem. The idea is to provide an artificial niche to the cells in the form of small cavities in a matrix engineered from biomaterials that aim at emulating the extra-cellular matrix of the real niche (Cordey et al., 2008). Also this assay is a single-cell assay which abstracts from the role of the other niche cells but is clearly a progress.

Assessing Multipotency

Multipotency practically means the presence of differentiating cells from two or three neural lineages

Routinely only generic, rather immature cell types are assessed within the neural lineages

Multipotency is routinely assessed after withdrawing growth factors EGF and FGF2 in stem cell cultures and instead adding factors such as retinoic acid, BDNF, serum, or simply nothing of this kind to the culture medium. The presence of cells from the three main neural lineages—neurons, astrocytes, and oligodendrocytes—is demonstrated by means of immunocytochemistry and taken as evidence of multipotency (Figs. 7–1 and 7–2). In most studies, the identification of neurons, astrocytes, and oligodendrocytes is based on the detection of key antigens, such as β-III-tubulin, Map-2, or NeuN for neurons; O4 for oligodendrocytes; and GFAP for astrocytes. Microglia has never been found in these cultures.

Tripotency for the three neural lineages of neurons, oligodendrocytes, and astrocytes is assessed as such common feature of neural precursor cells ex vivo and demanded to satisfy the definition of stemness criterion "self-renewal plus multipotency." Subspecification within these large categories is usually not performed, and functional maturation is not assessed, so that the cultures might gloss over existing differences. While such convergence to a generic differentiation spectrum can often be explained by the limited specificity of the readout, the culture conditions themselves may impose certain homogenizing influences (Gabay et al., 2003; Santa-Olalla et al., 2003). For adult hippocampal precursor cells, for example, modification to the culture protocols were required in order to allow the generation of granule cell-like neurons in culture and thus a neuronal subtype reflecting the in vivo competence (Babu et al., 2007).

The most widely used marker to detect neurons is β-III-tubulin, in lab jargon often referred to as "Tuj1" (which is neither the historic nor actual protein or gene name), a rather immature marker, weakly detectable even in some proliferating cells. β-III-tubulin is discussed further below with the other neuronal markers (p. 242), although in contrast to them, β-III-tubulin is almost exclusively used in vitro. The preferred marker is Map2ab.

Multipotency might be inducible by culture conditions

One important general concern in addition to the issues of marker sensitivity and specificity is the possibility that culture conditions themselves might induce multipotency in the precursor cells. Consequently, there is no true evidence that neural stem cells in vivo are multipotent as long as this conclusion is based only on an analogy drawn from in vitro studies, which might induce the multipotency in the first place.

Many conclusions depend on the specific culture conditions

In 1992, when Reynolds and Weiss first published data on EGF-dependent neural progenitor cells from the adult mouse brain (Reynolds et al., 1992; Reynolds and Weiss, 1992), the debate on culture conditions was further stimulated by the fact that the first results from rat hippocampi indicated that the precursor cells depended on FGF-2 (Palmer et al., 1995). On one hand this result could have reflected differences in populations of cells or among species, but on the other, concerns were raised that supposedly biological differences may have been introduced by the experimental conditions. This has been further exemplified by the finding that the question of whether or not cells that meet the more stringent criteria for stemness can be obtained from the hippocampus (Seaberg and van der Kooy, 2002; Bull and Bartlett, 2005) turned out to be at least partly dependent on the culture protocol (Babu et al., 2007).

Transplantation experiments have confirmed that the realization of the differentiation potential is dependent on cues in the host tissue (reviewed in Chap. 8, p. 286). On the other hand, the developmental potential is a function of the regionalization of the precursor cells and their own developmental history (see, for example, Chap. 4, p. 111, and Chap. 5, p. 172). This means that in the reductionistic assay aiming at assessing multipotency in vitro, we have no less a task than solving a "nature and nurture" riddle in the culture dish.

Induced pluripotency is the best example of how far the ex vivo manipulation towards multipotency can go

The finding that induced pluripotency is possible with small molecules and the manipulation of only a few genes cautions us about the influence culture conditions might exert on the potential of cells. Consequently, it remains a valid concern that criteria used to define aspects of stemness in truth mirror the conditions we apply to the cells (see also the discussion of induced pluripotency on p. 82, and the question of the developmental potential of mesenchymal stem cells, p. 87). One simple exemplification of this problem is the observation that cell growth in fetal precursor cell cultures is increased when cells are grown under lower levels of oxygen that are closer to the partial pressure in brain tissue than to that in the atmosphere (Studer et al., 2000).

Assessing Differences Between Precursor Cells

Precursor cell cultures cannot yet be adequately used to study regional differences ex vivo

Problematically, and despite much progress in recent years, the exact nature and identity of precursor cells from the adult brain and how closely they mirror in vivo characteristics remains an unresolved question. Even more problematic is the fact that the current tests still have an unknown sensitivity to identify differences between precursor cells. There are a few exceptions in which populations can be characterized based on a particular expression pattern of relevant transcription factors (Hack et al., 2004; Brill et al., 2008; Brill et al., 2009). Neither the many observed differences nor their apparent absence in other experiments allow final conclusions. There are competing interests: on one side we would like to have a generic test of "stemness," on the other hand we would like to detect small relevant differences between different precursor cells. Obviously, no single test can suit both needs equally well.

Neural precursor cells ex vivo show signs of regionalization and differential potentials

Nevertheless, precursor cells from different brain regions did not behave identically under identical culture conditions (Hitoshi et al., 2002b; Hitoshi et al., 2002a; Ostenfeld et al., 2002; Parmar et al., 2002; Parmar et al., 2003); nor do precursor cells from different areas of the SVZ (Merkle et al., 2007). This supports the idea that precursor cells in the adult brain are regionalized and show an inborn developmental potential that reflects the development that led to the germinative matrix from which they were derived. To assess this potential and to maintain it in vitro is a difficult task. In many contexts, however, the stakes are much lower, and rather general conclusions, most notably about self-renewal and multipotency, are sought. At this level, neural precursor cells behave so remarkably similarly, independent of their origin, that quite general criteria are often considered sufficient to identify a cell.

Both regional and intrinsic differences in vivo will be underestimated by in vitro assays. The homogenizing effect of cell culture conditions might also explain the otherwise surprisingly few differences in gene expression patterns between embryonic and adult neural precursor cells (Ramalho-Santos et al., 2002). Again, choosing a different approach to isolate the cells of interest for comparison led to a different outcome and revealed that a number of plausible genetic differences between pluripotent and multipotent stem cell populations exist (D'Amour and Gage, 2003).

The sequence from stem cell to progenitor cell to neuroblast has not yet been adequately recreated in vitro

In addition to an unsatisfying reflection of regional precursor cell heterogeneity in vivo, in vitro assays show difficulties in differentiating between transiently amplifying progenitor cells and stem cells and therefore do not allow unambiguous conclusions about the position the cultured cells would have in the stem cell hierarchy in vivo (Seaberg and van der Kooy, 2003). Given the greater number and greater activity of the intermediate precursor cells in both neurogenic zones in the adult brain, it seems plausible that the intermediate precursor cells might end up in culture with higher probability. The process-bearing radial glia–like primary stem cells might also be more sensitive to damage during the isolation process. Nevertheless, glial antigens are found on many precursor cells in vitro, but this is also true for intermediate progenitor cells in vivo, at least in the hippocampus.

By conventional standards, C cells in the SVZ and type-2 cells of the hippocampus are intermediate progenitor cells, and they might be the primary origin of precursor cell populations in vivo. But this has neither been proven nor it is likely that in fact the cultures represent only C or type-2 cells.

Lineage proportions are again dependent on culture conditions

Additional selection mechanisms might be in place. Most transcription factors are down-regulated in the growth phase of the neurosphere assay as well as in adherent cultures, with the notable exception of the Olig genes (Hack et al., 2004). Culture conditions thus seem to select for or induce the generation of Olig2-expressing precursor cells, which in the adult SVZ in vivo corresponds to a subset of C cells (Hack et al., 2004). Consequently, the contribution of Olig2-negative populations might be underestimated. The growing insight into the mechanisms generating the diversity of olfactory bulb interneurons brings up the question which of the potentially different progenitor cell types are actually found in a "normal" cell culture experiment from the SVZ.

Multipotency in vivo is difficult to assess, but appears more abundant ex vivo than in vivo

Again the homogenizing influences of the culture conditions are a concern. In both hippocampal and parenchymal precursor cell populations in the adult rat brain, FGF2 elicited the stem cell potential (Palmer et al., 1999). Exposure to epidermal growth factor transformed Dlx2-expressing C cells into multipotent stem cells (Doetsch et al., 2002). At least this is how the results have been interpreted. Our knowledge about multipotency in vivo, however, is still so limited that it is not clear which of the different precursor cells that can be distinguished in vivo are multipotent. The only in vivo lineage-tracing study to date was based on Sox2 expression, but again, Sox2 is expressed in both the radial and the non-radial precursor cells (Suh et al., 2007). At face value it seems that in many instances multipotency can be detected in cell populations ex vivo that do not seem to show it in vivo.

There are several options to explain these discrepancies: either (1) all of these cells are indeed multipotent in vivo and the in vivo assays are still too crude to detect this; or (2) their multipotency is actively suppressed by the environment in vivo; or (3) the apparent multipotency in vitro reflects a dedifferentiation and thus a non-physiological or even pathological state, with no valid relationship to the normal in vivo situation (but possibly to the situation in cancer); or (4) the in vitro assay does not adequately reflect the in vivo situation because in vitro and in vivo environments have different inductive effects on precursor cells. To date, no definitive answer to this important question has been found. Because in vivo experiments and developmental considerations indicate that heterogeneity among precursor cells and ex vivo experiments might suggest false homogeneity, the identity of precursor cells in the adult brain also remains unresolved.

Improving Neural Precursor Cell Cultures

The main future technical challenges in neural stem cell cultures are the following:

- To increase yield and homogeneity so that stem cell cultures really consist of stem cells. With respect to homogeneity, monolayer cultures clearly have an edge but require further improvement to increase the yield.
- Make ex vivo assays of stemness a valid indicator of the situation in vivo. Although yield of "neurospheres" is often taken as evidence of stem cells in vivo, validation of this equation has been very limited.
- Define the end-points of the in vitro studies. Validity also needs to be improved with respect to the exact nature of the cells that can be generated from the cultures. This relates to the methods used to determine the exact neuronal phenotypes with respect to both molecular characteristics and their electrophysiological properties. There is no such thing as a "generic" neuron, and neurons are certainly much more than β-III-tubulin positive cells with some processes.
- Let nature be the guide. As a consequence, all culture protocols must become more lifelike by more realistically mimicking the stem cell niche and overall making the protocols less artificial. But increasing complexity has then to be balanced with aspects of feasibility. Much of the appeal of the neurosphere assay stems from its ease of use, and this quality should preserved as much as possible. One type of culture might not suit all needs, but the requirements need to be graded and the expectations on quality should not be placed too low just because of feasibility.
- In general, standards need to be defined, and for many instances something like standard operating procedures (SOPs) have to be developed in order to allow comparability between laboratories and to allow researchers to distinguish biological differences from merely technical ones.

The determination of "sphere-forming units" is supposed to overcome the problem that the simple neurosphere assay overestimates the number of self-renewing cells

One first response to the critique of the neurosphere assay as it is in widespread use today (see p. 219 above) came from Reynolds and colleagues, who modified the neurosphere model as a stem cell detection assay by focusing on the colony-forming activity of the putative stem cells (Reynolds and Rietze, 2005; Rietze and Reynolds, 2006; Marshall et al., 2007). This strategy attempts to overcome the misconception that a one-to-one relationship between neurospheres and neural stem cells exists. It appears that the common neurosphere assay overestimated the frequency of stem cells by an order of magnitude. Reynolds and Rietze discovered for their conditions that in a clonal assay, 2.4% of the individualized cells of a sphere generated secondary spheres (Reynolds and Rietze, 2005). The production of secondary spheres alone has often been taken as sufficient evidence of self-renewal, and this value would hence indicate that 2.4% of the cells would be stem cells. The growth curves that can be observed when passaging the cultures are, however, less steep than one would have to predict from this stem cell frequency.

Reynolds and Rietze now proposed to calculate the number of long-term proliferating cells that are required to obtain the growth curves found in neurosphere assays upon serial propagation. The assay thus requires not just a single sub-cloning step but continued passaging. Two time-points are compared: one early (at passage 0), one late (e.g., at passage 10). The total number of cells at the late time-point is divided by the number of cells per sphere (which has to be determined independently) to obtain the number of "sphere-forming units (SFU)." This number divided by the number of cells at the early time-point, multiplied by 100, yields the percentage of SFUs in the cellular population at the early time-point. For their culture conditions this number was 0.16%, indeed substantially lower than 2.4% (Reynolds and Rietze, 2005).

Sphere size alone has also be proposed as indicator of stemness but remains questionable

Stemness of neurospheres might also be reflected in their size in that large spheres (greater than 150 μm) appear to be more reflective of stem cells than smaller spheres (Walker et al., 2008). Whether these spheres are larger because they contain stem cells or whether the likelihood to detect stem cells is greater just because the spheres contain more cells is not quite clear. However, excitation generated more larger spheres in the adult hippocampus, which was interpreted as indicating the presence of dormant stem cells, which could be activated by the excitatory stimulus (Walker et al., 2008). Still, given the tendency of neurospheres to fuse (Singec et al., 2006; Jessberger et al., 2007), sphere size alone is no relevant indicator of stemness and requires additional analyses.

Best single cell assays avoid fusion of spheres by separating cells in a matrix

Another approach to improving the neurosphere assay has been taken by Cordey and colleagues, who trapped individualized cells from adult SVZ neurospheres in hydrogel microcavities, which allowed the researchers to monitor their behavior by time-lapse videomicroscopy. No fusion could occur between the spheres in the different wells. The surprising finding was that, initially, almost half of the plated cells show some sphere-forming (or better: proliferative) behavior, but many cells die, and the effectively growing spheres vary greatly in size (Cordey et al., 2008). These results caution us that we might actually underestimate the number of stem cells over many passages because the cells behave heterogeneously in culture and might not find optimal conditions to elicit their potential.

Few initiatives have attempted to improve the assessment of differentiation potential

Whereas several initiatives have been aimed at improving the neurosphere protocol as a tool to semi-quantitatively prove the presence of stem cells (sphere- or colony-forming capacity), the problems related to neuronal differentiation have received somewhat less attention (Mokry et al., 2005; Pagani et al., 2006). The problem is that cells at the core of the sphere tend to differentiate. In this sense a neurosphere is similar to an embryoid body developing from ES cells in vitro. Cell–cell contacts, presumably originating from stochastic variations, lead to the initiation of development. On the other hand, neurospheres thereby offer a good opportunity to study the early steps of neuronal differentiation. These are interesting not only from the perspective of basic science, but also with potential biotechnological applications in mind. Early decisions in development limit the possible steps that follow.

Optimized culture protocols combine neurosphere and monolayer techniques

The complementary advantages of the neurosphere and adherent methodology of culturing neural precursor cells calls for strategies that actually make use of the best of both worlds. Prospectively isolated precursor cells might be expanded in neurospheres to make better use of the niche-like conditions prevailing in the cell clusters. After sufficient expansion and the establishment of reliable growth, enriched precursor cells could be transferred to monolayer conditions, where individual cells become accessible and observable. Skillful use of biomaterials will greatly improve the handling of the cultures and the homogeneity of the cell populations.

Adult Neurogenesis in Organotypic Slice Cultures

Adult neurogenesis has not yet been fully reproducible in organotypic slice culture models

Adult neurogenesis has also been replicated to some degree in organotypic slice cultures. Cell proliferation, migration, and maturation were shown for the hippocampus (Raineteau et al., 2004) and in an injury model for cell evasion from the SVZ to the ischemic striatum (Zhang et al., 2007). Generally, however, demonstrating adult neurogenesis in organotypic slice cultures has been remarkably difficult, and to date no truly satisfying and feasible methodology has found broader application. One problem is that standard slice preparations are generally from animals at an age at which "adult" neurogenesis has not yet been established. Most slice preparations are not obtained beyond P15 (Poulsen et al., 2005). For example, migration of progenitor cells from the adult SVZ in postnatal tissue has been followed after the injection of various tracer substances, and the cells could be followed over several days (De Marchis et al., 2001). But even in young cultures, results have been mixed. The cultures change over time and proliferation especially declines (Sadgrove et al., 2005; Sadgrove et al., 2006). And kainate, which robustly increases cell proliferation in the adult hippocampus, decreases proliferation in the slices corresponding to an age of P11 (as it actually happens in vivo at this age as well) (Sadgrove et al., 2005).

Acute slice preparations can be used for electrophysiological examination of precursor cells and newborn neurons

In contrast, numerous studies have used acute slice preparations to address the electrophysiological properties of precursor cells and their progeny as well as the network properties of the neurogenic zones. Maturation of new neurons has been so established in both the hippocampus (Song et al., 2002; van Praag et al., 2002; Filippov et al., 2003; Fukuda et al., 2003; Ambrogini et al., 2004; Esposito et al., 2005; Wang et al., 2005; Raineteau et al., 2006) and the SVZ/OB (Belluzzi et al., 2003; Wang et al., 2003; Platel et al., 2008; Liu et al., 2009).

Methodology In Vivo

Antigen-Based Identification of Precursor Cells

Precursor cells in vivo are routinely identified by their proliferative activity

Because stem cell properties cannot yet be effectively determined within the adult brain, the precursor cell nature in vivo is usually inferred only indirectly from cell culture experiments and by analogy. The fact that precursor cells are dividing allows us to use proliferation markers such as bromodeoxyuridine to identify precursor cells in vivo, but this strategy is obviously not specific (Fig. 7–3). If upon investigation at a certain time after division the BrdU-containing cells display, not only the proliferation marker, but also a marker for differentiated neurons or glial cells, the retrospective conclusion is made that the cell in question was a precursor cell at the time that BrdU was present. Conversely, cells that are BrdU-positive at late time-points after labeling but negative for all differentiation markers might constitute "label-retaining cells"; that is, cells that do not divide enough to dilute the label but also do not differentiate.

Proliferation markers plus antigens such as Sox2 or Nestin allow identification of putative precursor cells within the neurogenic regions

Constitutive proliferation markers such as cell cycle–associated antigens—for example, proliferating cell nuclear antigen (PCNA) or mKi67—can be combined with bona fide "precursor cell markers" such as nestin, Sox2, and BLPB to find further evidence of putative precursor cells in vivo (compare Figs. 5–1 and 6–6). Their biological properties are then inferred from other studies having shown precursor cell properties in cells with particular antigen profiles.

The glial nature of many neural precursor cells has complicated the search for unambiguous markers. GFAP alone is obviously insufficient,

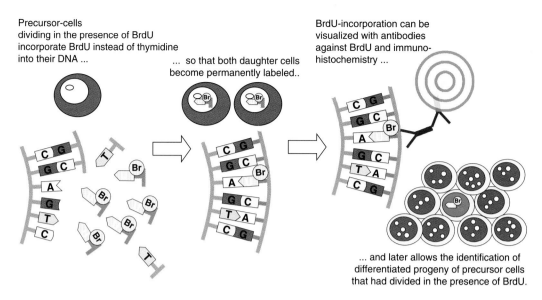

Precursor-cells dividing in the presence of BrdU incorporate BrdU instead of thymidine into their DNA ...

... so that both daughter cells become permanently labeled..

BrdU-incorporation can be visualized with antibodies against BrdU and immuno-histochemistry ...

... and later allows the identification of differentiated progeny of precursor cells that had divided in the presence of BrdU.

FIGURE 7–3 The bromodeoxyuridine method to permanently label dividing cells. BrdU is an exogenously applied false base that competes with endogenous thymidine for incorporation into the DNA.

because many GFAP-positive astrocytes are not stem cells. However, conversely, elimination of GFAP-expressing cells efficiently eliminated multipotent precursor cells in several assays (Morshead et al., 2003; Garcia et al., 2004b).

Sox2

Sox2 is a quintessential "stem cell gene" in ES cells and neuroectodermal precursor cells

Sox2 is member of the SRY-related HMG box (SOX) family of genes. Sox2 is constitutively expressed in embryonic stem cells, and Sox2 null-mutants fail to develop any ectodermal or trophoectodermal lineages (Avilion et al., 2003). Sox2 is expressed in neural stem cells throughout development (D'Amour and Gage, 2003; Ellis et al., 2004). Ablation of Sox2 reduced the number of neurospheres that could be obtained from the developing mouse forebrain (Miyagi et al., 2008). Adult neural stem cells in vivo express Sox2 (Komitova and Eriksson, 2004; Brazel et al., 2005; Steiner et al., 2006; Suh et al., 2007), but Sox2 does not identify a homogenous population of precursor cells (Brazel et al., 2005): In addition it is also expressed in a subset of S100β-expressing astrocytes (Steiner et al., 2006).

Sox2 has been used to identify putative precursor cells in vivo, and a large overlap with GFAP expression was found. Isolating precursor cells on the basis of their expression of Sox2, D'Amour and Gage (D'Amour and Gage, 2003) used gene expression profiling to compare the genetic profile of multipotent precursor cells with that of pluripotent stem cells from the embryo. They found that 158 genes from the chip were expressed only in neural stem cells, but not in embryonic stem cells. These genes included a number of plausible and useful candidates: Ascl1/Mash1, Hes5, Dlx1, Jun, Fgfr2, Notch1, Jak2, CyclinD2, Doublecortin, Dab1, and others. GFAP, by the way, was not among the candidates.

Sox2 is an extremely useful marker for neural precursor cells in vivo but should be combined with S100β in many circumstances, because some non-neurogenic astrocyte populations are Sox2-positive but in contrast to the precursor cells also express S100β.

Nestin

Nestin is expressed in many precursor cell populations but not limited to them

For many practical purposes, the intermediate filament *nestin* (Lendahl et al., 1990) is a useful precursor cell marker (Figs. 6–7 and 7–1). It is not, however, as one often reads, a "stem cell gene." Neither the sensitivity nor the specificity of nestin is particularly high. In vivo, for example, blood vessels in the neurogenic zones can express nestin (Palmer et al., 2000). On the other hand, relatively quiescent cells of the SVZ were found to be nestin-negative (Maslov et al., 2004). In vitro a population of nestin-negative precursor cells gave rise to neurons and glia (Kukekov et al., 1997). These unresolved issues notwithstanding, nestin has proven to be a useful marker in many circumstances, as long as its limitations are acknowledged.

Endogenous nestin expression is relatively low in the neurogenic regions of adult mice, and nestin immunohistochemistry often shows a cross-reaction with blood vessels. The development of transgenic reporter gene mice, which express green fluorescent protein (GFP) under neurally specific elements of the nestin promoter (Lothian and Lendahl, 1997), allowed the prospective isolation of precursor cells from the adult brain and the visualization of precursor cells in vivo (Yamaguchi et al., 2000; Sawamoto et al., 2001; Encinas et al., 2006). Nestin-GFP-positive cells can be found in essentially the entire brain. Nestin is expressed in astrocytes with no apparent neurogenic potential (Kronenberg et al., 2006) and is up-regulated in reactive astrocytes (Wilhelmsson et al., 2006). Reactive gliosis shows interesting relationships to neurogenesis, but not all activated astrocytes qualify as "neural stem cells" (Buffo et al., 2008).

Prominin1/CD133

Prominin1 is one of the best neural stem cell markers during development but has not yet been fully validated in the adult

Prominin1/CD133 is a membrane protein associated with neuroepithelial cells, where it plays an important role during asymmetrical division (Weigmann et al., 1997). It is a transmembrane protein located in microvilli, and in vitro is released during stem cell division with the so-called midbody (Dubreuil et al., 2007). Prominin1 is also expressed in several other precursor cell populations, most notably hematopoietic precursor cells (Corbeil et al., 2001), so it cannot serve as a neurally specific marker. Nestin-expressing cells isolated from the fetal mouse brain expressed Prominin1 (Sawamoto et al., 2001), and Prominin1 has been used to prospectively isolate neural precursor cells from the fetal human brain (Uchida et al., 2000). In the adult SVZ, Prominin1 is expressed by ependymal cells and B1 cells (Mirzadeh et al., 2008). Prominin1 expression by precursor cell populations of the adult hippocampus in vivo has not yet been unequivocally demonstrated.

Doublecortin

During adult neurogenesis, doublecortin is expressed from precursor cell stages to immature neurons but is not specific to neurogenesis

In vivo, the populations of the putative transiently amplifying progenitor cells in the neurogenic regions are negative for GFAP, but many of them are positive for doublecortin and the polysialylated form of the neural cell adhesion molecule (Brown et al., 2003; Rao and Shetty, 2004; Couillard-Despres et al., 2005). When adult neurogenesis is stimulated, expansion occurs in these intermediate populations (Kronenberg et al., 2003). DCX does not play a major role as precursor cell marker, because it is expressed in many post-mitotic immature neurons as well (see below, p. 224). In the absence of additional supportive evidence, DCX cannot be used to measure "neurogenesis" because it is not sensitive enough. This is an issue for studies on neurogenesis in the adult human brain, but with additional markers, specificity can be enhanced (Knoth et al., 2010). Some NG2 cells can be DCX-positive, and an overlap of nestin with DCX exists, although no further signs of neurogenesis are found (Dellarole and Grilli, 2008; Gomez-Climent et al., 2008; Xiong et al., 2008). Irrespective of this limitation for in vivo applications, DCX reporter gene mice can be used to isolate precursor cells from the adult mouse brain (Couillard-Despres et al., 2006). The specific promoter sequence has been identified (Karl et al., 2005). Because of the wide range of developmental stages during which DCX is expressed and the fact that most early post-mitotic cells die, the number of DCX-expressing cells is only a loose quantitative indicator of ongoing neurogenesis.

DCX is a poor antigen for use in vitro, because staining tends to be diffuse. Similarly, in embryonic and early postnatal neurogenesis, DCX immunoreaction is fuzzy, whereas in adult neurogenesis, the morphology of newborn neurons with their immature dendritic tree can be appreciated in great detail. DCX-positive precursor cells tend to show a homogenous and dense perinuclear staining.

NG2 (Neuron-Glia 2)

NG2 allows identification of white matter precursor cells but is not specific

The most interesting marker for parenchymal or white matter precursor cells is NG2 (Fig. 3–11). Although often equaled with "oligodendrocyte precursor cells," NG2 cells in fact mostly generate "NG2 cells." The use of the antigen as the name of the cell exposes researchers' insecurity about what these cells really are. But despite many open questions (or because of them), NG2 is an important marker in the context of adult neurogenesis.

NG2 was first described as neural lineage marker by William Stallcup in 1981 (Stallcup, 1981). NG2 stands for "neuron glia 2," which already hints at the nature of NG2 cells, lingering between glial and neuronal properties. The unusual features of NG2 cells have led to proposition that they represent an entire new class of brain cells (Butt et al., 2005). The issues of NG2 cell identity and function have been discussed in Chapter 3, p. 71. We are here concerned with the marker protein itself.

Whether NG2 itself plays a role in the potential precursor cell activity of NG2 cells is unclear

NG2 is a proteoglycan and might have several functions, none of which so far explains the extraordinary properties of NG2 cells. NG2 expression might still be an epiphenomenon in this situation. The cytoplasmic domain contains a PDZ-binding site that can bind to glutamate receptor interacting protein (GRIP), which in turn interacts with AMPA receptors (Stegmuller et al., 2003). NG2 cells have AMPA receptors and sense neuronal glutamate release (Bergles et al., 2000). This might be one pathway through which NG2-expressing cells are involved in synaptic or other forms of plasticity. NG2 cells might receive some synaptic input, the relevance of which is not quite clear (Gallo et al., 2008). NG2 cells might therefore play a particular role in neuron–glia communication underlying forms of plasticity. This function could be very subtle because NG2 null-mutant mice do not have an obvious phenotype (Grako et al., 1999).

NG2 increases cell migration in vitro (Fang et al., 1999), but in the adult brain, it does not appear to be expressed in migratory cells (Gensert and Goldman, 1996, 1997; Horner et al., 2002). NG2-expressing cells predominantly divide in loco.

NG2 inhibits axonal growth (Dou and Levine, 1994) by inducing a growth cone collapse (Ughrin et al., 2003). During development, NG2 is expressed in regions that should be avoided by growing axons—for example, the mesenchyma (Chen et al., 2002). If one interprets this finding from a precursor-cell perspective, as the other face of NG2 cells, it seems that NG2 cells are geared toward maintaining an immature stage and act against a maturing plasticity. If NG2 cells are indeed only one type of cell, they might normally modulate and stabilize neuronal function at synapses or nodes of Ranvier and prevent axonal sprouting. They are usually in a relatively repressed state, presumably by contact inhibition or factors secreted from neighboring cells (Miller, 1999). In cases of pathology, NG2 cells become disinhibited and generate new oligodendrocytes that remyelinate damage but at the same time are part of the glial scar, further preventing axonal regeneration.

NG2 is also expressed in pericytes and plays a role in angiogenesis (Ozerdem and Stallcup, 2003, 2004).

Taken together, despite some intriguing relations to neuronal development, NG2 cannot be viewed as an unambiguous marker of precursor cells, especially not with neurogenic potential. Although NG2 cells are involved in multiple types of plasticity, including cellular plasticity after damage, the few available markers that allow this conclusion indicate that only immature stages of neuronal development are reached. These stages coincide with a complex electrophysiological phenotype whose exact meaning is not yet clear but which is again associated with states of synaptic or cellular plasticity.

Persistent labeling of dividing precursor cells

Tritiated Thymidine

The core method in adult neurogenesis research is to permanently label a dividing precursor cell

In most publications, demonstration of adult neurogenesis is based on the "birth-marking" of cells with bromodeoxyuridine (BrdU). The underlying principle is straightforward: a permanent marker is brought into a cell of interest at the time-point of division, and the later fate of this cell is studied. Because the marker is persistent, it is possible to retrospectively conclude with confidence that a labeled cell must have undergone division at the time when the marker was injected (Fig. 7–3). Because neurons do not divide, the finding of a marked neuron signals that it must derive from a cell that once could divide. It is thus inferred that these proliferative cells are precursor cells. Initially, *precursor* was more or less a descriptive term without the connotations of today's stem cell biology. In his first complete description of adult hippocampal neurogenesis, Joseph Altman already assumed that "some precursor" cell might have been responsible for the "birth-dating" marker found in the granule cells (Altman and Das, 1965).

Radioactively labeled thymidine competes with endogenous thymidine and is stably incorporated into the DNA of dividing cells

The first widely used substance that allowed permanent labeling of cell divisions was tritiated thymidine ([3H]-thymidine; Fig. 2–4). In the 1950s it was discovered that radioactively labeled nucleotides such as 14 C-marked thymidine were incorporated into the DNA of dividing cells during the S phase of the cell cycle and could be detected autoradiographically (see, e.g., Friedkin and Wood, 1956; and compare also Chap. 11, p. 506). But 14 C thymidine was cumbersome to use and did not yield a resolution as high as that of tritiated thymidine, which was first applied to studies of brain by Messier and Leblond in 1958 (Messier et al., 1958). Thymidine, radioactively labeled with tritium, can be injected systemically. Once in the bloodstream, [3H]-thymidine competes with endogenous thymidine in the S phase of cell division and is permanently incorporated into the DNA. Labeled thymidine has a short half-life in vivo (although the exact value for the rodent brain is not known and presumably varies). Thus [3H]-thymidine labels all cells in the process of cell division when the label is injected. At a later time-point, tissue sections can be prepared and coated with a photo emulsion. The alpha radiation from the labeled thymidine molecules spotwise blackens the photo emulsion, thus making visible the typical grains of thymidine autoradiography.

Thymidine autoradiography does not allow combination with immunofluorescence

All studies about adult neurogenesis from 1962 to 1993 were based on thymidine autoradiography (e.g., Fig. 2–3). From today's perspective, the main disadvantage was that thymidine autoradiography could not be easily combined with the increasing arsenal of cell type–specific markers that would allow a deeper and three-dimensional analysis of the labeled cells. In particular, a combination of radiography with immunofluorescence is not possible. One advantage of [3H]-thymidine is that a relatively straightforward combination with electron-microscopy is possible (Kaplan and Hinds, 1977; Perez-Canellas and Garcia-Verdugo, 1996; Spassky et al., 2005). Very high doses of [3H]-thymidine can kill the cells that uptake the compound. In the context of adult neurogenesis research, this has been used by Cindi Morshead and colleagues to ablate neurogenesis in one of the early papers in the field ("thymidine kill": Morshead et al., 1994). A similar approach has been taken later by Gritti and colleagues (Gritti et al., 1999).

In 1993, Gould and Cameron used thymidine autoradiography together with immunohistochemistry against neuron-specific enolase to prove the existence of "new neurons" (Cameron et al., 1993). This technical advance enabled a description of adult neurogenesis that finally paved the way to wide acceptance of the phenomenon.

The spatial resolution of [3H]-thymidine is low; to relate the black grains that represent the positive signal to a particular cell requires a projection of information between two different focal planes. Therefore the method can only be used in rather thin sections, making three-dimensional reconstructions difficult. At the same time, double-labelings are very challenging and triple-labelings are virtually impossible. Because the autoradiographic and the immunohistochemical signal for the marker of interest are not in the same focal plane, confocal microscopy cannot be used, thus making the decision of whether a given cell is truly double-labeled difficult. On the other hand, autoradiography is very sensitive and, if done right, specific (Rakic, 2002). It even allows a certain quantification because the number of black grains per cell directly correlates with the amount of [3H]-thymidine incorporated. The lack of high spatial resolution and compatibility with confocal microscopy largely outweigh these advantages, however.

Bromodeoxyuridine (BrdU)

Today BrdU has replaced [3H]-thymidine

That same year (1993), the next methodological step was taken by Frank Corotto and coworkers, who first applied a nonradiographic technique, the BrdU-method, of birth-marking cells to the field of research on adult neurogenesis (Fig. 3–3; Corotto et al., 1993). The first larger study that made full use of the potential of the new method came from H. Georg Kuhn, then with Fred H. Gage at the Salk Institute in La Jolla (Kuhn et al., 1996).

Because BrdU is detected with antibodies, its fluorescent visualization can be combined with two or more other markers. (For a review of all one might possibly want to know about the history of BrdU antibodies and their use, see Leif et al., 2004). Analysis of newborn neurons via the confocal microscope became possible (Kuhn et al., 1996). Today, essentially all in vivo studies on adult neurogenesis rely on the BrdU method, thus we will cover it here in greater detail.

Like [3H]-thymidine, 5-bromo-2-deoxyuridine (BrdU, or in older reports, BrdUrd or BUrd) is an analog to thymidine that can be administered systemically. BrdU is a false base that competes with endogenous thymidine during the synthesis phase of the cell cycle. BrdU was initially developed as a antitumorigenic agent and used to sensitize tumor cells to radiotherapy (Sano et al., 1965; Sano et al., 1968). With the availability of more effective therapies, including other halogenated nucleotides such as fluorouracil with stronger antitumorigenic effects, this strategy was abandoned. That halogenated nucleotides can be incorporated into the DNA without interfering with transcription and the health of the cell was first described in the early 1950s (Dunn et al., 1954; Zamenhof and Gribiff, 1954). Nevertheless, very high doses of BrdU can kill dividing cells. Chromosome segregation during mitosis might also become impaired (Cortes et al., 2003). With the development of specific antibodies, BrdU could be detected immunohistochemically (Gratzner, 1982) and became a widely used marker for cell proliferation in vivo and in vitro (Dolbeare, 1995). BrdU was first applied to neurobiology by Miller and Nowakowski in 1988 (Miller and Nowakowski, 1988).

Despite its widespread use and relevance to the field, the pharmacokinetics and other properties of BrdU have not been fully investigated

BrdU allows the identification of a cohort of cells that underwent division at the time that BrdU was injected

The estimated bioavailability that is most widely found in the literature suggests that, in rodents, BrdU labels cells for a maximum of two hours after injection, but this estimation is rough and supposedly based on an old report dealing with a rather different situation (Packard et al., 1973). In contrast, Mandyam, Harburg, and Eisch estimated the bioavailability to be around 15 minutes for the adult mouse brain (Mandyam et al., 2007). This estimate was not based on pharmacokinetic measurements either, however. Also, the sensitivity of the BrdU method depends on the dose of BrdU injected and on the immunohistochemical technique used to visualize it (Cameron and McKay, 2001; Mandyam et al., 2007; Leuner et al., 2009), so that bioavailability is only one facet. The pharmacokinetics of BrdU in the brain appear to be largely inferred from those of thymidine (Thomas and Segal, 1997), including its uptake mechanism across the blood brain barrier (Thomas nee Williams and Segal, 1996).

In contrast to endogenous markers of cell division, such as Ki-67 or proliferating cell nuclear antigen (see below), BrdU and [3H]-thymidine identify cells dividing at the time of marker injection, not when the tissue specimen was obtained (see Eisch and Mandyam, 2007, for detailed review). BrdU allows the labeling of a cohort of cells undergoing division at the known time when BrdU was applied. It is this fundamental difference that is the basis for the identification of new neurons with the BrdU method. Between the time-point of birthmarking by BrdU injection and BrdU incorporation, and the collection of tissue and its analysis, there can be days, weeks, months, or even years. This interval gives the labeled cell sufficient time to differentiate and mature into a neuron. Endogenous cell cycle–related antigens do not represent such a memory of a past cell division. Because neurons are strictly post-mitotic, a BrdU-positive neuron that is detected weeks after BrdU injection must have originated from a cell that divided at exactly the time when BrdU was systemically available. The strength of the method lies in the fact that the experimenter can control the injection time-point and the length of interval to analysis. Despite these advantages, the BrdU method has a number of pitfalls and problems that can confound the results.

Combination of different halogenated thymidine analogs allows addressing cell cycle–related issues

There are other halogenated nucleotides that use iodine (IdU) or chlorine (CldU) instead of brome to mark the false base that competes with thymidine. Because antibodies exist that allow distinction of these molecules, double-labelings of cells with multiple S-phase markers are possible. These allow, for example, assessment of certain measures of cell cycle kinetics, of the question whether cells reenter the cell cycle, or whether precursor cells might produce different progeny at different times (Yokochi and Gilbert, 2007). The co-labeling method with two halogenated nucleotides was developed by J. Aten and colleagues (Aten et al., 1992; Manders et al., 1992; Aten et al., 1994), first used by Maslov, Pruitt, and colleagues to describe the properties of precursor cells in the adult SVZ (Maslov et al., 2004), and finally formed into a feasible and reliable protocol for use in the context of adult neurogenesis (Burns and Kuan, 2005; Vega and Peterson, 2005).

[3H] thymidine has been replaced for essentially all practical purposes

The question of whether to use BrdU or tritiated thymidine hardly poses itself to researchers any more, because BrdU has replaced tritiated thymidine as the method of choice for all but a very few special cases. In the years since 2000, there were fewer than 20 publications in the field of adult neurogenesis research that were based on tritiated thymidine. A few studies use both methods (Ormerod and Galea, 2001; Seri et al., 2004; Spassky et al., 2005), which also allows double-labeling precursor cells as with the newer IdU/CldU method.

Dosage of BrdU

BrdU is used in a wide range of doses and application protocols, not all of which are sensible

In vivo, BrdU is normally used at a concentration of between 50 and 250 mg per gram body weight. Its solubility in normal saline is pH-dependent and prefers a basic milieu. Addition of NaOH to the injection solution, however, makes it aggressive to the animal's skin and other tissues. It is thus preferred that a pH as neutral as possible be used and that BrdU be slowly dissolved by warming the solution to about 40°C. Afterwards, the solution should be cooled to room temperature, sterile filtered, and used immediately. The solution should not be placed in the refrigerator because the white crystals that fall out are very difficult to bring back into solution (except by adding NaOH). No good data exist on how long the BrdU solution is stable, so to be on the safe side it should be prepared fresh before the injection. Exposure to light and ultraviolet sources should be avoided.

Dosage and number of applications affect the result

The yield of labeled cells is dependent on the individual dose of BrdU administered (Cameron and McKay, 2001). However, more BrdU is not necessarily better because the total dose of BrdU and the period over which it is administered have to be taken into account. To increase the number of detectable BrdU-labeled cells for further analysis, the dose of the individual injection can be increased, or, at the price of a reduced temporal resolution, several injections can be spread out over consecutive days (cumulative labeling). Cumulative labeling was prevalent in most studies from the mid-to-late 1990s; today there is a tendency to use as few injections as possible. Cumulative paradigms remain necessary when only few proliferating cells are expected, such as, for example, in aged animals. There has been no systematic investigation of the tradeoff, but it seems that single injections at higher doses can be tolerated, as can multiple injections at lower doses.

|||||||||||||||||||||||||||||||||||||||||||||||||||||||

The recommended protocol is three injections of 50 µg/g body weight over 12 hours, which approximately saturates one S-phase

|||||||||||||||||||||||||||||||||||||||||||||||||||||||

For mice we recommend a single dose of 50 µg per gram body weight, given once per day over a period of one to three, maximally seven, days. Longer injection periods do not give additional benefit. For single-injection paradigms, the dose can be increased to 100 µg or possibly 200 µg per gram body weight but it is questionable whether this gives any real advantages. Mandyam, Harburg and Eisch recommend 150 µg/g body weight for mice (Mandyam et al., 2007), but cell proliferation could successfully be detected with the lower doses as well (Burns and Kuan, 2005). For rats, single doses of up to 250 mg per gram body weight appear to be feasible (Cameron and McKay, 2001), but in cumulative labeling paradigms one might want to reduce the dose to avoid overloading the cells and increasing toxicity. One might also resort to use different dosages for the assessment of proliferation (with a single injection at a higher dose with a short survival time of two to 24 hours) and of the phenotypes of surviving cells (with cumulative labeling at over a few days and a longer survival period of three to four weeks).

If the quantitative conclusions about the entire population of proliferative cells are intended, more than one injection has to be given. Because the S phase is much longer than the bioavailability of BrdU, several injections are needed to saturate the population of cells that undergo division at one time. Here, three injections over a period of 12 hours are sufficient.

|||||||||||||||||||||||||||||||||||||||||||||||||||||||

BrdU-positive cells have been followed up to several years after labeling

|||||||||||||||||||||||||||||||||||||||||||||||||||||||

At low doses, BrdU incorporation into the DNA does not seem to interfere with transcription, perhaps because only a low number of thymidine nucleotides are replaced, and endogenous DNA repair mechanisms can easily deal with the resulting mismatches within coding regions. Even one year after BrdU injection (in mice), two years (in rats), or a maximum of over two years in humans, BrdU-labeled neurons have been detectable. If one assumes that a neuron with severely disturbed transcription machinery would not be viable and be eliminated, this can be taken as an indication of relative cellular health. But no data exist to prove this assumption.

|||||||||||||||||||||||||||||||||||||||||||||||||||||||

Whether BrdU might become toxic to the cell depends on the dosage, the type of cell and other circumstances

|||||||||||||||||||||||||||||||||||||||||||||||||||||||

Cells can, however, be overloaded with BrdU and die. Ironically, developing neurons in vitro were particularly sensitive to BrdU toxicity (Caldwell et al., 2005). Massive cumulative labeling in vivo also reduces the number of BrdU-marked cells that can be detected at longer periods after BrdU administration, even though the numbers initially increase. For this reason, the application of BrdU via the drinking water is not recommended. Not only is the actual uptake not controllable, as rodents can spent considerable time without drinking and might consume very different amounts during a given time, the risk of overloading cells is also high. BrdU can also be applied intracerebroventricularly, which increases bioavailability in the brain and the number of BrdU-labeled cells (e.g., Kokoeva et al., 2007). Experience with this approach is limited, and the increased sensitivity of the method with this application route might have its price.

Pretreating Tissue for BrdU Immunohistochemistry

|||||||||||||||||||||||||||||||||||||||||||||||||||||||

BrdU-immunohistochemistry requires denaturation of the DNA

|||||||||||||||||||||||||||||||||||||||||||||||||||||||

To detect incorporated BrdU, the DNA needs to be denatured. Usually this is achieved by exposing the sections to 2N hydrochloric acid for 15 minutes at 37°C. Earlier protocols asked for an additional treatment with formamide at 65°C to precede the acid step, but it appears that the availability of new anti-BrdU antibodies has made this step dispensable. A direct comparison of different pretreatment paradigms has been published by Elisabeth Gould and colleagues confirming this change to the protocol (Leuner et al., 2009).

Exposure to hydrochloric acid is helpful in unmasking the BrdU antigen. However, it is potentially detrimental to other antigens of interest, particularly cell-surface antigens and receptors. The result that certain immunoreactions might appear weaker or are absent has to be taken into consideration when interpreting double- and triple-labelings. The use of proper controls (omission of pretreatment steps) helps assess the magnitude of this problem. As an alternative to pretreatment with hydrochloric acid, treatment of the sections with DNAse can be tried (approximately 10 U/ml; see Gonchoroff et al., 1986) but is not recommended.

Pretreatment for BrdU incorporation must be carefully adjusted to the antibodies used. Some antibody clones recognize methylated thymidine as well, so that under harsh pretreatment, essentially all nuclei can become positive. This problem might also underlie false-positive labeling in, for example, ischemic tissue.

Ethynyl Deoxyuridine (EdU)

EdU-immunohistochemistry
might solve the problem of
pretreatment damage, but
experience is limited so far

The main shortcoming of BrdU immunohistochemistry, that the pretreatment might damage the epitopes of interest on the BrdU-labeled cells, might also be overcome by a new labeling paradigm. Ethynyl deoxyuridine is incorporated into the DNA just like BrdU and can be used in the same way. However, it is visualized in a different way that avoids the problematic denaturation step. The visualization is based on "click chemistry," which stands for a copper-catalyzed binding between the alkenyl group of the EdU with either a BrdU azide probe, which can be recognized by any commercial anti-BrdU antibody, or an azide probe that is directly coupled to a fluorophore (Buck et al., 2008; Cappella et al., 2008; Warren et al., 2009). For the binding step, no unfolding of the DNA is necessary, so the helical structure is conserved. So far, in the context of neurogenesis research (albeit only up to P7), one study has been published (Chehrehasa et al., 2009).

Cell Cycle–Related Issues

For the interpretation of
BrdU-based results, cell cycle
measures matter

The main advantage of the BrdU method is that the "false base" is permanently incorporated into the DNA of the dividing cell and thus allows post-hoc analysis at late time-points after division (and after injection of BrdU). At short time periods after injection, however, BrdU can also be used to estimate the proliferative activity itself, but BrdU labels the S phase of the cell cycle only (Fig. 7–4). The efficiency of detecting all dividing cells thus depends, not only on the availability of BrdU, but also on cell cycle parameters. The S phase comprises only half or less of the entire cell cycle. Hayes and Nowakowski (2002) have pointed out that estimates of proliferation are strongly distorted by a mismatch between the duration of the S phase and the availability of BrdU (Hayes and Nowakowski, 2002). In tendency, the BrdU method will underestimate the number of cells that are in a cell cycle.

If one can assume that the influence of a given experimental manipulation on the length of the cell cycle is negligible, relative statements can be made, for example, when comparing groups within the same experiment. In a straightforward comparison of proliferation in different experimental groups, the impact of cell cycle issues will be of much less importance, unless, of course, the cell cycle itself is target of the experimental manipulation. In most cases incomplete comparisons will be sufficient (and the only feasible way), but it must be remembered that this estimate of proliferation is not identical to the absolute size of the proliferating cell population.

Caution is particularly necessary
if experimental manipulation
might affect cell cycle length

Issues of cell cycle kinetics do become potentially confounding, however, if estimates of the total number of dividing cells need to be made or if a strong influence of experimental manipulation on cell cycle parameters is suspected. Manipulating cell cycle length by slowing down or speeding up the G1 phase decreases or increases neurogenesis (Calegari et al., 2005; Lange et al., 2009). This remains to be shown for adult neurogenesis, but

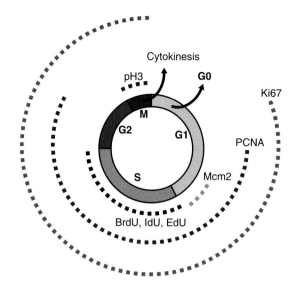

FIGURE 7–4 Cell cycle markers. Different cell cycle markers identify cells at different stages of the cell cycle. Because these divisions of the cell cycle differ in length, the chances of detecting a cell in shorter phases is lower, and the counts will hence be lower. M-phase marker pH3, for example, yields very low but highly reliable numbers.

the findings suffice to raise important caveats for some experimental situations. At the same time, the issue brought forth by Hayes and Nowakowski is complicated by the fact that dividing cells in the adult neurogenic zones are not homogeneous populations. In the dentate gyrus, at least seven morphologically and functionally distinct types of dividing cells can be distinguished (four types, or stages, of precursor cells plus microglia, endothelia, and NG2 cells), event though the vast majority of proliferation is accounted for by the transiently amplifying progenitor cells (type-2). But all existing estimates of cell cycle durations in the adult hippocampus (no such data exist yet for the adult sub-ventricular zone) relate to an average of all cell types that have incorporated BrdU (Nowakowski et al., 1989; Hayes and Nowakowski, 2002).

With these assumptions, the length of the cell cycle has been estimated to lie around 12 to 14 hours in the adult murine hippocampus, of which the S-phase would occupy roughly eight hours (Hayes and Nowakowski, 2002; Mandyam et al., 2007).

Does BrdU Detect DNA Repair and Cell Death?

The main issue for critics of the BrdU method is the question of whether BrdU might mistake dying or damaged cells for dividing cells

Cell-cycle activation might indeed precede apoptosis

The major argument brought forth against the BrdU method has been the suspicion that BrdU would not only label dividing cells but also pick up cell death or DNA repair (Cooper-Kuhn and Kuhn, 2002; Rakic, 2002). This is particularly relevant for the reported cases of regenerative neurogenesis after ischemic or hypoxic cell damage, which are associated with apoptotic cell death (Magavi et al., 2000; Nakatomi et al., 2002; Chen et al., 2004), but it might affect seemingly normal situations as well.

Induction of apoptosis can be preceded by an abortive attempt of the cell to enter into cell cycle. Cell-cycle entry of post-mitotic cells precedes their death (Herrup and Busser, 1995; Busser et al., 1998; Herrup et al., 2004). Damaged neurons might initiate abortive DNA synthesis (Copani et al., 2001; Katchanov et al., 2001; Kuan et al., 2004) but do not proceed to actual cell division. If DNA replication is completed in such a situation, this leads to aneuploidy, which could be measured. The process of entry into an abortive cell cycle appears to take place, for example, in Alzheimer disease, where mature neurons attempt to divide and activate the machinery for cell division (Neve et al., 2000; Lee et al., 2009). Because completion of the cell cycle is not possible in the post-mitotic cells, so the reasoning goes, the neurons instead die. This mechanism has not been unequivocally shown in animal studies, however, and it might also be that astrocytes rather than neurons show this behavior (Bondolfi et al., 2002; Gartner et al., 2003).

Theoretically it is difficult to exclude in such a scenario the possibility that the doomed neurons incorporate BrdU during the abortive S phase preceding their attempted division, which would severely confound supposed evidence of regenerative neurogenesis (Kuan et al., 2004). Kuan and colleagues found false BrdU incorporation into apoptotic cells only in paradigms combining ischemia and hypoxia but not in either paradigm alone. Several other studies similarly suggest that acute, single damaging events are not sufficient to cause significant uptake of BrdU (Katchanov et al., 2001; Tonchev et al., 2003).

Demonstrating "development" over different time-points after BrdU injection is not compatible with false-positive labeling due to damage

In any case, there is no indication that this process can be found in the normal brain. Also, the cells in the germinative matrices of the neurogenic zones that are identified by the BrdU method as proliferating are not mature neurons but fulfill immunohistochemical criteria of precursor cells. A BrdU signal in mature neurons at very early time points after the injection of BrdU would indeed raise the suspicion of a pathological event or a detection error. But note that as early as 24 hours after BrdU injection—i.e., with completion of the cell cycle—post-mitotic neuronal marker NeuN might be detectable. Some studies, however, have reported BrdU incorporation in neurons even sooner after BrdU application—too soon to allow differentiation from a precursor cell or a neurogenic division (Gu et al., 2000). The temporal pattern in many BrdU experiments, with progression through the expression of different markers compatible with increasing neuronal maturation (e.g., in the hippocampus from nestin to doublecortin, to calretinin, to calbindin), resolves this suspicion.

Theoretically BrdU might also be incorporated during DNA repair, but this does not seem to be of practical relevance, except in rather extreme models of pathology

Related to the issue of cell death is the argument that the BrdU method would actually detect DNA repair. During DNA repair, the DNA polymerase is active, and false nucleotides such as BrdU could be incorporated. Brain irradiation as well as irradiation of isolated precursor cells, which induces DNA repair mechanisms, led to reduced immunohistochemical detection of BrdU, not an increased number of marked cells (Palmer et al., 2000; Mizumatsu et al., 2003). This has also been assessed by the simultaneous application of BrdU immunohistochemistry and terminal deoxynucleotidyl transferase dUTP nick end labeling (TUNEL). At two hours after BrdU injection, no TUNEL/BrdU double-positive cells were found (Cooper-Kuhn and Kuhn, 2002). Kuan and colleagues found these cells only in the combined injury model (Kuan et al., 2004). Bauer and Patterson finally undertook an extensive examination of the BrdU/TUNEL paradigm in three models of brain injury: olfactory bulbectomy, irradiation, and kainate-induced seizures. No apoptotic BrdU-labeled cells were found, even when BrdU was directly infused into the brain. As final confirmation, sequential labeling with IdU (before irradiation) and CldU (after irradiation) was undertaken: no differences in the numbers of labeled cells were found. The counts, possibly confounded by cell death or DNA repair, which was targeted with the second label, were not higher than the estimate of cell proliferation from the first label (Bauer and Patterson, 2005).

Again, marker progression indicative of cellular differentiation serves as another argument against BrdU labeling of DNA repair. Related to this is the observation that the number of BrdU-positive cells approximately doubles in the first 24 hours after the injection of BrdU, and at that time BrdU-positive cells are often found in pairs. This is compatible with continuation of the cell cycle through the M and T phases and the appearance of two labeled daughter cells, but not with DNA repair.

A related issue is the possibility that BrdU might be released from dying cells and be taken up from neighboring cells (Burns et al., 2006). This is of primary relevance for transplantation studies, where BrdU has sometimes been used to pre-label cells (Gage et al., 1995; Suhonen et al., 1996). Theoretically, such marker transfer might also occur during physiological neurogenesis, but it would have effects on the temporal resolution only, if at all.

In summary, despite the valid concerns about potential confounds of the BrdU method (Breunig et al., 2007) and the necessity to remain skeptical of far-reaching claims based on a few potentially confounded data points, the great majority of studies is hardly touched by these issues. The critique of the BrdU methodology, which is as important as with any other method, cannot be considered as weakening the foundations of the entire field any more.

Absolute Cell Counts and Cell Cycle–Related Antigens

Large regulatory effects on adult neurogenesis might affect total granule cell numbers

BrdU-based results have also been confirmed by independent methods. Some questions can be addressed by measuring cell proliferation by other means. Net effects of neurogenesis on the brain structure of interest can best be assessed by measuring changes in the entire population of cells. The strongest confirmation of neurogenesis is the demonstration of an increase in total cell number in the neuronal population, to which adult neurogenesis adds. Such a demonstration can be achieved by use of stereological techniques, as described below. In a few studies, including Altman's inaugural paper from 1965, this has been done (Altman and Das, 1965; Kempermann et al., 1997; Bick-Sander et al., 2006).

Cell-cycle related antigens are often the only feasible way to assess putative precursor cell proliferation, if BrdU application is not possible

For some years in the 1980s, the BrdU-based method has been used to assess pathological cell proliferation and has been the method of choice for determining the mitotic activity of tumors. Peter Eriksson's demonstration of adult neurogenesis in humans was done on postmortem tissue from patients who had received BrdU injections during their lifetime to determine the proliferative index of surgically treated laryngeal carcinoma (Eriksson et al., 1998). Today this invasive method of tumor staging has been replaced by use of other markers that do not require the injection of substances into the patient before surgery. Assessment of cell proliferation in tissue is now based on the expression of cell cycle–related antigens detected immunohistochemically (Fig. 7–4). In situations where only proliferation needs to be assessed, cell cycle markers can be used instead of BrdU. However, the signal-to-background ratio of these markers can be lower than that of BrdU, which might justify the additional effort of administering BrdU.

Proliferating Cell Nuclear Antigen (PCNA)

PCNA is not specific enough for most purposes

PCNA, also called *cyclin*, is a 36 kDa protein. It is an auxiliary protein to the delta-DNA-polymerase and is expressed in G1 and S phase (Hall et al., 1990). It decreases in G2. In late S phase, PCNA is prominently found in the nucleoli. However, in formaldehyde-fixed tissue this does not become apparent, and the staining tends to be diffuse. PCNA staining requires an antigen retrieval procedure consisting of brief boiling in citrate buffer. Although PCNA has been used in a few studies in the context of adult neurogenesis, its application is rather cumbersome and it tends to yield a wide range of staining intensities that can be difficult to interpret. There is also evidence that PCNA remains "on" after the cells have become post-mitotic, which reduces temporal specificity (Mandyam et al., 2007). It is thus not recommended for routine use in rodents.

Ki67

Ki67 is the name of the original antibody clone that identifies a cell cycle–associated protein (mKi67) encoded on mouse chromosome 7. Although few functional data exist for mki67, it appears to be essential for cell cycle progression (Starborg et al., 1996; Endl and Gerdes, 2000). *Ki* stands for the Pathological Institute at the University of *Ki*el; the first antibody that could be used in paraffin-embedded sections was "made in Bostel," a place nearby, hence the mysterious synonymous acronym MIB. Ki67 antibodies identify cells in late G1, S, G2, and M phases (Scholzen and Gerdes, 2000). Thus only G0- and early G1-phase cells are not recognized. Mki67 is the broadest known cell cycle–associated antigen. Despite this wide specificity and presumably high sensitivity for identifying proliferating cells, in direct comparisons with analyses done shortly after BrdU injection, Ki67 immunohistochemistry yielded higher (Kee et al., 2002) or lower (McKeever et al., 1997) numbers. Most Ki67 antibodies require epitope retrieval procedures such as brief boiling in citrate buffer or microwaving.

Minichromosome Maintenance Complex Component 2 (Mcm2)

Mcm2 is involved in the initiation of chromosome replication and is part of the pre-replication process (Lei, 2005). Mcm2 has been used in a number of more recent studies on adult neurogenesis (Amrein et al., 2007; Fahrner et al., 2007; Sivilia et al., 2008). Mcm2 is a useful additional marker that is specific to replication, but its expression stands at the beginning of cell division. Mcm2 expression alone does not tell us about the completion of the cell cycle. As the available antibodies work very well, Mcm2 can offer welcome supportive evidence.

Phospho-histone 3 (pH3)

Phosphorylated histone 3 differs from the other mentioned proliferation markers in that it is specific for mitosis (M-phase). Due to this fact, it has a greater prognostic value in pathological examinations assessing tumor growth (Skaland et al., 2007). No detailed comparative study has been undertaken, but by analogy, the advantage in validity should also apply to adult neurogenesis. Note, however, that counts are substantially lower than with other markers because the chances to label a cell during its short M-phase are considerably lower than during S and G2. Several studies on adult neurogenesis have used pH3 (Komitova et al., 2005; Rodriguez et al., 2008) as a proliferation marker, but in combination with other cell cycle markers pH3 can also be used to estimate cell cycle kinetics (Mandyam et al., 2007). Antibodies to different phosphorylation sites are available (Ser10, Thr11, Ser28, etc.) but Ser10 is most commonly used. The great advantage of pH3 lies in the fact that it is not detectable in abortive DNA replication.

Identifying New Neurons: When Is a Cell Double- or Triple-Labeled?

General Considerations and Criteria

The immunohistochemical identification of new neurons is usually based on detection of BrdU in a cell that expresses neuronal markers. This straightforward approach poses two types of problems: (1) neuronal markers often cannot be unambiguously defined, and (2) determination of an intracellular marker overlap that constitutes a double- or triple-labeling is technically challenging.

The following criteria are consistent with the recommendations from several publications that have specifically dealt with the pitfalls of detecting adult neurogenesis (Peterson, 2004; Breunig et al., 2007; Gould, 2007; Kuhn and Peterson, 2008).

Dependencies on Markers

Markers are only as good as their validation for the context in which they are used: generalizations are problematic

For the neurogenic regions of the adult brain, the immunohistochemical detection of neuronal markers in BrdU-positive cells has been validated by numerous studies. Particular problems arise, however, in non-neurogenic zones, when markers from the neurogenic regions are applied and their identical meaning is tacitly assumed. Here the rule that extraordinary claims require extraordinary evidence should be followed, and independent, additional means of proving the neuronal identity of a cell need to be applied. But in the end, all markers will have to be validated exactly for the population of cells they are intended to be used for. Analogy alone is usually insufficient.

Brain pathology might alter marker specificity

Marker expression might not be stable. Pathological situations, most notably ischemia, pose particular challenges, and the possibility that damaged cells might show a fluctuation in antigen expression has to be taken into account. As a worst-case scenario, a cell might be damaged and be at the brink of dying when cell death–related mechanisms trigger BrdU incorporation, while at the same time proteins associated with neuronal maturity are down-regulated. If the cell recovers, its nucleus would contain BrdU and again the mature neuronal markers. This scenario has some plausibility and from a devil's advocate's point of view should be considered when difficult examples of evidence for neurogenesis are evaluated. As discussed above, however, BrdU uptake during DNA repair and cell death could not be confirmed in many studies, and without further substantiation must not be used as killer argument against new neurons.

Demonstration of development should lie at the heart of any description of neurogenesis

The best rule to support a claim of neurogenesis is to seek confirmation with at least one independent method, use many markers, and make every attempt to demonstrate development. Cells that only re-express a mature marker do not go through a developmentally plausible sequence of transcription factors or other developmental molecules. A good example of this strategy was used in a study by Andreas Arvidsson and colleagues, who showed regenerative neurogenesis in the ischemic striatum (Arvidsson et al., 2002). The emerging new neurons went through a stage of transient Pbx and Meis2 expression, two transcripts associated with the development of striatal interneurons (Fig. 12–7).

Neuron-Specific Enolase (NSE)

NSE is a historic neuronal marker but of little practical use today

Cameron and Gould's first double-labeling experiments to demonstrate adult hippocampal neurogenesis and Corotto's work in the olfactory bulb were based on the use of NSE (Cameron et al., 1993; Corotto et al., 1993). However, despite its name and researchers' original belief (Vinores et al., 1984; Schmechel, 1985), NSE is not entirely specific for neurons, especially if used in light microscopy. NSE immunoreaction is often diffuse and shows great variance in intensity even in neighboring cells. Immunoreaction in mice is lower than that in rats (Vinores et al., 1984). Currently, better and more practical neuronal markers are available and NSE should no longer be used for this purpose.

Neuronal Nuclei (NeuN)

The most widely accepted marker of neurons is NeuN, but NeuN identifies not all neurons and not only "mature" neurons

NeuN has become the most widely used marker to identify neurons. NeuN was first described by Richard J. Mullen in 1992 (Mullen et al., 1992). NeuN expression is restricted to post-mitotic neurons, but a few neuronal populations, including photoreceptors of the retina, cerebellar Purkinje cells, mitral cells of the olfactory bulb, and a population of neurons in the cochlear nucleus, are negative for NeuN. This list of exceptions may not be complete (Kim et al., 2009): an important factor to remember if NeuN is used to demonstrate extraordinary cases of adult neurogenesis outside the neurogenic regions. The three populations of neurons known to be generated in adult neurogenesis—hippocampal granule cells and glomerular and periglomerular interneurons of the olfactory bulb—are NeuN-positive. In no case has NeuN expression been noted in glial cells. *NeuN* stands for *Neu*ronal *N*uclei, and although NeuN is predominantly located in the nucleus, it can also be detected in purified cytoplasmic fractions from brain. Immunohistochemically, most neurons show a strong and reliable reaction in the nucleus and cytoplasm near the nucleus, occasionally reaching into some neurites.

NeuN is also expressed in neurons of the peripheral and autonomous nervous system. Neuroendocrine cells such as in the intestines, adrenal medulla, pituitary, and pineal gland are often negative for NeuN (Wolf et al., 1996). Also, sympathetic chain neurons do not show NeuN expression. One population of enteric neurons is particular in that it only shows cytoplasmic rather than the name-giving nuclear expression (Van Nassauw et al., 2005). The cytoplasmic NeuN largely corresponds to a 48 kDA-isoform of the protein, whereas the predominant nuclear form weighs 46 kDA (Lind et al., 2005).

NeuN is a splicing factor

Remarkably, despite the wide acceptance of NeuN as a neuronal marker, only a very few studies have addressed the properties of NeuN itself. Only in 2009, 17 years after the original description, was it reported that NeuN is the product of the Fox3 gene (Hrnbp3) and a splicing factor in neurons (Kim et al., 2009). Before, it had only been known that NeuN could bind to DNA and is switched on with the initiation of terminal differentiation (Mullen et al., 1992; Sarnat et al., 1998). The physical localization of the NeuN gene (still officially named "Neuna60") or Fox3 corresponds to locus D11Bwg0517e on murine chromosome 11. NeuN is a phosphoprotein, and the described two isoforms are thought to represent the phosphorylated versus the non-phosphorylated form (Lind et al., 2005). The antibody used to stain for NeuN requires the phosphorylated form.

Down-regulation of NeuN is not identical to a decrease in neuronal numbers

Pathological conditions such as ischemia can down-regulate NeuN expression (Collombet et al., 2006). The loss of immunoreactivity, though, does not necessarily relate to a loss of neurons (Unal-Cevik et al., 2004). Consequently, a reemergence of NeuN after recovery from a metabolic perturbation alone would not be indicative of newly generated neurons.

The main advantage of NeuN as a neuronal marker lies in its relatively high, and under normal conditions stable, expression in the nucleus, making it a good choice for confocal microscopic analyses in combination with BrdU. The presence of the two markers in the same compartment can be determined with high accuracy.

Contrary to the notion in many publications, NeuN is no marker specifically of mature neurons. In adult neurogenesis, NeuN is expressed very early after cell cycle-exit, when the new neurons are still very immature (Plumpe et al., 2006; Petrus et al., 2008). So NeuN alone is no maturation marker. For this, calbindin can be used in the hippocampus, the location together with neurotransmitter markers in the olfactory bulb.

β-III-tubulin

β-III-tubulin is a widely used in vitro but is expressed very early in the neuronal lineage so that even precursor cells might be labeled

β-III-tubulin is an isotype of tubulin, one of the major constituents of the cytoskeleton. Class β-III has historically been labeled the "minor neuronal" isotype associated with neurons in the central and peripheral nervous systems, in contrast to "major neuronal" isotype β-II, expressed in both neurons and glia. β-III-tubulin is often also referred to as "Tuj1," which is not a true synonym but the clonal designation for the originally used antibody against this antigen. β-III-tubulin has found the most widespread application as a neuronal marker in cell culture work. Some groups have also used β-III-tubulin as an indicator of neurogenesis in vivo. However, in vivo, staining against β-III-tubulin is unsatisfactory. Immature neurons in the neurogenic zones indeed seem to express β-III-tubulin, but nothing is known about re-expression in mature neurons, for example under stress. β-III-tubulin expression is switched on early in neuronal development and parallels NeuroD expression in vitro (Uittenbogaard and Chiaramello, 2002). β-III-tubulin is not strictly brain-specific and is also found in testis. In the neurogenic zones in vivo, no overlap between β-III-tubulin and glial markers has been found, but elsewhere some NG2 cells have been found to be β-III-tubulin-positive. The glial features of putative precursor cells that can generate neurons make it difficult to determine whether β-III-tubulin can be used as an indicator of neuronal lineage determination. In cell culture these ambiguities become a potentially severely confounding issue. β-III-tubulin has, for example, been found in retinal pigment epithelial cells after prolonged periods of time in culture, but not constitutively (Vinores et al., 1993). In cell cultures, low β-III-tubulin expression can also be found under proliferation conditions. The specificity of β-III-tubulin as a neuronal marker is further reduced by its expression in brain tumors of presumably glial origin (Ignatova et al., 2002). Although the latter point could also be taken as an argument that gliomas are in fact derived, not from differentiated glial cells, but from precursor cells, the point clearly requires further investigation. In any case, far-reaching conclusions about the neuronal nature of cells based solely on β-III-tubulin should be avoided.

TUC4 (Synonyms: TOAD-64, CRMP4)

TUC4 is an early post-mitotic neuronal marker, linked to axonal growth

TUC4 is expressed by post-mitotic neurons at the stage of initial differentiation and is associated with axonal outgrowth. It has been used in several studies on adult neurogenesis, because it labels early post-mitotic cells (Wang et al., 2000; Wen et al., 2002; Parent et al., 2006; Seki, 2002b) but due to some disadvantages is less often found today. Whereas β-III-tubulin is expressed along the entire axon, TUC4, or TOAD-64 (for Turned On After Division, 64 kDa), is found in the distal parts of the growth cones (Minturn et al., 1995). The TUC proteins are homologue to a *C. elegans* gene, *unc-33*, and like it presumably function as a collapsin-response mediator protein (CRMP) (Quinn et al., 1999). Collapsin belongs to the family of semaphorins, which are repulsive molecules active in axon pathfinding. TUC proteins mediate their effect by still unknown mechanisms.

In theory, because of its association with early post-mitotic neuronal development, TUC4 should be an ideal marker for adult neurogenesis research. There are, however, a few caveats. The first is that the highest TUC4 expression is not necessarily found in the cell soma but in the growth cone. This makes double-labeling with nuclear markers (such as BrdU) difficult. Also, the intensity of positive cells thus tends to be highly variable. For qualitative statements, TUC4 can be useful, although background staining is often strong. Quantification, especially in tissue treated for BrdU immunohistochemistry, is tricky. TUC4 might also be weakly expressed in NG2 cells throughout the brain parenchyma (Ricard et al., 2001).

TUC4 is expressed very soon after the cells have become post-mitotic, but it is not known how long it is expressed during maturation of the new cells. Because much of the quantitative regulation in hippocampal neurogenesis occurs during the post-mitotic stage, quantification of TUC4 immuno-histochemistry does not represent net neurogenesis but an intermediate stage of as-yet-undefined length. TUC4, however, shares this limitation with other markers, such as DCX and PSA-NCAM, which can be used as markers for neuronal development but not necessarily the quantity of net neurogenesis.

Like β-III-tubulin, TUC4 should not be used as the sole marker to prove adult hippocampal neurogenesis. TUC4 is not expressed in the SVZ, but only in the olfactory bulb (where the new neurons begin to extend their axons).

PSA-NCAM

PSA-NCAM shows an almost complete overlap with DCX

PSA-NCAM is the polysialyliated form of the neural cell adhesion mole-cule. Polysialic acid (PSA) is only found on NCAM, making it rather specific to the neuronal lineage (Tomasiewicz et al., 1993). At least in some brain regions, such as the hypothalamohypophysial system, PSA is also found on glial cells (Kiss et al., 2001), reducing its sensitivity in detecting potential neurogenesis outside the well-characterized neurogenic regions. For the dentate gyrus, the relationship between adult neurogenesis and PSA-NCAM expression was recognized and characterized early, most notably in the pioneering work by Tatsunori Seki (Seki and Arai, 1991a; Seki and Arai, 1993a, 1993b, 1999; Seki, 2002b). PSA-NCAM is widely considered to represent one key factor defining, albeit not specifically, the neurogenic regions (Bonfanti, 2006; Gascon et al., 2008).

In both the olfactory system and the hippocampus, PSA-NCAM is found on migratory neuronal cells. In these areas PSA-NCAM identifies type-2b and type-3 cells in the dentate gyrus and the C and A cells in the SVZ and olfactory bulb. NeuroD expression precedes PSA-NCAM expression (Seki, 2002b). The PSA residues reduce cell adhesion mediated by NCAM (Sadoul et al., 1983), thus PSA-NCAM is associated primarily with migratory stages. There is ex vivo evidence that glial progenitors after injury might require PSA-NCAM as well (Wang et al., 1996; Barral-Moran et al., 2003).

In addition, PSA-NCAM is still found on early post-mitotic stages of neuronal development and the cells that extend dendrites and axon. If PSA is removed enzymatically, aberrant mossy fiber growth is encountered (Seki and Rutishauser, 1998). In the hippocampus, developing neurons are often found in clusters, and the more immature cells are associated with the processes of the more mature cells, suggestive of a direct interaction (Seki, 2002a). The details of this interaction are not known, however. PSA-NCAM expression on the cell surface is activity-dependently regulated, and this pro-cess seems to be NMDA receptor-dependent (Wang et al., 1996).

PSA-NCAM expression is not limited to neurogenesis

PSA-NCAM is also associated with non-neurogenic regions (Varea et al., 2005; Mazzetti et al., 2007). In the piriform cortex, a region of par-ticularly strong expression is found, without any signs of neurogenesis (Seki and Arai, 1991b; Nacher et al., 2002; Nacher et al., 2004). Besides imma-ture neurons, PSA-NCAM also identifies a specific population of mature neurons in cortical regions (Luzzati et al., 2009). PSA-NCAM has func-tions independent of the adhesion mediated by NCAM, and these functions are beginning to be revealed. PSA plays a permissive role for axon pathfinding (Landmesser et al., 1990; Rutishauser and Landmesser, 1996) and in synaptogenesis (Seki and Rutishauser, 1998; Dityatev et al., 2004; Muller et al., 2008). PSA-NCAM is found on both the pre- and postsynaptic membrane (Muller et al., 1996) and is required for the structural plasticity following induction of long-term potentiation (LTP) in the hippocampus (Eckhardt et al., 2000).

PSA-NCAM might have a direct signaling function as well. For its function at synapses, activa-tion of Trk B receptors, the receptors for brain-derived neurotrophic factor, has been discussed (Muller et al., 2000).

In summary, PSA-NCAM is a very interesting molecule associated with embryonic and adult neurogenesis and other examples of plasticity. Its value as a marker is context-dependent. Within the neurogenic regions, PSA-NCAM immunohistochemistry allows identification of cells at a transient stage of neuronal development. Outside the neurogenic regions, this specificity of PSA-NCAM is not a given.

In the neurogenic zones, the time course of PSA-NCAM expression almost completely overlaps with doublecortin expression. Because PSA-NCAM is a surface molecule, immunohistochemical analysis can be difficult. Preservation of the glycoproteins and PSA residues depends on perfusion and fixation protocols. If the goal is only to identify the neuronal progenitor cells and their early post-mitotic progeny, doublecortin immunohistochemistry is preferable. Even more than doublecortin, however, PSA-NCAM staining enables visualization of cells with much of their dendritic morphology.

Doublecortin (DCX)

Doublecortin is a microtubule-associated protein enriched in migratory neuronal cells (Meyer et al., 2002), particularly in their leading processes (Schaar et al., 2004) and in the growth cones of neurites (Friocourt et al., 2003). DCX is also expressed in neuronally determined progenitor cells (see above, p. 229). In the neurogenic zones of the adult brain, DCX expression allows identification of type-2b and type-3 cells of the SGZ, and C and A cells of the SVZ and olfactory bulb. Like PSA-NCAM, it is transiently expressed identifying the phases of migration and neurite extension (Brandt et al., 2003; Brown et al., 2003; Kronenberg et al., 2003; Ambrogini et al., 2004). At the post-mitotic stage, DCX expression overlaps with calretinin expression (Brandt et al., 2003). An overlap between DCX and calretinin is also found in the developing brain (Meyer et al., 2002).

Doublecortin is the most widely used marker for neural progenitor cells and immature neurons in vivo but not specific for neurogenesis

Doublecortin was discovered because its mutation causes disturbed cortical lamination in humans. One of the resulting disorders is X-linked lissencephaly, associated with mental retardation and epilepsy (Gleeson et al., 1998; Gleeson et al., 1999a; Gleeson et al., 1999b). DCX appears to play a central role during the migration process (LoTurco and Bai, 2006). In lissencephaly a four-layered cortex is found instead of the normal six layers, and normal gyri and sulci are lacking. In other cases, some neurons do not reach their appropriate position in the cortex and accumulate in the white matter below a layered cortex. This particular structural abnormality is called "subcortical band heterotopia" or "double cortex," hence the name of the protein. Silencing of DCX in developing mice by the RNAi technique has produced a similar phenotype in mice (Bai et al., 2003) but, oddly, not the classical gene knock-out (Corbo et al., 2002). Similarly, the formation of heterotopias could be induced by RNA interference in rats but not in mice (Ramos et al., 2006). Re-expressing DCX via in utero electroporation in a related rat model of subcortical band heterotopia led to a partial resolution of the malformation, suggesting that DCX plays a major role in determining positioning of the neurons (Manent et al., 2009). During development, DCX extensively overlaps with reelin expression (Meyer et al., 2002).

Doublecortin has a function in neuronal migration

Function of DCX is tightly regulated by a balance of kinase and phosphatase activities (Schaar et al., 2004). It has thus been suggested that several signaling pathways might converge at DCX, thereby modulating the stability of microtubules and consequently migratory behavior (LoTurco, 2004). In vitro data indicate that DCX stabilizes microtubules (Moores et al., 2004), but there might be other DCX-dependent mechanisms involved in migration as well (Tanaka et al., 2004).

DCX is a very convenient and robust marker for defined stages of neuronal development in the neurogenic zones (Brown et al., 2003; Filippov et al., 2003; Rao and Shetty, 2004; Couillard-Despres et al., 2005; Knoth et al., 2010). Its cytoplasmic and nuclear expression make it ideally suited for double-labeling with nuclear antigens, such as proliferation marker BrdU. DCX is also detectable in the neurites and enables appreciation of dendritic morphology (Plumpe et al., 2006).

DCX completely overlaps with PSA-NCAM expression

Both inside and outside the neurogenic regions, DCX expression overlaps almost completely with PSA-NCAM (Nacher et al., 2001; Luzzati et al., 2009). Expression by mature astrocytes or oligodendrocytes has not been reported, but NG2 cells throughout the brain might express DCX (Tamura et al., 2007). As in the case of PSA-NCAM, sensitivity and specificity of DCX as a neurogenic marker in the non-neurogenic regions is reduced. Neither marker, therefore, can be used as sole indicator of neurogenesis.

Calretinin

Calretinin is transiently expressed at the early post-mitotic stage of adult neurogenesis in the hippocampus but is not specific to "neurogenesis"

Calretinin is a calcium-binding protein that is expressed during a post-mitotic stage in adult hippocampal neurogenesis (Brandt et al., 2003). All calretinin-positive cells are also Prox1-positive and NeuN-positive, identifying them as immature granule cells. A subset of the cells stains also for DCX, and dendritic morphology of the cells can be appreciated to some degree (Plumpe et al., 2006). About the function of Calretinin in newborn granule cells, nothing is known. The phase of Calretinin expression is particular in that some markers, e.g., the glucocorticoid receptor or Tis21 (Garcia et al., 2004a; Attardo et al., 2009), that are present at earlier and later stages are absent from the calretinin-positive cells. There is also no overlap with calbindin, which is expressed in mature granule cells (Brandt et al., 2003). The number of Calretinin-positive cells correlates well with adult the regulation of adult neurogenesis (Brandt et al., 2003), but it is still not fully clear whether all cells go through a calretinin stage. Calretinin is used as a marker of the immature post-mitotic stage of adult neurogenesis (Sun et al., 2004; Kronenberg et al., 2006; Chan et al., 2008; Petrus et al., 2008).

Surprisingly, calretinin expression does not seem to be found during adult hippocampal neurogenesis in rats. In the olfactory bulb, calretinin identifies a subpopulation of adult-generated neurons in the granule cell layer (Corotto et al., 1993; Bagley et al., 2007; Batista-Brito et al., 2008). Calretinin is also found in many populations of interneurons throughout the brain that are not generated in adulthood, so calretinin alone cannot serve as neurogenesis marker.

Calbindin

Calbindin is the most mature marker of adult-generated hippocampal granule cells

Mature hippocampal granule cells express the calcium-binding protein calbindin. In contrast to NeuN, calbindin is only expressed after the cells have matured, so together with BrdU it is a good and useful marker of "net neurogenesis" in the dentate gyrus (Kuhn et al., 1996; Brandt et al., 2003; Kempermann et al., 2004; Tozuka et al., 2005; Chan et al., 2008). Calbindin immunohistochemistry is often not straightforward because the antibodies tend to show poor penetration, so that cells in the center of the section might appear falsely negative.

Hu Protein

Hu is a marker of neuronal progenitor cells exiting the cell cycle and very immature neurons

Hu, officially *Elav* for "ELAV (embryonic lethal, abnormal vision, Drosophila)-like," is actually a family of genes that is evolutionarily highly conserved. Antibodies to Hu protein recognize one or more of several Hu isoforms, mostly HuC and HuD. The Hu proteins bind to mRNA in untranslated regions, for example of p21. The current hypothesis is that Hu proteins control the timing of the switch between precursor cell proliferation and neuronal differentiation. Given this mechanism it is clear that Hu is no marker for mature neurons. It has not been studied in any detail if Hu can also be expressed in NG2 cells, but some data suggest this (Dayer et al., 2005). Hu is also expressed in mesenchymal lineages (Gantt et al., 2004). Taken together, the current evidence indicates that Hu is not a suitable marker for mature neurons but rather identifies a progenitor cell stage, which might or might not lead to the presence of a newborn neuron.

Microtubule-Associated Protein 2ab (Map2ab), Tau Protein, and Neurofilament 200

Map2ab, Tau and Neurofilament 200 are useful rather mature markers in vitro but of limited use in vivo

Microtubule-associated protein 2ab, tau protein, and neurofilament 200 have good neuronal specificity and are widely used in vitro. In vivo, their applicability is limited because immunoreactions against these antigens visualize only neurites. This result makes it difficult to relate a BrdU-labeled nucleus to Map2ab- or neurofilament-positive processes or tau-positive axons. All three markers are expressed by mature neurons, which gives a very dense staining pattern. In vivo, dendrites of newly generated cells can be visualized more easily with DCX immunohistochemistry.

Pitfalls of Immunofluorescence and Confocal Microscopy

Confocal microscopy is the method of choice but can produce false-positive results

Analysis of double- and triple-stained fluorescent sections poses several challenges. Confocal microscopy has become the gold standard, and some problems of classical fluorescent microscopy can be avoided with this technique. On the other hand, confocal microscopy can also add errors. The three most important issues are:

- Problems due to "cross-talk" between excitation and detection channels
- Problems due to a neglect of three-dimensional information
- Problems due to incorrect detection sensitivity

Channel cross-talk is a serious confound but simple to prevent

In confocal microscopy, laser beams replace the light bulb as the source of light (Fig. 7–5). This allows one to focus the energy of the light very precisely. Filters (or acoustic-optical beam splitters) let only light of the desired wavelength pass through. Excitation wavelengths are chosen at which the fluorophores used can be excited. If hit by light of this particular wavelength, the fluorophores coupled to the secondary antibodies used in the immunohistochemical reaction send out light with a wavelength that is longer than the excitation wavelength. A detector blind to the excitation wavelength thus allows one to distinguish emission from excitation light. This emission spectrum is not a sharp line but rather a distribution with a peak and more or less steep sides. Consequently, emission spectra corresponding to different excitation wavelengths can overlap (Fig. 7–6). If more than one emission channel is analyzed at the same time, it is not possible to conclude from the recorded signal at which wavelength the excitation occurred and, consequently, which of the fluorophores was meant to be excited. This is the most common source of false-positive double-labelings. Confocal microscopy per se does not prevent this problem any more than a good fluorescent microscope does. The solution to this problem, commonly dubbed "bleeding," is to sequentially detect the different channels. For example, if the fluorophore FITC is excited at 488 nm, only the detection channel for FITC will be active; if TRTC is excited at 543 nm, only the detection channel for TRTC will be active. One advantage of confocal microscopy is that this sequential scanning is possible in a very straightforward way, and a computerized system can immediately superimpose the independently generated images to allow evaluation of double-stainings. However, automated imaging systems can now accomplish this principle for conventional fluorescent light microscopy as well, offering a cheaper alternative to a confocal microscope.

The advantage of confocal microscopy in gathering three-dimensional information has to be exploited

The true (and name-giving) advantage of confocal microscopy lies in its sophisticated optical construction, which differs from that of a normal microscope. Because the laser light produces a sharp focus yielding high energy exactly at the right wavelength and in a tiny area, the image projected to the detection system is very sharp, albeit very small. Whatever is in focus in the specimen is in focus in the detector, thus the name *confocal*. By positioning an aperture in front of a detector that can be adjusted to

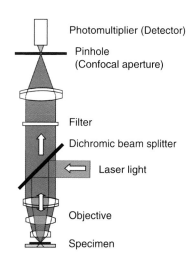

Confocal principle:

• Illuminate and detect from only a very small tissue volume and move the focal spot across the sample ("pixel by pixel")

• What is in the illumination focus is in the detection focus

• Pinhole blocks defracted light and allows optical sectioning

• Everything in focus appears bright, everything not in focus remains black

Photomultiplier (Detector)

Pinhole (Confocal aperture)

Filter

Dichromic beam splitter

Laser light

Objective

Specimen

FIGURE 7–5 The confocal principle. Confocal microscopy uses laser light to excite fluorescent dyes and generate digital microscopic images of high spatial resolution and strict separation of detection channels. The confocal principle allows placement of thin optical sections within a physical tissue section.

variable width, it is possible to exclude diffracted, out-of-focus light from the image, producing a sharp image of the optical plane of focus. The full image is collected by moving the focal point over the specimen pixel by pixel and line by line. The confocal image is thus always digital.

In a normal microscope, in contrast, the full image has to be taken at one single time-point, and the light is distributed over a much larger area. Consequently, diffracted light from one spot can

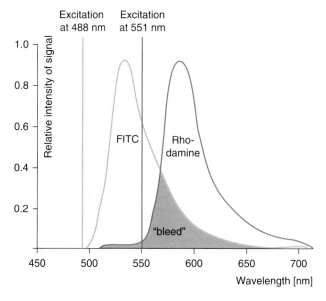

FIGURE 7–6 Excitation and emission spectra in immunofluorescence. Fluorescent dyes are excited at a wavelength that is lower than their emission spectrum (*the red and green curves*). For different fluorophores, these spectra might overlap (*gray area under the curves*) and cause false detection. Confocal microscopy allows sequential analysis of each dye separately to avoid this problem, which can otherwise lead to false-positive double-labeling.

interfere with emission of a neighboring spot. The spatial resolution of conventional light or fluorescent microscopy is thus much lower than that in confocal microscopy.

Because the confocal system works in all three dimensions, through confocal microscopy it is possible to place optical sections in the physical section. The laser beam can scan the specimen in x-, y-, and z-dimensions, producing a stack of images. This application is called a "z-series." Special imaging software can reconstruct three-dimensional renderings from such z-series. Confocal microscopy thus allows three-dimensional analysis of cells. For proving double-labelings, this approach is often the only feasible way.

Conventional microscopes in contrast produce essentially two-dimensional information. The true optical depth is compressed along the view axis. The three-dimensional specimen is projected, and the information in the z-axis is largely lost. To some degree, this can be overcome by manual focus through the section. Because of the limited spatial resolution, however, it can become difficult to distinguish closely neighboring structures (Peterson, 2004).

The misidentification of satellite cells as neurons can be avoided with confocal microscopy

Satellite cells are glial cells that, especially in cortical regions, are tightly attached to the soma of neurons (Fig. 7–7). They often engulf one side of the neuron like a cap. Depending on the direction of the observer, the satellite cells can completely cover the neuron. Because of the close proximity and low spatial resolution, the BrdU-labeled nucleus of the satellite cell can be mistaken as the nucleus of the neuron. The neuron would be falsely identified as BrdU-labeled and thus as new (Kuhn et al., 1997). The satellite phenomenon can even fool investigators who use a confocal microscope but do not carefully reconstruct the cell in question. An eccentric, kidney-shaped nucleus with dense chromatin found in a putative neuron should always raise suspicion.

Tweaking up the gain and opening the pinhole will evoke false-positive signals

Because in confocal microscopy the investigator can adjust both the energy of the excitation and the sensitivity of the detection, another problem is over-amplification of spurious signals. In the path of light of the microscope, the pinhole corresponds to the iris of a camera. Opening the iris increases the amount of light that falls through it but reduces spatial resolution. The wider the pinhole, the brighter the image, but more of the confocal effect is given away. Also, the detection sensitivity (gain) needs to

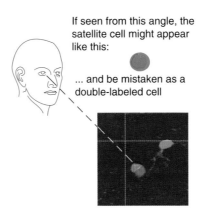

FIGURE 7–7 Satellite cells. Satellite glia are astrocytes that are tightly attached to neuronal cell bodies and often engulf the neuron like a cap. Depending on the viewing axis, satellite cells can be mistaken for the underlying neuron. BrdU incorporation in satellite glia might lead to the erroneous detection of neurogenesis. Z-series through the supposed new neuron can reveal that it actually consists of two cells, with BrdU incorporation in the nucleus of the satellite cell.

be adjusted carefully. If the gain is cranked up too far, more signal might seem present, but information is actually lost. Fine structures disappear, and the image becomes patchy. If the sensitivity is adjusted to a signal of too low intensity, even though other stronger signals at the same wavelength are found in the same image, all one is doing is amplifying the background. It will become increasingly difficult to distinguish signal from noise. No simple rules exist for the correct adjustment of excitation intensity and detection sensitivity, although modern microscopes do offer software tools that help determine the optimal settings.

How to Quantify Cell Numbers: The Issue of Stereology

To turn histology quantitative is a major challenge and source of errors

Reports on adult neurogenesis often contain quantitative statements reporting how much a given manipulation increased or decreased the production of new neurons. The parameter used to substantiate such claims is the "number of cells." Quantification becomes particularly relevant if far-reaching conclusions about regulation, function, and medical implications are made. Thus at one point or another in most studies on adult neurogenesis, cells are counted. Although nothing seems to be as straightforward as counting something, the procedure is not without pitfalls, because one simply cannot count all the cells that are there. Rather, cell counts are estimates based on an extrapolation from samples such as tissue sections.

The following forms of reporting cell numbers are found in the literature:

- Cells per area or section
- Cells per volume
- Cells per anatomically defined structure

Only the last one represents meaningful absolute numbers; the first two will always raise questions about their validity and are difficult to interpret.

The Problem of Treating Tissue Sections as Two-Dimensional

Cell counts must not be related to tissue areas

A tissue section under the microscope is essentially two-dimensional. One can add a third dimension by focusing through the section, but the spatial impression is limited because histological sections are thin. Thus in many practical situations histological sections are used as flat images. Their three-dimensional information is reduced to two dimensions, like a map. On a map it is simple to measure distances and areas as long as the scale is known. But for the third dimension, the heights of mountains, for example, one would have to turn to constructs such as altitude lines. Histological sections obviously do not come equipped with altitude lines.

When we count the number of cells in a histological section, we are counting colored dots ("profiles") in an area (although it is actually a small volume of tissue). But in biological contexts we are rarely interested in the numbers per area, because all organisms and their organs and cells are three-dimensional. The problem is that we usually do not know how the two-dimensional section represents the three-dimensional structure we are really interested in. After all, sections are not maps drawn with the idea of best possible representation. Sections are to a large degree random samples.

To display cell counts as cells per section is by its nature heavily confounded, even if we are seeking to "exactly match" sections between experimental groups. The slightest changes in orientation of a section according to the main axes can have large effects on resulting cell counts. More fundamentally, expression of cell counts as number per area is biologically almost meaningless and extremely difficult for the reader to visualize.

The Problem of Giving Quantities as Volume Densities

Cell counts must not be expressed as densities: the reference volume matters

Most frequently, cell counts are incorrectly expressed as cells per volume. This is not an optimal way to present quantitative data either, and not just because it usually would be rather easy to obtain absolute cell counts with the same tissue and equipment. When a measurement of "cells per volume" is mentioned, this volume is generally based on the volume of the samples in which the cells were counted. We are therefore depending entirely on the relationship between the sample volume and the volume of the complete structure of interest (reference volume) to determine whether this cell density is relevant and valid. The same cell density can refer to smaller or greater absolute numbers, depending on the size of the reference volume. If the size of the reference volume is not known, the reader cannot evaluate what a cell density tells us about the entire structure. In the hippocampus of adult mice, for example, cell densities differ dramatically across even a structure as homogenous as the dentate gyrus. This problem is called the "reference trap" (Braendgaard and Gundersen, 1986): it can completely destroy the sense to be made out of quantitative statements in morphological studies. One of the key rules of stereology, the method for circumventing these problems, is "never, ever do not measure the reference volume" (HJ Gundersen). A particularly interesting, albeit controversial, example is the neuroanatomical characterization of the p75 receptor null-mutant mouse. Depending on how the quantification was achieved, essentially opposite results were obtained by different groups (Van der Zee et al., 1996; Peterson et al., 1997).

Stereology as a Means of Obtaining Absolute Numbers

Quantitative histology always needs to consider rules of stereology, even if no full stereological analysis is possible or feasible

The only reasonable way to give relatively valid quantitative cell counts is to obtain absolute numbers, which are independent of the shape and size of the sample and relate to a meaningful anatomically defined structure. "Cells per dentate gyrus" tells us more about adult hippocampal neurogenesis than "cells per mm³." The art of obtaining such data is called "stereology" (Haug, 1986) and is incorrectly accused of being complicated and overly time consuming. Because of its sometimes mysterious rules that at first sight can only be followed by the initiated, the entire construct has sometimes been mocked as "the holy church of stereology," a world that is rather detached from the realities of everyday science. But if a stereological analysis is carefully designed and carried out, it will actually save time. As much as one does not have to know how a four-stroke engine works in order to drive a car, one does not have to understand the mathematics underlying stereology in order to use the method effectively. The principle is fairly simple.

Stereology is a method like others and should not be used as ideology

Stereology is often called an "unbiased" counting technique. The term is used to describe improved, design-based tissue sampling that is supposedly "assumption-free" in terms of size, shape, or distribution of the cell within the tissue, and free from methodological biases that could result from incorrect assumptions. However, no scientific method is free of assumptions, and thus bias remains possible. Stereology does not solve all problems, is no end in itself, and sometimes the full program is not feasible. But it is the best we can do, and the key considerations of stereology should always be taken into account.

Stereology is a way to safely estimate total counts in a structure from samples

The pitfalls of estimating cell counts from samples can be visualized with the following example (Figs. 7–8 and 7–9). Imagine a cylindrical hippocampus of 100 μm length, for the sake of simplicity, with only one neuron every 10 μm. Stereology would be a way to estimate the total number of neurons in the cylinder from counting the cells in a sample slice of this cylinder.

One 10-μm slice is taken from the middle of the hippocampus; in the example it contains one neuron (Fig. 7–4B). As stated previously, depending on the angle of the cut, a 10-μm slice can comprise different volumes and if looked at from the cutting surface, cover different areas.

Physical disector

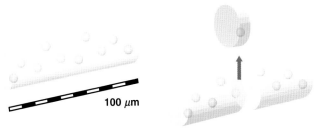

100 μm

1 cell per section x 10 = 10 cells

Physical fractionator

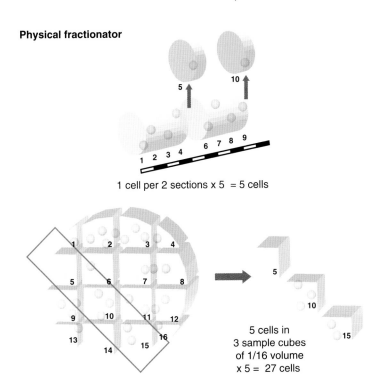

1 cell per 2 sections x 5 = 5 cells

5 cells in
3 sample cubes
of 1/16 volume
x 5 = 27 cells

FIGURE 7–8 Stereology 1. See text for explanation.

All of these different volumes or areas still would contain one neuron. If expressed as neuron per volume (or even worse, per area), the resulting densities will differ, although the true total hippocampal volume and the true total number of neurons within it remain the same. To extrapolate from the slice to the hippocampus would yield greatly varying results.

In the "fractionator" method, a known fraction of the tissue is analyzed

On the other hand, if one knew which part of the hippocampus the sample slice came from, one could easily calculate the total number of neurons. If the slice were exactly one-tenth of the hippocampus, and the assumption could be reasonably made that the neurons are as evenly distributed as they are in the example, one could calculate the total number of neurons by dividing the number of neurons in the sample slice by the ratio of the volume, 1/10. If the percentage of the volume occupied by the sample is known, the number of neurons in the entire brain can be calculated. In stereology, this principle is called the "fractionator."

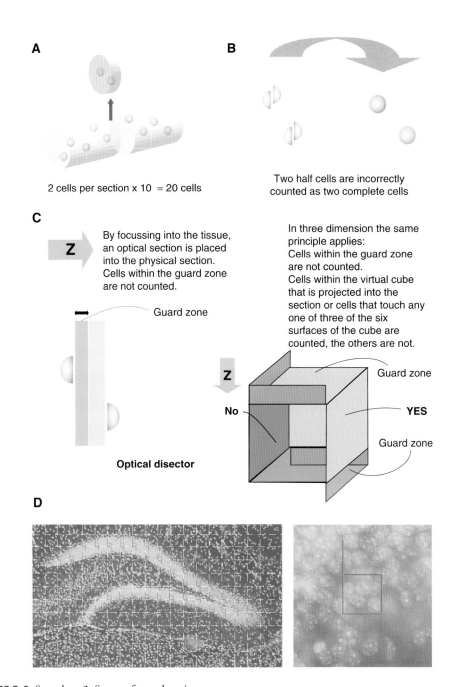

A

2 cells per section x 10 = 20 cells

B

Two half cells are incorrectly
counted as two complete cells

C

By focussing into the tissue,
an optical section is placed
into the physical section.
Cells within the guard zone
are not counted.

In three dimension the same
principle applies:
Cells within the guard zone
are not counted.
Cells within the virtual cube
that is projected into the
section or cells that touch any
one of three of the six
surfaces of the cube are
counted, the others are not.

Guard zone

Z

Guard zone

No

YES

Guard zone

Optical disector

D

FIGURE 7–9 Stereology 2. See text for explanation.

In neurobiology, the reference objects are complex spatial anatomical structures, such as the dentate gyrus or the olfactory bulb. Although they can be defined anatomically, neither their boundaries nor their three-dimensional extension are easily intelligible. This might be the reason why expressing cell counts as densities is so popular: a cube is a well-defined structure and easy to imagine. But biologically this cube needs to be related to the much more complicated three-dimensional representation of the anatomical structure of interest to convey the intended information. Densities are only a means of obtaining this absolute number, not an end in themselves.

With evenly distributed objects to count and knowledge of both the volume of the sample and of the total structure in which the objects are found, a very good estimate can be made with just one single sample. In reality this will never occur. The less evenly the objects are distributed, the more samples will need to be considered to lower the variance of the counted objects within each sample and to approximate the true mean.

Parameters for the fractionator need to be determined for each situation

For practical purposes, one should take series of slices (sections) and detect the number of neurons in every sixth or twelfth section (or any other useful regular interval). The estimate of total counts based on counts made in this series of physical sections is called the "physical fractionator." The more irregular the shape of the entire structure, the more sections will be needed to produce a good estimate of the total volume. For this estimate the areas of the sections are measured and added. The result is then multiplied by the distance between the sections. This method thus transforms every object of complex three-dimensional structure into the equivalent approximation of a cube or cylinder of the same volume: ground area times height yields volume.

In the example of the cylindrical hippocampus, the entire slice is examined for the presence of neurons. If the neurons were very small in relation to the slice to be examined, this would not be practical. One could, however, cut the slice into many cubes of identical size and reapply the fractionator principle (Fig. 7–8, bottom). For example, only every fifth or tenth tissue cube (or any other useful regular interval) would be examined. If the number of cubes per slice is known as well as the number of slices per hippocampus, the total number of neurons in the entire anatomical structure could be estimated. Again, the less evenly the objects to be counted are distributed in the slice, the more samples that need to be taken.

The optical fractionator avoids counting cells cut in half as two

If we return to the simple version of one slice through the cylindrical brain with one neuron per every 10 μm, we can visualize another pitfall (Fig. 7–9). Imagine that the two cuts made for removing the slice are accidentally made in such a way that not one neuron is caught but two are cut in half. Looking from rectangular to the cut, each of these incomplete neurons could easily be miscounted as complete, yielding an estimate of two instead of one neuron per sample slice. The total number of neurons per brain would be overestimated by 100%. The obvious solution is to count only one-half of the incomplete cells. When looking from one side onto the slice the most superficial cells are disregarded. In histology this is accomplished by focusing into the section and disregarding the objects to be counted in the uppermost focal plane. This focusing into the section effectively transforms the as-if-two-dimensional section (x- and y-axes) into a three-dimensional sample volume (x, y, and z), only part of which is used for counting. This method is called the *optical disector* (with only one *s*, as the term refers to two, hence *di*, sectors). The neglected uppermost part is called the *guard volume*. If we have counts of the sample volume analyzed by the optical disector, as well as the total volume of the structure being examined, the total number of neurons in the hippocampus can be estimated. Combining the optical disector principle with the fractionator principle gives the *optical fractionator*. If the volume of the sample volume placed into the physical section with the optical disector method represents a known fraction of the total structure volume, the total number of counts can be estimated by multiplying the counts per sample volume with the reverse of the fraction that the sample volume represents. Again, the less evenly the objects to be counted are distributed, the more sample volumes are needed. These sample volumes are spaced evenly throughout the entire volume—first by using series of physical sections, then by dissecting the section into (virtual) cubes.

Disregarding the uppermost focal plane solves only part (to be precise, one-third) of the problem. If objects are counted in a virtual cube placed into the physical section, the same effect of cutting objects in two can occur on any two opposing surfaces of the cube, not just the uppermost and lowermost surfaces. Accordingly, only objects that are found within the cube or touch any one out of three of the six surfaces of the cube may be counted. Objects that touch the other surfaces have to be disregarded.

To be exact, the bottom of the disector should have a guard zone as well to ensure that a questionable profile observed in the bottom focal plane that could be counted can be brought into focus to validate that it really was the top of a cell. This strategy actually becomes rather important when using confocal stereology (which does not really exist as yet) because the high resolution of the confocal microscope may make the investigator see signal in the bottom counting frame but not be sure that it is a real cell without focusing through further (Peterson, 1999).

Currently, stereology is normally done with a semi-automated system consisting of a motorized-stage microscope controlled by a computer. A video image is transferred to the monitor, and counting frames can be superimposed so that the actual counting can be done on the screen. The software keeps track of the section series and enables measurement of area and volume. The program allows projection of a grid onto the sections, and when focusing through the section depicts the forbidden surfaces of the virtual counting cube as red, and those at which objects should be counted as green. A simplified version that can be done without computerized setup is based on determining cell densities at many randomly determined spots (with a randomly placed grid, for example) and relating them to the total volume of the olfactory bulb, which can be measured efficiently with a technique called the "Cavalieri estimator."

In the case of heterogeneous distributions, it is important to assess the amount of variance. There are several statistical estimators that can be employed to provide a critical evaluation of the amount of variance within the sample. Unfortunately, there is no absolute rule about how much variance is acceptable. A good rule of thumb is to look at the raw numbers produced by sampling at each optical disector and see how consistent the number of counted cells are at each site. For example, if the number of cells consistently ranges between zero and five at all counting sites, then the distribution is relatively homogeneous and the variance is probably acceptably low. However, if the sites produce mostly no cells counted and there are a couple of sites where 20 to 30 cells are counted, then the cells are heterogeneous in their distribution and the variance may be unacceptably high. In this case, many more sites should be sampled to reduce the sampling variance.

Stereology solves some but not all problems related to estimating cell counts. Although stereology is much less laborious than often assumed, it is complicated to apply to some specific questions and can lead to technical overkill. If small differences (10% to 20%) are detected and proven, however, stereology is the only feasible way to go. If differences are huge, they are unlikely to be greatly distorted by volume effects, as long as the anatomical structures of interest do not show gross differences. However, the general rule is never to not consider the reference volume.

For counting BrdU-labeled cells in the SGZ, one can use a full series of sections covering the entire dentate gyrus in its rostrocaudal extension and exhaustively count all labeled cells in the SGZ except those in the uppermost focal plane. The resulting number is multiplied by the number of sections in each interval (i.e., six, if every sixth section is analyzed). This is not real stereology, but is an often-feasible compromise that still obeys some key stereological principles and avoids the reference trap.

Stereology usually relies on the semi-automated systems, which greatly simplify implementation

Quality control and the optimization process of the protocol include the assessment of variation

Stereology is the only safe way to detect small differences in cell counts

For counting cells in the SGZ, full stereology is not feasible, but a simplified version is highly practical and avoids serious pitfalls

The Genetic Toolbox of Research on Adult Neurogenesis

Genetic tools have greatly expanded the opportunities to study adult neurogenesis but these methods raise important methodological issues as well

Whereas immunohistological tools target the protein level, molecular biology has introduced a number of novel methods that assess the underlying genetic level of adult neurogenesis. It is often inferred that with this transition, research would also become less descriptive and more mechanistic, but this is not necessarily true. Neurobiology today lies under the spell of the molecular bias, which—largely due to a frame of mind driven by journal impact factors—pretends that any genetic approach is per se better than anything else. But no gene makes a mechanism, and there is something to be said in favor of a healthy phenomenology and description. Many genetic studies suffer from poor readouts, but at times where in reviewing processes it is considered a minus if a study is "exploratory," we should wonder if the virtues of the explorer, who used to be *the* scientist, are lost and we end with mechanistic bits and pieces that do not build a case—or a phenomenon, for that matter.

Despite this introductory remark against molecular overkill, there is no doubt that much of the progress in the field is owed to the invention of new gene-based methods. And it is clear that despite the impressive achievements so far, we are still at the beginning of this development.

The molecular tools such as reporter constructs, targeted mutagenesis, conditional single-cell knockdown, etc., are extremely powerful and have made possible tremendous progress in describing the details of adult neurogenesis and offered us the first glimpses of the mechanism of molecular control. But they are also, like most influential science, highly reductionistic. This is no disadvantage as long as the potential confounds of the methods are not let out of sight. Gene-based methods are not principly less confounded than other methods (for review, see, for example, Breunig et al., 2007). Today's genetic tools are still mostly based on single genes, very few on two. Constructs combining three genes are still exceedingly rare. This is important to note because the gene networks that supposedly interact to control complex phenotypes, in contrast, might contain hundreds to thousands of genes, and we currently do not begin to have a thorough understanding of how the individual genes contribute to the experienced complexity.

Viral Vectors

Viral vectors allow the expression of genes in non-transgenic animals

Viral vectors are used in many studies to transfer a gene of interest to either the precursor cells (in vitro or in vivo) or to the cellular microenvironment in SVZ and SGZ (for review, see, for example Suhonen et al., 2006). In the first case a modification of the intrinsic potential is sought, in the second a manipulation of the niche. The gene of interest can be a reporter gene such as EGFP, whose function is simply to allow better visualization of the cell itself or of a particular function or developmental step. If the precursor cells were, for example, transfected with an EGFP reporter that is driven by a neuronal promoter such as α-tubulin, synapsin I, calcium/calmodulin-dependent protein kinase II, neuron-specific enolase, platelet-derived growth factor beta chain or others (Liu et al., 2004; Hioki et al., 2007), their progeny would turn green, once they have differentiated far enough to turn on the promoter driving the transgene. The same principle can be applied to any gene of interest, other gene-based sensors of cellular function, and RNA interference. Sensitivity and specificity of the promoter is key to this approach. Three types of viral vectors are frequently used for studies of adult neurogenesis are: *retro-, lenti-,* and *adenoviruses.*

Retroviral vectors target dividing cells only and are thus relatively specific for the precursor cells

Retroviruses are special in that they require a full cell cycle for complete integration into the host genome. They can thus only infect dividing cells, which comes in very handy in the context of neurogenesis, when precursor cells should be targeted. Their most common application is to bring a reporter gene into the precursor cells in vivo, thereby allowing the full morphogical appreciation of their progeny (see below). Fred H. Gage and his colleagues have developed this method to maturity, and most of our present knowledge about neurite extension and maturation in adult neurogenesis is based on this strategy (Zhao et al., 2006; Toni et al., 2007; Jessberger et al., 2008; Toni et al., 2008). In addition, however, transduction of proliferating cells with a retrovirus expressing the green fluorescent protein (or enhanced GFP) allows visualization of living cells, for example in acute slice preparations. This retrovirus-based approach could therefore successfully be used in electrophysiological studies of newly generated neurons (van Praag et al., 2002; Esposito et al., 2005; Laplagne et al., 2006; Toni et al., 2008; Mongiat et al., 2009). On the basis of similar constructs, combined with an extrinsically regulatable element or with the Cre/Lox system, it is also possible to achieve a cell-specific knockdown or overexpression of genes of interest in developing neurons. A key example is the targeted acceleration of neuronal maturation by the overexpression of K+-coupled Cl– transporter KCC2 2, which turns the excitatory effects of GABA to become inhibitory as they are in mature neurons, thereby prematurely terminating maturation of newborn neurons in the adult hippocampus (Ge et al., 2006). Similarly, Tashiro and colleagues achieved cell-specific knockdown of the NMDA receptor NR1 by delivering a GFP/Cre fusion protein to excise the loxP-flanked target gene in a mouse line designed towards that end (Tashiro et al., 2006). Despite their dependency on full cell cycle completion to integrate into the host genome, retroviruses can possibly falsely label post-mitotic cells under pathological circumstances, presumably due to fusion processes with proliferative microglia (Ackman and LoTurco, 2006; Breunig et al., 2007).

Lentiviral vectors have a greater transfection efficiency but do not just target dividing cells

Lentiviral infection is a highly efficient means of transferring genes. In principle, lentiviral vectors, which were originally derived from the HIV (Naldini et al., 1996), should infect all cells. In reality, they tend to infect subsets of cells. This can be used, for example, to mark SVZ progenitor cells (Geraerts et al., 2006). Due to their high efficiency, lentiviral vectors are particularly suited for targeted overexpression or knockdown experiments. Lentivirus-mediated knockdown of Wnt function in the adult hippocampus, for example, is one effective way to suppress adult neurogenesis in order to assess the potential function of the new neurons (Jessberger et al., 2009).

Adeno- and adeno-associated viruses also do not target specifically precursor cells but allow efficient gene transfer

Several studies have used adenovirus (AV) to transfer genes to the adult brain in order to manipulate adult neurogenesis (Benraiss et al., 2001; Bauer and Patterson, 2006). An early groundbreaking transplantation study used AV to prelabel the precursor cells that were expanded in cell culture before their implantation into the dentate gyrus (Gage et al., 1995). To target neural precursor cells, AV have been combined with ligands that allow cell-specifc infection in vivo (Schmidt et al., 2007). Adeno-associated virus (AAV) has been used to achieve long-term conditional gene manipulations with the Cre/Lox system in the adult brain (Kaspar et al., 2002) and, for example, to mediate BDNF in order to induce adult neurogenesis in the SVZ/OB (Henry et al., 2007).

Virus-based techniques are technically challenging and introduce a number of potential confounds

As a word of caution, viral infections are not quantitative and might show varying tropisms. The efficiency of retrovirus-mediated reporter gene expression is also greatly dependent on the promoter used (Zhao et al., 2006). Finally, their application in vivo requires a stereotaxic injection, which causes tissue damage that in turn will elicit a response. Spread of the virus in the tissue is influenced by the amount of tissue damage but is usually limited within the parenchyma. Virus particle might, however, travel greater distances along fiber tracts. Although one can control for such problems by carefully chosing the injection site in relation to the site of analysis and with the injection of empty vectors as controls, the experimental setting remains challenging.

Transgenic Reporter Mice

Reporter gene mice allow the identification of cells of interest without immunohistochemistry and in cases where protein expression is poor

In the case of transgenic reporter mice, the reporter construct (e.g., EGFP or β-galactosidase) is stably expressed in transgenic mouse strains under the promoter of the marker gene of interest. This can be achieved via a knock-in of the gene, in which case the reporter gene would replace the target gene, resulting in a null-mutation. The signal from the reporter would then replace the endogenous gene function. If the transgene is placed after the endogenous gene, however, and, for example, coupled with an IRES (internal ribosomal entry site) or another linker, which allows the expression of both genes from the same promoter, the actual gene remains functional and is still coupled to the reporter signal. The ratio of the transcript of the target gene and the reporter gene is one to one, which leads to a usually weak signal that, if further endangered by the fact that the reporter comes only second after the endogenous gene, reduces the chance for correct transcription. In traditional transgenesis, in contrast, the reporter gene construct integrates randomly and in higher copy number into the genome. Unless this integration accidentally disrupts another gene, function of the cell is normal: the target gene is transcribed as normal, and the reporter gene is transcribed from the additionally integrated sequences with the same promoter.

The most widely used reporter mice in adult neurogenesis research are based on the nestin promoter

A very large number of reporter gene lines has been developed; for example, as BAC transgenics in the Gensat project (www.gensat.org). Few of these, however, have been validated. A growing number of lines from other sources has been published within the context of adult neurogenesis research, however, and essentially all stages of neuronal development in the adult can now be covered with reporter mice (Table 7–1). The most widely used reporter gene mice in adult neurogenesis research are nestin-GFP/ EGFP lines, of which a variety of versions exists. These lines are not identical in that they use different portions of the neural enhancer elements of the nestin promoter that target expression to the nervous system (Lothian and Lendahl, 1997; Lothian et al., 1999). Nevertheless, the expression patterns of the reporter in the neurogenic zones is very similar between the lines.

The reporter gene product persists longer in the cell than the actual activity of the promoter

A critical issue is the fact that presence of the reporter protein in the cell does not indicate that the promoter must be active at that time. GFP is degraded quickly, but the estimated halflife is still about one day (Corish and Tyler-Smith, 1999), enough to confound some experiments. Daughter cells might also inherit GFP from the mother cell, in which the promoter was active before or during division. This problem can occasionally be turned into an advantage, because it allows some very limited lineage tracing, "poor man's lineage tracing" (see, e.g., Steiner et al., 2006; Petrus et al., 2008). In any case, reporter gene lines have to be thoroughly validated before they can be routinely applied as research tools. But if used properly, they open an entirely new window.

Conditional and Inducible Genetic Tools

Conditional and inducible genetic tools allow us to temporally restrict transgene expression or knockdown/ knockout

The Cre/Lox system works as follows. A mouse line is engineered in which the gene of interest is flanked by two short sequences, the so-called pLox sites. Lab jargon for this situation is that the gene of interest is "floxed." The pLox sites are specifically cut by the Cre recombinase, which is not found in vertebrates. If Cre is applied extrinsically, however, the flanked gene is excised and a null-mutation is achieved (Fig. 7–10). Cre can be introduced by various means. In cell culture it can be applied directly as a protein; in vivo it is either transfected with viruses (e.g., Ide et al., 2008) or, more commonly, by crossing the floxed line into another transgenic line, in which Cre is expressed under a promoter of interest. This allows the researcher to target the excision of the floxed gene in a cell-specific manner resulting in a "conditional knockout." Except for the few cases in which a gene does not affect, or hardly affects, embryonic neurogenesis (e.g., Shi et al., 2004), conditional mutants are the only way to apply the knockout technique to studying the role of single genes in adult neurogenesis in vivo.

The CreERT2 construct allows regulatable conditional transgenesis and the generation of promoter-based permanent label

An ingenious refinement of this idea is to have a reporter gene be preceded by a floxed stop cassette, leading to the abortive expression under normal circumstances, but induced expression once the stop cassette is removed by Cre (Fig. 7–10). If the reporter gene is driven by a constitutive promoter it will remain "on." For this, usually the Rosa26 locus is chosen, which is ubiquitously and stably expressed, although its disruption by the inserted genes does not lead to any noticable phenotype. Mouse lines such as Nestin::CreERT2, which are based on this idea, allow the permanent

Table 7–1 Reporter mice in adult neurogenesis research

Gene	Reporter	Identifies…	Comment	Reference
Nestin	GFP	Precursor cells	Has been used as basis for the original distinction of type-1/2 precursor cells in the SGZ	(Yamaguchi et al., 2000; Filippov et al., 2003; Fukuda et al., 2003)
	GFP	Precursor cells	Used to propose a slightly different model of adult hippocampal neurogenesis than above, including ex vivo data	(Mignone et al., 2004)
	EGFP	Precurssor cells	Prospective isolation of precursor cells from adult brain	(Kawaguchi et al., 2001; Sawamoto et al., 2001)
Nestin-H2B	CFP	Precursor cells	Nuclear signal allows unambiguous identification of nuclei for counting purposes	(Encinas et al., 2006)
BLBP	EGFP, EYFP, dsRed2	Radial glia, astrocytes, precursor cells		(Schmid et al., 2006)
GFAP	EGFP	Astrocytes, precursor cells	Due to low specificity useful as precursor cell marker only in conjunction with other markers	(Nolte et al., 2001; Steiner et al., 2006)
DCX	promo-luciferase	Precursor cells, immature neurons	Allows in vivo imaging	(Couillard-Despres et al., 2008)
	EGFP, dsRed	Precursor cells, Immature neurons		(Couillard-Despres et al., 2006)
Sox2	EGFP	Precursor cells		(Ellis et al., 2004; Brazel et al., 2005)
POMC	GFP	Immature neurons about 2 weeks after precursor cell division	Unique in that a very late stage of development is specifically labeled	(Overstreet et al., 2004)
Tbr2	GFP	Intermediate progenitor cells in the hippocampus		(Hodge et al., 2008)
Tis21	EGFP	Late progenitor cells, late immature neurons	Identifies precursor cells before neurogenic division in fetal cortical development	(Haubensak et al., 2004; Attardo et al., 2009)

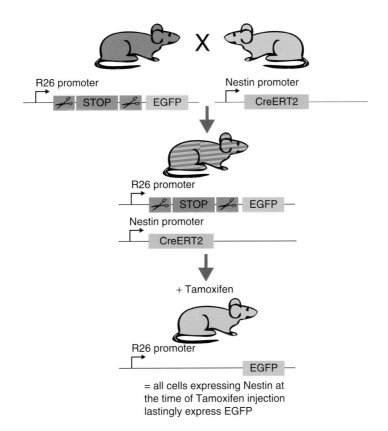

R26 promoter

Nestin promoter

R26 promoter

Nestin promoter

+ Tamoxifen

R26 promoter

= all cells expressing Nestin at
the time of Tamoxifen injection
lastingly express EGFP

FIGURE 7–10 The CreERT2 method. ERT2 is a genetic element that is activated by the binding of tamoxifen. If tamoxifen binds, the linked gene is transcribed. In the case of this method, the regulatable gene is Cre recombinase, which is a non-mammalian gene that excises at restriction sites ("floxed") which have been engineered before and after the target gene. The target gene is here a reporter gene under the ubiquitously expressed Rosa26 locus. As promoter for the CreERT2 construct a gene is chosen that is specific to the cell of interest; in the example, Nestin.

labeling of cells after the transfection with Cre. In this case. labeling is not constitutively targeted to nestin-expressing (bona fide) precursor cells but only after tamoxifen has been applied. Cohorts of cells with all their progeny can be permanently visualized after the system has been switched on. For adult hippocampal neurogenesis, a similar construct based on Glast::CreERT2 has been used to estimate the contribution of adult neurogenesis to the total population of granule cells (Ninkovic et al., 2007). With an analogous strategy, it was shown that type-2 progenitor cells of the adult hippocampus that transiently express Ascl1/Mash1 generate new neurons (Kim et al., 2008).

||||||||||||||||||||||||||||||||||||||||||||||||||||||||||||

Inducible conditional constructs also underlie strategies for targeted ablation of newborn neurons

||||||||||||||||||||||||||||||||||||||||||||||||||||||||||||

The same regulatable conditional approach can also be used to express "suicide genes" for targeted ablation of precursor cells or new neurons. Targeting of TrkB receptor in precursor cells with the Glast::CreERT2 construct, for example, impaired maturation of their progeny and suppressed neurogensis, which could be exploited for behavioral tests addressing the putative function of newborn neurons (Bergami et al., 2008). A similar approach involved the use of tetracycline-regulatable elements in a mouse line overexpressing the pro-apoptotic gene BAX under the nestin-promoter, targeting a death-inducing signal to nestin-positive cells (Dupret et al., 2008).

Regulatable bi- or trigenic constructs require complicated breeding and genotyping protocols and involve numerous controls, which are not always presented in the published studies. The general caveats of transgenic and knockout mice apply (see, e.g., Crusio, 1996; Gerlai and Clayton, 1999; Wolfer and Lipp, 2000; Wolfer et al., 2002). The absence of expected results, or counterintuitive findings are thus often difficult to interpret. The greater specificity achieved by targeting individual genes is paid for with complex experimental situations.

Genetic tools would often call for more controls than presented in the papers

One Word on Statistics: What Is in a P-Value?

The relevance of observed differences is usually underscored by the determination of their "statistical significance"

Because in adult neurogenesis research—as in most other research areas—often two (or more) experimental groups are compared, the paramount question is how valid the observed differences are. In the absence of knowledge about the truth (if this knowledge were available, we would not have to do the experiment in the first place), statistical tools are used to estimate the probability that the observed differences are real. This is at least the common perception of why we use hypothesis testing and assign p-values to our data. Unfortunately, the statistical procedures applied in such cases (a t-test or an ANOVA followed by a post hoc test such as Fisher's or Tukey's) do not really allow this conclusion. This common misunderstanding is actually well known (and discussed in numerous papers; see, e.g., Blume and Peipert, 2003; Goodman, 2003; Gigerenzer et al., 2004; Turkheimer et al., 2004) but might be the most disregarded fact in science. First of all, p-value statistics and hypothesis testing are conceptually different things that find themselves in a problematic marriage in everyday science.

The p-value is the conditional probability that an experiment yields the observed result under the assumption that in fact no difference exists

The p-value in the common statistical tests does not five the probability that the difference, to which it is assigned, is real, but the probability that the data would look as they do if in fact no difference exists. This means that the p-value does not tell us with which probability the "null-hypothesis" is correct. Polls among scientists who routinely use p-values reveal that almost none of them can correctly define p-values (Haller and Krauss, 2002).

Humans have intrinsic difficulties with grasping and interpreting probabilities in general, but conditional probabilities are clearly beyond our intuitive imagination (Krämer and Gigerenzer, 2005). To allow the conclusion that we actually would like to draw (i.e., the probability that an observation reflects the truth or is real), a completely different set of statistical tools would be necessary. These are called *Bayesian statistics*. Bayesian statistics cannot be as easily formalized as the regular forms of statistics we have become used to and cannot be pressed into "objective" schemes because they call for the inclusion of prior knowledge in the interpretation. This brings in a certain level of subjectivity, which science as a whole (and in the end for no good reason) seems to abhor.

Bayesian statistics would be more appropriate to gaining insight into the probability that an observed difference is real but are difficult to implement

The level of statistical significance at p = 0.05 is a more or less useful convention and not a law of nature

What we make out of the "objective" p-value from usual post-hoc tests is not free of assumptions either. The most important *a priori* is the threshold below which we consider a p-value to represent "statistical significance." For most purposes, this threshold is set to 0.05, but lower thresholds might be used. The intention of setting lower thresholds is to indicate greater "significance" and thereby greater confidence in the validity of the observed difference.

However, *N*, the number of samples, is part of the equation that generates p-values, so that p-values are strongly influenced by group sizes. This is the truth behind the often heard (and revealing) sentence in the presentation of preliminary data that N would have to be increased to "make the result significant." Problematically, with large enough samples, even the smallest differences can be made "significant." The most important question, however, is whether such "significant" difference is also biologically meaningful and relevant. In most cases, effect size does matter, and statistical significance alone does not tell us much about biological significance or relevance.

Invariably, the interpretation of effect sizes calls for the investigator with his hypothesis and assumptions, so the intended objectivity is lost to some degree at this stage. When, due to a small N and a large variance, a huge mean difference misses statistical significance, the observation might still be much more relevant than a "highly significant" 3% difference tickled out of a population of 25,000 subjects. But it is also not true that large differences would be necessarily more relevant. Transcription factors, for example, need only very small changes in expression to lead to massive changes in an organism. So it is true that many studies are too small to detect relevant differences. But the point is that we cannot decide "relevance" solely on the basis of that one p-value.

We all use the adventurous construct of conditional probability to get a grip on a difficult situation and provide some standards, which should help to make studies comparable. This is all very well, and for all practical purposes we probably need to continue doing so, but one should keep in mind that the common practice is far from ideal for this purpose. Staring a p-value is a ritual and not necessarily the best way of dealing with data (Gigerenzer et al., 2004). At places it might be outright counterproductive.

The absurd consequence of this ultimately unscientific behavior is that a great deal of data with potentially greatest relevance end up in the garbage because they missed conventional statistical significance. There is no inherent truth in p < 0.05, which in the words of one critique is the "threshold of decerebrate genuflection" (Gallagher, 1999).

It cannot be correct that a result at p = 0.049 goes to *Nature* but is irrelevant at p = 0.051. The probability that (under the assumption that there is no difference) 100 experiments would show a difference is just 2 per mille lower in the first than in the second case.

It is also not correct to conclude from any p > 0.05 that the two samples are not different, and even less so to use this as an argument to pool data from the two groups for another comparison (Gallagher, 1994). Similarly, the rejection of the null hypothesis does not necessarily mean that the alternative hypothesis is true, and in most cases no explicit alternative hypothesis is formulated anyway.

One general recommendation for solving this problem is to present actual p-values. This has already been suggested by Sir Ronald A. Fisher, the inventor of the p-value in this context, who moved away from the use of a "conventional level of statistical significance." Statements like "p > 0.05" or simply "n.s." should be avoided. The probability values are part of the data.

And data should be presented with error bars. Confidence intervals are a simple and more useful means to allow the visual assessment of potential differences (Gardner and Altman, 1986; Young and Lewis, 1997; Nakagawa and Cuthill, 2007). The International Committee of Medical Journal Editors called for the report of confidence intervals, stating "Avoid sole reliance on statistical hypothesis testing, such as the use of p-values, which fails to convey important quantitative information" (Bailar and Mosteller, 1988). Consequently, the automatic decision that a "significant" p-value equals a definite answer to the scientific question at hand is not good practice: what is possible is only a probabilistic approximation. This does not imply that we should water down any standards, but rather that we should introduce stricter ones: the pseudo-objectivity of the p-ritual should to be replaced with a thoughtful interpretation of the data based on a good hypothesis, the effect size, *N*, and the actual probabilities.

References

Ackman JB, LoTurco JJ (2006). The potential of endogenous neuronal replacement in developing cerebral cortex following hypoxic injury. *Exp Neurol* 199:5–9.

Altman J, Das GD (1965). Autoradiographic and histological evidence of postnatal hippocampal neurogenesis in rats. *J Comp Neurol* 124:319–335.

Ambrogini P, Lattanzi D, Ciuffoli S, Agostini D, Bertini L, Stocchi V, Santi S, Cuppini R (2004). Morpho-functional characterization of neuronal cells at different stages of maturation in granule cell layer of adult rat dentate gyrus. *Brain Res* 1017:21–31.

Amrein I, Dechmann DK, Winter Y, Lipp HP (2007). Absent or low rate of adult neurogenesis in the hippocampus of bats (*Chiroptera*). *PLoS ONE* 2:e455.

Arvidsson A, Collin T, Kirik D, Kokaia Z, Lindvall O (2002). Neuronal replacement from endogenous precursors in the adult brain after stroke. *Nat Med* 8:963–970.

Aten JA, Bakker PJ, Stap J, Boschman GA, Veenhof CH (1992). DNA double labelling with IdUrd and CldUrd for spatial and temporal analysis of cell proliferation and DNA replication. *Histochem J* 24:251–259.

Aten JA, Stap J, Hoebe R, Bakker PJ (1994). Application and detection of IdUrd and CldUrd as two independent cell-cycle markers. *Methods Cell Biol* 41:317–326.

Attardo A, Fabel K, Krebs J, Haubensak W, Huttner WB, Kempermann G (2009). Tis21 expression marks not only populations of neurogenic precursor cells but also new post-mitotic neurons in adult hippocampal neurogenesis. *Cereb Cortex* 20:304–14.

Avilion AA, Nicolis SK, Pevny LH, Perez L, Vivian N, Lovell-Badge R (2003). Multipotent cell lineages in early mouse development depend on Sox2 function. *Genes Dev* 17:126–140.

Babu H, Cheung G, Kettenmann H, Palmer TD, Kempermann G (2007). Enriched monolayer precursor cell cultures from micro-dissected adult mouse dentate gyrus yield functional granule cell–like neurons. *PLoS ONE* 2:e388.

Babu H, Ramirez-Rodriguez G, Fabel K, Bischofberger J, Kempermann G (2009). Synaptic network activity induces neuronal differentiation of adult hippocampal precursor cells through BDNF signaling. *Front. Neurosci.* **3**:49.

Bagley J, LaRocca G, Jimenez DA, Urban NN (2007). Adult neurogenesis and specific replacement of interneuron subtypes in the mouse main olfactory bulb. *BMC Neurosci* 8:92.

Bai J, Ramos RL, Ackman JB, Thomas AM, Lee RV, LoTurco JJ (2003). RNAi reveals doublecortin is required for radial migration in rat neocortex. *Nat Neurosci* 6:1277–1283.

Bailar JC, 3rd, Mosteller F (1988). Guidelines for statistical reporting in articles for medical journals. Amplifications and explanations. *Ann Intern Med* 108:266–273.

Barral-Moran MJ, Calaora V, Vutskits L, Wang C, Zhang H, Durbec P, Rougon G, Kiss JZ (2003). Oligodendrocyte progenitor migration in response to injury of glial monolayers requires the polysialic neural cell-adhesion molecule. *J Neurosci Res* 72:679–690.

Batista-Brito R, Close J, Machold R, Fishell G (2008). The distinct temporal origins of olfactory bulb interneuron subtypes. *J Neurosci* 28:3966–3975.

Bauer S, Patterson PH (2005). The cell cycle-apoptosis connection revisited in the adult brain. *J Cell Biol* 171: 641–650.

Bauer S, Patterson PH (2006). Leukemia inhibitory factor promotes neural stem cell self-renewal in the adult brain. *J Neurosci* 26:12089–12099.

Belluzzi O, Benedusi M, Ackman J, LoTurco JJ (2003). Electrophysiological differentiation of new neurons in the olfactory bulb. *J Neurosci* 23:10411–10418.

Benraiss A, Chmielnicki E, Lerner K, Roh D, Goldman SA (2001). Adenoviral brain-derived neurotrophic factor induces both neostriatal and olfactory neuronal recruitment from endogenous progenitor cells in the adult forebrain. *J Neurosci* 21:6718–6731.

Bergami M, Rimondini R, Santi S, Blum R, Gotz M, Canossa M (2008). Deletion of TrkB in adult progenitors alters newborn neuron integration into hippocampal circuits and increases anxiety-like behavior. *Proc Natl Acad Sci USA* 105:15570–15575.

Bergles DE, Roberts JD, Somogyi P, Jahr CE (2000). Glutamatergic synapses on oligodendrocyte precursor cells in the hippocampus. *Nature* 405:187–191.

Bick-Sander A, Steiner B, Wolf SA, Babu H, Kempermann G (2006). Running in pregnancy transiently increases postnatal hippocampal neurogenesis in the offspring. *Proc Natl Acad Sci USA* 103:3852–3857.

Blume J, Peipert JF (2003). What your statistician never told you about P-values. *J Am Assoc Gynecol Laparosc* 10:439–444.

Bondolfi L, Calhoun M, Ermini F, Kuhn HG, Wiederhold KH, Walker L, Staufenbiel M, Jucker M (2002). Amyloid-associated neuron loss and gliogenesis in the neocortex of amyloid precursor protein transgenic mice. *J Neurosci* 22:515–522.

Bonfanti L (2006). PSA-NCAM in mammalian structural plasticity and neurogenesis. *Prog Neurobiol* 80:129–164.

Braendgaard H, Gundersen HJ (1986). The impact of recent stereological advances on quantitative studies of the nervous system. *J Neurosci Methods* 18:39–78.

Brandt MD, Jessberger S, Steiner B, Kronenberg G, Reuter K, Bick-Sander A, von der Behrens W, Kempermann G (2003). Transient calretinin expression defines early post-mitotic step of neuronal differentiation in adult hippocampal neurogenesis of mice. *Mol Cell Neurosci* 24:603–613.

Brazel CY, Limke TL, Osborne JK, Miura T, Cai J, Pevny L, Rao MS (2005). Sox2 expression defines a heterogeneous population of neurosphere-forming cells in the adult murine brain. *Aging Cell* 4:197–207.

Breunig JJ, Arellano JI, Macklis JD, Rakic P (2007). Everything that glitters isn't gold: A critical review of postnatal neural precursor analyses. *Cell Stem Cell* 1:612–627.

Brill MS, Snapyan M, Wohlfrom H, Ninkovic J, Jawerka M, Mastick GS, Ashery-Padan R, Saghatelyan A, Berninger B, Gotz M (2008). A Dlx2- and Pax6-dependent transcriptional code for periglomerular neuron specification in the adult olfactory bulb. *J Neurosci* 28:6439–6452.

Brill MS, Ninkovic J, Winpenny E, Hodge RD, Ozen I, Yang R, Lepier A, Gascon S, Erdelyi F, Szabo G, Parras C, Guillemot F, Frotscher M, Berninger B, Hevner RF, Raineteau O, Gotz M (2009). Adult generation of glutamatergic olfactory bulb interneurons. *Nat Neurosci* 12:1524–33.

Brown JP, Couillard-Despres S, Cooper-Kuhn CM, Winkler J, Aigner L, Kuhn HG (2003). Transient expression of doublecortin during adult neurogenesis. *J Comp Neurol* 467:1–10.

Buck SB, Bradford J, Gee KR, Agnew BJ, Clarke ST, Salic A (2008). Detection of S-phase cell cycle progression using 5-ethynyl-2′-deoxyuridine incorporation with click chemistry, an alternative to using 5-bromo-2′-deoxyuridine antibodies. *Biotechniques* 44:927–929.

Buffo A, Rite I, Tripathi P, Lepier A, Colak D, Horn AP, Mori T, Gotz M (2008). Origin and progeny of reactive gliosis: A source of multipotent cells in the injured brain. *Proc Natl Acad Sci USA* 105:3581–3586.

Bull ND, Bartlett PF (2005). The adult mouse hippocampal progenitor is neurogenic but not a stem cell. *J Neurosci* 25:10815–10821.

Burns KA, Kuan CY (2005). Low doses of bromo- and iododeoxyuridine produce near-saturation labeling of adult proliferative populations in the dentate gyrus. *Eur J Neurosci* 21:803–807.

Burns TC, Ortiz-Gonzalez XR, Gutierrez-Perez M, Keene CD, Sharda R, Demorest ZL, Jiang Y, Nelson-Holte M, Soriano M, Nakagawa Y, Luquin MR, Garcia-Verdugo JM, Prosper F, Low WC, Verfaillie CM (2006). Thymidine analogs are transferred from prelabeled donor to host cells in the central nervous system after transplantation: a word of caution. *Stem Cells* 24:1121–1127.

Busser J, Geldmacher DS, Herrup K (1998). Ectopic cell cycle proteins predict the sites of neuronal cell death in Alzheimer's disease brain. *J Neurosci* 18:2801–2807.

Butt AM, Hamilton N, Hubbard P, Pugh M, Ibrahim M (2005). Synantocytes: The fifth element. *J Anat* 207:695–706.

Caldwell MA, He X, Svendsen CN (2005). 5-Bromo-2′-deoxyuridine is selectively toxic to neuronal precursors in vitro. *Eur J Neurosci* 22:2965–2970.

Calegari F, Haubensak W, Haffner C, Huttner WB (2005). Selective lengthening of the cell cycle in the neurogenic subpopulation of neural progenitor cells during mouse brain development. *J Neurosci* 25:6533–6538.

Cameron HA, McKay RD (2001). Adult neurogenesis produces a large pool of new granule cells in the dentate gyrus. *J Comp Neurol* 435:406–417.

Cameron HA, Woolley CS, McEwen BS, Gould E (1993). Differentiation of newly born neurons and glia in the dentate gyrus of the adult rat. *Neuroscience* 56:337–344.

Capela A, Temple S (2002). LeX/ssea-1 is expressed by adult mouse CNS stem cells, identifying them as non-ependymal. *Neuron* 35:865–875.

Cappella P, Gasparri F, Pulici M, Moll J (2008). A novel method based on click chemistry, which overcomes limitations of cell cycle analysis by classical determination of BrdU incorporation, allowing multiplex antibody staining. *Cytometry A* 73:626–36.

Chan JP, Cordeira J, Calderon GA, Iyer LK, Rios M (2008). Depletion of central BDNF in mice impedes terminal differentiation of new granule neurons in the adult hippocampus. *Mol Cell Neurosci* 39:372–383.

Chehrehasa F, Meedeniya AC, Dwyer P, Abrahamsen G, Mackay-Sim A (2009). EdU, a new thymidine analogue for labelling proliferating cells in the nervous system. *J Neurosci Methods* 177:122–130.

Chen J, Magavi SS, Macklis JD (2004). Neurogenesis of corticospinal motor neurons extending spinal projections in adult mice. *Proc Natl Acad Sci USA* 101:16357–16362.

Chen ZJ, Negra M, Levine A, Ughrin Y, Levine JM (2002). Oligodendrocyte precursor cells: Reactive cells that inhibit axon growth and regeneration. *J Neurocytol* 31:481–495.

Collombet JM, Masqueliez C, Four E, Burckhart MF, Bernabe D, Baubichon D, Lallement G (2006). Early reduction of NeuN antigenicity induced by soman poisoning in mice can be used to predict delayed neuronal degeneration in the hippocampus. *Neurosci Lett* 398:337–342.

Cooper-Kuhn CM, Kuhn HG (2002). Is it all DNA repair? Methodological considerations for detecting neurogenesis in the adult brain. *Brain Res Dev Brain Res* 134:13–21.

Copani A, Uberti D, Sortino MA, Bruno V, Nicoletti F, Memo M (2001). Activation of cell-cycle-associated proteins in neuronal death: A mandatory or dispensable path? *Trends Neurosci* 24:25–31.

Corbeil D, Roper K, Fargeas CA, Joester A, Huttner WB (2001). Prominin: A story of cholesterol, plasma membrane protrusions and human pathology. *Traffic* (Copenhagen, Denmark) 2:82–91.

Corbo JC, Deuel TA, Long JM, LaPorte P, Tsai E, Wynshaw-Boris A, Walsh CA (2002). Doublecortin is required in mice for lamination of the hippocampus but not the neocortex. *J Neurosci* 22:7548–7557.

Cordey M, Limacher M, Kobel S, Taylor V, Lutolf MP (2008). Enhancing the reliability and throughput of neurosphere culture on hydrogel microwell arrays. *Stem Cells* 26:2586–2594.

Corish P, Tyler-Smith C (1999). Attenuation of green fluorescent protein half-life in mammalian cells. *Protein Eng* 12:1035–1040.

Corotto FS, Henegar JA, Maruniak JA (1993). Neurogenesis persists in the subependymal layer of the adult mouse brain. *Neurosci Lett* 149:111–114.

Cortes F, Pastor N, Mateos S, Dominguez I (2003). The nature of DNA plays a role in chromosome segregation: Endoreduplication in halogen-substituted chromosomes. *DNA Repair* 2:719–726.

Couillard-Despres S, Winner B, Schaubeck S, Aigner R, Vroemen M, Weidner N, Bogdahn U, Winkler J, Kuhn HG, Aigner L (2005). Doublecortin expression levels in adult brain reflect neurogenesis. *Eur J Neurosci* 21:1–14.

Couillard-Despres S, Winner B, Karl C, Lindemann G, Schmid P, Aigner R, Laemke J, Bogdahn U, Winkler J, Bischofberger J, Aigner L (2006). Targeted transgene expression in neuronal precursors: Watching young neurons in the old brain. *Eur J Neurosci* 24:1535–1545.

Couillard-Despres S, Finkl R, Winner B, Ploetz S, Wiedermann D, Aigner R, Bogdahn U, Winkler J, Hoehn M, Aigner L (2008). In vivo optical imaging of neurogenesis: Watching new neurons in the intact brain. *Mol Imaging* 7:28–34.

Crusio WE (1996). Gene-targeting studies: New methods, old problems. *Trends Neurosci* 19:186–187; discussion 188–189.

D'Amour KA, Gage FH (2003). Genetic and functional differences between multipotent neural and pluripotent embryonic stem cells. *Proc Natl Acad Sci USA* 100 Suppl 1:11866–11872.

Dayer AG, Cleaver KM, Abouantoun T, Cameron HA (2005). New GABAergic interneurons in the adult neocortex and striatum are generated from different precursors. *J Cell Biol* 168:415–427.

De Marchis S, Fasolo A, Shipley M, Puche A (2001). Unique neuronal tracers show migration and differentiation of SVZ progenitors in organotypic slices. *J Neurobiol* 49:326–338.

Dellarole A, Grilli M (2008). Adult dorsal root ganglia sensory neurons express the early neuronal fate marker doublecortin. *J Comp Neurol* 511:318–328.

Dityatev A, Dityateva G, Sytnyk V, Delling M, Toni N, Nikonenko I, Muller D, Schachner M (2004). Polysialylated neural cell adhesion molecule promotes remodeling and formation of hippocampal synapses. *J Neurosci* 24:9372–9382.

Doetsch F, Caille I, Lim DA, Garcia-Verdugo JM, Alvarez-Buylla A (1999). Subventricular zone astrocytes are neural stem cells in the adult mammalian brain. *Cell* 97:703–716.

Doetsch F, Petreanu L, Caille I, Garcia-Verdugo JM, Alvarez-Buylla A (2002). EGF converts transit-amplifying neurogenic precursors in the adult brain into multipotent stem cells. *Neuron* 36:1021–1034.

Dolbeare F (1995). Bromodeoxyuridine: A diagnostic tool in biology and medicine. Part I: Historical perspectives, histochemical methods and cell kinetics. *Histochem J* 27:339–369.

Dou CL, Levine JM (1994). Inhibition of neurite growth by the NG2 chondroitin sulfate proteoglycan. *J Neurosci* 14:7616–7628.

Dubreuil V, Marzesco AM, Corbeil D, Huttner WB, Wilsch-Brauninger M (2007). Midbody and primary cilium of neural progenitors release extracellular membrane particles enriched in the stem cell marker prominin-1. *J Cell Biol* 176:483–495.

Dunn DB, Smith JD, Zamenhof S, Griboff G (1954). Incorporation of halogenated pyrimidines into the deoxyribonucleic acids of *Bacterium coli* and its bacteriophages. *Nature* 174:305–307.

Dupret D, Revest JM, Koehl M, Ichas F, De Giorgi F, Costet P, Abrous DN, Piazza PV (2008). Spatial relational memory requires hippocampal adult neurogenesis. *PLoS ONE* 3:e1959.

Eckhardt M, Bukalo O, Chazal G, Wang L, Goridis C, Schachner M, Gerardy-Schahn R, Cremer H, Dityatev A (2000). Mice deficient in the polysialyltransferase ST8SiaIV/PST-1 allow discrimination of the roles of neural cell adhesion molecule protein and polysialic acid in neural development and synaptic plasticity. *J Neurosci* 20:5234–5244.

Eisch AJ, Mandyam CD (2007). Adult neurogenesis: Can analysis of cell cycle proteins move us "Beyond BrdU"? *Curr Pharm Biotechnol* 8:147–165.

Ellis P, Fagan BM, Magness ST, Hutton S, Taranova O, Hayashi S, McMahon A, Rao M, Pevny L (2004). Sox2, a persistent marker for multipotential neural stem cells derived from embryonic stem cells, the embryo or the adult. *Dev Neurosci* 26:148–165.

Encinas JM, Vaahtokari A, Enikolopov G (2006). Fluoxetine targets early progenitor cells in the adult brain. *Proc Natl Acad Sci USA* 103:8233–8238.

Endl E, Gerdes J (2000). The Ki-67 protein: Fascinating forms and an unknown function. *Exp Cell Res* 257: 231–237.

Eriksson PS, Perfilieva E, Bjork-Eriksson T, Alborn AM, Nordborg C, Peterson DA, Gage FH (1998). Neurogenesis in the adult human hippocampus. *Nat Med* 4:1313–1317.

Esposito MS, Piatti VC, Laplagne DA, Morgenstern NA, Ferrari CC, Pitossi FJ, Schinder AF (2005). Neuronal differentiation in the adult hippocampus recapitulates embryonic development. *J Neurosci* 25:10074–10086.

Fahrner A, Kann G, Flubacher A, Heinrich C, Freiman TM, Zentner J, Frotscher M, Haas CA (2007). Granule cell dispersion is not accompanied by enhanced neurogenesis in temporal lobe epilepsy patients. *Exp Neurol* 203:320–332.

Fang X, Burg MA, Barritt D, Dahlin-Huppe K, Nishiyama A, Stallcup WB (1999). Cytoskeletal reorganization induced by engagement of the NG2 proteoglycan leads to cell spreading and migration. *Mol Biol Cell* 10: 3373–3387.

Filippov V, Kronenberg G, Pivneva T, Reuter K, Steiner B, Wang LP, Yamaguchi M, Kettenmann H, Kempermann G (2003). Subpopulation of nestin-expressing progenitor cells in the adult murine hippocampus shows electrophysiological and morphological characteristics of astrocytes. *Mol Cell Neurosci* 23:373–382.

Friedkin M, Wood HI (1956) Utilization of thymidine-C14 by bone marrow cells and isolated thymus nuclei. J Biol Chem 220:639–651.

Friocourt G, Koulakoff A, Chafey P, Boucher D, Fauchereau F, Chelly J, Francis F (2003). Doublecortin functions at the extremities of growing neuronal processes. *Cereb Cortex* 13:620–626.

Fukuda S, Kato F, Tozuka Y, Yamaguchi M, Miyamoto Y, Hisatsune T (2003). Two distinct subpopulations of nestin-positive cells in adult mouse dentate gyrus. *J Neurosci* 23:9357–9366.

Gabay L, Lowell S, Rubin LL, Anderson DJ (2003). Deregulation of dorsoventral patterning by FGF confers trilineage differentiation capacity on CNS stem cells in vitro. *Neuron* 40:485–499.

Gage FH, Coates PW, Palmer TD, Kuhn HG, Fisher LJ, Suhonen JO, Peterson DA, Suhr ST, Ray J (1995). Survival and differentiation of adult neuronal progenitor cells transplanted to the adult brain. *Proc Natl Acad Sci USA* 92:11879–11883.

Gallagher EJ (1994). No proof of a difference is not equivalent to proof of no difference. *J Emerg Med* 12: 525–527.

Gallagher EJ (1999). p < 0.05: Threshold for decerebrate genuflection. *Acad Emerg Med* 6:1084–1087.

Gallo V, Mangin JM, Kukley M, Dietrich D (2008). Synapses on NG2-expressing progenitors in the brain: Multiple functions? *J Physiol* 586:3767–3781.

Gantt KR, Jain RG, Dudek RW, Pekala PH (2004). HuB localizes to polysomes and alters C/EBP-beta expression in 3T3-L1 adipocytes. *Biochem Biophys Res Commun* 313:619–622.

Garcia A, Steiner B, Kronenberg G, Bick-Sander A, Kempermann G (2004a). Age-dependent expression of glucocorticoid- and mineralocorticoid receptors on neural precursor cell populations in the adult murine hippocampus. *Aging Cell* 3:363–371.

Garcia AD, Doan NB, Imura T, Bush TG, Sofroniew MV (2004b). GFAP-expressing progenitors are the principal source of constitutive neurogenesis in adult mouse forebrain. *Nat Neurosci* 7:1233–1241.

Gardner MJ, Altman DG (1986). Confidence intervals rather than P-values: Estimation rather than hypothesis testing. *BMJ* (Clinical research ed.) 292:746–750.

Gartner U, Bruckner MK, Krug S, Schmetsdorf S, Staufenbiel M, Arendt T (2003). Amyloid deposition in APP23 mice is associated with the expression of cyclins in astrocytes but not in neurons. *Acta Neuropathol* 106:535–544.

Gascon E, Vutskits L, Kiss JZ (2008). The Role of PSA-NCAM in adult neurogenesis. Adv *Exp Med Biol* 663: 127–36.

Ge S, Goh EL, Sailor KA, Kitabatake Y, Ming GL, Song H (2006). GABA regulates synaptic integration of newly generated neurons in the adult brain. *Nature* 439:589–593.

Gensert JM, Goldman JE (1996). In vivo characterization of endogenous proliferating cells in adult rat subcortical white matter. *Glia* 17:39–51.

Gensert JM, Goldman JE (1997). Endogenous progenitors remyelinate demyelinated axons in the adult CNS. *Neuron* 19:197–203.

Geraerts M, Eggermont K, Hernandez-Acosta P, Garcia-Verdugo JM, Baekelandt V, Debyser Z (2006). Lentiviral vectors mediate efficient and stable gene transfer in adult neural stem cells in vivo. *Hum Gene Ther* 17: 635–650.

Gerlai R, Clayton NS (1999). Analysing hippocampal function in transgenic mice: An ethological perspective. *Trends Neurosci* 22:47–51.

Gigerenzer G, Krauss S, Vitouch O (2004). The Null Ritual: What you always wanted to know about significance testing but were afraid to ask. In: *The Sage Handbook of Quantitative Methodology for the Social Sciences* (Kaplan D, ed.), pp. 391–408. Thousand Oaks, California: Sage.

Gleeson JG, Allen KM, Fox JW, Lamperti ED, Berkovic S, Scheffer I, Cooper EC, Dobyns WB, Minnerath SR, Ross ME, Walsh CA (1998). Doublecortin, a brain-specific gene mutated in human X-linked lissencephaly and double cortex syndrome, encodes a putative signaling protein. *Cell* 92:63–72.

Gleeson JG, Lin PT, Flanagan LA, Walsh CA (1999a). Doublecortin is a microtubule-associated protein and is expressed widely by migrating neurons. *Neuron* 23:257–271.

Gleeson JG et al. (1999b). Characterization of mutations in the gene doublecortin in patients with double cortex syndrome. *Ann Neurol* 45:146–153.

Gomez-Climent MA, Castillo-Gomez E, Varea E, Guirado R, Blasco-Ibanez JM, Crespo C, Martinez-Guijarro FJ, Nacher J (2008). A population of prenatally generated cells in the rat paleocortex maintains an immature neuronal phenotype into adulthood. *Cereb Cortex* 18:2229–2240.

Gonchoroff NJ, Katzmann JA, Currie RM, Evans EL, Houck DW, Kline BC, Greipp PR, Loken MR (1986). S-phase detection with an antibody to bromodeoxyuridine. Role of DNase pretreatment. *J Immunol Methods* 93:97–101.

Goodman S (2003). Commentary: The P-value, devalued. *Int J Epidemiol* 32:699–702.

Gould E (2007). How widespread is adult neurogenesis in mammals? *Nat Rev Neurosci* 8:481–488.

Grako KA, Ochiya T, Barritt D, Nishiyama A, Stallcup WB (1999). PDGF (alpha)-receptor is unresponsive to PDGF-AA in aortic smooth muscle cells from the NG2 knockout mouse. *J Cell Sci* 112 (Pt 6):905–915.

Gratzner HG (1982). Monoclonal antibody to 5-bromo- and 5-iododeoxyuridine: A new reagent for detection of DNA replication. *Science* 218:474–475.

Gritti A, Frolichsthal-Schoeller P, Galli R, Parati EA, Cova L, Pagano SF, Bjornson CR, Vescovi AL (1999). Epidermal and fibroblast growth factors behave as mitogenic regulators for a single multipotent stem cell-like population from the subventricular region of the adult mouse forebrain. *J Neurosci* 19:3287–3297.

Gu W, Brannstrom T, Wester P (2000). Cortical neurogenesis in adult rats after reversible photothrombotic stroke. *J Cereb Blood Flow Metab* 20:1166–1173.

Hack MA, Sugimori M, Lundberg C, Nakafuku M, Gotz M (2004). Regionalization and fate specification in neurospheres: The role of Olig2 and Pax6. *Mol Cell Neurosci* 25:664–678.

Hall PA, Levison DA, Woods AL, Yu CC, Kellock DB, Watkins JA, Barnes DM, Gillett CE, Camplejohn R, Dover R, et al. (1990). Proliferating cell nuclear antigen (PCNA) immunolocalization in paraffin sections: An index of cell proliferation with evidence of deregulated expression in some neoplasms. *J Pathol* 162: 285–294.

Haller H, Krauss S (2002). Misinterpretation of significance: A problem students share with their teachers? *Methods of Psychological Research* 7:1–20.

Haubensak W, Attardo A, Denk W, Huttner WB (2004). Neurons arise in the basal neuroepithelium of the early mammalian telencephalon: A major site of neurogenesis. *Proc Natl Acad Sci USA* 101:3196–3201.

Haug H (1986). History of neuromorphometry. *J Neurosci Methods* 18:1–17.

Hayes NL, Nowakowski RS (2002). Dynamics of cell proliferation in the adult dentate gyrus of two inbred strains of mice. *Brain Res Dev Brain Res* 134:77–85.

Henry RA, Hughes SM, Connor B (2007). AAV-mediated delivery of BDNF augments neurogenesis in the normal and quinolinic acid-lesioned adult rat brain. *Eur J Neurosci* 25:3513–3525.

Herrup K, Busser JC (1995). The induction of multiple cell cycle events precedes target-related neuronal death. *Development* 121:2385–2395.

Herrup K, Neve R, Ackerman SL, Copani A (2004). Divide and die: Cell cycle events as triggers of nerve cell death. *J Neurosci* 24:9232–9239.

Hioki H, Kameda H, Nakamura H, Okunomiya T, Ohira K, Nakamura K, Kuroda M, Furuta T, Kaneko T (2007). Efficient gene transduction of neurons by lentivirus with enhanced neuron-specific promoters. *Gene Ther* 14:872–882.

Hitoshi S, Tropepe V, Ekker M, van der Kooy D (2002a). Neural stem cell lineages are regionally specified, but not committed, within distinct compartments of the developing brain. *Development* 129:233–244.

Hitoshi S, Alexson T, Tropepe V, Donoviel D, Elia AJ, Nye JS, Conlon RA, Mak TW, Bernstein A, van der Kooy D (2002b). Notch pathway molecules are essential for the maintenance, but not the generation, of mammalian neural stem cells. *Genes Dev* 16:846–858.

Hodge RD, Kowalczyk TD, Wolf SA, Encinas JM, Rippey C, Enikolopov G, Kempermann G, Hevner RF (2008). Intermediate progenitors in adult hippocampal neurogenesis: Tbr2 expression and coordinate regulation of neuronal output. *J Neurosci* 28:3707–3717.

Horner PJ, Thallmair M, Gage FH (2002). Defining the NG2-expressing cell of the adult CNS. *J Neurocytol* 31:469–480.

Ide Y, Fujiyama F, Okamoto-Furuta K, Tamamaki N, Kaneko T, Hisatsune T (2008). Rapid integration of young newborn dentate gyrus granule cells in the adult hippocampal circuitry. *Eur J Neurosci* 28:2381–2392.

Ignatova TN, Kukekov VG, Laywell ED, Suslov ON, Vrionis FD, Steindler DA (2002). Human cortical glial tumors contain neural stem-like cells expressing astroglial and neuronal markers in vitro. *Glia* 39:193–206.

Jensen JB, Parmar M (2006). Strengths and limitations of the neurosphere culture system. *Mol Neurobiol* 34: 153–161.

Jessberger S, Clemenson GD, Jr., Gage FH (2007). Spontaneous fusion and nonclonal growth of adult neural stem cells. *Stem Cells* 25:871–874.

Jessberger S, Aigner S, Clemenson GD, Toni N, Lie DC, Karalay O, Overall R, Kempermann G, Gage FH (2008). Cdk5 regulates accurate maturation of newborn granule cells in the adult hippocampus. *PLoS Biol* 6:e272.

Jessberger S, Clark RE, Broadbent NJ, Clemenson GD, Jr., Consiglio A, Lie DC, Squire LR, Gage FH (2009). Dentate gyrus-specific knockdown of adult neurogenesis impairs spatial and object recognition memory in adult rats. *Learn Mem* 16:147–154.

Kaplan MS, Hinds JW (1977). Neurogenesis in the adult rat: Electron microscopic analysis of light radioautographs. *Science* 197:1092–1094.

Karl C, Couillard-Despres S, Prang P, Munding M, Kilb W, Brigadski T, Plotz S, Mages W, Luhmann H, Winkler J, Bogdahn U, Aigner L (2005). Neuronal precursor-specific activity of a human doublecortin regulatory sequence. *J Neurochem* 92:264–282.

Kaspar BK, Vissel B, Bengoechea T, Crone S, Randolph-Moore L, Muller R, Brandon EP, Schaffer D, Verma IM, Lee KF, Heinemann SF, Gage FH (2002). Adeno-associated virus effectively mediates conditional gene modification in the brain. *Proc Natl Acad Sci USA* 99:2320–2325.

Katchanov J, Harms C, Gertz K, Hauck L, Waeber C, Hirt L, Priller J, von Harsdorf R, Bruck W, Hortnagl H, Dirnagl U, Bhide PG, Endres M (2001). Mild cerebral ischemia induces loss of cyclin-dependent kinase inhibitors and activation of cell cycle machinery before delayed neuronal cell death. *J Neurosci* 21:5045–5053.

Kawaguchi A, Miyata T, Sawamoto K, Takashita N, Murayama A, Akamatsu W, Ogawa M, Okabe M, Tano Y, Goldman SA, Okano H (2001). Nestin-EGFP transgenic mice: Visualization of the self-renewal and multipotency of CNS stem cells. *Mol Cell Neurosci* 17:259–273.

Kee N, Sivalingam S, Boonstra R, Wojtowicz JM (2002). The utility of Ki-67 and BrdU as proliferative markers of adult neurogenesis. *J Neurosci Methods* 115:97–105.

Kempermann G, Kuhn HG, Gage FH (1997). More hippocampal neurons in adult mice living in an enriched environment. *Nature* 386:493–495.

Kempermann G, Jessberger S, Steiner B, Kronenberg G (2004). Milestones of neuronal development in the adult hippocampus. *Trends Neurosci* 27:447–452.

Kim EJ, Battiste J, Nakagawa Y, Johnson JE (2008). Ascl1 (Mash1) lineage cells contribute to discrete cell populations in CNS architecture. *Mol Cell Neurosci* 38:595–606.

Kim KK, Adelstein RS, Kawamoto S (2009). Identification of neuronal nuclei (NeuN) as Fox-3, a new member of the Fox-1 gene family of splicing factors. *J Biol Chem* 284:31052–31061.

Kim M, Morshead CM (2003). Distinct populations of forebrain neural stem and progenitor cells can be isolated using side-population analysis. *J Neurosci* 23:10703–10709.

Kiss JZ, Troncoso E, Djebbara Z, Vutskits L, Muller D (2001). The role of neural cell adhesion molecules in plasticity and repair. *Brain Res Brain Res Rev* 36:175–184.

Knoth R, Singec I, Ditter M, Pantazis G, Capetian P, Meyer RP, Horvat V, Volk B, Kempermann G (2010). Murine features of neurogenesis in the human hippocampus across the lifespan from 0 to 100 years. *PLoS ONE* 5:e8809.

Kokoeva MV, Yin H, Flier JS (2007). Evidence for constitutive neural cell proliferation in the adult murine hypothalamus. *J Comp Neurol* 505:209–220.

Komitova M, Eriksson PS (2004). Sox-2 is expressed by neural progenitors and astroglia in the adult rat brain. *Neurosci Lett* 369:24–27.

Komitova M, Mattsson B, Johansson BB, Eriksson PS (2005). Enriched environment increases neural stem/progenitor cell proliferation and neurogenesis in the subventricular zone of stroke-lesioned adult rats. *Stroke* 36:1278–1282.

Krämer W, Gigerenzer G (2005). How to confuse with statistics, or: The use and misuse of conditional probabilities. *Statistical Science* 20:223–230.

Kronenberg G, Reuter K, Steiner B, Brandt MD, Jessberger S, Yamaguchi M, Kempermann G (2003). Subpopulations of proliferating cells of the adult hippocampus respond differently to physiologic neurogenic stimuli. *J Comp Neurol* 467:455–463.

Kronenberg G, Bick-Sander A, Bunk E, Wolf C, Ehninger D, Kempermann G (2006). Physical exercise prevents age-related decline in precursor cell activity in the mouse dentate gyrus. *Neurobiol Aging* 27:1505–1513.

Kuan CY, Schloemer AJ, Lu A, Burns KA, Weng WL, Williams MT, Strauss KI, Vorhees CV, Flavell RA, Davis RJ, Sharp FR, Rakic P (2004). Hypoxia-ischemia induces DNA synthesis without cell proliferation in dying neurons in adult rodent brain. *J Neurosci* 24:10763–10772.

Kuhn HG, Peterson DA (2008). Detection and phenotypic characterization of adult neurogenesis. In: *Adult Neurogenesis* (Gage FH, Kempermann G, Song H, eds.), pp. 25–47. Cold Spring Harbor: Cold Spring Harbor Laboratory Press.

Kuhn HG, Dickinson-Anson H, Gage FH (1996). Neurogenesis in the dentate gyrus of the adult rat: Age-related decrease of neuronal progenitor proliferation. *J Neurosci* 16:2027–2033.

Kuhn HG, Winkler J, Kempermann G, Thal LJ, Gage FH (1997). Epidermal growth factor and fibroblast growth factor-2 have different effects on neural progenitors in the adult rat brain. *J Neurosci* 17:5820–5829.

Kukekov VG, Laywell ED, Thomas LB, Steindler DA (1997). A nestin-negative precursor cell from the adult mouse brain gives rise to neurons and glia. *Glia* 21:399–407.

Landmesser L, Dahm L, Tang JC, Rutishauser U (1990). Polysialic acid as a regulator of intramuscular nerve branching during embryonic development. *Neuron* 4:655–667.

Lange C, Huttner WB, Calegari F (2009). Cdk4/cyclinD1 overexpression in neural stem cells shortens G1, delays neurogenesis, and promotes the generation and expansion of basal progenitors. *Cell Stem Cell* 5:320–331.

Laplagne DA, Esposito MS, Piatti VC, Morgenstern NA, Zhao C, van Praag H, Gage FH, Schinder AF (2006). Functional convergence of neurons generated in the developing and adult hippocampus. *PLoS Biol* 4:e409.

Lee HG, Casadesus G, Zhu X, Castellani RJ, McShea A, Perry G, Petersen RB, Bajic V, Smith MA (2009). Cell cycle reentry mediated neurodegeneration and its treatment role in the pathogenesis of Alzheimer's disease. *Neurochem Int* 54:84–88.

Lei M (2005). The MCM complex: Its role in DNA replication and implications for cancer therapy. *Current Cancer Drug Targets* 5:365–380.

Leif RC, Stein JH, Zucker RM (2004). A short history of the initial application of anti-5-BrdU to the detection and measurement of S phase. *Cytometry A* 58:45–52.

Lendahl U, Zimmerman LB, McKay RDG (1990). CNS Stem cells express a new class of intermediate filament protein. *Cell* 60:585–595.

Leuner B, Glasper ER, Gould E (2009). Thymidine analog methods for studies of adult neurogenesis are not equally sensitive. *J Comp Neurol* 517:123–133.

Lind D, Franken S, Kappler J, Jankowski J, Schilling K (2005). Characterization of the neuronal marker NeuN as a multiply phosphorylated antigen with discrete subcellular localization. *J Neurosci Res* 79:295–302.

Liu BH, Wang X, Ma YX, Wang S (2004). CMV enhancer/human PDGF-beta promoter for neuron-specific transgene expression. *Gene Ther* 11:52–60.

Liu XS, Chopp M, Zhang XG, Zhang RL, Buller B, Hozeska-Solgot A, Gregg SR, Zhang ZG (2009). Gene profiles and electrophysiology of doublecortin-expressing cells in the subventricular zone after ischemic stroke. *J Cereb Blood Flow Metab* 29:297–307.

Lothian C, Lendahl U (1997). An evolutionarily conserved region in the second intron of the human nestin gene directs gene expression to CNS progenitor cells and to early neural crest cells. *Eur J Neurosci* 9:452–462.

Lothian C, Prakash N, Lendahl U, Wahlstrom GM (1999). Identification of both general and region-specific embryonic CNS enhancer elements in the nestin promoter. *Exp Cell Res* 248:509–519.

LoTurco J (2004). Doublecortin and a tale of two serines. *Neuron* 41:175–177.

LoTurco JJ, Bai J (2006). The multipolar stage and disruptions in neuronal migration. *Trends Neurosci* 29:407–413.

Luzzati F, Bonfanti L, Fasolo A, Peretto P (2009). DCX and PSA-NCAM expression identifies a population of neurons preferentially distributed in associative areas of different pallial derivatives and vertebrate species. *Cereb Cortex* 19:1028–1041.

Magavi S, Leavitt B, Macklis J (2000). Induction of neurogenesis in the neocortex of adult mice. *Nature* 405:951–955.

Manders EM, Stap J, Brakenhoff GJ, van Driel R, Aten JA (1992). Dynamics of three-dimensional replication patterns during the S-phase, analysed by double labelling of DNA and confocal microscopy. *J Cell Sci* 103:857–862.

Mandyam CD, Harburg GC, Eisch AJ (2007). Determination of key aspects of precursor cell proliferation, cell cycle length and kinetics in the adult mouse subgranular zone. *Neuroscience* 146:108–122.

Manent JB, Wang Y, Chang Y, Paramasivam M, LoTurco JJ (2009). DCX reexpression reduces subcortical band heterotopia and seizure threshold in an animal model of neuronal migration disorder. *Nat Med* 15:84–90.

Marshall GP, 2nd, Laywell ED, Zheng T, Steindler DA, Scott EW (2006). In vitro-derived "neural stem cells" function as neural progenitors without the capacity for self-renewal. *Stem Cells* 24:731–738.

Marshall GP, 2nd, Reynolds BA, Laywell ED (2007). Using the neurosphere assay to quantify neural stem cells in vivo. *Curr Pharm Biotechnol* 8:141–145.

Maslov AY, Barone TA, Plunkett RJ, Pruitt SC (2004). Neural stem cell detection, characterization, and age-related changes in the subventricular zone of mice. *J Neurosci* 24:1726–1733.

Mazzetti S, Ortino B, Inverardi F, Frassoni C, Amadeo A (2007). PSA-NCAM in the developing and mature thalamus. *Brain Res Bull* 71:578–586.

McKeever PE, Ross DA, Strawderman MS, Brunberg JA, Greenberg HS, Junck L (1997). A comparison of the predictive power for survival in gliomas provided by MIB-1, bromodeoxyuridine and proliferating cell nuclear antigen with histopathologic and clinical parameters. *J Neuropathol Exp Neurol* 56:798–805.

Merkle FT, Mirzadeh Z, Alvarez-Buylla A (2007). Mosaic organization of neural stem cells in the adult brain. *Science* 317:381–384.

Messier B, Leblond CP, Smart I (1958). Presence of DNA synthesis and mitosis in the brain of young adult mice. *Exp Cell Res* 14:224–226.

Meyer G, Perez-Garcia CG, Gleeson JG (2002). Selective expression of doublecortin and LIS1 in developing human cortex suggests unique modes of neuronal movement. *Cereb Cortex* 12:1225–1236.

Mignone JL, Kukekov V, Chiang AS, Steindler D, Enikolopov G (2004). Neural stem and progenitor cells in nestin-GFP transgenic mice. *J Comp Neurol* 469:311–324.

Miller MW, Nowakowski RS (1988). Use of bromodeoxyuridine-immunohistochemistry to examine the proliferation, migration and time of origin of cells in the central nervous system. *Brain Res* 457:44–52.

Miller RH (1999). Contact with central nervous system myelin inhibits oligodendrocyte progenitor maturation. *Dev Biol* 216:359–368.

Minturn JE, Geschwind DH, Fryer HJ, Hockfield S (1995). Early post-mitotic neurons transiently express TOAD-64, a neural specific protein. *J Comp Neurol* 355:369–379.

Mirzadeh Z, Merkle FT, Soriano-Navarro M, Garcia-Verdugo JM, Alvarez-Buylla A (2008). Neural stem cells confer unique pinwheel architecture to the ventricular surface in neurogenic regions of the adult brain. *Cell Stem Cell* 3:265–278.

Miyagi S, Masui S, Niwa H, Saito T, Shimazaki T, Okano H, Nishimoto M, Muramatsu M, Iwama A, Okuda A (2008). Consequence of the loss of Sox2 in the developing brain of the mouse. *FEBS Lett* 582: 2811–2815.

Mizumatsu S, Monje ML, Morhardt DR, Rola R, Palmer TD, Fike JR (2003). Extreme sensitivity of adult neurogenesis to low doses of X-irradiation. *Cancer Res* 63:4021–4027.

Mokry J, Karbanova J, Filip S (2005). Differentiation potential of murine neural stem cells in vitro and after transplantation. *Transplant Proc* 37:268–272.

Mongiat LA, Esposito MS, Lombardi G, Schinder AF (2009). Reliable activation of immature neurons in the adult hippocampus. *PLoS ONE* 4:e5320.

Moores CA, Perderiset M, Francis F, Chelly J, Houdusse A, Milligan RA (2004). Mechanism of microtubule stabilization by doublecortin. *Mol Cell* 14:833–839.

Morshead CM, Reynolds BA, Craig CG, McBurney MW, Staines WA, Morassutti D, Weiss S, van der Kooy D (1994). Neural stem cells in the adult mammalian forebrain: A relatively quiescent subpopulation of subependymal cells. *Neuron* 13:1071–1082.

Morshead CM, Garcia AD, Sofroniew MV, van Der Kooy D (2003). The ablation of glial fibrillary acidic protein-positive cells from the adult central nervous system results in the loss of forebrain neural stem cells but not retinal stem cells. *Eur J Neurosci* 18:76–84.

Mullen RJ, Buck CR, Smith AM (1992). NeuN, a neuronal specific nuclear protein in vertebrates. *Development* 116:201–211.

Muller D, Wang C, Skibo G, Toni N, Cremer H, Calaora V, Rougon G, Kiss JZ (1996). PSA-NCAM is required for activity-induced synaptic plasticity. *Neuron* 17:413–422.

Muller D, Djebbara-Hannas Z, Jourdain P, Vutskits L, Durbec P, Rougon G, Kiss JZ (2000). Brain-derived neurotrophic factor restores long-term potentiation in polysialic acid-neural cell adhesion molecule-deficient hippocampus. *Proc Natl Acad Sci USA* 97:4315–4320.

Muller D, Mendez P, De Roo M, Klauser P, Steen S, Poglia L (2008). Role of NCAM in spine dynamics and synaptogenesis. *Neurochem Res*.

Nacher J, Crespo C, McEwen BS (2001). Doublecortin expression in the adult rat telencephalon. *Eur J Neurosci* 14:629–644.

Nacher J, Alonso-Llosa G, Rosell D, McEwen B (2002). PSA-NCAM expression in the piriform cortex of the adult rat. Modulation by NMDA receptor antagonist administration. *Brain Res* 927:111–121.

Nacher J, Pham K, Gil-Fernandez V, McEwen BS (2004). Chronic restraint stress and chronic corticosterone treatment modulate differentially the expression of molecules related to structural plasticity in the adult rat piriform cortex. *Neuroscience* 126:503–509.

Nakagawa S, Cuthill IC (2007). Effect size, confidence interval and statistical significance: A practical guide for biologists. *Biol Rev Camb Philos Soc* 82:591–605.

Nakatomi H, Kuriu T, Okabe S, Yamamoto S, Hatano O, Kawahara N, Tamura A, Kirino T, Nakafuku M (2002). Regeneration of hippocampal pyramidal neurons after ischemic brain injury by recruitment of endogenous neural progenitors. *Cell* 110:429–441.

Naldini L, Blomer U, Gallay P, Ory D, Mulligan R, Gage FH, Verma IM, Trono D (1996). In vivo gene delivery and stable transduction of nondividing cells by a lentiviral vector [see comments]. *Science* 272:263–267.

Neve RL, McPhie DL, Chen Y (2000). Alzheimer's disease: A dysfunction of the amyloid precursor protein(1). *Brain Res* 886:54–66.

Ninkovic J, Mori T, Gotz M (2007). Distinct modes of neuron addition in adult mouse neurogenesis. *J Neurosci* 27:10906–10911.

Nolte C, Matyash M, Pivneva T, Schipke CG, Ohlemeyer C, Hanisch UK, Kirchhoff F, Kettenmann H (2001). GFAP promoter-controlled EGFP-expressing transgenic mice: A tool to visualize astrocytes and astrogliosis in living brain tissue. *Glia* 33:72–86.

Nowakowski RS, Lewin SB, Miller MW (1989). Bromodeoxyuridine immunohistochemical determination of the lengths of the cell cycle and the DNA-synthetic phase for an anatomically defined population. *J Neurocytol* 18:311–318.

Ormerod BK, Galea LA (2001). Reproductive status influences cell proliferation and cell survival in the dentate gyrus of adult female meadow voles: A possible regulatory role for estradiol. *Neuroscience* 102:369–379.

Ostenfeld T, Joly E, Tai YT, Peters A, Caldwell M, Jauniaux E, Svendsen CN (2002). Regional specification of rodent and human neurospheres. *Brain Res Dev Brain Res* 134:43–55.

Overstreet LS, Hentges ST, Bumaschny VF, de Souza FS, Smart JL, Santangelo AM, Low MJ, Westbrook GL, Rubinstein M (2004). A transgenic marker for newly born granule cells in dentate gyrus. *J Neurosci* 24:3251–3259.

Ozerdem U, Stallcup WB (2003). Early contribution of pericytes to angiogenic sprouting and tube formation. *Angiogenesis* 6:241–249.

Ozerdem U, Stallcup WB (2004). Pathological angiogenesis is reduced by targeting pericytes via the NG2 proteoglycan. *Angiogenesis* 7:269–276.

Packard DS, Jr., Menzies RA, Skalko RG (1973). Incorportaiton of thymidine and its analogue, bromodeoxyuridine, into embryos and maternal tissues of the mouse. *Differentiation* 1:397–404.

Pagani F, Lauro C, Fucile S, Catalano M, Limatola C, Eusebi F, Grassi F (2006). Functional properties of neurons derived from fetal mouse neurospheres are compatible with those of neuronal precursors in vivo. *J Neurosci Res* 83:1494–1501.

Palmer TD, Ray J, Gage FH (1995). FGF-2-responsive neuronal progenitors reside in proliferative and quiescent regions of the adult rodent brain. *Mol Cell Neurosci* 6:474–486.

Palmer TD, Markakis EA, Willhoite AR, Safar F, Gage FH (1999). Fibroblast growth factor-2 activates a latent neurogenic program in neural stem cells from diverse regions of the adult CNS. *J Neurosci* 19:8487–8497.

Palmer TD, Willhoite AR, Gage FH (2000). Vascular niche for adult hippocampal neurogenesis. *J Comp Neurol* 425:479–494.

Parent JM, von dem Bussche N, Lowenstein DH (2006). Prolonged seizures recruit caudal subventricular zone glial progenitors into the injured hippocampus. *Hippocampus* 16:321–328.

Parmar M, Skogh C, Bjorklund A, Campbell K (2002). Regional specification of neurosphere cultures derived from subregions of the embryonic telencephalon. *Mol Cell Neurosci* 21:645–656.

Parmar M, Sjoberg A, Bjorklund A, Kokaia Z (2003). Phenotypic and molecular identity of cells in the adult subventricular zone. in vivo and after expansion in vitro. *Mol Cell Neurosci* 24:741–752.

Perez-Canellas MM, Garcia-Verdugo JM (1996). Adult neurogenesis in the telencephalon of a lizard: A [3H] thymidine autoradiographic and bromodeoxyuridine immunocytochemical study. *Brain Res Dev Brain Res* 93:49–61.

Peterson DA (1999). Quantitative histology using confocal microscopy: Implementation of unbiased stereology procedures. *Methods* 18:493–507.

Peterson DA (2004). The use of fluorescent probes in cell-counting procedures. In: *Quantitative Methods in Neuroscience—A Neuroanatomical Approach* (Evans ME, Janson AM, Nyengaard JR, eds.), pp. 85–115. Oxford: Oxford University Press.

Peterson DA, Leppert JT, Lee KF, Gage FH (1997). Basal forebrain neuronal loss in mice lacking neurotrophin receptor p75. *Science* 277:837–839.

Petrus DS, Fabel K, Kronenberg G, Winter C, Steiner B, Kempermann G (2008). NMDA and benzodiazepine receptors have synergistic and antagonistic effects on precursor cells in adult hippocampal neurogenesis. *Eur J Neurosci*: 29:244–52.

Platel JC, Heintz T, Young S, Gordon V, Bordey A (2008). Tonic activation of GLUK5 kainate receptors decreases neuroblast migration in whole-mounts of the subventricular zone. *J Physiol* 586:3783–3793.

Plumpe T, Ehninger D, Steiner B, Klempin F, Jessberger S, Brandt M, Romer B, Rodriguez GR, Kronenberg G, Kempermann G (2006). Variability of doublecortin-associated dendrite maturation in adult hippocampal neurogenesis is independent of the regulation of precursor cell proliferation. *BMC Neurosci* 7:77.

Poulsen FR, Blaabjerg M, Montero M, Zimmer J (2005). Glutamate receptor antagonists and growth factors modulate dentate granule cell neurogenesis in organotypic, rat hippocampal slice cultures. *Brain Res* 1051: 35–49.

Quinn CC, Gray GE, Hockfield S (1999). A family of proteins implicated in axon guidance and outgrowth. *J Neurobiol* 41:158–164.

Raineteau O, Rietschin L, Gradwohl G, Guillemot F, Gahwiler BH (2004). Neurogenesis in hippocampal slice cultures. *Mol Cell Neurosci* 26:241–250.

Raineteau O, Hugel S, Ozen I, Rietschin L, Sigrist M, Arber S, Gahwiler BH (2006). Conditional labeling of newborn granule cells to visualize their integration into established circuits in hippocampal slice cultures. *Mol Cell Neurosci* 32:344–355.

Rakic P (2002). Neurogenesis in adult primate neocortex: An evaluation of the evidence. *Nat Rev Neurosci* 3: 65–71.

Ramalho-Santos M, Yoon S, Matsuzaki Y, Mulligan RC, Melton DA (2002). "Stemness": Transcriptional profiling of embryonic and adult stem cells. *Science* 298:597–600.

Ramos RL, Bai J, LoTurco JJ (2006). Heterotopia formation in rat but not mouse neocortex after RNA interference knockdown of DCX. *Cereb Cortex* 16:1323–1331.

Rao MS, Shetty AK (2004). Efficacy of doublecortin as a marker to analyse the absolute number and dendritic growth of newly generated neurons in the adult dentate gyrus. *Eur J Neurosci* 19:234–246.

Reynolds BA, Weiss S (1992). Generation of neurons and astrocytes from isolated cells of the adult mammalian central nervous system. *Science* 255:1707–1710.

Reynolds BA, Rietze RL (2005). Neural stem cells and neurospheres—re-evaluating the relationship. *Nat Methods* 2:333–336.

Reynolds BA, Tetzlaff W, Weiss S (1992). A multipotent EGF-responsive striatal embryonic progenitor cell produces neurons and astrocytes. *J Neurosci* 12:4565–4574.

Ricard D, Rogemond V, Charrier E, Aguera M, Bagnard D, Belin MF, Thomasset N, Honnorat J (2001). Isolation and expression pattern of human Unc-33–like phosphoprotein 6/collapsin response mediator protein 5 (Ulip6/CRMP5): Coexistence with Ulip2/CRMP2 in Sema3a-sensitive oligodendrocytes. *J Neurosci* 21:7203–7214.

Rietze RL, Reynolds BA (2006). Neural stem cell isolation and characterization. *Methods Enzymol* 419:3–23.

Rodriguez JJ, Jones VC, Tabuchi M, Allan SM, Knight EM, LaFerla FM, Oddo S, Verkhratsky A (2008). Impaired adult neurogenesis in the dentate gyrus of a triple transgenic mouse model of Alzheimer's disease. *PLoS ONE* 3:e2935.

Roy NS, Wang S, Harrison-Restelli C, Benraiss A, Fraser RA, Gravel M, Braun PE, Goldman SA (1999). Identification, isolation, and promoter-defined separation of mitotic oligodendrocyte progenitor cells from the adult human subcortical white matter. *J Neurosci* 19:9986–9995.

Roy NS, Benraiss A, Wang S, Fraser RA, Goodman R, Couldwell WT, Nedergaard M, Kawaguchi A, Okano H, Goldman SA (2000a). Promoter-targeted selection and isolation of neural progenitor cells from the adult human ventricular zone. *J Neurosci Res* 59:321–331.

Roy NS, Wang S, Jiang L, Kang J, Benraiss A, Harrison-Restelli C, Fraser RA, Couldwell WT, Kawaguchi A, Okano H, Nedergaard M, Goldman SA (2000b). In vitro neurogenesis by progenitor cells isolated from the adult human hippocampus. *Nat Med* 6:271–277.

Rutishauser U, Landmesser L (1996). Polysialic acid in the vertebrate nervous system: A promoter of plasticity in cell–cell interactions. *Trends Neurosci* 19:422–427.

Sadgrove MP, Chad JE, Gray WP (2005). Kainic acid induces rapid cell death followed by transiently reduced cell proliferation in the immature granule cell layer of rat organotypic hippocampal slice cultures. *Brain Res* 1035:111–119.

Sadgrove MP, Laskowski A, Gray WP (2006). Examination of granule layer cell count, cell density, and single-pulse BrdU incorporation in rat organotypic hippocampal slice cultures with respect to culture medium, septotemporal position, and time in vitro. *J Comp Neurol* 497:397–415.

Sadoul R, Hirn M, Deagostini-Bazin H, Rougon G, Goridis C (1983). Adult and embryonic mouse neural cell adhesion molecules have different binding properties. *Nature* 304:347–349.

Sano K, Sato F, Hoshino T, Nagai M (1965). Experimental and clinical studies of radiosensitizers in brain tumors, with special reference to BUdR-antimetabolite continuous regional infusion-radiation therapy (BAR therapy). *Neurol Med Chir* (Tokyo) 7:51–72.

Sano K, Hoshino T, Nagai M (1968). Radiosensitization of brain tumor cells with a thymidine analogue (bromouridine). *J Neurosurg* 28:530–538.

Santa-Olalla J, Baizabal JM, Fregoso M, del Carmen Cardenas M, Covarrubias L (2003). The in vivo positional identity gene expression code is not preserved in neural stem cells grown in culture. *Eur J Neurosci* 18: 1073–1084.

Sarnat HB, Nochlin D, Born DE (1998). Neuronal nuclear antigen (NeuN): A marker of neuronal maturation in early human fetal nervous system. *Brain Dev* 20:88–94.

Sawamoto K, Yamamoto A, Kawaguchi A, Yamaguchi M, Mori K, Goldman SA, Okano H (2001). Direct isolation of committed neuronal progenitor cells from transgenic mice coexpressing spectrally distinct fluorescent proteins regulated by stage-specific neural promoters. *J Neurosci Res* 65:220–227.

Schaar BT, Kinoshita K, McConnell SK (2004). Doublecortin microtubule affinity is regulated by a balance of kinase and phosphatase activity at the leading edge of migrating neurons. *Neuron* 41:203–213.

Schmechel DE (1985). Gamma-subunit of the glycolytic enzyme enolase: Nonspecific or neuron specific? *Lab Invest* 52:239–242.

Schmid RS, Yokota Y, Anton ES (2006). Generation and characterization of brain lipid-binding protein promoter-based transgenic mouse models for the study of radial glia. *Glia* 53:345–351.

Schmidt A, Haas SJ, Hildebrandt S, Scheibe J, Eckhoff B, Racek T, Kempermann G, Wree A, Putzer BM (2007). Selective targeting of adenoviral vectors to neural precursor cells in the hippocampus of adult mice: New prospects for in situ gene therapy. *Stem Cells* 25:2910–2918.

Scholzen T, Gerdes J (2000). The Ki-67 protein: From the known and the unknown. *J Cell Physiol* 182: 311–322.

Seaberg RM, van der Kooy D (2002). Adult rodent neurogenic regions: The ventricular subependyma contains neural stem cells, but the dentate gyrus contains restricted progenitors. *J Neurosci* 22:1784–1793.

Seaberg RM, van der Kooy D (2003). Stem and progenitor cells: The premature desertion of rigorous definitions. *Trends Neurosci* 26:125–131.

Seki T (2002a). Hippocampal adult neurogenesis occurs in a microenvironment provided by PSA-NCAM-expressing immature neurons. *J Neurosci Res* 69:772–783.

Seki T (2002b). Expression patterns of immature neuronal markers PSA-NCAM, CRMP-4 and NeuroD in the hippocampus of young adult and aged rodents. *J Neurosci Res* 70:327–334.

Seki T, Arai Y (1991a). The persistent expression of a highly polysialylated NCAM in the dentate gyrus of the adult rat. *Neurosci Res* 12:503–513.

Seki T, Arai Y (1991b). Expression of highly polysialylated NCAM in the neocortex and piriform cortex of the developing and the adult rat. *Anat Embryol* 184:395–401.

Seki T, Arai Y (1993a). Distribution and possible roles of the highly polysialylated neural cell adhesion molecule (NCAM-H) in the developing and adult central nervous system. *Neurosci Res* 17:265–290.

Seki T, Arai Y (1993b). Highly polysialylated neural cell adhesion molecule (NCAM-H) is expressed by newly generated granule cells in the dentate gyrus of the adult rat. *J Neurosci* 13:2351–2358.

Seki T, Rutishauser U (1998). Removal of polysialic acid-neural cell adhesion molecule induces aberrant mossy fiber innervation and ectopic synaptogenesis in the hippocampus. *J Neurosci* 18:3757–3766.

Seki T, Arai Y (1999). Temporal and spacial relationships between PSA-NCAM-expressing, newly generated granule cells, and radial glia–like cells in the adult dentate gyrus. *J Comp Neurol* 410:503–513.

Seri B, Garcia-Verdugo JM, McEwen BS, Alvarez-Buylla A (2001). Astrocytes give rise to new neurons in the adult mammalian hippocampus. *J Neurosci* 21:7153–7160.

Seri B, Garcia-Verdugo JM, Collado-Morente L, McEwen BS, Alvarez-Buylla A (2004). Cell types, lineage, and architecture of the germinal zone in the adult dentate gyrus. *J Comp Neurol* 478:359.

Shi Y, Chichung Lie D, Taupin P, Nakashima K, Ray J, Yu RT, Gage FH, Evans RM (2004). Expression and function of orphan nuclear receptor TLX in adult neural stem cells. *Nature* 427:78–83.

Singec I, Knoth R, Meyer RP, Maciaczyk J, Volk B, Nikkhah G, Frotscher M, Snyder EY (2006). Defining the actual sensitivity and specificity of the neurosphere assay in stem cell biology. *Nat Methods* 3:801–806.

Sivilia S, Giuliani A, Del Vecchio G, Giardino L, Calza L (2008). Age-dependent impairment of hippocampal neurogenesis in chronic cerebral hypoperfusion. *Neuropathol Appl Neurobiol* 34:52–61.

Skaland I, Janssen EA, Gudlaugsson E, Klos J, Kjellevold KH, Soiland H, Baak JP (2007). Phosphohistone H3 expression has much stronger prognostic value than classical prognosticators in invasive lymph node-negative breast cancer patients less than 55 years of age. *Mod Pathol* 20:1307–1315.

Song HJ, Stevens CF, Gage FH (2002). Neural stem cells from adult hippocampus develop essential properties of functional CNS neurons. *Nat Neurosci* 5:438–445.

Spassky N, Merkle FT, Flames N, Tramontin AD, Garcia-Verdugo JM, Alvarez-Buylla A (2005). Adult ependymal cells are post-mitotic and are derived from radial glial cells during embryogenesis. *J Neurosci* 25: 10–18.

Stallcup WB (1981). The NG2 antigen, a putative lineage marker: Immunofluorescent localization in primary cultures of rat brain. *Dev Biol* 83:154–165.

Starborg M, Gell K, Brundell E, Hoog C (1996). The murine Ki-67 cell proliferation antigen accumulates in the nucleolar and heterochromatic regions of interphase cells and at the periphery of the mitotic chromosomes in a process essential for cell cycle progression. *J Cell Sci* 109 (Pt 1):143–153.

Stegmuller J, Werner H, Nave KA, Trotter J (2003). The proteoglycan NG2 is complexed with alpha-amino-3-hydroxy-5-methyl-4-isoxazolepropionic acid (AMPA) receptors by the PDZ glutamate receptor interaction protein (GRIP) in glial progenitor cells. Implications for glial-neuronal signaling. *J Biol Chem* 278:3590–3598.

Steiner B, Klempin F, Wang L, Kott M, Kettenmann H, Kempermann G (2006). Type-2 cells as link between glial and neuronal lineage in adult hippocampal neurogenesis. *Glia* 54:805–814.

Studer L, Csete M, Lee SH, Kabbani N, Walikonis J, Wold B, McKay R (2000). Enhanced proliferation, survival, and dopaminergic differentiation of CNS precursors in lowered oxygen. *J Neurosci* 20:7377–7383.

Suh H, Consiglio A, Ray J, Sawai T, D'Amour KA, Gage FH (2007). In vivo fate analysis reveals the multipotent and self-renewal capacities of Sox2+ neural stem cells in the adult hippocampus. *Cell Stem Cell* 1:515–528.

Suhonen JO, Peterson DA, Ray J, Gage FH (1996). Differentiation of adult hippocampus-derived progenitors into olfactory neurons in vivo. *Nature* 383:624–627.

Suhonen J, Ray J, Blomer U, Gage FH, Kaspar B (2006). Ex vivo and in vivo gene delivery to the brain. *Curr Protoc Hum Genet* Chapter 13: Unit 13, p. 13.

Sun W, Winseck A, Vinsant S, Park OH, Kim H, Oppenheim RW (2004). Programmed cell death of adult-generated hippocampal neurons is mediated by the proapoptotic gene Bax. *J Neurosci* 24:11205–11213.

Tamura Y, Kataoka Y, Cui Y, Takamori Y, Watanabe Y, Yamada H (2007). Multi-directional differentiation of doublecortin- and NG2-immunopositive progenitor cells in the adult rat neocortex in vivo. *Eur J Neurosci* 25:3489–3498.

Tanaka T, Serneo FF, Higgins C, Gambello MJ, Wynshaw-Boris A, Gleeson JG (2004). Lis1 and doublecortin function with dynein to mediate coupling of the nucleus to the centrosome in neuronal migration. *J Cell Biol* 165:709–721.

Tashiro A, Sandler VM, Toni N, Zhao C, Gage FH (2006). NMDA-receptor-mediated, cell-specific integration of new neurons in adult dentate gyrus. *Nature* 442:929–933.

Thomas nee Williams SA, Segal MB (1996). Identification of a saturable uptake system for deoxyribonucleosides at the blood-brain and blood-cerebrospinal fluid barriers. *Brain Res* 741:230–239.

Thomas SA, Segal MB (1997). Saturation kinetics, specificity and NBMPR sensitivity of thymidine entry into the central nervous system. *Brain Res* 760:59–67.

Tomasiewicz H, Ono K, Yee D, Thompson C, Goridis C, Rutishauser U, Magnuson T (1993). Genetic deletion of a neural cell adhesion molecule variant (N-CAM-180) produces distinct defects in the central nervous system. *Neuron* 11:1163–1174.

Tonchev AB, Yamashima T, Zhao L, Okano HJ, Okano H (2003). Proliferation of neural and neuronal progenitors after global brain ischemia in young adult macaque monkeys. *Mol Cell Neurosci* 23:292–301.

Toni N, Teng EM, Bushong EA, Aimone JB, Zhao C, Consiglio A, van Praag H, Martone ME, Ellisman MH, Gage FH (2007). Synapse formation on neurons born in the adult hippocampus. *Nat Neurosci* 10:727–734.

Toni N, Laplagne DA, Zhao C, Lombardi G, Ribak CE, Gage FH, Schinder AF (2008). Neurons born in the adult dentate gyrus form functional synapses with target cells. *Nat Neurosci* 11:901–907.

Tozuka Y, Fukuda S, Namba T, Seki T, Hisatsune T (2005). GABAergic excitation promotes neuronal differentiation in adult hippocampal progenitor cells. *Neuron* 47:803–815.

Turkheimer FE, Aston JA, Cunningham VJ (2004). On the logic of hypothesis testing in functional imaging. *Eur J Nucl Med Mol Imaging* 31:725–732.

Uchida N, Buck DW, He D, Reitsma MJ, Masek M, Phan TV, Tsukamoto AS, Gage FH, Weissman IL (2000). Direct isolation of human central nervous system stem cells. *Proc Natl Acad Sci USA* 97:14720–14725.

Ughrin YM, Chen ZJ, Levine JM (2003). Multiple regions of the NG2 proteoglycan inhibit neurite growth and induce growth cone collapse. *J Neurosci* 23:175–186.

Uittenbogaard M, Chiaramello A (2002). Constitutive overexpression of the basic helix-loop-helix Nex1/MATH-2 transcription factor promotes neuronal differentiation of PC12 cells and neurite regeneration. *J Neurosci Res* 67:235–245.

Unal-Cevik I, Kilinc M, Gursoy-Ozdemir Y, Gurer G, Dalkara T (2004). Loss of NeuN immunoreactivity after cerebral ischemia does not indicate neuronal cell loss: A cautionary note. *Brain Res* 1015:169–174.

Van der Zee CE, Ross GM, Riopelle RJ, Hagg T (1996). Survival of cholinergic forebrain neurons in developing p75NGFR-deficient mice. *Science* 274:1729–1732.

Van Nassauw L, Wu M, De Jonge F, Adriaensen D, Timmermans JP (2005). Cytoplasmic, but not nuclear, expression of the neuronal nuclei (NeuN) antibody is an exclusive feature of Dogiel type II neurons in the guinea-pig gastrointestinal tract. *Histochem Cell Biol* 124:369–377.

Van Praag H, Schinder AF, Christie BR, Toni N, Palmer TD, Gage FH (2002). Functional neurogenesis in the adult hippocampus. *Nature* 415:1030–1034.

Varea E, Nacher J, Blasco-Ibanez JM, Gomez-Climent MA, Castillo-Gomez E, Crespo C, Martinez-Guijarro FJ (2005). PSA-NCAM expression in the rat medial prefrontal cortex. *Neuroscience* 136:435–443.

Vega CJ, Peterson DA (2005). Stem cell proliferative history in tissue revealed by temporal halogenated thymidine analog discrimination. *Nat Methods* 2:167–169.

Vinores SA, Herman MM, Rubinstein LJ, Marangos PJ (1984). Electron microscopic localization of neuron-specific enolase in rat and mouse brain. *J Histochem Cytochem* 32:1295–1302.

Vinores SA, Herman MM, Hackett SF, Campochiaro PA (1993). A morphological and immunohistochemical study of human retinal pigment epithelial cells, retinal glia, and fibroblasts grown on Gelfoam matrix in an organ culture system. A comparison of structural and nonstructural proteins and their application to cell type identification. *Graefes Arch Clin Exp Ophthalmol* 231:279–288.

Walker TL, White A, Black DM, Wallace RH, Sah P, Bartlett PF (2008). Latent stem and progenitor cells in the hippocampus are activated by neural excitation. *J Neurosci* 28:5240–5247.

Wang C, Pralong WF, Schulz MF, Rougon G, Aubry JM, Pagliusi S, Robert A, Kiss JZ (1996). Functional N-methyl-D-aspartate receptors in O-2A glial precursor cells: A critical role in regulating polysialic acid-neural cell adhesion molecule expression and cell migration. *J Cell Biol* 135:1565–1581.

Wang DD, Krueger DD, Bordey A (2003). GABA depolarizes neuronal progenitors of the postnatal subventricular zone via GABAA receptor activation. *J Physiol* 550:785–800.

Wang LP, Kempermann G, Kettenmann H (2005). A subpopulation of precursor cells in the mouse dentate gyrus receives synaptic GABAergic input. *Mol Cell Neurosci* 29:181–189.

Wang S, Scott BW, Wojtowicz JM (2000). Heterogenous properties of dentate granule neurons in the adult rat. *J Neurobiol* 42:248–257.

Warren M, Puskarczyk K, Chapman SC (2009). Chick embryo proliferation studies using EdU labeling. *Dev Dyn* 238:944–949.

Weigmann A, Corbeil D, Hellwig A, Huttner WB (1997). Prominin, a novel microvilli-specific polytopic membrane protein of the apical surface of epithelial cells, is targeted to plasmalemmal protrusions of non-epithelial cells. *Proc Natl Acad Sci USA* 94:12425–12430.

Wen PH, Friedrich VL, Jr., Shioi J, Robakis NK, Elder GA (2002). Presenilin-1 is expressed in neural progenitor cells in the hippocampus of adult mice. *Neurosci Lett* 318:53–56.

Wilhelmsson U, Bushong EA, Price DL, Smarr BL, Phung V, Terada M, Ellisman MH, Pekny M (2006). Redefining the concept of reactive astrocytes as cells that remain within their unique domains upon reaction to injury. *Proc Natl Acad Sci USA* 103:17513–17518.

Wolf HK, Buslei R, Schmidt-Kastner R, Schmidt-Kastner PK, Pietsch T, Wiestler OD, Bluhmke I (1996). NeuN: A useful neuronal marker for diagnostic histopathology. *J Histochem Cytochem* 44:1167–1171.

Wolfer DP, Lipp HP (2000). Dissecting the behaviour of transgenic mice: Is it the mutation, the genetic background, or the environment? *Exp Physiol* 85:627–634.

Wolfer DP, Crusio WE, Lipp HP (2002). Knockout mice: Simple solutions to the problems of genetic background and flanking genes. *Trends Neurosci* 25:336–340.

Xiong K, Luo DW, Patrylo PR, Luo XG, Struble RG, Clough RW, Yan XX (2008). Doublecortin-expressing cells are present in Layer II across the adult guinea pig cerebral cortex: Partial colocalization with mature interneuron markers. *Exp Neurol* 211:271–282.

Yamaguchi M, Saito H, Suzuki M, Mori K (2000). Visualization of neurogenesis in the central nervous system using nestin promoter-GFP transgenic mice. *Neuroreport* 11:1991–1996.

Yokochi T, Gilbert DM (2007). Replication labeling with halogenated thymidine analogs. *Curr Protoc Cell Biol* Chapter 22: Unit 22, p. 10.

Young KD, Lewis RJ (1997). What is confidence? Part 1: The use and interpretation of confidence intervals. *Ann Emerg Med* 30:307–310.

Zamenhof S, Gribiff G (1954). *E. coli* containing 5-bromouracil in its deoxyribonucleic acid. *Nature* 174: 307–308.

Zhang RL, LeTourneau Y, Gregg SR, Wang Y, Toh Y, Robin AM, Zhang ZG, Chopp M (2007). Neuroblast division during migration toward the ischemic striatum: A study of dynamic migratory and proliferative characteristics of neuroblasts from the subventricular zone. *J Neurosci* 27:3157–3162.

Zhao C, Teng EM, Summers RG, Jr., Ming GL, Gage FH (2006). Distinct morphological stages of dentate granule neuron maturation in the adult mouse hippocampus. *J Neurosci* 26:3–11.

8

Neurogenic and Non-neurogenic Regions

The adult rodent and primate brains have two "canonical" neurogenic regions in which adult neurogenesis occurs

THE NEUROGENIC REGIONS of the adult brain are characterized by the presence of a germinative matrix with a significant production of new neurons throughout life. As far as we know, in the adult rodent and primate brain, only the subventricular zone (SVZ)/olfactory bulb and the subgranular zone (SGZ) in the hippocampus qualify as neurogenic regions according to this definition. Consequently, the remainder of the brain would be the non-neurogenic regions. There are, however, numerous reports on neurogenesis outside the classical neurogenic regions even in rodents and primates (Fig. 8–1). Although many of these reports have raised occasionally substantial methodological doubts (the reason for this chapter's following the one on methods), today the widely held opinion is that under particular circumstances, exceptional neurogenesis can occur in non-neurogenic regions even if it does not do so under physiological conditions (Arvidsson et al., 2002; Magavi and Macklis, 2002b; Ohira et al., 2010). In such cases the SVZ might also serve as a reservoir for neuronal precursor cells for other brain regions than the olfactory bulb, which somewhat blurs the picture. Pathology in the cortex, for example, a tumor, has been seen to divert SVZ precursor cells towards the damage and has led to a concomitant decrease in neurogenesis in the olfactory bulb although no new neurons developed at the lesion cite (Walzlein et al., 2008). Another example is the production of interneurons for cortical Layer VI, which at least persists postnatally and into adolescence (if not into full adulthood) and with characteristics sufficiently different from fetal brain development (Dayer et al., 2005; Inta et al., 2008). But it has also been suggested that "new" precursor cell populations, e.g., in cortical Layer I, may underlie a neurogenic response of the neocortex to ischemia (Ohira et al., 2010).

The conventional distinction of neurogenic versus non-neurogenic regions refers to physiological conditions

The claim that neurogenesis happens in non-neurogenic regions, however, seems to be a contradiction in terms. Consequently, some researchers refute the distinction between neurogenic and non-neurogenic regions altogether. Elisabeth Gould has rightly argued that the history of research on adult neurogenesis is full of premature conclusions that had to be revised once better methodology became available, and it is hard to ultimately prove the absence of neurogenesis just on the basis of the current status in methodology (Gould, 2007). The fact that damage can elicit a neurogenic response where otherwise none is found suggests that in the end the adult brain might be somewhat more neurogenic than the tale of the two neurogenic regions suggests. Huge species differences also warn researchers against basing all their reasoning on a too simple concept. On the other hand, the

275

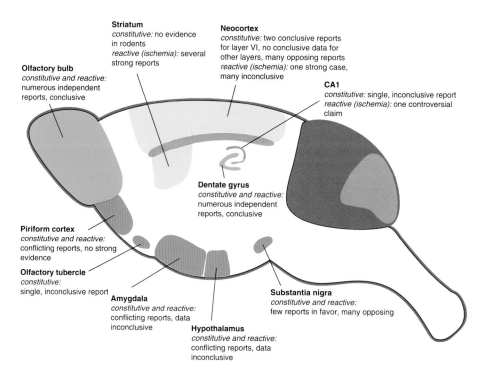

Striatum
constitutive: no evidence in rodents
reactive (ischemia): several strong reports

Neocortex
constitutive: two conclusive reports for layer VI, no conclusive data for other layers, many opposing reports
reactive (ischemia): one strong case, many inconclusive

Olfactory bulb
constitutive and reactive: numerous independent reports, conclusive

CA1
constitutive: single, inconclusive report
reactive (ischemia): one controversial claim

Dentate gyrus
constitutive and reactive: numerous independent reports, conclusive

Piriform cortex
constitutive and reactive: conflicting reports, no strong evidence

Olfactory tubercle
constitutive: single, inconclusive report

Amygdala
constitutive and reactive: conflicting reports, data inconclusive

Substantia nigra
constitutive and reactive: few reports in favor, many opposing

Hypothalamus
constitutive and reactive: conflicting reports, data inconclusive

FIGURE 8–1 Neurogenic and non-neurogenic regions in rodents. The "canonical" neurogenic regions of the adult rodent brain are the hippocampus and the olfactory bulb (*green*), in which sizable, well-characterized adult neurogenesis occurs. For several other regions, only anecdotal or very controversial evidence exists, so that adult neurogenesis in these regions cannot be taken as proven at the present time (*red*). In the neocortex, good suggestive evidence for very low levels of adult neurogenesis exists, especially in Layer VI (see Fig. 8–4) and after ischemia, but the description is not complete yet and several controversial issues remain (*yellow*). The distinction of neurogenic and non-neurogenic regions is a useful convention, based on the theoretical concept outlined in the text, but it is highly species-dependent and in some flux.

conservative and parsimonious approach that sets the physiological neurogenesis in the adult SVZ/OB and SGZ apart from the other examples has its merits and also helps prevent premature hype about biomedical implications. The concept of neurogenic permissiveness as a regional identity directs our gaze to the particular anatomical structures and molecular characteristics of the two "canonical" neurogenic regions, SVZ/OB and SGZ, that are not present in the other regions. Presence of these structures alone does not provide an iron-clad argument: adult neurogenesis might be achievable by more than one means, and already SVZ/OB and SGZ differ considerably. But SVZ/OB and SGZ also stand out because neurogenesis can be detected here almost with the "naked eye," whereas all other reported cases (if documented well) rely on exquisitely rare examples. This makes the stakes for proving a new case of adult neurogenesis extremely high: not only must a very rare event be confirmed, it must also be based on a mechanism that is distinct from the by-now-well-described processes in SVZ/OB and SGZ. This difference is not per se an argument against the existence of other cases of neurogenesis (as the example of adult neurogenesis in the zebrafish with its 16 considerably distinct neurogenic zones aptly shows: Grandel et al., 2006; Kaslin et al., 2009), but that neurogenesis outside SVZ/OB and SGZ cannot be simply confirmed by analogy.

Neurogenic Permissiveness

Neurogenic regions are characterized by precursor cells plus a neurogenically permissive microenvironment or niche

There is one litmus test for a neurogenic region in this sense: a neural precursor cell implanted in a neurogenic region should develop into a neuron, and when grafted onto a non-neurogenic region, it should become a glial cell or die. Thus the definition of "neurogenicity" is based on a general and physiological neurogenic permissiveness. It is not based on the presence or absence of neural precursor cells alone, and it is not influenced by the possibility that such a promotion could be induced under certain unusual conditions. The key question becomes, then, what makes a neurogenic region neurogenic? How is neurogenic permissiveness defined on a molecular and cellular level?

Taken together, *neurogenic zones* are defined by (1) the presence of neural precursor cells *plus* (2) the presence of a microenvironment, the niche, consisting of cell–cell contacts and diffusible factors, which promotes neuronal development from intrinsic precursor cells and whose neurogenic potential can be tested by the implantation of neural precursor cells into this region. Consequently, non-neurogenic regions may contain precursor cells but lack distinctive germinative cell clusters as well as the permissive microenvironment that under physiological conditions would promote neurogenesis from local or implanted precursor cells. In the end, the new neurons might end up outside their germinative matrix, especially if lured there as in cases of pathology. Hence a neurogenic region is somewhat less defined by the net result than by its potential and the structure that allows this potential to be realized.

Neurogenic regions vary between animal species

What is a non-neurogenic region in one species might very well be neurogenic in another. As a tendency, with increasing brain complexity along the course of evolution, the number of neurogenic sites and the extent of adult neurogenesis seems to decrease, but this idea actually rests on very sketchy experimental evidence. The olfactory system is almost always among the neurogenic zones, however. The following paragraphs refer to the mammalian brain, and within that order and with few exceptions, only to rodents and primates, including humans. We will return to the comparative and evolutionary perspectives in greater detail in Chapter 11.

The Neurogenic Niche

Stem cells and their niche form a functional unit

In 2000, Theo D. Palmer applied the niche concept of stem cell biology (Schofield, 1983) to adult neurogenesis and linked the process of adult neurogenesis to the vasculature within that niche (Fig. 8–2) (Palmer et al., 2000; Palmer, 2002). He recognized that the situation in the niche is more than just a set of particular cues affecting stem cells (Fabel et al., 2003a). He and coauthor Klaus Fabel wrote at the outset of their article: "As the instruments used to observe adult neurogenesis become more sophisticated, the concept of a discrete competent 'stem cell' has become less concrete" (Fabel et al., 2003a). They argued that the concept needs to be replaced with one that sees the stem cells in functional unity with their surroundings. We have discussed the elusive entity of adult neural stem cells in Chapter 3, p. 89. Fabel and Palmer described the hippocampus as an exemplary yet unique "cosmos of cellular interactions" resulting in neurogenesis. For them it is not "here the stem cells and there the permissive factors in the environment," but a "constellation" of various partners. In fact, from all we know, the precursor cells themselves are important niche factors (see, for example, Song et al., 2002; Filippov et al., 2003; Shapiro et al., 2005; Kunze et al., 2009). Cells defined by the potential they harbor and the permissive environment, nature and nurture, form one unit. In this view, regulation is a change in this constellation, affecting both precursor cells and their partners. The example of recent developments in the field of

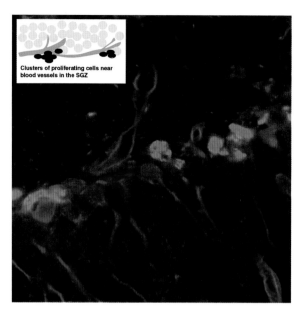

Clusters of proliferating cells near
blood vessels in the SGZ

FIGURE 8–2 The vascular niche. Precursor cell activity in the subgranular zone occurs in close proximity to blood vessels. Vasculature is an integrative part of the germinative niche (Palmer et al., 2000). Clusters of BrdU-positive cells are found in close proximity to blood vessels. Vessels in the SGZ are visualized with tomatolectin (*green*), proliferating cells with BrdU (*red*). Picture kindly provided by Klaus Fabel, Dresden.

spermatogenic stem cells shows that "stem cell systems" might be much more flexible than the theoretical concept and the knowledge for example from *Drosophila* models suggested (Yoshida, 2008).

Niches are first of all anatomical structures

At the first level of investigation strikes the hour of the anatomists. For the SVZ, Arturo Alvarez-Buylla and colleagues have over the years advanced our understanding of the three-dimensional structural complexity (Alvarez-Buylla et al., 2008). In comparison, our knowledge about the SGZ is still somewhat more limited, but here, too, an amazing complexity begins to shine through (see, for example, Palmer et al., 2000; Mercier et al., 2002; Filippov et al., 2003; Shapiro et al., 2005; Breunig et al., 2008). These cells engage in numerous interactions, employing various channels of communication. Radial glia–like precursor cells in the dentate gyrus, for example, are coupled, allowing the direct flow of information between them (Kunze et al., 2009). Besides direct contacts between the different cell types, autocrine, paracrine, and short- and long-range signaling mechanisms with numerous molecules establish a dynamic high-dimensional pattern of cues. For the stem cell niche in the aging bone marrow, a first exemplary concept has been developed that sees the issues of stem cell senescence neither in the stem cells nor in the niche, but in their ability to interact (Geiger et al., 2007). The interaction between precursor cells and the other niche elements is bi-directional. One problem with the ablation experiments that are used to study the possible functional relevance of adult-born neurons is that they disrupt this relationship, presumably doing more to the system than just removing one partner would.

The neurogenic niches resemble the microenvironment of synapses

In a review of the subject, Angelique Bordey has stressed the surprising observation that cell–cell interactions in the neurogenic niches of the adult brain have much in common with the niche-like environment of a synapse and much less with niches as we know them from development (Bordey, 2006). This applies in particular to the role of astrocytes in the niche.

If in the following paragraphs we take apart this complexity, we do so only for systematic and didactic reasons. We are only getting the first glimpses of what a neurogenic niche is. But the concept of a functional unit, however ill-defined it must be at present, is the key idea.

Cells as Niche Factors

Astrocytes and Astrocyte-like Precursor Cells

Neurogenesis is tightly linked to gliogenesis and dependent on glial cells

In the context of adult hippocampal neurogenesis, a population of astrocytes is generated: gliogenesis to some degree parallels neurogenesis (Kuhn et al., 1996; Kempermann et al., 1997a). These adult-generated astrocytes differ from the astrocyte-like cells acting as precursor cells in that they are S100β-positive, non-proliferative, and non-radial (Steiner et al., 2004). It is not clear if and how these S100β-positive astrocytes contribute to the stem cell niche. But astrocytes play a central role in germinative niches of the brain.

Precursor cells are glia-like cells

First, radial glia–like, astrocyte-like cells are the precursor cells of these regions (Doetsch, 2003). As discussed in greater detail in Chapter 3, p. 68, the use of the term *astrocytes* for both entities of cells is problematic in the absence of further qualification and is often misleading, although there is no doubt that lineage relationships exist and the cells share many features.

Astrocytes from neurogenic zones promote neurogenesis

Second, certain populations of astrocytes seem to have a region-specific potential to establish neurogenic permissiveness. Astrocytes from the dentate gyrus promote precursor cell activity in hippocampal precursor cells, whereas cortical astrocytes do not (Song et al., 2002). Conversely, astrocytes transplanted from the SGZ and SVZ to cortical regions could induce some signs of neuronal differentiation (Jiao and Chen, 2008). Irrespective of whether full neurogenesis might indeed occur, factors released by the astrocytes, including Sonic hedgehog (Shh) could obviously efficiently convert cell fate.

The close interaction between the developing neurons and S100β-negative astrocytes in the neurogenic zones, which are partly identical to the precursor cells of these regions, indicates a dual function of astrocytes or astrocyte-like cells in this context: they function as precursor cells and form part of the niche. Confusingly, in a double-knockout mouse model for both GFAP and vimentin, which would presumably affect astrocyte integrity (but the phenotype of the model is surprisingly mild), neurogenic permissiveness in the hippocampus was actually increased for implanted progenitor cells (Widestrand et al., 2007). The observation might, however, have to do with the reactive gliosis (see below) after transplantation and not generally reflect the situation in the niche.

Astrocytes act pro-neurogenic through cell–cell contacts as well as independent of contacts

To some degree astrocytes seem to "cradle" the newborn cells in the dentate gyrus (Fig. 8–3) (Shapiro et al., 2005; Plumpe et al., 2006). It sometimes seems that astrocytic processes shield the cells towards the hilus, while radial processes provide guidance structures (Seri et al., 2001; Shapiro et al., 2005). Such one-to-one relationship, however, was not seen in another study (Plumpe et al., 2006).

Astrocytes might thus provide the cell–cell interaction necessary for precursor cell function. However, as astrocytes in neurogenic regions appear to be specialized and as they even encompass the stem cells themselves, the argument becomes somewhat circular. Not all astrocytes are stem cells, though, and not all astrocytes are neurogenic. At present it also remains unclear by which means astrocytes induce neurogenic permissiveness. To some degree, cell–cell contact is required (Lim and Alvarez-Buylla, 1999), but astrocytes also exert contact-independent effects (Song et al., 2002; Jiao and Chen, 2008).

Glia-like precursor cells are coupled by gap-junctions

Astrocytes form functional networks through gap junctions, and calcium waves can travel through this network. It is plausible to assume that astrocytes as precursor cells take part in such networks, but this remains to be shown in detail. Radial (and possibly non-radial) precursor cells are coupled, and these contacts are necessary for adult neurogenesis to occur

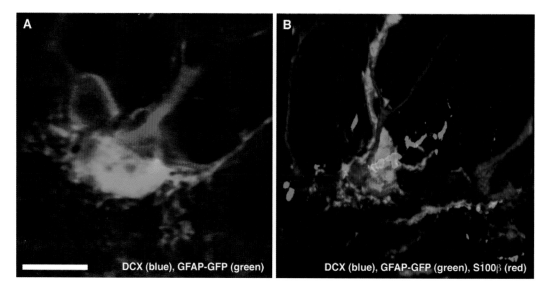

FIGURE 8–3 Cradling astrocyte. GFAP-positive astrocytes (here in a GFAP-GFP reporter mouse, *green*) often cradle newborn neurons (doublecortin-positive, *blue*) and seem to shield them towards the hilus. For rats even a 1:1 relationship between the astrocytes and the new neurons has been suggested, by Shapiro et al. (2005), but it has not been found confirmed for mice: See Plumpe et al. (2006).

(Kunze et al., 2009). The nature of the communication between the cells, however, has not yet been elucidated and might comprise multiple mechanisms. During development, calcium waves in radial glia modulate the proliferative activity of the precursor cells (Weissman et al., 2004). Similarly, precursor cell activity in the adult brain is calcium-dependent (Deisseroth et al., 2004) but here the link to a gap junction–mediated process has yet to be made.

Neurogenic astrocytes might confer neurogenesis to non-neurogenic regions

Hippocampal astrocytes are also able to elicit a neurogenic potential in precursor cells from the spinal cord, a non-neurogenic region, whereas spinal cord astrocytes have no effect. The same pattern applies to astrocytic effects on precursor cells isolated from the substantia nigra (Lie et al., 2002). Astrocytes from the neurogenic region elicited a potential for neuronal development, whereas astrocytes from non-neurogenic regions did not. Wnt-3 signaling has been suggested as one of the key mediators of this glial effect on precursor cells (Lie et al., 2005). In an ex vivo neurosphere experiment, astrocytes from the neurogenic regions conferred neurogenic permissiveness to precursor cells isolated from throughout the brain, and here Sonic hedgehog (Shh) was identified a primary mediator (Jiao and Chen, 2008). Transplantation of astrocytes from the SGZ into the neocortex elicited "neurogenesis" in these regions (Jiao and Chen, 2008), but the analysis was based only on Hu and DCX expression and only one single example of a tiny NeuN-positive cell, which alone cannot dispel the necessary skepticism. The observation is clearly important but requires further studies.

Astrocytes modulate neurotransmitter action in neurogenic zones

At least in the SVZ, astrocytes play a prominent role in modulating the influence of GABA on neuronal development. Astrocytes express both GABA receptors and the GABA transporters (Fraser et al., 1995) and ensheathe synapses like they seem to contact precursor cells in the niche. Activity of the GABA transporter balances the availability of GABA at the synapse and thereby presumably influences GABAA receptor activity at the precursor cells (analogous to Dingledine and Korn, 1985). Given the known role of GABA in controlling adult neurogenesis, such a mechanism seems plausible but remains to be proven (Bordey, 2006).

Ependymal Cells

In the SVZ, B1 precursor cells are engulfed by ependymal cells

Ependymal cells are niche cells in the SVZ but not the SGZ, which lies detached from the ventricular surface. The relationship between precursor cells and ependymal cells is so close and the parameters to distinguish them are often so difficult to apply that a longstanding controversy about an identity between the two populations arose. This identity has now been largely dismissed, but open questions remain, and the relationship remains particular. For a more detailed discussion of this issue, refer to Chapter 3, p. 69, and Chapter 5, p. 164. Irrespective of this problem, ependymal cells also serve as niche constituents. Structurally, they form an intriguing pattern surrounding the astrocyte-like B1 precursor cells extending to the ventricular surface (Fig. 5–4) (Mirzadeh et al., 2008).

Ependymal cells secrete noggin, which antagonizes the action of BMPs in the neurogenic niche of the SVZ and thus inhibits gliogenesis and promotes neurogenesis (Lim et al., 2000). Another example of a regulatory factor of ependymal origin, which is however of lesser cellular specificity, is pigment epithelium-derived factor (PEDF). PEDF promotes self-renewal and proliferation of SVZ precursor cells (Ramirez-Castillejo et al., 2006).

Vasculature

Neurogenic niches are vascularized

A key cellular component of the germinative zones is blood vessels (Fig. 8–1) (Palmer et al., 2000; Palmer, 2002). The definition of a vascular niche for neural stem cells raises the question of whether proximity to the blood vessels is required to allow neurogenesis. Endothelia or vascular smooth-muscle cells might be important cell types involved in building the germinative matrix. Type-1 cells in the SGZ have vascular endfeet characteristic of astrocytes, and these endfeet rest on the basal membrane of the endothelial cells (Filippov et al., 2003).

A vascular niche has also been described anatomically and functionally for the adult SVZ (Tonchev et al., 2007; Shen et al., 2008). B cells come into contact with both the ventricular wall and basal membrane invaginations associated with blood vessels. The vascular basal membrane from the vessels in the subependymal tissue engulfs the cells of the stem cell niche. The concept of a vascular niche for precursor cells has also been applied to embryonic development of the cortex, and a similar spatial relationship for intermediate progenitor cells and the vasculature has been described (Javaherian and Kriegstein, 2009; Stubbs et al., 2009). Ex vivo, endothelial cells promote the expansion and neuronal differentiation of precursor cells from the embryonic brain (Shen et al., 2004).

At least neurogenesis in the hippocampus is associated with angiogenesis

Precursor cell proliferation in the adult subgranular zone is paralleled by a proliferation of endothelial cells (Palmer et al., 2000). Induction of precursor cell proliferation can be accompanied by an induction of endothelial proliferation (Hellsten et al., 2004). And endothelial cells release neurogenic factors that both maintain precursor cells and promote neuronal development (Shen et al., 2004). A link between angiogenesis and neurogenesis is further supported by the finding that vascular endothelial growth factor (VEGF) has strong effects on adult neurogenesis (Cao et al., 2004; Jin et al., 2004; Schanzer et al., 2004). Interestingly, however, the effect in vivo was survival-promoting, whereas in vitro, a proproliferative effect was seen (Schanzer et al., 2004). It seems that in vivo, doublecortin-positive (type-2 and -3) progenitor cells of the SGZ express VEGF receptor 2, flk1 (Jin et al., 2002). Inhibition of VEGF action prevents the exercise-induced up-regulation of adult hippocampal neurogenesis (Fabel et al., 2003b). Despite its name and the suggestive spatial situation, however, VEGF action in neurogenic zones might be essentially independent of the vasculature.

There is also a conspicuous link between the stem cell hypothesis of gliomas and the vascular niche, in that the malformed vascular bed of glioblastomas might actually provide an abnormal stem cell niche that can sustain the tumor stem cells (Gilbertson and Rich, 2007). VEGF promoted cancerous growth from tumor stem cells, not in vitro but in vivo, and led to the existence of highly vascularized GBM, indicating that VEGF might promote tumor growth through the induction of angiogenesis (Oka et al., 2007).

In a co-transplantation model of neural precursor cells and endothelial cells in a stroke model, proliferation and neuronal differentiation from the precursor cells was increased, further supporting the ability of endothelia to convey neurogenic permissiveness (Nakagomi et al., 2009).

Speculation about possible lineage relationships between endothelial and neural precursor cells has not yet been resolved

Irritatingly, the link between neurogenesis and angiogenesis might be even closer than just interdependency: in the presence of endothelial cells, one study showed neural precursor cells transdifferentiating into endothelial cells (Wurmser et al., 2004). This report has not seen follow-ups, though, so far-reaching conclusions should be avoided. The study went to great lengths to avoid the many potential pitfalls of other transdifferentiation studies, but questions remain. The possibility of reverse transdifferentiation, from endothelia to neurons, however, that has been hypothesized by some investigators has not been found so far, although one other report has suggested precursor cell properties in adventitia cells (Yamashima et al., 2004). A link might exist through cells of the vessel wall, which are of neural crest origin and the issue of precursor cell properties of pericytes, discussed in Chapter 3, p. 73 and 88.

In both SGZ and SVZ, precursor cells have historic roots at the ventricular wall

Arturo Alvarez-Buylla has proposed that both neurogenic niches in the adult brain represent displaced neuroepithelium, even though this seems less obvious in the case of the SVZ, which is still adjacent to the ventricle (Alvarez-Buylla and Lim, 2004). During early embryonic brain development, neural precursor cells span the entire thickness of the future brain, coming into contact with both the ventricular lumen and the basal membrane at the pial surface. These radial glia–like stem cells produce neurons, very likely through expanding progenitor cell populations, and function as a guidance structure for migrating neurons that build up the layered cortex. Postnatally, the radial glia–like precursor cells in the SGZ (type-1) are shortened, spanning only the distance between the SGZ and the inner molecular layer. In the SVZ they lose their radial glia–like appearance and become specialized astrocytes (B cells) that maintain contact with the ventricular surface and the basal membrane around the blood vessels. When the cortex thickens during development, the basal lamina around these blood vessels maintain contact between the astrocyte-like precursor cells in the SVZ and the pial surface. In the hippocampus this continuity is less obvious, because the germinative matrix of the SGZ is even further displaced from the original primary and secondary germinative zones in the ventricular wall. The basal membrane around blood vessels of the SGZ provides the basal contact for the radial elements, but in the inner molecular layer, no counterpart is obvious. But the orientation remains towards the pial surface. There, the cells branch and terminate between the fiber tracts.

Microglia and Immune Cells

Microglia is a brain cell type of ultimately extra-cerebral origin exerting immunological functions

Besides the notorious NG2 cells, microglia might be the most enigmatic population of brain cells. Microglia are ultimately of extracerebral origin and derive from the monocytic lineage in the bone marrow. These precursors are Ly-6c-positive and express CCR2 (Mildner et al., 2007). At a given time, microglia in the brain is of mixed origin, one part stemming from the periphery, one from cell divisions in loco. Proliferation rate of microglia is generally low but besides by pathological stimuli could also be stimulated by physical activity (Ehninger and Kempermann, 2003). Nevertheless, controversial issues remain, and how microglia provides a link to the periphery is not yet fully clear (Davoust et al., 2008). Microglia exerts immunological functions in the brain but there is still no clear idea of the function of microglia in the absence of pathology. This is important with respect to the role of microglia as potential niche cells, because here we would have to assume a relevance that is not restricted to pathological situations. Under pathological conditions, microglia become activated, change their morphology, and might secrete numerous factors.

Microglia always has both positive and negative functions

Microglia can be found in both neurogenic niches of the adult brain, albeit in low numbers. The effect of microglia on adult neurogenesis is Janus-faced (Kempermann and Neumann, 2003; Ekdahl et al., 2009). Acute microglial activation impairs adult neurogenesis under both pathological and otherwise physiological conditions (Ekdahl et al., 2003; Monje et al., 2003). Similarly, microglia inhibited neuronal differentiation in vitro (Cacci et al., 2005), but the exact mechanism is still not known. The main candidates are interferon-γ (Ben-Hur et al., 2003), interleukin 1β (Koo and Duman, 2008), tumor necrosis factor α (Ben-Hur et al., 2003; Cacci et al., 2005), and particularly interleukin 6 (Vallieres et al., 2002; Monje et al., 2003). Most of these data have been obtained in the context of pathology, mostly ischemia or autoimmunity. So it is not always clear whether the same factors also act as inhibitory niche factors under physiological conditions. On the other hand, microglia might also secrete permissive factors and support neurons (Bessis et al., 2007). Adrenalectomy led to an increase in neurogenesis that was correlated to the concomitant increase in microglial proliferation and levels of TGFβ (Battista et al., 2006).

Most notably, microglia can secrete BDNF in vitro (Nakajima et al., 2001; Nakajima et al., 2002) and do so after stimulation with conditioned medium from neurons (Nakajima et al., 2007), suggesting a paracrine feedback loop.

These results are not simple to generalize, but they seem to suggest that microglia might in fact also be involved in balancing adult neurogenesis via different, partly opposing regulatory pathways. Microglia and microglia-conditioned medium, which prevented neuronal differentiation (Cacci et al., 2005), nevertheless promoted precursor cell proliferation over prolonged passages (Morgan et al., 2004; Walton et al., 2006), together possibly indicating a role of microglia in stem cell maintenance in the niche. However, partially conflicting data in favor of a positive influence of microglia on neuronal development also exist (Aarum et al., 2003) as well as data on pro-gliogenic influences (Nakanishi et al., 2007).

The idea of a balanced influence is consistent with results from other disease models and with respect to reactive neurogenesis in the ischemic striatum. Here, at late time-points after the insult (up to more than one year), accumulation of microglia corresponded to the maintained ability to form neurons (Thored et al., 2009).

Neurons

Precursor cells can sense surrounding neuronal activity

Neurons are not generally counted as niche cells, but in the SGZ, the niche is so closely surrounded by neurons that this separation is somewhat difficult. Immature neurons are found within the boundaries of the niche (if these are defined by, e.g., the basal membrane: Mercier et al., 2002) but how and when newborn cells leave these boundaries is not clear. Developing new granule cells are highly sensitive to excitatory signals, and much of the control of "neurogenesis" is dependent on synaptic input. This input goes beyond direct afferents but rather involves a sensitivity of the precursor cells and immature neurons to the surrounding network activity. How this information is translated to the new cells has not been fully explored, but BDNF that is activity-dependently released from the granule cells is a prime candidate (Babu et al., 2009).

In the SVZ itself, no neurons are found and the distance to the closest neurons in the striatum is variable. There is also no evidence that neurogenesis in the olfactory bulb would be dependent on neuronal activity in the striatum. How neuronal activity in the olfactory bulb might influence precursor cell activity in the niche of the SVZ remains to be shown. One conceivable pathway is via the dopaminergic fibers from the ventral tegmental area (VTA). In any case, with the exception of the neuroblasts (A cells), no neurons seem to be part of the stem cell niche of the SVZ.

Extracellular Matrix

Non-cellular niche components form a matrix that regulates availability of soluble factors and directly interacts with the precursor cells

Extracellular matrix (ECM) molecules are likely to play an important role in defining neurogenic conduciveness, but relatively little is known about the actual contributions and mechanisms. In precursor cell cultures, substrates such as laminin and fibronectin influence precursor cell survival and behavior, and integrins, which are receptors to laminin, influence the migration of neurosphere cells in vitro (Jacques et al., 1998). The adult neurogenic regions contain several ECM molecules such as tenascin-C, laminin, netrins, and different types of proteoglycans (Gates et al., 1995;

Mercier et al., 2002; Kerever et al., 2007). Composition of the ECM is complex and regionally diverse, but few details are known. In the SVZ, for example, no expression of laminin α1 was found, which is a ECM factor elsewhere (Kerever et al., 2007). Table 8–1 summarizes ECM molecules, for which information exists in the context of neurogenic niches in the adult brain. Essentially all our knowledge about ECM comes from work in the SVZ; there have been hardly any studies on ECM and adult hippocampal neurogenesis. One notable exception even addressing functionality is a report by Imbeault and colleagues, who tried to recapitulated the sequence of cell types (type-1 to -3) that is found in the adult dentate gyrus in vivo (see Chap. 6, p. 192, for details) in a neurosphere-type culture in vitro, and studied the effects of ECM molecule laminin on the expression of gap junction proteins in these cells (Imbeault et al., 2009).

In cell culture experiments, the coating of dishes with laminin, poly-D-lysin, fibronectin, or the like serves the purpose of mimicking ECM. Here it can be shown how different substrates influence either precursor cell maintenance or neuronal differentiation (Goetz et al., 2006).

The extracellular matrix provides no static environment but plays important roles in modulating, for example, the activity of growth factors, for which the ECM can serve as a reservoir. Most notably, heparan sulfate proteoglycans might bind to extracellular signaling molecules and thereby modify their action. This has, for example, been shown for FGF2 and ECM structures of the adult SVZ (Kerever et al., 2007). Novel substrates for cell culture are now constructed based on heparan sulfates that allow the controlled release of growth factors, bringing this "active" aspect of ECM function into culture models (Oezyuerek et al., 2009).

Table 8–1 ECM molecules in adult neurogenic niches

Molecule	Function	Expression	Reference
Tenascin C		SVZ (B cells)	(Kazanis et al., 2007)
		Olfactory bulb	(Miragall et al., 1990)
		SVZ Neurospheres	(Kukekov et al., 1999)
	Knockout leads to delayed onset of olfaction, while neurogenesis is unimpaired	SVZ	(de Chevigny et al., 2006; Kazanis et al., 2007)
Netrin-4	Forms complex with laminin γ1 that controls proliferation and migration; known relevance for axon guidance	SVZ astrocytes	(Staquicini et al., 2009)
N-sulfate heparan sulfate proteoglycan	Binds FGF-2	SVZ	(Kerever et al., 2007)
Laminin	Favors precursor cells, influences expression of gap junction proteins	SGZ neurospheres	(Imbeault et al., 2009)
Laminin β1		SVZ	(Kerever et al., 2007)
Collagen I		SVZ	(Mercier et al., 2002)
Collagen IV		SVZ	(Kerever et al., 2007)
Nidogen		SVZ	(Kerever et al., 2007)
Perlecan		SVZ	(Kerever et al., 2007)

	At the same time, cells often carry receptors for ECM constituents. One example is Netrin4, which forms a complex with laminin γ1: "neuroblasts" of the SVZ carry netrin receptors, which signal through integrins, here specifically α6β1 integrin, directing proliferation (Staquicini et al., 2009) and migration (Emsley and Hagg, 2003).

Some ECM factors are ligands to cellular receptors

Cells modify the ECM, for example through degrading enzymes like metalloproteinases

At the same time, cells often carry receptors for ECM constituents. One example is Netrin4, which forms a complex with laminin γ1: "neuroblasts" of the SVZ carry netrin receptors, which signal through integrins, here specifically α6β1 integrin, directing proliferation (Staquicini et al., 2009) and migration (Emsley and Hagg, 2003).

Cells interact with the ECM in a reciprocal way. ECM is modulated by a number of factors, most notably metalloproteinases, which can be expressed by many niche cells, including the precursor cells themselves (Lu et al., 2008). When adult neurogenesis is stimulated by electroconvulsive seizures, gene expression profiling indicates a parallel increase in growth factors, angiogenic factors, and metalloproteinases (Newton et al., 2003). In the adult SVZ, specifically a disintegrin and metalloproteinase 21 (ADAM21) was on one side found to be associated with ependyma, SVZ cells with long basal processes, and early postnatally also with radial glial elements, and on the other side with terminally differentiating cells in the olfactory bulb (Yang et al., 2005). This suggests a dual relevance for precursor cell maintenance and final, presumably axonal, integration. ADAMs exert multiple functions, for example in the cleavage-dependent activation of factors like Notch, amyloid precursor protein (APP), and transforming growth factor alpha (TGFα), can bind integrins and have thus implications for a great number of plasticity processes (Yang et al., 2006). One might expect that more implications for adult neurogenesis will be revealed in the future.

In the pancreas, metalloproteinase-9 increased VEGF action and transformed normal tissue into angiogenic foci (Bergers et al., 2000). In bone marrow, metalloproteinase-9 increased the expression of soluble ligand to the kit-receptor and thereby promoted the recruitment of ckit-positive precursor cells from the stem cell niche (Heissig et al., 2002).

Secreted Niche Factors

Niche signaling includes numerous secreted factors promoting or inhibiting neurogenesis

Fabienne Agasse, Michel Roger, and Valérie Coronas from the University of Poitiers in France described how proteins secreted in an autocrine or paracrine fashion by SVZ cells promote neurogenesis in vitro, whereas protein factors from cortex actively inhibit neurogenesis (Agasse et al., 2004). Intriguingly, if apoptosis was induced in the cortical explants used in the co-culture systems of these studies, the inhibitory effect was reversed and cell proliferation in the SVZ samples increased. These data indicate that beyond cell–cell interaction and extracellular matrix effects, neurogenic permissiveness is determined by both permissive and inhibitory diffusible cues. Non-neurogenic regions actively block neurogenesis, whereas neurogenic regions actively promote it. The interesting question is whether the removal of the blocking effects in non-neurogenic regions is sufficient to allow an intrinsic neurogenic program to unfold, or whether below the normal inhibitory effect of non-neurogenic regions a pro-neurogenic effect awaits and is up-regulated when the inhibition ceases. This question is of central importance for cases of targeted or induced regenerative neurogenesis. Does the tissue damage do more than remove the negative influence blocking neurogenic permissiveness? Can apoptosis actively induce pro-neurogenic factors in non-neurogenic regions?

Some of the regulatory molecules involved in the early stages of brain development might also influence adult neurogenesis. Bone morphogenic proteins (BMPs), for example, actively antagonize neurogenesis in non-neurogenic regions. When Daniel Lim and Arturo Alvarez-Buylla used noggin to inhibit BMP activity in the adult SVZ, they found neurogenesis in the otherwise non-neurogenic striatum (Lim et al., 2000). Similar observations have been made for ephrins (Holmberg et al., 2005) and Sonic hedgehog (Jiao and Chen, 2008). See Chapter 9 for more information on the humoral control of adult neurogenesis.

Towards the Systems Biology of the Niche

The complexity of the niche–
precursor cell interaction calls
for a systems biology approach
and modeling

The niche concept is a prime target for systems biology. Some attempts on both the transcriptome and the proteome level have targeted the precursor cells (Gurok et al., 2004; Maurer et al., 2008), but overall, the state of knowledge is rather modest. Proteomics especially will open a whole new take on how precursor cells are controlled (Paulson et al., 2007) and how cells interact in the niche. Gene profiling studies in vivo have generated first data sets (e.g., Zhang et al., 2007, for the SVZ) but to date no model exists that would allow making best use of this information. An example of what lies ahead has been provided in a mathematical model of cellular dynamics and development in the crypts of the colon (Johnston et al., 2007b; van Leeuwen et al., 2009), apparently a popular system among modelers (Johnston et al., 2007a). Bringing kinetic data about precursor cells together with information about gene and protein expression and data about the relative contributions of the different niche cells will allow us to generate more complete hypotheses on how stem cells and their niche make new neurons.

As part of a pioneering study to use co-expression patterns to identify cell populations and functional units in the brain that was based on data-driven analysis tools rather than the application of predefined classifications, Oldham and colleagues among other populations also identified the SVZ and its neurogenic astrocytes (Oldham et al., 2008). The approach clearly awaits refinement and confirmation, but the idea to identify "modules" out of the complex "brain soup" (Mirnics, 2008) rather than reverse-building the niche concepts from the pieces we know is a major leap forward.

Transplantation Studies

Transplantation studies support
the relevance of neurogenic
permissiveness in the target
region for neuronal
differentiation to occur

To some extent the distinction between neurogenic and non-neurogenic region is based on a small (and in the end insufficient) number of fundamental transplantation studies. Implantation of cultured precursor cells from the adult hippocampus back into the hippocampus of adult rats demonstrated that the grafted cells incorporated adequately only into the granule cell layer and nowhere else (Gage et al., 1995). When the same hippocampal precursor cells were implanted in the rostral migratory stream (RMS), they generated olfactory interneurons (Suhonen et al., 1996). Implantation into the cerebellum did not result in neuronal differentiation, confirming that the implanted cells developed according to the local cues present in the neurogenic zones.

SVZ precursor cells implanted in the striatum, cortex, and olfactory bulb showed integration and neuronal differentiation only in the olfactory bulb but not in the other regions (Herrera et al., 1999). Another study, however, described neuronal differentiation in the striatum (Zhang et al., 2003). Also, neonatally isolated precursor cells differentiated into neurons when transplanted into the same region (Zigova et al., 1998).

Precursor cells from the non-neurogenic spinal cord and substantia nigra also developed into granule cell neurons when placed in the hippocampus (Shihabuddin et al., 2000; Lie et al., 2002). However, when spinal cord precursor cells were placed in the neocortex, they failed to differentiate into neurons. Moreover, when substantia nigra precursor cells were grafted back into the substantia nigra, they did not result in neurogenesis (Lie et al., 2002). In contrast to these findings, early postnatal cerebellar precursor cells implanted in the SVZ and RMS failed to differentiate into olfactory neurons and migrated only a brief distance into the RMS (Jankovski and Sotelo, 1996).

When hippocampal precursor cells were implanted in the retina, many stages of photoreceptor development were seen, including the highly characteristic morphology of photoreceptors. However, no terminal differentiation occurred (Takahashi et al., 1998). This finding indicates that neurogenic permissiveness might be graded and that the composition of neurogenic factors in different brain regions might vary. It is thus likely that many factors have to act together to make a brain region fully neurogenic.

To some degree, neurogenic permissiveness could be transplanted together with astrocytes from neurogenic regions (Jiao and Chen, 2008), but the evidence in support of "neurogenesis" in that study did not yet meet all necessary standards. The finding is suggestive, however, and awaits confirmation.

Taken together, these results argue that extrinsic factors are more important than cell-intrinsic parameters in neurogenesis, but more transplantation studies would be necessary to fill the gaps. For example, neural precursor cells from the SVZ have not yet been implanted in the hippocampus.

Differences Between SVZ and SGZ

The two neurogenic regions differ with respect to most parameters

The two neurogenic regions in the adult rodent and primate brain are not created equal. Although the transplantation studies have indicated that the neurogenic niches are somewhat promiscuous in accepting precursor cells from other brain regions, the precursor cells themselves show differences, and neither the niche composition nor the molecular machinery that is expressed are identical. Some factors such as Mash1/Ascl1 have different functions in both regions (Parras et al., 2007; Jessberger et al., 2008; Kim et al., 2008).

The sets of factors determining neurogenic permissiveness in the two regions are still poorly defined

At present it is not clear to what degree the factors that define the stem cell niche are identical to those that determine neurogenic permissiveness. These two sets of parameters do not completely overlap, as suggested by the distribution of different stages of neuronal development in adult olfactory bulb neurogenesis over the distance between the SVZ, the RMS, and the olfactory bulb. This distinction would mirror the situation in the developing brain, where factors that determine positional identity need not be identical to those controlling neuronal differentiation. As the positional identity differs between the two neurogenic regions, so might the conditions that build on this identity to foster further neuronal development (Table 8–2).

Cell Genesis Outside the Neurogenic Regions

Cell genesis is much more widespread than neurogenesis in the adult brain

There is extensive cell proliferation going on throughout the adult mammalian brain. Precursor cells outside the canonical neurogenic regions have been introduced and discussed in Chapter 3, p. 74. Under physiological conditions, essentially all of these dividing cells express NG2 and were historically thought to represent oligodendrocyte progenitor cells or the elusive O2A progenitor cell in vivo. This association to the oligodendrocytic lineage, however, is only the partial truth. The nature of NG2 cells with glial and some neuronal features has made it clear that their function probably goes beyond being mere progenitor cells for oligodendroglia. We have discussed these properties of NG2 cells in detail in Chapter 3.

NG2 cells might show some neuronal markers but are no "neuroblasts"

Most studies used NG2 immunoreactivity, often in combination with the BrdU technology, to assess the behavior and fate of these cells. NG2 cells might be migratory cells originating from the SVZ, or resident cells. The latter cells have a suggestive territorial distribution and might otherwise exhibit properties of protoplasmic astrocytes. The migratory cells might co-express doublecortin, which has led to the erroneous expectation that these were migratory "neuroblast" capable of neurogenesis throughout

Table 8–2 Differences between SGZ and SVZ

	SGZ	SVZ
Positional identity	Dorsal forebrain	Ventral forebrain; possibly with dorsal subpopulations
Presumed origin	Ectopic tertiary germinative matrix derived from the hippocampal hem at the ventricular wall (Altman and Bayer, 1990b, a)	Presumably direct origin from lateral ganglionic eminences and dorsal SVZ of the embryo (Stenman et al., 2003; Merkle et al., 2004; Kohwi et al., 2007; Young et al., 2007)
Putative stem cell	Radial glia–like (Type-1) with apical contact to blood vessels (Seri et al., 2001; Filippov et al., 2003; Seri et al., 2004; Breunig et al., 2008; Kunze et al., 2009)	Astrocyte-like (B cell) with apical contact to ventricular surface (Doetsch et al., 1997; Doetsch et al., 1999b; Doetsch et al., 1999a; Mirzadeh et al., 2008)
Self-renewal and multipotency ex vivo?	Yes (Palmer et al., 1997; Babu et al., 2007)	Yes (Reynolds and Weiss, 1992)
	(Conflicting reports (Seaberg and van der Kooy, 2002; Bull and Bartlett, 2005))	
Migration	Very limited horizontal migration within SGZ Radial migration into granule cell layer	No radial migration in SVZ. Extensive chain migration within SVZ and through rostral migratory stream to olfactory bulb
Lineage potential in vivo	Glutamatergic granule cells, astrocytes, very few oligodendrocytes	2-6 types of inhibitory interneurons, 1 type of glutamatergic interneuron in the olfactory bulb; astrocytes; separate oligodendrocytic lineage with migration to cortical areas
Key transcription factors	Pax6, Tbr2, Tbr1 (Hodge et al., 2008), Ngn2 (Raineteau et al., 2006; Ozen et al., 2007) Prox1 (Pleasure et al., 2000; Steiner et al., 2006a)	Dlx1/2 (Doetsch et al., 2002), Er81 (Stenman et al., 2003) Olig2 (Hack et al., 2004; Menn et al., 2006)
Niche composition	Radial glia–like precursor cells Intermediate progenitor cells Astrocytes Blood vessels Few microglia, very few NG2 cells Neurons	Astrocyte-like precursor cells Ependymal cells Astrocytes Blood vessels No or few neurons
Regulatory input of key neurotransmitter systems	Glutamate: NMDA dependent GABA: promotes differentiation and synaptic integration (Tozuka et al., 2005; Ge et al., 2006)	Glutamate: No NMDA effects GABA: promotes proliferation (Liu et al., 2005) and migration (Bolteus and Bordey, 2004) and favors dopaminergic differentiation (Nakagomi et al., 2009)

the cortex. A large proportion of these cells is found in and directly below the corpus callosum, and a distinct "fountain" of migration leads from the SVZ anteriorly into the cortex.

Unfortunately, no reporter gene mouse for the NG2 promoter exists, and two surrogates have been used. Vittorio Gallo and colleagues studied the NG2 cells based on the activity of the Cnp (2′,3′-cyclic nucleotide 3′ phosphodiesterase, CNPase) promoter. They find a high overlap between the CNP-EGFP signal and NG2. Some aspects of this approach remain unclear, because Cnp is involved in cytoskeletal rearrangement during the process outgrowth of oligodendrocytes (Lee et al., 2005a). Cnp protein is often used as marker for oligodendrocytes, although it was noted early that Cnp is also found in "glioblast-like" cells of the SVZ (Braun et al., 1988). Consequently, the concern would be that Cnp might not appropriately identify undetermined NG2 cells. The data from the Cnp-EGFP reporter mouse, however, suggest the opposite and even include a neurogenic potential, for which there is comparatively little evidence from other resources (Belachew et al., 2003; Aguirre and Gallo, 2004). The reporter shows an overlap with Dlx2, which would explain such neurogenic potential and the assumed relationship to C cells of the SVZ (Aguirre et al., 2004; Chittajallu et al., 2007) but might argue against the idea that it reliably identifies the NG2 cells as we find them in the cortex.

Olig2-positive lineages of adult-generated cells are not limited to oligodendrocytes but also comprise post-mitotic NG2 cells

NG2 expression also shows a large overlap with Olig2, but despite its suggestive name (oligodendrocytes transcription factor 2), Olig2 is not specific to the oligodendrocyte lineage (Rowitch et al., 2002; Ligon et al., 2006). For example, reactive astrocytes were found to originate from Olig2-positive cells (Tatsumi et al., 2008), strengthening the idea of a precursor between the oligodendrocytic and astrocytic lineage. But in other contexts, Olig2 is also involved in other fate choice decisions; for example, in balance with Ngn2 in motor-neuron development in the spinal cord (Novitch et al., 2001; Lee et al., 2005b). Although the exact function of Olig2 in NG2 cells is thus not fully understood, roles other than in the generation of oligodendrocytes seem plausible, and its expression is not contradicting other potentials.

Leda Dimou, Magdalena Götz, and colleagues from Ludwigs Maximilians Universität in Munich, Germany, provided an extensive mapping study of adult cortical cell genesis using fate mapping based on tamoxifen-inducible Cre-recombination in the Olig2 locus. See Chapter 7 for the details on how this approach works (p. 257). Throughout the adult brain, cells that expressed the reporter gene were also proliferative at early time-points after induction of the recombination and expressed NG2. At later time-points, however, the progeny of the marked cells turned out to be oligodendrocytes in the white matter, but largely NG2 cells in the gray matter (Dimou et al., 2008). These NG2 cells presumably turned post-mitotic and could not be re-stimulated for example by injury. This result clearly lends further support to the idea that NG2 cells constitute a separate "mature" entity of glial cells in the adult brain, separate from the NG2 cells, with precursor cell functions (see, for example, Butt et al., 2005; as well as Chap. 3, p. 71). Consistent with other reports, there was no evidence of neurogenesis from the Olig2-positive cells in the adult brain, even under traumatic conditions (stab wound).

Reactive Gliosis Reconsidered

Astrocytes also respond to damage with proliferation, which might reflect a kind of precursor cell activity

While astrocytes show little proliferative activity under normal conditions, if at all, they respond to brain damage with a characteristic yet nonspecific response called "reactive gliosis." Phenomenologically, the astrocytes react by hypertrophy and an up-regulation of GFAP. The nature of this response as "friend or foe" has been a matter of debate for decades, and reactive gliosis is still seen as a two-edged sword (Ridet et al., 1997; Pekny and Nilsson, 2005). On one side, the response plays a role in limiting damage; on the other side, it might contribute in particular to the long-term consequences of lesion-induced changes (Pekny et al., 2007). A central part of the reactive gliosis is a proliferation of astrocytes, often leading to what is called a "glial scar." Can all kinds of astrocytes produce the gliotic response? Is a reactive gliosis a stem cell activity? Fate-mapping in vivo has revealed that the proliferative response in fact originated from quiescent astrocytes in loco, which ex vivo exhibited stem cell properties in a neurosphere assay (Buffo et al., 2008). These cells did not seem to be related to the NG2 cells of the adult murine neocortex, suggesting that the insult can elicit stem cell properties in astrocytes proper. This potential is not used in vivo, and the cells remain in the astrocytic lineage. The obvious question arises of whether this potential might be tapped to induced neurogenesis in otherwise non-neurogenic regions. The observation that overexpression of Pax6 was sufficient to induce a neurogenic potential in astrocytes from non-neurogenic areas in vitro supports this idea (Heins et al., 2002). A long road would have to be traveled to turn this approach into a route for functionally relevant restorative neurogenesis in non-neurogenic regions. Nevertheless, with such observations, "reactive gliosis" is not the same as it used to be.

Neurogenesis Outside the Neurogenic Regions

Numerous reports have claimed neurogenesis outside the mammalian SGZ and SVZ, but very few of them have been confirmed

Since the days of Altman and Das and the work of Kaplan in the 1970s, there have been attempts to discover neurogenesis in more regions than the hippocampal dentate gyrus and the SVZ and olfactory bulb. Whereas widespread adult neurogenesis is common in some lower vertebrates, in most mammalian species investigated, adult neurogenesis is essentially limited to the two now-canonical neurogenic regions (but rabbits appear to show more: Bonfanti and Ponti, 2008). We will return to the evolutionary perspectives in Chapter 11. Multiple studies have claimed adult neurogenesis outside the canonical neurogenic regions of rodents and primates, but in most cases the evidence supplied in the reports is weak. Very often only single ambiguous examples are shown, and sometimes the published images even overtly contradict the claim. The methodological problems and pitfalls have been discussed in Chapter 7. To date, none of these claims has been proven beyond doubt.

The reports should nevertheless be taken seriously, because some of the individual data are suggestive, even if they are not yet fully convincing, and the question should consequently be further addressed with an open attitude. The chances that a constitutively neurogenic region in the adult brain of rodents and primates has been overlooked, however, are slim. The situation is somewhat different in cases of pathology, where a few convincing examples of regenerative neurogenesis exist. A particular case is the report that previously uncharacterized precursor cells in cortical Layer I with an as-yet-unknown origin and regional identity might produce new cortical interneurons after ischemia (see below) (Ohira et al., 2010). Even in studies that go the experimental extra mile to prove their case, independent confirmation by others will be needed and does not yet exist for this report. It is generally problematic that isolated negative results tend to remain unpublished so that one finds few experimental data arguing for the absence of a phenomenon. Species and technical differences could be partially responsible for the discrepancies, without closing the case. As in the other controversial cases, the pro side has the burden of proof beyond a reasonable doubt, whereas the contra side has the fundamental problem of proving the complete absence of a trait.

Hippocampus

Reports on adult neurogenesis in CA1 were not confirmed

Whereas one study reported new neurons in region CA1 of adult rat hippocampus (Rietze et al., 2000), others did not find evidence for adult neurogenesis in CA1 of mice (Kempermann et al., 1997b; Nakatomi et al., 2002). Neurogenesis in CA1 might be inducible by ischemia and growth factor infusion (Nakatomi et al., 2002), a situation rather removed from physiological conditions. In another study, in the absence of pathology and additional manipulations, no migration of putative precursor cells from the SVZ into the CA1 region was found, and cell genesis was restricted to astrocytes and NG2 cells (Kronenberg et al., 2007).

Remarkably, however, this cell genesis could be enhanced by environmental enrichment and voluntary wheel-running (Kronenberg et al., 2007).

Olfactory Tubercle

One study reported that, in squirrel monkeys, migrating neuroblasts in the rostral migratory stream might divert from their path to the olfactory bulb and reach the olfactory tubercle (Bedard et al., 2002). The olfactory tubercle is an allocortical brain area at the base of the cerebral hemispheres. It receives input from the olfactory bulb, the substantia nigra, and the reticular formation, and projects to the hypothalamus. In the report, one single example of a BrdU/NeuN-doublepositive cell was shown in which the area of NeuN staining was far larger than the BrdU-positive nucleus, but no 3D-reconstruction was presented despite these circumstances.

Striatum

Neurogenesis in the adult striatum has primarily been described for pathological situations

In the lateral ganglionic eminence, Islet1 identifies the precursor cells for striatal projection neurons, in contrast to the Er81-positive progenitor cells of interneurons for the olfactory bulb (Stenman et al., 2003). In the postnatal SVZ, no Islet1-positive cells remain, but overexpression of Islet-1 together with Neurogenin2 induced a diversion of cells towards the striatum (Rogelius et al., 2006; Rogelius et al., 2008). Limited but long-lasting neurogenesis in the adult striatum has been observed after ischemia (Arvidsson et al., 2002; Thored et al., 2006) and is discussed further below.

These studies, too, did not observe any constitutive adult striatal neurogenesis. Dayer et al., in contrast, whose study is best known for their report on neurogenesis in Layer VI of the neocortex, also observed the constitutive production of new calretinin-expressing interneurons in a small sub-region of the dorso-lateral striatum (Dayer et al., 2005). The degree of sophistication in this analysis goes beyond most other reports on adult neurogenesis outside the neurogenic regions, but the examples presented in the study are not all equally convincing. The finding will have to be confirmed by others.

Neocortex

Claims of constitutive neurogenesis in the mammalian cortex have attracted much attention but remain essentially unconfirmed

Non-neurogenic regions withhold cues necessary for neurogenic development not only of endogenous precursor cells but also of implanted cells, which would turn into neurons when placed in the hippocampus or olfactory system. This definition of neurogenicity is thus independent of issues related to precursor cell heterogeneity. Unfortunately, only a few transplantation studies have been carried out (see above), and thus not all presumably non-neurogenic brain regions have been tested by this vigorous standard. Consequently, the debate about the stringency of the dichotomy of neurogenic versus non-neurogenic regions has repeatedly been questioned and considered biased (Gould, 2007). In the end, a more graded distinction will be necessary.

This is particularly true for the cerebral cortex. Altman's initial descriptions of adult neurogenesis, as well as several studies by Kaplan in the 1980s, reported neurogenesis in the neocortex (Altman, 1963; Kaplan, 1981). Unlike their reports on neurogenesis in the hippocampus and olfactory system, these claims were not confirmed by others. In 1998 a study claimed widespread neurogenesis in the cingulate and retrosplenial cortex, the lateral preoptic area, and the central gray of the hamster (Huang et al., 1998). Despite the major claim, the paper actually contains only one single example of a BrdU/NeuN double-labeled cell in the cortex, and this example of a small oval nucleus is not typical of cortical neurons and not particularly convincing. In 1999, Elisabeth Gould and colleagues described large numbers of new neurons' being found in the neocortex of adult macaque monkeys (Gould et al., 1999).

A similar finding has been reported for the cortex adjoining the amygdala (Bernier et al., 2002). Presumably, these cells originated from the SVZ, branched away from the RMS, and migrated into prefrontal, parietal, and temporal areas. The Gould study contained only a few examples and no three-dimensional reconstructions. There was also no description of neuronal development in the sense that different stages of neuronal maturation in new neurons could be demonstrated. The study received much attention but also raised substantial criticism (Nowakowski and Hayes, 2000; Rakic, 2002).

In series of experiments involving a total of 127 macaque monkeys up to 17 years of age, Pasko Rakic and his coworkers at Harvard University and Yale University had not found any new neurons in the adult neocortex (see Rakic, 2002; and Chap. 7 for the technical issues).

The examples of "transient" cortical neurogenesis might simply represent neuronal markers in NG2 cells

What struck many investigators as problematic regarding the reports on cortical neurogenesis in adult primates was the fact that the many groups working in rodents had never found evidence of adult cortical neurogenesis in the cortices of rats or mice. Although the possibility could not be categorically denied that primates had the ability of adult cortical neurogenesis and rodents did not, this seemed very unlikely, given the prevailing idea of an overall reduction in neurogenesis with increasing brain complexity during evolution. Somewhat in line with these doubts, Gould and colleagues (2001) later reported that new neurons in the cortex of adult monkeys had only a transient existence, and now showed examples of similar cortical "neurogenesis" in rodents (Gould et al., 2001).

The current interpretation of these data is that these cells are indeed of SVZ origin but, similar to NG2 cells (although in those studies their immunoreactivity for NG2 had not been shown), these new cells never reached a stage of complete neuronal maturation. As described elsewhere (Chap. 3, p. 71) and above, NG2 cells might share characteristics with neurons and might therefore easily be mistaken as neurons. The detection of immature neurons in the cortex might be equivalent to other reports indicating that the earliest (but not more mature) stages of neuronal development might in fact be found in non-neurogenic regions of the adult brain. If this is true, the lack of neurogenic permissiveness in non-neurogenic regions would affect primarily the stages of neuronal maturation, not the (short-term) maintenance of neuronal progenitor cells. Many NG2-expressing cells also express nestin and show (an often weak) co-expression of DCX. DCX was not studied in the experiments by Gould and coworkers, but TUC4 was. TUC4 expression, however, is not sufficient evidence of neuronal maturation, because it can also be found in the putative oligodendrocyte precursor cells of the adult brain (Ricard et al., 2001). It thus seems possible that presumed signs of adult cortical neurogenesis actually reflect activity on the level of NG2 cells that did not lead to completed neurogenesis. It is questionable whether such aborted neuronal development should be termed *neurogenesis*. This interpretation of aborted neurogenesis is entirely based on marker analogies to neuronal development in the neurogenic zones. These analogies might not be justified, because the specificity of the markers is not known. A more parsimonious interpretation is that in the adult brain, renewing cells can be found that show marker expression from both the glial and neuronal lineages, whose functional significance is not known at present.

This interpretation is not yet refuted but put into question by the observation that parenchymal precursor cells in cortical Layer I might be responsible for limited neurogenesis after stroke (Ohira et al., 2010).

Most studies searching for adult cortical neurogenesis did not detect it

Several studies have searched further for completed neurogenesis in the adult primate and rodent brain and found no evidence of it (Magavi et al., 2000; Kornack and Rakic, 2001; Ehninger and Kempermann, 2003; Koketsu et al., 2003). The methodological issues that have been brought forth against the reports claiming neurogenesis in the adult cortex are essentially those reviewed in Chapter 7: the pitfalls of BrdU immunohistochemistry, the lack of demonstration of development, and the problems of unambiguous neuronal markers. The 14C method that can be applied to study potential adult neurogenesis in humans also did not give evidence for neocortical neurogenesis (Bhardwaj et al., 2006). See Chapter 11 for more details on this method and the results of the studies that used it.

A particular case of neurogenesis in the adult cortex is the reports of new interneurons in Layer IV of the neocortex by Alexandre Dayer, Heather Cameron and colleagues (Fig. 8–4) (Dayer et al., 2005; Cameron and Dayer, 2008). The histological evidence is of high quality, but among a respectable 7,624 BrdU-labeled cells analyzed, only 33 were positive (equaling 0.43%), which inevitably raises general and unanswerable questions about the error rate of the method. The study obtained support, however, by an independent report based on an entirely different method. Lineage tracing in a transgenic reporter mouse for the serotonin receptor 3 revealed migration of precursor cells from the embryonic and postnatal SVZ to the cortex (Inta et al., 2008). In adulthood (P90) these cells were largely restricted to "lower cortical layers." Here, too, numbers were very low. Together these observations clearly deserve to be followed. It might be too early to settle for the available level of evidence, but the published information stands out from other reports. Dayer and Cameron provided an insightful discussion of what their findings might mean and how so few neurons could be meaningful.

Retina

The retina has been a brain region of particularly avid searches for neurogenesis—almost exclusively, however, in the context of cell transplantation. This limitation is somewhat surprising, given the fact that lifelong retinal neurogenesis is common in many species. In fish, amphibians, and birds, a ring of retinal precursor cells lies along the junction between the retina and the iris (Fig. 8–5). These precursor cells generate new retinal neurons throughout life, and adult retinal neurogenesis is stimulated after retinal cell loss (Hollyfield, 1968; Johns, 1977; Johns and Easter, 1977; Reh and Constantine-Paton, 1983; Marcus et al., 1999; Fischer and Reh, 2000).

In frogs, these precursor cells are multipotent and can generate all cell types in the retina (Wetts and Fraser, 1988). Many vertebrate species can reconstitute their retinae with high efficiency, but this ability has been lost in mammals (Lamba et al., 2008).

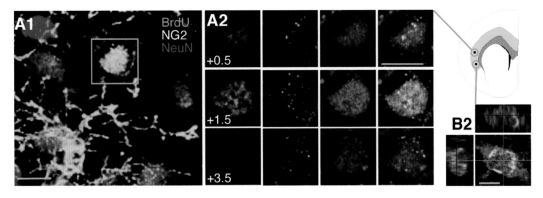

FIGURE 8–4 New cortical interneurons. (A): At 4–5 weeks after labeling with BrdU, a few cells express neuronal marker NeuN (*green* in A1). In A2, z-planes through the boxed area in A1 are shown with channel separation, demonstrating the co-localization of BrdU with NeuN but not with NG2. The location of the cell shown in (A) is circled in the small diagram of a coronal section, and the analyzed region of cortex is marked yellow on the diagram. The subcortical white matter that was used as a boundary for the analysis is indicated in blue. (B): A 4–5 week-old BrdU-positive (NG2–negative) neuron labeled with the early neuronal marker HuC/D is shown in orthogonal views. Images are taken from Dayer et al. (2005) and were kindly provided by Heather Cameron, Bethesda. Reprinted with permission from the authors and the publisher Rockefeller University Press.

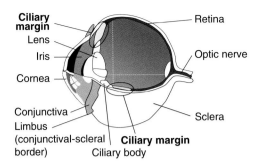

Ciliary margin — Retina

Lens

Iris — Optic nerve

Cornea —

Conjunctiva —

Limbus — Sclera

(conjunctival-scleral **Ciliary margin**

border) Ciliary body

FIGURE 8–5 Anatomy of the eye. The adult mammalian eye appears to contain three different populations of precursor cells: one in the limbic margin and one in the retina itself. There is still controversy about the relationship between these cell types and their potential. In addition, there is a non-neural precursor cell population in the limbus that replaces corneal epithelium.

In mammals, retinal neurogenesis ceases early postnatally (Young, 1985), and to date no reports on spontaneous adult neurogenesis in the retina have been published. But the case of the eye is somewhat particular. The ciliary body (or ciliary marginal zone, CMZ) contains multipotent precursor cells (Ahmad et al., 2000; Tropepe et al., 2000). Proliferation of cells in the murine ciliary body could be induced by growth factor infusion, but not damage to the eye, and the proliferating cells adopted an immature phenotype consistent with a potential role as precursor cells (Fischer and Reh, 2003). Precursor cells from the ciliary body could be induced ex vivo to generate cells with markers of retinal pigment epithelium, but no further differentiation was observed (Vossmerbaeumer et al., 2008). As yet, it has not been possible to generate all kinds of required functional retinal cells for transplantation, although retinal precursor cells have been successfully derived from embryonic stem cells (Lamba et al., 2006). Most notably, the generation of photoreceptors has only been possible with pre-differentiated cells isolated around birth, when photoreceptors are formed physiologically (MacLaren et al., 2006; Bartsch et al., 2008). Otherwise the potential of retinal precursor cells appears to be limited to glial cells or retinal ganglion cells (Canola et al., 2007). The specific precursor cell for photoreceptors, for example, has not yet been identified molecularly, and the exact sequence of events during the complex production of all different cell types in the retina has not been fully elucidated. A "unifying model" of photoreceptor development is not yet available (Adler and Raymond, 2008). Nevertheless, remarkable progress has been made to generate rod- and cone-like cells from ES cells in vitro by sequential cell culture protocols (Osakada et al., 2008), suggesting that the final gaps to truly functional cells can ultimately be closed (Klassen and Reubinoff, 2008). But the final step is not trivial. Rod-like cells have also been generated from adult-derived iris (Akagi et al., 2005), further complicating the picture.

Precursor cell populations are found in the ciliary body and the retinal cell layers themselves, there possibly including Müller glia

The precursor cells for photoreceptors are found in the pars plana of the ciliary body (Nishiguchi et al., 2008). They seem to "disappear" once retinal development is completed but are stimulated in response to injury (Nishiguchi et al., 2008) and degeneration (Nishiguchi et al., 2009). No mice older than 30 days have been studied, and it is not clear how functional the newly generated cells are, but these reports are as close to "adult neurogenesis" in vivo as we have gotten so far in the case of the retina.

Besides cells in the ciliary body, precursor cells with neurogenic potential are found in the retina itself, and these cells share a number of markers with precursor cells from the SVZ. They express putative precursor cell markers nestin, Blbp, Pax6, Flk-1, Hes1, and Musashi, and neurospheres could be derived ex vivo (Engelhardt et al., 2004; Engelhardt et al., 2005). Ex vivo, retinal precursor cells, like their ontogenetic predecessors in the wall of the developing optic vesicle, are bipotent and generate neurons and astrocytes, but no oligodendrocytes (Alexiades and Cepko, 1997; Ahmad et al., 1999; Engelhardt et al., 2004). Precursor cells for photoreceptors were not found when the cells were isolated later than around birth (Bartsch et al., 2008).

Müller cells, which are the radial glial elements of the eye, might also act as precursor cells in the adult mammalian retina (Fischer and Reh, 2001). Remarkably, Müller cells specifically formed amacrine cells in vivo (Karl et al., 2008), another important neuronal population in the retina besides the photoreceptors that attracted the most attention. This finding suggests that different precursor cell populations, some of which persist into adulthood, generate the different cell types in the eye.

After implantation of precursor cells from other brain regions, for example the hippocampus, the retina seems to allow an astonishing degree of neuronal development, but no full maturation (Tamada et al., 1995; Takahashi et al., 1998; Young et al., 2000; Warfvinge et al., 2001; Akita et al., 2002; Chacko et al., 2003; Mellough et al., 2004). This support of multi-lineage integration sets the retina apart from the other non-neurogenic regions and makes the retinal precursor cell niche particularly interesting.

Overall it is not clear which kinds of precursor cells, if any, persist in the adult retina and what their lineage relationship is. One study found putative precursor cells to be concentrated in the periphery of the retina and argued that these cells might be remnants from embryonic development (Zhao et al., 2005). The nature of the quiescent cells is as yet elusive. Accordingly, next to nothing is known about the potential functional role of precursor cells in the adult eye, including precursor cell functions of Müller glia, for normal and diseased physiology of the eye. But as in other brain regions, many efforts are directed at finding ways to tap the regenerative potential of the precursor cells of the eye (Boulton and Albon, 2004; Lamba et al., 2008).

In addition, the eye contains yet another important population of precursor cells that are not part of the central nervous system: limbal stem cells continuously maintain the corneal epithelium (Avunduk and Tekelioglu, 2006).

Substantia Nigra

Because of the important role in the dopaminergic system and thus its relevance for Parkinson disease, the adult substantia nigra has attracted particular interest from stem cell researchers. Whereas neural precursor cells could indeed be found in the substantia nigra (Lie et al., 2002), the question of whether physiological adult neurogenesis occurs in the substantia nigra was less clear for a while. One group first reported adult neurogenesis (Zhao et al., 2003), but many other studies have argued against it (Lie et al., 2002; Cooper and Isacson, 2004; Frielingsdorf et al., 2004; Yoshimi et al., 2005; Steiner et al., 2006b). One report even included the infusion of PDGF-BB or BDNF and still could not find BrdU-labeled cells expressing tyrosine-hydroxylase, the key enzyme for dopamine synthesis and a key marker for dopaminergic neurons in the substantia nigra (Mohapel et al., 2005). By vote this is a clear situation, but the situation was unusual in that Ming Zhao, Jonas Frisén, and coworkers from Karolinska Institute in Stockholm had used several methods to support their initial claim (Zhao et al., 2003). Taking the side of their study is a report that chronic intraventricular administration of 7-hydroxy-N,N-di-n-propyl-2-aminotetralin, an agonist of the dopaminergic D3 receptor, can stimulate neurogenesis in the substantia nigra, especially after a 7-OH-DPAT lesion, another model of Parkinson disease (Van Kampen and Robertson, 2005). And a similar claim has been made in a rather extensive study on the basis of a nestin-LacZ reporter mouse, in which co-localization of LacZ with NeuN was reported and interpreted as evidence of adult neurogenesis in the substantia nigra (Shan et al., 2006). But without conditional regulation (e.g., with the ERT2 system, see p. 257) and without persistent reporter expression, for example, from the Rosa26 locus, Nestin reporter-mice cannot be used for lineage tracing. Furthermore, the claims for co-localization of LacZ with NeuN cannot be reproduced from the published figures, so that this report must not be counted in support of neurogenesis in the adult substantia nigra.

Hypothalamus

The adult hypothalamus contains precursor cells that are neurogenic ex vivo but not in vivo

The hypothalamus is another physiologically non-neurogenic region that contains neural precursor cells that can be isolated and propagated in vitro (Markakis et al., 2004). These precursor cells produce neuroendocrine phenotypes after initiation of differentiation. Surprisingly, they behaved almost indistinguishably from hippocampal cultures maintained in parallel. When the cell culture protocol was adjusted so that hypothalamic precursor cells produced neuroendocrine phenotypes in vitro, it turned out that hippocampal precursor cells could do the same. Consequently, precursor cells in the adult brain, no matter which region they are from, share a wide spectrum of developmental potential. Obviously, this is not identical to the realization of this potential as it occurs under physiological conditions. In addition, only the initial stages of neuronal differentiation could be reproduced in vitro; beyond this stage, the cells displayed signs of multiple phenotypes in parallel.

The hypothalamus is rich in PSA-NCAM-expressing cells, but evidence of neurogenesis is weak

That the hypothalamus is a site of ongoing plasticity had been supported early by the observation that PSA-NCAM remains expressed there throughout adulthood (Bonfanti et al., 1992; Seki and Arai, 1993a, 1993b). Indeed, cell proliferation is routinely detectable in the adult hypothalamus, although the exact nature of the proliferating cells is not yet known. At least some of them appear to express NG2. A possible neurogenic potential of these cells was first indicated by a study in which BDNF was infused intracerebroventricularly, resulting in increased numbers of BrdU-labeled cells in the hypothalamus (Pencea et al., 2001). These had a homogenous distribution, and despite some higher density near the ventricle, no preference for the wall of the third ventricle was seen. At the time the finding was interpreted as "neurogenesis," but Map2 was used as marker, and the two examples of new neurons shown in the publication do not fully convince. Similarly, Fowler and colleagues reported BrdU-labeled cells in the hypothalamus of untreated prairie voles (Fowler et al., 2002). The level of evidence that neurogenesis occurred was still not sufficient: β-III-tubulin was used as neuronal marker, and in the figure no double-labeling can actually be seen. Still, the report was important because, interestingly, the number of BrdU-labeled cells decreased in a social-stress paradigm (Fowler et al., 2002). Similarly, some effect of the duration of the daily exposure to light on cell proliferation in the hypothalamus of hamsters was reported (Huang et al., 1998). Here the presumed occurrence of neurogenesis was exemplified by one highly condensed, irregular shaped BrdU and NeuN-positive nucleus: again not sufficient to make a general claim.

Another study found proliferating nestin-positive cells in the wall of the third ventricle, supposedly tanycytes (compare Chap. 5, p. 168) that migrated into the hypothalamus and differentiated into orexin-positive neurons (Xu et al., 2005).

Infusion of CNTF boosted the presence of cells with immature neuronal markers, which appeared to be of direct functional relevance

In 2005, however, Maia Kokoeva and colleagues from Boston showed that infusion of CNTF (ciliary neurotrophic factor) into the mouse brain boosted the production of BrdU-labeled cells that were positive for neuronal marker Hu and doublecortin (Kokoeva et al., 2005). About the shortcomings of Hu and DCX as markers of neurogenesis, see Chapter 7, pp. 244 and 245. Unfortunately, therefore, neither other and more specific markers were used, nor were developmental stages investigated. But the study was very remarkable in that it showed that, when cell proliferation and genesis were inhibited by the infusion of cytostatic agent cytosine arabinoside, the physiological long-term reduction of the body weight in response to CNTF disappeared, suggesting that the new cells are normally involved in that response. Kokoeva, together with colleagues Huali Yin and Jeffrey Fliers, went on to examine whether the presumed adult neurogenesis that was found after CNTF could also be detected physiologically, if the sensitivity of the detection was increased. BrdU was infused intracerebroventricularly, which greatly increased the number of labeled cells compared to intraperitoneal injections (Kokoeva et al., 2005). But again, besides DCX and Hu, only β-III-tubulin was used. The cells were negative for NeuN, and the authors speculate that the newborn cells might belong to the neuronal populations that are negative for NeuN (Kokoeva et al., 2007; compare

Mullen et al., 1992; and Weyer and Schilling, 2003). This of course could be possibly shown, and the absence of NG2 staining in these cells would also be instructive in this context. But only the demonstration of developmental stages of neurogenesis would dispel the doubts about constitutive and inducible adult hypothalamic neurogenesis that are still justified.

Amygdala

Several studies reported adult neurogenesis in the amygdala, but the level of evidence is still low

Because of its role in anxiety and other emotions as well as its involvement in depression, the amygdala has drawn considerable interest as a potential neurogenic region. Like in many of the other brain regions for which conflicting data on adult neurogenesis exist, the amygdala (and the neighboring piriform cortex; see below) show high expression of doublecortin and PSA-NCAM (Nacher et al., 2002; Nacher et al., 2004).

But the question of neurogenesis in the adult amygdala has not been conclusively answered and is rather similar to the situation in the hypothalamus. Some studies actually cover both areas and were thus prone to the same methodological issues discussed above (Fowler et al., 2002; Fowler et al., 2005). One report found evidence of new neurons in the adult amygdala in monkeys (Bernier et al., 2002), whereas others could not detect it in naïve rodents (Park et al., 2006).

In the case of seizure-induced neurogenesis in the amygdala of rats, the argument is again based on DCX expression in BrdU-labeled cells (Park et al., 2006). In addition, two tiny BrdU/NeuN-double-labeled profiles were shown, which were only a fraction of the size of the surrounding neuronal nuclei and revealed a very dense and homogenous staining pattern. Removal of the olfactory bulb, which is a bona fide model of depression in rodents, decreased cell proliferation in both SGZ and SVZ but increased cell proliferation in the basolateral amygdala (Keilhoff et al., 2006). Neurogenesis was assessed on the basis of NeuN detection in BrdU-positive cells, and the two cells shown as example of new neurons show nucleolus-sized BrdU immunoreaction in large NeuN-positive nuclei. This pattern seems too unusual to justify the conclusion of adult neurogenesis at this stage. Another example in the literature shows a similar size difference between the BrdU and NeuN immunoreaction, and here the orthogonal reconstruction reveals that the two also do not fully match in the xz- and yz-planes (Shapiro et al., 2009).

Piriform Cortex

The piriform cortex is the next neuronal relay station after the olfactory bulb, and several reports have claimed adult neurogenesis there as well

The existence of a ventral migratory stream from the RMS to the piriform cortex has been proposed

The piriform cortex represents a large proportion of the olfactory cortex and receives massive input from the olfactory bulb. Given the extreme neuronal turnover in the olfactory epithelium (see below) and the substantial and lifelong generation of new interneurons in the olfactory bulb, the assumption seemed reasonable that in the next relay station neurogenesis could be found. A number of studies have indeed suggested so (Bernier et al., 2002; Luzzati et al., 2003; Pekcec et al., 2006; Shapiro et al., 2007a; Shapiro et al., 2007b; Shapiro et al., 2009). Most examples of putative neurogenesis were based on DCX immunohistochemistry, but besides Tuc4, NeuN was also used (Pekcec et al., 2006). As in the examples from the hypothalamus that were discussed above, however, NeuN staining was cytoplasmic (Shapiro et al., 2007a), and no long-term survival of BrdU/NeuN-double-positive cells was observed (Pekcec et al., 2006). But interestingly, the number of the newly generated cells, even if they most likely were not neurons, was increased after olfactory enrichment (Shapiro et al., 2007a).

During embryonic brain development, the lateral cortical stream originates between the forming cortex and the striatum. Via this structure, Pax6-positive precursor cells reach what becomes the piriform cortex as early as E 11.5 in mice, followed by Dlx2-positive cells two days later. Both populations populate basomedial brain structures. Migration of Pax6-positive cells followed radial glia, whereas the Dlx population showed a pattern more resembling chain migration (Carney et al., 2006). These authors stated (but did not show) that this type of migration ceased postnatally. In contrast, based on intraventricular marker injection, De Marchis and colleagues described the existence of a ventral migratory stream originating from the "knee" of the postnatal rostral migratory stream. Along this stream, new cells reach the piriform cortex as well as other ventral cortical regions (De Marchis et al., 2004). The few examples of NeuN-immunostaining in the cell tracker-filled cells were not unambiguous, the study did not go beyond postnatal animals, and no details of development with marker progression were demonstrated. For both monkeys and rabbits, a similar route of migration, however, has been described that originates earlier from the RMS than at the knee (Bernier et al., 2002; Luzzati et al., 2003). Lee Shapiro, Charles Ribak, and colleagues from the University of California in Irvine investigated that same pathway in the adult mouse (the exact age was not reported) and confirmed the existence of DCX-positive cells migrating towards the piriform cortex (Shapiro et al., 2007b). Time course studies after BrdU injection confirmed that the cells did not proliferate in loco but migrated in. They might originate from a ventral portion of the SVZ, which, as Shapiro et al. discuss, hardly contributes to neurogenesis in the olfactory bulb (Merkle et al., 2004). This finding supports the idea that the adult SVZ contributes cells outside the oligodendrocytic lineage to numerous brain regions. Presumably these cells originate from Dlx2-positive cells, but for the ventral forebrain structures this remains to be shown. In this context, the possible connection between certain C cells in the SVZ and the NG2 cells should be remembered (Aguirre et al., 2004). Whether these can include neurons remains to be shown by the same standard that applies to the neurogenic regions. The phenomenon is interesting in any case, and invites further investigation.

Nestin-expressing radial cells and DCX-positive cells are found in the adult piriform cortex

This migration from the SVZ notwithstanding, the piriform cortex also contains a population of radial glia–like cells, which as far as is known, do not serve as precursor cells. They are clearly visible in Nestin-GFP mice (Chen et al., 2009). Similarly, both DCX and PSA-NCAM are expressed, in particular in Layer II of the piriform cortex (Seki and Arai, 1991; Nacher et al., 2002; Nacher et al., 2004), further suggesting that the piriform cortex is a particular structure of presumably high plasticity. Nestin expression was up-regulated by seizure activity (albeit in numerous cell types) in the piriform cortex (Nakagawa et al., 2004). Chronic restraint stress increased the number of DCX and PSA-NCAM-positive cells, but the chronic administration of corticosterone had the opposite effect (Nacher et al., 2004).

In summary, the piriform cortex might be one of the most interesting non-neurogenic regions, because it shows an intriguing degree of cellular and other plasticity, which might actually be linked to the regulation of neurogenesis elsewhere. The available results from the piriform cortex indicate that neurogenesis should not be considered in isolation but is part of more general plastic phenomena in the adult brain.

Reactive and Regenerative Adult Neurogenesis

Reactive and regenerative neurogenesis is adult neurogenesis in response to damage where no physiological adult neurogenesis occurs

Several studies have suggested that the lack of neurogenic permissiveness in some brain regions can be overcome by a pathological stimulus. This response does not necessarily have to lead to "regeneration" but might be purely reactive, without functional consequences. Examples have already been mentioned in the above list of brain areas with disputed reports of adult neurogenesis. Two conditions have to be fulfilled to make regenerative neurogenesis possible: neural precursor cells must be present at the site of regeneration or be attracted to it, and the local microenvironment must be switched to neurogenic permissiveness. Key triggers for this response appear to be hypoxia and ischemia.

Targeted induced cell death caused limited local neurogenesis

Jeffrey Macklis, Constance Scharff, and colleagues pioneered this work with a study on stimulating neurogenesis in the bird brain by inducing a very small and precisely positioned ischemic lesion (Scharff et al., 2000). The method consists of the stereotaxic application of a phototoxic drug into the area to which the cells of interest project. The cells take up the compound, transport it retrogradely, and become vulnerable to light. To induce cell death, laser light is shone onto the brain region of interest. Only the loaded neurons die. Because they die by apoptosis, the tissue reaction is minimal or absent. The great advantage of this experimental model is that the lesion is precisely on the level of single cells from a targeted population.

In 2000, Sanjay S. Magavi and Jeffrey D. Macklis at Harvard University applied the same method to the rodent brain (Figs. 8–6 and 8–7). They selectively killed corticothalamic neurons in Layer VI of the anterior cortex (Magavi et al., 2000) and observed that putative precursor cells, some of them DCX positive, migrated toward the lesion site. For at least 28 weeks after the lesion, they detected new neurons with the BrdU method. By retrograde tracing, they were able to confirm that the new cells adequately projected to the thalamus. This study suggested that adult neurogenesis could be induced in adult mammalian non-neurogenic regions, and that cell death is sufficient to trigger it. The key requirement might be that the lesion has to be small enough to prevent a general reaction of the surrounding tissue. In the absence of an inflammatory response, for example, neurogenesis might become possible. The damage, however, would have to fundamentally change the level of neurogenic permissiveness. This was also shown by implanting neural progenitor cells in the cortex of rats in which targeted apoptosis had been induced; the transplanted cells developed into neurons (Sheen and Macklis, 1995; Snyder et al., 1997; Shin et al., 2000). The model was extended in a study in which corticospinal projection neurons were regenerated after targeted photolysis (Chen et al., 2004a). There have not been more follow-ups, and consequently many questions remain unanswered.

Cell death is not generally a trigger for neurogenesis

The degree to which more extensive cell death can trigger adult neurogenesis has been debated for many years (Gould and Cameron, 1996). The data are ambiguous. On one hand, the induction of highly selective cell death in individual cells could trigger neurogenesis (Magavi and Macklis, 2002a). On the other hand, however, cell death under pathological conditions can be associated with an antineurogenic inflammatory response (Monje et al., 2003). It might thus be that apoptotic cell death is compatible with neurogenesis and even conducive, whereas the necrotic forms are not. There are no indications, however, that neurogenesis is necessarily linked to cell death, because adult neurogenesis in the neurogenic zones is not primarily a regenerative event. In non-neurogenic regions, this is fundamentally different, and neurogenesis in non-neurogenic regions might thus be more affected by death signals.

Intraocular NMDA injection caused degeneration of neurons in the inner retinal layers and stimulated the proliferation of Müller glia and caused the production of new amacrine cells (Karl et al., 2008), the first example of regenerative neurogenesis in the adult rodent retina.

Hypoxia might be a potent stimulus of neurogenesis in the striatum

In a middle cerebral artery occlusion model in which transient blockage of the blood flow to the brain mimics the situation of focal cerebral ischemia in human patients, the size of the infarct can be titrated by varying the ischemic interval. Andreas Arvidsson from Olle Lindvall's group in Lund, Sweden, showed that when the ischemia damaged only the striatum, a very small number of degenerating medium spiny interneurons in the striatum were replaced from precursor cells migrating in from the neighboring SVZ (Arvidsson et al., 2002). An independent parallel study by Jack Parent and colleagues from Ann Arbor showed very similar results (Parent et al., 2002). This attraction of SVZ precursor cells is dependent on the C-X-C chemokine receptor 4 (CXCR4) and its ligand, the chemokine stromal cell-derived factor-1α (SDF-1α) (Thored et al., 2006). Migration into the ischemic striatum occurred along the newly forming microvasculature as part of an angiogenic response to the insult (Thored et al., 2007). Because other studies have shown that the overexpression of BDNF in the ventricular wall can lead to striatal neurogenesis as well (Benraiss et al., 2001; Chmielnicki et al., 2004), striatal damage might thus result in a shift in the neurotrophic balance toward neurogenic

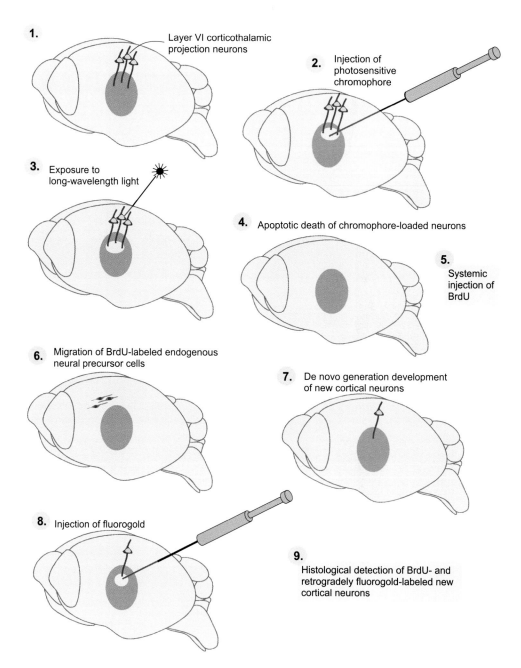

FIGURE 8–6 Targeted neurogenesis in the adult murine neocortex. Magavi et al. (2000) reported cortical neurogenesis after the selective phototoxic ablation of corticothalamic neurons in the anterior cortex of adult mice. Integration of new BrdU-labeled neurons was confirmed by retrograde fluorogold tracing from the thalamus (see Fig. 8–7).

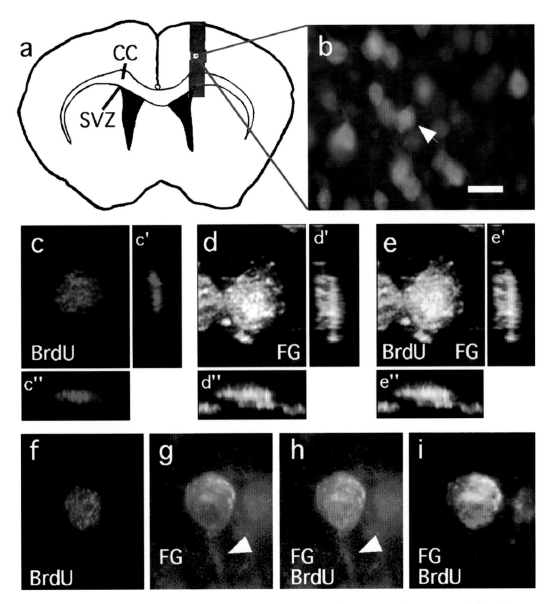

FIGURE 8–7 Targeted neurogenesis in the adult murine neocortex. In the experiment by Magavi et al. (2000) explained in Figure 8–6, new cortical neurons made appropriate target connections in the thalamus, here visualized with retrograde fluorogold tracing (FG) in BrdU-positive cells. (A and B): Camera lucida drawing and photo of a BrdU-positive (*red*)/FG (*white*) retrogradely labeled neuron. CC, corpus callosum; SVZ, subventricular zone. Arrowhead indicates labeled neuron. Scale bar: 20mm. (C–E): Confocal three-dimensional reconstruction of the neuron. (F): BrdU-positive (*red*) nucleus of a new FG-positive corticothalamic neuron. (G–I): FG-positive (*blue*) cell body with labeled axon (*arrowheads*). Reprinted from the cited publication with kind permission by Jeffrey D. Macklis, Cambridge, Massachusetts, and Macmillan Publishers (Nature), Copyright 2000.

permissiveness in the adult striatum. This increased neurogenic permissiveness could be linked to the presence of pro-neurogenic microglia (Thored et al., 2009) and showed remarkably little decrease with age (Darsalia et al., 2005). The number of new neurons in the ischemic striatum was extremely low, though, and thus unlikely to lead to a substantial functional recovery. On the other hand, the neurogenic response persisted for more than a year after the insult (Thored et al., 2006). In the controls no new neurons were detected. In this sense, the Arvidsson study could serve as further proof of principle that regenerative neurogenesis is possible in non-neurogenic regions. The Swedish group was able to show that, over weeks after the damage, the new cells progressed through developmental stages typical for medium spiny interneurons and identifiable by the expression of transcription factors Pbx and Meis2 (Arvidsson et al., 2002).

Growth factor infusion aided reactive neurogenesis after ischemia in CA1, but the finding remains unconfirmed

Hirofumi Nakatomi, Masato Nakafuko, and colleagues from Tokyo used a similar ischemia model in which ischemia caused damage of hippocampal region CA1 (Nakatomi et al., 2002). After the ischemic stimulus, and boosted by massive intraventricular infusions of growth factors EGF and FGF-2, they claimed that migratory precursor cells from the posterior SVZ contributed to an impressive structural reconstitution of CA1. This finding seemed to suggest that, in addition to interneurons in the olfactory bulb and possibly the striatum, SVZ precursor cells might also generate excitatory principal neurons in CA1. This range of developmental potential would reflect precursor cell heterogeneity in the SVZ (for example, in the sense of a dorsal vs. ventral positional identity of the precursor cells; Merkle et al., 2007). But this was the only report so far including the hippocampus in this spectrum of developmental options. The proposed route of migration of precursor cells has not been found under physiological conditions (Kronenberg et al., 2007).

The unprecedented extent of the presumably precursor cell-based regeneration also surprised many researchers. One skeptical question is whether BrdU-incorporation might have labeled dying neurons (Kuan et al., 2004), which survived as "zombies" without normal function and antigenic properties (e.g., the down-regulation of NeuN; see Chap. 7, p. 241) but could be rescued by the massive growth factor application.

In the majority of studies, there is no good evidence for reactive neurogenesis after ischemia in the cortex

Numerous reports claimed regenerative neurogenesis in the adult neocortex after stroke (Gu et al., 2000; Jiang et al., 2001; Zhang et al., 2001; Zhang et al., 2004; Hoehn et al., 2005; Zhang et al., 2005; Jin et al., 2006; Leker et al., 2007; Wang et al., 2007; Li et al., 2008; Ling et al., 2008; Abrahams et al., 2009; Gu et al., 2009; Kuge et al., 2009; Wang et al., 2009) but unfortunately in none of these cases does the evidence in favor of adult neurogenesis yet meet the most stringent criteria (for example, as outlined in Chap. 7, p. 239). Most of these publications are based on BrdU. Given the number of publications over almost a decade, the fact that no independent confirmation with additional methods (retroviral labeling, dense time-course studies, etc.) has been possible is worrying. Persistent expression of NG2, for example, was rarely excluded, and in many reports the conclusion of neurogenesis was based on DCX, β-III-tubulin or Map2 expression. In some of these cases the reported (questionable) neurogenesis could be boosted by the application of exogenous factors such as VEGF (Wang et al., 2009), kallikrein (Ling et al., 2008), FGF2 (Leker et al., 2007), indomethacin (Hoehn et al., 2005), or BDNF (Keiner et al., 2009), but these stimulating effects do not per se strengthen the argument in favor of adult neurogenesis, because the production of non-neuronal cells (and especially NG2 cells) might easily be stimulated as well.

One study suggests the neurogenic response of Layer I precursor cells to global ischemia

In the beginning of 2010, this interpretation was questioned by a very extensive study that found a proliferative response of a previously undescribed population of precursor cells in cortical Layer I of rats to global ischemia and the neogenesis of interneurons in underlying cortical layers (Ohira et al., 2010). The study was amazingly broad in its methodology and made a strong case. The proliferating cells were identified as Nkx2.1 and MafB expressing, which would suggest their origin from the ganglionic eminences, but they were Nestin-negative and also did not show signs of an oligodendrocytic or microglial lineage. On the other hand, such an unprecedented claim raises many questions and, not surprisingly, not all parts of the story are already equally convincing. But even assessment of cellular functionality was attempted. Cell genesis from the pial surface down (as in the cerebellum), however, is opposite to the prevailing directionality during cortical development, which makes this story, if confirmed, particularly interesting.

The cellular response to ischemia requires further investigation

But even if post-stroke neurogenesis in the penumbra has generally not yet been proven beyond reasonable doubt, in summary the available data clearly indicate an unexpected and as yet largely unexplored amount of cell genesis in response to ischemia, involving several types of cells with precursor cell properties in different regions. Given the substantial evidence for this cellular response in the penumbra, the next experimental steps have to be made to understand this process. The study on precursor cells in Layer I has made important steps into this direction (Ohira et al., 2010). But in most reports, the questionable new neurons were usually found in the peri-infarct area; that is, in the area of secondary ischemic damage. This predisposition has been linked to the ongoing neovascularization and angiogenesis in this zone (Ohab et al., 2006; Ohab and Carmichael, 2008).

Cell cycle marker expression in neurons near the infarct might be indicative of death rather than division

Besides the dominating explanation that precursor cells migrating in from the SVZ or local precursor cells are the origin of stroke-induced cortical neurogenesis, the idea has been proposed that neurons themselves would divide and give rise to the new neuronal daughter cells (Jiang et al., 2001; Gu et al., 2009). In some analogy to the activation of cell cycle machinery in neurons in Alzheimer disease (see p. 534), neurons predisposed to die from ischemia might enter the cell cycle and attempt cell divisions. The presence of cell cycle markers in NeuN-positive cells in the peri-infarct region is unexpected and unusual but would be consistent with this idea. But from these data alone, one cannot conclude that neurogenesis from neurons occurred.

Regenerative striatal neurogenesis has also been reported for excitotoxic lesions, the quinolic acid model of Huntington disease (Tattersfield et al., 2004; Collin et al., 2005). One study suggested that new DCX-expressing cells were recruited only within two days after the lesion, whereas afterwards only glial cells migrated from the SVZ into the striatum (Gordon et al., 2007). BDNF potently increased regenerative striatal neurogenesis in this model (Henry et al., 2007).

Functional relevance of the limited striatal neurogenesis after ischemia is unknown

Although the presented level of evidence for neurogenesis to occur in post-ischemic striatum has been quite good, the limited number of reports and the many potential pitfalls still caution us not to prematurely close the case. Regenerative neurogenesis remains an extraordinary claim, which as we all know, requires extraordinary evidence. The available information clearly points to the fact that, in terms of cellular plasticity after injury, much more is going on than previously thought possible. On the other hand, the functional consequences of these events remain essentially unknown and many details will have to be unraveled until "regenerative neurogenesis" is beyond doubt. Some healthy skepticism, but together with the openness to changing concepts, is still appropriate.

Exogenously Induced Adult Neurogenesis

Adult neurogenesis might be inducible with extrinsically introduced factors

Closely related to the question of regenerative neurogenesis is "targeted" or "induced neurogenesis," which can but does not have to take place in the context of pathology. Here the idea is to elicit a neurogenic response by some extrinsic manipulation; e.g., the injection or overexpression of a stimulating factor. Mediating pro-neurogenic molecules are described in Chapter 9 in greater detail. Table 8–3 provides an overview in the present context. The same serious methodological caveats apply than with the other putative cases of neurogenesis in non-neurogenic regions.

Table 8–3 Induced adult neurogenesis in physiologically non-neurogenic regions

Region	Mediator	Method	Type of neuron	Reference
Striatum	BDNF	Adenoviral gene transfer	Interneurons	(Benraiss et al., 2001; Chmielnicki et al., 2004; Bedard et al., 2006; Cho et al., 2007)]
	BDNF	AAV gene transfer in normal and quinolic acid lesioned rats	Interneurons	(Henry et al., 2007)
	BDNF, PDGF-BB	i.c.v. infusion in 6-OH-dopamine lesioned rats	Interneurons	(Mohapel et al., 2005)
	D3 receptor stimulation	7-hydroxy-N,N-di-n-propyl-2-aminotetralin	Interneurons	(Van Kampen et al., 2004)
Substantia nigra	D3 receptor stimulation	7-hydroxy-N,N-di-n-propyl-2-aminotetralin	TH-positive neurons	(Van Kampen and Robertson, 2005)

Adult Neurogenesis in the Olfactory Epithelium

Neuronal Development in the Olfactory Epithelium

Massive adult neurogenesis occurs in the olfactory epithelium, which is part of the peripheral nervous system

Neurogenesis in the adult olfactory epithelium replaces receptor neurons

The process of neuronal development in the olfactory epithelium is well described

Neurogenesis in the adult olfactory epithelium generates new olfactory receptor neurons. The first description of neurogenesis in the adult mammalian olfactory epithelium was reported by Nagahara in 1940 (Nagahara, 1940). Adult neurogenesis in the olfactory epithelium was rediscovered in the early 1970s (Graziadei, 1973; Graziadei and DeHan, 1973; Graziadei and Graziadei, 1979a; Graziadei and Graziadei, 1979b). Since then it has been described for humans and nonhuman primates (Graziadei et al., 1980) and is conserved in rodents into old age (Loo et al., 1996). Olfactory receptor neurons reside within the olfactory epithelium in the roof of the nasal cavity (Fig. 8–8). At least in mice, there are large regional differences in the amount of turnover and the kinetics of neurogenesis, so that subregions of the epithelium can be identified (Vedin et al., 2009). Adult neurogenesis is particularly low in dorsomedial regions (of the mouse), an area where during development distinct types of radial glia–like precursor cells can be identified that do not exist elsewhere in the epithelium (Murdoch and Roskams, 2008). Consequently, the olfactory epithelium is no homogenous structure.

Overall, the amount of neurogenesis in the adult olfactory epithelium is the highest of all known neurogenic regions. It has been speculated that one of the possible reasons for this high neuronal turnover in the olfactory epithelium is that olfactory receptor neurons are exposed to and particularly sensitive to noxious chemical agents. By virtue of the necessity to recognize chemical molecules, supposedly they are particularly vulnerable to the neurotoxic effects of these substances and thus need to be replaced constantly.

Development of olfactory receptor neurons has been studied in considerable detail (Kawauchi et al., 2004) and originates from the precursor cells, which are at the most basal level of the olfactory epithelium. Differentiating cells migrate vertically to more superficial layers while they mature. It seems that horizontal migration occurs only before neuronal

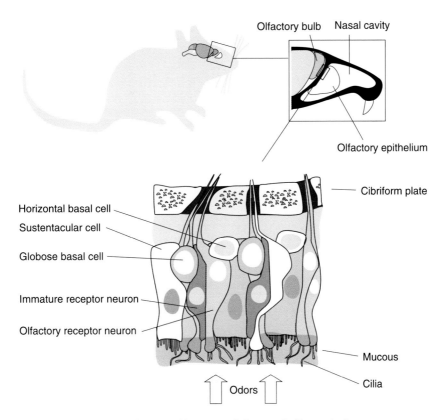

FIGURE 8–8 Anatomical relationship between olfactory epithelium and olfactory bulb.

differentiation begins on the level of precursor cells. This causes a columnar organization of neuronal development. The developing neurons begin to express a sequence of neuronal markers, from immature to mature. Immature neuronal markers, for example, transcription factors like NeuroD, and β-III-tubulin are expressed in the layer just above the basal globose cells. Mature markers are the olfactory-specific G-protein subunit (Golf) and olfactory marker protein (OMP).

The new neurons have to send their axons to the olfactory bulb. Early during their rise, the new neurons also express GAP43, a protein associated with axon elongation. The distance between olfactory epithelium and the olfactory bulb is therefore a region of extremely high axonal growth in the adult. A specialized local population of glia cells, called "olfactory ensheathing glia," promotes axonal growth.

Neuronal Precursor Cells in the Adult Olfactory Epithelium

Globose basal cells are the most prominent precursor cells in the olfactory epithelium, whereas horizontal basal cells might represent more quiescent stem-like cells

The precursor cell population for adult neurogenesis in the olfactory epithelium resides in the olfactory epithelium itself (Fig. 8–9).

The putative stem cells are found in the basal layer, express keratin, and are called *globose basal cells*. When globose basal cells are missing, no reconstitution of the epithelium is possible any more (Jang et al., 2003). Although widely equated with *the* stem cells in this system, it is not quite clear whether the globose cells actually are the most stemlike cells in the olfactory epithelium. The other candidates are "horizontal basal cells" and sustentacular cells. In any case, not all globose basal cells are stem cells, but

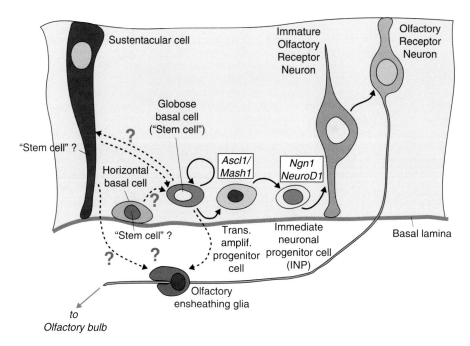

FIGURE 8–9 The olfactory epithelium. Olfactory receptor neurons are the first neurons of the olfactory system. They show lifelong turnover. The highest-ranking stem cells of the olfactory epithelium appear to be the sustentacular cells, which correspond to radial glial cells of the central nervous system. The globose basal cells are the transiently amplifying progenitor cells of this system. They give rise to immature receptor neurons. Differentiated receptor neurons carry cilia, which are embedded in a layer of mucous in which odorants dissolve.

they seem to account for most of the ongoing cell divisions (Huard and Schwob, 1995). Subpopulations express numerous stem cell markers, such as Pax6, Sox2 and members of the Notch system (see overview in Murdoch and Roskams, 2007). The globose basal cells can be identified by a specific antigen called GBC-1 (Goldstein and Schwob, 1996). Another population of (albeit slowly) proliferative cells in the epithelium are horizontal basal cells, which however do not express any stem cell markers and are of unclear lineage relationship. There is some evidence, though, that they might nevertheless represent the relatively quiescent stem cell of the olfactory epithelium (Iwai et al., 2008). In any case, the putative stem cell gives rise to an Ascl1/Mash1-expressing transient amplifying progenitor cell, which in turn generates an "intermediate neuronal progenitor cell" (INP) that expresses Neurogenin 1 (Davis and Reed, 1996).

Ex vivo, precursor cells from the olfactory epithelium are multipotent in that they generate neurons and glia (Chen et al., 2004b; Murrell et al., 2005; Iwai et al., 2008). Neuronal precursor cells have also been isolated from the human olfactory mucosa (Murrell et al., 2005; Zhang et al., 2006). A potential for transdifferentiation has been reported but deserves the same skepticism applicable to all claims for transdifferentiation (Murrell et al., 2005). In addition, at least during embryonic development, a type of mesenchymal precursor cell has also been isolated from the nasal mucosa (Tome et al., 2009).

Sustentacular cells share some similarities with radial glia and also might have precursor cell functions

Intermediate filament nestin is expressed in many CNS precursor cells. In the olfactory epithelium, however, it appears that the nestin-expressing cells are not the basal globose cells but the neighboring sustentacular cells. Sustentacular cells, in turn, might have a function similar to that of radial glia in the CNS. As we have seen in Chapter 3, radial glia–like cells play a dual role in adult neurogenesis. They serve as stem cells and scaffold for the ensuing neuronal migration and differentiation. A similar function of sustentacular cells has not yet been demonstrated for the olfactory epithelium but also not been ruled out (Murdoch and Roskams, 2007). The sustentacular cells express transcription factor Pax6 (Davis and Reed, 1996) and stem cell factor Sox2 (Murdoch and Roskams, 2007). Pax6 is one of the master genes in neurogenesis and affects lineage choices between glial and neuronal differentiation.

The intermediate progenitor cells give rise to immature neurons, which have a biopolar shape and extend their neurites. From these, the cilia-bearing mature olfactory receptor neurons develop and make functional contact with their appropriate target glomerulum in the olfactory bulb.

Regulation of Neurogenesis in the Adult Olfactory Epithelium

Neurogenesis in the adult olfactory epithelium is triggered by behavior and noxae

Neurogenesis in the adult olfactory epithelium can be influenced by numerous manipulations. A change in behavior leads to changes on a systems level that in turn affect communication between cells, leading to effects on the transcriptional control of genes responsible for neurogenesis (Nicolay et al., 2006). For example, exploratory behavior might set the stage for alertness in the olfactory system, which leads to an increase in neurotrophin 3 (NT3) that in turn stimulates the PI3K/Akt pathway promoting neuronal survival via the activation of pro-neurogenic transcription factors.

The means by which neurogenesis in the olfactory epithelium is regulated in response to behavior is not well studied. Whether the act of olfaction itself has consequences for neurogenesis is not clear. It is rather believed that an increase in olfactory behavior will lead to greater exposure to chemical noxae and thus to more damage to olfactory epithelium, which in turn should provoke increased cellular turnover (Loseva et al., 2009). The olfactory epithelium would not behave differently from other epithelia, which are constantly replaced (Graziadei, 1973). The turnover in response to noxae has in fact been demonstrated and is well conserved evolutionarily (Schmidt, 2007).

Direct damage to the epithelium stimulates neurogenesis

Tampering with the olfactory system induces neurogenesis of the olfactory epithelium. Already some of the earliest reports on adult olfactory receptor neurogenesis identified damage as a neurogenic stimulus. In 1979, Graziadei reported that damage to the axons of the olfactory receptor neurons through olfactory bulbectomy causes severe retrograde degeneration that essentially depletes the olfactory epithelium of receptor neurons (Graziadei and Graziadei, 1979a). Macrophages clear the area of damage. Macrophages also up-regulate leukocyte inhibitory factor (LIF), which can stimulate neurogenesis at the early stages of progenitor cell proliferation (Getchell et al., 2002). This fits with the hypothesis that immune cells and mediators are important regulators of adult neurogenesis (see Chap. 9, p. 375). When macrophages were depleted from the epithelium, this process was interrupted, and adult neurogenesis was not increased (Borders et al., 2007a; Borders et al., 2007b). Similarly, when scavenger receptor A was absent, which is required to attract the macrophages, neurogenesis was similarly not induced (Getchell et al., 2006). The gene expression in the olfactory epithelium has been studied in a dense time course after bulbectomy so that first concepts about the sequence of molecular events after lesion-induced neurogenesis become possible (Getchell et al., 2005; Getchell et al., 2006).

About one week after the damage, a strong proliferative response of basal cells can be detected, and after roughly one month, the olfactory epithelium is reconstituted to normal strengths. A detailed first description of developmental stages in olfactory receptor neurogenesis has already been given in the Graziadeis' admirable 1979 study, in which the basal cells were identified as the precursor cells driving adult neurogenesis (Graziadei and Graziadei, 1979a; Graziadei and Graziadei, 1979b). Many studies have elaborated on these initial findings. For example, it was found that the receptors of the reconstituted epithelium indeed become functional and can respond to odors (Costanzo, 1985).

Olfactory deprivation reduces neurogenesis

Closing one naris is a less invasive way to deprive olfactory receptor neurons. Naris occlusion on postnatal day one in rats acutely depressed cell proliferation in the olfactory epithelium. Reopening of the naris 20 days later resulted in a sharp rebound and reconstitution of the normal thickness of the epithelium within days (Farbman et al., 1988; Mirich and Brunjes, 2001).

Anne Calof has hypothesized that the precursor cells of the olfactory epithelium "sense" the neuronal density in the epithelium and that a local regulatory mechanism thus allows maintenance of the olfactory epithelium at a defined receptor density.

Many growth and neurotrophic factors support neurogenesis

On the level of growth factors, the regulators of neurogenesis in the adult olfactory epithelium include Fibroblast growth factor-2 (FGF-2) (Nakamura et al., 2002), Epidermal growth factor (EGF) (Barraud et al., 2007), Insulin-like growth factor-1 (IGF-1) (Mathonnet et al., 2001), Brain-derived neurotrophic factor (BDNF) (Isoyama et al., 2004), NT3 (Bianco et al., 2004), Transforming growth factor-α (TGF-α) (Herzog and Otto, 2002), and Leukemia-inhibiting factor (LIF) (Bauer et al., 2003). All of these have been shown to be able to influence neurogenesis. In vitro, FGF-2 acted as strong mitogen on the globose basal cells, TGF-β induced neuronal differentiation, and PDGF acted as a survival-promoting factor (Newman et al., 2000). This might be indicative of a sequential role of these factors. In explant cultures, however, only FGF-2 had an effect on neuronal outgrowth and differentiation (MacDonald et al., 1996). In vivo, LIF was the only factor that was measurably induced before the onset of precursor cell proliferation when ablation of the olfactory bulb was used to trigger neurogenesis in the olfactory epithelium (Bauer et al., 2003).

In female mice, prolactin induced cell proliferation in the olfactory epithelium and so did pregnancy (Orita et al., 2009).

In vitro, mechanical stress to rodent olfactory epithelial cell cultures resulted in increased neurogenesis even in the absence of any added growth factors (Feron et al., 1999b). However, these cultures were not clonal precursor cell cultures, thus still allowing for the possibility that endogenous growth factors participate in the proneurogenic "talk" between different cell types of cultures. These examples show how difficult it is to characterize the role of a single communication molecule in the regulation of adult neurogenesis.

Highly proliferative cells express transcription factor Ascl1/Mash1

Many relevant transcription factors are accordingly up-regulated after bulbectomy (Shetty et al., 2005).

Ascl1/Mash1 is a key transcription factor in this system. Ascl1/Mash1 is expressed in the highly proliferative progenitor cells of the olfactory epithelium and can be used to identify the precursor cells (Vedin et al., 2009). During development, Ascl1/Mash1 is also required to establish the precursor cell population of the medial ganglionic eminence. Accordingly, Ascl1/Mash1-mutant mice show neuronal losses in the olfactory epithelium and the autonomous system, both of which arise from the medial ganglionic eminence (Casarosa et al., 1999). Olfactory progenitor cells exposed to bone morphogenic proteins (BMPs), known as strong inhibitors of neurogenesis, lose Ascl1/Mash1 expression (Shou et al., 1999).

Transcription factor Gdf11 suppresses neurogenesis at the proliferation level; FoxG1 increases it

The second well-studied transcription factor that can be used as precursor cell marker is FoxG1, which antagonizes the effects of Gdf11 (Kawauchi et al., 2009). Gdf11 in turn is the key factor in an intriguing auto-regulatory loop that presumably is central to controlling neurogenesis. At least in vitro Gdf11 inhibits neurogenesis by inducing reversible cell cycle arrest through its effects on p27(Kip1) (Wu et al., 2003). Gdf11 knockout mice consequently have more precursor cells and hence more neurons in the olfactory epithelium, whereas mice that lack another Gdf11 antagonist, follistatin, had decreased neurogenesis (Wu et al., 2003).

During cortical development, FoxG1 plays a similar role: impaired FoxG1 function results in decreased basal progenitor cells (Siegenthaler et al., 2008). At least in postnatal hippocampal neurogenesis, haploinsufficiency for FoxG1 similarly reduced precursor cell function and a smaller dentate gyrus (Shen et al., 2006).

For Hes1 and Hes5 only developmental data are available (Cau et al., 2000). They indicate that Hes1 regulates Mash1 transcription, first to define regional domains in which neurogenesis is to occur in the olfactory placode, and second to control the number of precursor cells within this domain. Hes1 is a suppressor of transcription, thus acting as a negative regulator. In double mutants for Hes1 and Hes5, Neurogenin1, downstream of Ascl1 (Mash1) signaling, is up-regulated, inducing neurogenesis. It is not yet clear whether Hes1 has this role during adult neurogenesis. But Notch1, which signals through Hes5, is expressed in a subset of Ascl1/Mash1-negative basal cells, and Notch ligand is expressed in neighboring cells (Schwarting et al., 2007). Hes6 belonged to the neurogenesis-associated transcription factors up-regulated by bulbectomy (Shetty et al., 2005) and Hes6 promotes neurogenesis during cortical development (Jhas et al., 2006).

Adult Neurogenesis in the Olfactory Epithelium as a Window to the Central Nervous System

|||||||||||||||||||||||||||||||||||||||||||||||||||||||

Neurogenesis in the olfactory epithelium might allow diagnostic access to "neuronal development" in the adult

|||||||||||||||||||||||||||||||||||||||||||||||||||||||

In the general scientific view, neurogenesis in the olfactory epithelium might be somewhat undervalued. What makes adult olfactory neurogenesis particularly appealing is that it can serve as an accessible system in which to study neuronal development in adult humans. All the various steps, from precursor cell proliferation over migration to axon elongation and synaptogenesis, occur within a small area that lies outside the skull. Rett syndrome, for example, is an inherited neurodevelopmental disorder due to a mutation in the gene that encodes methyl CpG-binding protein 2 (Mecp2), which in turn is a repressor of gene transcription, primarily in neurons. Most neuronal populations are inaccessible even for more invasive routine diagnostics. But Rett syndrome has been successfully diagnosed from biopsies of olfactory epithelium (Ronnett et al., 2003)—probably not with good enough validity for clinical use, but nevertheless impressive. Many more disturbances of neuronal development could be diagnosed from the olfactory epithelium, provided the problem affects the olfactory cells as much as the central neurons.

This potential clinical applicability makes further comparative studies worthwhile. An understanding of adult neurogenesis in the olfactory epithelium and its relationship to neurogenesis in the olfactory bulb might help us learn about neurogenesis within the CNS generally. Some researchers have even attempted to study the cellular basis of psychiatric disorders such as schizophrenia from the olfactory epithelium (Arnold et al., 1998; Smutzer et al., 1998; Feron et al., 1999a; Cascella et al., 2007). Although the immense complexity of schizophrenia and the vast number of unknowns in this disorder require that caution be used in this approach, the strategy is intriguing and might lead to novel ways of diagnosing deficits in plasticity underlying neuropsychiatric disorders. The degree to which plasticity in the olfactory epithelium reflects plasticity within the CNS, however, must first be established.

|||||||||||||||||||||||||||||||||||||||||||||||||||||||

Because of their origin in the peripheral nervous system, use of olfactory epithelium precursor cells for cell therapy is problematic

|||||||||||||||||||||||||||||||||||||||||||||||||||||||

Beyond experimental and diagnostic aspects, some researchers are convinced that adult neurogenesis in the olfactory epithelium might provide a resource for cell therapy (Marshall et al., 2006). Autologous neural precursor cells in the olfactory epithelium could be harvested through biopsy from the nasal cavity. The open issue of precursor cell heterogeneity in the adult CNS, however, will have even greater relevance if stem cells from the periphery are included. Whether neural precursor cells from the olfactory epithelium could be used as a source to replace lost neurons in the CNS is not known at present. Most research focuses on the use of olfactory ensheathing glia.

Olfactory ensheathing glia, the specialized glia of the olfactory epithelium that promotes axonal growth, is also of special interest with regard to reconstitutive efforts after axonal damage (Barnett and Chang, 2004; Boyd et al., 2005; Barnett and Riddell, 2007). For example, in spinal cord injury, local glial cells prevent regenerative axonal growth; thus one could try to transplant permissive olfactory ensheathing glia to enable the regrowing axons to bridge the area of damage (Bunge and Pearse, 2003; Ramer et al., 2004).

Vomeronasal Organ

Neurogenesis also occurs in the specialized olfactory epithelium of the vomeronasal organ, which is of relevance for pheromone detection

The vomeronasal organ is a specialized part of the olfactory epithelium, consisting of two liquid-filled small recesses in the upper part of the nasal septum. The chemosensors in the vomeronasal organ are by far the most sensitive olfactory receptor cells (Leinders-Zufall et al., 2000). In many species the vomeronasal organ is involved in the detection of pheromones (Zufall et al., 2002). The pathways from the vomeronasal organ to its own cortical representations are separate from the other olfactory projections. The epithelium of the vomeronasal organ lacks the distinct structure of the olfactory epithelium proper. Its precursor cells reside in the center, and the progeny migrate vertically to contribute to neurogenesis in this region (Jia and Halpern, 1998; Martinez-Marcos et al., 2000a; Martinez-Marcos et al., 2000b). Additional proliferative cells of unknown function reside in the periphery of the epithelium (Martinez-Marcos et al., 2005). The vertically migrating cells in the center contribute to adult neurogenesis as evidenced by BrdU-co-localization with OMP as a marker of new neurons (Martinez-Marcos et al., 2000b; Martinez-Marcos et al., 2005). The relationship between adult neurogenesis in the olfactory epithelium proper and in the vomeronasal organ is not known.

Adult Neurogenesis in the Peripheral Nervous System Outside the Olfactory Epithelium

Dorsal Root Ganglia

Dorsal root ganglia contain neural crest precursor cells, which might allow some neurogenesis that, however, does not seem to take place physiologically

Dorsal root ganglia have attracted some attention as potential sites of adult neurogenesis. The reason is that after lesion of the peripheral nerve, the ganglia degenerate, and a subpopulation of dorsal root neurons undergoes apoptosis, reducing their number by about one third (Hu and McLachlan, 2003) but recovering over a period of a few months (Groves et al., 2003). Neural crest stem cells, from which dorsal root ganglia are derived, can be isolated from dorsal root ganglia of the adult rat (Li et al., 2007) and show hallmarks of stem cells (Singh et al., 2009). They might correspond to satellite glial cells in vivo and express nestin (Arora et al., 2007). In explant cultures, stimulation with a growth factor cocktail consisting of NGF, BDNF, NT-3, and the Neurogenin1 isoform GGF2 induced the generation of mature neurons that expressed the characteristic marker PGP9.5 as well as neurofilament 200 and acquired distinctive neuronal morphologies (Li et al., 2007). They expressed substance P, calcitonin gene-related peptide (CGRP), and p75NTR, as do their counterparts in intact dorsal root ganglia in vivo (Li et al., 2007). Similarly, neurogenesis from nestin-positive satellite cells ex vivo has been shown early postnatally (Arora et al., 2007). Taken together, no strong case in favor of adult neurogenesis in dorsal root ganglia can be made at present, but the available data encourage further studies.

Enteric Nervous System

Adult neurogenesis in the plexus of the gut has been demonstrated with a good level of evidence

The enteric nervous system is derived from the neural crest, and the precursor cells generating the myenteric plexus of the intestines differ from other stem cell populations building the peripheral nervous system (Bixby et al., 2002). In mice, neurogenesis in the myenteric plexus was thought to cease around three weeks after birth. If the migration and differentiation of these precursor cells fails, Hirschsprung disease, the absence of enteric ganglia, ensues (Iwashita et al., 2003; Heanue and Pachnis, 2007). Some of these stem cells persist into the adult phase, but there the potential appears reduced (Kruger et al., 2002). One early report suggested that both constitutively and after lesion, new enteric ganglion cells could develop from these precursor cells (Hanani et al., 2003). The report was based on time-course studies and careful histological analysis, including electronmicroscopy, only. Adult neurogenesis in the enteric nervous system was confirmed, however, with the BrdU method and numerous markers in 2009 (Liu et al., 2009). That study also revealed that the precursor cells reside in a niche near the longitudinal muscle of the gut and the developing neurons migrate a short distance into the intestinal ganglia. In addition, it was found that the serotonin receptor 4 is crucial in regulating intestinal neurogenesis (Liu et al., 2009).

Carotid Bodies

In the carotid body adult neurogenesis physiologically responds to hypoxia

The carotid bodies lie in the bifurcation of the carotid artery on each side of the body and sense changes in arterial oxygen tension (Fig. 8–10) (Lopez-Barneo et al., 2008). Ventilation is increased when hypoxia is encountered. The carotid bodies are of neural crest origin. The neurons of the carotid bodies are dopaminergic, and there is hence increasing interest in exploring if cells from the carotid bodies might become a source for cell therapy in Parkinson disease (Lopez-Barneo et al., 2009). Given the minute size of these organs, this seemed unlikely until the presence of neural precursor cells in the carotid bodies was demonstrated (Pardal et al., 2007). These precursor cells have glia-like properties, express GFAP, and as sustentacular cells surround the clusters of dopaminergic neurons. That they originate from neural crest was shown with Wnt1-based lineage tracing. These precursor cells are multipotent and generate new neurons ex vivo and, remarkably, also in vivo (Pardal et al., 2007). Functionally the newly generated cells ex vivo faithfully reflected the phenotype in vivo. They, for example, strongly express GDNF, a potent survival factor for dopaminergic neurons (Villadiego et al., 2005). Adult neurogenesis in carotid bodies is (not surprisingly) strongly triggered by hypoxia, explaining the amazing plasticity of this structure, which can increase its size manifold times under chronic hypoxia; for example, when acclimatizing to high altitudes. The pioneering report on adult neurogenesis in carotid bodies came from Ricardo Pardal, José López-Barneo and colleagues from the University of Sevilla, Spain, and with respect to the provided evidence made an exceptionally strong case (Pardal et al., 2007). Eszsebet Kokovay and Sally Temple wrote in their commentary on this extraordinary study that it would take

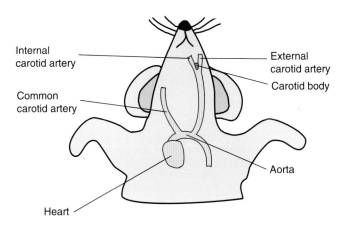

FIGURE 8–10 Adult neurogenesis in the carotid body. The carotid body is a small organ in the bifurcation of the common cerebral artery, and it senses oxygen pressure in the blood stream. Hypoxia was found to stimulate neurogenesis in the carotid bodies (Pardal et al. 2007), which are known to grow with altitude training.

"neural crest cells to new heights" and mused that when Edmund Hillary and Tenzing Norgay for the first time successfully reached the very thin air at the summit of Mt. Everest adult neurogenesis unexpectedly had its share in the success (Kokovay and Temple, 2007).

References

Aarum J, Sandberg K, Haeberlein SL, Persson MA (2003). Migration and differentiation of neural precursor cells can be directed by microglia. *Proc Natl Acad Sci USA* 100:15983–15988.

Abrahams JM, Lenart CJ, Tobias ME (2009). Temporal variation of induction neurogenesis in a rat model of transient middle cerebral artery occlusion. *Neurol Res* 31:528–533.

Adler R, Raymond PA (2008). Have we achieved a unified model of photoreceptor cell fate specification in vertebrates? *Brain Res* 1192:134–150.

Agasse F, Roger M, Coronas V (2004). Neurogenic and intact or apoptotic non-neurogenic areas of adult brain release diffusible molecules that differentially modulate the development of subventricular zone cell cultures. *Eur J Neurosci* 19:1459–1468.

Aguirre A, Gallo V (2004). Postnatal neurogenesis and gliogenesis in the olfactory bulb from NG2-expressing progenitors of the subventricular zone. *J Neurosci* 24:10530–10541.

Aguirre AA, Chittajallu R, Belachew S, Gallo V (2004). NG2-expressing cells in the subventricular zone are type C-like cells and contribute to interneuron generation in the postnatal hippocampus. *J Cell Biol* 165: 575–589.

Ahmad I, Dooley CM, Thoreson WB, Rogers JA, Afiat S (1999). In vitro analysis of a mammalian retinal progenitor that gives rise to neurons and glia. *Brain Res* 831:1–10.

Ahmad I, Tang L, Pham H (2000). Identification of neural progenitors in the adult mammalian eye. *Biochem Biophys Res Commun* 270:517–521.

Akagi T, Akita J, Haruta M, Suzuki T, Honda Y, Inoue T, Yoshiura S, Kageyama R, Yatsu T, Yamada M, Takahashi M (2005). Iris-derived cells from adult rodents and primates adopt photoreceptor-specific phenotypes. *Invest Ophthalmol Vis Sci* 46:3411–3419.

Akita J, Takahashi M, Hojo M, Nishida A, Haruta M, Honda Y (2002). Neuronal differentiation of adult rat hippocampus-derived neural stem cells transplanted into embryonic rat explanted retinas with retinoic acid pretreatment. *Brain Res* 954:286–293.

Alexiades MR, Cepko CL (1997). Subsets of retinal progenitors display temporally regulated and distinct biases in the fates of their progeny. *Development* 124:1119–1131.

Altman J (1963). Autoradiographic investigation of cell proliferation in the brains of rats and cats. *Anat Rec* 145:573–591.

Altman J, Bayer SA (1990a). Mosaic organization of the hippocampal neuroepithelium and the multiple germinal sources of dentate granule cells. *J Comp Neurol* 301:325–342.

Altman J, Bayer SA (1990b). Migration and distribution of two populations of hippocampal progenitors during the perinatal and postnatal periods. *J Comp Neurol* 301:365–381.

Alvarez-Buylla A, Lim DA (2004). For the long run: Maintaining germinal niches in the adult brain. *Neuron* 41:683–686.

Alvarez-Buylla A, Kohwi M, Nguyen TM, Merkle FT (2008). The heterogeneity of adult neural stem cells and the emerging complexity of their niche. *Cold Spring Harbor Symposia on Quantitative Biology* 73:357–365.

Arnold SE, Smutzer GS, Trojanowski JQ, Moberg PJ (1998). Cellular and molecular neuropathology of the olfactory epithelium and central olfactory pathways in Alzheimer's disease and schizophrenia. *Ann NY Acad Sci* 855:762–775.

Arora DK, Cosgrave AS, Howard MR, Bubb V, Quinn JP, Thippeswamy T (2007). Evidence of postnatal neurogenesis in dorsal root ganglion: Role of nitric oxide and neuronal restrictive silencer transcription factor. *J Mol Neurosci* 32:97–107.

Arvidsson A, Collin T, Kirik D, Kokaia Z, Lindvall O (2002). Neuronal replacement from endogenous precursors in the adult brain after stroke. *Nat Med* 8:963–970.

Avunduk AM, Tekelioglu Y (2006). Therapeutic use of limbal stem cells. *Curr Stem Cell Res Ther* 1:231–238.

Babu H, Cheung G, Kettenmann H, Palmer TD, Kempermann G (2007). Enriched monolayer precursor cell cultures from micro-dissected adult mouse dentate gyrus yield functional granule cell-like neurons. *PLoS ONE* 2:e388.

Babu H, Ramirez-Rodriguez G, Fabel K, Bischofberger J, Kempermann G (2009). Synaptic network activity induces neuronal differentiation of adult hippocampal precursor cells through BDNF signaling. *Front. Neurosci.* 3:49.

Barnett SC, Chang L (2004). Olfactory ensheathing cells and CNS repair: Going solo or in need of a friend? *Trends Neurosci* 27:54–60.

Barnett SC, Riddell JS (2007). Olfactory ensheathing cell transplantation as a strategy for spinal cord repair— what can it achieve? *Nat Clin Pract Neurol* 3:152–161.

Barraud P, He X, Zhao C, Ibanez C, Raha-Chowdhury R, Caldwell MA, Franklin RJ (2007). Contrasting effects of basic fibroblast growth factor and epidermal growth factor on mouse neonatal olfactory mucosa cells. *Eur J Neurosci* 26:3345–3357.

Bartsch U, Oriyakhel W, Kenna PF, Linke S, Richard G, Petrowitz B, Humphries P, Farrar GJ, Ader M (2008). Retinal cells integrate into the outer nuclear layer and differentiate into mature photoreceptors after subretinal transplantation into adult mice. *Exp Eye Res* 86:691–700.

Battista D, Ferrari CC, Gage FH, Pitossi FJ (2006). Neurogenic niche modulation by activated microglia: Transforming growth factor beta increases neurogenesis in the adult dentate gyrus. *Eur J Neurosci* 23:83–93.

Bauer S, Rasika S, Han J, Mauduit C, Raccurt M, Morel G, Jourdan F, Benahmed M, Moyse E, Patterson PH (2003). Leukemia inhibitory factor is a key signal for injury-induced neurogenesis in the adult mouse olfactory epithelium. *J Neurosci* 23:1792–1803.

Bedard A, Gravel C, Parent A (2006). Chemical characterization of newly generated neurons in the striatum of adult primates. *Exp Brain Res* 170:501–512.

Bedard A, Levesque M, Bernier PJ, Parent A (2002). The rostral migratory stream in adult squirrel monkeys: Contribution of new neurons to the olfactory tubercle and involvement of the antiapoptotic protein Bcl-2. *Eur J Neurosci* 16:1917–1924.

Belachew S, Chittajallu R, Aguirre AA, Yuan X, Kirby M, Anderson S, Gallo V (2003). Postnatal NG2 proteoglycan-expressing progenitor cells are intrinsically multipotent and generate functional neurons. *J Cell Biol* 161:169–186.

Ben-Hur T, Ben-Menachem O, Furer V, Einstein O, Mizrachi-Kol R, Grigoriadis N (2003). Effects of proinflammatory cytokines on the growth, fate, and motility of multipotential neural precursor cells. *Mol Cell Neurosci* 24:623–631.

Benraiss A, Chmielnicki E, Lerner K, Roh D, Goldman SA (2001). Adenoviral brain-derived neurotrophic factor induces both neostriatal and olfactory neuronal recruitment from endogenous progenitor cells in the adult forebrain. *J Neurosci* 21:6718–6731.

Bergers G, Brekken R, McMahon G, Vu TH, Itoh T, Tamaki K, Tanzawa K, Thorpe P, Itohara S, Werb Z, Hanahan D (2000). Matrix metalloproteinase-9 triggers the angiogenic switch during carcinogenesis. *Nat Cell Biol* 2:737–744.

Bernier PJ, Bedard A, Vinet J, Levesque M, Parent A (2002). Newly generated neurons in the amygdala and adjoining cortex of adult primates. *Proc Natl Acad Sci USA* 99:11464–11469.

Bessis A, Bechade C, Bernard D, Roumier A (2007). Microglial control of neuronal death and synaptic properties. *Glia* 55:233–238.

Bhardwaj RD, Curtis MA, Spalding KL, Buchholz BA, Fink D, Bjork-Eriksson T, Nordborg C, Gage FH, Druid H, Eriksson PS, Frisen J (2006). From the cover: Neocortical neurogenesis in humans is restricted to development. *Proc Natl Acad Sci USA* 103:12564–12568.

Bianco JI, Perry C, Harkin DG, Mackay-Sim A, Feron F (2004). Neurotrophin 3 promotes purification and proliferation of olfactory ensheathing cells from human nose. *Glia* 45:111–123.

Bixby S, Kruger GM, Mosher JT, Joseph NM, Morrison SJ (2002). Cell-intrinsic differences between stem cells from different regions of the peripheral nervous system regulate the generation of neural diversity. *Neuron* 35:643–656.

Bolteus AJ, Bordey A (2004). GABA release and uptake regulate neuronal precursor migration in the postnatal subventricular zone. *J Neurosci* 24:7623–7631.

Bonfanti L, Ponti G (2008). Adult mammalian neurogenesis and the New Zealand white rabbit. *Vet J* 175: 310–331.

Bonfanti L, Olive S, Poulain DA, Theodosis DT (1992). Mapping of the distribution of polysialylated neural cell adhesion molecule throughout the central nervous system of the adult rat: An immunohistochemical study. *Neuroscience* 49:419–436.

Borders AS, Getchell ML, Etscheidt JT, van Rooijen N, Cohen DA, Getchell TV (2007a). Macrophage depletion in the murine olfactory epithelium leads to increased neuronal death and decreased neurogenesis. *J Comp Neurol* 501:206–218.

Borders AS, Hersh MA, Getchell ML, van Rooijen N, Cohen DA, Stromberg AJ, Getchell TV (2007b) Macrophage-mediated neuroprotection and neurogenesis in the olfactory epithelium. *Physiol Genomics* 31:531–543.

Bordey A (2006). Adult neurogenesis: Basic concepts of signaling. *Cell Cycle* 5:722–728.

Boulton M, Albon J (2004). Stem cells in the eye. *Int J Biochem Cell Biol* 36:643–657.

Boyd JG, Doucette R, Kawaja MD (2005). Defining the role of olfactory ensheathing cells in facilitating axon remyelination following damage to the spinal cord. *Faseb J* 19:694–703.

Braun PE, Sandillon F, Edwards A, Matthieu JM, Privat A (1988). Immunocytochemical localization by electron microscopy of 2′3′-cyclic nucleotide 3′-phosphodiesterase in developing oligodendrocytes of normal and mutant brain. *J Neurosci* 8:3057–3066.

Breunig JJ, Sarkisian MR, Arellano JI, Morozov YM, Ayoub AE, Sojitra S, Wang B, Flavell RA, Rakic P, Town T (2008). Primary cilia regulate hippocampal neurogenesis by mediating sonic hedgehog signaling. *Proc Natl Acad Sci USA* 105:13127–13132.

Buffo A, Rite I, Tripathi P, Lepier A, Colak D, Horn AP, Mori T, Gotz M (2008). Origin and progeny of reactive gliosis: A source of multipotent cells in the injured brain. *Proc Natl Acad Sci USA* 105:3581–3586.

Bull ND, Bartlett PF (2005). The adult mouse hippocampal progenitor is neurogenic but not a stem cell. *J Neurosci* 25:10815–10821.

Bunge MB, Pearse DD (2003). Transplantation strategies to promote repair of the injured spinal cord. *J Rehabil Res Dev* 40:55–62.

Butt AM, Hamilton N, Hubbard P, Pugh M, Ibrahim M (2005). Synantocytes: The fifth element. *J Anat* 207:695–706.

Cacci E, Claasen JH, Kokaia Z (2005). Microglia-derived tumor necrosis factor-alpha exaggerates death of newborn hippocampal progenitor cells in vitro. *J Neurosci Res* 80:789–797.

Cameron HA, Dayer AG (2008). New interneurons in the adult neocortex: Small, sparse, but significant? *Biol Psychiatry* 63:650–655.

Canola K, Angenieux B, Tekaya M, Quiambao A, Naash MI, Munier FL, Schorderet DF, Arsenijevic Y (2007). Retinal stem cells transplanted into models of late stages of retinitis pigmentosa preferentially adopt a glial or a retinal ganglion cell fate. *Invest Ophthalmol Vis Sci* 48:446–454.

Cao L, Jiao X, Zuzga DS, Liu Y, Fong DM, Young D, During MJ (2004). VEGF links hippocampal activity with neurogenesis, learning and memory. *Nat Genet* 36:827–835.

Carney RS, Alfonso TB, Cohen D, Dai H, Nery S, Stoica B, Slotkin J, Bregman BS, Fishell G, Corbin JG (2006). Cell migration along the lateral cortical stream to the developing basal telencephalic limbic system. *J Neurosci* 26:11562–11574.

Casarosa S, Fode C, Guillemot F (1999). Mash1 regulates neurogenesis in the ventral telencephalon. *Development* 126:525–534.

Cascella NG, Takaki M, Lin S, Sawa A (2007). Neurodevelopmental involvement in schizophrenia: The olfactory epithelium as an alternative model for research. *J Neurochem* 102:587–594.

Cau E, Gradwohl G, Casarosa S, Kageyama R, Guillemot F (2000). Hes genes regulate sequential stages of neurogenesis in the olfactory epithelium. *Development* 127:2323–2332.

Chacko DM, Das AV, Zhao X, James J, Bhattacharya S, Ahmad I (2003). Transplantation of ocular stem cells: The role of injury in incorporation and differentiation of grafted cells in the retina. *Vision Res* 43:937–946.

Chen J, Magavi SS, Macklis JD (2004a). Neurogenesis of corticospinal motor neurons extending spinal projections in adult mice. *Proc Natl Acad Sci USA* 101:16357–16362.

Chen J, Kwon CH, Lin L, Li Y, Parada LF (2009). Inducible site-specific recombination in neural stem/progenitor cells. *Genesis* 47:122–131.

Chen X, Fang H, Schwob JE (2004b). Multipotency of purified, transplanted globose basal cells in olfactory epithelium. *J Comp Neurol* 469:457–474.

Chittajallu R, Kunze A, Mangin JM, Gallo V (2007). Differential synaptic integration of interneurons in the outer and inner molecular layers of the developing dentate gyrus. *J Neurosci* 27:8219–8225.

Chmielnicki E, Benraiss A, Economides AN, Goldman SA (2004). Adenovirally expressed noggin and brain-derived neurotrophic factor cooperate to induce new medium spiny neurons from resident progenitor cells in the adult striatal ventricular zone. *J Neurosci* 24:2133–2142.

Cho SR, Benraiss A, Chmielnicki E, Samdani A, Economides A, Goldman SA (2007). Induction of neostriatal neurogenesis slows disease progression in a transgenic murine model of Huntington disease. *J Clin Invest* 117:2889–2902.

Collin T, Arvidsson A, Kokaia Z, Lindvall O (2005). Quantitative analysis of the generation of different striatal neuronal subtypes in the adult brain following excitotoxic injury. *Exp Neurol* 195:71–80.

Cooper O, Isacson O (2004). Intrastriatal transforming growth factor alpha delivery to a model of Parkinson's disease induces proliferation and migration of endogenous adult neural progenitor cells without differentiation into dopaminergic neurons. *J Neurosci* 24:8924–8931.

Costanzo RM (1985). Neural regeneration and functional reconnection following olfactory nerve transection in hamster. *Brain Res* 361:258–266.

Darsalia V, Heldmann U, Lindvall O, Kokaia Z (2005). Stroke-induced neurogenesis in aged brain. *Stroke* 36:1790–1795.

Davis JA, Reed RR (1996). Role of Olf-1 and Pax-6 transcription factors in neurodevelopment. *J Neurosci* 16:5082–5094.

Davoust N, Vuaillat C, Androdias G, Nataf S (2008). From bone marrow to microglia: Barriers and avenues. *Trends Immunol* 29:227–234.

Dayer AG, Cleaver KM, Abouantoun T, Cameron HA (2005). New GABAergic interneurons in the adult neocortex and striatum are generated from different precursors. *J Cell Biol* 168:415–427.

de Chevigny A, Lemasson M, Saghatelyan A, Sibbe M, Schachner M, Lledo PM (2006). Delayed onset of odor detection in neonatal mice lacking tenascin-C. *Mol Cell Neurosci* 32:174–186.

De Marchis S, Fasolo A, Puche AC (2004). Subventricular zone-derived neuronal progenitors migrate into the subcortical forebrain of postnatal mice. *J Comp Neurol* 476:290–300.

Deisseroth K, Singla S, Toda H, Monje M, Palmer TD, Malenka RC (2004). Excitation-neurogenesis coupling in adult neural stem/progenitor cells. *Neuron* 42:535–552.

Dimou L, Simon C, Kirchhoff F, Takebayashi H, Gotz M (2008). Progeny of Olig2-expressing progenitors in the gray and white matter of the adult mouse cerebral cortex. *J Neurosci* 28:10434–10442.

Dingledine R, Korn SJ (1985). Gamma-aminobutyric acid uptake and the termination of inhibitory synaptic potentials in the rat hippocampal slice. *J Physiol* 366:387–409.

Doetsch F (2003). The glial identity of neural stem cells. *Nat Neurosci* 6:1127–1134.

Doetsch F, Garcia-Verdugo JM, Alvarez-Buylla A (1997). Cellular composition and three-dimensional organization of the subventricular germinal zone in the adult mammalian brain. *J Neurosci* 17:5046–5061.

Doetsch F, Garcia-Verdugo JM, Alvarez-Buylla A (1999a). Regeneration of a germinal layer in the adult mammalian brain. *Proc Natl Acad Sci USA* 96:11619–11624.

Doetsch F, Caille I, Lim DA, Garcia-Verdugo JM, Alvarez-Buylla A (1999b). Subventricular zone astrocytes are neural stem cells in the adult mammalian brain. *Cell* 97:703–716.

Doetsch F, Petreanu L, Caille I, Garcia-Verdugo JM, Alvarez-Buylla A (2002). EGF converts transit-amplifying neurogenic precursors in the adult brain into multipotent stem cells. *Neuron* 36:1021–1034.

Ehninger D, Kempermann G (2003). Regional effects of wheel running and environmental enrichment on cell genesis and microglia proliferation in the adult murine neocortex. *Cereb Cortex* 13:845–851.

Ekdahl CT, Kokaia Z, Lindvall O (2009). Brain inflammation and adult neurogenesis: The dual role of microglia. *Neuroscience* 158:1021–1029.

Ekdahl CT, Claasen JH, Bonde S, Kokaia Z, Lindvall O (2003). Inflammation is detrimental for neurogenesis in adult brain. *Proc Natl Acad Sci USA* 100:13632–13637.

Emsley JG, Hagg T (2003). Alpha6beta1 integrin directs migration of neuronal precursors in adult mouse forebrain. *Exp Neurol* 183:273–285.

Engelhardt M, Wachs FP, Couillard-Despres S, Aigner L (2004). The neurogenic competence of progenitors from the postnatal rat retina in vitro. *Exp Eye Res* 78:1025–1036.

Engelhardt M, Bogdahn U, Aigner L (2005). Adult retinal pigment epithelium cells express neural progenitor properties and the neuronal precursor protein doublecortin. *Brain Res* 1040:98–111.

Fabel K, Toda H, Palmer T (2003a). Copernican stem cells: Regulatory constellations in adult hippocampal neurogenesis. *J Cell Biochem* 88:41–50.

Fabel K, Fabel K, Tam B, Kaufer D, Baiker A, Simmons N, Kuo CJ, Palmer TD (2003b). VEGF is necessary for exercise-induced adult hippocampal neurogenesis. *Eur J Neurosci* 18:2803–2812.

Farbman AI, Brunjes PC, Rentfro L, Michas J, Ritz S (1988). The effect of unilateral naris occlusion on cell dynamics in the developing rat olfactory epithelium. *J Neurosci* 8:3290–3295.

Feron F, Perry C, Hirning MH, McGrath J, Mackay-Sim A (1999a). Altered adhesion, proliferation and death in neural cultures from adults with schizophrenia. *Schizophr Res* 40:211–218.

Feron F, Mackay-Sim A, Andrieu JL, Matthaei KI, Holley A, Sicard G (1999b) Stress induces neurogenesis in non-neuronal cell cultures of adult olfactory epithelium. *Neuroscience* 88:571–583.

Filippov V, Kronenberg G, Pivneva T, Reuter K, Steiner B, Wang LP, Yamaguchi M, Kettenmann H, Kempermann G (2003). Subpopulation of nestin-expressing progenitor cells in the adult murine hippocampus shows electrophysiological and morphological characteristics of astrocytes. *Mol Cell Neurosci* 23:373–382.

Fischer AJ, Reh TA (2000). Identification of a proliferating marginal zone of retinal progenitors in postnatal chickens. *Dev Biol* 220:197–210.

Fischer AJ, Reh TA (2001). Muller glia are a potential source of neural regeneration in the postnatal chicken retina. *Nat Neurosci* 4:247–252.

Fischer AJ, Reh TA (2003). Growth factors induce neurogenesis in the ciliary body. *Dev Biol* 259:225–240.

Fowler CD, Liu Y, Ouimet C, Wang Z (2002). The effects of social environment on adult neurogenesis in the female prairie vole. *J Neurobiol* 51:115–128.

Fowler CD, Johnson F, Wang Z (2005). Estrogen regulation of cell proliferation and distribution of estrogen receptor-alpha in the brains of adult female prairie and meadow voles. *J Comp Neurol* 489:166–179.

Fraser DD, Duffy S, Angelides KJ, Perez-Velazquez JL, Kettenmann H, MacVicar BA (1995). GABAA/benzodi-azepine receptors in acutely isolated hippocampal astrocytes. *J Neurosci* 15:2720–2732.

Frielingsdorf H, Schwarz K, Brundin P, Mohapel P (2004). No evidence for new dopaminergic neurons in the adult mammalian substantia nigra. *Proc Natl Acad Sci USA* 101:10177–10182.

Gage FH, Coates PW, Palmer TD, Kuhn HG, Fisher LJ, Suhonen JO, Peterson DA, Suhr ST, Ray J (1995). Survival and differentiation of adult neuronal progenitor cells transplanted to the adult brain. *Proc Natl Acad Sci USA* 92:11879–11883.

Gates MA, Thomas LB, Howard EM, Laywell ED, Sajin B, Faissner A, Gotz B, Silver J, Steindler DA (1995). Cell and molecular analysis of the developing and adult mouse subventricular zone of the cerebral hemispheres. *J Comp Neurol* 361:249–266.

Ge S, Goh EL, Sailor KA, Kitabatake Y, Ming GL, Song H (2006). GABA regulates synaptic integration of newly generated neurons in the adult brain. *Nature* 439:589–593.

Geiger H, Koehler A, Gunzer M (2007). Stem cells, aging, niche, adhesion and Cdc42: A model for changes in cell–cell interactions and hematopoietic stem cell aging. *Cell Cycle* 6:884–887.

Getchell ML, Li H, Vaishnav RA, Borders AS, Witta J, Subhedar N, de Villiers W, Stromberg AJ, Getchell TV (2006). Temporal gene expression profiles of target-ablated olfactory epithelium in mice with disrupted expression of scavenger receptor A: Impact on macrophages. *Physiol Genomics* 27:245–263.

Getchell TV, Shah DS, Partin JV, Subhedar NK, Getchell ML (2002). Leukemia inhibitory factor mRNA expression is upregulated in macrophages and olfactory receptor neurons after target ablation. *J Neurosci Res* 67:246–254.

Getchell TV, Liu H, Vaishnav RA, Kwong K, Stromberg AJ, Getchell ML (2005). Temporal profiling of gene expression during neurogenesis and remodeling in the olfactory epithelium at short intervals after target ablation. *J Neurosci Res* 80:309–329.

Gilbertson RJ, Rich JN (2007). Making a tumour's bed: Glioblastoma stem cells and the vascular niche. *Nature Rev* 7:733–736.

Goetz AK, Scheffler B, Chen HX, Wang S, Suslov O, Xiang H, Brustle O, Roper SN, Steindler DA (2006). Temporally restricted substrate interactions direct fate and specification of neural precursors derived from embryonic stem cells. *Proc Natl Acad Sci USA* 103:11063–11068.

Goldstein BJ, Schwob JE (1996). Analysis of the globose basal cell compartment in rat olfactory epithelium using GBC-1, a new monoclonal antibody against globose basal cells. *J Neurosci* 16:4005–4016.

Gordon RJ, Tattersfield AS, Vazey EM, Kells AP, McGregor AL, Hughes SM, Connor B (2007). Temporal profile of subventricular zone progenitor cell migration following quinolinic acid-induced striatal cell loss. *Neuroscience* 146:1704–1718.

Gould E (2007). How widespread is adult neurogenesis in mammals? *Nat Rev Neurosci* 8:481–488.

Gould E, Cameron HA (1996). Regulation of neuronal birth, migration and death in the rat dentate gyrus. *Dev Neurosci* 18:22–35.

Gould E, Reeves AJ, Graziano MS, Gross CG (1999). Neurogenesis in the neocortex of adult primates. *Science* 286:548–552.

Gould E, Vail N, Wagers M, Gross CG (2001). Adult-generated hippocampal and neocortical neurons in macaques have a transient existence. *Proc Natl Acad Sci USA* 98:10910–10917.

Grandel H, Kaslin J, Ganz J, Wenzel I, Brand M (2006). Neural stem cells and neurogenesis in the adult zebrafish brain: Origin, proliferation dynamics, migration and cell fate. *Dev Biol* 295:263–277.

Graziadei G, Graziadei P (1979a). Neurogenesis and neuron regeneration in the olfactory system of mammals. II. Degeneration and reconstitution of the olfactory neurons after axotomy. *J Neurocytol* 8:197–213.

Graziadei P, Graziadei G (1979b). Neurogenesis and neuron regeneration in the olfactory system of mammals. I. Morphological aspects of differentiation and structural organization of the olfactory sensory neurons. *J Neurocytol* 8:1–18.

Graziadei PP (1973). Cell dynamics in the olfactory mucosa. *Tissue Cell* 5:113–131.

Graziadei PP, DeHan RS (1973). Neuronal regeneration in frog olfactory system. *J Cell Biol* 59:525–530.

Graziadei PP, Karlan MS, Graziadei GA, Bernstein JJ (1980). Neurogenesis of sensory neurons in the primate olfactory system after section of the fila olfactoria. *Brain Res* 186:289–300.

Groves MJ, Schanzer A, Simpson AJ, An SF, Kuo LT, Scaravilli F (2003). Profile of adult rat sensory neuron loss, apoptosis and replacement after sciatic nerve crush. *J Neurocytol* 32:113–122.

Gu W, Brannstrom T, Wester P (2000). Cortical neurogenesis in adult rats after reversible photothrombotic stroke. *J Cereb Blood Flow Metab* 20:1166–1173.

Gu W, Brannstrom T, Rosqvist R, Wester P (2009). Cell division in the cerebral cortex of adult rats after photo-thrombotic ring stroke. *Stem Cell Res* 2:68–77.

Gurok U, Steinhoff C, Lipkowitz B, Ropers HH, Scharff C, Nuber UA (2004). Gene expression changes in the course of neural progenitor cell differentiation. *J Neurosci* 24:5982–6002.

Hack MA, Sugimori M, Lundberg C, Nakafuku M, Gotz M (2004). Regionalization and fate specification in neurospheres: The role of Olig2 and Pax6. *Mol Cell Neurosci* 25:664–678.

Hanani M, Ledder O, Yutkin V, Abu-Dalu R, Huang TY, Hartig W, Vannucchi MG, Faussone-Pellegrini MS (2003). Regeneration of myenteric plexus in the mouse colon after experimental denervation with benzalkonium chloride. *J Comp Neurol* 462:315–327.

Heanue TA, Pachnis V (2007). Enteric nervous system development and Hirschsprung's disease: Advances in genetic and stem cell studies. *Nat Rev Neurosci* 8:466–479.

Heins N, Malatesta P, Cecconi F, Nakafuku M, Tucker KL, Hack MA, Chapouton P, Barde YA, Gotz M (2002). Glial cells generate neurons: The role of the transcription factor Pax6. *Nat Neurosci* 5:308–315.

Heissig B, Hattori K, Dias S, Friedrich M, Ferris B, Hackett NR, Crystal RG, Besmer P, Lyden D, Moore MA, Werb Z, Rafii S (2002). Recruitment of stem and progenitor cells from the bone marrow niche requires MMP-9 mediated release of kit-ligand. *Cell* 109:625–637.

Hellsten J, Wennstrom M, Bengzon J, Mohapel P, Tingstrom A (2004). Electroconvulsive seizures induce endothelial cell proliferation in adult rat hippocampus. *Biol Psychiatry* 55:420–427.

Henry RA, Hughes SM, Connor B (2007). AAV-mediated delivery of BDNF augments neurogenesis in the normal and quinolinic acid-lesioned adult rat brain. *Eur J Neurosci* 25:3513–3525.

Herrera DG, Garcia-Verdugo JM, Alvarez-Buylla A (1999). Adult-derived neural precursors transplanted into multiple regions in the adult brain. *Ann Neurol* 46:867–877.

Herzog CD, Otto T (2002). Administration of transforming growth factor-alpha enhances anatomical and behavioral recovery following olfactory nerve transection. *Neuroscience* 113:569–580.

Hodge RD, Kowalczyk TD, Wolf SA, Encinas JM, Rippey C, Enikolopov G, Kempermann G, Hevner RF (2008). Intermediate progenitors in adult hippocampal neurogenesis: Tbr2 expression and coordinate regulation of neuronal output. *J Neurosci* 28:3707–3717.

Hoehn BD, Palmer TD, Steinberg GK (2005). Neurogenesis in rats after focal cerebral ischemia is enhanced by indomethacin. *Stroke* 36:2718–2724.

Hollyfield JG (1968). Differential addition of cells to the retina in Rana pipiens tadpoles. *Dev Biol* 18:163–179.

Holmberg J, Armulik A, Senti KA, Edoff K, Spalding K, Momma S, Cassidy R, Flanagan JG, Frisen J (2005). Ephrin-A2 reverse signaling negatively regulates neural progenitor proliferation and neurogenesis. *Genes Dev* 19:462–471.

Hu P, McLachlan EM (2003). Selective reactions of cutaneous and muscle afferent neurons to peripheral nerve transection in rats. *J Neurosci* 23:10559–10567.

Huang L, DeVries GJ, Bittman EL (1998). Photoperiod regulates neuronal bromodeoxyuridine labeling in the brain of a seasonally breeding mammal. *J Neurobiol* 36:410–420.

Huard JM, Schwob JE (1995). Cell cycle of globose basal cells in rat olfactory epithelium. *Dev Dyn* 203:17–26.

Imbeault S, Gauvin LG, Toeg HD, Pettit A, Sorbara CD, Migahed L, DesRoches R, Menzies AS, Nishii K, Paul DL, Simon AM, Bennett SA (2009). The extracellular matrix controls gap junction protein expression and function in postnatal hippocampal neural progenitor cells. *BMC Neurosci* 10:13.

Inta D, Alfonso J, von Engelhardt J, Kreuzberg MM, Meyer AH, van Hooft JA, Monyer H (2008). Neurogenesis and widespread forebrain migration of distinct GABAergic neurons from the postnatal subventricular zone. *Proc Natl Acad Sci USA* 105:20994–20999.

Isoyama K, Nagata H, Shino Y, Isegawa N, Arimoto Y, Koda M, Kumahara K, Okamoto Y, Shirasawa H (2004). Effects of adenoviral vector-mediated BDNF expression on the bulbectomy-induced apoptosis of olfactory receptor neurons. *Brain Res Mol Brain Res* 129:88–95.

Iwai N, Zhou Z, Roop DR, Behringer RR (2008). Horizontal basal cells are multipotent progenitors in normal and injured adult olfactory epithelium. *Stem Cells* 26:1298–1306.

Iwashita T, Kruger GM, Pardal R, Kiel MJ, Morrison SJ (2003). Hirschsprung disease is linked to defects in neural crest stem cell function. *Science* 301:972–976.

Jacques TS, Relvas JB, Nishimura S, Pytela R, Edwards GM, Streuli CH, ffrench-Constant C (1998). Neural precursor cell chain migration and division are regulated through different beta1 integrins. *Development* 125:3167–3177.

Jang W, Youngentob SL, Schwob JE (2003). Globose basal cells are required for reconstitution of olfactory epithelium after methyl bromide lesion. *J Comp Neurol* 460:123–140.

Jankovski A, Sotelo C (1996). Subventricular zone-olfactory bulb migratory pathway in the adult mouse: Cellular composition and specificity as determined by heterochronic and heterotopic transplantation. *J Comp Neurol* 371:376–396.

Javaherian A, Kriegstein A (2009). A stem cell niche for intermediate progenitor cells of the embryonic cortex. *Cereb Corte* 19 Suppl 1:i70–7.

Jessberger S, Toni N, Clemenson Jr GD, Ray J, Gage FH (2008). Directed differentiation of hippocampal stem/progenitor cells in the adult brain. *Nat Neurosci* 11:888–893.

Jhas S, Ciura S, Belanger-Jasmin S, Dong Z, Llamosas E, Theriault FM, Joachim K, Tang Y, Liu L, Liu J, Stifani S (2006). Hes6 inhibits astrocyte differentiation and promotes neurogenesis through different mechanisms. *J Neurosci* 26:11061–11071.

Jia C, Halpern M (1998). Neurogenesis and migration of receptor neurons in the vomeronasal sensory epithelium in the opossum, *Monodelphis domestica. J Comp Neurol* 400:287–297.

Jiang W, Gu W, Brannstrom T, Rosqvist R, Wester P (2001). Cortical neurogenesis in adult rats after transient middle cerebral artery occlusion. *Stroke* 32:1201–1207.

Jiao J, Chen DF (2008). Induction of neurogenesis in nonconventional neurogenic regions of the adult central nervous system by niche astrocyte-produced signals. *Stem Cells* 26:1221–1230.

Jin K, Zhu Y, Sun Y, Mao XO, Xie L, Greenberg DA (2002). Vascular endothelial growth factor (VEGF) stimulates neurogenesis in vitro and in vivo. *Proc Natl Acad Sci USA* 99:11946–11950.

Jin K, Galvan V, Xie L, Mao XO, Gorostiza OF, Bredesen DE, Greenberg DA (2004). Enhanced neurogenesis in Alzheimer's disease transgenic (PDGF-APPSw,Ind) mice. *Proc Natl Acad Sci USA* 101:13363–13367.

Jin K, Wang X, Xie L, Mao XO, Zhu W, Wang Y, Shen J, Mao Y, Banwait S, Greenberg DA (2006). Evidence for stroke-induced neurogenesis in the human brain. *Proc Natl Acad Sci USA* 103:13198–13202.

Johns PR (1977). Growth of the adult goldfish eye. III. Source of the new retinal cells. *J Comp Neurol* 176: 343–357.

Johns PR, Easter SS, Jr. (1977). Growth of the adult goldfish eye. II. Increase in retinal cell number. *J Comp Neurol* 176:331–341.

Johnston MD, Edwards CM, Bodmer WF, Maini PK, Chapman SJ (2007a). Examples of mathematical modeling: Tales from the crypt. *Cell Cycle* 6:2106–2112.

Johnston MD, Edwards CM, Bodmer WF, Maini PK, Chapman SJ (2007b). Mathematical modeling of cell population dynamics in the colonic crypt and in colorectal cancer. *Proc Natl Acad Sci USA* 104: 4008–4013.

Kaplan MS (1981). Neurogenesis in the 3-month-old rat visual cortex. *J Comp Neurol* 195:323–338.

Karl MO, Hayes S, Nelson BR, Tan K, Buckingham B, Reh TA (2008). Stimulation of neural regeneration in the mouse retina. *Proc Natl Acad Sci USA* 105:19508–19513.

Kaslin J, Ganz J, Geffarth M, Grandel H, Hans S, Brand M (2009). Stem cells in the adult zebrafish cerebellum: Initiation and maintenance of a novel stem cell niche. *J Neurosci* 29:6142–6153.

Kawauchi S, Beites CL, Crocker CE, Wu HH, Bonnin A, Murray R, Calof AL (2004). Molecular signals regulating proliferation of stem and progenitor cells in mouse olfactory epithelium. *Dev Neurosci* 26:166–180.

Kawauchi S, Kim J, Santos R, Wu HH, Lander AD, Calof AL (2009). Foxg1 promotes olfactory neurogenesis by antagonizing Gdf11. *Development* 136:1453–1464.

Kazanis I, Belhadi A, Faissner A, Ffrench-Constant C (2007). The adult mouse subependymal zone regenerates efficiently in the absence of tenascin-C. *J Neurosci* 27:13991–13996.

Keilhoff G, Becker A, Grecksch G, Bernstein HG, Wolf G (2006). Cell proliferation is influenced by bulbectomy and normalized by imipramine treatment in a region-specific manner. *Neuropsychopharmacology* 31:1165–1176.

Keiner S, Witte OW, Redecker C (2009). Immunocytochemical detection of newly generated neurons in the perilesional area of cortical infarcts after intraventricular application of brain-derived neurotrophic factor. *J Neuropathol Exp Neurol* 68:83–93.

Kempermann G, Neumann H (2003). Neuroscience. Microglia: The enemy within? *Science* 302:1689–1690.

Kempermann G, Kuhn HG, Gage FH (1997a). Genetic influence on neurogenesis in the dentate gyrus of adult mice. *Proc Natl Acad Sci USA* 94:10409–10414.

Kempermann G, Kuhn HG, Gage FH (1997b). More hippocampal neurons in adult mice living in an enriched environment. *Nature* 386:493–495.

Kerever A, Schnack J, Vellinga D, Ichikawa N, Moon C, Arikawa-Hirasawa E, Efird JT, Mercier F (2007). Novel extracellular matrix structures in the neural stem cell niche capture the neurogenic factor fibroblast growth factor 2 from the extracellular milieu. *Stem Cells* 25:2146–2157.

Kim EJ, Battiste J, Nakagawa Y, Johnson JE (2008). Ascl1 (Mash1) lineage cells contribute to discrete cell populations in CNS architecture. *Mol Cell Neurosci* 38:595–606.

Klassen H, Reubinoff B (2008). Stem cells in a new light. *Nat Biotechnol* 26:187–188.

Kohwi M, Petryniak MA, Long JE, Ekker M, Obata K, Yanagawa Y, Rubenstein JL, Alvarez-Buylla A (2007). A subpopulation of olfactory bulb GABAergic interneurons is derived from Emx1- and Dlx5/6-expressing progenitors. *J Neurosci* 27:6878–6891.

Koketsu D, Mikami A, Miyamoto Y, Hisatsune T (2003). Nonrenewal of neurons in the cerebral neocortex of adult macaque monkeys. *J Neurosci* 23:937–942.

Kokoeva MV, Yin H, Flier JS (2005). Neurogenesis in the hypothalamus of adult mice: Potential role in energy balance. *Science* 310:679–683.

Kokoeva MV, Yin H, Flier JS (2007). Evidence for constitutive neural cell proliferation in the adult murine hypothalamus. *J Comp Neurol* 505:209–220.

Kokovay E, Temple S (2007). Taking neural crest stem cells to new heights. *Cell* 131:234–236.

Koo JW, Duman RS (2008). IL-1beta is an essential mediator of the antineurogenic and anhedonic effects of stress. *Proc Natl Acad Sci USA* 105:751–756.

Kornack DR, Rakic P (2001). Cell proliferation without neurogenesis in adult primate neocortex. *Science* 294:2127–2130.

Kronenberg G, Wang LP, Geraerts M, Babu H, Synowitz M, Vicens P, Lutsch G, Glass R, Yamaguchi M, Baekelandt V, Debyser Z, Kettenmann H, Kempermann G (2007). Local origin and activity-dependent generation of nestin-expressing protoplasmic astrocytes in CA1. *Brain Structure & Function* 212:19–35.

Kruger GM, Mosher JT, Bixby S, Joseph N, Iwashita T, Morrison SJ (2002). Neural crest stem cells persist in the adult gut but undergo changes in self-renewal, neuronal subtype potential, and factor responsiveness. *Neuron* 35:657–669.

Kuan CY, Schloemer AJ, Lu A, Burns KA, Weng WL, Williams MT, Strauss KI, Vorhees CV, Flavell RA, Davis RJ, Sharp FR, Rakic P (2004). Hypoxia-ischemia induces DNA synthesis without cell proliferation in dying neurons in adult rodent brain. *J Neurosci* 24:10763–10772.

Kuge A, Takemura S, Kokubo Y, Sato S, Goto K, Kayama T (2009). Temporal profile of neurogenesis in the subventricular zone, dentate gyrus and cerebral cortex following transient focal cerebral ischemia. *Neurol Res* 31:969–76.

Kuhn HG, Dickinson-Anson H, Gage FH (1996). Neurogenesis in the dentate gyrus of the adult rat: Age-related decrease of neuronal progenitor proliferation. *J Neurosci* 16:2027–2033.

Kukekov VG, Laywell ED, Suslov O, Davies K, Scheffler B, Thomas LB, O'Brien TF, Kusakabe M, Steindler DA (1999). Multipotent stem/progenitor cells with similar properties arise from two neurogenic regions of adult human brain. *Exp Neurol* 156:333–344.

Kunze A, Congreso MR, Hartmann C, Wallraff-Beck A, Huttmann K, Bedner P, Requardt R, Seifert G, Redecker C, Willecke K, Hofmann A, Pfeifer A, Theis M, Steinhauser C (2009). Connexin expression by radial glia–like cells is required for neurogenesis in the adult dentate gyrus. *Proc Natl Acad Sci USA* 106:11336–1.

Lamba DA, Karl MO, Ware CB, Reh TA (2006). Efficient generation of retinal progenitor cells from human embryonic stem cells. *Proc Natl Acad Sci USA* 103:12769–12774.

Lamba D, Karl M, Reh T (2008). Neural regeneration and cell replacement: A view from the eye. *Cell Stem Cell* 2:538–549.

Lee J, Gravel M, Zhang R, Thibault P, Braun PE (2005a). Process outgrowth in oligodendrocytes is mediated by CNP, a novel microtubule assembly myelin protein. *J Cell Biol* 170:661–673.

Lee SK, Lee B, Ruiz EC, Pfaff SL (2005b). Olig2 and Ngn2 function in opposition to modulate gene expression in motor neuron progenitor cells. *Genes Dev* 19:282–294.

Leinders-Zufall T, Lane AP, Puche AC, Ma W, Novotny MV, Shipley MT, Zufall F (2000). Ultrasensitive phero-mone detection by mammalian vomeronasal neurons. *Nature* 405:792–796.

Leker RR, Soldner F, Velasco I, Gavin DK, Androutsellis-Theotokis A, McKay RD (2007). Long-lasting regen-eration after ischemia in the cerebral cortex. *Stroke* 38:153–161.

Li HY, Say EH, Zhou XF (2007). Isolation and characterization of neural crest progenitors from adult dorsal root ganglia. *Stem Cells* 25:2053–2065.

Li WL, Yu SP, Ogle ME, Ding XS, Wei L (2008). Enhanced neurogenesis and cell migration following focal ischemia and peripheral stimulation in mice. *Dev Neurobiol* 68:1474–1486.

Lie DC, Dziewczapolski G, Willhoite AR, Kaspar BK, Shults CW, Gage FH (2002). The adult substantia nigra contains progenitor cells with neurogenic potential. *J Neurosci* 22:6639–6649.

Lie DC, Colamarino SA, Song HJ, Desire L, Mira H, Consiglio A, Lein ES, Jessberger S, Lansford H, Dearie AR, Gage FH (2005). Wnt signalling regulates adult hippocampal neurogenesis. *Nature* 437:1370–1375.

Ligon KL, Fancy SP, Franklin RJ, Rowitch DH (2006). Olig gene function in CNS development and disease. *Glia* 54:1–10.

Lim DA, Alvarez-Buylla A (1999). Interaction between astrocytes and adult subventricular zone precursors stimu-lates neurogenesis. *Proc Natl Acad Sci USA* 96:7526–7531.

Lim DA, Tramontin AD, Trevejo JM, Herrera DG, Garcia-Verdugo JM, Alvarez-Buylla A (2000). Noggin antag-onizes BMP signaling to create a niche for adult neurogenesis. *Neuron* 28:713–726.

Ling L, Hou Q, Xing S, Yu J, Pei Z, Zeng J (2008). Exogenous kallikrein enhances neurogenesis and angiogenesis in the subventricular zone and the peri-infarction region and improves neurological function after focal cortical infarction in hypertensive rats. *Brain Res* 1206:89–97.

Liu MT, Kuan YH, Wang J, Hen R, Gershon MD (2009). 5–HT4 receptor-mediated neuroprotection and neu-rogenesis in the enteric nervous system of adult mice. *J Neurosci* 29:9683–9699.

Liu X, Wang Q, Haydar TF, Bordey A (2005). Nonsynaptic GABA signaling in postnatal subventricular zone controls proliferation of GFAP-expressing progenitors. *Nat Neurosci* 8:1179–1187.

Loo AT, Youngentob SL, Kent PF, Schwob JE (1996). The aging olfactory epithelium: Neurogenesis, response to damage, and odorant-induced activity. *Int J Dev Neurosci* 14:881–900.

Lopez-Barneo J, Ortega-Saenz P, Pardal R, Pascual A, Piruat JI (2008). Carotid body oxygen sensing. *Eur Respir J* 32:1386–1398.

Lopez-Barneo J, Pardal R, Ortega-Saenz P, Duran R, Villadiego J, Toledo-Aral JJ (2009). The neurogenic niche in the carotid body and its applicability to anti-parkinsonian cell therapy. *J Neural Transm* 116:975–982.

Loseva E, Yuan TF, Karnup S (2009). Neurogliogenesis in the mature olfactory system: A possible protective role against infection and toxic dust. *Brain Res Rev* 59:374–387.

Lu L, Tonchev AB, Kaplamadzhiev DB, Boneva NB, Mori Y, Sahara S, Ma D, Nakaya MA, Kikuchi M, Yamashima T (2008). Expression of matrix metalloproteinases in the neurogenic niche of the adult monkey hippocampus after ischemia. *Hippocampus* 18:1074–1084.

Luzzati F, Peretto P, Aimar P, Ponti G, Fasolo A, Bonfanti L (2003). Glia-independent chains of neuroblasts through the subcortical parenchyma of the adult rabbit brain. *Proc Natl Acad Sci USA* 100:13036–13041.

MacDonald KP, Murrell WG, Bartlett PF, Bushell GR, Mackay-Sim A (1996). FGF2 promotes neuronal differentiation in explant cultures of adult and embryonic mouse olfactory epithelium. *J Neurosci Res* 44:27–39.

MacLaren RE, Pearson RA, MacNeil A, Douglas RH, Salt TE, Akimoto M, Swaroop A, Sowden JC, Ali RR (2006). Retinal repair by transplantation of photoreceptor precursors. *Nature* 444:203–207.

Magavi S, Leavitt B, Macklis J (2000). Induction of neurogenesis in the neocortex of adult mice. *Nature* 405:951–955.

Magavi SS, Macklis JD (2002a). Induction of neuronal type-specific neurogenesis in the cerebral cortex of adult mice: Manipulation of neural precursors in situ. *Brain Res Dev Brain Res* 134:57–76.

Magavi SS, Macklis JD (2002b). Immunocytochemical analysis of neuronal differentiation. *Methods Mol Biol* 198:291–297.

Marcus RC, Delaney CL, Easter SS, Jr. (1999). Neurogenesis in the visual system of embryonic and adult zebrafish (*Danio rerio*). *Vis Neurosci* 16:417–424.

Markakis EA, Palmer TD, Randolph-Moore L, Rakic P, Gage FH (2004). Novel neuronal phenotypes from neural progenitor cells. *J Neurosci* 24:2886–2897.

Marshall CT, Lu C, Winstead W, Zhang X, Xiao M, Harding G, Klueber KM, Roisen FJ (2006). The therapeutic potential of human olfactory-derived stem cells. *Histol Histopathol* 21:633–643.

Martinez-Marcos A, Ubeda-Banon I, Halpern M (2000a). Cell turnover in the vomeronasal epithelium: Evidence for differential migration and maturation of subclasses of vomeronasal neurons in the adult opossum. *J Neurobiol* 43:50–63.

Martinez-Marcos A, Ubeda-Banon I, Deng L, Halpern M (2000b). Neurogenesis in the vomeronasal epithelium of adult rats: Evidence for different mechanisms for growth and neuronal turnover. *J Neurobiol* 44:423–435.

Martinez-Marcos A, Jia C, Quan W, Halpern M (2005). Neurogenesis, migration, and apoptosis in the vomeronasal epithelium of adult mice. *J Neurobiol* 63:173–187.

Mathonnet M, Comte I, Lalloue F, Ayer-Le Lievre C (2001). Insulin-like growth factor I induced survival of axotomized olfactory neurons in the chick. *Neurosci Lett* 308:67–70.

Maurer MH, Feldmann RE, Jr., Burgers HF, Kuschinsky W (2008). Protein expression differs between neural progenitor cells from the adult rat brain subventricular zone and olfactory bulb. *BMC Neurosci* 9:7.

Mellough CB, Cui Q, Spalding KL, Symons NA, Pollett MA, Snyder EY, Macklis JD, Harvey AR (2004). Fate of multipotent neural precursor cells transplanted into mouse retina selectively depleted of retinal ganglion cells. *Exp Neurol* 186:6–19.

Menn B, Garcia-Verdugo JM, Yaschine C, Gonzalez-Perez O, Rowitch D, Alvarez-Buylla A (2006). Origin of oligodendrocytes in the subventricular zone of the adult brain. *J Neurosci* 26:7907–7918.

Mercier F, Kitasako JT, Hatton GI (2002). Anatomy of the brain neurogenic zones revisited: Fractones and the fibroblast/macrophage network. *J Comp Neurol* 451:170–188.

Merkle FT, Tramontin AD, Garcia-Verdugo JM, Alvarez-Buylla A (2004). Radial glia give rise to adult neural stem cells in the subventricular zone. *Proc Natl Acad Sci USA* 101:17528–17532.

Merkle FT, Mirzadeh Z, Alvarez-Buylla A (2007). Mosaic organization of neural stem cells in the adult brain. *Science* 317:381–384.

Mildner A, Schmidt H, Nitsche M, Merkler D, Hanisch UK, Mack M, Heikenwalder M, Bruck W, Priller J, Prinz M (2007). Microglia in the adult brain arise from Ly-6ChiCCR2+ monocytes only under defined host conditions. *Nat Neurosci* 10:1544–1553.

Miragall F, Kadmon G, Faissner A, Antonicek H, Schachner M (1990). Retention of J1/tenascin and the polysialylated form of the neural cell adhesion molecule (N-CAM) in the adult olfactory bulb. *J Neurocytol* 19:899–914.

Mirich JM, Brunjes PC (2001). Activity modulates neuronal proliferation in the developing olfactory epithelium. *Brain Res Dev Brain Res* 127:77–80.

Mirnics K (2008). What is in the brain soup? *Nat Neurosci* 11:1237–1238.

Mirzadeh Z, Merkle FT, Soriano-Navarro M, Garcia-Verdugo JM, Alvarez-Buylla A (2008). Neural stem cells confer unique pinwheel architecture to the ventricular surface in neurogenic regions of the adult brain. *Cell Stem Cell* 3:265–278.

Mohapel P, Frielingsdorf H, Haggblad J, Zachrisson O, Brundin P (2005). Platelet-derived growth factor (PDGF-BB) and brain-derived neurotrophic factor (BDNF) induce striatal neurogenesis in adult rats with 6-hydroxydopamine lesions. *Neuroscience* 132:767–776.

Monje ML, Toda H, Palmer TD (2003). Inflammatory blockade restores adult hippocampal neurogenesis. *Science* 302:1760–1765.

Morgan SC, Taylor DL, Pocock JM (2004). Microglia release activators of neuronal proliferation mediated by activation of mitogen-activated protein kinase, phosphatidylinositol-3–kinase/Akt and delta-Notch signalling cascades. *J Neurochem* 90:89–101.

Mullen RJ, Buck CR, Smith AM (1992). NeuN, a neuronal specific nuclear protein in vertebrates. *Development* 116:201–211.

Murdoch B, Roskams AJ (2007). Olfactory epithelium progenitors: Insights from transgenic mice and in vitro biology. *J Mol Histol* 38:581–599.

Murdoch B, Roskams AJ (2008). A novel embryonic nestin-expressing radial glia–like progenitor gives rise to zonally restricted olfactory and vomeronasal neurons. *J Neurosci* 28:4271–4282.

Murrell W, Feron F, Wetzig A, Cameron N, Splatt K, Bellette B, Bianco J, Perry C, Lee G, Mackay-Sim A (2005). Multipotent stem cells from adult olfactory mucosa. *Dev Dyn* 233:496–515.

Nacher J, Alonso-Llosa G, Rosell D, McEwen B (2002). PSA-NCAM expression in the piriform cortex of the adult rat. Modulation by NMDA receptor antagonist administration. *Brain Res* 927:111–121.

Nacher J, Pham K, Gil-Fernandez V, McEwen BS (2004). Chronic restraint stress and chronic corticosterone treatment modulate differentially the expression of molecules related to structural plasticity in the adult rat piriform cortex. *Neuroscience* 126:503–509.

Nagahara Y (1940). Experimentelle studien uber die histologischen Veranderungen des Geruchsorgans nach der Olfactoriusdurchschneidung. Beitrage zur Kenntnis des feineren Baus der Geruchsorgans. *Japan J Med Sci Pathol* 5:165–199.

Nakagawa T, Miyamoto O, Janjua NA, Auer RN, Nagao S, Itano T (2004). Localization of nestin in amygdaloid kindled rat: An immunoelectron microscopic study. *Can J Neurol Sci* 31:514–519.

Nakagomi N, Nakagomi T, Kubo S, Nakano-Doi A, Saino O, Takata M, Yoshikawa H, Stern DM, Matsuyama T, Taguchi A (2009). Endothelial cells support survival, proliferation and neuronal differentiation of transplanted adult ischemia-induced neural stem/progenitor cells after cerebral infarction. *Stem Cells*. 27:2185–95

Nakajima K, Honda S, Tohyama Y, Imai Y, Kohsaka S, Kurihara T (2001). Neurotrophin secretion from cultured microglia. *J Neurosci Res* 65:322–331.

Nakajima K, Tohyama Y, Kohsaka S, Kurihara T (2002). Ceramide activates microglia to enhance the production/ secretion of brain-derived neurotrophic factor (BDNF) without induction of deleterious factors in vitro. *J Neurochem* 80:697–705.

Nakajima K, Tohyama Y, Maeda S, Kohsaka S, Kurihara T (2007). Neuronal regulation by which microglia enhance the production of neurotrophic factors for GABAergic, catecholaminergic, and cholinergic neurons. *Neurochem Int* 50:807–820.

Nakamura H, Higuchi Y, Kondoh H, Obata M, Takahashi S (2002). The effect of basic fibroblast growth factor on the regeneration of guinea pig olfactory epithelium. *Eur Arch Otorhinolaryngol* 259:166–169.

Nakanishi M, Niidome T, Matsuda S, Akaike A, Kihara T, Sugimoto H (2007). Microglia-derived interleukin-6 and leukaemia inhibitory factor promote astrocytic differentiation of neural stem/progenitor cells. *Eur J Neurosci* 25:649–658.

Nakatomi H, Kuriu T, Okabe S, Yamamoto S, Hatano O, Kawahara N, Tamura A, Kirino T, Nakafuku M (2002). Regeneration of hippocampal pyramidal neurons after ischemic brain injury by recruitment of endogenous neural progenitors. *Cell* 110:429–441.

Newman MP, Feron F, Mackay-Sim A (2000). Growth factor regulation of neurogenesis in adult olfactory epithelium. *Neuroscience* 99:343–350.

Newton SS, Collier EF, Hunsberger J, Adams D, Terwilliger R, Selvanayagam E, Duman RS (2003). Gene profile of electroconvulsive seizures: Induction of neurotrophic and angiogenic factors. *J Neurosci* 23:10841–10851.

Nicolay DJ, Doucette JR, Nazarali AJ (2006). Transcriptional regulation of neurogenesis in the olfactory epithelium. *Cell Mol Neurobiol* 26:803–821.

Nishiguchi KM, Kaneko H, Nakamura M, Kachi S, Terasaki H (2008). Identification of photoreceptor precursors in the pars plana during ocular development and after retinal injury. *Invest Ophthalmol Vis Sci* 49:422–428.

Nishiguchi KM, Kaneko H, Nakamura M, Kachi S, Terasaki H (2009). Generation of immature retinal neurons from proliferating cells in the pars plana after retinal histogenesis in mice with retinal degeneration. *Mol Vis* 15:187–199.

Novitch BG, Chen AI, Jessell TM (2001). Coordinate regulation of motor neuron subtype identity and pan-neuronal properties by the bHLH repressor Olig2. *Neuron* 31:773–789.

Nowakowski RS, Hayes NL (2000). New neurons: Extraordinary evidence or extraordinary conclusion? *Science* 288:771.

Oezyuerek Z, Franke K, Nitschke M, Schulze R, Simon F, Eichhorn KJ, Pompe T, Werner C, Voit B (2009). Sulfated glyco-block copolymers with specific receptor and growth factor binding to support cell adhesion and proliferation. *Biomaterials* 30:1026–1035.

Ohab JJ, Carmichael ST (2008). Poststroke neurogenesis: Emerging principles of migration and localization of immature neurons. *Neuroscientist* 14:369–380.

Ohab JJ, Fleming S, Blesch A, Carmichael ST (2006). A neurovascular niche for neurogenesis after stroke. *J Neurosci* 26:13007–13016.

Ohira K, Furuta T, Hioki H, Nakamura KC, Kuramoto E, Tanaka Y, Funatsu N, Shimizu K, Oishi T, Hayashi M, Miyakawa T, Kaneko T, Nakamura S (2010). Ischemia-induced neurogenesis of neocortical layer 1 progenitor cells. *Nat Neurosci* 13:173–179.

Oka N, Soeda A, Inagaki A, Onodera M, Maruyama H, Hara A, Kunisada T, Mori H, Iwama T (2007). VEGF promotes tumorigenesis and angiogenesis of human glioblastoma stem cells. *Biochem Biophys Res Commun* 360:553–559.

Oldham MC, Konopka G, Iwamoto K, Langfelder P, Kato T, Horvath S, Geschwind DH (2008). Functional organization of the transcriptome in human brain. *Nat Neurosci* 11:1271–1282.

Orita S, Yoshinobu J, Orita Y, Tsujigiwa H, Kakiuchi M, Nagatsuka H, Nomiya S, Nagai N, Nishizaki K (2009). Prolactin may stimulate proliferation in the olfactory epithelium of the female mouse. *Am J Rhinol Allergy* 23:135–138.

Osakada F, Ikeda H, Mandai M, Wataya T, Watanabe K, Yoshimura N, Akaike A, Sasai Y, Takahashi M (2008). Toward the generation of rod and cone photoreceptors from mouse, monkey and human embryonic stem cells. *Nat Biotechnol* 26:215–224.

Ozen I, Galichet C, Watts C, Parras C, Guillemot F, Raineteau O (2007). Proliferating neuronal progenitors in the postnatal hippocampus transiently express the proneural gene Ngn2. *Eur J Neurosci* 25:2591–2603.

Palmer TD (2002). Adult neurogenesis and the vascular Nietzsche. *Neuron* 34:856–858.

Palmer TD, Takahashi J, Gage FH (1997). The adult rat hippocampus contains premordial neural stem cells. *Mol Cell Neurosci* 8:389–404.

Palmer TD, Willhoite AR, Gage FH (2000). Vascular niche for adult hippocampal neurogenesis. *J Comp Neurol* 425:479–494.

Pardal R, Ortega-Saenz P, Duran R, Lopez-Barneo J (2007). Glia-like stem cells sustain physiologic neurogenesis in the adult mammalian carotid body. *Cell* 131:364–377.

Parent JM, Vexler ZS, Gong C, Derugin N, Ferriero DM (2002). Rat forebrain neurogenesis and striatal neuron replacement after focal stroke. *Ann Neurol* 52:802–813.

Park JH, Cho H, Kim H, Kim K (2006). Repeated brief epileptic seizures by pentylenetetrazole cause neurodegeneration and promote neurogenesis in discrete brain regions of freely moving adult rats. *Neuroscience* 140:673–684.

Parras CM, Hunt C, Sugimori M, Nakafuku M, Rowitch D, Guillemot F (2007). The proneural gene Mash1 specifies an early population of telencephalic oligodendrocytes. *J Neurosci* 27:4233–4242.

Paulson L, Eriksson PS, Curtis MA (2007). Defining primary and secondary progenitor disorders in the brain: Proteomic approaches for analysis of neural progenitor cells. *Curr Pharm Biotechnol* 8:117–125.

Pekcec A, Loscher W, Potschka H (2006). Neurogenesis in the adult rat piriform cortex. *Neuroreport* 17: 571–574.

Pekny M, Nilsson M (2005). Astrocyte activation and reactive gliosis. *Glia* 50:427–434.

Pekny M, Wilhelmsson U, Bogestal YR, Pekna M (2007). The role of astrocytes and complement system in neural plasticity. *Int Rev Neurobiol* 82:95–111.

Pencea V, Bingaman KD, Wiegand SJ, Luskin MB (2001). Infusion of brain-derived neurotrophic factor into the lateral ventricle of the adult rat leads to new neurons in the parenchyma of the striatum, septum, thalamus, and hypothalamus. *J Neurosci* 21:6706–6717.

Pleasure SJ, Collins AE, Lowenstein DH (2000). Unique expression patterns of cell fate molecules delineate sequential stages of dentate gyrus development. *J Neurosci* 20:6095–6105.

Plumpe T, Ehninger D, Steiner B, Klempin F, Jessberger S, Brandt M, Romer B, Rodriguez GR, Kronenberg G, Kempermann G (2006). Variability of doublecortin-associated dendrite maturation in adult hippocampal neurogenesis is independent of the regulation of precursor cell proliferation. *BMC Neurosci* 7:77.

Raineteau O, Hugel S, Ozen I, Rietschin L, Sigrist M, Arber S, Gahwiler BH (2006). Conditional labeling of newborn granule cells to visualize their integration into established circuits in hippocampal slice cultures. *Mol Cell Neurosci* 32:344–355.

Rakic P (2002). Neurogenesis in adult primate neocortex: An evaluation of the evidence. *Nat Rev Neurosci* 3: 65–71.

Ramer LM, Au E, Richter MW, Liu J, Tetzlaff W, Roskams AJ (2004). Peripheral olfactory ensheathing cells reduce scar and cavity formation and promote regeneration after spinal cord injury. *J Comp Neurol* 473:1–15.

Ramirez-Castillejo C, Sanchez-Sanchez F, Andreu-Agullo C, Ferron SR, Aroca-Aguilar JD, Sanchez P, Mira H, Escribano J, Farinas I (2006). Pigment epithelium-derived factor is a niche signal for neural stem cell renewal. *Nat Neurosci* 9:331–339.

Reh TA, Constantine-Paton M (1983). Qualitative and quantitative measures of plasticity during the normal development of the *Rana pipiens* retinotectal projection. *Brain Res* 312:187–200.

Reynolds BA, Weiss S (1992). Generation of neurons and astrocytes from isolated cells of the adult mammalian central nervous system. *Science* 255:1707–1710.

Ricard D, Rogemond V, Charrier E, Aguera M, Bagnard D, Belin MF, Thomasset N, Honnorat J (2001). Isolation and expression pattern of human Unc-33–like phosphoprotein 6/collapsin response mediator protein 5 (Ulip6/CRMP5): Coexistence with Ulip2/CRMP2 in Sema3a-sensitive oligodendrocytes. *J Neurosci* 21:7203–7214.

Ridet JL, Malhotra SK, Privat A, Gage FH (1997). Reactive astrocytes: Cellular and molecular cues to biological function. *Trends Neurosci* 20:570–577.

Rietze R, Poulin P, Weiss S (2000). Mitotically active cells that generate neurons and astrocytes are present in multiple regions of the adult mouse hippocampus. *J Comp Neurol* 424:397–408.

Rogelius N, Jensen JB, Lundberg C, Parmar M (2006). Retrovirally delivered Islet-1 increases recruitment of Ng2 expressing cells from the postnatal SVZ into the striatum. *Exp Neurol* 201:388–398.

Rogelius N, Hebsgaard JB, Lundberg C, Parmar M (2008). Reprogramming of neonatal SVZ progenitors by islet-1 and neurogenin-2. *Mol Cell Neurosci* 38:453–459.

Ronnett GV, Leopold D, Cai X, Hoffbuhr KC, Moses L, Hoffman EP, Naidu S (2003). Olfactory biopsies demonstrate a defect in neuronal development in Rett's syndrome. *Ann Neurol* 54:206–218.

Rowitch DH, Lu QR, Kessaris N, Richardson WD (2002). An 'oligarchy' rules neural development. *Trends Neurosci* 25:417–422.

Schanzer A, Wachs FP, Wilhelm D, Acker T, Cooper-Kuhn C, Beck H, Winkler J, Aigner L, Plate KH, Kuhn HG (2004). Direct stimulation of adult neural stem cells in vitro and neurogenesis in vivo by vascular endothelial growth factor. *Brain Pathol* 14:237–248.

Scharff C, Kirn JR, Grossman M, Macklis JD, Nottebohm F (2000). Targeted neuronal death affects neuronal replacement and vocal behavior in adult songbirds. *Neuron* 25:481–492.

Schmidt M (2007). The olfactory pathway of decapod crustaceans—an invertebrate model for life-long neurogenesis. *Chem Senses* 32:365–384.

Schofield R (1983). The stem cell system. *Biomed Pharmacother* 37:375–380.

Schwarting GA, Gridley T, Henion TR (2007). Notch1 expression and ligand interactions in progenitor cells of the mouse olfactory epithelium. *J Mol Histol* 38:543–553.

Seaberg RM, van der Kooy D (2002). Adult rodent neurogenic regions: The ventricular subependyma contains neural stem cells, but the dentate gyrus contains restricted progenitors. *J Neurosci* 22:1784–1793.

Seki T, Arai Y (1991). Expression of highly polysialylated NCAM in the neocortex and piriform cortex of the developing and the adult rat. *Anat Embryol* 184:395–401.

Seki T, Arai Y (1993a). Distribution and possible roles of the highly polysialylated neural cell adhesion molecule (NCAM-H) in the developing and adult central nervous system. *Neurosci Res* 17:265–290.

Seki T, Arai Y (1993b). Distribution and possible roles of the highly polysialylated neural cell adhesion molecule (NCAM-H) in the developing and adult central nervous system. *Neurosci Res* 17:265–290.

Seri B, Garcia-Verdugo JM, McEwen BS, Alvarez-Buylla A (2001). Astrocytes give rise to new neurons in the adult mammalian hippocampus. *J Neurosci* 21:7153–7160.

Seri B, Garcia-Verdugo JM, Collado-Morente L, McEwen BS, Alvarez-Buylla A (2004). Cell types, lineage, and architecture of the germinal zone in the adult dentate gyrus. *J Comp Neurol* 478:359.

Shan X, Chi L, Bishop M, Luo C, Lien L, Zhang Z, Liu R (2006). Enhanced de novo neurogenesis and dopaminergic neurogenesis in the substantia nigra of 1-methyl-4-phenyl-1,2,3,6-tetrahydropyridine-induced Parkinson's disease-like mice. *Stem Cells* 24:1280–1287.

Shapiro LA, Korn MJ, Shan Z, Ribak CE (2005). GFAP-expressing radial glia–like cell bodies are involved in a one-to-one relationship with doublecortin-immunolabeled newborn neurons in the adult dentate gyrus. *Brain Res* 1040:81–91.

Shapiro LA, Ng KL, Zhou QY, Ribak CE (2007a). Olfactory enrichment enhances the survival of newly born cortical neurons in adult mice. *Neuroreport* 18:981–985.

Shapiro LA, Ng KL, Kinyamu R, Whitaker-Azmitia P, Geisert EE, Blurton-Jones M, Zhou QY, Ribak CE (2007b). Origin, migration and fate of newly generated neurons in the adult rodent piriform cortex. *Brain Structure & Function* 212:133–148.

Shapiro LA, Ng K, Zhou QY, Ribak CE (2009). Subventricular zone-derived, newly generated neurons populate several olfactory and limbic forebrain regions. *Epilepsy Behav* 14 Suppl 1:74–80.

Sheen VL, Macklis JD (1995). Targeted neocortical cell death in adult mice guides migration and differentiation of transplanted embryonic neurons. *J Neurosci* 15:8378–8392.

Shen L, Nam HS, Song P, Moore H, Anderson SA (2006). FoxG1 haploinsufficiency results in impaired neurogenesis in the postnatal hippocampus and contextual memory deficits. *Hippocampus* 16:875–890.

Shen Q, Goderie SK, Jin L, Karanth N, Sun Y, Abramova N, Vincent P, Pumiglia K, Temple S (2004). Endothelial cells stimulate self-renewal and expand neurogenesis of neural stem cells. *Science* 304:1338–1340.

Shen Q, Wang Y, Kokovay E, Lin G, Chuang SM, Goderie SK, Roysam B, Temple S (2008). Adult SVZ stem cells lie in a vascular niche: A quantitative analysis of niche cell–cell interactions. *Cell Stem Cell* 3:289–300.

Shetty RS, Bose SC, Nickell MD, McIntyre JC, Hardin DH, Harris AM, McClintock TS (2005). Transcriptional changes during neuronal death and replacement in the olfactory epithelium. *Mol Cell Neurosci* 30:90–107.

Shihabuddin LS, Horner PJ, Ray J, Gage FH (2000). Adult spinal cord stem cells generate neurons after transplantation in the adult dentate gyrus. *J Neurosci* 20:8727–8735.

Shin JJ, Fricker-Gates RA, Perez FA, Leavitt BR, Zurakowski D, Macklis JD (2000). Transplanted neuroblasts differentiate appropriately into projection neurons with correct neurotransmitter and receptor phenotype in neocortex undergoing targeted projection neuron degeneration. *J Neurosci* 20:7404–7416.

Shou J, Rim PC, Calof AL (1999). BMPs inhibit neurogenesis by a mechanism involving degradation of a transcription factor. *Nat Neurosci* 2:339–345.

Siegenthaler JA, Tremper-Wells BA, Miller MW (2008). Foxg1 haploinsufficiency reduces the population of cortical intermediate progenitor cells: Effect of increased p21 expression. *Cereb Cortex* 18:1865–1875.

Singh RP, Cheng YH, Nelson P, Zhou FC (2009). Retentive multipotency of adult dorsal root ganglia stem cells. *Cell Transplant* 18:55–68.

Smutzer G, Lee VM, Trojanowski JQ, Arnold SE (1998). Human olfactory mucosa in schizophrenia. *Ann Otol Rhinol Laryngol* 107:349–355.

Snyder EY, Yoon C, Flax JD, Macklis JD (1997). Multipotent neural precursors can differentiate toward replacement of neurons undergoing targeted apoptotic degeneration in adult mouse neocortex. *Proc Natl Acad Sci USA* 94:11663–11668.

Song H, Stevens CF, Gage FH (2002). Astroglia induce neurogenesis from adult neural stem cells. *Nature* 417:39–44.

Staquicini FI, Dias-Neto E, Li J, Snyder EY, Sidman RL, Pasqualini R, Arap W (2009). Discovery of a functional protein complex of netrin-4, laminin gamma1 chain, and integrin alpha6beta1 in mouse neural stem cells. *Proc Natl Acad Sci USA* 106:2903–2908.

Steiner B, Kronenberg G, Jessberger S, Brandt MD, Reuter K, Kempermann G (2004). Differential regulation of gliogenesis in the context of adult hippocampal neurogenesis in mice. *Glia* 46:41–52.

Steiner B, Klempin F, Wang L, Kott M, Kettenmann H, Kempermann G (2006a). Type-2 cells as link between glial and neuronal lineage in adult hippocampal neurogenesis. *Glia* 54:805–814.

Steiner B, Winter C, Hosman K, Siebert E, Kempermann G, Petrus DS, Kupsch A (2006b). Enriched environment induces cellular plasticity in the adult substantia nigra and improves motor behavior function in the 6-OHDA rat model of Parkinson's disease. *Exp Neurol* 199:291–300.

Stenman J, Toresson H, Campbell K (2003). Identification of two distinct progenitor populations in the lateral ganglionic eminence: Implications for striatal and olfactory bulb neurogenesis. *J Neurosci* 23:167–174.

Stubbs D, Deproto J, Nie K, Englund C, Mahmud I, Hevner R, Molnar Z (2009). Neurovascular congruence during cerebral cortical development. *Cereb Cortex* 19 Suppl 1:i32–41.

Suhonen JO, Peterson DA, Ray J, Gage FH (1996). Differentiation of adult hippocampus-derived progenitors into olfactory neurons in vivo. *Nature* 383:624–627.

Takahashi M, Palmer TD, Takahashi J, Gage FH (1998). Widespread integration and survival of adult-derived neural progenitor cells in the developing optic retina. *Mol Cell Neurosci* 12:340–348.

Tamada A, Shirasaki R, Murakami F (1995). Floor plate chemoattracts crossed axons and chemorepels uncrossed axons in the vertebrate brain. *Neuron* 14:1083–1093.

Tatsumi K, Takebayashi H, Manabe T, Tanaka KF, Makinodan M, Yamauchi T, Makinodan E, Matsuyoshi H, Okuda H, Ikenaka K, Wanaka A (2008). Genetic fate mapping of Olig2 progenitors in the injured adult cerebral cortex reveals preferential differentiation into astrocytes. *J Neurosci Res* 86:3494–3502.

Tattersfield AS, Croon RJ, Liu YW, Kells AP, Faull RL, Connor B (2004). Neurogenesis in the striatum of the quinolinic acid lesion model of Huntington's disease. *Neuroscience* 127:319–332.

Thored P, Arvidsson A, Cacci E, Ahlenius H, Kallur T, Darsalia V, Ekdahl CT, Kokaia Z, Lindvall O (2006). Persistent production of neurons from adult brain stem cells during recovery after stroke. *Stem Cells* 24:739–747.

Thored P, Wood J, Arvidsson A, Cammenga J, Kokaia Z, Lindvall O (2007). Long-term neuroblast migration along blood vessels in an area with transient angiogenesis and increased vascularization after stroke. *Stroke* 38:3032–3039.

Thored P, Heldmann U, Gomes-Leal W, Gisler R, Darsalia V, Taneera J, Nygren JM, Jacobsen SE, Ekdahl CT, Kokaia Z, Lindvall O (2009). Long-term accumulation of microglia with proneurogenic phenotype concomitant with persistent neurogenesis in adult subventricular zone after stroke. *Glia* 57:835–849.

Tome M, Lindsay SL, Riddell JS, Barnett SC (2009). Identification of non-epithelial multipotent cells in the embryonic olfactory mucosa. *Stem Cells* 27:2196–208.

Tonchev AB, Yamashima T, Guo J, Chaldakov GN, Takakura N (2007). Expression of angiogenic and neurotrophic factors in the progenitor cell niche of adult monkey subventricular zone. *Neuroscience* 144:1425–1435.

Tozuka Y, Fukuda S, Namba T, Seki T, Hisatsune T (2005). GABAergic excitation promotes neuronal differentiation in adult hippocampal progenitor cells. *Neuron* 47:803–815.

Tropepe V, Coles BL, Chiasson BJ, Horsford DJ, Elia AJ, McInnes RR, van der Kooy D (2000). Retinal stem cells in the adult mammalian eye. *Science* 287:2032–2036.

Vallieres L, Campbell IL, Gage FH, Sawchenko PE (2002). Reduced hippocampal neurogenesis in adult transgenic mice with chronic astrocytic production of interleukin-6. *J Neurosci* 22:486–492.

Van Kampen JM, Robertson HA (2005). A possible role for dopamine D3 receptor stimulation in the induction of neurogenesis in the adult rat substantia nigra. *Neuroscience* 136:381–386.

Van Kampen JM, Hagg T, Robertson HA (2004). Induction of neurogenesis in the adult rat subventricular zone and neostriatum following dopamine D3 receptor stimulation. *Eur J Neurosci* 19:2377–2387.

van Leeuwen IM et al. (2009). An integrative computational model for intestinal tissue renewal. *Cell Prolif* 42:617–36.

Vedin V, Molander M, Bohm S, Berghard A (2009). Regional differences in olfactory epithelial homeostasis in the adult mouse. *J Comp Neurol* 513:375–384.

Villadiego J, Mendez-Ferrer S, Valdes-Sanchez T, Silos-Santiago I, Farinas I, Lopez-Barneo J, Toledo-Aral JJ (2005). Selective glial cell line-derived neurotrophic factor production in adult dopaminergic carotid body cells in situ and after intrastriatal transplantation. *J Neurosci* 25:4091–4098.

Vossmerbaeumer U, Kuehl S, Kern S, Kluter H, Jonas JB, Bieback K (2008). Induction of retinal pigment epithelium properties in ciliary margin progenitor cells. *Clin Experiment Ophthalmol* 36:358–366.

Walton NM, Sutter BM, Laywell ED, Levkoff LH, Kearns SM, Marshall GP, 2nd, Scheffler B, Steindler DA (2006). Microglia instruct subventricular zone neurogenesis. *Glia* 54:815–825.

Walzlein JH, Synowitz M, Engels B, Markovic DS, Gabrusiewicz K, Nikolaev E, Yoshikawa K, Kaminska B, Kempermann G, Uckert W, Kaczmarek L, Kettenmann H, Glass R (2008). The antitumorigenic response of neural precursors depends on subventricular proliferation and age. *Stem Cells* 26:2945–2954.

Wang Y, Jin K, Greenberg DA (2007). Neurogenesis associated with endothelin-induced cortical infarction in the mouse. *Brain Res* 1167:118–122.

Wang YQ, Cui HR, Yang SZ, Sun HP, Qiu MH, Feng XY, Sun FY (2009). VEGF enhance cortical newborn neurons and their neurite development in adult rat brain after cerebral ischemia. *Neurochem Int* 55(7): 629-36.

Warfvinge K, Kamme C, Englund U, Wictorin K (2001). Retinal integration of grafts of brain-derived precursor cell lines implanted subretinally into adult, normal rats. *Exp Neurol* 169:1–12.

Weissman TA, Riquelme PA, Ivic L, Flint AC, Kriegstein AR (2004). Calcium waves propagate through radial glial cells and modulate proliferation in the developing neocortex. *Neuron* 43:647–661.

Wetts R, Fraser SE (1988). Multipotent precursors can give rise to all major cell types of the frog retina. *Science* 239:1142–1145.

Weyer A, Schilling K (2003). Developmental and cell type-specific expression of the neuronal marker NeuN in the murine cerebellum. *J Neurosci Res* 73:400–409.

Widestrand A, Faijerson J, Wilhelmsson U, Smith PL, Li L, Sihlbom C, Eriksson PS, Pekny M (2007). Increased neurogenesis and astrogenesis from neural progenitor cells grafted in the hippocampus of GFAP-/- Vim-/- mice. *Stem Cells* 25:2619–2627.

Wu HH, Ivkovic S, Murray RC, Jaramillo S, Lyons KM, Johnson JE, Calof AL (2003). Autoregulation of neurogenesis by GDF11. *Neuron* 37:197–207.

Wurmser AE, Nakashima K, Summers RG, Toni N, D'Amour KA, Lie DC, Gage FH (2004). Cell fusion-independent differentiation of neural stem cells to the endothelial lineage. *Nature* 430:350–356.

Xu Y, Tamamaki N, Noda T, Kimura K, Itokazu Y, Matsumoto N, Dezawa M, Ide C (2005). Neurogenesis in the ependymal layer of the adult rat third ventricle. *Exp Neurol* 192:251–264.

Yamashima T, Tonchev AB, Vachkov IH, Popivanova BK, Seki T, Sawamoto K, Okano H (2004). Vascular adventitia generates neuronal progenitors in the monkey hippocampus after ischemia. *Hippocampus* 14:861–875.

Yang P, Baker KA, Hagg T (2005). A disintegrin and metalloprotease 21 (ADAM21) is associated with neurogenesis and axonal growth in developing and adult rodent CNS. *J Comp Neurol* 490:163–179.

Yang P, Baker KA, Hagg T (2006). The ADAMs family: Coordinators of nervous system development, plasticity and repair. *Prog Neurobiol* 79:73–94.

Yoshida S (2008). Spermatogenic stem cell system in the mouse testis. *Cold Spring Harbor Symposia on Quantitative Biology* 73:25–32.

Yoshimi K, Ren YR, Seki T, Yamada M, Ooizumi H, Onodera M, Saito Y, Murayama S, Okano H, Mizuno Y, Mochizuki H (2005). Possibility for neurogenesis in substantia nigra of Parkinsonian brain. *Ann Neurol* 58: 31–40.

Young KM, Fogarty M, Kessaris N, Richardson WD (2007). Subventricular zone stem cells are heterogeneous with respect to their embryonic origins and neurogenic fates in the adult olfactory bulb. *J Neurosci* 27: 8286–8296.

Young MJ, Ray J, Whiteley SJ, Klassen H, Gage FH (2000). Neuronal differentiation and morphological integration of hippocampal progenitor cells transplanted to the retina of immature and mature dystrophic rats. *Mol Cell Neurosci* 16:197–205.

Young RW (1985). Cell proliferation during postnatal development of the retina in the mouse. *Brain Res* 353: 229–239.

Zhang PB, Liu Y, Li J, Kang QY, Tian YF, Chen XL, Zhao JJ, Shi QD, Song TS, Qian YH (2005). Ependymal/subventricular zone cells migrate to the peri-infarct region and differentiate into neurons and astrocytes after focal cerebral ischemia in adult rats. *Di Yi Jun Yi Da Xue Xue Bao* 25:1201–1206.

Zhang R, Zhang Z, Wang L, Wang Y, Gousev A, Zhang L, Ho KL, Morshead C, Chopp M (2004). Activated neural stem cells contribute to stroke-induced neurogenesis and neuroblast migration toward the infarct boundary in adult rats. *J Cereb Blood Flow Metab* 24:441–448.

Zhang RL, Zhang ZG, Zhang L, Chopp M (2001). Proliferation and differentiation of progenitor cells in the cortex and the subventricular zone in the adult rat after focal cerebral ischemia. *Neuroscience* 105:33–41.

Zhang RL, Zhang L, Zhang ZG, Morris D, Jiang Q, Wang L, Zhang LJ, Chopp M (2003). Migration and differentiation of adult rat subventricular zone progenitor cells transplanted into the adult rat striatum. *Neuroscience* 116:373–382.

Zhang RL, Zhang ZG, Chopp M (2007). Gene profiles within the adult subventricular zone niche: Proliferation, differentiation and migration of neural progenitor cells in the ischemic brain. *Curr Mol Med* 7:459–462.

Zhang X, Klueber KM, Guo Z, Cai J, Lu C, Winstead WI, Qiu M, Roisen FJ (2006). Induction of neuronal differentiation of adult human olfactory neuroepithelial-derived progenitors. *Brain Res* 1073–1074:109–119.

Zhao M, Momma S, Delfani K, Carlen M, Cassidy RM, Johansson CB, Brismar H, Shupliakov O, Frisen J, Janson AM (2003). Evidence for neurogenesis in the adult mammalian substantia nigra. *Proc Natl Acad Sci USA* 100:7925–7930.

Zhao X, Das AV, Soto-Leon F, Ahmad I (2005). Growth factor-responsive progenitors in the postnatal mammalian retina. *Dev Dyn* 232:349–358.

Zigova T, Pencea V, Betarbet R, Wiegand SJ, Alexander C, Bakay RA, Luskin MB (1998). Neuronal progenitor cells of the neonatal subventricular zone differentiate and disperse following transplantation into the adult rat striatum. *Cell Transplant* 7:137–156.

Zufall F, Kelliher KR, Leinders-Zufall T (2002). Pheromone detection by mammalian vomeronasal neurons. *Microsc Res Tech* 58:251–260.

9

———

Regulation

Adult neurogenesis is regulated
by surprisingly many different
factors

Most original publications on adult neurogenesis are in one form or another concerned with its regulation. Factors as diverse as neurotrophins, stress, opioids, seizures, odors, diabetes, gingko biloba, physical activity, acupuncture, ischemia, learning, nitric oxide, alcohol, electromagnetic waves, and many others have been reported to "regulate" neurogenesis. Often it remains undecided whether influence actually reflects a direct effect. Many regulators might share a common mechanism or be entirely nonspecific. Since the first edition of this book, the body of literature on the regulation of adult neurogenesis has expanded by several hundred publications, and a book is not the place to review this literature in its entirety; the overview would be neither timely nor complete.

In this chapter, we will rather focus on the principles behind the regulation of adult neurogenesis. A first impression from the large variety of studies on this topic might be that the regulation of adult neurogenesis seems to be seismographically sensitive to almost any stimulus. Sometimes one would have been more surprised if a study had shown convincingly that "factor X" had indeed no influence on adult neurogenesis, rather than adding factor X to the long list of factors that "regulate" neurogenesis.

Regulation Versus Control

Regulation and control are not
the same: regulation means
departure from the controlled
baseline

A distinction can be made between "regulation" and "control," and although the boundary between these terms has become blurred, it is helpful to consider the original idea of a clearer separation. In molecular biology, for example, we speak of a gene as being *controlled* by its promoter but being *regulated* by extrinsic influences, ultimately through this promoter. Regulation is in this sense achieved on the basis of alterations on the level of one or more control units. In other words: regulation is macroscopic; control, microscopic. Regulation is a departure from baseline, control maintains baseline. Regulation is relatively more extrinsic, control relatively more intrinsic. This is also the terminology used in this book.

But this understanding is different from the same pair of terms, for example, in physiology. There regulation takes place when a system maintains a variable against outside forces; and control, in contrast, refers to the situation in which the system would adjust to this outside force by changing this variable over time. That meaning is, for example, also found in psychology, where self-regulation is opposed to the extrinsic control of a trait. Or in political science, where the (e.g., legal) regulation of a system is not the same as actively controlling it (as an executive function).

In the reality of experimental situations, regulation and control might often be addressed in an identical way, but the closer we move towards the role of transcription factors and epigenetic mechanisms, the more we learn about control rather than regulation. Consequently, demonstrating the involvement of a transcription factor known from embryonic neurogenesis also in adult neurogenesis might tell us much about neuronal development in general, but not necessarily about how overall regulation of *adult* neurogenesis is actually achieved. Conversely, demonstrating regulation alone does not necessarily lead to meaningful mechanistic conclusions, because we are only confronted with the "usual suspects." The involvement of, for example, Notch signaling in adult neurogenesis does not come as a surprise, and lacks all specificity for adulthood. Nevertheless, Notch signaling is indispensable, and the devil will lie in the details.

Regulation of adult neurogenesis is usually measured as change in cell numbers

Studying regulation usually means addressing a change in quantity, but only rarely, changes in quality. Therefore, important aspects of regulation might be missed because only selected parameters were measured. In most studies, regulation of adult neurogenesis is equivalent to an increase or decrease in the number of newly generated cells. The readout of these studies is the number of cells labeled with bromodeoxyuridine before or during the experimental manipulation. As we have seen, the number of BrdU-positive cells is only a good indicator of the number of new neurons if long-enough survival times are allowed after the injection of BrdU. Therefore, many studies that report changes in adult neurogenesis but base their claim only on an acute measurement of BrdU incorporation (or some other measure of proliferation; see Chap. 7) do not necessarily deliver a meaningful quantitative assessment of regulation as far as the resulting number of new neurons is concerned. The same applies (by analogy) to the use of surrogate marker doublecortin to measure "neurogenesis."

The growing awareness of this pitfall and increasing knowledge about the details of adult neuronal development are changing this situation. The use of electrophysiological methods, the characterization of complex regulatory situations by gene or protein expression profiling, and the tool box of systems biology allow us to define regulation on more levels than previously possible. Our concepts of how adult neurogenesis is both regulated and controlled are thus evolving at a dramatic speed.

Regulation occurs in many different steps in the course of neuronal development

Although "regulation" of adult neurogenesis that deserves this name in the stricter sense should have an effect on net neurogenesis—that is, on the quantity (and possibly the quality) of new neurons—the target of regulation in the course of adult neuronal development can vary. Theoretically, any identifiable step during neuronal development can be subject to regulatory influences resulting in the net effect. Consequently, the degree to which the results of an experiment are meaningful depends on the specificity with which the readout is adjusted to the needs of the regulatory mechanism of interest. The impression that adult neurogenesis boldly and nonspecifically reacts to many different stimuli in an apparently identical way might be a problem of methodology. An identical net effect on adult neurogenesis can result from many different combinations of effects on individual stages of development. Increased proliferation with decreased survival, for example, could lead to no net change at all. Decreased proliferation with much increased survival of the reduced progeny could still result in more new neurons at the end of the day. One thus cannot extrapolate information from effects on individual stages of neuronal development and apply this to net effects. Nevertheless, discrete and specific effects on neuronal development are relevant and conceptually might provide more useful information than the report of a generic net effect. Precise language needs to be used in such discussions. For the uninitiated reader, an *effect* on adult neurogenesis implies a measurable net effect on the number (or function) of new neurons. Subeffects during development with absent or unknown consequences for net neurogenesis need to be clearly identified as such.

Regulation Occurs on Different Conceptual Levels

Regulation can occur during many different aspects of neuronal development and thus mechanistically can mean very different things. We can distinguish the following three levels:

- *Regulation on the level of systems and behavior.* How does adult neurogenesis respond to the animal's behavior? Specifically, how do changes in the functional states of the olfactory system or the hippocampus that are related to certain behaviors affect adult neurogenesis? At this level, regulation is most closely linked to function of the new neurons.
- *Regulation on the level of cell–cell signaling.* Independent of the way the functional system detects the need for new neurons, how is the message conveyed to a precursor cell or a differentiating young neuron that it should divide, live or die, migrate, extend an axon and dendrites, or form synapses? What kind of extracellular signaling is involved in the regulation of adult neurogenesis?
- *Regulation within the individual cell.* What is the nature of the intracellular signaling involved in the regulation of precursor cell activity and adult neurogenesis? How do the many different signal transduction cascades interact in determining the next step in adult neurogenesis for the given cell? What are the transcriptional regulators? What are the epigenetic mechanisms that define the state of a cell and its genetic potential?

For the full picture, one level cannot be understood without the others. Systematic distinctions might be necessary, but they are artificial. Exposure to a complex environment, for example, stimulates adult hippocampal neurogenesis, but the ephemeral cognitive stimuli induced by the environment do not directly interfere with precursor cells. Voluntary wheel-running in mice, whose execution by itself requires preceding brain activity, activates a multitude of systemic responses in bodily systems and the central nervous system, reaches the germinative niches, and only through a variety of cell-to-cell interactions that find continuation in intracellular signaling pathways, finally effects transcriptional control of the relevant genes (Fig. 9–1). We can get first impressions of how regulation proceeds from the behavioral level down to the action of a single transcription factor. There is a network of regulatory pathways whose activation over many intermediate steps, supposedly in many different brain regions, ultimately leads to a particular change in the cellular niche engulfing the precursor cell. Such influences can come from neurotransmitters, hormones, growth factors, cell–cell contacts, extracellular matrix components, and others. Through intracellular second-messenger machinery, cells integrate across a large number of concurrent stimuli. These regulatory intracellular networks link activity at the many different receptors a cell carries as well as other intracellular events with the effects of transcription factors that interact with the DNA. Induction of proliferation is thus the consequence of a cell's sensing a condition that is "mitogenic" and reacting to it accordingly. The same principle applies to the regulation of all consecutive stages of adult neurogenesis.

Reductionistic experimental settings tend to suggest that single, identifiable stimuli are necessary and sufficient to elicit certain responses in biological systems, although this is usually nothing more than a direct consequence of the experimental design. Epidermal growth factor, for example, is a strong mitogen in vitro and in vivo, but that does not mean that EGF is the one and only factor responsible for control of proliferation. Excess of EGF can override other stimuli, and lack of EGF can block proliferation. Control, however, is more than all-or-nothing decisions. "Control" means integrating all available relevant stimuli. Nonspecific stimuli might override the complex pathways of specific regulation. Both kainic acid (KA) receptor activation and EGF receptor activation cause an increase in hippocampal precursor cell proliferation but involve different cascades: many regulatory pathways are redundant.

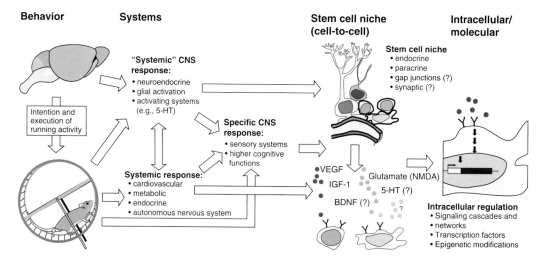

Behavior | **Systems** | **Stem cell niche (cell-to-cell)** | **Intracellular/molecular**

FIGURE 9–1 Regulation occurs on many conceptual levels. The example of physical activity inducing adult hippocampal neurogenesis is used here as an example to demonstrate that *regulation* is a multifaceted term that encompasses effects on levels from behavior, to single transcription factors. Note that the behavior that causes the regulation within the brain is itself dependent on brain activity. BDNF, brain-derived neurotrophic factor; IGF-1, insulin-like growth factor-1; VEGF, vascular endothelial growth factor; 5-HT, serotonin.

On a transcriptional level and on the level of regulating transcriptional control, many pathways in the net regulation of neurogenesis may share identical components. Although we have to study the small pieces of regulatory networks one by one, in reality they compete and interact with each other and are interdependent. The complex challenge is thus to understand, not only the mechanisms of regulation, but also their hierarchy and interdependencies. Net regulation is the result of integration of all these influences and is high-dimensional. From a phenomenological point of view, the net result of regulation (e.g., "seizures lead to more new neurons") might be predictable, but describing, understanding, and manipulating the state of the precursor cell or the immature neuron during this process remains a tremendous challenge. This situation underscores the importance of reductionism and the underestimated value of phenomenology in biological research. But the relevance of individually identified factors should not be overemphasized.

Natural Variation in Adult Hippocampal Neurogenesis

Adult neurogenesis in mice is a polygenic trait with huge natural variation

Comparison of adult hippocampal neurogenesis in different strains of inbred (Kempermann et al., 1997b; Kempermann and Gage, 2002b, a; Kempermann et al., 2006) and wild (Amrein et al., 2004a; Amrein et al., 2004b) strains of mice has revealed a large, natural variation in this trait.

Recombinant inbred (RI) strains of mice are essentially inbred progeny of F2 generations from a cross of two genetically defined parental strains such as C57BL/6 and DBA/2 (Fig. 9–2). The RI breeding paradigm leads to different strains within one set of strains (called "BXD" in the case of C57BL/6 and DBA/2) that contain roughly 50% genome from each parental strain in various compositions. Because they are inbred, RI strains are homologous at every locus and can be used for linkage studies (Korstanje and Paigen, 2002; Abiola et al., 2003). Figure 9–2 displays how the different mix of parental genomes in such strains results in large variations in the genetically determined baseline level of adult neurogenesis (data from Kempermann et al., 2006). In this example, the baseline difference in adult neurogenesis

Recombinant inbred breeding panel

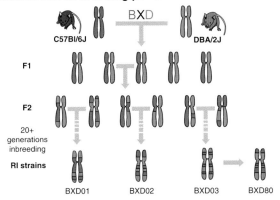

Natural variation in adult hippocampal neurogenesis

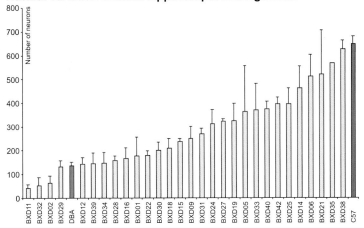

FIGURE 9–2 Natural variation in adult hippocampal neurogenesis. Adult hippocampal neurogenesis in mice is to a large degree genetically controlled and shows large strain differences. Strains C57BL/6 and DBA/2 show a fourfold difference in the baseline rate of neurogenesis (*number of new neurons, open bars*). The other strains are recombinant inbred strains of mice derived from these two parental strains and are essentially inbred F2 generation animals. The strains thus differ only in the genetic mix, but all have an approximately half C57BL/6 and half DBA/2 genome. C57BL/6 has been set as 100%. The data show how extensive the influence of genetic background is on adult neurogenesis. Data are taken from Kempermann and Gage (2002).

between BXD5 and BXD38 is more than twenty-fold. This difference is only due to a different 50:50 mix of the parental genome and considerably larger than any single-gene effects in transgenic or knockout studies (unless the mutations did not abolish neurogenesis altogether).

Many other parameters such as total granule cell number (Wimer and Wimer, 1989) and hippocampal weight (Peirce et al., 2003) and the size of the mossy fiber connection (Lassalle et al., 1994) show a similar variability. Especially in inbred mice, this variation mainly reflects the influence of genetic background. The data are not devoid of all environmental influence, because animals might react differently to an identical environment, but the environment can be held constant. The genetic contribution to the variation in total granule cell number, for example, was 86% (Abusaad et al., 1999). In the BXD panel, heritability of adult hippocampal neurogenesis was around 70% (Kempermann et al., 2006). Heritability is calculated as the ratio of between-strain variance over total variance (between-strain plus within-strain variance).

Regulation occurs at different stages of neuronal development

The first strain-difference studies also revealed that regulation of adult neurogenesis does not follow a simple on–off pattern but is differentially regulated at different stages of development, such as cell proliferation, rate of survival, and the relative contributions to neurogenesis and gliogenesis (Kempermann et al., 1997b; Kempermann and Gage, 2002a; Kempermann et al., 2006). This observation can be interpreted as a differential contribution of different sets of genes to the regulation of adult neurogenesis. Adult neurogenesis is a polygenic trait, and numerous genes will interact in determining the rate of neurogenesis and its regulation in response to external or internal neurogenic stimuli. Consequently, results from studies manipulating single genes erroneously treat adult neurogenesis as a monogenic Mendelian trait, which limits the conclusions that can be drawn from such experiments for understanding neurogenic regulation as a whole.

Strain differences have also been found in rats

In rats, no extensive strain-difference studies have been conducted, but the available data point to a similar heterogeneity (Perfilieva et al., 2001; Kronenberg et al., 2007a; Alahmed and Herbert, 2008). A particularly interesting case has been rat strains SHR and SHR-SP, which both show increased adult hippocampal neurogenesis compared to their genetic control strain Wistar Kyoto but achieved this increase by different means: the intermediate stages of precursor cells as identified by doublecortin expression were differently involved (Kronenberg et al., 2007a).

Finally, there was a first direct comparison of adult hippocampal neurogenesis in rats (mostly Sprague-Dawley) and in mice (C57BL/6 and in one sub-experiment CD1) stating that rats had more new neurons, which developed faster, were more likely to be recruited into functional connections, and showed greater contribution to establishing memory of fear (Snyder et al., 2009).

Linkage analysis allows identification of key gene regions associated with the control of adult neurogenesis

Natural variation of a phenotype in a genetic reference population such as BXD mice can be exploited to identify key genes associated with the trait. This can be done by conventional linkage analysis. The linkage map for the trait "new neurons" showed a great number of peaks, and although a few regions stand out, no locus will control much of the variation (Kempermann et al., 2006). This is not unexpected for a polygenic trait, but the linkage map aptly illustrates just how polygenic adult neurogenesis is going to be.

Given this general problem in analyzing highly polygenic traits, the RI breeding panels reveal their true power in combination with other techniques, for example with gene-expression profiling. In that situation, the expression level of every gene is treated as a phenotype of its own. The genetics of gene expression is called *expression genetics* or *genomical genetics*. Besides providing cues about how genes control genes, one might also correlate gene expression data with phenotypes such as *adult neurogenesis*. From many such data, one can construct covariance networks that allow a first glimpse at the molecular complexity underlying adult neurogenesis. Besides a few classical stem cell genes such as Prominin1, Musashi, or CamK4, the first of these analyses yielded a number of suggestive new candidates for genes that represent hubs in such networks of gene interaction (Kempermann et al., 2006). Examples were Ssbp1, which in myeloid cells is associated with the induction of differentiation; or Rapgef6, a member of the Ras cascade and, interestingly, the only high-ranking gene linking neurogenesis and gliogenesis in the adult hippocampus. There is still some leap of faith to be made between the recognition of covariance and the conclusion of functional causality in this link, but statistical methods allow us to control for mere random associations, and additional information can be incorporated into the analysis to strengthen a link. In principle, the hypotheses derived from these studies are directly testable, although, given the large number of interacting partners, the experiments will not always be straightforward.

Regulation of the Balance Between Cell Production and Cell Death

Adult neurogenesis is primarily regulated at the level of survival, not the proliferation of the precursor cells

The strain comparison studies in the BXD set also revealed that cell proliferation in the dentate gyrus explains only roughly 19% of the variance in the number of new neurons, whereas "survival" explained 85% (Kempermann et al., 2006). This indicates that adult neurogenesis is largely regulated at the level of cell survival. There is nevertheless an obvious influence of precursor cell proliferation in that the cells that might survive must first be generated, and increasing this potential for neurogenesis surprisingly also resulted in greater realization of this potential (Fabel et al., 2009); but it is the survival, not the production of new neurons that determines net neurogenesis. This insight is of great relevance for the consideration of the activity-dependent regulation of adult neurogenesis, where appropriate learning stimuli elicit a survival-promoting effect on newborn neurons.

Pro-proliferative factors do not need to be directly mitogenic

A large number of factors have been identified that influence the expansion phase (*proliferation*). This does not mean that such factors would necessarily be mitogens; their pro-proliferative action might be indirect. Regulation of precursor cell proliferation is largely considered to be rather nonspecific. Nevertheless, loss-of-function experiments with E2F, a transcription factor involved in the control of cell cycle progression, led to a reduction in adult neurogenesis without grossly affecting prenatal neurogenesis (Cooper-Kuhn et al., 2002). Similarly, loss of cyclin D2 (but not D1) reduced adult neurogenesis (Kowalczyk et al., 2004). Cell cycle duration is a key determinant in regulating neurogenesis; speeding up or slowing down cell cycle speed was sufficient to increase or decrease cortical development (Calegari et al., 2005; Lange et al., 2009).

New neurons are recruited from a surplus of immature progeny of the progenitor cells

Analogous to the prevailing mechanism during embryonic brain development, there is a surplus of new neurons generated in adult neurogenesis. From these cells, only those survive that make useful functional connections. Over a period of several weeks, the new cells mature in responsive to synaptic activation (Jessberger and Kempermann, 2003) and become electrophysiologically indistinguishable from older granule cells (van Praag et al., 2002). This sensitive period coincides with the phase of elimination during neuronal development in the adult (Gould et al., 1999; Kempermann et al., 2004b; Ge et al., 2007; Kee et al., 2007; Tashiro et al., 2007). The primary principle underlying the regulation of adult neurogenesis thus seems to be a Hebbian mechanism. The cells that are functionally beneficial survive, while the others die.

Adult hippocampal neurogenesis does not contribute to a appreciable cellular turnover

An early hypothesis in the field has been the idea that adult hippocampal neurogenesis would replace lost neurons in the adult hippocampus and cell death thus be a primary trigger of adult neurogenesis (Gould et al., 1991a, 1991b; Gould and McEwen, 1993; Cameron and Gould, 1994). The underlying concept was that adult neurogenesis would maintain a homeostasis. This idea was also later incorporated into a first dynamic, theoretical model of adult neurogenesis (Butz et al., 2006).

There is no evidence that adult neurogenesis would contribute to neuronal turnover in the hippocampus. Rather, new neurons are added to the existing networks, and the dentate gyrus grows throughout life, although in a rodent this absolute growth becomes minute and not measurable after the first weeks of life (Ben Abdallah et al., 2008). Lineage tracing based on GLAST-ERT2 transgenic mice indicated a certain turnover (Ninkovic et al., 2007), but Imayoshi and colleagues found that ablating adult neurogenesis did not cause a reduction in total granule cell count (Imayoshi et al., 2008). Evidence in support of limited turnover and a primary additive mode also comes from the observation that newborn neurons have a long-term existence (Kempermann et al., 2003). Neurogenesis in old age is so low that any potential increase in granule cell number is smaller than the interindividual variance. Obviously, because single granule cells might die over the course of life, on a population basis adult neurogenesis would nevertheless "replace" neurons.

In the olfactory bulb, the situation is more complex; both turnover and long-term addition of new neurons seem to occur (Winner et al., 2002; Imayoshi et al., 2008).

Superfluous cells are eliminated by programmed cell death

Of all regions in the adult brain, the neurogenic zones show the highest incidence of apoptotic cell death, being about a hundred times higher than in the rest of the brain (Fig. 9–3) (Blaschke et al., 1996; Biebl et al., 2000; Biebl et al., 2005). In the dentate gyrus, the highest number of apoptotic cells is found in the subgranular zone (SGZ); in the olfactory system, approximately 80% are detected in the olfactory bulb, the remainder in the rostral migratory stream (RMS) and the subventricular zone (SVZ). In both the SVZ and SGZ, the counts of BrdU-labeled cells at different times after BrdU reflect a strong initial expansion followed by a dramatic decrease in the number of newly generated cells (Winner et al., 2002; Kempermann et al., 2003). This elimination of cells is achieved by the programmed cell death of immature neurons. Consequently, measures of cell proliferation and cell death are related; in a model of chronic stress that resulted in a large decrease in cell proliferation in the dentate gyrus (see below), the number of apoptotic cells was reduced accordingly (Heine et al., 2004b). Manipulating cell death consequently changed adult neurogenesis (Biebl et al., 2005; Kuhn et al., 2005; Gemma et al., 2007).

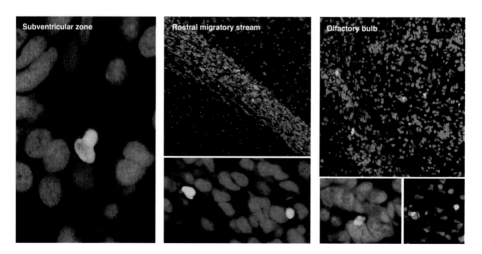

FIGURE 9–3 Apoptotic cell death during adult neurogenesis. New cells are generated in excess, and most of them are eliminated by programmed cell death. Eighty percent of cell death within the neurogenic region of the olfactory bulb occurs in the olfactory bulb, but the density of cell death is highest in the rostral migratory system. Apoptotic cell death is visualized with the (terminal deoxynucleotidyl transferase–mediated [dUPT] nick end labeling–TUNEL) method. Images by H. Georg Kuhn and Christiana Cooper-Kuhn, Gothenburg.

Specificity is brought into the net regulation by means of a selection process accompanied by the specific elimination of cells. At the highest conceptual level, benefit from adult neurogenesis would be measured as an improvement in cognitive function as a consequence of the addition of new neurons. Consistently with this concept, cognitive stimuli such as environmental enrichment and learning stimuli affect primarily the survival period during adult neurogenesis, although this distinction is not clear-cut (Kempermann et al., 1997a, 1998a; Gould et al., 1999; Dobrossy et al., 2003; Leuner et al., 2004). For many factors that can influence adult neurogenesis, it is not clear how their net effect on adult neurogenesis is achieved. Many paths might lead to the same net result. Cell proliferation during the expansion phase and survival of newly generated cells during the selection phase might be independently affected. However, some factors might act indirectly and influence both aspects through different pathways.

Appropriate survival-promoting factors include behavior, learning, and neuronal activity

Table 9–1 summarizes genes coding for cell-cycle factors with demonstrated function in adult neurogenesis. Interesting is the case of cyclin D2. In adult neurogenesis, cyclins are nonredundant, as they are elsewhere in the brain and the body. Knockout of cyclin D2 thus efficiently wipes out adult neurogenesis (Kowalczyk et al., 2004). Table 9–2 lists known apoptotic genes involved in the programmed elimination of new immature neurons generated in excess. Not unexpectedly, knockout of pro-apoptotic genes and overexpression of anti-apoptotic genes increased neurogenesis.

Related to the idea of a balance between cell birth and death is the concept of *quiescence*: precursor cells can withdraw from the cell cycle and remain in a dormant, nonproliferative stage from which they can be recruited. Protein p27kip1 (gene symbol: Cdkn1b) is associated with a transitional stage of cell cycle arrest in precursor cells, albeit not classical "quiescence": both neurogenic zones of the adult brain show p27kip1 expression (Doetsch et al., 2002b; Heine et al., 2004a). When p27kip1 function was abolished, proliferation in the SVZ increased (Doetsch et al., 2002b). A chronic stress model resulting in reduced adult neurogenesis caused a relative increase in p27kip1-positive cells. This finding indicates that the experimental manipulation caused an arrest of further neuronal development (Heine et al., 2004a).

In the literature, the distinction between neurogenesis phenotypes related to "proliferation" versus "survival" largely reflects the balance between precursor cell expansion and selective survival and integration.

Aging

Aging is a variable influencing living organisms like no other. In some sense, aging is just the passing of time. The neurobiology of aging is the neurobiology of very long time-scales. Key to this (rather simplistic approach) is the insight that development is unidirectional, and everything that happens builds on what happened before. The overall lifespan of any species is more or less set and characteristic for that species. Turtles get very old; *Drosophila,* not. Aging means gains and losses, but overall, losses prevail after adolescence. It is these losses, not the potential gain in wisdom, that makes aging a problem for modern societies. Aging is no disease, but even in the absence of neurodegeneration, cognitive impairments are manifold. Age-dependent changes are often seen as antithetical to "development." Plasticity and adult neurogenesis are on the plus side of this balance. We have argued that they provide a "reserve" to cope with age-dependent losses (Kempermann, 2008).

Age is a unique cofactor of all other regulatory mechanisms

The first edition of this book still stated: "The strongest known negative regulator of adult hippocampal neurogenesis is age." This is actually not correct. There is no doubt that neurogenesis greatly decreases with time, but this decrease is almost hyperbolic, not linear; indicating that at different ages, the passing of time has very different effects on adult neurogenesis. Joseph Altman's first complete description of adult hippocampal neurogenesis in 1965 already contained an account of this decrease over the

Adult neurogenesis decreases with age, but does so only in childhood and adolescence and is relatively stable thereafter

Table 9–1 Cell-cycle related genes in adult neurogenesis

Protein name	Gene Symbol	Description	Effect on adult neurogenesis	
Ataxia telangiectasia mutated homolog (human)	Atm	Serine protein kinase with role in cell cycle regulation and DNA repair	increased proliferation in ATM–/– mice but decreased neurogenesis (Allen et al., 2001)	SGZ
Cyclin D1	Ccnd1	see Ccnd2	Absent from precursor cells in the adult but not postnatal SGZ (Kowalczyk et al., 2004)	SGZ
Cyclin D2	Ccnd2	Regulates Cyclin-dependent kinases 4 and 6, which control G1/S transition of the cell cycle	Only cyclin expressed in precursor cells in the adult dentate gyrus. In Ccnd2 –/– cell proliferation was very low and neurogenesis, even on the level of immature neurons absent; environmental enrichment did not rescue this phenotype (Kowalczyk et al., 2004)	SGZ
Cyclin D3	Ccnd3	see Ccnd2	Absent from precursor cells in the adult but not postnatal SGZ (Kowalczyk et al., 2004)	SGZ
Cyclin-dependent kinase 2	Cdk2	Kinase presumably involved in G1/S transition during cell cycle	No differences between Cdk–/– and wildtype controls (Vandenbosch et al., 2007)	SGZ
Cyclin-dependent kinase inhibitor 1A (p21)	Cdkn1a	Regulates cell cycle progression at G1 and mediates p53-dependent cell cycle arrest	Expressed in DCX+ cells, reduced precursor cell proliferation and number of DCX+ cells in p21–/– mice based on FACS (Pechnick et al., 2008)	SGZ
Cyclin-dependent kinase inhibitor 1B (p27)	Cdkn1b	Reducing or stopping cell cycle progression at G1	In p27Kip1 –/– no effect on the number of B cells, increased numbers of C cells and reduced number of A cells (Doetsch et al., 2002b)	SVZ
			Stress increased p27 expression and reduced adult neurogenesis (Heine et al., 2004a)	SGZ
Cell cycle exit and neuronal differentiation 1 (BM88)	Cend1		Low expression in precursor cells, high expression in post-mitotic OB interneurons. Ex vivo overexpression favored differentiation; knockdown reduced differentiation (neurosphere assay) (Katsimpardi et al., 2008).	SVZ

Table 9–1 (Cont'd) Cell-cycle related genes in adult neurogenesis

Protein name	Gene Symbol	Description	Effect on adult neurogenesis	
Citron	Cit	Citron kinase is expressed during cytokinesis and possibly involved in neurogenic divisions	The flathead mutant rat (citron –/–) shows dysgenesis of the dentate gyrus, reduced precursor cell population and net adult neurogenesis (Ackman et al., 2007)	SGZ
E2F transcription factor 1	E2F1	Transcription factor with role in cell cycle control	Decreased proliferation and net neurogenesis in E2F1 –/–	SGZ, SVZ
p21	see Cdkn1a			
p27kip1	see Cdkn1b			

first months in the life of rodents (Altman and Das, 1965). Adult neurogenesis hits the highest level in early adulthood, during and after puberty, and quickly decreases thereafter (Fig. 9–4) (Kuhn et al., 1996; Cameron and McKay, 1999; Bondolfi et al., 2004; Ben Abdallah et al., 2010). Importantly, however, even in very old age, adult neurogenesis can be detected, although the levels are very low (Kempermann et al., 2002). The finding that adult neurogenesis declines but does not completely disappear with age apparently also applies to humans, in whom adult hippocampal neurogenesis was detected in postmortem samples from 72-year-old individuals (Eriksson et al., 1998). After the initial decrease, aging does not seem to exert a further, similarly strong influence. Consequently, one must

Table 9–2 Apoptosis-related genes in adult neurogenesis

Protein name	Gene Symbol	Description	Effect on adult neurogenesis	
Bcl2-associated X protein	Bax	Pro-apoptotic factor	Bax–/– mice showed reduced proliferation, increased ectopic migration of DCX cells, and increased numbers of granule cells (Sun et al., 2004)	SGZ
			No change in proliferation and size of the olfactory bulb but impaired migration of precursor cells and aberrant maturation in the SVZ (Kim et al., 2007)	SVZ
B-cell leukemia/ lymphoma 2	Bcl2	Anti-apoptotic factor	Overexpression of human Bcl2 increased survival, neurogenesis and granule cell numbers by inhibiting apoptosis (Kuhn et al., 2005); effect more pronounced under ischemia (Sasaki et al., 2003)	SGZ
Caspase 1	Casp1	Pro-apoptotic enzyme	Inhibition of Casp1 with Ac-YVAD-CMK increased neurogenesis (Gemma et al., 2007)	SGZ
Caspase 3	Casp3	Pro-apoptotic enzyme	Inhibition increased neurogenesis (Ekdahl et al., 2001)	SGZ

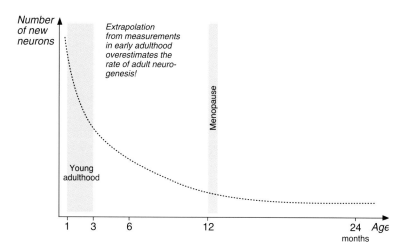

FIGURE 9–4 Age-related decline of adult neurogenesis in mice. The schematic graph highlights that the well-described decrease in adult neurogenesis actually takes place in young adulthood and adolescence, leading to low but rather constant levels for most part of the remaining life. Quantitative estimates obtained from young samples have to take into account this shape of the curve in order to avoid overestimation of cell numbers.

question whether aging per se is really a regulator at all. One might better call "aging" the possibly most important covariable in the regulation of adult neurogenesis. Aging might affect all other types of regulation. This makes aging research essentially interdisciplinary. A full appreciation of how neurogenesis changes over the course of life calls for the integration of results from many disciplines. To date, there is only one study that attempts such synthetic view, placing adult hippocampal neurogenesis in the rat into the context of other parameters of hippocampal aging, including water maze performance (Driscoll et al., 2006). Others have related the decrease in (precursor) cell proliferation to other variables describing adult neurogenesis, including the relative distribution of the different precursor cells and total granule cell counts (Ben Abdallah et al., 2008). What is still lacking is a detailed cohort study and (this is problematic due to the postmortem techniques used to measure adult neurogenesis) a truly longitudinal observation.

Adult neurogenesis is regulated in old age

As discussed in greater detail further below (p. 340), environmental enrichment stimulates adult hippocampal neurogenesis and is also effective in old age, suggesting that physiological regulation remains in place (Kempermann et al., 1998a; Kempermann et al., 2002). The same was shown for the other physiological stimulus, physical activity (van Praag et al., 2005). In relative terms, the effect of enrichment was even stronger in old than in young animals (Kempermann et al., 1998a). It seems that the functional stimulus provided by environmental complexity previously unknown to an animal challenged adult neurogenesis to deliver the maximum number of new neurons possible.

The physiological age-dependent decline can be partly prevented by activity

Even in the absence of significant inductions of cell proliferation, in most experiments with enriched environments, cell counts in the enriched animals have tended to be somewhat higher than those in controls. In mice exposed to an enriched environment for three months and returned afterwards to standard housing for an additional three months, this difference became significant (Kempermann and Gage, 1999). It appears that early stimulation of adult neurogenesis as a result of experiencing an enriched environment maintained neurogenesis at levels corresponding roughly to those of this much earlier age, and the stimulation thus counteracted the age effect. Even withdrawal for three months after environmental enrichment did not allow these animals to catch up with the age-related decline in precursor cell proliferation witnessed in the control mice. The hypothesis was confirmed in a later study demonstrating that voluntary wheel-running between the ages of six weeks and nine months (thus encompassing the physiological decline in adult neurogenesis) maintained precursor cell proliferation at the juvenile level, but, presumably in the absence of the appropriate cognitive stimulus, did not result in similarly increased net neurogenesis (Kronenberg et al., 2006).

This finding was confirmed when adult hippocampal neurogenesis was studied in mice that lived the second half of their life, from age 10 to 20 months, in an enriched environment (Kempermann et al., 2002). Although the environment might have been more complex than that for controls, the aspect of novelty disappeared during this chronic paradigm. These long-term-enriched mice not only had less age pigment in their dentate granule cells (and thus presumably a biologically younger hippocampus), their rate of adult hippocampal neurogenesis also corresponded to that of a much younger age. The baseline level was five times as high as that in controls. Both sustained physical and cognitive activity might underlie this effect.

Age affects different individual aspects of neurogenesis

There are conflicting data as to whether neuronal maturation in adult neurogenesis is slowed in the aged brain, but most studies tend to indicate that the changes are absent to minimal (van Praag et al., 2005; Couillard-Despres et al., 2006; Rao et al., 2006; Ahlenius et al., 2009). In learning-impaired old rats, however, there was reduced neuronal differentiation compared to well-learning rats (Nyffeler et al., 2008). And in old mice, the up-regulatory effects of fluoxetine on adult neurogenesis disappeared (Couillard-Despres et al., 2009). Despite changes in numerous parameters describing adult neurogenesis, spine density on the newborn neurons as a measure of the glutamatergic input they receive was unaltered in aged mice (Morgenstern et al., 2008), suggesting that, despite certain changes in some parameters, the potential for functionality might essentially remain unaltered.

Aging alters precursor cells and their niche

Nevertheless, and despite the relative shortage in detailed data, the neurogenic niches show obvious age-related changes. These alterations affect both the precursor cells and the niche factors themselves. Under baseline conditions, the number of precursor cells that can be isolated from the adult brain decreased with increasing age (Maslov et al., 2004), although a contradicting report had not supported this finding (Goldman et al., 1997). The precursor cell pool is maintained by self-renewal and the possible quiescence of some of the cells. This maintenance requires the specific conditions of the stem cell niche as well as cell-intrinsic properties.

Senescence of the precursor cells has nevertheless been investigated in only one study, which showed that the age-related loss of precursor cells in the SVZ (but not the dentate gyrus) was associated with an increase in p16INK4a, a cyclin-dependent kinase that has been linked to cellular senescence (Molofsky et al., 2006). As an extreme position it has been hypothesized that the decrease in precursor cell activity in the aging hippocampus was the consequence of increased niche-induced quiescence of precursor cells and not their reduced numbers (Hattiangady and Shetty, 2006).

Telomerase activity counteracts senescence of precursor cells

Long-term self-renewal in stem cells requires the activity of the enzyme *telomerase*, which counteracts the shortening of chromosome ends, the telomeres, that occurs during every cell division because the DNA polymerase misses the first nucleotides when synthesizing the new strand. Without telomerase activity, this increasing loss of telomeres would limit the number of possible divisions without chromosomal damage. Telomerase activity is thus a characteristic of stem cells. Telomerases add repetitive noncoding DNA sequence to the ends of chromosomes. Telomerase activity has been detected in the SVZ and remained present at low levels even after cytostatic treatment, a finding consistent with the idea that relatively quiescent stem cells can survive such treatment. Telomerase activity has also been found to be regulated in association with the regulation of neurogenesis (Caporaso et al., 2003). Whereas telomerase activity is widespread during development, it is restricted to the regions of high proliferative activity in the adult brain. Adult neural stem cells, however, seem to rely on telomerase activity even more than stem cells from the embryonic brain, possibly because redundant protection mechanisms exist in the embryo (Ferron et al., 2004). In vitro, telomerase activity was down-regulated when differentiation of primary neuronal precursor cells was initiated (Haik et al., 2000; Ostenfeld et al., 2000). In regenerative tissues of telomerase null-mutants, cell proliferation was not abolished but reduced, and cell death increased. This result further suggests that telomerase activity is necessary to maintain precursor cell proliferation over long periods of time and across generations (Lee et al., 1998; Herrera et al., 1999a; Herrera et al., 1999b).

Age-related hormonal changes might determine the decrease in neurogenesis

The neuroprotective properties of estrogens, their possible role in regulating adult neurogenesis (see below), and the possible link between estrogen deficiency and cognitive dysfunction in menopause have stimulated the idea of a hormonal dependency of the regulation of adult neurogenesis. However, no consistent association of hormone replacement and an improvement in cognitive function has been found (Hogervorst et al., 2002; Low and Anstey, 2006). Exogenous application of estradiol resulted in only a transient increase in cell proliferation in the hippocampus (Ormerod et al., 2003), although physiologically, proliferation peaks during estrogen-high pro-estrous (Tanapat et al., 1999). The situation is amazingly complex, however, given the high abundance of steroid hormones and their receptors and their intricate control by feedback mechanisms (Pawluski et al., 2009). But in any case, the age-dependent decrease of neurogenesis is essentially over well before menopause. So it is rather the hormonal changes during puberty than at midlife that might affect neurogenesis. Related to this question is the issue that, during puberty, hormones apparently control the "addition of neurons" to various sexually dimorphic brain structures, thereby shaping the female vs. the male (rat) brain (Ahmed et al., 2008). The study raised questions of how far the observed changes actually reflect adolescent or adult neurogenesis, but the impact of hormonal changes is appealing.

Glucocorticoids might play an ambiguous role in the age-related decrease in neurogenesis

Reduced neurogenesis in aging might also be causally linked to chronically increased levels of glucocorticoids in the elderly due to a dysregulation of the HPA axis (Koehl et al., 1999; Sapolsky, 1999). Acute treatment with glucocorticoid receptor antagonist mifepristone increased adult neurogenesis in old rats (Mayer et al., 2006). Preventing glucocorticoid action in the aging hippocampus maintained adult neurogenesis at a level corresponding to a much younger age (Cameron and McKay, 1999; Montaron et al., 2006). However, as to the singular role of glucocorticoids in this context, conflicting data exist, because the age-dependent decline is not necessarily associated with increased corticosterone levels (Heine et al., 2004b). With increasing age, the expression of corticosteroid receptors in the course of neuronal development shifts to more immature stages, which suggests that in older age, adult neurogenesis might become more sensitive to corticocorticoid action (Garcia et al., 2004a).

Neurogenic growth factors show age-related changes

Neurotrophic factors (see below) are other key candidates for mediators of cell-extrinsic mechanisms. Their receptors are abundant during development and decline with age (Shetty et al., 2005; Chadashvili and Peterson, 2006), although the precursor cells in the aging murine hippocampus retain the ability to respond to these signals (Lichtenwalner et al., 2001; Jin et al., 2003). Some factors such as cell cycle regulator transforming growth factor β1 (TFGβ1) might also increase with aging: overexpression of TGFβ1 in aged mice prevented proliferation of the precursor cells (Buckwalter et al., 2006).

Insulin-like growth factor-1 is one of the possible key mediators in the effects of physical activity on adult hippocampal neurogenesis (Carro et al., 2000; Trejo et al., 2001). Similarly, exogenous application of IGF1 to restore endogenous levels of IGF1 associated with younger age acutely counteracted the age-dependent decrease in adult hippocampal neurogenesis (Lichtenwalner et al., 2001).

Activity-Dependent Regulation

Enriched Environment and Learning

Exposure to a complex environment increases adult hippocampal neurogenesis

When it was reported in 1997 that mice living in an enriched environment had more new hippocampal granule cells than those of controls (Kempermann et al., 1997a) (Fig. 9–5), this was the first report of a positive regulator of adult hippocampal neurogenesis. Intriguingly, the up-regulation could thus be linked to hippocampal function. The same effect was later

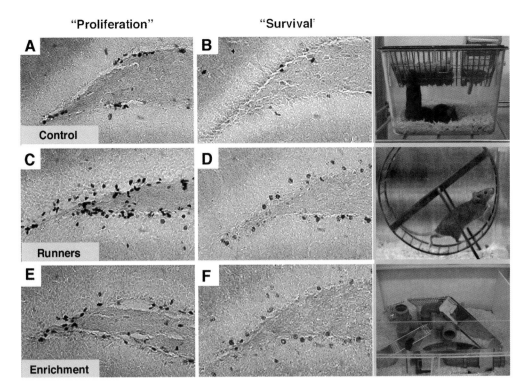

"Proliferation" **"Survival"**

A B
Control

C D
Runners

E F
Enrichment

FIGURE 9–5 Environmental enrichment increases adult hippocampal neurogenesis. Exposure to an enriched environment has a survival-promoting effect on newly generated cells in the adult dentate gyrus; physical activity also induces precursor cell proliferation. The survival-promoting effect of environmental enrichment affects primarily young post-mitotic neurons, whereas the pro-proliferative effect of wheel running has a strong effect on intermediate progenitor cells in the dentate gyrus. Neither manipulation affects neurogenesis in the adult olfactory bulb.

reported for rats as well (Nilsson et al., 1999). Environmental enrichment has a strong neurogenic effect on the hippocampus, but no effect on adult olfactory bulb neurogenesis (Brown et al., 2003a). These results are the opposite of effects of olfactory enrichment (Rochefort et al., 2002). The effects of environmental enrichment on adult neurogenesis were maintained in old age: here, the relative increase was even larger than in younger age, although the absolute numbers of new neurons were very low (Kempermann et al., 1998a; Kempermann et al., 2002). If the complexity of the enriched environment was experienced in early adulthood, when adult hippocampal neurogenesis still generated a very high number of neurons, this increase in adult hippocampal neurogenesis led to a measurable increase in the absolute number of granule cells (Kempermann et al., 1997a).

Environmental enrichment exerts a survival-promoting effect

In C57Bl/6 mice this effect was due to a survival-promoting effect on the progeny of the dividing precursor cells in the SGZ (calretinin stage) but not to increased divisions of the precursor cells (Brandt et al., 2003; Kronenberg et al., 2003). Consequently, environmental enrichment reduced apoptotic elimination of newborn cells in the dentate gyrus (Young et al., 1999).

Genetic factors determine the neurogenic response to environmental enrichment

However, the specificity of environmental enrichment for survival was dependent on the genetic background. In contrast to C57BL/6, mouse strain 129/SvJ showed a strong induction of cell proliferation after environmental enrichment (Kempermann et al., 1998b). Because 129/SvJ mice have very low endogenous levels of adult neurogenesis, this resulted in a net induction that was similar in the two strains. In 129/SvJ mice, environmental enrichment also affected cell proliferation. Inheritable traits thus influence the mechanism by which adult neurogenesis is regulated. It seems that, when faced with a challenging situation, the hippocampus can activate its neurogenic resources to increase adult neurogenesis and can achieve this by different means. The survival-promoting effect of environmental enrichment was found to be diminished in forebrain presenilin-1 null-mutants (Feng et al., 2001), in mice with mutations in the presenilin gene (Choi et al., 2008), in heterozygous BDNF knockouts (Rossi et al., 2006), and after blockade of VEGF-signaling (Cao et al., 2004). The common basis of these results with their apparent partly contradictory specificity is not yet known.

Environmental enrichment also affects the precursor cells and might promote their maintenance as well

Even in the absence of a pro-proliferative effect, environmental enrichment affected the number of nestin-GFP-positive cells that also expressed transcription factor Prox1, which is expressed early during granule cell development and stays expressed in all mature granule cells. The number of Prox1-positive new cells was significantly increased (Kronenberg et al., 2003). Acute exposure to an enriched environment similarly resulted in an increase in late precursor cells (type-3) (Steiner et al., 2008) suggesting that the pro-neurogenic effect of environmental enrichment in fact reaches rather early stages of neuronal development well before functional integration has occurred. Environmental enrichment is not purely, but is predominantly, survival-promoting. The observed apparent increase in cell proliferation in long-term experiments (Kempermann and Gage, 1999; Kempermann et al., 2002) in fact represents an increased survival of precursor cells, similar to the situation seen after prolonged voluntary exercise (see below). It is not clear if and how environmental enrichment affects late stages of neuronal development, after the cells have become functionally integrated. At that time a survival-promoting effect is postulated (and can be observed in certain experimental situations: Ge et al., 2007; Kee et al., 2007; Tashiro et al., 2007), but overall numbers are so low by that time after birth of the cells that it has not yet been possible to study this effect in the enrichment paradigm.

Environmental enrichment is a very straightforward experimental concept that historically has taught us much about activity-dependent brain plasticity (Chap. 2). An enriched environment typically consists of a larger cage, larger groups of animals, toys, and, for example, a tunnel system that is rearranged frequently. It is not known which of the many potential components of an enriched environment are responsible for the neurogenic effects. When locomotion and physical activity were controlled for, it turned out that they have an effect on adult neurogenesis of their own, in that they increase neurogenesis via stimulating precursor cell proliferation (Van Praag et al., 1999a; Kronenberg et al., 2003; Steiner et al., 2008). Physical activity thus does not explain the effects of environmental enrichment. The social component is suspected of playing a major role in mediating the effects of environmental enrichment, but specific studies (beyond investigating isolation and social dominance: see below) are missing.

Partial novelty, environmental complexity, and the challenge to learn might be main stimuli for neurogenesis

We hypothesize that the key stimuli are complexity and novelty (Kempermann, 2002; Kempermann et al., 2004a). In this sense, experiencing an enriched environment could be a form of continued learning. In fact, more or less isolated learning stimuli have been reported to have a survival-promoting effect on adult hippocampal neurogenesis similar to that of environmental enrichment (Gould et al., 1999; Leuner et al., 2004). In other modifications of the experimental settings, however, other groups could not detect such effects (Van Praag et al., 1999a; Ambrogini et al., 2004; Ehninger and Kempermann, 2006). Although there are clear differences in the experimental design between these studies, in particular with respect to a sensitive period of the new neurons, the question of whether learning per se can directly recruit new neurons into function in the end remains open.

Long-term potentiation increases adult neurogenesis

In favor of this hypothesis speaks the observation that the induction of long-term potentiation as the presumed electrophysiological correlate of learning induces adult hippocampal neurogenesis in vitro (Babu et al., 2009) and in vivo (Bruel-Jungerman et al., 2006). In vitro, this induction was dependent on BDNF-signaling from the excited neurons (Babu et al., 2009). Thus, although precursor cells can immediately sense excitation (see below), the LTP-dependent mechanism might be indirect. Outside such reductionist approach, the central problem is that learning as a stimulus cannot be presented in a pure form and always is embedded in situations that by themselves might influence neurogenesis by additional means. Whether learning is a form of environmental enrichment or whether the experience of enrichment is a form of learning remains undecided. The idea that learning could have a survival-promoting effect on newly generated cells, however, is in line with current concepts about the functional relevance of adult hippocampal neurogenesis (see Chap. 10).

Learning-induced survival might act back on the proliferating precursor cells

One study found a complex effect on late stages of the downward-sloping survival curve of new hippocampal neurons in rats and a small but specific increase in cell proliferation, which was interpreted as a some sort of feedback mechanism by which the precursor cells "learn" about the activity-induced integration of new neurons down the line (Dobrossy et al., 2003). The mechanism of this feedback remains elusive. Functionality of this regulation involves more than the promotion of survival and such potential feedback to the proliferating precursor cells. The cells not required for function also need to be eliminated (Dupret et al., 2007). There is therefore an intricate balance between the activity-dependent integration of new cells and the elimination of others.

Olfactory discrimination learning promotes survival of newborn cells during a critical period after their birth

In the olfactory bulb, the situation is less well studied. General environmental enrichment did not affect neurogenesis in the SVZ/olfactory bulb (Brown et al., 2003a), but olfactory enrichment did (Rochefort et al., 2002). More specifically, exposure to an olfactory discrimination task increased the survival of newly generated cells (Alonso et al., 2006). One other study, however, found that an associative learning task rather reduced the survival of newborn cells, except in situations of very closely related odors, where no reduction occurred (Mandairon et al., 2006). Consequently, despite their differences, both latter studies indicated that appropriate stimuli elicit a response that is analogous in both neurogenesis regions. The apparent discrepancy was resolved by a later study, in which Pierre-Marie Lledo and colleagues discovered that, depending on the time after birth of the cell, an olfactory discrimination challenge can either increase or decrease its survival (Mouret et al., 2008). There is consequently a critical time window for survival-promoting effects in adult olfactory bulb neurogenesis as well (Yamaguchi and Mori, 2005).

Physical Activity

Wheel-running induces adult hippocampal neurogenesis

Rodents provided with a running wheel in their cage make extensive use of this opportunity to exercise during their active period of the day. As nocturnal animals, mice run as much as three to eight kilometers a night, an amount that has been estimated to reflect natural physical activity. Nevertheless, not surprisingly one study claimed that mice caught in the wild did not show an increase in adult neurogenesis after exposure to a running wheel in their new habitat in captivity (Hauser et al., 2009). Otherwise, voluntary physical activity has been a very robust inducer of adult neurogenesis at all ages (Van Praag et al., 1999a; Van Praag et al., 1999b; Kronenberg et al., 2003; Van Praag et al., 2005; Bick-Sander et al., 2006; Steiner et al., 2008; Wu et al., 2008). Surprisingly, voluntary physical activity does not influence neurogenesis in the olfactory system (Brown et al., 2003a). Cell proliferation in other brain regions without evidence of adult neurogenesis is regulated by exercise, but the patterns are complex and not yet fully understood (Ehninger and Kempermann, 2003; Kronenberg et al., 2007b).

Wheel-running primarily increases the proliferation of precursor cells

Running has an acute and strong effect on cell proliferation that wears off over a number of days and weeks (Fig. 9–5). Proliferation peaked at four to 10 days and leveled out after four weeks. This pro-proliferative effect affected primarily type-2 progenitor cells in the hippocampus (Kronenberg et al., 2003; Steiner et al., 2008). There is a presumably not-independent effect on subsequent stages of neuronal development, most importantly a survival-promoting effect. Even after the effect on cell proliferation had returned to baseline, the population of doublecortin-positive cells continued to increase (Kronenberg et al., 2003).

Physical activity prevents the age-dependent decline in precursor cell proliferation

Like environmental enrichment (Kempermann et al., 2002), prolonged physical activity maintained precursor cells proliferation on a higher level and thus slowed the age-dependent decline (Kronenberg et al., 2006). Presumably, in the absence of appropriate survival-promoting stimuli, this increased potential was not translated into an increase in net neurogenesis, however. There are two different, activity-dependent effects: one acute effect that affects cell proliferation and is transient, and one that affects the stem cell niche as a whole and is long-lasting. There is a certain dose-dependency in this effect. The number of DCX-positive cells in the dentate gyrus correlated with the amount of running (Aberg et al., 2008). Similarly, four hours of exercise per night had greater effects than two hours per night, but there were no further significant differences after longer daily exercise, suggesting a ceiling effect (Holmes et al., 2004). Accordingly, rewarding mice for running increased running but did not further enhance neurogenesis (Klaus et al., 2009). The mRNA expression of some potential key regulators such as BDNF and VEGF was greater in animals with lower exercise intensity than in those with higher intensity (Lou et al., 2008), further supporting the idea that the dose–response relationship is not linear.

The effects of physical activity and environmental enrichment are additive

Voluntary wheel-running increases adult neurogenesis but even more so increases the potential for neurogenesis as embodied by the greater number of precursor cells. The initial increase in proliferation was much greater than the resulting increase in neurogenesis. And in the chronic model, the prevented decline in proliferation of the precursor cells did not lead to more new neurons at all (Kronenberg et al., 2006). Sequentially combining the potential-increasing effects of physical activity with the survival-promoting effect of environmental enrichment revealed that the two effects are indeed additive (Fabel et al., 2009). A stimulated potential yielded even more new neurons than the baseline potential, despite the fact that only a subset of new cells survives anyway.

Wheel-running behavior is influenced by cofactors

Why rodents voluntarily run so much is not quite clear, though. The wheel-running paradigm is sometimes also used as a model of obsessive compulsive behavior (Woods-Kettelberger et al., 1997), and despite being voluntary (although there is admittedly not much else to do in the cage), might be experienced as stressful by the animals. The stress hormone corticosterone is reliably increased in exercising mice (Van Praag et al., 1999a; Chang et al., 2008). Although exercise has antidepressant effects in humans, wheel running increased anxiety-like behavior in mice, cautioning against simplistic extrapolations (Fuss et al., 2009). When it was reported that exercise of single-caged mice did not result in increased neurogenesis, whereas running in larger groups did (Stranahan et al., 2006), this finding was prematurely extrapolated to the benefits of running groups for humans. The report was also not confirmed by others: there, wheel-running after restraint stress (which down-regulates neurogenesis; see below) was only effective in inducing adult neurogenesis in the socially isolated mice (Kannangara et al., 2009).

The increase in adult hippocampal neurogenesis is accompanied by increased long-term potentiation in the dentate gyrus and improved performance in the hippocampus-dependent learning task of the Morris water maze (Van Praag et al., 1999b). Suppression of adult neurogenesis by irradiation of the dentate gyrus abolished the beneficial effect of exercise on spatial memory (but not contextual fear conditioning: see Clark et al., 2008).

Effects of exercise on neurogenesis are transmissible from running dams to their offspring

Wheel-running during pregnancy delayed intrauterine development, and pups were born with a lower birth weight than from sedentary mothers. Cell proliferation in the hippocampal area was reduced, but increased to two-fold levels early postnatally, resulting in a larger dentate gyrus before returning to normal levels (Bick-Sander et al., 2006).

Some of the mechanisms underlying the effects of physical activity on adult neurogenesis have been unraveled. Growth factors, most notably insulin-like growth factor-1, have been discussed as key mediators, because IGF1, like other growth factors (Gomez-Pinilla et al., 1997), is increased in running animals, and the running-induced increase in cell proliferation can be blocked by scavenging circulating IGF1, and because it is missing in IGF1 null-mutants (Carro et al., 2000; Trejo et al., 2001). Similarly, but also contradicting the solitary role of IGF1, blockade of vascular endothelial growth factor prevents the induction of cell proliferation and adult neurogenesis (Fabel et al., 2003). Possibly consistent with a "runner's high" in mice (and opposing the idea of exercise only as stressor), β-endorphin signaling was found to be required for exercise-induced effects on adult neurogenesis as well (Koehl et al., 2008). Given the potentially interfering action of stress, the role of glucocorticoid signaling is of particular interest. Unexpectedly, treadmill training reduced expression of the mineralocorticoid receptor (and left the glucocorticoid receptor unchanged) (Chang et al., 2008).

> **Circulating growth factors might be involved in mediating the pro-neurogenic effect of exercise**

Consistent with the idea of a primarily proliferation-related effect of exercise, retinoic acid-dependent signaling, which is rather involved in differentiation, was dispensable (Aberg et al., 2008).

Voluntary wheel-running counteracted the consequences of a number of pathological conditions that otherwise down-regulate adult neurogenesis: irradiation (Naylor et al., 2008), ovariectomy (Jin et al., 2008), kainic-acid lesions to the hippocampus (Chen et al., 2006). In others, for example the Huntington model R6/2, no such rescue was observed (Kohl et al., 2007). Voluntary wheel-running after binge alcohol exposure in adolescence rescued the down-regulation of cell proliferation but not net neurogenesis (Helfer et al., 2009). These "beneficial" effects do not imply that exercise would alter the impact of the pathology itself. Rather, running might stimulate a compensation.

> **Exercise-induced neurogenesis can sometimes compensate for decreased neurogenesis in pathology**

If voluntary wheel-running is interpreted less as exercise and more as a way to measure the level of intrinsic activity (as the paradigm is widely used in research on circadian rhythms), an additional aspect becomes apparent. There is an activity-dependent component in the regulation of adult hippocampal neurogenesis that is not related to physical exercise itself, because circadian phase, and thus activity in a more general sense, also correlates with levels of adult hippocampal neurogenesis (Holmes et al., 2004). At least under acute conditions, forced physical activity on a treadmill seems to have similar effects on cell proliferation, but long-term effects have not been studied (Trejo et al., 2001; Ra et al., 2002; Kim et al., 2004).

> **Behavioral activity in a more general sense influences adult neurogenesis**

Beyond the reductionist setting of wheel-running in a cage, a conceptual problem is how to define "activity." Like *function*, which will be considered in the next chapter, *activity* has different (although often related) meanings on different conceptual levels. Equating the physical activity of wheel-running rodents with exercise in humans might be a misleading simplification. Arguably, most aspects of cognition in a rodent are inseparable from physical activity. Exploration, spatial navigation, and most types of learning accessible in a rodent are based on physical activity. Only language would allow a separation of motor and cognitive activity and something like pure thought (but even executed language is arguably a motor activity). The finding of running-induced neurogenesis appears less counterintuitive if one regards physical activity as a basis for cognition. In line with this speculation, it is very likely that neuronal activity and neurotransmitter-based signaling through, most notably, N-methyl-D-asparatate (NMDA) receptors contribute to the proneurogenic effect of physical activity. In vitro, NMDA receptor–mediated pathways are involved in determining the state of activation of neural precursor cells and consequently influence the cell's behavior and potential (Deisseroth et al., 2004; Babu et al., 2009). Walking has a noted effect on LTP, the electrophysiological correlate of learning (Leung et al., 2003), and it is thought that the induction of theta rhythms in the brain, which is promoted by repetitive regular movements, might underlie this finding. The direct link between theta waves (or other endogenous oscillations) and the regulation of adult neurogenesis has not yet been made. But in co-cultures of precursor cells with neurons in vitro, induction of LTP caused oscillating synchronous synaptic activity in the neurons, consequently inducing the precursor cells to differentiate into neurons (Babu et al., 2009).

Pereira et al. have claimed that cerebral blood flow volume in response to exercise would provide a correlate of adult neurogenesis that is measurable by MRI and thus also accessible in humans (Pereira et al., 2007). Although there appears to be a link a between angiogenesis and neurogenesis in the hippocampus and both are affected by exercise, evidence how this change at the capillary level might translate into measurable changes in blood volume is entirely missing. Cerebral blood flow is also highly variable between subjects, so that the method could not be calibrated. Most important, and efficiently undermining their own argumentation, the Pereira study suggested an inverted U shape for the relationship between cerebral blood volume and neurogenesis. Very low and very high levels of cerebral blood flow might thus be related to the same level of adult neurogenesis, eliminating all predictive power of the measurement.

Sleep

Acute short-term sleep deprivation increases, but chronic sleep deprivation decreases, adult neurogenesis

In contrast to common popular belief, sleep is a period of high brain activity. The prevailing hypothesis goes that it is the brain rather than any other organ that requires sleep: but not for resting. Sleep is important for memory formation and thus is of particular relevance to the hippocampus. Sleep is required for memory consolidation and presumably other aspects of memory formation (Walker and Stickgold, 2004). Disturbed sleep results in cognitive deficits (Muzur et al., 2002). Sleep deprivation decreases adult neurogenesis (Guzman-Marin et al., 2005; Mirescu et al., 2006). Sleep deprivation is stressful and raises glucocorticoid levels, which might be one of the mechanisms by which the down-regulation of neurogenesis is evoked (Mirescu et al., 2006). However, acute sleep deprivation for only one night (a form of sleep deprivation that is also used to acutely treat depression) actually increased hippocampal neurogenesis (Grassi Zucconi et al., 2006) and the recovery that occurs after more chronic sleep deprivation was found to be independent of both the restoration of normal sleep patterns as well as glucocorticoid levels (Mirescu et al., 2006). This independence was confirmed in studies comparing adrenalectomized rats with normal controls (Guzman-Marin et al., 2007; Mueller et al., 2008). Adult neurogenesis was also decreased after sleep fragmentation (Guzman-Marin et al., 2007). When the animals were woken up automatically when the EEG indicated the onset of rapid eye movement (REM) sleep, cell proliferation was also reduced (Guzman-Marin et al., 2008). The endogenous sleep hormone melatonin, however, does not influence cell proliferation but was found to be the first highly specific survival factor for newborn neurons (Ramirez-Rodriguez et al., 2009).

Odors

Odors are important carriers of information for many animals. For dogs, smelling is the primary way of extracting information from the outside world. Rodents, too, rely very much on their olfactory sense, although they complement it with tactile input from their whiskers. In experimental contexts, manipulation of the experience of odors is a paradigm that is relatively close to the animal's physiological range of experience.

Predator odor is stressful to rodents and can down-regulate adult neurogenesis

Certain odors can evoke stress in rodents. When rodents were exposed to fox odor and thus the smell of a natural enemy that preys upon them, severe stress reactions resulted that consequently led to a down-regulation of cell proliferation in the hippocampus (Tanapat et al., 2001). Another study could not replicate this observation, suggesting that details matter and this stress-response response is tightly controlled by other factors (Thomas et al., 2006). The response might also be a matter of the time window investigated (see below) (Petreanu and Alvarez-Buylla, 2002).

Adult neurogenesis in the olfactory bulb is dependent on the input from the olfactory epithelium

Most odors, however, relate to situations more within the physiological range of experiences. For rodents, olfaction is the leading sensory input system. Neurogenesis in the olfactory bulb has sometimes been conceptualized as a consequence of high cellular turnover in the olfactory epithelium. Accordingly, the continuous exchange of primary olfactory neurons would call for a consecutive turnover in downstream neurons within the neuronal network of the olfactory system. There are indeed data suggesting that this turnover is beneficial for the efficiency of olfactory discrimination (Mouret et al., 2009). The second (projection) neuron remains constant, and adult neurogenesis generates new interneurons modulating their activity.

This entire process is clearly input dependent: damage to the olfactory epithelium has negative consequences for olfactory bulb neurogenesis. Acutely, there is an increase in cell proliferation in the SVZ, the RMS, and especially the olfactory bulb that later gives way to a long-lasting decrease if the damage in the epithelium persists. It is tempting to believe that this negative regulation is due to sensory deprivation, a hypothesis supported by a number of arguments, but the lesion itself with its axonal damage might also have a lasting effect.

The net effect of decreased olfactory bulb neurogenesis is due to an interesting combination of effects on different stages of development. The axotomy of olfactory receptor neurons increased, not only cell division in the SVZ, but also cell death of the migrating neuroblasts. The two effects canceled each other out. Structural input to the olfactory bulb in the form of axons of the olfactory receptor neurons appears to be necessary to control the balance of cell birth and death in the regulation of adult olfactory bulb neurogenesis (Mandairon et al., 2003).

Similar effects can be observed in less invasive paradigms. Closure of one naris, which deprives the receptor neurons of olfactory stimuli, down-regulated adult neurogenesis in the olfactory epithelium on the same side and secondarily decreased olfactory bulb neurogenesis (Corotto et al., 1994). This finding supports the hypothesis that sensory activity itself is important for regulating adult olfactory bulb neurogenesis. However, sensory deprivation has been found to induce apoptosis in the olfactory glomeruli. Reopening of the closed naris caused a rebound in adult olfactory bulb neurogenesis (Cummings et al., 1997) and decreased cell death in the olfactory bulb (Fiske and Brunjes, 2001).

Olfactory stimuli recruit new neurons during a critical time window after birth of the cell

Even more compelling than these negative and compensatory effects are direct positive effects of odors on adult olfactory bulb neurogenesis. When Christelle Rochefort, Pierre-Marie Lledo, and colleagues from the Institut Pasteur in Paris exposed mice to a large number of odors over several days, they found that this enriched odor experience resulted in an increased number of BrdU-labeled neurons in the olfactory bulb at three weeks after BrdU injections (Rochefort et al., 2002). This result was interpreted as a survival-promoting effect, similar to the effects of environmental enrichment on adult hippocampal neurogenesis (see above), and was confirmed in a study that also found evidence of the involvement of a typical "survival pathway" (extracellular signal-regulated kinase 1/2, Erk1/2/mitogen-activated protein kinase, MAPK) (Miwa and Storm, 2005). New neurons respond particularly to new odors, but this response to novelty itself does not seem to be the specific survival-promoting stimulus (Magavi et al., 2005). Cauterization of the olfactory epithelium during different times after labeling newborn neurons revealed that a critical time window exists for this effect (Yamaguchi and Mori, 2005). There was no effect of odor enrichment on cell proliferation in the SVZ, and there was also no effect on adult neurogenesis in the hippocampus. Whereas olfactory enrichment was associated with improved olfactory learning, there was no effect on hippocampus-dependent learning (Rochefort et al., 2002). Interestingly, however, the survival-promoting effect was short-lived, indicating that in contrast to the hippocampus, olfactory function is tightly connected to persistent neurogenesis (Rochefort and Lledo, 2005). In anosmic mice, during the critical time-window for an odor-induced promotion of survival, no such increase was found, although survival was normal outside this time window (Petreanu and Alvarez-Buylla, 2002).

Pheromones induce neurogenesis in both neurogenic regions through different signaling mechanisms

In female mice, male pheromones induced neurogenesis (Larsen et al., 2008). This occurred only when the males were dominant, not subordinate, was found in the SVZ and the SGZ, and was independently mediated in both regions: by prolactin in the SVZ and by luteinizing hormone in the SGZ (Mak et al., 2007).

Nutrition

Nutrition effects on adult neurogenesis are underexplored

In some of the early experiments on the effect of environmental enrichment on adult neurogenesis, changes in diet were also part of the experimental program (because it was assumed that cheese or apples would be most appropriate for providing mice with an enriched experience). This led some readers to believe that dietary factors were the main mediators of the effect of an enriched environment. However, when identical food was given to enriched animals and controls, the effects of enrichment on adult neurogenesis were maintained (Kempermann et al., 1998a; Kempermann et al., 1998b). Nevertheless, diet can influence adult hippocampal neurogenesis, although very few studies exist.

Caloric restriction increased cell proliferation in the dentate gyrus

At a first level, nutrition means the provision of calories. Caloric restriction in turn is a classic experimental paradigm in which rodents receive about one-third less food than what they would eat on their own and is the only manipulation known to prolong life in many species. Caloric restriction induced cell proliferation in the dentate gyrus (Lee et al., 2000a; Lee et al., 2002a; Kumar et al., 2009). Caloric restriction might be one of the mild stressors that have a positive effect on adult neurogenesis (see paragraph on stress below, p. 349), but the mechanism might be more complicated. Given the effects on life span, it was thought that caloric restriction would slow the age-dependent decrease in adult neurogenesis, but this did not turn out to be the case (Bondolfi et al., 2004). Conversely, dams with food-induced obesity had offspring with impaired hippocampal neurogenesis (Tozuka et al., 2009).

"Healthy" food components like omega III fatty acids or flavonoids have positive effects on adult neurogenesis

Among the particular food components with a positive effect on adult neurogenesis, omega III fatty acids (known to have a neuroprotective effect) stand out (Kawakita et al., 2006), as do examples from the large class of compounds called flavonoids or polyphenols. A combination of a diet with poly-unsaturated fatty acids plus polyphenols stimulated neurogenesis in both hippocampus and SVZ (Valente et al., 2009). High-fat diet, in contrast, decreased neurogenesis (Lindqvist et al., 2006). Flavonoids are found in plants and had merely been considered antioxidants for a long time. They do have numerous additional effects, however, and some flavonoids have been shown to be not only neuroprotective but also anti-tumorigenic. The latter effects has also brought them into the focus of stem cell researchers. The best-known example is resveratrol, which is found in grape skins and consequently red wine. Resveratrol directly influenced epigenetic processes in fetal neural precursor cells via the expression of Sirt1 (Prozorovski et al., 2008). The analogous analysis for adult neurogenesis is still lacking. Justified by the fact that the effective component is in the red color, but presumably also tongue-in-cheek, the authors used white wine as a control (Wallenborg et al., 2009). The related substance epigallo catechin galleat (EGCG) and curcumin, a popular spice in Asian cuisine, increased adult neurogenesis in vivo (Kim et al., 2008b; Yoo et al., 2009). A somewhat preliminary experiment with blueberries, fruits particularly rich in flavonoids, did not yield a clear result, however (Casadesus et al., 2004). But the similarly flavonoid-rich extract from the Chinese plant Xiaobuxin-Tang rescued decreased levels in cell proliferation after exposure to a stressful experience (An et al., 2008).

Food deficient in vitamins, except for E, reduced proliferation

Dietary deficiency in vitamins A (Bonnet et al., 2008), B1 (thiamine) (Zhao et al., 2008; Zhao et al., 2009) and folate (Kruman et al., 2005; Kronenberg et al., 2008), as well as zinc (Corniola et al., 2008) reduced cell proliferation, and so did caffeine at "physiological concentrations" (Wentz and Magavi, 2009). In contrast, vitamin E supplementation decreased cell proliferation (Cuppini et al., 2001; Ciaroni et al., 2002; Ferri et al., 2003).

Chewing increases cell proliferation

A favorite in Journal Clubs is the report that impaired chewing after removal of molars reduced adult hippocampal neurogenesis by decreasing survival (Mitome et al., 2005). Relatedly, other studies showed reduced cell proliferation when rats were fed with a softer diet than normally (Aoki et al., 2005; Yamamoto et al., 2009). Possibly, this is a particular example of physical activity inducing neurogenesis, and one is tempted to speculate about the evolutionary role of this adaptation.

Nutrition and food are not the same

Overall, these data do not yet convey a coherent picture, but it seems that what is considered "good for you" for other reasons is also beneficial with respect to adult neurogenesis, at least on cell proliferation and in the highly reductionist settings of the these experiments. But feeding is a behavior with massive impact on higher organisms beyond satisfying caloric and certain nutritional needs.

In many species it has a strong social and communicative component. "Nutritionism" fails to recognize this complexity, which is likely to be important in our context as well.

Regulation by Stress

The stress response is an essential survival mechanism but can overreact

Stress is a term for the physical and mental reaction to real or imagined challenges and threats to an organism. In the very broad classical definition by the endocrinologist Hans Seyle, *stress* even encompasses essentially all body responses to any kind of external demand. This in-the-end not very practical definition is nevertheless implicit in many biological concepts and is in some contradiction to the popular use of the word, which focuses on negative psychological and somatic consequences of encountered challenges. But the stress response is in fact a system that is central for appropriate interaction with any environment, and as such, nothing negative. Acute stress puts the individual into a state of increased alertness and preparedness to react ("fight or flight") and is a life-saving mechanism, which still can turn negative if the response is not appropriate and the situation is uncontrollable. Chronic stress usually has only negative connotations. The stress response is thus Janus-faced. Some stress is necessary, but too much quickly turns against the organism.

"Stress" is widely known to down-regulate adult neurogenesis, but the true picture is more complex

Stress might be the most notorious negative regulator of adult hippocampal neurogenesis. Historically, the first research on adult neurogenesis that captured the imagination of a wider circle of scientists had to do with the effects of stress on the brain. Elizabeth Gould and Bruce McEwen's groundbreaking work on adult neurogenesis originated from the question of why stress did not lead to smaller hippocampal dentate gyri, even though many dying cells could be detected (Gould et al., 1992; Gould and McEwen, 1993). Gould reasoned that the new neurons generated in adult hippocampal neurogenesis could be involved in maintaining this cellular balance. On the basis of the idea that the stress hormone cortisol or corticosterone mediates stress effects, Gould published a first study in 1994 on the up-regulating effects of adrenalectomy and down-regulating effects of exogenous glucocorticoids on adult hippocampal neurogenesis (Cameron et al., 1993b; Cameron and Gould, 1994). Two influential studies on the effects of psychosocial stress on adult neurogenesis in primates followed (Fig. 9–6) (Gould et al., 1997; Gould et al., 1998).

Stress that is detrimental to adult neurogenesis is usually excessive in strength or duration

Since these first reports, numerous studies have confirmed the initial central finding: strong acute stress down-regulates cell proliferation in the dentate gyrus and the consecutive stages of neuronal development, although much less is known about this latter stage. From the available data, it can be concluded that severe, acute stress dramatically decreases cell proliferation in the adult dentate gyrus. The experimental models that have been used to demonstrate this link have included psychosocial stress—for example, the resident-intruder model in territorial tree shrews (Gould et al., 1997; Czeh et al., 2002) (Fig. 9–3), predator odor (Tanapat et al., 2001), restraint (Pham et al., 2003), social isolation (Lu et al., 2003b), or electrical foot shocks in rodents (Malberg and Duman, 2003; Vollmayr et al., 2003). Not only the strength of the stress response but also its duration is relevant. Chronic mild stress also has down-regulating effects on cell proliferation (Alonso et al., 2004). Stress also affected cell proliferation in the SVZ (Hitoshi et al., 2007).

Importantly, these experimental paradigms usually consist of unusual, non-physiological stress situations with respect either to strength or to duration. From the negative effects of severe stress on adult neurogenesis, some researches have implicitly concluded that down-regulating effects on adult neurogenesis are induced by any measure that is remotely stressful.

The HPA-axis plays a central role in mediating stress effects on adult neurogenesis

Although stress is a systemic response involving numerous organs and mediators, the role of the hypothalamic-pituitary-adrenal (HPA) axis is central. The key biological parameter for measuring stress in an organism is the serum level of the glucocorticoid hormones cortisol (in humans and nonhuman primates) or corticosterone (in rodents), which are released by the adrenal glands. The mediator regulating the corticosterone-release from

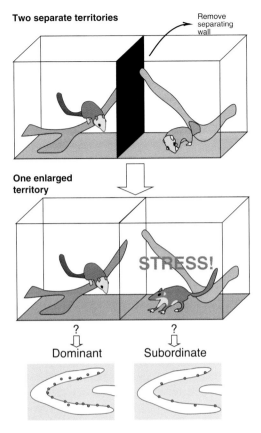

FIGURE 9–6 Resident-intruder model of stress. In the resident-intruder model of psychosocial stress, territorial tree shrews are confronted with another male in an enlarged territory. A dominant–subordinate relationship rapidly evolves that results in greater stress in the subordinate tree shrew, with elevated cortisol levels and reduced adult hippocampal neurogenesis (Gould et al., 1997).

the adrenal gland is adrenocorticotropin (ACTH), which in turn is controlled by hypothalamic secretion of corticotrophin-releasing factor (CRF). Blockade of the CRF receptor with a specific antagonist counteracted the stress-induced reduction in cell proliferation (Alonso et al., 2004).

High levels of stress lead to chronically increased levels of glucocorticoids and over time to a failure of feedback mechanisms that control glucocorticoid secretion on one hand, and glucocorticoid receptor expression on the other. Major depression (see Chap. 12, p. 541) is associated with disturbed regulation of cortisol levels, resulting in an abnormal circadian pattern of hormone secretion and chronically elevated glucocorticoid levels. Despite these suggestive findings, however, the relationship between corticosterone levels and adult neurogenesis is not linear. In some stress models the impact on corticosterone and neurogenesis dissociated. Many situations associated with mild to moderate and more chronic stress actually seem to increase adult neurogenesis. Voluntary physical activity and environmental enrichment are two prime examples of such conditions. Although detailed experimental analysis is still lacking, it seems that the dose–response relationship between corticosterone levels and adult neurogenesis probably follows an inverted-U curve. At low stress levels, experienced more as challenge, there is a positive effect; at high levels, a negative effect is seen. But the lack of any challenge is problematic as well. In vitro, this relationship was indeed found in one study

(Wolf et al., 2009a). Conversely, a preliminary report suggests that blocking adult neurogenesis increases an HPA-axis response (Schloesser et al., 2009). At present it remains undecided whether this is just a consequence of the irradiation procedure or, more intriguingly, a consequence of a feedback loop between hypothalamus and hippocampus. Given the involvement of the dentate gyrus in affective behavior, this idea is less farfetched as it might first seem. Glucocorticoids as regulators of adult neurogenesis are discussed further below.

Presumably, there are other glucocorticoid-independent mechanism mediating the stress response on adult neurogenesis. Myosin VI, for example, a protein motor involved in endocytosis, is expressed in neurons and precursor cells and is up-regulated in stress models (Tamaki et al., 2008). Overexpression reduced clonal size in a neuronal precursor cell line (Takarada et al., 2009).

Prenatal stress has lasting effects on adult neurogenesis

A particular form of stress is prenatal stress, because these early life events can have impact throughout adulthood. Prenatal stress plays an important role in attempts to explain the development of several psychiatric disorders, most notably schizophrenia and autism. Prenatal stress has long-lasting effects on adult neurogenesis and lowers the baseline level of adult hippocampal neurogenesis (Lemaire et al., 2000; Coe et al., 2003). It is not known how persistent the effects of stress in adulthood are, but some negative effects of prolonged stress even on adult hippocampal neurogenesis are reversible (Heine et al., 2004c).

Stress might directly affect the precursor cell pool

Stress also affects the number of sphere-forming cells that can be derived from the brain, which serves as an estimate of the number of neural precursor cells in the SVZ (Kippin et al., 2004; Hitoshi et al., 2007). In one study, whereas prenatal stress decreased the number of spheres that could be generated at both postnatal day one and the age of 14 months, postnatal handling (as a paradigm with well-known positive effects on reducing stress responses: Meaney et al., 1991) reversed this decline. Postnatal handling alone increased the yield of sphere-forming cells (Kippin et al., 2004). Derivation of neurospheres from the adult SVZ was reduced after corticosterone treatment but enhanced by serotonin (Hitoshi et al., 2007).

Stress plays a central role in the neurogenesis hypothesis of major depression

Impaired adult neurogenesis due to chronic stress plays a central role in the neurogenesis hypothesis of major depression (Dranovsky and Hen, 2006; Lucassen et al., 2006; Pittenger and Duman, 2008). Some of the relatively best animal models of depression are stress models; for example, either chronic mild stress or chronic unpredictable stress. The link between stress and levels of adult neurogenesis in these models is not linear, however. In the learned helplessness model, for example, adult neurogenesis was not impaired (Vollmayr et al., 2003). But in such studies, timing is critical, and general conclusions are difficult to draw. The neurogenesis theory of depression is discussed in Chapter 12, p. 541. A dysregulation of adult neurogenesis in response to stressors and involving impaired HPA axis signaling is a key idea of this theory.

Numerous studies have shown that antidepressants of various kinds counteract the downregulating effects of stress on adult neurogenesis. Stress-induced down-regulation of neurogenesis could be counteracted by environmental enrichment (Veena et al., 2009). In contrast to this finding, the similarly stress-protective paradigm of preweaning enrichment did not have lasting effects on neurogenesis in the adult hippocampus in vivo (Kohl et al., 2002). Most studies have of course focused on pharmacological interventions, most notably serotonin-reuptake inhibitors (Czeh et al., 2001; Malberg and Duman, 2003; Santarelli et al., 2003; Jayatissa, 2006; and see review in Dranovsky and Hen, 2006). Stress induced the down-regulation of a number of genes linked to neuronal differentiation, and clomipramine treatment counteracted this effect (Alfonso et al., 2004).

Loss of proliferating cells in the SVZ could be restored by treatment with antidepressants even weeks after the damage (Hitoshi et al., 2007). Olfactory bulbectomy is a severe stress model in rodents (Song and Leonard, 2005) with the obvious disadvantage that the direct consequences of the lesion and the indirect effects from the stress mix in way that is hard to untangle. In any case, imipramine treatment reversed the decrease in cell proliferation (Keilhoff et al., 2006).

Cell–Cell Communication: Signaling Mechanisms

Cell-to-cell signaling occurs over various distances

Cell–cell signaling can be roughly divided into three major categories: direct communication based on cell–cell contact (juxtracrine), short-range signaling (paracrine), and systemic long-range signaling (endocrine). In reality, the boundaries between these categories are sometimes blurred, and some factors might, for example, have both paracrine and endocrine effects. Figure 9–7 summarizes major signaling systems and their intra-cellular convergence.

Cell–Cell Contacts

Cell Coupling via Gap Junctions

Coupling of progenitor cells with gap junctions plays an important role during brain development

During development, cell–cell contact between progenitor or immature cells plays an important role in the communication between cells. For example, in *Drosophila*, cell communication through adherens junctions controls the proliferation of stem cells in the germline (Spradling et al., 2001). Because adherens junctions contain β-catenin, which is an important downstream part of Wnt signaling (see below), a similar role in maintenance of the precursor cell niche in the adult brain is conceivable.

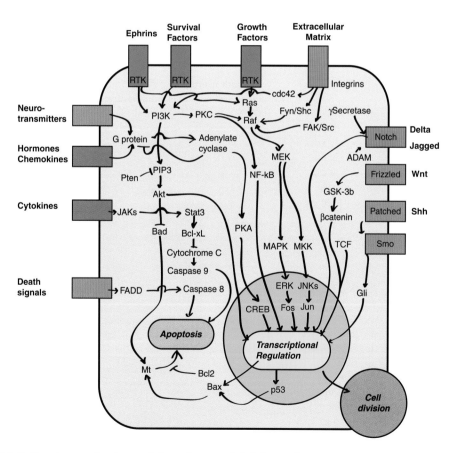

FIGURE 9–7 Overview over key intra-cellular pathways and their interactions.

Gap junctions and their key
constituents, the connexins, are
found in the adult neurogenic
zones

Gap junctions allow communication between neural cells before the appearance of chemical synapses. Small molecules and electrical currents can cross the cell membrane between coupled cells. In the developing neocortex, immature cells form a functional syncytium—a nonsynaptic system for neuronal communication. There are many examples from embryology in which precursor cells can be coupled (Bittman et al., 1997), including hippocampal progenitor cells in vitro (Rozental et al., 1995; Rozental et al., 1998). In the postnatal SVZ of rats, cell coupling has been visualized by dye effusion from one cell to the other (Menezes et al., 2000), although the range of coupled cell types and the reasons for the varying degree of coupling have not yet been determined. Connexins are molecules involved in the formation of gap junctions and are expressed in adult neurogenic zones (Peretto et al., 2005; Kunze et al., 2009). As one example, in the SGZ, NG2-positive oligodendrocyte precursor cells but not GFAP-positive cells expressed connexin 32. In the connexin 32 null-mutant mouse, the turnover of NG2-positive cells in the SGZ was increased (Melanson-Drapeau et al., 2003). Connexins are also found in the radial cells lining the central canal of the adult spinal cord (Russo et al., 2008).

Coupling of hippocampal type-1
cells is required for adult
neurogenesis

Christian Steinhäuser and their colleagues from the University of Bonn, Germany, demonstrated that radial glia–like type-1 cells of the adult dentate gyrus are coupled and express connexin 43. Virally mediated downregulation of connexins decreased proliferation. In mice lacking both connexins 43 and 30, proliferation of type-1 cells almost completely ceased: the number of radial cells and consequently of granule cells went down (Kunze et al., 2009).

Notch

Notch signaling has multiple functions during brain development (Figs 4–7 and 9–7). Most information is available for Notch1, which is expressed in both neurogenic regions (Stump et al., 2002). In the SVZ at least B cells as well as doublecortin-positive cells express Notch1 (Givogri et al., 2006; Wang et al., 2009). In the hippocampus, Notch1 is expressed in type-1 cells as well as presumably type-2a cells, where there is an overlap with Ascl1/Mash1, and in immature neurons. Here no expression in doublecortin-positive type-3 cells was found (Breunig et al., 2007). In textbooks the mechanism of action of Notch is usually explained by the so-called lateral inhibition, by which a cell might promote its becoming or staying different from its neighbor. In detail, lateral inhibition is difficult to understand and even harder to apply to the situation in the adult neurogenic zones.

Nevertheless, activation of Notch1 appears to maintain cells in a precursor cell state and prevent neuronal differentiation (Chojnacki et al., 2003). In Notch1 null-mutants, neural stem cells were depleted, although at early stages of development they could be generated independently of Notch signaling (Hitoshi et al., 2002). When key downstream regulators of Notch signaling, such as presenilin 1, which cleaves Notch after ligand binding, were mutated, a similar situation occurred (Hitoshi et al., 2002). See Table 9–4 and Chapter 12 p. 536 for more details on presenilin 1. Knockdown of Notch1 targeted to GFAP-expressing cells decreased proliferation and increased cell-cycle exit and neurogenesis but with less-complex dendrites; analogous overexpression increased proliferation and maintained the stem cells but also increased dendrite complexity (Breunig et al., 2007). In other contexts, however, Notch1 has been reported to actively induce gliogenesis. In neural crest stem cells, even a transient overexpression of Notch1 caused a switch from neurogenesis to gliogenesis (Morrison et al., 2000). The apparent conflict between maintenance of the precursor cell state and active promotion of gliogenesis might be resolved, if the astrocytic nature of neural precursor cells in the adult SVZ and SGZ turns out to be a sufficient explanation. This possibility is supported by the finding that when Notch1 and a marker gene were overexpressed retrovirally in the embryonic SVZ, radial glia became labeled in the fetal brain, and many of the transduced cells became SVZ astrocytes postnatally (Gaiano and Fishell, 2002).

Other Signaling Related to the Cell Membrane

A few membrane molecules with detectable impact on adult neurogenesis still defy unambiguous classification. Prominin1 is an exemplary case (see Chapter 7, p. 229). Table 9–3 also contains these genes, among them some receptors.

Extracellular Matrix

Extracellular matrix components fall into and between many categories. The activity of many growth factors as well as other signaling molecules is modulated by the extracellular matrix, which serves as a reservoir and intrinsic release system. At the same time, many factors within the extracellular matrix directly signal, for example, through members of the integrin receptor family. The extracellular matrix has been discussed in Chapter 8, p. 283, as part of the neurogenic niche concept.

Neurotransmitters

Neurotransmitters do more than relay excitation across the synaptic cleft

Neurotransmitters primarily act at synapses, where they are released from the presynaptic terminal and activate a postsynaptic receptor. Besides this receptor, responsible for transmitting the signal from one cell to the other, there might be also receptors at the presynaptic side as well as extrasynaptic receptors with other functions. The synaptic (and sometimes the extrasynaptic) action of neurotransmitters belongs to the paracrine mode of signaling. In addition, there are often local ambient effects. Such ambient activity is often referred to as "tonic" and contrasted to the "phasic" action potential–triggered release at the synapse.

Some neurotransmitters have also systemic effects: norepinephrine, for example, is produced in the adrenal gland and affects cells throughout the body.

The dentate gyrus receives input involving essentially all major neurotransmitter systems

The dentate gyrus receives input from a variety of brain regions, and axonal terminals using many different neurotransmitter classes project to the dentate gyrus (Fig. 6–3). The innervation of the SVZ has not been studied in sufficient detail, but is less dense (as it contains no neurons).

- *Glutamate.* The main afferent to the dentate gyrus is the perforant path from the entorhinal cortex. The synapses of perforant path projections on the dendrites of the granule cells are glutamatergic and thus excitatory. They occupy primarily the outer molecular layer.
- *Acetylcholine.* In the inner molecular layer, commissural projections from the contralateral hippocampus end. These axons use acetylcholine as a neurotransmitter. A second acetylcholinergic input comes from the septum and the nucleus basalis Meynert (NBM).
- *Serotonin (5-HT).* Serotonergic afferents project from the raphe nuclei in the brain stem to the SGZ.
- *Dopamine.* From the ventral tegmental area, dopaminergic fibers reach the SVZ. Direct dopaminergic innervation of the SGZ in contrast is very sparse. Fibers from the substantia nigra reach the striatum, immediately adjoining the SVZ and possibly the SVZ itself.
- *Catecholamines.* These include adrenaline and noradrenaline. Noradrenergic projections reach the SGZ from the locus coeruleus.
- *GABA.* Local networks of interneurons represent GABAergic-inhibitory systems necessary to balance excitatory activity. Neuronal activity in the granule cell layer is heavily blocked by interneurons. The architecture of this system is not completely understood. In the dentate gyrus, at least seven classes of interneurons can be found (Freund and Buzsaki, 1996; Houser, 2007). Their distinguishing characteristics are their morphology and location, their pattern of connectivity, and their expression of different calcium-binding proteins (calretinin, calbindin, parvalbumin) and other markers (VIP, NPY, STR).

Table 9-3 Genes in juxtacrine signaling and signaling related to not otherwise classifiable membrane proteins

Protein name	Gene Symbol	Description	Effect on adult neurogenesis	
Abca2 ATP-binding cassette, sub-family A (ABC1), member 2	Abca2	Membrane transporter, presumably for sterols	Expressed in DCX expressing cells (Broccardo et al., 2006)	SGZ, SVZ
Adenylate cyclase activating polypeptide 1 (PACAP)	Adcyap1	Neuropeptide binding to the vasoactive intestinal peptide receptor	Expressed in neurogenic zones; increases proliferation (Mercer et al., 2004)	SGZ, SVZ
Amyloid beta (A4) precursor protein	App	Membrane protein with high abundance in synapses	I.c.v. infusion increased number of EGF-responsive cells ex vivo (neurosphere assay) (Caille et al., 2004)	SVZ
Aquaporin 4	Aqp4	Membrane channel for water	Possibly expressed in precursor cells; in Agp4–/– mice the fluoxetine-induced increase in neurogenesis was abolished; no baseline effect (Kong et al., 2009);	SGZ
			Ex vivo expression in SVZ precursor cells (La Porta et al., 2006)	SVZ
Bystin-like	Bysl	Membrane protein mediating initial attachment during nidation of the embryo	Expressed in B cells (Ma et al., 2006)	SVZ
CD24a antigen	Cd24a	highly glycosylated molecule on differentiating neurons	In Cd24a–/– increased proliferation in SGZ and SVZ; increased apoptosis; no change in OB neurogenesis (Belvindrah et al., 2002)	SGZ, SVZ
Free fatty acid receptor 1 (GPR40)	Ffar1	Receptor for polyunsaturated free fatty acids (PUFA) which play regulatory role in neuronal development	Expressed in nestin+ cells, endothelia and neurons of the DG and in S100b+ cells of the SGZ but not elsewhere (Ma et al., 2008)	SGZ
Fucosyltransferase 4 (LeX, Ssea1)	Fut4	Adhesion molecule	Expressed on GFAP-positive radial cells; ex vivo stemness proof for Fut4+ cells from the SVZ (Capela and Temple, 2002)	SGZ, SVZ

(Continued)

Table 9–3 (Cont'd) Genes in juxtacrine signaling and signaling related to not otherwise classifiable membrane proteins

Protein name	Gene Symbol	Description	Effect on adult neurogenesis	
Jagged 1	Jag1	Ligand to the Notch receptor	Expressed in SVZ (but not unequivocally in the SGZ) (Stump et al., 2002). Enriched in the SGZ in DCX+ and Gfap+ cells (Breunig et al., 2007). Expressed in astrocytes of SVZ and RMS (Givogri et al., 2006; Wang et al., 2009)	SGZ, SVZ
see Slc12a5	Kcc2			
Potassium voltage-gated channel, subfamily Q, member 2	Kcnq2	Potassium channel strongly expressed in the mossy fiber tract and linked to plasticity	Expression covaried significantly with adult neurogenesis (Kempermann et al., 2006)	SGZ
Kinesin family member 3A	Kif3a	Motor protein in primary cilia, which is essential for Shh signaling and Smo function	Targeted ablation in GFAP+ cells eliminates type-1 cells (but not other astrocytes), reduced proliferation and the number of Ascl1/Mash1+ cells (Han et al., 2008)	SGZ
see Fut4	LeX			
Notch gene homologue 1 (Drosophila)	Notch1	Transmembrane receptor with important functions in neurogenesis, most notably fate choice between glial and neuronal lineages	Expressed in SGZ and SVZ (Stump et al., 2002). In the SVZ expressed at least in B and A cells (Givogri et al., 2006) or (in another report) in DCX+ cells (Wang et al., 2009). Expressed in type-1 cells (here overlap with Ascl1/Mash1) and immature neurons but absent from type-3 cells. Knockdown of Notch1 targeted to GFAP-expressing cells decreased proliferation and increased cell-cycle exit and neurogenesis but with less complex dendrites; analogous overexpression increased proliferation and maintained the stem cells but also increased dendrite complexity (Breunig et al., 2007).	SGZ, SVZ

Prominin 1	Prom1	Protein in the apical membrane of neuroepithelial cells, presumably involved in asymmetric divisions	Expression covaried with adult hippocampal neurogenesis (Kempermann et al., 2006). SVZ precursor cells can be prospectively isolated based on Prom1 expression (Corti et al., 2007). Expressed on cells along the ventricular extension in the human OB, the equivalent to the rodent RMS (Kam et al., 2009).	SGZ, SVZ
Presenilin 1	Psen1	Transmembrane protein that is part of the gamma-secretase protease complex; involved in control of neurotransmitter release and induction of long-term potentiation; mutations can cause early-onset Alzheimer disease	Expressed from type-2 cells to immature neurons (Wen et al., 2002a). Transgenic overexpression of mutant Psen1 (P117L) did not affect proliferation but reduced neurogenesis (Wen et al., 2004). Transgenic overexpression of wt Psen1 increased Tuc4+ cells but did not affect net neurogenesis (Wen et al., 2002b). Psen1 variants PS1delta9 or PS1M146L prevented increase in neurogenesis induced by environment enrichment (Choi et al., 2008) and so did the targeted deletion of Psen1 in excitatory forebrain neurons (Feng et al., 2001). Conditioned media from microglia of mutant mice decreased proliferation and neuronal differentiation in vitro (Choi et al., 2008).	SGZ
Solute carrier family 7 (cationic amino acid transporter, y+ system), member 11 (xCT)	Slc7a11	Cystine/glutamate exchanger (xCT) supplies intracellular cystine for the production of cellular anti-oxidant glutathione	Possibly slightly reduced proliferation in the SVZ but increased proliferation in the SGZ (Liu et al., 2007).	SGZ, SVZ
Sortilin-related receptor, LDLR class A repeats-containing (SORLA)	Sorl1	Receptor reducing the processing of amyloid precursor protein to soluble forms	Increased proliferation, survival and neurogenesis in Sorl1−/− mice (Rohe et al., 2008).	SGZ
Synapsin III	Syn3	Phosphoprotein at the cytoplasmic surface of synaptic vesicles	Expressed in immature neurons. Syn3−/− had reduced proliferation but possibly slight increase in neurogenesis (Kao et al., 2008).	SGZ
Trophinin	Tro	Membrane protein mediating initial attachment during nidation of the embryo	Expressed by C and A cells (Ma et al., 2006).	SVZ

Usually neurotransmitters act
pro-neurogenic

Numerous genes related to neurotransmitter effects on adult neurogenesis have been studied and are summarized in Table 9–4. In general, neurotransmitters promote or support adult neurogenesis. This does not exclude individual oppositional effects. Dependent on the receptors, several transmitters have divergent effects, which result in a balanced net effect.

Gaba

Genes (Table 9–4): Gabra1, Slc12a2, Slc12a5

GABA promotes neuronal
differentiation and maturation
from a progenitor cell onward

GABA plays an important role in the development of the new neurons in the dentate gyrus. In the course of development, its effect switches from excitatory to inhibitory. On progenitor cells (from type-2a onward: Wang et al., 2005) and immature neurons, GABA acts excitatory. Only after full integration into the neuronal network and the arrival of excitatory glutamatergic input does the effect of GABA switch to become inhibitory. Accelerating this switch by cell-targeted overexpression of Kcc2 (Slc12a5) led to an precocious integration and malfunction of the new neurons (see Chap. 6, p. 205) (Ge et al., 2006). Hippocampal granule cells are heavily inhibited and are sparsely firing. Under physiological conditions, the long-term potentiation that can be measured in the dentate gyrus is thought to arise almost exclusively from the newborn neurons, which are not yet inhibited (see Chap. 10, p. 447).

Because of the complexity of the interneuron network in the dentate gyrus, however (Freund and Buzsaki, 1996; Houser, 2007), it is not known which types of interneuron the ambient and synaptic GABA is coming from and how the inhibitory network as a whole is involved in controlling and regulating adult neurogenesis. Specific mechanical lesion studies of subtypes of interneurons have not been possible, and the hope is that targeted genetic models will change that. Inhibiting inhibition systemically by applying pentylenetetrazole (PTZ) induced hippocampal seizures and net adult neurogenesis (Jiang et al., 2003; see also Chap. 12, p. 528). Inhibitory circuits tend to counteract the pro-neurogenic effects of kainate-receptor activation and enforce the NMDA receptor–mediated balancing effects (see below).

The benzodiazepine receptor is a co-receptor to the GABA receptor. Chronic treatment with diazepam caused a reduction in the fraction of newly generated cells turning into neurons in the adult hippocampus (Deisseroth et al., 2004). A single injection, in contrast, enhanced adult neurogenesis and also counteracted some (but not all) of the other effects of NMDA-antagonist MK801 (Petrus et al., 2008). The as-yet-unproven hypothesis is that NMDA receptors on interneurons induce release of excitatory GABA.

Glutamate

Genes (Table 9–4): Gria1, Grin1, Grm1, Grm2, Grm5, Slc1a3

Glutamatergic fibers from the
entorhinal cortex are the main
excitatory input to the dentate
gyrus

Glutamatergic fibers from the entorhinal cortex reach the dentate gyrus through the perforant path. A tract from the entorhinal cortex (the perforant path) forms synapses at granule cell dendrites, as do connections from the contralateral hippocampus. Lesioning of the main excitatory input from the entorhinal cortex up-regulated cell proliferation in the dentate gyrus of adult rats (Gould, 1994; Cameron et al., 1995). From this finding, the hypothesis was developed that under physiological conditions, excitatory input would put a lid on adult neurogenesis. However, a later study found no such effect in mice; lesioning of the perforant path left cell proliferation unaffected but led to a transient increase in cell survival (Gama Sosa et al., 2004). The apparent increase might thus be nonspecific.

Glutamatergic signaling through
NMDA receptors promotes
neurogenesis

Glutamate acts through several receptor systems, each of which consists of different subtypes. The effects of these receptors are not homogenous but intricately balanced so that glutamate elicits a range of different responses in the same cell. Blockade of the NMDA receptors, as the main receptor type associated with "plasticity," with antagonist MK-801 increased

Table 9–4 Genes related to neurotransmitter signaling

Protein name	Gene Symbol	Description	Effect on adult neurogenesis	
Acetylcholin-esterase	Ache	Enzyme degrading acetylcholine	Blockade increased neurogenesis (Mohapel et al., 2005; Kotani et al., 2006). Conditional Ache−/− mice have increased proliferation, survival and neurogenesis; overexpression reduced neurogenesis (Cohen et al., 2008)	SGZ
			Blockade increased survival and neurogenesis (Kaneko et al., 2006a).	SGZ, SVZ
Cholinergic receptor, muscarinic 1, CNS	Chrm1	Mediates cholinergic signaling	Expression on 40% of BrdU-positive cells in the SGZ (Mohapel et al., 2005) and PSA-NCAM+ cells in SGZ and SVZ (Kaneko et al., 2006a). Reduced neurogenesis in DG and OB after cholinergic lesion (Cooper-Kuhn et al., 2004).	SGZ, SVZ
			Chrm blocker scopolamine did not affect proliferation but increased survival and neurogenesis (Kotani et al., 2006). Effect of exercise on proliferation preserved after cholinergic lesion but survival reduced (Ho et al., 2009)	SGZ
Cholinergic receptor, muscarinic 4	Chrm4	Mediates cholinergic signaling	Expression in BrdU+ cells, PSA-NCAM+ cells, and neurons (Mohapel et al., 2005; Kaneko et al., 2006a)	SGZ, SVZ
Cholinergic receptor, nicotinic, alpha polypeptide 7	Chrna7	Mediates cholinergic signaling	Expression on PSA-NCAM-positive cells	SGZ, SVZ
Cholinergic receptor, nicotinic, beta polypeptide 2	Chrnb2	Mediates cholinergic signaling	Expression on PSA-NCAM-positive cells	SGZ, SVZ
Dopamine receptor 2	Drb2	Neurotransmitter receptor	K.o. mice for beta2 subunit showed reduced precursor cell proliferation in adult but not in young or very old mice (Harrist et al., 2004)	SGZ
Dopamine receptor 3	Drb3	Neurotransmitter receptor	Agonist increased proliferation (presumably mediated by CNTF) (Yang et al., 2008); agonist restores reduced neurogenesis after dopaminergic deafferentiation (Hoglinger et al., 2004)	SVZ
			Agonist increased proliferation in SVZ and RMS (and neighboring striatum) (Van Kampen et al., 2004)	SVZ

(Continued)

Table 9–4 (Cont'd) Genes related to neurotransmitter signaling

Protein name	Gene Symbol	Description	Effect on adult neurogenesis	
Gamma-aminobutyric acid (GABA-A) receptor, subunit alpha 1	Gabra1	Neurotransmitter receptor	Treatment with GABA receptor agonists and antagonists had no effect on type-1 cells and regulated proliferation of type-2 cells; increase in BrdU+/Calb+ cells with agonist treatment. Conclusion is that GABA drives neuronal differentiation (Tozuka et al., 2005)	SGZ
Glutamate receptor, ionotropic, AMPA1 (alpha 1)	Gria1	Neurotransmitter receptor	Pharmacological potentiator LY451646 increased number of BrdU+ cells (Bai et al., 2003).	SGZ
Glutamate receptor, ionotropic, NMDA1 (zeta 1)	Grin1	Neurotransmitter receptor	NMDA and agonist decreased proliferation and neurogenesis; lesion of perforant path increased neurogenesis (Cameron et al., 1995). Treatment with NMDAR antagonist CGP43487 increased number of Nestin+ radial glia-like and PSA-NCAM-positive cells as well as cell proliferation between 2 and 7 days after treatment (Nacher et al., 2001). One single MK801 application increased precursor cell proliferation, cell cycle exit, and adult neurogenesis but not gliogenesis 4 weeks later; the effect was partly antagonized by diazepam treatment (Petrus et al., 2008)	SGZ
Glutamate receptor, metabotropic 1	Grm1	Neurotransmitter receptor	Antagonist reduced Tuc4+ cells in slice preparation (Baskys et al., 2005).	SGZ
Glutamate receptor, metabotropic 2	Grm2	Neurotransmitter receptor	Antagonist MGS0039 increased number of BrdU+ cells (Yoshimizu and Chaki, 2004).	SGZ
Glutamate receptor, metabotropic 5	Grm5	Neurotransmitter receptor	No effect of antagonist on Tuc4+ cells in slice preparation (Baskys et al., 2005).	SGZ
5-hydroxytryptamine (serotonin) receptor 1A	Htr1a	Neurotransmitter receptor	Agonist induced proliferation but not net neurogenesis, when endogenous serotonin had been depleted (Huang and Herbert, 2005a). No changes in proliferation and neurogenesis in Htr1a −/− (Santarelli et al., 2003). Agonist increased, antagonist decreased proliferation and neurogenesis (Banasr et al., 2004).	SGZ
5-hydroxytryptamine (serotonin) receptor 2A	Htr2a	Neurotransmitter receptor	No effect of agonist or antagonist on proliferation (Banasr et al., 2004).	SGZ

Full name	Gene	Function	Findings	Region
5-hydroxytryptamine (serotonin) receptor 2C	Htr2c	Neurotransmitter receptor	No effect of agonist or antagonist on proliferation and neurogenesis (Banasr et al., 2004).	SGZ
mGluR		see Grm		
Nkcc1		see Slc12a2		
Nmdar		see Grin		
Neuropeptide Y	Npy	Hypothalamic peptide neurotransmitter increasing food intake and decreasing physical activity	Ex vivo NPY increased proliferation and neuronal differentiation in hippocampal precursor cell cultures (Howell et al., 2005). I.c.v. infusion increased proliferation in the SVZ and BrdU/DCX + cells in RMS and OB (Decressac et al., 2009).	SGZ, SVZ
Neuropeptide Y receptor Y1	Npyry1	Neurotransmitter receptor	In the SGZ Npyry1−/− showed reduced proliferation but identical net neurogenesis (Howell et al., 2005) and reduced seizure induced increase in proliferation (Howell et al., 2007).	SGZ
			Npyry1−/− mice had reduced proliferation in the SVZ, reduced migration in the RMS and fewer new neurons in the OB (Hokfelt et al., 2008). I.c.v. injection of agonist increased proliferation and BrdU-+ DCX+ cells (Decressac et al., 2009).	SVZ
Neuropeptide Y receptor Y1	Npyry2	Neurotransmitter receptor	Npyry2−/− mice had reduced proliferation in the SVZ, reduced migration in the RMS and fewer new neurons in the OB (Hokfelt et al., 2008). No effect of i.c.v. injected agonist (Decressac et al., 2009).	SVZ
Solute carrier family 12, member 2 (Nkcc1)	Slc12a2	Chloride importer that causes input from GABAergic cells to be excitatory	Targeted knockdown in precursor cells and Nkcc1−/− resulted in impaired dendrite growth and synaptic integration of newborn neurons (Ge et al., 2006).	SGZ
Solute carrier family 12 (potassium-chloride transporter), member 5 (Kcc2)	Slc12a5	Chloride exporter that causes activity of GABAergic cells to be inhibitory	Targeted overexpression of Kcc2 in precursor cells caused premature abortion of neuronal differentiation (Ge et al., 2006).	SGZ

(Continued)

Table 9-4 (Cont'd) Genes related to neurotransmitter signaling

Protein name	Gene Symbol	Description	Effect on adult neurogenesis	
Solute carrier family 1 (glial high affinity glutamate transporter), member 3 (GLAST1)	Slc1a3	Astrocyte-specific glutamate transporter, expressed also in radial glia	Expressed by type-1 cells (Gubert et al., 2009). Expressed by precursor cells in the SVZ (Mori et al., 2006).	SGZ, SVZ
Solute carrier family 6 (neurotransmitter transporter, serotonin), member 4 (SERT)	Slc6a4	Mediates reuptake of serotonin into presynaptic terminals and thereby modulates serotonergic neurotransmission	In Slc6a4-/- reduced proliferation, survival and neurogenesis in 14.5 month-old mice but not at younger age (Schmitt et al., 2007).	SGZ
VGF nerve growth factor inducible	Vgf	NGF-inducible neuropeptide	Increased proliferation after infusion into hippocampus (Thakker-Varia et al., 2007).	SGZ

adult neurogenesis under physiological and pathological conditions (Gould, 1994; Cameron et al., 1995; Bernabeu and Sharp, 2000; Arvidsson et al., 2001; Nacher et al., 2001; Okuyama et al., 2004; Nacher and McEwen, 2006; Petrus et al., 2008). More-detailed analysis revealed that the effect was actually dual: MK-801 promoted differentiation as well as maintaining the proliferation of early progenitor cells. This implies that, conversely, glutamate signaling through NMDA receptors actually induces neurogenesis at the expense of precursor cell maintenance. Indeed, NMDA receptor–mediated "activity" induced neurogenesis from hippocampal precursor cells in vitro via L-type calcium channels, and directly changed transcription factor patterns to a neuronal program (Deisseroth et al., 2004; Babu et al., 2009).

Glutamatergic signaling through KA receptors increases precursor cell proliferation

Activation of kainate (KA) receptors as another class of glutamate receptors caused a strong induction of cell proliferation and net neurogenesis (Bengzon et al., 1997; Parent et al., 1997; and see Chap. 12, p. 528). The difference between the action mediated by NMDA- and KA-receptor activation suggests that excitatory input to the dentate gyrus has a balanced effect on adult neurogenesis, which is then modulated by other systems. KA-receptor activation would stimulate neurogenesis according to the incoming excitation, being equivalent to the data flow into the dentate gyrus. NMDA-receptor activation would limit this pro-neurogenic response at the precursor cell level and allow other systems to take part in fine-tuning it.

AMPA receptor activity also supports neurogenesis

AMPA receptors are responsible for the fast transmission of an excitation across the synaptic cleft. The effects of NMDA-receptor activation, in contrast, are slower and mediate the plastic structural alterations that might follow the excitation. Interestingly, when AMPA receptor–dependent transmission is enhanced by special pharmacological potentiators, the result is an effect on plasticity as well. BDNF mRNA, for example, is up-regulated in the adult rat hippocampus (Mackowiak et al., 2002). Upon chronic administration, an AMPA potentiator increased cell proliferation in the adult hippocampus and did so at a lower dose than had been necessary for a detectable induction of BDNF mRNA (Bai et al., 2003).

Glutamate signaling through metabotropic receptors also affects neurogenesis

In summary, glutamatergic input to the dentate gyrus is pro-neurogenic in the hippocampus. No information on the SVZ exists. In addition to the ionotropic NMDA, KA and AMPA receptor glutamate also acts in a more hormone-like fashion through so-called metabotropic receptors, for which limited data exist and no clear picture has yet emerged (gene symbols: Grm1-5; see Table 9–4).

Acetylcholine

Genes (Table 9–4): Ache, Chrm1, Chrm4, Charm7, Chrmb2

Cholinergic signaling supports survival

Both the hippocampus and the SVZ receive acetylcholinergic input from the septal region and the basal nucleus of Meynert. This input plays an important role in hippocampal function. In Alzheimer disease, cholinergic neurons in the basal forebrain are the primary targets of degeneration. The reduced cholinergic input to the hippocampus is made responsible for some parts of the loss of hippocampal function during the course of the disease (Leanza et al., 1996; Pizzo et al., 2002). Lesioning of the fimbria fornix, which contains the major cholinergic projections between the septum and the hippocampus, resulted in increased cell proliferation in the SGZ (Weinstein et al., 1996), but this may have been a nonspecific effect. In a more specific approach, lesioning of acetylcholinergic neurons in the basal forebrain, reliably achieved with an immunotoxin, caused a reduction in adult neurogenesis in both the hippocampus and the olfactory bulb (Cooper-Kuhn et al., 2004). The cholinergic system appears to exert a survival-promoting effect, because in the denervated rats, the numbers of apoptotic cells in the germinative regions increased. An interesting detail is that the small number of new neurons in the periglomerular layer of the olfactory bulb seemed to increase (although not significantly), despite the significant overall decrease in neurogenesis in the entire bulb. The meaning of this finding is not clear, but it might be one of the first examples of differential regulation of neurogenesis of the different types of new interneurons in the adult olfactory bulb (Cooper-Kuhn et al., 2004). Pharmacological increase of acetylcholine signaling by inhibiting its degradation through the enzyme acetylcholinesterase also increased neurogenesis in both neurogenic regions (gene symbol: Ache; Table 9–4).

Additional data on the acetylcholinergic regulation of adult neurogenesis have been gathered by manipulating the nicotinergic acetylcholine receptor in the brain with nicotine. These results are discussed together with results from the other drugs of abuse in Chapter 12, p. 548.

Serotonin

Genes (Table 9–4): Htr1a, Htr2a, Htr2c, Slc6a4

Serotonin acts generally pro-neurogenic

Serotonin plays a key role in controlling brain development, and it seems that a similar relevance is also found in adult neurogenesis. Serotonin function is multifaceted and is mediated through at least 15 receptors with different downstream effects (Gaspar et al., 2003). Serotonergic innervation is extremely widespread and reaches the entire brain. All serotonergic input to the brain arises from the raphe nuclei and the reticular formation of the brain stem. The number of serotonergic neurons is rather small (about 20,000 in a rat). Serotonergic neurons are generated very early during development. They are born between E10 to E12 in a mouse, thus preceding cortical (and hippocampal) development. These features make the serotonergic system an ideal candidate for providing relatively general and coordinating stimuli to brain development. In the mature brain, for example, the serotonergic system has effects on levels of wakefulness, attention, and alertness, on aggression, and on exploratory behavior. One could imagine the serotonergic system also as a sensor and announcer of activity in a very general sense. Because both brain development and brain plasticity (and thus adult neurogenesis) are strongly influenced by activity, a central role of serotonin in these processes makes intuitive sense.

In addition to brain stem serotonergic neurons, some developing neurons, mostly glutamatergic neurons in sensory systems, go through a transient stage of an incomplete serotonergic phenotype. They appear to borrow serotonin as a neurotransmitter; they can take up serotonin from their environment but cannot synthesize it. The exact functional consequences of this process are not yet understood, but it might be relevant to adult neurogenesis, because both the serotonin transporter (SERT) and Vmat2 (which packages serotonin into synaptic vesicles) remain expressed in the two neurogenic regions of the adult brain (Hansson et al., 1998; Lebrand et al., 1998).

Surprisingly, however, the patterns of serotonin effects on development are so complex but also redundant that many mutants of the serotonergic system have an almost normal phenotype. But the knockout mouse for the dominant brain isoform of the key enzyme in serotonin biosynthesis—tryptophan hydroxylase, Tph2—had a postnatal lethality of 50% and severe growth retardation (Alenina et al., 2009). Nevertheless, even if changes are not obvious, they might be highly relevant to brain functions. Such alterations have been connected with minute developmental changes hypothesized to underlie psychiatric disorders such as depression, schizophrenia, autism, and drug abuse.

Serotonergic input to the hippocampus and SVZ up-regulates adult neurogenesis (Brezun and Daszuta, 1999). This has been shown by lesioning the input fibers from the median raphe, causing a sharp decline in adult neurogenesis that could be rescued by transplanting serotonergic tissue into the lesion area (Brezun and Daszuta, 2000). Substances that increase the activity of serotonin, such as selective serotonin reuptake inhibitors (SSRI), which are effective antidepressants, increase adult hippocampal neurogenesis (Malberg et al., 2000; Encinas et al., 2006). In a knockout mouse for the 5-HT 1A receptor, in contrast, the effects on adult neurogenesis by fluoxetine (the most widely used SSRI, sold as Prozac or Fluctine) were abolished (Santarelli et al., 2003). Serotonin modified the corticosterone-induced down-regulation of neurogenesis, which is also relevant in the context of depression (Huang and Herbert, 2005b), while tampering with the diurnal regulation of glucocorticoid levels, a symptom often seen in depressed patients, reduced the proneurogenic effects of fluoxetine (Pinnock et al., 2009), together indicating a close interaction between other regulatory systems and serotonin. More information on the role of serotonin in the neurogenesis theory of depression is found in Chapter 12, p. 541.

Different serotonin receptors differently affect different aspects of adult neurogenesis

The 5-HT antagonist tianeptin, an "atypical" antidepressant because it inhibits serotonergic effects rather than stimulating them, also increased adult neurogenesis (Czeh et al., 2001). The explanation for this might lie in the activation of different 5-HT-receptor subtypes, leading to different net effects. Such differential effects at different receptor subtypes might also explain the apparent effect-latency of serotonin on cell proliferation in the dentate gyrus. Malberg and colleagues had related this delay to the effect-latency found in several clinical studies on the antidepressant action of SSRI (Malberg et al., 2000). As in the case of glutamate, serotonin-mediated regulation of adult neurogenesis consists of several balanced partial effects mediated by the various receptors, which appears to change in the course of neuronal development (Banasr et al., 2001).

Dopamine

Genes (Table 9–4): Drb2, Drb3

Together with acetylcholine and serotonin, the dopamine system belongs to the ubiquitous regulatory systems that reach out from a few control centers and allow fine-tuning and orchestrating of complex functions. Dopaminergic fibers from the ventral tegmental area reach almost the entire brain and are involved in mood functions. Dopaminergic input from the substantia nigra to the basal ganglia plays an important role in the extrapyramidal motor system. Conclusive data exist only for the SVZ.

Dopamine sustains neurogenesis in the adult SVZ

Gunter Höglinger, Etienne Hirsch, and colleagues at the Hôpital de la Salpetrière in Paris demonstrated that C cells of the SVZ expressed dopamine receptors, dopaminergic denervation caused reduced olfactory bulb neurogenesis in vivo, and, conversely, activation of the dopaminergic receptor D2 resulted in an increase in proliferation in vitro (Hoglinger et al., 2004). For the SGZ, the group also reported a negative effect on cell proliferation after dopaminergic denervation, although this effect might be indirect, as the presence of the receptors in the SGZ has not yet been convincingly shown. Dopamine receptor D2 and D3 agonists both increased cell proliferation in the SVZ (see details in Table 9–4).

In contrast to this report, another study claimed that damage to the dopamergic system by application of MPTP (1-methyl-4-phenyl-1,2,3,6-tetrahydropyridine) caused a reactive increase in the production of dopaminergic periglomerular interneurons in the olfactory bulb (Yamada et al., 2004).

When rodents were treated with dopamine antagonist haloperidol, the classical antipsychotic drug, as well as other antipsychotics, the results concerning neurogenesis were ambiguous; see Chapter 12, p. 547, for details.

Adrenaline (Epinephrine) and Noradrenaline (Norepinephrine)

In the only existing study on catecholamines, noradrenaline supported proliferation

Essentially no data exist yet on the role of the other catecholamines besides dopamine, adrenaline (epinephrine), and noradrenaline (norepinephrine), in the control of adult neurogenesis. There are extensive noradrenergic fiber connections into the dentate gyrus (Loy et al., 1980). In the dentate gyrus, depletion of norepinephrine by a selective neurotoxin decreased cell proliferation without affecting consecutive stages of neuronal development (Kulkarni et al., 2002). Given the overall abundance of adrenergic and noradrenergic effects throughout the body, it is likely that catecholamines can indeed affect precursor cell activity and adult neurogenesis, but this mechanism is tightly controlled by other systems and remains to be studied.

Neuropeptides

Genes (Table 9–4): Npy, Npyry1, Npyry2, Vgf

Neuropeptides are modulatory transmitters co-expressed with the classical neurotransmitters

Among the large number of peptide neurotransmitters that can be found in the brain, only two representatives have explicitly been studied in the context of adult neurogenesis: neuropeptide Y (NPY) and VGF, an NGF-inducible signaling molecule. Both systems act pro-neurogenic in SGZ and SVZ. Neuropeptides might mediate or modulate neuronal communication, and many neurons express several types of neuropeptides. Functionally, neuropeptides are extremely heterogeneous. The opioid system is usually counted among them, but here they are covered with the psychotropic systems of the brain (p. 388).

Activity-Dependent Regulation of Precursor Cell Proliferation

Precursor cells directly and indirectly sense neuronal activity and initiate neuronal development

A fundamental question is whether precursor cells are able to directly respond to local excitatory activity or whether such a reaction is always a function of the local network and the precursor cell niche. In a pioneering study, Karl Deisseroth, Robert Malenka, and colleagues at Stanford University showed that hippocampal precursor cells in vitro can directly sense excitation (Deisseroth et al., 2004) (Fig. 9–8). They cultured hippocampal precursor cells on either living or fixed hippocampal astrocyte cultures, an experimental paradigm that favors neuronal differentiation (Song et al., 2002). Under these conditions, even short-lasting and mild depolarization caused long-lasting increases in the number of newly generated neurons. This effect was specific in that cell proliferation was not broadly enhanced. Rather, in the proliferating cells, a fast up-regulation of NeuroD was found, accompanied by a down-regulation of transcription factors favoring glial phenotypes, such as Hes1 and Id2. Depolarization of post-mitotic cells had no effect. The NMDA receptor–dependent effect was mediated by calcium signaling through L-type calcium channels. Accordingly, in vivo, calcium-antagonist nifidipine decreased adult hippocampal neurogenesis, whereas treatment with a calcium channel agonist induced adult neurogenesis (Deisseroth et al., 2004).

Babu and colleagues later showed that the cell-intrinsic response to excitation is not the only mechanism and possibly not the most important one (Fig. 9–9) (Babu et al., 2009). Hippocampal precursor cells were cultured on a layer of primary hippocampal neurons. When under magnesium-free conditions glycine was added, an in vitro correlate of long-term potentiation was elicited. The neurons showed synchronous network activity, similar to the oscillations observed in vivo. In this situation the precursor cells were induced to differentiate into neurons. This stimulus was not active in cultures without neurons but it was again dependent on L-type calcium channels. The LTP-induced neuronal differentiation could be diminished by the addition of soluble TrkB receptor to the medium, which scavenges BDNF.

Extrinsic electrical stimulation increases at least cell proliferation in vivo

In vivo, the coupling of precursor cell activity to neuronal activity has only been studied quite indirectly, when the consequences of extrinsically applied currents were studied. Induction of LTP from implanted electrodes indeed increased adult neurogenesis (Bruel-Jungerman et al., 2006) and so did the kindling models of seizures and epilepsy (Parent et al., 1998; Scott et al., 1998; Fournier et al., 2009). Relatedly, in a model of deep brain stimulation, cell proliferation was increased (Toda et al., 2008), and the same was found after vagus nerve stimulation, a treatment option for major depression (Revesz et al., 2008).

Autocrine and Paracrine Niche Factors

The category of short-range signaling molecules in this section comprises very many and very different molecules (Table 9–5). They more or less represent the humoral or secreted factors determining the neurogenic niches.

Excitation increases neurogenesis

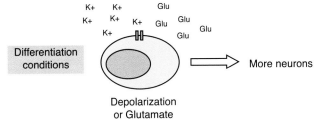

Excitation acts directly on neural precursor cells via L-type Calcium channels

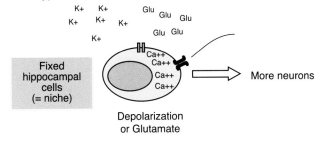

Bidirectional effects of NMDA- or L-type Calcium channel blockade or agonism on excitation-induced neurogenesis

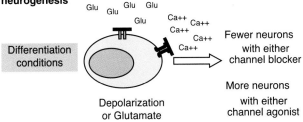

Excitation or Calcium channel activation cause rapid downregulation of anti-neurogenic transcription factors

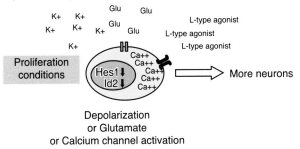

FIGURE 9–8 Direct effects of activity on precursor cells. Precursor cells in culture can react to excitatory stimuli via NMDA receptors on their surface. The drawing summarizes work by Deisseroth et al. (1994).

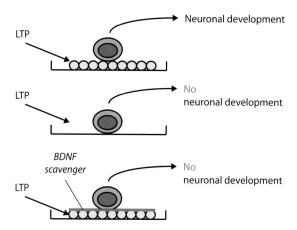

FIGURE 9–9 Precursor cells can sense neuronal excitation. Neuronal network activity (here the experimental induction of chemical LTP in a co-culture model of hippocampal neurons and adult neural precursor cells) is transmitted to the precursor cells through BDNF that is released from the neurons (Babu et al. 2009).

Nitric Oxide

Genes (Table 9–5): Nos1, Nos3

Nitric oxide is anti-proliferative and promotes differentiation

Nitric oxide (NO) is the smallest known signaling molecule and is in a class of its own. Among other mechanisms of action, it induces the synthesis of cGMP, and this step is a key restriction point in the regulation of neurogenesis in many invertebrates. In general, NO is antiproliferative and induces differentiation, but does so in concert with other factors. For example, NO inhibits the EGF receptor and thus reduces the mitogenic effects of EGF (Peranovich et al., 1995; Estrada et al., 1997). Among its numerous other functions, NO is also involved in the control of adult mammalian neurogenesis (Moreno-Lopez et al., 2000). External application of NO leads to an enhancement of regenerative neurogenesis after ischemia or trauma (Cheng et al., 2003; Lu et al., 2003a). Nitric oxide has a very short range of action and acts primarily at its site of production. The key enzyme is nitric oxide synthase (NOS). The major constitutive producers are neurons (nNOS, gene: Nos1) and endothelial cells (eNOS, gene: Nos3). These forms are involved in physiological pathways of regulation. In cases of pathology, an inducible isoform (iNOS) is up-regulated. Consequently, after damage, the first response involves primarily nNOS activity, which is followed by a later stage of iNOS effects. Null-mutants for iNOS failed to induce neurogenesis after ischemia (Zhu et al., 2003), whereas nNOS knockout mice showed constitutively increased proliferation and neurogenesis in nNOS-/- mice (Zhu et al., 2006; Fritzen et al., 2007).

In vitro, NO scavengers such as hemoglobin increase cell proliferation and reduce differentiation. NO donors do the reverse (Cheng et al., 2003). Thus, NO inhibits precursor cell proliferation and initiates differentiation. Many growth factors induce NOS in precursor cells; rising levels of NO will then lead to the exit from the cell cycle. Thus NO might serve as a built-in brake that prevents continued divisions.

In vivo, administration of an NO donor induced cell proliferation in the SVZ and SGZ as well as consecutive neuronal differentiation (Zhang et al., 2001; Cheng et al., 2003). BDNF caused an induction of NOS, which in turn is associated with the cessation of proliferation and the beginning of differentiation (Cheng et al., 2003). This does not necessarily prove that NO is an obligatory downstream factor of BDNF. Consistent with a role at the transition stage between proliferation phase and differentiation, NOS has been found in PSA-NCAM-positive cells of the dentate gyrus (Islam et al., 2003).

Table 9–5 Genes related to autocrine and paracrine signaling, except neurotransmitters

Protein name	Gene Symbol	Description	Effect on adult neurogenesis	
β-catenin		see Ctnnb1		
Bone morphogenetic protein 4	Bmp4	Paracrine signaling molecule, binds to the BMP receptor, which is prevented by Noggin (see there)	Bmp4 overexpression depleted SGZ from proliferating GFAP-positive cells and induced gliogenesis; Noggin-overexpression showed the reverse (Bonaguidi et al., 2005).	SGZ
			Expression associated with glial cells in SVZ, RMS and OB (Peretto et al., 2004); expression in B/C cells ex vivo (Lim et al., 2000). Overexpression inhibits neurogenesis ex vivo and in vivo (Lim et al., 2000).	SVZ
Catenin (cadherin associated protein), beta 1	Ctnnb1	Member of the Wnt pathway, can form activating complexes with many transcription factors	Expressed in PSA-NCAM+ cells of proliferative clusters in the SGZ (Seki et al., 2007); increased expression after antidepressant treatment (Mostany et al., 2008); ex vivo increased Ctnnb1 expression promoted neuronal differentiation (Wexler et al., 2008)	SGZ
			Expressed in the glial tube ensheathing the migrating cells of SVZ and RMS (Cleary et al., 2006).	SVZ
Cxcl12 chemokine (C-X-C motif) ligand 12 (SDF-1)	Cxcl12	Ligand for Cxcr4	Cxcl12-EGFP expression in PSA-NCAM+ cells (Tran et al., 2004); protein in granule cells (Berger et al., 2007); application of Cxcl12 (Cxcr4-ligand) increased excitability in immature granule cells (Kolodziej et al., 2008)	SGZ
Chemokine (C-X-C motif) receptor 4	Cxcr4	Chemokine receptor	Expression in proliferating cells (Lu et al., 2002), type-1 cells (Berger et al., 2007), DCX+, Cr+ and Cb+ cells (Tran et al., 2007); abnormal development of DG in Cxcr4-/- mice (Lu et al., 2002). Antagonist against Cxcr4 decreased survival of BrdU+/DCX+ and BrdU+/NeuN+ cells; agonist increased expression and BrdU+/DCX+ cells (Kolodziej et al., 2008).	SGZ
Ephrin A2	Efna2	Short-range signaling molecule, interacts with Ephrin A7	Epha2-/- showed reduced cell proliferation and neurogenesis, reduced cell cycle length, no effect on apoptosis, no migration defect; i.c.v. infusion of Ephrin A2-Fc increased proliferation (Holmberg et al., 2005)	SVZ
Ephrin B2	Efnb2	Short-range signaling molecule	I.c.v. injection of Ephrin B2-Fc increased proliferation in the SVZ (Conover et al., 2000)	SVZ

(Continued)

369

Table 9–5 (Cont'd) Genes related to autocrine and paracrine signaling, except neurotransmitters

Protein name	Gene Symbol	Description	Effect on adult neurogenesis	
Ephrin B3	Efnb3	Short-range signaling molecule	Efnb3–/– showed increase in proliferation and cell death in the SVZ (Furne et al., 2009). Infusing soluble ephrinB3-Fc molecules in Efnb3–/– suppressed cell proliferation to wild type levels (Ricard et al., 2006). In the DG of Efnb3–/– reduced precursor cell proliferation, resulted in ectopic precursor cells and impaired polarity of DCX+ cells (Chumley et al., 2007).	SGZ, SVZ
Eph receptor A4	Epha4	Ephrin receptor; upon removal of its ligand Ephrin B3, EphA4 triggers cell death	In Epha4–/– no effect on proliferation but reduced cell death in particular of neuroblasts (Furne et al., 2009).	SVZ
Eph receptor A7	Epha7	Ephrin receptor, interacts with ligand Ephrin A2	Eph A7–/– showed reduced cell proliferation and neurogenesis, no effect on apoptosis, no migration defect; i.c.v. infusion of Eph A7-Fc increased proliferation (Holmberg et al., 2005)	SVZ
Eph receptor B1	Ephb1	Ephrin receptor, primary receptor for Ephrin B3	Expressed in type-1 and type-2 precursor cells. Ephb1–/– showed decreased proliferation, ectopic proliferating cells in the granule cell layer and premature branching of dendrites in DCX+ cells. Compound mutants Ephb1 and Ephb2 had reduced volume of the granule cell layer and fewer precursor cells (Chumley et al., 2007).	SGZ, SVZ
Eph receptor B2	Ephb2	Ephrin receptor	Compound mutants Ephb1 and Ephb2 had reduced volume of the granule cell layer and fewer precursor cells (Chumley et al., 2007). I.c.v. injection of Ephb2-Fc increased proliferation in the SVZ (Conover et al., 2000)	SGZ, SVZ
V-erb-a erythroblastic leukemia viral oncogene homolog 4 (avian)	Erbb4	Receptor for Neuroregulin 2	Expressed mostly on PSA-NCAM-positive A cells; conditional Erbb4–/– showed impaired chain migration and placement and differentiation of new OB neurons (Anton et al., 2004), reduced number of B and A cells, whereas C cells where unchanged (Ghashghaei et al., 2006)	SVZ
Kit oncogene (c-kit)	Kit	Receptor for the cytokine stem cell factor (SCF)	Expression on proliferating cells of the monkey SVZ (Tonchev et al., 2007). I.c.v. infusion of stem cell factor increased proliferation of NeuroD expressing cells (Jin et al., 2002a).	SGZ, SVZ

Name	Symbol	Description	Details	Region
Multiple coagulation factor deficiency 2 (Sdnf)	Mcfd2	Neural stem cell derived neuronal survival protein	Molecule derived from hippocampal precursor cells, expressed in the SGZ. In vitro Mcdf2 had survival-promoting effect after Fgf2 withdrawal and maintained stemness properties without affecting mitotic activity (Toda et al., 2003).	SGZ
Noggin	Nog	Endogenous antagonist to BMP signaling (see also there)	Expressed in ependymal cells; antagonizes anti-neurogenic effect of Bmp4 in the SVZ (Lim et al., 2000). I.c.v. infusion decreased neurogenesis in the OB and increased oligodendrogenesis in the corpus callosum (Colak et al., 2008). Difference between results presumably dose-dependent.	SVZ
			Overexpression increased glia-like precursor cells in the SGZ (Bonaguidi et al., 2005); promoted self-renewal in vitro (Bonaguidi et al., 2008).	SGZ
nNOS; nitric oxide synthase 1, neuronal	Nos1	Enzyme synthesizing short range signaling molecule nitric oxide	Expressed in PSA-NCAM+ cells (Islam et al., 2003). Chronic inhibition of nNOS increased proliferation (Park et al., 2003) and number of pCREB/PSA-NCAM+ cells (Park et al., 2004). Increased proliferation and neurogenesis in nNos–/– mice (Zhu et al., 2006; Fritzen et al., 2007). Overexpression of nNOS in muscle rescued aberrant increase in neurogenesis induced by dystrophin (Dmd) mutation (Deng et al., 2009).	SGZ
			Expressed in neurons in the SVZ and OB (Moreno-Lopez et al., 2000). Decreased nNos expression after ischemia, when proliferation was increased (Zhang et al., 2006). NOS inhibitor L-NAME enhanced proliferation and inhibited neuronal differentiation (Cheng et al., 2003); did not alter cell proliferation in the SVZ or in the OB postnatally, but enhanced proliferation in adults (Romero-Grimaldi et al., 2008).	SVZ
eNOS; nitric oxide synthase 3, endothelial	Nos3	Enzyme synthesizing short range signaling molecule nitric oxide	Reduced proliferation in Nos3–/– (Reif et al., 2004). In Nos1/Nos3 –/– but not in Nos3–/– reduced survival (Fritzen et al., 2007).	SGZ
Prokineticin 2	Prok2	Chemoattractant	Expressed in RMS and OB; acts as chemoattractant on SVZ-derived precursor cells ex vivo; Prok2–/– mice have smaller OB and accumulation of neuroblasts in the RMS (Ng et al., 2005).	SVZ
Prokineticin receptor 2	Prokr2	G-protein coupled receptor mediating Prok2 effects	Expressed in C and A cells, decreased in Ascl1 (Mash1)–/– mutant mice (Puverel et al., 2009).	SVZ

(Continued)

Table 9–5 (Cont'd) Genes related to autocrine and paracrine signaling, except neurotransmitters

Protein name	Gene Symbol	Description	Effect on adult neurogenesis	
Reelin	Reln	Short-range signaling molecule involved in positioning migrating cortical neurons and hippocampal granule cells	rl/rl mice had reduced proliferation and number of DCX-positive cells with uncoordinated migration (Won et al., 2006). Neutralization of endogenous reelin resulted in increased aberrant migration (dispersion) after seizures (Heinrich et al., 2006).	SGZ
see Cxcl12	SDF-1			
see Mcfd2	SDNSF			
Sonic hedgehog	Shh	Secreted signaling molecule with multiple functions in neural development, including defining positional identity and precursor cell maintenance (binds to receptor Smoothened, Smo; see there)	In mutants, which lack primary cilia of type-1 cells, Shh signaling was abolished, resulting in a loss of type-1 cells, reduced self-renewal and increased cell cycle exit (Breunig et al., 2008; Han et al., 2008). Shh antagonist decreased baseline and seizure-induced proliferation (Banerjee et al., 2005). Shh increased neuronal differentiation in vitro (Babu et al., 2007). Agonist increased proliferation (Machold et al., 2003).	SGZ
			Shh increased formation of neurospheres from postnatal mice and increased proliferation as co-factor to EGF (Palma et al., 2005). Blockade of Shh signaling prevented carbamylated	SVZ
			Erythropoietin-enhanced proliferation and differentiation in vitro; Shh induced number of Ascl1/Mash1+ cells (Wang et al., 2007). Agonist increased proliferation (Machold et al., 2003).	
Smoothened homologue (Drosophila)	Smo	G protein-coupled receptor for Shh located in the primary cilium	Targeted deletion of Smo in astrocytes (hGFAP::Cre; Smof/flh) results in absence of type-1 cells and very few PSA-NCAM+ cells (Han et al., 2008).	SGZ
Wingless-related MMTV integration site 3	Wnt3	Member of a large family of secreted signaling molecules with numerous functions in (brain) development	Retrovirally applied dominant-negative Wnt3 reduced cell proliferation and number of new DCX+ cells; reverse effect with overexpression (Lie et al., 2005). Activates NeuroD signaling (Kuwabara et al., 2009). In vitro antagonist promoted neurogenesis; abolishing autocrine signaling decreased multipotent precursor cells (Wexler et al., 2009).	SGZ

Secreted Proteins

Several types of auto- and paracrine factors control precursor cell activity and other aspects of neuronal development in the neurogenic niches. The genes coding for these factors and their receptors are listed in Table 9–5, and key representatives are also discussed in the following paragraphs.

Bone Morphogenic Proteins and Noggin

Genes (Table 9–5): Bmp4, Nog

||||||||||||||||||||||||||||||||||||||||||||||||||

BMPs promote gliogenesis and inhibit neurogenesis at the precursor cell level

||||||||||||||||||||||||||||||||||||||||||||||||||

Bone morphogenic proteins are extracellular signaling molecules that play many different roles at various stages of neuronal development. The BMPs generally antagonize neurogenesis. Noggin disinhibits neurogenesis by attenuating BMP action (Wilson and Hemmati-Brivanlou, 1997). In the adult SVZ, ependymal cells secrete noggin, and by counteracting the antineurogenic effects of local BMP4, contribute to creating a region of neurogenic permissiveness. Radial glia–like cells appear to be the main targets. Overexpression of noggin consequently further promotes neurogenesis (Lim et al., 2000). In the adult SGZ, both noggin and BMP4 are expressed (Fan et al., 2003), and inhibition of noggin activity by the application of anti-sense oligonucleotides in vivo results in a decrease in cell proliferation (Fan et al., 2004).

Wnt

Genes (Table 9–5): Ctnnb1 (β-catenin), Wnt

||||||||||||||||||||||||||||||||||||||||||||||||||

Wnt comprises a large family of largely pro-neurogenic factors

||||||||||||||||||||||||||||||||||||||||||||||||||

Members of the Wnt (Wingless) family of signaling molecules play important roles in development. Wnt is, for example, involved in maintaining self-renewal in hematopoietic stem cells (van de Wetering et al., 2002; Reya et al., 2003; Willert et al., 2003) and in the induction of neural specification (Munoz-Sanjuan and Brivanlou, 2002; Muroyama et al., 2002). Wnt signaling occurs through two complex signal transduction pathways that are tightly controlled by a large number of interacting factors. This multifaceted regulation converges to β-catenin (official gene symbol: Ctnnb1) as downstream target (Figs. 4–7 and 9–7). In the presence of Wnt, the concentration of β-catenin in the cell rises because its degradation is inhibited. In the cell nucleus, β-catenin associates with transcription factors TCF/Lef and induces the transcription of Wnt-dependent genes. The loss of appropriate β-catenin signaling is thought to play an important role in cancer development (Morin, 1999), possibly including brain tumors, including medulloblastomas (Itoh et al., 1993; Zurawel et al., 1998).

||||||||||||||||||||||||||||||||||||||||||||||||||

During brain development, different Wnt molecules exert specific functions in specific neuronal populations

||||||||||||||||||||||||||||||||||||||||||||||||||

The effects of Wnt are cell type–specific and dependent on the Wnt family members involved. Different Wnt genes show different distinct expression patterns in the course of brain development. In embryonic stem cells in vitro, Wnt3 is sufficient to induce neuronal lineage commitment (Otero et al., 2004). Wnt1 and Wnt5a increase the number of dopaminergic neurons through two different mechanisms (Castelo-Branco et al., 2003). In midbrain development, Wnt3a stimulates the proliferation of ventral precursor cells. In neural precursor cells of the CNS, Wnt controls self-renewal, whereas in neural crest stem cells, Wnt additionally induces the generation of sensory neurons at the expense of other cell types (Ikeya et al., 1997; Lee et al., 2004). Wnt signaling seems to be particularly relevant for dorsal brain development (Dickinson et al., 1994; Liem et al., 1995; Muroyama et al., 2002), especially in the spinal cord. In vitro, too, a Wnt antagonist has promoted neurogenesis. At least in culture, Wnt is active in an autocrine fashion. The direct inhibition of autocrine Wnt signaling decreased the number of multipotent precursor cells (Wexler et al., 2009).

||||||||||||||||||||||||||||||||||||||||||||||||||

Wnt signaling is required as maintenance factor for adult hippocampal neurogenesis

||||||||||||||||||||||||||||||||||||||||||||||||||

Wnt3a signaling is also necessary for normal hippocampal development by controlling the expansion of neural precursor cells in the cortical SVZ (Lee et al., 2000b). In the adult hippocampus, proliferative (BrdU-incorporating) cells are associated with β-catenin immunoreactivity. When experimental seizures were induced by electroconvulsive seizures, β-catenin in BrdU-labeled cells and expression of Wnt2 in the dentate gyrus were induced (Madsen et al., 2003). Inhibiting Wnt signaling in the hippocampus decreased precursor cell proliferation and induced NeuroD1 expression (Lie et al., 2005). Wnt directly controls NeuroD signaling, and thus has a dual effect on maintaining precursor cells and preventing differentiation (Kuwabara et al., 2009). Overexpression of Wnt3 promoted precursor cell activity in the dentate gyrus (Lie et al., 2005). Overexpression of dominant-negative Wnt abolished adult hippocampal neurogenesis (Jessberger et al., 2009).

Dysfunctional Wnt signaling has been linked with the pathogenesis of mood disorders (Gould and Manji, 2002), Alzheimer disease (De Ferrari and Inestrosa, 2000), and schizophrenia (Kozlovsky et al., 2002).

Sonic Hedgehog
Genes (Table 9–5): Shh, Smo

During brain development Shh is a major patterning factor

Sonic hedgehog (Shh; Figs. 4–7 and 9–7) serves a multitude of functions during development and is required in the brain for patterning of the ventral brain (McMahon et al., 2003). Ectopic expression of Shh causes ventralization (Kohtz et al., 1998); in Shh null-mutants, the telencephalon is greatly reduced, dorsoventral patterning is disturbed, and precursor cell divisions throughout the brain are disturbed (Chiang et al., 1996). Using a conditional mutant based on the neural enhancer element of the nestin promoter, and restricting the loss of Shh function to nestin-expressing cells in the brain, Robert Machold, Gord Fishell, and colleagues studied the role of Shh signaling in neural precursor cells during development. In terms of the general patterning, the phenotype of these mice was minor. Those that reached adulthood had a much smaller brain and enlarged ventricles. In the postnatal neurogenic zones, the number of precursor cells was greatly reduced. Both the olfactory bulb and dentate gyrus were reduced in size (Machold et al., 2003). This postnatal phenotype was not independent of a role of Shh signaling during prior development, but the hypothesis of Shh as a self-renewal factor (Wechsler-Reya and Scott, 1999) could also be applied to the adult brain.

Shh is a stem cell maintenance and differentiation factor in the adult brain

Treatment with an Shh antagonist decreased both baseline and seizure-induced proliferation in the hippocampus (Banerjee et al., 2005), whereas an agonist increased proliferation (Machold et al., 2003). Shh also increased neuronal differentiation in vivo and in vitro (Lai et al., 2003; Babu et al., 2007). Shh also increased the formation of neurospheres from the SVZ of postnatal mice and increased proliferation as co-factor to EGF (Palma et al., 2005), as well as increasing the number of Ascl1/Mash1-expressing intermediate progenitor cells in vivo (Wang et al., 2007). As in the SGZ, agonist treatment increased proliferation (Machold et al., 2003).

Shh is a maintenance factor for type-1 cells

Neural stem cells in the adult neurogenic niches are cilia-bearing, and Shh is located close to the cilium. In mice, lacking primary cilia on type-1 cells, Shh signaling was abolished, resulting in a loss of type-1 cells, reduced self-renewal, and an increased cell cycle from the cell cycle exit (Breunig et al., 2008; Han et al., 2008). Similarly, deletion of Shh receptor Smoothened eliminated type-1 cells (Han et al., 2008).

Ephrins
Genes (Table 9–5): Ephb2

Ephrins exert multiple complex and tendentially positive effects on adult neurogenesis

One of the most complex signaling systems is the ephrin family of signaling factors, acting through various receptors of varying specificity (Fig. 9–7). Several ephrins and ephrin receptors have been studied in the context of adult neurogenesis; they are listed in Table 9–5. There is no simple picture emerging from these data. Ephrins act on precursor cells and their progeny, depending on the type of ephrin and the expression of the receptors. The entire ephrin system seems to be a means of tightly controlling homeostasis in neurogenesis. But little is known about the interactive mechanisms within this system (Martinez and Soriano, 2005; Lai and Ip, 2009).

Reelin

Gene (Table 9–5): Reln

|||||||||||||||||||||||||||||||||||||||||||||||||||||||||||||||||

Reelin supports the glial scaffold and regulates migration of newborn neurons in the hippocampus

|||||||||||||||||||||||||||||||||||||||||||||||||||||||||||||||||

Reelin is secreted by Cajal Retzius neurons during cortical development and is instrumental in neuronal positioning. Reelin binds to the VLDL and ApoE receptors. The Reeler mutant, a natural knockout for the reelin gene, has a characteristic disturbance in layer formation. Reelin has classically been considered a "stop signal" during migration, but its functions clearly go beyond that. Reeler mice show a similar layering defect in the dentate gyrus, and a mutation in the reelin gene might be responsible for the granule cell dispersion associated with temporal lobe epilepsy (Haas and Frotscher, 2010). In adult reeler mice, precursor cell proliferation and the number of DCX-positive cells were reduced. Migration appeared uncoordinated (Won et al., 2006). Neutralization of endogenous reelin resulted in increased aberrant migration (dispersion) after seizures (Heinrich et al., 2006). Exogenous reelin counteracted granule cell dispersion (Muller et al., 2009a). As induced seizures also disproportionately increase proliferation and migration of type-3 cells (Jessberger et al., 2005), it is tempting to speculate that seizures might also interfere with reelin signaling (rather than only reduced reelin being a prerequisite for seizures) and thereby impair adult neurogenesis. But this remains to be investigated.

Immune Cells and Inflammatory Cytokines

Inflammation as a systemic response is found in numerous pathological conditions. Secreted mediators and immune cells are involved, including microglia as the brain's own immune cells, which are activated by many pathological stimuli and can secrete factors associated with inflammation. Inflammation has negative as well as some positive effects on adult neurogenesis (Ekdahl et al., 2009), which raises the question of whether immunity in a more general sense might be involved in the regulation of adult neurogenesis. Genes associated with the immune system and known effects on adult neurogenesis are listed in Table 9–6.

Microglia

|||||||||||||||||||||||||||||||||||||||||||||||||||||||||||||||||

Microglia might harm or support neurogenesis, presumably depending on their activation status

|||||||||||||||||||||||||||||||||||||||||||||||||||||||||||||||||

Michelle Monje, Theo Palmer, and colleagues found that reduction in adult neurogenesis inflicted by irradiation or injection of bacterial lipopolysaccharide (LPS) could be restored to normal levels by treating the irradiated animals with a common anti-inflammatory drug, indomethacin (Monje et al., 2003) (Fig. 9–10). The number of activated microglia correlated with the amount of damage. Christine Ekdahl, Olle Lindvall, and colleagues from Lund University in Sweden showed that microglia activation after LPS injections is detrimental to adult hippocampal neurogenesis (Ekdahl et al., 2003). Inhibition of microglia prevented the damage. In cell culture, activated (but not resting) microglia inhibited neuronal development from neural precursor cells. On the other hand, there are also pro-neurogenic microglia that secrete factors supporting neurogenesis (Jakubs et al., 2008). Consequently, a complex function of microglia in controlling cellular plasticity seems likely (Kempermann and Neumann, 2003). Very little is known about the physiological function of microglia (in the absence of pathology), but, intriguingly, at least cortical microglia also responded to physical activity with increased proliferation (Ehninger and Kempermann, 2003).

Table 9–6 Factors related to immunity and inflammation

Protein name	Gene Symbol	Description	Effect on adult neurogenesis	
Cyclooxygenase 2	Cox2	see Ptgs2		
Interferon alpha	Ifna	Pro-inflammatory cytokine	IFN-α treatment reduced cell proliferation but did not affect neurogenesis (Kaneko et al., 2006b).	SGZ
Interleukin 1 beta	Il1b	Pro-inflammatory cytokine	Il1β antagonist blocked interferon alpha-induced reduction in cell proliferation (Kaneko et al., 2006b). Inhibition of Casp1, which cleaves Il1β, reduced neurogenesis (Gemma et al., 2007).	SGZ
Interleukin 2	Il2	Cytokine with key role in mediating the T cell response	Relatively more proliferation in male Il2–/– mice (but no differences in females and male wildtype animals showed lower cell proliferation than females) (Beck et al., 2005).	SGZ
Interleukin 6	Il6	Cytokine	Transgenic overexpression of Il6 in astrocytes caused reduction in proliferation, survival and neuronal differentiation (Vallieres et al., 2002).	SGZ
Leukemia inhibitory factor	Lif	Cytokine	Unaltered neurogenesis in Lif–/– (Muller et al., 2009b)	SGZ
Prostaglandin E receptor 1 (subtype EP1)	Ptger1	Prostaglandin receptor	Expression was necessary for Toll-like receptor 4 (Tlr4)-dependent LPS-induced depletion of Tbr2+ progenitor cells (Keene et al., 2009)	SGZ
Prostaglandin E receptor 2 (subtype EP2)	Ptger2	Prostaglandin receptor	Expression was necessary for Toll-like receptor 4 (Tlr4)-dependent LPS-induced depletion of Tbr2+ progenitor cells (Keene et al., 2009)	SGZ
Prostaglandin-endoperoxide synthase 2 (Cox2)	Ptgs2	Enzyme in the formation of prostaglandins and prostacyclins	Injected analog of prostaglandin E2, substrate of Ptgs2, increased proliferation (Uchida et al., 2002). Cox2 inhibitor decreased ischemia-induced increase in BrdU+ cells; no baseline change in Cox2–/– and after inhibitor infusion (Sasaki et al., 2003)	SGZ
Transforming growth factor beta 1	Tgfb1	Secreted signaling factor	Chronic transgenic overexpression in astrocytes reduced proliferation and number of DCX+ cells and thereby neurogenesis in the SGZ (Buckwalter et al., 2006). I.c.v. infusion reduced proliferation and neurogenesis in DG and OB (Wachs et al., 2006).	SGZ

Table 9–6 (Cont'd) Factors related to immunity and inflammation				
Protein name	**Gene Symbol**	**Description**	**Effect on adult neurogenesis**	
Toll-like receptor 2	Tlr2	Membrane receptor best studied on immune cells, acting on Nfkb mediated through Myd88	In Tlr2-deficient mice proliferation and survival were unaltered but gliogenesis favored over neurogenesis in the SGZ. In vitro overexpression of Tlr2 induced neuronal differentiation (Rolls et al., 2007).	SGZ
Toll-like receptor 4	Tlr4	See Tlr2	In Tlr4-deficient mice increased proliferation and neurogenesis (Rolls et al., 2007).	SGZ

Adaptive Versus Innate Immunity

Adaptive immune responses seem to be supportive of neurogenesis, whereas innate immunity tends to be anti-neurogenic

Besides evoking a microglial response, the LPS-triggered immune response—called an "innate immune response"—also led to a massive increase in the stress hormone corticosterone, which might be responsible for the decrease in adult neurogenesis, because blockade of glucocorticoid signaling prevented the effect (Wolf et al., 2009a). In contrast, an "adaptive" immune response after the induction of an experimental adjuvant-induced autoimmune or rheumatoid arthritis resulted in a much lower increase in corticosterone and an increase in adult neurogenesis. Besides shedding light on the influence of corticosterone on adult neurogenesis, the result also suggests considerable differences between innate and adaptive immune responses in their effect on adult neurogenesis.

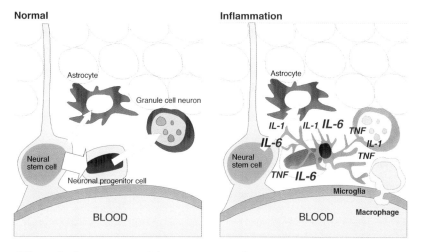

FIGURE 9–10 Effects of inflammation on adult neurogenesis. Inflammation inhibits adult neurogenesis through microglial activation and secreted inflammatory molecules. Gliogenesis is increased. Anti-inflammatory treatment and depletion of microglia prevented this effect. (Monje et al., 2003; Ekdahl et al., 2003). IL, interleukin; TNF, tumor necrosis factor.

T Lymphocytes

In a groundbreaking study, Yaniv Ziv, Michal Schwartz, and colleagues from the Weizmann Institute of Science in Rehovot, Israel, found that CNS-specific T cells in the EAE (experimental auto-immune encephalitis) model of multiple sclerosis supported adult hippocampal neurogenesis (Ziv et al., 2006). The general idea is that regulatory T cells are involved in maintaining homeostasis in the brain, thereby preventing degeneration (Kipnis and Schwartz, 2005). It later turned out that specificity for CNS antigens was not required for such support, and a similar observation could also be made after induction of adjuvant-induced rheumatoid arthritis as an example of a peripheral autoimmune disease (Wolf et al., 2009a).

Wolf and colleagues also found evidence for a role of innate immunity in controlling neurogenesis. Depleting CD4-positive (but not CD8-positive) lymphocytes by FAC sorting reduced adult hippocampal neurogenesis, and the same result was also found in transgenic mice devoid of T cells (RAG1-/-) (Wolf et al., 2009b). As essentially no T lymphocytes are present in the SGZ, this effect is presumably mediated by secreted factors, which, however, have not yet been identified.

Cytokines and Other Secreted Factors

Genes: see Table 9–6

|||||||||||||||||||||||||||||||||||||||||||||||||||||||||||||||

Interleukin 6 promotes gliogenesis

|||||||||||||||||||||||||||||||||||||||||||||||||||||||||||||||

Of the three interleukins that have so far been studied, Il-1b, Il-2, and Il-6, Il-1b might have a positive effect (although the data are not yet fully conclusive). Overexpression of Il-6 in astrocytes strongly decreased adult hippocampal neurogenesis, but left gliogenesis and glial cell counts unaffected (Vallieres et al., 2002). In vitro, Il-6 steered cortical precursor cells to an astrocytic fate (Bonni et al., 1997). Il-2 knockout mice had increased cell proliferation, but the effect was only observed in males (Beck et al., 2005).

|||||||||||||||||||||||||||||||||||||||||||||||||||||||||||||||

Prostaglandins might be involved in the up-regulation of cell proliferation after ischemia

|||||||||||||||||||||||||||||||||||||||||||||||||||||||||||||||

Another class of inflammatory molecules potentially involved in the regulation of adult neurogenesis is the prostaglandins (Uchida et al., 2002). The rate-limiting enzyme in prostaglandin synthesis in the brain is cyclo-oxygenase-2 (COX2). COX2 knockout mice and the pharmacological inhibition of COX2 caused a reduction in ischemia-induced up-regulation of cell proliferation in the SGZ, but did not have a significant effect under physiological conditions (Sasaki et al., 2003). Similarly, inhibition of prostaglandin synthesis with acetylsalicylic acid led to a reduction in ischemia-induced cell proliferation (Kumihashi et al., 2001).

Growth, Survival, and Neurotrophic Factors

|||||||||||||||||||||||||||||||||||||||||||||||||||||||||||||||

Growth factors are a heterogeneous class of short- to long-range peptide factors supporting neurogenesis

|||||||||||||||||||||||||||||||||||||||||||||||||||||||||||||||

Growth factors are peptides that affect the growth of cells through binding to a receptor on the cell surface. In the stricter sense, growth is often equated with proliferation and expansion. Historically, growth factors such as EGF and fibroblast growth factor have been distinguished from the neurotrophic factors. Neurotrophic factors such as nerve growth factor were considered primarily survival factors giving trophic support to neurons and thus maintaining the integrity of nervous tissue. However, in many contexts these distinctions seem arbitrary, and often the distinction between the biological net effects of growth factors and neurotrophic factors is not clear-cut. But in the end, there is no unambiguous and ubiquitously accepted distinction between the terms *growth factors*, *survival factors*, and *neurotrophic factors*. As a general rule, factors within this class act positively on adult neurogenesis. The term *neurotrophins*, however, is reserved for trophic factors signaling through tyrosine kinase receptors (Trk) and p75: NGF, BDNF, NT3, NT4/5. Genes for growth factors in the wider sense, or their receptors, for which an effect on adult neurogenesis has been described are listed in Table 9–7.

Table 9–7 Growth factor–dependent signaling of adult neurogenesis

Protein name	Gene Symbol	Description	Effect on adult neurogenesis	
Brain derived neurotrophic factor (BDNF)	Bdnf	Neurotrophin, binding to TrkB receptor (see also Nrtk2, TrkB)	Intrahippocampal infusion of Bdnf increased neurogenesis; ectopic new granule cells in the hilus (Scharfman et al., 2005). No changes in neurogenesis in Bdnf+/- mice but abolished up-regulating effect of environmental enrichment (Rossi et al., 2006). Reduced proliferation in Bdnf+/- mice (Lee et al., 2002b)	SGZ
			Intracerebroventricular infusion had no effect on neurogenesis in mice but decreased neurogenesis in rats (Galvao et al., 2008). Another rat study found an increase (Pencea et al., 2001a).	SVZ
Ciliary neurotrophic factor	Cntf	Neurotrophic cytokine promoting neuronal survival	I.c.v. injected CNTF increased proliferation and number of B cells, anti-CNTF antibodies decreased proliferation in the SVZ (Emsley and Hagg, 2003); reduced proliferation in CNTF−/− mice (Yang et al., 2008)	SVZ
			Reduced proliferation, number of type-1 cells and neurogenesis in CNTF−/− (Muller et al., 2009b)	SGZ
Granulocyte colony stimulating factor (G-CSF)	Csf3	Co-factor for FGF2	I.c.v. infusion of G-CSF increased neurogenesis (Schneider et al., 2005); additive enhancing effect with arm maze training (Diederich et al., 2009)	SGZ
Cystatin C	Cst3	Co-factor for FGF2	Increase in number of endogenous BrdU/beta-III-tubulin+ cells after implantation of Cst3 expressing progenitor cells. Expression in BrdU+/GFAP- cells (Taupin et al., 2000). Reduced neurogenesis in Cst3−/− (Pirttila et al., 2005).	SGZ
Epidermal growth factor	Egf	Growth factor	I.c.v. infusion increased proliferation, reduced neuronal differentiation and stimulated gliogenesis (Kuhn et al., 1997)	SGZ, SVZ
Epidermal growth factor receptor	Egfr	Growth factor receptor	Expressed on the cilia of B and A cells and ependyma and at the mitotic spindle (Danilov et al., 2009); ex vivo Egfr+/NG2- cells are transiently amplifying progenitor cells (Cesetti et al., 2009); interaction with Dlx2 (Suh et al., 2009)	SVZ
Fibroblast growth factor 1	Fgf1	Growth factor	Gadd45b (see there) binds to Fgf1b promoter; Fgf1 promotes precursor cell proliferation ex vivo (Ma et al., 2009b)	SGZ
Fibroblast growth factor 2	Fgf2	Growth factor	I.c.v. infusion of Fgf2 induced proliferation and neurogenesis in the SGZ in some (Jin et al., 2003; Rai et al., 2007) but not all studies (Kuhn et al., 1997); increased proliferation and neurogenesis in the SVZ/OB (Kuhn et al., 1997)	SGZ, SVZ

(Continued)

Table 9–7 (Cont'd) Growth factor–dependent signaling of adult neurogenesis

Protein name	Gene Symbol	Description	Effect on adult neurogenesis	
	Fgf2		Expressed by neural precursor cells ex vivo; overexpression in precursor cells inhibited neuronal differentiation (Li et al., 2008).	SGZ
Fibroblast growth factor receptor 1	Fgfr1	Growth factor receptor; primary receptor for Fgf2	Increased proliferation and neurogenesis in cond. Fgf1r–/– (Zhao et al., 2007)	SGZ
Fetal liver kinase-1, see Kdr	Flk1	VEGF receptor 2		
FMS-like tyrosine kinase 1	Flt1	VEGF receptor 1	Expression in immature neurons of the monkey SVZ (Tonchev et al., 2007) but not on proliferating cells in the SGZ; agonist placental growth factor nevertheless down-regulated proliferation (Cao et al., 2004).	
Glial cell line derived neurotrophic factor	Gdnf	Neurotrophic factor	Intrastriatal infusion increased cell proliferation and net neurogenesis (Chen et al., 2005)	SGZ
see Slc1a3	Glast1			
Heparin-binding EGF-like growth factor	Hbegf	Growth factor, member of the EGF family	I.c.v. infusion increased proliferation at 3 and 20 months of age (Jin et al., 2003)	SGZ, SVZ
Insulin-like growth factor 1	Igf1	Growth factor	Peripheral Igf1 infusion increased proliferation and neurogenesis (Aberg et al., 2000). I.c.v. infusion did so also in old age (Lichtenwalner et al., 2001). Ames dwarf mutation with constitutively low GH and compensatory high Igf1 had high proliferation and neurogenesis (Sun et al., 2005). Adult-onset depletion of Igf1 in Lewis dwarf rats reduced survival and neurogenesis (Lichtenwalner et al., 2006). Anti-Igf1-antibodies abolished exercise induced increase in neurogenesis (Trejo et al., 2001). No difference in adult neurogenesis in Igf1 overexpressing mice compared to wildtypes (He and Crews, 2007).	SGZ
Insulin-like growth factor binding protein 1	Igfbp1	Plasma protein that binds Igf1, increases bioavailability and modifies receptor interaction	No difference in adult neurogenesis in Igfbp1 overexpressing mice compared to wildtypes (He and Crews, 2007).	SGZ

Name	Symbol	Description		Region
Kinase insert domain protein receptor	Kdr	VEGF receptor 2 (VLDL receptor 2, flk1)	Gene transfer of a mutant KDR abolished up-regulatory effects of VEGF on neurogenesis (Cao et al., 2004). Antagonist prevented effects of selective serotonin reuptake inhibitor fluoxetine (Warner-Schmidt and Duman, 2007) and electroconvulsive seizures on neurogenesis (Segi-Nishida et al., 2008).	SGZ
see Ptprf	LAR			
Nerve growth factor receptor (TNFR superfamily, member 16 (p75))	Ngfr		Expressed in GFAP- and PSA-NCAM-, but Nestin+ cells (Giuliani et al., 2004); expressed in BDNF-responsive subpopulation of SVZ cells ex vivo, activation stimulates neuronal development in neurospheres (Young et al., 2007)	SVZ
Neuregulin 1	Nrg1	Member of EGF family of signaling molecules, binding e.g. to ErbB4 (see there)	Expressed in A cells (PSA-NCAM+); i.c.v. infusion of Nrg1 caused accumulation of Dlx+ cells and impaired migration into the RMS (Ghashghaei et al., 2006).	SVZ
Neuregulin 2	Nrg2	Member of EGF family of signaling molecules, binding e.g. to ErbB4 (see there)	I.c.v. infusion of NRG2 increased Sox2+ and GFAP+ cells, A cells migrating in the RMS and net neurogenesis in the OB (Ghashghaei et al., 2006).	SVZ
Neurotrophin 3 (NT3)	Ntf3	Neurotrophic factor binding to TrkB (Ntrk2) and C (Ntrk3)	Conditional ablation of NT3 in Nestin+ cells did not affect proliferation but reduced survival and neurogenesis and impaired spatial learning and synaptic plasticity (Shimazu et al., 2006).	SGZ
neurotrophin 5 (NT 4/5)	Ntf5	Neurotrophic factor binding to TrkC (Ntrk3)	Ntf5-/- showed no difference in net neurogenesis to wildtypes under both baseline and enriched conditions (increase) (Rossi et al., 2006).	SGZ
Neurotrophic tyrosine kinase, receptor, type 1 (TrkA)	Ntrk1	Neurotrophin receptor for ligand NGF	Expressed on immature neurons of the monkey SVZ (Tonchev et al., 2007).	SVZ
Neurotrophic tyrosine kinase, receptor, type 2 (TrkB)	Ntrk2	Neurotrophin receptor for ligand BDNF and Ntf3	Expressed on type-1 cells and DCX-expressing cells; absent from highly cycling cells; increased expression with maturation (Donovan et al., 2008). Fusion-protein increased BrdU+/NeuN+ cells 6 weeks after ischemia; no unlesioned group (Gustafsson et al., 2003a)	SGZ

(Continued)

Table 9–7 (Cont'd) Growth factor–dependent signaling of adult neurogenesis

Protein name	Gene Symbol	Description	Effect on adult neurogenesis	
			Expressed in SVZ astrocytes and ependyma, not in neuroblasts; reduced number of dopaminergic periglomerular interneurons in TrkB–/– mice; other populations unaffected (Galvao et al., 2008). Expressed on migratory RMS cells; ex vivo stimulation of migration by BDNF (Chiaramello et al., 2007). Expressed on A cells; decreased TrkB phosphorylation in mice with the BDNF variant Val66Met (Bath et al., 2008)	SVZ
p75		see Ngfr		
Placental growth factor	Pgf	Growth factor signaling through VEGFR1 (Flt1)	Virus-mediated overexpression reduced proliferation (Cao et al., 2004).	SGZ
Protein tyrosine phosphatase, receptor type, F (LAR)	Ptprf	Protein tyrosine phosphatase counteracting tyrosine kinase receptor signaling (growth factors)	Ptprf–/– and siRNA knockdown of Ptprf increased proliferation, neuronal differentiation, net neurogenesis and total granule cell counts (Bernabeu et al., 2006).	SGZ
Vascular endothelial growth factor A	Vegfa	Angiogenic and neurogenic growth factor, signaling through Kdr and Flt1	Expressed by granule cells, increased after cAMP-dependent stimulation (Lee et al., 2009). I.c.v. infusion left proliferation unchanged but increased survival (Schanzer et al., 2004). Increased proliferation after i.c.v. infusion (Segi-Nishida et al., 2008). Gene transfer increased proliferation and neurogenesis; knockdown with shRNA prevented enrichment-induced increase in neurogenesis (Cao et al., 2004). Vegf-scavenging antibodies prevented running-induced increase in neurogenesis (Fabel et al., 2003).	SGZ
	Vegfa		I.c.v. infusion left proliferation unchanged but increased survival (Schanzer et al., 2004).	SVZ
Vascular endothelial growth factor B	Vegfb	Angiogenic growth factor with large homology to VEGFA, signaling through Kdr only	Reduced proliferation in Vegfb–/–; i.c.v. infusion increased neurogenesis (Sun et al., 2006).	SGZ SVZ

Epidermal Growth Factor

EGF promotes proliferation of neural precursor cells and is pro-gliogenic

In vitro, neural precursor cells grow in the presence of EGF (Reynolds and Weiss, 1992), which has a strong mitogenic effect. Upon withdrawal of EGF, the neural precursor cells are induced to differentiate.

The precursor cells differ, however, with respect to their expression of EGF receptors, and the general picture is far from clear. Some cells express EGF receptors only, some express only FGF receptors, and some express both. This distribution is species-, region-, and time-dependent (Represa et al., 2001). While it is thus undisputed that SVZ precursor cells express the EGF receptor, expression and function in individual cells might vary. According to one detailed study, in the SVZ, the EGF receptor is expressed on the cilia of B and A cells and ependyma and at the mitotic spindle (Danilov et al., 2009). Others hold that it is the C cells in the SVZ that express the EGF receptor and can be isolated based on this expression (Ciccolini et al., 2005). This is supported by the interaction between the EGF receptor and Dlx1/2, which is expressed in C cells. When EGF was infused into the ventricles, proliferation in the SVZ increased dramatically (Fig. 9–11) and could lead to hyperplasias that protruded into the ventricle but resolved after discontinuation of the infusion (Kuhn et al., 1997). A similar effect was achieved when a form of EGF was administered intranasally (Jin et al., 2003). EGF promoted glial phenotypes over neuronal determination, presumably including the glia-like precursor cells.

EGF diverts precursor cells into the normal and ischemic striatum

There was an increased number of BrdU-labeled cells in the striatum, reflecting cell migration from the SVZ or the induction of local responsive cells. Under normal conditions, no neuronal differentiation of these EGF-induced striatal cells was found. EGF (and its relative, heparin-binding EGF, or HB-EGF) has been used to boost regenerative neurogenesis in the ischemic striatum (Teramoto et al., 2003; Jin et al., 2004b; Baldauf and Reymann, 2005; Sugiura et al., 2005).

EGF promotes precursor cell division in vitro and is pro-gliogenic in vivo

The EGF receptor is also expressed on proliferative cells in the SGZ (Okano et al., 1996). Intracerebroventricular infusion of EGF in rats did not induce proliferation in the SGZ but favored gliogenesis over neurogenesis ((Kuhn et al., 1997). Murine hippocampal precursor cells ex vivo require EGF (Babu et al., 2007), whereas corresponding cells from rats do not (Palmer et al., 1995).

Fibroblast Growth Factor-2

FGF2 is sufficient to maintain adult neural stem cells

In vitro at least, the hippocampal precursor cells are FGF2-dependent (Palmer et al., 1995; Palmer et al., 1999; Babu et al., 2007), and to this date it has not been fully sorted out how far this dependency is a distinguishing feature between precursor cells from SVZ and SGZ. FGF2 is also sufficient to successfully elicit a neurogenic potential in precursor cells isolated from non-neurogenic regions in rats (Palmer et al., 1999). Murine precursor cells, independent of their origin, are routinely cultured in FGF2 and EGF. To some degree, EGF and FGF2 appear to have redundant functions in precursor cells—they are both mitogenic. However, the mitogenic effect of FGF2 seems lower than that of EGF. The finding that FGF2 induces telomerase activity in neural precursor cells might indicate that FGF2 also plays a role in maintaining precursor cell function (Haik et al., 2000).

In FGF2 knockout mice, no increase in hippocampal cell proliferation was found in response to KA-induced seizures or ischemia, normally two strong inducers of precursor cell division (Yoshimura et al., 2001). This effect could be rescued when FGF2 expression was reintroduced with a viral vector. This observation suggests that FGF2 might play a similarly central role for murine precursor cells as it does in rat.

FGF2 promotes neuronal differentiation

When infused into the ventricle, FGF2, unlike EGF, not only expanded the dividing cells of the SVZ, but also induced net neurogenesis in the olfactory bulb (Kuhn et al., 1997), suggesting a pro-neurogenic component. This finding shares an unexplained link with data from embryogenesis,

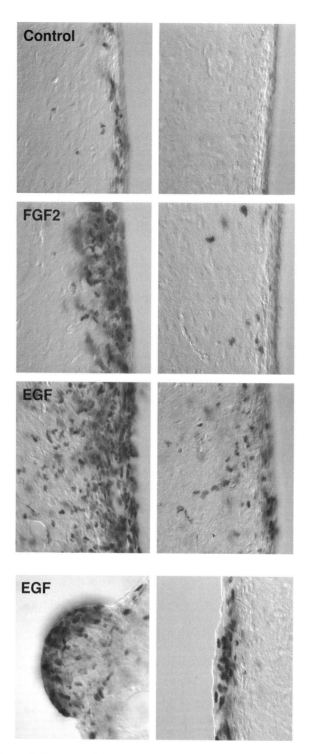

FIGURE 9–11 Effects of growth factors on adult neurogenesis. Cell proliferation in the subventricular zone was differentially affected by the infusion of growth factors epidermal growth factor and fibroblast growth factor-2 into the ventricles of adult rats. EGF had a stronger effect and resulted in increased numbers of BrdU-labeled cells, not only in the SVZ but also in the adjacent striatum. EGF-induced cell proliferation resulted in protrusions into the ventricle that resolved after discontinuation of the treatment (Kuhn et al., 1997). Image reproduced with kind permission of H. Georg Kuhn, Gothenburg. Reprinted with permission from The Society for Neuroscience.

where microinjected FGF2 induced neurogenesis at embryonic day 15.5 (E15.5), but induced gliogenesis at E20.5 (Vaccarino et al., 1999). Like EGF, FGF2 can be applied systemically and crosses the blood–brain barrier (Wagner et al., 1999), which suggests that it can act like a hormone. Intraventricular infusion of FGF2 induced proliferation and neurogenesis in the SGZ in some (Jin et al., 2003; Rai et al., 2007) but not all studies (Kuhn et al., 1997).

Growth factors like FGF2 might require autocrine and paracrine cofactors for their action on neural stem cells in vivo and in vitro; cystatin C is the first identified example in conditioned media from hippocampal precursor cells (Taupin et al., 2000; Dahl et al., 2003; Dahl et al., 2004). Cystatin C is up-regulated in many pathological situations (Nagai et al., 2008) and in such situations is thought to induce cell death (Nagai et al., 2002). On the other hand, cystatin C is considered a neuroprotective factor in that it can bind amyloid beta and reduce its aggregation (Mi et al., 2007). In cystatin C–knockout mice, adult hippocampal neurogenesis was reduced (Pirttila et al., 2005).

> Cystatin C is a co-factor for FGF2 in adult neurogenesis

Insulin-Like Growth Factor-1

> IGF1 is a systemically acting pleiotropic growth factor

Whereas most of the trophic support for neurons originates from the local microenvironment, glia, and other neurons, IGF1, which shows large sequence homologies to insulin, is an example of a systemically acting factor. Most IGF1 in the body is secreted from the liver, and IGF1 affects many body systems. There is restricted production of IGF1 in the brain as well (Werther et al., 1990). Growth hormone induces IGF1 levels, which in turn increase the action of BDNF. Thus in addition to other potential direct mechanisms, IGF1 essentially mediates BDNF action. IGF1 is therefore regarded as a "survival factor." IGF1 signals through two membrane receptors, which primarily signal through the Akt pathway. The daf-2 gene codes for a homologue receptor in *C. elegans*. Mutations of the daf-2 gene can double the lifespan of the worm (Kimura et al., 1997; Apfeld and Kenyon, 1998).

> IGF1 induces precursor cell proliferation and neurogenesis and might mediate the neurogenic response to physical activity

When infused peripherally or intracerebroventricularly, IGF1 induced both cell proliferation and net neurogenesis (Aberg et al., 2000; Lichtenwalner et al., 2001; Aberg et al., 2003). During exercise, IGF1 is secreted from muscle, presumably representing one of the key anabolic stimuli associated with physical activity (Berg and Bang, 2004). Not surprisingly, IGF1 has thus been associated with the increase in adult neurogenesis in response to physical activity. With physical activity, IGF1 accumulates in neurons, and at least some of the exercise-induced effects on neurons can be inhibited by blocking the uptake of IGF1 into the brain (Carro et al., 2000; Trejo et al., 2001). Administration of an antibody that blocked IGF1 activity prevailed many positive effects of physical activity on the brain (Carro et al., 2000). This shows that IGF1 is necessary, but not that it is sufficient, to elicit activity-dependent effects. Nevertheless, the finding supports the idea that lack of activity (a sedentary life) causes a decrease in systemic trophic support that neurons receive. This tonic effect (to which not only IGF1, but also FGF2 and the neurotrophins contribute) has been called "neuroprotective surveillance" (Torres-Aleman, 2000). Like adrenalectomy (Cameron and McKay, 1999), FGF2, HB-EGF (Jin et al., 2003), and notably exercise (Kronenberg et al., 2006), IGF1 administration can prevent some of the age-related decline in adult hippocampal neurogenesis (Lichtenwalner et al., 2001).

> IGF1 induces precursor cell proliferation and neurogenesis and might mediate the neurogenic response to physical activity

The exact function of IGF1 is not clear: it has a comparatively mild mitogenic effect that is physiologically modulated by specific IGF-binding proteins (Clemmons, 1991; Hill and Han, 1991). Interaction with estrogen signaling is evident in that IGF1-mediated increase in cell proliferation can be prevented by blocking the estrogen receptor (Perez-Martin et al., 2003). Both IGF1 and its binding proteins are found in the CSF (Ferry et al., 1999) and the blood. Besides the effect on neuronal development, IGF1 preferentially induces oligodendrocyte differentiation from hippocampal precursor cells both in vitro and in vivo (Hsieh et al., 2004).

Vascular Endothelial Growth Factor (VEGF)

Genes (Table 9–7): Vegfa, Vegfb, Kdr, Flt1

VEGF is a key neurogenic factor in songbird, linking neurogenesis and angiogenesis

Vascular endothelial growth factor a (VEGFa) was first identified as a hypoxia-induced growth factor primarily thought to be involved in angiogenesis. However, in the songbird brain, angiogenesis and adult neurogenesis are coordinated (Louissaint et al., 2002). Testosterone increased neurogenesis in the HVC of songbirds and induced BDNF expression in endothelial cells in the adult HVC. This BDNF expression was preceded by an induced expression of both VEGF proteins and the endothelial VEGF receptor. When the receptor was blocked, both the angiogenic and the neurogenic response was abolished.

VEGF promotes neurogenesis by acting on neuronally determined progenitor cells

In the adult neurogenic zone of rodents, VEGF receptor 2 (Flk1) is expressed on cells expressing DCX (Jin et al., 2002b). This finding suggests that VEGF acts on late progenitor cells in vivo. However, flk-1 was also expressed on clonal precursor cells from the rat brain (Schanzer et al., 2004). Blockage of Flk1 inhibited the VEGF effect on cultured precursor cells.

After infusion into the lateral ventricle, VEGF did not increase precursor cell proliferation but acted as a survival-promoting factor in the hippocampus and the SVZ/OB (Schanzer et al., 2004). In conflict with this result, another study that used viral gene transfer found an induction of cell proliferation in vivo as well (Cao et al., 2004), and so did one other study based on intracerebroventricular injection (Segi-Nishida et al., 2008). In contrast, stimulation of VEGF receptor 1 (Flt1) with its specific agonist, placental growth factor, reduced hippocampal neurogenesis (Cao et al., 2004). Gene transfer of a mutant VEGF receptor 2, KDR, abolished any up-regulatory effects of VEGF on adult neurogenesis (Cao et al., 2004).

VEGF is involved in the regulation of adult neurogenesis above baseline; e.g., in physical activity or environmental enrichment

Physiologically, VEGF is expressed by granule cells, and this expression is increased after cAMP-dependent stimulation, suggesting a potential involvement in activity-dependent regulation (Lee et al., 2009). Appropriately, thus, a knockdown with shRNA prevented the increase in adult hippocampal neurogenesis in response to environmental enrichment (Cao et al., 2004). Blocking of peripheral action of VEGF in vivo inhibited the induction of adult neurogenesis in response to physical activity (Fabel et al., 2003), just as was shown for IGF1 earlier (Trejo et al., 2001). Inhibition of VEGF had no effect on baseline neurogenesis, however. The positive and negative effects of signaling through the VEGF receptors were paralleled by a suggestive change in hippocampal learning parameters (Cao et al., 2004). Similarly, treatment with a KDR antagonist prevented the pro-neurogenic effects of selective serotonin reuptake inhibitor fluoxetine (Warner-Schmidt and Duman, 2007) and of electroconvulsive seizures on adult neurogenesis (Segi-Nishida et al., 2008).

Neurotrophic Factors

Genes (Table 9–7): Bdnf, Ngfr, Nrf3, Nrf5, Ntrk1, Ntrk2, Ntrk3

Neurotrophic factors are survival factors for neurons (and neural precursor cells)

Neurotrophic factors are extracellular signaling molecules that exert a wide range of functions in brain development and maintenance (Lewin and Barde, 1996). Several classes of neurotrophic factors can be distinguished: (1) neurotrophins—that is, NGF, BDNF, and the other factors that act through the trk receptors and p75; (2) the glial cell–derived neurotrophic factor; (3) the hepatocyte growth factor families of neurotrophic factors; and (4) the neurotrophic cytokines such as ciliary neurotrophic factor, Il-6, and others. The neurotrophin hypothesis proposes that a lack of this trophic support causes neurodegeneration. This trophic support can be mediated through retrograde transport from the innervated target, through anterograde transport from afferent regions (Altar et al., 1997), or in autocrine loops as has been found for developing retinal ganglion cells (Wright et al., 1992). Genes associated with the effects of neurotrophic factors on adult neurogenesis are listed in Table 9–7.

Brain-Derived Neurotrophic Factor (BDNF)

BDNF is by far the best-studied neurotrophin, and even if one concedes a certain bias in research, BDNF seems in fact to be the most relevant representative. BDNF is often considered *the* secreted factor modulating brain plasticity. For example, BDNF plays a major role in synaptic plasticity during learning; LTP is BDNF-dependent (Korte et al., 1995; Minichiello et al., 2002), and there is abundant literature on the potential role of BDNF in mental disorders (Hashimoto et al., 2004; Duman and Monteggia, 2006). Both physical activity and environmental enrichment increase BDNF expression (Zajac et al., 2009). BDNF might thus be responsible for the inductive effects of physical activity on LTP (Farmer et al., 2004). Proliferating cells in both neurogenic zones express the low affinity neurotrophin receptor p75 and the high affinity receptor TrkB, but the pattern is complex and variable (see below) (Okano et al., 1996; Young et al., 2007).

BDNF is a survival and maturation factor in adult hippocampal neurogenesis

Baseline hippocampal neurogenesis was reduced in heterozygous BDNF knockout mice through an effect on cell survival (Lee et al., 2002b; Sairanen et al., 2005). In heterozygous BDNF knockout mice, the survival-promoting effect as response to the exposure to an enriched environment was abolished (Rossi et al., 2006). Overexpression of a mutant TrkB receptor (Sairanen et al., 2005) or direct knockdown of BDNF in the dentate gyrus (but not CA3) did the same (Taliaz et al., 2010). The conditional knockdown of TrkB in precursor cells (based on a GLAST-CreERT2 construct) impaired survival, differentiation, dendritic maturation, and functional synaptic integration of newborn neurons (Bergami et al., 2008). TrkB is little expressed on proliferating precursor cells, except for the radial glial–like cells, but increases with maturation (Donovan et al., 2008). In vitro, the LTP-induced release of BDNF from hippocampal neurons was responsible for the observed stimulation of neuronal differentiation (Babu et al., 2009). Intrahippocampal infusion of BDNF increased adult hippocampal neurogenesis but also caused the appearance of ectopic new granule cells in the hilus (Scharfman et al., 2005).

BDNF has also attracted interest because of the hypothesis that BDNF might play a pivotal role in the association between adult neurogenesis and major depression (D'Sa and Duman, 2002).

Under pathological conditions, BDNF might have negative effects on adult neurogenesis

Because BDNF sticks to cellular surfaces, the effectiveness of infusion studies has been limited. BDNF also does not cross the blood–brain barrier. Virus-mediated overexpression of BDNF in the ischemic hippocampus dampened the normal endogenous neurogenic response to ischemia (Larsson et al., 2002), but peripheral application of BDNF resulted in an increase (Schabitz et al., 2007). In line with the first result and further complicating the picture, when endogenous BDNF was inhibited by intraventricular infusion of a BDNF blocker, induction of hippocampal neurogenesis by ischemia was stimulated (Gustafsson et al., 2003b). Although the picture is not quite clear, it seems that under pathological conditions, BDNF might have an effect opposite its normal action and block neuronal differentiation.

In the SVZ BDNF is involved in the control of migration and survival during migration

One early study suggested that infusion of BDNF into the ventricles induced neurogenesis originating from the SVZ (Pencea et al., 2001b), and virally mediated overexpression of BDNF caused long-lasting induction of olfactory bulb neurogenesis (Benraiss et al., 2001; Henry et al., 2007). Peripheral injection of BDNF after ischemia also enhanced neurogenesis in the striatum (Schabitz et al., 2007).

A study based on intrinsic BDNF activity, however, could not replicate this finding for baseline conditions and observed no effects in rats, and even a decrease in mice (Galvao et al., 2008). In the proliferating precursor cells, only a truncated, inactive version of the TrkB receptor was found. An ex vivo study suggested that p75 expression might identify the stem cell population in the SVZ (Young et al., 2007), but this claim is in some conflict with the other expression studies in vivo, suggesting p75 to be found on C cells and the data from the p75 knockout mice, which did not show any changes in proliferation and differentiation (Bath et al., 2008). BDNF, however, seems to be involved in the migration of SVZ-precursor cells along blood vessels (Chiaramello et al., 2007; Snapyan et al., 2009), and BDNF is also critical for the survival during migration (Bath et al., 2008). The migrating cells express p75 and GABA released from the migrating A cells induced TrkB expression on surrounding astrocytes, which in turn removed BDNF from the extracellular space, thereby limiting migration (Snapyan et al., 2009). This finding linked the role of GABA in precursor cell migration (Bolteus and Bordey, 2004) with the effects of BDNF, and suggested a mechanism by which migration could be regulated by activity.

In vitro, BDNF is a differentiation factor

In vitro, BDNF is a differentiation factor that can down-regulate precursor cell proliferation (Cheng et al., 2003; Babu et al., 2007). Purified precursor cells that have been expanded with FGF2 can be primed to become responsive to BDNF by the application of retinoic acid (RA) (Takahashi et al., 1999). Upon contact with RA, hippocampal progenitor cells in vitro up-regulated trk receptors and p75, expressed p21 as a molecule associated with exit from the cell cycle, and expressed NeuroD, indicating neuronal determination.

Other Neurotrophins

Nerve growth factor binds to tyrosine kinase receptors trk-A and trk-B. Consequently, its effects should overlap somewhat with those of BDNF. Like BDNF and NT3, NGF is highly expressed in the adult hippocampus (Maisonpierre et al., 1990). Despite its suggestive name, the role of NGF in regulating adult neurogenesis appears to be limited. Intracerebroventricular infusion of NGF did not have any effect on SVZ or SGZ precursor cells or on net neurogenesis (Kuhn et al., 1997).

Similarly, little is known about the effects of the NTs on adult neurogenesis. Both BDNF and NT3 are induced by dietary restriction (Lee et al., 2002b), an experimental paradigm that not only has robust positive effects on life expectancy in rodents, but also induces cell proliferation in the SGZ. In vitro, RA induced expression not only of primary BDNF receptor trkB, but also of trkC, to which NT3 preferentially binds. Retinoic acid also sensitized cultured precursor cells to the differentiation-inducing effects of NT3 (Takahashi et al., 1999). NT3 increased neuronal differentiation of adult hippocampal precursor cells in culture (Babu et al., 2007). During embryonic neural development, NT3 is the most highly expressed neurotrophic factor, but this does not seem to be the case in adult neurogenesis. Nevertheless, expression of NT3 mRNA covaried with adult neurogenesis phenotypes (Kempermann et al., 2006). In NT4 knockout mice, adult neurogenesis and its up-regulation after environmental enrichment was unimpaired (Rossi et al., 2006).

Endogenous Psychotropic Systems

Cannabinoids

Cannabinoid receptor 1 activation probably promotes survival of newly generated neurons

The endocannabinoid system plays multiple modulatory roles in the brain, and its receptors are widely expressed. Its hormone-like functions include effects as different as analgesia, antiemesis, bronchodilation, and neuroprotection, but the modulatory influences on synaptic transmission and its psychotropic effects are the best studied. The reported effects on adult neurogenesis are partly contradictory, but the bottom line appears to be that activation of the CB1 receptor (gene symbol: Cnr1) increases cell survival and net neurogenesis but reduces cell proliferation. The confusing picture is partly explained by the background strain issues, by differences in the experimental design, and by readout parameters for adult neurogenesis.

Precursor cell proliferation tends to be decreased by cannabinoids

In the SGZ, all cells involved in adult neurogenesis express the CB1 receptor, but expression levels increase with neuronal maturation (Aguado et al., 2006); the pattern for CB2 is not known. There was increased precursor cell proliferation and adult neurogenesis in CB1-/- mice (Palazuelos et al., 2006; Aguado et al., 2007), but only on a C57BL/6 background: on a CD1 background, a decrease was found (Jin et al., 2004a). CB2 mutants, however, showed reduced proliferation and neurogenesis (Palazuelos et al., 2006).

A CB1 antagonist did not change hippocampal cell proliferation in wild types but paradoxically quadrupled it in the knockouts (Jin et al., 2004a). In the SVZ, the pattern was similar but less strong, except that an effect on cell proliferation in wild-type mice was seen. Others found a slight phenotypic shift in adult hippocampal neurogenesis of rats treated with a CB1 antagonist to relatively more glial cells (Rueda et al., 2002) or a stimulation of proliferation with reduced neurogenesis (Kim et al., 2006). An agonist stimulated neurogenesis ex vivo (Aguado et al., 2005). The synthetic agonist HU210 decreased the number of PSA-NCAM-positive cells (Mackowiak et al., 2007) but promoted neuronal differentiation in a different study (Jiang et al., 2005). Ex vivo, the activation of CB1 increased the generation of neurospheres, which was absent in CB1 knockouts (Aguado et al., 2005; Aguado et al., 2006; Aguado et al., 2007) but the endocannabinoid anandamide (AEA) inhibited precursor cell proliferation in vitro in another study (Rueda et al., 2002). A knockout for fatty acid amide hydroxylase (FAAH), which breaks down cannabinoids, increased neurogenesis in the hippocampus (Aguado et al., 2005).

Endorphins

Opioid receptor agonists decrease adult neurogenesis in most studies

Expression of the m-opioid receptor during development correlates with regions and times of neurogenesis, suggesting a functional role of the endorphins in the regulation of neurogenesis (Tong et al., 2000). In addition, opioids enhanced long-term potentiation in the dentate gyrus, which might be indicative of a modulatory role in plasticity (Drake et al., 2007). The general impression from the available data, however, is that opioids decrease adult neurogenesis. Overall the picture is not yet conclusive.

Adult hippocampal precursor cells in culture release β-endorphin and express the m- and a-opioid receptors. Endogenous ligands are also the peptides dynorphin and enkephalin. Exogenous activation of the receptors caused a decrease in cell proliferation. Under differentiation conditions, receptor activation led to increased neuronal but decreased glial differentiation (Persson et al., 2003). Chronic application of morphine and heroin decreased adult hippocampal neurogenesis in vivo (Eisch et al., 2000). Upon morphine withdrawal a strong rebound was observed (Kahn et al., 2005). In knockout mice, for the m-receptor no change in cell proliferation was found but an increase in survival and differentiation, resulting in a larger granule cell layer (Harburg et al., 2007).

Acute morphine treatment decreased S phase entry of precursor cells, presumably primarily at the type-2b stage, and progression through consecutive stages (Arguello et al., 2008). There was a compensatory, yet insufficient up-regulation of neurogenic factor VEGF in chronically treated mice (Arguello et al., 2009). Treatment with a receptor antagonist only mildly reduced the exercise-induced increase in cell proliferation in the SHR rat strain but increased proliferation in the controls (Persson et al., 2004).

Systemic Mediators: Hormones

There is no sharp line between short-range and long-range signaling. An intuitive boundary is the blood–brain barrier, but the fact that the brain can, for example, synthesize steroid hormones ("neurosteroids") and many other "peripheral" factors blurs this distinction as well. We have thus assembled in this category factors that conventionally would be considered "hormones." They are listed in Table 9–8.

Table 9-8 Genes in the hormonal regulation of adult neurogenesis

Protein name	Gene Symbol	Description	Effect on adult neurogenesis	
Erythropoietin	Epo	Peptide hormone	I.c.v. infusion decreased proliferation, increased migration to the OB and increased neurogenesis; anti-EPO antibodies did the reverse (Shingo et al., 2001). I.p. injections of Epo increased proliferation of precursor cells but did not affect net neurogenesis (Ransome and Turnley, 2007).	SVZ
Estrogen receptor 2 (beta)	Esr2	Nuclear receptor for estrogen	Expression in DCX+ cells (Herrick et al., 2006)	SGZ
Growth hormone	Gh	Polypeptide hormone from the anterior pituitary gland	Lewis dwarf mutation as natural null-mutant requiring lifelong GH substitution; induction of GH/Igf1 deficiency by GH withdrawal resulted in decreased net neurogenesis in the absence of altered proliferation (Lichtenwalner et al., 2006)	SGZ
GR		see Nr3c1		
Leptin	Lep	Adipocyte-derived cytokine	Chronic leptin treatment increased cell proliferation but not survival and neurogenesis (Garza et al., 2008).	SGZ
Luteinizing hormone beta	Lhb	Anterior pituitary gland hormone triggers ovulation in females and testosterone production in males	Subcutaneous application of LH increased proliferation, number of DCX+ cells and neurogenesis in SGZ and SVZ (Mak et al., 2007).	SGZ, SVZ
Luteinizing hormone/ choriogonadotropin receptor	Lhcgr	Receptor for Lhb	Expressed in the SGZ; female Lhcgr−/− showed no baseline difference in proliferation in SGZ and SVZ but in the SGZ no induction by exposure to male pheromones (Mak et al., 2007).	SGZ, SVZ

MR	see Nr3c2			
Nuclear receptor subfamily 3, group C, member 1, GR	Nr3c1	Nuclear receptor mediating glucocorticoid action	Expressed in type-1, type-2b, type-3 cells and mature granule cells but not in type2b and early after cell cycle exit (Garcia et al., 2004a). No effect on net neurogenesis in Nr3c1+/− except under stress conditions (Kronenberg et al., 2009). Unaltered proliferation in Nestin-Cre::Nr3c1−/− mice (Gass et al., 2000). Antagonist Mifepriston had no baseline effect but reversed CORT-induced decrease in proliferation of adrenalectomized rats (Wong and Herbert, 2005).	SGZ
Nuclear receptor subfamily 3, group C, member 2; MR	Nr3c2	Nuclear receptor mediating glucocorticoid action	Expressed in mature neurons and few proliferating cells in older age (Garcia et al., 2004a). MR−/− reduced proliferation (Gass et al., 2000). Antagonist Spironolacton increased proliferation in adrenalectomized corticosterone-supplemented rats (Wong and Herbert, 2005).	SGZ
Prolactin	Prl	Anterior pituitary gland hormone with multiple effects including the stimulation of lactation and as mediator of sexual satisfaction	I.c.v. or s.c. application of prolactin increased SVZ proliferation and OB neurogenesis but not in the hippocampus of ovariectomized female mice (as did pregnancy) (Shingo et al., 2003). S.c. injections in males counteracted down-regulating effects of stress on proliferation and neurogenesis in the SGZ (Torner et al., 2009).	SGZ, SVZ
Prolactin receptor	Prlr	Transmembrane cytokine receptor acting on the JAK-STAT pathway.	Expressed in the SVZ; female Prlr−/− showed no baseline difference in proliferation in SGZ and SVZ, but in the SVZ lacked induced proliferation by exposure to male pheromones like it occurred in the SGZ (Mak et al., 2007).	SGZ, SVZ

391

Corticosteroid Hormones

Genes (Table 9–8): Nr3c1, Nr3c2

Glucocorticoids or corticosteroid hormones, corticosterone in rodents and cortisol in primates and humans, play a complex role in regulating adult neurogenesis. They were the first modulators of adult neurogenesis that were described in the early 1990s, and they gained a notorious reputation as being negative regulators of adult neurogenesis. Indeed, the acute stress response (see above, p. 349), which is associated with the activation of the HPA axis system and increased levels of corticosterone, robustly down-regulates precursor cell proliferation and adult neurogenesis. And so did the exogenous application of corticosterone analogs, including dexamethasone. In major depression (see Chap. 12, p. 541), a circadian dysregulation of cortisol is found: often but not always with temporarily increased levels of the hormone. The neurogenesis hypothesis of depression builds on the effects of glucocorticoids for mood regulation and their negative effects on adult neurogenesis. Not all of the stress response on adult neurogenesis is mediated by glucocorticoids, however (Mueller et al., 2008).

Glucocorticoids are the main mediators of the stress response and tend to decrease adult neurogenesis

Conversely, removing corticosterone by adrenalectomy, a radical way of doing so that requires substitution of low levels of the essential hormone, increased neurogenesis. In aging, corticosteroid levels increase, and neurogenesis goes down. Adrenalectomy restored neurogenesis to levels corresponding to a younger age (Cameron and McKay, 1999), but another study could not find different rates of neurogenesis between groups with lifelong reduction of glucocorticoid levels (Brunson et al., 2005). Adrenalectomy also abolished stress-induced down-regulation of neurogenesis (Tanapat et al., 2001). Adrenalectomy after cell division increased the survival of BrdU-labeled cells, whereas exogenous corticosterone decreased it. Adrenalectomy before cell division, in contrast, decreased cell survival (Wong and Herbert, 2004).

Adrenalectomy usually increases neurogenesis

As suggested by the nonlinear effects of stress on adult neurogenesis, however, the link between glucocorticoids and adult neurogenesis is also more complex. Both experimental paradigms of physiological activity, environmental enrichment, and physical activity are associated with increased adult neurogenesis and increased corticosterone levels (Brunson et al., 2005). This suggested that the dose–response curve for glucocorticoids might have an inverted U-shape. At too low and too high levels, glucocorticoids would down-regulate adult neurogenesis, but at an intermediate range, the hormone would sustain neurogenesis. A first hint that this relationship is probably true came from the observation that low-level activation of glucocorticoid signaling was associated with increased neurogenesis, while high-level activation caused a decrease. In vitro this was confirmed: the different effects on the precursor cells could be abolished by transfecting the cells with a construct that rendered the glucocorticoid receptor inactive (Wolf et al., 2009a).

High levels of corticosterone decrease, low levels support adult neurogenesis

The action of glucocorticoids is mediated by two types of receptors, the low-affinity receptor GR (or glucocorticoid receptor I, gene symbol: Nr3c1) and the high-affinity receptor MR (or glucocorticoid receptor II, gene symbol: Nr3c2). The GR is abundantly expressed throughout the brain but enriched in the hippocampus; the MR is found primarily in the hippocampus. Both receptors are expressed on the nuclear membrane. In hippocampal granule cells (and some other neuronal populations), GR and MR co-localize. The effect of MR action is called "tonic" and stable, largely due to the fact that the MR receptors tend to be saturated so that little actual regulation occurs. The GR, in contrast, mediates the acute and fast effects of glucocorticoids, including most of the stress response. This simple dichotomy is complicated by the facts that expression of the receptors underlies a tight feedback control and that the subunits of the two receptor types can form heterodimers, blurring the distinction.

Corticosterone action is mediated by two receptors: GR and MR

In the dentate gyrus, the GR is expressed in type-1, type-2a, and type-3 cells and mature granule cells, but not in type2b; i.e., at the time of neuronal determination, and early after cell cycle exit (Garcia et al., 2004a). Consequently, the proliferative, the developing, and the mature differ in their glucocorticoid receptor expression and thus presumably in their

The GR is involved in above- or below-baseline regulation of neurogenesis and affects proliferation

sensitivity to corticosterone (Cameron et al., 1993a; Garcia et al., 2004a). In knockout mice for the GR, no effect on net neurogenesis was found, except under stress conditions (Kronenberg et al., 2009). This result is consistent with the traditional general view of GR function. Similarly, there was no change in proliferation in Nestin-Cre::Nr3c1 knockout mice, further supporting the idea that baseline neurogenesis is not controlled by GR (Gass et al., 2000). Consequently, GR-antagonist mifepristone had no effect on baseline neurogenesis either, but reversed the decrease in proliferation that is observed after a corticosterone challenge of adrenalectomized rats (Wong and Herbert, 2005).

The MR exerts a survival-promoting effect and presumably maintains precursor cells

The MR, in contrast, is expressed in mature granule cells and few proliferating cells. With age, this relative distribution changes, in that younger cells begin to express the MR (Garcia et al., 2004a). Conditional knockout of the MR reduced cell proliferation in the dentate gyrus, reduced net neurogenesis, and resulted in lower granule cell counts (Gass et al., 2000). Similar to the observation for the GR, the MR antagonist spironolacton increased proliferation in adrenalectomized corticosterone-supplemented rats (Wong and Herbert, 2005). Treatment with MR agonist aldosterone was associated with increased cell survival, albeit in a model of adrenalectomy-induced cell death (Montaron et al., 1999). This suggests that MR activation sustains adult neurogenesis and exerts a survival-promoting effect. In a treadmill-running paradigm (as a presumably primarily pro-proliferative stimulus), the MR but not the GR was found to be down-regulated (Chang et al., 2008).

Glucocorticoid effects on adult neurogenesis might be indirect and are modified by other regulatory systems

Despite these results, it is not clear whether glucocorticoid-dependent regulation of adult neurogenesis is due to direct effects on precursor cells. Indirect pathways, for example involving neuronal populations, might be involved as well, and even predominate. The effects of glucocorticoids on cell proliferation are dependent on NMDA receptor activity. If the NMDA receptor is blocked, no decrease in response to glucocorticoid treatment is found (Cameron et al., 1998). Other hormonal systems might modify corticosterone effects. Dehydroepiandrosterone (DHEA) is a steroid hormone from the adrenal gland that can modulate and even counteract glucocorticoid activity. For example, DHEA levels in serum and cerebrospinal fluid (CSF) decrease, not only with age (Orentreich et al., 1992; Guazzo et al., 1996), but also in major depression (Michael et al., 2000). In the adult rodent hippocampus, DHEA on its own had a significantly inductive effect on cell proliferation and adult neurogenesis, which could dose-dependently counteract the down-regulating effect of corticosterone (Karishma and Herbert, 2002). Fmr1 knockout mice, a model of fragile X syndrome, show a diminished glucocorticoid response to stress but reduced neurogenesis (Eadie et al., 2009). Other known regulatory interaction partners are, for example, nitric oxide, BDNF (Pinnock and Herbert, 2008), and serotonin (Huang and Herbert, 2005b). Genetic differences in glucocorticoid signaling will contribute to strain differences in adult neurogenesis, although the study that specifically suggested a very strong link based its conclusion on only two strains (Alahmed and Herbert, 2008).

Sex Hormones

Different effects of female and male steroid hormones, but no overall gender differences in adult neurogenesis

There is no evidence of strong gender differences in adult neurogenesis. Nevertheless, sex hormones influence adult neurogenesis, and both female and male hormones seem to do so (Galea, 2008). In most situations, divergent actions of different hormone systems seem to be balanced so that no gender-specific pattern arises. Female hormones tend to have an acute pro-proliferative effect, whereas male hormones mostly enhance cell survival (Spritzer and Galea, 2007).

Sex hormones in adult neurogenesis were first studied in the songbird studies of Fernando Nottebohm, because song behavior is sexually dimorphic. In songbirds, testosterone is a positive regulator of neurogenesis in the adult higher vocal center, the brain region responsible for song learning (Rasika et al., 1999; Alvarez-Borda and Nottebohm, 2002; Louissaint et al., 2002). However, the HVC is close to a subventricular layer of cells that express estrogen receptors. Estrogen also induced cell proliferation, supported survival of the new neurons, and resulted in a higher number of neurons in the HVC (Nordeen and Nordeen, 1989; Hidalgo et al., 1995).

In the female rodent hippocampus, cell proliferation has been found to peak in proestrous, when estrogen levels are high, and to be reduced in the phases when estrogen is low (Tanapat et al., 1999). During pregnancy, when estrogens are down-regulated and gestagens are high, adult neurogenesis is decreased (Rolls et al., 2008), and this decrease is maintained in the early postpartum period (Pawluski and Galea, 2007). The latter effect is, however, thought to be mediated by elevated glucocorticoid levels (Leuner et al., 2007). Fittingly, in meadow voles a seasonal influence (which is tightly correlated with hormonal status) on cell proliferation in the dentate gyrus was found. In winter, high cell proliferation and survival coincided with reproductive inactivity (Galea and McEwen, 1999).

> In rodents, estrogens increase neurogenesis at the level of cell proliferation, but the effect is not lasting

Exogenously applied estradiol acutely increased cell proliferation in vivo but then led to a consecutive decrease within 48 hours, probably reflecting a feedback loop activating other adrenal hormone pathways (Ormerod et al., 2003). In male meadow voles, estradiol had survival-promoting effects when administered six to 10 days after BrdU, a finding suggesting a sensitive period in neuronal development that is possibly associated with axon elongation (Ormerod et al., 2004). Many studies on the effects of hormone levels in the ovary cycle have been conducted in female prairie voles, because in this species, the entry into estrus can be provoked by exposure to the odor of a male. The number of proliferating cells in the SVZ and RMS almost doubled in female voles exposed to a male, compared to those exposed to another female (Smith et al., 2001). The effect disappeared in ovariectomized animals but could be rescued when estrogen was injected. Increased estrogen levels during pregnancy did not induce cell proliferation in the dentate gyrus, but the total number of PSA-NCAM-positive cells increased (Smith et al., 2001).

> Prolactin induces neurogenesis in the SVZ/olfactory bulb

While pregnancy reduced adult hippocampal neurogenesis, an increase in olfactory bulb neurogenesis has been reported, which is mediated by prolactin (Shingo et al., 2003). Speculatively, this finding has been connected with the increased sense of smell experienced by many pregnant women. The story is not yet fully conclusive, because throughout most of pregnancy, prolactin levels are actually rather low and increase only after birth. That prolactin induces neurogenesis, however, has been confirmed, and it was found that male pheromones can stimulate this effect (Larsen et al., 2008). Female knockout mice for the receptor show no baseline difference in proliferation in the SGZ and SVZ, and in the SVZ lacked induced proliferation by exposure to male pheromones as it occurred in the SGZ, where the pheromone-induced increase in neurogenesis is independently mediated by luteinizing hormone (Mak et al., 2007). During the lactation period, the offspring of exercising mice had greatly increased hippocampal neurogenesis after a slight growth retardation in utero, resulting in a larger dentate gyrus compared to pups from sedentary dams (Bick-Sander et al., 2006).

Rat neural precursor cells in vitro express receptors for estrogen, and treatment with estrogen results in reduced mitogenic effectiveness of EGF. Embryonic precursor cells, by contrast, were induced to proliferate by estradiol (Brannvall et al., 2002). Just as corticosteroids might require NMDA receptor activity to exert their effect on adult neurogenesis, estrogen might depend on activation of the serotonergic system (Banasr et al., 2001). However, blockade of the estrogen receptor inhibited the effect of IGF1 on adult hippocampal neurogenesis (Perez-Martin et al., 2003), which further illustrates the complex interdependence of the many regulatory systems.

Intracellular and Transcriptional Regulation

Structure and Integrity of the Cell

Shape and function of cells are tightly connected. Cell division, migration, dendritic maturation, and synaptogenesis are all dependent on changes in the cytoskeleton. Many key molecules in neurogenesis such as nestin or doublecortin are cytoskeleton-associated factors. The conventional "neuronal marker" in cell culture, β-III-tubulin, is a cytoskeletal protein, and so is the quintessential glial marker GFAP. Despite this predominance as markers, there is actually only limited knowledge about their immediate functional contribution to neurogenesis. This is particularly remarkable in the case of doublecortin, which is thought to increase cytoskeletal flexibility for migration and neurite extension. Intermediate filament nestin is involved in the complex remodeling processes during the division of neural precursor cells. Table 9–9 lists genes in this category so far investigated in the context of adult neurogenesis. For nestin, doublecortin, and β-III-tubulin, see also Chapter 7, pp. 229, 242, and 244.

Intracellular Signaling

In this category are members of classical signaling cascades, which usually do not have specific roles in adult neurogenesis

From the gigantic spectrum of intracellular signaling molecules and cascades (see also Fig. 9–7 for a rough overview of examples), only a few examples have been studied in the context of adult neurogenesis. Overall knowledge is still limited, but the literature is growing rapidly. Usually the action of these intermediate signaling molecules is not specific to neurogenesis. The tacit assumption is usually that signaling will not be fundamentally different in adult neural precursor cells than in fetal precursor cells or actually in any given other cell. Many signaling cascades are evolutionarily well conserved. Most general knowledge about these pathways comes from research on *Saccharomyces cerevisiae* to immortalized cell lines, so that similar assumptions have to be made in other contexts than neural stem cell biology as well. What might be cell-specific, though, is the cross-talk between signaling pathways. The integration of different signal transduction cascades arguably determines the identity of the cells, if the signaling mechanisms per se lack specificity. This is the field of systems biology.

The current knowledge is summarized in Table 9–10 and obviously sketchy. Most major pathways have not yet been studied. All molecules listed in Table 9–10 are enzymes, mostly kinases.

Not all intracellular factors associated with the regulation of adult neurogenesis are classical signaling molecules and enzymes like kinases. Table 9–11 gives an overview over genes and factors in this very heterogenous group.

Tailless (TLX, Official Gene Symbol: Nr2e1)

Tailless (TLX) is a nuclear orphan receptor; that is, a receptor without known ligand. The mechanism of action of TLX is not known.

TLX has a particular function for neuronal determination in adult neurogenesis, which is distinct from its role during embryonic cortical development

In the case of TLX, researchers on adult neurogenesis were in an unusually advantageous situation, because the role of TLX during development appears to shift and becomes more pronounced in late neurogenesis than during embryonic brain development. TLX is not necessary for intrauterine survival and general formation of the brain, although it is primarily expressed in the ventricular walls from which neurons and glia originate (Monaghan et al., 1997). In TLX null-mutants, cortical development was disturbed and led to a premature specification of cortical layers (Land and Monaghan, 2003; Roy et al., 2004). Adult TLX null-mutants showed hypotrophic limbic structures, notably including the postnatally developing brain regions such as the dentate gyrus and olfactory bulb, partially because the late progenitor cells have a prolonged cell cycle time (Roy et al., 2004). Although TLX function is thus not specific to adult neurogenesis and is not independent of a developmental phenotype, relevance of TLX to precursor cell function appears to increase in postnatal and adult neurogenesis. TLX-expressing cells derived from wild-type mice were self-renewing and multipotent, but TLX–/– mice lacked precursor cells in adult neurogenic zones (Fig. 9–12) (Shi et al., 2004). TLX also down-regulates the expression of glial proteins such as GFAP, which suggests that TLX is expressed, not in the radial glia–like cells, but in their progeny, the intermediate progenitor cells of the neurogenic zones. In partial contrast to this assumption, TLX has been

Table 9–9 Regulating factors related to cytoskeleton and other structural intracellular elements

Protein name	Gene Symbol	Description	Effect on adult neurogenesis	
Collapsin response mediator protein	Crmp	See Dpysl		
Doublecortin	Dcx	Cytoskeleton-associated	Expressed in lineage-determined progenitor cells (but not radial glia-like stem cells) and immature new neurons (but not mature neurons) (Winner et al., 2002; Brown et al., 2003b; Rao and Shetty, 2004)	SGZ, SVZ
Dystrophin	Dmd	Actin-binding protein in the wall of muscle cells; mutated in Duchenne's muscular dystrophy; loss down-regulates muscular nitric oxide synthase	Mutation disrupted adult neurogenesis by increasing cell proliferation and suppressing neuronal differentiation; phenotype could be rescued by overexpression of muscular nitric oxide synthase (Deng et al., 2009)	SGZ
Ezrin	Ezr	Protein linking membrane and cytoskeleton; involved in cell motility and cell-cell communication	Expressed by glial cells ensheathing migrating cells in SVZ and RMS (Cleary et al., 2006).	SVZ
Glial fibrillary acidic protein	Gfap	Intermediate filament characteristic for astrocytes	Expressed in astrocyte-like and radial glia-like precursor cells (type-1 and B) in vivo (Doetsch et al., 1999; Seri et al., 2001; Filippov et al., 2003); cell ablation based on GFAP-promoter abolishes adult neurogenesis (Doetsch et al., 1999; Seri et al., 2001; Garcia et al., 2004b) and eliminates stem cells ex vivo (Morshead et al., 2003)	SGZ, SVZ
Lis1		see Pafah1b1		
Nestin	Nes	Intermediate filament	Expressed in precursor cells of SGZ and SVZ: type-1 and B, type-2 and C (Yamaguchi et al., 2000; Sawamoto et al., 2001; Filippov et al., 2003; Mignone et al., 2004; Seri et al., 2004).	SGZ, SVZ
p35		see Cdk5r1		
Platelet-activating factor acetylhydrolase, isoform 1b, beta1 subunit	Pafah1b1	Non-catalytic subunit of an enzyme that when mutated or lost causes lissencephaly; gene more commonly known as Lis1	Lis1 +/- mice show dispersed adult neurogenesis with reduced numbers of radial astrocytes, reduced survival and ectopic migration (Wang and Baraban, 2007).	SGZ
T alpha 1 tubulin	Tuba1	Neuron-specific cytoskeletal molecule	Expressed in Sox2 negative proliferating cells; promoter activity precedes Map2 and NeuN (Coksaygan et al., 2006)	SGZ, SVZ

Table 9–10 Genes for enzymes in intracellular signaling in adult neurogenesis

Protein name	Gene Symbol	Description	Effect on adult neurogenesis	
Thymoma viral proto-oncogene (Protein kinase B)	Akt	Serine/threonine kinase involved in several signaling pathways downstream of growth factor receptors	Increased expression after venlafaxin-induced proliferation (Mostany et al., 2008). Overexpressing Akt resulted in soma hypertrophy and ectopic positioning and increased dendritic branching (Kim et al., 2009). Blocking PI3K/Akt prevented exercise-induced increase in adult neurogenesis (Bruel-Jungerman et al., 2009).	SGZ
Alkaline phosphatase, liver/bone/kidney	Alpl	Enzyme generating the P1 nucleotide receptor agonist adenosine from extracellular ATP	Expressed by B, A and at least subsets of C cells of the SVZ and throughout the RMS; expressed in all proliferating cells (Langer et al., 2007).	SVZ
Calcium/calmodulin-dependent protein kinase IV	Camk4	Kinase controlling transcriptional regulation	Expression covaries significantly with adult neurogenesis (Kempermann et al., 2006)	SGZ
Cyclin-dependent kinase 5	Cdk5	Despite its name no function in cell cycle but in neuronal plasticity	Expression absent from ventricular wall but maintained in the adult dentate gyrus (Zheng et al., 1998)	SGZ, SVZ
		(see also Cdk5r1, p35)	Single-cell-specific knockdown of in newborn cells caused aberrant growth of dendritic processes and migration pattern of newborn cells but Cdk5 is dispensable for precursor cell proliferation and differentiation in vitro (Jessberger et al., 2008b). Conditional knockout in mature granule cells as well as in precursor cells caused reduction in number of immature new neurons without impairing precursor cell proliferation (Lagace et al., 2008).	SGZ
Cyclin-dependent kinase 5, regulatory subunit 1 (p35)\	Cdk5r1		Expression absent from ventricular wall but maintained in the adult dentate gyrus (Zheng et al., 1998)	SGZ, SVZ
Dihydrolipoamide dehydrogenase	Dld	Mitochondrial flavoprotein enzyme with possible role in neurodegeneration	Decreased number of Dcx+ cells in the SGZ (Calingasan et al., 2008)	SGZ
Dihydropyrimidinase-like 1	Dpysl1	Cytosolic phosphoprotein in the semaphorin pathway	Expressed throughout adult neurogenesis from SVZ precursor cells to mature OB interneurons (Veyrac et al., 2005).	SVZ

(Continued)

Table 9–10 (Cont'd) Genes for enzymes in intracellular signaling in adult neurogenesis

Protein name	Gene Symbol	Description	Effect on adult neurogenesis	
Dihydropyrimidinase-like 2	Dpysl2	Cytosolic phosphoprotein in the semaphorin pathway	Expressed throughout adult neurogenesis from SVZ precursor cells to mature OB interneurons (Veyrac et al., 2005).	SVZ
Dihydropyrimidinase-like 3 (Crmp4, Tuc4)	Dpysl3	Cytosolic phosphoprotein in the semaphorin pathway	Expressed in immature post-mitotic neurons in the SGZ (Seki, 2002)	SGZ
Dihydropyrimidinase-like 5 (Crmp5)	Dpysl5	Cytosolic phosphoprotein in the semaphorin pathway	Co-localizes with GAP43 in newborn neurons in the OB, down-regulated in mature interneurons (Veyrac et al., 2005)	SVZ
Ectonucleoside triphosphate diphosphohydrolase 2	Entpd2	Enzyme controlling ligand availability at nucleotide 2 receptors	Expressed specifically in B cells in the SVZ (Langer et al., 2007). Expressed by proliferating cells, type-1 cells, and Dcx+ cells in the SGZ (type-2 not explicitly mentioned) (Shukla et al., 2005).	SVZ
Extracellular signal-regulated kinase 1/2	Erk1/2	see Mapk1/3		
Growth associated protein 43	Gap43	Cytoplasmic protein involved in axon growth	Expressed in immature neurons in the olfactory bulb (Veyrac et al., 2005)	SVZ
Mitogen activated protein kinase 1/3 (Erk1/2)	Mapk1/3	Protein kinase with key role at the G1- to S-phase transition	Ex vivo, activation of Erk1/2 is necessary and sufficient for FGF2 effects on the self-renewal of hippocampal precursor cells from adult rat (Ma et al., 2009a).	SGZ
Musashi homologue 1 (Drosophila)	Msi1	Neural RNA binding protein in embryonic neural stem cells	Expressed in proliferating non-glial cells of the SGZ after ischemia (Yagita et al., 2002). Expressed in non-radial precursor cells of the SGZ (Suh et al., 2007). Expression covaried significantly with adult neurogenesis (Kempermann et al., 2006). Increased numbers of Msi1+ cells that co-labeled with nestin but rarely with Dcx and not with mature markers in specimen from humans with temporal lobe epilepsy (Crespel et al., 2005).	SGZ
PDZ binding kinase	Pbk	Mitogen-activated protein kinase that phosphorylates p38	Expressed in C cells; inhibition of Pbk in SVZ precursor cells in vitro impaired self-renewal (Dougherty et al., 2005).	SVZ
Phospholipase C, gamma 1	Plcg1	Class of enzymes that cleave phospholipids	Ex vivo in adult hippocampal rat precursor cells Plcg1, downstream of Fgf2 signaling, was required for stem cell maintenance and self-renewal (Ma et al., 2009a)	SGZ

Table 9–11 Other intracellular molecules involved in adult neurogenesis

Protein	Gene Symbol	Description	Effect on adult neurogenesis	
Brain lipid binding protein	Blbp	see Fabp7		
Calbindin	Calb1	Calcium-binding protein	Expressed in mature newborn granule cells (Kuhn et al., 1996)	SGZ
Calretinin	Calb2	Calcium-binding protein	Transiently in early post-mitotic neurons in the SGZ (Brandt et al., 2003) and in a subset of newborn olfactory interneurons (not the periglomerular TH-positive cells) (Li et al., 2002).	SGZ, SVZ
Disrupted in schizophrenia 1	Disc1	Protein with a wide range of functions in neuronal development, mutation associated with some but not all cases of schizophrenia	Cell-specific down-regulation of DISC1 accelerated neuronal integration, enhanced dendritic growth with aberrant morphology, premature synaptic integration, greater excitability and misplaced new dentate granule cells (Duan et al., 2007). Disc1 acts via Akt in precursor cells (Kim et al., 2009).	SGZ
Fatty acid binding protein 7, brain (BLBP)	Fabp7	Cytosolic protein with high affinity to long-chain fatty acids, characteristic expression in radial glia	Expressed in type-1 and some type-2a cells (Steiner et al., 2006)	SGZ
Fragile X mental retardation syndrome 1 homologue	Fmr1	Codes for a protein of unknown exact function (but involvement in protein synthesis) whose loss causes mental retardation	Fmr1–/– showed normal proliferation but reduced survival in the ventral part of the dentate gyrus, which is more related to affective function (Eadie et al., 2009).	SGZ
NADH dehydrogenase (ubiquinone) Fe-S protein 2	Ndufs2	Protein in mitochondrial complex I	Expression covaried inversely with adult neurogenesis (Kempermann et al., 2006)	SGZ
NeuN	Neuna60	Nuclear protein expressed in most neurons, identical to splicing factor Fox3	Expressed in granule cells and OB interneurons, including new ones (Kuhn et al., 1996; Kuhn et al., 1997). Can be expressed within hours after cell cycle exit (Petrus et al., 2008).	SGZ, SVZ
Nuclear receptor subfamily 2, group E, member 1 (Tailless, Tlx)	Nr2e1	Nuclear orphan receptor	Hippocampal precursor cells can be isolated based on Tlx expression. Tlx-/- precursor cells did not self-renew, re-introduced Tlx rescued self-renewal. Tlx–/– mice had reduced proliferation. Tlx silenced GFAP in precursor cells (Shi et al., 2004). Targeted ablation of Tlx+ cells reduced neurogenesis and impaired spatial learning (Zhang et al., 2008).	SGZ

(Continued)

Table 9–11 (Cont'd) Other intracellular molecules involved in adult neurogenesis

Protein	Gene Symbol	Description	Effect on adult neurogenesis	
Myeloid differentiation primary response gene 88	Myd88	Adapter protein mediating between Toll-like receptors (Tlr) and Nfkb	In Myd88-deficient mice increased proliferation and neurogenesis (Rolls et al., 2007).	SGZ
Nuclear distribution gene E-like homologue 1 (A. nidulans)	Ndel1	Binding partner to Disc1 (see also there)	ShRNA knockdown of Ndel1 induces aberrant dendritic growth and ectopic position of new cells, similar to Disc1 phenotype (Duan et al., 2007).	SGZ
Pro-opiomelanocortin-alpha	Pomc	Precursor polypeptide for various bioactive factors with many functions, including b-endorphin	Transgenic (but not endogenous) promoter activity in immature neurons of the late Dcx+ stage, absent in mature neurons and precursor cells (Overstreet et al., 2004). In b-endorphin-deficient mice (with mutant POMC allele POMC*4) baseline neurogenesis was unaffected but exercise induced increase in proliferation and neurogenesis was abolished (Koehl et al., 2008).	SGZ
S100 protein, beta polypeptide, neural	S100b	Cytoplasmic protein in astrocytes with growth factor functions promoting survival	I.c.v. infusion of S100beta increased proliferation and neurogenesis (Kleindienst et al., 2005).	SGZ
Synuclein, alpha	Snca	Presynaptic protein involved in synaptic remodeling; aggregates intracellularly in the Lewy bodies, e.g. in Parkinson disease	Normal proliferation and reduced survival and neurogenesis in hSnca overexpressing mice (Winner et al., 2004).	SGZ, SVZ
Tailless	Tlx	see Nr2e1	Expressed in B cells; targeted mutation of Tlx abolished neurogenesis in the OB (Liu et al., 2008).	SVZ

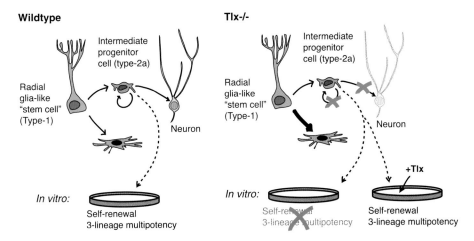

Wildtype

Radial glia-like "stem cell" (Type-1)

Intermediate progenitor cell (type-2a)

Neuron

In vitro:

Self-renewal
3-lineage multipotency

Tlx-/-

Radial glia-like "stem cell" (Type-1)

Intermediate progenitor cell (type-2a)

Neuron

+Tlx

In vitro:

Self-renewal
3-lineage multipotency

Self-renewal
3-lineage multipotency

FIGURE 9–12 Tailless (Tlx). Tlx is an orphan receptor involved in the balance between precursor cell expansion and differentiation. If Tlx is knocked out, gliogenesis is promoted, precursor cell numbers are reduced, and neurogenesis is prevented. If Tlx is added back to cultured precursor cells, self-renewal and multipotency are reconstituted (Shi et al., 2004).

described as being expressed in B cells of the SVZ. Targeted mutation also abolished neurogenesis in the olfactory bulb (Liu et al., 2008).

These findings suggests that TLX could serve a relatively specific function in adult neurogenesis and might be involved in determining precursor cells in neuronal lineage versus their maintenance as glia-like precursor cells. The relative specificity of TLX for the adult neurogenic regions has been exploited to ablate adult neurogenesis on the basis of the TLX promotor (see Chap. 10, Table 10–2, p. 456) (Zhang et al., 2008).

Transcription Factors

Transcription factors brain regulatory events to the DNA level

As long as regulation of neurogenesis involves differential regulation of genes, transcription factors are the home stretch of regulation: the target factor directly binding to the promoter of the gene in question onto which various signaling pathways might merge. Transcription factors rarely act alone; they require co-factors whose combination might determine how the gene is transcribed. This explains how one transcription factor can exert different effects depending on context and time. Within one context, transcription factors can often serve as specific signature markers, such as Dlx1/2 for C cells in the SVZ, Pax6 for the radial glia–like cells, or Tbr2 for late intermediate progenitors in the SGZ. Table 9–12 summarizes the state of knowledge about transcription factors in adult neurogenesis.

Pax6

In the SVZ, Pax6 maintains astrocyte-like stem cells and promotes development of dopaminergic periglomerular interneurons in the olfactory bulb

Paired box gene 6 is expressed in radial glia during development and in the radial glia–like cells of the adult neurogenic zones. Pax6 has many functions in neurogenesis (Osumi et al., 2008). In embryonic cortical astrocytes, forced overexpression of Pax6 is sufficient to induce a neuronal fate in the glial cells (Berninger et al., 2007). In the SVZ, Pax6 is expressed by B cells (Gubert et al., 2009). Pax6 has a dual function in maintaining these precursor cells and in promoting the dopaminergic lineage in the adult olfactory bulb (Hack et al., 2005; Brill et al., 2008). Transplanted mutant Pax6(Sey/Sey) progenitor cells produced neuroblasts that migrated into the OB but

Table 9–12 Transcription factors in adult neurogenesis

Protein name	Gene Symbol	Description	Effect on adult neurogenesis	
Achaete-scute complex homologue 1 (Drosophila); Mash1	Ascl1	Transcription factor in fate choice decision	Expressed in some type-2 cells, which respond to exercise (Uda et al., 2007; Kim et al., 2008a); overexpression generates oligodendrocytes (Jessberger et al., 2008a).	SGZ
			Upstream of Dlx1/2 and induces migratory A cells (neuroblast) (Fode et al., 2000); is also expressed in postnatal SVZ precursor cells, which generate NG2 cells and oligodendrocytes (Parras et al., 2007) (no data yet from adult!)	SVZ
Bmi1 polycomb ring finger oncogene	Bmi1	Self-renewal factor in hematopoietic stem cells	Overexpression increased neurogenesis in vitro, presumably by acting through p16(Ink4a) and p19(Arf) but had no effect in vivo (He et al., 2009)	SVZ
Bm88		see Cend1		
cAMP responsive element binding protein 1	Creb1	Transcription factor that upon phosphorylation mediates neuronal survival	Phosphorylated Creb1 in PSA-NCAM/TUC4-positive cells (Nakagawa et al., 2002a); conditional Creb1−/− show reduced proliferation; Rolipram (inhibits cAMP breakdown) increased proliferation (Nakagawa et al., 2002b) and branching and dendrite length (Fujioka et al., 2004). Cell-specific CREB loss-of-function reduced survival, decreased NeuroD and Dcx expression and impaired dendritic development; rescue by the gain-of-function vector (Jagasia et al., 2009).	SGZ
			Most Creb expression in A cells and immature neurons of SVZ, RMS and OB; most phosphorylation at shift from tangential to radial migration in OB; no defect in cond. Cerb−/− mice but in double mutants with Crem−/− (Giachino et al., 2005)	SVZ
cAMP responsive element modulator	Crem		No change in Crem−/− mice but reduced neurogenesis in double-mutants with cond. Creb−/− (Giachino et al., 2005)	SVZ
Distal-less homeobox 2	Dlx2	Transcription factor	Expressed in SVZ but not SGZ (Saino-Saito et al., 2003); expressed in C cells (Doetsch et al., 2002a); necessary for neurogenesis in the OB, together with Pax6 additional role in specification of new dopaminergic periglomerular neurons (Brill et al., 2008); promotes transition from B to C cell and increases sensitivity to EGF (Suh et al., 2009)	SVZ

Name	Symbol	Type	Description	Location
Empty spiracles homologue 1	Emx1	Homeobox transcription factor	Emx–/– mice had smaller DG, reduced proliferation and fewer DCX+ cells; they also showed a lack of learning-induced increase in BrdU+ and DCX+ cells; no effect in SVZ (Hong et al., 2007).	SGZ, SVZ
T-box brain gene 2, Tbr2	Eomes	Transcription factor	Study of Tbr2-GFP reporter mice revealed expression in typea2a, type-2b and type-3 cells but no overlap with Calretinin (Hodge et al., 2008).	SGZ
Forkhead box G1	FoxG1	Transcription factor	FoxG1 +/- mice showed reduced proliferation in relation to the reduced granule cell number only at older age. Relative survival, including CR-positive cells, was reduced earlier (Shen et al., 2006).	SGZ
GLI-Kruppel family member GLI1	Gli1	Transcription factor, downstream of Shh signaling (see there)	Expressed in GFAP-expressing type-1 and B cells and intermediate progenitor cells (Ahn and Joyner, 2005; Palma et al., 2005). Barely detectable in the SGZ of mice with abolished Shh signaling (Breunig et al., 2008).	SGZ, SVZ
Hairy and enhancer of split 5 (Drosophila)	Hes5	Transcription factor	Expressed in SVZ and SGZ (Stump et al., 2002). Increased after ischemia (Kawai et al., 2005).	SGZ, SVZ
HOP homeobox	Hopx	Atypical homeodomain protein and tumor suppressor gene	Expressed in type-1 cells. Deletion or down-regulation decreased apoptosis of type-1 cells without altering proliferation, and increased neurogenesis (De Toni et al., 2008)	SGZ
Inhibitor of DNA binding 1	Id1	Transcriptional regulator inhibiting basic helix-loop-helix transcription factors	Id1 is expressed in B1 cells and required for self renewal (Nam and Benezra, 2009)	SVZ
Inhibitor of DNA binding 2	Id2	Transcriptional regulator inhibiting basic helix-loop-helix transcription factors	Id2–/– had reduced OB size with very few dopaminergic interneurons because migratory A cells prematurely differentiated into astrocytes; no change in B and C cells (Havrda et al., 2008).	SVZ
Insulinoma-associated 1	Insm1	Zinc finger protein associated with neurogenesis and expressed in basal progenitor cells	Expressed in proliferating cells of the SGZ and proliferating and migrating cells in SVZ and RMS (Duggan et al., 2008).	SGZ, SVZ

(Continued)

Table 9–12 (Cont'd) Transcription factors in adult neurogenesis

Protein name	Gene Symbol	Description	Effect on adult neurogenesis	
see Ascl1	Mash1			
MYST histone acetyltransferase monocytic leukemia 4 (Querkopf, Qkf)	Myst4	Transcriptional co-activator	Myst4–/– mice have fewer A cells and olfactory bulb interneurons as well as fewer neurosphere-forming cells ex vivo (Merson et al., 2006).	SVZ
Neurogenic differentiation 1	Neurod1	Basic-helix-loop-helix (bHLH) transcription factor	Expressed in type-2b and type-3 cells (Steiner et al., 2006). Inducible deletion of Neurod1 reduced neurogenesis (Gao et al., 2009). Overexpression promoted neurogenesis (Roybon et al., 2009b). Wnt induced Neurod1 in SGZ (Kuwabara et al., 2009).	SGZ
	Neurod1		Ex vivo, silencing of Clock or Bmal1 in SVZ precursor cells by RNAi decreased the percentages of neuronal marker Map2-positive cells and expression levels of NeuroD1 mRNA (Kimiwada et al., 2009). Inducible deletion of Neurod1 reduced neurogenesis (Gao et al., 2009)	SVZ
Neurogenin 2	Neurog2	Proneural basic-helix-loop-helix (bHLH) protein whose deletion causes malformation of the dentate gyrus	In the SGZ expressed in type-2a cells; overexpression promoted neurogenesis (Roybon et al., 2009b). Overlap with Dcx expression (Ozen et al., 2007). In the SVZ expressed in A cells, downstream of Ascl1 (Mash1), co-localizing with Tbr2 and Tbr1 (Roybon et al., 2009a).	SGZ, SVZ
Nuclear factor of kappa light polypeptide gene enhancer in B-cells 1, p50	Nfkb1	Transcription factor involved in the cellular response to multiple extrinsic stimuli	Expression in SVZ precursor cells (Denis-Donini et al., 2005). In the SGZ p50–/– led to unaltered proliferation but reduced survival and differentiation (Denis-Donini et al., 2008).	SVZ
NK2 transcription factor related, locus 2 (Drosophila)	Nkx2-2	Oligodendrogenic transcription factor	Co-localized expression with Olig2 in the SVZ (Kuhlmann et al., 2008). Expressed in proliferating cells of the adult monkey SVZ, induced by ischemia (Tonchev et al., 2007).	SVZ

Name	Symbol	Function	Details	Region
Neuronal PAS domain protein 3	Npas3	bHLH-PAS transcription factors with possible link to schizophrenia	Npas3 −/− had reduced proliferation (Pieper et al., 2005).	SGZ
Oligodendrocyte lineage transcription factor 2	Olig2	Basic helix-loop-helix transcription factor with key function in oligodendrogenesis	Expressed by some C-cells (Menn et al., 2006) and PSA-NCAM+ cells of the SVZ (Marshall et al., 2005). Expression of normal and dominant-interfering Olig2 in vivo prevented neuronal differentiation and induced astrocytic and oligodendrocytic fates (Marshall et al., 2005).	SVZ
Paired box gene 6	Pax6	Key transcription factor in brain development with many functions, some of which are linked to radial glia	Expressed in B cells (Gubert et al., 2009). Transplanted mutant Pax6(Sey/Sey) progenitor cells produced neuroblasts that migrated into the OB but failed to generate dopaminergic periglomerular and superficial granule cells (Kohwi et al., 2005). In vivo knockdown of Pax6 prevented Dlx2 induced specification of dopaminergic periglomerular neurons (Brill et al., 2008).	SVZ
			Expressed in type-1 and type-2a cells (in mice) (Hodge et al., 2008). Expressed in precursor cells (in rats) (Nacher et al., 2005). In Pax6-deficient rSey2/+ rats overall decreased number of proliferating cells with relatively more late progenitor cells at the expense of the radial glia-like cells (Maekawa et al., 2005).	SGZ
Prospero-related homeobox 1	Prox1	Transcription factor with specific role in granule cell development	Expressed from type-2b cells and persistently in all granule cells (Steiner et al., 2008).	SGZ
RE1-silencing transcription factor	Rest	Transcriptional suppressor of neuronal fate choice	In hippocampal precursor cells in vitro double-stranded piece of DNA (NRSE), contained in the promoter of many neuronal genes, binds to Rest, which then releases its suppression of neuronal differentiation (Kuwabara et al., 2004; Kuwabara et al., 2005).	SGZ
	Smad4	Transcription factor mediating BMP signaling	Targeted deletion in Glast1+ astrocyte-like cells caused increased oligodendrogenesis in the corpus callosum and decreased neurogenesis in the OB but did not affect precursor cell proliferation (Colak et al., 2008).	SVZ
SRY-box containing gene 1	Sox1	Transcription factor involved in precursor cell maintenance	Expressed in proliferating cells of the SGZ (Limke et al., 2003).	SGZ
SRY-box containing gene 2	Sox2	Transcription factor involved in the maintenance of ES cells and neural stem cells	Expressed in proliferating GFAP+ cells (Komitova and Eriksson, 2004). Expressed in type-1 and type-2a cells (Steiner et al., 2006; Suh et al., 2007). Mutation and targeted ablation of Sox2 reduced number of type-1 cells, proliferation and net neurogenesis (Ferri et al., 2004; Favaro et al., 2009).	SGZ

(Continued)

Table 9-12 (Cont'd) Transcription factors in adult neurogenesis

Protein name	Gene Symbol	Description	Effect on adult neurogenesis	
			Expressed in proliferating GFAP+ cells (Komitova and Eriksson, 2004). Mutation reduced number of astrocyte-like precursor cells and adult neurogenesis (Ferri et al., 2004). Knockdown reduced number of new interneurons in the OB (Cavallaro et al., 2008).	SVZ
SRY-box containing gene 3	Sox3	Transcription factor in precursor cells	Expressed in proliferating cells, most of them GFAP+, and early post-mitotic neurons (Wang et al., 2006).	SGZ
SRY-box containing gene 9	Sox9	Transcription factor involved in early stages of neuronal development	Expressed in C cells and involved in the micro RNA mi124-dependent transition to A cells; overexpression decreased, while knockdown increased neurogenesis (Cheng et al., 2009).	SVZ
SRY-box containing gene 11	Sox 11	Transcription factor involved in early stages of neuronal development	Expressed in Dcx+ cells but not in uncommitted precursor cells (no overlap with Sox2); ex vivo overexpression of Sox11 reduced Dcx expression (Haslinger et al., 2009).	SGZ, SVZ
Signal Transducer and Activator of Transcription 3	Stat3	Transcription factor activated by numerous cytokines and growth factors	Conditional ablation in GFAP+ cells lead to a phenotype similar to CNTF−/−, i.e. reduced number of radial glia-like cells, proliferation and neurogenesis (Muller et al., 2009b).	SGZ
T-box brain gene 1	Tbr1	Transcription factor whose disruption leads to a reeler-like cortical malformation	Expressed by immature neurons and granule cells (Hodge et al., 2008).	SGZ
Tbr2		see Eomes		
Transcription factor 12	Tcf12	bHLH transcription factor, often co-localizing with Hes1	Expressed in SVZ (Uittenbogaard and Chiaramello, 2002).	SVZ
Transformation related protein 53	Trp53	Transcription factor and tumor suppressor gene	Less pronounced reductions in proliferating cells in p53−/− (Uberti et al., 2001; Limoli et al., 2004).	SGZ

failed to generate dopaminergic periglomerular and superficial granule cells (Kohwi et al., 2005). In vivo knockdown of Pax6 prevented Dlx2-induced specification of dopaminergic periglomerular neurons (Brill et al., 2008).

In the SGZ Pax6 is expressed in type-1 and type-2a cells (in mice) (Nacher et al., 2005; Hodge et al., 2008). In Pax6-deficient rSey2/+ rats, there was an overall decrease in the number of proliferating cells in the SGZ and a relative increase in advanced progenitor cells at the expense of the radial glia–like cells (Maekawa et al., 2005).

Sox Genes

The Sox (SRY-box containing) genes are a large class of transcription factors characterized by the name-giving motif of the "sex-determining region of the Y chromosome" (SRY) box, which was first described in testis and has close homology to the DNA-binding motif named HMG. Sox genes have numerous functions during development, and most of them play critical roles at the stem cell level.

Sox1 is involved in fate choice decision and the initiation of neuronal differentiation. Sox1 is first expressed during formation of the neural plate (Collignon et al., 1996). It later becomes restricted to neural precursor cells. Null-mutants for Sox1 show a loss of ventral brain structures (Malas et al., 2003). In neural progenitor cell lines and neurosphere cultures from the embryonic brain, Sox1 induced neurogenesis by blocking Notch activity and by attenuating Wnt signaling (Kan et al., 2004). Sox1 is also expressed in the adult SGZ but has not yet been studied functionally in this context (Limke et al., 2003).

Sox2 is a transcription factor that controls the development of the nervous system from its earliest stages (Gubbay et al., 1990; Uwanogho et al., 1995). Sox2 is expressed in embryonic stem cells in the inner cell mass of the blastocyst and persists in many multipotent cell lineages. In the neurogenic zones of the adult brain, Sox2 is expressed by the radial glia–like stem cells (B cells/type-1 cells) and their progenitor progeny (C cells/type-2 cells), but not all cells within these populations express Sox2 (Ferri et al., 2004; Komitova and Eriksson, 2004; Steiner et al., 2006). Adult neurogenesis was decreased in mice in which the neural enhancer element of the Sox2 promoter had been deleted, thus preventing embryonic lethality. Type-1 cells were greatly decreased (Ferri et al., 2004; Favaro et al., 2009). In the SVZ, the picture was similar: mutation of Sox2 reduced the number of B cells and adult neurogenesis in the olfactory bulb (Ferri et al., 2004; Cavallaro et al., 2008).

On the basis of Sox2 expression, both pluripotent embryonic stem cells and multipotent neural precursor cells could be isolated from the mouse brain (D'Amour and Gage, 2003). This isolation technique allowed comparison of the two precursor cell populations. Sox2-positive multipotent neural precursor cells differed from Sox2-positive embryonic stem cells in the expression of a total of about 270 identified genes, which were present in one and absent in the other population. This finding suggests that Sox2 exerts its function in precursor cells whose positional identity has already been determined. In this determination, the neural enhancer element of the Sox2 promoter is involved, which suggests that, despite its early expression in embryonic stem cells, Sox2 in adult neural precursor cells is controlled by yet another factor. Nevertheless, to date, Sox2 is the one transcription factor that is most closely associated with "stemness" in adult neurogenesis.

Ascl1/Mash1

Achaete-scute complex homologue (Ascl1) is still better known by its former name Mash1 and plays a key role in neuronal commitment. During development, Ascl1 is expressed by precursor cells migrating from the SVZ to the site of the developing dentate gyrus (Pleasure et al., 2000). What Ascl1 does in the adult brain is not yet fully clear. Ascl1 seems to play a central role in fate choice decisions and is expressed in type-2a cells in the SGZ (Uda et al., 2007; Kim et al., 2008a). Fittingly, exercise and ischemia induced Ascl1 in the dentate gyrus (Kawai et al., 2005; Kim et al., 2008a). Retrovirally mediated overexpression, however, induced the generation of oligo-dendrocytes (Jessberger et al., 2008a). In the SVZ, Ascl1 is expressed before Dlx1/2 and reportedly induces A cells, also suggesting that Ascl1 is in the neuronal lineage (Fode et al., 2000). But again, in partial conflict with this observation, in the postnatal SVZ, Ascl1 was also expressed in cells that would generate NG2 cells and oligodendrocytes (Parras et al., 2007). Ascl1 might thus play different roles in the two lineages.

cAMP Response Element Binding Protein (CREB)

CREB is the transcription factor mediating, among others, BDNF action

CREB has received particular attention as transcription factor controlling adult neurogenesis, because phosphorylation allows CREB to bind to the BDNF promoter and thus modulate BDNF action. The consequences of CREB signaling, however, seem to go beyond mere BDNF effects. Main target cells are type-2b and type-3 progenitor cells. Phosphorylated CREB highly co-localizes with PSA-NCAM in the dentate gyrus (Nakagawa et al., 2002a). Hippocampus-specific expression of a dominant negative mutant of CREB led to reduced cell proliferation (Nakagawa et al., 2002b). Similarly, blocking the metabolism of cAMP by application of the drug rolipram increased both the number of dividing cells and the number of surviving newly generated cells (Nakagawa et al., 2002a; Nakagawa et al., 2002b). Consequently, in a situation of induced neurogenesis, such as in response to focal ischemia, CREB expression is increased as well (Zhu et al., 2004). Blockade of CREB prevented the induction of adult neurogenesis, whereas overexpression of CREB further stimulated it (Zhu et al., 2004). The same pattern applied to physiological situations where targeted knockdown of CREB in hippocampal precursor cells reduced NeuroD and Dcx expression, decreased cell survival, and impaired dendritic development, all of which could be rescued by the corresponding gain-of-function vector (Jagasia et al., 2009).

RNA as Signaling Molecules

MicroRNA

MicroRNA are RNA molecules that do not code for proteins but exert complex signaling functions by binding to promoters of other genes much like transcription factors do. They show regionally and temporally distinct expression patterns, suggesting that networks of microRNA coordinate gene expression (Barbato et al., 2008; Liu and Zhao, 2009; Shen and Temple, 2009). Because the binding motifs are shared among many promoters, one microRNA can have pleiotropic effects throughout the genome. This is an efficient way of coordinating gene expression in developmental programs.

miR-124 promotes neurogenesis in the SVZ

In the adult SVZ the particular role of miR-124 has been identified, because miR-124 is the most abundant microRNA in the adult brain. Through target gene Sox9, miR-124 promoted neurogenesis at the transition between C and A cells. Knockdown of miR-124 maintained the precursor cells in their self-renewal; overexpression increased differentiation (Cheng et al., 2009). MiR-124 level are highest in mature neurons.

smRNA

Transcription factor NRSF/REST suppresses neurogenesis

Numerous genes involved in neuronal development contain a highly conserved DNA-responsive element, NRSE/RE1, to which a key transcription factor NRSF/REST (neuronal-restricted silencing factor/RE-1 silencing transcription factor) can bind. NRSF/REST binding suppresses activation of the neuronal genes and is thus antineurogenic (Ballas and Mandel, 2005; Lunyak and Rosenfeld, 2005). The system plays an important role in preventing the expression of neuronal genes in non-neuronal cells (Schoenherr and Anderson, 1995).

Tomoko Kuwabara and colleagues from Fred H. Gage's group at the Salk Institute discovered that a small noncoding piece of small modulatory double-stranded RNA (smRNA) binds to NRSF/REST in hippocampal precursor cells in vitro, and by suppressing the inhibitory action of NRSF/REST, induces a fate choice decision toward neuronal differentiation (Kuwabara et al., 2004). In the presence of NRSE/RE1, smRNA NRSF/REST acted as an activator of neuronal genes. If the uptake of NRSE dsRNA into the nucleus was blocked, no neuronal differentiation of hippocampal precursor cells occurred (Fig. 3–16).

Small modulatory double stranded RNA molecules inhibit NRSF/REST and depress neurogenesis

This finding helps explain how neural precursor cells switch to activation from the suppression of neuronal genes associated with the stemness state when differentiation is induced (Kuwabara et al., 2005). Similar to the situation with micro RNAs, because the NRSE/RE1 sequence is part of so many neuronal genes, a great number of genes are efficiently controlled at the same time.

Epigenetics

Epigenetic mechanisms control accessibility of promoters for transcription factors

Transcription factor binding is controlled by additional mechanisms. Epigenetic regulation controls gene activity by altering chromatin structure. DNA methylation and acetylation in promoter regions, for example, make them unavailable for transcription factor binding (see Fisher and Merkenschlager, 2002; Hsieh and Gage, 2004; Namihira et al., 2008; and Sanosaka et al., 2009, for review). This type of control might bundle many different genes and help coordinate regulation in complex developmental processes.

Control of adult neurogenesis by epigenetic mechanisms is still under-explored. Table 9–13 lists the three genes that have been described in this context so far. Expression of additional genes (Mll, Bmi1, Smarcad1, Baf53a, and Hat1) involved in chromatin remodeling as another epigenetic mechanism has been discovered in the adult SVZ (Lim et al., 2006). For Bmi1, see Table 9–12.

Treatment with histone deacetylase inhibitor valproic acid prevented seizure-induced neurogenesis

Valproic acid is an anti-epileptic drug that also blocks the activity of histone deacetylase (HDAC), which acetyl groups from histones, thereby increasing the high-affinity binding between the histones and DNA, which in turn condenses the DNA structure and prevents transcription. HDAC inhibitors thus increase transcription. In seizure models, treatment with HDAC inhibitor valproic acid prevented seizure-induced neurogenesis in the dentate gyrus (Jessberger et al., 2007). In vitro and in vivo, the acetylated histone H4 was, among others, associated with the promotor of neurogenin1 (Ngn1), suggesting a possible downstream target (Yu et al., 2009).

Retrotransposons

Retrotransposons individualize genomes

The common perception is that environment- or activity-dependent regulation acts upon a stable genome, which differs between individuals but not between cells. But the genome harbors an incredibly large number of so-called retrotransposons, or "jumping genes," which might accomplish exactly this: the individualization of cells. The most abundant retrotransposon (and the only one active in humans) is LINE 1 (long interspersed nuclear element 1) or short L1. When retrotransposon genes are transcribed, they make an mRNA copy of themselves, but they also code for the genetic machinery that allows that mRNA copy to revert to DNA. Retrotransposons are thus genes that can make copies of themselves, which reintegrate into the genome. By this process, the genome is altered, even though retrotransposon genes do not code for any other meaningful or useful protein. Because the retrotransposition is largely (but as we will see, not entirely) random, the resulting genome is individualized. Retrotransposition will have

Table 9–13 Genes involved in the epigenetic control of adult neurogenesis

Protein name	Gene Symbol	Description	Effect on adult neurogenesis	
Growth arrest and DNA-damage-inducible, beta	Gadd45b	Gene responds to environmental stress by activation of the p38/JNK pathway	Immediate early gene, up-regulated in the dentate gyrus by neuronal activity; is necessary for DNA demethylation on genes required for adult neurogenesis, including FGF1 and BDNF (Ma et al., 2009b)	SGZ
Methyl-CpG binding domain protein 1	Mbd1	Protein binding to methylation at CpG sites of the DNA as epigenetic mechanism to suppress transcription	Normal cell proliferation but reduced survival and neurogenesis in Mbd1–/–; reduction in dentate gyrus LTP characteristic for lacking new neurons; ex vivo reduced neuronal differentiation (Zhao et al., 2003) and inhibited Fgf2 expression in precursor cells (Li et al., 2008).	SGZ
Mixed-lineage leukemia 1	Mll1	Histone methyltransferase that acts as positive epigenetic regulator of gene transcription; interacts e.g. with CREB binding protein	Mll1–/– showed normal proliferation, survival and glial differentiation but impaired neuronal differentiation. In Mll1-deficient cells, early proneural Ascl1 (Mash1) and Olig2 expression were preserved, but downstream Dlx2 was not expressed (Lim et al., 2009).	SVZ

effects on the expression of other genes only, when these are hit and if the insertion into one copy of the gene is sufficient to affect gene expression. The activity of that gene can be either disrupted or increased.

L1 retrotransposon, more often than expected from randomness, integrate near neural genes

Most retrotransposons are inactive through truncation, but it turned out that they are highly expressed in neural stem cells, in rats as well as in humans, and their insertion tended to be somewhat more likely to occur near or in (neuronally expressed) genes than expected from the assumptions of sheer randomness (Muotri et al., 2005; Coufal et al., 2009). Alyson Muotri, Fred H. Gage, and their colleagues could now show that the forced expression of L1 in hippocampal precursor cells in vitro promoted neuronal differentiation through overexpression of Psd93 (official symbol: Dlg2), a plasticity-related factor associated with NMDA receptors and synapses (Muotri et al., 2005). L1 was found to be turned on when Sox2 was down-regulated, suggesting that the retrotransposition does not take place in the "stem cells" but in intermediate progenitor cells, presumably type-2b and type-3. Wnt3a caused an increase in NeuroD1 together with the down-regulation of Sox2 and required an overlapping promoter site (Sox/LEF), which is also found on L1 elements (Kuwabara et al., 2009). The intriguing aspect is thus that Wnt is apparently both upstream and downstream of the L1-dependent effects.

Exercise induced somatic mosaicism in the hippocampus

Muotri and colleagues also generated a transgenic mouse that expressed reporter gene EGFP under the ubiquitously active CMV promoter inside a L1 retrotransposon. Consequently, wherever L1 would insert and become active in a cell, EGFP would be turned on. With this mouse they could reveal a "somatic mosaicism" in neuronal populations throughout the brain. Most remarkably, wheel-running increased the number of cells expressing the reporter gene in the hippocampus (Muotri et al., 2009), suggesting that activity could directly influence reshaping of the genome towards a state that is potentially pro-neurogenic.

The entire story of genetic somatic mosaicism in neuronal progenitor cells as source for neuronal individualization still has many speculative elements, and it is difficult to estimate how relevant the relatively unlikely event of L1 inserting into a neurogenically active gene is for adult neurogenesis (Ostertag and Kazazian, 2005; Martin, 2009). Some healthy skepticism is certainly still in order, and many more experiments are needed. How different are granule cells in the adult, and how relevant is their diversity? Much depends on the answer to the question of how random the process really is and whether there is any indication that, in the end, there is benefit to be gained from the process rather than a disadvantage, due to the disruption of gene function. But the idea is appealing that a mechanism exists to give neurons a certain individuality and that blurs the boundary between nature and nurture (Muotri and Gage, 2006; Muotri et al., 2007).

References

Aberg E, Perlmann T, Olson L, Brene S (2008). Running increases neurogenesis without retinoic acid receptor activation in the adult mouse dentate gyrus. *Hippocampus* 18:785–792.

Aberg MA, Aberg ND, Hedbacker H, Oscarsson J, Eriksson PS (2000). Peripheral infusion of IGF-I selectively induces neurogenesis in the adult rat hippocampus. *J Neurosci* 20:2896–2903.

Aberg MA, Aberg ND, Palmer TD, Alborn AM, Carlsson-Skwirut C, Bang P, Rosengren LE, Olsson T, Gage FH, Eriksson PS (2003). IGF-I has a direct proliferative effect in adult hippocampal progenitor cells. *Mol Cell Neurosci* 24:23-40.

Abiola O et al. (2003). The nature and identification of quantitative trait loci: A community's view. *Nat Rev Genet* 4:911–916.

Abusaad I, MacKay D, Zhao J, Stanford P, Collier DA, Everall IP (1999). Stereological estimation of the total number of neurons in the murine hippocampus using the optical disector. *J Comp Neurol* 408:560–566.

Ackman JB, Ramos RL, Sarkisian MR, Loturco JJ (2007). Citron kinase is required for postnatal neurogenesis in the hippocampus. *Dev Neurosci* 29:113–123.

Aguado T, Monory K, Palazuelos J, Stella N, Cravatt B, Lutz B, Marsicano G, Kokaia Z, Guzman M, Galve-Roperh I (2005). The endocannabinoid system drives neural progenitor proliferation. *Faseb J* 19:1704–1706.

Aguado T, Palazuelos J, Monory K, Stella N, Cravatt B, Lutz B, Marsicano G, Kokaia Z, Guzman M, Galve-Roperh I (2006). The endocannabinoid system promotes astroglial differentiation by acting on neural progenitor cells. *J Neurosci* 26:1551–1561.

Aguado T, Romero E, Monory K, Palazuelos J, Sendtner M, Marsicano G, Lutz B, Guzman M, Galve-Roperh I (2007). The CB1 cannabinoid receptor mediates excitotoxicity-induced neural progenitor proliferation and neurogenesis. *J Biol Chem* 282:23892–23898.

Ahlenius H, Visan V, Kokaia M, Lindvall O, Kokaia Z (2009). Neural stem and progenitor cells retain their potential for proliferation and differentiation into functional neurons despite lower number in aged brain. *J Neurosci* 29:4408–4419.

Ahmed EI, Zehr JL, Schulz KM, Lorenz BH, DonCarlos LL, Sisk CL (2008). Pubertal hormones modulate the addition of new cells to sexually dimorphic brain regions. *Nat Neurosci* 11:995–997.

Ahn S, Joyner AL (2005). In vivo analysis of quiescent adult neural stem cells responding to Sonic hedgehog. *Nature* 437:894–897.

Alahmed S, Herbert J (2008). Strain differences in proliferation of progenitor cells in the dentate gyrus of the adult rat and the response to fluoxetine are dependent on corticosterone. *Neuroscience* 157:677–682.

Alenina N, Kikic D, Todiras M, Mosienko V, Qadri F, Plehm R, Boye P, Vilianovitch L, Sohr R, Tenner K, Hortnagl H, Bader M (2009). Growth retardation and altered autonomic control in mice lacking brain serotonin. *Proc Natl Acad Sci USA* 106:10332–10337.

Alfonso J, Pollevick GD, Van Der Hart MG, Flugge G, Fuchs E, Frasch AC (2004). Identification of genes regulated by chronic psychosocial stress and antidepressant treatment in the hippocampus. *Eur J Neurosci* 19:659–666.

Allen DM, van Praag H, Ray J, Weaver Z, Winrow CJ, Carter TA, Braquet R, Harrington E, Ried T, Brown KD, Gage FH, Barlow C (2001). Ataxia telangiectasia mutated is essential during adult neurogenesis. *Genes Dev* 15:554–566.

Alonso M, Viollet C, Gabellec MM, Meas-Yedid V, Olivo-Marin JC, Lledo PM (2006). Olfactory discrimination learning increases the survival of adult-born neurons in the olfactory bulb. *J Neurosci* 26:10508–10513.

Alonso R, Griebel G, Pavone G, Stemmelin J, Le Fur G, Soubrie P (2004). Blockade of CRF(1) or V(1b) receptors reverses stress-induced suppression of neurogenesis in a mouse model of depression. *Mol Psychiatry* 9: 278–286, 224.

Altar CA, Cai N, Bliven T, Juhasz M, Conner JM, Acheson AL, Lindsay RM, Wiegand SJ (1997). Anterograde transport of brain-derived neurotrophic factor and its role in the brain. *Nature* 389:856–860.

Altman J, Das GD (1965). Autoradiographic and histological evidence of postnatal hippocampal neurogenesis in rats. *J Comp Neurol* 124:319–335.

Alvarez-Borda B, Nottebohm F (2002). Gonads and singing play separate, additive roles in new neuron recruitment in adult canary brain. *J Neurosci* 22:8684–8690.

Ambrogini P, Orsini L, Mancini C, Ferri P, Ciaroni S, Cuppini R (2004). Learning may reduce neurogenesis in adult rat dentate gyrus. *Neurosci Lett* 359:13–16.

Amrein I, Slomianka L, Lipp HP (2004a). Granule cell number, cell death and cell proliferation in the dentate gyrus of wild-living rodents. *Eur J Neurosci* 20:3342–3350.

Amrein I, Slomianka L, Poletaeva, II, Bologova NV, Lipp HP (2004b). Marked species and age-dependent differences in cell proliferation and neurogenesis in the hippocampus of wild-living rodents. *Hippocampus* 14:1000–10

An L, Zhang YZ, Yu NJ, Liu XM, Zhao N, Yuan L, Chen HX, Li YF (2008). The total flavonoids extracted from Xiaobuxin-Tang up-regulate the decreased hippocampal neurogenesis and neurotrophic molecules expression in chronically stressed rats. *Prog Neuropsychopharmacol Biol Psychiatry* 32:1484–1490.

Anton ES, Ghashghaei HT, Weber JL, McCann C, Fischer TM, Cheung ID, Gassmann M, Messing A, Klein R, Schwab MH, Lloyd KC, Lai C (2004). Receptor tyrosine kinase ErbB4 modulates neuroblast migration and placement in the adult forebrain. *Nat Neurosci* 7:1319–1328.

Aoki H, Kimoto K, Hori N, Toyoda M (2005). Cell proliferation in the dentate gyrus of rat hippocampus is inhibited by soft diet feeding. *Gerontology* 51:369–374.

Apfeld J, Kenyon C (1998). Cell nonautonomy of *C. elegans* daf-2 function in the regulation of diapause and life span. *Cell* 95:199–210.

Arguello AA, Harburg GC, Schonborn JR, Mandyam CD, Yamaguchi M, Eisch AJ (2008). Time course of morphine's effects on adult hippocampal subgranular zone reveals preferential inhibition of cells in S phase of the cell cycle and a subpopulation of immature neurons. *Neuroscience* 157:70–79.

Arguello AA, Fischer SJ, Schonborn JR, Markus RW, Brekken RA, Eisch AJ (2009). Effect of chronic morphine on the dentate gyrus neurogenic microenvironment. *Neuroscience* 159:1003–1010.

Arvidsson A, Kokaia Z, Lindvall O (2001). N-methyl-D-aspartate receptor-mediated increase of neurogenesis in adult rat dentate gyrus following stroke. *Eur J Neurosci* 14:10–18.

Babu H, Cheung G, Kettenmann H, Palmer TD, Kempermann G (2007). Enriched monolayer precursor cell cultures from micro-dissected adult mouse dentate gyrus yield functional granule cell-like neurons. *PLoS ONE* 2:e388.

Babu H, Ramirez-Rodriguez G, Fabel K, Bischofberger J, Kempermann G (2009). Synaptic network activity induces neuronal differentiation of adult hippocampal precursor cells through BDNF signaling. *Front. Neurosci.* 3:49

Bai F, Bergeron M, Nelson DL (2003). Chronic AMPA receptor potentiator (LY451646) treatment increases cell proliferation in adult rat hippocampus. *Neuropharmacology* 44:1013–1021.

Baldauf K, Reymann KG (2005). Influence of EGF/bFGF treatment on proliferation, early neurogenesis and infarct volume after transient focal ischemia. *Brain Res* 1056:158–167.

Ballas N, Mandel G (2005). The many faces of REST oversee epigenetic programming of neuronal genes. *Curr Opin Neurobiol* 15:500–506.

Banasr M, Hery M, Brezun JM, Daszuta A (2001). Serotonin mediates oestrogen stimulation of cell proliferation in the adult dentate gyrus. *Eur J Neurosci* 14:1417–1424.

Banasr M, Hery M, Printemps R, Daszuta A (2004). Serotonin-induced increases in adult cell proliferation and neurogenesis are mediated through different and common 5-HT receptor subtypes in the dentate gyrus and the subventricular zone. *Neuropsychopharmacology* 29:450–460.

Banerjee SB, Rajendran R, Dias BG, Ladiwala U, Tole S, Vaidya VA (2005). Recruitment of the Sonic hedgehog signalling cascade in electroconvulsive seizure-mediated regulation of adult rat hippocampal neurogenesis. *Eur J Neurosci* 22:1570–1580.

Barbato C, Giorgi C, Catalanotto C, Cogoni C (2008). Thinking about RNA? MicroRNAs in the brain. *Mamm Genome* 19:541–551.

Baskys A, Bayazitov I, Fang L, Blaabjerg M, Poulsen FR, Zimmer J (2005). Group I metabotropic glutamate receptors reduce excitotoxic injury and may facilitate neurogenesis. *Neuropharmacology* 49 Suppl 1:146–156.

Bath KG, Mandairon N, Jing D, Rajagopal R, Kapoor R, Chen ZY, Khan T, Proenca CC, Kraemer R, Cleland TA, Hempstead BL, Chao MV, Lee FS (2008). Variant brain-derived neurotrophic factor (Val66Met) alters adult olfactory bulb neurogenesis and spontaneous olfactory discrimination. *J Neurosci* 28:2383–2393.

Beck RD, Jr., Wasserfall C, Ha GK, Cushman JD, Huang Z, Atkinson MA, Petitto JM (2005). Changes in hippocampal IL-15, related cytokines, and neurogenesis in IL-2 deficient mice. *Brain Res* 1041:223–230.

Belvindrah R, Rougon G, Chazal G (2002). Increased neurogenesis in adult mCD24–deficient mice. *J Neurosci* 22:3594–3607.

Ben Abdallah N, Slomianka L, Vyssotski AL, Lipp HP (2010). Early age-related changes in adult hippocampal neurogenesis in C57 mice. *Neurobiol Aging* 31:151–61.

Bengzon J, Kokaia Z, Elmér E, Nanobashvili A, Kokaia M, Lindvall O (1997). Apoptosis and proliferation of dentate gyrus neurons after single and intermittent limbic seizures. *Proc Natl Acad Sci USA* 94:10432–10437.

Benraiss A, Chmielnicki E, Lerner K, Roh D, Goldman SA (2001). Adenoviral brain-derived neurotrophic factor induces both neostriatal and olfactory neuronal recruitment from endogenous progenitor cells in the adult forebrain. *J Neurosci* 21:6718–6731.

Berg U, Bang P (2004). Exercise and circulating insulin-like growth factor I. *Horm Res* 62 Suppl 1:50–58.

Bergami M, Rimondini R, Santi S, Blum R, Gotz M, Canossa M (2008). Deletion of TrkB in adult progenitors alters newborn neuron integration into hippocampal circuits and increases anxiety-like behavior. *Proc Natl Acad Sci USA* 105:15570–15575.

Berger O, Li G, Han SM, Paredes M, Pleasure SJ (2007). Expression of SDF-1 and CXCR4 during reorganization of the postnatal dentate gyrus. *Dev Neurosci* 29:48–58.

Bernabeu R, Sharp FR (2000). NMDA and AMPA/kainate glutamate receptors modulate dentate neurogenesis and CA3 synapsin-I in normal and ischemic hippocampus. *J Cereb Blood Flow Metab* 20:1669–1680.

Bernabeu R, Yang T, Xie Y, Mehta B, Ma SY, Longo FM (2006). Downregulation of the LAR protein tyrosine phosphatase receptor is associated with increased dentate gyrus neurogenesis and an increased number of granule cell layer neurons. *Mol Cell Neurosci* 31:723–738.

Berninger B, Costa MR, Koch U, Schroeder T, Sutor B, Grothe B, Gotz M (2007). Functional properties of neurons derived from in vitro reprogrammed postnatal astroglia. *J Neurosci* 27:8654–8664.

Bick-Sander A, Steiner B, Wolf SA, Babu H, Kempermann G (2006). Running in pregnancy transiently increases postnatal hippocampal neurogenesis in the offspring. *Proc Natl Acad Sci USA* 103:3852–3857.

Biebl M, Cooper CM, Winkler J, Kuhn HG (2000). Analysis of neurogenesis and programmed cell death reveals a self-renewing capacity in the adult rat brain. *Neurosci Lett* 291:17–20.

Biebl M, Winner B, Winkler J (2005). Caspase inhibition decreases cell death in regions of adult neurogenesis. *Neuroreport* 16:1147–1150.

Bittman K, Owens DF, Kriegstein AR, LoTurco JJ (1997). Cell coupling and uncoupling in the ventricular zone of developing neocortex. *J Neurosci* 17:7037–7044.

Blaschke AJ, Staley K, Chun J (1996). Widespread programmed cell death in proliferative and postmitotic regions of the fetal cerebral cortex. *Development* 122:1165–1174.

Bolteus AJ, Bordey A (2004). GABA release and uptake regulate neuronal precursor migration in the postnatal subventricular zone. *J Neurosci* 24:7623–7631.

Bonaguidi MA, McGuire T, Hu M, Kan L, Samanta J, Kessler JA (2005). LIF and BMP signaling generate separate and discrete types of GFAP-expressing cells. *Development* 132:5503–5514.

Bonaguidi MA, Peng CY, McGuire T, Falciglia G, Gobeske KT, Czeisler C, Kessler JA (2008). Noggin expands neural stem cells in the adult hippocampus. *J Neurosci* 28:9194–9204.

Bondolfi L, Ermini F, Long JM, Ingram DK, Jucker M (2004). Impact of age and caloric restriction on neurogenesis in the dentate gyrus of C57BL/6 mice. *Neurobiol Aging* 25:333–340.

Bonnet E, Touyarot K, Alfos S, Pallet V, Higueret P, Abrous DN (2008). Retinoic acid restores adult hippocampal neurogenesis and reverses spatial memory deficit in vitamin A deprived rats. *PLoS ONE* 3:e3487.

Bonni A, Sun Y, Nadal-Vicens M, Bhatt A, Frank DA, Rozovsky I, Stahl N, Yancopoulos GD, Greenberg ME (1997). Regulation of gliogenesis in the central nervous system by the JAK-STAT signaling pathway. *Science* 278:477–483.

Brandt MD, Jessberger S, Steiner B, Kronenberg G, Reuter K, Bick-Sander A, von der Behrens W, Kempermann G (2003). Transient calretinin expression defines early postmitotic step of neuronal differentiation in adult hippocampal neurogenesis of mice. *Mol Cell Neurosci* 24:603–613.

Brannvall K, Korhonen L, Lindholm D (2002). Estrogen-receptor-dependent regulation of neural stem cell proliferation and differentiation. *Mol Cell Neurosci* 21:512–520.

Breunig JJ, Silbereis J, Vaccarino FM, Sestan N, Rakic P (2007). Notch regulates cell fate and dendrite morphology of newborn neurons in the postnatal dentate gyrus. *Proc Natl Acad Sci USA* 104:20558–20563.

Breunig JJ, Sarkisian MR, Arellano JI, Morozov YM, Ayoub AE, Sojitra S, Wang B, Flavell RA, Rakic P, Town T (2008). Primary cilia regulate hippocampal neurogenesis by mediating sonic hedgehog signaling. *Proc Natl Acad Sci USA* 105:13127–13132.

Brezun JM, Daszuta A (1999). Depletion in serotonin decreases neurogenesis in the dentate gyrus and the subventricular zone of adult rats. *Neuroscience* 89:999–1002.

Brezun JM, Daszuta A (2000). Serotonin may stimulate granule cell proliferation in the adult hippocampus, as observed in rats grafted with foetal raphe neurons. *Eur J Neurosci* 12:391–396.

Brill MS, Snapyan M, Wohlfrom H, Ninkovic J, Jawerka M, Mastick GS, Ashery-Padan R, Saghatelyan A, Berninger B, Gotz M (2008). A dlx2– and pax6–dependent transcriptional code for periglomerular neuron specification in the adult olfactory bulb. *J Neurosci* 28:6439–6452.

Broccardo C, Nieoullon V, Amin R, Masmejean F, Carta S, Tassi S, Pophillat M, Rubartelli A, Pierres M, Rougon G, Nieoullon A, Chazal G, Chimini G (2006). ABCA2 is a marker of neural progenitors and neuronal subsets in the adult rodent brain. *J Neurochem* 97:345–355.

Brown J, Cooper-Kuhn CM, Kempermann G, Van Praag H, Winkler J, Gage FH, Kuhn HG (2003a). Enriched environment and physical activity stimulate hippocampal but not olfactory bulb neurogenesis. *Eur J Neurosci* 17:2042–2046.

Brown JP, Couillard-Despres S, Cooper-Kuhn CM, Winkler J, Aigner L, Kuhn HG (2003b). Transient expression of doublecortin during adult neurogenesis. *J Comp Neurol* 467:1–10.

Bruel-Jungerman E, Davis S, Rampon C, Laroche S (2006). Long-term potentiation enhances neurogenesis in the adult dentate gyrus. *J Neurosci* 26:5888–5893.

Bruel-Jungerman E, Veyrac A, Dufour F, Horwood J, Laroche S, Davis S (2009). Inhibition of PI3K-Akt signaling blocks exercise-mediated enhancement of adult neurogenesis and synaptic plasticity in the dentate gyrus. *PLoS ONE* 4:e7901.

Brunson KL, Baram TZ, Bender RA (2005). Hippocampal neurogenesis is not enhanced by lifelong reduction of glucocorticoid levels. *Hippocampus* 15:491–501.

Buckwalter MS, Yamane M, Coleman BS, Ormerod BK, Chin JT, Palmer T, Wyss-Coray T (2006). Chronically increased transforming growth factor-beta1 strongly inhibits hippocampal neurogenesis in aged mice. *Am J Pathol* 169:154–164.

Butz M, Lehmann K, Dammasch IE, Teuchert-Noodt G (2006). A theoretical network model to analyse neurogenesis and synaptogenesis in the dentate gyrus. *Neural Netw* 19:1490–1505.

Caille I, Allinquant B, Dupont E, Bouillot C, Langer A, Muller U, Prochiantz A (2004). Soluble form of amyloid precursor protein regulates proliferation of progenitors in the adult subventricular zone. *Development* 131:2173–2181.

Calegari F, Haubensak W, Haffner C, Huttner WB (2005). Selective lengthening of the cell cycle in the neurogenic subpopulation of neural progenitor cells during mouse brain development. *J Neurosci* 25:6533–6538.

Calingasan NY, Ho DJ, Wille EJ, Campagna MV, Ruan J, Dumont M, Yang L, Shi Q, Gibson GE, Beal MF (2008). Influence of mitochondrial enzyme deficiency on adult neurogenesis in mouse models of neurodegenerative diseases. *Neuroscience* 153:986–996.

Cameron HA, Gould E (1994). Adult neurogenesis is regulated by adrenal steroids in the dentate gyrus. *Neuroscience* 61:203–209.

Cameron HA, McKay RD (1999). Restoring production of hippocampal neurons in old age. *Nat Neurosci* 2:894–897.

Cameron HA, Woolley CS, Gould E (1993a). Adrenal steroid receptor immunoreactivity in cells born in the adult rat dentate gyrus. *Brain Res* 611:342–346.

Cameron HA, Woolley CS, McEwen BS, Gould E (1993b). Differentiation of newly born neurons and glia in the dentate gyrus of the adult rat. *Neuroscience* 56:337–344.

Cameron HA, McEwen BS, Gould E (1995). Regulation of adult neurogenesis by excitatory input and NMDA receptor activation in the dentate gyrus. *J Neurosci* 15:4687–4692.

Cameron HA, Tanapat P, Gould E (1998). Adrenal steroids and N-methyl-D-aspartate receptor activation regulate neurogenesis in the dentate gyrus of adult rats through a common pathway. *Neuroscience* 82:349–354.

Cao L, Jiao X, Zuzga DS, Liu Y, Fong DM, Young D, During MJ (2004). VEGF links hippocampal activity with neurogenesis, learning and memory. *Nat Genet* 36:827–835.

Capela A, Temple S (2002). LeX/ssea-1 is expressed by adult mouse CNS stem cells, identifying them as nonependymal. *Neuron* 35:865–875.

Caporaso GL, Lim DA, Alvarez-Buylla A, Chao MV (2003). Telomerase activity in the subventricular zone of adult mice. *Mol Cell Neurosci* 23:693–702.

Carro E, Nunez A, Busiguina S, Torres-Aleman I (2000). Circulating insulin-like growth factor I mediates effects of exercise on the brain. *J Neurosci* 20:2926–2933.

Casadesus G, Shukitt-Hale B, Stellwagen HM, Zhu X, Lee HG, Smith MA, Joseph JA (2004). Modulation of hippocampal plasticity and cognitive behavior by short-term blueberry supplementation in aged rats. *Nutr Neurosci* 7:309–316.

Castelo-Branco G, Wagner J, Rodriguez FJ, Kele J, Sousa K, Rawal N, Pasolli HA, Fuchs E, Kitajewski J, Arenas E (2003). Differential regulation of midbrain dopaminergic neuron development by Wnt-1, Wnt-3a, and Wnt-5a. *Proc Natl Acad Sci USA* 100:12747–12752.

Cavallaro M, Mariani J, Lancini C, Latorre E, Caccia R, Gullo F, Valotta M, DeBiasi S, Spinardi L, Ronchi A, Wanke E, Brunelli S, Favaro R, Ottolenghi S, Nicolis SK (2008). Impaired generation of mature neurons by neural stem cells from hypomorphic Sox2 mutants. *Development* 135:541–557.

Cesetti T, Obernier K, Bengtson CP, Fila T, Mandl C, Holzl-Wenig G, Worner K, Eckstein V, Ciccolini F (2009). Analysis of stem cell lineage progression in the neonatal subventricular zone identifies EGFR+/NG2– cells as transit-amplifying precursors. *Stem Cells* 27:1443–1454.

Chadashvili T, Peterson DA (2006). Cytoarchitecture of fibroblast growth factor receptor 2 (FGFR-2) immunore-activity in astrocytes of neurogenic and non-neurogenic regions of the young adult and aged rat brain. *J Comp Neurol* 498:1–15.

Chang YT, Chen YC, Wu CW, Yu L, Chen HI, Jen CJ, Kuo YM (2008). Glucocorticoid signaling and exercise-induced down-regulation of the mineralocorticoid receptor in the induction of adult mouse dentate neurogenesis by treadmill running. *Psychoneuroendocrinology* 33:1173–1182.

Chen L, Gong S, Shan LD, Xu WP, Zhang YJ, Guo SY, Hisamitsu T, Yin QZ, Jiang XH (2006). Effects of exercise on neurogenesis in the dentate gyrus and ability of learning and memory after hippocampus lesion in adult rats. *Neuroscience Bulletin* 22:1–6.

Chen Y, Ai Y, Slevin JR, Maley BE, Gash DM (2005). Progenitor proliferation in the adult hippocampus and substantia nigra induced by glial cell line–derived neurotrophic factor. *Exp Neurol* 196:87–95.

Cheng A, Wang S, Cai J, Rao MS, Mattson MP (2003). Nitric oxide acts in a positive feedback loop with BDNF to regulate neural progenitor cell proliferation and differentiation in the mammalian brain. *Dev Biol* 258:319–333.

Cheng LC, Pastrana E, Tavazoie M, Doetsch F (2009). miR-124 regulates adult neurogenesis in the subventricular zone stem cell niche. *Nat Neurosci* 12:399–408.

Chiang C, Litingtung Y, Lee E, Young KE, Corden JL, Westphal H, Beachy PA (1996). Cyclopia and defective axial patterning in mice lacking Sonic hedgehog gene function. *Nature* 383:407–413.

Chiaramello S, Dalmasso G, Bezin L, Marcel D, Jourdan F, Peretto P, Fasolo A, De Marchis S (2007). BDNF/TrkB interaction regulates migration of SVZ precursor cells via PI3–K and MAP-K signalling pathways. *Eur J Neurosci* 26:1780–1790.

Choi SH, Veeraraghavalu K, Lazarov O, Marler S, Ransohoff RM, Ramirez JM, Sisodia SS (2008). Non-cell-autonomous effects of presenilin 1 variants on enrichment-mediated hippocampal progenitor cell proliferation and differentiation. *Neuron* 59:568–580.

Chojnacki A, Shimazaki T, Gregg C, Weinmaster G, Weiss S (2003). Glycoprotein 130 signaling regulates Notch1 expression and activation in the self-renewal of mammalian forebrain neural stem cells. *J Neurosci* 23:1730–1741.

Chumley MJ, Catchpole T, Silvany RE, Kernie SG, Henkemeyer M (2007). EphB receptors regulate stem/progenitor cell proliferation, migration, and polarity during hippocampal neurogenesis. *J Neurosci* 27:13481–13490.

Ciaroni S, Cecchini T, Ferri P, Cuppini R, Ambrogini P, Santi S, Benedetti S, Del Grande P, Papa S (2002). Neural precursor proliferation and newborn cell survival in the adult rat dentate gyrus are affected by vitamin E deficiency. *Neurosci Res* 44:369–377.

Ciccolini F, Mandl C, Holzl-Wenig G, Kehlenbach A, Hellwig A (2005). Prospective isolation of late development multipotent precursors whose migration is promoted by EGFR. *Dev Biol* 284:112–125.

Clark PJ, Brzezinska WJ, Thomas MW, Ryzhenko NA, Toshkov SA, Rhodes JS (2008). Intact neurogenesis is required for benefits of exercise on spatial memory but not motor performance or contextual fear conditioning in C57BL/6J mice. *Neuroscience* 155:1048–1058.

Cleary MA, Uboha N, Picciotto MR, Beech RD (2006). Expression of ezrin in glial tubes in the adult subventricular zone and rostral migratory stream. *Neuroscience* 143:851–861.

Clemmons DR (1991). Insulin-like growth factor binding proteins: Roles in regulating IGF physiology. *J Dev Physiol* 15:105–110.

Coe CL, Kramer M, Czeh B, Gould E, Reeves AJ, Kirschbaum C, Fuchs E (2003). Prenatal stress diminishes neurogenesis in the dentate gyrus of juvenile rhesus monkeys. *Biol Psychiatry* 54:1025–1034.

Cohen JE, Zimmerman G, Melamed-Book N, Friedman A, Dori A, Soreq H (2008). Transgenic inactivation of acetylcholinesterase impairs homeostasis in mouse hippocampal granule cells. *Hippocampus* 18:182–192.

Coksaygan T, Magnus T, Cai J, Mughal M, Lepore A, Xue H, Fischer I, Rao MS (2006). Neurogenesis in Talpha-1 tubulin transgenic mice during development and after injury. *Exp Neurol* 197:475–485.

Colak D, Mori T, Brill MS, Pfeifer A, Falk S, Deng C, Monteiro R, Mummery C, Sommer L, Gotz M (2008). Adult neurogenesis requires Smad4–mediated bone morphogenic protein signaling in stem cells. *J Neurosci* 28:434–446.

Collignon J, Sockanathan S, Hacker A, Cohen-Tannoudji M, Norris D, Rastan S, Stevanovic M, Goodfellow PN, Lovell-Badge R (1996). A comparison of the properties of Sox-3 with Sry and two related genes, Sox-1 and Sox-2. *Development* 122:509–520.

Conover JC, Doetsch F, Garcia-Verdugo JM, Gale NW, Yancopoulos GD, Alvarez-Buylla A (2000). Disruption of Eph/ephrin signaling affects migration and proliferation in the adult subventricular zone. *Nat Neurosci* 3:1091–1097.

Cooper-Kuhn CM, Vroemen M, Brown J, Ye H, Thompson MA, Winkler J, Kuhn HG (2002). Impaired adult neurogenesis in mice lacking the transcription factor E2F1. *Mol Cell Neurosci* 21:312–323.

Cooper-Kuhn CM, Winkler J, Kuhn HG (2004). Decreased neurogenesis after cholinergic forebrain lesion in the adult rat. *J Neurosci Res* 77:155–165.

Corniola RS, Tassabehji NM, Hare J, Sharma G, Levenson CW (2008). Zinc deficiency impairs neuronal precursor cell proliferation and induces apoptosis via p53-mediated mechanisms. *Brain Res* 1237:52–61.

Corotto FS, Henegar JR, Maruniak JA (1994). Odor deprivation leads to reduced neurogenesis and reduced neuronal survival in the olfactory bulb of the adult mouse. *Neuroscience* 61:739–744.

Corti S, Nizzardo M, Nardini M, Donadoni C, Locatelli F, Papadimitriou D, Salani S, Del Bo R, Ghezzi S, Strazzer S, Bresolin N, Comi GP (2007). Isolation and characterization of murine neural stem/progenitor cells based on Prominin-1 expression. *Exp Neurol* 205:547–562.

Coufal NG, Garcia-Perez JL, Peng GE, Yeo GW, Mu Y, Lovci MT, Morell M, O'Shea KS, Moran JV, Gage FH (2009). L1 retrotransposition in human neural progenitor cells. *Nature* 460:1127–1131.

Couillard-Despres S, Winner B, Karl C, Lindemann G, Schmid P, Aigner R, Laemke J, Bogdahn U, Winkler J, Bischofberger J, Aigner L (2006). Targeted transgene expression in neuronal precursors: Watching young neurons in the old brain. *Eur J Neurosci* 24:1535–1545.

Couillard-Despres S, Wuertinger C, Kandasamy M, Caioni M, Stadler K, Aigner R, Bogdahn U, Aigner L (2009). Ageing abolishes the effects of fluoxetine on neurogenesis. *Mol Psychiatry* 14:856–864.

Crespel A, Rigau V, Coubes P, Rousset MC, de Bock F, Okano H, Baldy-Moulinier M, Bockaert J, Lerner-Natoli M (2005). Increased number of neural progenitors in human temporal lobe epilepsy. *Neurobiol Dis* 19: 436–450.

Cummings DM, Henning HE, Brunjes PC (1997). Olfactory bulb recovery after early sensory deprivation. *J Neurosci* 17:7433–7440.

Cuppini R, Ciaroni S, Cecchini T, Ambrogini P, Ferri P, Del Grande P, Papa S (2001). Alpha-tocopherol controls cell proliferation in the adult rat dentate gyrus. *Neurosci Lett* 303:198–200.

Czeh B, Michaelis T, Watanabe T, Frahm J, de Biurrun G, van Kampen M, Bartolomucci A, Fuchs E (2001). Stress-induced changes in cerebral metabolites, hippocampal volume, and cell proliferation are prevented by antidepressant treatment with tianeptine. *Proc Natl Acad Sci USA* 98:12796–12801.

Czeh B, Welt T, Fischer AK, Erhardt A, Schmitt W, Muller MB, Toschi N, Fuchs E, Keck ME (2002). Chronic psychosocial stress and concomitant repetitive transcranial magnetic stimulation: Effects on stress hormone levels and adult hippocampal neurogenesis. *Biol Psychiatry* 52:1057–1065.

D'Amour KA, Gage FH (2003). Genetic and functional differences between multipotent neural and pluripotent embryonic stem cells. *Proc Natl Acad Sci USA* 100 Suppl 1:11866–11872.

D'Sa C, Duman RS (2002). Antidepressants and neuroplasticity. *Bipolar Disord* 4:183–194.

Dahl A, Eriksson PS, Davidsson P, Persson AI, Ekman R, Westman-Brinkmalm A (2004). Demonstration of multiple novel glycoforms of the stem cell survival factor CCg. *J Neurosci Res* 77:9–14.

Dahl A, Eriksson PS, Persson AI, Karlsson G, Davidsson P, Ekman R, Westman-Brinkmalm A (2003). Proteome analysis of conditioned medium from cultured adult hippocampal progenitors. *Rapid Commun Mass Spectrom* 17:2195–2202.

Danilov AI, Gomes-Leal W, Ahlenius H, Kokaia Z, Carlemalm E, Lindvall O (2009). Ultrastructural and antigenic properties of neural stem cells and their progeny in adult rat subventricular zone. *Glia* 57: 136–152.

De Ferrari GV, Inestrosa NC (2000). Wnt signaling function in Alzheimer's disease. *Brain Res Brain Res Rev* 33:1–12.

De Toni A, Zbinden M, Epstein JA, Ruiz i Altaba A, Prochiantz A, Caille I (2008). Regulation of survival in adult hippocampal and glioblastoma stem cell lineages by the homeodomain-only protein HOP. *Neural Dev* 3:13.

Decressac M, Prestoz L, Veran J, Cantereau A, Jaber M, Gaillard A (2009). Neuropeptide Y stimulates proliferation, migration and differentiation of neural precursors from the subventricular zone in adult mice. *Neurobiol Dis* 34:441–449.

Deisseroth K, Singla S, Toda H, Monje M, Palmer TD, Malenka RC (2004). Excitation-neurogenesis coupling in adult neural stem/progenitor cells. *Neuron* 42:535–552.

Deng B, Glanzman D, Tidball JG (2009). Nitric oxide generated by muscle corrects defects in hippocampal neurogenesis and neural differentiation caused by muscular dystrophy. *J Physiol* 587:1769–1778.

Denis-Donini S, Caprini A, Frassoni C, Grilli M (2005). Members of the NF-kappaB family expressed in zones of active neurogenesis in the postnatal and adult mouse brain. *Brain Res Dev Brain Res* 154:81–89.

Denis-Donini S, Dellarole A, Crociara P, Francese MT, Bortolotto V, Quadrato G, Canonico PL, Orsetti M, Ghi P, Memo M, Bonini SA, Ferrari-Toninelli G, Grilli M (2008). Impaired adult neurogenesis associated with short-term memory defects in NF-kappaB p50–deficient mice. *J Neurosci* 28:3911–3919.

Dickinson ME, Krumlauf R, McMahon AP (1994). Evidence for a mitogenic effect of Wnt-1 in the developing mammalian central nervous system. *Development* 120:1453–1471.

Diederich K, Schabitz WR, Kuhnert K, Hellstrom N, Sachser N, Schneider A, Kuhn HG, Knecht S (2009). Synergetic effects of granulocyte-colony stimulating factor and cognitive training on spatial learning and survival of newborn hippocampal neurons. *PLoS One* 4:e5303.

Dobrossy MD, Drapeau E, Aurousseau C, Le Moal M, Piazza PV, Abrous DN (2003). Differential effects of learning on neurogenesis: Learning increases or decreases the number of newly born cells depending on their birth date. *Mol Psychiatry* 8:974–982.

Doetsch F, Caille I, Lim DA, Garcia-Verdugo JM, Alvarez-Buylla A (1999). Subventricular zone astrocytes are neural stem cells in the adult mammalian brain. *Cell* 97:703–716.

Doetsch F, Petreanu L, Caille I, Garcia-Verdugo JM, Alvarez-Buylla A (2002a). EGF converts transit-amplifying neurogenic precursors in the adult brain into multipotent stem cells. *Neuron* 36:1021–1034.

Doetsch F, Verdugo JM, Caille I, Alvarez-Buylla A, Chao MV, Casaccia-Bonnefil P (2002b). Lack of the cell-cycle inhibitor p27Kip1 results in selective increase of transit-amplifying cells for adult neurogenesis. *J Neurosci* 22:2255–2264.

Donovan MH, Yamaguchi M, Eisch AJ (2008). Dynamic expression of TrkB receptor protein on proliferating and maturing cells in the adult mouse dentate gyrus. *Hippocampus* 18:435–439.

Dougherty JD, Garcia AD, Nakano I, Livingstone M, Norris B, Polakiewicz R, Wexler EM, Sofroniew MV, Kornblum HI, Geschwind DH (2005). PBK/TOPK, a proliferating neural progenitor-specific mitogen-activated protein kinase kinase. *J Neurosci* 25:10773–10785.

Drake CT, Chavkin C, Milner TA (2007). Opioid systems in the dentate gyrus. *Prog Brain Res* 163:245–263.

Dranovsky A, Hen R (2006). Hippocampal neurogenesis: Regulation by stress and antidepressants. *Biol Psychiatry* 59:1136–1143.

Driscoll I, Howard SR, Stone JC, Monfils MH, Tomanek B, Brooks WM, Sutherland RJ (2006). The aging hippocampus: A multi-level analysis in the rat. *Neuroscience* 139:1173–1185.

Duan X, Chang JH, Ge S, Faulkner RL, Kim JY, Kitabatake Y, Liu XB, Yang CH, Jordan JD, Ma DK, Liu CY, Ganesan S, Cheng HJ, Ming GL, Lu B, Song H (2007). Disrupted-In-Schizophrenia 1 regulates integration of newly generated neurons in the adult brain. *Cell* 130:1146–1158.

Duggan A, Madathany T, de Castro SC, Gerrelli D, Guddati K, Garcia-Anoveros J (2008). Transient expression of the conserved zinc finger gene INSM1 in progenitors and nascent neurons throughout embryonic and adult neurogenesis. *J Comp Neurol* 507:1497–1520.

Duman RS, Monteggia LM (2006). A neurotrophic model for stress-related mood disorders. *Biol Psychiatry* 59:1116–1127.

Dupret D, Fabre A, Dobrossy MD, Panatier A, Rodriguez JJ, Lamarque S, Lemaire V, Oliet SH, Piazza PV, Abrous DN (2007). Spatial learning depends on both the addition and removal of new hippocampal neurons. *PLoS Biol* 5:e214.

Eadie BD, Zhang WN, Boehme F, Gil-Mohapel J, Kainer L, Simpson JM, Christie BR (2009). Fmr1 knockout mice show reduced anxiety and alterations in neurogenesis that are specific to the ventral dentate gyrus. *Neurobiol Dis* 36:361–73.

Ehninger D, Kempermann G (2003). Regional effects of wheel running and environmental enrichment on cell genesis and microglia proliferation in the adult murine neocortex. *Cereb Cortex* 13:845–851.

Ehninger D, Kempermann G (2006). Paradoxical effects of learning the Morris water maze on adult hippocampal neurogenesis in mice may be explained by a combination of stress and physical activity. *Genes Brain Behav* 5:29–39.

Eisch AJ, Barrot M, Schad CA, Self DW, Nestler EJ (2000). Opiates inhibit neurogenesis in the adult rat hippocampus. *Proc Natl Acad Sci USA* 97:7579–7584.

Ekdahl CT, Mohapel P, Elmer E, Lindvall O (2001). Caspase inhibitors increase short-term survival of progenitor-cell progeny in the adult rat dentate gyrus following status epilepticus. *Eur J Neurosci* 14:937–945.

Ekdahl CT, Claasen JH, Bonde S, Kokaia Z, Lindvall O (2003). Inflammation is detrimental for neurogenesis in adult brain. *Proc Natl Acad Sci USA* 100:13632–13637.

Ekdahl CT, Kokaia Z, Lindvall O (2009). Brain inflammation and adult neurogenesis: The dual role of microglia. *Neuroscience* 158:1021–1029.

Emsley JG, Hagg T (2003). Endogenous and exogenous ciliary neurotrophic factor enhances forebrain neurogenesis in adult mice. *Exp Neurol* 183:298–310.

Encinas JM, Vaahtokari A, Enikolopov G (2006). Fluoxetine targets early progenitor cells in the adult brain. *Proc Natl Acad Sci USA* 103:8233–8238.

Eriksson PS, Perfilieva E, Bjork-Eriksson T, Alborn AM, Nordborg C, Peterson DA, Gage FH (1998). Neurogenesis in the adult human hippocampus. *Nat Med* 4:1313–1317.

Estrada C, Gomez C, Martin-Nieto J, De Frutos T, Jimenez A, Villalobo A (1997). Nitric oxide reversibly inhibits the epidermal growth factor receptor tyrosine kinase. *Biochem J* 326 (Pt 2):369–376.

Fabel K, Fabel K, Tam B, Kaufer D, Baiker A, Simmons N, Kuo CJ, Palmer TD (2003). VEGF is necessary for exercise-induced adult hippocampal neurogenesis. *Eur J Neurosci* 18:2803–2812.

Fabel K, Wolf SA, Ehninger D, Babu H, Galicia PL, Kempermann G (2009). Additive effects of physical exercise and environmental enrichment on adult hippocampal neurogenesis in mice. *Front. Neurosci* 3:50

Fan X, Xu H, Cai W, Yang Z, Zhang J (2003). Spatial and temporal patterns of expression of Noggin and BMP4 in embryonic and postnatal rat hippocampus. *Brain Res Dev Brain Res* 146:51–58.

Fan XT, Xu HW, Cai WQ, Yang H, Liu S (2004). Antisense Noggin oligodeoxynucleotide administration decreases cell proliferation in the dentate gyrus of adult rats. *Neurosci Lett* 366:107–111.

Farmer J, Zhao X, van Praag H, Wodtke K, Gage FH, Christie BR (2004). Effects of voluntary exercise on synaptic plasticity and gene expression in the dentate gyrus of adult male Sprague-Dawley rats in vivo. *Neuroscience* 124:71–79.

Favaro R, Valotta M, Ferri AL, Latorre E, Mariani J, Giachino C, Lancini C, Tosetti V, Ottolenghi S, Taylor V, Nicolis SK (2009). Hippocampal development and neural stem cell maintenance require Sox2–dependent regulation of Shh. *Nat Neurosci* 12:1248–56.

Feng R, Rampon C, Tang YP, Shrom D, Jin J, Kyin M, Sopher B, Martin GM, Kim SH, Langdon RB, Sisodia SS, Tsien JZ (2001). Deficient neurogenesis in forebrain-specific presenilin-1 knockout mice is associated with reduced clearance of hippocampal memory traces. *Neuron* 32:911–926.

Ferri AL, Cavallaro M, Braida D, Di Cristofano A, Canta A, Vezzani A, Ottolenghi S, Pandolfi PP, Sala M, DeBiasi S, Nicolis SK (2004). Sox2 deficiency causes neurodegeneration and impaired neurogenesis in the adult mouse brain. *Development* 131:3805–3819.

Ferri P, Cecchini T, Ciaroni S, Ambrogini P, Cuppini R, Santi S, Benedetti S, Pagliarani S, Del Grande P, Papa S (2003). Vitamin E affects cell death in adult rat dentate gyrus. *J Neurocytol* 32:1155–1164.

Ferron S, Mira H, Franco S, Cano-Jaimez M, Bellmunt E, Ramirez C, Farinas I, Blasco MA (2004). Telomere shortening and chromosomal instability abrogates proliferation of adult but not embryonic neural stem cells. *Development* 131:4059–4070.

Ferry RJ, Jr., Katz LE, Grimberg A, Cohen P, Weinzimer SA (1999). Cellular actions of insulin-like growth factor binding proteins. *Horm Metab Res* 31:192–202.

Filippov V, Kronenberg G, Pivneva T, Reuter K, Steiner B, Wang LP, Yamaguchi M, Kettenmann H, Kempermann G (2003). Subpopulation of nestin-expressing progenitor cells in the adult murine hippocampus shows electrophysiological and morphological characteristics of astrocytes. *Mol Cell Neurosci* 23:373–382.

Fisher AG, Merkenschlager M (2002). Gene silencing, cell fate and nuclear organisation. *Curr Opin Genet Dev* 12:193–197.

Fiske BK, Brunjes PC (2001). Cell death in the developing and sensory-deprived rat olfactory bulb. *J Comp Neurol* 431:311–319.

Fode C, Ma Q, Casarosa S, Ang SL, Anderson DJ, Guillemot F (2000). A role for neural determination genes in specifying the dorsoventral identity of telencephalic neurons. *Genes Dev* 14:67–80.

Fournier NM, Andersen DR, Botterill JJ, Sterner EY, Lussier AL, Caruncho HJ, Kalynchuk LE (2009). The effect of amygdala kindling on hippocampal neurogenesis coincides with decreased reelin and DISC1 expression in the adult dentate gyrus. *Hippocampus* 20:659-71.

Freund TF, Buzsaki G (1996). Interneurons of the hippocampus. *Hippocampus* 6:347–470.

Fabel K, Wolf SA, Ehninger D, Babu H, Galicia PL, Kempermann G (2009). Additive effects of physical exercise and environmental enrichment on adult hippocampal neurogenesis in mice. *Front. Neurosci* **3**:50.

Fritzen S, Schmitt A, Koth K, Sommer C, Lesch KP, Reif A (2007). Neuronal nitric oxide synthase (NOS-I) knockout increases the survival rate of neural cells in the hippocampus independently of BDNF. *Mol Cell Neurosci* 35:261–271.

Fujioka T, Fujioka A, Duman RS (2004). Activation of cAMP signaling facilitates the morphological maturation of newborn neurons in adult hippocampus. *J Neurosci* 24:319–328.

Furne C, Ricard J, Cabrera JR, Pays L, Bethea JR, Mehlen P, Liebl DJ (2009). EphrinB3 is an anti-apoptotic ligand that inhibits the dependence receptor functions of EphA4 receptors during adult neurogenesis. *Biochim Biophys Acta* 1793:231–238.

Fuss J, Ben Abdallah NM, Vogt MA, Touma C, Pacifici PG, Palme R, Witzemann V, Hellweg R, Gass P (2009). Voluntary exercise induces anxiety-like behavior in adult C57BL/6J mice correlating with hippocampal neurogenesis. *Hippocampus* 20:364–76.

Gaiano N, Fishell G (2002). The role of notch in promoting glial and neural stem cell fates. *Annu Rev Neurosci* 25:471–490.

Galea LA (2008). Gonadal hormone modulation of neurogenesis in the dentate gyrus of adult male and female rodents. *Brain Res Rev* 57:332–341.

Galea LA, McEwen BS (1999). Sex and seasonal differences in the rate of cell proliferation in the dentate gyrus of adult wild meadow voles. *Neuroscience* 89:955–964.

Galvao RP, Garcia-Verdugo JM, Alvarez-Buylla A (2008). Brain-derived neurotrophic factor signaling does not stimulate subventricular zone neurogenesis in adult mice and rats. *J Neurosci* 28:13368–13383.

Gama Sosa MA, Wen PH, De Gasperi R, Perez GM, Senturk E, Friedrich VL, Jr., Elder GA (2004). Entorhinal cortex lesioning promotes neurogenesis in the hippocampus of adult mice. *Neuroscience* 127:881–891.

Gao Z, Ure K, Ables JL, Lagace DC, Nave KA, Goebbels S, Eisch AJ, Hsieh J (2009). Neurod1 is essential for the survival and maturation of adult-born neurons. *Nat Neurosci* 12:1090–1092.

Garcia A, Steiner B, Kronenberg G, Bick-Sander A, Kempermann G (2004a). Age-dependent expression of glucocorticoid- and mineralocorticoid receptors on neural precursor cell populations in the adult murine hippocampus. *Aging Cell* 3:363–371.

Garcia AD, Doan NB, Imura T, Bush TG, Sofroniew MV (2004b). GFAP-expressing progenitors are the principal source of constitutive neurogenesis in adult mouse forebrain. *Nat Neurosci* 7:1233–1241.

Garza JC, Guo M, Zhang W, Lu XY (2008). Leptin increases adult hippocampal neurogenesis in vivo and in vitro. *J Biol Chem* 283:18238–18247.

Gaspar P, Cases O, Maroteaux L (2003). The developmental role of serotonin: News from mouse molecular genetics. *Nat Rev Neurosci* 4:1002–1012.

Gass P, Kretz O, Wolfer DP, Berger S, Tronche F, Reichardt HM, Kellendonk C, Lipp HP, Schmid W, Schutz G (2000). Genetic disruption of mineralocorticoid receptor leads to impaired neurogenesis and granule cell degeneration in the hippocampus of adult mice. *EMBO Rep* 1:447–451.

Ge S, Goh EL, Sailor KA, Kitabatake Y, Ming GL, Song H (2006). GABA regulates synaptic integration of newly generated neurons in the adult brain. *Nature* 439:589–593.

Ge S, Yang CH, Hsu KS, Ming GL, Song H (2007). A critical period for enhanced synaptic plasticity in newly generated neurons of the adult brain. *Neuron* 54:559–566.

Gemma C, Bachstetter AD, Cole MJ, Fister M, Hudson C, Bickford PC (2007). Blockade of caspase-1 increases neurogenesis in the aged hippocampus. *Eur J Neurosci* 26:2795–2803.

Ghashghaei HT, Weber J, Pevny L, Schmid R, Schwab MH, Lloyd KC, Eisenstat DD, Lai C, Anton ES (2006). The role of neuregulin-ErbB4 interactions on the proliferation and organization of cells in the subventricular zone. *Proc Natl Acad Sci USA* 103:1930–1935.

Giachino C, De Marchis S, Giampietro C, Parlato R, Perroteau I, Schutz G, Fasolo A, Peretto P (2005). cAMP response element-binding protein regulates differentiation and survival of newborn neurons in the olfactory bulb. *J Neurosci* 25:10105–10118.

Giuliani A, D'Intino G, Paradisi M, Giardino L, Calza L (2004). p75(NTR)-immunoreactivity in the subventricular zone of adult male rats: Expression by cycling cells. *J Mol Histol* 35:749–758.

Givogri MI, de Planell M, Galbiati F, Superchi D, Gritti A, Vescovi A, de Vellis J, Bongarzone ER (2006). Notch signaling in astrocytes and neuroblasts of the adult subventricular zone in health and after cortical injury. *Dev Neurosci* 28:81–91.

Goldman SA, Kirschenbaum B, Harrison-Restelli C, Thaler HT (1997). Neuronal precursors of the adult rat subependymal zone persist into senescence, with no decline in spatial extent or response to BDNF. *J Neurobiol* 32:554–566.

Gomez-Pinilla F, Dao L, So V (1997). Physical exercise induces FGF2 and its mRNA in the hippocampus. *Brain Res* 764:1–8.

Gould E (1994). The effects of adrenal steroids and excitatory input on neuronal birth and survival. *Ann NY Acad Sci* 743:73–92; discussion, 92–73.

Gould E, McEwen BS (1993). Neuronal birth and death. *Curr Opin Neurobiol* 3:676–682.

Gould E, Woolley CS, McEwen BS (1991a). Naturally occurring cell death in the developing dentate gyrus of the rat. *J Comp Neurol* 304:408–418.

Gould E, Woolley CS, McEwen BS (1991b). Adrenal steroids regulate postnatal development of the rat dentate gyrus: I. Effects of glucocorticoids on cell death. *J Comp Neurol* 313:479–485.

Gould E, Cameron HA, Daniels DC, Woolley CS, McEwen BS (1992). Adrenal hormones suppress cell division in the adult rat dentate gyrus. *J Neurosci* 12:3642–3650.

Gould E, McEwen BS, Tanapat P, Galea LA, Fuchs E (1997). Neurogenesis in the dentate gyrus of the adult tree shrew is regulated by psychosocial stress and NMDA receptor activation. *J Neurosci* 17:2492–2498.

Gould E, Tanapat P, McEwen BS, Flügge G, Fuchs E (1998). Proliferation of granule cell precursors in the dentate gyrus of adult monkeys is diminished by stress. *Proc Natl Acad Sci USA* 95:3168–3171.

Gould E, Beylin A, Tanapat P, Reeves A, Shors TJ (1999). Learning enhances adult neurogenesis in the hippocampal formation. *Nat Neurosci* 2:260–265.

Gould TD, Manji HK (2002). The Wnt signaling pathway in bipolar disorder. *Neuroscientist* 8:497–511.

Grassi Zucconi G, Cipriani S, Balgkouranidou I, Scattoni R (2006). "One night" sleep deprivation stimulates hippocampal neurogenesis. *Brain Res Bull* 69:375–381.

Guazzo EP, Kirkpatrick PJ, Goodyer IM, Shiers HM, Herbert J (1996). Cortisol, dehydroepiandrosterone (DHEA), and DHEA sulfate in the cerebrospinal fluid of man: Relation to blood levels and the effects of age. *J Clin Endocrinol Metab* 81:3951–3960.

Gubbay J, Collignon J, Koopman P, Capel B, Economou A, Munsterberg A, Vivian N, Goodfellow P, Lovell-Badge R (1990). A gene mapping to the sex-determining region of the mouse Y chromosome is a member of a novel family of embryonically expressed genes. *Nature* 346:245–250.

Gubert F, Zaverucha-do-Valle C, Pimentel-Coelho PM, Mendez-Otero R, Santiago MF (2009). Radial glia–like cells persist in the adult rat brain. *Brain Res* 1258:43–52.

Gustafsson E, Lindvall O, Kokaia Z (2003a). Intraventricular infusion of TrkB-Fc fusion protein promotes ischemia-induced neurogenesis in adult rat dentate gyrus. *Stroke* 34:2710–2715.

Gustafsson E, Andsberg G, Darsalia V, Mohapel P, Mandel RJ, Kirik D, Lindvall O, Kokaia Z (2003b). Anterograde delivery of brain-derived neurotrophic factor to striatum via nigral transduction of recombinant adeno-associated virus increases neuronal death but promotes neurogenic response following stroke. *Eur J Neurosci* 17:2667–2678.

Guzman-Marin R, Suntsova N, Methippara M, Greiffenstein R, Szymusiak R, McGinty D (2005). Sleep deprivation suppresses neurogenesis in the adult hippocampus of rats. *Eur J Neurosci* 22:2111–2116.

Guzman-Marin R, Bashir T, Suntsova N, Szymusiak R, McGinty D (2007). Hippocampal neurogenesis is reduced by sleep fragmentation in the adult rat. *Neuroscience* 148:325–333.

Guzman-Marin R, Suntsova N, Bashir T, Nienhuis R, Szymusiak R, McGinty D (2008). Rapid eye movement sleep deprivation contributes to reduction of neurogenesis in the hippocampal dentate gyrus of the adult rat. *Sleep* 31:167–175.

Haas CA, Frotscher M (2010). Reelin deficiency causes granule cell dispersion in epilepsy. *Exp Brain Res* 200:141–149.

Hack MA, Saghatelyan A, de Chevigny A, Pfeifer A, Ashery-Padan R, Lledo PM, Gotz M (2005). Neuronal fate determinants of adult olfactory bulb neurogenesis. *Nat Neurosci* 8:865–872.

Haik S, Gauthier LR, Granotier C, Peyrin JM, Lages CS, Dormont D, Boussin FD (2000). Fibroblast growth factor 2 up-regulates telomerase activity in neural precursor cells. *Oncogene* 19:2957–2966.

Han YG, Spassky N, Romaguera-Ros M, Garcia-Verdugo JM, Aguilar A, Schneider-Maunoury S, Alvarez-Buylla A (2008). Hedgehog signaling and primary cilia are required for the formation of adult neural stem cells. *Nat Neurosci* 11:277–284.

Hansson SR, Cabrera-Vera TM, Hoffman BJ (1998). Infraorbital nerve transection alters serotonin transporter expression in sensory pathways in early postnatal rat development. *Brain Res Dev Brain Res* 111:305–314.

Harburg GC, Hall FS, Harrist AV, Sora I, Uhl GR, Eisch AJ (2007). Knockout of the mu opioid receptor enhances the survival of adult-generated hippocampal granule cell neurons. *Neuroscience* 144:77–87.

Harrist A, Beech RD, King SL, Zanardi A, Cleary MA, Caldarone BJ, Eisch A, Zoli M, Picciotto MR (2004). Alteration of hippocampal cell proliferation in mice lacking the beta 2 subunit of the neuronal nicotinic acetylcholine receptor. *Synapse* 54:200–206.

Hashimoto K, Shimizu E, Iyo M (2004). Critical role of brain-derived neurotrophic factor in mood disorders. *Brain Res Brain Res Rev* 45:104–114.

Haslinger A, Schwarz TJ, Covic M, Chichung Lie D (2009). Expression of Sox11 in adult neurogenic niches suggests a stage-specific role in adult neurogenesis. *Eur J Neurosci* 29:2103–2114.

Hattiangady B, Shetty AK (2006). Aging does not alter the number or phenotype of putative stem/progenitor cells in the neurogenic region of the hippocampus. *Neurobiol Aging* 29:129-47.

Hauser T, Klaus F, Lipp HP, Amrein I (2009). No effect of running and laboratory housing on adult hippocampal neurogenesis in wild caught long-tailed wood mouse. *BMC Neurosci* 10:43.

Havrda MC, Harris BT, Mantani A, Ward NM, Paolella BR, Cuzon VC, Yeh HH, Israel MA (2008). Id2 is required for specification of dopaminergic neurons during adult olfactory neurogenesis. *J Neurosci* 28:14074–14086.

He J, Crews FT (2007). Neurogenesis decreases during brain maturation from adolescence to adulthood. *Pharmacol Biochem Behav* 86:327–333.

He S, Iwashita T, Buchstaller J, Molofsky AV, Thomas D, Morrison SJ (2009). Bmi-1 over-expression in neural stem/progenitor cells increases proliferation and neurogenesis in culture but has little effect on these functions in vivo. *Dev Biol* 328:257–272.

Heine VM, Maslam S, Joels M, Lucassen PJ (2004a). Increased P27KIP1 protein expression in the dentate gyrus of chronically stressed rats indicates G1 arrest involvement. *Neuroscience* 129:593–601.

Heine VM, Maslam S, Joels M, Lucassen PJ (2004b). Prominent decline of newborn cell proliferation, differentiation, and apoptosis in the aging dentate gyrus, in absence of an age-related hypothalamus-pituitary-adrenal axis activation. *Neurobiol Aging* 25:361–375.

Heine VM, Maslam S, Zareno J, Joels M, Lucassen PJ (2004c). Suppressed proliferation and apoptotic changes in the rat dentate gyrus after acute and chronic stress are reversible. *Eur J Neurosci* 19:131–144.

Heinrich C, Nitta N, Flubacher A, Muller M, Fahrner A, Kirsch M, Freiman T, Suzuki F, Depaulis A, Frotscher M, Haas CA (2006). Reelin deficiency and displacement of mature neurons, but not neurogenesis, underlie the formation of granule cell dispersion in the epileptic hippocampus. *J Neurosci* 26:4701–4713.

Helfer JL, Goodlett CR, Greenough WT, Klintsova AY (2009). The effects of exercise on adolescent hippocampal neurogenesis in a rat model of binge alcohol exposure during the brain growth spurt. *Brain Res* 1294:1–11.

Henry RA, Hughes SM, Connor B (2007). AAV-mediated delivery of BDNF augments neurogenesis in the normal and quinolinic acid-lesioned adult rat brain. *Eur J Neurosci* 25:3513–3525.

Herrera E, Samper E, Blasco MA (1999a). Telomere shortening in mTR-/- embryos is associated with failure to close the neural tube. *Embo J* 18:1172–1181.

Herrera E, Samper E, Martin-Caballero J, Flores JM, Lee HW, Blasco MA (1999b). Disease states associated with telomerase deficiency appear earlier in mice with short telomeres. *Embo J* 18:2950–2960.

Herrick SP, Waters EM, Drake CT, McEwen BS, Milner TA (2006). Extranuclear estrogen receptor beta immunoreactivity is on doublecortin-containing cells in the adult and neonatal rat dentate gyrus. *Brain Res* 1121:46–58.

Hidalgo A, Barami K, Iversen K, Goldman SA (1995). Estrogens and non-estrogenic ovarian influences combine to promote the recruitment and decrease the turnover of new neurons in the adult female canary brain. *J Neurobiol* 27:470–487.

Hill DJ, Han VK (1991). Paracrinology of growth regulation. *J Dev Physiol* 15:91–104.

Hitoshi S, Alexson T, Tropepe V, Donoviel D, Elia AJ, Nye JS, Conlon RA, Mak TW, Bernstein A, van der Kooy D (2002). Notch pathway molecules are essential for the maintenance, but not the generation, of mammalian neural stem cells. *Genes Dev* 16:846–858.

Hitoshi S, Maruta N, Higashi M, Kumar A, Kato N, Ikenaka K (2007). Antidepressant drugs reverse the loss of adult neural stem cells following chronic stress. *J Neurosci Res* 85:3574–3585.

Ho NF, Han SP, Dawe GS (2009). Effect of voluntary running on adult hippocampal neurogenesis in cholinergic lesioned mice. *BMC Neurosci* 10:57.

Hodge RD, Kowalczyk TD, Wolf SA, Encinas JM, Rippey C, Enikolopov G, Kempermann G, Hevner RF (2008). Intermediate progenitors in adult hippocampal neurogenesis: Tbr2 expression and coordinate regulation of neuronal output. *J Neurosci* 28:3707–3717.

Hogervorst E, Yaffe K, Richards M, Huppert F (2002). Hormone replacement therapy for cognitive function in postmenopausal women. *Cochrane Database Syst Rev* CD003122.

Hoglinger GU, Rizk P, Muriel MP, Duyckaerts C, Oertel WH, Caille I, Hirsch EC (2004). Dopamine depletion impairs precursor cell proliferation in Parkinson disease. *Nat Neurosci* 7:726–735.

Hokfelt T, Stanic D, Sanford SD, Gatlin JC, Nilsson I, Paratcha G, Ledda F, Fetissov S, Lindfors C, Herzog H, Johansen JE, Ubink R, Pfenninger KH (2008). NPY and its involvement in axon guidance, neurogenesis, and feeding. *Nutrition (Burbank, Los Angeles County, Calif.)* 24:860–868.

Holmberg J, Armulik A, Senti KA, Edoff K, Spalding K, Momma S, Cassidy R, Flanagan JG, Frisen J (2005). Ephrin-A2 reverse signaling negatively regulates neural progenitor proliferation and neurogenesis. *Genes Dev* 19:462–471.

Holmes MM, Galea LA, Mistlberger RE, Kempermann G (2004). Adult hippocampal neurogenesis and voluntary running activity: Circadian and dose-dependent effects. *J Neurosci Res* 76:216–222.

Hong SM, Liu Z, Fan Y, Neumann M, Won SJ, Lac D, Lum X, Weinstein PR, Liu J (2007). Reduced hippocampal neurogenesis and skill reaching performance in adult Emx1 mutant mice. *Exp Neurol* 206:24–32.

Houser CR (2007). Interneurons of the dentate gyrus: An overview of cell types, terminal fields and neurochemical identity. *Prog Brain Res* 163:217–232.

Howell OW, Doyle K, Goodman JH, Scharfman HE, Herzog H, Pringle A, Beck-Sickinger AG, Gray WP (2005). Neuropeptide Y stimulates neuronal precursor proliferation in the post-natal and adult dentate gyrus. *J Neurochem* 93:560–570.

Howell OW, Silva S, Scharfman HE, Sosunov AA, Zaben M, Shatya A, McKhann G, 2nd, Herzog H, Laskowski A, Gray WP (2007). Neuropeptide Y is important for basal and seizure-induced precursor cell proliferation in the hippocampus. *Neurobiol Dis* 26:174–188.

Hsieh J, Gage FH (2004). Epigenetic control of neural stem cell fate. *Curr Opin Genet Dev* 14:461–469.

Hsieh J, Aimone JB, Kaspar BK, Kuwabara T, Nakashima K, Gage FH (2004). IGF-I instructs multipotent adult neural progenitor cells to become oligodendrocytes. *J Cell Biol* 164:111–122.

Huang GJ, Herbert J (2005a). The role of 5-HT1A receptors in the proliferation and survival of progenitor cells in the dentate gyrus of the adult hippocampus and their regulation by corticoids. *Neuroscience* 135:803–813.

Huang GJ, Herbert J (2005b). Serotonin modulates the suppressive effects of corticosterone on proliferating progenitor cells in the dentate gyrus of the hippocampus in the adult rat. *Neuropsychopharmacology* 30:231–241.

Ikeya M, Lee SM, Johnson JE, McMahon AP, Takada S (1997). Wnt signalling required for expansion of neural crest and CNS progenitors. *Nature* 389:966–970.

Imayoshi I, Sakamoto M, Ohtsuka T, Takao K, Miyakawa T, Yamaguchi M, Mori K, Ikeda T, Itohara S, Kageyama R (2008). Roles of continuous neurogenesis in the structural and functional integrity of the adult forebrain. *Nat Neurosci* 11:1153–1161.

Islam AT, Kuraoka A, Kawabuchi M (2003). Morphological basis of nitric oxide production and its correlation with the polysialylated precursor cells in the dentate gyrus of the adult guinea pig hippocampus. *Anat Sci Int* 78:98–103.

Itoh H, Hirata K, Ohsato K (1993). Turcot's syndrome and familial adenomatous polyposis associated with brain tumor: Review of related literature. *Int J Colorectal Dis* 8:87–94.

Jagasia R, Steib K, Englberger E, Herold S, Faus-Kessler T, Saxe M, Gage FH, Song H, Lie DC (2009). GABA-cAMP response element-binding protein signaling regulates maturation and survival of newly generated neurons in the adult hippocampus. *J Neurosci* 29:7966–7977.

Jayatissa MN, Bisgaard C, Tingstrom A, Papp M, Wiborg O (2006). Hippocampal cytogenesis correlates to escitalopram-mediated recovery in a chronic mild stress rat model of depression. *Neuropsychopharmacology* 31:2395–2404.

Jakubs K, Bonde S, Iosif RE, Ekdahl CT, Kokaia Z, Kokaia M, Lindvall O (2008). Inflammation regulates functional integration of neurons born in adult brain. *J Neurosci* 28:12477–12488.

Jessberger S, Kempermann G (2003). Adult-born hippocampal neurons mature into activity-dependent responsiveness. *Eur J Neurosci* 18:2707–2712.

Jessberger S, Romer B, Babu H, Kempermann G (2005). Seizures induce proliferation and dispersion of doublecortin-positive hippocampal progenitor cells. *Exp Neurol* 196:342–351.

Jessberger S, Nakashima K, Clemenson GD, Jr., Mejia E, Mathews E, Ure K, Ogawa S, Sinton CM, Gage FH, Hsieh J (2007). Epigenetic modulation of seizure-induced neurogenesis and cognitive decline. *J Neurosci* 27:5967–5975.

Jessberger S, Toni N, Clemenson Jr GD, Ray J, Gage FH (2008a). Directed differentiation of hippocampal stem/progenitor cells in the adult brain. *Nat Neurosci* 11:888–893.

Jessberger S, Aigner S, Clemenson GD, Toni N, Lie DC, Karalay O, Overall R, Kempermann G, Gage FH (2008b). Cdk5 regulates accurate maturation of newborn granule cells in the adult hippocampus. *PLoS Biol* 6:e272.

Jessberger S, Clark RE, Broadbent NJ, Clemenson GD, Jr., Consiglio A, Lie DC, Squire LR, Gage FH (2009). Dentate gyrus-specific knockdown of adult neurogenesis impairs spatial and object recognition memory in adult rats. *Learn Mem* 16:147–154.

Jiang W, Wan Q, Zhang ZJ, Wang WD, Huang YG, Rao ZR, Zhang X (2003). Dentate granule cell neurogenesis after seizures induced by pentylenetrazol in rats. *Brain Res* 977:141–148.

Jiang W, Zhang Y, Xiao L, Van Cleemput J, Ji SP, Bai G, Zhang X (2005). Cannabinoids promote embryonic and adult hippocampus neurogenesis and produce anxiolytic- and antidepressant-like effects. *J Clin Invest* 115:3104–3116.

Jin J, Jing H, Choi G, Oh MS, Ryu JH, Jeong JW, Huh Y, Park C (2008). Voluntary exercise increases the new cell formation in the hippocampus of ovariectomized mice. *Neurosci Lett* 439:260–263.

Jin K, Mao XO, Sun Y, Xie L, Greenberg DA (2002a). Stem cell factor stimulates neurogenesis in vitro and in vivo. *J Clin Invest* 110:311–319.

Jin K, Zhu Y, Sun Y, Mao XO, Xie L, Greenberg DA (2002b). Vascular endothelial growth factor (VEGF) stimulates neurogenesis in vitro and in vivo. *Proc Natl Acad Sci USA* 99:11946–11950.

Jin K, Sun Y, Xie L, Batteur S, Mao XO, Smelick C, Logvinova A, Greenberg DA (2003). Neurogenesis and aging: FGF2 and HB-EGF restore neurogenesis in hippocampus and subventricular zone of aged mice. *Aging Cell* 2:175–183.

Jin K, Xie L, Kim SH, Parmentier-Batteur S, Sun Y, Mao XO, Childs J, Greenberg DA (2004a). Defective adult neurogenesis in CB1 cannabinoid receptor knockout mice. *Mol Pharmacol* 66:204–208.

Jin K, Minami M, Xie L, Sun Y, Mao XO, Wang Y, Simon RP, Greenberg DA (2004b). Ischemia-induced neurogenesis is preserved but reduced in the aged rodent brain. *Aging Cell* 3:373–377.

Kahn L, Alonso G, Normand E, Manzoni OJ (2005). Repeated morphine treatment alters polysialylated neural cell adhesion molecule, glutamate decarboxylase-67 expression and cell proliferation in the adult rat hippocampus. *Eur J Neurosci* 21:493–500.

Kam M, Curtis MA, McGlashan SR, Connor B, Nannmark U, Faull RL (2009). The cellular composition and morphological organization of the rostral migratory stream in the adult human brain. *J Chem Neuroanat* 37:196–205.

Kan L, Israsena N, Zhang Z, Hu M, Zhao LR, Jalali A, Sahni V, Kessler JA (2004). Sox1 acts through multiple independent pathways to promote neurogenesis. *Dev Biol* 269:580–594.

Kaneko N, Okano H, Sawamoto K (2006a). Role of the cholinergic system in regulating survival of newborn neurons in the adult mouse dentate gyrus and olfactory bulb. *Genes Cells* 11:1145–1159.

Kaneko N, Kudo K, Mabuchi T, Takemoto K, Fujimaki K, Wati H, Iguchi H, Tezuka H, Kanba S (2006b). Suppression of cell proliferation by interferon-alpha through interleukin-1 production in adult rat dentate gyrus. *Neuropsychopharmacology* 31:2619–2626.

Kannangara TS, Webber A, Gil-Mohapel J, Christie BR (2009). Stress differentially regulates the effects of voluntary exercise on cell proliferation in the dentate gyrus of mice. *Hippocampus* 19:889–897.

Kao HT, Li P, Chao HM, Janoschka S, Pham K, Feng J, McEwen BS, Greengard P, Pieribone VA, Porton B (2008). Early involvement of synapsin III in neural progenitor cell development in the adult hippocampus. *J Comp Neurol* 507:1860–1870.

Karishma KK, Herbert J (2002). Dehydroepiandrosterone (DHEA) stimulates neurogenesis in the hippocampus of the rat, promotes survival of newly formed neurons and prevents corticosterone-induced suppression. *Eur J Neurosci* 16:445–453.

Katsimpardi L, Gaitanou M, Malnou CE, Lledo PM, Charneau P, Matsas R, Thomaidou D (2008). BM88/Cend1 expression levels are critical for proliferation and differentiation of subventricular zone-derived neural precursor cells. *Stem Cells* 26:1796–1807.

Kawai T, Takagi N, Nakahara M, Takeo S (2005). Changes in the expression of Hes5 and Mash1 mRNA in the adult rat dentate gyrus after transient forebrain ischemia. *Neurosci Lett* 380:17–20.

Kawakita E, Hashimoto M, Shido O (2006). Docosahexaenoic acid promotes neurogenesis in vitro and in vivo. *Neuroscience* 139:991–997.

Kee N, Teixeira CM, Wang AH, Frankland PW (2007). Preferential incorporation of adult-generated granule cells into spatial memory networks in the dentate gyrus. *Nat Neurosci* 10:355–362.

Keene CD, Chang R, Stephen C, Nivison M, Nutt SE, Look A, Breyer RM, Horner PJ, Hevner R, Montine TJ (2009). Protection of hippocampal neurogenesis from toll-like receptor 4–dependent innate immune activation by ablation of prostaglandin E2 receptor subtype EP1 or EP2. *Am J Pathol* 174:2300–2309.

Keilhoff G, Becker A, Grecksch G, Bernstein HG, Wolf G (2006). Cell proliferation is influenced by bulbectomy and normalized by imipramine treatment in a region-specific manner. *Neuropsychopharmacology* 31:1165–1176.

Kempermann G (2002). Why new neurons? Possible functions for adult hippocampal neurogenesis. *J Neurosci* 22:635–638.

Kempermann G (2008). The neurogenic reserve hypothesis: what is adult hippocampal neurogenesis good for? *Trends Neurosci* 31:163–169.

Kempermann G, Gage FH (1999). Experience-dependent regulation of adult hippocampal neurogenesis: Effects of long-term stimulation and stimulus withdrawal. *Hippocampus* 9:321–332.

Kempermann G, Gage FH (2002a). Genetic influence on phenotypic differentiation in adult hippocampal neurogenesis. *Brain Res Dev Brain Res* 134:1–12.

Kempermann G, Gage FH (2002b). Genetic determinants of adult hippocampal neurogenesis correlate with acquisition, but not probe trial performance in the water maze task. *Eur J Neurosci* 16:129–136.

Kempermann G, Neumann H (2003). Neuroscience. Microglia: The enemy within? *Science* 302:1689–1690.

Kempermann G, Kuhn HG, Gage FH (1997a). More hippocampal neurons in adult mice living in an enriched environment. *Nature* 386:493–495.

Kempermann G, Kuhn HG, Gage FH (1997b). Genetic influence on neurogenesis in the dentate gyrus of adult mice. *Proc Natl Acad Sci USA* 94:10409–10414.

Kempermann G, Kuhn HG, Gage FH (1998a). Experience-induced neurogenesis in the senescent dentate gyrus. *J Neurosci* 18:3206–3212.

Kempermann G, Brandon EP, Gage FH (1998b). Environmental stimulation of 129/SvJ mice causes increased cell proliferation and neurogenesis in the adult dentate gyrus. *Curr Biol* 8:939–942.

Kempermann G, Gast D, Gage FH (2002). Neuroplasticity in old age: Sustained fivefold induction of hippocampal neurogenesis by long-term environmental enrichment. *Ann Neurol* 52:135–143.

Kempermann G, Gast D, Kronenberg G, Yamaguchi M, Gage FH (2003). Early determination and long-term persistence of adult-generated new neurons in the hippocampus of mice. *Development* 130:391–399.

Kempermann G, Wiskott L, Gage FH (2004a). Functional significance of adult neurogenesis. *Curr Opin Neurobiol* 14:186–191.

Kempermann G, Jessberger S, Steiner B, Kronenberg G (2004b). Milestones of neuronal development in the adult hippocampus. *Trends Neurosci* 27:447–452.

Kempermann G, Chesler EJ, Lu L, Williams RW, Gage FH (2006). Natural variation and genetic covariance in adult hippocampal neurogenesis. *Proc Natl Acad Sci USA* 103:780–785.

Kim EJ, Battiste J, Nakagawa Y, Johnson JE (2008a). Ascl1 (Mash1) lineage cells contribute to discrete cell populations in CNS architecture. *Mol Cell Neurosci* 38:595–606.

Kim JY, Duan X, Liu CY, Jang MH, Guo JU, Pow-anpongkul N, Kang E, Song H, Ming GL (2009). DISC1 regulates new neuron development in the adult brain via modulation of AKT-mTOR signaling through KIAA1212. *Neuron* 63:761–773.

Kim SH, Won SJ, Mao XO, Ledent C, Jin K, Greenberg DA (2006). Role for neuronal nitric-oxide synthase in cannabinoid-induced neurogenesis. *J Pharmacol Exp Ther* 319:150–154.

Kim SJ, Son TG, Park HR, Park M, Kim MS, Kim HS, Chung HY, Mattson MP, Lee J (2008b). Curcumin stimulates proliferation of embryonic neural progenitor cells and neurogenesis in the adult hippocampus. *J Biol Chem* 283:14497–14505.

Kim WR, Kim Y, Eun B, Park OH, Kim H, Kim K, Park CH, Vinsant S, Oppenheim RW, Sun W (2007). Impaired migration in the rostral migratory stream but spared olfactory function after the elimination of programmed cell death in Bax knock-out mice. *J Neurosci* 27:14392–14403.

Kim YP, Kim H, Shin MS, Chang HK, Jang MH, Shin MC, Lee SJ, Lee HH, Yoon JH, Jeong IG, Kim CJ (2004). Age-dependence of the effect of treadmill exercise on cell proliferation in the dentate gyrus of rats. *Neurosci Lett* 355:152–154.

Kimiwada T, Sakurai M, Ohashi H, Aoki S, Tominaga T, Wada K (2009). Clock genes regulate neurogenic transcription factors, including NeuroD1, and the neuronal differentiation of adult neural stem/progenitor cells. *Neurochem Int* 54:277–285.

Kimura KD, Tissenbaum HA, Liu Y, Ruvkun G (1997). daf-2, an insulin receptor-like gene that regulates longevity and diapause in *Caenorhabditis elegans*. *Science* 277:942–946.

Kipnis J, Schwartz M (2005). Controlled autoimmunity in CNS maintenance and repair: Naturally occurring CD4+CD25+ regulatory T-Cells at the crossroads of health and disease. *Neuromol Med* 7:197–206.

Kippin TE, Cain SW, Masum Z, Ralph MR (2004). Neural stem cells show bidirectional experience-dependent plasticity in the perinatal mammalian brain. *J Neurosci* 24:2832–2836.

Klaus F, Hauser T, Slomianka L, Lipp HP, Amrein I (2009). A reward increases running-wheel performance without changing cell proliferation, neuronal differentiation or cell death in the dentate gyrus of C57BL/6 mice. *Behav Brain Res* 204:175–181.

Kleindienst A, McGinn MJ, Harvey HB, Colello RJ, Hamm RJ, Bullock MR (2005). Enhanced hippocampal neurogenesis by intraventricular S100B infusion is associated with improved cognitive recovery after traumatic brain injury. *J Neurotrauma* 22:645–655.

Koehl M, Darnaudery M, Dulluc J, Van Reeth O, Le Moal M, Maccari S (1999). Prenatal stress alters circadian activity of hypothalamo-pituitary-adrenal axis and hippocampal corticosteroid receptors in adult rats of both gender. *J Neurobiol* 40:302–315.

Koehl M, Meerlo P, Gonzales D, Rontal A, Turek FW, Abrous DN (2008). Exercise-induced promotion of hippocampal cell proliferation requires beta-endorphin. *Faseb J* 22:2253–2262.

Kohl Z, Kuhn HG, Cooper-Kuhn CM, Winkler J, Aigner L, Kempermann G (2002). Preweaning enrichment has no lasting effects on adult hippocampal neurogenesis in four-month-old mice. *Genes Brain Behav* 1:46–54.

Kohl Z, Kandasamy M, Winner B, Aigner R, Gross C, Couillard-Despres S, Bogdahn U, Aigner L, Winkler J (2007). Physical activity fails to rescue hippocampal neurogenesis deficits in the R6/2 mouse model of Huntington's disease. *Brain Res* 1155:24–33.

Kohtz JD, Baker DP, Corte G, Fishell G (1998). Regionalization within the mammalian telencephalon is mediated by changes in responsiveness to Sonic hedgehog. *Development* 125:5079–5089.

Kohwi M, Osumi N, Rubenstein JL, Alvarez-Buylla A (2005). Pax6 is required for making specific subpopulations of granule and periglomerular neurons in the olfactory bulb. *J Neurosci* 25:6997–7003.

Kolodziej A, Schulz S, Guyon A, Wu DF, Pfeiffer M, Odemis V, Hollt V, Stumm R (2008). Tonic activation of CXC chemokine receptor 4 in immature granule cells supports neurogenesis in the adult dentate gyrus. *J Neurosci* 28:4488–4500.

Komitova M, Eriksson PS (2004). Sox-2 is expressed by neural progenitors and astroglia in the adult rat brain. *Neurosci Lett* 369:24–27.

Kong H, Sha LL, Fan Y, Xiao M, Ding JH, Wu J, Hu G (2009). Requirement of AQP4 for antidepressive efficiency of fluoxetine: Implication in adult hippocampal neurogenesis. *Neuropsychopharmacology* 34:1263–1276.

Korstanje R, Paigen B (2002). From QTL to gene: The harvest begins. *Nat Genet* 31:235–236.

Korte M, Carroll P, Wolf E, Brem G, Thoenen H, Bonhoeffer T (1995). Hippocampal long-term potentiation is impaired in mice lacking brain-derived neurotrophic factor. *Proc Natl Acad Sci USA* 92:8856–8860.

Kotani S, Yamauchi T, Teramoto T, Ogura H (2006). Pharmacological evidence of cholinergic involvement in adult hippocampal neurogenesis in rats. *Neuroscience* 142:505–514.

Kowalczyk A, Filipkowski RK, Rylski M, Wilczynski GM, Konopacki FA, Jaworski J, Ciemerych MA, Sicinski P, Kaczmarek L (2004). The critical role of cyclin D2 in adult neurogenesis. *J Cell Biol* 167:209–213.

Kozlovsky N, Belmaker RH, Agam G (2002). GSK-3 and the neurodevelopmental hypothesis of schizophrenia. *Eur Neuropsychopharmacol* 12:13–25.

Kronenberg G, Reuter K, Steiner B, Brandt MD, Jessberger S, Yamaguchi M, Kempermann G (2003). Subpopulations of proliferating cells of the adult hippocampus respond differently to physiologic neurogenic stimuli. *J Comp Neurol* 467:455–463.

Kronenberg G, Bick-Sander A, Bunk E, Wolf C, Ehninger D, Kempermann G (2006). Physical exercise prevents age-related decline in precursor cell activity in the mouse dentate gyrus. *Neurobiol Aging* 27:1505–1513.

Kronenberg G, Lippoldt A, Kempermann G (2007a). Two genetic rat models of arterial hypertension show different mechanisms by which adult hippocampal neurogenesis is increased. *Dev Neurosci* 29:124–133.

Kronenberg G, Wang LP, Geraerts M, Babu H, Synowitz M, Vicens P, Lutsch G, Glass R, Yamaguchi M, Baekelandt V, Debyser Z, Kettenmann H, Kempermann G (2007b). Local origin and activity-dependent generation of nestin-expressing protoplasmic astrocytes in CA1. *Brain Structure & Function* 212:19–35.

Kronenberg G, Harms C, Sobol RW, Cardozo-Pelaez F, Linhart H, Winter B, Balkaya M, Gertz K, Gay SB, Cox D, Eckart S, Ahmadi M, Juckel G, Kempermann G, Hellweg R, Sohr R, Hortnagl H, Wilson SH, Jaenisch R, Endres M (2008). Folate deficiency induces neurodegeneration and brain dysfunction in mice lacking uracil DNA glycosylase. *J Neurosci* 28:7219–7230.

Kronenberg G, Kirste I, Inta D, Chourbaji S, Heuser I, Endres M, Gass P (2009). Reduced hippocampal neurogenesis in the GR(+/-) genetic mouse model of depression. *Eur Arch Psychiatry Clin Neurosci* 259:499-504.

Kruman, II, Mouton PR, Emokpae R, Jr., Cutler RG, Mattson MP (2005). Folate deficiency inhibits proliferation of adult hippocampal progenitors. *Neuroreport* 16:1055–1059.

Kuhlmann T, Miron V, Cuo Q, Wegner C, Antel J, Bruck W (2008). Differentiation block of oligodendroglial progenitor cells as a cause for remyelination failure in chronic multiple sclerosis. *Brain* 131:1749–1758.

Kuhn HG, Dickinson-Anson H, Gage FH (1996). Neurogenesis in the dentate gyrus of the adult rat: Age-related decrease of neuronal progenitor proliferation. *J Neurosci* 16:2027–2033.

Kuhn HG, Winkler J, Kempermann G, Thal LJ, Gage FH (1997). Epidermal growth factor and fibroblast growth factor-2 have different effects on neural progenitors in the adult rat brain. *J Neurosci* 17:5820–5829.

Kuhn HG, Biebl M, Wilhelm D, Li M, Friedlander RM, Winkler J (2005). Increased generation of granule cells in adult Bcl-2–overexpressing mice: A role for cell death during continued hippocampal neurogenesis. *Eur J Neurosci* 22:1907–1915.

Kulkarni VA, Jha S, Vaidya VA (2002). Depletion of norepinephrine decreases the proliferation, but does not influence the survival and differentiation, of granule cell progenitors in the adult rat hippocampus. *Eur J Neurosci* 16:2008–2012.

Kumar S, Parkash J, Kataria H, Kaur G (2009). Interactive effect of excitotoxic injury and dietary restriction on neurogenesis and neurotrophic factors in adult male rat brain. *Neurosci Res* 65:367–374.

Kumihashi K, Uchida K, Miyazaki H, Kobayashi J, Tsushima T, Machida T (2001). Acetylsalicylic acid reduces ischemia-induced proliferation of dentate cells in gerbils. *Neuroreport* 12:915–917.

Kunze A, Congreso MR, Hartmann C, Wallraff-Beck A, Huttmann K, Bedner P, Requardt R, Seifert G, Redecker C, Willecke K, Hofmann A, Pfeifer A, Theis M, Steinhauser C (2009). Connexin expression by radial glia–like cells is required for neurogenesis in the adult dentate gyrus. *Proc Natl Acad Sci USA* 106:11336–41

Kuwabara T, Hsieh J, Nakashima K, Taira K, Gage FH (2004). A small modulatory dsRNA specifies the fate of adult neural stem cells. *Cell* 116:779–793.

Kuwabara T, Hsieh J, Nakashima K, Warashina M, Taira K, Gage FH (2005). The NRSE smRNA specifies the fate of adult hippocampal neural stem cells. *Nucleic Acids Symposium Series* (2004):87–88.

Kuwabara T, Hsieh J, Muotri A, Yeo G, Warashina M, Lie DC, Moore L, Nakashima K, Asashima M, Gage FH (2009). Wnt-mediated activation of NeuroD1 and retro-elements during adult neurogenesis. *Nat Neurosci* 12:1097–1105.

La Porta CA, Gena P, Gritti A, Fascio U, Svelto M, Calamita G (2006). Adult murine CNS stem cells express aquaporin channels. *Biol Cell* 98:89–94.

Lagace DC, Benavides DR, Kansy JW, Mapelli M, Greengard P, Bibb JA, Eisch AJ (2008). Cdk5 is essential for adult hippocampal neurogenesis. *Proc Natl Acad Sci USA* 105:18567–18571.

Lai KO, Ip NY (2009). Synapse development and plasticity: Roles of ephrin/Eph receptor signaling. *Curr Opin Neurobiol* 19:275–283.

Land PW, Monaghan AP (2003). Expression of the transcription factor, tailless, is required for formation of superficial cortical layers. *Cereb Cortex* 13:921–931.

Lange C, Huttner WB, Calegari F (2009). Cdk4/cyclinD1 overexpression in neural stem cells shortens G1, delays neurogenesis, and promotes the generation and expansion of basal progenitors. *Cell Stem Cell* 5:320–331.

Langer D, Ikehara Y, Takebayashi H, Hawkes R, Zimmermann H (2007). The ectonucleotidases alkaline phosphatase and nucleoside triphosphate diphosphohydrolase 2 are associated with subsets of progenitor cell populations in the mouse embryonic, postnatal and adult neurogenic zones. *Neuroscience* 150:863–879.

Larsen CM, Kokay IC, Grattan DR (2008). Male pheromones initiate prolactin-induced neurogenesis and advance maternal behavior in female mice. *Horm Behav* 53:509–517.

Larsson E, Mandel RJ, Klein RL, Muzyczka N, Lindvall O, Kokaia Z (2002). Suppression of insult-induced neurogenesis in adult rat brain by brain-derived neurotrophic factor. *Exp Neurol* 177:1–8.

Lassalle JM, Halley H, Roullet P (1994). Analysis of behavioral and hippocampal variation in congenic albino and pigmented BALB mice. *Behav Genet* 24:161–169.

Leanza G, Muir J, Nilsson OG, Wiley RG, Dunnett SB, Bjorklund A (1996). Selective immunolesioning of the basal forebrain cholinergic system disrupts short-term memory in rats. *Eur J Neurosci* 8:1535–1544.

Lebrand C, Cases O, Wehrle R, Blakely RD, Edwards RH, Gaspar P (1998). Transient developmental expression of monoamine transporters in the rodent forebrain. *J Comp Neurol* 401:506–524.

Lee HW, Blasco MA, Gottlieb GJ, Horner JW, 2nd, Greider CW, DePinho RA (1998). Essential role of mouse telomerase in highly proliferative organs. *Nature* 392:569–574.

Lee HY, Kleber M, Hari L, Brault V, Suter U, Taketo MM, Kemler R, Sommer L (2004). Instructive role of Wnt/beta-catenin in sensory fate specification in neural crest stem cells. *Science* 303:1020–1023.

Lee J, Duan W, Long JM, Ingram DK, Mattson MP (2000a). Dietary restriction increases the number of newly generated neural cells, and induces BDNF expression, in the dentate gyrus of rats. *J Mol Neurosci* 15:99–108.

Lee J, Seroogy KB, Mattson MP (2002a). Dietary restriction enhances neurotrophin expression and neurogenesis in the hippocampus of adult mice. *J Neurochem* 80:539–547.

Lee J, Duan W, Mattson MP (2002b). Evidence that brain-derived neurotrophic factor is required for basal neurogenesis and mediates, in part, the enhancement of neurogenesis by dietary restriction in the hippocampus of adult mice. *J Neurochem* 82:1367–1375.

Lee JS, Jang DJ, Lee N, Ko HG, Kim H, Kim YS, Kim B, Son J, Kim SH, Chung H, Lee MY, Kim WR, Sun W, Zhuo M, Abel T, Kaang BK, Son H (2009). Induction of neuronal vascular endothelial growth factor expression by cAMP in the dentate gyrus of the hippocampus is required for antidepressant-like behaviors. *J Neurosci* 29:8493–8505.

Lee SM, Tole S, Grove E, McMahon AP (2000b). A local Wnt-3a signal is required for development of the mammalian hippocampus. *Development* 127:457–467.

Lemaire V, Koehl M, Le Moal M, Abrous DN (2000). Prenatal stress produces learning deficits associated with an inhibition of neurogenesis in the hippocampus. *Proc Natl Acad Sci USA* 97:11032–11037.

Leuner B, Mendolia-Loffredo S, Kozorovitskiy Y, Samburg D, Gould E, Shors TJ (2004). Learning enhances the survival of new neurons beyond the time when the hippocampus is required for memory. *J Neurosci* 24:7477–7481.

Leuner B, Mirescu C, Noiman L, Gould E (2007). Maternal experience inhibits the production of immature neurons in the hippocampus during the postpartum period through elevations in adrenal steroids. *Hippocampus* 17:434–442.

Leung LS, Shen B, Rajakumar N, Ma J (2003). Cholinergic activity enhances hippocampal long-term potentiation in CA1 during walking in rats. *J Neurosci* 23:9297–9304.

Lewin GR, Barde YA (1996). Physiology of the neurotrophins. *Annu Rev Neurosci* 19:289–317.

Li X, Barkho BZ, Luo Y, Smrt RD, Santistevan NJ, Liu C, Kuwabara T, Gage FH, Zhao X (2008). Epigenetic regulation of the stem cell mitogen FGF2 by Mbd1 in adult neural stem/progenitor cells. *J Biol Chem* 283:27644–27652.

Li Z, Kato T, Kawagishi K, Fukushima N, Yokouchi K, Moriizumi T (2002). Cell dynamics of calretinin-immunoreactive neurons in the rostral migratory stream after ibotenate-induced lesions in the forebrain. *Neurosci Res* 42:123–132.

Lichtenwalner RJ, Forbes ME, Bennett SA, Lynch CD, Sonntag WE, Riddle DR (2001). Intracerebroventricular infusion of insulin-like growth factor-I ameliorates the age-related decline in hippocampal neurogenesis. *Neuroscience* 107:603–613.

Lichtenwalner RJ, Forbes ME, Sonntag WE, Riddle DR (2006). Adult-onset deficiency in growth hormone and insulin-like growth factor-I decreases survival of dentate granule neurons: Insights into the regulation of adult hippocampal neurogenesis. *J Neurosci Res* 83:199–210.

Lie DC, Colamarino SA, Song HJ, Desire L, Mira H, Consiglio A, Lein ES, Jessberger S, Lansford H, Dearie AR, Gage FH (2005). Wnt signalling regulates adult hippocampal neurogenesis. *Nature* 437:1370–1375.

Liem KF, Jr., Tremml G, Roelink H, Jessell TM (1995). Dorsal differentiation of neural plate cells induced by BMP-mediated signals from epidermal ectoderm. *Cell* 82:969–979.

Lim DA, Tramontin AD, Trevejo JM, Herrera DG, Garcia-Verdugo JM, Alvarez-Buylla A (2000). Noggin antagonizes BMP signaling to create a niche for adult neurogenesis. *Neuron* 28:713–726.

Lim DA, Suarez-Farinas M, Naef F, Hacker CR, Menn B, Takebayashi H, Magnasco M, Patil N, Alvarez-Buylla A (2006). In vivo transcriptional profile analysis reveals RNA splicing and chromatin remodeling as prominent processes for adult neurogenesis. *Mol Cell Neurosci* 31:131–148.

Lim DA, Huang YC, Swigut T, Mirick AL, Garcia-Verdugo JM, Wysocka J, Ernst P, Alvarez-Buylla A (2009). Chromatin remodelling factor Mll1 is essential for neurogenesis from postnatal neural stem cells. *Nature* 458:529–533.

Limke TL, Cai J, Miura T, Rao MS, Mattson MP (2003). Distinguishing features of progenitor cells in the late embryonic and adult hippocampus. *Dev Neurosci* 25:257–272.

Limoli CL, Giedzinski E, Rola R, Otsuka S, Palmer TD, Fike JR (2004). Radiation response of neural precursor cells: Linking cellular sensitivity to cell cycle checkpoints, apoptosis and oxidative stress. *Radiat Res* 161:17–27.

Lindqvist A, Mohapel P, Bouter B, Frielingsdorf H, Pizzo D, Brundin P, Erlanson-Albertsson C (2006). High-fat diet impairs hippocampal neurogenesis in male rats. *Eur J Neurol* 13:1385–1388.

Liu C, Zhao X (2009). MicroRNAs in adult and embryonic neurogenesis. *Neuromolecular Med* 11:141–152.

Liu HK, Belz T, Bock D, Takacs A, Wu H, Lichter P, Chai M, Schutz G (2008). The nuclear receptor tailless is required for neurogenesis in the adult subventricular zone. *Genes Dev* 22:2473–2478.

Liu RR, Brown CE, Murphy TH (2007). Differential regulation of cell proliferation in neurogenic zones in mice lacking cystine transport by xCT. *Biochem Biophys Res Commun* 364:528–533.

Lou SJ, Liu JY, Chang H, Chen PJ (2008). Hippocampal neurogenesis and gene expression depend on exercise intensity in juvenile rats. *Brain Res* 1210:48–55.

Louissaint A, Jr., Rao S, Leventhal C, Goldman SA (2002). Coordinated interaction of neurogenesis and angiogenesis in the adult songbird brain. *Neuron* 34:945–960.

Low LF, Anstey KJ (2006). Hormone replacement therapy and cognitive performance in postmenopausal women—a review by cognitive domain. *Neurosci Biobehav Rev* 30:66–84.

Loy R, Koziell DA, Lindsey JD, Moore RY (1980). Noradrenergic innervation of the adult rat hippocampal formation. *J Comp Neurol* 189:699–710.

Lu D, Mahmood A, Zhang R, Copp M (2003a). Upregulation of neurogenesis and reduction in functional deficits following administration of DEtA/NONOate, a nitric oxide donor, after traumatic brain injury in rats. *J Neurosurg* 99:351–361.

Lu L, Bao G, Chen H, Xia P, Fan X, Zhang J, Pei G, Ma L (2003b). Modification of hippocampal neurogenesis and neuroplasticity by social environments. *Exp Neurol* 183:600–609.

Lu M, Grove EA, Miller RJ (2002). Abnormal development of the hippocampal dentate gyrus in mice lacking the CXCR4 chemokine receptor. *Proc Natl Acad Sci USA* 99:7090–7095.

Lucassen PJ, Heine VM, Muller MB, van der Beek EM, Wiegant VM, De Kloet ER, Joels M, Fuchs E, Swaab DF, Czeh B (2006). Stress, depression and hippocampal apoptosis. *CNS Neurol Disord Drug Targets* 5:531–546.

Lunyak VV, Rosenfeld MG (2005). No rest for REST: REST/NRSF regulation of neurogenesis. *Cell* 121:499–501.

Ma D, Lu L, Boneva NB, Warashina S, Kaplamadzhiev DB, Mori Y, Nakaya MA, Kikuchi M, Tonchev AB, Okano H, Yamashima T (2008). Expression of free fatty acid receptor GPR40 in the neurogenic niche of adult monkey hippocampus. *Hippocampus* 18:326–333.

Ma DK, Ponnusamy K, Song MR, Ming GL, Song H (2009a). Molecular genetic analysis of FGFR1 signalling reveals distinct roles of MAPK and PLCgamma1 activation for self-renewal of adult neural stem cells. *Mol Brain* 2:16.

Ma DK, Jang MH, Guo JU, Kitabatake Y, Chang ML, Pow-Anpongkul N, Flavell RA, Lu B, Ming GL, Song H (2009b). Neuronal activity-induced Gadd45b promotes epigenetic DNA demethylation and adult neurogenesis. *Science* 323:1074–1077.

Ma L, Yin M, Wu X, Wu C, Yang S, Sheng J, Ni H, Fukuda MN, Zhou J (2006). Expression of trophinin and bystin identifies distinct cell types in the germinal zones of adult rat brain. *Eur J Neurosci* 23:2265–2276.

Machold R, Hayashi S, Rutlin M, Muzumdar MD, Nery S, Corbin JG, Gritli-Linde A, Dellovade T, Porter JA, Rubin LL, Dudek H, McMahon AP, Fishell G (2003). Sonic hedgehog is required for progenitor cell maintenance in telencephalic stem cell niches. *Neuron* 39:937–950.

Mackowiak M, O'Neill MJ, Hicks CA, Bleakman D, Skolnick P (2002). An AMPA receptor potentiator modulates hippocampal expression of BDNF: An in vivo study. *Neuropharmacology* 43:1–10.

Mackowiak M, Chocyk A, Markowicz-Kula K, Wedzony K (2007). Acute activation of CB1 cannabinoid receptors transiently decreases PSA-NCAM expression in the dentate gyrus of the rat hippocampus. *Brain Res* 1148:43–52.

Madsen TM, Newton SS, Eaton ME, Russell DS, Duman RS (2003). Chronic electroconvulsive seizure up-regulates beta-catenin expression in rat hippocampus: Role in adult neurogenesis. *Biol Psychiatry* 54:1006–1014.

Maekawa M, Takashima N, Arai Y, Nomura T, Inokuchi K, Yuasa S, Osumi N (2005). Pax6 is required for production and maintenance of progenitor cells in postnatal hippocampal neurogenesis. *Genes Cells* 10:1001–1014.

Magavi SS, Mitchell BD, Szentirmai O, Carter BS, Macklis JD (2005). Adult-born and preexisting olfactory granule neurons undergo distinct experience-dependent modifications of their olfactory responses in vivo. *J Neurosci* 25:10729–10739.

Maisonpierre PC, Belluscio L, Friedman B, Alderson RF, Wiegand SJ, Furth ME, Lindsay RM, Yancopoulos GD (1990). NT-3, BDNF, and NGF in the developing rat nervous system: Parallel as well as reciprocal patterns of expression. *Neuron* 5:501–509.

Mak GK, Enwere EK, Gregg C, Pakarainen T, Poutanen M, Huhtaniemi I, Weiss S (2007). Male pheromone-stimulated neurogenesis in the adult female brain: Possible role in mating behavior. *Nat Neurosci* 10:1003–1011.

Malas S, Postlethwaite M, Ekonomou A, Whalley B, Nishiguchi S, Wood H, Meldrum B, Constanti A, Episkopou V (2003). Sox1-deficient mice suffer from epilepsy associated with abnormal ventral forebrain development and olfactory cortex hyperexcitability. *Neuroscience* 119:421–432.

Malberg JE, Duman RS (2003). Cell proliferation in adult hippocampus is decreased by inescapable stress: Reversal by fluoxetine treatment. *Neuropsychopharmacology* 28:1562–1571.

Malberg JE, Eisch AJ, Nestler EJ, Duman RS (2000). Chronic antidepressant treatment increases neurogenesis in adult rat hippocampus. *J Neurosci* 20:9104–9110.

Mandairon N, Jourdan F, Didier A (2003). Deprivation of sensory inputs to the olfactory bulb up-regulates cell death and proliferation in the subventricular zone of adult mice. *Neuroscience* 119:507–516.

Mandairon N, Sacquet J, Garcia S, Ravel N, Jourdan F, Didier A (2006). Neurogenic correlates of an olfactory discrimination task in the adult olfactory bulb. *Eur J Neurosci* 24:3578–3588.

Marshall CA, Novitch BG, Goldman JE (2005). Olig2 directs astrocyte and oligodendrocyte formation in post-natal subventricular zone cells. *J Neurosci* 25:7289–7298.

Martin SL (2009). Developmental biology: Jumping-gene roulette. *Nature* 460:1087–1088.

Martinez A, Soriano E (2005). Functions of ephrin/Eph interactions in the development of the nervous system: Emphasis on the hippocampal system. *Brain Res Brain Res Rev* 49:211–226.

Maslov AY, Barone TA, Plunkett RJ, Pruitt SC (2004). Neural stem cell detection, characterization, and age-related changes in the subventricular zone of mice. *J Neurosci* 24:1726–1733.

Mayer JL, Klumpers L, Maslam S, de Kloet ER, Joels M, Lucassen PJ (2006). Brief treatment with the glucocorticoid receptor antagonist mifepristone normalises the corticosterone-induced reduction of adult hippocampal neurogenesis. *J Neuroendocrinol* 18:629–631.

McMahon AP, Ingham PW, Tabin CJ (2003). Developmental roles and clinical significance of hedgehog signaling. *Curr Top Dev Biol* 53:1–114.

Meaney MJ, Aitken DH, Bhatnagar S, Sapolsky RM (1991). Postnatal handling attenuates certain neuroendocrine, anatomical, and cognitive dysfunctions associated with aging in female rats. *Neurobiol Aging* 12:31–38.

Melanson-Drapeau L, Beyko S, Dave S, Hebb AL, Franks DJ, Sellitto C, Paul DL, Bennett SA (2003). Oligodendrocyte progenitor enrichment in the connexin32 null-mutant mouse. *J Neurosci* 23:1759–1768.

Menezes JR, Froes MM, Moura Neto V, Lent R (2000). Gap junction-mediated coupling in the postnatal anterior subventricular zone. *Dev Neurosci* 22:34–43.

Menn B, Garcia-Verdugo JM, Yaschine C, Gonzalez-Perez O, Rowitch D, Alvarez-Buylla A (2006). Origin of oligodendrocytes in the subventricular zone of the adult brain. *J Neurosci* 26:7907–7918.

Mercer A, Ronnholm H, Holmberg J, Lundh H, Heidrich J, Zachrisson O, Ossoinak A, Frisen J, Patrone C (2004). PACAP promotes neural stem cell proliferation in adult mouse brain. *J Neurosci Res* 76:205–215.

Merson TD, Dixon MP, Collin C, Rietze RL, Bartlett PF, Thomas T, Voss AK (2006). The transcriptional coactivator Querkopf controls adult neurogenesis. *J Neurosci* 26:11359–11370.

Mi W, Pawlik M, Sastre M, Jung SS, Radvinsky DS, Klein AM, Sommer J, Schmidt SD, Nixon RA, Mathews PM, Levy E (2007). Cystatin C inhibits amyloid-beta deposition in Alzheimer's disease mouse models. *Nat Genet* 39:1440–1442.

Michael A, Jenaway A, Paykel ES, Herbert J (2000). Altered salivary dehydroepiandrosterone levels in major depression in adults. *Biol Psychiatry* 48:989–995.

Mignone JL, Kukekov V, Chiang AS, Steindler D, Enikolopov G (2004). Neural stem and progenitor cells in nestin-GFP transgenic mice. *J Comp Neurol* 469:311–324.

Minichiello L, Calella AM, Medina DL, Bonhoeffer T, Klein R, Korte M (2002). Mechanism of TrkB-mediated hippocampal long-term potentiation. *Neuron* 36:121–137.

Mirescu C, Peters JD, Noiman L, Gould E (2006). Sleep deprivation inhibits adult neurogenesis in the hippocampus by elevating glucocorticoids. *Proc Natl Acad Sci USA* 103:19170–19175.

Mitome M, Hasegawa T, Shirakawa T (2005). Mastication influences the survival of newly generated cells in mouse dentate gyrus. *Neuroreport* 16:249–252.

Miwa N, Storm DR (2005). Odorant-induced activation of extracellular signal-regulated kinase/mitogen-activated protein kinase in the olfactory bulb promotes survival of newly formed granule cells. *J Neurosci* 25:5404–5412.

Mohapel P, Leanza G, Kokaia M, Lindvall O (2005). Forebrain acetylcholine regulates adult hippocampal neurogenesis and learning. *Neurobiol Aging* 26:939–946.

Molofsky AV, Slutsky SG, Joseph NM, He S, Pardal R, Krishnamurthy J, Sharpless NE, Morrison SJ (2006). Increasing p16INK4a expression decreases forebrain progenitors and neurogenesis during ageing. *Nature* 443:448–452.

Monaghan AP, Bock D, Gass P, Schwager A, Wolfer DP, Lipp HP, Schutz G (1997). Defective limbic system in mice lacking the tailless gene. *Nature* 390:515–517.

Monje ML, Toda H, Palmer TD (2003). Inflammatory blockade restores adult hippocampal neurogenesis. *Science* 302:1760–1765.

Montaron MF, Petry KG, Rodriguez JJ, Marinelli M, Aurousseau C, Rougon G, Le Moal M, Abrous DN (1999). Adrenalectomy increases neurogenesis but not PSA-NCAM expression in aged dentate gyrus. *Eur J Neurosci* 11:1479–1485.

Montaron MF, Drapeau E, Dupret D, Kitchener P, Aurousseau C, Le Moal M, Piazza PV, Abrous DN (2006). Lifelong corticosterone level determines age-related decline in neurogenesis and memory. *Neurobiol Aging* 27:645–654.

Moreno-Lopez B, Noval JA, Gonzalez-Bonet LG, Estrada C (2000). Morphological bases for a role of nitric oxide in adult neurogenesis. *Brain Res* 869:244–250.

Morgenstern NA, Lombardi G, Schinder AF (2008). Newborn granule cells in the ageing dentate gyrus. *J Physiol* 586:3751–3757.

Mori T, Tanaka K, Buffo A, Wurst W, Kuhn R, Gotz M (2006). Inducible gene deletion in astroglia and radial glia–a valuable tool for functional and lineage analysis. *Glia* 54:21–34.

Morin PJ (1999). Beta-catenin signaling and cancer. *Bioessays* 21:1021–1030.

Morrison SJ, Perez SE, Qiao Z, Verdi JM, Hicks C, Weinmaster G, Anderson DJ (2000). Transient Notch activation initiates an irreversible switch from neurogenesis to gliogenesis by neural crest stem cells. *Cell* 101:499–510.

Morshead CM, Garcia AD, Sofroniew MV, van Der Kooy D (2003). The ablation of glial fibrillary acidic protein-positive cells from the adult central nervous system results in the loss of forebrain neural stem cells but not retinal stem cells. *Eur J Neurosci* 18:76–84.

Mostany R, Valdizan EM, Pazos A (2008). A role for nuclear beta-catenin in SNRI antidepressant-induced hippocampal cell proliferation. *Neuropharmacology* 55:18–26.

Mouret A, Gheusi G, Gabellec MM, de Chaumont F, Olivo-Marin JC, Lledo PM (2008). Learning and survival of newly generated neurons: When time matters. *J Neurosci* 28:11511–11516.

Mouret A, Lepousez G, Gras J, Gabellec MM, Lledo PM (2009). Turnover of newborn olfactory bulb neurons optimizes olfaction. *J Neurosci* 29:12302–12314.

Mueller AD, Pollock MS, Lieblich SE, Epp JR, Galea LA, Mistlberger RE (2008). Sleep deprivation can inhibit adult hippocampal neurogenesis independent of adrenal stress hormones. *Am J Physiol* 294:R1693–1703.

Muller MC, Osswald M, Tinnes S, Haussler U, Jacobi A, Forster E, Frotscher M, Haas CA (2009a). Exogenous reelin prevents granule cell dispersion in experimental epilepsy. *Exp Neurol* 216:390–397.

Muller S, Chakrapani BP, Schwegler H, Hofmann HD, Kirsch M (2009b). Neurogenesis in the dentate gyrus depends on ciliary neurotrophic factor and signal transducer and activator of transcription 3 signaling. *Stem Cells* 27:431–441.

Munoz-Sanjuan I, Brivanlou AH (2002). Neural induction, the default model and embryonic stem cells. *Nat Rev Neurosci* 3:271–280.

Muotri AR, Gage FH (2006). Generation of neuronal variability and complexity. *Nature* 441:1087–1093.

Muotri AR, Chu VT, Marchetto MC, Deng W, Moran JV, Gage FH (2005). Somatic mosaicism in neuronal precursor cells mediated by L1 retrotransposition. *Nature* 435:903–910.

Muotri AR, Marchetto MC, Coufal NG, Gage FH (2007). The necessary junk: New functions for transposable elements. *Hum Mol Genet* 16 Spec No. 2:R159–167.

Muotri AR, Zhao C, Marchetto MC, Gage FH (2009). Environmental influence on L1 retrotransposons in the adult hippocampus. *Hippocampus* 19:1002–1007.

Muroyama Y, Fujihara M, Ikeya M, Kondoh H, Takada S (2002). Wnt signaling plays an essential role in neuronal specification of the dorsal spinal cord. *Genes Dev* 16:548–553.

Muzur A, Pace-Schott EF, Hobson JA (2002). The prefrontal cortex in sleep. *Trends Cogn Sci* 6:475–481.

Nacher J, McEwen BS (2006). The role of N-methyl-D-asparate receptors in neurogenesis. *Hippocampus* 16: 267–270.

Nacher J, Rosell DR, Alonso-Llosa G, McEwen BS (2001). NMDA receptor antagonist treatment induces a long-lasting increase in the number of proliferating cells, PSA-NCAM-immunoreactive granule neurons and radial glia in the adult rat dentate gyrus. *Eur J Neurosci* 13:512–520.

Nacher J, Varea E, Blasco-Ibanez JM, Castillo-Gomez E, Crespo C, Martinez-Guijarro FJ, McEwen BS (2005). Expression of the transcription factor Pax 6 in the adult rat dentate gyrus. *J Neurosci Res* 81:753–761.

Nagai A, Ryu JK, Kobayash S, Kim SU (2002). Cystatin C induces neuronal cell death in vivo. *Ann NY Acad Sci* 977:315–321.

Nagai A, Terashima M, Sheikh AM, Notsu Y, Shimode K, Yamaguchi S, Kobayashi S, Kim SU, Masuda J (2008). Involvement of cystatin C in pathophysiology of CNS diseases. *Front Biosci* 13:3470–3479.

Nakagawa S, Kim JE, Lee R, Chen J, Fujioka T, Malberg J, Tsuji S, Duman RS (2002a). Localization of phospho-rylated cAMP response element-binding protein in immature neurons of adult hippocampus. *J Neurosci* 22:9868–9876.

Nakagawa S, Kim JE, Lee R, Malberg JE, Chen J, Steffen C, Zhang YJ, Nestler EJ, Duman RS (2002b). Regulation of neurogenesis in adult mouse hippocampus by cAMP and the cAMP response element-binding protein. *J Neurosci* 22:3673–3682.

Nam HS, Benezra R (2009). High levels of Id1 expression define B1 type adult neural stem cells. *Cell Stem Cell* 5:515–526.

Namihira M, Kohyama J, Abematsu M, Nakashima K (2008). Epigenetic mechanisms regulating fate specification of neural stem cells. *Philos Trans R Soc Lond B Biol Sci* 363:2099–2109.

Naylor AS, Bull C, Nilsson MK, Zhu C, Bjork-Eriksson T, Eriksson PS, Blomgren K, Kuhn HG (2008). Voluntary running rescues adult hippocampal neurogenesis after irradiation of the young mouse brain. *Proc Natl Acad Sci USA* 105:14632–14637.

Ng KL, Li JD, Cheng MY, Leslie FM, Lee AG, Zhou QY (2005). Dependence of olfactory bulb neurogenesis on prokineticin 2 signaling. *Science* 308:1923–1927.

Nilsson M, Perfilieva E, Johansson U, Orwar O, Eriksson P (1999). Enriched environment increases neurogen-esis in the adult rat dentate gyrus and improves spatial memory. *J Neurobiol* 39:569–578.

Ninkovic J, Mori T, Gotz M (2007). Distinct modes of neuron addition in adult mouse neurogenesis. *J Neurosci* 27:10906–10911.

Nordeen EJ, Nordeen KW (1989). Estrogen stimulates the incorporation of new neurons into avian song nuclei during adolescence. *Brain Res Dev Brain Res* 49:27–32.

Nyffeler M, Yee BK, Feldon J, Knuesel I (2008). Abnormal differentiation of newborn granule cells in age-related working memory impairments. *Neurobiol Aging.*

Okano HJ, Pfaff DW, Gibbs RB (1996). Expression of EGFR-, p75NGFR-, and PSTAIR (cdc2)-like immunore-activity by proliferating cells in the adult rat hippocampal formation and forebrain. *Dev Neurosci* 18:199–209.

Okuyama N, Takagi N, Kawai T, Miyake-Takagi K, Takeo S (2004). Phosphorylation of extracellular-regulating kinase in NMDA receptor antagonist-induced newly generated neurons in the adult rat dentate gyrus. *J Neurochem* 88:717–725.

Orentreich N, Brind JL, Vogelman JH, Andres R, Baldwin H (1992). Long-term longitudinal measurements of plasma dehydroepiandrosterone sulfate in normal men. *J Clin Endocrinol Metab* 75:1002–1004.

Ormerod BK, Lee TT, Galea LA (2003). Estradiol initially enhances but subsequently suppresses (via adrenal steroids) granule cell proliferation in the dentate gyrus of adult female rats. *J Neurobiol* 55:247–260.

Ormerod BK, Lee TT, Galea LA (2004). Estradiol enhances neurogenesis in the dentate gyri of adult male meadow voles by increasing the survival of young granule neurons. *Neuroscience* 128:645–654.

Ostenfeld T, Caldwell MA, Prowse KR, Linskens MH, Jauniaux E, Svendsen CN (2000). Human neural precursor cells express low levels of telomerase in vitro and show diminishing cell proliferation with extensive axonal outgrowth following transplantation. *Exp Neurol* 164:215–226.

Ostertag EM, Kazazian HH (2005). Genetics: LINEs in mind. *Nature* 435:890–891.

Osumi N, Shinohara H, Numayama-Tsuruta K, Maekawa M (2008). Concise review: Pax6 transcription factor contributes to both embryonic and adult neurogenesis as a multifunctional regulator. *Stem Cells* 26:1663–1672.

Otero JJ, Fu W, Kan L, Cuadra AE, Kessler JA (2004). Beta-catenin signaling is required for neural differentiation of embryonic stem cells. *Development* 131:3545–3557.

Overstreet LS, Hentges ST, Bumaschny VF, de Souza FS, Smart JL, Santangelo AM, Low MJ, Westbrook GL, Rubinstein M (2004). A transgenic marker for newly born granule cells in dentate gyrus. *J Neurosci* 24:3251–3259.

Ozen I, Galichet C, Watts C, Parras C, Guillemot F, Raineteau O (2007). Proliferating neuronal progenitors in the postnatal hippocampus transiently express the proneural gene Ngn2. *Eur J Neurosci* 25:2591–2603.

Palazuelos J, Aguado T, Egia A, Mechoulam R, Guzman M, Galve-Roperh I (2006). Non-psychoactive CB2 can-nabinoid agonists stimulate neural progenitor proliferation. *Faseb J* 20:2405–2407.

Palma V, Lim DA, Dahmane N, Sanchez P, Brionne TC, Herzberg CD, Gitton Y, Carleton A, Alvarez-Buylla A, Ruiz i Altaba A (2005). Sonic hedgehog controls stem cell behavior in the postnatal and adult brain. *Development* 132:335–344.

Palmer TD, Ray J, Gage FH (1995). FGF2–responsive neuronal progenitors reside in proliferative and quiescent regions of the adult rodent brain. *Mol Cell Neurosci* 6:474–486.

Palmer TD, Markakis EA, Willhoite AR, Safar F, Gage FH (1999). Fibroblast growth factor-2 activates a latent neurogenic program in neural stem cells from diverse regions of the adult CNS. *J Neurosci* 19:8487–8497.

Parent JM, Yu TW, Leibowitz RT, Geschwind DH, Sloviter RS, Lowenstein DH (1997). Dentate granule cell neurogenesis is increased by seizures and contributes to aberrant network reorganization in the adult rat hip-pocampus. *J Neurosci* 17:3727–3738.

Parent JM, Janumpalli S, McNamara JO, Lowenstein DH (1998). Increased dentate granule cell neurogenesis following amygdala kindling in the adult rat. *Neurosci Lett* 247:9–12.

Park C, Sohn Y, Shin KS, Kim J, Ahn H, Huh Y (2003). The chronic inhibition of nitric oxide synthase enhances cell proliferation in the adult rat hippocampus. *Neurosci Lett* 339:9–12.

Park C, Shin KS, Ryu JH, Kang K, Kim J, Ahn H, Huh Y (2004). The inhibition of nitric oxide synthase enhances PSA-NCAM expression and CREB phosphorylation in the rat hippocampus. *Neuroreport* 15:231–234.

Parras CM, Hunt C, Sugimori M, Nakafuku M, Rowitch D, Guillemot F (2007). The proneural gene Mash1 specifies an early population of telencephalic oligodendrocytes. *J Neurosci* 27:4233–4242.

Pawluski JL, Galea LA (2007). Reproductive experience alters hippocampal neurogenesis during the postpartum period in the dam. *Neuroscience* 149:53–67.

Pawluski JL, Brummelte S, Barha CK, Crozier TM, Galea LA (2009). Effects of steroid hormones on neurogenesis in the hippocampus of the adult female rodent during the estrous cycle, pregnancy, lactation and aging. *Front Neuroendocrinol* 30:343–357.

Pechnick RN, Zonis S, Wawrowsky K, Pourmorady J, Chesnokova V (2008). p21Cip1 restricts neuronal proliferation in the subgranular zone of the dentate gyrus of the hippocampus. *Proc Natl Acad Sci USA* 105:1358–1363.

Peirce JL, Chesler EJ, Williams RW, Lu L (2003). Genetic architecture of the mouse hippocampus: Identification of gene loci with selective regional effects. *Genes Brain Behav* 2:238–252.

Pencea V, Bingaman KD, Wiegand SJ, Luskin MB (2001a). Infusion of brain-derived neurotrophic factor into the lateral ventricle of the adult rat leads to new neurons in the parenchyma of the striatum, septum, thalamus, and hypothalamus. *J Neurosci* 21:6706–6717.

Pencea V, Bingaman KD, Freedman LJ, Luskin MB (2001b). Neurogenesis in the subventricular zone and rostral migratory stream of the neonatal and adult primate forebrain. *Exp Neurol* 172:1–16.

Peranovich TM, da Silva AM, Fries DM, Stern A, Monteiro HP (1995). Nitric oxide stimulates tyrosine phosphorylation in murine fibroblasts in the absence and presence of epidermal growth factor. *Biochem J* 305 (Pt 2):613–619.

Pereira AC, Huddleston DE, Brickman AM, Sosunov AA, Hen R, McKhann GM, Sloan R, Gage FH, Brown TR, Small SA (2007). An in vivo correlate of exercise-induced neurogenesis in the adult dentate gyrus. *Proc Natl Acad Sci USA* 104:5638–5643.

Peretto P, Dati C, De Marchis S, Kim HH, Ukhanova M, Fasolo A, Margolis FL (2004). Expression of the secreted factors noggin and bone morphogenetic proteins in the subependymal layer and olfactory bulb of the adult mouse brain. *Neuroscience* 128:685–696.

Peretto P, Giachino C, Aimar P, Fasolo A, Bonfanti L (2005). Chain formation and glial tube assembly in the shift from neonatal to adult subventricular zone of the rodent forebrain. *J Comp Neurol* 487:407–427.

Perez-Martin M, Azcoitia I, Trejo JL, Sierra A, Garcia-Segura LM (2003). An antagonist of estrogen receptors blocks the induction of adult neurogenesis by insulin-like growth factor-I in the dentate gyrus of adult female rat. *Eur J Neurosci* 18:923–930.

Perfilieva E, Risedal A, Nyberg J, Johansson BB, Eriksson PS (2001). Gender and strain influence on neurogenesis in dentate gyrus of young rats. *J Cereb Blood Flow Metab* 21:211–217.

Persson AI, Thorlin T, Bull C, Zarnegar P, Ekman R, Terenius L, Eriksson PS (2003). Mu- and delta-opioid receptor antagonists decrease proliferation and increase neurogenesis in cultures of rat adult hippocampal progenitors. *Eur J Neurosci* 17:1159–1172.

Persson AI, Naylor AS, Jonsdottir IH, Nyberg F, Eriksson PS, Thorlin T (2004). Differential regulation of hippocampal progenitor proliferation by opioid receptor antagonists in running and non-running spontaneously hypertensive rats. *Eur J Neurosci* 19:1847–1855.

Petreanu L, Alvarez-Buylla A (2002). Maturation and death of adult-born olfactory bulb granule neurons: Role of olfaction. *J Neurosci* 22:6106–6113.

Petrus DS, Fabel K, Kronenberg G, Winter C, Steiner B, Kempermann G (2008). NMDA and benzodiazepine receptors have synergistic and antagonistic effects on precursor cells in adult hippocampal neurogenesis. *Eur J Neurosci* 29:244–252.

Pham K, Nacher J, Hof PR, McEwen BS (2003). Repeated restraint stress suppresses neurogenesis and induces biphasic PSA-NCAM expression in the adult rat dentate gyrus. *Eur J Neurosci* 17:879–886.

Pieper AA, Wu X, Han TW, Estill SJ, Dang Q, Wu LC, Reece-Fincanon S, Dudley CA, Richardson JA, Brat DJ, McKnight SL (2005). The neuronal PAS domain protein 3 transcription factor controls FGF-mediated adult hippocampal neurogenesis in mice. *Proc Natl Acad Sci USA* 102:14052–14057.

Pinnock SB, Herbert J (2008). Brain-derived neurotropic factor and neurogenesis in the adult rat dentate gyrus: Interactions with corticosterone. *Eur J Neurosci* 27:2493–2500.

Pinnock SB, Lazic SE, Wong HT, Wong IH, Herbert J (2009). Synergistic effects of dehydroepiandrosterone and fluoxetine on proliferation of progenitor cells in the dentate gyrus of the adult male rat. *Neuroscience* 158:1644–1651.

Pirttila TJ, Lukasiuk K, Hakansson K, Grubb A, Abrahamson M, Pitkanen A (2005). Cystatin C modulates neurodegeneration and neurogenesis following status epilepticus in mouse. *Neurobiol Dis* 20:241–253.

Pittenger C, Duman RS (2008). Stress, depression, and neuroplasticity: A convergence of mechanisms. *Neuropsychopharmacology* 33:88–109.

Pizzo DP, Thal LJ, Winkler J (2002). Mnemonic deficits in animals depend upon the degree of cholinergic deficit and task complexity. *Exp Neurol* 177:292–305.

Pleasure SJ, Collins AE, Lowenstein DH (2000). Unique expression patterns of cell fate molecules delineate sequential stages of dentate gyrus development. *J Neurosci* 20:6095–6105.

Prozorovski T, Schulze-Topphoff U, Glumm R, Baumgart J, Schroter F, Ninnemann O, Siegert E, Bendix I, Brustle O, Nitsch R, Zipp F, Aktas O (2008). Sirt1 contributes critically to the redox-dependent fate of neural progenitors. *Nat Cell Biol* 10:385–394.

Puverel S, Nakatani H, Parras C, Soussi-Yanicostas N (2009). Prokineticin receptor 2 expression identifies migrating neuroblasts and their subventricular zone transient-amplifying progenitors in adult mice. *J Comp Neurol* 512:232–242.

Ra SM, Kim H, Jang MH, Shin MC, Lee TH, Lim BV, Kim CJ, Kim EH, Kim KM, Kim SS (2002). Treadmill running and swimming increase cell proliferation in the hippocampal dentate gyrus of rats. *Neurosci Lett* 333:123–126.

Rai KS, Hattiangady B, Shetty AK (2007). Enhanced production and dendritic growth of new dentate granule cells in the middle-aged hippocampus following intracerebroventricular FGF2 infusions. *Eur J Neurosci* 26:1765–1779.

Ramirez-Rodriguez G, Klempin F, Babu H, Benitez-King G, Kempermann G (2009). Melatonin modulates cell survival of new neurons in the hippocampus of adult mice. *Neuropsychopharmacology* 34:2180–2191.

Ransome MI, Turnley AM (2007). Systemically delivered Erythropoietin transiently enhances adult hippocampal neurogenesis. *J Neurochem* 102:1953–1965.

Rao MS, Shetty AK (2004). Efficacy of doublecortin as a marker to analyse the absolute number and dendritic growth of newly generated neurons in the adult dentate gyrus. *Eur J Neurosci* 19:234–246.

Rao MS, Hattiangady B, Shetty AK (2006). The window and mechanisms of major age-related decline in the production of new neurons within the dentate gyrus of the hippocampus. *Aging Cell* 5:545–558.

Rasika S, Alvarez-Buylla A, Nottebohm F (1999). BDNF mediates the effects of testosterone on the survival of new neurons in an adult brain. *Neuron* 22:53–62.

Reif A, Schmitt A, Fritzen S, Chourbaji S, Bartsch C, Urani A, Wycislo M, Mossner R, Sommer C, Gass P, Lesch KP (2004). Differential effect of endothelial nitric oxide synthase (NOS-III) on the regulation of adult neurogenesis and behaviour. *Eur J Neurosci* 20:885–895.

Revesz D, Tjernstrom M, Ben-Menachem E, Thorlin T (2008). Effects of vagus nerve stimulation on rat hippocampal progenitor proliferation. *Exp Neurol* 214:259–265.

Reya T, Duncan AW, Ailles L, Domen J, Scherer DC, Willert K, Hintz L, Nusse R, Weissman IL (2003). A role for Wnt signalling in self-renewal of haematopoietic stem cells. *Nature* 423:409–414.

Reynolds BA, Weiss S (1992). Generation of neurons and astrocytes from isolated cells of the adult mammalian central nervous system. *Science* 255:1707–1710.

Ricard J, Salinas J, Garcia L, Liebl DJ (2006). EphrinB3 regulates cell proliferation and survival in adult neurogenesis. *Mol Cell Neurosci* 31:713–722.

Rochefort C, Lledo PM (2005). Short-term survival of newborn neurons in the adult olfactory bulb after exposure to a complex odor environment. *Eur J Neurosci* 22:2863–2870.

Rochefort C, Gheusi G, Vincent JD, Lledo PM (2002). Enriched odor exposure increases the number of newborn neurons in the adult olfactory bulb and improves odor memory. *J Neurosci* 22:2679–2689.

Rohe M, Carlo AS, Breyhan H, Sporbert A, Militz D, Schmidt V, Wozny C, Harmeier A, Erdmann B, Bales KR, Wolf S, Kempermann G, Paul SM, Schmitz D, Bayer TA, Willnow TE, Andersen OM (2008). Sortilin-related receptor with A-type repeats (SORLA) affects the amyloid precursor protein-dependent stimulation of ERK signaling and adult neurogenesis. *J Biol Chem* 283:14826–14834.

Rolls A, Shechter R, London A, Ziv Y, Ronen A, Levy R, Schwartz M (2007). Toll-like receptors modulate adult hippocampal neurogenesis. *Nat Cell Biol* 9:1081–1088.

Rolls A, Schori H, London A, Schwartz M (2008). Decrease in hippocampal neurogenesis during pregnancy: A link to immunity. *Mol Psychiatry* 13:468–469.

Romero-Grimaldi C, Moreno-Lopez B, Estrada C (2008). Age-dependent effect of nitric oxide on subventricular zone and olfactory bulb neural precursor proliferation. *J Comp Neurol* 506:339–346.

Rossi C, Angelucci A, Costantin L, Braschi C, Mazzantini M, Babbini F, Fabbri ME, Tessarollo L, Maffei L, Berardi N, Caleo M (2006). Brain-derived neurotrophic factor (BDNF) is required for the enhancement of hippocampal neurogenesis following environmental enrichment. *Eur J Neurosci* 24:1850–1856.

Roy K, Kuznicki K, Wu Q, Sun Z, Bock D, Schutz G, Vranich N, Monaghan AP (2004). The Tlx gene regulates the timing of neurogenesis in the cortex. *J Neurosci* 24:8333–8345.

Roybon L, Deierborg T, Brundin P, Li JY (2009a). Involvement of Ngn2, Tbr and NeuroD proteins during post-natal olfactory bulb neurogenesis. *Eur J Neurosci* 29:232–243.

Roybon L, Hjalt T, Stott S, Guillemot F, Li JY, Brundin P (2009b). Neurogenin2 directs granule neuroblast production and amplification while NeuroD1 specifies neuronal fate during hippocampal neurogenesis. *PLoS One* 4:e4779.

Rozental R, Mehler MF, Morales M, Andrade-Rozental AF, Kessler JA, Spray DC (1995). Differentiation of hippocampal progenitor cells in vitro: Temporal expression of intercellular coupling and voltage- and ligand-gated responses. *Dev Biol* 167:350–362.

Rozental R, Morales M, Mehler MF, Urban M, Kremer M, Dermietzel R, Kessler JA, Spray DC (1998). Changes in the properties of gap junctions during neuronal differentiation of hippocampal progenitor cells. *J Neurosci* 18:1753–1762.

Rueda D, Navarro B, Martinez-Serrano A, Guzman M, Galve-Roperh I (2002). The endocannabinoid anandamide inhibits neuronal progenitor cell differentiation through attenuation of the Rap1/B-Raf/ERK pathway. *J Biol Chem* 277:46645–46650.

Russo RE, Reali C, Radmilovich M, Fernandez A, Trujillo-Cenoz O (2008). Connexin 43 delimits functional domains of neurogenic precursors in the spinal cord. *J Neurosci* 28:3298–3309.

Saino-Saito S, Berlin R, Baker H (2003). Dlx-1 and Dlx-2 expression in the adult mouse brain: Relationship to dopaminergic phenotypic regulation. *J Comp Neurol* 461:18–30.

Sairanen M, Lucas G, Ernfors P, Castren M, Castren E (2005). Brain-derived neurotrophic factor and antidepressant drugs have different but coordinated effects on neuronal turnover, proliferation, and survival in the adult dentate gyrus. *J Neurosci* 25:1089–1094.

Sanosaka T, Namihira M, Nakashima K (2009). Epigenetic mechanisms in sequential differentiation of neural stem cells. *Epigenetics* 4:89–92.

Santarelli L, Saxe M, Gross C, Surget A, Battaglia F, Dulawa S, Weisstaub N, Lee J, Duman R, Arancio O, Belzung C, Hen R (2003). Requirement of hippocampal neurogenesis for the behavioral effects of antidepressants. *Science* 301:805–809.

Sapolsky RM (1999). Glucocorticoids, stress, and their adverse neurological effects: Relevance to aging. *Exp Gerontol* 34:721–732.

Sasaki T, Kitagawa K, Sugiura S, Omura-Matsuoka E, Tanaka S, Yagita Y, Okano H, Matsumoto M, Hori M (2003). Implication of cyclooxygenase-2 on enhanced proliferation of neural progenitor cells in the adult mouse hippocampus after ischemia. *J Neurosci Res* 72:461–471.

Sawamoto K, Yamamoto A, Kawaguchi A, Yamaguchi M, Mori K, Goldman SA, Okano H (2001). Direct isolation of committed neuronal progenitor cells from transgenic mice coexpressing spectrally distinct fluorescent proteins regulated by stage-specific neural promoters. *J Neurosci Res* 65:220–227.

Schabitz WR, Steigleder T, Cooper-Kuhn CM, Schwab S, Sommer C, Schneider A, Kuhn HG (2007). Intravenous brain-derived neurotrophic factor enhances poststroke sensorimotor recovery and stimulates neurogenesis. *Stroke* 38:2165–2172.

Schanzer A, Wachs FP, Wilhelm D, Acker T, Cooper-Kuhn C, Beck H, Winkler J, Aigner L, Plate KH, Kuhn HG (2004). Direct stimulation of adult neural stem cells in vitro and neurogenesis in vivo by vascular endothelial growth factor. *Brain Pathol* 14:237–248.

Scharfman H, Goodman J, Macleod A, Phani S, Antonelli C, Croll S (2005). Increased neurogenesis and the ectopic granule cells after intrahippocampal BDNF infusion in adult rats. *Exp Neurol* 192:348–356.

Schloesser RJ, Manji HK, Martinowich K (2009). Suppression of adult neurogenesis leads to an increased hypothalamo-pituitary-adrenal axis response. *Neuroreport* 20:553–557.

Schmitt A, Benninghoff J, Moessner R, Rizzi M, Paizanis E, Doenitz C, Gross S, Hermann M, Gritti A, Lanfumey L, Fritzen S, Reif A, Hamon M, Murphy DL, Vescovi A, Lesch KP (2007). Adult neurogenesis in serotonin transporter deficient mice. *J Neural Transm* 114:1107–1119.

Schneider A, Kruger C, Steigleder T, Weber D, Pitzer C, Laage R, Aronowski J, Maurer MH, Gassler N, Mier W, Hasselblatt M, Kollmar R, Schwab S, Sommer C, Bach A, Kuhn HG, Schabitz WR (2005). The hematopoietic factor G-CSF is a neuronal ligand that counteracts programmed cell death and drives neurogenesis. *J Clin Invest* 115:2083–2098.

Schoenherr CJ, Anderson DJ (1995). The neuron-restrictive silencer factor (NRSF): A coordinate repressor of multiple neuron-specific genes. *Science* 267:1360–1363.

Scott BW, Wang S, Burnham WM, De Boni U, Wojtowicz JM (1998). Kindling-induced neurogenesis in the dentate gyrus of the rat. *Neurosci Lett* 248:73–76.

Segi-Nishida E, Warner-Schmidt JL, Duman RS (2008). Electroconvulsive seizure and VEGF increase the proliferation of neural stem-like cells in rat hippocampus. *Proc Natl Acad Sci USA* 105:11352–11357.

Seki T (2002). Expression patterns of immature neuronal markers PSA-NCAM, CRMP-4 and NeuroD in the hippocampus of young adult and aged rodents. *J Neurosci Res* 70:327–334.

Seki T, Namba T, Mochizuki H, Onodera M (2007). Clustering, migration, and neurite formation of neural precursor cells in the adult rat hippocampus. *J Comp Neurol* 502:275–290.

Seri B, Garcia-Verdugo JM, McEwen BS, Alvarez-Buylla A (2001). Astrocytes give rise to new neurons in the adult mammalian hippocampus. *J Neurosci* 21:7153–7160.

Seri B, Garcia-Verdugo JM, Collado-Morente L, McEwen BS, Alvarez-Buylla A (2004). Cell types, lineage, and architecture of the germinal zone in the adult dentate gyrus. *J Comp Neurol* 478:359.

Shen L, Nam HS, Song P, Moore H, Anderson SA (2006). FoxG1 haploinsufficiency results in impaired neurogenesis in the postnatal hippocampus and contextual memory deficits. *Hippocampus* 16:875–890.

Shen Q, Temple S (2009). Fine control: MicroRNA regulation of adult neurogenesis. *Nat Neurosci* 12:369–370.

Shetty AK, Hattiangady B, Shetty GA (2005). Stem/progenitor cell proliferation factors FGF2, IGF1, and VEGF exhibit early decline during the course of aging in the hippocampus: Role of astrocytes. *Glia* 51:173–186.

Shi Y, Chichung Lie D, Taupin P, Nakashima K, Ray J, Yu RT, Gage FH, Evans RM (2004). Expression and function of orphan nuclear receptor TLX in adult neural stem cells. *Nature* 427:78–83.

Shimazu K, Zhao M, Sakata K, Akbarian S, Bates B, Jaenisch R, Lu B (2006). NT-3 facilitates hippocampal plasticity and learning and memory by regulating neurogenesis. *Learn Mem* 13:307–315.

Shingo T, Sorokan ST, Shimazaki T, Weiss S (2001). Erythropoietin regulates the in vitro and in vivo production of neuronal progenitors by mammalian forebrain neural stem cells. *J Neurosci* 21:9733–9743.

Shingo T, Gregg C, Enwere E, Fujikawa H, Hassam R, Geary C, Cross JC, Weiss S (2003). Pregnancy-stimulated neurogenesis in the adult female forebrain mediated by prolactin. *Science* 299:117–120.

Shukla V, Zimmermann H, Wang L, Kettenmann H, Raab S, Hammer K, Sevigny J, Robson SC, Braun N (2005). Functional expression of the ecto-ATPase NTPDase2 and of nucleotide receptors by neuronal progenitor cells in the adult murine hippocampus. *J Neurosci Res* 80:600–610.

Smith MT, Pencea V, Wang Z, Luskin MB, Insel TR (2001). Increased number of BrdU-labeled neurons in the rostral migratory stream of the estrous prairie vole. *Horm Behav* 39:11–21.

Snapyan M, Lemasson M, Brill MS, Blais M, Massouh M, Ninkovic J, Gravel C, Berthod F, Gotz M, Barker PA, Parent A, Saghatelyan A (2009). Vasculature guides migrating neuronal precursors in the adult mammalian forebrain via brain-derived neurotrophic factor signaling. *J Neurosci* 29:4172–4188.

Snyder JS, Choe JS, Clifford MA, Jeurling SI, Hurley P, Brown A, Kamhi JF, Cameron HA (2009). Adult-born hippocampal neurons are more numerous, faster maturing, and more involved in behavior in rats than in mice. *J Neurosci* 29:14484–14495.

Song C, Leonard BE (2005). The olfactory bulbectomised rat as a model of depression. *Neurosci Biobehav Rev* 29:627–647.

Song H, Stevens CF, Gage FH (2002). Astroglia induce neurogenesis from adult neural stem cells. *Nature* 417:39–44.

Spradling A, Drummond-Barbosa D, Kai T (2001). Stem cells find their niche. *Nature* 414:98–104.

Spritzer MD, Galea LA (2007). Testosterone and dihydrotestosterone, but not estradiol, enhance survival of new hippocampal neurons in adult male rats. *Dev Neurobiol* 67:1321–1333.

Steiner B, Klempin F, Wang L, Kott M, Kettenmann H, Kempermann G (2006). Type-2 cells as link between glial and neuronal lineage in adult hippocampal neurogenesis. *Glia* 54:805–814.

Steiner B, Zurborg S, Horster H, Fabel K, Kempermann G (2008). Differential 24 h responsiveness of Prox1–expressing precursor cells in adult hippocampal neurogenesis to physical activity, environmental enrichment, and kainic acid-induced seizures. *Neuroscience* 154:521–529.

Stranahan AM, Khalil D, Gould E (2006). Social isolation delays the positive effects of running on adult neurogenesis. *Nat Neurosci* 9:526–533.

Stump G, Durrer A, Klein AL, Lutolf S, Suter U, Taylor V (2002). Notch1 and its ligands Delta-like and Jagged are expressed and active in distinct cell populations in the postnatal mouse brain. *Mech Dev* 114:153–159.

Sugiura S, Kitagawa K, Tanaka S, Todo K, Omura-Matsuoka E, Sasaki T, Mabuchi T, Matsushita K, Yagita Y, Hori M (2005). Adenovirus-mediated gene transfer of heparin-binding epidermal growth factor-like growth factor enhances neurogenesis and angiogenesis after focal cerebral ischemia in rats. *Stroke* 36:859–864.

Suh H, Consiglio A, Ray J, Sawai T, D'Amour KA, Gage FH (2007). In vivo fate analysis reveals the multipotent and self-renewal capacities of Sox2+ neural stem cells in the adult hippocampus. *Cell Stem Cell* 1:515–528.

Suh Y, Obernier K, Holzl-Wenig G, Mandl C, Herrmann A, Worner K, Eckstein V, Ciccolini F (2009). Interaction between DLX2 and EGFR in the regulation of proliferation and neurogenesis of SVZ precursors. *Mol Cell Neurosci* 42:308–14.

Sun LY, Evans MS, Hsieh J, Panici J, Bartke A (2005). Increased neurogenesis in dentate gyrus of long-lived Ames dwarf mice. *Endocrinology* 146:1138–1144.

Sun W, Winseck A, Vinsant S, Park OH, Kim H, Oppenheim RW (2004). Programmed cell death of adult-generated hippocampal neurons is mediated by the proapoptotic gene Bax. *J Neurosci* 24:11205–11213.

Sun Y, Jin K, Childs JT, Xie L, Mao XO, Greenberg DA (2006). Vascular endothelial growth factor-B (VEGFB) stimulates neurogenesis: Evidence from knockout mice and growth factor administration. *Dev Biol* 289:329–335.

Takahashi J, Palmer TD, Gage FH (1999). Retinoic acid and neurotrophins collaborate to regulate neurogenesis in adult-derived neural stem cell cultures. *J Neurobiol* 38:65–81.

Takarada T, Tamaki K, Takumi T, Ogura M, Ito Y, Nakamichi N, Yoneda Y (2009). A protein-protein interaction of stress-responsive myosin VI endowed to inhibit neural progenitor self-replication with RNA binding protein, TLS, in murine hippocampus. *J Neurochem* 110:1457–1468.

Taliaz D, Stall N, Dar DE, Zangen A (2010). Knockdown of brain-derived neurotrophic factor in specific brain sites precipitates behaviors associated with depression and reduces neurogenesis. *Mol Psychiatry* 15:80–92.

Tamaki K, Kamakura M, Nakamichi N, Taniura H, Yoneda Y (2008). Upregulation of Myo6 expression after traumatic stress in mouse hippocampus. *Neurosci Lett* 433:183–187.

Tanapat P, Hastings NB, Reeves AJ, Gould E (1999). Estrogen stimulates a transient increase in the number of new neurons in the dentate gyrus of the adult female rat. *J Neurosci* 19:5792–5801.

Tanapat P, Hastings NB, Rydel TA, Galea LA, Gould E (2001). Exposure to fox odor inhibits cell proliferation in the hippocampus of adult rats via an adrenal hormone-dependent mechanism. *J Comp Neurol* 437:496–504.

Tashiro A, Makino H, Gage FH (2007). Experience-specific functional modification of the dentate gyrus through adult neurogenesis: A critical period during an immature stage. *J Neurosci* 27:3252–3259.

Taupin P, Ray J, Fischer WH, Suhr ST, Hakansson K, Grubb A, Gage FH (2000). FGF2–responsive neural stem cell proliferation requires CCg, a novel autocrine/paracrine cofactor. *Neuron* 28:385–397.

Teramoto T, Qiu J, Plumier JC, Moskowitz MA (2003). EGF amplifies the replacement of parvalbumin-expressing striatal interneurons after ischemia. *J Clin Invest* 111:1125–1132.

Thakker-Varia S, Krol JJ, Nettleton J, Bilimoria PM, Bangasser DA, Shors TJ, Black IB, Alder J (2007). The neuropeptide VGF produces antidepressant-like behavioral effects and enhances proliferation in the hippocampus. *J Neurosci* 27:12156–12167.

Thomas RM, Urban JH, Peterson DA (2006). Acute exposure to predator odor elicits a robust increase in corticosterone and a decrease in activity without altering proliferation in the adult rat hippocampus. *Exp Neurol* 201:308–315.

Toda H, Hamani C, Fawcett AP, Hutchison WD, Lozano AM (2008). The regulation of adult rodent hippocampal neurogenesis by deep brain stimulation. *J Neurosurg* 108:132–138.

Toda H, Tsuji M, Nakano I, Kobuke K, Hayashi T, Kasahara H, Takahashi J, Mizoguchi A, Houtani T, Sugimoto T, Hashimoto N, Palmer TD, Honjo T, Tashiro K (2003). Stem cell-derived neural stem/progenitor cell supporting factor is an autocrine/paracrine survival factor for adult neural stem/progenitor cells. *J Biol Chem* 278:35491–35500.

Tonchev AB, Yamashima T, Guo J, Chaldakov GN, Takakura N (2007). Expression of angiogenic and neurotrophic factors in the progenitor cell niche of adult monkey subventricular zone. *Neuroscience* 144:1425–1435.

Tong Y, Chabot JG, Shen SH, O'Dowd BF, George SR, Quirion R (2000). Ontogenic profile of the expression of the mu opioid receptor gene in the rat telencephalon and diencephalon: An in situ hybridization study. *J Chem Neuroanat* 18:209–222.

Torner L, Karg S, Blume A, Kandasamy M, Kuhn HG, Winkler J, Aigner L, Neumann ID (2009). Prolactin prevents chronic stress-induced decrease of adult hippocampal neurogenesis and promotes neuronal fate. *J Neurosci* 29:1826–1833.

Torres-Aleman I (2000). Serum growth factors and neuroprotective surveillance: Focus on IGF1. *Mol Neurobiol* 21:153–160.

Tozuka Y, Wada E, Wada K (2009). Diet-induced obesity in female mice leads to peroxidized lipid accumulations and impairment of hippocampal neurogenesis during the early life of their offspring. *Faseb J* 23:1920–1934.

Tozuka Y, Fukuda S, Namba T, Seki T, Hisatsune T (2005). GABAergic excitation promotes neuronal differentiation in adult hippocampal progenitor cells. *Neuron* 47:803–815.

Tran PB, Ren D, Veldhouse TJ, Miller RJ (2004). Chemokine receptors are expressed widely by embryonic and adult neural progenitor cells. *J Neurosci Res* 76:20–34.

Tran PB, Banisadr G, Ren D, Chenn A, Miller RJ (2007). Chemokine receptor expression by neural progenitor cells in neurogenic regions of mouse brain. *J Comp Neurol* 500:1007–1033.

Trejo JL, Carro E, Torres-Aleman I (2001). Circulating insulin-like growth factor I mediates exercise-induced increases in the number of new neurons in the adult hippocampus. *J Neurosci* 21:1628–1634.

Uberti D, Piccioni L, Cadei M, Grigolato P, Rotter V, Memo M (2001). p53 is dispensable for apoptosis but controls neurogenesis of mouse dentate gyrus cells following gamma-irradiation. *Brain Res Mol Brain Res* 93:81–89.

Uchida K, Kumihashi K, Kurosawa S, Kobayashi T, Itoi K, Machida T (2002). Stimulatory effects of prostaglandin E2 on neurogenesis in the dentate gyrus of the adult rat. *Zoolog Sci* 19:1211–1216.

Uda M, Ishido M, Kami K (2007). Features and a possible role of Mash1–immunoreactive cells in the dentate gyrus of the hippocampus in the adult rat. *Brain Res* 1171:9–17.

Uittenbogaard M, Chiaramello A (2002). Expression of the bHLH transcription factor Tcf12 (ME1) gene is linked to the expansion of precursor cell populations during neurogenesis. *Brain Res Gene Expr Patterns* 1:115–121.

Uwanogho D, Rex M, Cartwright EJ, Pearl G, Healy C, Scotting PJ, Sharpe PT (1995). Embryonic expression of the chicken Sox2, Sox3 and Sox11 genes suggests an interactive role in neuronal development. *Mech Dev* 49:23–36.

Vaccarino FM, Schwartz ML, Raballo R, Nilsen J, Rhee J, Zhou M, Doetschman T, Coffin JD, Wyland JJ, Hung YT (1999). Changes in cerebral cortex size are governed by fibroblast growth factor during embryogenesis. *Nat Neurosci* 2:246–253.

Valente T, Hidalgo J, Bolea I, Ramirez B, Angles N, Reguant J, Morello JR, Gutierrez C, Boada M, Unzeta M (2009). A diet enriched in polyphenols and polyunsaturated fatty acids, LMN diet, induces neurogenesis in the subventricular zone and hippocampus of adult mouse brain. *J Alzheimers Dis* 18:849–65.

Vallieres L, Campbell IL, Gage FH, Sawchenko PE (2002). Reduced hippocampal neurogenesis in adult transgenic mice with chronic astrocytic production of interleukin-6. *J Neurosci* 22:486–492.

Van de Wetering M, de Lau W, Clevers H (2002). WNT signaling and lymphocyte development. *Cell* 109 Suppl:S13–19.

Van Kampen JM, Hagg T, Robertson HA (2004). Induction of neurogenesis in the adult rat subventricular zone and neostriatum following dopamine D3 receptor stimulation. *Eur J Neurosci* 19:2377–2387.

Van Praag H, Kempermann G, Gage FH (1999a). Running increases cell proliferation and neurogenesis in the adult mouse dentate gyrus. *Nat Neurosci* 2:266–270.

Van Praag H, Christie BR, Sejnowski TJ, Gage FH (1999b). Running enhances neurogenesis, learning and long-term potentiation in mice. *Proc Natl Acad Sci USA* 96:13427–13431.

Van Praag H, Schinder AF, Christie BR, Toni N, Palmer TD, Gage FH (2002). Functional neurogenesis in the adult hippocampus. *Nature* 415:1030–1034.

Van Praag H, Shubert T, Zhao C, Gage FH (2005). Exercise enhances learning and hippocampal neurogenesis in aged mice. *J Neurosci* 25:8680–8685.

Vandenbosch R, Borgs L, Beukelaers P, Foidart A, Nguyen L, Moonen G, Berthet C, Kaldis P, Gallo V, Belachew S, Malgrange B (2007). CDK2 is dispensable for adult hippocampal neurogenesis. *Cell Cycle* 6:3065–3069.

Veena J, Srikumar BN, Mahati K, Bhagya V, Raju TR, Shankaranarayana Rao BS (2009). Enriched environment restores hippocampal cell proliferation and ameliorates cognitive deficits in chronically stressed rats. *J Neurosci Res* 87:831–843.

Veyrac A, Giannetti N, Charrier E, Reymond-Marron I, Aguera M, Rogemond V, Honnorat J, Jourdan F (2005). Expression of collapsin response mediator proteins 1, 2 and 5 is differentially regulated in newly generated and mature neurons of the adult olfactory system. *Eur J Neurosci* 21:2635–2648.

Vollmayr B, Simonis C, Weber S, Gass P, Henn F (2003). Reduced cell proliferation in the dentate gyrus is not correlated with the development of learned helplessness. *Biol Psychiatry* 54:1035–1040.

Wachs FP, Winner B, Couillard-Despres S, Schiller T, Aigner R, Winkler J, Bogdahn U, Aigner L (2006). Transforming growth factor-beta1 is a negative modulator of adult neurogenesis. *J Neuropathol Exp Neurol* 65:358–370.

Wagner JP, Black IB, DiCicco-Bloom E (1999). Stimulation of neonatal and adult brain neurogenesis by subcutaneous injection of basic fibroblast growth factor. *J Neurosci* 19:6006–6016.

Walker MP, Stickgold R (2004). Sleep-dependent learning and memory consolidation. *Neuron* 44:121–133.

Wallenborg K, Vlachos P, Eriksson S, Huijbregts L, Arner ES, Joseph B, Hermanson O (2009). Red wine triggers cell death and thioredoxin reductase inhibition: Effects beyond resveratrol and SIRT1. *Exp Cell Res* 315:1360–1371.

Wang L, Zhang ZG, Gregg SR, Zhang RL, Jiao Z, LeTourneau Y, Liu X, Feng Y, Gerwien J, Torup L, Leist M, Noguchi CT, Chen ZY, Chopp M (2007). The Sonic hedgehog pathway mediates carbamylated erythropoietin-enhanced proliferation and differentiation of adult neural progenitor cells. *J Biol Chem* 282:32462–32470.

Wang LP, Kempermann G, Kettenmann H (2005). A subpopulation of precursor cells in the mouse dentate gyrus receives synaptic GABAergic input. *Mol Cell Neurosci* 29:181–189.

Wang TW, Stromberg GP, Whitney JT, Brower NW, Klymkowsky MW, Parent JM (2006). Sox3 expression identifies neural progenitors in persistent neonatal and adult mouse forebrain germinative zones. *J Comp Neurol* 497:88–100.

Wang X, Mao X, Xie L, Greenberg DA, Jin K (2009). Involvement of Notch1 signaling in neurogenesis in the subventricular zone of normal and ischemic rat brain in vivo. *J Cereb Blood Flow Metab* 29:1644–54.

Wang Y, Baraban SC (2007). Granule cell dispersion and aberrant neurogenesis in the adult hippocampus of an LIS1 mutant mouse. *Dev Neurosci* 29:91–98.

Warner-Schmidt JL, Duman RS (2007). VEGF is an essential mediator of the neurogenic and behavioral actions of antidepressants. *Proc Natl Acad Sci USA* 104:4647–4652.

Wechsler-Reya RJ, Scott MP (1999). Control of neuronal precursor proliferation in the cerebellum by Sonic hedgehog. *Neuron* 22:103–114.

Weinstein DE, Burrola P, Kilpatrick TJ (1996). Increased proliferation of precursor cells in the adult rat brain after targeted lesioning. *Brain Res* 743:11–16.

Wen PH, Friedrich VL, Jr., Shioi J, Robakis NK, Elder GA (2002a). Presenilin-1 is expressed in neural progenitor cells in the hippocampus of adult mice. *Neurosci Lett* 318:53–56.

Wen PH, Shao X, Shao Z, Hof PR, Wisniewski T, Kelley K, Friedrich VL, Jr., Ho L, Pasinetti GM, Shioi J, Robakis NK, Elder GA (2002b). Overexpression of wild type but not an FAD mutant presenilin-1 promotes neurogenesis in the hippocampus of adult mice. *Neurobiol Dis* 10:8–19.

Wen PH, Hof PR, Chen X, Gluck K, Austin G, Younkin SG, Younkin LH, DeGasperi R, Gama Sosa MA, Robakis NK, Haroutunian V, Elder GA (2004). The presenilin-1 familial Alzheimer disease mutant P117L impairs neurogenesis in the hippocampus of adult mice. *Exp Neurol* 188:224–237.

Wentz CT, Magavi SS (2009). Caffeine alters proliferation of neuronal precursors in the adult hippocampus. *Neuropharmacology* 56:994–1000.

Werther GA, Abate M, Hogg A, Cheesman H, Oldfield B, Hards D, Hudson P, Power B, Freed K, Herington AC (1990). Localization of insulin-like growth factor-I mRNA in rat brain by in situ hybridization—relationship to IGF-I receptors. *Mol Endocrinol* 4:773–778.

Wexler EM, Geschwind DH, Palmer TD (2008). Lithium regulates adult hippocampal progenitor development through canonical Wnt pathway activation. *Mol Psychiatry* 13:285–292.

Wexler EM, Paucer A, Kornblum HI, Plamer TD, Geschwind DH (2009). Endogenous Wnt signaling maintains neural progenitor cell potency. *Stem Cells* 27:1130–1141.

Willert K, Brown JD, Danenberg E, Duncan AW, Weissman IL, Reya T, Yates JR, 3rd, Nusse R (2003). Wnt proteins are lipid-modified and can act as stem cell growth factors. *Nature* 423:448–452.

Wilson PA, Hemmati-Brivanlou A (1997). Vertebrate neural induction: Inducers, inhibitors, and a new synthesis. *Neuron* 18:699–710.

Wimer CC, Wimer RE (1989). On the sources of strain and sex differences in granule cell number in the dentate area of house mice. *Brain Res Dev Brain Res* 48:167–176.

Winner B, Cooper-Kuhn CM, Aigner R, Winkler J, Kuhn HG (2002). Long-term survival and cell death of newly generated neurons in the adult rat olfactory bulb. *Eur J Neurosci* 16:1681–1689.

Winner B, Lie DC, Rockenstein E, Aigner R, Aigner L, Masliah E, Kuhn HG, Winkler J (2004). Human wild-type alpha-synuclein impairs neurogenesis. *J Neuropathol Exp Neurol* 63:1155–1166.

Wolf SA, Steiner B, Wengner A, Lipp M, Kammertoens T, Kempermann G (2009a). Adaptive peripheral immune response increases proliferation of neural precursor cells in the adult hippocampus. *Faseb J* 23:3121–3128.

Wolf SA, Steiner B, Akpinarli A, Kammertoens T, Nassenstein C, Braun A, Blankenstein T, Kempermann G (2009b). CD4-positive T lymphocytes provide a neuroimmunological link in the control of adult hippocampal neurogenesis. *J Immunol* 182:3979–3984.

Won SJ, Kim SH, Xie L, Wang Y, Mao XO, Jin K, Greenberg DA (2006). Reelin-deficient mice show impaired neurogenesis and increased stroke size. *Exp Neurol* 198:250–259.

Wong EY, Herbert J (2004). The corticoid environment: A determining factor for neural progenitors' survival in the adult hippocampus. *Eur J Neurosci* 20:2491–2498.

Wong EY, Herbert J (2005). Roles of mineralocorticoid and glucocorticoid receptors in the regulation of progenitor proliferation in the adult hippocampus. *Eur J Neurosci* 22:785–792.

Woods-Kettelberger A, Kongsamut S, Smith CP, Winslow JT, Corbett R (1997). Animal models with potential applications for screening compounds for the treatment of obsessive-compulsive disorder. *Expert Opin Investig Drugs* 6:1369–1381.

Wright EM, Vogel KS, Davies AM (1992). Neurotrophic factors promote the maturation of developing sensory neurons before they become dependent on these factors for survival. *Neuron* 9:139–150.

Wu CW, Chang YT, Yu L, Chen HI, Jen CJ, Wu SY, Lo CP, Kuo YM (2008). Exercise enhances the proliferation of neural stem cells and neurite growth and survival of neuronal progenitor cells in dentate gyrus of middle-aged mice. *J Appl Physiol* 105:1585–1594.

Yagita Y, Kitagawa K, Sasaki T, Miyata T, Okano H, Hori M, Matsumoto M (2002). Differential expression of Musashi1 and nestin in the adult rat hippocampus after ischemia. *J Neurosci Res* 69:750–756.

Yamada M, Onodera M, Mizuno Y, Mochizuki H (2004). Neurogenesis in olfactory bulb identified by retroviral labeling in normal and 1-methyl-4-phenyl-1,2,3,6-tetrahydropyridine-treated adult mice. *Neuroscience* 124:173–181.

Yamaguchi M, Mori K (2005). Critical period for sensory experience-dependent survival of newly generated granule cells in the adult mouse olfactory bulb. *Proc Natl Acad Sci USA* 102:9697–9702.

Yamaguchi M, Saito H, Suzuki M, Mori K (2000). Visualization of neurogenesis in the central nervous system using nestin promoter-GFP transgenic mice. *Neuroreport* 11:1991–1996.

Yamamoto T, Hirayama A, Hosoe N, Furube M, Hirano S (2009). Soft-diet feeding inhibits adult neurogenesis in hippocampus of mice. *Bull Tokyo Dent Coll* 50:117–124.

Yang P, Arnold SA, Habas A, Hetman M, Hagg T (2008). Ciliary neurotrophic factor mediates dopamine D2 receptor-induced CNS neurogenesis in adult mice. *J Neurosci* 28:2231–2241.

Yoo KY, Choi JH, Hwang IK, Lee CH, Lee SO, Han SM, Shin HC, Kang IJ, Won MH (2009). (-)-epigallocatechin-3-gallate increases cell proliferation and neuroblasts in the subgranular zone of the dentate gyrus in adult mice. *Phytother Res*.

Yoshimizu T, Chaki S (2004). Increased cell proliferation in the adult mouse hippocampus following chronic administration of group II metabotropic glutamate receptor antagonist, MGS0039. *Biochem Biophys Res Commun* 315:493–496.

Yoshimura S, Takagi Y, Harada J, Teramoto T, Thomas SS, Waeber C, Bakowska JC, Breakefield XO, Moskowitz MA (2001). FGF2 regulation of neurogenesis in adult hippocampus after brain injury. *Proc Natl Acad Sci USA* 98:5874–5879.

Young D, Lawlor PA, Leone P, Dragunow M, During MJ (1999). Environmental enrichment inhibits spontaneous apoptosis, prevents seizures and is neuroprotective. *Nat Med* 5:448–453.

Young KM, Merson TD, Sotthibundhu A, Coulson EJ, Bartlett PF (2007). p75 neurotrophin receptor expression defines a population of BDNF-responsive neurogenic precursor cells. *J Neurosci* 27:5146–5155.

Yu IT, Park JY, Kim SH, Lee JS, Kim YS, Son H (2009). Valproic acid promotes neuronal differentiation by induction of proneural factors in association with H4 acetylation. *Neuropharmacology* 56:473–480.

Zajac MS, Pang TY, Wong N, Weinrich B, Leang LS, Craig JM, Saffery R, Hannan AJ (2009). Wheel running and environmental enrichment differentially modify exon-specific BDNF expression in the hippocampus of wild-type and pre-motor symptomatic male and female Huntington's disease mice. *Hippocampus* 20:621–36.

Zhang CL, Zou Y, He W, Gage FH, Evans RM (2008). A role for adult TLX-positive neural stem cells in learning and behavior. *Nature* 2008.

Zhang P, Liu Y, Li J, Kang Q, Tian Y, Chen X, Shi Q, Song T (2006). Cell proliferation in ependymal/subventricular zone and nNOS expression following focal cerebral ischemia in adult rats. *Neurol Res* 28:91–96.

Zhang R, Zhang L, Zhang Z, Wang Y, Lu M, Lapointe M, Chopp M (2001). A nitric oxide donor induces neurogenesis and reduces functional deficits after stroke in rats. *Ann Neurol* 50:602–611.

Zhao M, Li D, Shimazu K, Zhou YX, Lu B, Deng CX (2007). Fibroblast growth factor receptor-1 is required for long-term potentiation, memory consolidation, and neurogenesis. *Biol Psychiatry* 62:381–390.

Zhao N, Zhong C, Wang Y, Zhao Y, Gong N, Zhou G, Xu T, Hong Z (2008). Impaired hippocampal neurogenesis is involved in cognitive dysfunction induced by thiamine deficiency at early pre-pathological lesion stage. *Neurobiol Dis* 29:176–185.

Zhao X, Ueba T, Christie BR, Barkho B, McConnell MJ, Nakashima K, Lein ES, Eadie BD, Willhoite AR, Muotri AR, Summers RG, Chun J, Lee KF, Gage FH (2003). Mice lacking methyl-CpG binding protein 1 have deficits in adult neurogenesis and hippocampal function. *Proc Natl Acad Sci USA* 100:6777–6782.

Zhao Y, Pan X, Zhao J, Wang Y, Peng Y, Zhong C (2009). Decreased transketolase activity contributes to impaired hippocampal neurogenesis induced by thiamine deficiency. *J Neurochem* 111:537–546.

Zheng M, Leung CL, Liem RK (1998). Region-specific expression of cyclin-dependent kinase 5 (cdk5) and its activators, p35 and p39, in the developing and adult rat central nervous system. *J Neurobiol* 35:141–159.

Zhu DY, Liu SH, Sun HS, Lu YM (2003). Expression of inducible nitric oxide synthase after focal cerebral ischemia stimulates neurogenesis in the adult rodent dentate gyrus. *J Neurosci* 23:223–229.

Zhu DY, Lau L, Liu SH, Wei JS, Lu YM (2004). Activation of cAMP-response-element-binding protein (CREB) after focal cerebral ischemia stimulates neurogenesis in the adult dentate gyrus. *Proc Natl Acad Sci USA* 101:9453–9457.

Zhu XJ, Hua Y, Jiang J, Zhou QG, Luo CX, Han X, Lu YM, Zhu DY (2006). Neuronal nitric oxide synthase-derived nitric oxide inhibits neurogenesis in the adult dentate gyrus by down-regulating cyclic AMP response element binding protein phosphorylation. *Neuroscience* 141:827–836.

Ziv Y, Ron N, Butovsky O, Landa G, Sudai E, Greenberg N, Cohen H, Kipnis J, Schwartz M (2006). Immune cells contribute to the maintenance of neurogenesis and spatial learning abilities in adulthood. *Nat Neurosci* 9:268–275.

Zurawel RH, Chiappa SA, Allen C, Raffel C (1998). Sporadic medulloblastomas contain oncogenic beta-catenin mutations. *Cancer Res* 58:896–899.

10

Function

Specific function is the key to the relevance of adult neurogenesis

ADULT NEUROGENESIS GAINS all its relevance from the fact that the new neurons are functional. Most people intuitively assume that new neurons in the adult brain are beneficial and might provide a means to actively improve brain function and cognition. It seems almost trivial to say that a neuron is not a true neuron unless it functions as one.

Functional new neurons should lead to altered functionality of the hippocampus and the olfactory bulb, which in turn should have behavioral consequences. The issue that *function* means very different things on these different levels deserves our attention and is discussed below. From the perspective of such "compound function," there are profound differences between the two adult neurogenic regions (Table 10–1). Table 10–1 should be seen in the context of the related Table 8–2, p. 288, which makes a similar comparison with regard to the niche and key anatomical properties of adult neurogenesis.

This chapter will focus entirely on the functional relevance of adult neurogenesis in the rodent brain. For information on neurogenesis in the vocal system of songbirds, see Chapter 11, p. 491.

In this chapter there is no speculation about potential functionality and functional relevance of regenerative or induced neurogenesis outside the hippocampus and olfactory bulb. The available experimental evidence is still too limited. In the end, to prove function will be crucial to demonstrate the clinical relevance of regenerative neurogenesis. But so far, very little has been achieved in this regard.

The Stability–Plasticity Dilemma

Do complex brains tend to have less neurogenesis than simpler ones?

The brain is no silicone-chip–based computer with a clear distinction between hardware and software. The very nature of the brain is that it is plastic: structure and function are intricately interwoven and can be considered two faces of the same coin. There is no such thing as functional plasticity without structural changes. These changes can occur at different scales, however, ranging from sub-synaptic adaptations to adult neurogenesis. As we will discuss in Chapter 11, p. 484, many researchers believe that over the course of evolution there is so-called phylogenetic reduction in neurogenesis. They claim that, during the course of evolution, the ability to undergo adult neurogenesis decreased with growing

	Hippocampus	Olfactory bulb
Table 10–1 Differences in neurogenesis-dependent function between hippocampus and olfactory bulb		
Adult neurogenesis generates ...	One type of principle neuron (granule cell)	Several types of interneurons
New neurons are ...	Excitatory	Inhibitory (plus one type of glutamatergic interneuron: Brill et al., 2009)
	Largely additive	Largely turnover
Functional context on the systems level	Memory formation	Sensory input
	Affective behavior	(Affective behavior)
Functionality is ...	Probably dual: transient during acute processing plus long-lasting network adaptations	Not known
Putative key function of the new neurons	1. Pattern separation and avoidance of catastrophic interference 2. Temporal and contextual encoding 3. Long-term memory vs. extinction 4. Accelerating processing and clearance time 5. Flexible relearning and advanced functionality	1. Olfactory discrimination 2. Retaining olfactory memory

brain complexity. From this observation, if it is true, one might be tempted to conclude that the most complex brain functions as found in primates and humans are not compatible with the ability to produce new neurons and integrate them into the functional networks. However, it is difficult to evaluate whether function of the spinal cord in lizards, which shows robust and extensive adult neurogenesis, is in a fundamental sense less complex than in a human. And as discussed in Chapter 11, as a "law of nature," the idea of phylogenetic reduction rests on shaky grounds.

Adult-generated neurons were thought to disrupt existing memories

Nevertheless, in 1985, Pasko Rakic hypothesized that, during the course of evolution, adult neurogenesis might have been reduced and was not found in the primate brain (Rakic, 1985). One of his arguments was based on the stability–plasticity dilemma, a fundamental issue in network theory. A neuronal network that is very plastic will learn very well, but what has been learned will not last. Any new information will overwrite the previously learned items. A stable network can remember for long times and is resistant to this damage due to new information. But the price it has to pay is that it cannot learn new things. Consequently, a balance between the maintenance of stability and the allowance of plasticity has to be found. Rakic argued that new neurons would be disruptive to existing networks. This argument, which is plausible at face value, does not hold up, however, for all types of networks. Modeling studies show that neuronal networks exist (e.g., the so-called cascade correlation models and several others) that not only accommodate new neurons but actually call for them. In biological neuronal networks, plasticity is not limitless but regulated and controlled, dependent on the computational needs.

The stability–plasticity dilemma needs to be solved in any network that is supposed to learn

The issue, here brought up by Rakic, is central to the considerations about the functions of adult neurogenesis. But in contrast to his conclusions at the time, adult neurogenesis might now be considered a particular means to solve the stability–plasticity dilemma (Wiskott et al., 2006). Independent of the degree to which different species rely on adult neurogenesis and why, there is probably a balance between the benefits that can be gained from new neurons and the problems the integration of new cells causes for the network structure. In the two neurogenic regions of the adult mammalian brain, the benefits seem to outweigh the disadvantages. Given the fact that we generally still know relatively little about the exact function of individual brain parts (and this even applies to the heavily studied hippocampus), and much less is known about their precise contribution to cognition, the role of new neurons within these functions is difficult to determine.

What Does "Function" Mean?

Function can be assessed on the cellular, network, and systems level and beyond

The term *function* has theoretical and practical meanings on different conceptual levels. To speak of *function* therefore often remains vague. We are left with the considerable challenge of developing concepts of what function should mean in the context of adult neurogenesis. To date, adult neurogenesis has been examined on the levels of individual cells, neuronal networks, and neuronal systems (Fig. 10–1). Function can be attributed to all of these levels. In the olfactory bulb, the *cellular* level refers to the physiology of the new granule cells, the *network* level to the integration of these cells into the circuitry of the olfactory bulb, and the *systems* level to the question of how the new neurons might contribute to olfaction. Analogously, in the hippocampus, study at the cellular level involves the functional properties of individual new granule cells; at the network level, analysis of the means by which the new granule cells are integrated into

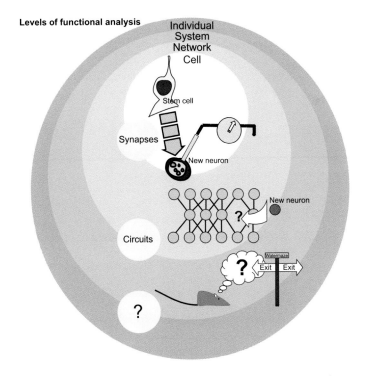

FIGURE 10–1 Levels of function in adult neurogenesis. Just as regulation occurs on several conceptual levels (Fig. 9–1), *function* is an elusive term that can be addressed with different meanings, from single-cell level to complex behaviors. Figure from Kempermann et al. (2004); reprinted with kind permission from Elsevier, copyright 2004.

the neuronal circuits of the dentate gyrus and the mossy fiber tract; and at the systems level, investigation of ways in which new granule cells might contribute to hippocampus-dependent learning and memory. Tests of function are very different at the different conceptual levels. Thus *function* will mean very different things, depending on the experimental situation.

In addition, the conceptual levels overlap. They are helpful constructs but are not strictly separated in reality. Neurons form synapses to build networks. Integration of different local networks leads to the establishment of the complex circuitry in functional systems, such as the hippocampus or the olfactory system. Neither synapses, neurons, local networks, nor systems are functionally self-sustained islands. The higher the conceptual level, the more its functions rely on the integrity of function at the levels below.

Action potentials can be detected in neurons in vitro, but olfaction, which clearly depends on action potential in many neurons, cannot be fully represented in single cells. This is not a trivial point, because the mechanism of synaptic strengthening, long-term potentiation that can be studied in living hippocampal slices (and thus on a network level) is thought to be the synaptic mechanism underlying learning. When *learning* is implicitly used in a behavioral sense (which is not the same as the learning of the neuronal networks of computational neurosciences), this tacit transfer can become problematic. Consequently, whether LTP *is* learning has been debated for decades, despite very good evidence that a profound link between them exists. But the fundamental problem is that the similar language and concepts used at the different levels are not identical; implications vary with the concepts.

Matters become even more complicated when a functional significance of the new neurons is inferred from the three levels of cells, networks, and systems, leading to a fourth level, the cognitive and behavioral level. *Cognition* and *behavior* in the laboratory sense tend to be far away from what the same terms mean under feral conditions. This is the level of the individual as it behaves and is integrated into its psychological, historical, and social contexts. Extrapolations from reductionistic studies on animal models can provide insight into the broader picture of neuronal function, but they nevertheless usually remain reduced in their perspective.

Functions on different conceptual levels are not independent from each other

Experimental assessments of function allow only approximations of cognitive and behavioral function under real-life conditions

Signs of Neuronal Function in Individual New Neurons

Neuronal features like sodium currents and excitability to generate action potential can be assessed in isolated neurons in vitro

The essence of neuronal function is communication. Consequently, only limited conclusions about function can be drawn from individual neurons. Nevertheless, in many experimental contexts, single cells are the only available objects of functional investigations, and it is thus useful to consider the function of single new neurons. The need to prove neuronal differentiation by these means is most evident in studies in which neurons are grown from precursor cells in vitro (for example: Berninger et al., 2007; Song et al., 2007; Erceg et al., 2008; Husseini et al., 2008). Critical properties are the development of sodium currents, a neuronal membrane potential, and excitability leading to the generation of action potentials. Pharmacological interventions are used to demonstrate the involvement of particular receptor systems.

Precursor cells from the adult neurogenic zones generated appropriate functional neuronal phenotypes in vitro

In cultures of adult neural precursor cells, the maturation of the appropriate granule cell-like neuronal phenotype from isolated hippocampal precursor cells was demonstrated by electrophysiological measurements (Song et al., 2002; Babu et al., 2007). The same applies to the olfactory bulb, but due to the more complex lineage relationships and the greater variety of interneurons, the result is not as homogenous (Brill et al., 2009).

In vivo there is a sequence of
electrophysiological maturation
stages in adult hippocampal
neurogenesis

In vivo, the membrane properties and currents of individual identified cells can be studied in the course of neuronal development with patch-clamp techniques. For both the olfactory system and the hippocampus, sequences of "electrophysiological maturation" have been described (Carlen et al., 2002; Filippov et al., 2003; Fukuda et al., 2003; Ambrogini et al., 2004b) (Fig. 10–2). The radial glia–like precursor cells of the subgranular zone (SGZ) and the subventricular zone (SVZ) share, not only morphological characteristics, but also electrophysiological features with astrocytes: they have passive membrane properties and show potassium currents (Filippov et al., 2003; Fukuda et al., 2003). In addition, type-1 (and presumably type-2a) cells are coupled by gap junctions (Kunze et al., 2009). Many type-2 progenitor cells of the SGZ have electrophysiological properties that have been named "complex" and originally were described for glial precursor cells (Steinhauser et al., 1994). Nevertheless, the transiently amplifying type-2 progenitor cells, identified by their nestin-expression, contribute to the neuronal lineage because early neuronal markers and sodium currents can be found in a small portion of type-2 cells (Filippov et al., 2003). Presumably, this distinction relates to the difference between the glia-like type-2a and the neuronally committed type-2b cells.

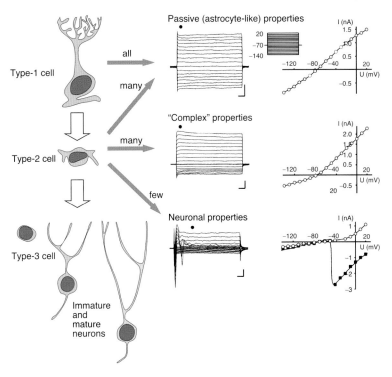

FIGURE 10–2 Membrane properties of nestin green fluorescent protein-expressing precursor cells in the dentate gyrus. Radial glia–like type-1 cells show the electrophysiological characteristics of astrocytes: passive membrane properties and potassium currents. Type-2 cells can have astrocytic or "complex" properties. The complex pattern was originally described for oligodendrocyte precursor cells (Steinhauser et al., 1994) and appears to be a shared feature of many nestin- and NG2-expressing cells throughout the brain. A small percentage of the nestin-GFP-expressing cells show neuronal characteristics with sodium currents, indicative of the first stages of neuronal differentiation. The presented data are based on patch-clamp recordings by Li-Ping Wang and Helmut Kettenmann, Berlin.

| Responsiveness to GABA is critical during the differentiation and maturation stage | After a first study of electrophysiological differentiation and maturation in the adult hippocampus by Ambrogini and colleagues from Urbino, who used an immunohistochemical post-hoc identification of the patch-clamped cells filled with biocytin and β-III-tubulin expression (Ambrogini et al., 2004b), Esposito and colleagues from Alejandro Schinder's group in Buenos Aires provided a very detailed account of adult granule cell maturation based on a retroviral labeling method (Esposito et al., 2005). Both |

studies show that in the course of their development, the new neurons became increasingly hyperpolarized, progressed from low to high capacitance, and from receiving GABAergic input to glutamatergic input. While input resistance decreased over a maturation period of about four weeks, excitability increased. GABAergic input, initially ambient, later synaptic, reaches the cells as early as at the type-2a stage (Wang et al., 2005). This input was excitatory until connections had been made to the entorhinal cortex and CA3, when GABA became inhibitory (Esposito et al., 2005; Ge et al., 2006; Karten et al., 2006). During the phase of excitatory GABA, GABA signaling promoted maturation (Tozuka et al., 2005; Ge et al., 2006). See also Chapter 6, p. 198.

| Co-culture models reveal the influence of the niche on developing neuronal functionality | Co-culture systems and organotypic slice cultures add a moment of cellular interaction to in vitro studies and cross the border into studies on the network level. Hongjun Song and colleagues from the Salk Institute co-cultured adult hippocampal precursor cells with astrocytes from the same region and showed that this induced the development of electrically active neurons that formed synaptic networks in the culture dish (Song et al., 2002). This was the first demonstration that, within the highly reductionistic system in vitro, a meaningful level of neuronal function can be |

achieved by specifically introducing a natural cellular environment. Similar observations could be made with neuronal co-cultures (Deisseroth et al., 2004; Babu et al., 2007). When neurons derived from embryonic stem cells in vitro were implanted in early postnatal hippocampal slice cultures, over time the new cells expressed voltage-gated ion channels and a mature profile of receptors and synaptically integrated into the host tissue (Benninger et al., 2003).

Local Network Integration of New Neurons In Vivo

Anatomical Integration

| Precursor cells in the SGZ receive synaptic input | Functional integration on a network level in vivo requires the establishment of synapses. A small number of nestin-expressing cells and many β-III-tubulin- or doublecortin (DCX)-positive cells in the dentate gyrus receive synaptic input (van Praag et al., 2002; Filippov et al., 2003; Ambrogini et al., 2004b; Schmidt-Hieber et al., 2004; Esposito et al., 2005; Wang et al., 2005). These findings imply that the first signs of synaptic |

activity and thus network integration are found at a precursor cell stage. The very first synapses have not yet been unambiguously visualized. Precursor cells also respond to ambient GABA, independent of synapses. Because new neurons appear to serve functions before of their full maturation, relevance of their first synaptic integration will go beyond the modulation of further development.

| New neurons in the adult dentate gyrus function as granule cells | Henriette van Praag and colleagues were the first to succeed in a prospective electrophysiological characterization of newly generated granule cells (van Praag et al., 2002). Their study provided the proof that adult-generated neurons become electrophysiologically functional in vivo and are integrated as granule cells. The new neurons were not identified post hoc, but dividing cells were labeled with a retrovirus expressing green fluorescent protein in vivo. Their living progeny could thus be identified under the |

fluorescent microscope. After a maturation period of about seven weeks, the new granule cells showed electrophysiological properties very similar to those of the older granule cells. The finding of increased synaptic plasticity during the course of granule cell development (Wang et al., 2000; Snyder et al., 2001; Schmidt-Hieber et al., 2004) suggests that the cautious statement of "similarity" opens opportunities for further research. The current consensus seems to be, however, that newborn neurons ultimately become indistinguishable from their older siblings.

Functional maturation of newborn neurons from the SVZ takes place after migration to the olfactory bulb

During neuronal development in the adult olfactory bulb with its long migratory phase, this functional maturation appears delayed until the new cells have reached the olfactory bulb. In a groundbreaking study, Alan Carleton, Pierre-Marie Lledo, and colleagues from Paris applied the same experimental approach as chosen by van Praag for the hippocampus and followed migrating enhanced green fluorescent protein (EGFP)–marked neuroblasts from the SVZ through the rostral migratory stream (RMS) to the olfactory bulb (Carleton et al., 2003). Here, too, the expression of GABA receptors preceded the appearance of glutamate receptors. Five distinct stages of neuronal development could be distinguished, beginning with the migratory neuroblasts of the RMS that show immature properties but can already express first glutamate receptors. See below, p. 474, for more details.

Within days after exit from cell cycle, new neurons establish axonal contacts to CA3

For fully functional integration, the new neurons have to extend dendrites and axons and form dendritic and axonal synapses in the appropriate target regions. Using retrograde labeling, Stanfield and Trice were the first to show that the new granule cells in the adult dentate gyrus extend axons along the mossy fiber tract (Stanfield and Trice, 1988). This study was truly pioneering because it predated the general interest in this question by several years. Axonal projections of new granule cells to CA3 were later confirmed (Hastings and Gould, 1999; Markakis and Gage, 1999). Retrograde labeling consists of injecting a dye, usually fluorogold, into the target region to which the neurons in question project. Neurons will take up the dye through their axons. Tracers such as fluorogold are retrogradely transported along the axon to the cell soma, where they accumulate. Consequently, all regions that project to the injection site will be visualized with fluorogold. To prove that new hippocampal granule cells project along the mossy fiber tract, animals received BrdU injection to label the new neurons. Several weeks later, fluorogold was injected into CA3. The dye was taken up into the mossy fiber terminals and retrogradely transported to the granule cell layer, where the cells that had incorporated fluorogold became fluorescent. Immunohistochemistry for BrdU clarified whether these cells were newly generated, and in fact a small percentage of them were. The labeling study by Elizabeth Gould's group provided the additional piece of information that the extension of the new axon is very fast and can occur within days after labeling with BrdU (Hastings and Gould, 1999). After leaving the cell cycle, terminal differentiation is initiated; it seems that axon elongation is an early part of this. This phase of increased plasticity is also characterized by the transient expression of growth cone–associated protein TUC4 (see Chap. 7). The formation of synapses between new neurons and CA3 pyramidal neurons is described in Chapter 6, p. 204. In targeted regenerative neurogenesis in the murine neocortex after a directed ablation of corticothalamic neurons, retrograde labeling similarly confirmed that the new (BrdU-positive) neurons had normal projections to their physiological target in the thalamus (Magavi et al., 2000).

Synaptic integration of new neurons can also be demonstrated with pseudorabies virus

Jonas Frisén's group in Stockholm proved synaptic integration with pseudorabies virus–expressing EGFP. The particular property of rabies viruses is that they are propagated strictly transsynaptically. After injection into the piriform cortex, EGFP expression was found in BrdU-labeled periglomerular neurons. Because the interneurons do not project to the piriform cortex, whereas the principal neurons, onto which the interneurons synapse, do, EGFP in the periglomerular cells demonstrated integration of the new neurons into the synaptic network. Similarly, injection into CA1 led to EGFP-positive new neurons of the dentate gyrus and pyramidal neurons of CA3 to which the granule cells projected, suggesting that the new granule cells became part of the synaptic network (Carlen et al., 2002).

Functional Responsiveness of New Neurons

The retrovirus-based morphological maturation studies have greatly expanded our knowledge about the functional integration of new neurons (Zhao et al., 2006). These studies have raised two questions. First, does an entire cohort of new cells become integrated in parallel? The retrovirus-based experiments identified only individual cells. What can be said about functional integration across the entire population of newly generated cells? Second, does functional integration also imply that the new cells would respond to physiological stimuli known to activate the respective neuronal populations into which the new cells are born? Both questions have been addressed with experimental paradigms that make use of the fact that synaptic activation induces fast up-regulation of immediate early genes such as c-fos, zif268, or Homer1A. Their transcribed proteins can be detected immunohistochemically. This allows assessment of the responsiveness of all cells visible in a tissue section. Also, the analysis can be made quantitative by determining the numbers of activated cells and be combined with BrdU-immunohistochemistry to determine the ratio of new versus old activated neurons.

Immediate early gene regulation can be used to visualize neuronal activation after specific stimuli. Training in the hippocampus-dependent learning task of the Morris water maze induced c-fos expression in single granule cells, including newly generated cells (Jessberger and Kempermann, 2003). In an experimental *tour de force,* Paul Frankland and colleagues showed that such activated neurons become preferentially recruited into networks activated by the learning task (Fig. 10–3) (Kee et al., 2007). With time after their birth, the likelihood of becoming recruited first increased until reaching a plateau (and presumably decreasing thereafter). There was no transient responsiveness in one-week-old neurons, at least not on the basis of c-fos expression. As no response was detected before the neurons were two weeks old, the initiation of responsiveness coincided with the establishment of glutamatergic input to the new neurons, which occurs approximately at this time (Kee et al., 2007).

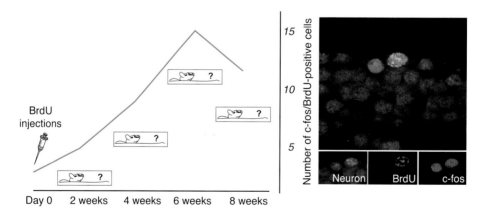

FIGURE 10–3 Sensitive period during adult neurogenesis. New hippocampal neurons go through a period during which they are particularly responsive to a learning stimulus, here the Morris water maze. New neurons were labeled with BrdU, and their activation by the learning stimulus was assessed by c-fos immunohistochemistry. Kee et al. (2007), whose results are presented here in a schematic rendering, also showed that the activated neurons become preferentially integrated into the network.

More evidence for this critical time window came from a study on the recruitment of new neurons after exposure to an enriched environment. Environmental enrichment increased not only the total number of new neurons but also the number of new neurons that specifically responded after re-exposure to the same environment. That increase was only seen after exposure to the same environment at about two to three weeks after they had been born, but not to a different experience and at different times (Tashiro et al., 2007). Finally, that time window was also observed with a direct induction of long-term potentiation via electrodes and the expression of immediate early gene zif-268 (Bruel-Jungerman et al., 2006).

Most new neurons become responsive to excitatory stimuli over a period of seven weeks

Systemic injections of kainic acid led to a generalized synaptic activation of hippocampal granule cells that in turn caused a detectable up-regulation of immediate early proteins (c-fos, zif-268, and homer 1A) in about 80% of the cells (Jessberger and Kempermann, 2003). When this type of general activation was applied at different time-points after labeling the newborn cells with BrdU, it was found that over a period of weeks, the new cells matured into full activity-dependent responsiveness. At two weeks after BrdU labeling, kainic acid injections did not lead to any measurable immediate early gene regulation in BrdU-labeled cells. After about four weeks, roughly half of the new cells responded, and after seven weeks, the rate of 80%, found also in the older granule cells, was detected (Jessberger and Kempermann, 2003).

In the olfactory bulb, Jonas Frisén's group showed that olfactory stimuli evoked an up-regulation of c-fos gene expression in newborn neurons, similarly supporting the functional synaptic integration (Carlen et al., 2002).

The Contribution of New Neurons to Long-Term Potentiation in the Hippocampus

Unless the general network inhibition is released, LTP in the dentate gyrus originates from the synaptic plasticity of new neurons

Sabrina Wang and Martin Wojtowicz from the University of Toronto were the first to compare the properties of immature neurons in the dentate gyrus with those of older granule cells (Wang et al., 2000). The younger cells were identified by morphology because of their sparser dendritic tree, and post hoc by immunohistochemistry against TUC4. The intriguing result was that new neurons showed a higher degree of synaptic plasticity. Long-term potentiation was induced at a lower threshold, and the cells were entirely insensitive to $GABA_A$ inhibition (Wang et al., 2000). This is consistent with the idea that GABA acts excitatory at the newly generated cells. Consequently, if new neurons are less inhibited than the more mature cells, LTP induction under superfusion with artificial cerebrospinal fluid should reflect LTP by the new cells, whereas treatment with a blocker of GABAergic synapses should elicit a response from young and old cells. Indeed, LTP induction through stimulation of the medial perforant path showed the facilitated LTP induction supposedly attributable to the immature cells under physiological conditions, and an increase in LTP under GABA blockade that required more stimuli and higher frequency (Snyder et al., 2001). On the basis of post-hoc immunohistochemistry against PSA-NCAM, Christoph Schmidt-Hieber and coworkers (2004) found that such enhanced synaptic plasticity was due to isolated Ca^{2+} spikes in young neurons that promote fast action potentials (Schmidt-Hieber et al., 2004). Low-dose irradiation of the brain prior to the electrophysiological examination in vitro reduced adult hippocampal neurogenesis and abolished the LTP that can be induced by stimulating the medial perforant path (Snyder et al., 2001; Saxe et al., 2006); and the same observation was made after cytostatic ablation of adult neurogenesis (Garthe et al., 2009) and in genetic ablation models (Bergami et al., 2008). Based on these observations, the now widely accepted opinion is that the LTP that can be detected under aCSF conditions in the adult dentate gyrus is due to the contribution of the newborn neurons (Fig. 10–4).

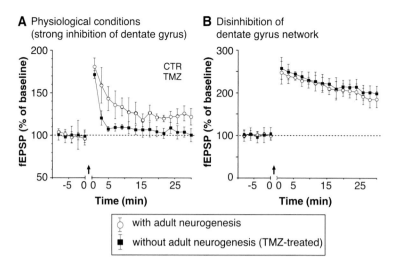

A Physiological conditions
(strong inhibition of dentate gyrus)

B Disinhibition of
dentate gyrus network

with adult neurogenesis

without adult neurogenesis (TMZ-treated)

FIGURE 10–4 Neurogenesis in the adult hippocampus is responsible for LTP in the dentate gyrus. Network activity in the dentate gyrus is physiologically inhibited by local interneurons. (A): Electrophysiological recordings in the presence of artificial cerebrospinal fluid thus detect only LTP from neurons that have not yet become inhibited (*open circles*). These are the new neurons, which GABA excites for a transient period during their development. If adult neurogenesis is suppressed (e.g., with cytostatic drug temozolomide [TMZ]), this type of LTP is eliminated (*filled boxes*). (B): If the network is disinhibited, normal LTP is elicited from both treated and untreated hippocampi. Recordings were performed by Alexander Garthe and Joachim Behr; the figure is modified from Garthe et al., 2009.

Concepts of Hippocampal Function

‖‖‖

The hippocampus consolidates declarative contents for long-term memory

‖‖‖

The hippocampus is metaphorically referred to as the "gateway to memory." A widely accepted idea is that the hippocampus *consolidates memory*, meaning that it processes information for long-term storage. These functions include pattern separation, so that units of information remain distinct; compression, so that memory load is decreased; and annotation, so that information is associated with temporal, spatial, or emotional contexts. Not all types of memory formation require a hippocampal contribution, however. Procedural learning—i.e., the acquisition of motor skills; or priming—i.e., types of instinctive learning, are independent of the hippocampus. One type of learning that requires processing in the hippocampus is the learning of declarative contents. For humans *declarative* can be quite literally be explained as information that can be "declared" or put in words. A more general definition of declarative memory is knowledge of facts and events: knowledge about facts is called "semantic memory," and knowledge about events is called "episodic memory."

‖‖‖

The hippocampus also plays a role in memory retrieval

‖‖‖

To achieve memory consolidation, the hippocampus provides means to link information to contexts and to place it into internal coordinate systems. The hippocampus might also help identify consistencies across many experiences and thereby prepare the grounds for future association (Eichenbaum, 2003; Wirth et al., 2003). It seems that all these maps, however, are not stored in the hippocampus itself (Eichenbaum et al., 1999). Rather, the processed information is stored primarily elsewhere in the cortex and, together with their coordinates, becomes independent of the hippocampus. The process of consolidation can take days to weeks, and during this time the information to be stored remains sensitive to modification. The hippocampus is also involved in the retrieval of stored information. During recall, the stored information can again become vulnerable to alterations, further suggesting that the hippocampus plays an active role in this process.

The temporospatial information associated with the stored contents is the basis of episodic memory (Fortin et al., 2002). In addition, the hippocampus allows us to deal with onetime experiences (Nakazawa et al., 2003): in a constantly changing world, most events occur only once. The integration of the hippocampus into the circuitry of the limbic system with its role in emotion is thought to play a central role in the evaluation of facts or events. For episodic memory, one-trial learning is indispensable. One-trial learning in turn requires at least transient storage of information in the hippocampus. This transient memory may be located in CA3 (Nakazawa et al., 2003).

Network Theories of the Dentate Gyrus and Hippocampus

The hippocampus has a network architecture that is relatively well understood. It consists of an essentially tri-synaptic backbone (Fig. 10–5). The first relay station within this core circuit is the dentate gyrus; the second, CA3; the third, CA1. Excitation follows this path coming from the entorhinal cortex, in which input from many other brain regions, including the sensory areas, converges. From CA1, the flow of information goes via the subiculum back to the entorhinal cortex and from the entorhinal cortex to cortical association areas. Several shortcuts and side paths complicate the general pattern, but for most considerations these complications are usually ignored. Within the tri-synaptic path, adult neurogenesis occurs only in the dentate gyrus and thus has a direct influence on only the number of granule cells in the dentate gyrus and the mossy fiber connections built by the axons of the granule cells and linking the dentate gyrus with CA3.

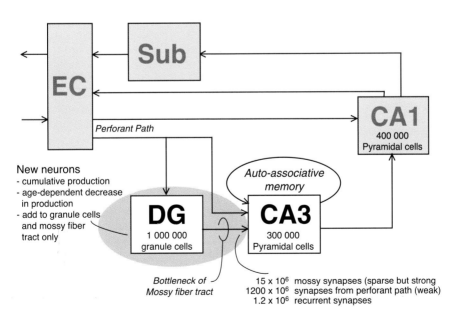

FIGURE 10–5 The main circuitry of the hippocampus. The input into hippocampal formation from the entorhinal cortex (EC) and dentate gyrus (DG) is a bottleneck within the network of the hippocampus. The numbers are based on a review by Treves and Rolls (1994). Sub, subiculum.

Adult neurogenesis occurs at a bottleneck in the hippocampal network

The hippocampus is a bottleneck structure into which numerous inputs converge and from which a wide range of outputs emerges. The entrance structure for the hippocampus, formed by the entorhinal cortex and the dentate gyrus, requires a massive reduction in input channels as well as a reduction in information complexity. The dentate gyrus thus seems to be involved in compressing data. While a divergence is found between the entorhinal cortex and the dentate gyrus, there is a strong convergence between the dentate gyrus and CA3. New neurons are thus found at a bottleneck within a bottleneck. Firing patterns in the dentate gyrus are sparse, which is in apparent contrast to the massive flow of information through this structure and the greater cell numbers in the dentate gyrus compared to the entorhinal cortex. But the sparseness of activity is consistent with the idea that the dentate gyrus stabilizes neuronal activity flooding into the hippocampus. The underlying assumption is that, for complex cognitive functions to be performed by the hippocampus, a reduction in noise and fluctuations on very short time scales is necessary.

CA3 is an auto-associative short-term memory and has to avoid "catastrophic interference"

Judging from the network architecture, CA3, the second station, serves as a temporary auto-associative memory. The general idea is that the function of the dentate gyrus is to encode information to make it usable by CA3. Data compression might be one aspect of this processing; the avoidance of catastrophic interference a second. *Catastrophic interference* is the problem that would ensue if a constant flow of information made it impossible to keep together the pieces of information that belong to each other, leading to a catastrophic accumulation of useless contents. New contents would constantly overwrite older associations. The dentate gyrus might reduce the overlap—that is, the number of commonly active neurons—between different input patterns by generating a sparse representation of that information. Since sparse patterns share fewer active neurons at a given time, they result in less interference, or "cross-talk." The individual bouts of information are "orthogonalized"; that is, if expressed as vectors, they share as few directions as possible. The avoidance of catastrophic interference would improve the reliability of pattern storage and retrieval in CA3. In contrast to CA3, the dentate gyrus does not have the architectural properties of a memory network. There is an ongoing debate about what degree information is actually stored within the hippocampus and not the cortical regions. Clearly, the overall storage capacity of the hippocampus is low compared to that of the cortex, and within the hippocampus, only CA3 has the network characteristics of a memory device. This property is thought to be relevant for the one-trial learning associated with episodic memory (Nakazawa et al., 2003). Consequently, adult neurogenesis in the dentate gyrus would add neurons only to a structure that just feeds into memory, but does not seem to be directly involved in building or modifying a structure that can store information.

The key function of CA1 is the transfer of memory for cortical long-term storage

Formation of long-term memory, so-called memory consolidation, requires the transfer or integration of memory contents into the cortex and is mainly dependent on CA1. This conclusion is based mainly on electrophysiological data showing that memory correlates best with LTP as the electrophysiological equivalent of learning in CA1. Episodic memory, a particular type of long-term memory, is dependent on glutamate action in CA1 (Day et al., 2003). Nightly sleep plays a central role during consolidation in that the hippocampus replays contents to be transferred during REM sleep (Walker and Stickgold, 2004, 2006). Contents destined for long-term memory might have a hippocampal transition time of variable length, and during the consolidation period, the information is dependent on the hippocampus and vulnerable to hippocampal damage. Emotional contexts strongly influence the passage time. With the transfer, the information is cleared from the hippocampus; which plays, however, a role during the recall of the information after consolidation as well. This clearance is sometimes casually referred to as "forgetting," but that term is misleading (see below, p. 464). The finding that adult neurogenesis occurs two synaptic relay stations before this consolidation step supports the idea that neurogenesis functionally contributes to data processing essentially independently of the actual storage into the cortex. Appropriately, adult hippocampal neurogenesis correlated better with parameters describing the acquisition of a hippocampal learning task but not with measures of retention or recall of the stored information (Kempermann and Gage, 2002). The contribution of new neurons to the variance in "learning" was relatively low, ranging between 10% and 20% in that study. Many factors, not just new neurons, contribute to hippocampal function.

The unidirectional flow of information through the hippocampus and the position of adult neurogenesis at the network bottleneck suggest that many downstream sub-functions within the hippocampus might be influenced by neurogenesis, and direct and indirect effects will be very difficult to untangle. But it is reasonable to assume that the primary functional role of adult-generated granule cells should be found within the contribution of the dentate gyrus to overall hippocampal function.

The key to the function of adult hippocampal neurogenesis lies in the function of the dentate gyrus

Adult neurogenesis occurs only in the dentate gyrus. Consequently, the function of new neurons must be tightly linked to the function and the evolution of the dentate gyrus. In a seminal review, Allessandro Treves, Ayumu Tashiro, Menno Witter, and Edvard Moser have worked out the conceptual framework in which to place the function of the dentate gyrus, including the contribution by new neurons (Treves et al., 2008). The dentate might be part of the "classical" tri-synaptic backbone of the hippocampus, but they state that in evolutionary terms, it is an invention by mammals only (compare also Chap. 11, p. 485). Whatever this function exactly is, they argue, other classes of animals must have solved the underlying problem either not at all or by different means. While the putative functions of the dentate gyrus include compression, pattern separation, and temporospatial annotation, its exact functional role remains to be unraveled. As new neurons provide new mossy fibers connecting to CA3, the functional relevance of adult neurogenesis must be seen within the context of what is known about the functional relevance of mossy fibers as well. But these lines of research are only beginning to merge.

Computational Modeling

Computational models are powerful tools for testing hypotheses at the network level

Considerable progress for our understanding of how new neurons might contribute to hippocampal function has been made by the development of computational models in which the consequences of neurogenesis on the network's performance could be tested. A common critique of this approach is that the model is only as good as its design, and that one might be able to construct models for any expected outcome. Although there are intrinsic and important caveats with network modeling, the critique largely misses the point: experimental systems need to be reductionistic, and the quality of a model does not per se depend on the level of abstraction. The model needs to be correctly designed, and the limits need to be identified, but this applies to all experimental models, independently of whether they take place in vitro, in vivo, or "in silico."

Different models have focused on different aspects of adult neurogenesis

The computational models for adult neurogenesis proposed to date differ by their closeness to a realistic representation of the hippocampal network situation. Although ultimately superior, a more complex model mimicking more of the real connections is not necessarily better suited to generating useful theories in the beginning because of the sheer number of assumptions that have to be made. There is a balance to be found between the level of abstraction and the usefulness of the readout and the relevance of the conclusions. A key difference between different models is, for example, the underlying assumption of whether adult neurogenesis is additive or contributes to a turnover of neurons (see below, p. 470).

In Table 10–2, the key properties and conclusions from the modeling studies are summarized. Across the different approaches it has become obvious that, under many different circumstances, neurogenesis is beneficial and not detrimental for the performance of the network. This observation alone refutes the idea that adult neurogenesis would in principle be useless or damaging for an advanced network. Validation and confirmation are still required, and some of the discrepancies between different models have to be resolved. But with the help of the computational models, much

Table 10–2 Computational models of adult hippocampal neurogenesis

Reference	Main key word	Model features	Main conclusions or predictions
Becker, 2005	Pattern separation, catastrophic interference	Reductionistic turnover model	A turnover improves recall of information
Wiskott et al., 2006	Catastrophic interference	Reductionistic additive model	New neurons reduce errors in situations of novelty
Appleby and Wiskott, 2009	Catastrophic interference	Sparsely coding network; comparison of turnover and additive models	Additive neurogenesis is superior to turnover models
Becker et al., 2009	Catastrophic interference	Largely turnover model	New neurons reduce the overlap between memories and associate contextual information
Aimone et al., 2006, 2009	Pattern separation and temporal encoding	Largely additive model including many features of hippocampal circuitry	Neurogenesis improves encoding and the distinction between familiar and novel contexts
Weisz and Argibay, 2009	Memory retrieval	Additive model based on "realistic" hippocampal circuitry	With neurogenesis there is improved retrieval of recent memories
Butz et al., 2006; Butz et al., 2008	Homeostasis of neuronal activity	Highly theoretical turnover model centered around the proliferation of precursor cells and synaptogenesis	There is an optimal rate of cell proliferation above which new cells destabilize the network; neurogenesis and synaptic rewiring are inversely linked
Deisseroth et al., 2004; Meltzer et al., 2005	Circuit homeostasis	Reductionistic turnover model	Balanced integration of neurogenesis avoids negative effects of too many new neurons

more precise behavioral tests might be developed. The model by Wiskott and colleagues (Wiskott et al., 2006), for example, generated a hypothesis for modifying the classical water maze protocol to test for the hippocampal contribution to flexible re-learning in the reversal phase of that test (Garthe et al., 2009). The model by Aimone and colleagues focusing on temporal encoding and pattern separation (Aimone et al., 2006, 2009) sparked the Clelland and coworker's study testing that hypothesis directly in vivo (Clelland et al., 2009).

The two models by Butz et al. (2006, 2008) and Meltzer et al. (2005) are different from the others in that they focus on the question of whether there is an optimal rate of turnover and neurogenesis. In the case of the Meltzer study, the question is even how potentially deleterious effects of new neurons for the network due to "runaway excitation" might be avoided. The model thus makes an important contribution to our understanding of how adult neurogenesis contributes to solving the stability–plasticity dilemma (Meltzer et al., 2005). In this context it is worth remembering the idea that, in temporal lobe epilepsy, newborn neurons might actually contribute to maintaining excitational homeostasis in the dentate gyrus (Jakubs et al., 2006; Kempermann, 2006).

A computational model for olfactory bulb neurogenesis, which was the very first computational model in neurogenesis research, is described further below, p. 477 (Cecchi et al., 2001).

Behavioral Relevance of Adult-Generated Neurons in the Hippocampus

Behavioral Tests

The central conclusions on the functional relevance of adult-born neurons to date essentially rest on two behavioral tests: the Morris water maze, which assesses spatial learning; and fear conditioning, which measures emotional memory. Both tests exist in hippocampus-dependent and hippocampus-independent versions.

The Morris Water Maze

Spatial navigation assessed with the water maze is a widely used, yet selective, measure of hippocampal function

Spatial navigation is an example of learning that not only is dependent on hippocampal integrity but also exemplifies the need to place information in a set of temporal and spatial coordinates. A path through a new environment is a sequence of temporospatial information. In patients with Alzheimer disease, the hippocampus is affected early, and disorientation and lack of spatial memory are typical early symptoms. Spatial navigation can be tested in rodents and has become the key surrogate measure for assessing declarative memory in animals. This does not imply, however, that the hippocampus is "just for place" (Eichenbaum et al., 1999).

In the water maze, rodents learn to navigate to an escape platform hidden in the pool

One of the most widely used tests of spatial memory as a measure of hippocampal function is the Morris water maze (Morris et al., 1982; Morris, 1984; Wolfer et al., 2001). Rodents are placed in a circular pool of water and learn to navigate to a small escape platform hidden just below the surface of the water (Fig. 10–6). The time taken (latency) to find the platform and the distance the animals swim to it can be plotted as a learning curve to describe acquisition of the task. The steeper the slope of the learning curve, the faster the animals have learned the task. Over few days of training, most normal rodents find the platform quickly and with a short swimming distance. If position of the platform is changed in a reversal, most rodents adapt to the new position quickly. The first trial of the reversal can be used as the so-called probe-trial. Here, the perseverance of the animals in finding the platform at the position where it used to be is taken as a measure of how well the animal learned the task and retained the information or can recall it. Although there are some conceptual caveats with this interpretation, the probe-trial measures different aspects of learning than the parameters that describe the acquisition period of the task. If after the acquisition of the task, the platform is moved to a new position, the animals will after a brief period of irritation (as assessed by the probe trial) normally learn the new position quickly. This reversal situation reveals how well the animal can cope with an important partial change in the learned representation.

An animal's ability to navigate the hidden version of the water maze with high efficiency is dependent on the integrity of the hippocampus, but rodents with hippocampal damage might be able to learn the task to a certain degree (e.g., by "circling"; that is, swimming at a fixed distance from the wall in which the platform was encountered). If the hidden platform is made visible, the test becomes independent of the hippocampus.

Advanced analysis of the water maze by including search strategies increases the sensitivity and specificity of the test

Including the analysis of search strategies, which are applied in a characteristic sequence by the animals, greatly enhances the readout from the water maze task (Wolfer et al., 2001; Garthe et al., 2009). Hippocampus-dependent and -independent aspects can be distinguished, and subtler differences in performance quality can be identified.

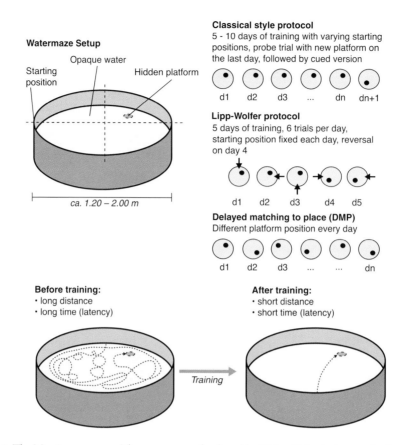

Watermaze Setup

Starting position

Opaque water

Hidden platform

ca. 1.20 – 2.00 m

Classical style protocol
5 - 10 days of training with varying starting positions, probe trial with new platform on the last day, followed by cued version

d1 d2 d3 ... dn dn+1

Lipp-Wolfer protocol
5 days of training, 6 trials per day, starting position fixed each day, reversal on day 4

d1 d2 d3 d4 d5

Delayed matching to place (DMP)
Different platform position every day

d1 d2 d3 ... ... dn

Before training:
• long distance
• long time (latency)

After training:
• short distance
• short time (latency)

Training

FIGURE 10–6 The Morris water maze. The water maze, developed by Richard Morris, is a standard tool to assess hippocampal function in rodents and consists of a pool of water with an escape platform hidden under the surface of the water. The animals learn to navigate from varying starting positions to the platform by orienting themselves according to cues in the room. The time and swim path needed is recorded as a measure of test performance and improves over days of training. Different protocols can be distinguished.

One confounding problem is that learning the water maze places animals under a considerable amount of stress, and the situation is rather artificial for dry-land animals. Ethologically more-relevant tests of hippocampal functions have to be developed that allow us to relate adult neurogenesis more closely to hippocampal function. These tests will have to go beyond the assessment of spatial navigation and appreciate the complexity of hippocampal function and the role of the dentate gyrus within it.

‖‖

Adult hippocampal neurogenesis shows a clear association with specific aspects of water maze performance

‖‖

In mice, there was a small but significant correlation between parameters describing the acquisition phase of the water maze and the genetically determined baseline level of adult hippocampal neurogenesis (Kempermann and Gage, 2002). Similarly, in old rats, water-maze performance predicted levels of adult neurogenesis (Drapeau et al., 2003). On the other hand, elimination of precursor cell activity in the dentate gyrus did not affect water-maze learning in some of the ablation study (Shors et al., 2002; Saxe et al., 2006; Bergami et al., 2008; Jaholkowski et al., 2009). It turned out that this discrepancy was largely due to the test protocol used. In a training protocol first allowing the establishment of a robust representation of maze and target, reducing adult neurogenesis was associated with specific impairments in the water maze task (Dupret et al., 2008; Garthe et al., 2009; Wolf et al., 2009).

Contextual Fear Conditioning

Contextual fear conditioning tests hippocampus-dependent emotional memory

Contextual fear conditioning belongs to the tests for conditioned emotional responses and addresses "emotional memory." The animals learn to associate a neutral event (tone) and environment (context) with a fearful experience, usually a food shock (Fig. 10–7). What is actually measured in the test is the time of immobility ("freezing") of the animal, when it is re-exposed to the conditioned stimulus or context. Longer freezing means more fear and better learning. Fear conditioning is primarily based on the function of the amygdala, but especially the contextual version is also hippocampus-dependent (Phillips and LeDoux, 1992). In cued fear conditioning, the trained animal is transferred to a different environment but exposed to the same conditioned tone: this version of the test is independent of the hippocampus.

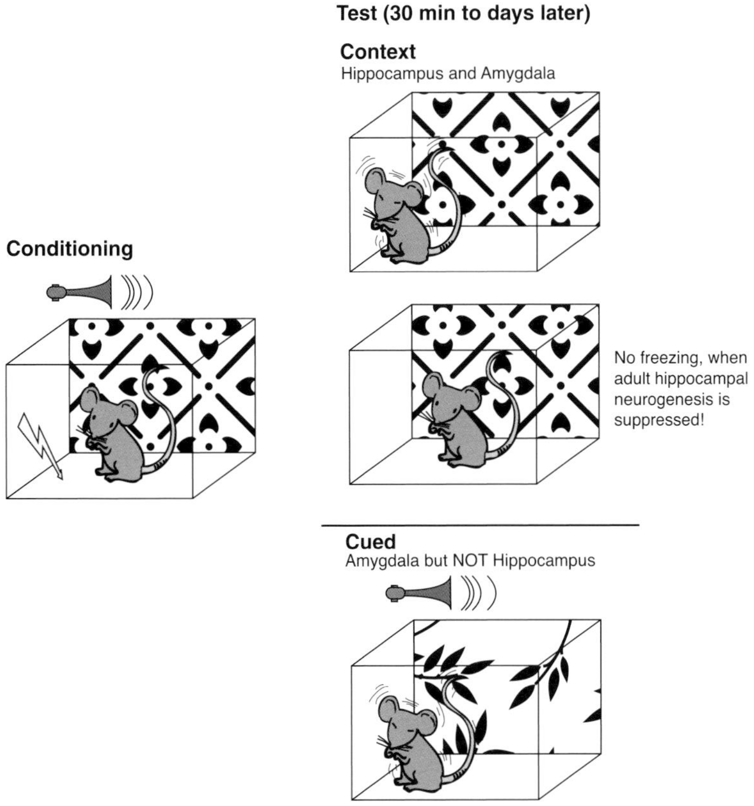

FIGURE 10–7 Contextual fear conditioning. Mice are conditioned to associate an unpleasant cessation (food shock) with a tone and an environment (context). In the test situation they are either exposed to the environment (context version of the task) or to the tone but in a different environment (cued version). If learning was successful, the mice will show freezing behavior under these conditions, because they fear the coming shock. Better learning thus means more freezing and more fear. The context version response is hippocampus-dependent and eliminated in mice with suppressed adult neurogenesis, whereas the cued version response is not.

Most studies support a role for
new neurons in contextual fear
conditioning

As with the water maze, the results with fear conditioning after ablation of adult neurogenesis were heterogeneous, but the studies supporting
a functional link outnumber the negative observations (see Table 10–3).

Fear conditioning measures hippocampus-dependent memory as does
the water maze but has a strongly emotional component and is not dependent on navigation. Consequently, they form a bridge between cognitive
and affective functions of the hippocampus. The question is thus what the
specifically emotional contribution of the hippocampus to affective behavior is. For the greater
picture, one might argue that the hippocampal contribution to emotional behavior is the aspect of
learned association, not the emotionality itself, because learning is what the hippocampus does best.
The emotional connotation might vary across the range of possible experiences, but the principle
underlying learning does not.

In contextual fear conditioning, "extinction" stands for the observation that with time the
learned association begins to weaken, and the context or conditioned stimulus become less effective
in eliciting the response. Contextual fear conditioning is more prone to extinction than cued fear
conditioning (Phillips and LeDoux, 1992).

Blocking Adult Neurogenesis to Study the Role of New Neurons in Hippocampal Function

The classical loss-of-function
experiment to assess the
function of adult neurogenesis is
based on the targeted
elimination of new neurons by
cytostatic drugs, irradiation, or
conditionally transgenic ablation

A straightforward way to study the functional contribution of adult neurogenesis to hippocampal function is to block adult neurogenesis altogether.
Most researchers who have chosen this approach have aimed at eliminating
the precursor cells as the cellular source for new neurons. This has been
achieved by either cytostatic drugs or irradiation as they are known from
cancer therapy. Joseph Altman was a pioneer in this regard as well. He irradiated the dentate gyrus of rats early postnatally (thus not addressing adult
neurogenesis) and found lasting damage to the dentate gyrus and reduced
performance in the T-maze (Bayer and Altman, 1975; Gazzara and Altman,
1981). Irradiation that is targeted specifically to the brain area containing
the hippocampus has been the method of choice in many later studies
(Table 10–3). Cytostatic drugs vary in their specificity, but the most modern versions provide an
experimentally simple and straightforward way to eliminate precursor cell activity (Garthe et al.,
2009). In the past few years a variety of different genetic models was introduced in which, through
conditional transgenesis, precursor cells or new neurons are either eliminated or rendered dysfunctional (Saxe et al., 2006; Bergami et al., 2008; Dupret et al., 2008; Imayoshi et al., 2008; Jessberger
et al., 2009; Kitamura et al., 2009). Finally, there are a few cases, such as conditional Presinilin-1
knockouts or constitutive Cyclin D2 or Tailless knockouts, which show reduced neurogenesis and
have been used to elucidate aspects of the function of adult-born hippocampal neurons.

Specificity of the ablation is an
important issue also applicable
to the genetic models

The general critique of all ablation studies aims at the supposedly
questionable specificity of the effect. Does the manipulation indeed only
abolish adult neurogenesis, or are there side effects that might explain the
behavioral outcome that is ascribed to the lack of new neurons? The
assumption that the genetic models are solely by virtue of the underlying
genetic hypothesis more selective and thus more specific is wishful thinking
(see Chap. 7, p. 259). In the end, very similar caveats apply to pharmaco-
logical, x-ray–based, or genetic models. The data collected so far across the different models indicate
that the method of ablation has limited consequences for the result. Across all studies, a complex but
increasingly coherent picture emerges. For the bigger picture, the advantages and disadvantages of the
various methods should have cancelled each other out. Consequently, there is no reason to demand
the extremely expensive and, due to their complex breeding schemes, technically demanding genetic
studies, especially given that most experiments of this kind, often involving double transgenics or
conditional and regulatable constructs, cannot contain all the possible control groups that would

Table 10-3 The ablation studies for adult hippocampal neurogenesis

Method	Species	Behavioral test	Result	Conclusion	Reference
Chemical ablation (MAM)	Rats	Eye blink conditioning (Trace; hippocampus-dependent)	Impaired	Adult neurogenesis is required to associate stimuli in time	Shors et al., 2001
		Eye blink conditioning (Delay; hippocampus-independent)	Not impaired		
Conditional genetic ablation based on Presenilin 1 expression eliminated enrichment-induced increase only	Mice	Contextual fear conditioning	Increased fear with lacking increase in neurogenesis after enrichment	Adult neurogenesis involved in clearance of the hippocampus for new memories	Feng et al., 2001
Chemical ablation (MAM)	Rats	Trace fear conditioning	Impaired		Shors et al., 2002
		Contextual fear conditioning	Not impaired		
		Morris water maze	Not impaired		
		Elevated plus maze	Not impaired		
Radiation	Mice	Novelty-suppressed feeding	Baseline performance not impaired, but anxiolytic effect of antidepressants on test performance was lost	Adult neurogenesis required for action of antidepressants	Santarelli et al., 2003
Radiation	Rats	Morris water maze	Impaired	Adult neurogenesis is involved in formation of spatial memory	Snyder et al., 2005
Radiation/ Genetic/toxic ablation of GFAP-positive cells	Mice	Contextual fear conditioning	Impaired	Adult neurogenesis is relevant for certain aspects of contextual learning	Saxe et al., 2006
		Cued fear conditioning Morris water maze Delayed matching to place Y maze Elevated Plus maze Light-Dark-Choice Test	Not impaired		

(Continued)

Table 10–3 (Cont'd) The ablation studies for adult hippocampal neurogenesis

Method	Species	Behavioral test	Result	Conclusion	Reference
Radiation	Rats	Contextual fear conditioning	Impaired	Adult neurogenesis is relevant for certain aspects of contextual learning	Winocur et al., 2006
		Non-matching-to-sample task (conditional rule learning)	Impaired, if intervals were long		
Radiation	Mice	Radial arm maze: High memory load/no interference task	Not impaired	Adult neurogenesis constrains certain types of working memory	Saxe et al., 2007
		Radial arm maze: Low memory load/high interference task	Improved, if delay longer than 30s		
Radiation/ Chemical ablation based on conditional transgenic sensitivity to ganciclovir in GFAP expression cells (GFAP-tk)	Mice	Radial arm maze: Low memory load/limited interference task	Improved, if delay longer than 35s		Saxe et al., 2007
Conditional genetic ablation of Tlx-expressing cells	Mice	Morris water maze		Adult neurogenesis is involved in formation of spatial memory	Zhang et al., 2007
		Contextual fear conditioning	Not impaired		
Prevention of cell death during memory stabilization	Rats	Morris water maze	Impaired acquisition and retention	Adult neurogenesis, including the specific elimination of new neurons improves spatial memory	Dupret et al., 2007
Conditional genetic ablation based on Bax expression	Mice	Habituation to new environment Contextual fear conditioning Water maze: fixed starting point	Not impaired	Adult neurogenesis contributes to flexibility in spatial learning	Dupret et al., 2008

Manipulation	Species	Task	Result	Conclusion	Reference
Conditional regulatable genetic ablation based on Nestin expression	Mice	Water maze: novel starting point	Impaired acquisition and probe trial	Adult neurogenesis is required for spatial learning	Imayoshi et al., 2008
		Water maze	Reduced acquisition and impaired retention (probe trial)		
		Contextual fear conditioning	Impaired	Adult neurogenesis is involved in the association of contexts	
Lentivirally mediated dominant negative Wnt, dentate-gyrus specific	Rats	Water maze	Dose-dependent impairment in probe trial performance	Adult neurogenesis is involved in long-term memory formation	Jessberger et al., 2009
		Socially transmitted food preference	Not impaired		
Radiation	Rats	Morris water maze, T maze	Not impaired		Hernandez-Rabaza et al., 2009
		Contextual fear conditioning	Impaired	Adult neurogenesis involved in the rapid acquisition of emotionally relevant contextual information	
Radiation, dominant-negative Wnt	Mice	Spatial radial arm maze task	Impaired learning, when locations are similar, not distinct	New neurons involved in pattern separation and spatial discrimination	Clelland et al., 2009
Radiation	Mice	Spatial touchscreen task without navigation	Impaired learning	New neurons involved in pattern separation and spatial discrimination	Clelland et al., 2009
Chemical ablation based on conditional transgenic sensitivity to ganciclovir in nestin expression cells (Nestin-tk)	Mice	Water maze	No effect on acquisition but on short- and long-term retention at 1 week after training but not later	New neurons are required for long-term memory during critical time window	Deng et al., 2009

(Continued)

Table 10–3 (Cont'd) The ablation studies for adult hippocampal neurogenesis

Method	Species	Behavioral test	Result	Conclusion	Reference
Cytostatic ablation (MAM) Radiation	Mice	Contextual fear conditioning	Impaired only with high-dose radiation but no effect on extinction	Adult neurogenesis not involved in extinction	Ko et al., 2009
			Prolonged period of hippocampus-dependency to associate fear	Adult neurogenesis is involved in memory clearance from the hippocampus	Kitamura et al., 2009
Cytostatic ablation (Temozolomide)	Mice	Morris water maze	Slightly reduced acquisition, impaired reversal	Adult neurogenesis necessary for "re-learning" in old contexts and flexible use of learning strategies	Garthe et al., 2009

In contextual fear conditioning, impaired performance means reduced fear (less freezing)!

theoretically be necessary. Success of the ablation studies in identifying the function of new neurons is more dependent on the quality of the behavioral readout than on the method of ablation.

|||||||||||||||||||||||||||||||||||||||||||||||||||||||||

The first ablation study
suggested a role of new neurons
in trace eye-blink conditioning

|||||||||||||||||||||||||||||||||||||||||||||||||||||||||

The first ablation study for adult neurogenesis was published by Tracey Shors and Elizabeth Gould in 2001 and stirred substantial controversy (Shors et al., 2001). Dividing cells were eliminated by a treatment with methylazoxymethanol acetate (MAM), an antiproliferative drug that acts by methylating the DNA. Because the DNA is exposed, cells that undergo division are sensitive to the effects of cytostatic drugs such as MAM. Cytostatics prevent successful completion of the cell cycle, which causes the death of the dividing cell. They are used in cancer therapy because tumor cells are highly proliferative, but so are neural precursor cells.

In the experiment, MAM blocked precursor cell divisions in the dentate gyrus and led to a down-regulation of adult hippocampal neurogenesis. In the treated animals, performance on a hippocampus-dependent learning task (trace conditioning) was disturbed, whereas that on a hippocampus-independent version of the same task was spared (Shors et al., 2001) (Fig. 10–8). The conclusion was that new neurons were needed to mediate the hippocampal contribution to the task. Eye-blink conditioning is a classical conditioning task in which the unconditioned stimulus and the conditioned stimulus can be separated by different time spans. If the two stimuli overlap (and co-terminate) the version of the test is called "delay." The delay version of the test is independent of the hippocampus. The information of the unconditioned stimulus remains present for this brief time and can be associated with the conditioned stimulus. In the "trace" version of the same test, the interval between unconditioned and conditioned stimulus is 500 ms, and therefore the hippocampus is required to associate unconditioned and conditioned stimulus across the interval. Although the trace version of this task has a hippocampal component, eye-blink conditioning does not adequately represent the complexity of hippocampal function, including higher cognitive processes. It is often used as a test assessing cerebellar function (McCormick et al., 1982; Freeman and Nicholson, 2000). There is a long-standing debate about the information that conditioning tasks provide about hippocampal function, and the answer largely depends on the definition of "function." Other points of concern are the range of side effects of MAM, which might be impossible to control sufficiently. In fact, the effects of MAM on presumably neurogenesis-dependent hippocampal learning were not detectable or ambiguous on other hippocampal tests, including the Morris water maze (Shors, 2004). In hindsight, these issues mostly fall into place, and the study turns out to have assessed the potential function of new neurons in temporal associations.

|||||||||||||||||||||||||||||||||||||||||||||||||||||||||

The ablation studies clearly
indicate functional relevance of
adult neurogenesis, but no
unifying theory has yet been
achieved

|||||||||||||||||||||||||||||||||||||||||||||||||||||||||

Table 10–3 summarizes the results from the by-now-long list of ablation studies that vary in their method of down-regulating adult neurogenesis and their spectrum of functional readouts. Across these studies it cannot remain disputable that adult-generated neurons contribute to hippocampal function. The interpretation of the entire set of experiments does not yet allow one coherent and elegant synthesis, but, if the speed of development in the years 2007 to 2009 is any indicator of further progress, that will only be a matter of, presumably, not a very long time.

Roles of New Hippocampal Neurons in Learning and Memory

|||||||||||||||||||||||||||||||||||||||||||||||||||||||||

From the ablation studies, five
overlapping key functions of
new neurons have emerged

|||||||||||||||||||||||||||||||||||||||||||||||||||||||||

Together with a number of additional pieces of experimental evidence, data from the computational models and theoretical considerations the ablation studies allowed the identification of five types of potential functions of new neurons. Neither conceptually nor in the experimental reality can these functions often be clearly separated, but the distinction brings some order into the complicated picture. Table 10–4 attempts an abstraction where a true synthesis is not yet possible. The prerequisites for a true synthesis are outlined further below (p. 468).

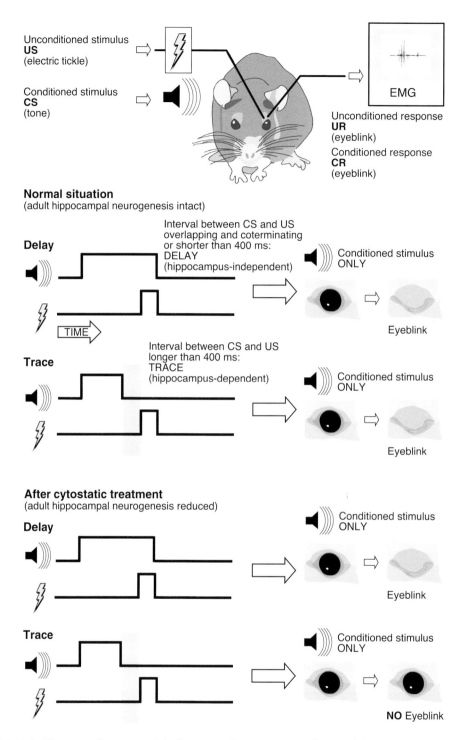

FIGURE 10–8 Hippocampal neurogenesis in formation of trace memories. Shors et al. (2001) reported that treating rats with cytostatic agent MAM resulted in a defect in acquisition of a hippocampus-dependent version of a learning task (delay eyeblink conditioning), whereas the hippocampus-independent version of the same task (trace eyeblink conditioning) was spared. The finding has been interpreted as evidence for a role of new neurons in learning the hippocampus-dependent task.

Table 10–4 Possible functions of new neurons in learning and memory

Functional quality	Arguments
Improving temporal resolution, pattern separation and avoiding catastrophic interference	• Relevance of new neurons for trace vs. delayed eye blink conditioning (Shors et al., 2001; Shors et al., 2002) • Impaired spatial pattern separation after ablation of adult neurogenesis (Clelland et al., 2009) • Impairment in reversal task (Garthe et al., 2009) • Results from computational modeling studies (Becker, 2005; Aimone et al., 2006; Wiskott et al., 2006; Aimone et al., 2009; Appleby and Wiskott, 2009; Becker et al., 2009)
Increasing the ability to add temporal, emotional and other contextual information	• Requirement of new neurons for anxiety-like behavior: e.g. novelty-suppressed feeding (Santarelli et al., 2003; Bergami et al., 2008; Revest et al., 2009) • Relevance of new neurons in contextual fear conditioning (Saxe et al., 2006; Winocur et al., 2006; Hernandez-Rabaza et al., 2009) • Results from computational modeling studies (Becker et al., 2009)
Improving long-term memory and retention or extinction	• Improved probe-trial performance in the water maze (Snyder et al., 2005; Imayoshi et al., 2008; Deng et al., 2009; Garthe et al., 2009; Jessberger et al., 2009) • Extinction in fear conditioning positively correlates with adult neurogenesis (Feng et al., 2001; Kitamura et al., 2009)
Determining processing speed and clearance time	• Correlation between adult neurogenesis and acquisition speed in the water maze across mouse strains (Kempermann and Gage, 2002) • Results from computational modeling (Deisseroth et al., 2004k; Meltzer et al., 2005) • Critical time window after cell ablation for behavioral consequences (Feng et al., 2001; Dupret et al., 2007; Deng et al., 2009; Kitamura et al., 2009)
Improving the formation of allocentric representations and increasing the flexibility to integrate new relevant details to existing cognitive maps	• New neurons improve only hippocampus-dependent aspects of water maze performance (Dupret et al., 2008) • New neurons improve reversal learning and the flexible use of advanced spatial strategies (Garthe et al., 2009)

Pattern Separation, Prevention of Catastrophic Interference

Different lines of experimental evidence support the idea of new neurons in pattern separation and avoidance of catastrophic interference

Closest to the specific function of the dentate gyrus in the hippocampal network and its presumed contribution to hippocampal function are the conclusions from the ablation studies that deal with pattern separation and the avoidance of catastrophic interference, both clear predictions from computational models (see the section on the computational models above, p. 451). The first experimental evidence came from the classical eye-blink conditioning studies by Shors and Gould (Fig. 10–8) (Shors et al., 2001; Shors et al., 2002). Clelland and colleagues addressed the question of spatial pattern separation directly in a cleverly designed study, including the use of a touch-screen test for rodents, which allowed the researchers to address spatial memory in mice without navigation and locomotion in a three-dimensional environment (Clelland et al., 2009). In that study, learning was more impaired for similar cues than for distinct cues. Garthe and colleagues showed that, after ablation of adult neurogenesis, a specific deficit occurs in the reversal test of the water maze paradigm (where the platform that has been stable so far is moved to a new position and the acquisition of the new platform location is assessed) (Garthe et al., 2009). Here, the mice have to integrate a new but highly relevant element (the new position of the platform) into an existing representation of the environment, which has not changed.

Contextual, Emotional, and Temporal Connotations

The second set of results deals with another circumscribed function of the hippocampus: its relevance for placing information into a context and a coordinate system that allows a temporospatial order, for example in form of a "time stamp" (Aimone et al., 2006). This order is the prerequisite for memory, especially episodic memory and, for humans, autobiographical memory. At the same time, a content is also associated with an emotional coloring, which decides about its priority for further processing. This affective component is arguably the strongest version of contextual information to go with a memory. Results from fear conditioning studies and the behavioral experiments involving tests of affective behavior suggest that new neurons might be relevant to assign to contexts.

The distinction from pattern separation, etc., is actually not very sharp and presumably largely a matter of the time scale considered:

> At much shorter time scales, within the span of a single memory episode, rather than contributing to pattern separation, neurogenesis might play more of an integrative role in mediating contextual associative learning. Consistent with this, empirical evidence from animal studies suggests a role for the new neurons in forming complex event memories that bridge across time delays. (Becker et al., 2009)

One might argue, though, that contextual and temporospatial association occur at time scales both shorter and longer than is relevant for pattern separation. To separate patterns but still link them temporally as connected in an orderly way is a particularly challenging task. New neurons might help solve this challenge.

Context is, of course, more than temporospatial coordinates and an ad hoc emotional evaluation. To grasp the essence of a context and assign it to a memory trace requires abstraction from the incoming information. Contexts are defined by key cues, not true representation of the entire circumstances. The dentate gyrus with its sparse firing and its position at the structural bottleneck of the network seems destined to perform this task. How exactly new neurons are involved in this associative compression remains a future challenge.

Long-Term Memory vs. Extinction

Time is a critical issue for this third type of possible function as well, but now the longest time scales are concerned. Several studies have suggested that adult neurogenesis is associated either with long-term memory (as measured by probe trial performance in the water maze: Snyder et al., 2005; Imayoshi et al., 2008; Deng et al., 2009; Jessberger et al., 2009) or the extinction of memories (as measured by repeated exposure to the context in contextual fear conditioning: Feng et al., 2001; Kitamura et al., 2009). The discrepancy at the conceptual level (long-term memory vs. "forgetting") might obviously be primarily due to the particularities of the different tasks (water maze vs. fear conditioning), but on a mechanistic level, the two sets of studies might not be as far apart as it seems. What is considered long-term memory might be the outcome of the neurogenesis-dependent processing on a shorter time scale. It might depend on the strength and the coloring (positive or negative) of the emotional connotation that determines whether upon repeated exposure to a stimulus in the absence of further negative reinforcement, a habituation or extinction occurs, or whether, as in the case of the lack of extinction in the water maze (Deng et al., 2009), the relevance of the learned information remains unchanged.

Extinction is the weakening of a response despite repeated exposure to the conditioned stimulus or context

Important in this context is the question whether new neurons are involved in the extinction of memories. The concept of "extinction" is not identical to "forgetting," although some popular connotations will suggest so. In the nomenclature of the classical conditioning literature, forgetting is not permanent, whereas extinction is. Forgetting is largely a matter of time, during which no reinforcement occurs. In contextual fear conditioning, "extinction" means the weakening of the association over time despite repeated exposure to the conditioned stimulus or context.

A turnover of new neurons might help clear the hippocampus for new memory traces

The first study to suggest that new neurons might be needed for extinction in fear conditioning was published in 2001 (Feng et al., 2001). Conditional Presinilin-1 knockout mice had normal baseline neurogenesis but lacked the increase in adult neurogenesis in response to an enriched environment. Nevertheless, they showed greater fear memory than enriched controls (when enrichment was applied after the learning experience). In this context it has to be remembered that in fear conditioning experiments more fear stands for better memory of the trained association. In essence this was a combined loss- and gain-of-function experiment, in which only an increase was eliminated, and hence the specific behavioral change due to this increase could be evaluated. At the time there were no other ablation studies, and even today this report is a solitaire in its unique combination of experimental increase and decrease.

In the end, the interpretation of its findings are complicated. One cannot directly conclude from these data that new neurons would be required to actively eliminate memory traces from the hippocampus or even beyond CA1. The fact that the result was only seen, when enrichment occurred after the learning also complicates the discussion. The study has also been criticized because, in the mutant mice, environmental enrichment might affect fear conditioning in more than one interacting way. A later ablation study contained a detailed analysis of critical time windows for learning and extinction in fear conditioning and also included a group in which adult neurogenesis was increased by running and in which the hippocampus-dependency of the association was decreased (Kitamura et al., 2009). This study strongly suggested an association of test performance with the time the learned association is dependent on neurogenesis. There is a dissenting report contending that only substantial reduction in adult neurogenesis reduced freezing as indicator of the learned association. Even then, no effect on extinction was detectable, and there was no simple correlation between extinction and ongoing neurogenesis (Ko et al., 2009). In addition, suppressing adult neurogenesis did not lead to enhanced extinction in the water maze task (Deng et al., 2009). Taken together, the arguments in support of an effect of adult neurogenesis on processing time in the hippocampus are quite strong for contextual fear conditioning, but whether indeed "extinction" in the sense of active clearance is affected, or the observed phenotype is simply a side effect of increased processing speed, requires further investigation.

Processing Speed and Clearance Time

More new neurons might decrease the time required for memory acquisition

First evidence that adult neurogenesis might have influence on the speed of learning came from a strain comparison study, in which a significant correlation between the slope of the learning curve in the water maze and the rate of adult neurogenesis was found (Kempermann and Gage, 2002).

Elegant time-course studies in which the behavioral assessment was placed at different times in relation to the birth (or elimination) of the new neuron allowed two closely related types of conclusions. First, the identification of critical time windows suggested that new neurons might actually have a transient function (Kee et al., 2007; Tashiro et al., 2007). One consequence of this thought is that information processing is directly dependent on the availability of new neurons. The rate of adult neurogenesis would, in turn, have a quantitative impact on learning. Several studies suggest that this is indeed the case. In a particularly extensive study, Dupret and colleagues from Nora Abrous's group in Bordeaux have shown that the particular functional relevance during such time windows also acts back on the regulation of neurogenesis itself (Dupret et al., 2007). Not only was cell survival affected (as predicted from many other studies; see Chap. 9, p. 340) but also cell proliferation, which was increased in a kind of feedback mechanism to the precursor cells, when more cells were needed.

Second, new neurons might directly influence memory consolidation in that more new neurons would reduce the time period during which learning is dependent on the hippocampus (Kitamura et al., 2009).

Quality and Flexibility of Allocentric Map Formation

New neurons increase flexibility in learning and the use of more advanced hippocampus-dependent learning strategies

So far, all functions considered were those related to the act of information processing. A very urgent question, therefore, is whether there are any "higher" cognitive consequences of adult neurogenesis. The first study to suggest that this higher relevance exists demonstrated that new neurons are relevant to the hippocampus-dependent versions of the water maze test but not the hippocampus-independent ones (Dupret et al., 2008). The cued version (with the platform visible) can be learned by rodents with a damaged hippocampus. Reducing neurogenesis had no impact on performance in the cued version but affected acquisition and retention in the hidden version. Most important, there was a difference between performance in a test paradigm where the starting position was kept constant and when it was moved to a new position. Mice with suppressed neurogenesis failed to cope with a novel starting position (Dupret et al., 2008).

In the course of learning the hidden version of the water maze, a key transition from hippocampus-independent to hippocampus-dependent learning occurs. This switch is associated with the transition from an egocentric representation of the environment (that is, the outer world is represented with reference to the learning individual) to an allocentric representation (that is, the representation has become independent of the position of the learner, like a printed map). It is the allocentric representations that are hippocampus-dependent. There are other strategies for learning the water maze that do not require a true allocentric map, which explains the paradox that animals with an impaired hippocampus can still learn the water maze. While the exact nature of the representation has be inferred, the transition from hippocampus-independent to hippocampus-dependent strategies can be directly assessed by analyzing the search patterns of the animals in the water maze. These swim patterns follow a characteristic sequence (see above, p. 453). With suppressed neurogenesis, a selective impairment of advanced, hippocampus-dependent search strategies, presumably associated with the formation of an allocentric representation, was found (Garthe et al., 2009). This impairment was particularly prominent in a reversal situation, when the animals lacking new neurons stuck with the old representation and did not seem to be able to adjust to the new one (Fig. 10–9). It took them longer to return to the more advanced strategies without reaching the same levels of proficiency as the mice with normal neurogenesis.

Discrepancies and Paradoxes

New neurons increase flexibility in learning and the use of more advanced hippocampus-dependent learning strategies

The ablation studies and the experiments based on a different methodology begin to allow us to draw an increasingly coherent picture, but they are not free of contradictions. One paradox has already been identified: the observation that neurogenesis is involved in extinction points into the opposite direction from the finding that new neurons are good for long-term retention. The type of test and the difference in emotional connotation might explain these differences.

Some other discrepancies in Table 10–3 can be explained by strain and species differences. When the hippocampus of both rats and mice was irradiated to suppress neurogenesis, a phenotype in contextual fear conditioning was observed only in rats but (in that study) not in mice (Snyder et al., 2005).

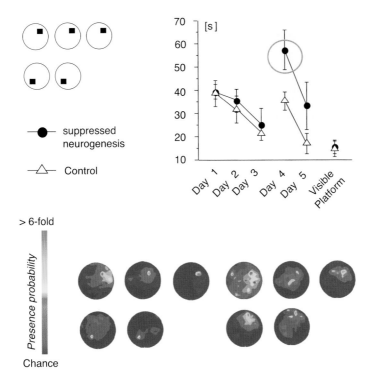

FIGURE 10–9 Effect of suppressed neurogenesis on water maze performance. Animal with suppressed neurogenesis learn the water maze well but show reduced performance in the reversal situation, when the platform is moved to another quadrant (*red circle*). The heat maps indicate the probability that an animal is found at a position in the pool. Red colors refer to high probabilities. Mice with adult neurogenesis quickly show a strong place preference for the appropriate target position and quickly adjust in the reversal. If adult neurogenesis is suppressed, however, the preference is more diffuse during the acquisition phase, and upon reversal, the mice show a perseverance in swimming to the previous platform location. From this the hypothesis is derived that new neurons are needed for flexible adaptations, when in a learned context important information changes (here, the platform position). Data for the figure are taken from Wolf et al. (2009) and Garthe et al. (2009) and combined here for didactic purposes.

The details of the protocols for behavioral testing also matter. Different water maze protocols differ with respect to the aspects of learning they actually test. Performance in the delayed matching to place (DMP) paradigm of the Morris water maze, for example, was not sensitive to a disruption of adult neurogenesis (Saxe et al., 2006). In DMP the platform position is constantly changed. According to the results from studies based on the Lipp-Wolfer protocol, with intensive training and a fixed platform position, one might assume that in DMP no strong representation of the environment is formed, so that the challenge of integrating new information into an established framework is not tested by this paradigm (Garthe et al., 2009).

Even more striking is the working memory paradox: irradiated animals showed improved working memory in a particular form of the radial arm maze (Saxe et al., 2007). Working memory per se is not hippocampus-dependent and thus is unlikely to be directly dependent on neurogenesis, irrespective of the direction of the effect. The surprising effect might thus be compensatory: neurogenesis might change the network in such a way that working memory is inhibited. On the other hand, an animal lacking the advantages of new neurons might simply have to rely more on its working memory. Interestingly, the improvement in working memory was found only in situations in which the animals had to ignore information from previous trials. In this respect, the situation is somewhat similar to the reversal task in the water maze, where a specific deficit has been described, too (Garthe et al., 2009).

Caution is also necessary with regard to the studies aiming at finding the link between functional relevance and activity-dependent regulation of adult neurogenesis (see Chap. 9, p. 366). Despite the overwhelming evidence that learning induces the survival (and functional integration) of newborn neurons, at least two studies showed the clear opposite (Ambrogini et al., 2004a; Ehninger and Kempermann, 2006). The apparent down-regulation might be the result of the stressing test situation. Habituation to the test environment at least abolished the decrease in neurogenesis (Ehninger and Kempermann, 2006). The test clearly influences the result in more than one way, so that misinterpretations are possible.

Remaining Challenges and Open Questions

The unifying theory of the function of new neurons will help us identify the evolutionarily unique and advantageous network function of the dentate gyrus

Taken together, the ablation studies show that identifying the ultimate function of new neurons for cognition will depend less on the paradigm used to manipulate adult neurogenesis than on the cleverness of the behavioral test. The hypothesis must be specific, and the causal link to the function of the dentate gyrus needs to be attempted. Being able to confidently state that new neurons are relevant for memory formation is clearly wonderful progress over only ten years ago, but no more sufficient. What is the *specific* contribution of adult neurogenesis? Why did, instead of adult neurogenesis, no solution based on synaptic plasticity evolve, which is so much easier to achieve and if comparably effective would have all the evolutionary advantages on its side? There is very likely a unique computational problem for which new neurons are the better solution. This is why the question of the functional relevance of adult hippocampal neurogenesis cannot be solved by practical experimentalists alone, and computational modeling plays such an increasingly important role. Conversely, thus, with this original and novel approach, adult neurogenesis research will now help us unravel the old mystery of what the mammalian dentate gyrus is good for.

The available data suggest that time plays a particular role in this function. Pattern separation, processing speed, long-term memory, critical time window, time stamps: all these keywords relate to aspects of time. The key question for the coming years will be how functions at different time scales relate to each other.

The second, even larger, challenge is the conceptual synthesis. Are the more complex functions, such as in the case of learning flexibility, only consequences of more basic functions, or do they constitute distinct, super-additive quality? Will there be a grand unifying theory? Probably yes, but we are still lacking a few key data. The hypotheses are still too much bound by our knowledge about what the hippocampus does, rather than what it might theoretically do in addition (if we only had better tools to measure it).

The unifying theory will also have to incorporate two further aspects of "function" that have not yet been covered. These are the potential function of adult neurogenesis in affective behavior and across the lifespan.

Adult Neurogenesis and Affective Behavior

The depression hypothesis of adult neurogenesis has implied that new neurons might be involved in affective behavior

One of the most stimulating ideas in research on adult hippocampal neurogenesis has been the hypothesis that failure of adult hippocampal neurogenesis might help explain the pathogenesis (of at least the hippocampal symptoms) of major depression (see Chap. 12, p. 541). Independently of whether this theory will ultimately hold true, it has certainly provoked a different perspective on the possible function of adult-born neurons by expanding the spectrum beyond cognition and learning. Depression is a mood disorder that includes cognitive symptoms of hippocampal origin.

The central issue is the link between the cognitive contents and emotional contexts that the hippocampus is thought to provide.

Affective and cognitive functions are linked

Input from other limbic structures that mediate the emotional contexts of information to be processed in the hippocampus converge at the dentate gyrus and supposedly modulate information processing. Snyder and colleagues discovered that more of the newborn neurons responding with immediate early gene expression to a learning stimulus were found in the ventral dentate gyrus (which plays a larger role in affective behavior), whereas more of the older granule cells responded in the dorsal hippocampus with its greater involvement in cognition (Snyder et al., 2009a; Snyder et al., 2009b). This might suggest that, even within functions subsumed under cognition, neurogenesis might contribute to those aspects relating to affect. At the same time, human emotional memory is associated with enhanced episodic memory, another hippocampal function (Richardson et al., 2004). The question thus arises as to how distinct the two aspects really are.

Ablation of adult neurogenesis increases anxiety-like behavior

The idea of a specific role for adult hippocampal neurogenesis in hippocampus-related affective behavior was first presented when Luca Santarelli, René Hen, and colleagues at Columbia University used irradiation to knockdown adult hippocampal neurogenesis and tested the animals in an anxiety test ("novelty-suppressed feeding") (Santarelli et al., 2003). The group showed that the effectiveness of antidepressant drugs in improving their performance in this test depended on adult neurogenesis. Antidepressants have a stimulating effect on adult hippocampal neurogenesis (Malberg et al., 2000), and this effect was absent in the irradiated animals. The study was a major breakthrough in that it extended the question of the functional relevance of adult neurogenesis to include hippocampal function beyond the strict hippocampal learning paradigms. (See Chapter 12, p. 541, for more information on the neurogenesis hypothesis of depression and the different studies from the Hen group.)

The involvement of adult neurogenesis in affective behavior has found support by other ablation studies. In a model based on the conditional deletion of the TrkB receptor in radial glia–like (GLAST-positive) cells, anxiety was increased, while with the tests used here, no effect on cognition was detected (Bergami et al., 2008). A very detailed study using another genetic model further supported this idea (Revest et al., 2009). On the other end of the spectrum, physical activity, which increases adult neurogenesis, has also anxiolytic and anti-depressant effects (Bjornebekk et al., 2005; Zheng et al., 2006; Duman et al., 2008; Marais et al., 2009). One dissenting study found an increase in anxiety-like behavior after wheel-running and despite increased adult neurogenesis (Fuss et al., 2009).

Anxiety, fear, and depression might share a common pathology for which new neurons are relevant

In common theory, fear is concrete while anxiety is diffuse, but phobias are concrete and count under anxiety disorders, as do panic episodes, which are not. Animal models of "anxiety" often actually test fear, and the results from the fear conditioning tasks might simply for this reason relate to what is operationally defined as "anxiety" but in reality is not. On the other hand, there is no true biological evidence that, mechanistically speaking, fear and anxiety are fundamentally different (Charney, 2003; Marcin and Nemeroff, 2003). At the same time the clinical definitions of *anxiety disorders* and *depression* merge (see Chap. 12, p. 541). There is thus presumably a continuum of symptoms and underlying causes across the spectrum spanned by anxiety, fear, and depression. Available data indicate that adult neurogenesis might be linked to the entire spectrum, but whether the new neurons are thus a common denominator or only a shared side aspect remains to be shown.

Patients with symptoms of depression showed impairment in cognitive tasks predicted from a neurogenesis-dependent model of hippocampal function

The discussion of adult neurogenesis in complex emotional behavior and functional pathology, as in the case of major depression, requires an interpretation on the network *and* the systems level and also beyond. Consequently, the study of adult neurogenesis in the highly reductionistic animal models of depression has not yet helped much to clarify the situation (Vollmayr et al., 2003). Sue Becker and Martin Wojtowicz, who in 2007 had already presented a theoretical framework for the contribution of adult-born neurons to the hippocampal involvement in mood disorders

(Becker and Wojtowicz, 2007), used their computational model to predict test performance of human subjects with pre-clinical stages of depression (Becker et al., 2009). Patients who had higher score in a depression scale indeed showed reduced performance in a delayed matching-to-sample (DMS) task.

The positive affective response to tickling is associated with increased adult neurogenesis

Gunter Höglinger's group in Marburg, Germany, delivered unusual evidence in support of the role of adult neurogenesis in emotions by demonstrating that affect in turn might regulate neurogenesis. As good doctors do, he and his colleagues—sort of—listened to the animals directly. Rats emit vocalization at different frequencies depending on whether they are in appetitive or aversive situations. Rats like tickling, and produce calls at 50 Hz in response. Cell proliferation in the hippocampus was correlated with calls at this frequency, but not with 22 Hz calls associated with aversive affect (Wohr et al., 2009). How direct this effect is remains to be seen, but the study certainly deserves a bonus for originality.

Acute Transient Versus Lasting Adaptive Effects of Adult Neurogenesis

Adult neurogenesis might affect cognition differently at different time scales

Also, from yet another perspective, the field has long struggled with concepts about the functions of the new neurons in the adult hippocampus. Two positions were held: (1) that new neurons have a transient function in the process of learning itself (Gould et al., 1999a), or (2) that adult neurogenesis allows long-term network adaptations that are an adaptive investment for future challenges (Kempermann, 2002). Both ideas were never mutually exclusive, but there was so little experimental support for either of the ideas that the argument remained rather theoretical. As demonstrated by Table 10–2, this situation has dramatically changed. Table 10–5 gives an overview of the arguments. Although clear proof in one experiment covering both aspects is still lacking, the current idea is that new neurons have both transient functions in the process of acquisition and lasting functions in network adaptation. The latter might, but need not necessarily, have to be a consequence of the former.

The transient function of new neurons was first hypothesized by Elizabeth Gould in 1999 and is now supported by many experiments

In 1999, in what is the first concise theory paper on the subject, Elizabeth Gould (Gould et al., 1999a) suggested:

A rapidly changing population of adult-generated neurons would be particularly suitable as a substrate for such a transient role of the hippocampal formation in memory storage. Neurons produced in adulthood might play a role in information processing related to memory storage during a discrete time after their generation. These cells might then degenerate or undergo changes in connectivity, gene expression, or both, coincident with the end of hippocampal storage of that particular memory.

Table 10–5 Summary of arguments for turnover vs. cumulative network adaptations	
Pro turnover	**Pro long-term network adaptation**
Critical time window for learning-induced recruitment	Hippocampus is the gateway to memory, not the site of memory storage
Period of increased synaptic plasticity suggests distinct function during this period	Long-term survival of newborn neurons beyond critical time window
Adult neurogenesis influences period during which memories are dependent on the hippocampus	Relative growth of the dentate gyrus according to cumulative lineage tracing studies
Effects of neurogenesis on processing speed	Very low rate of neurogenesis in old age might indicate reduced needs due to previous adaptation processes

The idea is that new neurons are not produced on demand but that the hippocampus (and olfactory bulb) provide a constant flow of immature cells that could be used in an acute situation. The findings that learning stimuli and experience might specifically recruit new neurons in the hippocampus and induce their long-term survival (Kempermann et al., 1997; Gould et al., 1999b; Leuner et al., 2004; Tashiro et al., 2006; Tashiro et al., 2007; Steiner et al., 2008) and that immature new granule cells show signs of increased synaptic plasticity (Wang et al., 2000; Snyder et al., 2001; Schmidt-Hieber et al., 2004; Garthe et al., 2009), as well as theoretical considerations (Feng et al., 2001; Deisseroth et al., 2004; Becker, 2005; Aimone et al., 2009), indicate that adult neurogenesis might be beneficial to hippocampal activity through a functional contribution of precursor cells and immature neurons themselves (Fig. 10–10). The new cells might exert a specific function in an immature state, just after maturation or even at a lineage-determined progenitor cell stage. During the consolidation process of memory formation, the immature new neurons might form temporary memory that is passed on to long-term storage and might activity-dependently add to the processing network (Deisseroth et al., 2004). After consolidation, the function of the new cells would become obsolete, and they would be replaced by the next generation of neurons. The early neuronal responsiveness might be a relevant property in and of itself and be distinct from what could be perceived as mature neuronal function. The experiment to directly test this hypothesis was done only ten years after Gould's foresight and showed that with fewer new neurons, the time for which new memories remain hippocampus-dependent is increased (Kitamura et al., 2009).

However, the idea that the "time-limited existence [of new neurons] might be related to transient processes thought to be involved in memory" (Gross, 2000) is challenged by the finding that new neurons, including those generated in response to an acute learning stimulus, tend to survive for long periods of time (Kempermann et al., 2003; Leuner et al., 2004), and no good evidence of a general neuronal turnover has been found in the adult dentate gyrus (Bayer, 1982, 1985; Crespo et al., 1986; Ben Abdallah et al., 2010; Imayoshi et al., 2008). This makes it unlikely that new neurons would have only a transient function.

Net growth, however, could still be accounted for by turnover and just a lower cell death rate than the production rate of new neurons, but it seems that it is the surplus of new cells that die, not the older granule cells. This net increase has consequences for functional hypotheses at a network and systems level. During adult hippocampal neurogenesis, cell death occurs primarily at the level of early post-mitotic cells, presumably at the same stage that shows increased synaptic plasticity. Once the cells have become integrated into the local network and survived a Hebbian selection process, the new neurons are likely to persist for long periods of time (Kempermann et al., 2003; Leuner et al., 2004). This early determination and long-term persistence are also reflected in the distribution of new neurons in the adult dentate gyrus. The new cells find their position within the granule cell layer early, and the location within the granule cell layer barely changes over time (Kempermann et al., 2003).

A turnover largely limited to the population of intermediate cells, in contrast, seems more plausible. Consequently, the two concepts of transient and lasting functions are not necessarily mutually exclusive. On one side, the new immature neuron might perform a specific function during memory acquisition; on the other side, the specific and distinct early functions of new neurons might be a preparatory step necessary for persistent integration and mature functioning in the long-term altered network. The increased synaptic plasticity of immature cells might then be necessary to recruit the new neurons into the network but also allows improved acute processing. Strictly speaking, these two groups of new neurons need not be identical, and it is actually conceivable that immature cells with transient function and those destined for long-term function might not be the same. As a first hint, the expression of immediate early gene Arc/Arg3.1, which is classically induced by "learning," was detectable in very young new neurons although it was not inducible by LTP at this stage. "Spontaneous" Arc/Arg3.1 expression was rather associated with the survival of new cells at a somewhat later stage (Kuipers et al., 2009).

In a complex conditional transgenic model, in which new neurons were not ablated but impaired in their differentiation and maturation, new neurons up to an age of three to four weeks were required for the acquisition of new spatial memories (Farioli-Vecchioli et al., 2008). The idea was confirmed and extended with a different model and different tasks by another group: mice that lack immature neurons but not new mature neurons were impaired in the water maze task and fear conditioning (Deng et al., 2009). This is the application of the idea of a critical time window that exists for the recruitment of new neurons (see Chap. 6, p. 199) to the function of new neurons.

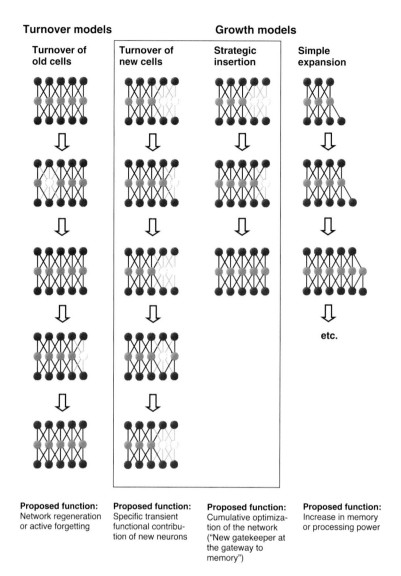

Turnover models

Growth models

Turnover of old cells	Turnover of new cells	Strategic insertion	Simple expansion

etc.

Proposed function:
Network regeneration or active forgetting

Proposed function:
Specific transient functional contribution of new neurons

Proposed function:
Cumulative optimization of the network ("New gatekeeper at the gateway to memory")

Proposed function:
Increase in memory or processing power

FIGURE 10–10 Theoretical concepts about the function of new neurons in the hippocampus. Concepts can be roughly divided in to turnover models and growth models. There is no experimental evidence for plain and general growth of the hippocampus. Adult neurogenesis affects only the granule cell layer. In turnover models, two forms can be distinguished. In the first, old cells are replaced because they have been lost or have become useless. There is little evidence for this idea of a regenerative turnover, but elimination due to lack of activity might exist. Many current theoretical concepts, however, argue in favor of a turnover of newly generated cells, which in these concepts would have a particular and transient function. Thus, a version of a turnover model is imaginable in which both old and new cells are replaced because the network function requires the turnover of certain neurons. Clearance of the hippocampus for new information or active forgetting might be such functions. The strongest argument against a dominance of this mechanism is the very low level of adult hippocampal neurogenesis in old animals, which still have good hippocampal function. For long-term effects of adult neurogenesis, we favor a strategic growth model, the model of "new gatekeepers" at the gateway to memory. This theory states that adult hippocampal neurogenesis strategically inserts new neurons to adjust the hippocampal network for coping with levels of complexity and novelty encountered by an individual (Kempermann, 2002; Kempermann and Wiskott, 2004). This long-term effect would work alongside a transient function of the new neurons (*box*).

The key problem with this interpretation is that in such a model, learning requires a steady stream of newborn neurons. There is, however, the strong age-dependent decline in adult neurogenesis that prevents the availability of greater numbers of immature neurons throughout most of a rodent's life. This decrease in neurogenesis leads to extremely low numbers of new neurons in aged subjects, which nevertheless learn many new hippocampal tasks considerably well (Merrill et al., 2003). New experiments would have to be more specific here. It might well be that aged animals indeed lose certain key functions attributable to the transient function of new neurons, while still benefiting from the long-term network changes achieved with the help of new neurons earlier in life.

New neurons might help to lastingly optimize the strength of the mossy fiber connection

The strategic insertion of new neurons at a narrow, key spot in the neuronal network of the hippocampus is the basis of a theory that places more emphasis on the long-term benefits of new neurons (Fig. 10–10) (Kempermann, 2002; Kempermann and Wiskott, 2004; Kempermann et al., 2004). The assumption behind the idea that new neurons are "new gatekeepers at the gateway to memory" is that it is beneficial for the mossy fiber connection to be as sparse as possible, while at the same time as strong as necessary. Adult hippocampal neurogenesis would thereby provide a means of optimizing the mossy fiber system to allow efficient processing at a level of information complexity and novelty frequently encountered by the individual. The benefits of an adaptation of the mossy fiber system are cumulative and an investment for the future. This is consistent with the known slow macroscopic changes of the mossy fiber tract over life and their strong dependence on genetic factors. The "new gatekeeper theory" would also explain why adult neurogenesis can decrease with increasing age. Normally, younger animals encounter relatively more novel experiences, whereas older animals have seen it all and thus require less potential for adaptation. If older individuals, however, are challenged by a novel experience, a relatively much stronger induction of adult neurogenesis is possible in the aged brain than in the young brain (Kempermann et al., 1998). Consistent with this theory is the finding that the absolute level of adult neurogenesis at the oldest age alone did not predict performance in the recall phase of the Morris water maze (Bizon and Gallagher, 2003). The new gatekeeper theory links adult hippocampal neurogenesis to the lifelong capability of coping with novelty and complexity. Rats bred for strong reaction to novelty have low levels of adult neurogenesis, presumably because their novelty-seeking behavior compensates for their inability to efficiently and quietly cope with novelty (Lemaire et al., 1999).

Functional Relevance of Adult Neurogenesis Across the Lifespan: The Neurogenic Reserve Hypothesis

Adult neurogenesis as a whole might provide a functional advantage over the lifespan

Therefore, the potential contributions of adult neurogenesis to cognition might be found in long-term adaptations of the hippocampal circuitry as well as in acute benefits. In that sense, the network modification would be immediate but not restricted to that initial period. The new neurons would be recruited during a particularly sensitive phase of their late development and in response to challenges of complexity and novelty in order to cumulatively adapt the network to cope with similar challenges in the future. During their critical recruitment phase, they would contribute other related or unrelated functional benefits to the network. As adult neurogenesis declines with age, one might predict, therefore, that the acute functions should be more sensitive to aging than the cumulative network adaptations. Depending on the preceding experience and the resulting neurogenesis-dependent network adaptations, there should be, by tendency, an improvement in these functions with increasing age.

The idea of "neural reserves" explains the discrepancy between signs of degeneration and functional performance

There is no strict correlation between cognitive performance and neuropathological signs of degeneration with age. From this observation, the theory of *neural* or *cognitive reserves* has been developed. The idea is that different brains might be differently equipped to cope with pathology and age-related changes and would have a different range of options for compensation. This compensatory capacity is malleable and represents a particular form of plasticity. Epidemiological studies indicate that leading a physically and cognitively active life with high levels of education and lifelong learning reduce the risk of dementia and different types of neurodegeneration. The personal risk profile is clearly determined by genetic factors, but there does not seem to be any genetic predisposition under which an individual would not benefit from activity. Activity would thus be the natural means to develop and increase the cognitive reserve.

The neurogenic reserve theory attempts to explain how activity-dependent regulation of adult neurogenesis might contribute to greater compensatory capacity and cognitive flexibility with aging

The neurogenic reserve hypothesis integrates adult neurogenesis into this concept (Fig. 10–11) (Kempermann, 2008). As for an animal, locomotion and cognitive challenges are inseparable, "activity" in a broader sense will always signal a greater chance of coming cognitive challenges to the brain. Adult neurogenesis is high in young animals and allows optimal adaptation of the hippocampal network to this individual level of activity and exposure to cognitive challenges. Over the course of time, this prevailing high level of activity slows the age-related cognitive decline in precursor cell activity (Kronenberg et al., 2006). The hippocampus remains prepared for new adaptational needs and builds a "neurogenic reserve." If a great cognitive challenge arises, it can draw from a greater pool of precursor cells and immature neurons and better cope with the situation. The same applies to compensation in the face of neurodegeneration. The increased pool can cope with more losses and still remain functional. In contrast, if life is not challenging and activity is low in youth, the potential for adult neurogenesis rapidly decreases with age and less adaptation is possible, should an unforeseen challenge arise in advanced age. Adult neurogenesis would thus be a means of optimizing the hippocampal network according to the level of activity, novelty, and complexity usually encountered by the individual and would maintain the network prepared for challenges at this level in the future. At the extreme, this might mean that beyond one point an optimally adapted hippocampus might require (and have) less neurogenesis than one that remains "challengeable." But within the physiological range (however defined), more activity would mean more plasticity in the future.

Functional Relevance of Adult-Generated Neurons in the Olfactory Bulb

The Network Situation in the Olfactory Bulb

Neurogenesis in the olfactory bulb modifies the network of a key sensory relay station

Neurogenesis in the olfactory bulb shows a few distinctions from its counterpart in the dentate gyrus (Table 10–1; and see also Table 8–2, Chap. 8, p. 288), including its potential functional relevance. As a first hypothesis, adult neurogenesis allows an activity-dependent adaptation of GABAergic inhibition in the olfactory bulb (Carleton et al., 2002). The olfactory bulb is a major sensory relay center. The first synapse in the olfactory path is in the olfactory bulb, where the olfactory receptor neurons synapse onto the dendrites of mitral and tufted cells (M/T cells). The synaptic contact between the two cells occurs in the glomeruli of the olfactory bulb (Fig. 4–11). Each glomerulum represents one odor, in the sense that all receptors of one type project to the same glomerulum. Periglomerular cells provide modulating contacts between different glomeruli. Granule cell interneurons provide modulating contacts between different M/T cells.

The general network structure of the olfactory bulb and its integration into the olfactory system has been identified (Hopfield, 1991), but knowledge about the electrophysiology of the olfactory bulb is still more limited than that of the hippocampus.

Adult neurogenesis produces a variety of different interneurons in the olfactory bulb

Adult neurogenesis generates two major classes of interneurons with several subtypes in the adult olfactory bulb, most of them in the granule cell layer, and a few percent in the periglomerular regions (see Chap. 5, Table 5–1). Based on neurotransmitter types, there is at least a total of six inhibitory GABAergic new neurons and possibly one type of glutamatergic interneuron. Interneurons have multiple functions in a neuronal network,

A Adult neurogenesis
allows adaptation to new experiences

young-adult high levels
of adult
neurogenesis

B Lack of adult neurogenesis
results in reduced functional plasticity

young-adult low levels
of adult
neurogenesis

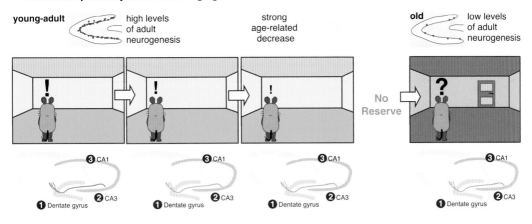

C Lack of exposure and training in younger age lead to reduced cellular and
functional plasticity with increasing age

young-adult high levels
of adult
neurogenesis

strong
age-related
decrease

old low levels
of adult
neurogenesis

No
Reserve

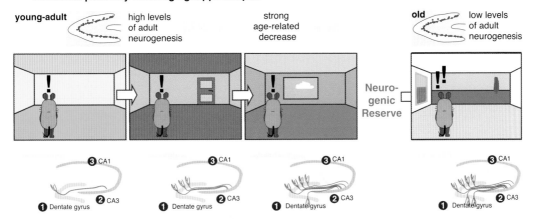

D Exposure and training in younger age build a neurogenic reserve for cellular and
functional plasticity in the aging hippocampus

young-adult high levels
of adult
neurogenesis

strong
age-related
decrease

old low levels
of adult
neurogenesis

Neuro-
genic
Reserve

FIGURE 10–11 The neurogenic reserve hypothesis. Summarizing cartoon from (Kempermann, 2008), reprinted with kind permission from Elsevier, copyright 2008. 'Die Maus' appears with kind permission of copyright holder Westdeutscher Rundfunk Köln. I. Schmitt-Menzel/WDR mediagroup licensing GmbH/Die Sendung mit der Maus, WDR.

and together these functions go far beyond a mere balance of excitation. In the olfactory bulb, the interneurons form the oscillatory network activity, which is important in representing odors in a temporally and spatially precise manner (Lledo and Lagier, 2006). This theoretical aspect would suggest a potential function of new neurons in forming such representations and would place particular emphasis on a temporal domain.

Functional Maturation of Adult-Born Olfactory Interneurons

Neuronal maturation is delayed until the migrating cells reach the olfactory bulb

Migratory neuroblasts (A cells) reach the olfactory bulb approximately 14 days after they are born. Their functional maturation is delayed until this point. They switch from a tangential migratory pattern in the stream to radial migration toward the granule cell layer and the glomeruli. Pierre-Marie Lledo and colleagues from Paris have used retrovirally expressed GFP to study the fine structure of the incoming new neurons and perform electrophysiological recordings. First spines can be found on migrating cells, indicating that these cells might already receive synapses. The synapses are first GABAergic, and later glutamatergic (Panzanelli et al., 2009). No later than after having arrived in their target location, the cells show signs of advanced maturation with the expression of neurotransmitter receptor channels and the generation of spontaneous action potentials (Carleton et al., 2003). Intuitively it makes sense that fully functional maturation should be delayed until the cells have entered the local network structure in which they ultimately have to integrate. During the final period of maturation, the new cells transiently showed increased long-term potentiation, which suggests that they go through a critical period of enhanced responsiveness (Nissant et al., 2009). Due to this greater plasticity, they might successfully compete with the older neurons for input and sustained survival. Such an interval was also found when olfactory training sessions were performed at various times after the birth of the new neurons. Interestingly, the consequences of the exposure to the discrimination challenge not only favors survival, at other times it might actively foster the elimination of cells (Mouret et al., 2008). This is again similar to analogous studies in the adult hippocampus (Dobrossy et al., 2003). Stimulation of the olfactory nerve resulted in action potentials in newly integrated interneurons in the adult olfactory bulb (Belluzzi et al., 2003). Based on immediate early gene expression, the population of newly integrated neurons was more responsive to novel odors than the older neurons, and when the animals were familiarized with the test odors, the responsiveness of the new neurons was long-lasting but unaltered in the older cells. That familiarization (but possibly the immediate responsiveness itself) did not have an effect on the survival of the newborn neurons (Magavi et al., 2005).

Systems- and Behavior-Level Function of Adult Olfactory Neurogenesis

Exposure to odors increases adult neurogenesis

The ethological relevance of olfaction is immense, particularly in rodents. Appropriately, olfaction plays a major role in regulating neurogenesis in the adult olfactory bulb (Petreanu and Alvarez-Buylla, 2002). Exposure to an enriched odor environment increased adult olfactory bulb neurogenesis but left hippocampal neurogenesis unaffected (Rochefort et al., 2002). Conversely, physiological activators of adult hippocampal neurogenesis, such as voluntary physical activity and exposure to a complex environment, did not affect neurogenesis in the olfactory bulb (Brown et al., 2003).

Ablation studies indicate a function of adult neurogenesis in odor discrimination

Most insight about functional relevance of adult-generated interneurons in the olfactory bulb comes from ablation studies as well. As expected, when adult neurogenesis was ablated, mitral cells received reduced inhibitory input (Breton-Provencher et al., 2009). Reduced olfactory bulb neurogenesis in mutants for NCAM was associated with disturbed odor discrimination (Gheusi et al., 2000). Most later studies have focused on this aspect of odor discrimination (Enwere et al., 2004; Mandairon et al., 2006). Olfactory discrimination in turn also increases the survival of new neurons (Alonso et al., 2006).

A slightly different twist was added to the picture by a study showing that adult-generated olfactory bulb interneurons might be needed for "perceptual learning," a particular aspect of olfactory discrimination. Perceptual learning is a form of implicit memory and independent of associations between the stimuli. The discrimination of new stimuli is improved after exposure to other non-related stimuli of the same kind. Improved discrimination was associated with increased survival of new neurons, whereas ablation with AraC abolished perceptual learning (Moreno et al., 2009).

In one irradiation-based ablation study, fear conditioning was examined and found to be impaired only for olfactory but not auditory cues, but this might be explained simply by the impaired olfaction, not necessarily a specific contribution to emotional learning (Valley et al., 2009).

In addition, a functional relevance of new neurons in olfactory memory has been proposed. In the studies on olfactory enrichment, the animals not only showed increased olfactory neurogenesis, but also sustained and more stable memory of odors (Rochefort et al., 2002). Irradiation of the SVZ impaired neurogenesis in the olfactory bulb and caused a lasting deficit in long-term memory, while leaving the distinction between odors and short-term memory unaffected (Lazarini et al., 2009). In an ablation study with cytostatic agent cytosine arabinoside (AraC), in contrast, short-term olfactory memory was impaired (Breton-Provencher et al., 2009). In the irradiation study, the identification of socially relevant olfactory cues was also unaltered. That study thus stands in some contrast to other ablation experiments, both with regard to a role of new neurons in olfactory discrimination and to its relevance for social behavior. Exposure of female mice to the odor of pheromones from dominant males increased neurogenesis in the olfactory bulb and in the hippocampus; elimination of the new neurons by the infusion of AraC into the ventricle abolished the preference of the females for the dominant males (Mak et al., 2007). To conclude from this that the new neurons are mediating this preference will apply a certain stretch to the data, but behind this observation might lie an important aspect of a functional contribution of new neurons to ethologically highly relevant olfaction-dependent behavior.

Most evidence indicates that adult neurogenesis in the olfactory bulb results in a turnover. Studies based on permanent genetic lineage labeling indicate this (Ninkovic et al., 2007; Imayoshi et al., 2008), although there is also strong evidence that many adult-generated neurons survive for long times (Winner et al., 2002). The essence of a turnover is that the elimination of cells would be as relevant as the addition of new ones. Genetically ablating adult neurogenesis over time led to a decrease in the size of the granule cell layer of the olfactory bulb and an impairment in olfactory discrimination (Imayoshi et al., 2008). Conversely, when cell death was pharmacologically blocked in the olfactory bulb, odor discrimination was reduced, suggesting that an optimal amount of cell death is required (Mouret et al., 2009). In BAX knockout mice, in which cell death is also abolished, however, no such deficit was found, despite impaired migration and neurogenesis (Kim et al., 2007). The method also could not distinguish between the death of new versus the death of older neurons, but in the olfactory bulb this distinction might be of less relevance than in the adult hippocampus. First, the rate of neurogenesis is much higher in the olfactory bulb than in the granule cell layer, and, second, the position of the new interneurons in the neuronal network indicates a substantially different situation. Plasticity in the olfactory bulb depends on the extreme and undisputed turnover in the olfactory epithelium and its function (Petreanu and Alvarez-Buylla, 2002). The mechanism by which the input-dependent turnover is controlled, especially in the face of potentially unpredictable variation in supply from the SVZ, is not yet known (Valley et al., 2009).

On the basis of the network structure, the general pattern of regulation and the types of neurons that generate the contribution of new neurons to olfactory bulb function might thus be fundamentally different from their counterparts in the dentate gyrus. A first computational model has been developed to explain how new neurons contribute to function of the olfactory bulb (Cecchi et al., 2001). It is based on a simplified network with receptor neurons that project to the M/T cells and with inhibitory granule

Data on a role of adult olfactory neurogenesis in other behaviors are still inconsistent

Neurogenesis in the adult olfactory bulb contributes to a neuronal turnover

A computational model supports the role of adult neurogenesis in odor discrimination

cell neurons that make pair-wise connections with the M/T cells. Consistent with the general ideas applicable to the hippocampus, the underlying assumption was that neural production proceeds at a constant and unregulated rate, whereas cell survival is selective and activity-dependent. Within the network model, only interneurons were replaced, because adult olfactory neurogenesis affects only interneurons. Survival of the new interneurons was linked to their activity, making more active neurons more likely to survive. Learning in this model was unsupervised—that is, no external instance adjusted key parameters in the course of the training. Even this comparatively simple model revealed that neurogenesis allowed the network to maximize the discrimination of odors (Cecchi et al., 2001).

References

Aimone JB, Wiles J, Gage FH (2006). Potential role for adult neurogenesis in the encoding of time in new memories. *Nat Neurosci* 9:723–727.

Aimone JB, Wiles J, Gage FH (2009). Computational influence of adult neurogenesis on memory encoding. *Neuron* 61:187–202.

Alonso M, Viollet C, Gabellec MM, Meas-Yedid V, Olivo-Marin JC, Lledo PM (2006). Olfactory discrimination learning increases the survival of adult-born neurons in the olfactory bulb. *J Neurosci* 26:10508–10513.

Ambrogini P, Orsini L, Mancini C, Ferri P, Ciaroni S, Cuppini R (2004a). Learning may reduce neurogenesis in adult rat dentate gyrus. *Neurosci Lett* 359:13–16.

Ambrogini P, Lattanzi D, Ciuffoli S, Agostini D, Bertini L, Stocchi V, Santi S, Cuppini R (2004b). Morpho-functional characterization of neuronal cells at different stages of maturation in granule cell layer of adult rat dentate gyrus. *Brain Res* 1017:21–31.

Appleby PA, Wiskott L (2009). Additive neurogenesis as a strategy for avoiding interference in a sparsely coding dentate gyrus. *Network* 20:137–161.

Babu H, Cheung G, Kettenmann H, Palmer TD, Kempermann G (2007). Enriched monolayer precursor cell cultures from micro-dissected adult mouse dentate gyrus yield functional granule cell-like neurons. *PLoS ONE* 2:e388.

Bayer SA (1982). Changes in the total number of dentate granule cells in juvenile and adult rats: A correlated volumetric and 3H-thymidine autoradiographic study. *Exp Brain Res* 46:315–323.

Bayer SA (1985). Neuron production in the hippocampus and olfactory bulb of the adult rat brain: Addition or replacement? *Ann NY Acad Sci* 457:163–172.

Bayer SA, Altman J (1975). The effects of X-irradiation on the postnatally-forming granule cell populations in the olfactory bulb, hippocampus, and cerebellum of the rat. *Exp Neurol* 48:167–174.

Becker S (2005). A computational principle for hippocampal learning and neurogenesis. *Hippocampus* 15:722–738.

Becker S, Wojtowicz JM (2007). A model of hippocampal neurogenesis in memory and mood disorders. *Trends Cogn Sci* 11:70–76.

Becker S, Macqueen G, Wojtowicz JM (2009). Computational modeling and empirical studies of hippocampal neurogenesis-dependent memory: Effects of interference, stress and depression. *Brain Res* 1299:45–54.

Belluzzi O, Benedusi M, Ackman J, LoTurco JJ (2003). Electrophysiological differentiation of new neurons in the olfactory bulb. *J Neurosci* 23:10411–10418.

Ben Abdallah N, Slomianka L, Vyssotski AL, Lipp HP (2010). Early age-related changes in adult hippocampal neurogenesis in C57 mice. *Neurobiol Aging* 31:151-161.

Benninger F, Beck H, Wernig M, Tucker KL, Brustle O, Scheffler B (2003). Functional integration of embryonic stem cell-derived neurons in hippocampal slice cultures. *J Neurosci* 23:7075–7083.

Bergami M, Rimondini R, Santi S, Blum R, Gotz M, Canossa M (2008). Deletion of TrkB in adult progenitors alters newborn neuron integration into hippocampal circuits and increases anxiety-like behavior. *Proc Natl Acad Sci USA* 105:15570–15575.

Berninger B, Costa MR, Koch U, Schroeder T, Sutor B, Grothe B, Gotz M (2007). Functional properties of neurons derived from in vitro reprogrammed postnatal astroglia. *J Neurosci* 27:8654–8664.

Bizon JL, Gallagher M (2003). Production of new cells in the rat dentate gyrus over the lifespan: Relation to cognitive decline. *Eur J Neurosci* 18:215–219.

Bjornebekk A, Mathe AA, Brene S (2005). The antidepressant effect of running is associated with increased hippocampal cell proliferation. *Int J Neuropsychopharmacol* 8:357–368.

Breton-Provencher V, Lemasson M, Peralta MR, 3rd, Saghatelyan A (2009). Interneurons produced in adulthood are required for the normal functioning of the olfactory bulb network and for the execution of selected olfactory behaviors. *J Neurosci* 29:15245–15257.

Brill MS, Ninkovic J, Winpenny E, Hodge RD, Ozen I, Yang R, Lepier A, Gascon S, Erdelyi F, Szabo G, Parras C, Guillemot F, Frotscher M, Berninger B, Hevner RF, Raineteau O, Gotz M (2009). Adult generation of glutamatergic olfactory bulb interneurons. *Nat Neurosci* 12:1524–1533.

Brown J, Cooper-Kuhn CM, Kempermann G, Van Praag H, Winkler J, Gage FH, Kuhn HG (2003). Enriched environment and physical activity stimulate hippocampal but not olfactory bulb neurogenesis. *Eur J Neurosci* 17:2042–2046.

Bruel-Jungerman E, Davis S, Rampon C, Laroche S (2006). Long-term potentiation enhances neurogenesis in the adult dentate gyrus. *J Neurosci* 26:5888–5893.

Butz M, Lehmann K, Dammasch IE, Teuchert-Noodt G (2006). A theoretical network model to analyse neurogenesis and synaptogenesis in the dentate gyrus. *Neural Netw* 19:1490–1505.

Butz M, Teuchert-Noodt G, Grafen K, van Ooyen A (2008). Inverse relationship between adult hippocampal cell proliferation and synaptic rewiring in the dentate gyrus. *Hippocampus* 18:879–898.

Carlen M, Cassidy RM, Brismar H, Smith GA, Enquist LW, Frisen J (2002). Functional integration of adult-born neurons. *Curr Biol* 12:606–608.

Carleton A, Rochefort C, Morante-Oria J, Desmaisons D, Vincent JD, Gheusi G, Lledo PM (2002). Making scents of olfactory neurogenesis. *J Physiol Paris* 96:115–122.

Carleton A, Petreanu LT, Lansford R, Alvarez-Buylla A, Lledo PM (2003). Becoming a new neuron in the adult olfactory bulb. *Nat Neurosci* 6:507–518.

Cecchi GA, Petreanu LT, Alvarez-Buylla A, Magnasco MO (2001). Unsupervised learning and adaptation in a model of adult neurogenesis. *J Comput Neurosci* 11:175–182.

Charney DS (2003). Neuroanatomical circuits modulating fear and anxiety behaviors. *Acta Psychiatr Scand* Suppl :38–50.

Clelland CD, Choi M, Romberg C, Clemenson GD, Jr., Fragniere A, Tyers P, Jessberger S, Saksida LM, Barker RA, Gage FH, Bussey TJ (2009). A functional role for adult hippocampal neurogenesis in spatial pattern separation. *Science* 325:210–213.

Crespo D, Stanfield BB, Cowan WM (1986). Evidence that late-generated granule cells do not simply replace earlier formed neurons in the rat dentate gyrus. *Exp Brain Res* 62:541–548.

Day M, Langston R, Morris RG (2003). Glutamate-receptor-mediated encoding and retrieval of paired-associate learning. *Nature* 424:205–209.

Deisseroth K, Singla S, Toda H, Monje M, Palmer TD, Malenka RC (2004). Excitation-neurogenesis coupling in adult neural stem/progenitor cells. *Neuron* 42:535–552.

Deng W, Saxe MD, Gallina IS, Gage FH (2009). Adult-born hippocampal dentate granule cells undergoing maturation modulate learning and memory in the brain. *J Neurosci* 29:13532–13542.

Dobrossy MD, Drapeau E, Aurousseau C, Le Moal M, Piazza PV, Abrous DN (2003). Differential effects of learning on neurogenesis: Learning increases or decreases the number of newly born cells depending on their birth date. *Mol Psychiatry* 8:974–982.

Drapeau E, Mayo W, Aurousseau C, Le Moal M, Piazza PV, Abrous DN (2003). Spatial memory performances of aged rats in the water maze predict levels of hippocampal neurogenesis. *Proc Natl Acad Sci USA* 100: 14385–14390.

Duman CH, Schlesinger L, Russell DS, Duman RS (2008). Voluntary exercise produces antidepressant and anxiolytic behavioral effects in mice. *Brain Res* 1199:148–158.

Dupret D, Fabre A, Dobrossy MD, Panatier A, Rodriguez JJ, Lamarque S, Lemaire V, Oliet SH, Piazza PV, Abrous DN (2007). Spatial learning depends on both the addition and removal of new hippocampal neurons. *PLoS Biol* 5:e214.

Dupret D, Revest JM, Koehl M, Ichas F, De Giorgi F, Costet P, Abrous DN, Piazza PV (2008). Spatial relational memory requires hippocampal adult neurogenesis. *PLoS ONE* 3:e1959.

Ehninger D, Kempermann G (2006). Paradoxical effects of learning the Morris water maze on adult hippocampal neurogenesis in mice may be explained by a combination of stress and physical activity. *Genes Brain Behav* 5:29–39.

Eichenbaum H (2003). How does the hippocampus contribute to memory? *Trends Cogn Sci* 7:427–429.

Eichenbaum H, Dudchenko P, Wood E, Shapiro M, Tanila H (1999). The hippocampus, memory, and place cells: Is it spatial memory or a memory space? *Neuron* 23:209–226.

Enwere E, Shingo T, Gregg C, Fujikawa H, Ohta S, Weiss S (2004). Aging results in reduced epidermal growth factor receptor signaling, diminished olfactory neurogenesis, and deficits in fine olfactory discrimination. *J Neurosci* 24:8354–8365.

Erceg S, Lainez S, Ronaghi M, Stojkovic P, Perez-Arago MA, Moreno-Manzano V, Moreno-Palanques R, Planells-Cases R, Stojkovic M (2008). Differentiation of human embryonic stem cells to regional specific neural precursors in chemically defined medium conditions. *PLoS ONE* 3:e2122.

Esposito MS, Piatti VC, Laplagne DA, Morgenstern NA, Ferrari CC, Pitossi FJ, Schinder AF (2005). Neuronal differentiation in the adult hippocampus recapitulates embryonic development. *J Neurosci* 25:10074–10086.

Farioli-Vecchioli S, Saraulli D, Costanzi M, Pacioni S, Cina I, Aceti M, Micheli L, Bacci A, Cestari V, Tirone F (2008). The timing of differentiation of adult hippocampal neurons is crucial for spatial memory. *PLoS Biol* 6:e246.

Feng R, Rampon C, Tang YP, Shrom D, Jin J, Kyin M, Sopher B, Martin GM, Kim SH, Langdon RB, Sisodia SS, Tsien JZ (2001). Deficient neurogenesis in forebrain-specific Presenilin-1 knockout mice is associated with reduced clearance of hippocampal memory traces. *Neuron* 32:911–926.

Filippov V, Kronenberg G, Pivneva T, Reuter K, Steiner B, Wang LP, Yamaguchi M, Kettenmann H, Kempermann G (2003). Subpopulation of nestin-expressing progenitor cells in the adult murine hippocampus shows electro-physiological and morphological characteristics of astrocytes. *Mol Cell Neurosci* 23:373–382.

Fortin NJ, Agster KL, Eichenbaum HB (2002). Critical role of the hippocampus in memory for sequences of events. *Nat Neurosci* 5:458–462.

Freeman JH, Jr., Nicholson DA (2000). Developmental changes in eye-blink conditioning and neuronal activity in the cerebellar interpositus nucleus. *J Neurosci* 20:813–819.

Fukuda S, Kato F, Tozuka Y, Yamaguchi M, Miyamoto Y, Hisatsune T (2003). Two distinct subpopulations of nestin-positive cells in adult mouse dentate gyrus. *J Neurosci* 23:9357–9366.

Fuss J, Ben Abdallah NM, Vogt MA, Touma C, Pacifici PG, Palme R, Witzemann V, Hellweg R, Gass P (2009). Voluntary exercise induces anxiety-like behavior in adult C57BL/6J mice correlating with hippocampal neurogenesis. *Hippocampus* 20:364–376.

Garthe A, Behr J, Kempermann G (2009). Adult-generated hippocampal neurons allow the flexible use of spatially precise learning strategies. *PLoS ONE* 4:e5464.

Gazzara RA, Altman J (1981). Early postnatal x-irradiation of the hippocampus and discrimination learning in adult rats. *J Comp Physiol Psychol* 95:484–495.

Ge S, Goh EL, Sailor KA, Kitabatake Y, Ming GL, Song H (2006). GABA regulates synaptic integration of newly generated neurons in the adult brain. *Nature* 439:589–593.

Gheusi G, Cremer H, McLean H, Chazal G, Vincent JD, Lledo PM (2000). Importance of newly generated neurons in the adult olfactory bulb for odor discrimination. *Proc Natl Acad Sci USA* 97:1823–1828.

Gould E, Tanapat P, Hastings NB, Shors TJ (1999a). Neurogenesis in adulthood: A possible role in learning. *Trends Cogn Sci* 3:186–192.

Gould E, Beylin A, Tanapat P, Reeves A, Shors TJ (1999b). Learning enhances adult neurogenesis in the hippocampal formation. *Nat Neurosci* 2:260–265.

Gross CG (2000). Neurogenesis in the adult brain: Death of a dogma. *Nat Rev Neurosci* 1:67–73.

Hastings NB, Gould E (1999). Rapid extension of axons into the CA3 region by adult-generated granule cells. *J Comp Neurol* 413:146–154.

Hernandez-Rabaza V, Llorens-Martin M, Velazquez-Sanchez C, Ferragud A, Arcusa A, Gumus HG, Gomez-Pinedo U, Perez-Villalba A, Rosello J, Trejo JL, Barcia JA, Canales JJ (2009). Inhibition of adult hippocampal neurogenesis disrupts contextual learning but spares spatial working memory, long-term conditional rule retention and spatial reversal. *Neuroscience* 159:59–68.

Hopfield JJ (1991). Olfactory computation and object perception. *Proc Natl Acad Sci USA* 88:6462–6466.

Husseini L, Schmandt T, Scheffler B, Schroder W, Seifert G, Brustle O, Steinhauser C (2008). Functional analysis of embryonic stem cell-derived glial cells after integration into hippocampal slice cultures. *Stem Cells Dev* 17:1141–1152.

Imayoshi I, Sakamoto M, Ohtsuka T, Takao K, Miyakawa T, Yamaguchi M, Mori K, Ikeda T, Itohara S, Kageyama R (2008). Roles of continuous neurogenesis in the structural and functional integrity of the adult forebrain. *Nat Neurosci* 11:1153–1161.

Jaholkowski P, Kiryk A, Jedynak P, Ben Abdallah NM, Knapska E, Kowalczyk A, Piechal A, Blecharz-Klin K, Figiel I, Lioudyno V, Widy-Tyszkiewicz E, Wilczynski GM, Lipp HP, Kaczmarek L, Filipkowski RK (2009). New hippocampal neurons are not obligatory for memory formation; cyclin D2 knockout mice with no adult brain neurogenesis show learning. *Learn Mem* 16:439–451.

Jakubs K, Nanobashvili A, Bonde S, Ekdahl CT, Kokaia Z, Kokaia M, Lindvall O (2006). Environment matters: synaptic properties of neurons born in epileptic brain develop to reduce excitability. *Neuron* 52:1047–59

Jessberger S, Kempermann G (2003). Adult-born hippocampal neurons mature into activity-dependent responsiveness. *Eur J Neurosci* 18:2707–2712.

Jessberger S, Clark RE, Broadbent NJ, Clemenson GD, Jr., Consiglio A, Lie DC, Squire LR, Gage FH (2009). Dentate gyrus-specific knockdown of adult neurogenesis impairs spatial and object recognition memory in adult rats. *Learn Mem* 16:147–154.

Karten YJ, Jones MA, Jeurling SI, Cameron HA (2006). GABAergic signaling in young granule cells in the adult rat and mouse dentate gyrus. *Hippocampus* 16:312–320.

Kee N, Teixeira CM, Wang AH, Frankland PW (2007). Preferential incorporation of adult-generated granule cells into spatial memory networks in the dentate gyrus. *Nat Neurosci* 10:355–362.

Kempermann G (2002). Why new neurons? Possible functions for adult hippocampal neurogenesis. *J Neurosci* 22:635–638.

Kempermann G (2006). They are not too excited: The possible role of adult-born neurons in epilepsy. *Neuron* 52:935–937.

Kempermann G (2008). The neurogenic reserve hypothesis: What is adult hippocampal neurogenesis good for? *Trends Neurosci* 31:163–169.

Kempermann G, Gage FH (2002). Genetic determinants of adult hippocampal neurogenesis correlate with acquisition, but not probe trial performance in the water maze task. *Eur J Neurosci* 16:129–136.

Kempermann G, Wiskott L (2004). What is the functional role of new neurons in the adult dentate gyrus? In: *Stem Cells in the Nervous System: Function and Clinical Implications* (Gage F, Björklund A, Prochiatz A, Christen Y, eds.), pp 57–65. Berlin and Heidelberg: Springer.

Kempermann G, Kuhn HG, Gage FH (1997). More hippocampal neurons in adult mice living in an enriched environment. *Nature* 386:493–495.

Kempermann G, Kuhn HG, Gage FH (1998). Experience-induced neurogenesis in the senescent dentate gyrus. *J Neurosci* 18:3206–3212.

Kempermann G, Gast D, Kronenberg G, Yamaguchi M, Gage FH (2003). Early determination and long-term persistence of adult-generated new neurons in the hippocampus of mice. *Development* 130:391–399.

Kempermann G, Wiskott L, Gage FH (2004). Functional significance of adult neurogenesis. *Curr Opin Neurobiol* 14:186–191.

Kim WR, Kim Y, Eun B, Park OH, Kim H, Kim K, Park CH, Vinsant S, Oppenheim RW, Sun W (2007). Impaired migration in the rostral migratory stream but spared olfactory function after the elimination of programmed cell death in Bax knock-out mice. *J Neurosci* 27:14392–14403.

Kitamura T, Saitoh Y, Takashima N, Murayama A, Niibori Y, Ageta H, Sekiguchi M, Sugiyama H, Inokuchi K (2009). Adult neurogenesis modulates the hippocampus-dependent period of associative fear memory. *Cell* 139:814–827.

Ko HG, Jang DJ, Son J, Kwak C, Choi JH, Ji YH, Lee YS, Son H, Kaang BK (2009). Effect of ablated hippocampal neurogenesis on the formation and extinction of contextual fear memory. *Mol Brain* 2:1.

Kronenberg G, Bick-Sander A, Bunk E, Wolf C, Ehninger D, Kempermann G (2006). Physical exercise prevents age-related decline in precursor cell activity in the mouse dentate gyrus. *Neurobiol Aging* 27:1505–1513.

Kuipers SD, Tiron A, Soule J, Messaoudi E, Trentani A, Bramham CR (2009). Selective survival and maturation of adult-born dentate granule cells expressing the immediate early gene Arc/Arg3.1. *PLoS ONE* 4:e4885.

Kunze A, Congreso MR, Hartmann C, Wallraff-Beck A, Huttmann K, Bedner P, Requardt R, Seifert G, Redecker C, Willecke K, Hofmann A, Pfeifer A, Theis M, Steinhauser C (2009). Connexin expression by radial glia–like cells is required for neurogenesis in the adult dentate gyrus. *Proc Natl Acad Sci USA* 106:11336–11341.

Lazarini F, Mouthon MA, Gheusi G, de Chaumont F, Olivo-Marin JC, Lamarque S, Abrous DN, Boussin FD, Lledo PM (2009). Cellular and behavioral effects of cranial irradiation of the subventricular zone in adult mice. *PLoS ONE* 4:e7017.

Lemaire V, Aurousseau C, Le Moal M, Abrous DN (1999). Behavioural trait of reactivity to novelty is related to hippocampal neurogenesis. *Eur J Neurosci* 11:4006–4014.

Leuner B, Mendolia-Loffredo S, Kozorovitskiy Y, Samburg D, Gould E, Shors TJ (2004). Learning enhances the survival of new neurons beyond the time when the hippocampus is required for memory. *J Neurosci* 24:7477–7481.

Lledo PM, Lagier S (2006). Adjusting neurophysiological computations in the adult olfactory bulb. *Semin Cell Dev Biol* 17:443–453.

Magavi S, Leavitt B, Macklis J (2000). Induction of neurogenesis in the neocortex of adult mice. *Nature* 405: 951–955.

Magavi SS, Mitchell BD, Szentirmai O, Carter BS, Macklis JD (2005). Adult-born and preexisting olfactory granule neurons undergo distinct experience-dependent modifications of their olfactory responses in vivo. *J Neurosci* 25:10729–10739.

Mak GK, Enwere EK, Gregg C, Pakarainen T, Poutanen M, Huhtaniemi I, Weiss S (2007). Male pheromone-stimulated neurogenesis in the adult female brain: Possible role in mating behavior. *Nat Neurosci* 10:1003–1011.

Malberg JE, Eisch AJ, Nestler EJ, Duman RS (2000). Chronic antidepressant treatment increases neurogenesis in adult rat hippocampus. *J Neurosci* 20:9104–9110.

Mandairon N, Sacquet J, Garcia S, Ravel N, Jourdan F, Didier A (2006). Neurogenic correlates of an olfactory discrimination task in the adult olfactory bulb. *Eur J Neurosci* 24:3578–3588.

Marais L, Stein DJ, Daniels WM (2009). Exercise increases BDNF levels in the striatum and decreases depressive-like behavior in chronically stressed rats. *Metab Brain Dis* 24:587–97.

Marcin MS, Nemeroff CB (2003). The neurobiology of social anxiety disorder: The relevance of fear and anxiety. *Acta Psychiatr Scand* Suppl: 51–64.

Markakis E, Gage FH (1999). Adult-generated neurons in the dentate gyrus send axonal projections to the field CA3 and are surrounded by synaptic vesicles. *J Comp Neurol* 406:449–460.

McCormick DA, Clark GA, Lavond DG, Thompson RF (1982). Initial localization of the memory trace for a basic form of learning. *Proc Natl Acad Sci USA* 79:2731–2735.

Meltzer LA, Yabaluri R, Deisseroth K (2005). A role for circuit homeostasis in adult neurogenesis. *Trends Neurosci* 28:653–660.

Merrill DA, Karim R, Darraq M, Chiba AA, Tuszynski MH (2003). Hippocampal cell genesis does not correlate with spatial learning ability in aged rats. *J Comp Neurol* 459:201–207.

Moreno MM, Linster C, Escanilla O, Sacquet J, Didier A, Mandairon N (2009). Olfactory perceptual learning requires adult neurogenesis. *Proc Natl Acad Sci USA* 106:17980–17985.

Morris R (1984). Developments of a water-maze procedure for studying spatial learning in the rat. *J Neurosci Methods* 11:47–60.

Morris RG, Garrud P, Rawlins JN, O'Keefe J (1982). Place navigation impaired in rats with hippocampal lesions. *Nature* 297:681–683.

Mouret A, Gheusi G, Gabellec MM, de Chaumont F, Olivo-Marin JC, Lledo PM (2008). Learning and survival of newly generated neurons: When time matters. *J Neurosci* 28:11511–11516.

Mouret A, Lepousez G, Gras J, Gabellec MM, Lledo PM (2009). Turnover of newborn olfactory bulb neurons optimizes olfaction. *J Neurosci* 29:12302–12314.

Nakazawa K, Sun LD, Quirk MC, Rondi-Reig L, Wilson MA, Tonegawa S (2003). Hippocampal CA3 NMDA receptors are crucial for memory acquisition of one-time experience. *Neuron* 38:305–315.

Ninkovic J, Mori T, Gotz M (2007). Distinct modes of neuron addition in adult mouse neurogenesis. *J Neurosci* 27:10906–10911.

Nissant A, Bardy C, Katagiri H, Murray K, Lledo PM (2009). Adult neurogenesis promotes synaptic plasticity in the olfactory bulb. *Nat Neurosci* 12:728–730.

Panzanelli P, Bardy C, Nissant A, Pallotto M, Sassoe-Pognetto M, Lledo PM, Fritschy JM (2009). Early synapse formation in developing interneurons of the adult olfactory bulb. *J Neurosci* 29:15039–15052.

Petreanu L, Alvarez-Buylla A (2002). Maturation and death of adult-born olfactory bulb granule neurons: Role of olfaction. *J Neurosci* 22:6106–6113.

Phillips RG, LeDoux JE (1992). Differential contribution of amygdala and hippocampus to cued and contextual fear conditioning. *Behav Neurosci* 106:274–285.

Rakic P (1985). Limits of neurogenesis in primates. *Science* 227:1054–1056.

Revest JM, Dupret D, Koehl M, Funk-Reiter C, Grosjean N, Piazza PV, Abrous DN (2009). Adult hippocampal neurogenesis is involved in anxiety-related behaviors. *Mol Psychiatry* 14:959–967.

Richardson MP, Strange BA, Dolan RJ (2004). Encoding of emotional memories depends on amygdala and hippocampus and their interactions. *Nat Neurosci* 7:278–285.

Rochefort C, Gheusi G, Vincent JD, Lledo PM (2002). Enriched odor exposure increases the number of newborn neurons in the adult olfactory bulb and improves odor memory. *J Neurosci* 22:2679–2689.

Santarelli L, Saxe M, Gross C, Surget A, Battaglia F, Dulawa S, Weisstaub N, Lee J, Duman R, Arancio O, Belzung C, Hen R (2003). Requirement of hippocampal neurogenesis for the behavioral effects of antidepressants. *Science* 301:805–809.

Saxe MD, Battaglia F, Wang JW, Malleret G, David DJ, Monckton JE, Garcia AD, Sofroniew MV, Kandel ER, Santarelli L, Hen R, Drew MR (2006). Ablation of hippocampal neurogenesis impairs contextual fear conditioning and synaptic plasticity in the dentate gyrus. *Proc Natl Acad Sci USA* 103:17501–17506.

Saxe MD, Malleret G, Vronskaya S, Mendez I, Garcia AD, Sofroniew MV, Kandel ER, Hen R (2007). Paradoxical influence of hippocampal neurogenesis on working memory. *Proc Natl Acad Sci USA* 104:4642–4646.

Schmidt-Hieber C, Jonas P, Bischofberger J (2004). Enhanced synaptic plasticity in newly generated granule cells of the adult hippocampus. *Nature* 429:184–187.

Shors TJ (2004). Memory traces of trace memories: Neurogenesis, synaptogenesis and awareness. *Trends Neurosci* 27:250–256.

Shors TJ, Miesegaes G, Beylin A, Zhao M, Rydel T, Gould E (2001). Neurogenesis in the adult is involved in the formation of trace memories. *Nature* 410:372–376.

Shors TJ, Townsend DA, Zhao M, Kozorovitskiy Y, Gould E (2002). Neurogenesis may relate to some but not all types of hippocampal-dependent learning. *Hippocampus* 12:578–584.

Snyder JS, Kee N, Wojtowicz JM (2001). Effects of adult neurogenesis on synaptic plasticity in the rat dentate gyrus. *J Neurophysiol* 85:2423–2431.

Snyder JS, Hong NS, McDonald RJ, Wojtowicz JM (2005). A role for adult neurogenesis in spatial long-term memory. *Neuroscience* 130:843–852.

Snyder JS, Radik R, Wojtowicz JM, Cameron HA (2009a). Anatomical gradients of adult neurogenesis and activity: Young neurons in the ventral dentate gyrus are activated by water maze training. *Hippocampus* 19:360–370.

Snyder JS, Ramchand P, Rabbett S, Radik R, Wojtowicz JM, Cameron HA (2009b). Septo-temporal gradients of neurogenesis and activity in 13-month-old rats. *Neurobiol Aging*.

Song HJ, Stevens CF, Gage FH (2002). Neural stem cells from adult hippocampus develop essential properties of functional CNS neurons. *Nat Neurosci* 5:438–445.

Song S, Zhang H, Cuevas J, Sanchez-Ramos J (2007). Comparison of neuron-like cells derived from bone marrow stem cells to those differentiated from adult brain neural stem cells. *Stem Cells Dev* 16:747–756.

Stanfield BB, Trice JE (1988). Evidence that granule cells generated in the dentate gyrus of adult rats extend axonal projections. *Exp Brain Res* 72:399–406.

Steiner B, Zurborg S, Horster H, Fabel K, Kempermann G (2008). Differential 24 h responsiveness of Prox1-expressing precursor cells in adult hippocampal neurogenesis to physical activity, environmental enrichment, and kainic acid-induced seizures. *Neuroscience* 154:521–529.

Steinhauser C, Kressin K, Kuprijanova E, Weber M, Seifert G (1994). Properties of voltage-activated Na+ and K+ currents in mouse hippocampal glial cells in situ and after acute isolation from tissue slices. *Pflugers Arch* 428:610–620.

Tashiro A, Sandler VM, Toni N, Zhao C, Gage FH (2006). NMDA-receptor-mediated, cell-specific integration of new neurons in adult dentate gyrus. *Nature* 442:929–933.

Tashiro A, Makino H, Gage FH (2007). Experience-specific functional modification of the dentate gyrus through adult neurogenesis: A critical period during an immature stage. *J Neurosci* 27:3252–3259.

Tozuka Y, Fukuda S, Namba T, Seki T, Hisatsune T (2005). GABAergic excitation promotes neuronal differentiation in adult hippocampal progenitor cells. *Neuron* 47:803–815.

Treves A, Tashiro A, Witter ME, Moser EI (2008). What is the mammalian dentate gyrus good for? *Neuroscience* 154:1155–1172.

Valley MT, Mullen TR, Schultz LC, Sagdullaev BT, Firestein S (2009). Ablation of mouse adult neurogenesis alters olfactory bulb structure and olfactory fear conditioning. *Front. Neurosci.* 3:51.

Van Praag H, Schinder AF, Christie BR, Toni N, Palmer TD, Gage FH (2002). Functional neurogenesis in the adult hippocampus. *Nature* 415:1030–1034.

Vollmayr B, Simonis C, Weber S, Gass P, Henn F (2003). Reduced cell proliferation in the dentate gyrus is not correlated with the development of learned helplessness. *Biol Psychiatry* 54:1035–1040.

Walker MP, Stickgold R (2004). Sleep-dependent learning and memory consolidation. *Neuron* 44:121–133.

Walker MP, Stickgold R (2006). Sleep, memory, and plasticity. *Annu Rev Psychol* 57:139–166.

Wang LP, Kempermann G, Kettenmann H (2005). A subpopulation of precursor cells in the mouse dentate gyrus receives synaptic GABAergic input. *Mol Cell Neurosci* 29:181–189.

Wang S, Scott BW, Wojtowicz JM (2000). Heterogenous properties of dentate granule neurons in the adult rat. *J Neurobiol* 42:248–257.

Weisz VI, Argibay PF (2009). A putative role for neurogenesis in neuro-computational terms: Inferences from a hippocampal model. *Cognition* 112:229–240.

Winner B, Cooper-Kuhn CM, Aigner R, Winkler J, Kuhn HG (2002). Long-term survival and cell death of newly generated neurons in the adult rat olfactory bulb. *Eur J Neurosci* 16:1681–1689.

Winocur G, Wojtowicz JM, Sekeres M, Snyder JS, Wang S (2006). Inhibition of neurogenesis interferes with hippocampus-dependent memory function. *Hippocampus* 16:296–304.

Wirth S, Yanike M, Frank LM, Smith AC, Brown EN, Suzuki WA (2003). Single neurons in the monkey hippocampus and learning of new associations. *Science* 300:1578–1581.

Wiskott L, Rasch MJ, Kempermann G (2006). A functional hypothesis for adult hippocampal neurogenesis: Avoidance of catastrophic interference in the dentate gyrus. *Hippocampus* 16:329–343.

Wohr M, Kehl M, Borta A, Schanzer A, Schwarting RK, Hoglinger GU (2009). New insights into the relationship of neurogenesis and affect: Tickling induces hippocampal cell proliferation in rats emitting appetitive 50-kHz ultrasonic vocalizations. *Neuroscience* 163:1024–1030.

Wolf SA, Steiner B, Akpinarli A, Kammertoens T, Nassenstein C, Braun A, Blankenstein T, Kempermann G (2009). CD4-positive T lymphocytes provide a neuroimmunological link in the control of adult hippocampal neurogenesis. *J Immunol* 182:3979–3984.

Wolfer DP, Madani R, Valenti P, Lipp HP (2001). Extended analysis of path data from mutant mice using the public domain software Wintrack. *Physiol Behav* 73:745–753.

Zhang C, McNeil E, Dressler L, Siman R (2007). Long-lasting impairment in hippocampal neurogenesis associated with amyloid deposition in a knock-in mouse model of familial Alzheimer's disease. *Exp Neurol* 204:77–87.

Zhao C, Teng EM, Summers RG, Jr., Ming GL, Gage FH (2006). Distinct morphological stages of dentate granule neuron maturation in the adult mouse hippocampus. *J Neurosci* 26:3–11.

Zheng H, Liu Y, Li W, Yang B, Chen D, Wang X, Jiang Z, Wang H, Wang Z, Cornelisson G, Halberg F (2006). Beneficial effects of exercise and its molecular mechanisms on depression in rats. *Behav Brain Res* 168:47–55.

11

Adult Neurogenesis in Different Animal Species

Adult neurogenesis has hardly
been studied beyond the classical
laboratory species

LIFELONG GENERATION of new neurons, something comparable to adult neurogenesis in rodents and primates, is not rare across the animal kingdom and even appears to be more the rule than the exception. A comparative approach to studying adult neurogenesis will thus help us avoid seeing the subject from a strictly mammalian point of view and allow us to reach a profounder understanding of the phenomenon. To rely for this purpose only on the classical laboratory animals might severely limit our ability to see the general principles. Our knowledge about the evolutionary side of adult neurogenesis and adult neurogenesis in different species is still very sketchy, though. More problematical even, despite the explosion in publications on adult neurogenesis (and with the exception of adult neurogenesis in zebrafish), is the fact that recent progress in this particular area of neurogenesis research has been slow.

It is questionable whether a
"phylogenetic reduction" in
adult neurogenesis best describes
the differences between different
animal species

It is often written that, as a general rule, organisms with more primitive nervous systems would tend to have a higher level of persisting neurogenesis than those with more complex brains. As we will see, however, this "rule" is rather loose, and it is questionable whether a "phylogenetic restriction" exists et all. *Drosophila*, for example, does not seem to show adult neurogenesis, and among the mammalian species studied so far, many seem to have two neurogenic regions (rodents and primates), some have more (rabbits) and some fewer (bats).

The hypothesis of the "phylogenetic restriction" proposes that simpler brains can accommodate more adult neurogenesis, and that a loss of the ability to create abundant adult neurogenesis has been the price paid for higher "processing power." With increasing brain complexity, the dilemma between plasticity and stability might have forced a decision in favor of more stability. This has been the key thought in Pasko Rakic's famous article "Limit to Neurogenesis in Primates," written in 1985 (Rakic, 1985).

Was a reduction in adult
neurogenesis the price to pay for
evolving a more complex brain?

In rodents and primates, both neurogenic regions are part of the limbic system. The fact that the environment for both neurogenic regions is phylogenetically old has stimulated the idea that only a few highly specialized, yet fundamental and evolutionarily conserved, functions might require adult neurogenesis, whereas others evolved away from it. Just as with increasing brain complexity during the course of evolution the capability of adult neurogenesis supposedly has decreased, a similar pattern might also be visible within each species: some older brain parts retained lifelong neurogenesis, whereas phylogenetically younger regions did not.

There are several problems with this reasoning. The most interesting point is that only mammals seem to have evolved a dentate gyrus, and the equivalent to the hippocampus in reptiles and birds shows considerable differences from the situation in mammals (Treves et al., 2008). Although part of a phylogenetically old brain region, the hippocampal dentate gyrus is something new. Treves and colleagues hypothesized that the mammalian dentate gyrus is the answer to a very particular cognitive challenge and as such, an example of a very advanced adaptation (Treves et al., 2008).

Continued neurogenesis in a phylogenetically new structure means that this structure is also ontogenetically late in developing; at least in the sense that its development does not seem to cease until well into adulthood and possibly continues lifelong. In Chapter 9, we have discussed the hypothesis that the new neurons do not serve a primordial function but make a highly specific and advanced contribution to the dentate gyrus. The dentate gyrus adds complex functionality to the hippocampus and sets mammals apart from other classes. One can argue that adult hippocampal neurogenesis would thus take place in a brain region that is high in the hierarchy. Persisting neurogenesis in a structure that evolved late might consequently imply that here the existing tool of adult neurogenesis was adopted to solve a new problem that is relevant for the high position of the dentate gyrus in the functional hierarchy of brain regions. Adult hippocampal neurogenesis might thus be at the top rather than the bottom. In the hippocampus, there might have been evolution towards adult neurogenesis rather than away from it.

It is true that many lower animals show more adult neurogenesis than mammals, but in the end, mammals might be special in that they evolved a new use for adult neurogenesis, and that it was this step that brought them their particular advantage in the struggle for survival of the fittest.

Comparative studies on adult neurogenesis therefore should not just focus on the question of why mammals (and humans) have so little of it, but should also consider the idea that in terms of quality, we might actually have more of it. The evolutionary perspective also cautions that adult neurogenesis in the hippocampus and in the olfactory bulb might be fundamentally different.

What Is "Adult" in Different Species?

One fundamental difficulty in using a comparative approach to studying adult neurogenesis is the not-so-trivial task of determining what is "adult" in different organisms (Rakic, 2002; Abrous et al., 2005; Lindsey and Tropepe, 2006; Ricklefs, 2006). One can apply a concept of life periods such as childhood, youth, adulthood, and senescence to all organisms, independent of the actual life span. But although longevity can be assessed in flies, the study of aging is hampered by the fact that it also depends on absolute time scales and not only on those relative to the organism's life expectancy. Even in the oldest fly, the cells are infants compared to longer-living organisms. Organisms age at very different rates, and the relative share of the distinctive life periods across the lifespan can differ dramatically. Knowledge about the genetic bases of these fundamental differences is still relatively scarce (Finch, 1994).

Adulthood is often equaled with sexual maturation, although most biologists would not like to concede that definition to their own children. Nevertheless, this seems to be the most straightforward and feasible biological indicator. (Refer to Chapter 1, p. 11, for additional information on this subject.)

It seems that only mammals have evolved a dentate gyrus

Brain structures that add complexity to basic functions develop late

Adult hippocampal neurogenesis might provide an evolutionary advantage rather than being an atavism

"Adult" usually means that the animal can reproduce

For practical purposes, and in the reality of mouse research, *adult* often means not more than "after weaning"—that is, after postnatal day 21, when the pups can be removed from their mothers without problems. To some degree, this early attribution of adulthood is justified in the context of adult neurogenesis because by that time the germinative matrices have undergone the profound changes that lead to the structure that persists for the rest of the organism's life. Operational definitions of adulthood are useful and often necessary, but are limited if extrapolations are made from one species to another. To apply quantitative information from a four-week-old mouse to an adult human is difficult, if not impossible. A thorough comparative consideration can put such disparities into perspective.

> The age of the animals studied always needs to be considered in the interpretation of research data on neurogenesis

In species in which adult neurogenesis has been less extensively studied than in rodents, these questions become even more difficult. To avoid the word *adult* in these problematic contexts, Myriam Cayre and colleagues (2002) have proposed using the term *secondary neurogenesis*, which indeed brings some semantic advantage but does not solve the problem of determining whether a continuity between primary and secondary neurogenesis exists (Cayre et al., 2002).

Adult Mammalian Neurogenesis Is Mostly Adolescent Neurogenesis

> The majority of studies on adult neurogenesis addresses young adulthood and puberty

In this context, it is worth reconsidering how much in mammals adult neurogenesis decreases with age. From the onset there is a steep decline in adult hippocampal neurogenesis that had already been acknowledged by Joseph Altman in his inaugural study in 1965 and has been confirmed many times since (Altman and Das, 1965; Kuhn et al., 1996; Kempermann et al., 1998; Cameron and McKay, 1999; Kronenberg et al., 2006; Ben Abdallah et al., 2010; Knoth et al., 2010). The rate of this decrease is different in different species. It is important to note, however, that adult neurogenesis seems to level out in rats and mice and remains detectable even in oldest age. About the question of at which age this persisting low level is reached, different estimates exist, presumably reflecting strain and species differences. But irrespective of the exact slope of the curve, high levels of adult neurogenesis are only found in young animals. Arguably, thus, with respect to quantities, "adult" neurogenesis is rather "adolescent" or even "juvenile" neurogenesis. (See Chapter 9, p. 335, for details on age as "regulator" of neurogenesis.)

> Possibly adult neurogenesis is of particular relevance during puberty

In an article from 1973, Joseph Altman argued that this temporal pattern would be indicative of a role of adult neurogenesis during the rewiring of the brain in puberty (Altman et al., 1973). This idea has been extended and adjusted to the information gathered since this time by Irmgard Amrein and Hans-Peter Lipp (Amrein and Lipp, 2009). The idea is that neurogenesis might contribute to "transforming juvenile unpredictable to a predictable behavior, typically characterizing mammalian behavior once reproductive competence has been attained." A slightly different explanation, favored here, is that at a young age, there is more need for adaptation and integration of novelty than at an older age. After all, neurogenesis does not sharply drop after puberty, but linearly or hyperbolically declines across that decisive age. Rather, the level of activity, exposure to complexity, and likelihood of encountering novelty might determine levels of adult neurogenesis. This idea is also consistent with observations made in other vertebrate species where ongoing neurogenesis has been related to sensory experience. These and related thoughts have led to the "neurogenic reserve hypothesis," outlined in Chapter 10, p. 473. In terms of a comparative approach to adult neurogenesis, this implies that the idea of "adolescent neurogenesis" should not be applied too strictly. After all, the quantity of adult neurogenesis tells us little about the functional significance of the new neurons, and we have argued (in Chapter 10, p. 451) why and how very few new neurons can make relevant functional contributions.

Non-Vertebrates

No evidence for adult
neurogenesis in Drosophila

Many but not all insect species
show adult neurogenesis in the
mushroom bodies

In the cricket, adult
neurogenesis has been linked to
learning

The mushroom body equivalent
of crustacean species shows adult
neurogenesis but is otherwise
not widespread

The niche structure resembles
the situation in rodents

Of all insects, the most interesting species in which to study adult neurogenesis is the dew fly, *Drosophila melanogaster*, because *Drosophila* is one of the genetically best-characterized organisms. The potential of studying the molecular basis of adult neurogenesis in *Drosophila* is vast. Unfortunately, there is no evidence of continued neurogenesis in the adult fly brain. There is, however, an interesting example of "adult" neurogenesis occurring in the wing sensory cells when the flies emerge from their pupae (Ben Rokia-Mille et al., 2008).

Adult neurogenesis has been found in the mushroom bodies—the main processing unit of sensory information in the insect brain in many species (Cayre et al., 1994; Cayre et al., 1996; Mashaly et al., 2008), but (as reviewed in Cayre et al., 2002) not in *Drosophila* (Ito and Hotta, 1992), bees (Fahrbach et al., 1995), monarch butterflies (Nordlander and Edwards, 1968), or the migratory locust (Cayre et al., 1996). Especially the latter species, with their fascinating navigational abilities, seemed to be good candidates for activity-dependently regulated adult neurogenesis as we know it from rodents and birds. Insect species that show adult neurogenesis in the mushroom bodies include milkweed bugs, moths, beetles, and crickets (Bieber and Fuldner, 1979; Cayre et al., 1994; Cayre et al., 1996; Malaterre et al., 2003; Dufour and Gadenne, 2006; Mashaly et al., 2008). The variability of adult neurogenesis between different insect species and within a structure that serves very similar if not identical purposes in all of these species remains surprising.

Environmental enrichment had a stimulating effect on neurogenesis in mushroom bodies of adult crickets (Scotto Lomassese et al., 2000), whereas unilateral sensory deprivation down-regulated adult neurogenesis ipsilaterally to the damage (Scotto-Lomassese et al., 2002) (Fig. 11–1). Precursor cells form clusters in the calices of the mushroom bodies and generate interneurons in the cortex of the mushroom bodies, the Kenyon cells. This process is not homogenous across the mushroom body, indicating that there are different clusters of Kenyon cells with different properties, also with respect to neurogenesis (Mashaly et al., 2008). It has been estimated that about one fifth of Kenyon cells of crickets are produced during adulthood (Malaterre et al., 2003). Because the effect could be stimulated unilaterally, the hypothesis that this regulation is primarily hormonal could be ruled out. Positive knowledge about how adult neurogenesis is regulated in the cricket brain is lacking, however. Ex vivo, precursor cells from adult crickets were inducible by insulin but not fibroblast growth factor-2 (Malaterre et al., 2003).

When neurogenesis was suppressed by irradiation, the crickets exhibited delayed learning and reduced memory retention in an operant associative learning task, when olfactory cues were used (Scotto-Lomassese et al., 2003; Cayre et al., 2007). For a detailed review on neurogenesis in the adult insect brain, see Cayre et al., 2007.

Crustaceans have received considerable attention in terms of adult neurogenesis (for a very detailed review, see Schmidt, 2007). Adult neurogenesis is here restricted to two sites: the olfactory pathways and the optic lobe (Schmidt, 1997; Sullivan and Beltz, 2005a). Neurogenesis in the olfactory system has been studied in a total of nine species, including crabs, shrimp, and lobsters (Schmidt, 1997; Harzsch et al., 1999; Schmidt, 2001, 2007). Neurogenesis in the optic lobe occurs in the eyestalk ganglia but has been found in only two species so far (Schmidt, 1997; Sullivan and Beltz, 2005a). Here, new neurons are generated in the hemiellipsoid body, which can be considered the equivalent to the mushroom body of insects (Schmidt, 1997).

The precursor cells have glial properties and reside in a niche-like multicellular proliferation zone that is relatively compact and has a constant location in the cluster near the inner surface and is in close contact to the neurites of the neurons of that cluster (Sullivan et al., 2007a; Sullivan et al., 2007b). The niche shows a remarkable resemblance to the structures we see

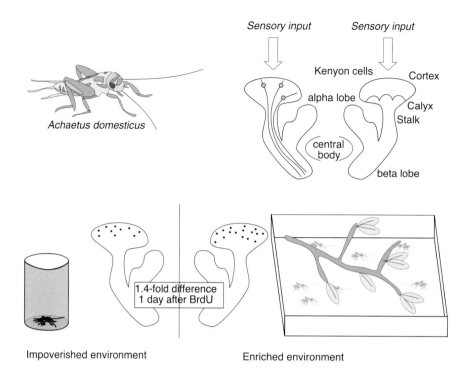

FIGURE 11–1 Activity-dependent regulation of adult neurogenesis in crickets. Environmental enrichment increased the generation of interneurons in the mushroom bodies of crickets (Scotto-Lomassese et al., 2000). Mushroom bodies are the main integration sites of sensory information in insects. Enrichment consisted of a larger, well-lit cage with more animals and leaves and other objects, whereas impoverished crickets were held in isolation and darkness.

in rodents, including, for example, the association with the vasculature (Zhang et al., 2009). Serotonin has been identified as a key regulatory molecule here, too, and this control is exerted by a single dorsal giant neuron (Sandeman et al., 2009). In crayfish, adult neurogenesis produces two types of interneurons in the olfactory lobe: the functional homologue of the vertebrate olfactory bulb, and the accessory lobe. These two populations form two distinct clusters lateral and medial to the lobes, which are referred to as clusters 10 and 9. Within these regions, cell proliferation is restricted to two regions, which are called the lateral (LPZ for cluster 10) and medial (MPZ for cluster 9) proliferation zones (Sullivan and Beltz, 2005a). In crabs and crayfish, olfactory neurogenesis leads to an almost linear increase and doubling in neuron numbers across the lifespan (Schmidt, 1997; Sandeman and Sandeman, 2000).

Neurogenesis in lobsters is regulated by both circadian (Goergen et al., 2002) and seasonal influences (Hansen and Schmidt, 2001). Signs of dependency of adult neurogenesis on the sensory input have been found in lobsters (Hansen and Schmidt, 2001) and crayfish (Sandeman and Sandeman, 2000). But neurogenesis in the crustacean olfactory bulb is interesting insofar as it takes place in the absence of any neuronal turnover in the olfactory epithelium (Sullivan and Beltz, 2005b).

On a side note, one of the first known scientific works in history that described regeneration of lost body parts is an account by René-Antoine Ferchault de Réaumur, who in 1712 reported, "*Sur les diverses reproductions qui se font dans les écrivesses, les omars, les crabes, etc. et entr'autres sur celles de leurs jambs et de leurs écailles*" ("On the diverse reproductions, which are found in crayfish, lobsters, crabs, etc., and among others in their legs and their shells"), to the French Royal Academy.

We have relatively detailed information on lifelong neurogenesis in hydrozoans (a complex class of animals that also includes jellyfish, corals, and the sweetwater hydra) but this comes from, to date, only two studies (Sakaguchi et al., 1996; Miljkovic-Licina et al., 2007). As Lindsey and Tropepe point out in their excellent review on adult neurogenesis across the animal kingdom, this neglect of hydrozoa is somewhat surprising, given the fact that the phylum Cnidaria presumably represents the first organisms with a true CNS, which would make hydrozoans the natural starting point of any exploration of adult neurogenesis from a comparative and evolutionary perspective (Lindsey and Tropepe, 2006)—although the CNS in this case is an enteric nervous system rather than a brain, and determining adulthood in these organisms is an art of its own.

> **Hydrozoans, the first animals with a CNS, also show adult neurogenesis**

Adult neurogenesis in more primitive invertebrates such as nematodes and flatworms does not seem to have been studied yet.

Non-Mammalian Vertebrates

In many vertebrate species, including fish, amphibians, and reptiles, the brain shows lifelong growth together with an increase in total body size. This brain growth is not proportional for all brain regions and cell populations, and the resulting shifts in relative size presumably reflect an adaptation to the particular needs of that species in their habitat (Kaslin et al., 2008). The current hypothesis is that adult neurogenesis preferentially affects sensory input systems and sensory cortices in order to adjust these systems to computational needs (Leonard et al., 1978; Brandstatter and Kotrschal, 1990). The fact that this differential "adult" neurogenesis decreases in older age supports this idea and resembles the situation in mammals, including a similar functional explanation. But for most species, the underlying mechanisms, populations of precursor cells, and routes of migrations are not yet known.

> **Adult neurogenesis in many vertebrate species is linked to plasticity of sensory systems**

Reptiles and Amphibians

Lizards show an amazing degree of brain plasticity and after injury can regrow entire parts of the brain. This was noted by Aristotle as early as three centuries B.C. (see (Odelberg, 2004)).

> **Lizards show very widespread adult neurogenesis**

Adult neurogenesis has been reported for most parts of the lizard forebrain in at least four different lizard species (Lopez-Garcia et al., 1988; Perez-Canellas and Garcia-Verdugo, 1996; Font et al., 1997), one of which, a tropical species, differed from the others in that the new neurons matured very slowly and in very low numbers (Marchioro et al., 2005). Cell proliferation occurs in the walls of the ventricles and apparently generates only neurons. The ventricular zone of the adult lizard is subdivided into four sulci (Schulz, 1969) that contain glial fibrillary acidic protein (GFAP)–positive radial glial cells, which can undergo divisions (Font et al., 1995). Some of the newborn cells migrate along the radial fibers into cortical regions. Analogous to the findings in the adult rodent subventricular zone, it is assumed that these radial glial cells serve as the precursor cells for adult cortical neurogenesis in lizards (Garcia-Verdugo et al., 2002; Weissman et al., 2003; Romero-Aleman et al., 2004). Between the germinative sulci, the ventricular walls show a simple epithelium without the characteristics of a neurogenic region. A population of migratory cells has also been identified in lizards, and as in rodents, these cells migrate toward the olfactory bulb, where they turn into new interneurons (Perez-Canellas and Garcia-Verdugo, 1996). Other cells from the ventricular zone migrate to the medial cerebral cortex, whose granule cell layer can be considered a homologue to the mammalian dentate gyrus (Lopez-Garcia et al., 1990). Most of the cells seem to survive for long times, suggesting that adult neurogenesis here indeed contributes to the lifelong growth of the brain region (Font et al., 2001).

Turtles at least have adult
neurogenesis in the olfactory
system

Because turtles can get very old, they might seem to be particularly promising targets of studies on lifelong brain plasticity. Indeed, neurogenesis has been found in the olfactory system of red-eared slider turtles in a pattern similar to that in lizards (Perez-Canellas et al., 1997). Continued neurogenesis in turtles argues against the idea that adult neurogenesis is preferentially a feature of short-lived animals and that with increasing lifespan, a decrease in adult neurogenesis will occur (Rakic, 2004). Turtles also show a great regenerative capacity in the spinal cord that presumably originates from a population of neural precursor cells near the central canal (Russo et al., 2004; Russo et al., 2008).

Adult neurogenesis in frogs and
other amphibians has hardly
been investigated

There is only a little information about adult neurogenesis in amphibians, although some were investigated already in the late 1960s. Early reports notably included the olfactory epithelium (Graziadei and Metcalf, 1971). Diverse regions with ongoing proliferation have been described and include telencephalon, midbrain, cerebellum, and hypothalamus and thalamus (Richter and Kranz, 1981; Chetverukhin and Polenov, 1993; Polenov and Chetverukhin, 1993; Margotta et al., 2005), but only a more recent study went into greater detail to characterize these regions in the adult frog brain (again, unfortunately only based on PCNA immunohistochemistry: Raucci et al., 2006). Finally, one report used TUC4 together with BrdU to give more evidence of adult neurogenesis in the bullfrog, confirming widespread adult "neurogenesis," including, in particular, the hypothalamus (Simmons et al., 2008).

The axolotl, *Ambystoma mexicanum*, is one of the prime model organisms in regeneration research because after amputation it can regrow its entire tail, including the bony vertebral column and the spinal cord. Nevertheless, information about constitutive adult neurogenesis in this species is scant and comprises just one historic report on cell proliferation (Richter and Kranz, 1981). Neurogenesis in the vomeronasal organ, with its particular role in detecting prey odor and pheromones, has also been the subject of the sole study on neurogenesis in adult snakes (Wang and Halpern, 1988) and a salamander species (Dawley et al., 2006).

Zebrafish

Zebrafish (*Danio rerio*) is a vertebrate species that is used for large-scale genetic screens in mutagenesis experiments similar to those done in *Drosophila,* and many genetic tools such as transgenic reporter lines, genetic knockdown, and overexpression are very straightforward. Quite surprisingly, reports on adult neurogenesis in fish covered numerous species before finally showing that adult neurogenesis occurs in the olfactory bulbs of the adult zebrafish (Byrd and Brunjes, 2001). Other regions of sustained proliferative activity were described (Oehlmann et al., 2004), but only more recently have detailed accounts of neurogenesis in the adult zebrafish brain been presented (Zupanc et al., 2005; Grandel et al., 2006). The new cells survive for very long times: about half of the cells present at 10 days after labeling with BrdU were still present more than a year later (Hinsch and Zupanc, 2007). Michael Brand and his colleagues mapped a total of 16 neurogenic zones in the adult zebrafish and began to characterize them in detail (Fig. 10–2)(Grandel et al., 2006; Kaslin et al., 2009). It soon became clear that these regions differ greatly with regard to the precursor cells they possess, the composition of the stem cell niches, the routes of migration the newborn cells take, and finally, the fate of the new neurons. In the cerebellum, for example, unique precursor cells with a neuroepithelial phenotype exist, not resembling the radial glia–like precursor cells prevalent in many other forms of adult neurogenesis (Kaslin et al., 2009). Adult cerebellar neurogenesis in zebrafish is more prevalent in males than in females (Ampatzis and Dermon, 2007).

It seems that neurogenesis in the adult zebrafish is associated with the production of surprisingly great numbers of cells with nuclear aberrations during mitosis, many of which persists for long times (Zupanc et al., 2009). The significance of this observation is not yet clear.

Zebrafish are particularly useful
for generating genetic models

Different kinds of transgenic reporter lines have already been developed to allow further investigations of the precursor cells in these niches (Kassen et al., 2008; Lam et al., 2009) and their regulation (Thummel et al., 2006; Molina et al., 2007). Among the factors regulating adult zebrafish neurogenesis, Fgf has received particular attention. Knockdown of Fgf2 decreased neurogenesis in the cerebellum (Kaslin et al., 2009). A detailed analysis of many Fgf isoforms in the ventricular wall, however, revealed that Fgf was primarily associated with the radial glia–like precursor cells and not specifically with precursor cell proliferation, suggesting a broader function (Topp et al., 2008).

The zebrafish retina shows
regenerative neurogenesis after
damage

Also among the constitutively neurogenic regions of the zebrafish is the retina, where regenerative neurogenesis occurs after light-induced degeneration (Vihtelic and Hyde, 2000). Regeneration originates from Müller cells in the inner nuclear layer, which give rise to intermediate progenitor cells in the outer nuclear layer, from which both rod and cone photoreceptors are derived (Thummel et al., 2008). In contrast, one study claimed that activated Müller cells would generate only new cone photoreceptors, whereas new rods come from an independent population of precursor cells in the outer nuclear layer (Morris et al., 2008b). This seems unlikely, however, given that blockade of Müller cell proliferation (directly by morpholino-mediated inhibition of PCNA) abolished all photoreceptor regeneration (Kassen et al., 2007). Stat3 seems to play a key role in inducing the proliferation of Müller cells in this process (Kassen et al., 2007). Nevertheless, the exact nature of rod photoreceptor cells is not quite clear yet, and new studies outside the setting of acute lesions have been proposed to solve the issues (Morris et al., 2008a).

Other Fish

Other teleost fish species also
show widespread neurogenesis

The extensive work by Gunther Zupanc and coworkers focused on the mildly electric gymnotiform fish (like zebrafish another teleost species). The proliferative zones containing the putative precursor cells (for which no ex vivo data exist yet) surround the brain ventricles (Zupanc and Zupanc, 1992), and neurogenesis decreases with age (Kranz and Richter, 1975). This, together with directed migration of newborn neurons and cell death in the neurogenic zones, might suggest a similar mechanism of function-dependent selection of newborn cells to that in adult mammalian neurogenesis (Soutschek and Zupanc, 1995, 1996; Zupanc, 1999). Another example of possibly activity-regulated neurogenesis in the adult fish brain is the generation of new neurons in the goldfish tectum, which seems to depend on input from the optic nerves (Raymond and Easter, 1983; Raymond et al., 1983).

Electric fish communicate by the exchange of electric signals. This kind of social interaction increased neurogenesis in the periventricular region and the number of radial glia–like cells (Dunlap et al., 2008).

Another early study reported increasing neuron numbers in the guppy (Birse et al., 1980), and a study by Alonso and colleagues added goldfish, salmon, carp, and barb to the list (Alonso et al., 1989). In trout, cell proliferation was reported for the pineal gland (Omura, 2007).

Neurogenesis in the goldfish retina was described as early as 1977 (Johns, 1977). The adult goldfish retina produces new cone photoreceptors (Wu et al., 2001).

All of the mentioned fish belong to the class of Teleostei, which comprises most fish we encounter today. With the exception of the parasitic lamprey, often not even considered a fish at all, no other, more ancient class of fish has been studied (Vidal Pizarro et al., 2004).

Birds

Neurogenesis in adult birds has been described for many species: canaries (Goldman and Nottebohm, 1983), zebra finches (Kirn and DeVoogd, 1989; Barkan et al., 2007), chickadees (Barnea and

Nottebohm, 1994; Barnea and Nottebohm, 1996), sparrows (Thompson and Brenowitz, 2009), and ring doves (Ling et al., 1997; Chen and Cheng, 2007).

Adult Neurogenesis in the Song Learning System

In male canary birds song-learning is associated with neurogenesis in the adult higher vocal center

Steven A. Goldman and Fernando Nottebohm's first (1983) description of newly generated neurons in the telencephalon of canaries was a true breakthrough and brought research on adult neurogenesis the long-awaited acceptance of a wider scientific audience (Goldman and Nottebohm, 1983). From an anthropocentric point of view, this attention might at first seem surprising because the phylogenetic relationship between humans and birds is much less close than that between humans and rodents. Adult neurogenesis had long been reported in rodents with no lesser evidence, but without being able to lose the stigma of being a curiosity. However, in songbirds, adult neurogenesis was found in the higher vocal center, the brain region responsible for song learning, and the production of new neurons correlated with the seasons in which the birds learn their songs. The researchers had been searching for mechanisms underlying the parallel seasonal changes found in the songs and the volume of the HVC. Because adult neurogenesis added a considerable amount of new neurons to the well-defined neuronal circuit of the HVC, Nottebohm and colleagues were able to verify the birth of new neurons with a variety of methods, ranging from microscopical techniques to retrograde tracing and electrophysiological recordings (Kirn et al., 1991; Alvarez-Buylla and Kirn, 1997). The newly generated neurons in the HVC can survive for long periods (Kirn et al., 1991).

In the HVC, only neurons projecting to the motor nucleus are generated in adulthood

The HVC projects to two different brain regions, the motor nucleus in the archistriatum (RA) and the basal ganglia–like area X (Vates and Nottebohm, 1995) (Fig. 11–2). These downstream regions also vary in size across seasons (but no neurogenesis occurs there), and this process seems to be dependent on BDNF release from the HVC (Wissman and Brenowitz, 2009). In the HVC, both types of projection neurons are intermingled with interneurons. Only the type of RA-projecting neurons are born in

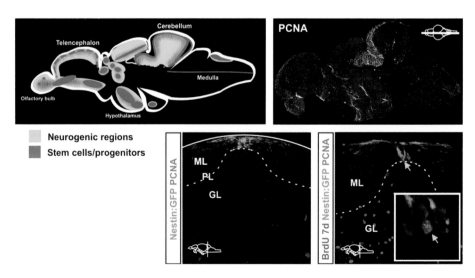

FIGURE 11–2 The adult zebrafish brain has 16 regions with constitutive cell proliferation and neurogenesis. The structure of the niches differs considerably between these regions. The schematic drawing is based on the work by Grandel et al. The figure was kindly provided by Michael Brand and Jan Kaslin.

adulthood (Fig. 11–3). Constance Scharff and colleagues from Fernando Nottebohm's group applied a technique developed by Jeffrey Macklis (Macklis, 1993; Madison and Macklis, 1993) to show that selective elimination of the RA projections, but not of the area-X projections, induced adult neurogenesis (Scharff et al., 2000). This method, called "chromophore-targeted neuronal degeneration," is described in Figure 8–3. It allows the specific elimination of projection neurons and leads to the disappearance of the neuron without a reaction of the surrounding tissue. If area-X projections were killed in young birds, however, neurogenesis was induced, but only RA-projecting neurons were produced. Zebra finches, which in contrast to canaries learn their songs only in their youth, lost their songs after ablation of the RA-projecting neurons. With recovery of the HVC due to adult neurogenesis, the lost songs returned (Scharff et al., 2000). This indicates that the HVC neurons are not involved in storing song information but are required to use that information. In canaries, the number of new neurons in the HVC-RA projection is so high that the entire system might turn over in one year (Nottebohm and Alvarez-Buylla, 1993). In the targeted ablation studies, there was a curious variability in the degree to which the birds lost their songs and recovered them. This finding might indicate that specific projections must be hit to obtain the full effect, and that even with induced adult neurogenesis, recovery might not be optimal.

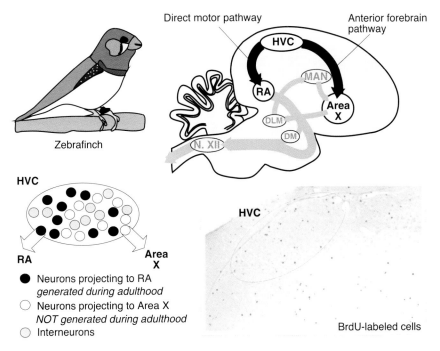

FIGURE 11–3 Adult neurogenesis in songbirds. In male zebra finches and other songbirds, adult neurogenesis is found in the higher vocal center, a brain region centrally involved in song learning. The HVC contains three types of neurons. One projects to the robust nucleus of the archistriatum (RA) and constitutes the output pathway to the motor systems involved in singing. The second population projects to the anterior forebrain, area X. As a third neuronal population, the HVC contains interneurons. Only the neurons that project to RA are replaced in adult neurogenesis. The photomicrograph at the bottom shows BrdU-positive cells in the HVC. BrdU-positive cells and cell genesis, however, can be found throughout the entire adult bird brain; the HVC is not highlighted by a particular density of newly generated cells. DLM, dorsolateral thalamic nucleus; DM, dorsomedial nucleus of the intercollicular complex; MAN, magnucellular nucleus of the anterior neostriatum. The schematic drawing is based on Vates et al. (1995). Photomicrograph by Alexander Garthe, Dresden, and Constance Scharff, Berlin.

Precursor Cells

The precursor cells are found in the SVZ equivalent at the ventricular wall

The stem cells have radial glia properties and also serve as guidance structure

Precursor cells generating new HVC neurons reside in the subependymal zone, similar to the situation in the rodent SVZ. The dividing cells are not evenly distributed but cluster in proliferative "hot spots" in the ventricular wall (Alvarez-Buylla et al., 1990). Similar to the lizard brain, these different domains differ in their cytoarchitecture. The germinative zones show a pseudostratified epithelium (Garcia-Verdugo et al., 2002).

Newborn cells migrate to their target location largely along the processes of radial glia that terminate at blood vessels (Alvarez-Buylla and Nottebohm, 1988, Louissaint et al., 2002). Both the radial scaffold and the neurogenic progenitor cell derive from a common precursor cell in the SVZ (Goldman et al., 1996). The radial glia–like cells serve as stem cells in this system, comparable to the B cells in the rodent SVZ, generating both the progenitor cells (comparable to A cells) and new radial glia. The greatest proliferation occurs in B-like cells, which divide with a mitotic spindle perpendicular to the ventricular surface (Alvarez-Buylla et al., 1998). Their nuclei show interkinetic movements, with the mitosis occurring close to the ventricular wall. Cell division thus occurs in a highly orchestrated way and differs considerably from that of rodents, despite the existing similarities in the cytoarchitecture of the germinative matrix. From the lateral ventricle, the new neurons disperse widely throughout the brain and migrate through the brain parenchyma. The overall pattern is thus remarkably similar between the species (Garcia-Verdugo et al., 2002). There does not seem to be a true equivalent to the A cells in rodents, however.

A common marker for cells in the subependymal zone is N-cadherin, which is down-regulated when the cells leave this area (Barami et al., 1994). The radial cells express BLBP (Rousselot et al., 1997). Hu protein is expressed soon after the cells became committed to the neuronal lineage (Barami et al., 1995).

No good in vitro models of avian precursor cells exist, but a few studies have successfully used explant cultures that allowed to address migratory patterns (Goldman, 1990; Goldman and Nedergaard, 1992; Goldman et al., 1993).

Regulation

The recruitment of new neurons apparently occurs in an activity-dependent manner: the neurons produced in the spring, when no songs are learned, have a lower chance of survival than those generated in the fall, when song learning peaks (Nottebohm et al., 1994).

In sparrows, HVC neurogenesis is affected by seasonal influences

Sparrows are also seasonally breeding birds. Within days after transition into the non-breeding state (which is achieved by manipulating the light period during the day), their HVC shrinks by a quarter in neuron numbers. This elimination occurs by apoptosis. When apoptosis is blocked, no degeneration occurs (Thompson and Brenowitz, 2008); however, fewer new neurons are recruited, suggesting that here apoptosis provides a necessary trigger for adult neurogenesis (Thompson and Brenowitz, 2009). The entire process of the seasonal growth of the song system is highly autonomous and not dependent on auditory feedback and input: it occurs as well in deafened birds (Brenowitz et al., 2007). Deafening, however, still reduces adult neurogenesis, suggesting that not only neurogenesis explains the seasonal fluctuations in size (Wang et al., 1999). Migration of newborn neurons is strongly regulated by IGF-1, which is expressed by the radial glial cells (Jiang et al., 1998).

Only male canaries sing and have adult neurogenesis, which is regulated by testosterone

The brain centers responsible for song learning are sexually dimorphic in songbirds. It is usually the males that sing the complex songs, and seasonal neurogenesis in the HVC of canaries occurs only in males. Neurogenesis is induced by testosterone (Goldman and Nottebohm, 1983; Rasika et al., 1994).Testosterone withdrawal immediately triggers degeneration of the song system (Thompson et al., 2007), but it also exerts direct effects on the precursor cells, which carry the androgen (as well as estrogen) receptors

(Gahr et al., 1996; Gahr and Metzdorf, 1997). Female canaries can be masculinized in many aspects of their song behavior by the application of testosterone. Testosterone does not have an effect on precursor cell proliferation, but increases survival of newly generated cells, presumably through brain-derived neurotrophic factor (Dittrich et al., 1999; Rasika et al., 1999; Li et al., 2000). In addition, testosterone induces angiogenesis in the HVC, possibly by increasing vascular endothelial growth factor. Testosterone also increased BDNF secretion from HVC endothelial cells, a result suggesting that, through this interaction, testosterone coordinates angiogenesis and neurogenesis in the HVC (Louissaint et al., 2002). In addition, testosterone induced the expression and release of metalloproteinases from endothelial cells: when these were inhibited, recruitment of new neurons to the song nuclei was reduced (Kim et al., 2008).

Function

Not all adult neurogenesis in birds is related to song learning

Many songbird species, such as zebra finches, learn their songs only once during a critical period in their youth, but some species such as canaries continue to learn songs every year. When a seasonal fluctuation of neurogenesis in the HVC of canaries was found, adult neurogenesis was placed in a context of learning and memory. Intriguingly, song learning is a trait that moves songbirds much closer to humans than rodents, with which humans share larger parts of the genome. Song learning shows many similarities to speech acquisition in humans, so songbirds are a good model organism in which to study this particular type of learning (Brainard and Doupe, 2002). Song learning in birds is as close to the human use of language as can be found anywhere in the animal kingdom. Only dolphins, whales, and bats show something similar. No dog, for example, has to learn how to bark. However, the learned vocalizations are not *language* in that they lack the ability to convey symbolic contents. But in songbirds, as in humans, vocal learning depends on critical periods during which the birds have to hear the songs from adults. The birds also have to hear themselves to adequately learn their songs.

How new neurons contribute to song learning is not clear

Despite these suggestive findings, it is still not known what the exact functional relevance of new neurons is for song learning. In starlings, for example, which are open-ended learners like canaries, an age-dependent decline in HVC neurogenesis has been noted (Absil et al., 2003), suggesting that the quantity of required new neurons might not be constant. Adult HVC neurogenesis is also constitutively found in zebra finches, although this species does not have to learn new songs in adulthood, so that in contrast to canaries, a functional interpretation is less obvious in this species. Zebra finches nevertheless show a seasonal fluctuation in incorporating new neurons into the HVC (Scharff and Nottebohm, 1991). There is a correlation between a decrease in neurogenesis and increasing stereotypy in the song repertoire, consistent with the idea that the new neurons still somehow contribute to plasticity (Pytte et al., 2007). Causality remains to be proven. Zebra finches do rely on persistent auditory feedback of their own songs in order to maintain their song repertoire. This might indicate that new neurons in the adult HVC serve other functions than the mere storage of new songs. This is supported by the finding that, in wild canaries, in contrast to the domesticated birds of the other experiments, song learning had seasonal patterns, whereas the neuron numbers and HVC volume did not (Leitner et al., 2001). Consequently, in wild canaries, as in zebra finches, a link between new neurons and learning is neither linear nor mandatory. As an alternative hypothesis, neurogenesis in the HVC might be driven by singing itself and thus represent a form of activity-dependent cellular plasticity (Ball et al., 2002). This hypothesis might help explain the seasonal changes by making neurogenesis a phenomenon secondary to seasonally fluctuating behavior. But it does not address the question of the function the new neurons serve. Although adult neurogenesis is much more abundant in birds than in rodents, and at first glance the functional relevance of new neurons seems much more compelling in songbirds than in mammals, adult neurogenesis in the HVC still raises as many functional questions as neurogenesis in the adult hippocampus or olfactory system of mammals. Vocal learning might provide a psychologically plausible explanation for why humans are particularly interested in neurogenesis in adult songbirds. However, so far there is no evidence that adult neurogenesis occurs in brain areas involved in speech generation in humans.

Avian Adult Neurogenesis Outside the Song System

Adult neurogenesis in birds also comprises the structure corresponding to the mammalian hippocampus

Adult neurogenesis in birds is not limited to the HVC but includes the different parts of the striatum, parolfactory lobe, and the equivalent to the hippocampal formation. Nevertheless, in the end, all bird neurogenesis seems to originate from a single neurogenic niche in the ventricular wall with their "hot spots" of proliferation (Alvarez-Buylla et al., 1990). The greatest proliferation is found in the medial telencephalon, which is thought to be the structure corresponding to the mammalian hippocampus (Alvarez-Buylla and Lois, 1995; Garcia-Verdugo et al., 2002).

In food-caching birds, neurogenesis in the "hippocampus" correlated with the food-storing activity

As in the HVC, activity-dependent regulation of neurogenesis has been described for the avian hippocampal formation. In chickadees, a navigational challenge appears to be the neurogenic stimulus: adult hippocampal neurogenesis is correlated with seasons in which food-caching chickadees memorize their food storage sites for the upcoming winter (Barnea and Nottebohm, 1994; Barnea and Nottebohm, 1996). It has thus been concluded that spatial learning induces adult hippocampal neurogenesis in food-caching birds (Patel et al., 1997). Consistent with this hypothesis, in one comparison, food-storing birds tended to have more neurogenesis than non–food-storing birds (Hoshooley and Sherry, 2007). However, as in the case of song learning, not all food-storing bird species—for example, woodpeckers—show a close correlation (Volman et al., 1997). These species differences suggest that the adaptive benefit provided by new neurons can be obtained by other means as well. One theory sees memory load as the main stimulus for adult neurogenesis (Barnea, 2009).

Socially dominant mountain chickadees showed more proliferation in the SVZ near the hippocampal formation than did subordinates, and this difference correlated with spatial memory (Pravosudov and Omanska, 2005).

Neurogenesis in the parolfactory lobe of birds corresponds to some degree to the mammalian olfactory system, and as in mammals, a large number of new neurons is generated here (Alvarez-Buylla et al., 1994).

Mammals

Adult Neurogenesis in Wild Populations

To study adult neurogenesis in feral species is technically demanding

As long as adult neurogenesis had only been described for laboratory rodents, there was some concern that these new neurons might represent something like a laboratory artifact or even sign of degeneration not applicable to feral situations (Barker et al., 2005; Amrein and Lipp, 2009). This concern turned out to be unjustified, although quantitative comparisons are difficult. To study adult neurogenesis in wild-living animals poses considerable technical challenges. Few experimental parameters can be controlled, the age of the animals is often difficult to determine, and experimental interventions, including BrdU-injections, are either impossible or raise difficult consequences such as the need to recapture the animals. Stress due to capture, injection, and recapture represents a confounding variable whose impact is difficult to assess.

Laboratory housing conditions remove the strong and variable influence of natural habitats

Irrespective of these concerns, to study adult neurogenesis in natural populations allows us to "*deal with animals that live out their lives in the context of what they were actually selected to do*" (Amrein et al., 2008). The reasoning goes that such consideration might provide deeper insight into the regulation and function of adult neurogenesis. For most other mammals, C57BL/6 mice might be very poor representatives, given the extreme variation in habitats, size, lifespan, behavior, nutrition, etc., among mammals.

Already in a direct comparison with Sprague Dawley rats (not a fair competition anyway), C57 fell short (Snyder et al., 2009). Accordingly, all studies on adult neurogenesis that venture beyond the boundaries of laboratory species confirm that no simple pattern exists. The simplistic notion of a "phylogenetic reduction" in adult neurogenesis with increasing brain complexity is questioned also by the extreme variability that is found between different mammalian species.

An age-related decrease in neurogenesis appears to be a common phenomenon

All studies that detected signs of adult neurogenesis in wild-living animals and studied different ages confirmed the age-dependent decline in cell proliferation (see also Fig. 9–4) (Amrein et al., 2004a; Barker et al., 2005).

There are hints at an activity-dependent regulation also in the wild

A few studies in voles suggested a link between the seasonal variation in reproductive state and cell proliferation (and laboratory studies confirm that sex hormones can influence adult neurogenesis), but other studies investigating seasonal influences, such as variation in caching activity, on adult neurogenesis could not detect such a link in squirrels (Lavenex et al., 2000). Between different species, for example, comparing food-caching squirrels with chipmunks, different aspects of adult neurogenesis such as proliferation or the total number of doublecortin-positive cells were not affected uniformly, but animals covering larger territories tended to have at least more cell proliferation (Barker et al., 2005). Similarly, wild-living wood mice showed more proliferation than bank voles, who control much smaller territories (Amrein et al., 2004a). Reportedly, there was also a weak correlation between these parameters and learning that was assessed with an automated learning task in the cage ("Intellicage"), into which the captured animals were placed (Galsworthy et al., 2005). Yellow-necked wood mice appear to be the species that among all species studied has by far the highest rate of cell proliferation. Amrein et al. estimated that the number of proliferating cells amounted to as much as 1.1% of the total granule cell count, which is at least three to four times higher than in other species (Amrein et al., 2004a). There is no information on net hippocampal neurogenesis in yellow-necked wood mice, but the high rate of proliferation was associated with considerable cell death. Between different wild strains and like in laboratory mice there is, however, no clear relationship between the rate of proliferation, cell death, or survival, and presumably adult neurogenesis (Kempermann et al., 1997a; Amrein et al., 2004a; Kempermann et al., 2006; Amrein et al., 2008).

Bats

Most bat species investigated so far have adult neurogenesis only in the SVZ but not the hippocampus

Bats have achieved a notorious reputation in the adult neurogenesis literature because in a large study they did not show adult hippocampal neurogenesis, or (in a few species) very little of it, whereas cell proliferation in the SVZ revealed much less variation (Amrein et al., 2007). Measurements of adult neurogenesis in bats have been based on Ki67 immunohistochemistry together with markers such as doublecortin, so these data are highly suggestive but might not be the final word. Additional methodology would be required. Nevertheless, the observations are interesting and might reflect fundamental differences in the functional demands on the hippocampus in the different classes of mammals. Contrary to common belief (and the misleading terminology in some languages; e.g., *Fledermaus* in German) bats are not flying rodents but rather distant relatives, about as far from both rodents and primates as are cats and dogs. As bats are masters of spatial navigation, the putative absence of adult hippocampal neurogenesis in bats might indicate that adult neurogenesis has little relevance for navigation. But the hippocampus is "not just for place" anyway, and if adult neurogenesis is indeed an indication of high adaptability to different environments, the lack of adult neurogenesis in bats might adequately reflect their extreme specialization to very particular ecological niches. Bats can also reach very old age, so that the apparent absence of "adult neurogenesis" might also be simply indicative of an earlier age-dependent decrease.

Rabbits

Rabbits probably have more
neurogenic regions than rodents

Together with the order Rodentia, the order Lagomorpha, to which rabbits belong, represents the Glires, which are "small, gnawing placental mammals" (Douzery and Huchon, 2004).

Adult hippocampal neurogenesis in rabbits was first reported in 1982 (Gueneau et al., 1982). This study was amazingly complete, given the technical limitations of the time, and described a developmental sequence with a putative precursor cell, an intermediate neuroblast, and the new mature neuron. It was thus the first study covering the temporal aspects of adult neurogenesis. Neuronal development reached what was considered the neuroblast stage at about four days after the initial division, which is remarkably close to the estimates in mice.

If bats have less adult neurogenesis than mice, rats, and primates, rabbits seem to have more of it. There is considerable evidence for adult neurogenesis in the rabbit SVZ/OB (Fasolo et al., 2002), striatum (Luzzati et al., 2007), caudate nucleus (Luzzati et al., 2006), and cerebellum (Ponti et al., 2008; see Bonfanti and Ponti, 2008, for review). Many questions remain open, but the available body of data is very suggestive.

The structure of the rabbit SVZ
is similar to that of the human
brain

Rabbits share with humans (and for example cows: Rodriguez-Perez et al., 2003) an SVZ with a "hypocellular gap" between the ependyma and a band of astrocytes. Like in humans, the "neuroblasts" do not show the strict chain-migratory pattern but can be encountered individually and in the absence of an ensheathing glial tube linking the ventricular region with the olfactory bulb. In rabbits, but not in adult rodents, there is a ventricular extension into the olfactory bulb, the olfactory ventricle. This structure obliterates in rodents and primates but remains open in rabbits and appears to serve as the guidance structure for the migrating neuroblasts (Luzzati et al., 2003; Ponti et al., 2006). As discussed below, the remnants of the olfactory ventricle, which appear to remain present in many more cases than previously thought, might also have that same function in the human brain, making rabbits a particularly interesting model. Rabbits show particularly extensive parenchymal migration of "neuroblasts," showing some similarity to the migration of doublecortin-positive cells—e.g., to the corpus callosum in rodents—but suggesting a different degree of cellular plasticity.

Rabbits, like humans but unlike rodents, also possess a relatively large temporal lobe with another ventricular extension, the inferior horn. Quite surprisingly, adult neurogenesis in the temporal lobe of rabbits, especially the hippocampus, has not been investigated beyond the one initial study (Gueneau et al., 1982).

Other Mammals

Mammalian species other than
mice and rats have not been
systematically studied, but
results are indicative of adult
neurogenesis

After Altman and Das had first described adult neurogenesis in rats (Altman and Das, 1965), the next animal they turned to was the guinea pig (Altman and Das, 1967)—not the mouse, which was added to the list only in 1997 (Kempermann et al., 1997b). The reason for the choice of the guinea pig lay in the fact that, in contrast to other rodents, guinea pigs do not have prolonged postnatal brain development and are born with comparatively mature brains. Consequently, the detection of newly generated cells in the adult guinea pig hippocampus supported the idea of truly adult neurogenesis instead of only delayed postnatal brain development. Single studies exist on gerbils (Dawirs et al., 1998; Teuchert-Noodt et al., 2000) and voles (Ormerod and Galea, 2001; Smith et al., 2001; Fowler et al., 2002; Amrein et al., 2004a; Amrein et al., 2004b). Voles have attracted particular interest in the context of addressing the hormonal influences on adult neurogenesis because of their social structure and their seasonal behaviors (Ormerod and Galea, 2001; Fowler et al., 2002; Ormerod et al., 2003, 2004).

Not unexpectedly, the golden hamster as primary research animal in chronobiology has been used to study the effects of the photoperiod on adult neurogenesis. The data from one single study are not yet fully conclusive but suggest a situation closer to that of rabbits than to other rodent species in that neurogenesis might be more abundant (Huang et al., 1998).

There is not much literature about adult neurogenesis in carnivores. About cats we do not have even rudimentary knowledge, although they were studied very early by Joseph Altman and included in the 1963 paper predating the first true proof of adult neurogenesis (Altman, 1963). A more recent study claimed reactive neurogenesis in the vestibular nuclei after deafferentation (Tighilet et al., 2007).

Comparatively good data, but only from two studies so far, exist for adult neurogenesis in the hippocampus of dogs. Sadly, only BrdU-labeled cells and doublecortin expression were studied, but a strong age-related decrease in these parameters was found (Hwang et al., 2007; Siwak-Tapp et al., 2007). In one of the studies, the effects of antioxidant-fortified food and of environmental enrichment, both for over 2.8 years, were examined but did not yield differences in the numbers of BrdU- or doublecortin-positive cells (Siwak-Tapp et al., 2007). However, in the enriched groups, some interesting correlations were found between doublecortin-positive cells and spatial memory, size, or black-and-white discrimination and, notably, the reversal of these latter performances. The oldest dogs studied so far were 15-year-old beagles (Siwak-Tapp et al., 2007).

Two reports have addressed neurogenesis in explants from the bovine ventricular wall, which argues in favor of ongoing neurogenesis and an anatomical situation with a hypocellular zone similar to that of rabbits and humans (Perez-Martin et al., 2003; Rodriguez-Perez et al., 2003). Otherwise, most questions remain open.

In sheep, the number of BrdU-labeled cells in the dentate gyrus roughly doubled after exposure of female animals to unfamiliar males (Hawken et al., 2009). Unfortunately, no further phenotypical analysis of the cells and no investigation of neurogenesis proper has been done.

Marsupials

Adult marsupialian neurogenesis resembles neurogenesis in mammals

Initially there had been a small, not widely appreciated, controversy over the question of whether marsupials showed adult hippocampal neurogenesis as well (Reynolds et al., 1985; Harman, 1997), but in a classical tritiated thymidine study, Alison Harman and coworkers showed that in a small marsupial, the fat-tailed dunnart, hippocampal granule cells are produced in adulthood (Harman et al., 2003). In contrast to mice, putting the animals in an enriched environment reduced the survival of the new cells, leaving the proliferation unchanged, an effect possibly attributable to stress.

Meanwhile we have a BrdU-based study for the gray short-tailed opossum at the ages of 10 and 30 months. These animals have a lifespan of approximately three years. At both ages, treatment with serotonin receptor 1A agonist buspirone increased cell survival and neurogenesis (Grabiec et al., 2009a). Conversely, treatment with an antagonist decreased olfactory bulb neurogenesis, which was associated with longer searches for food, interpreted as impaired olfaction (Grabiec et al., 2009b).

Nonhuman Primates

Adult hippocampal neurogenesis has been found in Old and New World monkeys

The first signs of DNA synthesis in SVZ cells of adult rhesus monkeys were reported by Michael Kaplan in 1983. However, the level of labeling with tritiated thymidine was very low, and the evidence remained controversial (Kaplan, 1983).

Other early autoradiographic studies based on tritiated thymidine injections failed to detect adult hippocampal neurogenesis in rhesus monkeys beyond the age of three years, which equals postpuberty (Eckenhoff and Rakic, 1988). But in 1997, Elizabeth Gould, Eberhard Fuchs, and colleagues found adult neurogenesis in the dentate gyrus of the adult tree shrew, a species of New World monkey (Gould et al., 1997).

Similar to what the group had found earlier in rats, a stress-induced and NMDA receptor–dependent down-regulation of cell proliferation was described, making this study the first to report, not only adult neurogenesis in primates (although, precisely, tree shrews are considered half-primates, phylogenetically located between insectivores and primates), but also its regulation in a species other than rodents, thus linking the study to the large body of previous evidence (Gould et al., 1997). In 1999, hippocampal neurogenesis in Old World monkeys (macaques), an undisputed primate species, was independently described by Elizabeth Gould and coworkers and by David Kornack and Pasko Rakic, who used bromodeoxyuridine immunohistochemistry and colocalization of BrdU and neuronal markers (Gould et al., 1999b; Kornack and Rakic, 1999). Neurogenesis was found even in 23-year-old animals (Gould et al., 1999b), but overall levels were reported to be very low (Kornack and Rakic, 1999). In marmosets, the age-dependent decrease in neurogenesis in the dentate gyrus and olfactory bulb was almost linear between young (1.5 years) and middle age (7 years) (Leuner et al., 2007). Adult hippocampal neurogenesis was also found in young rhesus monkeys, and prenatal stress had long-lasting effects on hippocampal neurogenesis in these monkeys (Coe et al., 2003). Hippocampal neurogenesis in adult rhesus monkeys was also confirmed by electronmicroscopy (Ngwenya et al., 2008). A time-course study with different time points after BrdU-injection between two hours and 98 days combining immunohistochemistry and electronmicroscopy revealed a pattern of neuronal development in the adult hippocampus that was similar to rodents', although maturation appeared to be slower (Ngwenya et al., 2006).

Adult neurogenesis is also found in the monkey olfactory bulb

Olfactory bulb neurogenesis, too, was detected in macaque and rhesus monkeys in patterns similar to that known from rodents (Kornack and Rakic, 2001a; Pencea et al., 2001) (Fig. 11–4). In addition, olfactory bulb neurogenesis was found in squirrel monkeys, a New World species (Bedard et al., 2002). Interestingly, in macaques, the fine anatomical structure of the SVZ was similar to the situation reported for humans and rabbits, showing for example the same characteristic hypocellular gap and the band of astrocytes, whereas the migratory patterns of neuroblasts towards the olfactory bulb resembled more the migratory chains in rodents (Gil-Perotin et al., 2009).

Ischemia increased neurogenesis in the neurogenic regions

Focal ischemia induced neurogenesis in the olfactory bulb and SVZ of macaque monkeys (Koketsu et al., 2003). Global ischemia at least induced the proliferation of Musahi- and nestin-positive cells in the dentate gyrus and the SVZ of macaques (Tonchev et al., 2003a; Tonchev et al., 2003b). Both findings suggested a responsiveness of the precursor cells similar to the situation found in rodents. Matrixmetalloproteinases were up-regulated in newborn hippocampal neurons approximately two weeks after ischemia (Lu et al., 2008), but no further mechanistic insight is available.

With respect to functional implications, only correlational evidence yet exists

In aging macaque monkeys, Aizawa and colleagues from Tatsuhiro Hisatsunes group at the University of Tokyo observed a positive correlation between cell proliferation in the dentate gyrus and performance in a visual pattern discrimination task (Aizawa et al., 2009). The study did not control for many potential covariants and confounding factors and did not measure neurogenesis proper. The report will thus have to be taken with some caution, but it nevertheless represents the first attempt to assess the functional contribution of new neurons in the primate hippocampus.

Neurogenesis in the cortex of adult non-human primates remained unconfirmed

The controversy over reports on neocortical neurogenesis in macaques was covered in Chapter 8. The initial report about a robust and steady stream of new cortical neurons originating from the SVZ (Gould et al., 1999a) was not confirmed by other groups (Kornack and Rakic, 2001b; Koketsu et al., 2003); mostly methodological concerns have been raised (Rakic, 2002). The observed cells might correspond to NG2 cells showing some neuronal features but never terminally differentiating into functional neurons. A number of additional studies have implied that neurogenesis could occur outside the canonical neurogenic regions of primates—for example, in the amygdala and adjacent temporal cortex (Bernier et al., 2002), olfactory tubercle (Bedard et al., 2002), and spinal cord (Vessal et al., 2007).

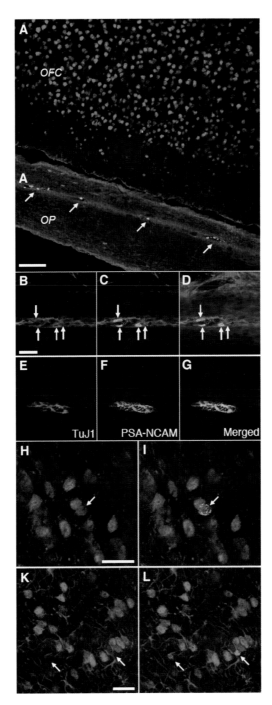

FIGURE 11–4 Neurogenesis in the olfactory bulb of adult monkeys. Kornack and Rakic (2001) reported newly generated neurons in the olfactory bulb of macaque monkeys. (A): Below the orbitofrontal cortex (OFC), the rostral migratory stream (*arrows*) is visible in the olfactory peduncle (OP). BrdU-labeled cells (*green*) are found in the GFAP-positive (*blue*) pathway. (B–D): β-III-tubulin-expressing cells are in the stream (B, *red*), a few of which are BrdU-labeled (C, *green*); (D) again shows the GFAP-positive environment (*blue*). (E–G): β-III-tubulin-labeling coincides with expression of PSA-NCAM. (H and I): A newly generated neuron in the olfactory bulb of a macaque monkey is shown, 97 days after the last injection of BrdU (NeuN, *red*; BrdU, *green*; GFAP, *blue*). (K and L): An example of a newly generated, non-neuronal cell in the same region. Scale bars, 100 mm for (A); 25 mm for (B–G); 20 mm for (H–L). Reprinted with kind permission of the authors and Copyright 2001 National Academy of Sciences, U.S.A.

Neither study could conclusively dispel the concerns regarding the validity of marker co-localization or of the markers themselves.

With respect to ex vivo evidence of neural stem cells in the adult monkey, we so far have only one neurosphere study, which confirms the existence of precursor cells as we know them from the rodent literature (Yue et al., 2006).

Humans

Adult neurogenesis in humans remains a key question for adult neurogenesis research

From a justifiable if not somewhat superficial perspective, the value of research on adult neurogenesis rises and falls with the question how much of what is described in this book is relevant to the human situation. This question is important because research on adult neurogenesis is to a large degree justified to funding agencies and the tax-paying public by the belief that adult neurogenesis occurs, or can occur, in humans, and that its exploration will lead to progress in medicine.

Neurogenesis in the Adult Human Dentate Gyrus

One seminal BrdU study confirmed adult neurogenesis in five humans up to 72 years of age

Adult neurogenesis can be found in humans. The proof of principle, in terms of the standards set by research in rodents, has been led by Peter Eriksson in a 1998 landmark study (Eriksson et al., 1998). Eriksson, a neurologist in Gothenburg, Sweden, identified a group of patients suffering from throat carcinomas who, as part of their treatment regimen, had received a single BrdU infusion to allow staging of the tumor after its surgical removal. Eriksson reasoned that the BrdU would have labeled, not only the dividing tumor cells, but also the proliferating precursor cells in the brain. He obtained informed consent from the patients to perform a brain autopsy after their death. The five patients died at the ages of 57, 58, 67, 68, and 72 years, between two weeks and two years after the BrdU injection. The same method that had been established for the analysis of adult hippocampal neurogenesis in rodents was applied to the examination of these patients' brains. Because no complete series through the entire dentate gyrus could be obtained, no absolute counts of BrdU-labeled neurons could be generated. The numbers of BrdU-marked neurons per cubic millimeter ranged between about two and 40 cells. Given that the patients in this study were older, that the human dentate gyrus contains 50 times as many granule cells as in a mouse (15 million vs. 300,000), and that, as in rodents, human adult hippocampal neurogenesis is likely to decline with increasing age, this detected rate of adult hippocampal neurogenesis in the five patients appeared very substantial.

All other studies rely on markers and marker combinations indicative of neurogenesis validated in other species

Eriksson's study was based on the unique availability of brain tissue after BrdU incorporation in humans. As BrdU is not needed for tumor staging any more, and because its potentially cancerogenic and mutagenic properties prevent its application just for research purposes, it cannot be expected that another study will confirm or extend Eriksson's findings with the same methodology. In surgical specimens from epilepsy surgery as well as post-mortem samples from subjects without temporal lobe epilepsy, proliferating (PCNA-labeled) doublecortin-positive cells and doublecortin/NeuN-double-positive cells were detected (Liu et al., 2008). The seizure samples consistently revealed greater numbers of Doublecorti-positive cells: on average, twice as many per volume. Although there are potential confounds due to a lack of information on the reference volume (see Chap. 7, p. 249), there is no general indication in the literature that epilepsy would lead to a smaller dentate gyrus. Consistent with this observation is the detection of nestin-expressing Ki67-labeled cells in resected hippocampi after epilepsy surgery (Blumcke et al., 2001).

Rolf Knoth and colleagues from the Department of Neuropathology at the University of Freiburg in Germany undertook a systematic survey of the expression of markers that in rodents are associated with adult neurogenesis in the human brain across the lifespan between birth and 100 years (Knoth et al., 2010). Doublecortin was taken as lead antigen, and co-localizations with a total of 21 other markers were investigated. Essentially all marker combinations known from rodent data were also found in the human brain, but not equally at all ages. There was a decrease both in the number of doublecortin-positive cells as well as co-localization with other markers, but at least until about the mid-40s, DCX overlapped with Ki67, Mcm2, Sox2, nestin, Prox1, PSA-NCAM, calretinin, NeuN, and others. Some key markers such as nestin, Sox2, and Prox1 remained co-expressed into oldest age (Knoth et al., 2010).

Neurogenesis in the Adult Human SVZ/Olfactory Bulb

No BrdU data exist for adult neurogenesis in the adult olfactory bulb, and one extensive study speaks against it

Neurogenesis in the olfactory bulb was not covered in the original Eriksson study of 1998, because the olfactory bulbs had not been available for investigation. To this date, adult neurogenesis in the human olfactory bulb has not yet been proven with the same degree of evidence as in the hippocampus. Two studies that searched for markers associated with neurogenesis in the adult human olfactory bulb found dividing cells that expressed nestin, doublecortin, NeuroD, vimentin, calretinin, or β-III-tubulin (Liu and Martin, 2003; Bedard and Parent, 2004). Nevertheless, in an extensive study from 2004, Arturo Alvarez-Buylla and his colleagues found SVZ precursor cells, but only a very few migrating neuroblasts were found in vivo, and those that were identified did not show the characteristic pattern of chain migration (Sanai et al., 2004). They concluded that their "*findings raise the unexpected possibility that migration from the SVZ to the olfactory bulb does not take place in humans or, if it does, precursors migrate as individual cells.*" (Sanai et al., 2004).

Humans do not rely much on smell and thus might require less adult neurogenesis in the olfactory bulb

The question was, however, how much neurogenesis could be expected. After all, in rodents, the olfactory bulb comprises about one fifth of the brain weight, compared to 0.07% in humans. This relationship notwithstanding, the distance between SVZ and olfactory bulb is about two orders of magnitude longer in humans than in rodents. In addition to these issues, the fact that humans do not rely heavily on olfaction (they are a microsmatic species and might thus have relatively less adult olfactory neurogenesis) and that an age-dependent decline in olfactory neurogenesis is to be expected might account for a considerable dilution of detectable cells on their migratory route to the human olfactory bulb. Only very few cells might actually migrate and still maintain a low but reasonable level of adult olfactory neurogenesis.

Migratory "neuroblasts" but no migrating chains have been identified in humans

In 2007, Curtis and colleagues from Peter Eriksson's group in Göteborg re-addressed this issue in a meticulous anatomical study. They found ultrastructural and immunohistochemical evidence for migratory "neuroblasts" despite the absence of chain migration as it is found in rodents (Curtis et al., 2007).

The adult human SVZ shows a characteristic hypocellular gap below the ependyma

The adult human SVZ is organized into four layers (Sanai et al., 2004; Curtis et al., 2007; Kam et al., 2009). Layer I essentially corresponds to the ependyma, Layer II contains very few cells and constitutes a "hypocellular gap," Layer III is heterogenous in terms of the cell types and sizes and included many Prominin1-positive cells and the cell bodies of astrocytes, and Layer IV is rich in nerve fibers interspersed with astrocytes. The presumably migratory PSA-NCAM-positive neuroblasts were found in Layers III and IV (Kam et al., 2009).

In the descending limb of the RMS, Layers I to III apparently merge, so that a bi-layered structure ensues for the rostral part of the RMS (Kam et al., 2009). There is, however, considerable variation between different sections of the SVZ and RMS, although the overall pattern does not differ much. Interspersed longitudinal cavities in the RMS were interpreted as remnants of the olfactory ventricle. Cells lining these cavities bore cilia, possibly suggesting a structure similar to the SVZ. Curtis and colleagues suggest that the ventriculo-olfactory extension (VOE) represents something like an SVZ extended towards the bulb (Curtis et al., 2007; Kam et al., 2009).

> Remnants of the olfactory ventricle that exists during development might serve as guidance structure towards the bulb

During fetal brain development the olfactory bulb forms from the wall of an anterior extension of the lateral ventricle (Bossy, 1980). This extension is called the *olfactory ventricle* and remains open in many species, including rabbits, but was thought to generally disappear in humans around birth. Accordingly, the idea that new neurons might migrate along remnants of this structure was heavily criticized (Sanai et al., 2007). It seems, however, that the presence or absence of cavities in the olfactory nerve and bulb has never been studied systematically, so that single case reports in which a persisting olfactory ventricle was linked to pathology ("rhinocele") dominate the literature (Roy et al., 1987; Andrews and Hulette, 1993). The often cited study by McFarland was actually on dolphins (McFarland et al., 1969). More recent MRI studies suggest that residuals of the olfactory ventricle are much more common than previously thought and do not generally relate to pathology (Smitka et al., 2009). Like on the histological side, however, there are at least partially contradicting observations here as well (Schneider and Floemer, 2009).

> Open olfactory ventricles are probably much more common than previously thought

Neural Precursor Cells in the Adult Human Brain

Several ex vivo studies confirmed that the adult human brain contains neural precursor cells. Most of these experiments were done on brain tissue that had been surgically removed. The indication for the operation was usually pharmacologically intractable epilepsy, for which a unilateral temporal lobectomy (removal of the part of the temporal lobe that contains the hippocampus) was performed. A unilateral resection of the hippocampus can be compensated for functionally; only a bilateral removal would lead to complete anterograde amnesia. Other samples were taken from resections that had become necessary to decompress a swelling brain after hemorrhage and stroke. Samples from the ventricular wall were usually obtained from ventricular biopsies or shunt operations in case of a hydrocephalus.

> Neural precursor cells can be isolated from the adult human brain

In any case, most of the presently available evidence about precursor cells in the adult human brain has been obtained from unhealthy brains. However, the concern that precursor cells in the adult human brain are only a consequence of brain pathology seems unlikely to be true, given the evidence from animal studies, including primates. Nevertheless, the caveat that the pathology might have changed the cells' properties has to be taken seriously until precursor cells can be directly derived from normal adult human brain. To date, the oldest documented human individual from whom precursor cells were isolated in the absence of brain pathology, but postmortem, was only 11 months old (Palmer et al., 2001).

> Most samples come from epilepsy surgery, a few from biopsies of the ventricular wall

A series of detailed experiments about neural precursor cells in the adult human brain came from Steven Goldman's group at Cornell University in New York. Their first study in 1994 indicated that dividing cells from the human ventricular wall could give rise to neurons in vitro (Kirschenbaum et al., 1994). In a similar study from 1998, the same group used explant cultures from the adult human SVZ to show that, over a period of weeks, new neurons would grow out of the explant, and that this outgrowth could be boosted by the application of BDNF and FGF-2. This study also contained histological analyses of the human SVZ, where putative precursor cells were identified by their expression of Musashi and Hu antigens (Pincus et al., 1998).

In a series of elegant experiments, N.S. Roy and colleagues from Steven Goldman's laboratory gave further evidence of precursor cells in the human hippocampus (Roy et al., 2000b) and SVZ (Roy et al., 2000a). In these studies, precursor cells were isolated from the surgical specimens by transfecting the cultured cells with plasmid DNA encoding the green fluorescent protein under precursor cell–relevant promoters such as Alpha1-tubulin or nestin. Precursor cells from the human (as well as rodent) SVZ showed some regional differences in growth and differentiation parameters in vitro (Bernier et al., 2000; Ostenfeld et al., 2002).

Arturo Alvarez-Buylla and colleagues investigated cells from the adult human SVZ in the most extensive study thus far (Sanai et al., 2004). They examined a total of 110 surgical and postmortem samples that allowed good spatial resolution of different regions of the SVZ. The precursor cells isolated from the adult human SVZ had astrocytic properties. Through transfection of isolated cells with GFP under the GFAP promoter and sorting for GFP-positive cells, a population of multipotent cells was identified. Precursor cells have also been isolated from the adult human olfactory bulb (Pagano et al., 2000; Sanai et al., 2004).

Many of these in vitro studies indicated that neural precursor cells from the adult human brain behaved differently than their rodent counterparts. For example, human precursor cells showed a preference for laminin as coating of the surface on which the cells were plated out for differentiation. Also, expansion of human precursor cells can be promoted by the presence of leukemia inhibiting factor (LIF), which has no effect in rodents. Human neural precursor cells can differentiate in the presence of growth factors, whereas in mice, growth factors have to be withdrawn to induce proliferation. The growth factor effects themselves showed differences as well. For example, platelet-derived growth factor increases oligodendrocytic differentiation in rodents, but had no such effect on human precursor cells (Arsenijevic et al., 2001).

There is conflicting evidence about neural precursor cells in the human cortex. Whereas two ex vivo studies detected precursor cells in cortical regions (Arsenijevic et al., 2001; Palmer et al., 2001), others did not (Kirschenbaum et al., 1994; Sanai et al., 2004). The human corpus callosum contains parenchymal precursor cells that could generate neurons ex vivo (Roy et al., 1999). Neuronal nature was determined not only immunocytochemically but also electrophysiologically, and the production of neurons was even found shortly after the isolation of precursor cells from the intraoperative specimens. This immediacy is important because it minimizes the chance that the neurogenic potential is induced only by culture conditions. The neurogenic potential was further confirmed after xenografting the cells into the developing rat brain, where they incorporated into the neurogenic regions and formed neurons.

The first experiments that included assays to determine stemness in cells isolated from human brain samples came from Dennis A. Steindler and colleagues, who in 1999 showed that both the SVZ and hippocampus contained precursor cells that formed primary and secondary neurospheres after cultivation at clonal density (Kukekov et al., 1999). The spheres also expressed transcription factor Pax6, which in rodents is found in astrocyte-like precursor cells both in vivo and in vitro. In this study, the individuals were 24 to 57 years old.

Precursor cells from the adult human brain could not be propagated indefinitely in vitro, prohibiting the use of the term *stem cell* for them. After a maximum of about 30 doublings, the cells appear to become senescent. To date it is not clear whether this result reflects a technical problem or a fundamental biological property. It is conceivable that all studies were successful in expanding, not true stem cells, but selective populations of progenitor cells with a limited capability for expansion.

On the other hand, expandability of fetal human progenitor cells could be greatly enhanced by maintaining cell–cell contacts in the cultures. Clive Svendsen, Maeve Caldwell, and colleagues from Cambridge "chopped" the neurospheres in quarters instead of triturating them to single-cell suspensions. This treatment increased, not only the number of possible cell doublings (Svendsen et al., 1998), but also the number of neurons per sphere (Caldwell et al., 2001). The proposed explanation was that cell–cell contacts were maintained through this method.

Adult neurogenesis in the human olfactory epithelium has hardly been studied yet

Neuronal outgrowth from dividing cells in biopsies taken from the olfactory epithelium of adult humans confirmed the existence of neurogenesis (Murrell et al., 1996) to the same degree as had been done by the pioneering 1994 study by Kirschenbaum, Goldman, and colleagues for the human SVZ. Surprisingly, further detailed studies on precursor cell properties (most notably clonal expansion) in cells isolated from the adult human olfactory epithelium are still lacking, and so is an in vivo study. Despite these shortcomings, however, the fact that human neuronal development in the olfactory epithelium is accessible by comparatively simple biopsy in the adult has raised a number of interesting speculations (see Chap. 8).

Adult Human Neurogenesis and Brain Pathology

Pathology affected cells expressing neurogenesis markers in the adult human brain

Besides the increase in signs of adult neurogenesis in the brains of patients with temporal lobe epilepsy that were mentioned above, a few reports dealt with other diseases. In accordance with similar observations in murine models of Alzheimer disease, Jin and colleagues reported that, in postmortem samples from Alzheimer patients, they had detected increased numbers of doublecortin-positive cells (Jin et al., 2004). A partially conflicting paper claimed that the "neuronal maturation marker" Map2ab was decreased in hippocampal samples from Alzheimer brains, whereas total Map and "immature marker" Map2c were not altered (Li et al., 2008). As this lack of maturation was not related directly to newborn cells, the interpretation that adult neurogenesis would fail in Alzheimer disease still rests on shaky grounds.

Höglinger and colleagues found decreased numbers of proliferating cells and nestin-positive cells (many of which co-expressed PSA-NCAM) in the SVZ of patients with Parkinson disease. For the hippocampus, a similar claim for "decreased neurogenesis" was made on the basis of a reduced number of nestin- and β-III-tubulin-positive cells (Hoglinger et al., 2004).

In patients with hypoxic-ischemic encephalopathy, an often lethal condition, for example after failed resuscitation, the number of PCNA-, TUC4-, and calretinin-positive cells in the dentate gyrus was increased, and so was the number of apoptotic nuclei (Mattiesen et al., 2009). Again, this does not prove a responsive increase in adult neurogenesis (as hypothesized from the rodent data on neurogenesis after ischemia) but adds important circumstantial information both about adult neurogenesis in humans in general and about its potential response to pathology.

Bomb Carbon

Studies based on the incorporation of atmospheric Carbon 14 allowed the estimation of cellular turnover in human tissues, including brain

In their search for new ways to investigate adult neurogenesis in humans, Kirsty Spalding and Jonas Frisén from the Karolinska Institute in Stockholm turned to a particularly intriguing solution: birth-dating based on carbon from atomic bomb tests (Arlotta and Macklis, 2005; Eisenstein, 2005; Spalding et al., 2005b; Spalding et al., 2005a).

Besides the prevailing isotope 12, there is a small, relatively constant level of carbon isotope 14 (14C) in the atmosphere. For any biosynthesis carbon is needed, and this is ultimately (via plants) derived from atmospheric carbon dioxide. There is an equilibrium between atmospheric 14C and the biomass. Between the late 1940s until their ban in 1963, atomic bomb tests were conducted aboveground and released unprecedented levels of 14C into the atmosphere. Consequently, 14C levels also rose in living cells. Since the ban, atmospheric 14C is declining again, leaving behind an impressive 14C pulse. Spalding and Frisén reasoned that this peak might provide some sort of endogenous BrdU-like labeling, if one were able to measure 14C in the DNA of neurons and compare it to other cell types. This turned out to be a major technical challenge. Extremely sensitive mass spectrometry was needed, as well as a strategy to reliably isolate cell type–specific DNA. When all the methodological hurdles were finally cleared, the group could first show that while cortical neurons are generally as old as the individual, its glial cells are not (Fig. 11–5) (Bhardwaj et al., 2006). Most people had not, however, been overly concerned that humans might have too much neurogenesis, so the reverse question still seems more pressing. As of this writing, the world is unfortunately still awaiting confirmation of adult neurogenesis with this method for the human olfactory bulb and the hippocampus, but with the extremely low numbers of newborn cells, the challenges to the sensitivity of mass spectroscopy are particularly daunting.

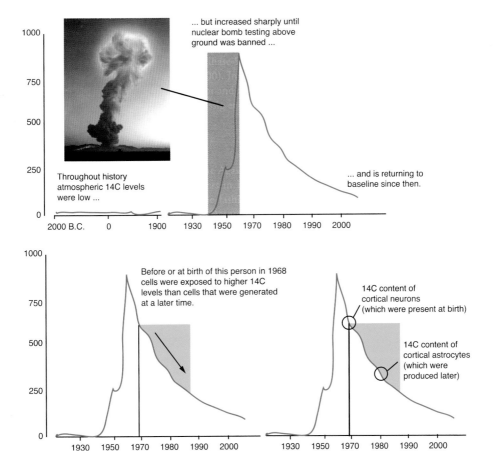

FIGURE 11–5 Birth-dating of cells with bomb carbon. Atmospheric 14C was taken up by cells during the period of above-surface nuclear bomb tests and the still-ongoing decline phase after the test ban treaty in 1963. Measuring 14C in cell nuclei allows estimation of their birth date relative to the birth date of the individual from whom the cells were taken (Spalding, et al., 2005, Bhardwaj et al., 2006). Details are explained in the text. Cloud photo courtesy of National Nuclear Security Administration / Nevada Site Office; August 31, 1957 (SMOKY).

References

Abrous DN, Koehl M, Le Moal M (2005). Adult neurogenesis: From precursors to network and physiology. *Physiol Rev* 85:523–569.

Absil P, Pinxten R, Balthazart J, Eens M (2003). Effect of age and testosterone on autumnal neurogenesis in male European starlings (Sturnus vulgaris). *Behav Brain Res* 143:15–30.

Aizawa K, Ageyama N, Yokoyama C, Hisatsune T (2009). Age-dependent alteration in hippocampal neurogenesis correlates with learning performance of macaque monkeys. *Exp Anim* 58:403–407.

Alonso JR, Lara J, Vecino E, Covenas R, Aijon J (1989). Cell proliferation in the olfactory bulb of adult freshwater teleosts. *J Anat* 163:155–163.

Altman J (1963). Autoradiographic investigation of cell proliferation in the brains of rats and cats. *Anat Rec* 145:573–591.

Altman J, Das GD (1965). Autoradiographic and histological evidence of postnatal hippocampal neurogenesis in rats. *J Comp Neurol* 124:319–335.

Altman J, Das GD (1967). Postnatal neurogenesis in the guinea-pig. *Nature* 214:1098–1101.

Altman J, Brunner RL, Bayer SA (1973). The hippocampus and behavioral maturation. *Behav Biol* 8:557–596.

Alvarez-Buylla A, Nottebohm F (1988). Migration of young neurons in adult avian brain. *Nature* 335: 353–354.

Alvarez-Buylla A, Lois C (1995). Neuronal stem cells in the brain of adult vertebrates. *Stem Cells* (Dayt) 13: 263–272.

Alvarez-Buylla A, Kirn JR (1997). Birth, migration, incorporation, and death of vocal control neurons in adult songbirds. *J Neurobiol* 33:585–601.

Alvarez-Buylla A, Theelen M, Nottebohm F (1990). Proliferation "hot spots" in adult avian ventricular zone reveal radial cell division. *Neuron* 5:101–109.

Alvarez-Buylla A, Ling CY, Yu WS (1994). Contribution of neurons born during embryonic, juvenile, and adult life to the brain of adult canaries: Regional specificity and delayed birth of neurons in the song-control nuclei. *J Comp Neurol* 347:233–248.

Alvarez-Buylla A, Garcia-Verdugo JM, Mateo AS, Merchant-Larios H (1998). Primary neural precursors and intermitotic nuclear migration in the ventricular zone of adult canaries. *J Neurosci* 18:1020–1037.

Ampatzis K, Dermon CR (2007). Sex differences in adult cell proliferation within the zebrafish (Danio rerio) cerebellum. *Eur J Neurosci* 25:1030–1040.

Amrein I, Lipp HP (2009). Adult hippocampal neurogenesis of mammals: Evolution and life history. *Biol Lett* 5:141–144.

Amrein I, Slomianka L, Lipp HP (2004a). Granule cell number, cell death and cell proliferation in the dentate gyrus of wild-living rodents. *Eur J Neurosci* 20:3342–3350.

Amrein I, Slomianka L, Poletaeva, II, Bologova NV, Lipp HP (2004b). Marked species and age-dependent differences in cell proliferation and neurogenesis in the hippocampus of wild-living rodents. *Hippocampus*. 14:1000–1010.

Amrein I, Dechmann DK, Winter Y, Lipp HP (2007). Absent or low rate of adult neurogenesis in the hippocampus of bats (*Chiroptera*). *PLoS ONE* 2:e455.

Amrein I, Lipp HP, Boonstra R, Wojtowicz JM (2008). Adult hippocampal neurogenesis in natural populations of mammals. In: *Adult Neurogenesis* (Gage F, Kempermann G, Son H, eds.), pp. 645–659. Cold Spring Harbor, New York: Cold Spring Harbor Laboratory Press.

Andrews PI, Hulette CM (1993). An infant with macrocephaly, abnormal neuronal migration and persistent olfactory ventricles. *Clin Neuropathol* 12:13–18.

Arlotta P, Macklis JD (2005). Archeo-cell biology: Carbon dating is not just for pots and dinosaurs. *Cell* 122:4–6.

Arsenijevic Y, Villemure JG, Brunet JF, Bloch JJ, Deglon N, Kostic C, Zurn A, Aebischer P (2001). Isolation of multipotent neural precursors residing in the cortex of the adult human brain. *Exp Neurol* 170:48–62.

Ball GF, Riters LV, Balthazart J (2002). Neuroendocrinology of song behavior and avian brain plasticity: Multiple sites of action of sex steroid hormones. *Front Neuroendocrinol* 23:137–178.

Barami K, Kirschenbaum B, Lemmon V, Goldman SA (1994). N-cadherin and Ng-CAM/8D9 are involved serially in the migration of newly generated neurons into the adult songbird brain. *Neuron* 13:567–582.

Barami K, Iversen K, Furneaux H, Goldman SA (1995). Hu protein as an early marker of neuronal phenotypic differentiation by subependymal zone cells of the adult songbird forebrain. *J Neurobiol* 28:82–101.

Barkan S, Ayali A, Nottebohm F, Barnea A (2007). Neuronal recruitment in adult zebra finch brain during a reproductive cycle. *Dev Neurobiol* 67:687–701.

Barker JM, Wojtowicz JM, Boonstra R (2005). Where's my dinner? Adult neurogenesis in free-living food-storing rodents. *Genes Brain Behav* 4:89–98.

Barnea A (2009). Interactions between environmental changes and brain plasticity in birds. *Gen Comp Endocrinol* 163:128–134.

Barnea A, Nottebohm F (1994). Seasonal recruitment of hippocampal neurons in adult free-ranging black-capped chickadees. *Proc Natl Acad Sci USA* 91:11217–11221.

Barnea A, Nottebohm F (1996). Recruitment and replacement of hippocampal neurons in young and adult chickadees: An addition to the theory of hippocampal learning. *Proc Natl Acad Sci USA* 93:714–718.

Bedard A, Parent A (2004). Evidence of newly generated neurons in the human olfactory bulb. *Brain Res Dev Brain Res* 151:159–168.

Bedard A, Levesque M, Bernier PJ, Parent A (2002). The rostral migratory stream in adult squirrel monkeys: Contribution of new neurons to the olfactory tubercle and involvement of the antiapoptotic protein Bcl-2. *Eur J Neurosci* 16:1917–1924.

Ben Abdallah N, Slomianka L, Vyssotski AL, Lipp HP (2010). Early age-related changes in adult hippocampal neurogenesis in C57 mice. *Neurobiol Aging* 31:151–161.

Ben Rokia-Mille S, Tinette S, Engler G, Arthaud L, Tares S, Robichon A (2008). Continued neurogenesis in adult *Drosophila* as a mechanism for recruiting environmental cue-dependent variants. *PLoS One* 3:e2395.

Bernier PJ, Vinet J, Cossette M, Parent A (2000). Characterization of the subventricular zone of the adult human brain: Evidence for the involvement of Bcl-2. *Neurosci Res* 37:67–78.

Bernier PJ, Bedard A, Vinet J, Levesque M, Parent A (2002). Newly generated neurons in the amygdala and adjoining cortex of adult primates. *Proc Natl Acad Sci USA* 99:11464–11469.

Bhardwaj RD, Curtis MA, Spalding KL, Buchholz BA, Fink D, Bjork-Eriksson T, Nordborg C, Gage FH, Druid H, Eriksson PS, Frisen J (2006). From the cover: Neocortical neurogenesis in humans is restricted to development. *Proc Natl Acad Sci USA* 103:12564–12568.

Bieber M, Fuldner D (1979). Brain growth during the adult stage of a holometabolous insect. *Naturwissenschaften* 66:426.

Birse SC, Leonard RB, Coggeshall RE (1980). Neuronal increase in various areas of the nervous system of the guppy, *Lebistes*. *J Comp Neurol* 194:291–301.

Blumcke I, Schewe JC, Normann S, Brustle O, Schramm J, Elger CE, Wiestler OD (2001). Increase of nestin-immunoreactive neural precursor cells in the dentate gyrus of pediatric patients with early-onset temporal lobe epilepsy. *Hippocampus* 11:311–321.

Bonfanti L, Ponti G (2008). Adult mammalian neurogenesis and the New Zealand white rabbit. *Vet J* 175:310–331.

Bossy J (1980). Development of olfactory and related structures in staged human embryos. *Anat Embryol* (Berl) 161:225–236.

Brainard MS, Doupe AJ (2002). What songbirds teach us about learning. *Nature* 417:351–358.

Brandstatter R, Kotrschal K (1990). Brain growth patterns in four European cyprinid fish species (Cyprinidae, Teleostei): Roach (*Rutilus rutilus*), bream (*Abramis brama*), common carp (*Cyprinus carpio*) and sabre carp (*Pelecus cultratus*). *Brain Behav Evol* 35:195–211.

Brenowitz EA, Lent K, Rubel EW (2007). Auditory feedback and song production do not regulate seasonal growth of song control circuits in adult white-crowned sparrows. *J Neurosci* 27:6810–6814.

Byrd CA, Brunjes PC (2001). Neurogenesis in the olfactory bulb of adult zebrafish. *Neuroscience* 105:793–801.

Caldwell MA, He X, Wilkie N, Pollack S, Marshall G, Wafford KA, Svendsen CN (2001). Growth factors regulate the survival and fate of cells derived from human neurospheres. *Nat Biotechnol* 19:475–479.

Cameron HA, McKay RD (1999). Restoring production of hippocampal neurons in old age. *Nat Neurosci* 2:894–897.

Cayre M, Strambi C, Strambi A (1994). Neurogenesis in an adult insect brain and its hormonal control. *Nature* 368:57–59.

Cayre M, Strambi C, Charpin P, Augier R, Meyer MR, Edwards JS, Strambi A (1996). Neurogenesis in adult insect mushroom bodies. *J Comp Neurol* 371:300–310.

Cayre M, Malaterre J, Scotto-Lomassese S, Strambi C, Strambi A (2002). The common properties of neurogenesis in the adult brain: From invertebrates to vertebrates. *Comp Biochem Physiol B Biochem Mol Biol* 132:1–15.

Cayre M, Scotto-Lomassese S, Malaterre J, Strambi C, Strambi A (2007). Understanding the regulation and function of adult neurogenesis: Contribution from an insect model, the house cricket. *Chem Senses* 32:385–395.

Chen G, Cheng MF (2007). Inhibition of lesion-induced neurogenesis impaired behavioral recovery in adult ring doves. *Behav Brain Res* 177:358–363.

Chetverukhin VK, Polenov AL (1993). Ultrastructural radioautographic analysis of neurogenesis in the hypothalamus of the adult frog, Rana temporaria, with special reference to physiological regeneration of the preoptic nucleus. I. Ventricular zone cell proliferation. *Cell Tissue Res* 271:341–350.

Coe CL, Kramer M, Czeh B, Gould E, Reeves AJ, Kirschbaum C, Fuchs E (2003). Prenatal stress diminishes neurogenesis in the dentate gyrus of juvenile rhesus monkeys. *Biol Psychiatry* 54:1025–1034.

Curtis MA, Kam M, Nannmark U, Anderson MF, Axell MZ, Wikkelso C, Holtas S, van Roon-Mom WM, Bjork-Eriksson T, Nordborg C, Frisen J, Dragunow M, Faull RL, Eriksson PS (2007). Human neuroblasts migrate to the olfactory bulb via a lateral ventricular extension. *Science* 315:1243–1249.

Dawirs RR, Hildebrandt K, Teuchert-Noodt G (1998). Adult treatment with haloperidol increases dentate granule cell proliferation in the gerbil hippocampus. *J Neural Transm* 105:317–127.

Dawley EM, Nelsen M, Lopata A, Schwartz J, Bierly A (2006). Cell birth and survival following seasonal periods of cell proliferation in the chemosensory epithelia of red-backed salamanders, *Plethodon cinereus*. *Brain Behav Evol* 68:26–36.

Dittrich F, Feng Y, Metzdorf R, Gahr M (1999). Estrogen-inducible, sex-specific expression of brain-derived neurotrophic factor mRNA in a forebrain song control nucleus of the juvenile zebra finch. *Proc Natl Acad Sci USA* 96:8241–8246.

Douzery EJ, Huchon D (2004). Rabbits, if anything, are likely Glires. *Mol Phylogenet Evol* 33:922–935.

Dufour MC, Gadenne C (2006). Adult neurogenesis in a moth brain. *J Comp Neurol* 495:635–643.

Dunlap KD, McCarthy EA, Jashari D (2008). Electrocommunication signals alone are sufficient to increase neurogenesis in the brain of adult electric fish, *Apteronotus leptorhynchus*. *Dev Neurobiol* 68:1420–1428.

Eckenhoff MF, Rakic P (1988). Nature and fate of proliferative cells in the hippocampal dentate gyrus during the life span of the rhesus monkey. *J Neurosci* 8:2729–2747.

Eisenstein M (2005). Positive fallout from the bomb. *Nat Methods* 2:638–639.

Eriksson PS, Perfilieva E, Bjork-Eriksson T, Alborn AM, Nordborg C, Peterson DA, Gage FH (1998). Neurogenesis in the adult human hippocampus. *Nat Med* 4:1313–1317.

Fahrbach SE, Strande JL, Robinson GE (1995). Neurogenesis is absent in the brains of adult honey bees and does not explain behavioral neuroplasticity. *Neurosci Lett* 197:145–148.

Fasolo A, Peretto P, Bonfanti L (2002). Cell migration in the rostral migratory stream. *Chem Senses* 27:581–582.

Finch CE (1994). *Longevity, Senescence, and the Genome*. Chicago: University of Chicago Press.

Font E, Garcia-Verdugo JM, Desfilis E, Perez-Canellas M (1995). Neuron-glia interrelations during 3-acetylpyridine-induced degeneration and regeneration in the adult lizard brain. In: *Neuron-Glia Interrelations During Phylogeny* (Vernadakis A, Roots B, eds.), pp. 275–302. Totowa: Humana.

Font E, Desfilis E, Perez-Canellas M, Alcantara S, Garcia-Verdugo JM (1997). 3-Acetylpyridine-induced degeneration and regeneration in the adult lizard brain: A qualitative and quantitative analysis. *Brain Res* 754:245–259.

Font E, Desfilis E, Perez-Canellas MM, Garcia-Verdugo JM (2001). Neurogenesis and neuronal regeneration in the adult reptilian brain. *Brain Behav Evol* 58:276–295.

Fowler CD, Liu Y, Ouimet C, Wang Z (2002). The effects of social environment on adult neurogenesis in the female prairie vole. *J Neurobiol* 51:115–128.

Gahr M, Metzdorf R (1997). Distribution and dynamics in the expression of androgen and estrogen receptors in vocal control systems of songbirds. *Brain Res Bull* 44:509–517.

Gahr M, Metzdorf R, Aschenbrenner S (1996). The ontogeny of the canary HVC revealed by the expression of androgen and oestrogen receptors. *Neuroreport* 8:311–315.

Galsworthy MJ, Amrein I, Kuptsov PA, Poletaeva, II, Zinn P, Rau A, Vyssotski A, Lipp HP (2005). A comparison of wild-caught wood mice and bank voles in the Intellicage: Assessing exploration, daily activity patterns and place learning paradigms. *Behav Brain Res* 157:211–217.

Garcia-Verdugo JM, Ferron S, Flames N, Collado L, Desfilis E, Font E (2002). The proliferative ventricular zone in adult vertebrates: A comparative study using reptiles, birds, and mammals. *Brain Res Bull* 57:765–775.

Gil-Perotin S, Duran-Moreno M, Belzunegui S, Luquin MR, Garcia-Verdugo JM (2009). Ultrastructure of the subventricular zone in *Macaca fascicularis* and evidence of a mouse-like migratory stream. *J Comp Neurol* 514:533–554.

Goergen EM, Bagay LA, Rehm K, Benton JL, Beltz BS (2002). Circadian control of neurogenesis. *J Neurobiol* 53:90–95.

Goldman SA (1990). Neuronal development and migration in explant cultures of the adult canary forebrain. *J Neurosci* 10:2931–2939.

Goldman SA, Nottebohm F (1983). Neuronal production, migration and differentiation in a vocal control nucleus of the adult female canary brain. *Proc Natl Acad Sci USA* 80:2390–2394.

Goldman SA, Nedergaard M (1992). Newly generated neurons of the adult songbird brain become functionally active in long-term culture. *Brain Res Dev Brain* Res 68:217–223.

Goldman SA, Lemmon V, Chin SS (1993). Migration of newly generated neurons upon ependymally derived radial guide cells in explant cultures of the adult songbird forebrain. *Glia* 8:150–160.

Goldman SA, Zukhar A, Barami K, Mikawa T, Niedzwiecki D (1996). Ependymal/subependymal zone cells of postnatal and adult songbird brain generate both neurons and nonneuronal siblings in vitro and in vivo. *J Neurobiol* 30:505–520.

Gould E, McEwen BS, Tanapat P, Galea LA, Fuchs E (1997). Neurogenesis in the dentate gyrus of the adult tree shrew is regulated by psychosocial stress and NMDA receptor activation. *J Neurosci* 17: 2492–2498.

Gould E, Reeves AJ, Graziano MS, Gross CG (1999a). Neurogenesis in the neocortex of adult primates. *Science* 286:548–552.

Gould E, Reeves AJ, Fallah M, Tanapat P, Gross CG, Fuchs E (1999b). Hippocampal neurogenesis in adult Old World primates. *Proc Natl Acad Sci USA* 96:5263–5267.

Grabiec M, Turlejski K, Djavadian RL (2009a). The partial 5-HT1A receptor agonist buspirone enhances neurogenesis in the opossum (*Monodelphis domestica*). *Eur Neuropsychopharmacol* 19:431–439.

Grabiec M, Turlejski K, Djavadian R (2009b). Reduction of the number of new cells reaching olfactory bulbs impairs olfactory perception in the adult opossum. *Acta Neurobiol Exp* (Warsz) 69:168–176.

Grandel H, Kaslin J, Ganz J, Wenzel I, Brand M (2006). Neural stem cells and neurogenesis in the adult zebrafish brain: Origin, proliferation dynamics, migration and cell fate. *Dev Biol* 295:263–277.

Graziadei PP, Metcalf JF (1971). Autoradiographic and ultrastructural observations on the frog's olfactory mucosa. *Z Zellforsch Mikrosk Anat* 116:305–318.

Gueneau G, Privat A, Drouet J, Court L (1982). Subgranular zone of the dentate gyrus of young rabbits as a secondary matrix. A high-resolution autoradiographic study. *Dev Neurosci* 5:345–358.

Hansen A, Schmidt M (2001). Neurogenesis in the central olfactory pathway of the adult shore crab *Carcinus maenas* is controlled by sensory afferents. *J Comp Neurol* 441:223–233.

Harman A, Meyer P, Ahmat A (2003). Neurogenesis in the hippocampus of an adult marsupial. *Brain Behav Evol* 62:1–12.

Harman AM (1997). Development and cell generation in the hippocampus of a marsupial, the quokka wallaby (*Setonix brachyurus*). *Brain Res Dev Brain Res* 104:41–54.

Harzsch S, Miller J, Benton J, Beltz B (1999). From embryo to adult: Persistent neurogenesis and apoptotic cell death shape the lobster deutocerebrum. *J Neurosci* 19:3472–3485.

Hawken PA, Jorre TJ, Rodger J, Esmaili T, Blache D, Martin GB (2009). Rapid induction of cell proliferation in the adult female ungulate brain (*Ovis aries*) associated with activation of the reproductive axis by exposure to unfamiliar males. *Biol Reprod* 80:1146–1151.

Hinsch K, Zupanc GK (2007). Generation and long-term persistence of new neurons in the adult zebrafish brain: A quantitative analysis. *Neuroscience* 146:679–696.

Hoglinger GU, Rizk P, Muriel MP, Duyckaerts C, Oertel WH, Caille I, Hirsch EC (2004). Dopamine depletion impairs precursor cell proliferation in Parkinson disease. *Nat Neurosci* 7:726–735.

Hoshooley JS, Sherry DF (2007). Greater hippocampal neuronal recruitment in food-storing than in non–food-storing birds. *Dev Neurobiol* 67:406–414.

Huang L, DeVries GJ, Bittman EL (1998). Photoperiod regulates neuronal bromodeoxyuridine labeling in the brain of a seasonally breeding mammal. *J Neurobiol* 36:410–420.

Hwang IK, Yoo KY, Li H, Choi JH, Kwon YG, Ahn Y, Lee IS, Won MH (2007). Differences in doublecortin immunoreactivity and protein levels in the hippocampal dentate gyrus between adult and aged dogs. *Neurochem Res* 32:1604–1609.

Ito K, Hotta Y (1992). Proliferation pattern of postembryonic neuroblasts in the brain of *Drosophila melanogaster*. *Dev Biol* 149:134–148.

Jiang J, McMurtry J, Niedzwiecki D, Goldman SA (1998). Insulin-like growth factor-1 is a radial cell–associated neurotrophin that promotes neuronal recruitment from the adult songbird edpendyma/subependyma. *J Neurobiol* 36:1–15.

Jin K, Peel AL, Mao XO, Xie L, Cottrell BA, Henshall DC, Greenberg DA (2004). Increased hippocampal neurogenesis in Alzheimer's disease. *Proc Natl Acad Sci USA* 101:343–347.

Johns PR (1977). Growth of the adult goldfish eye. III. Source of the new retinal cells. *J Comp Neurol* 176: 343–357.

Kam M, Curtis MA, McGlashan SR, Connor B, Nannmark U, Faull RL (2009). The cellular composition and morphological organization of the rostral migratory stream in the adult human brain. *J Chem Neuroanat* 37:196–205.

Kaplan MS (1983). Proliferation of subependymal cells in the adult primate CNS: Differential uptake of DNA labelled precursors. *J Hirnforsch* 24:23–33.

Kaslin J, Ganz J, Brand M (2008). Proliferation, neurogenesis and regeneration in the non-mammalian vertebrate brain. *Philos Trans R Soc Lond B Biol Sci* 363:101–122.

Kaslin J, Ganz J, Geffarth M, Grandel H, Hans S, Brand M (2009). Stem cells in the adult zebrafish cerebellum: Initiation and maintenance of a novel stem cell niche. *J Neurosci* 29:6142–6153.

Kassen SC, Ramanan V, Montgomery JE, C TB, Liu CG, Vihtelic TS, Hyde DR (2007). Time course analysis of gene expression during light-induced photoreceptor cell death and regeneration in albino zebrafish. *Dev Neurobiol* 67:1009–1031.

Kassen SC, Thummel R, Burket CT, Campochiaro LA, Harding MJ, Hyde DR (2008). The Tg(ccnb1:EGFP) transgenic zebrafish line labels proliferating cells during retinal development and regeneration. *Mol Vis* 14: 951–963.

Kempermann G, Kuhn HG, Gage FH (1997a). Genetic influence on neurogenesis in the dentate gyrus of adult mice. *Proc Natl Acad Sci USA* 94:10409–10414.

Kempermann G, Kuhn HG, Gage FH (1997b). More hippocampal neurons in adult mice living in an enriched environment. *Nature* 386:493–495.

Kempermann G, Kuhn HG, Gage FH (1998). Experience-induced neurogenesis in the senescent dentate gyrus. *J Neurosci* 18:3206–3212.

Kempermann G, Chesler EJ, Lu L, Williams RW, Gage FH (2006). Natural variation and genetic covariance in adult hippocampal neurogenesis. *Proc Natl Acad Sci USA* 103:780–785.

Kim DH, Lilliehook C, Roides B, Chen Z, Chang M, Mobashery S, Goldman SA (2008). Testosterone-induced matrix metalloproteinase activation is a checkpoint for neuronal addition to the adult songbird brain. *J Neurosci* 28:208–216.

Kirn JR, DeVoogd TJ (1989). Genesis and death of vocal control neurons during sexual differentiation in the zebra finch. *J Neurosci* 9:3176–3187.

Kirn JR, Alvarez-Buylla A, Nottebohm F (1991). Production and survival of projection neurons in a forebrain vocal center of adult male canaries. *J Neurosci* 11:1756–1762.

Kirschenbaum B, Nedergaard M, Preuss A, Barami K, Fraser RA, Goldman SA (1994). In vitro neuronal production and differentiation by precursor cells derived from the adult human forebrain. *Cereb Cortex* 6:576–589.

Knoth R, Singec I, Ditter M, Pantazis G, Capetian P, Meyer RP, Horvat V, Volk B, Kempermann G (2010). Murine features of neurogenesis in the human hippocampus across the lifespan from 0 to 100 years. *PLoS ONE* 5:e8809.

Koketsu D, Mikami A, Miyamoto Y, Hisatsune T (2003). Nonrenewal of neurons in the cerebral neocortex of adult macaque monkeys. *J Neurosci* 23:937–942.

Kornack DR, Rakic P (1999). Continuation of neurogenesis in the hippocampus of the macaque monkey. *Proc Natl Acad Sci USA* 96:5768–5773.

Kornack DR, Rakic P (2001a). The generation, migration, and differentiation of olfactory neurons in the adult primate brain. *Proc Natl Acad Sci USA* 98:4752–4757.

Kornack DR, Rakic P (2001b). Cell proliferation without neurogenesis in adult primate neocortex. *Science* 294:2127–2130.

Kranz VD, Richter W (1975). [Neurogenesis and regeneration in the brain of teleosts in relation to age. (Autoradiographic studies)]. *Z Alternsforsch* 30:371–382.

Kronenberg G, Bick-Sander A, Bunk E, Wolf C, Ehninger D, Kempermann G (2006). Physical exercise prevents age-related decline in precursor cell activity in the mouse dentate gyrus. *Neurobiol Aging* 27:1505–1513.

Kuhn HG, Dickinson-Anson H, Gage FH (1996). Neurogenesis in the dentate gyrus of the adult rat: Age-related decrease of neuronal progenitor proliferation. *J Neurosci* 16:2027–2033.

Kukekov VG, Laywell ED, Suslov O, Davies K, Scheffler B, Thomas LB, O'Brien TF, Kusakabe M, Steindler DA (1999). Multipotent stem/progenitor cells with similar properties arise from two neurogenic regions of adult human brain. *Exp Neurol* 156:333–344.

Lam CS, Marz M, Strahle U (2009). Gfap and nestin reporter lines reveal characteristics of neural progenitors in the adult zebrafish brain. *Dev Dyn* 238:475–486.

Lavenex P, Steele MA, Jacobs LF (2000). The seasonal pattern of cell proliferation and neuron number in the dentate gyrus of wild adult Eastern grey squirrels. *Eur J Neurosci* 12:643–648.

Leitner S, Voigt C, Garcia-Segura LM, Van't Hof T, Gahr M (2001). Seasonal activation and inactivation of song motor memories in wild canaries is not reflected in neuroanatomical changes of forebrain song areas. *Horm Behav* 40:160–168.

Leonard RB, Coggeshall RE, Willis WD (1978). A documentation of an age related increase in neuronal and axonal numbers in the stingray, *Dasyatis sabina*, Leseuer. *J Comp Neurol* 179:13–21.

Leuner B, Kozorovitskiy Y, Gross CG, Gould E (2007). Diminished adult neurogenesis in the marmoset brain precedes old age. *Proc Natl Acad Sci USA* 104:17169–17173.

Li B, Yamamori H, Tatebayashi Y, Shafit-Zagardo B, Tanimukai H, Chen S, Iqbal K, Grundke-Iqbal I (2008). Failure of neuronal maturation in Alzheimer disease dentate gyrus. *J Neuropathol Exp Neurol* 67:78–84.

Li XC, Jarvis ED, Alvarez-Borda B, Lim DA, Nottebohm F (2000). A relationship between behavior, neurotrophin expression, and new neuron survival. *Proc Natl Acad Sci USA* 97:8584–8589.

Lindsey BW, Tropepe V (2006). A comparative framework for understanding the biological principles of adult neurogenesis. *Prog Neurobiol* 80:281–307.

Ling C, Zuo M, Alvarez-Buylla A, Cheng MF (1997). Neurogenesis in juvenile and adult ring doves. *J Comp Neurol* 379:300–312.

Liu YW, Curtis MA, Gibbons HM, Mee EW, Bergin PS, Teoh HH, Connor B, Dragunow M, Faull RL (2008). Doublecortin expression in the normal and epileptic adult human brain. *Eur J Neurosci* 28:2254–2265.

Liu Z, Martin LJ (2003). Olfactory bulb core is a rich source of neural progenitor and stem cells in adult rodent and human. *J Comp Neurol* 459:368–391.

Lopez-Garcia C, Molowny A, Garcia-Verdugo JM, Ferrer I (1988). Delayed postnatal neurogenesis in the cerebral cortex of lizards. *Brain Res* 471:167–174.

Lopez-Garcia C, Molowny A, Garcia-Verdugo JM, Perez-Sanchez F, Martinez-Guijarro FJ (1990). Postnatal neurogenesis in the brain of the lizard *Podarcis hispanica*. In: *The Forebrain in Non-Mammals: New Aspects of Structure and Development* (Schwerdtfeger WK, Germroth P, eds.), pp. 103–117. Berlin: Springer.

Louissaint A, Jr., Rao S, Leventhal C, Goldman SA (2002). Coordinated interaction of neurogenesis and angiogenesis in the adult songbird brain. *Neuron* 34:945–960.

Lu L, Tonchev AB, Kaplamadzhiev DB, Boneva NB, Mori Y, Sahara S, Ma D, Nakaya MA, Kikuchi M, Yamashima T (2008). Expression of matrix metalloproteinases in the neurogenic niche of the adult monkey hippocampus after ischemia. *Hippocampus* 18:1074–1084.

Luzzati F, Peretto P, Aimar P, Ponti G, Fasolo A, Bonfanti L (2003). Glia-independent chains of neuroblasts through the subcortical parenchyma of the adult rabbit brain. *Proc Natl Acad Sci USA* 100:13036–13041.

Luzzati F, De Marchis S, Fasolo A, Peretto P (2006). Neurogenesis in the caudate nucleus of the adult rabbit. *J Neurosci* 26:609–621.

Luzzati F, De Marchis S, Fasolo A, Peretto P (2007). Adult neurogenesis and local neuronal progenitors in the striatum. *Neuro-Degenerative Diseases* 4:322–327.

Macklis JD (1993). Transplanted neocortical neurons migrate selectively into regions of neuronal degeneration produced by chromophore-targeted laser photolysis. *J Neurosci* 13:3848–3863.

Madison RD, Macklis JD (1993). Noninvasively induced degeneration of neocortical pyramidal neurons in vivo: Selective targeting by laser activation of retrogradely transported photolytic chromophore. *Exp Neurol* 121:153–159.

Malaterre J, Strambi C, Aouane A, Strambi A, Rougon G, Cayre M (2003). Effect of hormones and growth factors on the proliferation of adult cricket neural progenitor cells in vitro. *J Neurobiol* 56:387–397.

Marchioro M, Nunes JM, Ramalho AM, Molowny A, Perez-Martinez E, Ponsoda X, Lopez-Garcia C (2005). Postnatal neurogenesis in the medial cortex of the tropical lizard *Tropidurus hispidus*. *Neuroscience* 134:407–413.

Margotta V, Morelli A, Caronti B (2005). Expression of PCNA positivity in the brain of normal adult heterothermic vertebrates: Further observations. *Ital J Anat Embryol* 110:59–74.

Mashaly A, Winkler M, Frambach I, Gras H, Schurmann FW (2008). Sprouting interneurons in mushroom bodies of adult cricket brains. *J Comp Neurol* 508:153–174.

Mattiesen WR, Tauber SC, Gerber J, Bunkowski S, Bruck W, Nau R (2009). Increased neurogenesis after hypoxic-ischemic encephalopathy in humans is age related. *Acta Neuropathol* 117:525–534.

McFarland WL, Morgane PJ, Jacobs MS (1969). Ventricular system of the brain of the dolphin, *Tursiops truncatus*, with comparative anatomical observations and relations to brain specializations. *J Comp Neurol* 135:275–368.

Miljkovic-Licina M, Chera S, Ghila L, Galliot B (2007). Head regeneration in wild-type hydra requires de novo neurogenesis. *Development* 134:1191–1201.

Molina GA, Watkins SC, Tsang M (2007). Generation of FGF reporter transgenic zebrafish and their utility in chemical screens. *BMC Dev Biol* 7:62.

Morris AC, Scholz T, Fadool JM (2008a). Rod progenitor cells in the mature zebrafish retina. *Adv Exp Med Biol* 613:361–368.

Morris AC, Scholz TL, Brockerhoff SE, Fadool JM (2008b). Genetic dissection reveals two separate pathways for rod and cone regeneration in the teleost retina. *Dev Neurobiol* 68:605–619.

Murrell W, Bushell GR, Livesey J, McGrath J, MacDonald KP, Bates PR, Mackay-Sim A (1996). Neurogenesis in adult human. *Neuroreport* 7:1189–1194.

Ngwenya LB, Peters A, Rosene DL (2006). Maturational sequence of newly generated neurons in the dentate gyrus of the young adult rhesus monkey. *J Comp Neurol* 498:204–216.

Ngwenya LB, Rosene DL, Peters A (2008). An ultrastructural characterization of the newly generated cells in the adult monkey dentate gyrus. *Hippocampus* 18:210–220.

Nordlander RH, Edwards JS (1968). Morphology of the larval and adult brains of the monarch butterfly, *Danaus plexippus plexippus*, L. *J Morphol* 126:67–94.

Nottebohm F, Alvarez-Buylla A (1993). Neurogenesis and neuronal replacement in adult birds. In: *Neuronal Cell Death and Repair* (Cuello AC, ed.), pp. 227–236. Amsterdam: Elsevier.

Nottebohm F, O'Loughlin B, Gould K, Yohay K, Alvarez-Buylla A (1994). The life span of new neurons in a song control nucleus of the adult canary brain depends on time of year when these cells are born. *Proc Natl Acad Sci USA* 91:7849–7853.

Odelberg SJ (2004). Unraveling the molecular basis for regenerative cellular plasticity. *PLoS Biol* 2:E232.

Oehlmann VD, Berger S, Sterner C, Korsching SI (2004). Zebrafish beta tubulin 1 expression is limited to the nervous system throughout development, and in the adult brain is restricted to a subset of proliferative regions. *Gene Expr Patterns* 4:191–198.

Omura Y (2007). The distribution of proliferating cell nuclear antigen-immunoreactive cells in the pineal organ of the rainbow trout *Oncorhynchus mykiss. Arch Histol Cytol* 70:225–234.

Ormerod BK, Galea LA (2001). Reproductive status influences cell proliferation and cell survival in the dentate gyrus of adult female meadow voles: A possible regulatory role for estradiol. *Neuroscience* 102:369–379.

Ormerod BK, Lee TT, Galea LA (2003). Estradiol initially enhances but subsequently suppresses (via adrenal steroids) granule cell proliferation in the dentate gyrus of adult female rats. *J Neurobiol* 55:247–260.

Ormerod BK, Lee TT, Galea LA (2004). Estradiol enhances neurogenesis in the dentate gyri of adult male meadow voles by increasing the survival of young granule neurons. *Neuroscience* 128:645–654.

Ostenfeld T, Joly E, Tai YT, Peters A, Caldwell M, Jauniaux E, Svendsen CN (2002). Regional specification of rodent and human neurospheres. *Brain Res Dev Brain Res* 134:43–55.

Pagano SF, Impagnatiello F, Girelli M, Cova L, Grioni E, Onofri M, Cavallaro M, Etteri S, Vitello F, Giombini S, Solero CL, Parati EA (2000). Isolation and characterization of neural stem cells from the adult human olfactory bulb. *Stem Cells* 18:295–300.

Palmer TD, Schwartz PH, Taupin P, Kaspar B, Stein SA, Gage FH (2001). Cell culture. Progenitor cells from human brain after death. *Nature* 411:42–43.

Patel SN, Clayton NS, Krebs JR (1997). Spatial learning induces neurogenesis in the avian brain. *Behav Brain Res* 89:115–128.

Pencea V, Bingaman KD, Freedman LJ, Luskin MB (2001). Neurogenesis in the subventricular zone and rostral migratory stream of the neonatal and adult primate forebrain. *Exp Neurol* 172:1–16.

Perez-Canellas MM, Garcia-Verdugo JM (1996). Adult neurogenesis in the telencephalon of a lizard: A [3H] thymidine autoradiographic and bromodeoxyuridine immunocytochemical study. *Brain Res Dev Brain Res* 93:49–61.

Perez-Canellas MM, Font E, Garcia-Verdugo JM (1997). Postnatal neurogenesis in the telencephalon of turtles: Evidence for nonradial migration of new neurons from distant proliferative ventricular zones to the olfactory bulbs. *Brain Res Dev Brain Res* 101:125–137.

Perez-Martin M, Cifuentes M, Grondona JM, Bermudez-Silva FJ, Arrabal PM, Perez-Figares JM, Jimenez AJ, Garcia-Segura LM, Fernandez-Llebrez P (2003). Neurogenesis in explants from the walls of the lateral ventricle of adult bovine brain: Role of endogenous IGF-1 as a survival factor. *Eur J Neurosci* 17:205–211.

Pincus DW, Keyoung HM, Harrison-Restelli C, Goodman RR, Fraser RA, Edgar M, Sakakibara S, Okano H, Nedergaard M, Goldman SA (1998). Fibroblast growth factor-2/brain-derived neurotrophic factor-associated maturation of new neurons generated from adult human subependymal cells. *Ann Neurol* 43:576–585.

Polenov AL, Chetverukhin VK (1993). Ultrastructural radioautographic analysis of neurogenesis in the hypothalamus of the adult frog, *Rana temporaria*, with special reference to physiological regeneration of the preoptic nucleus. II. Types of neuronal cells produced. *Cell Tissue Res* 271:351–362.

Ponti G, Aimar P, Bonfanti L (2006). Cellular composition and cytoarchitecture of the rabbit subventricular zone and its extensions in the forebrain. *J Comp Neurol* 498:491–507.

Ponti G, Peretto P, Bonfanti L (2008). Genesis of neuronal and glial progenitors in the cerebellar cortex of peripuberal and adult rabbits. *PLoS ONE* 3:e2366.

Pravosudov VV, Omanska A (2005). Dominance-related changes in spatial memory are associated with changes in hippocampal cell proliferation rates in mountain chickadees. *J Neurobiol* 62:31–41.

Pytte CL, Gerson M, Miller J, Kirn JR (2007). Increasing stereotypy in adult zebra finch song correlates with a declining rate of adult neurogenesis. *Dev Neurobiol* 67:1699–1720.

Rakic P (1985). Limits of neurogenesis in primates. *Science* 227:1054–1056.

Rakic P (2002). Neurogenesis in adult primate neocortex: An evaluation of the evidence. *Nat Rev Neurosci* 3:65–71.

Rakic P (2004). Neuroscience: Immigration denied. *Nature* 427:685–686.

Rasika S, Nottebohm F, Alvarez-Buylla A (1994). Testosterone increases the recruitment and/or survival of new high vocal center neurons in adult female canaries. *Proc Natl Acad Sci USA* 91:7854–7858.

Rasika S, Alvarez-Buylla A, Nottebohm F (1999). BDNF mediates the effects of testosterone on the survival of new neurons in an adult brain. *Neuron* 22:53–62.

Raucci F, Di Fiore MM, Pinelli C, D'Aniello B, Luongo L, Polese G, Rastogi RK (2006). Proliferative activity in the frog brain: A PCNA-immunohistochemistry analysis. *J Chem Neuroanat* 32:127–142.

Raymond PA, Easter SS, Jr. (1983). Postembryonic growth of the optic tectum in goldfish. I. Location of germinal cells and numbers of neurons produced. *J Neurosci* 3:1077–1091.

Raymond PA, Easter SS, Jr., Burnham JA, Powers MK (1983). Postembryonic growth of the optic tectum in goldfish. II. Modulation of cell proliferation by retinal fiber input. *J Neurosci* 3:1092–1099.

Reynolds ML, Cavanagh ME, Dziegielewska KM, Hinds LA, Saunders NR, Tyndale-Biscoe CH (1985). Postnatal development of the telencephalon of the tammar wallaby (*Macropus eugenii*). An accessible model of neocortical differentiation. *Anat Embryol* (Berl) 173:81–94.

Richter W, Kranz D (1981). [Autoradiographic investigations on postnatal proliferative activity of the telencephalic and diencephalic matrix-zones in the axolotl (*Ambystoma mexicanum*), with special references to the olfactory organ (author's trans.)]. *Z Mikrosk Anat Forsch* 95:883–904.

Ricklefs RE (2006). Embryo development and ageing in birds and mammals. *Proceedings* 273:2077–2082.

Rodriguez-Perez LM, Perez-Martin M, Jimenez AJ, Fernandez-Llebrez P (2003). Immunocytochemical characterisation of the wall of the bovine lateral ventricle. *Cell Tissue Res* 314:325–335.

Romero-Aleman MM, Monzon-Mayor M, Yanes C, Lang D (2004). Radial glial cells, proliferating periventricular cells, and microglia might contribute to successful structural repair in the cerebral cortex of the lizard *Gallotia galloti*. *Exp Neurol* 188:74–85.

Rousselot P, Heintz N, Nottebohm F (1997). Expression of brain lipid binding protein in the brain of the adult canary and its implications for adult neurogenesis. *J Comp Neurol* 385:415–426.

Roy EP, 3rd, Frost JL, Schochet SS, Jr. (1987). Persistent olfactory bulb ventricle. *Clin Neuropathol* 6:86–87.

Roy NS, Wang S, Harrison-Restelli C, Benraiss A, Fraser RA, Gravel M, Braun PE, Goldman SA (1999). Identification, isolation, and promoter-defined separation of mitotic oligodendrocyte progenitor cells from the adult human subcortical white matter. *J Neurosci* 19:9986–9995.

Roy NS, Benraiss A, Wang S, Fraser RA, Goodman R, Couldwell WT, Nedergaard M, Kawaguchi A, Okano H, Goldman SA (2000a). Promoter-targeted selection and isolation of neural progenitor cells from the adult human ventricular zone. *J Neurosci* Res 59:321–331.

Roy NS, Wang S, Jiang L, Kang J, Benraiss A, Harrison-Restelli C, Fraser RA, Couldwell WT, Kawaguchi A, Okano H, Nedergaard M, Goldman SA (2000b). In vitro neurogenesis by progenitor cells isolated from the adult human hippocampus. *Nat Med* 6:271–277.

Russo RE, Fernandez A, Reali C, Radmilovich M, Trujillo-Cenoz O (2004). Functional and molecular clues reveal precursor-like cells and immature neurones in the turtle spinal cord. *J Physiol* 560:831–838.

Russo RE, Reali C, Radmilovich M, Fernandez A, Trujillo-Cenoz O (2008). Connexin 43 delimits functional domains of neurogenic precursors in the spinal cord. *J Neurosci* 28:3298–3309.

Sakaguchi M, Mizusina A, Kobayakawa Y (1996). Structure, development, and maintenance of the nerve net of the body column in Hydra. *J Comp Neurol* 373:41–54.

Sanai N, Tramontin AD, Quinones-Hinojosa A, Barbaro NM, Gupta N, Kunwar S, Lawton MT, McDermott MW, Parsa AT, Manuel-Garcia Verdugo J, Berger MS, Alvarez-Buylla A (2004). Unique astrocyte ribbon in adult human brain contains neural stem cells but lacks chain migration. *Nature* 427:740–744.

Sanai N, Berger MS, Garcia-Verdugo JM, Alvarez-Buylla A (2007). Comment on "Human neuroblasts migrate to the olfactory bulb via a lateral ventricular extension." *Science* 318:393; author reply, 393.

Sandeman DC, Benton JL, Beltz BS (2009). An identified serotonergic neuron regulates adult neurogenesis in the crustacean brain. *Dev Neurobiol* 69:530–545.

Sandeman R, Sandeman D (2000). "Impoverished" and "enriched" living conditions influence the proliferation and survival of neurons in crayfish brain. *J Neurobiol* 45:215–226.

Scharff C, Nottebohm F (1991). A comparative study of the behavioral deficits following lesions of various parts of the zebra finch song system: Implications for vocal learning. *J Neurosci* 11:2896–2913.

Scharff C, Kirn JR, Grossman M, Macklis JD, Nottebohm F (2000). Targeted neuronal death affects neuronal replacement and vocal behavior in adult songbirds. *Neuron* 25:481–492.

Schmidt M (1997). Continuous neurogenesis in the olfactory brain of adult shore crabs, *Carcinus maenas*. *Brain Res* 762:131–143.

Schmidt M (2001). Neuronal differentiation and long-term survival of newly generated cells in the olfactory midbrain of the adult spiny lobster, *Panulirus argus*. *J Neurobiol* 48:181–203.

Schmidt M (2007). The olfactory pathway of decapod crustaceans—an invertebrate model for life-long neurogenesis. *Chem Senses* 32:365–384.

Schneider JF, Floemer F (2009). Maturation of the olfactory bulbs: MR imaging findings. *AJNR* 30: 1149–1152.

Scotto-Lomassese S, Strambi C, Strambi A, Charpin P, Augier R, Aouane A, Cayre M (2000). Influence of environmental stimulation on neurogenesis in the adult insect brain. *J Neurobiol* 45:162–171.

Scotto-Lomassese S, Strambi C, Aouane A, Strambi A, Cayre M (2002). Sensory inputs stimulate progenitor cell proliferation in an adult insect brain. *Curr Biol* 12:1001–1005.

Scotto-Lomassese S, Strambi C, Strambi A, Aouane A, Augier R, Rougon G, Cayre M (2003). Suppression of adult neurogenesis impairs olfactory learning and memory in an adult insect. *J Neurosci* 23:9289–9296.

Simmons AM, Horowitz SS, Brown RA (2008). Cell proliferation in the forebrain and midbrain of the adult bullfrog, *Rana catesbeiana*. *Brain Behav Evol* 71:41–53.

Siwak-Tapp CT, Head E, Muggenburg BA, Milgram NW, Cotman CW (2007). Neurogenesis decreases with age in the canine hippocampus and correlates with cognitive function. *Neurobiol Learn Mem* 88:249–259.

Smith MT, Pencea V, Wang Z, Luskin MB, Insel TR (2001). Increased number of BrdU-labeled neurons in the rostral migratory stream of the estrous prairie vole. *Horm Behav* 39:11–21.

Smitka M, Abolmaali N, Witt M, Gerber JC, Neuhuber W, Buschhueter D, Puschmann S, Hummel T (2009). Olfactory bulb ventricles as a frequent finding in magnetic resonance imaging studies of the olfactory system. *Neuroscience* 162:482–485.

Snyder JS, Choe JS, Clifford MA, Jeurling SI, Hurley P, Brown A, Kamhi JF, Cameron HA (2009). Adult-born hippocampal neurons are more numerous, faster maturing, and more involved in behavior in rats than in mice. *J Neurosci* 29:14484–14495.

Soutschek J, Zupanc GK (1995). Apoptosis as a regulator of cell proliferation in the central posterior/prepacemaker nucleus of adult gymnotiform fish, *Apteronotus leptorhynchus*. *Neurosci Lett* 202:133–136.

Soutschek J, Zupanc GK (1996). Apoptosis in the cerebellum of adult teleost fish, *Apteronotus leptorhynchus*. *Brain Res Dev Brain Res* 97:279–286.

Spalding KL, Buchholz BA, Bergman LE, Druid H, Frisen J (2005a). Forensics: Age written in teeth by nuclear tests. *Nature* 437:333–334.

Spalding KL, Bhardwaj RD, Buchholz BA, Druid H, Frisen J (2005b). Retrospective birth dating of cells in humans. *Cell* 122:133–143.

Sullivan JM, Beltz BS (2005a). Newborn cells in the adult crayfish brain differentiate into distinct neuronal types. *J Neurobiol* 65:157–170.

Sullivan JM, Beltz BS (2005b). Adult neurogenesis in the central olfactory pathway in the absence of receptor neuron turnover in *Libinia emarginata*. *Eur J Neurosci* 22:2397–2402.

Sullivan JM, Sandeman DC, Benton JL, Beltz BS (2007a). Adult neurogenesis and cell cycle regulation in the crustacean olfactory pathway: From glial precursors to differentiated neurons. *J Mol Histol* 38:527–542.

Sullivan JM, Benton JL, Sandeman DC, Beltz BS (2007b). Adult neurogenesis: A common strategy across diverse species. *J Comp Neurol* 500:574–584.

Svendsen CN, ter Borg MG, Armstrong RJ, Rosser AE, Chandran S, Ostenfeld T, Caldwell MA (1998). A new method for the rapid and long term growth of human neural precursor cells. *J Neurosci Methods* 85:141–152.

Teuchert-Noodt G, Dawirs RR, Hildebrandt K (2000). Adult treatment with methamphetamine transiently decreases dentate granule cell proliferation in the gerbil hippocampus. *J Neural Transm* 107:133–143.

Thompson CK, Brenowitz EA (2008). Caspase inhibitor infusion protects an avian song control circuit from seasonal-like neurodegeneration. *J Neurosci* 28:7130–7136.

Thompson CK, Brenowitz EA (2009). Neurogenesis in an adult avian song nucleus is reduced by decreasing caspase-mediated apoptosis. *J Neurosci* 29:4586–4591.

Thompson CK, Bentley GE, Brenowitz EA (2007). Rapid seasonal-like regression of the adult avian song control system. *Proc Natl Acad Sci USA* 104:15520–15525.

Thummel R, Burket CT, Hyde DR (2006). Two different transgenes to study gene silencing and re-expression during zebrafish caudal fin and retinal regeneration. *The Scientific World Journal* 6 Suppl 1:65–81.

Thummel R, Kassen SC, Enright JM, Nelson CM, Montgomery JE, Hyde DR (2008). Characterization of Muller glia and neuronal progenitors during adult zebrafish retinal regeneration. *Exp Eye Res* 87:433–444.

Tighilet B, Brezun JM, Sylvie GD, Gaubert C, Lacour M (2007). New neurons in the vestibular nuclei complex after unilateral vestibular neurectomy in the adult cat. *Eur J Neurosci* 25:47–58.

Tonchev AB, Yamashima T, Zhao L, Okano H (2003a). Differential proliferative response in the postischemic hippocampus, temporal cortex, and olfactory bulb of young adult macaque monkeys. *Glia* 42:209–224.

Tonchev AB, Yamashima T, Zhao L, Okano HJ, Okano H (2003b). Proliferation of neural and neuronal progenitors after global brain ischemia in young adult macaque monkeys. *Mol Cell Neurosci* 23:292–301.

Topp S, Stigloher C, Komisarczuk AZ, Adolf B, Becker TS, Bally-Cuif L (2008). Fgf signaling in the zebrafish adult brain: Association of Fgf activity with ventricular zones but not cell proliferation. *J Comp Neurol* 510:422–439.

Treves A, Tashiro A, Witter ME, Moser EI (2008). What is the mammalian dentate gyrus good for? *Neuroscience* 154:1155–1172.

Vates GE, Nottebohm F (1995). Feedback circuitry within a song-learning pathway. *Proc Natl Acad Sci USA* 92:5139–5143.

Vessal M, Aycock A, Garton MT, Ciferri M, Darian-Smith C (2007). Adult neurogenesis in primate and rodent spinal cord: Comparing a cervical dorsal rhizotomy with a dorsal column transection. *Eur J Neurosci* 26:2777–2794.

Vidal Pizarro I, Swain GP, Selzer ME (2004). Cell proliferation in the lamprey central nervous system. *J Comp Neurol* 469:298–310.

Vihtelic TS, Hyde DR (2000). Light-induced rod and cone cell death and regeneration in the adult albino zebrafish (*Danio rerio*) retina. *J Neurobiol* 44:289–307.

Volman SF, Grubb TC, Jr., Schuett KC (1997). Relative hippocampal volume in relation to food-storing behavior in four species of woodpeckers. *Brain Behav Evol* 49:110–120.

Wang N, Aviram R, Kirn JR (1999). Deafening alters neuron turnover within the telencephalic motor pathway for song control in adult zebra finches. *J Neurosci* 19:10554–10561.

Wang RT, Halpern M (1988). Neurogenesis in the vomeronasal epithelium of adult garter snakes: 3. Use of H3-thymidine autoradiography to trace the genesis and migration of bipolar neurons. *Am J Anat* 183: 178–185.

Weissman T, Noctor SC, Clinton BK, Honig LS, Kriegstein AR (2003). Neurogenic radial glial cells in reptile, rodent and human: From mitosis to migration. *Cereb Cortex* 13:550–559.

Wissman AM, Brenowitz EA (2009). The role of neurotrophins in the seasonal-like growth of the avian song control system. *J Neurosci* 29:6461–6471.

Wu DM, Schneiderman T, Burgett J, Gokhale P, Barthel L, Raymond PA (2001). Cones regenerate from retinal stem cells sequestered in the inner nuclear layer of adult goldfish retina. *Invest Ophthalmol Vis Sci* 42:2115–2124.

Yue F, Chen B, Wu D, Dong K, Zeng SE, Zhang Y (2006). Biological properties of neural progenitor cells isolated from the hippocampus of adult cynomolgus monkeys. *Chin Med J* (Engl) 119:110–116.

Zhang Y, Allodi S, Sandeman DC, Beltz BS (2009). Adult neurogenesis in the crayfish brain: Proliferation, migration, and possible origin of precursor cells. *Dev Neurobiol* 69:415–436.

Zupanc GK (1999). Neurogenesis, cell death and regeneration in the adult gymnotiform brain. *J Exp Biol* 202 (Pt 10):1435–1446.

Zupanc GK, Zupanc MM (1992). Birth and migration of neurons in the central posterior/prepacemaker nucleus during adulthood in weakly electric knifefish (*Eigenmannia sp.*). *Proc Natl Acad Sci USA* 89:9539–9543.

Zupanc GK, Hinsch K, Gage FH (2005). Proliferation, migration, neuronal differentiation, and long-term survival of new cells in the adult zebrafish brain. *J Comp Neurol* 488:290–319.

Zupanc GK, Wellbrock UM, Sirbulescu RF, Rajendran RS (2009). Generation, long-term persistence, and neuronal differentiation of cells with nuclear aberrations in the adult zebrafish brain. *Neuroscience* 159: 1338–1348.

12

Medicine

Medical relevance of adult
neurogenesis lies in a new
perspective on brain plasticity
and the pathogenesis and course
of several neurological and
psychiatric disorders

RESEARCH ON ADULT NEUROGENESIS and neural stem cells is often justified to funding agencies, policy makers, and taxpayers by its potential benefit to medicine. The general tenor of most of these justifications is the aim of replacing lost neurons in cases of neuronal loss. In this chapter, we will review what is known about adult neurogenesis in the context of pathology and therapy, and we will see that, despite the huge medical implications of adult neurogenesis, in many circumstances the concrete goals of cell replacement are at odds with the realities of adult neurogenesis. Adult neurogenesis might rather offer profoundly new perspectives on the pathogenesis, course, and possible therapy of a number of neuropsychiatric disorders, rather than providing an easy fix to these problems. Adult neurogenesis is no "magic bullet"; its applications are not limitless and generic but are specific to those functions adult neurogenesis relates to in the first place. The potential might occasionally be stretched (as in the case of new striatal neurons after ischemia), and the caveat thus does not categorically exclude medical relevance of the examples of "regenerative" or "targeted" neurogenesis discussed in Chapter 8, p. 298. But as impressive and important as these example are, their obvious limitations still warn us that adult neurogenesis is not the straightforward way to turn the brain into a regenerative organ. This judgment is less pessimistic than it might appear, however. The medical implications of adult neurogenesis will just be very different than most people (and the introductory statements to numerous papers) imagine.

Adult neurogenesis is no
primary means of endogenous
cell replacement

There is presently no reasonable evidence that adult mammalian neurogenesis is a mechanism of endogenous cell replacement after damage. Adult neurogenesis is not primarily regenerative and reactive, but contributes to brain plasticity in an activity-dependent way. Adult neurogenesis represents a special case of structural adaptation, and the situations evoking this plasticity include pathology. But all realistic medical implications must in some way relate to the function of the new neurons that can be generated. Adult neurogenesis does not, on demand, produce types of neurons other than those that lie within the developmental potential of the neurogenic precursor cells. It is certainly possible to envision manipulations that extend this potential, especially within the context of induced pluripotency (see Chap. 3, p. 82). The case is thus somewhat different for adult neurogenesis from exogenously applied neural stem cells or their progeny. Here a wider development spectrum is conceivable, and the urgent question becomes whether the microenvironment into which the cells are transplanted will support the survival and functionality of the new neurons. After much enthusiasm in the 1980s, however, it has turned considerably quieter around neurotransplantation. Many expectations and promises have been reduced.

Assessment of medical relevance of adult neurogenesis is limited by the lack of knowledge about new neurons in the adult human brain

In the previous chapter we have also seen that our present knowledge about adult neurogenesis in humans is still limited. Neural precursor cells can, however, be isolated from the human SVZ and subgranular zone, and adult hippocampal neurogenesis has been directly shown to occur in humans. These findings allow subdued enthusiasm and call for much further research, as the presently available data certainly do raise more questions about the applicability of rodent data to the human situation than they answer. With respect to regulation and function, two key aspects of adult neurogenesis, we know next to nothing about the specific situation in humans. This lack of knowledge does not disqualify the hopes for human therapy based on results obtained in rodents (or occasionally primate) species, but it cautions us to consider the animal data as what they are—models. It remains to be established how good these models are. The fact that neural stem cells and adult neurogenesis generally exist in the adult human brain justifies the use of the models. On the other hand, the data obtained in rodents have to be carefully evaluated for their relevance to the human situation if they are to be used in clinical applications.

Adult neurogenesis does not render the brain a "regenerative organ"

This caveat is important because, after what has been perceived as the dissolution of the "no-new-neurons dogma" (see Chap. 1 on the problems with this term), the common perception sometimes seems to have swung to the other extreme. The few signs of adult human neurogenesis, together with the much more suggestive "therapeutic" findings in rodents, have led to the popular misconception that adult neurogenesis per se makes the impossible possible. Quite surprisingly, neural stem cells have sometimes been greeted as the last missing link in an otherwise apparently coherent picture of what would constitute successful neuroregeneration. However, the adult brain regenerates as poorly after the discovery of adult neurogenesis as it did before. But it does so *despite* the presence of neural stem cells and not, as was previously thought, because of their absence.

Neural Precursor Cells and the Therapy of Neuro-Psychiatric Disorders

Cell replacement therapies aim for adult neurogenesis from exogenous neural precursor cells or neurons

Therapeutic options related to adult neurogenesis need to be seen in the greater context of cell-based therapies of neuropsychiatric disorders. Figure 12–1 gives an overview. A classic concept states that precursor cells could contribute to neuroregenerative therapies in the following two ways (Bjorklund and Lindvall, 2000):

- Replace lost cells
- Provide a pro-regenerative microenvironment for other cells.

This concept includes therapeutic strategies based on neurotransplantation. In that context, neural stem cell biology is seen primarily as a tool for generating transplantable cells. Most strategies rely on predifferentiating donor cells in vitro before implantation. Nevertheless, the cellular environment in the host brain will influence graft survival and function. Insights from research on adult neurogenesis in vivo can help us understand the neurogenic requirements in the host brain for promoting graft function. We will cover examples of neurogenesis from implanted precursor cells and their progeny below (p. 520).

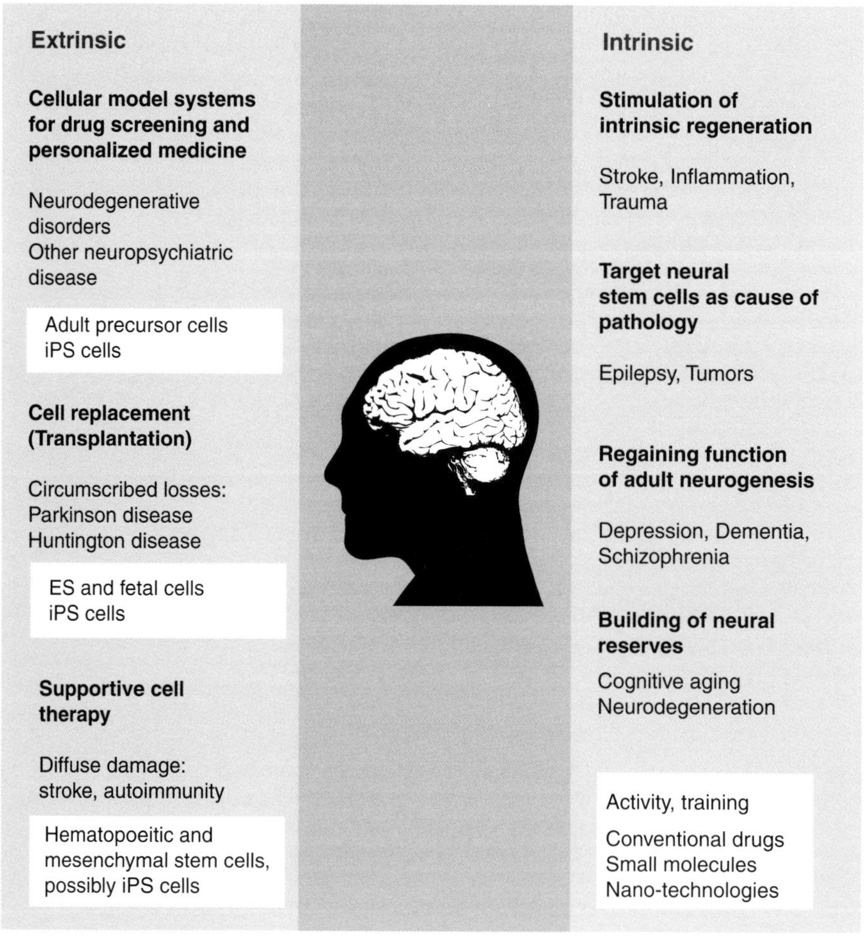

Extrinsic

Cellular model systems for drug screening and personalized medicine

Neurodegenerative disorders
Other neuropsychiatric disease

Adult precursor cells
iPS cells

Cell replacement (Transplantation)

Circumscribed losses:
Parkinson disease
Huntington disease

ES and fetal cells
iPS cells

Supportive cell therapy

Diffuse damage:
stroke, autoimmunity

Hematopoeitic and mesenchymal stem cells, possibly iPS cells

Intrinsic

Stimulation of intrinsic regeneration

Stroke, Inflammation, Trauma

Target neural stem cells as cause of pathology

Epilepsy, Tumors

Regaining function of adult neurogenesis

Depression, Dementia, Schizophrenia

Building of neural reserves

Cognitive aging
Neurodegeneration

Activity, training

Conventional drugs
Small molecules
Nano-technologies

FIGURE 12–1 Therapeutic options based on neural stem cells and adult neurogenesis.

Neurotransplantation for Lost Neuronal or Glial Populations

Cell replacement therapies in the brain are in principle possible, but actual success has been limited

Precursor cells have been successfully transplanted into a number of animal models of neurological disease, including traumatic brain injury (Riess et al., 2002), spinal cord injury (McDonald et al., 1999; Ogawa et al., 2002), and Parkinson disease (PD) (Bjorklund et al., 2002). In these studies, integration of the graft into the host tissue was shown, functionality of the grafted cells could be visualized, and improvements in functional tests could be correlated with the engraftment.

Together, these findings were the first to show that precursor cell–based cell therapy is generally possible in the brain. In humans there is much experience with the transplantation of fetal mesencaphalic tissue into patients with PD. Some of these patients showed extraordinary improvements and have not required any PD medication for many years (Lindvall et al., 1987; Wenning et al., 1997). Double-blind studies of cell therapy in PD have indicated, however, that not all patients benefit from transplantation, and that side effects such as dyskinesias can occur—that is, a hypermotility typically seen after long treatment with dopamine (Freed et al., 2001; Hagell et al., 2002; Olanow et al., 2003). Clinical trials are also under way for Huntington disease, which shares with PD the trait that a relatively circumscribed population of neurons is primarily affected, in this case medium spiny interneurons in the striatum (Bachoud-Levi et al., 2000; Keene et al., 2007). There are many problems associated with cell therapy in this case, too. Not only did the transplants succumb to the original disease (Cicchetti et al., 2009), in another case the unfortunate patient developed a mass lesion (Keene et al., 2009).

The use of stem cells as a source for transplantable cells is highly desirable and thus at the focus of research in many laboratories worldwide. Stem cells would alleviate the logistic and ethical problems associated with the use of fetal cells that have to be obtained from aborted human fetuses.

Precursor Cell Therapy for Diffuse Cell Losses

Neural transplantation for diffuse brain pathology is problematic

Implantation of wild-type embryonic stem cells into genetic models of disease—for example, an inherited demyelinating disorder—demonstrated that, particularly if introduced during development, grafted stem cells can widely distribute and lead to almost complete structural reconstitution (Brustle et al., 1999). Because multiple sclerosis has a strong inflammatory component and is not a developmental disorder, though, cell-based treatment strategies face considerable technical difficulties in this instance. Neurospheres injected into mice with chronic EAE, however, caused remyelination and even functional recovery (Pluchino et al., 2003), albeit through a completely different mechanism (see below, p. 558).

One theoretical solution to the problem of directed migration to the sites of pathology is the diffuse application of exogenous precursor cells. The simplest strategy is to distribute precursor cells through the blood stream, although in this case the cells have to cross the blood–brain barrier. Only a few reporter gene–labeled bone marrow or blood cells have been detected in the intact brain after intravenous infusion (Priller et al., 2001). In stroke, however, or at other sites of pathology, the blood–brain barrier is open, allowing the transition of blood cells into the brain parenchyma. In such cases, most of the cells found after infusion of marked bone marrow cells are microglia, which originate from blood monocytes.

Cell-Based Therapies Beyond Cell Replacement

Implanted cells might also provide trophic support for endogenous cells at risk

Infusion of bone marrow cells into rodent models of ischemia were nevertheless associated with some therapeutic benefit (Li et al., 2001; Li et al., 2002a; Kurozumi et al., 2004). Infusion of bone marrow cells into patients with myocardial infarction resulted in a small clinical improvement, also in larger clinical trials (Dill et al., 2009; Herbots et al., 2009), but overall the data nevertheless remained ambiguous. One study found that the beneficial effect occurred independently of whether the peripherally administered precursor cells entered the brain (Borlongan et al., 2004). The analogous application of stem cells for stroke therapy has also been tested in clinical trials and yielded similarly ambiguous results (Kawada et al., 2006; Suarez-Monteagudo et al., 2009; Zawadzka et al., 2009). In animal research, a low number of new cells with neuronal properties were initially taken as signs of transdifferentiation of bone marrow cells but are today largely explained by cell fusion (see Chap. 3, p. 80). There is no conclusive experimental evidence showing that bone marrow–derived cells would incorporate into an ischemic area and thereby lead to structural restoration. Rather, the infused cells seem to exert a trophic effect or induce angiogenesis. As precursor cells they escape the immune surveillance, and this might be the reason why "stem cells" provide advantages in this approach (Haas et al., 2005; Parr et al., 2007; Li and Chopp, 2009).

Other types of neural cell therapy are conceivable

A related yet much more specific approach has been taken by implanting un-modified neural precursor cells into a mouse model of infantile neuronal ceroid lipofuscinosis (Batten or Spielmeyer-Vogt disease), a storage disorder due to the lack of the enzyme lysosomal palmitoyl protein thioesterase-1 (PPT1). The transplanted precursor cells who had the enzyme exerted a neuroprotective effect on the host neurons who lacked it (Tamaki et al., 2009). Again, no differentiation into neurons occurred, but that had also not been the goal.

False Promises and Hopes

Many unrealistic or outright dangerous promises are made for neural stem cell therapies

Other, often aggressively advertised "cell therapies," in contrast, are not backed by experimental evidence (and sometimes not even based on a plausible hypothesis) and entail the infusion or injection of various cell types to "cure" numerous neurological disorders, including chronically progressive diseases like amyotrophic lateral sclerosis (Lou Gehrig's disease). These experimental treatments usually take place outside professional and academic medicine, and their proponents use legal loopholes or the absence of regulations or their enforcement in some countries to do their business. The only gain from such manipulations is in the bank account of the provider. The risk for the patient can be very high. Occasionally the appeal of adult neurogenesis is exploited in these schemes to attract the victims. There is of course no reason to believe that these "business people" know secrets of adult neurogenesis that have escaped the appreciation of the entire, extremely dynamic, field of research.

Cell therapy within official medicine has seen some worrying examples of questionable strategies, too. There has been a clinical trial for neuronal replacement in stroke patients with cells derived from a human teratocarcinoma cell line. Luckily, in this attempt no patients developed tumors, but a thorough risk assessment and understanding of the biology of these cells appears difficult to achieve (Kleppner et al., 1995; Kondziolka et al., 2000). The strategy was also not therapeutically successful.

Stem cells, including neural stem cells, might also be used as vehicles for gene therapy. In these cases, no neuronal differentiation is desired, only the ability of the precursor cells to express the transgene. An ill-designed trial in gene therapy that resulted in the death of the victim has paralyzed progress in many other, thoroughly planned attempts to bring gene therapy to the clinic. This is only slowly beginning to change again.

Besides the imminent dangers in premature clinical applications, a second threat to cell-based therapy in neurology and psychiatry arises from those for whom development is still not fast enough. There are voices that categorically question the value of stem cell research in general because of the present lack of convincing clinical success and because of the general difficulties in transferring promising results from rodents to the human situation. Both those who leap into clinic and the pessimistic critics, however, miss the point that the relevance of adult neurogenesis and neural stem cell biology go far beyond cell replacement strategies.

Targeted Neurogenesis

Can cell replacement be achieved by activating endogenous potentials for neurogenesis?

It is often stated that stimulation of an endogenous potential for regeneration is the great alternative to transplanting neurons to the brain. In reality, evidence for this claim is scant, partly because our knowledge about the potential of cell replacement therapies in the adult brain is generally rather limited.

Visions of making medical use of the endogenous potential for neurogenesis tend to leave the boundaries of the neurogenic zones. In the remaining section of this chapter, we will see that adult neurogenesis can be linked with several disorders that affect the neurogenic zones, primarily the hippocampus. In this context it is straightforward to envision therapeutic strategies by trying to reverse diminished neurogenesis as a potential basis of major depression, or diminish disturbed neurogenesis in temporal lobe epilepsy. Nevertheless, therapeutic visions based on endogenous neural precursor cells and adult neurogenesis have been most prominently put forth for situations in which a pathogenic involvement of precursor cells and neurogenesis is much less obvious. In the absence of a clear link to normal precursor cell function, the proposed therapies thus become more utopian than in the case of tumor, epilepsy, dementia, or depression.

In such therapies, precursor cells are primarily considered a resource for endogenous cell replacement. However, both endogenously recruited and exogenously expanded and implanted cells face the environment of a diseased brain that in general is more or less adverse to neuronal regeneration and neurogenesis. In non-neurogenic regions, the key issue is the lack of neurogenic permissiveness, not the absence (or promoted presence) of neural precursor cells. Regenerative neurogenesis in non-neurogenic regions has to overcome this inhibition. In some cases it seems that the pathology itself induces a pattern change. In models of striatal ischemia, for example, a low number of new neurons has been found (Arvidsson et al., 2002; Parent et al., 2002b).

New replacement neurons might succumb to the same pathology as the lost original cells

It also seems that inhibitory and promoting stimuli are competing, with the promoting factors being easily overwhelmed by the inhibitory influences. The experiments by Sanjay S. Magavi, Jeffrey D. Macklis, and colleagues from Harvard University have demonstrated that targeted lesions of individual neuronal populations (as described in detail in Chap. 8), leading to the apoptotic death of single cells without reaction from the local environment, can induce neuronal replacement, probably from resident cortical precursor cells (Magavi et al., 2000; Chen et al., 2004). In some sense, these studies embody the quintessential goal of all attempts to achieve regenerative neurogenesis. The finding that death of the cell was sufficient to trigger neurogenesis is ambivalent. On the encouraging side, it shows that conducive clues can be expressed by the adult brain; on the discouraging side, it indicates that unless the lesion is so discrete that it does not trigger an adverse tissue response, the competing antineurogenic signals are stronger. That the apoptotic elimination of single neurons can indeed induce changes that locally transform the non-neurogenic adult cortex into a neurogenic zone is supported by the finding that, under the condition of targeted apoptosis, exogenous precursor cells implanted into the cortex also differentiate into neurons (Snyder et al., 1992; Sheen and Macklis, 1995; Shin et al., 2000).

Targeted restorative neurogenesis has been possible under specific experimental conditions

Targeted neurogenesis remains a fascinating strategy (Chap. 8, p. 298), and one day its therapeutic potential might indeed be tapped (Arlotta et al., 2003; Mitchell et al., 2004; Okano and Sawamoto, 2008). But the obstacles to inducing regenerative neurogenesis remain high; at present the available experimental evidence from animal models provides impressive proofs of principle but no direct access to clinical solutions. However, this perspective in the end reduces adult neurogenesis to just an endogenous variant of cell replacement. This is not the most radical orientation possible.

Plasticity-Based Therapies

Adult neurogenesis is neuronal development in the adult brain. We have emphasized before that the rise of stem cell biology has led to a fundamental paradigm shift: the insight that brain development never ends, and precursor cells with their accessible genetic potential persist in the adult brain and make lifelong contributions to brain plasticity. This is the potential that can be tapped.

The neurogenic zones are brain regions of exceptional structural plasticity

The neurogenic regions are privileged regions in that they have the cells and the permissive microenvironment that makes such plasticity possible. Therefore, plasticity of whole cells (in addition to the other types of plasticity affecting synapses and neurites) will play a larger role in these areas than in the non-neurogenic regions, where this plasticity is limited to glial responses. Considering thus the facts that brain development never ends, and precursor cells are carriers of cellular brain plasticity, the following two related aspects can be added to the classical perspective. Neural precursor cells might also be of clinical and ultimately therapeutic relevance because:

- a failure of normal precursor cell activity might underlie or contribute to brain pathology, and
- prevention and therapy strategies might target precursor cells to maintain or reconstitute normal brain plasticity.

These two aspects are two sides of the same coin, but both together are considerably different from a direct cell-replacement strategy. Yet another type of medical relevance arises from the use of neural precursor cells as model systems for neuropsychiatric disease (see Chap. 3, p. 83).

Adult Neurogenesis in Brain Pathology

Adult neurogenesis might be involved in the pathogenesis and course of neuropsychiatric disease

Acute pathology tends to stimulate cell proliferation in the neurogenic zones

When considering adult neurogenesis in the context of brain pathology, the two mentioned aspects blend. Failing adult neurogenesis might contribute to the disease, and the regulatory response at the level of adult neurogenesis might in turn influence the course of the disease. The disease will act upon adult neurogenesis and be shaped by the plastic response by the new neurons. This is "plasticity" in the context of pathology.

Pathologies in general tend to have a strong, acute effect on cell proliferation in both the SGZ and the SVZ. As in the case of the deafferentation of the olfactory bulb, an initial increase in cell proliferation is usually followed by a lasting decrease (Graziadei and Monti Graziadei, 1980; Corotto et al., 1994; Jankovski et al., 1998; Mandairon et al., 2003). Thus, acute effects have to be distinguished from chronic consequences of pathology. The principle seems to be the following: acute damage causes a (transient) effect on cell proliferation; chronic damage results in a lasting decrease in adult neurogenesis. There might also be dose effects such as in the case of corticosterone signaling. Moderately increased levels of corticosterone are proneurogenic (at least in vitro), but high doses disrupt neurogenesis (Wolf et al., 2009a). Neuronal excitation stimulates neurogenesis; over-excitation in seizures strongly induces proliferation; but continued seizures in epilepsy might reduce neurogenesis.

Consequently, pathological regulation needs to be seen in the context of physiological regulation, and there is no simple and constant one-to-one relationship between pathology and the effect on adult neurogenesis.

Brain Tumors

Brain tumors might arise from neurogenic precursor cells

Neural precursor cells interact with brain tumor cells and exert anti-tumorigenic effects

Because precursor cells are proliferating cells with developmental potential, control of both proliferation and development might fail and lead to tumors (Noble and Dietrich, 2004). The potential role of neural precursor cells in tumor formation has been discussed in Chapter 3, p. 83, in the context of general neural stem cell biology.

Experimental tumors—gliomas implanted into the murine neocortex—attracted precursor cells from the SVZ to migrate towards the lesion (Glass et al., 2005; Walzlein et al., 2008). Intriguingly, in one study the implanted cells even appeared to "track down" migrating tumor cells (Aboody et al., 2000). Fewer cells arrived in the olfactory bulb after the diversion towards the tumor, and in a genetic model with reduced adult neurogenesis, this response was lacking (Walzlein et al., 2008). The precursor cells seemed to exert an anti-tumorigenic action: At young age, many precursor cells surrounded the tumor, which was smaller than in older age, where with reduced adult neurogenesis, fewer cells reached the tumor, which remained bigger . When precursor cells from young mice were transplanted into aged mice with tumors, the survival of the mice was improved (Glass et al., 2005). The anti-tumorigenic effect could also be confirmed in vitro, but the actual mediator is not yet known (Walzlein et al., 2008).

Similar anti-tumorigenic observations have been made in studies using transplanted neural precursor cells into rodent models of brain cancer, mostly in the form of immortalized cell lines, in which the precursor cells were used to introduce an additional therapeutic intervention such as immune manipulations, the insertion of suicide genes, or the induction of angiogenesis (Aboody et al., 2000; Benedetti et al., 2000; Herrlinger et al., 2000; Barresi et al., 2003; Kim et al., 2005; Danks et al., 2007).

Chemotherapy

Both radiation and chemotherapy as anti-cancer treatments can be associated with cognitive dysfunction (Kramer et al., 1992; Crossen et al., 1994; Abayomi, 1996; Dietrich et al., 2008; Welzel et al., 2008). Overall, the literature concentrates heavily on side effects of treatment of brain tumors, although systemic therapy for many types of cancer might affect the brain. Children who received high-dose radiation prior to bone marrow transplantation for acute myeloid leukemia often have learning difficulties later (Nathan et al., 2007). In one study, children who underwent chemotherapy for similar diagnoses took up an academic profession less often than their healthy peers (Kingma et al., 2000). A relatively large literature exists for the treatment of breast cancer (Falleti et al., 2005; Reid-Arndt et al., 2009). The cognitive and mental side effects of chemotherapy are often referred to as "chemo-brain," a rather unfortunate term that has rightfully irritated patients in need of a therapy (Hede, 2008).

Both chemotherapy and radiation target proliferating cells, and common side effects consequently affect highly proliferating tissues, most notably bone marrow. Consequently, cognitive side effects might be due to the ablation of highly proliferative precursor cells. Results from all studies are very consistent.

Methotrexate (MTX) is a cytostatic drug used as adjuvant compound for chemotherapy of breast cancer and is associated with cognitive impairment in some patients. MTX treatment lastingly reduced cell proliferation in the dentate gyrus with some delay after the treatment and also impaired water maze performance (Seigers et al., 2008; Seigers et al., 2009).

Intracerebroventricularly infused cytosine arabinoside has been used experimentally to reduce precursor cell activity in the adult brain (Doetsch et al., 1999; Seri et al., 2001; Zhang et al., 2004; Kokoeva et al., 2005; Mak et al., 2007; Lau et al., 2009). Similarly, the clinically less relevant methylazoxymethanol, a drug with many side effects, acutely diminished cell proliferation in the dentate gyrus. This decrease was associated with impaired performance on some but not other hippocampal learning tasks. Again, hippocampus-independent cognitive abilities were spared (Shors et al., 2001; Shors, 2004). Given the many side effects of these substances, in most of the more recent studies, the experimental suppression of adult neurogenesis has been achieved by radiation or genetic tools. But Garthe and colleagues used Temozolomide, the first-in-line anticancer agent in the therapy of glioma to efficiently suppress adult hippocampal neurogenesis in the absence of gross side effects (Garthe et al., 2009).

The dentate gyrus can recover from acute and short-term treatment with cytostatics quite well. Presumably, the rarely dividing stem cells are not hit by the antiproliferative drug and can repopulate the germinative matrix of the SGZ (Seri et al., 2001; Ciaroni et al., 2002). If the damage occurs early, compensation from this source can be complete (Ciaroni et al., 2002). The apparent difference between the long-term effects of radiation, which did not allow recovery even after four months (Tada et al., 2000), and chemotherapy is most likely not a principal difference but a matter of dosage.

Radiation

Neural precursor cell dysfunction underlying cognitive impairment after radiation might be an as-yet-underestimated side effect of cancer treatment (Monje et al., 2002; Raber et al., 2004; Rola et al., 2004a; Monje, 2008). Precursor cells in the adult SVZ and SGZ are more sensitive to radiation than other brain cells. In dividing cells, gamma rays and X-rays damage

DNA so extensively that endogenous repair mechanisms are overwhelmed and apoptotic elimination is induced (Limoli et al., 2004).

Radiation ablates proliferating cells

In vivo, exposure to X-rays efficiently wiped out cell proliferation from the SGZ (Shinohara et al., 1997; Parent et al., 1999; Peissner et al., 1999; Monje et al., 2002; Santarelli et al., 2003; Raber et al., 2004). Signs of such reduction have also been found in postmortem samples from humans (Monje et al., 2007). The sensitivity of the precursor cells in vivo was dose-dependent and caused long-lasting damage (Mizumatsu et al., 2003; Rola et al., 2004b). A similar, dose-dependent reduction was also found after the exposure to ionized particles (Rola et al., 2004b; Casadesus et al., 2005). In one study, precursor cell proliferation in the dentate gyrus decreased by up to 95% at all doses above 1 Gy and led to a long-lasting reduction in neurogenesis but, interestingly, not ingliogenesis (Mizumatsu et al., 2003). Radiation therapy early in life caused a long-term reduction in adult neurogenesis (Zhu et al., 2009). In cell culture, radiation blocked proliferation of neural precursor cells (Palmer et al., 2000).

Radiation is associated with inflammation, which affects adult neurogenesis

Radiation was associated with a strong inflammatory response in the hippocampus. Treatment of the irradiated animals concomitantly with anti-inflammatory agents rescued adult hippocampal neurogenesis, a finding suggesting that part of the precursor cell damage is secondary and due to the inflammatory reaction (Monje et al., 2003).

Radiation not only ablated cell proliferation but also caused gliosis in the stem cell niche of the adult hippocampus (Monje et al., 2002). In addition, angiogenesis in the SGZ, normally associated with adult neurogenesis, was diminished, whereas the number of microglia increased.

Radiation is a common strategy to ablate adult neurogenesis for experimental purposes

Radiation has been used to block adult hippocampal neurogenesis for experimental purposes (Parent et al., 1999; Santarelli et al., 2003; Winocur et al., 2006; Wojtowicz, 2006; Clark et al., 2008; Iwata et al., 2008; Ko et al., 2009). The effects of antidepressants, for example, on both anxiety-related behavior and adult hippocampal neurogenesis were eliminated in mice in which the hippocampus had been irradiated (Santarelli et al., 2003), although this finding was partly dependent on the strain (Holick et al., 2008) and the drug that was used (Surget et al., 2008). Fractionated radiation also reduced performance in a hippocampus-dependent learning task, whereas hippocampus-independent learning was spared (Madsen et al., 2003). In another study, contextual learning was impaired, while spatial working memory was unaffected (Hernandez-Rabaza et al., 2009). Irradiation of the SVZ abolished neurogenesis in the olfactory bulb and reduced olfactory long-term memory (Lazarini et al., 2009).

Recovery of neurogenesis after radiation depends on whether or not the stem cells have been hit

Because highly proliferative cells are more sensitive to radiation damage than post-mitotic and quiescent cells, radiation should primarily affect the transiently amplifying progenitor cells, and only to a lesser degree, the stem cells (Mizumatsu et al., 2003; Limoli et al., 2004). Consequently, a reconstitution should be principally possible. However, in one study, within nine weeks, recovery from a comparable drop in precursor cell proliferation in both neurogenic regions was only observed in the SVZ, suggesting a differential sensitivity (Hellstrom et al., 2009). In contrast, good recovery within a week was found in another study (Ben Abdallah et al., 2007). The difference probably reflects a dose effect (6 versus 4 Gy). In the high-dose study it was specifically reported that type-1 precursor cells decreased by 90%, indicating a loss of the stem cells that would allow reconstitution (Hellstrom et al., 2009). Treatment with pro-survival factor melatonin (Ramirez-Rodriguez et al., 2009) before the radiation saved proliferating cells (Manda et al., 2009). Physical exercise three months after moderate radiation brought proliferation back to normal levels (Naylor et al., 2008), and so did environmental enrichment for two months after radiation (Fan et al., 2007).

The proposed mechanism of radiation-induced precursor cell dysfunction, besides inflammation, is an increased sensitivity to oxidative stress as well as an increase in the phosphorylation of cell cycle–associated protein transformation-related protein 53 (Trp53), which leads to higher rates of apoptosis (Limoli et al., 2004). Consequently, precursor cells in Trp53 null-mutant mice were less sensitive to radiation than wild types.

Radiation of the SVZ, probably quite surprisingly to many, led to improved engraftment of implanted exogenous neural precursor cells (Marshall et al., 2005), but this priming of the host tissue actually resembles the situation in bone marrow transplantation, where the host marrow has to be killed in order to allow engraftment of the transplanted cells.

Infection

Most infections tend to decrease precursor cell function and neurogenesis

Infection with human herpes virus 6 is detectable in 70% of all brain samples that come to autopsy (Sanders et al., 1996). In most cases the infection is benign and does not lead to a clinical phenotype. However, herpes virus 6 infection can also lead to leukencephalopathy with severe demyelination. In vitro, herpes virus 6 infected human glial precursor cells and caused cell cycle arrest and premature oligodendrocytic differentiation (Dietrich et al., 2004). On one hand, this finding raises the possibility that failing precursor cell function could be part of the pathogenesis of manifest herpes virus 6 infection. Classical herpes encephalitis can affect the temporal lobes and thus might exert some of its signs of hippocampal dysfunction through damage to adult neurogenesis. Precursor cells are also susceptible to cytomegalovirus (CMV) infection throughout life (Tsutsui et al., 2008).

Neonatal infection with the lymphocytic choriomeningitis virus (LCMV) caused a reduction in granule cell numbers, a decrease in proliferating cells and Mash1/Ascl1–positive cells in the SGZ, and a lasting reduction in adult neurogenesis (Sharma et al., 2002). Infection with *Escherichia coli* early postnatally made the brain more susceptible to an immune challenge with bacterial endotoxin lipopolysaccharide, including a more marked reduction in cell proliferation in the dentate gyrus (Bland et al., 2009). LPS injection to pregnant dams caused a postnatal decrease in precursor cell proliferation, which recovered within two months (Cui et al., 2009). As also discussed in the following paragraph on immunity, the mediator of this response might be the stress hormone corticosterone. By means of treatment with ibuprofen, the authors of this study could exclude LPS-induced fever and the concomitant increase in eicosanoids as mediators (Cui et al., 2009).

Meningitis and scrapie infection increase cell proliferation

After bacterial meningeal infection, in contrast, cell proliferation was increased in the dentate gyrus of mice and rabbits. At later time points, and after treatment, the number of cells expressing immature neuronal markers remained elevated (Gerber et al., 2003; Tauber et al., 2009). Neurogenesis was also increased in mice infected with scrapie, a prion disorder, leading to neurodegeneration (Na et al., 2009).

Impaired adult neurogenesis in HIV models

The human immunodeficiency virus HIV inhibited the proliferation of neural precursor cells by binding of HIV coat proteins to the C-X-C receptors 3 or 4 (Krathwohl and Kaiser, 2004; see also Chap. 9, p. 369; and Tran and Miller, 2005). Studies in a genetic mouse model overexpressing the HIV-envelope glycoprotein gp120 revealed that gp120 arrested cell-cycle progression of hippocampal precursor cells in vivo and in vitro before the G1/S phase transition (Okamoto et al., 2007). The speculation that impaired adult neurogenesis might thus contribute to the AIDS dementia complex is tempting but has not yet been investigated in sufficient detail (Venkatesan et al., 2007).

Immunodeficiency and Autoimmunity

Immunity tends to support adult neurogenesis

In the case of HIV, the consequence of the infection would be immune deficiency, which in the presently available studies on adult neurogenesis and HIV has not been specifically addressed. In a genetic model of immunodeficiency, the SCID mice, which lack PRKDC (protein kinase, DNA activated, catalytic polypeptide), an enzyme involved in DNA repair, and do not have a functional immune system, showed reduced adult neurogenesis (Ziv et al., 2006).

Activation of the adaptive immune response up-regulates adult neurogenesis

That finding in SCID mice was part of a landmark study by Michal Schwartz's group at the Weizman Institute in Tel Aviv, which above all revealed that CNS-specific T cells induced in an animal model of multiple sclerosis called *experimental autoimmune encephalitis* (EAE) sustained adult neurogenesis (see also Chap. 9, p. 375). The authors concluded that too little activity of the immune system as, for example, in the immune deficiency syndromes, or too high levels of immune activity such as in severe inflammatory disease would lead to impaired hippocampal neurogenesis (Ziv and Schwartz, 2008). There would thus be an optimal level of immune activity. Whereas a challenge of the innate immune system with LPS caused high levels of corticosterone and a reduction in adult neurogenesis, stimulation of the innate immune response in a model of adjuvant-induced rheumatoid arthritis (AIA) resulted in only moderately increased corticosterone levels and more neurogenesis (Wolf et al., 2009a).

CD4-positive T cells sustain adult neurogenesis

The finding has to be placed in the context, however, that CD4-positive T cells, independently of CNS specificity, sustain adult neurogenesis. Their selective elimination decreased adult neurogenesis (Wolf et al., 2009b). The mediators are not yet known.

Seizures and Epilepsy

Epilepsies are disorders with seizures

Seizures result from synchronized hyperexcitation in neuronal networks. *Epilepsy* is an umbrella term for disorders with seizures. One seizure does not constitute epilepsy. Mechanisms that provoke seizures ("ictogenic") might not be identical to those that cause epilepsy ("epileptogenic"). But many species and strains used in animal experiments (except for C57BL/6 mice) develop spontaneously recurring seizures after the injection of the excitation-enhancing compound (see, e.g., Sharma et al., 2008).

Temporal lobe epilepsy affects the hippocampus and is associated with sclerosis, ectopic granule cells, and persisting Cajal Retzius cells

The temporal lobe is frequently the site of origin for epileptic seizures, typically psychomotor seizures, with their wide spectrum of symptoms adequately reflecting the range of functions of the temporal lobe. Temporal lobe epilepsy is associated with sclerosis of the hippocampus in about 90% of operated cases and a characteristic dispersion of granule cells in the dentate gyrus in about 40% of cases (Houser, 1990; Thom et al., 2002). In 10% of cases, the granule cell dispersion shows a bilaminar pattern similar to that of doublecortex syndrome, a developmental disorder caused by mutations in the doublecortin gene. Whether doublecortin expression in the migrating, newly generated granule cells is responsible for this coincidence or not has to be further examined. The organization of the granule cell layer itself can thus be profoundly disturbed in cases of human temporal lobe epilepsy (Houser, 1990; Lurton et al., 1997; El Bahh et al., 1999). It is not clear, however, whether a preexisting hippocampal abnormality is responsible for the development of the seizures (Blumcke et al., 2002; and see also the discussion below). Magnetic resonance imaging studies in families with febrile convulsions suggest this causal link (Fernandez et al., 1998). It has also been postulated that Cajal-Retzius neurons persist in the epileptic dentate gyrus, which might be indicative of disturbed development (Blumcke et al., 1999; Garbelli et al., 2001; Thom et al., 2002). Severer cases of epilepsy tend to have fewer Cajal-Retzius cells (which secrete migratory stop signal reelin) in the molecular layer of the adult dentate gyrus than in milder cases but still more than controls (Thom et al., 2002). Reelin in turn has been connected with the development of neuropathological features of temporal lobe epilepsy (Haas and Frotscher, 2010; Kobow et al., 2009). Clinically, several classes of seizures can be distinguished, depending on the site of origin and the symptoms. In the context of adult hippocampal neurogenesis, generalized seizures as models of temporal lobe epilepsy are of particular relevance. In an experimental setting, seizures robustly increase adult neurogenesis in both neurogenic regions (Bengzon et al., 1997; Parent et al., 1997; Parent et al., 1998; Parent et al., 2002a;

Jessberger et al., 2005; Parent et al., 2006; Jessberger et al., 2007b). The induction of cell proliferation is massive and has been shown in several models of epilepsy. The first group of models contains chemoconvulsant-dependent seizures—that is, seizures that are pharmacologically induced. Injections of kainic acid and pilocarpine elicit seizures via excitation of glutamate receptors (Bengzon et al., 1997; Parent et al., 1997; Gray and Sundstrom, 1998; Covolan et al., 2000), whereas pentylene tetrazole acts by inhibiting GABAergic inhibition (Jiang et al., 2003; Yin et al., 2007). Seizures can also be induced directly by external electrodes in animal models of electroconvulsive therapy (Madsen et al., 2000; Banerjee et al., 2005; Ekstrand et al., 2008), and by kindling; i.e., by subthreshold stimulation from an intraparenchymal electrode (Parent et al., 1998; Scott et al., 1998; Nakagawa et al., 2000; Auvergne et al., 2002; Ferland et al., 2002; Fournier et al., 2009).

The common principle is massive overactivity of excitatory synapses. Kindling in which an inserted needle provides a sort of epileptic focus might be closest to the situation in patients with temporal lobe epilepsy. In general, the models will only reflect single aspects of the disease. Most importantly, with respect to the open question of whether the seizures or the neurogenic dysregulation comes first, these models necessarily take a biased stand.

Seizures robustly increase precursor cells and adult neurogenesis

All seizure models lead to a similar response in adult hippocampal neurogenesis. Cell proliferation increases eight- to tenfold within a few days after the seizures (Fig. 12–2). Mild seizures were, however, less effective than strong seizures and status epilepticus (Yang et al., 2008a). Labeling proliferative cells prior to the induction of seizures first revealed that it is not a quiescent population that responds to the stimulus, but one that is highly dividing (Parent et al., 1999). Seizures induced proliferation of

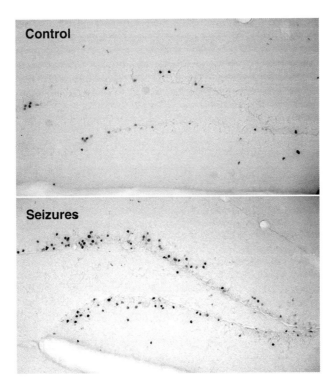

FIGURE12–2 Epileptic seizures induce cell proliferation in the adult dentate gyrus. Systemic application of glutamate receptor agonist kainic acid strongly up-regulates cell proliferation in the subgranular zone. Image by Sebastian Jessberger, Zürich.

type-1 cells in the hippocampus and all precursor cell stages thereafter (Huttmann et al., 2003; Steiner et al., 2008). There was an over-proportional increase in the number of dividing type-3 cells (Jessberger et al., 2005). In human autopsy samples from epilepsy patients, the number of doublecortin-positive cells also positive for a proliferation marker and β-III-tubulin was markedly increased (Liu et al., 2008). In seizures elicited by kindling in the amygdala, the total number of PSA-NCAM-expressing cells in the dentate gyrus increased (Saegusa et al., 2004). This increase correlated with an increase in PSA-NCAM-positive mossy fibers in CA3. The long-term effects of seizures on adult hippocampal neurogenesis depends on the severity of the seizures. Both partial and full status epilepticus caused a similar increase in cell proliferation, but in full status epilepticus, survival of the newly generated neurons was reduced (Mohapel et al., 2004). This result might be due to increased damage to the permissive microenvironment in the SGZ in this condition. Chronic seizures resulted in a lasting decrease in adult neurogenesis months after induction of the seizures (Hattiangady et al., 2004). In old age, the effect on net neurogenesis in response to seizures was reduced or absent, although proliferation remained increased (Rao et al., 2008).

Seizures also affect maturation stages of adult neurogenesis

In addition, seizures seem to interfere with other aspects of adult neurogenesis, especially polarity and dendrite development (Jessberger et al., 2007a) and changes at the ultrastructural level, for example in spines and synapses (Kraev et al., 2009), although the general maturation of doublecortin-expressing cells and the speed of this development were not affected (Plumpe et al., 2006). But in one study, stronger seizures induced neuronal differentiation of relatively more cells (Yang et al., 2008a). It also seems that seizure-induced new neurons are generally functional (Smith et al., 2006). Their activity, however, might be somewhat reduced, and the interesting hypothesis has been raised that their tamed synaptic plasticity might represent a compensatory mechanism (Jakubs et al., 2006; Kempermann, 2006). In the end, only a clearer understanding of the physiological and pathological circuitry will allow researchers to test this hypothesis. Suppressing adult neurogenesis by radiation, however, not only prevented the seizure-induced increase in adult neurogenesis but also enhanced excitability (Raedt et al., 2007).

Ectopic new neurons can be found in the hilus of rats and mice with seizure-induced neurogenesis. Trains of migrating cells have been described as leading from the SGZ into the hilus (Parent et al., 1999), suggesting that these ectopic cells originated from the dividing precursor cells of the SGZ. Such ectopic hilar granule cells fire with CA3 pyramidal neurons and not with granule cells, further supporting their incorrect integration (Scharfman et al., 2000; McCloskey et al., 2006). Mossy cells synapse onto these ectopic granule cells (Pierce et al., 2007). In specimens from human patients with chronic temporal lobe epilepsy, ectopic granule-like cells in hilus and CA3 were found in about 18% of cases (Thom et al., 2002). Increased numbers of proliferative cells have been found in several studies on human autopsy tissue (Takei et al., 2007).

At four weeks after BrdU and KA injections, a significantly increased number of calretinin-expressing cells was found (Brandt et al., 2003), and these cells were more widely distributed across the granule cell layer. Such increased migration was also apparent at the level of type-3 progenitor cells (Jessberger et al., 2005). This effect might be relevant to explaining the so-called granule cell dispersion found in many patients with temporal lobe epilepsy. Normal migration as well as development of this dispersion has been linked to reelin signaling (Gong et al., 2007), suggesting that in seizures, reelin signaling to the migratory precursor cells is impaired. Intrahippocampal injection of KA increased cell proliferation but not neurogenesis and was followed by complete cessation of adult neurogenesis. At that time, granule cell dispersion was still increasing, which led to the interpretation that granule cell dispersion were generally independent of neurogenesis (Heinrich et al., 2006). However, the one mechanism does obviously not exclude the other. In another model (amygdala kindling), the seizures reduced reelin expression in the dentate gyrus (Fournier et al., 2009).

Does seizure-induced adult neurogenesis contribute to the development of epilepsy?

Despite the suggestive link with regard to granule cell dispersion, which however, is no mandatory finding in temporal lobe epilepsy and is of unclear pathogenic relevance, the degree to which disturbed adult hippocampal neurogenesis is responsible for propagating seizures and thus the development of epilepsy is not clear. Taken together, the available evidence rather supports the hypothesis that aberrant neurogenesis is both consequence and cause of seizures and epilepsy and not a mere epiphenomenon (Parent, 2007; Zhao and Overstreet-Wadiche, 2008).

Newborn neurons are sensitive to the induction of axonal sprouting during a vulnerable period

In their 1997 report on seizure-induced neurogenesis (Parent et al., 1997), Jack Parent and Daniel Lowenstein speculated that the seizure-induced new neurons might be responsible for the aberrant axonal sprouting found in temporal lobe epilepsy (Houser, 1990; Lurton et al., 1997) and its animal models (Tauck and Nadler, 1985; Mello et al., 1993). They reported that axons of newly generated granule cells contributed to mossy fiber sprouting in both CA3 and the inner molecular layer. However, they also found that inhibiting neurogenesis through radiation did not prevent axonal sprouting, suggesting that sprouting might originate from both new and old cells (Parent et al., 1999). But the occurrence of ectopic granule cells, which fired with CA3 cells as well as the altered synaptic drive and plasticity from seizure-induced neurons, suggests impairment to the circuitry. Treatment with endoneuraminidase, which sheds the polysialylation from PSA-NCAM and thereby presumably interferes with the function of PSA-NCAM-positive cells, reduced the proliferative response to seizures but did not prevent the development of epilepsy (Pekcec et al., 2008). As the seizures affect other precursor cells as well, and the neuraminidase action is transient, this finding does not allow strong conclusions, however.

New neurons in seizure models have more basal dendrites

Often the ectopic new granule cells retain a basal dendrite that is normally absent in mature granule cells (Ribak et al., 2000). Developing granule cells in the adult go through a period during which they possess basal dendrites (Ribak et al., 2004). In the ectopic granule cells, the number of excitatory synapses on these dendrites was increased, further supporting the idea that these cells are involved in the generation or propagation of seizures (Ribak et al., 2000). The development of these aberrant neurites was prevented by blockade of microglial action, suggesting that inflammation might play a key role in this process (Yang et al., 2009). About half of seizure-induced new granule cells showed aberrant basal dendrites, and there is a critical time window during which seizures can induce the sprouting of these dendrites. After single bouts of seizures, it is not the progeny of cells whose proliferation is induced by the seizures that produces the aberrant dendrites, but cells that have been born a few weeks before (Walter et al., 2007). Visualizing newborn neurons with retrovirally expressed GFP revealed massive changes (Fig. 12–3) (Jessberger et al., 2007a).

Seizure-induced neurogenesis is associated with angiogenesis

In a model of electroconvulsive seizures (with externally applied electrodes), the induction of precursor cell proliferation was paralleled by an induction of endothelial proliferation in the SGZ (Hellsten et al., 2004). Other components of the precursor cell niche have not yet been studied in detail, but it is likely that, after seizures, profound structural alterations in the stem cell niche are found. Pretreatment with corticosterone (see below) did not counteract the seizure-induced increase in precursor cell proliferation, although it prevented the increase in angiogenesis (Ekstrand et al., 2008). The dose-dependency of corticosterone action as well as of the seizure effects themselves need to be taken into account when interpreting this finding.

Only limited information about seizure-induced neurogenesis in the SVZ

Prolonged seizures also affected neurogenesis in the SVZ and olfactory bulb. Most of the newly generated cells took the normal path of migration to the olfactory bulb; a certain percentage, however, prematurely left the RMS and were found in other forebrain regions (Parent et al., 2002a). Subcallosal oligodendrocyte precursor cells also respond to seizures with an increase in proliferation (Laskowski et al., 2007).

Direct glutamate signaling, not cell death, is the main trigger of seizure-induced neurogenesis

It was initially thought by many that cell death elicited by the seizures would be prominently involved in triggering the neurogenic response. However, in C57Bl/6 mice, kainic acid does not induce cell death but strongly increases neurogenesis. Also, in amygdala kindling, tissue damage is minimized, compared to that in other models, without losing a strong induction of neurogenesis (Ferland et al., 2002). The proliferative response is dose-dependent, and single seizures are still able to induce division of the

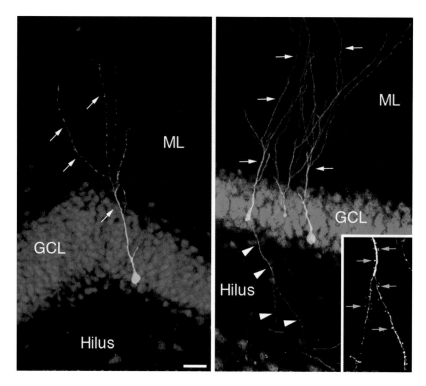

FIGURE 12–3 Seizures affect the dendritic morphology of newborn neurons.

precursor cells (Ferland et al., 2002) and a long-lasting response after status epilepticus. Today it is thought that the pro-proliferative effect is directly driven by glutamatergic signaling to the precursor cells or the niche (Deisseroth et al., 2004; Babu et al., 2009).

Anti-epileptic drug valproic acid supports adult neurogenesis through epigenetic modifications

The question of whether modulation of adult neurogenesis is a reasonable strategy to treat temporal lobe epilepsy is unresolved (Shetty and Hattiangady, 2007; Scharfman and McCloskey, 2009). Valproic acid, a classical antiepileptic drug, not only prevents seizures but also has direct effects on precursor cells. Valproic acid is a histone deacetylase inhibitor, indicating that seizures might cause epigenetic changes in precursor cells, which are prevented by the valproic acid treatment (Jessberger et al., 2007b). In the absence of seizures, valproic acid promoted the differentiation of neurons and suppressed a glial cell fate through the induction of NeuroD (Hsieh et al., 2004).

Neurodegenerative Disease

Neurodegenerative diseases share the accumulation of misfolded proteins

Neurodegenerative disorders in the stricter sense of the name are caused by the misfolding and accumulation of proteins such as a-synuclein in Parkinson disease, amyloid β in Alzheimer disease, and huntingtin in Huntington disease. In a broader sense, *neurodegenerative disorders* covers all diseases with a primary or even secondary degeneration of neurons. Although the limitation to a pathogenesis through the aggregation of misfolded proteins seems sometimes too narrow, the broader definition, which might encompass stroke and multiple sclerosis, makes this class of disorder difficult to handle. In one sense or another, almost all neurological disorders are "neurodegenerative," and in that sense the term is used rather metaphorically (Steiner et al., 2006a). The important issue with respect to adult neurogenesis is the question of whether the pathology affects the precursor cells themselves or only causes a relatively nonspecific response from the precursor cells. A tempting speculation has been that neurodegeneration might be only the flip side of neuroregeneration and neuronal development. The key idea would be that neurodegenerative disease would be the consequence rather of lacking plasticity and regeneration than of degeneration per se. Cell birth and death would be in an equilibrium that in the case of pathology might be disturbed and result in the manifestation of the disease. There is no evidence that this idea would be correct in a fundamental sense. Different diseases differ by their specific pathology. However, interesting continua between seemingly very different disorders exist; for example between Alzheimer and Parkinson disease, with Lewy body disease taking the middle ground (Fig. 12–4). At present there is no information about adult neurogenesis in Lewy body disease, but given the relationship to both other disorders, it would be interesting to study.

Alzheimer Disease

Alzheimer disease (AD) is the most common cause of dementia in the elderly and is characterized by the development of extracellular deposits of β-amyloid (Aβ) in the form of the characteristic plaques and intraneuronal fibrillary tangles. The deposition of Aβ is the result of the disturbed degradation of the larger transmembrane amyloid precursor protein, and begins characteristically in the hippocampus. Plaque formation is a primarily glial

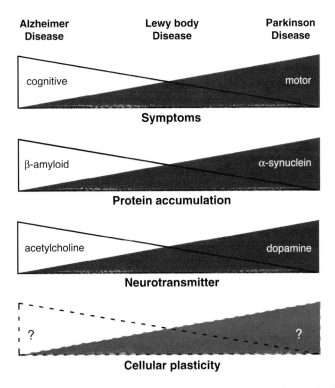

FIGURE 12–4 Continuity between Alzheimer and Parkinson disease. Dementia with Lewy bodies takes a middle position with respect to several parameters. How this spectrum of neurodegenerative disorders relates to plasticity and adult neurogenesis is not quite clear yet. Preliminary data suggest that Aβ pathology might increase neurogenesis, while lack of dopaminergic input decreases the production of new neurons. For Lewy body disease, no good animal model exists.

issue, which suggests that the glia-like neural precursor cells might be affected, too. The formation of neurofibrillary tangles, in contrast, is due to an incorrect processing of tau protein. The initial cognitive symptoms of AD often concern the hippocampus in that patients have learning and memory problems and become disoriented in space and time. The cognitive symptoms correlate with the tau pathology, not with the plaques. The relationship between the two aspects of the pathology is still not really understood.

Mouse models of AD are based on mutations known from hereditary forms of the disease

Early-onset forms of AD are caused by mutations in the APP genes or Presenilin-1 or 2 (Price et al., 1998). Presenilin-1 codes for the enzyme γ-secretase that cuts APP. Most cases of AD, however, are sporadic and usually do not show the mutations that are characteristic of the familial forms. Available mouse models of AD almost exclusively rely on the genotypes known from the familial cases, and most concern the amyloid pathology. Usually the mutated protein is overexpressed transgenically. In a transgenic mouse model, overexpressing a mutated form of the amyloid precursor protein, the enzymatic breakdown of APP leads to the presence of β-amyloid, which accumulates and leads to an AD-like plaque load (Sturchler-Pierrat et al., 1997). Double- and triple-transgenics combining the various described mutations in the APP and presenilin genes, sometimes also including tau pathology, exist. Strictly speaking, these mice are no models of AD and mimic only a very few aspects of the disease. One main problem is that pathology develops much faster in the mutant mice than it does in humans.

The literature on adult neurogenesis in AD models is confusing and becomes even more so, if one looks at the effects of exercise or environmental enrichment on adult neurogenesis in these models. There are effects of the genotype (that is the model chosen), the age of the mice, the duration of the intervention, and the readout parameters chosen.

Increased cell cycle markers in AD

In the late 1990s it was noted that cell cycle–dependent markers are expressed in the hippocampus of AD patients (Nagy et al., 1997b; Nagy et al., 1997a; Nagy, 2000). This observation led to the hypothesis that in the course of AD, as in other cases of neuronal damage such as ischemia (Katchanov et al., 2001; Kuan et al., 2004), post-mitotic neurons might be induced to enter into a cell cycle that cannot be completed, which results in their death (Nagy, 2000; Herrup and Yang, 2001; Yang et al., 2001; Yang et al., 2003; Herrup and Yang, 2007). Besides this neuronal activation of the cell cycle machinery, there is also an effect on proliferation of the neurogenic precursor cells in the dentate gyrus. Chromosomal missegregation and aneuplody have been observed in AD. These would not be readily explainable by abortive cell cycle events on the level of neurons, but they could be due to impaired adult neurogenesis (Zekanowski and Wojda, 2009).

Most AD-like animal models of Aβ pathology increase parameters of adult neurogenesis

Consistent with the idea that the pathology of AD involves an up-regulatory effect on cell division, increased hippocampal neurogenesis has been reported in several studies, mostly based on Aβ pathology (Jin et al., 2004a; Lopez-Toledano and Shelanski, 2007; Rohe et al., 2008; Mirochnic et al., 2009), whereas others reported no effect (Zhang et al., 2007). In one study it was suggested that the apparent increase in proliferation rather reflected vasculogenesis than neural precursor cell activity (Boekhoorn et al., 2006). Ex vivo, wildtype APP that had been injected into the ventricle also elicited increased proliferation from SVZ precursor cells in the neurosphere assay (Caille et al., 2004). In vitro, APP was able to stimulate the proliferation of neural precursor cells in one study (Ohsawa et al., 1999), but not in another (Haughey et al., 2002b). SORLA is a membrane receptor whose activation reduces the processing of amyloid precursor protein to soluble forms (gene symbol: Sorl1). In knockout mice for Sorl1, increased proliferation, survival, and neurogenesis were found (Rohe et al., 2008). Aβ is also not generally inhibitory or toxic to precursor cells. On the contrary, a dose-dependent increase in the number of neuronal progeny was found in an in vitro assay (Lopez-Toledano and Shelanski, 2004). Relatedly, in old transgenic mice, an up to sixfold increase in cortical cell genesis was found, but none of the new cells exhibited a neuronal phenotype (Bondolfi et al., 2002), and most of the new cells were microglia. In the APP23 mouse model, only a mild overall effect on neurogenesis was seen that included an increase in proliferation despite a decrease of the type-2b and -3 progenitor cells (Mirochnic et al., 2009). In other studies, however, overexpression of mutated APP interfered with neurogenesis in both the SGZ and SVZ (Haughey et al., 2002a; Haughey et al., 2002b).

Double and triple mutants including presenilin and tau pathology show reduced neurogenesis

In contrast to these observations, double-transgenics for mutated APP and Presenilin-1 (see below) and triple transgenics for APP, Presenilin and Tau revealed a down-regulation of adult neurogenesis, suggesting an overall minor role of APP in this regulation (Zhang et al., 2007; Niidome et al., 2008; Rodriguez et al., 2008; Pietropaolo et al., 2009).

There is relatively little information on the possible specific mechanisms underlying these effects on adult neurogenesis. Tau phosphorylation itself might be a pre-requisite for adult neurogenesis, in that the posphorylation is specifically present in doublecortin- and NeuroD1-positive cells (Hong et al., 2009). In AD Tau becomes hyperphosphorylated, which might impair neurogenesis. Noggin/BMP4 signaling has been implicated (Tang et al., 2009) as well as (of course) an involvement of Wnt (Toledo et al., 2008). Cortical (!) progenitor cells derived post mortem from AD patients and healthy controls showed reduced self-renewal and "neurogenesis" in vitro, which correlated with increased posphorylation of β-catenin (He and Shen, 2009).

Environmental enrichment improves learning test performance despite stable plaque load

Education as well as cognitive and physical activity reduce the risk of developing dementia, including AD. The "neural" or "cognitive reserve" hypothesis tries to explain the inconsistency between the amount of neuropathological alterations and the degree of cognitive impairment often observed in subjects with AD. Even apparently healthy people can die with a brain full of signs of AD pathology. This hypothesis can be applied to the animal models by exposing the mutated mice to an enriched environment or physical activity. See Chapter 10, p. 473, for the "neurogenic reserve hypothesis," which introduces adult neurogenesis to such considerations. Four months of environmental enrichment in 16-month-old mice carrying the Swedish double-mutation (K670N, M671L) improved learning and memory despite stable plaque load (Arendash et al., 2004). In line with that result, enrichment between the ages of two and 13 months improved water maze performance and increased measures of adult neurogenesis, also with unaltered plaque load (Wolf et al., 2006). Social isolation, in contrast, did not affect neurogenesis in a triple transgenic model (Pietropaolo et al., 2009).

Physical exercise was less effective in both regards. But age affected both the consequences of control and exercise and enrichment conditions. Exercise had a reduced effect on cell proliferation compared to wildtype mice at younger age, but at old age, both interventions were effective (Mirochnic et al., 2009). In one study the effect of exercise was gender-specific for females (Pietropaolo et al., 2008).

Effects of environmental enrichment and exercise on Aβ pathology are ambiguous

Although no effect on plaque load was found in these studies, enrichment increased, while exercise decreased the expression of endogenous (murine) APP, and both interventions shifted the ratio between the non-amyloidogenic form of Ab (Aβ1-40) and the amyloidogenic form (Aβ1-42) towards the non-amyloidogenic form. This result suggested that the interventions might in fact directly interfere with the pathological process (Mirochnic et al., 2009). In some contrast, one earlier study had found that exercise even exacerbated pathology (Jankowsky et al., 2003), whereas others had reported that exposure to an enriched environment reduced the plaque load (Lazarov et al., 2005). A careful behavioral analysis revealed that wheel-running in the transgenic mice reduced stereotypical behavior, while being positively correlated with plaque load, possibly suggesting that wheel-running itself might have become a stereotypical behavior acquired by the impaired animals (Richter et al., 2008).

ApoE4 is a risk factor for AD. In ApoE4 mice, exercise improved learning performance as well as markers of hippocampal plasticity (although adult neurogenesis was not investigated), supporting the clinical impression that ApoE4 carriers might particularly benefit from exercise (Nichol et al., 2009).

Presenilin mutations prevent enrichment-induced increase in neurogenesis

Presenilin-1 (gene symbol: Psen1) is expressed from type-2 cells to immature neurons (Wen et al., 2002a). The transgenic overexpression of mutant Psen1 (mutation P117L) did not affect proliferation but reduced net neurogenesis (Wen et al., 2004). In contrast, the transgenic overexpression of wildtype Psen1 increased TUC4-positive post-mitotic cells without having an overall effect on neurogenesis (Wen et al., 2002b). Psen1 variants PS1Δ9 or PS1M146L prevented the increase in neurogenesis induced by environmental enrichment (Choi et al., 2008) and so did the targeted deletion of Psen1 in excitatory forebrain neurons (Feng et al., 2001). In the 2008 study, wildtype precursor cells had been implanted in a mutant hippocampus, so the result suggests that (in accordance with the idea of Aβ pathology's being primarily non-neuronal) the observed effects were not cell-autonomous (Choi et al., 2008). Conditioned media from microglia of mutant mice decreased proliferation and neuronal differentiation in vitro (Choi et al., 2008).

More DCX-positive cells in postmortem samples from human hippocampus

Consistently with some of the mouse data, in human postmortem samples, doublecortin immunohistochemistry suggested that adult hippocampal neurogenesis might be increased in AD patients (Jin et al., 2004a). However, quantification is difficult in postmortem samples (see Chap. 7) because of potential volume effects on the granule cell layer, consistent with the atrophy found in AD. A reduced volume with the same or even a decreased number of cells might lead to an increased cell count because the cells are closer to each other and thus more likely to be visible in one section plane. Western blot analysis confirmed the presence of markers associated with adult neurogenesis such as doublecortin, PSA-NCAM, and NeuroD. The immature cells had a condensed morphology, though, and it was speculated that they might undergo apoptotic cell death before maturing into granule cells.

Adult neurogenesis might contribute to building a "neurogenic reserve" in AD

Overall, it seems likely that precursor cell plasticity in AD models is only moderately impaired unless double and triple mutations are used, and neurogenesis parameters might even be up-regulated. Regulation of adult neurogenesis is usually maintained in the models. Given the fact that is unknown how closely the combined mutants reflect pathogenesis in this scenario, disturbed adult neurogenesis still might or might not primarily contribute to the hippocampal cognitive deficits found in AD patients. The course of the disease and its duration are certainly different between the murine model and human patients. The data from the intervention studies in the admittedly imperfect mouse models paint the relatively optimistic picture that, because of the lack of strong primary impairment, adult neurogenesis might help build a neural or neurogenic reserve at pre-manifestation stages in order to cope with the disease-related impairment.

Besides environmental enrichment and exercise, a number of strategies have used to increase adult neurogenesis in AD models. Lithium, for example, not only lowers GSK3b levels (Mendes et al., 2009), which in turn reduce tau hyperphosphorylation (Plattner et al., 2006), but also, possibly via the same mediator, increases neurogenesis (Silva et al., 2008; Boku et al., 2009). Glatiramer acetate is another candidate neuroprotective drug that might have pro-neurogenic effects (Butovsky et al., 2006)

Central to the cognitive pathology in AD is an impairment of cholinergic signaling from the basal forebrain to the hippocampus. Lesioning this modulatory input caused a reduction in adult neurogenesis (Cooper-Kuhn et al., 2004), suggesting that the cholinergic input sustains adult neurogenesis. Conversely, treatment with cholinergic drugs increased neurogenesis (see Chap. 9, p. 363).

Parkinson Disease

Parkinson disease is characterized by dopaminergic cell loss and intracellular deposition of α-synuclein

In contrast to AD, the brain damage in Parkinson disease (PD) is not as generalized. Although many brain areas are afflicted during the course of the disease, the primary degenerating neurons are dopaminergic neurons in the substantia nigra of the midbrain. This relatively circumscribed cell loss has made PD the prime case for attempts at transplanting neurons. Several genes have been identified that are responsible for familial cases of PD; among them the gene coding for a-synuclein (gene symbol: Snca).

A-synuclein is a presynaptic protein involved in synaptic remodeling. In PD, a-synuclein aggregates intracellularly in the Lewy bodies, which are a neuropathological hallmark of the disease and not limited to the substantia nigra. In mice, overexpressing human Snca cell proliferation in SGZ and SVZ was normal, but survival and neurogenesis were reduced (Winner et al., 2004).

The question of whether physiological neurogenesis can be found in the adult substantia nigra has been controversial for years. Both preventive and therapeutic approaches based on a stimulation of endogenous precursor cells would be conceivable. However, most found no evidence of new neurons in the adult substantia nigra (Lie et al., 2002; Frielingsdorf et al., 2004). The initial positive report by Zhao et al. (2003) was unusual in its range of methods, so the exact reasons for the discrepancy are unclear (Zhao et al., 2003) (Chap. 8, p. 295).

The classical animal models of PD are lesions of the medial forebrain bundle, the axonal projection of substantia nigra neurons to the striatum, by injections of 6-hydroxydopamine, and the degeneration of substantia nigra neurons by the toxin MPTP. Neither of these models has been unambiguously shown to induce regenerative neurogenesis, either in the substantia nigra or the striatum.

The 6-hydroxdopamine lesion induced cell proliferation in the substantia nigra, but only glial cells developed (Lie et al., 2002). Another study found a reduction in glial differentiation (Steiner et al., 2006b). Environmental enrichment increased the number of NG2-positive and of GFAP-positive cells in the substantia nigra and improved apomorphine-induced rotations, the typical behavioral readout in this type of study (Steiner et al., 2006b). Precursor cells could, however, be isolated from the substantia nigra and were able to generate neurons in vitro and after implantation into a neurogenic region (Lie et al., 2002). Thus not all pathological stimuli seem to be sufficient to elicit degrees of neurogenic permissiveness. In the debated report of neurogenesis in the substantia nigra, MPTP induced a strong increase in the generation of new dopaminergic neurons (Zhao et al., 2003).

The 6-hydroxydopamine lesion also caused directed migration of SVZ precursor cells into the striatum, and early neuronal differentiation (to the DCX stage) could be boosted by the infusion of transforming growth factor-a, either alone (Fallon et al., 2000; Cooper and Isacson, 2004) or in combination with noggin (de Chevigny et al., 2008), or with liver growth factor (Gonzalo-Gobernado et al., 2009). In the absence of further manipulations, these cells migrating from the SVZ into the striatum differentiated into astrocytes (Aponso et al., 2008); with the manipulation, they more readily adopted a "neuroblast-like" phenotype (but see Chap. 7, p. 244, for discussion of DCX as marker in this context). Intracerebroventricularly infused FGF2 and EGF are effective in promoting olfactory bulb neurogenesis and also enhance migration into the striatum. In the case of dopaminergic deafferentation, however, this migration was greatly enhanced (Winner et al., 2008a). Apomorphine-induced rotations, the typical behavioral readout in this type of study, was improved in the TGF-α infused rats (Fallon et al., 2000). It seems that unlike the adult ischemic striatum, where a low level of regenerative neurogenesis was constitutively found (Arvidsson et al., 2002; Parent et al., 2002b), the appearance of supposedly neuronal markers such as DCX in the cells of SVZ origin in response to dopaminergic deafferentation could only be induced by the infusion of an appropriate growth factor in addition to the intrinsic effect of the pathogenic stimulus. The degree of true functional differentiation remains questionable in these cases. Neuronal damage itself seems to be a necessary trigger for regenerative neurogenesis and neuroregeneration, although additional support by exogenously applied factors is necessary to actually elicit a measurable response on the level of new neurons (Chap. 8, p. 298).

In the canonical neurogenic regions, reduced neurogenesis was found in models of PD and no signs of induced regeneration (Hoglinger et al., 2004). Dopaminergic denervation of the striatum led to a decrease in local precursor cell proliferation (Baker et al., 2004; Hoglinger et al., 2004) but there is a conflicting report that saw an increase (Aponso et al., 2008). In the SVZ, C cells have been identified as the targets of the dopaminergic innervation.

D2 and D3 receptors are expressed in the SVZ and co-localize with the EGF receptor and Pax6 (Winner et al., 2009). Dopamine agonist pramipexole promoted neurogenesis, including dopaminergic differentiation in the SVZ/olfactory bulb (Winner et al., 2009). In the MPTP model, migrating neuroblasts (A cells) showed greater vulnerability to the toxin than the neurons and were eliminated by apoptotic cell death (He et al., 2008). B cells showed some endangerment, but C cells were saved (He et al., 2006). In the 6-OH-model, the number of Pax6-expressing neural precursor cells actually increased in the SVZ after lesion. Subsequently, a decrease in the granule cell layer contrasted with a sustained increase of newly generated dopaminergic neurons in the periglomerular zone (Winner et al., 2006). A reduction of neurogenesis after dopaminergic deafferentation has also been reported for the SGZ (Hoglinger et al., 2004). Patients with PD who lack dopaminergic innervation might consequently face a loss of precursor cells in the SVZ and SGZ. Data from human postmortem samples suggested that patients with PD indeed have reduced levels of cells expressing precursor cell markers (Hoglinger et al., 2004). This technically challenging part of the study will have to be replicated with more samples and stereological methods, and with more markers. The influence of dopaminergic signaling on adult hippocampal neurogenesis is not fully clear and is less well understood than in the SVZ. With respect to a potential functional relevance of impaired adult neurogenesis in PD, in PD patients often a loss or alteration of smell is found (Berendse et al., 2001), and Parkinson dementia is a frequent yet underestimated problem.

α-Synuclein reduces adult neurogenesis

Overexpression of a-synuclein impaired survival of newly generated cells in the RMS and olfactory bulb without affecting cell proliferation (Winner et al., 2004). The α-synuclein mutant mice showed, however, a much more pronounced age-dependent decrease in adult neurogenesis in both neurogenic regions (Winner et al., 2008b).

A PD treatment has been the only case of successful cell therapy in the human brain (Lindvall et al., 1987; Wenning et al., 1997). The transplanted cells were immature mesencephalic dopaminergic neurons from aborted fetuses and were placed into the striatum. Two double-blind trials confirmed that this cell therapy can be successful (Freed et al., 2001; Hagell et al., 2002; Olanow et al., 2003), but cautioned that the patients have to be carefully selected (in that the transplantation worked better for tremor but less for rigor and akinesia, the three cardinal symptoms of PD) and that in some patients dyskinesias as signs of too much dopamine action developed. The transplantation of fetal cells nevertheless yielded a clear proof of principles. To overcome the ethical and logistical problems associated with the use of tissue from aborted fetuses, the use of stem cell–derived neurons is desirable. Several protocols for targeted development of dopaminergic neurons in vitro have been developed. Surprisingly, however, even the implantation of undifferentiated ES cells yielded positive results in a rat model, indicating that differentiating cues exist in the adult brain (Bjorklund et al., 2002). Implanted immortalized neural precursor cells also exerted a neuroprotective effect when transplanted at the time of the lesion (Yasuhara et al., 2006), which might not be the most realistic setting but supports the view that the role of stem cells in this context might not necessarily lie in cell replacement.

Huntington Disease

Huntington disease is an inherited triple-nucleotide repeat disease affecting primarily the striatum

Huntington disease is an autosomal-dominant disorder leading to the primary degeneration of medium spiny neurons in the striatum. Affected subjects show massive numbers of triple-nucleotide repeats (CAG) in the Huntingtin gene. The mutated proteins accumulate, but how this accumulation causes the death of striatal interneurons is still a matter of intense study. Injection of quinolinic acid into the striatum of rodents mimics some aspects of the pathological features of the disease, but not this complex pathogenesis and the Huntingtin aggregates.

In the quinolinic-acid model of Huntington disease, the lesion induced cell proliferation in the neighboring SVZ and migration of doublecortin-positive cells toward the lesion (Tattersfield et al., 2004). The study thus reflected a traumatic lesion to the striatum and its consequences on precursor cell activity in the SVZ, not specifically the situation in Huntington disease.

The R6 mouse model mimics the actual pathogenesis more closely in that it is based on the transgenic overexpression of the expanded Huntingtin gene (Li et al., 2005). In the R6/2 mice, one study found that adult neurogenesis in the dentate gyrus was altered on the level of cell proliferation and maturation of the cell, but not of survival and net neurogenesis. The pro-proliferative response to seizures was suppressed. No differences were seen in the SVZ and when precursor cells were analyzed ex vivo (Phillips et al., 2005). Another group found a stronger impairment of neurogenesis and observed that the reduced proliferation in the dentate gyrus was not increased in response to wheel running (Kohl et al., 2007). A more pronounced defect was also seen in the R6/1 line, where environmental enrichment rescued some of the problems in the dentate gyrus but not in the SVZ (Lazic et al., 2006). Interestingly, environmental enrichment rescued protein aggregates in the R6/1 model and improved the cognitive but not the motor symptoms (Spires et al., 2004). Wheel-running activity from pre-symptomatic ages onward delayed the onset of the motor symptoms, but without changing the protein aggregates (van Dellen et al., 2008). Wheel running also did not rescue the phenotype if started later in life. Reduced BDNF levels in the striatum but not the hippocampus were normalized by running (Pang et al., 2006). Adult neurogenesis was unfortunately not assessed in these latter studies.

When BDNF and Noggin were transgenically overexpressed in the SVZ to increase the number of DCX-expressing "neuroblasts" in the striatum, disease development and progression was reduced in R6/2 mice, and motor performance improved. The effect was abolished when the animals were treated intraventricularly with cytostatic drug cytosine arabinoside (Cho et al., 2007).

In the brain of deceased Huntington patients, increased levels of subventricular cell proliferation were detected, and the presence of β-III-tubulin in such cells was taken as sign of increased neurogenesis (Curtis et al., 2003; Curtis et al., 2005). Although this conclusion raises questions (see Chap. 7, p. 242), the finding fits in the context of other examples of signs of increased cellular plasticity in cases of human neurodegenerative disorders.

Because of the initially circumscribed pathology with the affecting of only one type of neuron, as well as the fatal consequences for the individuals who suffer from the disease, Huntington disease is one of the main targets in attempts at neural cell therapy. Various cell types have been explored (Kim and de Vellis, 2009) and clinical studies have been conducted with immature fetal neurons (in analogy to the trials in Parkinson disease) (Freeman et al., 2000; Hauser et al., 2002; Keene et al., 2007). Many animal studies found an improvement of the motor symptoms, but cognitive symptoms, which are difficult to mimic in the animal models, were not addressed. In the human studies, the transplants were afflicted by the disease as well (Cicchetti et al., 2009).

In genetic models of Huntington disease, adult neurogenesis tended to be impaired and could only be partially rescued by exercise and environmental enrichment

SVZ-derived precursor cells counteract disease development

Increased signs of proliferation in the SVZ of Huntington patients

Cell therapy studies for Huntington disease have shown mixed results

Amyotrophic Lateral Sclerosis

Amytrophic lateral sclerosis is the fatal ascending degeneration of motor neurons

The goal of most therapeutic studies is trophic support to the dying neurons

Amyotrophic lateral sclerosis (ALS; Lou Gehrig's disease) is the fatal primary degeneration of the lower motor neuron in the spinal cord and the upper motor neuron in the brainstem and cortex. ALS leads to progressive muscular weakness and death within a few years after onset. As there is no effective treatment available, great hopes rest on stem cell–based therapeutic options.

In ALS rats, a trophic effect supporting the damaged motor neurons has been achieved by infusing human embryonic germ cells into the cerebrospinal fluid (Kerr et al., 2003) or human cortical progenitor cells that were genetically engineered to release GDNF (Klein et al., 2005). Even though endangered motor neurons were saved by the GDNF treatment, their projection to the muscle was not, so that the overall value of this approach is still questionable (Suzuki et al., 2007). A more promising approach seemed to be to apply GDNF-secreting mesenchymal cells directly to the muscle, which rescued both neurons and their function (Suzuki et al., 2008).

In a phase I clinical trial, the infusion of autologous mesenchymal stem cells into the cerebrospinal fluid of ALS patients was found to be safe, but the therapeutic efficacy is not yet known (Mazzini et al., 2009). Neuroprotection is the goal of these attempts. Classical replacement transplantation of precursor cells is, however, in principle conceivable. Motor neurons have been generated in vitro with remarkable success, and the iPS technology will yield further advantages (Papadeas and Maragakis, 2009; Thonhoff et al., 2009). If and how such cells might integrate into the adult spinal cord and become functional, however, is not yet clear.

The best animal model is based on a mutation of superoxide dismutate 1

A human cellular disease model has been developed on the basis of human embryonic stem cells that carry a mutation of superoxide dismutase 1 (SOD1), a mutation found only in the rare familiar ALS cases but still the best molecular model of ALS. Neurons derived from these ES cells showed typical signs of ALS pathology (Karumbayaram et al., 2009). However, the pathology in ALS does not seem to be entirely cell-autonomous. Astrocytes might contribute to the damage as well. Inflammatory activation of human astrocytes carrying the mutation favored the death of wildtype ES cell–derived motor neurons (Nagai et al., 2007; Marchetto et al., 2008).

Limited endogenous proliferation response to the SOD mutation

In mice expressing mutated SOD, precursor cell proliferation in vivo and ex vivo was reduced (Liu and Martin, 2006). But precursor cell proliferation at the central canal was induced. Increased neurogenesis from these precursors was also claimed but not unambiguously proven (Chi et al., 2006). On a side note: adult hippocampal neurogenesis in SOD1-overexpressing mice was identical to that in wildtypes (Kamsler et al., 2007).

Cell therapy with neural precursor cells, often in combination with other manipulations, has also been studied, mostly in the SOD1-overexpressing mice, sometimes in another model in which motor neurons are damaged with the sindivius virus (Garbuzova-Davis et al., 2003; Nayak et al., 2006; Yan et al., 2006; Lepore et al., 2008; Park et al., 2009).

Psychiatric Disorders

Dementia and Mental Retardation

Dementias are acquired forms of cognitive impairment, whereas mental retardation is inborn

Dementia is the acquired irreversible loss of cognitive abilities. Pseudodementia, as found for example in major depression, is reversible. Alzheimer disease is the most common cause of dementia. Historically, scientists considered that vascular forms of dementia accounted for a large fraction of dementias. It turned out, however, that many of these forms share with AD the deposition of amyloid in the vessel walls. "Vascular dementia" might thus not be as distinct from AD as initially thought, and various transition forms might exist. The possible functional relevance of adult neurogenesis discussed in Chapter 10 needs to be reconsidered in the context of possible causes underlying hippocampal dysfunction. The current theories about how neurogenesis contributes to very specific yet broadly required cognitive functions (Kempermann, 2002) might also explain how disturbed adult hippocampal neurogenesis could lead to a maladaptation of hippocampal circuitry to cognitive challenges. It is unlikely that this theory will explain dementias completely. However, if considered in the larger contexts of normal and disturbed brain plasticity, it might be that adult neurogenesis plays a particularly prominent role in hippocampal dysfunction as seen in various forms of dementia. An interesting aspect is the increased incidence of depression in dementia, which might find a common causal ground in disturbed adult neurogenesis.

Frontotemporal lobar degeneration is a rare form of dementia and caused by a mutation in the progranulin gene (Gass et al., 2006). Progranulin (gene symbol: Pgrn) is a secreted protein with growth factor properties and pleiotropic functions. Interestingly, mutations in the Pgrn gene might also increase the susceptibility to other neurodegenerative disorders, so that wider implications of progranulin mutation exist than for the rare frontotemporal degeneration (Cruts and Van Broeckhoven, 2008). Scavenging progranulin with specific antibodies prevented the pro-proliferative effects of estrogen on adult hippocampal precursor cells in vitro (Chiba et al., 2007).

Mutation or dysregulation of the progranulin gene are found in many dementias

Adult neurogenesis is reduced in models of trisomy 21 (Down syndrome)

Mental retardation syndromes are forms of inborn cognitive impairment and as such are distinguished from the acquired dementias. Down syndrome (trisomy 21) causes impaired brain development and affects synapse formation (Wisniewski et al., 1984). The Ts65Dn mouse is a murine model of Down syndrome. The mice show defects in hippocampal development with reduced granule cell numbers and also decreased adult hippocampal neurogenesis (Lorenzi and Reeves, 2006), presumably through reduced cell-cycle speed (Contestabile et al., 2007). Signs of reduced proliferation in the hippocampus have also been found in human fetuses with Down syndrome (Guidi et al., 2008).

Fragile X syndrome is associated with a specific survival deficit of new cells in the ventral hippocampus

Fragile X syndrome is caused by either the loss of the Fmr1 gene or an expansion of a trinucleotide-repeat region (with the sequence CGG) between 40 and hundreds of copies. Because the Frm1 protein is involved in synaptic plasticity, both alterations cause mental retardation. A knockout mouse for Fmr1 showed normal cell proliferation in the dentate gyrus but a reduced survival especially in the ventral hippocampus, which is considered to be more involved in affective function than the dorsal hippocampus. Anxiety was increased in the Fmr1 mutants, whereas water maze performance was normal (Eadie et al., 2009).

Major Depression

"Major depression" is a defined mood disorder, while "depression" might refer to many conditions of low mood

Major depression is a mood disorder with generalized low mood, reduced self-esteem, and a loss of interest and pleasure in activities that are normally enjoyable ("anhedonia"). The qualifier "major" is used to distinguish the disease from many other conditions that might be referred to as "depression" but do not fulfill the diagnostic criteria as laid out, for example, in the *Diagnostic and Statistical Manual of Mental Disorders* (DSM-IV) classification of the American Psychiatric Association. Depression and anxiety are very closely related, and the fifth edition of the DSM will probably merge both entities. Many animal studies of depression measure anxiety.

Impaired adult neurogenesis might prepare the grounds for the complex pathogenesis of major depression

The hypothesis that a failure of adult hippocampal neurogenesis causes depression was first proposed by Barry Jacobs from Princeton University, who suggested that the "waxing and waning of adult hippocampal neurogenesis" could underlie the fluctuating nature of the disease (Jacobs et al., 2000). Today, the idea that impaired adult neurogenesis might be involved in the pathogenesis of major depression and other forms of depression is based on three different lines of reasoning: one empirical, one experimental, and one theoretical. First, patients with chronic depression have hippocampal symptoms and often show a reduced hippocampal volume. Second, all known antidepressants stimulate adult neurogenesis, and some antidepressants might not be effective in the absence of adult neurogenesis. Third, on the basis of functional interpretation of adult neurogenesis as outlined in Chapter 10, a theory can be developed that explains how a lack of adult neurogenesis might contribute to the clinical picture of major depression.

Since Jacobs's first proposal, a number of refinements have been made. Table 12–1 lists the arguments in favor of and against the hypothesis in its original form. Numerous review articles cover the

Table 12–1 The neurogenesis hypothesis of major depression	
Pro	**Contra**
• Reduced hippocampal volume in patients with depression	• Adult neurogenesis produces too few new neurons to explain the volume differences
• Stress and increased corticosterone levels as associated with depression decrease adult neurogenesis	• Role of stress and corticosterone on the regulation of adult neurogenesis is ambivalent
• All known anti-depressants up-regulate adult neurogenesis	• Anti-depressant effects on adult neurogenesis might be epiphenomena
• Adult neurogenesis is involved in affective behavior	• The dentate gyrus is primarily involved in learning and memory
• Depressed patients show hippocampus-dependent cognitive deficits	• Major depression is not primarily a hippocampal disorder
• Increased neurogenesis in models of depression is associated with behavioral improvement	• Animal models of depression poorly reflect the human disease and in many existing models of depression no change in adult neurogenesis is found
• Abolishment of adult neurogenesis renders anti-depressant fluoxetine ineffective	• The dependence of anti-depressant effects on adult neurogenesis is strain-dependent and relates to sub-functions only

underlying debate (Kempermann and Kronenberg, 2003; Becker and Wojtowicz, 2007; Sahay and Hen, 2007; Eisch et al., 2008; Kempermann et al., 2008; Thomas and Peterson, 2008; Lucassen et al., 2009).

It is thus very unlikely that (major) depression is a pure neurogenesis-disorder, but there is in fact increasing evidence that impaired adult neurogenesis might explain an important part of the disease. Failing cellular plasticity might prepare the ground for the developing pathology. The involvement of the hippocampus in the limbic system and the contribution of adult-born neurons to affective behavior play a major role in these refined concepts (see Chap. 10, p. 468).

Patients with chronic major depression have a reduced hippocampal volume

Meta-analyses over numerous MRI studies (Videbech and Ravnkilde, 2004; Geuze et al., 2005; McKinnon et al., 2009) have confirmed the observation reported by Yvette Sheline in 1996 that patients with chronic depression can have smaller hippocampi (Sheline et al., 1996; Sheline, 2000). That report had broken the taboo that major depression, as a purely mental disorder, could not have any biological correlate. The finding has been partially confirmed in postmortem samples from depressed patients (Rajkowska et al., 1999; Rajkowska, 2000). At first glance, this site of pathology indeed coincides well with the location of adult neurogenesis, but adult hippocampal neurogenesis affects only the dentate gyrus, and the low rates in the production of new neurons in adult humans are unlikely to generate enough neurons that their absence would become discernible by MRI. Neither in animal models of depression nor in human postmortem samples (in the latter, with certain caveats) was a reduction in cell proliferation found (Vollmayr et al., 2003; Reif et al., 2006). But the case should not be prematurely dismissed, and interesting speculations with respect to impaired cell-cycle parameters in depression and its treatment have been put forward (Chesnokova and Pechnick, 2008).

Failure of adult neurogenesis might therefore stand in the context of other changes to the hippocampus. Because major depression is, despite these morphological changes, not primarily a hippocampal disorder, but affects cognitive function in many brain regions, most notably but not exclusively structures in the (pre-)frontal cortex, a narrow pathogenetic interpretation of these data might not be justified. To accommodate the complexity of the disorder, it might be necessary to broaden the neurogenesis hypothesis of depression to a neuroplasticity hypothesis. Adult hippocampal neurogenesis might be the most visible aspect of a more general precursor cell–based plasticity that also includes glial cells in many brain regions, neurogenic or not (Kempermann and Kronenberg, 2003).

Corticosterone effects play a major role in explaining the involvement of adult neurogenesis in major depression

Because patients with major depression have a disturbed regulation of cortisol levels, chronically elevated and dysregulated serum cortisol might be responsible for both signs of atrophy in the entire hippocampus and the diminished level of adult hippocampal neurogenesis. In animal models of depression, however, this straightforward connection has not yet been unambiguously confirmed (Malberg and Duman, 2003; Vollmayr et al., 2003). Adult neurogenesis was also reduced in heterozygous knockout mice for the glucocorticoid receptor, GR (Kronenberg et al., 2009).

As discussed in Chapter 9, p. 349, the role of stress and stress hormone corticosterone in the regulation of adult neurogenesis is complex. Also, models such as learned helplessness, in which rodents are exposed to unpredictable stress that ultimately makes them resigned with many symptoms and biochemical parameters of depression, do not mirror the very long time scales of the human disease. Consequently, depression has been difficult to model in animals.

BDNF as a key regulatory molecule in depression might provide a molecular link to the control of adult neurogenesis

Many growth factors, most notably brain-derived neurotrophic factor, hormones, and neurotransmitters, have an effect on adult neurogenesis and also play an important role in current general hypotheses of depression (D'Sa and Duman, 2002). A neurotrophin hypothesis of depression preceded the neurogenesis hypothesis by several years. In adult neurogenesis, BDNF is a pro-survival factor that is presumably involved in the specific recruitment of newborn cells (see Chap. 9, p. 387). Ablation of TrkB reduced adult neurogenesis and abolished the effects not only of exercise on neurogenesis but also of antidepressants on performance in several anxiety-related behavioral tests (Li et al., 2008b).

Although not as widely discussed as a factor relevant for depression, a similar case can be made for VEGF (Lee et al., 2009), which also has an important function in regulating adult neurogenesis (see Chap. 9, p. 386).

In human postmortem samples, a reduction in hippocampal BDNF (and reelin) expression was found (Fatemi et al., 2000; Knable et al., 2004), but no change in the number of Ki67-positive cells (Reif et al., 2006).

Serotonin links regulation of adult neurogenesis with depression

Serotonin plays a central role in the pathogenesis of depression, and many antidepressants exert their effect through a manipulation of the serotonergic system. The most widely used class of antidepressants are selective serotonin reuptake inhibitors (SSRI), which increase the action of serotonin at the synapse (Cipriani et al., 2005; Deshauer et al., 2008). As outlined in Chapter 9, p. 364, serotonin is a potent positive regulator of adult neurogenesis. Most important in this context, serotonergic action is tonic and sustains baseline levels, rather than mediating the flexible regulation above this level. Antidepressants have little to no effect on healthy subjects, and acute effects of serotonin on neurogenesis are balanced due to oppositional effects of different serotonin receptors that affect the same system.

In depressed patients with their reduced serotonergic action, cellular plasticity, including adult hippocampal neurogenesis (but not limited to it), might thus be chronically and severely disturbed. Coping with the many situations of complexity and novelty in everyday life might constantly work at its limits, probably for a long time without clinical symptoms. In the brains of these patients, form can no longer follow function. A depressive episode might be a functional decompensation from this precious and increasingly fragile balance, to which a brain lacking mechanisms of plasticity cannot easily return.

SSRI such as fluoxetine enhance adult hippocampal neurogenesis

The finding that an SSRI such as fluoxetine increases neurogenesis is in line with the positive correlation of serotonergic input to the dentate gyrus with the level of adult neurogenesis (Malberg et al., 2000; Radley and Jacobs, 2002; Malberg and Duman, 2003; Santarelli et al., 2003;

Encinas et al., 2006; Pinnock et al., 2009). Fluoxetine enhancer drug Eszopiclone also boosted the effects of fluoxetine on adult neurogenesis (Su et al., 2009). The effect was diminished in aging, however (Cowen et al., 2008; Couillard-Despres et al., 2009) and was much reduced after stress in early life (Navailles et al., 2008). The bottom line of these many studies is that fluoxetine primarily seems to increase net neurogenesis without acute strong net effects on cell proliferation. The fluoxetine effects on adult neurogenesis were abolished in Aquaporin 4 knockout mice (Kong et al., 2009).

The negative effects of inescapable stress on cell proliferation in the adult hippocampus were reversed by fluoxetine (Malberg and Duman, 2003). Specifically, serotonin receptor 1A appears to mediate the effect: specific antagonists decreased cell proliferation in the adult dentate gyrus (Radley and Jacobs, 2002). In serotonin receptor 1A knockouts, the effect on adult neurogenesis and the behavioral effects of fluoxetine were abolished (Santarelli et al., 2003). The level of serotonin receptor 1a and the expression of its G(-1019) allele is considered a risk factor for depression and the reduced response to antidepressants (Le Francois et al., 2008).

Gender differences in fluoxetine effects on adult neurogenesis might reflect differences in susceptibility to depression

The effect of fluoxetine was not found in very young rats, and in that same study only in males (Hodes et al., 2009). Although this observation is put into perspective by positive findings in females (and young animals) in other studies, it might nevertheless point to the modulatory influence of other systems. But females are more at risk for major depression, and other studies have found gender differences with respect to findings relating to adult neurogenesis and behavioral alterations relevant to depression (Oomen et al., 2009). In males there was a suggestive interaction between dehydroepiandrosterone and fluoxetine-mediate serotonin effects on adult neurogenesis, supporting the idea of differential susceptibility (Pinnock et al., 2009).

Acute and chronic effects need to be clearly distinguished

In general, the interpretation of findings on SSRI effects on adult neurogenesis is often complicated, and the same can be said for the other antidepressants, because the temporal resolution on the cellular level is often not good enough. Chronic and acute effects are often mixed. A very detailed study, for example, reported that fluoxetine would affect symmetrical versus asymmetrical divisions of intermediate progenitor cells in the hippocampus, but the measurement of BrdU-positive cells was taken after chronic treatment (Encinas et al., 2006). At that time, a distinction of pro-proliferative effects on specific progenitor cell populations from a survival effect that over the course of weeks increased the population of these same progenitor cells would have required complicated additional experiments. Irrespective of this, the study provided important first evidence of a specificity of antidepressants for particular stages of adult neurogenesis.

All known antidepressants up-regulate adult neurogenesis

Antidepressants of various types, not only those with direct pharmacological interaction with serotonin, positively affect adult hippocampal neurogenesis. Substances of other classes than SSRI also increase adult neurogenesis, among them atypical antidepressant drugs such as tianeptine, which blocks serotonergic action rather than increasing it (Fuchs et al., 2002). Tianeptine robustly diminished the behavioral consequences of chronic psychosocial stress and rescued the parallel loss of cell proliferation in the dentate gyrus. The effect on cell proliferation, however, was found only in stressed, not in normal, animals (Czeh et al., 2001). In a different study, tianeptine was found to reduce apoptotic cell death in the dentate gyrus of normal and stressed tree shrews (Lucassen et al., 2004). Because essentially all dying cells in the normal dentate gyrus are newly generated neurons that are not recruited for long-term survival and functional integration, it is conceivable that tianeptine has some effects on cell survival.

Other antidepressant measures with effects on adult hippocampal neurogenesis are classical tricyclic antidepressants such as desipramine and imipramine (Malberg et al., 2000; Santarelli et al., 2003), electroconvulsive therapy (Madsen et al., 2000; Segi-Nishida et al., 2008), vagus nerve stimulation (Revesz et al., 2008; Biggio et al., 2009), short-term sleep deprivation (Grassi Zucconi et al., 2006), and general measures such as physical activity (Van Praag et al., 1999). However, one study found ambiguous results with respect to physical exercise, which in that study seemed to increase anxiety rather then relieving symptoms of depression (Bjornebekk et al., 2005; Fuss et al., 2009).

To date, only transcranial magnetic stimulation, which had good effects on reversing the consequences of chronic psychosocial stress and the associated disturbances in the hypothalamic-pituitary-adrenocortical system, did not disinhibit adult neurogenesis and had further negative effects on cell survival (Czeh et al., 2002).

The apparently delayed clinical onset of antidepressant activity cannot be easily linked to a latency in effect on adult neurogenesis

The psychopharmacological interventions showed the typical latency of two to four weeks to reach their full effectiveness on adult neurogenesis, a period of time coinciding with the latency known from antidepressant therapy in patients (Malberg et al., 2000). This argument has been widely popularized and has been taken as an argument that it is the newly matured neurons that mediate the effect on depression. However, the latency of the clinical effect might actually be an artifact, and has not been found in careful reanalyses and studies specifically directed at this issue (Posternak and Zimmerman, 2005; Katz et al., 2006a; Katz et al., 2006b; Mitchell, 2006). What is more, the Malberg study actually did not show a delayed increase in the number of mature neurons, but it did show an increase in proliferation four weeks after the initiation of treatment. The question thus arises of why a supposedly direct, pro-proliferative effect of serotonin on precursor cells (Encinas et al., 2006) would take several weeks to occur.

SSRI might require adult neurogenesis for their action

When adult hippocampal neurogenesis was blocked by radiation, the fluoxetine effects on behavior were prevented, a finding suggesting that adult neurogenesis is necessary for the action of fluoxetine (Fig. 12–5) (Santarelli et al., 2003). This result by Luca Santarelli, Rene Hen, and colleagues from Columbia University in New York City surprised observers by its suggestive simplicity and raised many questions. The group later had the stature to publish a study that partly relativized their own previous findings in that they reported that the observation that had originally been made in C57BL/6 mice could not be replicated in Balb/c (Holick et al., 2008). Most researchers would probably left it at that, but Hen and colleagues kept investigating and found that there are in fact neurogenesis-dependent and -independent aspects of fluoxetine effects on depression (David et al., 2009). The effects were also dependent on the specific antidepressants used (Surget et al., 2008).

Are, conversely, all pro-neurogenic compounds antidepressants?

The entire story has taken a surprising qualitative turn with numerous studies, which concluded from an effect of a compound on adult neurogenesis that it might be useful as antidepressant. This turned the original idea on its head. Although it might very well be that new antidepressants might be discovered through their effect on adult neurogenesis, there is no guarantee that one follows from the other, even if the neurogenesis-hypothesis turns out to be essentially correct. Nevertheless, for many substances that increase adult neurogenesis, a positive effect in tests of anxiety or "depression" was found: FGF2 (Perez et al., 2009) and a FGF-binding protein named FGL (Aonurm-Helm et al., 2008), Notch1 (Guo et al., 2009), and nitric oxide (Hua et al., 2008), and similar observations were made, for example, for nonpharmacological interventions such as environmental enrichment (Veena et al., 2009). Even cell therapy has been tried for depression-like states. The Flinders rat shows depression-like behavior. Implantation of mesenchymal stem cells not only increased adult hippocampal neurogenesis but also improved its performance in behavioral tests like the forced swim test (Tfilin et al., 2009).

The key to the neurogenesis hypothesis of depression lies in the contribution of new neurons to hippocampal function

Patients with major depression have a number of hippocampal symptoms that relate to both the cognitive and the affective functions of the hippocampus. Although the hippocampus is part of the limbic circuitry, these latter functions have long been neglected. As predominant cognitive symptom, depressed patients can show pseudodementia; that is, a reversible impairment of learning and memory. But they also show a reduced ability to assign emotional values to content that has to be learned. For depressed patients, the entire world is gray, and even simple cognitive tasks involving learning and memory become arduous. These and other symptoms do not make major depression a hippocampal disorder, though. To causally link major depression and adult hippocampal neurogenesis, not only would a central role of the hippocampus in major depression have to be shown, but also

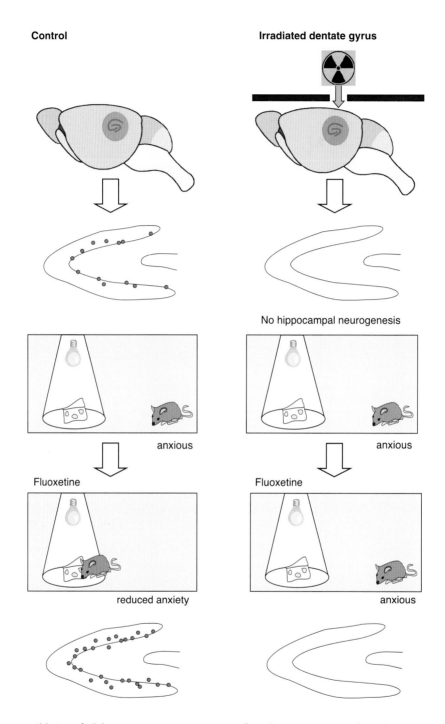

FIGURE 12–5 Ablation of adult neurogenesis impairs action of antidepressants. Santarelli et al. (2003) irradiated the hippocampal region of mice and thereby wiped out adult neurogenesis. In these mice, the antidepressant fluoxetine, a selective serotonin uptake inhibitor (SSRI), did not reduce anxiety about a task in which the rodents had to cross a brightly lit area to reach food, whereas it did induce it in the non-irradiated animals. In non-irradiated animals, fluoxetine induces an increase in adult hippocampal neurogenesis (compare with Malberg et al., 2000).

an indispensable contribution of new neurons to this relevant function (Kempermann and Kronenberg, 2003; Kempermann et al., 2008). This crucial prerequisite for the entire theory was considered to be lacking for some time (Lledo, 2009) but has now come within reach. Several studies have elucidated the involvement of adult neurogenesis in affective behavior (see Chap. 10, p. 468), including a suggestive model-based study on human patients (which, however, could not assess adult neurogenesis directly) (Becker et al., 2009).

Schizophrenia

Schizophrenia is a mental disorder of higher cognitive functions with several risk genes identified, but unknown cause

Schizophrenia is a severe chronic psychiatric disease with complex cognitive and affective symptoms. There are "positive" symptoms such as delusions, hallucinations, disorganized speech, and disorganized behavior; and "negative" symptoms (i.e., the absence of traits found in healthy people), such as a lack of interest, empathy, motivation, or the ability to express emotions or pursue a goal. Negative symptoms might resemble the anhedonia found in depression. In addition, there are cognitive deficits such as attention or memory problems.

Schizophrenia, like major depression, is often lethal because of a greatly increased risk for suicide. Because schizophrenia manifests itself usually in young adulthood, consequences of the disease for both the individual and society are massive. The underlying cause is not known, but there is a strong polygenic genetic component with several risk genes that have been identified without explaining much of the disease. Most of these genes suggest a link to "neurogenesis" (Le Strat et al., 2009). A neurogenesis hypothesis of schizophrenia seemed plausible (perhaps much more so than in the case of depression) because schizophrenia is now widely considered a neurodevelopmental disorder, which might, for example be triggered by maternal infection during pregnancy. There is also a vulnerable period during adolescence. In postmortem samples from humans with schizophrenia cell proliferation markers in the hippocampus were decreased (Reif et al., 2006). Relatedly, hippocampal volume is reduced in schizophrenic patients (Steen et al., 2006).

How failing adult neurogenesis might contribute to schizophrenia has been outlined in several theoretical papers (Reif et al., 2007; Toro and Deakin, 2007; Eisch et al., 2008; Kempermann et al., 2008), but the hypothesis is not yet particularly elaborated. In schizophrenia, altered adult hippocampal neurogenesis might reflect the late stages of subtle neurodevelopmental dysregulation, resulting in a particular set of hippocampal symptoms. Because of the connection between the prefrontal cortex with the ventral hippocampus, which is of particular relevance to concepts of schizophrenia (Tseng et al., 2008), observations about differential changes in adult neurogenesis in the ventral versus the dorsal hippocampus (compare Chap. 6, p. 189).

Animal models of schizophrenia and treatment with antipsychotic drugs were associated with ambiguous consequences for adult neurogenesis

Modeling schizophrenia in animals is extremely difficult, given the fact that many of the cognitive deficits relate to functions that are not found in animals. In one of the few established rodent models of schizophrenia, based on a transient toxic lesion of the developing hippocampus, no alteration in adult hippocampal neurogenesis was found (Lipska, 2004). The hypothesis of schizophrenia as triggered by a prenatal infection can be modeled by maternal polyriboinosinic-polyribocytidylic acid (poly I:C) injections, which cause a prenatal immune activation. In these mice, a long-term immune deficit was found, and adult neurogenesis was reduced (Cardon et al., 2009; Meyer et al., 2010). Treatment with Clozapine improved the deficits in working memory but did not return neurogenesis levels to normal (Meyer et al., 2010). Prenatal lipopolysaccharide injections, a bacterial toxin, also reduced hippocampal cell proliferation (Cui et al., 2009).

Phencyclidine injections are a pharmacological model of schizophrenia, in which adult neurogenesis is decreased (Liu et al., 2006; Maeda et al., 2007), also in the offspring of treated rats (Tanimura et al., 2009).

Antipsychotic drugs had
contradictory effects on adult
neurogenesis

The classic antipsychotic agent, haloperidol, a dopamine antagonist, did not affect adult hippocampal neurogenesis in some studies (Malberg et al., 2000; Halim et al., 2004) but did increase survival in one other (Keilhoff et al., 2009). In one report with a rather unorthodox way of quantification, a small stimulating effect on cell proliferation was seen (Dawirs et al., 1998). Atypical antipsychotics, in contrast, affected cell proliferation (Newton and Duman, 2007). Olanzapine and risperidon increased cell divisions in the SVZ (Wakade et al., 2002) and in the prefrontal cortex (Wang et al., 2004). Risperidon also promoted survival in the hippocampus (Keilhoff et al., 2009). Low doses of clozapine induced proliferation but did not increase net neurogenesis (Halim et al., 2004). The diversity of results might simply reflect the fact that common antipsychotic agents act on other receptor systems besides the dopamine receptors.

Overall, the plausibility of a neurogenesis hypothesis of schizophrenia was thus contested with contradictory experimental evidence. Nevertheless, and similar to the situation in depression research, models primarily aiming at impairing adult neurogenesis have been proposed as models of schizophrenia more or less solely based on this hypothesis (Iwata et al., 2008). Given the general problems with the animal models of schizophrenia, a genetic approach to the problem appeared more promising.

Disc1 (disrupted in schizophrenia 1) is one of the main candidate susceptibility genes for schizophrenia and is expressed in regions of neurogenesis during development and in adulthood (Austin et al., 2004; Schurov et al., 2004). Hongjun Song and colleagues from Johns Hopkins University in Baltimore found that Disc1 is important for the post-mitotic maturation of newborn neurons (Duan et al., 2007). This effect is dependent on signaling through the Akt/mTOR pathway, and Disc1 probably plays an important role in orchestrating Akt/mTOR signaling in developing neurons (Kim et al., 2009). In addition, Disc1 modulates precursor cell proliferation through the GSK3β/β-catenin pathway, which links Disc1 also to Wnt signaling (Mao et al., 2009). Mutation of GSK3β impairs precursor cell proliferation (Eom and Jope, 2009). Disc1 might thus play the role of a master regulator and is of particular interest because the Scottish family in which it was identified shows increased incidence not just of schizophrenia but also of major depression and bipolar disorder.

Mutations in other susceptibility
genes for schizophrenia also
reduce adult neurogenesis

Long-term amygdala kindling, a model of seizures (see above, p. 528), did not increase cell proliferation but altered dendritic morphology of doublecortin-positive cells and was associated with a decrease in hippocampal Disc1 and reelin expression (Fournier et al., 2009). Reelin is another candidate gene of susceptibility for schizophrenia. Mutations in the reelin gene reduce adult neurogenesis (Kim et al., 2002; Weiss et al., 2003; Won et al., 2006). Knockout mice for candidate susceptibility gene Npas3 (coding for the transcription factor neural PAS domain protein 3) also showed a reduction in adult neurogenesis (Pieper et al., 2005). Schizophrenia is also associated with the locus of the neuregulin gene. Impaired NRG1 signaling is thought to play a role in NMDA receptor dysfunction and thus contributes to the glutamate hypothesis of schizophrenia (Morrison and Pilowsky, 2007). NRG1 has not yet been studied in adult hippocampal neurogenesis, but is well described in the regulation of adult olfactory neurogenesis (Ghashghaei et al., 2006).

Drug Abuse and Addiction

Opioids, Cocaine, and Amphetamines

Cognitive and affective symptoms that might relate to adult neurogenesis are common in drug abuse. Exogenous ligands for endogenous psychotropic receptor systems (here the opioid and cannabinoid systems) are taken as drugs of abuse and highjack the endogenous systems with their widespread effects on brain function. Their impact on adult neurogenesis is outlined in

Chapter 9, p. 389. As a general rule, dosage matters. Low doses, as corresponding to the levels of endogenous activity, tended to increase adult neurogenesis, but higher doses and chronic applications usually impaired adult neurogenesis. Drugs of abuse, with the exception of the cannabinoids (see Chap. 9, p. 388) are listed in Table 12–2.

Cocaine and other psychostimulants directly interfere with the dopaminergic system and thus with the brain's reward center. Remarkably and perhaps counterintuitively, even under acute application these drugs did not stimulate adult neurogenesis. Only one study indicates that in fact one hour of intermittent access to methamphetamine self-administration increased proliferation, but net effects on neurogenesis were cancelled out by opposite effects on later stages of neuronal development. Hour-long daily exposure decreased cell proliferation (Mandyam et al., 2008). A similar observation was made for cell proliferation and gliogenesis in the prefrontal cortex (Mandyam et al., 2007).

For most drugs of abuse, the effects on adult neurogenesis have not yet been studied in sufficient detail to allow similar conclusions. Most reports only addressed cell proliferation, and functional considerations have hardly been made.

Alcohol

Alcoholism causes hippocampal dysfunction

In contrast to the drugs of abuse, which bind to specific receptors, the effect of alcohol is less specific. Alcohol does not serve a physiological regulatory function. But brain and body cells are equipped with the physiological machinery to deal with alcohol, and it is largely a matter of dosage that determines whether alcohol causes health problems. However, susceptibility to alcohol-related toxicity differs among various cell populations and brain regions, thus explaining the sequence of symptoms after acute alcohol intake. The same is true for the mechanism underlying addiction to alcohol, which is not primarily covered in this paragraph, because it has not yet been addressed in the context of adult neurogenesis.

There may be a link between alcohol and adult neurogenesis because chronic alcohol abuse in humans causes learning and memory problems and decreased olfaction in late stages of the disease. MRI studies in humans have shown fluctuations in hippocampal volume under alcohol abuse (White et al., 2000; Geuze et al., 2005; Wilhelm et al., 2008). Chronic alcoholism in mice results in atrophy of the dentate gyrus (Walker et al., 1980). A rather speculative hypothesis is thus that these consequences of alcohol abuse might be caused by alcohol-related negative effects on adult neurogenesis. While this has some plausibility for the cognitive deficits, the volume effects cannot be explained by the lack of a few adult-generated cells.

Alcohol reduces proliferation and survival in adult hippocampal neurogenesis

Acute alcohol consumption (binge drinking) of one or four days acutely reduced cell proliferation in the SGZ; the four-day treatment also reduced cell survival (Nixon and Crews, 2002). Physical exercise was able to counteract these effects to some degree (Crews et al., 2004). A later study, however, suggested that survival of the newborn cells might still be impaired (Helfer et al., 2009). Although there might thus be specific effects on neuronal survival (which are highly plausible), this observation is also consistent with the idea that voluntary exercise primarily stimulates cell proliferation, and that additional stimuli are need to recruit these cells (Fabel et al., 2009).

Adult neurogenesis might recover from alcohol-induced impairment only in relatively acute settings

Chronic alcohol intake for six weeks at moderate doses decreased the number of newly generated neurons in the adult dentate gyrus of rats but not in the olfactory bulb. This decrease was due to a reduction in cell survival, whereas cell proliferation was not impaired. Intake of Ebselen, an antioxidative agent, together with alcohol prevented the damage (Herrera et al., 2003; Crews et al., 2006), which indicates that the damage might be due to oxidative stress. Abstinence from alcohol after four days of binge alcohol intake caused a strong rebound in cell proliferation (Nixon and Crews, 2004). In more chronic models of alcoholism, such "recovery" did not occur, and the damage was lasting (Richardson et al., 2009). This was also true for single exposure to alcohol early in life, suggesting a particularly vulnerable period (Ieraci and Herrera, 2007). Precursor cell cultures from

Table 12–2 Effects of psychostimulatory drugs of abuse on adult hippocampal neurogenesis

Drug	Duration	Species	Mechanism	Effect on neurogenesis	Reference
Cocaine	acute to chronic	rat	Inhibits reuptake of serotonin, dopamine and norepinephrine	Reduced proliferation, unaltered survival	Yamaguchi et al., 2004; Yamaguchi et al., 2005; Dominguez-Escriba et al., 2006
Amphetamine	acute	rat	Increases action of dopamine and norepinephrine through different mechanisms	No change	Mao and Wang, 2001
"Ecstasy"*	acute	rat		Unchanged proliferation but reduced survival	Hernandez-Rabaza et al., 2006
Methamphetamine	acute	gerbil		Reduced proliferation	Teuchert-Noodt et al., 2000
	acute to chronic	rat		Reduced proliferation**	Mandyam et al., 2008
	acute	rat		Reduced proliferation***	Kochman et al., 2009
Morphine	acute	rat, mouse	Exerts direct effects after binding to opiod receptors	No change or reduced proliferation and survival	Eisch et al., 2000; Kahn et al., 2005
Heroin	chronic	rat		Reduced proliferation	Eisch et al., 2000

*3,4-methylenedioxymethamphetamine
**different effects dependent on dosage and duration of the application
***only when applied during inactive time (daytime, as opposed to night)

pups of mothers with chronic alcohol intake during pregnancy showed reduced neurogenic potential in vitro (Singh et al., 2009). In vivo, the reduction in cell proliferation after prenatal alcohol exposure was again rescuable by voluntary exercise, but here survival was not specifically studied (Redila et al., 2006). Environmental enrichment seems not to have been equally effective (Choi et al., 2005).

Forced abstinence from "voluntary" alcohol intake also reduced cell proliferation, but did not affect net neurogenesis (Stevenson et al., 2009). Interestingly, the abstinence also induced a depression-like state in the animals.

Alcohol thus seems to have an acute down-regulating effect on SGZ precursor cells that can be compensated for unless the damage is prolonged, which leads to a sustained net reduction in adult neurogenesis. In contrast to these observations, one study showed that moderate alcohol intake could actually increase precursor cell proliferation, and in this case the numbers returned to baseline levels after withdrawal (Aberg et al., 2005).

Only moderate acute intake of alcohol might increase adult neurogenesis

Does red wine have different effects on neural precursor cells than white wine?

On a side note, alcoholic beverages might affect neuronal plasticity beyond their actual alcohol content. In cell culture, red wine increased cell death of neural precursor cells independently of the alcohol content. White wine did not have this effect, which was found to be related to the inhibition of thioredoxin reductase by the red pigments (Wallenborg et al., 2009). This is in partial contrast to the idea that flavonoid resveratrol, which is found in the red pigment of red wine, would be actually beneficial for precursor cells (Prozorovski et al., 2008).

Nicotine

Nicotine impairs adult neurogenesis

Nora Abrous and colleagues from Bordeaux allowed rats self-administration of nicotine and found that cell death in the dentate gyrus increased, whereas adult neurogenesis decreased, both in a dose-dependent manner (Abrous et al., 2002). Similarly, injection of nicotine decreased the numbers of PSA-NCAM-expressing cells in the dentate gyrus in a dose-dependent manner (Shingo and Kito, 2005). Assuming a positive role of adult neurogenesis in hippocampal function, the finding conflicts with the claims heard from many smokers that nicotine would actually improve their cognitive performance—despite the fact that heavy smokers who quit show impaired cognition (Changeux et al., 1998). Indeed, one study using a lower dose and a more acute exposure found an increase in cell proliferation in the SVZ, possibly related to increased FGF2 expression (Mudo et al., 2007). A similar strong induction was also found in aged rats (Belluardo et al., 2008). Neurogenesis beyond precursor cell proliferation was unfortunately not studied. Nicotinic receptors are down-regulated in many neurodegenerative disorders, including Alzheimer disease and Parkinson disease (James and Nordberg, 1995). This finding has contributed to the controversial idea that smoking might actually be beneficial because it helps prevent neurodegeneration (Hernan et al., 2002; Allam et al., 2004). The Abrous study clarified this claim, indicating that such a generalized conclusion is at least not justified from a hippocampal perspective (Abrous et al., 2002). In addition, a range of negative effects of nicotine on other aspects of brain morphology are known (Aramakis et al., 2000). Generally, the abundance of nicotinic receptors in the brain, the multitude of effects of nicotine in different cell types and brain regions, the complexity of cognition, and the problem of different net effects of the same molecule on different time scales make it difficult to draw a clear picture. The interesting observation for our context is that precursor cells seem to directly respond to nicotine. This is also supported by in vitro findings in neural cell lines (Newman et al., 2002).

Addiction

The hippocampus is not the major brain region responsible for the development of addiction, but the many results that show effects of addictive compounds on adult neurogenesis have also raised the question of whether impairment of the new neurons might also somehow be involved in the development of addiction itself (Miller and Spear, 2006). The idea is that addiction is the uncontrollable craving for reward. Mood, it has been argued, is an intrinsic system to guide individuals to make use of their brains in the most efficient way (Bar, 2009), and the reward system is the central reinforcing mechanism. Reward is mediated by the dopaminergic system, and a reward system out of bounds is considered the basis of addiction.

Direct stimulation of reward systems increases adult neurogenesis

Rewarding mice for running, further increased running but did not increase adult neurogenesis beyond the level of unrewarded runners (Klaus et al., 2009). The study did not contain a "reward only" group (which is admittedly hard to devise), so it might tell more about the dose–effect relationship of exercise-induced neurogenesis than about reward effects, but the study has been the first attempt to address the question. But wheel running in rodents is sometimes interpreted as addictive or at least a compulsive behavior because it is rewarding. In a model of "pure" reward, three days of one hour of intracranial self-stimulation with an electrode increased adult neurogenesis in rats (Takahashi et al., 2009).

Cognitive Enhancement and Brain Doping

Most drugs used for cognitive enhancement seem to decrease adult neurogenesis

Pharmacological manipulation of adult neurogenesis might theoretically be attempted in order to boost cognitive performance in healthy people. The underlying rationale might not withstand critical reasoning in the light of what the new neurons are actually good for, but the idea has of course an intriguing face-value appeal. However, amphetamines and methamphetamines, classical psychostimulants, decreased neurogenesis rather than increasing it. They have been listed above with drugs of abuse. SSRI and other antidepressants, consumed by many students under pressure, are discussed on page 543, and tend to increase neurogenesis. It is unclear, however, whether a stimulated increase in neuronal production alone indeed has beneficial functional consequences. Clinically, antidepressants usually do not have effects on non-depressed patients.

Coffee-drinking is an accepted form of pharmacological cognitive enhancement. Studies on caffeine effects on adult hippocampal neurogenesis yielded surprising results in that low- to medium-high doses for seven days decreased cell proliferation, whereas supraphysiologically high doses increased cell proliferation and reduced survival. Acute caffeine had no effects (Wentz and Magavi, 2009). In another study, caffeine reduced cell proliferation if applied during the active period and had no effect during the inactive period (Kochman et al., 2009).

Methylphenidate (Ritalin), perhaps the quintessential drug in brain doping, reduced neurogenesis in adolescent rats (Lagace et al., 2006), but had no effects on hippocampal neurogenesis in adolescent gerbils (Schaefers et al., 2009).

Just like caffeine and methamphetamine, modafinil, a compound used (not only) in the military to keep people awake, reduced cell proliferation in rats during their active period (Kochman et al., 2009).

Stroke

The question of whether adult neurogenesis might contribute to restoration after stroke has great medical relevance

Given the clinical relevance of stroke, it is no surprise that a large number of experiments are aimed at identifying an endogenous neurogenic response to ischemic brain damage. Stroke generally leads to necrotic tissue loss that cannot be replaced. Transplantation with the goal of reconstructing the entire lost structure is an unlikely option because of the typical size of the lesion and the complexity of the destroyed brain structure. Cell therapy, however, might aid in retaining tissue that would otherwise succumb to secondary loss and might provide trophic or other support to the spared structures. In theory, this benefit could also be obtained from endogenous resources.

Consequently, a promising long-term therapeutic strategy might be to promote endogenous attempts at regeneration in order to improve outcome after stroke. Does the adult brain show a precursor cell–based response to ischemia that could be exploited to develop novel strategies to overcome persistent structural damage of brain tissue after stroke?

Clinically, two major forms of ischemic damage can be distinguished: stroke and hemorrhage. Stroke is the complete or partial interruption of cerebral blood flow; hemorrhage is due to bleeding into the parenchyma, for example, from a ruptured aneurysm or a subarachnoidal artery. If the entire carotid artery is occluded or blood flow is abruptly stopped after cardiac arrest, the result is global ischemia; if single cerebral arteries are occluded by a thrombus, focal ischemia is found. One fundamental difference between the forms of ischemic brain damage lies in the amount and distribution of spared tissue. Also, the kind and severity of cell damage and its distribution can greatly vary among different types of ischemia. This variation has important consequences for the interpretation of studies dealing with the effects of ischemic brain damage on adult neurogenesis.

> **Clinical and neuropathological manifestations of stroke vary dramatically**

On a phenomenological level, both global (Liu et al., 1998; Takagi et al., 1999; Iwai et al., 2001; Kee et al., 2001; Yagita et al., 2001) and focal (Arvidsson et al., 2001; Jin et al., 2001; Komitova et al., 2002; Takasawa et al., 2002) ischemia models robustly induce cell proliferation in the SGZ. Numerous studies have confirmed these initial reports (see, for example, Zhang et al., 2005; Lichtenwalner and Parent, 2006; Greenberg, 2007; and Liu et al., 2009b, for review). Net neurogenesis from these dividing cells varied but, not surprisingly, could be increased by anti-apoptotic measures (Doeppner et al., 2009). Adult hippocampal neurogenesis is also increased after photothrombotic ischemic lesions to the cortex (Kluska et al., 2005). Radial glia–like type-1 cells are part of this response (Kunze et al., 2006). After such damage to the motor cortex, enforced training of the impaired paw not only improved the behavioral outcome but also increased the neurogenic response (Zhao et al., 2009).

> **Unless the stem cell niche is damaged, ischemia increases adult neurogenesis**

In contrast, an acute decrease with consecutive recovery of cell proliferation in the SGZ and SVZ has been found after subarachnoid hemorrhage (Mino et al., 2003). In none of these cases was it resolved whether the newly generated cells actively participated in structural reconstitution. The increase in hippocampal neurogenesis also occurs in the absence of direct (at least visible) hippocampal damage (Arvidsson et al., 2001), suggesting a relatively nonspecific mechanism. There was also no sign of cell migration out of the SGZ toward ischemic regions. This is different in the SVZ.

In partial contrast to these results, a much-publicized study claimed that precursor cells migrating from the SVZ into the hippocampus can reconstitute ischemic damage in CA1 if this process is supported by massive growth factor infusions (Nakatomi et al., 2002). Unfortunately, the study lacked an absolute quantification of both the initial damage and the consecutive regeneration. The vague correlational link to an improvement in functional outcome has also been criticized. Methodological questions in this context have been discussed in Chapter 7, p. 236. There has not been a follow-up or confirmation study since the original publication. Under physiological conditions, no migration from the dorsal SVZ towards CA1 occurs (Kronenberg et al., 2007), although many DCX-positive cells are found in the sub-callosal region (Takemura, 2005; Kronenberg et al., 2007).

> **Does ischemia induce neurogenesis in CA1?**

The important problem of falsely identifying BrdU incorporation in dying neurons as evidence of neurogenesis applies to both non-neurogenic and neurogenic regions (Cooper-Kuhn and Kuhn, 2002; Kuan et al., 2004; Breunig et al., 2007; see also Chap. 7), but in non-neurogenic regions, proof of regenerative neurogenesis is much harder to find, because no undisputed baseline exist. There is little doubt, however, that ischemia and hypoxia can influence precursor cells and adult neurogenesis in neurogenic regions. Blockade of CREB, for example, inhibited ischemia-induced

> **Nonspecific uptake of BrdU or expression of cell cycle markers is a matter of particular concern in models of ischemia**

neurogenesis in the dentate gyrus of rats (Zhu et al., 2004), an effect that would not be observable in a case of falsely detected cell death. Otherwise, evidence of neuronal development, e.g., in the sense of a marker progression, can circumvent this problem because dying cells would not show signs of gradual maturation.

Ischemia induces regenerative neurogenesis in the adult striatum

In humans, stroke most frequently affects the internal capsule and striatal tissue. Most studies have thus focused on a neurogenic response to ischemia in the SVZ because it is close to the striatum.

Focal cerebral ischemia after middle cerebral artery occlusion (MCAO) caused an increase in cell proliferation in the SVZ (Jin et al., 2001; Zhang et al., 2001; Jin et al., 2004b), presumably primarily C and A cells, because immature neuronal markers were detected in the dividing cells (Jin et al., 2001). This response decreased with increasing age (Jin et al., 2004b) but is maintained into old age (as is the neurogenic response in the SGZ) (Darsalia et al., 2005). From these activated precursor cells some reactive neurogenesis in the striatum occurs (Arvidsson et al., 2002; Parent et al., 2002b; Kokaia et al., 2006).

A study by Andreas Arvidsson, Olle Lindvall, and colleagues from Lund University is particularly noteworthy in this context, because the ischemia-induced generation of new striatal neurons was followed over different stages of development. About two weeks after the insult, the attracted migrating cells expressed transcription factors Pbx and Meis2 (Figs. 12–6 and 12–7), which are characteristic for the development of medium spiny interneurons in the striatum (Toresson et al., 2000). About two or three weeks later, the new cells expressed DARPP-32, a marker of mature striatal neurons

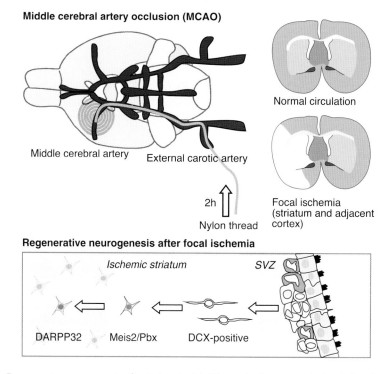

FIGURE 12–6 Regenerative neurogenesis after ischemia. Middle cerebral artery occlusion induced regenerative neurogenesis in the striatum (Arvidsson et al., 2002). Precursor cells' migration in from the subventricular zone resulted in a low level of neuronal replacement. The new cells went through an identifiable intermediate stage of development, depicted in Figure 12–7, a result arguing against the concern that neurons dying in response to ischemia had taken up BrdU and were falsely identified as new. DCX, doublecortin.

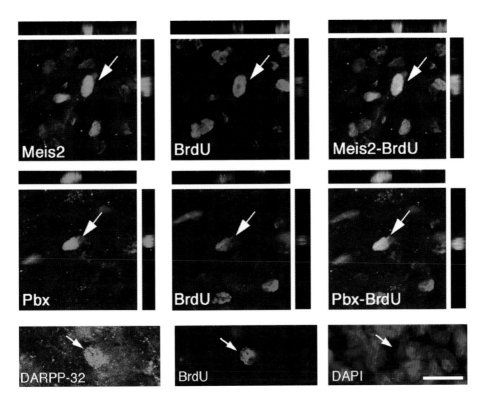

FIGURE 12–7 Regenerative neurogenesis after ischemia. During ischemia-induced regenerative neurogenesis in the adult striatum, the new neurons went through an intermediate transient stage at which they expressed transcription factors Pbx and Meis2. Mature neurons expressed DARP-32. See text and Figure 12–6 for details. Figure from Arvidsson et al. (2002), reprinted with kind permission of Olle Lindvall, Lund, and Macmillan Publishers *(Nature Medicine),* copyright 2002.

(Arvidsson et al., 2002; Parent et al., 2002b). This response to ischemia was long-lasting and remained visible even after one year (Thored et al., 2006). SVZ cells migrated along blood vessels (Thored et al., 2007). No such response was found after an excitotoxic lesion to the striatum (and cortex), suggesting that the exact mechanism of the lesion might matter (Deierborg et al., 2009). The response could be physiologically supported by environmental enrichment (Plane et al., 2008).

Whereas Nakatomi and colleagues had claimed extensive structural reconstitution of CA1, in the Arvidsson study, only 0.2% of the lost striatal neurons were replaced, making it unlikely that these cells alone would be able to support substantial functional recovery. The question is therefore whether the process is some sort of epiphenomenon and whether it can be stimulated to exert concrete functional benefits. There is first evidence, from a study labeling SVZ cells with an EGFP-expressing retrovirus and performing electrophysiological analyses at various time-points after ischemia, that the newborn cells acquire a functional neuronal phenotype of GABAergic and cholinergic neurons (Hou et al., 2008). Another study argued that, based on the expression of transcription factor Sp8 and calcium-binding protein Calretinin, the new striatal neurons would actually represent misplaced olfactory bulb neurons, rather than medium spiny interneurons (Liu et al., 2009a).

With this neuronal phenotype, the new cells in the striatum differ from those that migrate from SVZ to the olfactory bulb, where they become interneurons that express neither Pbx and Meis2 during development nor DARPP-32 in their mature condition. This finding suggests either that specific subpopulations of precursor cells are responsible for the regenerative neurogenesis after ischemia, or that an instructive influence alters the fate of identical precursor cells.

Is there cortical neurogenesis after stroke?

Several studies had reported specifically cortical neurogenesis after stroke (Jiang et al., 2001), but were contradicted by others that did not find new neurons in or around the ischemic cortex (Arvidsson et al., 2002; Parent et al., 2002b). Sometimes individual examples of a BrdU/NeuN double-labeled cell might indeed look convincing. But in the absence of additional data, that one cell does not constitute "neurogenesis," and healthy skepticism remains necessary. Many studies, including one on human stroke victims (Jin et al., 2006), were based on the expression of doublecortin or proliferation markers alone, which do not prove neurogenesis (see Chap. 7, p. 239). After ischemia to the barrel cortex, the representation area for sensory input from the whiskers in rodents, stimulation of the whiskers increased the number of doublecortin-positive cells near the lesion (Li et al., 2008a).

With an impressively expanded arsenal of methods, however, Ohira and colleagues reported the response of a precursor cell population in Layer I of the neocortex and adult neurogenesis of interneurons in the underlying layers after global ischemia (Ohira et al., 2010). The study described development of the new cells, including marker progression, and gave first hints at their functional integration.

A particular case is reports of induced cell division of neurons after stroke (Gu et al., 2000; Gu et al., 2009), which might relate to other cases of abortive cell cycle events prior to cell death rather than true division with cytokinesis.

Stroke-induced neurogenesis in the striatum depends on SVZ precursor cells

The response of SVZ cells is dependent on control of the local stem cell pool. Stem cell populations are maintained by Pten: conditional ablation of Pten increases migration to the olfactory bulb and also to a peri-infarct area. While in the olfactory bulb these cells turned into the neurons, no neurogenesis was observed near the stroke area (Gregorian et al., 2009). Ischemia also recruited ependymal cells, which are normally quiescent and kept in that stage by Notch1 signaling. Overexpression of Notch prevented this response (Carlen et al., 2009). Similar data relate to Notch1 controlling other SVZ precursor cells (presumably B cells) (Wang et al., 2009). A role for Notch1 has also been postulated in the ischemia-induced increase in adult hippocampal neurogenesis (Guo et al., 2009).

Inflammation modulates the neurogenic response to ischemia

Stroke causes an immune response in the brain, including the activation of microglia. Subsets of microglia are pro-neurogenic and accumulate within the neurogenic zone (Thored et al., 2009). What renders microglia permissive or supportive to neurogenesis is not fully clear, but IGF-1 is a candidate molecule expressed by ischemia-induced microglia. Microglia (like immunity in general) thus seems to have a dual role in this context, being responsible for both detrimental effects as well as playing a permissive role (Ekdahl et al., 2009). Tumor necrosis factor alpha is up-regulated after stroke and suppressed the neurogenic response (Iosif et al., 2008).

Spreading depression stimulates a neurogenic response after stroke

The increase of neurogenesis at least in the adult hippocampus that is found after stroke is also mediated by neuronal activity. The brain might respond to ischemia with a particular pattern of neuronal activity, detectable in EEG, and referred to as "cortical spreading depression" (not to be confused with the mental problem of "post-stroke depression"). Induction of spreading depression also increased neurogenesis, thus linking ischemia mechanistically to the induction of neurogenesis in epilepsy (Urbach et al., 2008).

Hypoxia increases pro-neurogenesis genes and factors

Hypoxia as a stimulus for gene expression is an evolutionarily conserved mechanism that in bacteria governs the switch from aerobic to anaerobic growth states. Certain promoter systems can directly respond to hypoxia. A key example of hypoxia-inducible genes is Hif1a, which quite surprisingly has not yet been studied in the context of adult neurogenesis. Hypoxia induces a number of other factors that might underlie the neurogenic response to ischemia, including erythropoietin (Shingo et al., 2001), VEGF (Jin et al., 2002b; Schanzer et al., 2004), stem cell factor (Jin et al., 2002a), and heparin-binding EGF (Jin et al., 2002c).

The receptors for all of these are expressed in the SVZ, making their involvement plausible. Their relative contributions and the necessity of their involvement in the regulation remain unclear, however.

As in other cases of pathology-induced plasticity, NMDA receptor activation seems to play a prominent role in mediating the neurogenic response to stroke. Blockade of NMDA receptors prevented the induction of hippocampal neurogenesis in both global and focal ischemia models (Bernabeu and Sharp, 2000; Arvidsson et al., 2001).

Nitric oxide (NO) might also take part in ischemia-induced induction of adult hippocampal neurogenesis (Zhu et al., 2003; Sehara et al., 2006; Zhu et al., 2006). Inhibition of inducible nitric oxide synthase (iNOS) or an iNOS null mutation blocked the increase in neurogenesis (Zhu et al., 2003). NO signaling leads to CREB phosphorylation (see Chap. 9, p. 408). In the hippocampus, neuronal NOS (nNOS) went down, while iNOS was increased, in response to ischemia (Luo et al., 2007). Activation of iNOS and thus the induction of neurogenesis is dependent on the activity of L-type calcium channels (Luo et al., 2005).

The ischemia-induced neurogenic response can be influenced by exogenous factors

Transforming growth factor α (gene symbol: Tgfa), a growth factor in the EGF family of signaling molecules, supported stroke-induced neurogenesis even if applied only four weeks after the insult (Guerra-Crespo et al., 2009). Intranasal application of transforming growth factor β 1 (Tgfb1) bypasses the blood–brain barrier and not only reduced stroke volume but also increased neurogenesis from the SVZ (Ma et al., 2008). Related evidence also exists for FGF2 (Wang et al., 2008 Wada, 2003 #5364). In two models of ischemia, infusion of hematopoeitic factor G-CSF increased adult neurogenesis by improving survival (Schneider et al., 2005).

Brain-derived neurotrophic factor is a likely candidate mediator of the neurogenic response to stroke. BDNF is induced by nitric oxide (Chen et al., 2005a). In the healthy rat, infusion of BDNF into the ventricle and overexpression of BDNF through use of an adenovirus were reported to not only increase neurogenesis in the olfactory bulb but also induce neurogenesis in the striatum (Zigova et al., 1998; Benraiss et al., 2001; Pencea et al., 2001). Under ischemic conditions, adenovirally transferred overexpression of BDNF in neurons projecting to the damaged striatum increased neurogenesis (Gustafsson et al., 2003b), and the peripheral application of BDNF increased neurogenesis in the dentate gyrus and promoted migration of cells into the striatum (Schabitz et al., 2007). It therefore remains to be discovered whether ischemia-induced tissue conditions counteract BDNF action. In heterozygous BDNF knockout mice, however, ischemia-induced migration of "neuroblasts" into the striatum was increased (Nygren et al., 2006). Blockage of BDNF action in vivo by means of a TrkB fusion protein (TrkB being the receptor for BDNF) increased neuronal differentiation after ischemia without affecting cell proliferation (Gustafsson et al., 2003a). Statins, which have a clinically relevant neuroprotective effect, increase neurogenesis and angiogenesis after stroke (Chen et al., 2003). This action might be mediated through BDNF and VEGF (Chen et al., 2005b). BDNF also increased the number of putative new neurons in the penumbra of cortical ischemic lesions (Keiner et al., 2009) (but see the general caveat above and in Chap. 7, p. 236). Striatal neurogenesis after stroke was also enhanced by infusion of GDNF into the ventricle (Kobayashi et al., 2006). Retinoic acid, as a molecule promoting differentiation, also increased neurogenesis after ischemia (Plane et al., 2008).

Stroke early in life stimulates neurogenesis but might also result in lasting impairment

Ischemia can also damage precursor cells. When neonatal rats were exposed to ischemia, proliferation initially increased (Plane et al., 2004; Ong et al., 2005; Spiegler et al., 2007), returned to baseline after about three weeks (Iwai et al., 2006), but might result in lasting damage (Kadam et al., 2008). Interestingly, oligodendrocyte precursor cells, which are very numerous at this stage of brain development, were particularly vulnerable to ischemia. This resulted in white matter defects and a reduced SVZ (Levison et al., 2001; Back et al., 2002). That precursor cells show such a marked sensitivity dependent on the developmental stage might underlie the neurological consequences of hypoxia in utero or during birth. Hyperbaric oxygen treatment reversed some of the resulting deficits in hippocampus and SVZ (Yang et al., 2008b).

Demyelinating Disorders

Multiple sclerosis is an autoimmune disease causing demyelination and neuronal death

Multiple sclerosis is characterized by autoimmunity against myelin, leading to foci of demyelination and consecutive axonal damage as well as to direct, T cell–mediated neuronal damage. Oligodendrocytes that could remyelinate demyelinated axons are in the focus of attempts to treat multiple sclerosis with cell therapy (Karussis and Kassis, 2008). Only more recently have approaches to deal with the neuronal problems received more widespread attention. In experimental autoimmune encephalitis, the standard experimental model of multiple sclerosis, autoimmune T cells sustained adult hippocampal neurogenesis (Ziv et al., 2006), suggesting that a neurogenic response to the disease is inseparable from the disease process (see also Chap. 9, p. 377, on the regulation of adult neurogenesis by the immune system).

Endogenous precursor cells outside the neurogenic regions respond to the autoimmune lesion

In the adult brain, new oligodendrocytes originate from specified oligodendrocyte precursor cells among the NG2 cells of the SVZ, of the brain parenchyma, the fiber tracts, and only sparsely the SGZ (see Chap. 3, p. 71, Chap. 5, p. 175, Chap. 6, p. 208, and Chap. 8, p. 287). NG2-positive cells migrate in large numbers from the SVZ, mostly in a sub-callosal stream. Not all NG2 cells seem to be oligodendrocyte progenitor cells. In the normal adult brain, only relatively few new oligodendrocytes are generated (McCarthy and Leblond, 1988; Ehninger and Kempermann, 2003), but after demyelinating injury, at least a subset of NG2 cells can give rise to new oligodendrocytes (Redwine and Armstrong, 1998; Levine et al., 2001). Oligodendrocyte precursor cells have also been found near plaques in multiple sclerosis (Scolding et al., 1998; Maeda et al., 2001; Petratos et al., 2004). But in advanced stages of the disease, which are probably associated with increasing axonal damage, they become ineffective in remyelinating the axons (Wolswijk, 1998). Demyelinating lesions in the corpus callosum attract precursor cells from the SVZ, but it is not clear to what degree they can contribute to remyelination (Nait-Oumesmar et al., 1999). With increasing age of the animals in the experiment, this response decreased but could be reconstituted by growth factor injections (Decker et al., 2002). Demyelination due to chemical injury rather than autoimmunity induced an SVZ response leading to astrogenesis, possibly indicating lineage-specific regulation by the different paradigms (Nait-Oumesmar et al., 1999).

In the spinal cord of EAE rats, precursor cells from the central canal migrated into the parenchyma and expressed neuronal markers, possibly indicating an example of regenerative neurogenesis, which awaits confirmation (Danilov et al., 2006). Demyelinated plaques also caused a suggestive cellular response in the adult human brain, which—as in similar reports for other disorders—were identified as "immature neurons" (see Chap. 8 for the applicable caveats) (Chang et al., 2008).

Because demyelination does not seem to attract oligodendrocyte precursor cells from faraway brain regions, and only precursors at the plaque site participate in remyelination, damage to the precursor cells efficiently prevents further endogenous repair (Gensert and Goldman, 1997). Consequently, to achieve increased remyelination from intrinsic oligodendrocyte progenitor or multipotent precursor cells, the rescuing cells have to be attracted to the lesion, protected from damage, and induced to remyelinate over prolonged periods of time, barring progression of the underlying disease.

Neural precursor cells change properties in response to the autoimmune lesion

Demyelination in EAE also directly induced changes in precursor cells. Under the influence of interferon-γ or tumor necrosis factor-α, the precursor cells expressed markers such as CD80 and CD86, which are normally found on pluripotent cells. Cross-linking of CD80 on isolated EAE-induced precursor cells caused apoptotic cell death, a result suggesting that inflammation in EAE might directly damage neural precursor cells (Imitola et al., 2004). This finding is in accordance with the others that inflammation is detrimental to precursor cells.

Implanted and systematically applied neural precursor cells exert immune-modulatory functions

Peripheral infusion of neural precursor cells (in the form of neuro-spheres) resulted in remarkable behavioral improvements of the treated EAE mice, but this effect was not due to a differentiation into oligodendro-cytes or neurons. Rather, the implanted cells somehow seemed to contrib-ute to a modulating immune response (Pluchino et al., 2005). A similar observation had also been made after injection of neurospheres into the ventricle (Einstein et al., 2003). Such a beneficial effect has also been seen with peripherally or intraventricularly applied human neural precursor cells in EAE in marmoset monkeys, where the implanted cells persisted up to three months in the brain and were also found in draining lymph nodes (Pluchino et al., 2009a). One idea is that the neural precursor cells somehow prevent T cell activation in the lymph nodes (Einstein et al., 2007) and the expression of inflammatory cytokines (Fainstein et al., 2008). Neural precursor cells might also inter-fere with antigen-presenting dendritic cells (Pluchino et al., 2009b).

Interferon beta is successfully used as a therapeutic agent in the treatment of multiple sclerosis. In vitro, neural precursor cells expressed the interferon beta receptor and responded to interferon treatment with unchanged baseline neurogenesis, but decreased cell death upon growth factor with-drawal (Hirsch et al., 2009).

Trauma

Traumatic brain injury induces migration of cells of SVZ origin into the damaged regions

Trauma is mechanical injury to the brain. Besides the direct damage by cuts, contusion, or shear force, traumatic brain injury (often abbreviated as TBI) includes indirect secondary damage from axonopathies, ischemia, hypoxia, and toxicity from free radicals and excitatory amino acids. Traumatic injury thus comprises several types of cellular damage, each of which might harm precursor cells. The overall net effect is determined by the relative contribution of the individual damaging stimuli. Consequently, plasticity in response to trauma is variable, and so far no unique reaction of precursor cells to trauma has been identified. The response to trauma in neurogenic vs. non-neurogenic regions differs, and in contrast to ischemia, in which a few new neurons become detectable in the striatum, no cases of trauma-induced neurogenesis outside the neurogenic regions have yet been described. As after isch-emia, migration of cells expressing immature neuronal markers into the damaged regions occurs (Lu et al., 2003; Sun et al., 2004; Ramaswamy et al., 2005; Sundholm-Peters et al., 2005).

Trauma induces cell proliferation in the parenchyma and in the neurogenic zones, but stimulated neurogenesis has only been found in the hippocampus

It has been noted early that the adult SVZ responds to ipsilateral corti-cal injury with increased cell proliferation (Altman, 1962; Reznikov, 1975). This increase includes both a microglial response (Tzeng and Wu, 1999) and an increase in PSA-NCAM-expressing cells (Szele and Chesselet, 1996). The response has not yet been specified with respect to the precursor cell types in the SVZ. In a fluid percussion model of traumatic brain injury, induction of cell proliferation in both the SVZ and hippocampus was found (Dash et al., 2001; Chirumamilla et al., 2002); in the hippocampus this led to a net increase in neurogenesis (Dash et al., 2001). The same observation was made with other models of traumatic brain injury (Sun et al., 2007).

Induction of proliferation by this contusion-type trauma was not limited to the neurogenic regions but included the damaged cortex surrounding the injury site as well. Most of the dividing cells expressed nestin or glial fibrillary acidic protein (GFAP) (Kernie et al., 2001). Whereas in the hip-pocampus cortical injury robustly induced the generation of neurons by a factors of four or five, cell genesis in the cortex produced mostly astrocytes. These results suggest that this type of proliferation builds the glial scar found late after injury (Kernie et al., 2001). Interestingly, the induction of hip-pocampal neurogenesis also occurred contralateral to the injury (Lu et al., 2003) and even predated the ipsilateral response (Kernie et al., 2001).

The long-term fate of trauma-induced newborn cells in the SVZ is not yet clear. Percussion trauma induces cell division in the SVZ, which leads to increased astrogenesis (Holmin et al., 1997). However, it is not known whether brain injury nonspecifically induces neurogenesis in the olfactory bulb as it does in the hippocampus. Damage to the olfactory bulb itself induces cell proliferation in the SVZ, which can only lead to abortive neurogenesis because the target of cell migration is missing (Kirschenbaum et al., 1999; Li et al., 2002b). Immature migratory cells build up in the RMS. Limited migration from the SVZ to the corpus callosum and into the striatum has also been reported (Lu et al., 2003).

The glial response would have been described only as reactive gliosis until quite recently, but this term requires reconsideration, given the role of astrocytes as precursor cells (Silver and Steindler, 2009) (see also Chap. 8, p. 289). This new appreciation does not imply that the response would lead to neurogenesis rather than gliogenesis but indicates a more complex picture than the old idea that was largely limited to scar formation. When neural precursor cells were implanted into the lesioned adult brain, the cells differentiated along the immature neuronal and astroglial lineage on the ipsilateral side. Surprisingly, in one study only immature neuronal but no astroglial differentiation was found on the contralateral side (Riess et al., 2002).

As in stroke, supportive transplantation of bone marrow cells has been attempted in models of traumatic brain injury and yielded neuroprotective effects (Mahmood et al., 2005; Tsyb et al., 2009). Direct implantation improved recovery, but with respect to the reported occurrence of a few transdifferentiated neurons, the usual caveats apply (see Chap. 8, p. 286) (Bonilla et al., 2009).

One of the central concepts in neurodegeneration is excitatory toxicity. An overload of glutamate (or under experimental conditions, its agonists) triggers cell death and specific receptor-dependent tissue responses. Increased glutamate concentrations in the parenchyma can be found in many pathological circumstances. The injection of NMDA receptor agonist ibutenic acid can be used to model strong excitatory toxicity. In a first step, it leads to the degeneration of neuronal populations sensitive to glutamate. In the dentate gyrus, ibutenic acid injections induced both cell death and cell proliferation (Gould and Tanapat, 1997). In the RMS, ibutenic acid down-regulated proliferation and the number of calretinin-expressing cells. This finding might indicate that the calretinin stage is particularly sensitive to excitatory damage. Two weeks after the injury, cell proliferation rebounded, suggesting an attempt to compensate (Li et al., 2002b).

The trauma-induced neurogenic response in the dentate gyrus is reduced in FGF-2-deficient mice and can be stimulated by gene transfer with FGF-2 (Yoshimura et al., 2003) or the infusion of FGF2 (Sun et al., 2009). Similarly, a nitric oxide donor induced the endogenous neurogenic response (Lu et al., 2003), which might involve a BDNF-dependent mechanism. Treatment with statins, which have a similar effect, increased BDNF and VEGF after traumatic brain injury and were associated with increased adult hippocampal neurogenesis (Wu et al., 2008). Expression of VEGF might also be directly induced in blood vessels in the SVZ, far from the lesion (Gotts and Chesselet, 2005). Several studies have addressed the supporting effect of erythropoietin on the cellular response to traumatic brain injury (Xiong et al., 2008; Xiong et al., 2009). Infusion of S100b protein was also beneficial of hippocampal neurogenesis after trauma (Kleindienst et al., 2005). The migration of precursor cells to the site of injury is mediated by stem cell factor (SCF) and dependent on SCF-receptor c-kit on the precursor cells (Sun et al., 2004). These findings support the idea that neurogenesis after trauma (and possibly other aspects of precursor cell–based plasticity) can be supported exogenously, thereby opening a window for therapeutic interventions.

Spinal Cord Injury

The adult spinal cord of rodents and primates is a non-neurogenic region

Spinal cord injury is a special case of brain trauma, in that its most prominent symptoms are due to white matter damage. Consequently, regeneration of the damaged fiber tracts has to originate from neurons whose cell bodies are at long distances from the injury site. For regeneration to occur, the local microenvironment must become primarily permissive to axonal regrowth. Neuronal cell replacement at the damage site is a lesser issue. To enable regeneration, it is necessary to prevent scar formation, support and promote axonal elongation and path-finding, and remyelinate the regenerating axons. In therapeutic strategies developed toward this goal, precursor cells play diverse roles, but few have focused on the contribution of endogenous precursor cells (Horner and Gage, 2000). Whereas in the normal spinal cord a number of oligodendrocytes are produced, no further strong induction of oligodendrocytes has been found after damage. Transplantation studies confirmed that the normal and injured adult spinal cord is not permissive for neuronal differentiation and only partially permissive for oligodendrocytic differentiation (Horner et al., 2000; Shihabuddin et al., 2000; Cao et al., 2002). Astrocytes isolated from the adult spinal cord were not able to induce neurogenesis, whereas astrocytes from the neurogenic regions were (Song et al., 2002). The hypothesis that it might be BMPs that inhibit neuronal differentiation from implanted precursor cells was dismissed after transfection of the cells with noggin did not lead to more neurons and even increased the damage (Enzmann et al., 2005).

Adult amphibians can regenerate the spinal cord

Amphibian species show an amazing degree of regeneration after amputation of the tail, including the lower part of the spinal cord. Axolotl regrow the entire tail with the spinal cord, and several types of progenitor cells are involved in this process (Schnapp et al., 2005; McHedlishvili et al., 2007). In comparative approaches, researchers try to understand why axolotl can do this, while other species cannot (Tanaka and Ferretti, 2009). Profound understanding of regenerative neurogenesis in lower vertebrates might give insight into how regeneration could be achieved in other species as well. The interplay of different specialized precursor cells, which are found at the lesion site (Kragl et al., 2009), suggests however, that the potential of the precursor cells is a major determinant.

The adult spinal cord contains neural precursor cells but no neurogenesis occurs

Although the adult rodent spinal cord contains multipotent precursor cells that can generate neurons after implantation into neurogenic regions (Shihabuddin et al., 1997; Shihabuddin et al., 2000), no constitutive adult neurogenesis is found in the spinal cord; it also appears to remain absent after injury. Precursor cells are found near the central canal and within the gray and white matter, particularly the substantia gelatinosa (Horner et al., 2000). Many of the cells at the central canal express BLBP, Pax6, and Connexin 43 and show electric coupling, which suggests that they belong to a population of radial glia–like cells, similar to the precursor cells in the neurogenic niches (Russo et al., 2008). Both areas responded to traumatic injury with increased proliferation, and progeny from proliferating ependymal cells were described as migrating towards a compression lesion (Ke et al., 2006). Proliferation could also be stimulated by physical exercise (Cizkova et al., 2009). Except for one report describing reactive neurogenesis (Ke et al., 2006) the other studies in rodents (Yamamoto et al., 2001; Takahashi et al., 2003) and rhesus monkeys (Yang et al., 2006) found only the generation of new astrocytes from these precursor cells.

Pro-Neurogenic Drugs and Compounds

Neural precursor cells and adult neurogenesis have been discovered as targets for pharmacological interventions. Both "Big Pharma" and biotechnology startup companies have adventured into the market of pro-neurogenic drugs. The disorders that are the primary targets of this research are major depression and neurodegenerative disorders, but others are conceivable. In the previous paragraphs as well as in Chapter 9, several compounds have been mentioned that are pro-neurogenic. Antidepressants have received particular attention, but numerous others are effective as well: NMDA receptor antagonists, acetylcholine esterase inhibitors, low-dose cortisol, melatonin, statins, nitric oxide donors, anti-inflammatory agents, vitamins, etc. The problem with these compounds is their lack of specificity, so that major research efforts are undertaken to identify more specific pro-neurogenic drugs.

The United States Patent Office has so far issued several patents for neurogenic compounds or their targets. These include Prolactin (see p. 347; U.S. Patent No. 7,393,830), agonists to the Prokineticin receptor (see p. 371; 7,323,334), nitric oxide donors (see p. 386; 7,135,498), deacetylase inhibitors (see p. 409; 7,229,963), calcitonin analogs (6,969,702), secretoneurin (7,427,597), and a few others. Roughly fifty applications were pending at the time of this writing, including factors like FGF2 (see p. 383; United States Patent Application No. 20090111748), renin inhibitors (20090197823), GSK3b (20080188457), modafinil (20080171750), melatonergic agents (20080167363), and 5-HT receptor-mediated mechanisms (20070270449). Companies like Brain-Cells in San Diego (*www.braincellsinc.com*) or NeuroNova in Stockholm (*www.neuronova.com*) essentially build their portfolio and hopes on the success of such approaches.

To date, these factors are all relatively conventional, but small-molecule strategies aiming at manipulating very specific neurogenic mechanisms are on the horizon. For many of the patents, issued or pending, the scientific evidence supporting the proposed approach, however, is still rather slim, even though the factor might influence neurogenesis in an experimental setting. There might be many economical reasons for securing patent protection for ideas around individual mechanisms. But the idea that any of the mentioned factors by itself would promote neurogenesis in a functionally and therapeutically relevant way seems rather bold. The reason is that neurogenesis is an extremely complex process, regulated and controlled at different levels by numerous factors, and—most importantly—dependent on activity and function. The irony might be that in one pharmaceutical company, groups of researchers try to make brain cells grow, while others dedicate their life to treating brain cancer. In the end it seems likely that only integrated approaches will succeed. For drug development in the field of adult neurogenesis, it will become expensive to ignore systems biology and behavioral neurobiology. Neurogenic drugs will be possible and will give new hope for many incurable brain disorders, but it is probably safe to bet that a coming "blockbuster" in this area will look different from what is anticipated now.

Outlook

Adult neurogenesis is interesting for medical reasons, but not only for medical reasons. Adult neurogenesis is primarily a fundamental, albeit unusual, mechanism of plasticity, presumably critical for higher cognition. This mechanism can fail and succumb to disease, but neurogenesis remains a mechanism involved in physiological brain function, not in the brain's response to pathology. Although some regenerative neurogenesis is possible under certain circumstances, adult neurogenesis does not seem to have evolved as intrinsic tool for brain repair. Too many articles about adult neurogenesis prominently contain the sentence that the research in question was conducted because it would help to discover "new ways to treat neurodegenerative disease." This might occasionally turn out to be true, but for most individual manipulations it is a justification very much based on speculation. There is no evidence that simply "more neurons would be better." The field should avoid being reduced or reducing itself to too simple promises and messages. The reality is much more complicated but also much more beautiful, and in the end even more promising. The new neurons in the adult brain touch very central concepts about ourselves and about how the brain works. Unless one wants to extent the meaning of "medicine" to all of physiology, which seems to be a certain tendency in Western societies, the reduction of our interest in adult neurogenesis to its medical usefulness misses the point. The true philosophical, societal, and—somewhat ironically also—medical relevance of research on adult neurogenesis will lie in a deeper understanding of how the brain works.

References

Abayomi OK (1996). Pathogenesis of irradiation-induced cognitive dysfunction. *Acta Oncol* 35:659–663.

Aberg E, Hofstetter CP, Olson L, Brene S (2005). Moderate ethanol consumption increases hippocampal cell proliferation and neurogenesis in the adult mouse. *Int J Neuropsychopharmacol* 8:557–567.

Aboody KS, Brown A, Rainov NG, Bower KA, Liu S, Yang W, Small JE, Herrlinger U, Ourednik V, Black PM, Breakefield XO, Snyder EY (2000). Neural stem cells display extensive tropism for pathology in adult brain: Evidence from intracranial gliomas. *Proc Natl Acad Sci USA* 97:12846–12851.

Abrous DN, Adriani W, Montaron MF, Aurousseau C, Rougon G, Le Moal M, Piazza PV (2002). Nicotine self-administration impairs hippocampal plasticity. *J Neurosci* 22:3656–3662.

Allam MF, Campbell MJ, Del Castillo AS, Fernandez-Crehuet Navajas R (2004). Parkinson's disease protects against smoking? *Behav Neurol* 15:65–71.

Altman J (1962). Autoradiographic study of degenerative and regenerative proliferation of neuroglia cells with tritiated thymidine. *Exp Neurol* 5:302–318.

Aonurm-Helm A, Jurgenson M, Zharkovsky T, Sonn K, Berezin V, Bock E, Zharkovsky A (2008). Depression-like behaviour in neural cell adhesion molecule (NCAM)-deficient mice and its reversal by an NCAM-derived peptide, FGL. *Eur J Neurosci* 28:1618–1628.

Aponso PM, Faull RL, Connor B (2008). Increased progenitor cell proliferation and astrogenesis in the partial progressive 6-hydroxydopamine model of Parkinson's disease. *Neuroscience* 151:1142–1153.

Aramakis VB, Hsieh CY, Leslie FM, Metherate R (2000). A critical period for nicotine-induced disruption of synaptic development in rat auditory cortex. *J Neurosci* 20:6106–6116.

Arendash GW, Garcia MF, Costa DA, Cracchiolo JR, Wefes IM, Potter H (2004). Environmental enrichment improves cognition in aged Alzheimer's transgenic mice despite stable beta-amyloid deposition. *Neuroreport* 15:1751–1754.

Arlotta P, Magavi SS, Macklis JD (2003). Induction of adult neurogenesis: Molecular manipulation of neural precursors in situ. *Ann NY Acad Sci* 991:229–236.

Arvidsson A, Kokaia Z, Lindvall O (2001). N-methyl-D-aspartate receptor-mediated increase of neurogenesis in adult rat dentate gyrus following stroke. *Eur J Neurosci* 14:10–18.

Arvidsson A, Collin T, Kirik D, Kokaia Z, Lindvall O (2002). Neuronal replacement from endogenous precursors in the adult brain after stroke. *Nat Med* 8:963–970.

Austin CP, Ky B, Ma L, Morris JA, Shughrue PJ (2004). Expression of Disrupted-In-Schizophrenia-1, a schizophrenia-associated gene, is prominent in the mouse hippocampus throughout brain development. *Neuroscience* 124:3–10.

Auvergne R, Lere C, El Bahh B, Arthaud S, Lespinet V, Rougier A, Le Gal La Salle G (2002). Delayed kindling epileptogenesis and increased neurogenesis in adult rats housed in an enriched environment. *Brain Res* 954:277–285.

Babu H, Ramirez-Rodriguez G, Fabel K, Bischofberger J, Kempermann G (2009). Synaptic network activity induces neuronal differentiation of adult hippocampal precursor cells through BDNF signaling. *Frontiers in Neurogenesis* 1.

Bachoud-Levi AC et al. (2000). Motor and cognitive improvements in patients with Huntington's disease after neural transplantation. *Lancet* 356:1975–1979.

Back SA, Han BH, Luo NL, Chricton CA, Xanthoudakis S, Tam J, Arvin KL, Holtzman DM (2002). Selective vulnerability of late oligodendrocyte progenitors to hypoxia-ischemia. *J Neurosci* 22:455–463.

Baker SA, Baker KA, Hagg T (2004). Dopaminergic nigrostriatal projections regulate neural precursor proliferation in the adult mouse subventricular zone. *Eur J Neurosci* 20:575–579.

Banerjee SB, Rajendran R, Dias BG, Ladiwala U, Tole S, Vaidya VA (2005). Recruitment of the Sonic hedgehog signalling cascade in electroconvulsive seizure-mediated regulation of adult rat hippocampal neurogenesis. *Eur J Neurosci* 22:1570–1580.

Bar M (2009). A cognitive neuroscience hypothesis of mood and depression. *Trends Cogn Sci* 13:456–463.

Barresi V, Belluardo N, Sipione S, Mudo G, Cattaneo E, Condorelli DF (2003). Transplantation of prodrug-converting neural progenitor cells for brain tumor therapy. *Cancer Gene Ther* 10:396–402.

Becker S, Wojtowicz JM (2007). A model of hippocampal neurogenesis in memory and mood disorders. *Trends Cogn Sci* 11:70–76.

Becker S, Macqueen G, Wojtowicz JM (2009). Computational modeling and empirical studies of hippocampal neurogenesis-dependent memory: Effects of interference, stress and depression. *Brain Res* 1299:45–54.

Belluardo N, Mudo G, Bonomo A, Di Liberto V, Frinchi M, Fuxe K (2008). Nicotine-induced fibroblast growth factor-2 restores the age-related decline of precursor cell proliferation in the subventricular zone of rat brain. *Brain Res* 1193:12–24.

Ben Abdallah NM, Slomianka L, Lipp HP (2007). Reversible effect of X-irradiation on proliferation, neurogenesis, and cell death in the dentate gyrus of adult mice. *Hippocampus* 17:1230–1240.

Benedetti S, Pirola B, Pollo B, Magrassi L, Bruzzone MG, Rigamonti D, Galli R, Selleri S, Di Meco F, De Fraja C, Vescovi A, Cattaneo E, Finocchiaro G (2000). Gene therapy of experimental brain tumors using neural progenitor cells. *Nat Med* 6:447–450.

Bengzon J, Kokaia Z, Elmér E, Nanobashvili A, Kokaia M, Lindvall O (1997). Apoptosis and proliferation of dentate gyrus neurons after single and intermittent limbic seizures. *Proc Natl Acad Sci USA* 94:10432–10437.

Benraiss A, Chmielnicki E, Lerner K, Roh D, Goldman SA (2001). Adenoviral brain-derived neurotrophic factor induces both neostriatal and olfactory neuronal recruitment from endogenous progenitor cells in the adult forebrain. *J Neurosci* 21:6718–6731.

Berendse HW, Booij J, Francot CM, Bergmans PL, Hijman R, Stoof JC, Wolters EC (2001). Subclinical dopaminergic dysfunction in asymptomatic Parkinson's disease patients' relatives with a decreased sense of smell. *Ann Neurol* 50:34–41.

Bernabeu R, Sharp FR (2000). NMDA and AMPA/kainate glutamate receptors modulate dentate neurogenesis and CA3 synapsin-I in normal and ischemic hippocampus. *J Cereb Blood Flow Metab* 20:1669–1680.

Biggio F, Gorini G, Utzeri C, Olla P, Marrosu F, Mocchetti I, Follesa P (2009). Chronic vagus nerve stimulation induces neuronal plasticity in the rat hippocampus. *Int J Neuropsychopharmacol* 12:1209–1221.

Bjorklund A, Lindvall O (2000). Cell replacement therapies for central nervous system disorders. *Nat Neurosci* 3:537–544.

Bjorklund LM, Sanchez-Pernaute R, Chung S, Andersson T, Chen IY, McNaught KS, Brownell AL, Jenkins BG, Wahlestedt C, Kim KS, Isacson O (2002). Embryonic stem cells develop into functional dopaminergic neurons after transplantation in a Parkinson rat model. *Proc Natl Acad Sci USA* 8:8.

Bjornebekk A, Mathe AA, Brene S (2005). The antidepressant effect of running is associated with increased hippocampal cell proliferation. *Int J Neuropsychopharmacol* 8:357–368.

Bland ST, Beckley JT, Young S, Tsang V, Watkins LR, Maier SF, Bilbo SD (2009). Enduring consequences of early-life infection on glial and neural cell genesis within cognitive regions of the brain. *Brain Behav Immun* 24(3):329–38.

Blumcke I, Thom M, Wiestler OD (2002). Ammon's horn sclerosis: A maldevelopmental disorder associated with temporal lobe epilepsy. *Brain Pathol* 12:199–211.

Blumcke I, Beck H, Suter B, Hoffmann D, Fodisch HJ, Wolf HK, Schramm J, Elger CE, Wiestler OD (1999). An increase of hippocampal calretinin-immunoreactive neurons correlates with early febrile seizures in temporal lobe epilepsy. *Acta Neuropathol* (Berl) 97:31–39.

Boekhoorn K, Joels M, Lucassen PJ (2006). Increased proliferation reflects glial and vascular-associated changes, but not neurogenesis in the presenile Alzheimer hippocampus. *Neurobiol Dis* 24:1–14.

Boku S, Nakagawa S, Masuda T, Nishikawa H, Kato A, Kitaichi Y, Inoue T, Koyama T (2009). Glucocorticoids and lithium reciprocally regulate the proliferation of adult dentate gyrus-derived neural precursor cells through GSK-3beta and beta-catenin/TCF pathway. *Neuropsychopharmacology* 34:805–815.

Bondolfi L, Calhoun M, Ermini F, Kuhn HG, Wiederhold KH, Walker L, Staufenbiel M, Jucker M (2002). Amyloid-associated neuron loss and gliogenesis in the neocortex of amyloid precursor protein transgenic mice. *J Neurosci* 22:515–522.

Bonilla C, Zurita M, Otero L, Aguayo C, Vaquero J (2009). Delayed intralesional transplantation of bone marrow stromal cells increases endogenous neurogenesis and promotes functional recovery after severe traumatic brain injury. *Brain Inj* 23:760–769.

Borlongan CV, Hadman M, Sanberg CD, Sanberg PR (2004). Central nervous system entry of peripherally injected umbilical cord blood cells is not required for neuroprotection in stroke. *Stroke* 35:2385–2389.

Brandt MD, Jessberger S, Steiner B, Kronenberg G, Reuter K, Bick-Sander A, von der Behrens W, Kempermann G (2003). Transient calretinin expression defines early postmitotic step of neuronal differentiation in adult hippocampal neurogenesis of mice. *Mol Cell Neurosci* 24:603–613.

Breunig JJ, Arellano JI, Macklis JD, Rakic P (2007). Everything that glitters isn't gold: A critical review of postnatal neural precursor analyses. *Cell Stem Cell* 1:612–627.

Brustle O, Jones KN, Learish RD, Karram K, Choudhary K, Wiestler OD, Duncan ID, McKay RD (1999). Embryonic stem cell-derived glial precursors: A source of myelinating transplants. *Science* 285:754–756.

Butovsky O, Koronyo-Hamaoui M, Kunis G, Ophir E, Landa G, Cohen H, Schwartz M (2006). Glatiramer acetate fights against Alzheimer's disease by inducing dendritic-like microglia expressing insulin-like growth factor 1. *Proc Natl Acad Sci USA* 103:11784–11789.

Caille I, Allinquant B, Dupont E, Bouillot C, Langer A, Muller U, Prochiantz A (2004). Soluble form of amyloid precursor protein regulates proliferation of progenitors in the adult subventricular zone. *Development* 131:2173–2181.

Cao QL, Howard RM, Dennison JB, Whittemore SR (2002). Differentiation of engrafted neuronal-restricted precursor cells is inhibited in the traumatically injured spinal cord. *Exp Neurol* 177:349–359.

Cardon M, Ron-Harel N, Cohen H, Lewitus GM, Schwartz M (2009). Dysregulation of kisspeptin and neuro-genesis at adolescence link inborn immune deficits to the late onset of abnormal sensorimotor gating in congenital psychological disorders. *Mol Psychiatry* 15:415–25.

Carlen M, Meletis K, Goritz C, Darsalia V, Evergren E, Tanigaki K, Amendola M, Barnabe-Heider F, Yeung MS, Naldini L, Honjo T, Kokaia Z, Shupliakov O, Cassidy RM, Lindvall O, Frisen J (2009). Forebrain ependymal cells are Notch-dependent and generate neuroblasts and astrocytes after stroke. *Nat Neurosci* 12:259–267.

Casadesus G, Shukitt-Hale B, Stellwagen HM, Smith MA, Rabin BM, Joseph JA (2005). Hippocampal neurogen-esis and PSA-NCAM expression following exposure to 56Fe particles mimics that seen during aging in rats. *Exp Gerontol* 40:249–254.

Chang A, Smith MC, Yin X, Fox RJ, Staugaitis SM, Trapp BD (2008). Neurogenesis in the chronic lesions of multiple sclerosis. *Brain* 131:2366–2375.

Changeux JP, Bertrand D, Corringer PJ, Dehaene S, Edelstein S, Lena C, Le Novere N, Marubio L, Picciotto M, Zoli M (1998). Brain nicotinic receptors: Structure and regulation, role in learning and reinforcement. *Brain Res Brain Res* Rev 26:198–216.

Chen J, Magavi SS, Macklis JD (2004). Neurogenesis of corticospinal motor neurons extending spinal projections in adult mice. *Proc Natl Acad Sci USA* 101:16357–16362.

Chen J, Zacharek A, Zhang C, Jiang H, Li Y, Roberts C, Lu M, Kapke A, Chopp M (2005a). Endothelial nitric oxide synthase regulates brain-derived neurotrophic factor expression and neurogenesis after stroke in mice. *J Neurosci* 25:2366–2375.

Chen J, Zhang C, Jiang H, Li Y, Zhang L, Robin A, Katakowski M, Lu M, Chopp M (2005b). Atorvastatin induc-tion of VEGF and BDNF promotes brain plasticity after stroke in mice. *J Cereb Blood Flow Metab* 25:281–290.

Chen J, Zhang ZG, Li Y, Wang Y, Wang L, Jiang H, Zhang C, Lu M, Katakowski M, Feldkamp CS, Chopp M (2003). Statins induce angiogenesis, neurogenesis, and synaptogenesis after stroke. *Ann Neurol* 53:743–751.

Chesnokova V, Pechnick RN (2008). Antidepressants and Cdk inhibitors: Releasing the brake on neurogenesis? *Cell Cycle* 7:2321–2326.

Chi L, Ke Y, Luo C, Li B, Gozal D, Kalyanaraman B, Liu R (2006). Motor neuron degeneration promotes neural progenitor cell proliferation, migration, and neurogenesis in the spinal cords of amyotrophic lateral sclerosis mice. *Stem Cells* 24:34–43.

Chiba S, Suzuki M, Yamanouchi K, Nishihara M (2007). Involvement of granulin in estrogen-induced neurogen-esis in the adult rat hippocampus. *The Journal of reproduction and development* 53:297–307.

Chirumamilla S, Sun D, Bullock MR, Colello RJ (2002). Traumatic brain injury induced cell proliferation in the adult mammalian central nervous system. *J Neurotrauma* 19:693–703.

Cho SR, Benraiss A, Chmielnicki E, Samdani A, Economides A, Goldman SA (2007). Induction of neostriatal neurogenesis slows disease progression in a transgenic murine model of Huntington disease. *J Clin Invest* 117:2889–2902.

Choi IY, Allan AM, Cunningham LA (2005). Moderate fetal alcohol exposure impairs the neurogenic response to an enriched environment in adult mice. *Alcohol Clin Exp Res* 29:2053–2062.

Choi SH, Veeraraghavalu K, Lazarov O, Marler S, Ransohoff RM, Ramirez JM, Sisodia SS (2008). Non-cell-autonomous effects of presenilin 1 variants on enrichment-mediated hippocampal progenitor cell proliferation and differentiation. *Neuron* 59:568–580.

Ciaroni S, Cecchini T, Ferri P, Ambrogini P, Cuppini R, Riccio M, Lombardelli G, Papa S, Del Grande P (2002). Impairment of neural precursor proliferation increases survival of cell progeny in the adult rat dentate gyrus. *Mech Ageing Dev* 123:1341–1352.

Cicchetti F, Saporta S, Hauser RA, Parent M, Saint-Pierre M, Sanberg PR, Li XJ, Parker JR, Chu Y, Mufson EJ, Kordower JH, Freeman TB (2009). Neural transplants in patients with Huntington's disease undergo disease-like neuronal degeneration. *Proc Natl Acad Sci USA* 106:12483–12488.

Cipriani A, Brambilla P, Furukawa T, Geddes J, Gregis M, Hotopf M, Malvini L, Barbui C (2005). Fluoxetine versus other types of pharmacotherapy for depression. *Cochrane Database Syst Rev*:CD004185.

Cizkova D, Nagyova M, Slovinska L, Novotna I, Radonak J, Cizek M, Mechirova E, Tomori Z, Hlucilova J, Motlik J, Sulla I, Jr., Vanicky I (2009). Response of ependymal progenitors to spinal cord injury or enhanced physical activity in adult rat. *Cell Mol Neurobiol* 29:999–1013.

Clark PJ, Brzezinska WJ, Thomas MW, Ryzhenko NA, Toshkov SA, Rhodes JS (2008). Intact neurogenesis is required for benefits of exercise on spatial memory but not motor performance or contextual fear conditioning in C57BL/6J mice. *Neuroscience* 155:1048–1058.

Contestabile A, Fila T, Ceccarelli C, Bonasoni P, Bonapace L, Santini D, Bartesaghi R, Ciani E (2007). Cell cycle alteration and decreased cell proliferation in the hippocampal dentate gyrus and in the neocortical germinal matrix of fetuses with Down syndrome and in Ts65Dn mice. *Hippocampus* 17:665–678.

Cooper O, Isacson O (2004). Intrastriatal transforming growth factor alpha delivery to a model of Parkinson's disease induces proliferation and migration of endogenous adult neural progenitor cells without differentiation into dopaminergic neurons. *J Neurosci* 24:8924–8931.

Cooper-Kuhn CM, Kuhn HG (2002). Is it all DNA repair? Methodological considerations for detecting neurogenesis in the adult brain. *Brain Res Dev Brain Res* 134:13–21.

Cooper-Kuhn CM, Winkler J, Kuhn HG (2004). Decreased neurogenesis after cholinergic forebrain lesion in the adult rat. *J Neurosci Res* 77:155–165.

Corotto FS, Henegar JR, Maruniak JA (1994). Odor deprivation leads to reduced neurogenesis and reduced neuronal survival in the olfactory bulb of the adult mouse. *Neuroscience* 61:739–744.

Couillard-Despres S, Wuertinger C, Kandasamy M, Caioni M, Stadler K, Aigner R, Bogdahn U, Aigner L (2009). Ageing abolishes the effects of fluoxetine on neurogenesis. *Mol Psychiatry* 14:856–864.

Covolan L, Ribeiro LT, Longo BM, Mello LE (2000). Cell damage and neurogenesis in the dentate granule cell layer of adult rats after pilocarpine- or kainate-induced status epilepticus. *Hippocampus* 10:169–180.

Cowen DS, Takase LF, Fornal CA, Jacobs BL (2008). Age-dependent decline in hippocampal neurogenesis is not altered by chronic treatment with fluoxetine. *Brain Res* 1228:14–19.

Crews F, Nixon K, Kim D, Joseph J, Shukitt-Hale B, Qin L, Zou J (2006). BHT blocks NF-kappaB activation and ethanol-induced brain damage. *Alcohol Clin Exp Res* 30:1938–1949.

Crews FT, Nixon K, Wilkie ME (2004). Exercise reverses ethanol inhibition of neural stem cell proliferation. *Alcohol* 33:63–71.

Crossen JR, Garwood D, Glatstein E, Neuwelt EA (1994). Neurobehavioral sequelae of cranial irradiation in adults: A review of radiation-induced encephalopathy. *J Clin Oncol* 12:627–642.

Cruts M, Van Broeckhoven C (2008). Loss of progranulin function in frontotemporal lobar degeneration. *Trends Genet* 24:186–194.

Cui K, Ashdown H, Luheshi GN, Boksa P (2009). Effects of prenatal immune activation on hippocampal neurogenesis in the rat. *Schizophr Res* 113:288–297.

Curtis MA, Penney EB, Pearson J, Dragunow M, Connor B, Faull RL (2005). The distribution of progenitor cells in the subependymal layer of the lateral ventricle in the normal and Huntington's disease human brain. *Neuroscience* 132:777–788.

Curtis MA, Penney EB, Pearson AG, van Roon-Mom WM, Butterworth NJ, Dragunow M, Connor B, Faull RL (2003). Increased cell proliferation and neurogenesis in the adult human Huntington's disease brain. *Proc Natl Acad Sci USA* 100:9023–9027.

Czeh B, Michaelis T, Watanabe T, Frahm J, de Biurrun G, van Kampen M, Bartolomucci A, Fuchs E (2001). Stress-induced changes in cerebral metabolites, hippocampal volume, and cell proliferation are prevented by antidepressant treatment with tianeptine. *Proc Natl Acad Sci USA* 98:12796–12801.

Czeh B, Welt T, Fischer AK, Erhardt A, Schmitt W, Muller MB, Toschi N, Fuchs E, Keck ME (2002). Chronic psychosocial stress and concomitant repetitive transcranial magnetic stimulation: Effects on stress hormone levels and adult hippocampal neurogenesis. *Biol Psychiatry* 52:1057–1065.

D'Sa C, Duman RS (2002). Antidepressants and neuroplasticity. *Bipolar Disord* 4:183–194.

Danilov AI, Covacu R, Moe MC, Langmoen IA, Johansson CB, Olsson T, Brundin L (2006). Neurogenesis in the adult spinal cord in an experimental model of multiple sclerosis. *Eur J Neurosci* 23:394–400.

Danks MK, Yoon KJ, Bush RA, Remack JS, Wierdl M, Tsurkan L, Kim SU, Garcia E, Metz MZ, Najbauer J, Potter PM, Aboody KS (2007). Tumor-targeted enzyme/prodrug therapy mediates long-term disease-free survival of mice bearing disseminated neuroblastoma. *Cancer Res* 67:22–25.

Darsalia V, Heldmann U, Lindvall O, Kokaia Z (2005). Stroke-induced neurogenesis in aged brain. *Stroke* 36:1790–1795.

Dash PK, Mach SA, Moore AN (2001). Enhanced neurogenesis in the rodent hippocampus following traumatic brain injury. *J Neurosci Res* 63:313–319.

David DJ, Samuels BA, Rainer Q, Wang JW, Marsteller D, Mendez I, Drew M, Craig DA, Guiard BP, Guilloux JP, Artymyshyn RP, Gardier AM, Gerald C, Antonijevic IA, Leonardo ED, Hen R (2009). Neurogenesis-dependent and -independent effects of fluoxetine in an animal model of anxiety/depression. *Neuron* 62:479–493.

Dawirs RR, Hildebrandt K, Teuchert-Noodt G (1998). Adult treatment with haloperidol increases dentate granule cell proliferation in the gerbil hippocampus. *J Neural Transm* 105:317–127.

de Chevigny A, Cooper O, Vinuela A, Reske-Nielsen C, Lagace DC, Eisch AJ, Isacson O (2008). Fate mapping and lineage analyses demonstrate the production of a large number of striatal neuroblasts after transforming growth factor alpha and noggin striatal infusions into the dopamine-depleted striatum. *Stem Cells* 26:2349–2360.

Decker L, Picard-Riera N, Lachapelle F, Baron-Van Evercooren A (2002). Growth factor treatment promotes mobilization of young but not aged adult subventricular zone precursors in response to demyelination. *J Neurosci Res* 69:763–771.

Deierborg T, Staflin K, Pesic J, Roybon L, Brundin P, Lundberg C (2009). Absence of striatal newborn neurons with mature phenotype following defined striatal and cortical excitotoxic brain injuries. *Exp Neurol* 219:363–367.

Deisseroth K, Singla S, Toda H, Monje M, Palmer TD, Malenka RC (2004). Excitation-neurogenesis coupling in adult neural stem/progenitor cells. *Neuron* 42:535–552.

Deshauer D, Moher D, Fergusson D, Moher E, Sampson M, Grimshaw J (2008). Selective serotonin reuptake inhibitors for unipolar depression: A systematic review of classic long-term randomized controlled trials. *CMAJ* 178:1293–1301.

Dietrich J, Blumberg BM, Roshal M, Baker JV, Hurley SD, Mayer-Proschel M, Mock DJ (2004). Infection with an endemic human herpesvirus disrupts critical glial precursor cell properties. *J Neurosci* 24:4875–4883.

Dietrich J, Monje M, Wefel J, Meyers C (2008). Clinical patterns and biological correlates of cognitive dysfunction associated with cancer therapy. *Oncologist* 13:1285–1295.

Dill T, Schachinger V, Rolf A, Mollmann S, Thiele H, Tillmanns H, Assmus B, Dimmeler S, Zeiher AM, Hamm C (2009). Intracoronary administration of bone marrow-derived progenitor cells improves left ventricular function in patients at risk for adverse remodeling after acute ST-segment elevation myocardial infarction: Results of the Reinfusion of Enriched Progenitor cells And Infarct Remodeling in Acute Myocardial Infarction study (REPAIR-AMI) cardiac magnetic resonance imaging substudy. *Am Heart J* 157:541–547.

Doeppner TR, Nagel F, Dietz GP, Weise J, Tonges L, Schwarting S, Bahr M (2009). TAT-Hsp70–mediated neuroprotection and increased survival of neuronal precursor cells after focal cerebral ischemia in mice. *J Cereb Blood Flow Metab* 29:1187–1196.

Doetsch F, Caille I, Lim DA, Garcia-Verdugo JM, Alvarez-Buylla A (1999). Subventricular zone astrocytes are neural stem cells in the adult mammalian brain. *Cell* 97:703–716.

Dominguez-Escriba L, Hernandez-Rabaza V, Soriano-Navarro M, Barcia JA, Romero FJ, Garcia-Verdugo JM, Canales JJ (2006). Chronic cocaine exposure impairs progenitor proliferation but spares survival and maturation of neural precursors in adult rat dentate gyrus. *Eur J Neurosci* 24:586–594.

Duan X, Chang JH, Ge S, Faulkner RL, Kim JY, Kitabatake Y, Liu XB, Yang CH, Jordan JD, Ma DK, Liu CY, Ganesan S, Cheng HJ, Ming GL, Lu B, Song H (2007). Disrupted-In-Schizophrenia 1 regulates integration of newly generated neurons in the adult brain. *Cell* 130:1146–1158.

Eadie BD, Zhang WN, Boehme F, Gil-Mohapel J, Kainer L, Simpson JM, Christie BR (2009). Fmr1 knockout mice show reduced anxiety and alterations in neurogenesis that are specific to the ventral dentate gyrus. *Neurobiol Dis* 36:361–73.

Ehninger D, Kempermann G (2003). Regional effects of wheel running and environmental enrichment on cell genesis and microglia proliferation in the adult murine neocortex. *Cereb Cortex* 13:845–851.

Einstein O, Karussis D, Grigoriadis N, Mizrachi-Kol R, Reinhartz E, Abramsky O, Ben-Hur T (2003). Intraventricular transplantation of neural precursor cell spheres attenuates acute experimental allergic encephalomyelitis. *Mol Cell Neurosci* 24:1074–1082.

Einstein O, Fainstein N, Vaknin I, Mizrachi-Kol R, Reihartz E, Grigoriadis N, Lavon I, Baniyash M, Lassmann H, Ben-Hur T (2007). Neural precursors attenuate autoimmune encephalomyelitis by peripheral immunosuppression. *Ann Neurol* 61:209–218.

Eisch AJ, Barrot M, Schad CA, Self DW, Nestler EJ (2000). Opiates inhibit neurogenesis in the adult rat hippocampus. *Proc Natl Acad Sci USA* 97:7579–7584.

Eisch AJ, Cameron HA, Encinas JM, Meltzer LA, Ming GL, Overstreet-Wadiche LS (2008). Adult neurogenesis, mental health, and mental illness: Hope or hype? *J Neurosci* 28:11785–11791.

Ekdahl CT, Kokaia Z, Lindvall O (2009). Brain inflammation and adult neurogenesis: The dual role of microglia. *Neuroscience* 158:1021–1029.

Ekstrand J, Hellsten J, Wennstrom M, Tingstrom A (2008). Differential inhibition of neurogenesis and angiogenesis by corticosterone in rats stimulated with electroconvulsive seizures. *Prog Neuropsychopharmacol Biol Psychiatry* 32:1466–1472.

El Bahh B, Lespinet V, Lurton D, Coussemacq M, Le Gal La Salle G, Rougier A (1999). Correlations between granule cell dispersion, mossy fiber sprouting, and hippocampal cell loss in temporal lobe epilepsy. *Epilepsia* 40:1393–1401.

Encinas JM, Vaahtokari A, Enikolopov G (2006). Fluoxetine targets early progenitor cells in the adult brain. *Proc Natl Acad Sci USA* 103:8233–8238.

Enzmann GU, Benton RL, Woock JP, Howard RM, Tsoulfas P, Whittemore SR (2005). Consequences of noggin expression by neural stem, glial, and neuronal precursor cells engrafted into the injured spinal cord. *Exp Neurol* 195:293–304.

Eom TY, Jope RS (2009). Blocked inhibitory serine-phosphorylation of glycogen synthase kinase-3alpha/beta impairs in vivo neural precursor cell proliferation. *Biol Psychiatry* 66:494–502.

Fabel K, Wolf SA, Ehninger D, Babu H, Galicia PL, Kempermann G (2009). Additive effects of physical exercise and environmental enrichment on adult hippocampal neurogenesis in mice. *Frontiers Neurosci* 3:50.

Fainstein N, Vaknin I, Einstein O, Zisman P, Ben Sasson SZ, Baniyash M, Ben-Hur T (2008). Neural precursor cells inhibit multiple inflammatory signals. *Mol Cell Neurosci* 39:335–341.

Falleti MG, Sanfilippo A, Maruff P, Weih L, Phillips KA (2005). The nature and severity of cognitive impairment associated with adjuvant chemotherapy in women with breast cancer: A meta-analysis of the current literature. *Brain Cogn* 59:60–70.

Fallon J, Reid S, Kinyamu R, Opole I, Opole R, Baratta J, Korc M, Endo TL, Duong A, Nguyen G, Karkehabadhi M, Twardzik D, Patel S, Loughlin S (2000). In vivo induction of massive proliferation, directed migration, and differentiation of neural cells in the adult mammalian brain. *Proc Natl Acad Sci USA* 97:14686–14691.

Fan Y, Liu Z, Weinstein PR, Fike JR, Liu J (2007). Environmental enrichment enhances neurogenesis and improves functional outcome after cranial irradiation. *Eur J Neurosci* 25:38–46.

Fatemi SH, Earle JA, McMenomy T (2000). Reduction in Reelin immunoreactivity in hippocampus of subjects with schizophrenia, bipolar disorder and major depression. *Mol Psychiatry* 5:654–663, 571.

Feng R, Rampon C, Tang YP, Shrom D, Jin J, Kyin M, Sopher B, Martin GM, Kim SH, Langdon RB, Sisodia SS, Tsien JZ (2001). Deficient neurogenesis in forebrain-specific Presenilin-1 knockout mice is associated with reduced clearance of hippocampal memory traces. *Neuron* 32:911–926.

Ferland RJ, Gross RA, Applegate CD (2002). Increased mitotic activity in the dentate gyrus of the hippocampus of adult C57BL/6J mice exposed to the flurothyl kindling model of epileptogenesis. *Neuroscience* 115: 669–683.

Fernandez G, Effenberger O, Vinz B, Steinlein O, Elger CE, Dohring W, Heinze HJ (1998). Hippocampal malformation as a cause of familial febrile convulsions and subsequent hippocampal sclerosis. *Neurology* 50:909–917.

Fournier NM, Andersen DR, Botterill JJ, Sterner EY, Lussier AL, Caruncho HJ, Kalynchuk LE (2009). The effect of amygdala kindling on hippocampal neurogenesis coincides with decreased reelin and DISC1 expression in the adult dentate gyrus. *Hippocampus* 20:659-671.

Freed CR, Greene PE, Breeze RE, Tsai WY, DuMouchel W, Kao R, Dillon S, Winfield H, Culver S, Trojanowski JQ, Eidelberg D, Fahn S (2001). Transplantation of embryonic dopamine neurons for severe Parkinson's disease. *N Engl J Med* 344:710–719.

Freeman TB, Cicchetti F, Hauser RA, Deacon TW, Li XJ, Hersch SM, Nauert GM, Sanberg PR, Kordower JH, Saporta S, Isacson O (2000). Transplanted fetal striatum in Huntington's disease: Phenotypic development and lack of pathology. *Proc Natl Acad Sci USA* 97:13877–13882.

Frielingsdorf H, Schwarz K, Brundin P, Mohapel P (2004). No evidence for new dopaminergic neurons in the adult mammalian substantia nigra. *Proc Natl Acad Sci USA* 101:10177–10182.

Fuchs E, Czeh B, Michaelis T, de Biurrun G, Watanabe T, Frahm J (2002). Synaptic plasticity and tianeptine: Structural regulation. *Eur Psychiatry* 17 Suppl 3:311–317.

Fuss J, Ben Abdallah NM, Vogt MA, Touma C, Pacifici PG, Palme R, Witzemann V, Hellweg R, Gass P (2009). Voluntary exercise induces anxiety-like behavior in adult C57BL/6J mice correlating with hippocampal neuro-genesis. *Hippocampus* 20: 364–376.

Garbelli R, Frassoni C, Ferrario A, Tassi L, Bramerio M, Spreafico R (2001). Cajal-Retzius cell density as marker of type of focal cortical dysplasia. *Neuroreport* 12:2767–2771.

Garbuzova-Davis S, Willing AE, Zigova T, Saporta S, Justen EB, Lane JC, Hudson JE, Chen N, Davis CD, Sanberg PR (2003). Intravenous administration of human umbilical cord blood cells in a mouse model of amyotrophic lateral sclerosis: Distribution, migration, and differentiation. *J Hematother Stem Cell Res* 12:255–270.

Garthe A, Behr J, Kempermann G (2009). Adult-generated hippocampal neurons allow the flexible use of spatially precise learning strategies. *PLoS ONE* 4:e5464.

Gass J et al. (2006). Mutations in progranulin are a major cause of ubiquitin-positive frontotemporal lobar degeneration. *Hum Mol Genet* 15:2988–3001.

Gensert JM, Goldman JE (1997). Endogenous progenitors remyelinate demyelinated axons in the adult CNS. *Neuron* 19:197–203.

Gerber J, Bottcher T, Bering J, Bunkowski S, Bruck W, Kuhnt U, Nau R (2003). Increased neurogenesis after experimental *Streptococcus pneumoniae* meningitis. *J Neurosci Res* 73:441–446.

Geuze E, Vermetten E, Bremner JD (2005). MR-based in vivo hippocampal volumetrics: 2. Findings in neuropsychiatric disorders. *Mol Psychiatry* 10:160–184.

Ghashghaei HT, Weber J, Pevny L, Schmid R, Schwab MH, Lloyd KC, Eisenstat DD, Lai C, Anton ES (2006). The role of neuregulin-ErbB4 interactions on the proliferation and organization of cells in the subventricular zone. *Proc Natl Acad Sci USA* 103:1930–1935.

Glass R, Synowitz M, Kronenberg G, Walzlein JH, Markovic DS, Wang LP, Gast D, Kiwit J, Kempermann G, Kettenmann H (2005). Glioblastoma-induced attraction of endogenous neural precursor cells is associated with improved survival. *J Neurosci* 25:2637–2646.

Gong C, Wang TW, Huang HS, Parent JM (2007). Reelin regulates neuronal progenitor migration in intact and epileptic hippocampus. *J Neurosci* 27:1803–1811.

Gonzalo-Gobernado R, Reimers D, Herranz AS, Diaz-Gil JJ, Osuna C, Asensio MJ, Baena S, Rodriguez-Serrano M, Bazan E (2009). Mobilization of neural stem cells and generation of new neurons in 6-OHDA-lesioned rats by intracerebroventricular infusion of liver growth factor. *J Histochem Cytochem* 57:491–502.

Gotts JE, Chesselet MF (2005). Vascular changes in the subventricular zone after distal cortical lesions. *Exp Neurol* 194:139–150.

Gould E, Tanapat P (1997). Lesion-induced proliferation of neuronal progenitor cells in the dentate gyrus of the adult rat. *Neuroscience* 80:427–436.

Grassi Zucconi G, Cipriani S, Balgkouranidou I, Scattoni R (2006). "One night" sleep deprivation stimulates hippocampal neurogenesis. *Brain Res Bull* 69:375–381.

Gray WP, Sundstrom LE (1998). Kainic acid increases the proliferation of granule cell progenitors in the dentate gyrus of the adult rat. *Brain Res* 790:52–59.

Graziadei PP, Monti Graziadei GA (1980). Neurogenesis and neuron regeneration in the olfactory system of mammals. III. Deafferentation and reinnervation of the olfactory bulb following section of the fila olfactoria in rat. *J Neurocytol* 9:145–162.

Greenberg DA (2007). Neurogenesis and stroke. *CNS Neurol Disord Drug Targets* 6:321–325.

Gregorian C, Nakashima J, Le Belle J, Ohab J, Kim R, Liu A, Smith KB, Groszer M, Garcia AD, Sofroniew MV, Carmichael ST, Kornblum HI, Liu X, Wu H (2009). Pten deletion in adult neural stem/progenitor cells enhances constitutive neurogenesis. *J Neurosci* 29:1874–1886.

Gu W, Brannstrom T, Wester P (2000). Cortical neurogenesis in adult rats after reversible photothrombotic stroke. *J Cereb Blood Flow Metab* 20:1166–1173.

Gu W, Brannstrom T, Rosqvist R, Wester P (2009). Cell division in the cerebral cortex of adult rats after photothrombotic ring stroke. *Stem Cell Res* 2:68–77.

Guerra-Crespo M, Gleason D, Sistos A, Toosky T, Solaroglu I, Zhang JH, Bryant PJ, Fallon JH (2009). Transforming growth factor-alpha induces neurogenesis and behavioral improvement in a chronic stroke model. *Neuroscience* 160:470–483.

Guidi S, Bonasoni P, Ceccarelli C, Santini D, Gualtieri F, Ciani E, Bartesaghi R (2008). Neurogenesis impairment and increased cell death reduce total neuron number in the hippocampal region of fetuses with Down syndrome. *Brain Pathol* 18:180–197.

Guo YJ, Zhang ZJ, Wang SH, Sui YX, Sun Y (2009). Notch1 signaling, hippocampal neurogenesis and behavioral responses to chronic unpredicted mild stress in adult ischemic rats. *Prog Neuropsychopharmacol Biol Psychiatry* 33:688–694.

Gustafsson E, Lindvall O, Kokaia Z (2003a). Intraventricular infusion of TrkB-Fc fusion protein promotes ischemia-induced neurogenesis in adult rat dentate gyrus. *Stroke* 34:2710–2715.

Gustafsson E, Andsberg G, Darsalia V, Mohapel P, Mandel RJ, Kirik D, Lindvall O, Kokaia Z (2003b). Anterograde delivery of brain-derived neurotrophic factor to striatum via nigral transduction of recombinant adeno-associated virus increases neuronal death but promotes neurogenic response following stroke. *Eur J Neurosci* 17:2667–2678.

Haas CA, Frotscher M (2010). Reelin deficiency causes granule cell dispersion in epilepsy. *Exp Brain Res*. 200:141–9.

Haas S, Weidner N, Winkler J (2005). Adult stem cell therapy in stroke. *Curr Opin Neurol* 18:59–64.

Hagell P, Piccini P, Bjorklund A, Brundin P, Rehncrona S, Widner H, Crabb L, Pavese N, Oertel WH, Quinn N, Brooks DJ, Lindvall O (2002). Dyskinesias following neural transplantation in Parkinson's disease. *Nat Neurosci* 5:627–628.

Halim ND, Weickert CS, McClintock BW, Weinberger DR, Lipska BK (2004). Effects of chronic haloperidol and clozapine treatment on neurogenesis in the adult rat hippocampus. *Neuropsychopharmacology* 29:1063–1069.

Hattiangady B, Rao MS, Shetty AK (2004). Chronic temporal lobe epilepsy is associated with severely declined dentate neurogenesis in the adult hippocampus. *Neurobiol Dis* 17:473–490.

Haughey NJ, Liu D, Nath A, Borchard AC, Mattson MP (2002a). Disruption of neurogenesis in the subventricular zone of adult mice, and in human cortical neuronal precursor cells in culture, by amyloid beta-peptide: Implications for the pathogenesis of Alzheimer's disease. *Neuromolecular Med* 1:125–135.

Haughey NJ, Nath A, Chan SL, Borchard AC, Rao MS, Mattson MP (2002b). Disruption of neurogenesis by amyloid beta-peptide, and perturbed neural progenitor cell homeostasis, in models of Alzheimer's disease. *J Neurochem* 83:1509–1524.

Hauser RA, Furtado S, Cimino CR, Delgado H, Eichler S, Schwartz S, Scott D, Nauert GM, Soety E, Sossi V, Holt DA, Sanberg PR, Stoessl AJ, Freeman TB (2002). Bilateral human fetal striatal transplantation in Huntington's disease. *Neurology* 58:687–695.

He P, Shen Y (2009). Interruption of beta-catenin signaling reduces neurogenesis in Alzheimer's disease. *J Neurosci* 29:6545–6557.

He XJ, Nakayama H, Dong M, Yamauchi H, Ueno M, Uetsuka K, Doi K (2006). Evidence of apoptosis in the subventricular zone and rostral migratory stream in the MPTP mouse model of Parkinson disease. *J Neuropathol Exp Neurol* 65:873–882.

He XJ, Yamauchi H, Uetsuka K, Nakayama H (2008). Neurotoxicity of MPTP to migrating neuroblasts: Studies in acute and subacute mouse models of Parkinson's disease. *Neurotoxicology* 29:413–420.

Hede K (2008). Chemobrain is real but may need new name. *J Natl Cancer Inst* 100:162–163, 169.

Heinrich C, Nitta N, Flubacher A, Muller M, Fahrner A, Kirsch M, Freiman T, Suzuki F, Depaulis A, Frotscher M, Haas CA (2006). Reelin deficiency and displacement of mature neurons, but not neurogenesis, underlie the formation of granule cell dispersion in the epileptic hippocampus. *J Neurosci* 26:4701–4713.

Helfer JL, Goodlett CR, Greenough WT, Klintsova AY (2009). The effects of exercise on adolescent hippocampal neurogenesis in a rat model of binge alcohol exposure during the brain growth spurt. *Brain Res* 1294:1–11.

Hellsten J, Wennstrom M, Bengzon J, Mohapel P, Tingstrom A (2004). Electroconvulsive seizures induce endothelial cell proliferation in adult rat hippocampus. *Biol Psychiatry* 55:420–427.

Hellstrom NA, Bjork-Eriksson T, Blomgren K, Kuhn HG (2009). Differential recovery of neural stem cells in the subventricular zone and dentate gyrus after ionizing radiation. *Stem Cells* 27:634–641.

Herbots L, D'Hooge J, Eroglu E, Thijs D, Ganame J, Claus P, Dubois C, Theunissen K, Bogaert J, Dens J, Kalantzi M, Dymarkowski S, Bijnens B, Belmans A, Boogaerts M, Sutherland G, Van de Werf F, Rademakers F, Janssens S (2009). Improved regional function after autologous bone marrow-derived stem cell transfer in patients with acute myocardial infarction: A randomized, double-blind strain rate imaging study. *Eur Heart J* 30:662–670.

Hernan MA, Takkouche B, Caamano-Isorna F, Gestal-Otero JJ (2002). A meta-analysis of coffee drinking, cigarette smoking, and the risk of Parkinson's disease. *Ann Neurol* 52:276–284.

Hernandez-Rabaza V, Dominguez-Escriba L, Barcia JA, Rosel JF, Romero FJ, Garcia-Verdugo JM, Canales JJ (2006). Binge administration of 3,4-methylenedioxymethamphetamine ("Ecstasy") impairs the survival of neural precursors in adult rat dentate gyrus. *Neuropharmacology* 51:967–973.

Hernandez-Rabaza V, Llorens-Martin M, Velazquez-Sanchez C, Ferragud A, Arcusa A, Gumus HG, Gomez-Pinedo U, Perez-Villalba A, Rosello J, Trejo JL, Barcia JA, Canales JJ (2009). Inhibition of adult hippocampal neurogenesis disrupts contextual learning but spares spatial working memory, long-term conditional rule retention and spatial reversal. *Neuroscience* 159:59–68.

Herrera DG, Yague AG, Johnsen-Soriano S, Bosch-Morell F, Collado-Morente L, Muriach M, Romero FJ, Garcia-Verdugo JM (2003). Selective impairment of hippocampal neurogenesis by chronic alcoholism: Protective effects of an antioxidant. *Proc Natl Acad Sci USA* 100:7919–7924.

Herrlinger U, Woiciechowski C, Sena-Esteves M, Aboody KS, Jacobs AH, Rainov NG, Snyder EY, Breakefield XO (2000). Neural precursor cells for delivery of replication-conditional HSV-1 vectors to intracerebral gliomas. *Mol Ther* 1:347–357.

Herrup K, Yang Y (2001). Pictures in molecular medicine: Contemplating Alzheimer's disease as cancer: A loss of cell-cycle control. *Trends Mol Med* 7:527.

Herrup K, Yang Y (2007). Cell cycle regulation in the postmitotic neuron: Oxymoron or new biology? *Nat Rev Neurosci* 8:368–378.

Hirsch M, Knight J, Tobita M, Soltys J, Panitch H, Mao-Draayer Y (2009). The effect of interferon-beta on mouse neural progenitor cell survival and differentiation. *Biochem Biophys Res Commun* 388:181–186.

Hodes GE, Yang L, Van Kooy J, Santollo J, Shors TJ (2009). Prozac during puberty: Distinctive effects on neurogenesis as a function of age and sex. *Neuroscience* 163:609–617.

Hoglinger GU, Rizk P, Muriel MP, Duyckaerts C, Oertel WH, Caille I, Hirsch EC (2004). Dopamine depletion impairs precursor cell proliferation in Parkinson disease. *Nat Neurosci* 7:726–735.

Holick KA, Lee DC, Hen R, Dulawa SC (2008). Behavioral effects of chronic fluoxetine in BALB/cJ mice do not require adult hippocampal neurogenesis or the serotonin 1A receptor. *Neuropsychopharmacology* 33:406–417.

Holmin S, Almqvist P, Lendahl U, Mathiesen T (1997). Adult nestin-expressing subependymal cells differentiate to astrocytes in response to brain injury. *Eur J Neurosci* 9:65–75.

Hong XP, Peng CX, Wei W, Tian Q, Liu YH, Yao XQ, Zhang Y, Cao FY, Wang Q, Wang JZ (2009). Essential role of tau phosphorylation in adult hippocampal neurogenesis. *Hippocampus*, Oct 8. [Epub ahead of print].

Horner PJ, Gage FH (2000). Regenerating the damaged central nervous system. *Nature* 407:963–970.

Horner PJ, Power AE, Kempermann G, Kuhn HG, Palmer TD, Winkler J, Thal LJ, Gage FH (2000). Proliferation and differentiation of progenitor cells throughout the intact adult rat spinal cord. *J Neurosci* 20:2218–2228.

Hou SW, Wang YQ, Xu M, Shen DH, Wang JJ, Huang F, Yu Z, Sun FY (2008). Functional integration of newly generated neurons into striatum after cerebral ischemia in the adult rat brain. *Stroke* 39:2837–2844.

Houser CR (1990). Granule cell dispersion in the dentate gyrus of humans with temporal lobe epilepsy. *Brain Res* 535:195–204.

Hsieh J, Nakashima K, Kuwabara T, Mejia E, Gage FH (2004). Histone deacetylase inhibition-mediated neuronal differentiation of multipotent adult neural progenitor cells. *Proc Natl Acad Sci USA* 101:16659–16664.

Hua Y, Huang XY, Zhou L, Zhou QG, Hu Y, Luo CX, Li F, Zhu DY (2008). DETA/NONOate, a nitric oxide donor, produces antidepressant effects by promoting hippocampal neurogenesis. *Psychopharmacology* (Berl) 200:231–242.

Huttmann K, Sadgrove M, Wallraff A, Hinterkeuser S, Kirchhoff F, Steinhauser C, Gray WP (2003). Seizures preferentially stimulate proliferation of radial glia–like astrocytes in the adult dentate gyrus: Functional and immunocytochemical analysis. *Eur J Neurosci* 18:2769–2778.

Ieraci A, Herrera DG (2007). Single alcohol exposure in early life damages hippocampal stem/progenitor cells and reduces adult neurogenesis. *Neurobiol Dis* 26:597–605.

Imitola J, Comabella M, Chandraker AK, Dangond F, Sayegh MH, Snyder EY, Khoury SJ (2004). Neural stem/progenitor cells express costimulatory molecules that are differentially regulated by inflammatory and apoptotic stimuli. *Am J Pathol* 164:1615–1625.

Iosif RE, Ahlenius H, Ekdahl CT, Darsalia V, Thored P, Jovinge S, Kokaia Z, Lindvall O (2008). Suppression of stroke-induced progenitor proliferation in adult subventricular zone by tumor necrosis factor receptor 1. *J Cereb Blood Flow Metab* 28:1574–1587.

Iwai M, Hayashi T, Zhang WR, Sato K, Manabe Y, Abe K (2001). Induction of highly polysialylated neural cell adhesion molecule (PSA-NCAM) in postischemic gerbil hippocampus mainly dissociated with neural stem cell proliferation. *Brain Res* 902:288–293.

Iwai M, Ikeda T, Hayashi T, Sato K, Nagata T, Nagano I, Shoji M, Ikenoue T, Abe K (2006). Temporal profile of neural stem cell proliferation in the subventricular zone after ischemia/hypoxia in the neonatal rat brain. *Neurol Res* 28:461–468.

Iwata Y, Suzuki K, Wakuda T, Seki N, Thanseem I, Matsuzaki H, Mamiya T, Ueki T, Mikawa S, Sasaki T, Suda S, Yamamoto S, Tsuchiya KJ, Sugihara G, Nakamura K, Sato K, Takei N, Hashimoto K, Mori N (2008). Irradiation in adulthood as a new model of schizophrenia. *PLoS One* 3:e2283.

Jacobs BL, Praag H, Gage FH (2000). Adult brain neurogenesis and psychiatry: A novel theory of depression. *Mol Psychiatry* 5:262–269.

Jakubs K, Nanobashvili A, Bonde S, Ekdahl CT, Kokaia Z, Kokaia M, Lindvall O (2006). Environment matters: Synaptic properties of neurons born in epileptic brain develop to reduce excitability. *Neuron* 52:1047–1059,

James JR, Nordberg A (1995). Genetic and environmental aspects of the role of nicotinic receptors in neurodegenerative disorders: Emphasis on Alzheimer's disease and Parkinson's disease. *Behav Genet* 25:149–159.

Jankovski A, Garcia C, Soriano E, Sotelo C (1998). Proliferation, migration and differentiation of neuronal progenitor cells in the adult mouse subventricular zone surgically separated from its olfactory bulb. *Eur J Neurosci* 10:3853–3868.

Jankowsky JL, Xu G, Fromholt D, Gonzales V, Borchelt DR (2003). Environmental enrichment exacerbates amyloid plaque formation in a transgenic mouse model of Alzheimer disease. *J Neuropathol Exp Neurol* 62:1220–1227.

Jessberger S, Romer B, Babu H, Kempermann G (2005). Seizures induce proliferation and dispersion of doublecortin-positive hippocampal progenitor cells. *Exp Neurol* 196:342–351.

Jessberger S, Zhao C, Toni N, Clemenson GD, Jr., Li Y, Gage FH (2007a). Seizure-associated, aberrant neurogenesis in adult rats characterized with retrovirus-mediated cell labeling. *J Neurosci* 27:9400–9407.

Jessberger S, Nakashima K, Clemenson GD, Jr., Mejia E, Mathews E, Ure K, Ogawa S, Sinton CM, Gage FH, Hsieh J (2007b). Epigenetic modulation of seizure-induced neurogenesis and cognitive decline. *J Neurosci* 27:5967–5975.

Jiang W, Gu W, Brannstrom T, Rosqvist R, Wester P (2001). Cortical neurogenesis in adult rats after transient middle cerebral artery occlusion. *Stroke* 32:1201–1207.

Jiang W, Wan Q, Zhang ZJ, Wang WD, Huang YG, Rao ZR, Zhang X (2003). Dentate granule cell neurogenesis after seizures induced by pentylenetrazol in rats. *Brain Res* 977:141–148.

Jin K, Minami M, Lan JQ, Mao XO, BattEur S, Simon RP, Greenberg DA (2001). Neurogenesis in dentate subgranular zone and rostral subventricular zone after focal cerebral ischemia in the rat. *Proc Natl Acad Sci USA* 98:4710–4715.

Jin K, Mao XO, Sun Y, Xie L, Greenberg DA (2002a). Stem cell factor stimulates neurogenesis in vitro and in vivo. *J Clin Invest* 110:311–319.

Jin K, Zhu Y, Sun Y, Mao XO, Xie L, Greenberg DA (2002b). Vascular endothelial growth factor (VEGF) stimulates neurogenesis in vitro and in vivo. *Proc Natl Acad Sci USA* 99:11946–11950.

Jin K, Mao XO, Sun Y, Xie L, Jin L, Nishi E, Klagsbrun M, Greenberg DA (2002c). Heparin-binding epidermal growth factor-like growth factor: Hypoxia-inducible expression in vitro and stimulation of neurogenesis in vitro and in vivo. *J Neurosci* 22:5365–5373.

Jin K, Peel AL, Mao XO, Xie L, Cottrell BA, Henshall DC, Greenberg DA (2004a). Increased hippocampal neurogenesis in Alzheimer's disease. *Proc Natl Acad Sci USA* 101:343–347.

Jin K, Minami M, Xie L, Sun Y, Mao XO, Wang Y, Simon RP, Greenberg DA (2004b). Ischemia-induced neurogenesis is preserved but reduced in the aged rodent brain. *Aging Cell* 3:373–377.

Jin K, Wang X, Xie L, Mao XO, Zhu W, Wang Y, Shen J, Mao Y, Banwait S, Greenberg DA (2006). Evidence for stroke-induced neurogenesis in the human brain. *Proc Natl Acad Sci USA* 103:13198–13202.

Kadam SD, Mulholland JD, McDonald JW, Comi AM (2008). Neurogenesis and neuronal commitment following ischemia in a new mouse model for neonatal stroke. *Brain Res* 1208:35–45.

Kahn L, Alonso G, Normand E, Manzoni OJ (2005). Repeated morphine treatment alters polysialylated neural cell adhesion molecule, glutamate decarboxylase-67 expression and cell proliferation in the adult rat hippocampus. *Eur J Neurosci* 21:493–500.

Kamsler A, Avital A, Greenberger V, Segal M (2007). Aged SOD overexpressing mice exhibit enhanced spatial memory while lacking hippocampal neurogenesis. *Antioxidants & Redox Signaling* 9:181–189.

Karumbayaram S, Kelly TK, Paucar AA, Roe AJ, Umbach JA, Charles A, Goldman SA, Kornblum HI, Wiedau-Pazos M (2009). Human embryonic stem cell-derived motor neurons expressing SOD1 mutants exhibit typical signs of motor neuron degeneration linked to ALS. *Disease Models & Mechanisms* 2:189–195.

Karussis D, Kassis I (2008). The potential use of stem cells in multiple sclerosis: An overview of the preclinical experience. *Clin Neurol Neurosurg* 110:889–896.

Katchanov J, Harms C, Gertz K, Hauck L, Waeber C, Hirt L, Priller J, von Harsdorf R, Bruck W, Hortnagl H, Dirnagl U, Bhide PG, Endres M (2001). Mild cerebral ischemia induces loss of cyclin-dependent kinase inhibitors and activation of cell cycle machinery before delayed neuronal cell death. *J Neurosci* 21:5045–5053.

Katz MM, Frazer A, Bowden CL (2006a). "Delay" hypothesis of onset of antidepressant action. *Br J Psychiatry* 188:586; author reply 587.

Katz MM, Bowden CL, Berman N, Frazer A (2006b). Resolving the onset of antidepressants' clinical actions: Critical for clinical practice and new drug development. *J Clin Psychopharmacol* 26:549–553.

Kawada H, Takizawa S, Takanashi T, Morita Y, Fujita J, Fukuda K, Takagi S, Okano H, Ando K, Hotta T (2006). Administration of hematopoietic cytokines in the subacute phase after cerebral infarction is effective for functional recovery facilitating proliferation of intrinsic neural stem/progenitor cells and transition of bone marrow-derived neuronal cells. *Circulation* 113:701–710.

Ke Y, Chi L, Xu R, Luo C, Gozal D, Liu R (2006). Early response of endogenous adult neural progenitor cells to acute spinal cord injury in mice. *Stem Cells* 24:1011–1019.

Kee NJ, Preston E, Wojtowicz JM (2001). Enhanced neurogenesis after transient global ischemia in the dentate gyrus of the rat. *Exp Brain Res* 136:313–320.

Keene CD, Sonnen JA, Swanson PD, Kopyov O, Leverenz JB, Bird TD, Montine TJ (2007). Neural transplantation in Huntington disease: Long-term grafts in two patients. *Neurology* 68:2093–2098.

Keene CD, Chang RC, Leverenz JB, Kopyov O, Perlman S, Hevner RF, Born DE, Bird TD, Montine TJ (2009). A patient with Huntington's disease and long-surviving fetal neural transplants that developed mass lesions. *Acta Neuropathol* 117:329–338.

Keilhoff G, Grecksch G, Bernstein HG, Roskoden T, Becker A (2009). Risperidone and haloperidol promote survival of stem cells in the rat hippocampus. *Eur Arch Psychiatry Clin Neurosci.* 260:151–162.

Keiner S, Witte OW, Redecker C (2009). Immunocytochemical detection of newly generated neurons in the perilesional area of cortical infarcts after intraventricular application of brain-derived neurotrophic factor. *J Neuropathol Exp Neurol* 68:83–93.

Kempermann G (2002). Why new neurons? Possible functions for adult hippocampal neurogenesis. *J Neurosci* 22:635–638.

Kempermann G (2006). They are not too excited: The possible role of adult-born neurons in epilepsy. *Neuron* 52:935–937.

Kempermann G, Kronenberg G (2003). Depressed new neurons—adult hippocampal neurogenesis and a cellular plasticity hypothesis of major depression. *Biol Psychiatry* 54:499–503.

Kempermann G, Krebs J, Fabel K (2008). The contribution of failing adult hippocampal neurogenesis to psychiatric disorders. *Curr Opin Psychiatry* 21:290–295.

Kernie SG, Erwin TM, Parada LF (2001). Brain remodeling due to neuronal and astrocytic proliferation after controlled cortical injury in mice. *J Neurosci Res* 66:317–326.

Kerr DA, Llado J, Shamblott MJ, Maragakis NJ, Irani DN, Crawford TO, Krishnan C, Dike S, Gearhart JD, Rothstein JD (2003). Human embryonic germ cell derivatives facilitate motor recovery of rats with diffuse motor neuron injury. *J Neurosci* 23:5131–5140.

Kim HM, Qu T, Kriho V, Lacor P, Smalheiser N, Pappas GD, Guidotti A, Costa E, Sugaya K (2002). Reelin function in neural stem cell biology. *Proc Natl Acad Sci USA* 99:4020–4025.

Kim JY, Duan X, Liu CY, Jang MH, Guo JU, Pow-anpongkul N, Kang E, Song H, Ming GL (2009). DISC1 regulates new neuron development in the adult brain via modulation of AKT-mTOR signaling through KIAA1212. *Neuron* 63:761–773.

Kim SK, Cargioli TG, Machluf M, Yang W, Sun Y, Al-Hashem R, Kim SU, Black PM, Carroll RS (2005). PEX-producing human neural stem cells inhibit tumor growth in a mouse glioma model. *Clin Cancer Res* 11:5965–5970.

Kim SU, de Vellis J (2009). Stem cell-based cell therapy in neurological diseases: A review. *J Neurosci Res* 87:2183–2200.

Kingma A, Rammeloo LA, van Der Does-van den Berg A, Rekers-Mombarg L, Postma A (2000). Academic career after treatment for acute lymphoblastic leukaemia. *Arch Dis Child* 82:353–357.

Kirschenbaum B, Doetsch F, Lois C, Alvarez-Buylla A (1999). Adult subventricular zone neuronal precursors continue to proliferate and migrate in the absence of the olfactory bulb. *J Neurosci* 19:2171–2180.

Klaus F, Hauser T, Slomianka L, Lipp HP, Amrein I (2009). A reward increases running-wheel performance without changing cell proliferation, neuronal differentiation or cell death in the dentate gyrus of C57BL/6 mice. *Behav Brain Res* 204:175–181.

Klein SM, Behrstock S, McHugh J, Hoffmann K, Wallace K, Suzuki M, Aebischer P, Svendsen CN (2005). GDNF delivery using human neural progenitor cells in a rat model of ALS. *Hum Gene Ther* 16:509–521.

Kleindienst A, McGinn MJ, Harvey HB, Colello RJ, Hamm RJ, Bullock MR (2005). Enhanced hippocampal neurogenesis by intraventricular S100B infusion is associated with improved cognitive recovery after traumatic brain injury. *J Neurotrauma* 22:645–655.

Kleppner SR, Robinson KA, Trojanowski JQ, Lee VM (1995). Transplanted human neurons derived from a teratocarcinoma cell line (NTera-2) mature, integrate, and survive for over 1 year in the nude mouse brain. *J Comp Neurol* 357:618–632.

Kluska MM, Witte OW, Bolz J, Redecker C (2005). Neurogenesis in the adult dentate gyrus after cortical infarcts: Effects of infarct location, N-methyl-D-aspartate receptor blockade and anti-inflammatory treatment. *Neuroscience* 135:723–735.

Knable MB, Barci BM, Webster MJ, Meador-Woodruff J, Torrey EF (2004). Molecular abnormalities of the hippocampus in severe psychiatric illness: Postmortem findings from the Stanley Neuropathology Consortium. *Mol Psychiatry* 9:609–620, 544.

Ko HG, Jang DJ, Son J, Kwak C, Choi JH, Ji YH, Lee YS, Son H, Kaang BK (2009). Effect of ablated hippocampal neurogenesis on the formation and extinction of contextual fear memory. *Mol Brain* 2:1.

Kobayashi T, Ahlenius H, Thored P, Kobayashi R, Kokaia Z, Lindvall O (2006). Intracerebral infusion of glial cell line-derived neurotrophic factor promotes striatal neurogenesis after stroke in adult rats. *Stroke* 37:2361–2367.

Kobow K, Jeske I, Hildebrandt M, Hauke J, Hahnen E, Buslei R, Buchfelder M, Weigel D, Stefan H, Kasper B, Pauli E, Blumcke I (2009). Increased reelin promoter methylation is associated with granule cell dispersion in human temporal lobe epilepsy. *J Neuropathol Exp Neurol* 68:356–364.

Kochman LJ, Fornal CA, Jacobs BL (2009). Suppression of hippocampal cell proliferation by short-term stimulant drug administration in adult rats. *Eur J Neurosci* 29:2157–2165.

Kohl Z, Kandasamy M, Winner B, Aigner R, Gross C, Couillard-Despres S, Bogdahn U, Aigner L, Winkler J (2007). Physical activity fails to rescue hippocampal neurogenesis deficits in the R6/2 mouse model of Huntington's disease. *Brain Res* 1155:24–33.

Kokaia Z, Thored P, Arvidsson A, Lindvall O (2006). Regulation of stroke-induced neurogenesis in adult brain—recent scientific progress. *Cereb Cortex* 16 Suppl 1:i162–167.

Kokoeva MV, Yin H, Flier JS (2005). Neurogenesis in the hypothalamus of adult mice: Potential role in energy balance. *Science* 310:679–683.

Komitova M, Perfilieva E, Mattsson B, Eriksson PS, Johansson BB (2002). Effects of cortical ischemia and postischemic environmental enrichment on hippocampal cell genesis and differentiation in the adult rat. *J Cereb Blood Flow Metab* 22:852–860.

Kondziolka D, Wechsler L, Goldstein S, Meltzer C, Thulborn KR, Gebel J, Jannetta P, DeCesare S, Elder EM, McGrogan M, Reitman MA, Bynum L (2000). Transplantation of cultured human neuronal cells for patients with stroke. *Neurology* 55:565–569.

Kong H, Sha LL, Fan Y, Xiao M, Ding JH, Wu J, Hu G (2009). Requirement of AQP4 for antidepressive efficiency of fluoxetine: Implication in adult hippocampal neurogenesis. *Neuropsychopharmacology* 34:1263–1276.

Kraev IV, Godukhin OV, Patrushev IV, Davies HA, Popov VI, Stewart MG (2009). Partial kindling induces neurogenesis, activates astrocytes and alters synaptic morphology in the dentate gyrus of freely moving adult rats. *Neuroscience* 162:254–267.

Kragl M, Knapp D, Nacu E, Khattak S, Maden M, Epperlein HH, Tanaka EM (2009). Cells keep a memory of their tissue origin during axolotl limb regeneration. *Nature* 460:60–65.

Kramer JH, Crittenden MR, Halberg FE, Wara WM, Cowan MJ (1992). A prospective study of cognitive functioning following low-dose cranial radiation for bone marrow transplantation. *Pediatrics* 90:447–450.

Krathwohl MD, Kaiser JL (2004). HIV-1 promotes quiescence in human neural progenitor cells. *J Infect Dis* 190:216–226.

Kronenberg G, Kirste I, Inta D, Chourbaji S, Heuser I, Endres M, Gass P (2009). Reduced hippocampal neurogenesis in the GR(+/-) genetic mouse model of depression. *Eur Arch Psychiatry Clin Neurosci*. 259:499–504.

Kronenberg G, Wang LP, Geraerts M, Babu H, Synowitz M, Vicens P, Lutsch G, Glass R, Yamaguchi M, Baekelandt V, Debyser Z, Kettenmann H, Kempermann G (2007). Local origin and activity-dependent generation of nestin-expressing protoplasmic astrocytes in CA1. *Brain Structure & Function* 212:19–35.

Kuan CY, Schloemer AJ, Lu A, Burns KA, Weng WL, Williams MT, Strauss KI, Vorhees CV, Flavell RA, Davis RJ, Sharp FR, Rakic P (2004). Hypoxia-ischemia induces DNA synthesis without cell proliferation in dying neurons in adult rodent brain. *J Neurosci* 24:10763–10772.

Kunze A, Grass S, Witte OW, Yamaguchi M, Kempermann G, Redecker C (2006). Proliferative response of distinct hippocampal progenitor cell populations after cortical infarcts in the adult brain. *Neurobiol Dis.* 21:324–332.

Kurozumi K, Nakamura K, Tamiya T, Kawano Y, Kobune M, Hirai S, Uchida H, Sasaki K, Ito Y, Kato K, Honmou O, Houkin K, Date I, Hamada H (2004). BDNF gene-modified mesenchymal stem cells promote functional recovery and reduce infarct size in the rat middle cerebral artery occlusion model. *Mol Ther* 9: 189–197.

Lagace DC, Yee JK, Bolanos CA, Eisch AJ (2006). Juvenile administration of methylphenidate attenuates adult hippocampal neurogenesis. *Biol Psychiatry* 60:1121–1130.

Laskowski A, Howell OW, Sosunov AA, McKhann G, Gray WP (2007). NPY mediates basal and seizure-induced proliferation in the subcallosal zone. *Neuroreport* 18:1005–1008.

Lau BW, Yau SY, Lee TM, Ching YP, Tang SW, So KF (2009). Intracerebroventricular infusion of cytosine-arabinoside causes prepulse inhibition disruption. *Neuroreport* 20:371–377.

Lazarini F, Mouthon MA, Gheusi G, de Chaumont F, Olivo-Marin JC, Lamarque S, Abrous DN, Boussin FD, Lledo PM (2009). Cellular and behavioral effects of cranial irradiation of the subventricular zone in adult mice. *PLoS One* 4:e7017.

Lazarov O, Robinson J, Tang YP, Hairston IS, Korade-Mirnics Z, Lee VM, Hersh LB, Sapolsky RM, Mirnics K, Sisodia SS (2005). Environmental enrichment reduces Abeta levels and amyloid deposition in transgenic mice. *Cell* 120:701–713.

Lazic SE, Grote HE, Blakemore C, Hannan AJ, van Dellen A, Phillips W, Barker RA (2006). Neurogenesis in the R6/1 transgenic mouse model of Huntington's disease: Effects of environmental enrichment. *Eur J Neurosci* 23:1829–1838.

Le Francois B, Czesak M, Steubl D, Albert PR (2008). Transcriptional regulation at a HTR1A polymorphism associated with mental illness. *Neuropharmacology* 55:977–985.

Le Strat Y, Ramoz N, Gorwood P (2009). The role of genes involved in neuroplasticity and neurogenesis in the observation of a gene-environment interaction (GxE) in schizophrenia. *Curr Mol Med* 9:506–518.

Lee JS, Jang DJ, Lee N, Ko HG, Kim H, Kim YS, Kim B, Son J, Kim SH, Chung H, Lee MY, Kim WR, Sun W, Zhuo M, Abel T, Kaang BK, Son H (2009). Induction of neuronal vascular endothelial growth factor expression by cAMP in the dentate gyrus of the hippocampus is required for antidepressant-like behaviors. *J Neurosci* 29:8493–8505.

Lepore AC, Rauck B, Dejea C, Pardo AC, Rao MS, Rothstein JD, Maragakis NJ (2008). Focal transplantation-based astrocyte replacement is neuroprotective in a model of motor neuron disease. *Nat Neurosci* 11:1294–1301.

Levine JM, Reynolds R, Fawcett JW (2001). The oligodendrocyte precursor cell in health and disease. *Trends Neurosci* 24:39–47.

Levison SW, Rothstein RP, Romanko MJ, Snyder MJ, Meyers RL, Vannucci SJ (2001). Hypoxia/ischemia depletes the rat perinatal subventricular zone of oligodendrocyte progenitors and neural stem cells. *Dev Neurosci* 23:234–247.

Li JY, Popovic N, Brundin P (2005). The use of the R6 transgenic mouse models of Huntington's disease in attempts to develop novel therapeutic strategies. *NeuroRx* 2:447–464.

Li WL, Yu SP, Ogle ME, Ding XS, Wei L (2008a). Enhanced neurogenesis and cell migration following focal ischemia and peripheral stimulation in mice. *Dev Neurobiol* 68:1474–1486.

Li Y, Chopp M (2009). Marrow stromal cell transplantation in stroke and traumatic brain injury. *Neurosci Lett* 456:120–123.

Li Y, Chen J, Wang L, Lu M, Chopp M (2001). Treatment of stroke in rat with intracarotid administration of marrow stromal cells. *Neurology* 56:1666–1672.

Li Y, Luikart BW, Birnbaum S, Chen J, Kwon CH, Kernie SG, Bassel-Duby R, Parada LF (2008b). TrkB regulates hippocampal neurogenesis and governs sensitivity to antidepressive treatment. *Neuron* 59:399–412.

Li Y, Chen J, Chen XG, Wang L, Gautam SC, Xu YX, Katakowski M, Zhang LJ, Lu M, Janakiraman N, Chopp M (2002a). Human marrow stromal cell therapy for stroke in rat: Neurotrophins and functional recovery. *Neurology* 59:514–523.

Li Z, Kato T, Kawagishi K, Fukushima N, Yokouchi K, Moriizumi T (2002b). Cell dynamics of calretinin-immunoreactive neurons in the rostral migratory stream after ibotenate-induced lesions in the forebrain. *Neurosci Res* 42:123–132.

Lichtenwalner RJ, Parent JM (2006). Adult neurogenesis and the ischemic forebrain. *J Cereb Blood Flow Metab* 26:1–20.

Lie DC, Dziewczapolski G, Willhoite AR, Kaspar BK, Shults CW, Gage FH (2002). The adult substantia nigra contains progenitor cells with neurogenic potential. *J Neurosci* 22:6639–6649.

Limoli CL, Giedzinski E, Rola R, Otsuka S, Palmer TD, Fike JR (2004). Radiation response of neural precursor cells: Linking cellular sensitivity to cell cycle checkpoints, apoptosis and oxidative stress. *Radiat Res* 161:17–27.

Lindvall O, Backlund EO, Farde L, Sedvall G, Freedman R, Hoffer B, Nobin A, Seiger A, Olson L (1987). Transplantation in Parkinson's disease: Two cases of adrenal medullary grafts to the putamen. *Ann Neurol* 22:457–468.

Lipska BK (2004). Using animal models to test a neurodevelopmental hypothesis of schizophrenia. *J Psychiatry Neurosci* 29:282–286.

Liu F, You Y, Li X, Ma T, Nie Y, Wei B, Li T, Lin H, Yang Z (2009a). Brain injury does not alter the intrinsic differentiation potential of adult neuroblasts. *J Neurosci* 29:5075–5087.

Liu J, Solway K, Messing RO, Sharp FR (1998). Increased neurogenesis in the dentate gyrus after transient ischemia in gerbils. *J Neurosci* 18:7768–7778.

Liu J, Suzuki T, Seki T, Namba T, Tanimura A, Arai H (2006). Effects of repeated phencyclidine administration on adult hippocampal neurogenesis in the rat. *Synapse* 60:56–68.

Liu YP, Lang BT, Baskaya MK, Dempsey RJ, Vemuganti R (2009b). The potential of neural stem cells to repair stroke-induced brain damage. *Acta Neuropathol* 117:469–80.

Liu YW, Curtis MA, Gibbons HM, Mee EW, Bergin PS, Teoh HH, Connor B, Dragunow M, Faull RL (2008). Doublecortin expression in the normal and epileptic adult human brain. *Eur J Neurosci* 28:2254–2265.

Liu Z, Martin LJ (2006). The adult neural stem and progenitor cell niche is altered in amyotrophic lateral sclerosis mouse brain. *J Comp Neurol* 497:468–488.

Lledo PM (2009). Dissecting the pathophysiology of depression with a Swiss army knife. *Neuron* 62:453–455.

Lopez-Toledano MA, Shelanski ML (2004). Neurogenic effect of beta-amyloid peptide in the development of neural stem cells. *J Neurosci* 24:5439–5444.

Lopez-Toledano MA, Shelanski ML (2007). Increased neurogenesis in young transgenic mice overexpressing human APP(Sw, Ind). *J Alzheimers Dis* 12:229–240.

Lorenzi HA, Reeves RH (2006). Hippocampal hypocellularity in the Ts65Dn mouse originates early in development. *Brain Res* 1104:153–159.

Lu D, Mahmood A, Zhang R, Copp M (2003). Upregulation of neurogenesis and reduction in functional deficits following administration of DEtA/NONOate, a nitric oxide donor, after traumatic brain injury in rats. *J Neurosurg* 99:351–361.

Lucassen PJ, Fuchs E, Czeh B (2004). Antidepressant treatment with tianeptine reduces apoptosis in the hippocampal dentate gyrus and temporal cortex. *Biol Psychiatry* 55:789–796.

Lucassen PJ, Meerlo P, Naylor AS, van Dam AM, Dayer AG, Fuchs E, Oomen CA, Czeh B (2009). Regulation of adult neurogenesis by stress, sleep disruption, exercise and inflammation: Implications for depression and antidepressant action. *Eur Neuropsychopharmacol* 20:1–17.

Luo CX, Zhu XJ, Zhang AX, Wang W, Yang XM, Liu SH, Han X, Sun J, Zhang SG, Lu Y, Zhu DY (2005). Blockade of L-type voltage-gated Ca channel inhibits ischemia-induced neurogenesis by down-regulating iNOS expression in adult mouse. *J Neurochem* 94:1077–1086.

Luo CX, Zhu XJ, Zhou QG, Wang B, Wang W, Cai HH, Sun YJ, Hu M, Jiang J, Hua Y, Han X, Zhu DY (2007). Reduced neuronal nitric oxide synthase is involved in ischemia-induced hippocampal neurogenesis by up-regulating inducible nitric oxide synthase expression. *J Neurochem* 103:1872–1882.

Lurton D, Sundstrom L, Brana C, Bloch B, Rougier A (1997). Possible mechanisms inducing granule cell dispersion in humans with temporal lobe epilepsy. *Epilepsy Res* 26:351–361.

Ma M, Ma Y, Yi X, Guo R, Zhu W, Fan X, Xu G, Frey WH, 2nd, Liu X (2008). Intranasal delivery of transforming growth factor-beta1 in mice after stroke reduces infarct volume and increases neurogenesis in the subventricular zone. *BMC Neurosci* 9:117.

Madsen TM, Treschow A, Bengzon J, Bolwig TG, Lindvall O, Tingstrom A (2000). Increased neurogenesis in a model of electroconvulsive therapy. *Biol Psychiatry* 47:1043–1049.

Madsen TM, Kristjansen PE, Bolwig TG, Wortwein G (2003). Arrested neuronal proliferation and impaired hippocampal function following fractionated brain irradiation in the adult rat. *Neuroscience* 119:635–642.

Maeda K, Sugino H, Hirose T, Kitagawa H, Nagai T, Mizoguchi H, Takuma K, Yamada K (2007). Clozapine prevents a decrease in neurogenesis in mice repeatedly treated with phencyclidine. *Journal of Pharmacological Sciences* 103:299–308.

Maeda Y, Solanky M, Menonna J, Chapin J, Li W, Dowling P (2001). Platelet-derived growth factor-alpha receptor-positive oligodendroglia are frequent in multiple sclerosis lesions. *Ann Neurol* 49:776–785.

Magavi S, Leavitt B, Macklis J (2000). Induction of neurogenesis in the neocortex of adult mice. *Nature* 405:951–955.

Mahmood A, Lu D, Qu C, Goussev A, Chopp M (2005). Human marrow stromal cell treatment provides long-lasting benefit after traumatic brain injury in rats. *Neurosurgery* 57:1026–1031; discussion 1026–1031.

Mak GK, Enwere EK, Gregg C, Pakarainen T, Poutanen M, Huhtaniemi I, Weiss S (2007). Male pheromone-stimulated neurogenesis in the adult female brain: Possible role in mating behavior. *Nat Neurosci* 10:1003–1011.

Malberg JE, Duman RS (2003). Cell proliferation in adult hippocampus is decreased by inescapable stress: Reversal by fluoxetine treatment. *Neuropsychopharmacology* 28:1562–1571.

Malberg JE, Eisch AJ, Nestler EJ, Duman RS (2000). Chronic antidepressant treatment increases neurogenesis in adult rat hippocampus. *J Neurosci* 20:9104–9110.

Manda K, Ueno M, Anzai K (2009). Cranial irradiation-induced inhibition of neurogenesis in hippocampal dentate gyrus of adult mice: Attenuation by melatonin pretreatment. *Journal of Pineal Research* 46:71–78.

Mandairon N, Jourdan F, Didier A (2003). Deprivation of sensory inputs to the olfactory bulb up-regulates cell death and proliferation in the subventricular zone of adult mice. *Neuroscience* 119:507–516.

Mandyam CD, Wee S, Eisch AJ, Richardson HN, Koob GF (2007). Methamphetamine self-administration and voluntary exercise have opposing effects on medial prefrontal cortex gliogenesis. *J Neurosci* 27:11442–11450.

Mandyam CD, Wee S, Crawford EF, Eisch AJ, Richardson HN, Koob GF (2008). Varied access to intravenous methamphetamine self-administration differentially alters adult hippocampal neurogenesis. *Biol Psychiatry* 64:958–965.

Mao L, Wang JQ (2001). Gliogenesis in the striatum of the adult rat: Alteration in neural progenitor population after psychostimulant exposure. *Brain Res Dev Brain Res* 130:41–51.

Mao Y, Ge X, Frank CL, Madison JM, Koehler AN, Doud MK, Tassa C, Berry EM, Soda T, Singh KK, Biechele T, Petryshen TL, Moon RT, Haggarty SJ, Tsai LH (2009). Disrupted in schizophrenia 1 regulates neuronal progenitor proliferation via modulation of GSK3beta/beta-catenin signaling. *Cell* 136:1017–1031.

Marchetto MC, Muotri AR, Mu Y, Smith AM, Cezar GG, Gage FH (2008). Non-cell-autonomous effect of human SOD1 G37R astrocytes on motor neurons derived from human embryonic stem cells. *Cell Stem Cell* 3:649–657.

Marshall GP, 2nd, Scott EW, Zheng T, Laywell ED, Steindler DA (2005). Ionizing radiation enhances the engraftment of transplanted in vitro-derived multipotent astrocytic stem cells. *Stem Cells* 23:1276–1285.

Mazzini L, Ferrero I, Luparello V, Rustichelli D, Gunetti M, Mareschi K, Testa L, Stecco A, Tarletti R, Miglioretti M, Fava E, Nasuelli N, Cisari C, Massara M, Vercelli R, Oggioni GD, Carriero A, Cantello R, Monaco F, Fagioli F (2009). Mesenchymal stem cell transplantation in amyotrophic lateral sclerosis: A Phase I clinical trial. *Exp Neurol.* 223:229–37.

McCarthy GF, Leblond CP (1988). Radioautographic evidence for slow astrocyte turnover and modest oligodendrocyte production in the corpus callosum of adult mice infused with 3H-thymidine. *J Comp Neurol* 271:589–603.

McCloskey DP, Hintz TM, Pierce JP, Scharfman HE (2006). Stereological methods reveal the robust size and stability of ectopic hilar granule cells after pilocarpine-induced status epilepticus in the adult rat. *Eur J Neurosci* 24:2203–2210.

McDonald JW, Liu XZ, Qu Y, Liu S, Mickey SK, Turetsky D, Gottlieb DI, Choi DW (1999). Transplanted embryonic stem cells survive, differentiate and promote recovery in injured rat spinal cord. *Nat Med* 5:1410–1412.

McHedlishvili L, Epperlein HH, Telzerow A, Tanaka EM (2007). A clonal analysis of neural progenitors during axolotl spinal cord regeneration reveals evidence for both spatially restricted and multipotent progenitors. *Development* 134:2083–2093.

McKinnon MC, Yucel K, Nazarov A, MacQueen GM (2009). A meta-analysis examining clinical predictors of hippocampal volume in patients with major depressive disorder. *J Psychiatry Neurosci* 34:41–54.

Mello LE, Cavalheiro EA, Tan AM, Kupfer WR, Pretorius JK, Babb TL, Finch DM (1993). Circuit mechanisms of seizures in the pilocarpine model of chronic epilepsy: Cell loss and mossy fiber sprouting. *Epilepsia* 34:985–995.

Mendes CT, Mury FB, de Sa Moreira E, Alberto FL, Forlenza OV, Dias-Neto E, Gattaz WF (2009). Lithium reduces Gsk3b mRNA levels: Implications for Alzheimer disease. *Eur Arch Psychiatry Clin Neurosci* 259:16–22.

Meyer U, Knuesel I, Nyffeler M, Feldon J (2010). Chronic clozapine treatment improves prenatal infection-induced working memory deficits without influencing adult hippocampal neurogenesis. *Psychopharmacology* (Berl) 208:531–543.

Miller MW, Spear LP (2006). The alcoholism generator. *Alcohol Clin Exp Res* 30:1466–1469.

Mino M, Kamii H, Fujimura M, Kondo T, Takasawa S, Okamoto H, Yoshimoto T (2003). Temporal changes of neurogenesis in the mouse hippocampus after experimental subarachnoid hemorrhage. *Neurol Res* 25:839–845.

Mirochnic S, Wolf S, Staufenbiel M, Kempermann G (2009). Age effects on the regulation of adult hippocampal neurogenesis by physical activity and environmental enrichment in the APP23 mouse model of Alzheimer disease. *Hippocampus* 19:1008–1018.

Mitchell AJ (2006). Two-week delay in onset of action of antidepressants: New evidence. *Br J Psychiatry* 188: 105–106.

Mitchell BD, Emsley JG, Magavi SS, Arlotta P, Macklis JD (2004). Constitutive and induced neurogenesis in the adult mammalian brain: Manipulation of endogenous precursors toward CNS repair. *Dev Neurosci* 26: 101–117.

Mizumatsu S, Monje ML, Morhardt DR, Rola R, Palmer TD, Fike JR (2003). Extreme sensitivity of adult neurogenesis to low doses of X-irradiation. *Cancer Res* 63:4021–4027.

Mohapel P, Ekdahl CT, Lindvall O (2004). Status epilepticus severity influences the long-term outcome of neurogenesis in the adult dentate gyrus. *Neurobiol Dis* 15:196–205.

Monje M (2008). Cranial radiation therapy and damage to hippocampal neurogenesis. *Journal of Pharmacological Sciences* 14:238–242.

Monje ML, Mizumatsu S, Fike JR, Palmer TD (2002). Irradiation induces neural precursor-cell dysfunction. *Nat Med* 8:955–962.

Monje ML, Toda H, Palmer TD (2003). Inflammatory blockade restores adult hippocampal neurogenesis. *Science* 302:1760–1765.

Monje ML, Vogel H, Masek M, Ligon KL, Fisher PG, Palmer TD (2007). Impaired human hippocampal neurogenesis after treatment for central nervous system malignancies. *Ann Neurol* 62:515–520.

Morrison PD, Pilowsky LS (2007). Schizophrenia: More evidence for less glutamate. *Expert Rev Neurother* 7:29–31.

Mudo G, Belluardo N, Mauro A, Fuxe K (2007). Acute intermittent nicotine treatment induces fibroblast growth factor-2 in the subventricular zone of the adult rat brain and enhances neuronal precursor cell proliferation. *Neuroscience* 145:470–483.

Na YJ, Jin JK, Lee YJ, Choi EK, Carp RI, Kim YS (2009). Increased neurogenesis in brains of scrapie-infected mice. *Neurosci Lett* 449:66–70.

Nagai M, Re DB, Nagata T, Chalazonitis A, Jessell TM, Wichterle H, Przedborski S (2007). Astrocytes expressing ALS-linked mutated SOD1 release factors selectively toxic to motor neurons. *Nat Neurosci* 10:615–622.

Nagy Z (2000). Cell cycle regulatory failure in neurones: Causes and consequences. *Neurobiol Aging* 21: 761–769.

Nagy Z, Esiri MM, Smith AD (1997a). Expression of cell division markers in the hippocampus in Alzheimer's disease and other neurodegenerative conditions. *Acta Neuropathol* (Berl) 93:294–300.

Nagy Z, Esiri MM, Cato AM, Smith AD (1997b). Cell cycle markers in the hippocampus in Alzheimer's disease. *Acta Neuropathol* (Berl) 94:6–15.

Nait-Oumesmar B, Decker L, Lachapelle F, Avellana-Adalid V, Bachelin C, Van Evercooren AB (1999). Progenitor cells of the adult mouse subventricular zone proliferate, migrate and differentiate into oligodendrocytes after demyelination. *Eur J Neurosci* 11:4357–4366.

Nakagawa E, Aimi Y, Yasuhara O, Tooyama I, Shimada M, McGeer PL, Kimura H (2000). Enhancement of progenitor cell division in the dentate gyrus triggered by initial limbic seizures in rat models of epilepsy. *Epilepsia* 41:10–18.

Nakatomi H, Kuriu T, Okabe S, Yamamoto S, Hatano O, Kawahara N, Tamura A, Kirino T, Nakafuku M (2002). Regeneration of hippocampal pyramidal neurons after ischemic brain injury by recruitment of endogenous neural progenitors. *Cell* 110:429–441.

Nathan PC, Patel SK, Dilley K, Goldsby R, Harvey J, Jacobsen C, Kadan-Lottick N, McKinley K, Millham AK, Moore I, Okcu MF, Woodman CL, Brouwers P, Armstrong FD (2007). Guidelines for identification of, advocacy for, and intervention in neurocognitive problems in survivors of childhood cancer: A report from the Children's Oncology Group. *Arch Pediatr Adolesc Med* 161:798–806.

Navailles S, Hof PR, Schmauss C (2008). Antidepressant drug-induced stimulation of mouse hippocampal neurogenesis is age-dependent and altered by early life stress. *J Comp Neurol* 509:372–381.

Nayak MS, Kim YS, Goldman M, Keirstead HS, Kerr DA (2006). Cellular therapies in motor neuron diseases. *Biochim Biophys Acta* 1762:1128–1138.

Naylor AS, Bull C, Nilsson MK, Zhu C, Bjork-Eriksson T, Eriksson PS, Blomgren K, Kuhn HG (2008). Voluntary running rescues adult hippocampal neurogenesis after irradiation of the young mouse brain. *Proc Natl Acad Sci USA* 105:14632–14637.

Newman MB, Kuo YP, Lukas RJ, Sanberg PR, Douglas Shytle R, McGrogan MP, Zigova T (2002). Nicotinic acetylcholine receptors on NT2 precursor cells and hNT (NT2–N) neurons. *Brain Res Dev Brain Res* 139 :73–86.

Newton SS, Duman RS (2007). Neurogenic actions of atypical antipsychotic drugs and therapeutic implications. *CNS Drugs* 21:715–725.

Nichol K, Deeny SP, Seif J, Camaclang K, Cotman CW (2009). Exercise improves cognition and hippocampal plasticity in APOE epsilon4 mice. *Alzheimers Dement* 5:287–294.

Niidome T, Taniuchi N, Akaike A, Kihara T, Sugimoto H (2008). Differential regulation of neurogenesis in two neurogenic regions of APPswe/PS1dE9 transgenic mice. *Neuroreport* 19:1361–1364.

Nixon K, Crews FT (2002). Binge ethanol exposure decreases neurogenesis in adult rat hippocampus. *J Neurochem* 83:1087–1093.

Nixon K, Crews FT (2004). Temporally specific burst in cell proliferation increases hippocampal neurogenesis in protracted abstinence from alcohol. *J Neurosci* 24:9714–9722.

Noble M, Dietrich J (2004). The complex identity of brain tumors: Emerging concerns regarding origin, diversity and plasticity. *Trends Neurosci* 27:148–154.

Nygren J, Kokaia M, Wieloch T (2006). Decreased expression of brain-derived neurotrophic factor in BDNF(+/-) mice is associated with enhanced recovery of motor performance and increased neuroblast number following experimental stroke. *J Neurosci Res* 84:626–631.

Ogawa Y, Sawamoto K, Miyata T, Miyao S, Watanabe M, Nakamura M, Bregman BS, Koike M, Uchiyama Y, Toyama Y, Okano H (2002). Transplantation of in vitro-expanded fetal neural progenitor cells results in neurogenesis and functional recovery after spinal cord contusion injury in adult rats. *J Neurosci Res* 69:925–933.

Ohira K, Furuta T, Hioki H, Nakamura KC, Kuramoto E, Tanaka Y, Funatsu N, Shimizu K, Oishi T, Hayashi M, Miyakawa T, Kaneko T, Nakamura S (2010). Ischemia-induced neurogenesis of neocortical layer 1 progenitor cells. *Nat Neurosci* 13:173–179.

Ohsawa I, Takamura C, Morimoto T, Ishiguro M, Kohsaka S (1999). Amino-terminal region of secreted form of amyloid precursor protein stimulates proliferation of neural stem cells. *Eur J Neurosci* 11:1907–1913.

Okamoto S, Kang YJ, Brechtel CW, Siviglia E, Russo R, Clemente A, Harrop A, McKercher S, Kaul M, Lipton SA (2007). HIV/gp120 decreases adult neural progenitor cell proliferation via checkpoint kinase-mediated cell-cycle withdrawal and G1 arrest. *Cell Stem Cell* 1:230–236.

Okano H, Sawamoto K (2008). Neural stem cells: Involvement in adult neurogenesis and CNS repair. *Philos Trans R Soc Lond B Biol Sci* 363:2111–2122.

Olanow CW, Goetz CG, Kordower JH, Stoessl AJ, Sossi V, Brin MF, Shannon KM, Nauert GM, Perl DP, Godbold J, Freeman TB (2003). A double-blind controlled trial of bilateral fetal nigral transplantation in Parkinson's disease. *Ann Neurol* 54:403–414.

Ong J, Plane JM, Parent JM, Silverstein FS (2005). Hypoxic-ischemic injury stimulates subventricular zone proliferation and neurogenesis in the neonatal rat. *Pediatr Res* 58:600–606.

Oomen CA, Girardi CE, Cahyadi R, Verbeek EC, Krugers H, Joels M, Lucassen PJ (2009). Opposite effects of early maternal deprivation on neurogenesis in male versus female rats. *PLoS One* 4:e3675.

Palmer TD, Willhoite AR, Gage FH (2000). Vascular niche for adult hippocampal neurogenesis. *J Comp Neurol* 425:479–494.

Pang TY, Stam NC, Nithianantharajah J, Howard ML, Hannan AJ (2006). Differential effects of voluntary physical exercise on behavioral and brain-derived neurotrophic factor expression deficits in Huntington's disease transgenic mice. *Neuroscience* 141:569–584.

Papadeas ST, Maragakis NJ (2009). Advances in stem cell research for Amyotrophic Lateral Sclerosis. *Curr Opin Biotechnol* 20:545–551.

Parent JM (2007). Adult neurogenesis in the intact and epileptic dentate gyrus. *Prog Brain Res* 163:529–540.

Parent JM, Yu TW, Leibowitz RT, Geschwind DH, Sloviter RS, Lowenstein DH (1997). Dentate granule cell neurogenesis is increased by seizures and contributes to aberrant network reorganization in the adult rat hippocampus. *J Neurosci* 17:3727–3738.

Parent JM, Janumpalli S, McNamara JO, Lowenstein DH (1998). Increased dentate granule cell neurogenesis following amygdala kindling in the adult rat. *Neurosci Lett* 247:9–12.

Parent JM, Tada E, Fike JR, Lowenstein DH (1999). Inhibition of dentate granule cell neurogenesis with brain irradiation does not prevent seizure-induced mossy fiber synaptic reorganization in the rat. *J Neurosci* 19:4508–4519.

Parent JM, Valentin VV, Lowenstein DH (2002a). Prolonged seizures increase proliferating neuroblasts in the adult rat subventricular zone-olfactory bulb pathway. *J Neurosci* 22:3174–3188.

Parent JM, Vexler ZS, Gong C, Derugin N, Ferriero DM (2002b). Rat forebrain neurogenesis and striatal neuron replacement after focal stroke. *Ann Neurol* 52:802–813.

Parent JM, von dem Bussche N, Lowenstein DH (2006). Prolonged seizures recruit caudal subventricular zone glial progenitors into the injured hippocampus. *Hippocampus* 16:321–328.

Park S, Kim HT, Yun S, Kim IS, Lee J, Lee IS, Park KI (2009). Growth factor-expressing human neural progenitor cell grafts protect motor neurons but do not ameliorate motor performance and survival in ALS mice. *Exp Mol Med* 41:487–500.

Parr AM, Tator CH, Keating A (2007). Bone marrow-derived mesenchymal stromal cells for the repair of central nervous system injury. *Bone Marrow Transplant* 40:609–619.

Peissner W, Kocher M, Treuer H, Gillardon F (1999). Ionizing radiation-induced apoptosis of proliferating stem cells in the dentate gyrus of the adult rat hippocampus. *Brain Res Mol Brain Res* 71:61–68.

Pekcec A, Fuest C, Muhlenhoff M, Gerardy-Schahn R, Potschka H (2008). Targeting epileptogenesis-associated induction of neurogenesis by enzymatic depolysialylation of NCAM counteracts spatial learning dysfunction but fails to impact epilepsy development. *J Neurochem* 105:389–400.

Pencea V, Bingaman KD, Wiegand SJ, Luskin MB (2001). Infusion of brain-derived neurotrophic factor into the lateral ventricle of the adult rat leads to new neurons in the parenchyma of the striatum, septum, thalamus, and hypothalamus. *J Neurosci* 21:6706–6717.

Perez JA, Clinton SM, Turner CA, Watson SJ, Akil H (2009). A new role for FGF2 as an endogenous inhibitor of anxiety. *J Neurosci* 29:6379–6387.

Petratos S, Gonzales MF, Azari MF, Marriott M, Minichiello RA, Shipham KA, Profyris C, Nicolaou A, Boyle K, Cheema SS, Kilpatrick TJ (2004). Expression of the low-affinity neurotrophin receptor, p75(NTR), is upregulated by oligodendroglial progenitors adjacent to the subventricular zone in response to demyelination. *Glia* 48:64–75.

Phillips W, Morton AJ, Barker RA (2005). Abnormalities of neurogenesis in the R6/2 mouse model of Huntington's disease are attributable to the in vivo microenvironment. *J Neurosci* 25:11564–11576.

Pieper AA, Wu X, Han TW, Estill SJ, Dang Q, Wu LC, Reece-Fincanon S, Dudley CA, Richardson JA, Brat DJ, McKnight SL (2005). The neuronal PAS domain protein 3 transcription factor controls FGF-mediated adult hippocampal neurogenesis in mice. *Proc Natl Acad Sci USA* 102:14052–14057.

Pierce JP, Punsoni M, McCloskey DP, Scharfman HE (2007). Mossy cell axon synaptic contacts on ectopic granule cells that are born following pilocarpine-induced seizures. *Neurosci Lett* 422:136–140.

Pietropaolo S, Sun Y, Li R, Brana C, Feldon J, Yee BK (2008). The impact of voluntary exercise on mental health in rodents: A neuroplasticity perspective. *Behav Brain Res* 192:42–60.

Pietropaolo S, Sun Y, Li R, Brana C, Feldon J, Yee BK (2009). Limited impact of social isolation on Alzheimer-like symptoms in a triple transgenic mouse model. *Behav Neurosci* 123:181–195.

Pinnock SB, Lazic SE, Wong HT, Wong IH, Herbert J (2009). Synergistic effects of dehydroepiandrosterone and fluoxetine on proliferation of progenitor cells in the dentate gyrus of the adult male rat. *Neuroscience* 158:1644–1651.

Plane JM, Liu R, Wang TW, Silverstein FS, Parent JM (2004). Neonatal hypoxic-ischemic injury increases forebrain subventricular zone neurogenesis in the mouse. *Neurobiol Dis* 16:585–595.

Plane JM, Whitney JT, Schallert T, Parent JM (2008). Retinoic acid and environmental enrichment alter subventricular zone and striatal neurogenesis after stroke. *Exp Neurol*.

Plattner F, Angelo M, Giese KP (2006). The roles of cyclin-dependent kinase 5 and glycogen synthase kinase 3 in tau hyperphosphorylation. *J Biol Chem* 281:25457–25465.

Pluchino S, Quattrini A, Brambilla E, Gritti A, Salani G, Dina G, Galli R, Del Carro U, Amadio S, Bergami A, Furlan R, Comi G, Vescovi AL, Martino G (2003). Injection of adult neurospheres induces recovery in a chronic model of multiple sclerosis. *Nature* 422:688–694.

Pluchino S, Zanotti L, Rossi B, Brambilla E, Ottoboni L, Salani G, Martinello M, Cattalini A, Bergami A, Furlan R, Comi G, Constantin G, Martino G (2005). Neurosphere-derived multipotent precursors promote neuroprotection by an immunomodulatory mechanism. *Nature* 436:266–271.

Pluchino S, Gritti A, Blezer E, Amadio S, Brambilla E, Borsellino G, Cossetti C, Del Carro U, Comi G, t Hart B, Vescovi A, Martino G (2009a). Human neural stem cells ameliorate autoimmune encephalomyelitis in non-human primates. *Ann Neurol* 66:343–354.

Pluchino S, Zanotti L, Brambilla E, Rovere-Querini P, Capobianco A, Alfaro-Cervello C, Salani G, Cossetti C, Borsellino G, Battistini L, Ponzoni M, Doglioni C, Garcia-Verdugo JM, Comi G, Manfredi AA, Martino G (2009b). Immune regulatory neural stem/precursor cells protect from central nervous system autoimmunity by restraining dendritic cell function. *PLoS ONE* 4:e5959.

Plumpe T, Ehninger D, Steiner B, Klempin F, Jessberger S, Brandt M, Romer B, Rodriguez GR, Kronenberg G, Kempermann G (2006). Variability of doublecortin-associated dendrite maturation in adult hippocampal neurogenesis is independent of the regulation of precursor cell proliferation. *BMC Neurosci* 7:77.

Posternak MA, Zimmerman M (2005). Is there a delay in the antidepressant effect? A meta-analysis. *J Clin Psychiatry* 66:148–158.

Price DL, Tanzi RE, Borchelt DR, Sisodia SS (1998). Alzheimer's disease: Genetic studies and transgenic models. *Annu Rev Genet* 32:461–493.

Priller J, Flugel A, Wehner T, Boentert M, Haas CA, Prinz M, Fernandez-Klett F, Prass K, Bechmann I, de Boer BA, Frotscher M, Kreutzberg GW, Persons DA, Dirnagl U (2001). Targeting gene-modified hematopoietic cells to the central nervous system: Use of green fluorescent protein uncovers microglial engraftment. *Nat Med* 7:1356–1361.

Prozorovski T, Schulze-Topphoff U, Glumm R, Baumgart J, Schroter F, Ninnemann O, Siegert E, Bendix I, Brustle O, Nitsch R, Zipp F, Aktas O (2008). Sirt1 contributes critically to the redox-dependent fate of neural progenitors. *Nat Cell Biol* 10:385–394.

Raber J, Fan Y, Matsumori Y, Liu Z, Weinstein PR, Fike JR, Liu J (2004). Irradiation attenuates neurogenesis and exacerbates ischemia-induced deficits. *Ann Neurol* 55:381–389.

Radley JJ, Jacobs BL (2002). 5-HT1A receptor antagonist administration decreases cell proliferation in the dentate gyrus. *Brain Res* 955:264–267.

Raedt R, Boon P, Persson A, Alborn AM, Boterberg T, Van Dycke A, Linder B, De Smedt T, Wadman WJ, Ben-Menachem E, Eriksson PS (2007). Radiation of the rat brain suppresses seizure-induced neurogenesis and transiently enhances excitability during kindling acquisition. *Epilepsia* 48:1952–1963.

Rajkowska G (2000). Postmortem studies in mood disorders indicate altered numbers of neurons and glial cells. *Biol Psychiatry* 48:766–777.

Rajkowska G, Miguel-Hidalgo JJ, Wei J, Dilley G, Pittman SD, Meltzer HY, Overholser JC, Roth BL, Stockmeier CA (1999). Morphometric evidence for neuronal and glial prefrontal cell pathology in major depression. *Biol Psychiatry* 45:1085–1098.

Ramaswamy S, Goings GE, Soderstrom KE, Szele FG, Kozlowski DA (2005). Cellular proliferation and migration following a controlled cortical impact in the mouse. *Brain Res* 1053:38–53.

Ramirez-Rodriguez G, Klempin F, Babu H, Benitez-King G, Kempermann G (2009). Melatonin modulates cell survival of new neurons in the hippocampus of adult mice. *Neuropsychopharmacology* 34:2180–2191.

Rao MS, Hattiangady B, Shetty AK (2008). Status epilepticus during old age is not associated with enhanced hippocampal neurogenesis. *Hippocampus* 18:931–944.

Redila VA, Olson AK, Swann SE, Mohades G, Webber AJ, Weinberg J, Christie BR (2006). Hippocampal cell proliferation is reduced following prenatal ethanol exposure but can be rescued with voluntary exercise. *Hippocampus* 16:305–311.

Redwine JM, Armstrong RC (1998). In vivo proliferation of oligodendrocyte progenitors expressing PDGFalphaR during early remyelination. *J Neurobiol* 37:413–428.

Reid-Arndt SA, Yee A, Perry MC, Hsieh C (2009). Cognitive and psychological factors associated with early post-treatment functional outcomes in breast cancer survivors. *Journal of Psychosocial Oncology* 27:415–434.

Reif A, Fritzen S, Finger M, Strobel A, Lauer M, Schmitt A, Lesch KP (2006). Neural stem cell proliferation is decreased in schizophrenia, but not in depression. *Mol Psychiatry* 11:514–522.

Reif A, Schmitt A, Fritzen S, Lesch KP (2007). Neurogenesis and schizophrenia: Dividing neurons in a divided mind? *Eur Arch Psychiatry Clin Neurosci* 257:290–299.

Revesz D, Tjernstrom M, Ben-Menachem E, Thorlin T (2008). Effects of vagus nerve stimulation on rat hippocampal progenitor proliferation. *Exp Neurol* 214:259–265.

Reznikov K (1975). [Incorporation of 3H-thymidine into glial cells of the parietal region and cells of the subependymal zone of two-week and adult mice under normal conditions and following brain injury]. *Ontogenez* 6:169–176.

Ribak CE, Tran PH, Spigelman I, Okazaki MM, Nadler JV (2000). Status epilepticus-induced hilar basal dendrites on rodent granule cells contribute to recurrent excitatory circuitry. *J Comp Neurol* 428:240–253.

Ribak CE, Korn MJ, Shan Z, Obenaus A (2004). Dendritic growth cones and recurrent basal dendrites are typical features of newly generated dentate granule cells in the adult hippocampus. *Brain Res* 1000:195–199.

Richardson HN, Chan SH, Crawford EF, Lee YK, Funk CK, Koob GF, Mandyam CD (2009). Permanent impairment of birth and survival of cortical and hippocampal proliferating cells following excessive drinking during alcohol dependence. *Neurobiol Dis* 36:1–10.

Richter H, Ambree O, Lewejohann L, Herring A, Keyvani K, Paulus W, Palme R, Touma C, Schabitz WR, Sachser N (2008). Wheel-running in a transgenic mouse model of Alzheimer's disease: Protection or symptom? *Behav Brain Res* 190:74–84.

Riess P, Zhang C, Saatman KE, Laurer HL, Longhi LG, Raghupathi R, Lenzlinger PM, Lifshitz J, Boockvar J, Neugebauer E, Snyder EY, McIntosh TK (2002). Transplanted neural stem cells survive, differentiate, and improve neurological motor function after experimental traumatic brain injury. *Neurosurgery* 51:1043–1052; discussion 1052–1044.

Rodriguez JJ, Jones VC, Tabuchi M, Allan SM, Knight EM, LaFerla FM, Oddo S, Verkhratsky A (2008). Impaired adult neurogenesis in the dentate gyrus of a triple transgenic mouse model of Alzheimer's disease. *PLoS One* 3:e2935.

Rohe M, Carlo AS, Breyhan H, Sporbert A, Militz D, Schmidt V, Wozny C, Harmeier A, Erdmann B, Bales KR, Wolf S, Kempermann G, Paul SM, Schmitz D, Bayer TA, Willnow TE, Andersen OM (2008). Sortilin-related receptor with A-type repeats (SORLA) affects the amyloid precursor protein-dependent stimulation of ERK signaling and adult neurogenesis. *J Biol Chem* 283:14826–14834.

Rola R, Raber J, Rizk A, Otsuka S, VandenBerg SR, Morhardt DR, Fike JR (2004a). Radiation-induced impairment of hippocampal neurogenesis is associated with cognitive deficits in young mice. *Exp Neurol* 188:316–330.

Rola R, Otsuka S, Obenaus A, Nelson GA, Limoli CL, VandenBerg SR, Fike JR (2004b). Indicators of hippocampal neurogenesis are altered by 56Fe-particle irradiation in a dose-dependent manner. *Radiat Res* 162:442–446.

Russo RE, Reali C, Radmilovich M, Fernandez A, Trujillo-Cenoz O (2008). Connexin 43 delimits functional domains of neurogenic precursors in the spinal cord. *J Neurosci* 28:3298–3309.

Saegusa T, Mine S, Iwasa H, Murai H, Seki T, Yamaura A, Yuasa S (2004). Involvement of highly polysialylated neural cell adhesion molecule (PSA-NCAM)-positive granule cells in the amygdaloid-kindling-induced sprouting of a hippocampal mossy fiber trajectory. *Neurosci Res* 48:185–194.

Sahay A, Hen R (2007). Adult hippocampal neurogenesis in depression. *Nat Neurosci* 10:1110–1115.

Sanders VJ, Felisan S, Waddell A, Tourtellotte WW (1996). Detection of herpesviridae in postmortem multiple sclerosis brain tissue and controls by polymerase chain reaction. *J Neurovirol* 2:249–258.

Santarelli L, Saxe M, Gross C, Surget A, Battaglia F, Dulawa S, Weisstaub N, Lee J, Duman R, Arancio O, Belzung C, Hen R (2003). Requirement of hippocampal neurogenesis for the behavioral effects of antidepressants. *Science* 301:805–809.

Schabitz WR, Steigleder T, Cooper-Kuhn CM, Schwab S, Sommer C, Schneider A, Kuhn HG (2007). Intravenous brain-derived neurotrophic factor enhances poststroke sensorimotor recovery and stimulates neurogenesis. *Stroke* 38:2165–2172.

Schaefers AT, Teuchert-Noodt G, Bagorda F, Brummelte S (2009). Effect of postnatal methamphetamine trauma and adolescent methylphenidate treatment on adult hippocampal neurogenesis in gerbils. *Eur J Pharmacol* 616:86–90.

Schanzer A, Wachs FP, Wilhelm D, Acker T, Cooper-Kuhn C, Beck H, Winkler J, Aigner L, Plate KH, Kuhn HG (2004). Direct stimulation of adult neural stem cells in vitro and neurogenesis in vivo by vascular endothelial growth factor. *Brain Pathol* 14:237–248.

Scharfman HE, McCloskey DP (2009). Postnatal neurogenesis as a therapeutic target in temporal lobe epilepsy. *Epilepsy Res* 85:150–161.

Scharfman HE, Goodman JH, Sollas AL (2000). Granule-like neurons at the hilar/CA3 border after status epilepticus and their synchrony with area CA3 pyramidal cells: Functional implications of seizure-induced neurogenesis. *J Neurosci* 20:6144–6158.

Schnapp E, Kragl M, Rubin L, Tanaka EM (2005). Hedgehog signaling controls dorsoventral patterning, blastema cell proliferation and cartilage induction during axolotl tail regeneration. *Development* 132:3243–3253.

Schneider A, Kruger C, Steigleder T, Weber D, Pitzer C, Laage R, Aronowski J, Maurer MH, Gassler N, Mier W, Hasselblatt M, Kollmar R, Schwab S, Sommer C, Bach A, Kuhn HG, Schabitz WR (2005). The hematopoietic factor G-CSF is a neuronal ligand that counteracts programmed cell death and drives neurogenesis. *J Clin Invest* 115:2083–2098.

Schurov IL, Handford EJ, Brandon NJ, Whiting PJ (2004). Expression of disrupted in schizophrenia 1 (DISC1) protein in the adult and developing mouse brain indicates its role in neurodevelopment. *Mol Psychiatry* 9:1100–1110.

Scolding N, Franklin R, Stevens S, Heldin CH, Compston A, Newcombe J (1998). Oligodendrocyte progenitors are present in the normal adult human CNS and in the lesions of multiple sclerosis. *Brain* 121 (Pt 12):2221–2228.

Scott BW, Wang S, Burnham WM, De Boni U, Wojtowicz JM (1998). Kindling-induced neurogenesis in the dentate gyrus of the rat. *Neurosci Lett* 248:73–76.

Segi-Nishida E, Warner-Schmidt JL, Duman RS (2008). Electroconvulsive seizure and VEGF increase the proliferation of neural stem-like cells in rat hippocampus. *Proc Natl Acad Sci USA* 105:11352–11357.

Sehara Y, Hayashi T, Deguchi K, Nagotani S, Zhang H, Shoji M, Abe K (2006). Distribution of inducible nitric oxide synthase and cell proliferation in rat brain after transient middle cerebral artery occlusion. *Brain Res* 1093:190–197.

Seigers R, Schagen SB, Beerling W, Boogerd W, van Tellingen O, van Dam FS, Koolhaas JM, Buwalda B (2008). Long-lasting suppression of hippocampal cell proliferation and impaired cognitive performance by methotrexate in the rat. *Behav Brain Res* 186:168–175.

Seigers R, Schagen SB, Coppens CM, van der Most PJ, van Dam FS, Koolhaas JM, Buwalda B (2009). Methotrexate decreases hippocampal cell proliferation and induces memory deficits in rats. *Behav Brain Res* 201:279–284.

Seri B, Garcia-Verdugo JM, McEwen BS, Alvarez-Buylla A (2001). Astrocytes give rise to new neurons in the adult mammalian hippocampus. *J Neurosci* 21:7153–7160.

Sharma A, Valadi N, Miller AH, Pearce BD (2002). Neonatal viral infection decreases neuronal progenitors and impairs adult neurogenesis in the hippocampus. *Neurobiol Dis* 11:246–256.

Sharma AK, Jordan WH, Reams RY, Hall DG, Snyder PW (2008). Temporal profile of clinical signs and histopathologic changes in an F-344 rat model of kainic acid-induced mesial temporal lobe epilepsy. *Toxicol Pathol* 36:932–943.

Sheen VL, Macklis JD (1995). Targeted neocortical cell death in adult mice guides migration and differentiation of transplanted embryonic neurons. *J Neurosci* 15:8378–8392.

Sheline YI (2000). 3D MRI studies of neuroanatomic changes in unipolar major depression: The role of stress and medical comorbidity. *Biol Psychiatry* 48:791–800.

Sheline YI, Wang PW, Gado MH, Csernansky JG, Vannier MW (1996). Hippocampal atrophy in recurrent major depression. *Proc Natl Acad Sci USA* 93:3908–3913.

Shetty AK, Hattiangady B (2007). Concise review: Prospects of stem cell therapy for temporal lobe epilepsy. *Stem Cells* 25:2396–2407.

Shihabuddin LS, Ray J, Gage FH (1997). FGF-2 is sufficient to isolate progenitors found in the adult mammalian spinal cord. *Exp Neurol* 148:577–586.

Shihabuddin LS, Horner PJ, Ray J, Gage FH (2000). Adult spinal cord stem cells generate neurons after transplantation in the adult dentate gyrus. *J Neurosci* 20:8727–8735.

Shin JJ, Fricker-Gates RA, Perez FA, Leavitt BR, Zurakowski D, Macklis JD (2000). Transplanted neuroblasts differentiate appropriately into projection neurons with correct neurotransmitter and receptor phenotype in neocortex undergoing targeted projection neuron degeneration. *J Neurosci* 20:7404–7416.

Shingo AS, Kito S (2005). Effects of nicotine on neurogenesis and plasticity of hippocampal neurons. *J Neural Transm* 112:1475–1478.

Shingo T, Sorokan ST, Shimazaki T, Weiss S (2001). Erythropoietin regulates the in vitro and in vivo production of neuronal progenitors by mammalian forebrain neural stem cells. *J Neurosci* 21:9733–9743.

Shinohara C, Gobbel GT, Lamborn KR, Tada E, Fike JR (1997). Apoptosis in the subependyma of young adult rats after single and fractionated doses of X-rays. *Cancer Res* 57:2694–2702.

Shors TJ (2004). Memory traces of trace memories: Neurogenesis, synaptogenesis and awareness. *Trends Neurosci* 27:250–256.

Shors TJ, Miesegaes G, Beylin A, Zhao M, Rydel T, Gould E (2001). Neurogenesis in the adult is involved in the formation of trace memories. *Nature* 410:372–376.

Silva R, Mesquita AR, Bessa J, Sousa JC, Sotiropoulos I, Leao P, Almeida OF, Sousa N (2008). Lithium blocks stress-induced changes in depressive-like behavior and hippocampal cell fate: The role of glycogen-synthase-kinase-3beta. *Neuroscience* 152:656–669.

Silver DJ, Steindler DA (2009). Common astrocytic programs during brain development, injury and cancer. *Trends Neurosci* 32:303–311.

Singh AK, Gupta S, Jiang Y, Younus M, Ramzan M (2009). In vitro neurogenesis from neural progenitor cells isolated from the hippocampus region of the brain of adult rats exposed to ethanol during early development through their alcohol-drinking mothers. *Alcohol Alcohol* 44:185–198.

Smith PD, McLean KJ, Murphy MA, Turnley AM, Cook MJ (2006). Functional dentate gyrus neurogenesis in a rapid kindling seizure model. *Eur J Neurosci* 24:3195–3203.

Snyder EY, Deitcher DL, Walsh C, Arnold-Aldea S, Hartwieg EA, Cepko CL (1992). Multipotent neural cell lines can engraft and participate in development of mouse cerebellum. *Cell* 68:33–51.

Song H, Stevens CF, Gage FH (2002). Astroglia induce neurogenesis from adult neural stem cells. *Nature* 417: 39–44.

Spiegler M, Villapol S, Biran V, Goyenvalle C, Mariani J, Renolleau S, Charriaut-Marlangue C (2007). Bilateral changes after neonatal ischemia in the P7 rat brain. *J Neuropathol Exp Neurol* 66:481–490.

Spires TL, Grote HE, Varshney NK, Cordery PM, van Dellen A, Blakemore C, Hannan AJ (2004). Environmental enrichment rescues protein deficits in a mouse model of Huntington's disease, indicating a possible disease mechanism. *J Neurosci* 24:2270–2276.

Steen RG, Mull C, McClure R, Hamer RM, Lieberman JA (2006). Brain volume in first-episode schizophrenia: Systematic review and meta-analysis of magnetic resonance imaging studies. *Br J Psychiatry* 188:510–518.

Steiner B, Wolf SA, Kempermann G (2006a). Adult neurogenesis and neurodegenerative disorders. *Regen Medicine* 1:15–28.

Steiner B, Winter C, Hosman K, Siebert E, Kempermann G, Petrus DS, Kupsch A (2006b). Enriched environment induces cellular plasticity in the adult substantia nigra and improves motor behavior function in the 6–OHDA rat model of Parkinson's disease. *Exp Neurol* 199:291–300.

Steiner B, Zurborg S, Horster H, Fabel K, Kempermann G (2008). Differential 24 h responsiveness of Prox1-expressing precursor cells in adult hippocampal neurogenesis to physical activity, environmental enrichment, and kainic acid-induced seizures. *Neuroscience* 154:521–529.

Stevenson JR, Schroeder JP, Nixon K, Besheer J, Crews FT, Hodge CW (2009). Abstinence following alcohol drinking produces depression-like behavior and reduced hippocampal neurogenesis in mice. *Neuropsychopharmacology* 34:1209–1222.

Sturchler-Pierrat C, Abramowski D, Duke M, Wiederhold KH, Mistl C, Rothacher S, Ledermann B, Burki K, Frey P, Paganetti PA, Waridel C, Calhoun ME, Jucker M, Probst A, Staufenbiel M, Sommer B (1997). Two amyloid precursor protein transgenic mouse models with Alzheimer disease-like pathology. *Proc Natl Acad Sci USA* 94:13287–13292.

Su XW, Li XY, Banasr M, Duman RS (2009). Eszopiclone and fluoxetine enhance the survival of newborn neurons in the adult rat hippocampus. *Int J Neuropsychopharmacol* 12:1421–1428.

Suarez-Monteagudo C et al. (2009). Autologous bone marrow stem cell neurotransplantation in stroke patients. An open study. *Restorative Neurology and Neuroscience* 27:151–161.

Sun D, McGinn MJ, Zhou Z, Harvey HB, Bullock MR, Colello RJ (2007). Anatomical integration of newly generated dentate granule neurons following traumatic brain injury in adult rats and its association to cognitive recovery. *Exp Neurol* 204:264–272.

Sun D, Bullock MR, McGinn MJ, Zhou Z, Altememi N, Hagood S, Hamm R, Colello RJ (2009). Basic fibroblast growth factor-enhanced neurogenesis contributes to cognitive recovery in rats following traumatic brain injury. *Exp Neurol* 216:56–65.

Sun L, Lee J, Fine HA (2004). Neuronally expressed stem cell factor induces neural stem cell migration to areas of brain injury. *J Clin Invest* 113:1364–1374.

Sundholm-Peters NL, Yang HK, Goings GE, Walker AS, Szele FG (2005). Subventricular zone neuroblasts emigrate toward cortical lesions. *J Neuropathol Exp Neurol* 64:1089–1100.

Surget A, Saxe M, Leman S, Ibarguen-Vargas Y, Chalon S, Griebel G, Hen R, Belzung C (2008). Drug-dependent requirement of hippocampal neurogenesis in a model of depression and of antidepressant reversal. *Biol Psychiatry* 64:293–301.

Suzuki M, McHugh J, Tork C, Shelley B, Klein SM, Aebischer P, Svendsen CN (2007). GDNF secreting human neural progenitor cells protect dying motor neurons, but not their projection to muscle, in a rat model of familial ALS. *PLoS One* 2:e689.

Suzuki M, McHugh J, Tork C, Shelley B, Hayes A, Bellantuono I, Aebischer P, Svendsen CN (2008). Direct muscle delivery of GDNF with human mesenchymal stem cells improves motor neuron survival and function in a rat model of familial ALS. *Mol Ther* 16:2002–2010.

Szele FG, Chesselet M-F (1996). Cortical lesions induce an increase in cell number and PSA-NCAM expression in the subventricular zone of adult rats. *J Comp Neurol* 368:439–454.

Tada E, Parent JM, Lowenstein DH, Fike JR (2000). X-irradiation causes a prolonged reduction in cell proliferation in the dentate gyrus of adult rats. *Neuroscience* 99:33–41.

Takagi Y, Nozaki K, Takahashi J, Yodoi J, Ishikawa M, Hashimoto N (1999). Proliferation of neuronal precursor cells in the dentate gyrus is accelerated after transient forebrain ischemia in mice. *Brain Res* 831:283–287.

Takahashi M, Arai Y, Kurosawa H, Sueyoshi N, Shirai S (2003). Ependymal cell reactions in spinal cord segments after compression injury in adult rat. *J Neuropathol Exp Neurol* 62:185–194.

Takahashi T, Zhu Y, Hata T, Shimizu-Okabe C, Suzuki K, Nakahara D (2009). Intracranial self-stimulation enhances neurogenesis in hippocampus of adult mice and rats. *Neuroscience* 158:402–411.

Takasawa K, Kitagawa K, Yagita Y, Sasaki T, Tanaka S, Matsushita K, Ohstuki T, Miyata T, Okano H, Hori M, Matsumoto M (2002). Increased proliferation of neural progenitor cells but reduced survival of newborn cells in the contralateral hippocampus after focal cerebral ischemia in rats. *J Cereb Blood Flow Metab* 22:299–307.

Takei H, Wilfong A, Yoshor D, Armstrong DL, Bhattacharjee MB (2007). Evidence of increased cell proliferation in the hippocampus in children with Ammon's horn sclerosis. *Pathol Int* 57:76–81.

Takemura NU (2005). Evidence for neurogenesis within the white matter beneath the temporal neocortex of the adult rat brain. *Neuroscience* 134:121–132.

Tamaki SJ, Jacobs Y, Dohse M, Capela A, Cooper JD, Reitsma M, He D, Tushinski R, Belichenko PV, Salehi A, Mobley W, Gage FH, Huhn S, Tsukamoto AS, Weissman IL, Uchida N (2009). Neuroprotection of host cells by human central nervous system stem cells in a mouse model of infantile neuronal ceroid lipofuscinosis. *Cell Stem Cell* 5:310–319.

Tanaka EM, Ferretti P (2009). Considering the evolution of regeneration in the central nervous system. *Nat Rev Neurosci* 10:713–723.

Tang J, Song M, Wang Y, Fan X, Xu H, Bai Y (2009). Noggin and BMP4 co-modulate adult hippocampal neurogenesis in the APP(swe)/PS1(DeltaE9) transgenic mouse model of Alzheimer's disease. *Biochem Biophys Res Commun* 385:341–345.

Tanimura A, Liu J, Namba T, Seki T, Matsubara Y, Itoh M, Suzuki T, Arai H (2009). Prenatal phencyclidine exposure alters hippocampal cell proliferation in offspring rats. *Synapse* 63:729–736.

Tattersfield AS, Croon RJ, Liu YW, Kells AP, Faull RL, Connor B (2004). Neurogenesis in the striatum of the quinolinic acid lesion model of Huntington's disease. *Neuroscience* 127:319–332.

Tauber SC, Bunkowski S, Ebert S, Schulz D, Kellert B, Nau R, Gerber J (2009). Enriched environment fails to increase meningitis-induced neurogenesis and spatial memory in a mouse model of pneumococcal meningitis. *J Neurosci Res* 87:1877–1883.

Tauck DL, Nadler JV (1985). Evidence of functional mossy fiber sprouting in hippocampal formation of kainic acid-treated rats. *J Neurosci* 5:1016–1022.

Teuchert-Noodt G, Dawirs RR, Hildebrandt K (2000). Adult treatment with methamphetamine transiently decreases dentate granule cell proliferation in the gerbil hippocampus. *J Neural Transm* 107:133–143.

Tfilin M, Sudai E, Merenlender A, Gispan I, Yadid G, Turgeman G (2009). Mesenchymal stem cells increase hippocampal neurogenesis and counteract depressive-like behavior. *Mol Psychiatry*.

Thom M, Sisodiya SM, Beckett A, Martinian L, Lin WR, Harkness W, Mitchell TN, Craig J, Duncan J, Scaravilli F (2002). Cytoarchitectural abnormalities in hippocampal sclerosis. *J Neuropathol Exp Neurol* 61:510–519.

Thomas RM, Peterson DA (2008). Even neural stem cells get the blues: Evidence for a molecular link between modulation of adult neurogenesis and depression. *Gene Expr* 14:183–193.

Thonhoff JR, Ojeda L, Wu P (2009). Stem cell-derived motor neurons: Applications and challenges in amyotrophic lateral sclerosis. *Curr Stem Cell Res Ther* 4:178–199.

Thored P, Arvidsson A, Cacci E, Ahlenius H, Kallur T, Darsalia V, Ekdahl CT, Kokaia Z, Lindvall O (2006). Persistent production of neurons from adult brain stem cells during recovery after stroke. *Stem Cells* 24: 739–747.

Thored P, Wood J, Arvidsson A, Cammenga J, Kokaia Z, Lindvall O (2007). Long-term neuroblast migration along blood vessels in an area with transient angiogenesis and increased vascularization after stroke. *Stroke* 38:3032–3039.

Thored P, Heldmann U, Gomes-Leal W, Gisler R, Darsalia V, Taneera J, Nygren JM, Jacobsen SE, Ekdahl CT, Kokaia Z, Lindvall O (2009). Long-term accumulation of microglia with proneurogenic phenotype concomitant with persistent neurogenesis in adult subventricular zone after stroke. *Glia* 57:835–849.

Toledo EM, Colombres M, Inestrosa NC (2008). Wnt signaling in neuroprotection and stem cell differentiation. *Prog Neurobiol* 86:281–296.

Toresson H, Parmar M, Campbell K (2000). Expression of Meis and Pbx genes and their protein products in the developing telencephalon: Implications for regional differentiation. *Mech Dev* 94:183–187.

Toro CT, Deakin JF (2007). Adult neurogenesis and schizophrenia: A window on abnormal early brain development? *Schizophr Res* 90:1–14.

Tran PB, Miller RJ (2005). HIV-1, chemokines and neurogenesis. *Neurotox Res* 8:149–158.

Tseng KY, Lewis BL, Hashimoto T, Sesack SR, Kloc M, Lewis DA, O'Donnell P (2008). A neonatal ventral hippocampal lesion causes functional deficits in adult prefrontal cortical interneurons. *J Neurosci* 28: 12691–12699.

Tsutsui Y, Kosugi I, Kawasaki H, Arai Y, Han GP, Li L, Kaneta M (2008). Roles of neural stem progenitor cells in cytomegalovirus infection of the brain in mouse models. *Pathol Int* 58:257–267.

Tsyb AF et al. (2009). Morphofunctional study of the therapeutic efficacy of human mesenchymal and neural stem cells in rats with diffuse brain injury. *Bull Exp Biol Med* 147:132–146.

Tzeng SF, Wu JP (1999). Responses of microglia and neural progenitors to mechanical brain injury. *Neuroreport* 10:2287–2292.

Urbach A, Redecker C, Witte OW (2008). Induction of neurogenesis in the adult dentate gyrus by cortical spreading depression. *Stroke* 39:3064–3072.

Van Dellen A, Cordery PM, Spires TL, Blakemore C, Hannan AJ (2008). Wheel running from a juvenile age delays onset of specific motor deficits but does not alter protein aggregate density in a mouse model of Huntington's disease. *BMC Neurosci* 9:34.

Van Praag H, Kempermann G, Gage FH (1999). Running increases cell proliferation and neurogenesis in the adult mouse dentate gyrus. *Nat Neurosci* 2:266–270.

Veena J, Srikumar BN, Raju TR, Shankaranarayana Rao BS (2009). Exposure to enriched environment restores the survival and differentiation of new born cells in the hippocampus and ameliorates depressive symptoms in chronically stressed rats. *Neurosci Lett* 455:178–182.

Venkatesan A, Nath A, Ming GL, Song H (2007). Adult hippocampal neurogenesis: Regulation by HIV and drugs of abuse. *Cell Mol Life Sci* 64:2120–2132.

Videbech P, Ravnkilde B (2004). Hippocampal volume and depression: A meta-analysis of MRI studies. *Am J Psychiatry* 161:1957–1966.

Vollmayr B, Simonis C, Weber S, Gass P, Henn F (2003). Reduced cell proliferation in the dentate gyrus is not correlated with the development of learned helplessness. *Biol Psychiatry* 54:1035–1040.

Wakade CG, Mahadik SP, Waller JL, Chiu FC (2002). Atypical neuroleptics stimulate neurogenesis in adult rat brain. *J Neurosci Res* 69:72–79.

Walker DW, Barnes DE, Zornetzer SF, Hunter BE, Kubanis P (1980). Neuronal loss in hippocampus induced by prolonged ethanol consumption in rats. *Science* 209:711–713.

Wallenborg K, Vlachos P, Eriksson S, Huijbregts L, Arner ES, Joseph B, Hermanson O (2009). Red wine triggers cell death and thioredoxin reductase inhibition: Effects beyond resveratrol and SIRT1. *Exp Cell Res* 315: 1360–1371.

Walter C, Murphy BL, Pun RY, Spieles-Engemann AL, Danzer SC (2007). Pilocarpine-induced seizures cause selective time-dependent changes to adult-generated hippocampal dentate granule cells. *J Neurosci* 27: 7541–7552.

Walzlein JH, Synowitz M, Engels B, Markovic DS, Gabrusiewicz K, Nikolaev E, Yoshikawa K, Kaminska B, Kempermann G, Uckert W, Kaczmarek L, Kettenmann H, Glass R (2008). The antitumorigenic response of neural precursors depends on subventricular proliferation and age. *Stem Cells* 26:2945–2954.

Wang HD, Dunnavant FD, Jarman T, Deutch AY (2004). Effects of antipsychotic drugs on neurogenesis in the forebrain of the adult rat. *Neuropsychopharmacology* 29:1230–1238.

Wang L, Chopp M, Zhang RL, Zhang L, Letourneau Y, Feng YF, Jiang A, Morris DC, Zhang ZG (2009). The Notch pathway mediates expansion of a progenitor pool and neuronal differentiation in adult neural progenitor cells after stroke. *Neuroscience* 158:1356–1363.

Wang ZL, Cheng SM, Ma MM, Ma YP, Yang JP, Xu GL, Liu XF (2008). Intranasally delivered bFGF enhances neurogenesis in adult rats following cerebral ischemia. *Neurosci Lett* 446:30–35.

Weiss KH, Johanssen C, Tielsch A, Herz J, Deller T, Frotscher M, Forster E (2003). Malformation of the radial glial scaffold in the dentate gyrus of reeler mice, scrambler mice, and ApoER2/VLDLR-deficient mice. *J Comp Neurol* 460:56–65.

Welzel G, Fleckenstein K, Schaefer J, Hermann B, Kraus-Tiefenbacher U, Mai SK, Wenz F (2008). Memory function before and after whole brain radiotherapy in patients with and without brain metastases. *Int J Radiat Oncol Biol Phys* 72:1311–1318.

Wen PH, Friedrich VL, Jr., Shioi J, Robakis NK, Elder GA (2002a). Presenilin-1 is expressed in neural progenitor cells in the hippocampus of adult mice. *Neurosci Lett* 318:53–56.

Wen PH, Shao X, Shao Z, Hof PR, Wisniewski T, Kelley K, Friedrich VL, Jr., Ho L, Pasinetti GM, Shioi J, Robakis NK, Elder GA (2002b). Overexpression of wild type but not an FAD mutant presenilin-1 promotes neurogenesis in the hippocampus of adult mice. *Neurobiol Dis* 10:8–19.

Wen PH, Hof PR, Chen X, Gluck K, Austin G, Younkin SG, Younkin LH, DeGasperi R, Gama Sosa MA, Robakis NK, Haroutunian V, Elder GA (2004). The presenilin-1 familial Alzheimer disease mutant P117L impairs neurogenesis in the hippocampus of adult mice. *Exp Neurol* 188:224–237.

Wenning GK, Odin P, Morrish P, Rehncrona S, Widner H, Brundin P, Rothwell JC, Brown R, Gustavii B, Hagell P, Jahanshahi M, Sawle G, Bjorklund A, Brooks DJ, Marsden CD, Quinn NP, Lindvall O (1997). Short- and long-term survival and function of unilateral intrastriatal dopaminergic grafts in Parkinson's disease. *Ann Neurol* 42:95–107.

Wentz CT, Magavi SS (2009). Caffeine alters proliferation of neuronal precursors in the adult hippocampus. *Neuropharmacology* 56:994–1000.

White AM, Matthews DB, Best PJ (2000). Ethanol, memory, and hippocampal function: A review of recent findings. *Hippocampus* 10:88–93.

Wilhelm J, Frieling H, Hillemacher T, Degner D, Kornhuber J, Bleich S (2008). Hippocampal volume loss in patients with alcoholism is influenced by the consumed type of alcoholic beverage. *Alcohol Alcohol* 43:296–299.

Winner B, Lie DC, Rockenstein E, Aigner R, Aigner L, Masliah E, Kuhn HG, Winkler J (2004). Human wild-type alpha-synuclein impairs neurogenesis. *J Neuropathol Exp Neurol* 63:1155–1166.

Winner B, Geyer M, Couillard-Despres S, Aigner R, Bogdahn U, Aigner L, Kuhn G, Winkler J (2006). Striatal deafferentation increases dopaminergic neurogenesis in the adult olfactory bulb. *Exp Neurol* 197:113–121.

Winner B, Couillard-Despres S, Geyer M, Aigner R, Bogdahn U, Aigner L, Kuhn HG, Winkler J (2008a). Dopaminergic lesion enhances growth factor-induced striatal neuroblast migration. *J Neuropathol Exp Neurol* 67:105–116.

Winner B, Rockenstein E, Lie DC, Aigner R, Mante M, Bogdahn U, Couillard-Despres S, Masliah E, Winkler J (2008b). Mutant alpha-synuclein exacerbates age-related decrease of neurogenesis. *Neurobiol Aging* 29:913–925.

Winner B, Desplats P, Hagl C, Klucken J, Aigner R, Ploetz S, Laemke J, Karl A, Aigner L, Masliah E, Buerger E, Winkler J (2009). Dopamine receptor activation promotes adult neurogenesis in an acute Parkinson model. *Exp Neurol* 219:543–552.

Winocur G, Wojtowicz JM, Sekeres M, Snyder JS, Wang S (2006). Inhibition of neurogenesis interferes with hippocampus-dependent memory function. *Hippocampus* 16:296–304.

Wisniewski KE, Laure-Kamionowska M, Wisniewski HM (1984). Evidence of arrest of neurogenesis and synaptogenesis in brains of patients with Down's syndrome. *N Engl J Med* 311:1187–1188.

Wojtowicz JM (2006). Irradiation as an experimental tool in studies of adult neurogenesis. *Hippocampus* 16:261–266.

Wolf SA, Kronenberg G, Lehmann K, Blankenship A, Overall R, Staufenbiel M, Kempermann G (2006). Cognitive and physical activity differently modulate disease progression in the amyloid precursor protein (APP)-23 model of Alzheimer's disease. *Biol Psychiatry* 60:1314–1323.

Wolf SA, Steiner B, Wengner A, Lipp M, Kammertoens T, Kempermann G (2009a). Adaptive peripheral immune response increases proliferation of neural precursor cells in the adult hippocampus. *Faseb J* 23:3121–3128.

Wolf SA, Steiner B, Akpinarli A, Kammertoens T, Nassenstein C, Braun A, Blankenstein T, Kempermann G (2009b). CD4-positive T lymphocytes provide a neuroimmunological link in the control of adult hippocampal neurogenesis. *J Immunol* 182:3979–3984.

Wolswijk G (1998). Chronic stage multiple sclerosis lesions contain a relatively quiescent population of oligodendrocyte precursor cells. *J Neurosci* 18:601–609.

Won SJ, Kim SH, Xie L, Wang Y, Mao XO, Jin K, Greenberg DA (2006). Reelin-deficient mice show impaired neurogenesis and increased stroke size. *Exp Neurol* 198:250–259.

Wu H, Lu D, Jiang H, Xiong Y, Qu C, Li B, Mahmood A, Zhou D, Chopp M (2008). Simvastatin-mediated upregulation of VEGF and BDNF, activation of the PI3K/Akt pathway, and increase of neurogenesis are associated with therapeutic improvement after traumatic brain injury. *J Neurotrauma* 25:130–139.

Xiong Y, Lu D, Qu C, Goussev A, Schallert T, Mahmood A, Chopp M (2008). Effects of erythropoietin on reducing brain damage and improving functional outcome after traumatic brain injury in mice. *J Neurosurg* 109:510–521.

Xiong Y, Mahmood A, Meng Y, Zhang Y, Qu C, Schallert T, Chopp M (2009). Delayed administration of erythropoietin reducing hippocampal cell loss, enhancing angiogenesis and neurogenesis, and improving functional outcome following traumatic brain injury in rats: comparison of treatment with single and triple dose. *J Neurosurg.*

Yagita Y, Kitagawa K, Ohtsuki T, Takasawa K, Miyata T, Okano H, Hori M, Matsumoto M (2001). Neurogenesis by progenitor cells in the ischemic adult rat hippocampus. *Stroke* 32:1890–1896.

Yamaguchi M, Suzuki T, Seki T, Namba T, Juan R, Arai H, Hori T, Asada T (2004). Repetitive cocaine administration decreases neurogenesis in adult rat hippocampus. *Ann NY Acad Sci* 1025:351–362.

Yamaguchi M, Suzuki T, Seki T, Namba T, Liu J, Arai H, Hori T, Shiga T (2005). Decreased cell proliferation in the dentate gyrus of rats after repeated administration of cocaine. *Synapse* 58:63–71.

Yamamoto S, Yamamoto N, Kitamura T, Nakamura K, Nakafuku M (2001). Proliferation of parenchymal neural progenitors in response to injury in the adult rat spinal cord. *Exp Neurol* 172:115–127.

Yan J, Xu L, Welsh AM, Chen D, Hazel T, Johe K, Koliatsos VE (2006). Combined immunosuppressive agents or CD4 antibodies prolong survival of human neural stem cell grafts and improve disease outcomes in amyotrophic lateral sclerosis transgenic mice. *Stem Cells* 24:1976–1985.

Yang F, Wang JC, Han JL, Zhao G, Jiang W (2008a). Different effects of mild and severe seizures on hippocampal neurogenesis in adult rats. *Hippocampus* 18:460–468.

Yang F, Liu ZR, Chen J, Zhang SJ, Quan QY, Huang YG, Jiang W (2009). Roles of astrocytes and microglia in seizure-induced aberrant neurogenesis in the hippocampus of adult rats. *J Neurosci Res* 88:519–29.

Yang H, Lu P, McKay HM, Bernot T, Keirstead H, Steward O, Gage FH, Edgerton VR, Tuszynski MH (2006). Endogenous neurogenesis replaces oligodendrocytes and astrocytes after primate spinal cord injury. *J Neurosci* 26:2157–2166.

Yang Y, Geldmacher DS, Herrup K (2001). DNA replication precedes neuronal cell death in Alzheimer's disease. *J Neurosci* 21:2661–2668.

Yang Y, Mufson EJ, Herrup K (2003). Neuronal cell death is preceded by cell cycle events at all stages of Alzheimer's disease. *J Neurosci* 23:2557–2563.

Yang YJ, Wang XL, Yu XH, Wang X, Xie M, Liu CT (2008b). Hyperbaric oxygen induces endogenous neural stem cells to proliferate and differentiate in hypoxic-ischemic brain damage in neonatal rats. *Undersea Hyperb Med* 35:113–129.

Yasuhara T, Matsukawa N, Hara K, Yu G, Xu L, Maki M, Kim SU, Borlongan CV (2006). Transplantation of human neural stem cells exerts neuroprotection in a rat model of Parkinson's disease. *J Neurosci* 26:12497–12511.

Yin J, Ma Y, Yin Q, Xu H, An N, Liu S, Fan X, Yang H (2007). Involvement of over-expressed BMP4 in pentylenetetrazol kindling-induced cell proliferation in the dentate gyrus of adult rats. *Biochem Biophys Res Commun* 355:54–60.

Yoshimura S, Teramoto T, Whalen MJ, Irizarry MC, Takagi Y, Qiu J, Harada J, Waeber C, Breakefield XO, Moskowitz MA (2003). FGF-2 regulates neurogenesis and degeneration in the dentate gyrus after traumatic brain injury in mice. *J Clin Invest* 112:1202–1210.

Zawadzka M, Lukasiuk K, Machaj EK, Pojda Z, Kaminska B (2009). Lack of migration and neurological benefits after infusion of umbilical cord blood cells in ischemic brain injury. *Acta Neurobiol Exp* (Warsz) 69:46–51.

Zekanowski C, Wojda U (2009). Aneuploidy, chromosomal missegregation, and cell cycle reentry in Alzheimer's disease. *Acta Neurobiol Exp* (Warsz) 69:232–253.

Zhang C, McNeil E, Dressler L, Siman R (2007). Long-lasting impairment in hippocampal neurogenesis associated with amyloid deposition in a knock-in mouse model of familial Alzheimer's disease. *Exp Neurol* 204:77–87.

Zhang R, Zhang Z, Wang L, Wang Y, Gousev A, Zhang L, Ho KL, Morshead C, Chopp M (2004). Activated neural stem cells contribute to stroke-induced neurogenesis and neuroblast migration toward the infarct boundary in adult rats. *J Cereb Blood Flow Metab* 24:441–448.

Zhang RL, Zhang ZG, Zhang L, Chopp M (2001). Proliferation and differentiation of progenitor cells in the cortex and the subventricular zone in the adult rat after focal cerebral ischemia. *Neuroscience* 105:33–41.

Zhang RL, Zhang ZG, Chopp M (2005). Neurogenesis in the adult ischemic brain: Generation, migration, survival, and restorative therapy. *Neuroscientist* 11:408–416.

Zhao C, Wang J, Zhao S, Nie Y (2009). Constraint-induced movement therapy enhanced neurogenesis and behavioral recovery after stroke in adult rats. *Tohoku J Exp Med* 218:301–308.

Zhao CS, Overstreet-Wadiche L (2008). Integration of adult generated neurons during epileptogenesis. *Epilepsia* 49 Suppl 5:3–12.

Zhao M, Momma S, Delfani K, Carlen M, Cassidy RM, Johansson CB, Brismar H, Shupliakov O, Frisen J, Janson AM (2003). Evidence for neurogenesis in the adult mammalian substantia nigra. *Proc Natl Acad Sci USA* 100:7925–7930.

Zhu C, Huang Z, Gao J, Zhang Y, Wang X, Karlsson N, Li Q, Lannering B, Björk-Eriksson T, Georg Kuhn H, Blomgren K (2009) Irradiation to the immature brain attenuates neurogenesis and exacerbates subsequent hypoxic-ischemic brain injury in the adult. *J Neurochem.* 111:1447–56.

Zhu DY, Liu SH, Sun HS, Lu YM (2003). Expression of inducible nitric oxide synthase after focal cerebral ischemia stimulates neurogenesis in the adult rodent dentate gyrus. *J Neurosci* 23:223–229.

Zhu DY, Lau L, Liu SH, Wei JS, Lu YM (2004). Activation of cAMP-response-element-binding protein (CREB) after focal cerebral ischemia stimulates neurogenesis in the adult dentate gyrus. *Proc Natl Acad Sci USA* 101:9453–9457.

Zhu XJ, Hua Y, Jiang J, Zhou QG, Luo CX, Han X, Lu YM, Zhu DY (2006). Neuronal nitric oxide synthase-derived nitric oxide inhibits neurogenesis in the adult dentate gyrus by down-regulating cyclic AMP response element binding protein phosphorylation. *Neuroscience* 141:827–836.

Zigova T, Pencea V, Wiegand SJ, Luskin MB (1998). Intraventricular administration of BDNF increases the number of newly generated neurons in the adult olfactory bulb. *Mol Cell Neurosci* 11:234–245.

Ziv Y, Schwartz M (2008). Immune-based regulation of adult neurogenesis: Implications for learning and memory. *Brain Behav Immun* 22:167–176.

Ziv Y, Ron N, Butovsky O, Landa G, Sudai E, Greenberg N, Cohen H, Kipnis J, Schwartz M (2006). Immune cells contribute to the maintenance of neurogenesis and spatial learning abilities in adulthood. *Nat Neurosci* 9:268–275.

Index

Note: Page references followed by "*f*" and "*t*" denote figures and tables, respectively.

Neuronally determined transient
amplifying progenitor cell,
193–94. *See also* Progenitor cell
Neuronal maturation
and aging, 339
marker, 506
Neuronal nuclei, *See* NeuN
Neuronal polarity, 131–32
Neuronal-restricted silencing factor/
RE-1 silencing transcription
factor (NRSF/REST), 408–9
Neuronal survival, 136–37
Neuron-glia 2 (NG2), 230–31, 287,
289, 292, 302
Neurons, 3, 24, 283
of cerebellum, 11
division of, 27–28
types of, 10*f*
Neuron-specific enolase (NSE), 40, 241
Neuropeptides, 366
Neuroplasticity, 30
Neuroprotective surveillance, 385
Neuro-psychiatric disorders, therapy of,
519–24
Neurosphere cultures. *See* Neurospheres,
Neurospheres, 77–79, 219–20,
219*f*, 294
critique of, 220
advantages of, 77
cultures vs. monolayer cultures,
comparison, 78*t*
Neurotransmitter signaling, 354,
358–66
genes related to, 359–62*t*
Neurotransplantation, 519, 520–21
Neurotrophic factors, 340, 386–88
for cell survival or death, 136–37
Neurotrophins (NTs), 137, 378, 388
New gatekeeper theory, 473
New neurons, 44
in adult brain, 5*f*
functional responsiveness of,
446–47, 446*f*
identification markers for, 240–46
in vivo, local network integration of,
444–53
to long-term potentiation in
hippocampus, contribution of,
447, 448*f*
role in hippocampal function,
448–49, 456, 461
signs of neuronal function in
individual, 442–44, 443*f*
Nfkb1, 404*t*
NG2 (Neuron-glia 2), 72*f*, 230–31,
287, 289, 292, 302
overlap with Doublecortin, 229
NG2 cells, 26, 67*t*, 71–74, 72*f*, 177
Ngfr, 381*t*
Ngn1, 122, 123, 306, 309, 409

Ngn2, 122, 123, 198
Niche, 121, 221–22, 277–86
astrocytes and astrocyte-like precursor
cells, 279–80
ependymal cells, 281
extracellular matrix, 283–85
factors, cells as, 279–86
microglia and immune cells, 282–83
neurogenic, 191
neurons, 283
secreted factors, 285
stem cell, 191
system biology of, 286
vascular, 191
vasculature, 281–82
Nicotine, 551–52
Nitric oxide (NO), 368
and ischemia-induced
neurogenesis, 557
Nitric oxide synthase (NOS), 368
NKCCl, 198
Nkx2-1, 138
Nkx2-2, 404*t*
NMDA (*N*-methyl-d-asparatate)
receptors, 135, 345, 393
Nodes of Ranvier, 72
Noggin, 281, 285, 371*t*, 373, 539
Nomenclature, 34
No-new-neurons dogma, 15–17
Nonhuman primates, adult
hippocampal neurogenesis in,
499–500, 501*f*
Non-mammalian vertebrates, adult
hippocampal neurogenesis in,
489–96
amphibians, 489–90
birds, 491–92
electric gymnotiform fish, 491
function, 495
precursor cells, 494
regulation, 494–95
reptiles, 489–90
song learning system, 492–93, 493*f*
zebrafish (*Danio rerio*), 490–91, 492*f*
Non-myelinating precursor cells, 177
Non-neurogenic regions, 7, 275, 276*f*
neural precursor cells in, 74–75
Non-vertebrates, adult hippocampal
neurogenesis in, 487–89, 488*f*
Noradrenaline, 354, 365
Norepinephrine, 354, 365
Nos1, 371*t*
Nos3, 371*t*
Notch1, 76, 117, 309, 353, 356*t*
Notch signaling, 124–26, 285, 306,
328, 353, 556
Nottebohm, Fernando, 39–40
Novelty-suppressed feeding, 469
Npas3, 405*t*
Npy, 361*t*, 366

Npyry1, 361*t*
Npyry2, 361*t*
Nr2e1, 395, 399*t*, 401
Nr3c1, 391*t*
Nr3c2, 391*t*
Nrg1, 381*t*, 548
Nrg2, 381*t*
NSC-34, 28
NT3, 308, 388
Ntf3, 381*t*
Ntf5, 381*t*
Ntrk1, 381*t*
Ntrk2, 381*t*
Nucleus basalis Meynert (NBM), 354
Numb, 117
Nutrition, 347–49

O2A precursor cell, 68
Ockham's razor, 240
Oct4, 82, 110
Odor, 346–47
Oken, Lorenz, 23
Olanzapine, 548
Olfaction, 346
Olfactory bulb (OB), 7, 286, 334
adult neurogenesis in, 158, 159
adult-generated interneurons in, 159*t*
adult-generated neurons in, 174*f*
adult-generated neurons in,
functional relevance of, 474–78
anatomy of, 305*f*
development of, 140*f*
environmental enrichment in, 343
network structure of, 141*f*
neurogenesis in adult human, 503–4
neurogenesis, and odor, 346–47
neurogenesis-dependent
function, 440*t*
neurogenic zones, of adult
brain, 139–42
new glutamatergic interneurons
in, 176*f*
precursor cells for, 159
and SVZ, 158
Olfactory bulbectomy, 351
Olfactory deprivation, 307
Olfactory ensheathing glia, 309
Olfactory epithelium, 11
adult neurogenesis in, 304–10
anatomy of, 305*f*
neurogenesis, regulation of, 307–9
neuronal development in, 304
neuronal precursor cells in,
305–7, 306*f*
vomeronasal organ, 310
Olfactory marker protein (OMP), 305
Olfactory neuroblastoma, 84
Olfactory receptor neurons, 158–59
Olfactory-specific G-protein subunit
(Golf), 305